Water Management in the Era of Climatic Change—Volume III

Water Management in the Era of Climatic Change—Volume III

Editors

Alban Kuriqi
Luis Garrote

MDPI • Basel • Beijing • Wuhan • Barcelona • Belgrade • Novi Sad • Cluj • Manchester

Editors
Alban Kuriqi
University of Lisbon
Lisbon, Portugal

Luis Garrote
Universidad Politécnica de Madrid
Madrid, Spain

Editorial Office
MDPI
St. Alban-Anlage 66
4052 Basel, Switzerland

This is a reprint of articles from the Topic published online in the open access journals *Water* (ISSN 2073-4441), *Agriculture* (ISSN 2077-0472), *International Journal of Environmental Research and Public Health* (ISSN 1660-4601), *Remote Sensing* (ISSN 2072-4292), and *Sustainability* (ISSN 2071-1050) (available at: https://www.mdpi.com/topics/water).

For citation purposes, cite each article independently as indicated on the article page online and as indicated below:

Lastname, A.A.; Lastname, B.B. Article Title. *Journal Name* **Year**, *Volume Number*, Page Range.

Volume III
ISBN 978-3-0365-9348-7 (Hbk)
ISBN 978-3-0365-9349-4 (PDF)
doi.org/10.3390/books978-3-0365-9349-4

Set
ISBN 978-3-0365-9342-5 (Hbk)
ISBN 978-3-0365-9343-2 (PDF)

Contents

About the Editors

Alban Kuriqi

Alban Kuriqi is a Research Scientist at CERIS–Civil Engineering Research and Innovation for Sustainability, University of Lisbon. He earned his Ph.D. in Civil Engineering from the University of Lisbon, specializing in River Restoration and Management. His extensive research interests and expertise encompass a wide array of areas, including renewable energy, with a particular focus on hydropower and complementary resources, as well as the impacts of hydropower and water resources' management-related issues.

Luis Garrote

Luis Garrote is Full Professor of Hydraulic Engineering at Universidad Politécnica de Madrid. His research focus is the application of hydrological and hydraulic models in water resources' planning and management, including floods, droughts, environmental constraints and reservoir operation, with a special emphasis on dealing with uncertainties, particularly those connected to global change.

Preface

Water is a crucial element on Earth for all living and non-living components. Climate change is an alarming issue for managing and sustaining life on Earth. Given climate change, water resources worldwide have been under drastically stressed conditions, as is evident from the uneven weather patterns, droughts, floods, and cloud bursts. Only three percent of the water resources on Earth are fresh, and two-thirds of the freshwater is locked up in ice caps and glaciers. Of the remaining one percent, a fifth is in remote, inaccessible areas. Much of the seasonal rainfall in monsoonal deluges and floods cannot be easily used. Only about 0.08 percent of all the world's freshwater is exploited by humankind, with an ever-increasing demand for sanitation, drinking, manufacturing, leisure, and agriculture. The ever-increasing water exploitation has intensively degraded freshwater ecosystems, notably rivers.

Furthermore, the climate extremes and water scarcity that are enhanced by climate change induce additional stress on the freshwater ecosystems and may stimulate conflicts among water users. In addition, we know that water is needed for several vital human activities, of which agricultural and industrial activities are the primary water consumers. In the context in which we observe more frequent droughts and incidences of water scarcity in the world, water systems' management requires the most advanced approaches and tools to rigorously address all of the dimensions involved in the sustainability of its development.

Therefore, this Topic Collection's main objective is to contribute to the understanding of water systems' management, and to provide science-based knowledge, new ideas/approaches, and solutions for water resources' management. Water demand for irrigation has been steadily increasing during in recent decades. However, other water users have simultaneously been competing with agricultural sectors for water resources. The conservation of freshwater ecosystems also needs special attention, such as the sufficient allocation of environmental flows. In addition, in terms of the projected climate change caused by warmer temperatures and shifting precipitation patterns, water availability is expected to decrease, and water demand to increase, in many areas of the world.

Consequently, soil productivity and, thus, crop production could be drastically reduced. These trends raise concerns highlighting the role of water and natural resources' management and their conservation to ensure the sustainability of irrigated agriculture. How well-irrigated agriculture adapts to water scarcity scenarios, particularly by increasing water use efficiency and better-estimating evapotranspiration, will directly affect the future and sustainability of the sector. The 89 papers published in this Topic Collection encompass a diverse range of critical issues and potential solutions concerning the sustainable management of water resources. We anticipate that this collection will serve as a source of inspiration for engineers, scientists, policymakers, and decision-makers worldwide, helping them to identify appropriate solutions and make informed decisions regarding their specific water-related challenges.

Alban Kuriqi and Luis Garrote
Editors

International Journal of
Environmental Research and Public Health

MDPI

Study Protocol

A Systematic Study Site Selection Protocol to Determine Environmental Flows in the Headwater Catchments of the Vhembe Biosphere Reserve

Anesu D. Gumbo [1,*], Evison Kapangaziwiri [2] and Fhumulani I. Mathivha [3]

1 Department of Geography and Environmental Sciences, University of Venda, Thohoyandou 0950, South Africa
2 Hydrosciences Research Group, Council for Scientific and Industrial Research (CSIR), Pretoria 0001, South Africa; ekapangaziwiri@csir.co.za
3 Department of Hydrology, University of Zululand, Empangeni 3886, South Africa; mathivhaf@unizulu.ac.za
* Correspondence: diongumbo@gmail.com

Abstract: Developing nations will be worst hit by the impacts of climate change because limited resources hinder the spatial reach of climate studies, effort, and subsequent implementation to help with the improvement of livelihoods. Therefore, finding the best-case study is an essential undertaking in environmental assessments. This study explains one systematic approach to selecting a study site for an environmental assessment project. A desktop review of relevant literature, a simple factor scoring assessment process, reliance on expert opinion, and a field survey for ground-truthing were conducted. The desktop review showed the most critical factors to site selection. The scoring of these factors selected those that were crucial for the study. Experts validated the results and suggested the best study site among the ones identified. While the design is simplified, the proposed approach selects the most appropriate study site for environmental assessments.

Keywords: climate change; case study; environmental assessment; data scarcity; science communication; selection protocol

Citation: Gumbo, A.D.; Kapangaziwiri, E.; Mathivha, F.I. A Systematic Study Site Selection Protocol to Determine Environmental Flows in the Headwater Catchments of the Vhembe Biosphere Reserve. *Int. J. Environ. Res. Public Health* **2022**, *19*, 6259. https://doi.org/10.3390/ijerph19106259

Academic Editors: Alban Kuriqi and Luis Garrote

Received: 28 February 2022
Accepted: 17 April 2022
Published: 21 May 2022

1. Introduction

It is imperative and apparent that understanding changes in climate, and subsequently their implications on society and the environment (especially the devastating adverse effects), is a necessary undertaking. Integrating several disciplines to inform mitigation and adaptation options to the unprecedented climate changes and their devastating impacts has become important [1]. The general scientific fact is that climate change will increase the occurrence, magnitude, and frequency of extreme weather events and potentially lead to loss of food production and livelihoods [2,3]. A crucial global message is that there is a need for robust responses to arrest the adverse impacts to save lives and livelihoods, particularly in the most vulnerable societies of the world [4].

According to [5], vulnerability to climate change is regarded as the degree to which a system is susceptible to and unable to cope with the adverse effects of climate change. How much a system is exposed, its sensitivity to the exposure, and its adaptive capacity provides an understanding of how vulnerable it is. Climate change impacts are distributed unevenly around the globe [6], based on geographical location, level of vulnerability, level of preparedness and access to necessary resources to implement mitigation and adaptation, and societal capacity to understand and perceive climate changes [7]. Area-specific climate change assessments are, thus, vital undertakings to address the global challenge at local levels to increase the adaptation capacity [8]. The global agenda has been placed on climate change mitigation and adaptation.

Developing countries, however, have low mitigation capacity, and adapting to these changes is ideal and many countries are developing their own National Adaptation Plans

(NAPs) [9]. These strategies to adapt to climate change are backed up by science which projects future possible scenarios. Southern Africa is one of the regions to experience some of the most adverse effects of climate change. Studies e.g., [10–13] indicate that the region's vulnerability is high, while its preparedness is low to nonexistent. Several issues such as poverty, corruption, political instability, limited access to climate information, pandemics, and other problems influence the responsive capacity of countries in southern Africa.

The past few decades have produced scientific reports on the projected magnitudes and directions of changes in the climate and the potential impacts on water resources' quality and quantity [14–20]. Improved projections are due to massive technological advancements [21,22] that have improved data collection and analysis methods and the efficiency and robustness of environmental assessment tools [23]. Such improvements have reduced some uncertainties related to environmental assessments and built more confidence and dependability in climate change projections to influence policy and decision making. High confidence in climate projections has made science a vital pillar to support the development of mitigation and adaptation strategies, approaches, and technologies necessary to combat the potential shocks of a changing climate [24]. These adaptation strategies are tested and modified in areas that have human–environment interactions. Several areas with these characteristics have been demarcated around the world and have been termed biospheres.

Biosphere reserves have been created as study sites to understand the climate change–human–environment interactions. The United Nations Educational, Scientific, and Cultural Organization (UNESCO) regards biosphere reserves as learning places for sustainable development [25]. They are sites for testing interdisciplinary approaches to understanding and managing changes and interactions between social and ecological systems, including conflict prevention and management of biodiversity. Consequently, there are 727 established biosphere reserves in 131 countries [25] whose purpose is to provide a learning environment for sustainable development in diverse and fragile but significant ecological regions. They utilize different socio-economic contexts to enhance the lives and livelihoods of the communities within them [26]. Local solutions to global challenges affecting sustainable development are developed and tested in these reserves. The value of these reserves in mitigating and adapting to climate change is significant, offering opportunities for case studies to be conducted as learning platforms. The Vhembe Biosphere Reserve (VBR) is a protected area in the Limpopo Province of South Africa. Its spatial extent limits the implementation of research work to only selected areas. These selected areas need to be carefully identified as they become the general representation of the biosphere and the results gathered from their study would need to be adopted by similar areas. Out of the 10 biospheres in South Africa, the VBR was selected as the area of study because of a high rural population vulnerable to climate change impacts. It is a data-deprived area that requires focus to generate information/data to aid informed decision making. Data availability and data quality are common problems in many areas in Africa [27] and this affects studies being carried out in such areas. Climate change impacts on water resources are affected by the lack of adequate and good quality time series of hydrometeorological data [28]. Creating protocols for carrying out research in such areas becomes critical to informing sustainable decision making.

Quantifying water resource availability against climate change is an important undertaking as water is central to sustainable development. Developing nations, such as the southern African regions, are mainly rural [29] whose socio-economic activities are closely tied to the availability of water resources. Many rivers in the southern African region have high flows during the rainy season and low to no flows during the dry season [29]. The dry season is persistent throughout the year and communities rely on the low flows to sustain their socio-economic activities and the environment. The South African Water Act of 1998 identifies the environment as a legitimate and important user of water resources in any sub-basin requiring that a determined amount of water be reserved. Hence, the environmental water requirements (popularly known as e-flows worldwide) are important to catchment water management. In the Water Act of 1998, the e-flows are referred to as the reserve. The quantification of these e-flows in South Africa is an ongoing undertaking of great importance to water resources' budgeting and accounting [30,31]. The task generally requires a water practitioner or user to provide a holistic assessment utilizing any one of the several acceptable approaches and tools. This is to fully understand the ecological functioning and value chain of a given sub-catchment and determine a sensible and viable e-flow requirements regime for the survival and sustenance of the riverine ecosystem. The choice of a representative area of study or river stretch should then be carefully considered to produce results that can inform decision making for the chosen area and its implementation in other areas of similar characteristics.

Ref. [32] explained that case studies are intense meticulous studies of a particular area to generate results that can be generalized over a more extensive set of units. They are used in exploratory research [33], where no past data or a few referral studies exist. Therefore, identifying a study site that can address one or several objectives of the overall aim is essential. The overall aim of the intended study for which the case study site(s) must be selected is to determine the e-flows for headwater catchments that are suitable and necessary to sustain riverine ecology within major selected parts of the Vhembe biosphere and assess the potential impacts of a changing climate on the availability and sustainability of these flow regimes. Several headwater catchments are potentially available for study within the reserve. However, because of several factors (time, funding, technical expertise, and best representative of areas in the VBR), the study will have to choose one or a few sub-basins to use. Results would then have to be extrapolated to the rest of the headwater catchments. Thus, representative study sites must be selected to generate and establish baseline information and methodologies that could be applied in the other areas. This paper proposes a step-by-step and objective approach to determine the ideal case study site(s) to determine e-flows within the VBR.

2. Study Area

The Vhembe District Municipality and the Blouberg Local Municipality (within the Capricorn District Municipality) together form the boundaries of the VBR. Figure 1 shows the location and spatial extent of the VBR. The VBR is the largest of the 10 biosphere reserves in South Africa [34]. It reaches 30,701 km^2, supporting a human population of approximately 1.5 million, with 97% of the population rural [35]. The Vhembe region holds a unique and extraordinary biological and cultural diversity in the Soutpansberg and Blouberg Mountains, together with the Mapungubwe World Heritage Site and northern Kruger National Park [35]. Local communities have a high, direct reliance on natural resources for livelihood, with 66% of households relying on firewood for fuel and subsistence agriculture [35]. Agriculture and tourism are the major socio-economic activities that sustain livelihoods in the reserve.

Several rivers within the VBR maintain diverse flora, fauna, and riverine ecosystems. These rivers include the Luvuvhu, Dzindi, Nzhelele, Tshipise, and several others, either tributary or pour directly into the Limpopo River. The transboundary nature of the Limpopo River brings about delicate management issues that need a coordinated effort for all riparian countries to benefit from the river's ecosystem values and functions. The VBR comprises three biomes: the Savannah, Grassland, and the Forest [35]. This implies the existence of rich biodiversity, some of which is endemic to the area. Such biodiversity requires coordinated and well-structured management strategies to ensure conservation viability. The general growth in population and dependence on the environment for energy and livelihood negatively impacts natural resources, and a changing climate is likely only to aggravate the situation.

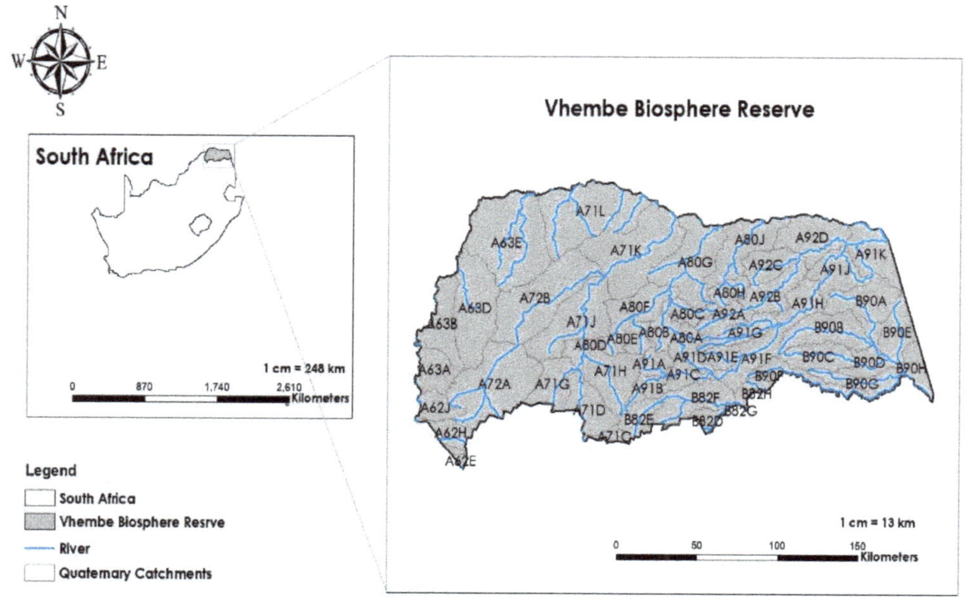

Figure 1. A map showing the spatial extent of the Vhembe Biosphere Reserve (VBR) and the catchments for potential study sites.

3. Materials and Methods

The study used several methods to determine which area(s) would be ideal for determining appropriate e-flows within the headwater catchments of the VBR. The study is based on a systematic review of relevant literature, a simple factor scoring assessment process, experts' opinions, and a field survey for ground-truthing.

3.1. Systematic Review

A detailed desktop scan of literature was carried out to determine the study sites used, especially the motivation for selecting the study sites. Literature on or referring to the VBR was reviewed to understand the previously carried out studies' nature and thrust. This also determined the potential gaps in scientific knowledge and understanding that require additional studies to address them successfully. As information is made available through research, these gaps are plugged, promoting sustainable management for this fragile ecoregion. Google Scholar identified journal articles using the keywords *'study site selection for environmental assessment'* and about 78,300 articles. A 10-year coverage of literature (2011–2021) was adopted for the study. Selection and rejection criteria (Table 1) were used to select the ideal research for review. The review only analyzed peer-reviewed work and excluded grey literature. Using the selection criteria, seven peer-reviewed journals were critically reviewed. A thematic analysis of the selected articles determined the significant factors that contributed to selecting the study site. The complete review of the selected literature led the study to adopt an exploratory factor analysis (EFA) to determine (without explicit ranking) the factors influential in selecting a study site.

Table 1. Literature selection criteria to determine factors that contribute to study site selection.

Search Protocol	Inclusion	Exclusion
Initial Google Scholar search	- English literature.	- Non-English.
	- Peer-reviewed.	- Gray area.
	- Any region.	
Title and abstract review	- Site selection of a study area/case study.	- Sight selection for project implementation, e.g., wind farm, landfill, solar.
	- Climate research.	- Any other study.
Study site description	- Field studies.	- Non-field studies.
	- Areas specific studies.	- Non-area-specific studies.

3.2. *Exploratory Factor Analysis*

Several factors influencing the selection of a study site were obtained from the desktop review and were used to inform the selection criteria for this study. These included accessibility, availability of funds, familiarity with the area, and research gap. The exploratory factor analysis (EFA) was implemented to reduce the identified factors to only those regarded as crucial to achieving the aim and objectives of the study. A factor-scoring criterion was developed to select the variables that could comprehensively address the aim of the study and the associated specific objectives. The variables were scored in importance from very important to not important.

3.3. *Expert Opinions*

Valuable expert opinion was solicited from knowledgeable practitioners with extensive work experience within the VBR. The definition of an expert, in this case, is rather loose and much more encompassing than the normal dictionary one. Experts included academics, researchers working or who had worked in the area and were assumed to possess valuable in-depth knowledge of the area's processes, functions, and importance. The definition was extended to include people who have resided in the confines of the biosphere and have a wealth of knowledge accumulated over many years of observation of relevant phenomena in the area and can therefore discuss and share valuable insights on the scientific queries.

3.4. *GIS-Based Assessment and Ground-Truthing*

Geographical Information Systems (GIS) were used to show all the potential sites selected to implement the study. This provided visual confirmation and display of the spatial distribution of the selected study sites in a way that could show their advantages over the other competing sites. Google Earth and available land use maps (e.g., the 2018 land use map of SA and its recent update from the Department of Forestry, Fisheries, and the Environment) were used to determine the land uses and land cover distributions, including the location of hydraulic structures, as well as vulnerability to possible degradation around the study site(s), among others. Site visits were then carried out as the last activity to familiarize with the area and confirm assumptions and GIS findings.

4. Results

4.1. *Systematic Review Results*

The literature review sought to determine what others considered important factors when selecting a study site for their environmental research. Information on where similar studies have been carried out and the gaps in research from such studies were identified. Table 2 shows the main factors informing study site selection, with those with an Asterix (*) being most common. A total of 12 factors were drawn from the literature search.

Table 2. Factors influencing the choice of a study site and their level of importance in the intended study. X marks the importance of the factor.

Factor	Importance			
	Very	Just	Neutral	None
Accessibility: ease of reach (near to the researcher). *	X			
Need for human presence at the study site or beneficiation from the headwater catchment/s.	X			
The presence of environmental phenomena to be quantified and understood. *	X			
The rich biodiversity in the area.	X			
Availability of hydrometeorological data of adequate quality and quantity. (*Headwater catchments with missing or no observed time series were more critical than those with).*			X	
The economic value of the area.	X			
Need to test a model or equipment (experimentation).			X	
Past similar work has been carried out in the area. *	X			
Availability of relevant literature.			X	
Influence of expert opinion.	X			
Communication barriers (with stakeholders).				X
Availability of (adequate) funding. *	X			

* Most typical reasons why researchers select a study site.

4.2. Factor Scoring Analysis Results

The factors derived from the desktop analysis of literature were listed and given a score based on how important they were in meeting the aim and objectives of the study. Table 2 shows the 12 factors used for this study and the factor score results. The factors considered very important for the study had the most influence on the site selection process. Of the 12 factors, 8 of them were regarded as significant. Table 3 shows the results of the selection criteria which indicate that the catchments A91A (Luvuvhu), A92C (Mbodi), and A91B (Sterkstroom) satisfy the criterion created.

Table 3. The factors that are important to site selection and the possible catchments of study. The tick (√) shows the quaternary catchments that meet the desired factor and could be chosen for study.

Quaternary Catchment Name.	Accessibility: Ease of Reach (Near to the Researcher).	Need for Human Presence at the Study Site or Beneficiation from the Headwater Catchment/s.	The Presence of Environmental Phenomena (to Be Quantified and Understood).	The Rich Biodiversity in the Area.	The Economic Value of the Area.	There Is No Similar Work Done in the Area.	Availability of Experts for the Area.	Availability of (Adequate) Funding.	No Streamflow Data Is Available.
A91A	√	√	√	√	√	√	√	√	√
A91B	√	√	√	√	√	√	√	√	√
A91D		√	√			√	√	√	
A91G		√	√			√	√	√	√
A92A		√	√			√	√	√	√
A92C	√	√	√	√	√	√	√	√	√

4.3. Expert Opinion Findings

From the desktop review, three experts were identified, an aquatic ecologist who has worked within the VBR and two hydrologists who have done extensive work in environmental modeling and climate change impact on streamflow availability. These experts gave guidance on the best site for the intended environmental study. Literature review and factor analysis resulted in selecting the headwaters of the Nzhelele and Luvuvhu River catchments as possible study sites as they met most of the criteria developed. These choices were then validated by experts whose opinions favored sites in the Luvuvhu River Catchment (LRC) as the most suitable.

4.4. Findings of GIS Assessment and Ground-Truthing

The selected quaternary catchments are shown in Figure 2. They are located at the uppermost part of the LRC, making them the headwater catchments for the basin. Several more headwater catchments join the LRC downstream of A91A and A91B. Most of the streamflow drained by the selected headwater catchments collects in the Albasini Dam, creating a different hydrological regime downstream of the dam. It was also confirmed that some parts are densely vegetated and deprived of human interference while others have human activities. The water collected in the dam supports domestic use, agricultural and fishing activities. This phase of the research managed to show the location of the selected study site within the VBR. It confirmed that the objectives of the comprehensive study to be implemented could be achieved.

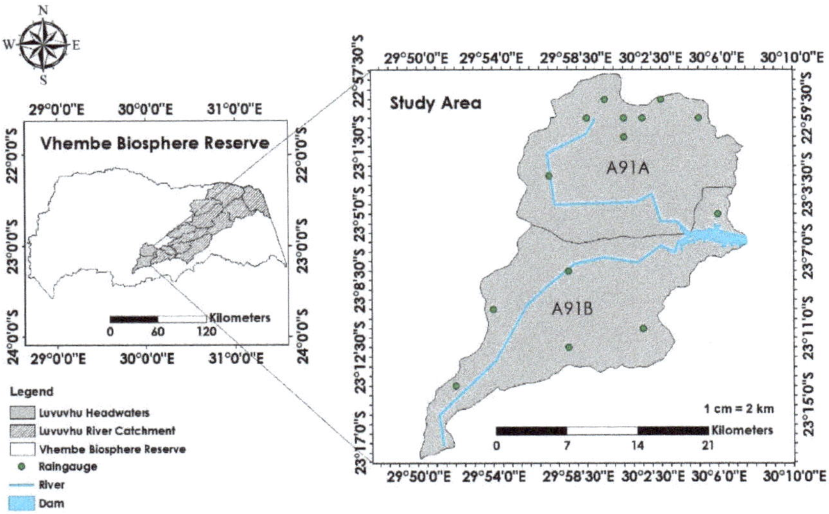

Figure 2. The selected headwater catchments of the Luvuvhu River Catchment in the Vhembe Biosphere Reserve.

5. Discussion

The assessment of the literature shed light on the reasons that drive researchers to select study sites. The need to address or understand an environmental problem appeared as the primary reason for site selection. The absence of similar work in an area can prompt a researcher to implement an environmental assessment within a region. This is particularly important where global challenges need to be addressed through sound scientific research [36]. Some regions will be worst affected by climatic changes, such as southern Africa, Asia, and Latin America, mainly due to low mitigation and adaptation capacity [37]. The knowledge of this future possibility requires more scientific inquiry in these regions. The need to contribute locally developed solutions to global challenges encourages researchers to implement case studies, especially in vulnerable regions. The availability of funds through donors, government, or private institutions significantly affects site selection [38]. Funders can influence a study and site selection based on institutional interests and agendas. However, this does not always give the best results in sustainable development due to the often top-down approach. The proximity of the researcher to the study area, and therefore accessibility, also plays a vital role in site selection. This is especially true in academic research as it is often limited in time, funding, and adequate supporting technical expertise. Several environmental assessments are based on the lack of adequate information on a specific environmental phenomenon that influences the direction of research in an area.

Environmental management and climate change mitigation and adaptation strategies have acknowledged the need for more research to be carried out to aid decision making and policy formulation [39,40]. The study site selection is not well documented, as shown by a lack of publications focused exclusively on this topic in environmental studies. This is not a popular undertaking or conscious decision in environmental studies. However, other disciplines such as health sciences have clear protocols to follow when selecting a study site [41]. They thoroughly analyze the research aim, objectives, and expected deliverables and select the best study area to achieve these. Adopting this approach in environmental assessments will be of great value and could account for common deterrents such as data unavailability. Prior knowledge of what the study requires and targets to achieve against what potential study sites can provide enables the researcher to design the best methodology. It becomes time-consuming and costly when a research design cannot be implemented because the chosen study site lacks the components that enable the methodology to be successfully implemented. This is important in data-scarce regions [42]. Alternative data sources, in such cases, may need to be used, which may require downscaling, interpolations, and extrapolations that the project might not have accounted for in the design phase. This inevitably leads to delays in project implementation and the invasion of additional uncertainties in the generated results, which may reduce the degree of confidence in these results.

The factor scoring results showed the likely catchments to implement the study. Data scarcity was regarded as a unique factor that, in its availability, would simplify the research and in its absence, presents an opportunity to generate it. Data of sufficient quality and quantity are always required in environmental studies, but often, they are the missing component. Data scarcity is a pandemic in environmental studies and limits research in some areas. However, data-scarce regions are usually the most vulnerable to climate change, and sustainable development of such areas can benefit from local scientific investigations. The selection protocol regarded the lack of observed hydro-climatological data (rainfall, streamflow, evaporation) as the desired factor. These areas are usually understudied because they lack observed historical data [43]. Therefore, it is imperative to carry out scientific investigations in these areas to generate reliable baseline information to support development initiatives. An opportunity arises to carry out a study in a data-scarce area to generate more data for various purposes, including decision and policymaking and creating a methodology that is transferrable to other sites. Designing a methodology that utilizes the bare minimum of data will give a more substantial basis for adoption on other sites. The current study needs to be carried out in headwater regions due to their unique provision of ecological goods and services essential to riverine integrity [44].

Due to their hard-to-reach and underdeveloped nature, headwater catchments are subject to limited studies [45]. In the African context, headwaters are data-scarce in terms of observed hydrometeorological data of sufficient quality and literature of previous studies carried on them [29]. Though insufficiently represented in research, these areas are home to a larger rural population that depends on natural resources for its livelihood. Water availability is one natural resource that is important in these communities and its availability is crucial to livelihoods. Southern Africa experiences a long dry period during the year. Low flows are crucial during this period to cater to the socio-economic activities that sustain the riparian rural communities. Understanding the minimum water requirements for the survival of riverine ecosystems is essential to river management, especially in a changing climate. Therefore, the headwater catchments must support rural communities benefiting from them. The study intends to improve rural resilience as the local climate changes. Building resilience through scientific approaches will also rest on the co-production of work to formulate frameworks that can be adopted at the local level using the resources available. The scoring criteria, thus, realized any potential benefits that would accrue to the rural communities in the study area to be of great importance in selecting a suitable study site.

Ref. [46] discuss the challenges expected in e-flow determination in the river catchment, though no specific e-flows quantities were recommended for the area. They acknowledge that this river is crucial to rural livelihoods as it provides a source of animal protein and ecosystems goods and services to the Kruger National Park downstream of the Luvuvhu River. The importance of the LRC was emphasized as well through the consultations carried out with experts who have worked in the biosphere and e-flow determination. The research utilized the information and recommendations of at least three experts. Literature review and factor scoring managed to further narrow down the site selection. Expert opinion was important to determine which catchments, from the assessments, would best address the overall objectives of the study and provide a good case study. The vast knowledge and understanding of the area, areas that still need researching, knowledge of research that was carried out, and the agreement of the experts led to the selection of the catchments A91A and A91B. According to the experts, the LRC offers an excellent opportunity to study most of the issues that arise in catchment management. The area is rich in biodiversity and carrying out this study will be of great importance to understanding the central role of water in this region. The area currently provides room for collaborative efforts in catchment studies as several projects are currently underway that deal with water and environmental management. Through networking and collaboration, the determination of e-flows in the LRC can be aided by the work being carried such as water quality and distribution of freshwater fish species. The headwaters of LRC have also been demarcated as part of the Strategic Water Source Areas of South Africa that have a unique ability to provide a substantial amount of water resources to the river basin than the other sections of the river [30]. The South African government has embarked on protecting these areas, and studies that demarcate and quantify the water resources potential of these areas are crucial [30]. One expert corroborated this and emphasized the need to carry out the e-flow determination in the LRC and its importance to the national agenda on WSAs. What the research drew from the consultations with experts was that though the criteria are important for determining study sites, input from people familiar with the characteristics of the area is invaluable.

6. Conclusions

Understanding the impacts of climate change on water resources and how these influence rural livelihoods is of great value to sustainable development. With the projections that rural populations are more vulnerable to the impacts of climate change, there is a need for science to aid adaptation in these communities. Biospheres provide the opportunity to test the man and environment relationship well. As such, this research sought to determine the best-case study to implement an e-flow determination on headwater catchments within the biosphere. The research, through limitations in funding, time, and human resources formulated a protocol to select the best-case study. Though simple, this case study can be modified accordingly and can be implemented in environmental studies in other areas. The study used a detailed review of relevant literature, a simple factor scoring approach, reliance on expert opinions, and ground-truthing to select the best study site to determine e-flows for the LRC. Several benefits are derived when using a systematic approach to site selection. Environmental research relies on data of sufficient quantity and quality. The availability or lack of this data enables researchers to formulate the most appropriate methodology to implement in their studies. A properly formulated methodology saves time in its implementation. Frequently, environmental research goes beyond the anticipated time as unprecedented hindrances are met during the study. This is particularly true for scholars who usually have limited time, a tight budget, and are on a learning curve. Implementing this systematic site selection approach before they begin their studies will inform them on what is and what is not achievable. At the project level, a proper site selection protocol saves on resources. Southern Africa is limited in its adaptation capacity because of limited resources. Each research carried out in this area should be within a small range on the uncertainty spectrum. Most countries in the region cannot afford to implement research work that drags on for longer than anticipated because of limited resources. The site selection protocol in this study selected headwater catchments that are populated by rural communities and have hydrometeorological data scarcity issues. This was undertaken mainly because most catchments in southern Africa face these challenges as alluded to by [47]. With the technical guidance and the knowledge to generate data, this study is expected to generate new information for the area and the methods used can be transferred to other similar areas in the region. As protocols to study marginalized areas become available, adaptation strategies that cater to all can be achieved. This article emphasizes the importance of carrying out a careful and systematic study site selection as the initial undertaking in any environmental assessment. Such would provide intelligence on the study's shared challenges and outcomes and, thus, provide a plausible methodology to overcome the challenges and achieve the expected outcomes.

Author Contributions: A.D.G.–Conceptualization A.D.G. and E.K.; Writing—original draft, A.D.G. and E.K.; Writing—review & editing, A.D.G., E.K. and F.I.M.; Supervision, E.K. and F.I.M. All authors have read and agreed to the published version of the manuscript.

Funding: This research was funded by South Africa-Flanders Climate Change Adaptation Research and Training Partnership: building the adaptation knowledge and capacity base.

Institutional Review Board Statement: Not applicable.

Informed Consent Statement: Not applicable.

Data Availability Statement: Not applicable.

Conflicts of Interest: The authors declare no conflict of interest.

References

1. De Bruin, W.B.; Morgan, M.G. Reflections on an interdisciplinary collaboration to inform public understanding of climate change, mitigation, and impacts. *Proc. Natl. Acad. Sci. USA* **2019**, *116*, 7676–7683. [CrossRef] [PubMed]
2. Meehl, G.A.; Zwiers, F.; Evans, J.; Knutson, T.; Mearns, L.; Whetton, P. Trends in Extreme Weather and Climate Events: Issues Related to Modeling Extremes in Projections of Future Climate Change. *Bull. Am. Meteorol. Soc.* **2000**, *81*, 427–436. [CrossRef]
3. Swain, D.L.; Wing, O.E.J.; Bates, P.D.; Done, J.M.; Johnson, K.A.; Cameron, D.R. Increased flood exposure due to climate change and population growth in the United States. *Earth's Future* **2020**, *8*, 1–17. [CrossRef]
4. Niang, I.; Ruppel, O.C.; Abdrabo, M.A.; Essel, A.; Lennard, C.; Padgham, J.; Urquhart, P. Africa. In *Climate Change 2014: Impacts, Adaptation, and Vulnerability. Part B: Regional Aspects. Contribution of Working Group II to the Fifth Assessment Report of the Intergovernmental Panel on Climate Change*; Cambridge University Press: Cambridge, UK, 2014.
5. International Panel on Climate Change (IPCC). Summary for policymakers. In *Climate Change 2014: Impacts, Adaptation, and Vulnerability. Part A: Global and Sectoral Aspects. Contribution of Working Group II to the Fifth Assessment Report of the Intergovernmental Panel on Climate Change*; Core Writing Team, Pachauri, R.K., Meyer, L.A., Eds.; Cambridge University Press: Cambridge, UK; New York, NY, USA, 2014; pp. 1–32.
6. Thomas, K.; Hardy, R.D.; Lazrus, H.; Mendez, M.; Orlove, B.; Rivera-Collazo, I.; Roberts, T.; Rockman, M.; Warner, B.P.; Winthrop, R. Explaining differential vulnerability to climate change: A social science review. *WIREs Clim. Change* **2019**, *10*, 1–18. [CrossRef]
7. Phuong, L.T.H.; Wals, A.; Sen, L.T.S.; Hoa, N.Q.; Phan Van Lu, P.V.; Biesbroek, R. Using a social learning configuration to increase Vietnamese smallholder farmers' adaptive capacity to respond to climate change. *Local Environ.* **2018**, *23*, 879–897. [CrossRef]
8. Baudoin, M.A.; Ziervogel, G. What role for local organisations in climate change adaptation? Insights from South Africa. *Reg. Environ. Change* **2017**, *17*, 691–702. [CrossRef]
9. McKenzie, M. Climate change education and communication in global review: Tracking progress through national submissions to the UNFCCC Secretariat. *Environ. Educ. Res.* **2021**, *27*, 631–651. [CrossRef]
10. Adenle, A.A.; Ford, J.D.; Morton, J.; Twomlow, S.; Alverson, K.; Cattaneo, A.; Cervigni, R.; Kurukulasuriya, P.; Huq, S.; Helfgott, A.; et al. Managing Climate Change Risks in Africa—A Global Perspective. *Ecol. Econ.* **2017**, *141*, 190–201. [CrossRef]

11. Tumushabe, J.T. Climate Change, Food Security and Sustainable Development in Africa. In *The Palgrave Handbook of African Politics, Governance and Development*; Oloruntoba, S., Falola, T., Eds.; Palgrave Macmillan: New York, NY, USA, 2018.
12. Ziervogel, G. Building transformative capacity for adaptation planning and implementation that works for the urban poor: Insights from South Africa. *Ambio* **2019**, *48*, 494–506. [CrossRef]
13. Filho, W.L.; Balogun, A.L.; Olayide, O.E.; Azeiteiro, U.M.; Ayal, D.Y.; Muñoz, D.C.; Nagy, G.J.; Bynoe, P.; Oguge, O.; Toamukum, N.Y.; et al. Assessing the impacts of climate change in cities and their adaptive capacity: Towards transformative approaches to climate change adaptation and poverty reduction in urban areas in a set of developing countries. *Sci. Total Environ.* **2019**, *692*, 1175–1190. [CrossRef]
14. Frederick, K.D.; Major, D.C. Climate Change and Water Resources. *Clim. Change* **1997**, *37*, 7–23. [CrossRef]
15. Anthony, F. Potential climate change impacts on water resources in the Auckland Region (New Zealand). *Clim. Res.* **1999**, *11*, 221–245.
16. Zhang, C.; Sun, L.; Han, T.; Wan, X.; Jiang, Y. Climate Change Impacts on Water Resources and Ecological Environment of Inland Watersheds in the Hexi Region of West Gansu, China. In Proceedings of the 2009 International Conference on Management and Service Science, Beijing, China, 20–22 September 2009; pp. 1–4.
17. Nóbrega, M.T.; Collischonn, W.; Tucci, C.E.M.; Paz, A.R. Uncertainty in climate change impacts on water resources in the Rio Grande Basin, Brazil. *Hydrol. Earth Syst. Sci.* **2011**, *15*, 585–595. [CrossRef]
18. Estrela, T.; Pérez-Martin, M.A.; Vargas, E. Impacts of climate change on water resources in Spain. *Hydrol. Sci. J.* **2012**, *57*, 1154–1167. [CrossRef]
19. Gayar, A.E.; Hamed, Y. Climate Change and Water Resources Management in Arab Countries. In *Recent Advances in Environmental Science from the Euro-Mediterranean and Surrounding Regions; Advances in Science, Technology & Innovation (IEREK Interdisciplinary Series for Sustainable Development)*; Kallel, A., Ksibi, M., Ben Dhia, H., Khélifi, N., Eds.; Springer: Cham, Switzerland, 2018.
20. Tramblay, Y.; Llasat, M.C.; Randin, C.; Coppola, E. Climate change impacts on water resources in the Mediterranean. *Reg. Environ. Change* **2020**, *20*, 1–3. [CrossRef]
21. Greenwood, D.; Zhang, K.; Hilton, H.; Thompson, A. Opportunities for improving irrigation efficiency with quantitative models, soil water sensors and wireless technology. *J. Agric. Sci.* **2010**, *148*, 1–16. [CrossRef]
22. Beven, K. How to make advances in hydrological modelling. *Hydrol. Res.* **2019**, *50*, 1481–1494. [CrossRef]
23. Chang, F.J.; Guo, S. Advances in hydrologic forecasts and water resource management. *Water* **2020**, *12*, 1819. [CrossRef]
24. Swart, R.; Biesbroek, R.; Lourenco, T.C. Science of adaptation to climate change and science for adaptation. *Front. Environ. Sci.* **2014**, *2*, 1–8. [CrossRef]
25. UNESCO. Vhembe Biosphere Reserve, South Africa. 2021. Available online: https://en.unesco.org/biosphere/africa/vhembe (accessed on 5 November 2021).
26. Hedden-Dunkhorst, B.; Schmitt, F. Exploring the potential and contribution of UNESCO Biosphere Reserves for landscape governance and management in Africa. *Land* **2020**, *9*, 237. [CrossRef]
27. Dinku, T. Challenges with availability and quality of climate data in Africa. In *Extreme Hydrology and Climate Variability*; Assefa, M.M., Wossenu, A., Gabriel, S., Eds.; Elsevier: Amsterdam, The Netherlands, 2019.
28. Roth, V.; Lemann, T.; Zeleke, G.; Subhatu, A.T.; Nigussie, T.K.; Hurni, H. Effects of climate change on water resources in the upper Blue Nile Basin of Ethiopia. *Heliyon* **2018**, *4*, e00771. [CrossRef] [PubMed]
29. Gumbo, A.D.; Kapangaziwiri, E.; Chikoore, H.; Pienaar, H.; Mathivha, F. Assessing water resources availability in headwater subcatchments of Pungwe River Basin in a changing climate. *J. Hydrol. Reg. Stud.* **2021**, *35*, 1–13.
30. Nel, J.; Colvin, C.; Le Maitre, D.; Smith, J.; Haines, I. *South Africa's Strategic Water Source Areas*; CSIR/NRE/ECOS/ER/2013/0031/A; WWF-South Africa: Pretoria, South Africa, 2013.
31. Nel, J.L.; Le Maitre, D.C.; Roux, D.J.; Colvin, C.; Smith, J.S.; Smith-Adao, L.B.; Maherry, A.; Sitas, N. Strategic water source areas for urban water security: Making the connection between protecting ecosystems and benefiting from their services. *Ecosyst. Serv.* **2017**, *28 Pt B*, 251–259. [CrossRef]
32. Crowe, S.; Cresswell, K.; Robertson, A.; Huby, G.; Avery, A.; Sheikh, A. The case study approach. *BMC Med. Res. Methodol.* **2011**, *11*, 100. [CrossRef] [PubMed]
33. Ponelis, S.R. Using interpretive qualitative case studies for exploratory research in doctoral studies: A case of information systems research in small and medium enterprise. *Int. J. Dr. Stud.* **2015**, *10*, 535–550. [CrossRef]
34. UNESCO. Biosphere Reserves in Africa. 2021. Available online: https://en.unesco.org/biosphere/africa (accessed on 8 November 2021).
35. VBR. Vhembe Biosphere. 2021. Available online: https://www.vhembebiosphere.org/ (accessed on 8 May 2021).
36. Oyinlola, M.; Whitehead, T.; Abuzeina, A.; Adefila, A.; Akinola, Y.; Fatai Anafi, F.; Farukh, F.; Jegede, O.; Kandan, K.; Kim, B.; et al. Bottle house: A case of transdisciplinary research for tackling challenges. *Habitat Int.* **2018**, *79*, 18–29. [CrossRef]
37. Wiebe, K.; Robinson, S.; Cattaneo, A. Climate Change, Agriculture and Food Security: Impacts and the Potential for Adaptation and Mitigation. In *Sustainable Food and Agricultur*; Campanhola, C., Pandey, S., Eds.; Elsevier Inc.: Amsterdam, Netherlands, 2019.
38. Heyard, R.; Hottenrott, H. The value of research funding for knowledge creation and dissemination: A study of SNSF Research Grants. *Humanit. Soc. Sci. Commun.* **2021**, *8*, 217. [CrossRef]
39. Crane, T.A.; Roncoli, C.; Hoogenboom, G. Adaptation to climate change and climate variability: The importance of understanding agriculture as performance. *NJAS—Wagening. J. Life Sci.* **2011**, *57*, 179–185. [CrossRef]

40. Malhi, Y.; Franklin, J.; Seddon, N.; Solan, M.; Turner, M.G.; Field, C.B.; Knowlton, N. Climate change and ecosystems: Threats, opportunities, and solutions. *Philos. Trans. R. Soc. B* **2020**, *375*, 1–8. [CrossRef]
41. Regmi, K.; Jones, L. A systematic review of the factors—Enablers and barriers—Affecting e-learning in health sciences education. *BMC Med. Educ.* **2020**, *20*, 91. [CrossRef]
42. Brocca, L.; Massari, C.; Pellarin, T.; Filippucci, P.; Ciabatta, L.; Camici, S.; Kerr, Y.H.; Fernandez-Preto, D. River flow prediction in data scarce regions: Soil moisture integrated satellite rainfall products outperform rain gauge observations in West Africa. *Sci. Rep.* **2020**, *10*, 1–14. [CrossRef]
43. Pandya, U.; Patel, D.P.; Singh, S.K. A flood assessment of data scarce region using an open source 2D hydrodynamic modeling and Google Earth Image: A case of Sabarmati flood, India. *Arab. J. Geosci.* **2021**, *14*, 1–18. [CrossRef]
44. De Paula, M., Jr.; Costa Silva, T.A.; Araújo, A.S.; Lacorte, G.A. Assessments of Bacterial Community Shifts in Sediments along the Headwaters of São Francisco River, Brazil. *Conservation* **2021**, *1*, 91–105. [CrossRef]
45. Koch, J.C.; Sjöberg, Y.; O'Donnell, J.A.; O'Donnell, J.A.; Carey, M.P.; Sullivan, P.F.; Terskaia, A. Sensitivity of headwater streamflow to thawing permafrost and vegetation change in a warming Arctic. *Environ. Res. Lett.* **2022**, *17*, 044074. [CrossRef]
46. Ramulifho, P.; Ndou, E.; Thifhulufhelwi, R.; Dalu, T. Challenges to Implementing an Environmental Flow Regime in the Luvuvhu River Catchment, South Africa. *Int. J. Environ. Res. Public Health* **2019**, *16*, 3694. [CrossRef] [PubMed]
47. Hughes, D.A.; Kapangaziwiri, E.; Sawunyama, T. Hydrological model uncertainty assessment in southern Africa. *J. Hydrol.* **2010**, *387*, 221–232. [CrossRef]

water

MDPI

Article

Assessment of Daily of Reference Evapotranspiration Using CLDAS Product in Different Climate Regions of China

Li-Feng Wu [1,2,†], Long Qian [3,†], Guo-Min Huang [1], Xiao-Gang Liu [3,*], Yi-Cheng Wang [2], Hua Bai [1] and Shao-Fei Wu [1]

1 School of Hydraulic and Ecological Engineering, Nanchang Institute of Technology, Nanchang 330099, China; china.sw@163.com (L.-F.W.); g.huang@nit.edu.cn (G.-M.H.); baihua1985@126.com (H.B.); sfw17@nit.edu.cn (S.-F.W.)
2 State Key Laboratory of Simulation and Regulation of Water Cycle in River Basin, China Institute of Water Resources and Hydropower Research, Beijing 100038, China; wangych@iwhr.com
3 Faculty of Agriculture Engineering, Kunming University of Science and Technology, Kunming 650500, China; qianlong@stu.kust.edu.cn
* Correspondence: liuxiaogangjy@126.com
† These authors contributed equally to this work.

Abstract: Reference Crop evapotranspiration (ET_0) datasets based on reanalysis products can make up for the time discontinuity and the spatial insufficiency of surface meteorological platform data, which is of great significance for water resources planning and irrigation system formulation. However, a rigorous evaluation must be conducted to verify if reanalysis products have application values. This study first evaluated the ability of the second-generation China Meteorological Administration Land Data Assimilation System (CLDAS) dataset for officially estimating ET_0 (the local meteorological station data is used as the reference dataset). The results suggest that the temperature data of CLDAS have high accuracy in all regions except the Qinghai Tibet Plateau (QTP) region. In contrast, the global solar radiation data accuracy is fair, and the relative humidity and wind speed data quality are poor. The overall accuracy of ET_0 is acceptable other than QTP, but there are also less than 15% (103) of stations with significant errors. In terms of seasons, the error is largest in summer and smallest in winter. Additionally, there are inter-annual differences in the ET_0 of this data set. Overall, the CLDAS dataset is expected to have good applicability in the Inner Mongolia Grassland area for estimating ET_0, Northeast Taiwan, the Semi Northern Temperate zone, the Humid and Semi Humid warm Temperate zone, and the subtropical region. However, there are certain risks in other regions. In addition, of all seasons, summer and spring have the slightest bias, followed by autumn and winter. From 2017 to 2020, bias in 2019 and 2020 are the smallest, and the areas with large deviation are south of climate zone 3, the coastal area of climate zone 6, and the boundary area of climate zone 7.

Keywords: raw reanalysis data; grid data; reference evapotranspiration; meteorological variables; Penman-Monteith equation

Citation: Wu, L.-F.; Qian, L.; Huang, G.-M.; Liu, X.-G.; Wang, Y.-C.; Bai, H.; Wu, S.-F. Assessment of Daily of Reference Evapotranspiration Using CLDAS Product in Different Climate Regions of China. *Water* **2022**, *14*, 1744. https://doi.org/10.3390/w14111744

Academic Editor: Alban Kuriqi

Received: 9 April 2022
Accepted: 26 May 2022
Published: 29 May 2022

1. Introduction

Reference Crop evapotranspiration (ET_0) is a critical factor for calculating crop evapotranspiration, the accurate estimation of which plays a vital role in irrigation engineering design and planning, water resources management, and climate change research [1–3]. Due to its large population and rapid economic development, China is facing a severe water shortage problem. The country's per capita water resource is only one-fourth of the world average level [4]. Therefore, an accurate estimation of ET_0 in this region would provide a scientific basis for rationally allocating water resources and minimizing the imbalance between water supply and demand [5]. Currently, the standard estimating method of ET_0 is the Penman-Monteith equation (FAO56 PM) recommended by the Food and Agriculture Organization of the United Nations (FAO) [6,7]. This method combines energy balance and the aerodynamic theory, which is strongly applicable under different climatic conditions. However, the main drawback of this method is that it requires a high quality of meteorological data, including air temperature, relative humidity or dew temperature, solar radiation, and wind speed [8,9]. In many regions of the world, there are not enough weather stations to monitor the meteorological factors. Additionally, high-quality, long-term observational data are lacking, especially in developing countries, which hinders the application of the PM method for ET_0 estimation on large spatial scales [10–12].

In recent years, reanalysis products have become one of the main grid data sources for water resource management research [13]. Reanalysis data are generated by running a numerical weather-predicting model that assimilates the observed atmospheric and surface data to reconstruct the past surface, ocean, and atmospheric state variables. Unlike geostatistical grid data derived from spatial interpolation, the spatial structure of weather variables (such as temperature and wind speed) synthesizes physical laws embedded in numerical models [8].

Nowadays, many reanalysis data sets have developed rapidly and are used in various fields. Baatz et al. (2020) [14] analyzed state-of-the-art methods, recent developments, and prospects of reanalysis for three subcomponents of the Earth system (atmosphere, ocean, and land), they points out the method's increasing computational capabilities, the growing availability of long-term satellite data with global coverage, and the advancements in model-data fusion methods such as variational and sequential data assimilation. In addition, the above paper discusses the increasing awareness of the drastic changes in the Earth system related to anthropogenic and climatic factors and the way they drive reanalysis development. Recently, networks of distributed in-situ sensors such as buoys and biogeochemical Argo floats [15], eddy covariance stations [16], surface water runoff observations [17], and meteorological station data [18] were used in the reanalysis of physical and biogeochemical Earth system processes. Munoz-Sabater et al. (2021) [19] presented the new global ERA5-Land reanalysis. The quality of ERA5-Land fields was evaluated by direct comparison to many in situ observations collected for the period 2001–2018, and for comparison to additional model or satellite-based global reference datasets. Overall, the water cycle was improved in ERA5-Land compared to ERA5 according to the different variables evaluated, whereas the energy cycle variables showed similar performances. Both ERA5 and ERA5-Land perform substantially better than ERA-Interim.

Reanalysis data have also been applied and compared to estimate evapotranspiration in different regions of the world. Boulard et al. (2016) [20] calculated ET_0 using the ERA-Interim reanalysis data and verified its accuracy in a water balance study in northeastern France. Srivastava et al. (2016) [21] found that ERA-Interim ET_0 was superior to NCEP/NCAR ET_0 in the UK. Pelosi et al. (2020) [22] also compared two reanalysis datasets for ET_0 estimation in southern Italy. Woldesenbet et al. (2021) [23] evaluated the ET_0 in the Omo-Gibetta watershed and achieved good prediction results. Song et al. (2015) [24] judged the spatiotemporal characteristics of ET_0 in the Shaanxi Province based on NCEP reanalysis data and made future predictions. Liu et al. (2019) [25] estimated the future ET_0 in the Poyang Lake basin based on the CMIP5 model. The results showed that the stepwise regression downscaling model established by the NCEP reanalysis data and the basin ET_0 had better simulation results. ET_0 was assessed in the Iberian Peninsula by Martins et al. (2016) [26]. The focus here is to use the National Center for Environmental Prediction/National Center for Atmospheric Research (NCEP/NCAR) hybrid reanalysis product and gridded dataset to calculate ET_0 with good simulation results. Raziei (2021) [27] used the National Center for Environmental Prediction/National Center for Atmospheric Research (NCEP/NCAR) reanalysis, combined with a gridded dataset, to calculate monthly ET_0 for 43 meteorological stations distributed across Iran. The results show that the ET_0 calculated by the mixed reanalysis had a better effect than the ET_0 calculated by the observations at most research stations. Milad and Mehdi (2022) [28] used reanalysis products to estimate ET_0 in areas with sparse data and showed that ERA5 provided more accurate estimates of daily and monthly ET_0. Some scholars have also compared satellite grid data with meteorological station values. Wang et al. (2019) [29] comprehensively evaluated and compared this newly released precipitation product (Integrated Multi-satellite Retrievals V05B) and its predecessor TRMM 3B42V7 based upon the ground-based observations under complex topographic and climatic conditions over the Hexi Region in the northwest arid region of China. Their results indicated that compared to ground-based observations, both IMERG and 3B42V7 showed good performance with slight overestimation. Prakash et al. (2016) [30] investigated the capabilities of the Tropical Rainfall Measuring Mission (TRMM), Multi-satellite Precipitation Analysis (TMPA), and the recently released Integrated Multi-satellitE Retrievals for GPM (IMERG) in detecting and estimating heavy rainfall across India. The results indicated that the multi-satellite product systematically overestimates its inter-annual variations. With continuous advances in numerical weather models, computing, information, and communication technology (ict) tools, and data assimilation techniques, along with continuous improvements in the quality of atmospheric and ground data obtained from satellites, the spatial and temporal resolution and reliability of reanalysis data have been gradually improved year after year.

The China Meteorological Administration Land Data Assimilation System (CLDAS) is the only real-time service system in land surface data assimilation systems in China. It uses a combined technology of integration and assimilation to fuse data from various sources, such as ground observation, satellite observation, and numerical model products [31]. The output of this system includes high spatial and temporal resolution land surface driving products such as temperature, air pressure, specific humidity, wind speed, precipitation, solar shortwave radiation, and soil moisture. These could be applied in agricultural drought monitoring, mountain flood geological disaster meteorological services, climate system model assessments, and spatial fine grid real data services. Although many studies have evaluated the quality of the CLDAS data, there are limited reports on the estimation of ET_0 by this dataset. In this paper, we used the meteorological reanalysis data of 689 surface meteorological stations in China from 2017 to 2020 and found four grid data points around each meteorological station through calculation and processing. We then calculated the value of the target station using the inverse distance weight method, compared it with the measured data of local meteorological stations and evaluated the accuracy of CLDAS data through statistical indicators. Therefore, this study aims to evaluate the accuracy of ET_0 simulation with CLDAS products for the first time by comparing meteorological data from 689 ground weather stations and to exploring a product that could provide accurate ET_0 for areas lacking meteorological data observation.

2. Materials and Methods

2.1. Introduction to CLDAS

CLDAS is an isolatitude and longitude mesh fusion analysis product covering the Asian region (0–65° N, 60–160° E) with a resolution of 0.0625° × 0.0625° and 1 h, and a spatial geometric distance of 9 km between the corresponding grid points, including six variables: 2 m air temperature, 2 m specific humidity, 10 m wind speed, surface pressure, precipitation, and shortwave radiation. This dataset is developed by using the Space and Time Mesoscale Analysis System (STMAS), optimal interpolation (OI), probability density function matching (CDF), physical inversion, terrain correction, and other techniques based on ground and satellite observations from a variety of sources. The dataset's quality and spatio-temporal resolution in China is better and higher than in the international market. The scientific goal of CLDAS is to use data fusion and assimilation technology, on the ground observation, satellite observations, numerical model products, and other sources of data fusion to obtain high space-time resolution and high-quality temperature, pressure, humidity, wind speed, precipitation, and radiation elements such as lattice data to drive the land surface model, obtain soil temperature and humidity, etc. The research focuses on processing and acquiring land surface driving data, realizing the operation and integration of multiple land surface models, and improving the underlying surface data, vegetation parameters, and atmospheric driving data.

2.1.1. Data Sources for CLDAS

(1) Ground observation data: hourly temperature, air pressure, humidity, wind speed, precipitation, and other data observed by more than 2400 national automatic weather stations and nearly 40,000 regional automatic weather stations after quality control.

(2) ECMWF (European Center for Mediumrange Weather Forecasts) numerical analysis/forecast products: global 3 h, 0.125° resolution 2 m temperature, 2 m humidity, 10 mU/V wind speed, surface pressure, and other data products released by EC (European Center).

(3) GFS numerical analysis/prediction products: NCEP released global ozone, atmospheric precipitation, surface pressure, and other data products with 3 h and 0.5°.

(4) Satellite precipitation products: FY2 precipitation estimation products (nominal disk chart) of the National Satellite Meteorological Center; East Asia Multi-Satellite Integrated Precipitation Data Product (EMSIP) with a resolution of 1 h and 0.0625° for the Asian region operated by the National Meteorological Information Center.

(5) Fusion precipitation product: the fusion product of FY2/CMORPH precipitation and automatic ground station precipitation with 1 h and 0.1° resolution in China operated by the National Meteorological Information Center.

(6) FY2 satellite entire disk nominal map: multi-channel geostationary satellite observation data with 1 h and 5 km resolution (subsatellite point) of the Service of National Satellite Meteorological Center (nominal disk map).

(7) DEM data: a global 30m spatial resolution topographic data product jointly measured by NASA (National Aeronautics and Space Administration) and NIMA (National Bureau of Surveying and Mapping of the Ministry of Defense) was used to re-sample DEM topographic data with a spatial resolution of 0.0625° in the Asian region using the area weight method.

2.1.2. CLDAS Data Processing Methods

The 2 m temperature, 2 m specific humidity, 10 m wind speed, and surface pressure products take ECMWF numerical analysis/forecast products as the background field. Topographic adjustment and multi-grid variational technology (STMAS) are used to integrate the observation data of automatic ground stations in China. The background field outside China is formed by topographic adjustment, variable diagnosis, and interpolation to the analysis grid.

The DISORT radiative transfer model used ozone, atmospheric precipitation, and surface pressure in GFS numerical analysis products as the dynamic input parameters for the radiative transfer model. Additionally, FY2E/G satellite VIS channel complete disk nominal map data inversion was used to form the short-wave radiation product.

The above information comes from the China Meteorological Data Sharing Network (https://data.cma.cn/, accessed on 1 May 2021). The data used in this study included temperature, global solar radiation, relative humidity, and wind speed, of which the height of wind speed was 10 m and the height of other meteorological variables was 2 m. The data spanned from 2017 to 2020.

2.2. *Reference Evapotranspiration*

According to the FAO56 PM equation [32], reference evapotranspiration (ET_0; mm d^{-1}) can be calculated as:

$$ET_0 = \frac{0.408(R_n - G) + \gamma \frac{900}{T_a+273} U(e_s - e_a)}{\Delta + \gamma(1 + 0.34U)} \quad (1)$$

where R_n is the net radiation at the crop surface, usually calculated by R_s (Global solar radiation); G is the soil heat flux density; T_a is the mean daily air temperature at 2 m height; U is the wind speed at 2 m height; e_s and e_a are the saturation and actual vapor pressure, respectively; Δ is the slope of vapor pressure curve, and γ is the air psychrometric constant. In daily time-step in this study G can be neglected [33,34].

2.3. Data Sources

To examine the performance of this dataset, meteorological data from 689 ground meteorological observation stations of the China Meteorological Administration (CMA) were collected, which included maximum and minimum temperatures at 2 m, global surface radiation or sunshine durations, relative humidity at 2 m, and wind speed at 10 m. If necessary, sunshine durations were converted into global radiation using a formula from a previous study [35]. The stations were divided into seven climate zones [36,37]. The specific distribution is shown in Figure 1, and the names are shown in Table 1.

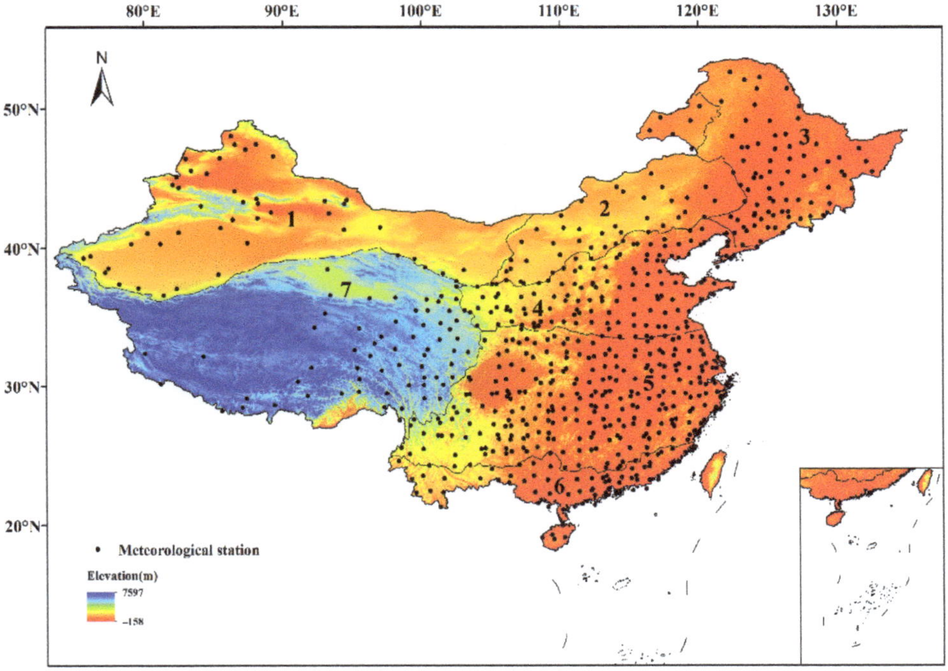

Figure 1. Climate zones of China and geographical distribution of 689 meteorological stations. (See Table 1 for the names of climatic zones 1–7).

Table 1. Names of the seven climate zones.

Zone	Area Name
1	Northwest desert zone
2	Inner Mongolia grassland zone
3	Northeast humid and semi humid temperate zone
4	Humid and semi humid warm temperate zone
5	Humid subtropical zone
6	Humid tropical zone
7	Qinghai Tibet Plateau zone

To obtain the daily reanalysis variables for Equation (1) (identified by subscript CLD), the following steps were taken: (a) daily T_{maxCLD} and T_{minCLD} were selected as the maximum and minimum of the 24 daily available 1-h values of the T_{max} and T_{min} sequences, respectively; (b) daily RH_{CLD} was obtained by calculating the 24-h average value of 24 RH values per day; (c) calculating the 24-h cumulative value of the 12-h R_s as the daily R_{sCLD} value; (d) wind speed at 10 m (U_{10CLD}) was calculated as the 24-h average of 24 1-h values, which were then converted to a height of 2 m (U_{CLD}) using Formula (2) as follows, respectively:

$$U = U_z \frac{4.87}{\ln(67.8z - 5.42)} \quad (2)$$

where z is the height of the wind speed observation instrument (in this paper, z is equal to 10) for each meteorological station. Grid data from four grid points around it were selected and interpolated to the station by the inverse distance weight (IDW) method. The formula is as follows:

$$V = \frac{\sum_{i=1}^{n} \frac{v_i}{D_i^2}}{\sum_{i=1}^{n} \frac{1}{D_i^2}} \quad (3)$$

where V is the inverse value, v_i is the value of the control point, and D_i is the weight coefficient.

2.4. Statistics Indicators

Three common statistical indicators, including the coefficient of determination (R^2), root mean square error (RMSE), mean absolute error (MAE), and percent bias (PBias) were chosen to evaluate the accuracy of the CLDAS meteorological variables and ET_0 in this study. The corresponding formulas are:

$$\mathrm{MAE} = \frac{1}{n}\sum_{i=1}^{n}|M_i - P_i| \quad (4)$$

$$\mathrm{RMSE} = \sqrt{\frac{1}{n}\sum_{i=1}^{n}(M_i - P_i)^2} \quad (5)$$

$$R^2 = \frac{\left[\sum_{i=1}^{n}\left(M_i - M_i\right)\left(P_i - P_i\right)\right]^2}{\sum_{i=1}^{n}(M_i - M_i)^2 \sum_{i=1}^{n}(P_i - P_i)^2} \quad (6)$$

$$\mathrm{PB} = \frac{\sum_{i=1}^{n}(P_i - M_i)}{\sum_{i=1}^{n} M_i} \quad (7)$$

where M_i is ET_0 calculated by meteorological station data, P_i is ET_0 calculated by the CLDAS gridded data, M_i is average ET_0 calculated by meteorological station data, P_i is average ET_0 calculated by the CLDAS gridded data, and n is the number corresponding to ET_0 data. Higher R^2 values (closer to 1) or lower RMSE and MAE values indicate a better estimation performance of the CLDAS dataset. The closer PB is to 0, the better the estimation performance of the CLDAS dataset.

3. Results

3.1. Meteorological Factors

3.1.1. Air Temperature

Table 2 shows the statistical indicators of maximum and minimum temperatures in the CLDAS data for the seven climate zones in China. Results indicated that the accuracy for the maximum and minimum temperatures differed in different climatic regions. For the maximum temperature, CLDAS data showed a high correlation with data from ground stations in the four northern climate zones (i.e., climate zones 1–4), with R^2 larger than 0.9. Climate zone 5 in the humid climate region also yielded a good correlation. In climate zones 6 and 7, the correlations between the two datasets were slightly worse when compared with other climate zones. However, climate zone 6 showed the smallest values in terms of statistical errors, with RMSE and MAE valued at 2.9 and 2.3 °C, respectively. This may be since the range of temperature changes in this region is not as large as that in other regions, and the area of this climate zone is significantly smaller than that in other climate zones, so the temperature change in this region is not as dramatic as that in other climate zones. The RMSE and MAE of the high-altitude climate zone (i.e., climate zone 7) were 6.55 °C and 5.83 °C, respectively. Figures 2–4 show the spatial error distribution of the maximum temperature in CLDAS. Overall, the errors at most stations were within a small range. However, in climate zone 7 and the north-central area of climate zone 1, there was a big error in the regions, with RMSE and MAE of many stations more significant than 10 °C, while R^2 was lower than 0.5. Such huge variations in model errors in these stations might be resulted from the regional climate model parameter variations and were unlikely caused by the overall overvalued or undervalued problem of models that may cause significant variation for ET_0 estimation.

Table 2. Statistical indicators of maximum and minimum temperatures in different climate zones of China.

Climate Zone	T_{max}			T_{min}		
	RMSE °C %	MAE °C %	R^2	RMSE °C %	MAE °C %	R^2
1	4.99	4.32	0.95	4.03	3.45	0.96
2	3.82	2.94	0.93	2.50	1.85	0.97
3	3.75	2.88	0.94	3.02	2.27	0.97
4	3.83	3.05	0.92	2.65	2.11	0.97
5	3.51	2.75	0.88	2.24	1.79	0.96
6	2.90	2.30	0.82	1.87	1.50	0.94
7	6.55	5.83	0.81	5.10	4.53	0.92

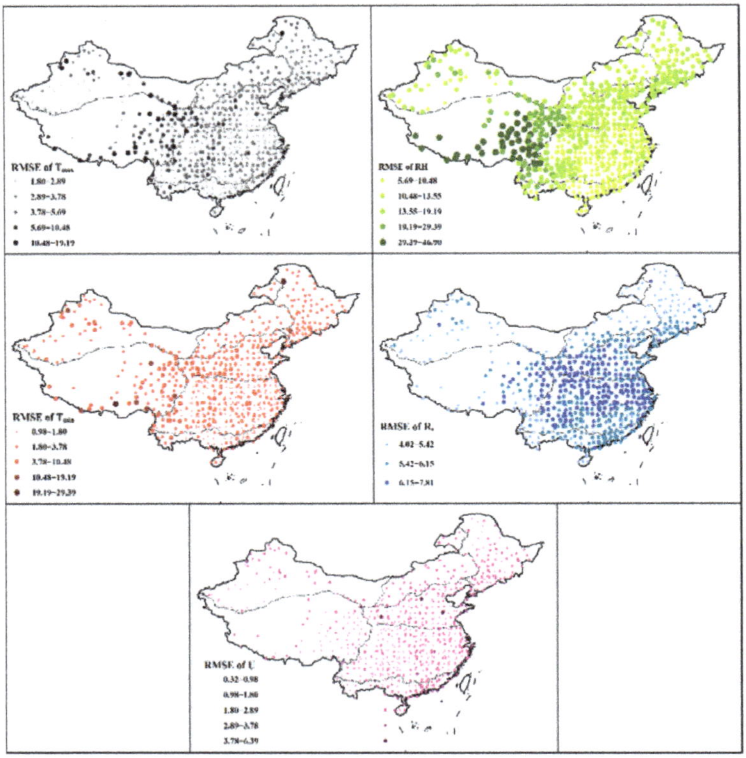

Figure 2. RMSE values of the five meteorological factors of in all stations.

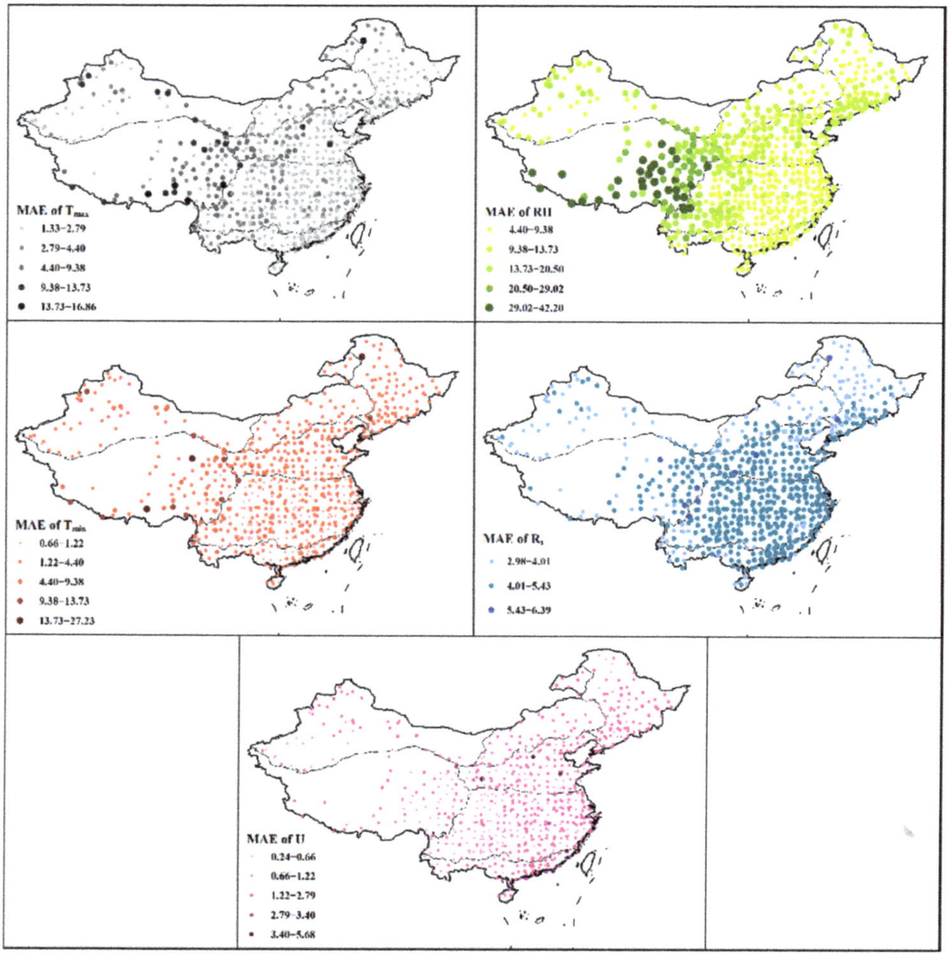

Figure 3. MAE values of the five meteorological factors of in all stations.

The minimum temperature behavior in the CLDAS data set was similar to the maximum temperature. However, compared with the maximum temperature, correlations between the minimum temperature of CLDAS and the station's temperature were higher in all seven climate zones, with all R^2 larger than 0.9. In climate zone 6, the minimum temperature error was the lowest among all climate zones, with RMSE and MAE valued at 1.87 °C and 1.5 °C, respectively. On the contrary, climate zone 7 had the highest error, with RMSE and MAE reaching 5.1 and 4.53 °C, respectively. In addition, the lowest temperature error in climate zone 1 was also relatively high, with RMSE and MAE valued at 4.03 °C and 3.45 °C, respectively, which might affect the accuracy of ET_0 estimation. As can be seen from Figures 2–4, the performance of T_{min} was similar to that of T_{max}. Therefore, the accuracy of most stations was within an acceptable range. However, some stations showed significant errors, which were mainly located in the middle of climate zone 1 and climate zone 7. These stations with high error in the minimum temperature had a high coincidence with corresponding high T_{max} error stations, indicating severe problems in the temperature simulation of the stations.

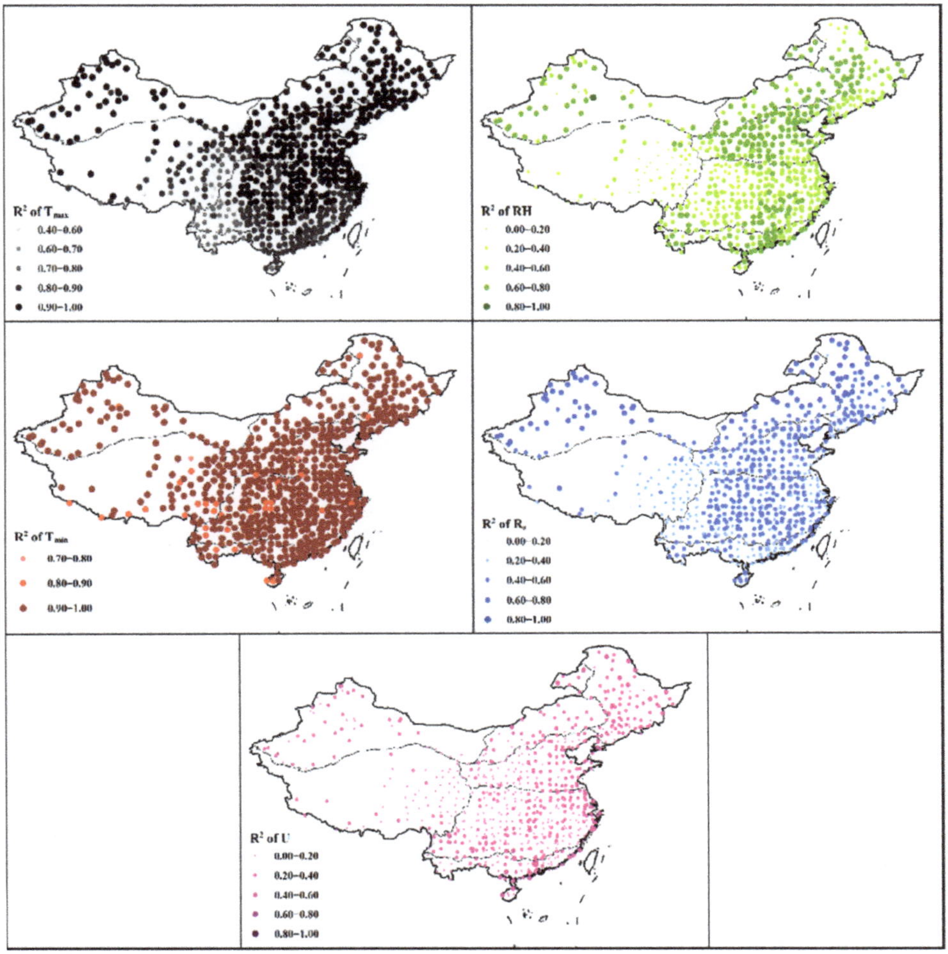

Figure 4. R^2 values of the five meteorological factors of in all stations.

3.1.2. Solar Radiation

Table 3 shows the statistical indicators of solar radiation (R_s) in the CLDAS data for the seven climate zones. Across climate zones, RMSE ranged from 5.18 to 6.21 MJ m^{-2} d^{-1}, and MAE ranged from 3.83 to 4.54 MJ m^{-2} d^{-1}. The differences in R_s errors among different climate regions were not as apparent as those in air temperature. However, the R^2 of climate zone 7 was significantly lower than that of other climatic regions. These results were similar to the results reported by Liu et al. (2009) [38]. However, their values were generally higher than that of the radiation model based on temperature, where the median RMSE was 3.3 MJ m^{-2} d^{-1} in humid regions of China (Fan et al., 2019) [39]. The above phenomenon indicated that the radiation data in the CLDAS data set did not perform well.

Figures 2–4 show the spatial distribution of *Rs* errors. Overall, the error of *Rs* in climate zones 1–3 was better than in other climate zones. The RMSE of most stations was more significant than 6 MJ m^{-2} d^{-1}. This might be due to the severe air pollution in the above areas [40], which would pose particular challenges to accurate simulation.

Table 3. Statistical indicators of solar radiation, relative humidity, and wind speed in different climate zones of China.

	R_s			RH			U		
Climate Zone	RMSE $MJ\ m^{-2}\ d^{-1}$	MAE $MJ\ m^{-2}\ d^{-1}$	R^2	RMSE %	MAE %	R^2	RMSE $m\ s^{-1}$	MAE $m\ s^{-1}$	R^2
1	5.18	3.83	0.55	14.93	12.11	0.59	1.18	0.88	0.22
2	5.41	3.93	0.65	12.60	9.79	0.55	1.32	1.01	0.29
3	5.33	3.93	0.55	12.81	9.88	0.50	1.28	1.00	0.30
4	6.21	4.51	0.60	13.62	10.73	0.44	1.26	1.01	0.20
5	6.13	4.54	0.48	11.90	9.72	0.42	1.05	0.85	0.21
6	5.54	4.19	0.59	9.19	7.43	0.39	1.38	1.13	0.22
7	5.80	4.43	0.28	31.29	27.62	0.33	1.03	0.81	0.14

3.1.3. Relative Humidity

Statistical indicators of CLDAS RH are shown in Table 3. Climate zone 7 had the most significant error among all regions, with RMSE and MAE reaching 31.29% and 27.62%, respectively, close to a random distribution. The consistency between CLDAS and site data in other climatic regions was also not high, with R^2 ranging from 0.39 to 0.59. However, the values of RMSE and MAE indicated that they were still within acceptable limits. Compared climate zone 1 with climate zone 6, the consistency of climate zone 1 was higher, but the RMSE and MAE of climate zone 6 are lower. This result could be attributed to the fact that climate zone 6 is within the humid region with high annual average relative humidity, while climate zone 1 is in the arid area where the relative humidity changes more sharply. Figures 2–4 showed that the overall error of this data set was larger in climate zone 7 than in other climate zones. In addition, compared with the northeast part of climate region 5, the error of RH in the western part of the same climate zone (i.e., areas bordering climate zone 7) was significantly larger. Although there was a relatively large error for RH in some regions, the estimation of ET_0 would unlikely be affected, as previous studies have found that RH would have a low contribution to ET_0 in most regions of China [41].

3.1.4. Wind Speed

Statistical indicators of CLDAS U are shown in Table 3. From the perspective of R^2, the consistency between CLDAS near-surface wind speed and station data was poor in all climate zones, while from the perspective of RMSE and MAE, their accuracies were acceptable. In addition, the mean difference between climate zones was within 30%. However, according to the spatial distribution of the error (Figures 2–4), the RMSE of some stations was more than 3 m s^{-1}, of which the errors of most stations in climate zone 7 were significant. Compared with the inland stations, the R^2 of the coastal stations was higher. However, RMSE and MAE were also higher, indicating a problem of overestimation or underestimation. The worldwide modeling for wind speed is challenging and often inaccurate. Similar results were obtained for ERA5 [22], NCEP/NCAR [27], and GLDAS [42]. This is mainly due to the complex terrain changes on the ground, and the wind speed is greatly affected by the roughness of the underlying surface. In addition, it is not easy to simulate the movement direction of winds accurately.

3.2. *Reference Evapotranspiration*

The statistical indicators of calculated ET_0 based on the CLDAS dataset are shown in Table 4. Among all climate zones, climate zone 1 had the best consistency (R^2 = 0.84) between CLDAS data and station data, while climate zone 3 showed the lowest errors (RMSE = 0.87 mm d^{-1} and MAE = 0.58 mm d^{-1}). For climate zone 7, the values of RMSE (1.37 mm d^{-1}) and MAE (1.19 mm d^{-1}) were higher than the corresponding values in any of the other climate zones. Figure 5 shows the spatial distribution of statistical indicators. Across climate zones, R^2 overall showed a decreasing trend from the north to the south, and the southernmost region (i.e., climate zone 6) had the lowest value of R^2. However, the spatial distributing patterns of RMSE and MAE were different from R^2. The stations with significant errors are mainly distributed west of climate zone 1, the coastal areas, and the boundary areas between climate zone 7 and other climate zones. This is mainly due to the more complex climate change between climate zones. In addition, the high wind speed error in the coastal areas often leads to a significant ET_0 error.

Table 4. Statistical indicators of reference evapotranspiration in different climate zones of China.

Zone	RMSE	MAE	R^2
1	1.10	0.74	0.84
2	0.94	0.63	0.80
3	0.87	0.58	0.74
4	1.03	0.75	0.71
5	0.99	0.72	0.64
6	1.08	0.83	0.52
7	1.37	1.19	0.62

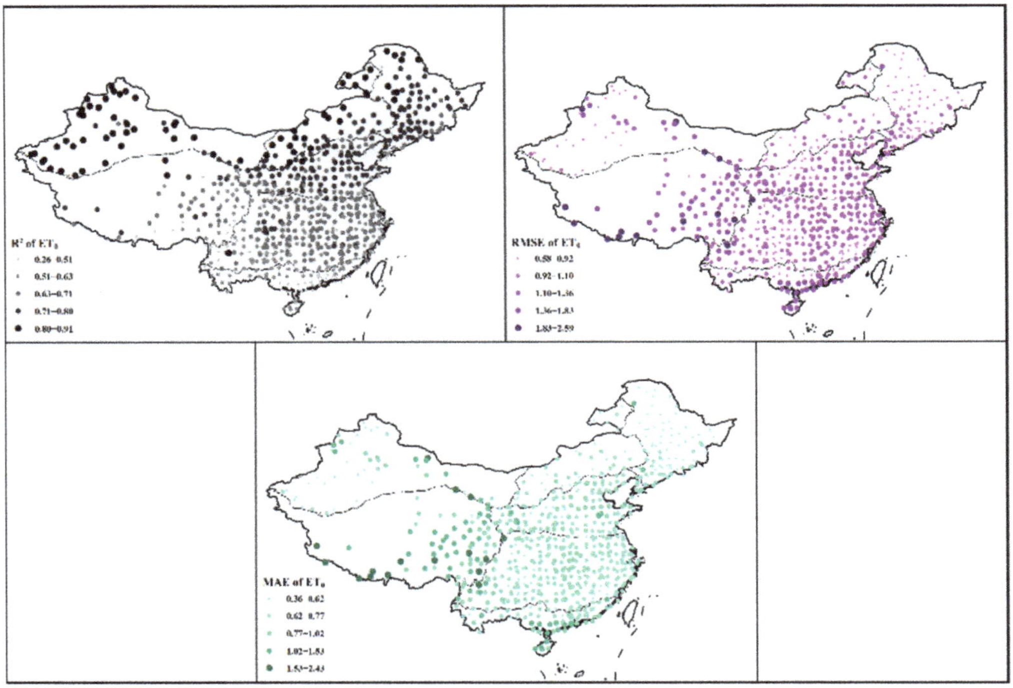

Figure 5. Spatial distribution of ET_0 statistical indicators.

To explore the differences in the CLDAS data in different climate regions, one station from each climate zone was randomly selected to fit the correlation between the calculated ET_0 based on CLDAS and the FAO56-PM ET_0 (Figure 6). Although there were a few outliers, the scatter points in climate zone 1 were more concentrated to the 1:1 line than those in other climate zones. Scatter points in climate region 2 were slightly more dispersed than in climate zone 1 and showed some obvious overestimations when ET_0 was more significant than 6. In climate zone 3, the accuracy was excellent when the value of ET_0 was low (<2 mm) but showed a decline as the following scatter points started to discretize. However, no overestimation or underestimation existed. In climate zone 4, the error was relatively large when ET_0 ranged from 3 mm to 6 mm. When the ET_0 of climate zone 5 was less than 2, the problem of underestimation appeared, and then the points were scattered in the 1:1 line for two measurements, but the distance from the 1:1 line was far. In climate zone 6, the error was significant when ET_0 was greater than 3 mm, and some scatter points were obviously overestimated or underestimated. Although the points were not as discrete as those in climate zones 4 and 5, the ET_0 of climate zone 7 showed a significant underestimation.

Figure 7 shows the ET_0 box diagram of a station randomly selected from each climate zone. From the median value, there are differences in the performance of different climate regions. Among them, the ET_0 prediction bias of climate region 2 and climate region 3 is slight. The bias of climate region 6 and 7 is large. In addition, from the extreme value, the bias of ET_0 estimated in climate zone 4 and climate zone 6 is small, and other regions have overestimated or underestimated in varying degrees. From the quartile line, there are significant differences in ET_0 estimation in climate regions 5, 6, and 7. The predicted ET_0 performance of climate zones 1, 2, 3, and 4 is relatively good.

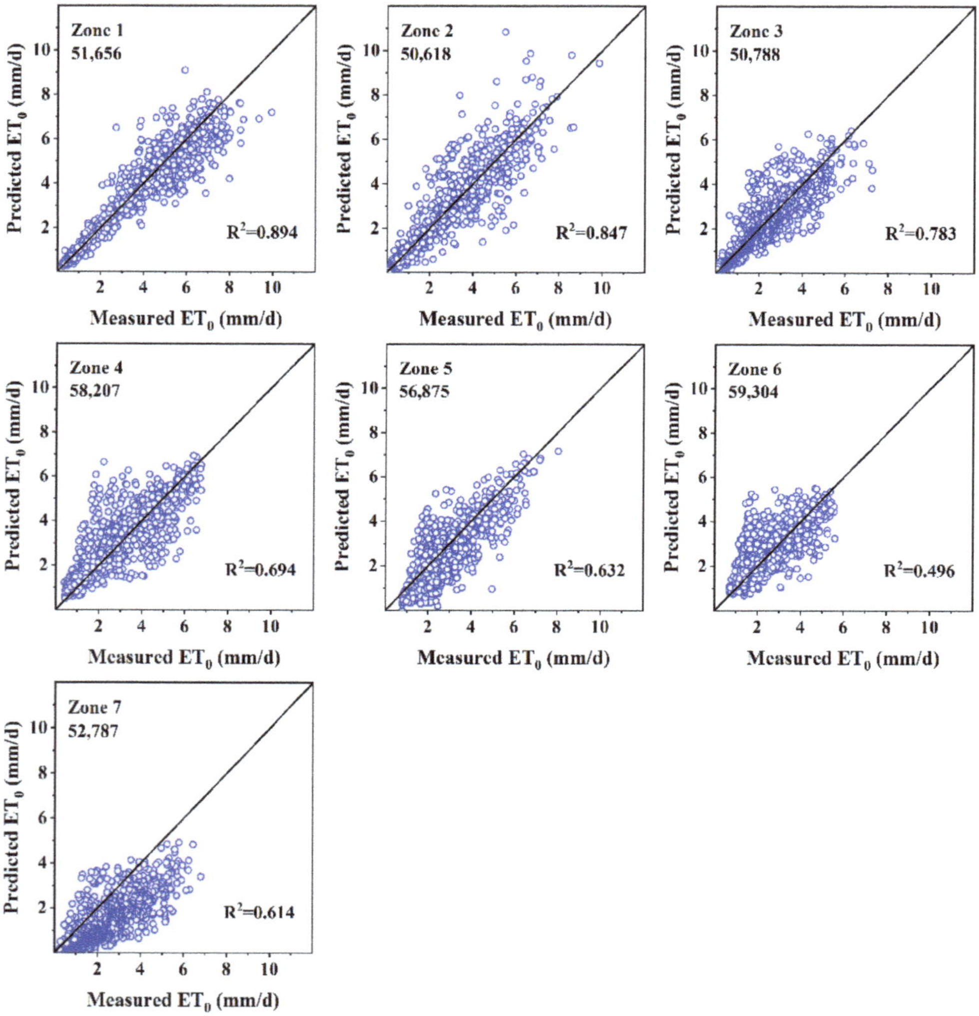

Figure 6. Scatter plots of CLDAS and FAO56 PM values of ET_0 in different climates.

3.3. Seasonal Performance of Reference Evapotranspiration from CLDAS

Since the demand for water resources varies significantly between seasons, it is necessary to assess the performance of the CLDAS dataset in different seasons. Figure 8 shows the RMSE performance of CLDAS ET_0 in the four seasons. In spring (March–May), stations with RMSE smaller than 1.5 mm d^{-1} accounted for more than 85% of all stations across China. The RMSE was lower in the south of climate zone 1 and the middle and north of climate zone 3, ranging from 0.5 to 1 mm d^{-1}. For most stations of climate zones 2, 4, 5, and 6, RMSE values ranged from 1–1.5 mm d^{-1}. Stations with errors greater than 1.5 mm d^{-1} are mainly located in climate zones 1 and 7.

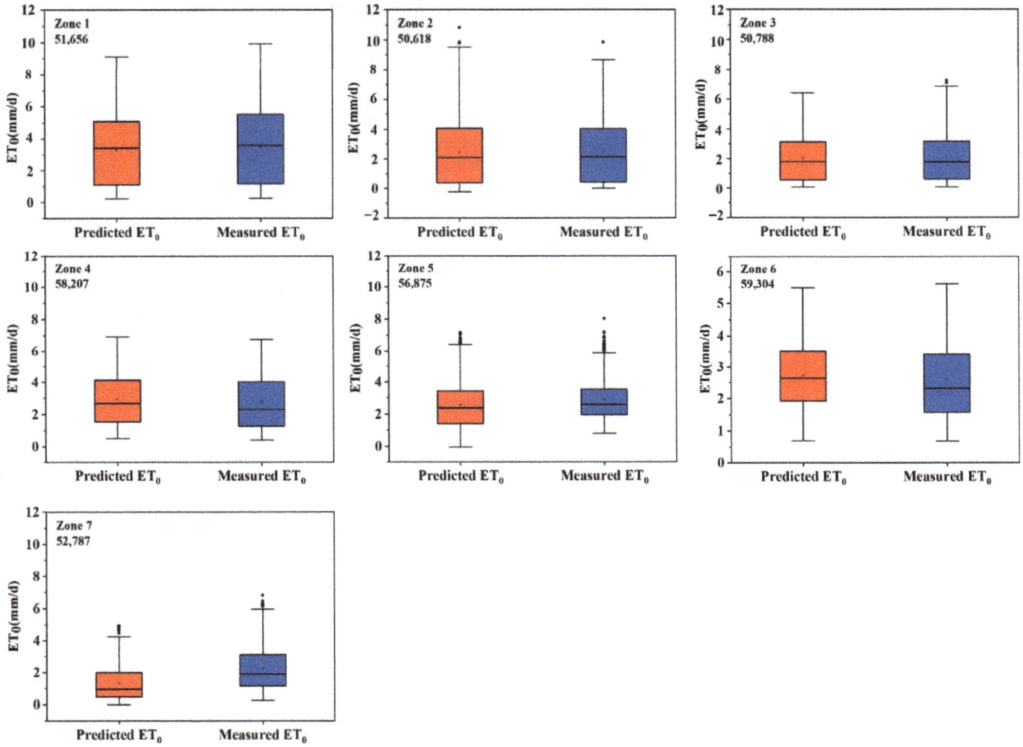

Figure 7. The box diagram of CLDAS and FAO56 PM values of ET_0 in different climates.

In summer (June–August), the RMSE of CLDAS ET_0 was generally higher than that of spring. More than 80% of the stations had RMSE ranging between 1 mm d^{-1} and 1.5 mm d^{-1}. Stations with RMSE smaller than 1 mm d^{-1} were mainly concentrated in the southern part of China and near the boundary area between climate zones 5 and 6. Stations with RMSE greater than 1.5 mm d^{-1} were distributed in all climatic regions, of which climate zone 7 had the largest RMSE, followed by climate zone 1. Especially for the southwest area of climate zone 7, stations in this area were sparse, and the error was relatively large, with the value of RMSE larger than 2 mm d^{-1}.

In autumn (September–November), RMSE was less than 1 mm d^{-1} in 80% of all stations, and stations with a significant error were still mainly concentrated in climate zone 7. It is worth mentioning that there were also many stations with RMSE greater than 1 mm d^{-1} in the coastal areas of climatic zone 6. This is mainly due to the relatively high temperature of this area in autumn, resulting in a relatively large RMSE.

In winter (December–February), RMSE in northern regions (i.e., climate zones 1–3) was lower than 0.5 mm d^{-1} due to the minimal ET_0 value. RMSE of most stations in climate zones 4 and 5 was less than 1 mm d^{-1}. However, the values of RMSE in the southern part of climate zone 7, the coastal part of climate zone 6, and the western part of climatic zone 5 were more outstanding than 1.5 mm d^{-1}.

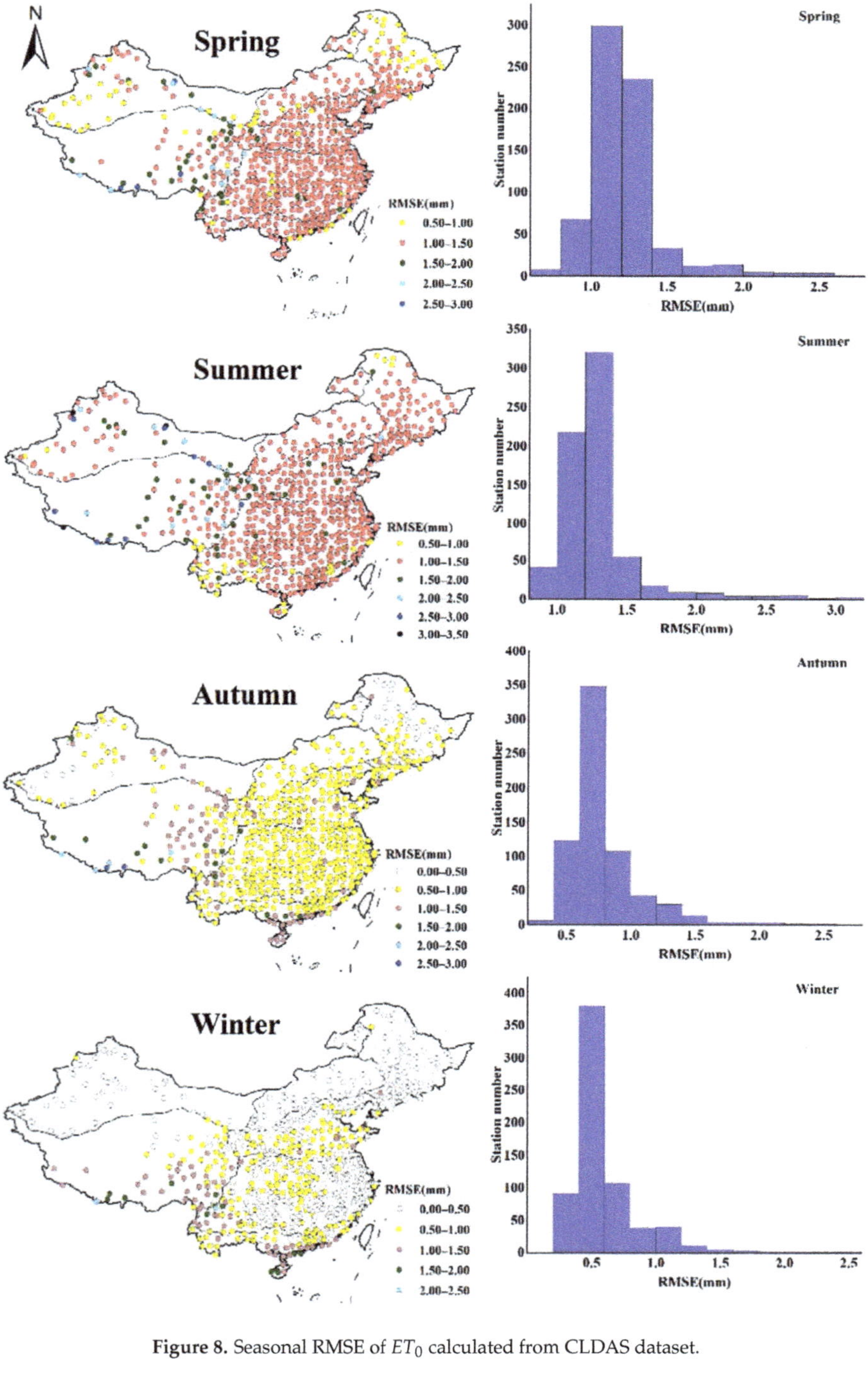

Figure 8. Seasonal RMSE of ET_0 calculated from CLDAS dataset.

Among all seasons, summer had the most significant RMSE error, followed in order by spring, autumn, and winter. The CLDAS dataset performed well in climate zones 2, 3, 4, and 5, but performed poorly in all seasons in climate zone 7. In addition, the coastal areas of climate zone 6 also did not perform well in autumn and winter.

Because the demand for water resources varies significantly in different seasons, it is also necessary to evaluate the specific overestimation or underestimation of the CLDAS dataset in different seasons. This provides a more detailed reference for practical production and life applications. Figure 9 shows the PBias distribution of ET_0 calculated by CLDAS in the four seasons. In spring, the sites with PBias between 0.2 and 0.2 accounted for about 70% of all sites in the country, and the overall forecast stability was good. The values of ET_{0CLD} in the southern regions of climate zone 1, climate zone 2, the southern part of climate zone 3, most of climate zone 4, and the central and northern parts of climate zone 5 are within 10% of the local station data. In climate zone 7 (Underestimated), numerical biases are generally greater than 30%. The prediction of ET_{0CLD} in the junction area of climate zone 7 and other climate zones is not very stable, and most of them are underestimated. In addition, the ET_{0CLD} in coastal areas will have a relatively large bias.

In summer, the bias of ET_0 calculated by CLDAS is generally smaller than that in spring, but some sites have large fluctuations (the bias is greater than 60%), and the PBias of more than 60% of the sites is between −10% and 10%. ET_{0CLD} is in the climate zone 3. The southeastern coastal areas (overestimated), the southern part of climate zone 5 (overestimated), and the western coastal areas of climate zone 6 (overestimated) have large biases from the local weather station data, with a gap of about 10% to 30%. Numerical bias with zone 7 (underestimation) is generally greater than 30%. The prediction of ET_{0CLD} for meteorological stations in the junction of climate zone 7 and other climate zones is not very stable, and most of them are underestimated.

In autumn, the bias of ET_0 calculated by CLDAS is generally larger than that in spring and autumn, and only about 50% of the sites have PBias between 10% and 10%. The western region (underestimated), the central and western regions of climate zone 5 (underestimated), the southern coastal region of climate zone 6 (underestimated), and the climate zone 7 (underestimated) have large biases from the data of local meteorological stations, with a gap of more than 30%. In addition, the prediction accuracy of the ET_{0CLD} of the meteorological stations at the junction of climate zone 3 and other climate zones decreased significantly. Most showed an overall underestimation.

In winter, the bias of ET_0 calculated by CLDAS is generally the largest, among which the bias of ET_{0CLD} from the local station data in the southern region of climate zone 1, the central region of climate zone 2, the central region of climate zone 3, and the central and eastern regions of climate zone 5 is 10%. Within %; ET_{0CLD} in the northern region of climate zone 1 (overestimated), the southern coastal region of climate zone 3 (overestimated), most of climate zone 4 (underestimated), the central and western regions of climate zone 5 (underestimated), the southern part of climate zone 6 Coastal areas (underestimated), and climate zone 7 have large biases from local weather station data, with a gap of more than 30%. In addition, the prediction accuracy for the ET_{0CLD} of the meteorological stations in the transition areas of different climatic zones will drop significantly, and both overestimation and underestimation exist.

Figure 10 shows a boxplot of the calculated PBias for the CLDAS dataset. From the median value of PBias, there are biass in the performance of different seasons. Among these, the estimated bias in spring and summer is smaller, the performance in autumn is second, and the performance in winter is the largest. From the quartile line (aside from winter), the estimated differences in the other three seasons were small. From the perspective of extreme values, the estimated maximum and minimum values in winter are not good. The performance in summer is the best, and the estimated bias is the smallest. In conclusion, of all seasons, summer and spring have the slightest bias, followed by autumn and winter. The CLDAS dataset performs well in climate zones 2, 3, 4, and 5 but not in all seasons in climate zone 7. In addition, the coastal areas of climate zone 3 and climate zone 6 also performed poorly in autumn and winter, and the performance at the interface of different climate zones was also relatively poor.

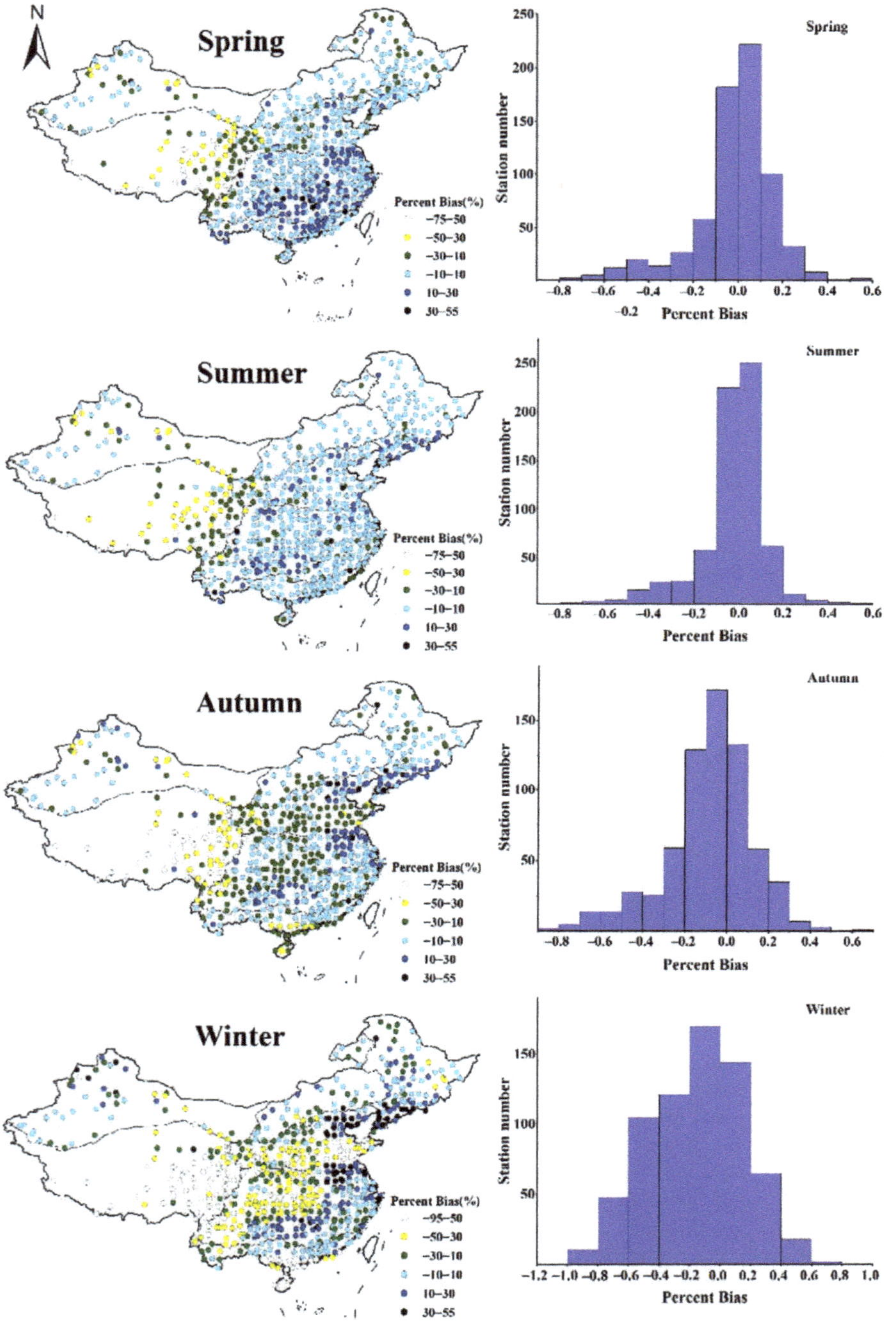

Figure 9. Seasonal PBias of ET_0 calculated from CLDAS dataset.

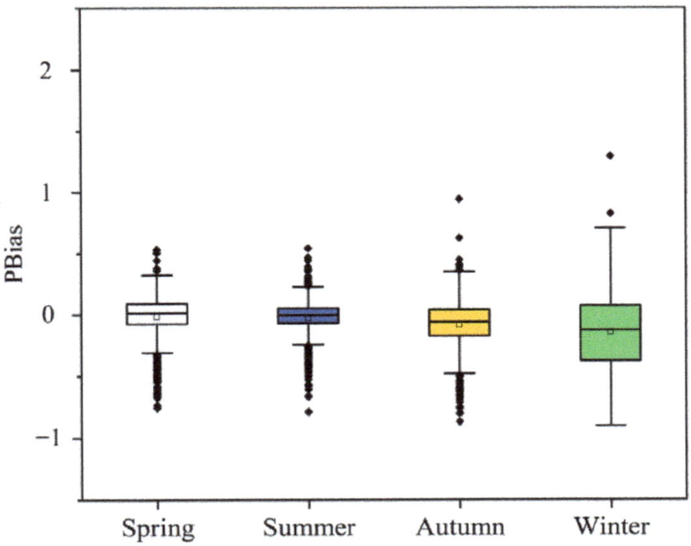

Figure 10. The box diagram of Seasonal PBias of ET_0 calculated from CLDAS dataset.

3.4. Annual Performance of Reference Evapotranspiration from CLDAS

China is a country with frequent droughts and floods. Water demand also varies widely between years. Therefore, it is necessary to evaluate the difference in CLDAS ET_0 error in different years. The RMSE of CLDAS ET_0 in 2017–2020 is shown in Figure 11. Overall, RMSE in 2019 was lower than that in other years, with more than 85% of all stations having a value less than 1 mm d^{-1}. In 2020, stations with RMSE less than 1 mm d^{-1} accounted for 60% of the total stations. Regarding the spatial distribution of errors, climate zones 3, 4, 5, and 6 overall performed better than other climate zones. Significant errors were in the southern part of climate zone 7 and the eastern part of climate zone 1, with RMSE generally more significant than 1.5 mm d^{-1}. This may be related to the special geographical location of these stations, such as at the boundary of different climate zones. The above results further confirmed that the data set had good performance in some regions. At the same time, there were also significant uncertainties in other regions, which could bring certain risks to the application.

Figure 12 shows the PB distribution of ET_0 calculated by CLDAS in 2019–2020. In 2017, the sites with PBias between 0.1 and 0.1 accounted for about 60% of all sites in the country, and the overall forecast stability was good. The values of ET_{0CLD} in the southern region of climate zone 1, climate zone 2, the central and northern regions of climate zone 3, the central region of climate zone 4, the central and northern regions of climate zone 5, and the central region of climate zone 6 are compared with local weather station data. The bias is within 10%; the bias of ET_{0CLD} from the local weather station data is larger in the northern area of climate zone 1, the southern coastal area of climate zone 3 (overestimated), and the southern coastal area of climate zone 6 (underestimated), with a gap of 10% to 30%. However, the numerical bias of climate zone 7 (underestimated) is generally greater than 30%. The prediction of ET_{0CLD} in the junction area between climate zone 7 and other climate zones is not very stable, and most of them are underestimated.

In 2018, the bias of ET_0 calculated by CLDAS was similar to that in 2017, with about 60% of sites having a PBias between 10% and 10%, and ET_{0CLD} in the central region of climate zone 1 and a few in the southern part of climate zone 3. The coastal areas (overestimated), parts of the southern part of climate zone 5 (overestimated), and the southern coastal areas of climate zone 6 (overestimated) have large biases from local weather station data, with a gap of about 10% to 30%. The (underestimated) numerical bias is generally greater than 30%. Similarly, the prediction of ET_{0CLD} for meteorological stations in the junction of climate zone 7 with other climate zones is not very stable, and most of them are underestimated.

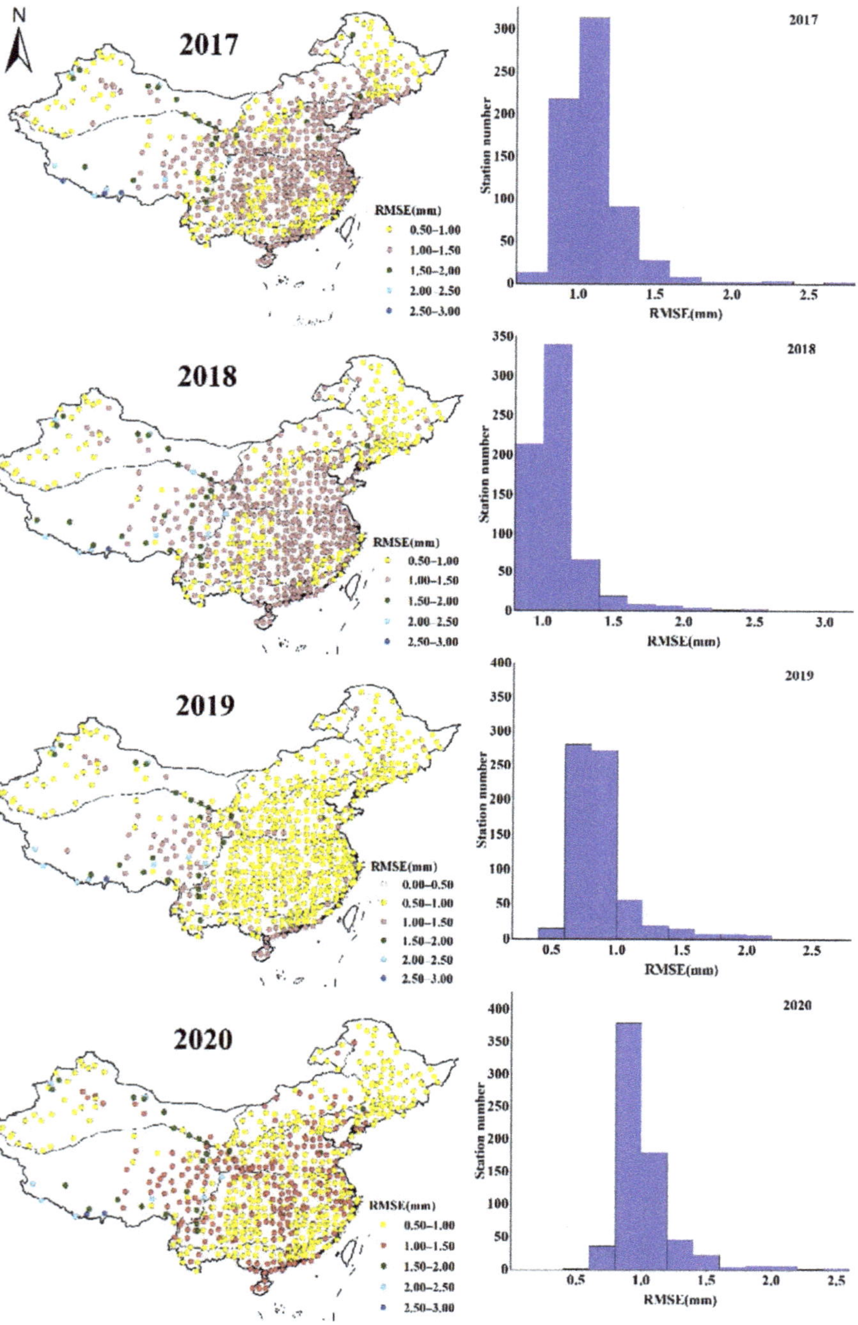

Figure 11. Annual RMSE of ET_0 calculated from CLDAS dataset.

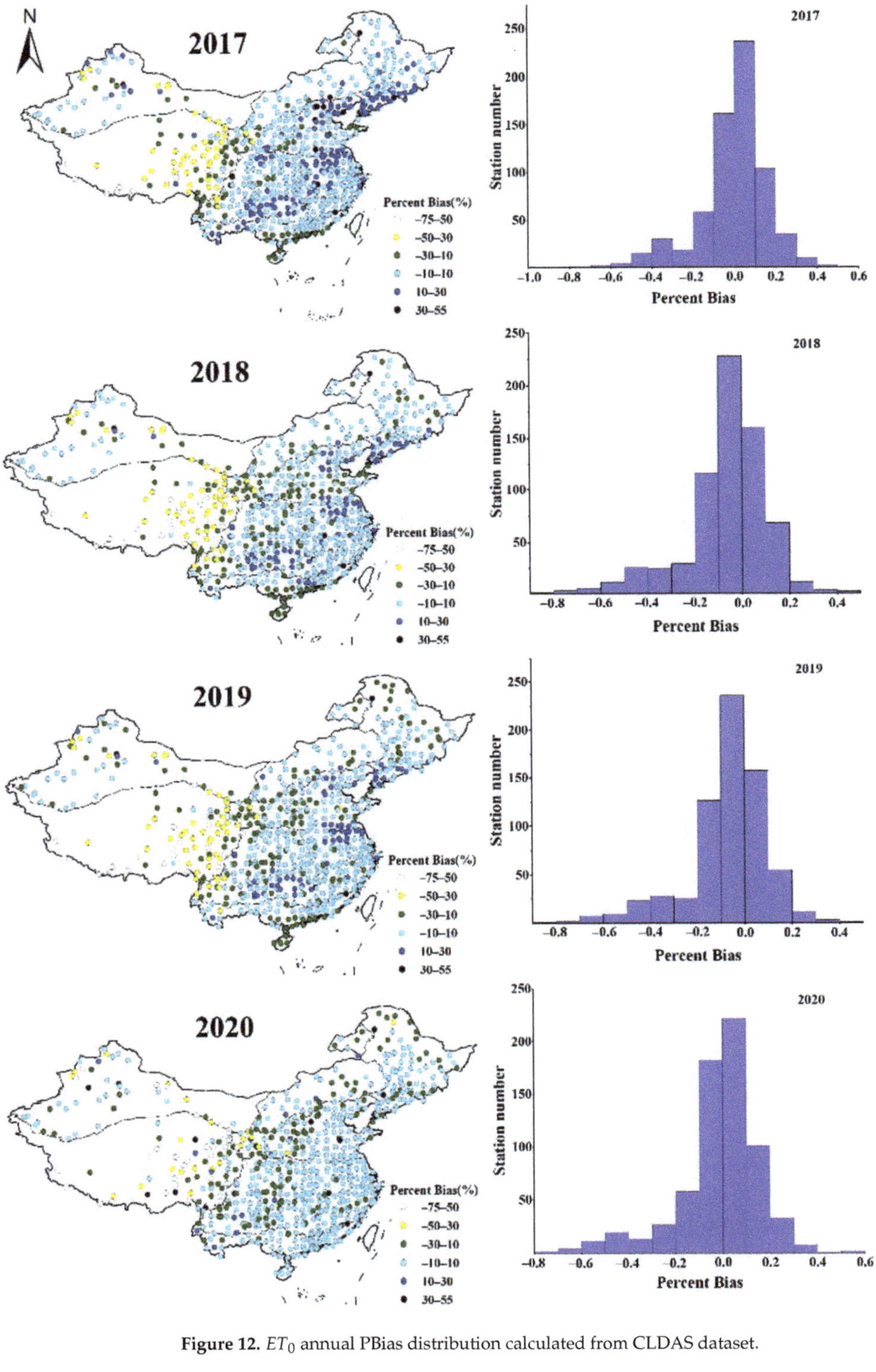

Figure 12. ET_0 annual PBias distribution calculated from CLDAS dataset.

In 2019, about 55% of the sites had PBias between −10% and 10%. The ET_{0CLD} in the southern part of climate zone 1, most of climate zone 2, the middle part of climate zone 3, small parts of the central region of climate zone 4, the central and eastern regions of climate zone 5, and the central region of climate zone 6 within 10% of the data from local weather stations. ET_{0CLD} is in the central region of climate zone 1, the southern coastal (overestimated) and northern regions (underestimated) of climate zone 3, the central and western regions of climate zone 4 (underestimated), and the southern part of climate zone 5 (overestimated). The bias between the southern coastal areas of climate zone 6 (overestimated) and the local weather station data is relatively large, about 10% to 30%. The numerical bias of climate zone 7 (underestimated) is generally greater than 30%.

In 2020, the bias of ET_0 calculated by CLDAS was generally the smallest, and about 60% of the sites have PBias between 10% and 10%. ET_{0CLD} is in the southern part of climate zone 1 and the southwest of climate zone 2. The central region of climate zone 3, the central and southern regions of climate zone 4, most of climate zone 5, and most of climate zone 6 were within 10% of data from local weather stations. ET_{0CLD} is in the central region of climate zone 1, a few southern coastal (overestimated) and northern regions (underestimated) of climate zone 3, the central and western regions of climate zone 4 (underestimated), and a small number of regions in climate zone 5 (underestimated). The bias of local weather station data is large, with a gap of about 10% to 30%. The numerical bias of climate zone 7 (underestimated) was generally greater than 30%. Similarly, the prediction of ET_{0CLD} for meteorological stations in the junction of climate zone 7 with other climate zones is not very stable, and most of them are underestimated.

Figure 13 shows a boxplot of annual PBias calculated for the CLDAS dataset. From the median value of PBias, there is a bias in the performance of different years. The estimated bias in 2020 and 2017 is smaller, and the performance of the other two years is relatively poor. From the quartile line, the estimated bias in 2020 is the smallest and more compact, and the estimated differences in other years have different degrees of fluctuation. From the perspective of extreme values, there was a clear overestimation in 2017 and a clear underestimation in 2018. Additionally, 2020 had the best performance, and the estimated bias was the smallest. In conclusion, from 2017 to 2020, bias in 2019 and 2020 was the smallest. The CLDAS dataset performs well in climate zones 2, 3, 4, and 5 but not in all seasons in climate zone 7. In addition, the coastal areas of climate zone 3 and climate zone 6 also performed poorly except in 2020. The performance of the boundary areas of different climate zones was also relatively poor.

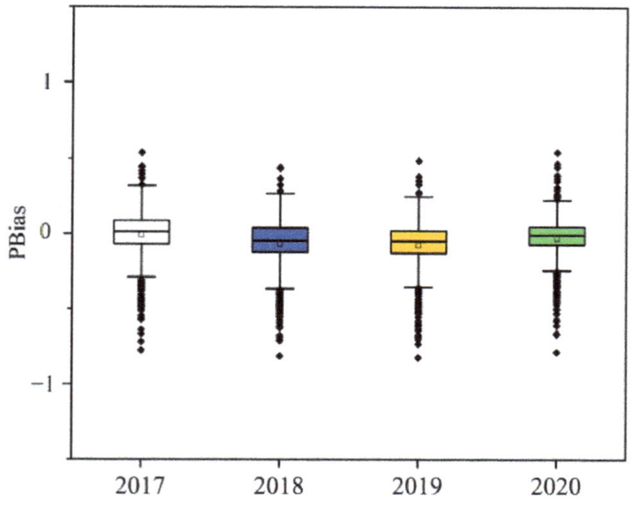

Figure 13. The box diagram of annual PBias of ET_0 calculated from CLDAS dataset.

3.5. Main Factors Affecting Reference Crop Evapotranspiration

The evapotranspiration process is affected by many factors [43], and its changes are mainly attributed to the changes in meteorological factors. The country's climate is complex and diverse. From a geographical point of view, the eastern part is mostly a monsoon climate zone with a complex and changeable climate. The air above it is severely polluted, affecting solar radiation and surface wind speed. Therefore, the performance of estimated ET_0 in coastal areas will decline. The northwest region is far from the sea, is a non-monsoon region, and belongs to a temperate continental climate. The ground topography in this region (the junction of climate zones 2, 4, and 5) is complex and changeable, and the wind speed is greatly affected by the roughness of the underlying surface. The direction of wind movement is accurately simulated, so the reduction in the accuracy of wind speed is likely to lead to a decline in the performance of estimating ET_0 in some areas. The Qinghai-Tibet Plateau belongs to the plateau climate zone. Due to its complex and changeable terrain, the climate itself on the Qinghai-Tibet Plateau will fluctuate depending on the region, which greatly affects the estimation of ET_0. To sum up, the closer are to inland areas (such as climate zones 1, 2, and 3), the higher the accuracy of ET_0 estimation will be. The performance of ET_0 estimation in coastal areas, the Qinghai-Tibet Plateau, and the junction of climatic zones will be negative effects [44–46]. From the seasonal point of view, the summer is affected by the warm and humid air from the ocean, with high temperature, humidity, and rain. The climate is oceanic, so the estimation error in summer is the largest. In winter, affected by the dry and cold airflow from the continent, the climate is cold, dry, and less rainy, and the climate is continental, and the estimation error will be relatively small [47,48]. In addition, specific regions need to be further analyzed according to the actual situation.

4. Discussion

The calculation of ET_0 is affected by a variety of climatic factors. Ma et al. (2010) [49] studied the influence of main climatic factors on ET_0 in mountainous plateau areas and found that the change of wind speed had the most significant impact on ET_0 at each site. Luo et al. (2010) [50] conducted a sensitivity analysis on ET_0 and main meteorological factors in the main agricultural areas of Tibet, and the results showed that ET_0 in the whole region had a declining trend over the past 50 years. The meteorological element that had the most significant impact on ET_0 was R_s. Similar results were obtained in our study, where the accuracy of ET_0 was affected by the error of R_s. Xie et al. (2017) [51] analyzed the impact of changes in meteorological factors on ET_0 in China's main grain-producing areas from 1961 to 2013, in which ET_0 showed a saw-tooth decline. The changing characteristics of main meteorological factors in the study area and the response of ET_0 are similar to the results of our study, showing regional and seasonal variations. Overall, our study suggests that the errors of meteorological factors in the Qinghai-Tibet Plateau region and the boundary region of the climate zones are more significant than in other regions, with the highest errors observed in summer.

Due to the incomplete understanding of the physical mechanism of weather changes and limited observational data, there is still a specific error in the reanalysis data [52], and the magnitude of this error tends to vary with different climatic factors. Temperature is a meteorological variable with minor errors, usually less than 10% [53,54]. Similar results were found in our study, in which the R^2 of T_{max} and T_{min} are generally greater than 0.9 in the seven climatic zones. Due to the influence of topography, the errors of wind speed and relative humidity are usually large [26], and similar results were obtained in our study.

It is worth mentioning that the weather stations in our study are affiliated with the China Meteorological Administration. The ground of the weather station is usually covered with short grass under adequate irrigation conditions. However, areas in the grid system do not necessarily have lush vegetation. Therefore, there might be some differences in the environmental factors between the two types of systems, especially for the radiant energy (i.e., $R^2 < 0.65$ in the seven climate zones for the R_{sCLD} estimation in our study).

This may lead to the problem of overestimating or underestimating the reanalysis data, which indirectly explains why the estimated ET_0 values in some areas fluctuate severely in our study. In addition, the variation of wind speed is greatly affected by the terrain and the type of underlying surface, and it is not easy to obtain the average wind speed in a specific area. Similar results were obtained in our study, where the overall U_{CLA} accuracy is not satisfactory.

Finally, this study can provide an idea for economically underdeveloped countries and contribute to improving the reanalysis data set and the accuracy of ET_0 estimation. Therefore, when other developing countries establish regional climate models, they can consider their own terrain and climate characteristics and establish a more local model.

5. Conclusions

ET_0 data set based on reanalysis products can make up for the time discontinuity and spatial insufficiency of surface meteorological platform data, which is significant for water resources planning and irrigation system formulation. However, a rigorous evaluation of reanalysis products must be carried out to see if they have value in application. This study evaluated the ability of the CLDAS dataset officially published by the Chinese meteorological system for ET_0 estimation. Results indicate that the temperature data of CLDAS have high accuracy in all regions except the Qinghai Tibet Plateau (QTP) region. In contrast, the accuracy of the total radiation data is average, and the quality of relative humidity and wind speed data is poor. The overall accuracy of ET_0 is acceptable except for QTP, but there are many stations with large errors. Among seasons, RMSE is the largest in summer and smallest in winter. There are also inter-annual differences in the ET_0 of this data set. Overall, the CLDAS dataset is expected to have good applicability in the Inner Mongolia Grassland area, Northeast Taiwan, the Semi-Northern Temperate zone, the Humid and Semi Humid warm Temperate zone, and the subtropical region. However, there are certain risks in other regions. In addition, of all seasons, summer and spring have the slightest bias, followed by autumn and winter. From 2017 to 2020, bias in 2019 and 2020 are the smallest, and the areas with large deviation are in the south of climate zone 3, the coastal area of climate zone 6, and the boundary area of climate zone 7.

Author Contributions: Conceptualization, L.-F.W. and L.Q.; methodology, L.-F.W., G.-M.H. and L.Q.; data curation, L.Q. and X.-G.L.; writing—original draft preparation, L.-F.W., Y.-C.W., G.-M.H. and L.Q.; writing—review and editing, L.-F.W., H.B., X.-G.L. and L.Q.; project administration, L.-F.W. and S.-F.W. All authors have read and agreed to the published version of the manuscript.

Funding: This research was funded by the Science and Technology Project of Jiangxi Provincial Department of Education (GJJ180925), National Natural Science Foundation of China (51979133 and 51769010) and Natural Science Foundation of Jiangxi province in China (20181BBG78078 and 20212BDH80016).

Institutional Review Board Statement: Not applicable.

Informed Consent Statement: Not applicable.

Data Availability Statement: Not applicable.

Conflicts of Interest: The authors declare no conflict of interest. The funders had no role in the design of the study; in the collection, analysis, or interpretation of data; in the writing of the manuscript, or in the decision to publish the results.

Abbreviations

Meaning of main acronyms:

CLDAS: China Meteorological Administration Land Data Assimilation System; ET_0: Reference Crop evapotranspiration; FAO: Food and Agriculture Organization of the United Nations; ECMWF: European Center for Medium Weather Forecasting; NCEP: National Centers for Environmental Prediction; R_s: Global solar radiation; U: wind speed at 2 m;

RH: relative humidity; T_{max}: maximum temperature; T_{min}: minimum temperature. (When the subscript CLD exists in these meteorological data, it is the corresponding CLDAS meteorological data); RMSE: Root Mean Square Error; MAE: Mean Absolute Error; PBias: percent bias; R^2: coefficient of determination.

References

1. Mao, Y.P.; Fang, S.Z. Research of reference evapotranspiration's simulation based on machine learning. *J. Geo-Inf. Sci.* **2020**, *22*, 1692–1701. [CrossRef]
2. Feng, Y.; Peng, Y.; Cui, N.; Gong, D.; Zhang, K. Modeling reference evapotranspiration using extreme learning machine and generalized regression neural network only with temperature data. *Comput. Electron. Agric.* **2017**, *136*, 71–78. [CrossRef]
3. Feng, Y.; Jia, Y.; Cui, N.; Zhao, L.; Li, C.; Gong, D. Calibration of Hargreaves model for reference evapotranspiration estimation in Sichuan basin of southwest China. *Agric. Water Manag.* **2017**, *181*, 1–9. [CrossRef]
4. Wang, Y.; Sheng, L.X.; Li, K.; Sun, H.Y. Analysis of present situation of water resources and countermeasures for sustainble development in China. *J. Water Resour. Water Eng.* **2008**, *3*, 10–14.
5. Jia, Y.; Cui, N.; Wei, X.; Gong, D.; Hu, X. Applicability evaluation of different algorithms for reference crop evapotranspiration in Yangtze River Basin based on inverse distance weighted method. *Trans. Chin. Soc. Agric. Eng.* **2016**, *32*, 130–138. [CrossRef]
6. Lei, G.; Zeng, W.; Zhu, J.; Zha, Y.; Fang, Y.; Song, Y.; Chen, M.; Qian, Y.; Wu, J.; Huang, J. Quantification of leaf growth, height increase, and compensatory root water uptake of sunflower in heterogeneous saline soils. *Agron. J.* **2019**, *111*, 1010–1027. [CrossRef]
7. Li, L.A.; Qiu, R.J.; Liu, C.W. Comparison and improvement of estimation models for the reference evapotranspiration using temperature data. *Trans. Chin. Soc. Agric. Eng.* **2021**, *37*, 123–130. [CrossRef]
8. Paredes, P.; Martins, D.S.; Pereira, L.S.; Cadima, J.; Pires, C. Accuracy of daily estimation of grass reference evapotranspiration using ERA-Interim reanalysis products with assessment of alternative bias correction schemes. *Agric. Water Manag.* **2018**, *210*, 340–353. [CrossRef]
9. Li, C. Performance of revised Hargreaves models in hilly area of Central Sichuan Basin. *Water Sav. Irrig.* **2021**, *11*, 88–96. [CrossRef]
10. Dong, J.H.; Liu, X.G.; Wu, L.F.; Huang, G.M.; Yang, Q.L. Cross-station adaptability of ET_0 based on machine learning. *J. Northwest Agric. For. Univ. (Nat. Sci. Ed.)* **2021**, *49*, 144–154. [CrossRef]
11. Zhu, B.; Feng, Y.; Gong, D.Z.; Jiang, S.Z.; Zhao, L.; Cui, N.B. Hybrid particle swarm optimization with extreme learning machine for daily reference evapotranspiration prediction from limited climatic data. *Comput. Electron. Agric.* **2020**, *173*, 105430. [CrossRef]
12. Huang, C.X.; Zhao, D.M.; Wang, B.F. A comparative study on calculation methods of reference crop evapotranspiration in the semi-arid region of Central Eastern Gansu. *Agric. Res. Arid Areas* **2018**, *36*, 41–47. [CrossRef]
13. Mahto, S.S.; Mishra, V. Does ERA-5 outperform other reanalysis products for hydrologic applications in India? *J. Geophys. Res. Atmos.* **2019**, *124*, 9423–9441. [CrossRef]
14. Baatz, R.; Franssen, H.J.H.; Euskirchen, E.; Sihi, D.; Vereecken, H. Reanalysis in Earth System Science: Toward Terrestrial Ecosystem Reanalysis. *Rev. Geophys.* **2021**, *59*, e2020RG000715. [CrossRef]
15. Fennel, K.; Gehlen, M.; Brasseur, P.; Brown, C.W.; Ciavatta, S.; Cossarini, G. Advancing marine biogeochemical and ecosystem reanalyses and forecasts as tools for monitoring and managing ecosystem health. *Front. Mar. Sci.* **2019**, *6*, 89. [CrossRef]
16. Peylin, P.; Bacour, C.; MacBean, N.; Leonard, S.; Rayner, P.; Kuppel, S. A new stepwise carbon cycle data assimilation system using multiple data streams to constrain the simulated land surface carbon cycle. *Geosci. Model Dev.* **2016**, *9*, 3321–3346. [CrossRef]
17. Hobeichi, S.; Abramowitz, G.; Evans, J.; Beck, H.E. Linear optimal runoff aggregate (LORA): A global gridded synthesis runoff product. *Hydrol. Earth Syst. Sci.* **2019**, *23*, 851–870. [CrossRef]
18. Hersbach, H.; Bell, B.; Berrisford, P.; Hirahara, S.; Horanyi, A.; Munoz-Sabater, J. The ERA5 global reanalysis. *Q. J. R. Meteorol. Soc.* **2020**, *146*, 1999–2049. [CrossRef]
19. Sabater, J.M.; Dutra, E.; Panareda, A.; Albergel, C.; Thepaut, J.N. ERA5-Land: A state-of-the-art global reanalysis dataset for land applications. *Earth Syst. Sci. Data* **2021**, *13*, 4349–4383. [CrossRef]
20. Boulard, D.; Castel, T.; Camberlin, P.; Sergent, A.S.; Bréda, N.; Badeau, V.; Rossi, A.; Pohl, B. Capability of a regional climate model to simulate climate variables requested for water balance computation: A case study over northeastern France. *Clim. Dyn.* **2016**, *46*, 2689–2716. [CrossRef]
21. Srivastava, P.K.; Han, D.; Islam, T.; Petropoulos, G.P.; Gupta, M.; Dai, Q. Seasonal evaluation of evapotranspiration fluxes from MODIS satellite and mesoscale model downscaled global reanalysis datasets. *Theor. Appl. Climatol.* **2016**, *124*, 461–473. [CrossRef]
22. Pelosi, A.; Terribile, F.; D'Urso, G.; Chirico, G.B. Comparison of ERA5-Land and UERRA MESCAN-SURFEX reanalysis data with spatially interpolated weather observations for the regional assessment of reference evapotranspiration. *Water* **2020**, *12*, 1669. [CrossRef]
23. Woldesenbet, T.; Elagib, N. Spatial-temporal evaluation of different reference evapotranspiration methods based on the climate forecast system reanalysis data. *Hydrol. Process.* **2021**, *35*, e14239. [CrossRef]
24. Song, Y.; Su, X.L.; Niu, J.P.; Cui, C.F. Temporal and spatial characteristics and forecasting of reference crop evaporation in Shaanxi. *J. Northwest Agric. For. Univ. (Nat. Sci. Ed.)* **2015**, *43*, 225–234. [CrossRef]

25. Liu, Z.; Lu, J.Z.; Huang, J.; Chen, X.; Zhang, L.; Sheng, Y. Prediction and trend of future reference crop evapotranspiration in the Poyang Lake Basin based on CMIP5 Models. *J. Lake Sci.* **2019**, *31*, 1685–1697. [CrossRef]
26. Martins, D.S.; Paredes, P.; Raziei, T.; Pires, C.; Cadima, J.; Pereira, L.S. Assessing reference evapotranspiration estimation from reanalysis weather products. An application to the Iberian Peninsula. *Int. J. Climatol.* **2017**, *37*, 2378–2397. [CrossRef]
27. Raziei, T.; Parehkar, A. Performance evaluation of NCEP/NCAR reanalysis blended with observation-based datasets for estimating reference evapotranspiration across Iran. *Theor. Appl. Climatol.* **2021**, *144*, 885–903. [CrossRef]
28. Milad, N.; Mehdi, H. Reference crop evapotranspiration for data-sparse regions using reanalysis products–ScienceDirect. *Agric. Water Manag.* **2022**, *262*, 107319. [CrossRef]
29. Wang, X.; Ding, Y.; Zhao, C.; Wang, J. Similarities and improvements of GPM IMERG upon TRMM 3B42 precipitation product under complex topographic and climatic conditions over Hexi region, Northeastern Tibetan Plateau. *Atmos. Res.* **2019**, *218*, 347–363. [CrossRef]
30. Prakash, S.; Mitra, A.; Pai, D.S.; Aghakouchak, A. From TRMM to GPM: How well can heavy rainfall be detected from space? *Adv. Water Resour.* **2016**, *88*, 1–7. [CrossRef]
31. Liu, J.; Shi, C.; Sun, S.; Liang, J.; Yang, Z.L. Improving land surface hydrological simulations in China using CLDAS meteorological forcing data. *J. Meteorol. Res.* **2019**, *33*, 1194–1206. [CrossRef]
32. Allen, R.G.; Pereira, L.S.; Raes, D.; Smith, M. *Crop Evapotranspiration: Guidelines for Computing Crop Water Requirements*; FAO Irrigation and Drainage Paper; FAO: Rome, Italy, 1998.
33. Mobilia, M.; Longobardi, A. Prediction of Potential and Actual Evapotranspiration Fluxes Using Six Meteorological Data-Based Approaches for a Range of Climate and Land Cover Types. *Int. J. Geo-Inf.* **2021**, *10*, 192. [CrossRef]
34. Takakura, T.; Kubata, C.; Sase, S.; Hayashi, M.; Ishii, M.; Takayama, K.; Nishina, H.; Kurata, K.; Giacomelli, G.A. Measurement of evapotranspiration rate in a single-span greenhouse using the energy¬balance equation. *Biosyst. Eng.* **2009**, *702*, 298–304. [CrossRef]
35. Fan, J.; Wu, L.; Zhang, F.; Cai, H.; Wang, X.; Lu, X.; Xiang, Y. Evaluating the effect of air pollution on global and diffuse solar radiation prediction using support vector machine modeling based on sunshine duration and air temperature. *Renew. Sustain. Energy Rev.* **2018**, *94*, 732–747. [CrossRef]
36. Yin, Y.H.; Li, B.Y. A New Scheme for Climate Regionalization in China. *Acta Geogr. Sin.* **2010**, *65*, 3–12. [CrossRef]
37. Zheng, J.Y.; Bian, J.J.; Ge, Q.S.; Hao, Z.X.; Yin, Y.H. Liao, Y.M. The climate regionalization in China for 1981–2010. *Geogr. Res.* **2013**, *32*, 987–997. [CrossRef]
38. Liu, X.; Mei, X.; Li, Y.; Wang, Q.; Jensen, J.R.; Zhang, Y.; Porter, J.R. Evaluation of temperature-based global solar radiation models in China. *Agric. For. Meteorol.* **2009**, *149*, 1433–1446. [CrossRef]
39. Fan, J.; Wu, L.; Zhang, F.; Cai, H.; Zeng, W.; Wang, X. Empirical and machine learning models for predicting daily global solar radiation from sunshine duration: A review and case study in China. *Renew. Sustain. Energy Rev.* **2019**, *100*, 186–212. [CrossRef]
40. Wang, Y.Q.; Zhang, X.Y.; Sun, J.Y.; Zhang, X.C.; Che, H.Z.; Li, Y.J.A.C. Spatial and temporal variations of the concentrations of PM 10, PM 2.5 and PM 1 in China. *Atmos. Chem. Phys.* **2015**, *15*, 13585–13598. [CrossRef]
41. Wang, Z.; Xie, P.; Lai, C.; Chen, X.; Wu, X.; Zeng, Z.; Li, J. Spatiotemporal variability of reference evapotranspiration and contributing climatic factors in China during 1961–2013. *J. Hydrol.* **2017**, *544*, 97–108. [CrossRef]
42. Blankenau, P.A.; Kilic, A.; Allen, R. An evaluation of gridded weather data sets for the purpose of estimating reference evapotranspiration in the United States. *Agric. Water Manag.* **2020**, *242*, 106376. [CrossRef]
43. Zhang, Y.S.; Zhao, X.Q.; Zhao, S.X.; Feng, C.B. Correlation between Evapotranspiration and Climate Factors in Warm Steppe in Source Region of Yangtze, Yellow and Yalu Tsangpo Rivers. *J. Desert Res.* **2010**, *30*, 363–368.
44. Li, S.C.; Liu, F.Y. Climate Complexity and Spatial Variation in China. *Clim. Environ. Res.* **2008**, *13*, 1. [CrossRef]
45. Dong, Y.; Hu, J.L.; Wang, J.; Chen, X. Study of Temporal and Spatial Variation of the Reference Crop Evapotranspiration in Xinjiang Uygur Autonomous Region During the Period from 1961 to 2013. *Res. Soil Water Conserv.* **2016**, *23*, 304–308 + 313. [CrossRef]
46. Huo, Z.L.; Shi, H.B.; Chen, Y.X.; Wei, Z.M.; Qu, Z.Y. Spatio-temporal variation and dependence analysis of ET_0 in north arid and cold region. *Trans. Chin. Soc. Agric. Eng.* **2004**, *6*, 60–63. [CrossRef]
47. Fu, J.; Qin, J.X.; Li, Z.X.; Zhang, Z.B.; Hu, S.S. Changing Reference Evapotranspiration and Effects of Climatic Factors. *Resour. Environ. Yangtze Basin* **2018**, *27*, 7.
48. Hu, Q.; Dong, B.; Pan, X.B.; Jiang, H.F.; Pan, Z.H.; Qiao, Y.; Shao, C.X.; Ding, M.L.; Yin, M.L.; Hu, L.T. Spatiotemporal variation and causes analysis of dry-wet climate over period of 1961–2014 in China. *Trans. Chin. Soc. Agric. Eng.* **2017**, *33*, 124–132. [CrossRef]
49. Ma, X.; Li, J.; Gu, S.X.; Wang, J.; Liu, T.; Duan, S.; Luo, S. Research on the Impact of Main Climatic Factors on ET_0 in Plateau Areas. *China Rural Water Hydropower* **2010**, *10*, 9–12.
50. Luo, H.; Cui, Y.; Duan, Z. Analysis of the Sensitivity of ET0 and the Main Meteorological Factors in Major Agricultural Regions in Tibet. In *Modern Water-Saving and Efficient Agriculture and Ecological Irrigation Area Construction*; Yunnan University Press: Kunming, China; pp. 130–136.
51. Xie, R.; Cui, N.; Li, Z.; Zhao, L.; Hu, X.; Gong, D. Spatiotemporal Variation of Main Meteorological Factors and Their Impact on Reference Crop Evapotranspiration in Main Agricultural Production Areas in China. *J. Irrig. Drain.* **2017**, *36*, 81–89. [CrossRef]

52. Nacar, S.; Kankal, M.; Okkan, U. Evaluation of the suitability of NCEP/NCAR, ERA-Interim and, ERA5 reanalysis data sets for statistical downscaling in the Eastern Black Sea Basin, Turkey. *Meteorol. Atmos. Phys.* **2022**, *134*, 39. [CrossRef]
53. Yan, S.; Wu, L.; Fan, J.; Zhang, F.; Zou, Y.; Wu, Y. A novel hybrid WOA-XGB model for estimating daily reference evapotranspiration using local and external meteorological data: Applications in arid and humid regions of China. *Agric. Water Manag.* **2021**, *244*, 106594. [CrossRef]
54. Tarek, M.; Brissette, F.; Arsenault, R. Evaluation of the ERA5 reanalysis as a potential reference dataset for hydrological modelling over North America. *Hydrol. Earth Syst. Sci.* **2020**, *24*, 2527–2544. [CrossRef]

water

MDPI

Article

Estimating Canopy-Scale Evapotranspiration from Localized Sap Flow Measurements

James Solum † and Bwalya Malama *

Department of Natural Resources Management & Environmental Sciences, College of Agriculture, Food & Environmental Sciences, California Polytechnic State University, San Luis Obispo, CA 93407, USA; james.solum@yahoo.com

* Correspondence: bmalama@calpoly.edu; Tel.: +1-805-756-2971

† Current affiliation: US Geological Survey, California Water Science Center, Santa Maria Field Office, Santa Maria, CA 93455, USA.

Abstract: The results reported in this work are based in part on measurements of sap flow in a few select trees on a representative riparian forest plot coupled with a forest-wide randomized sampling of tree sapwood area in a watershed located along the Pacific coast in Santa Cruz County, California. These measurements were upscaled to estimate evapotranspiration (ET) across the forest and to quantify groundwater usage by dominant phreatophyte vegetation. Canopy cover in the study area is dominated by red alder (*Alnus rubra*) and arroyo willow (*Salix lasiolepis*), deciduous phreatophyte trees from which a small sample was selected for instrumentation with sap flow sensors on a single forest plot. These localized sap flow measurements were then upscaled to the entire riparian forest to estimate forest ET using data from a survey of sapwood area on six plots scattered randomly across the entire forest. The estimated canopy-scale ET was compared to reference ET and NDVI based estimates. The results show positive correlation between sap flow based estimates and those of the other two methods, though over the winter months, sap flow-based ET values were found to significantly underestimate ET as predicted by the other two methods. The results illustrate the importance of ground-based measurements of sap flow for calibrating satellite based methods and for providing site-specific estimates and to better characterize the ET forcing in groundwater flow models.

Keywords: evapotranspiration; sapflow; phreatophyte; riparian; groundwater

Citation: Solum, J.; Malama, B. Estimating Canopy-Scale Evapotranspiration from Localized Sap Flow Measurements. *Water* **2022**, *14*, 1812. https://doi.org/10.3390/w14111812

Academic Editors: Luis Garrote and Alban Kuriqi

Received: 22 April 2022
Accepted: 30 May 2022
Published: 4 June 2022

1. Introduction

Prolonged drought conditions in California and the associated increased reliance on groundwater resources for irrigation in coastal areas, necessitates a re-examination of agricultural groundwater use in riparian corridors, particularly the impacts of groundwater pumping on instream flows. Minimum flow requirements in coastal creeks are a source of serious concern for riparian forest and land managers, fisheries biologists, and agencies assigned to evaluate sustainable instream flow requirements. Prior works in coastal riparian systems (e.g., [1]) have focused entirely on groundwater pumping for irrigation, with only cursory attention given to consumptive groundwater use by riparian vegetation. An accurate understanding of the impacts of groundwater pumping for irrigation, requires a consideration and characterization of all the components (inputs, outputs, and storage) of watershed-scale water budgets, including the poorly understood consumptive groundwater use by phreatophytic vegetation.

Riparian forests are among the most productive natural ecosystems and perform such ecological functions as filtering agricultural runoff of sediment, nutrients, and other solutes, thereby minimizing non-point source contamination of streams and groundwater. They help maintain the stability of stream banks as well as the quality and quantity of groundwater recharge [2–4]. In addition to ecological functions, riparian forests also play a central role in the earth's strongly coupled energy and hydrologic cycles through consumptive water use from evapotranspiration (ET) and effects on surface roughness and surface reflectivity (albedo). Direct measurement of tree sap flow to better characterize the ET forcing on groundwater flow could lead to improved understanding of the ET component of the hydrologic cycle attributable to consumptive groundwater use by phreatophytic vegetation. Water requirements of riparian vegetation are usually fulfilled by soil moisture and groundwater [5]. However, riparian forests also often contain phreatophytic species, which depend primarily on groundwater for long-term survival [6,7]. The root systems of such species extend to the capillary fringe, the water-table and the underlying saturated zone [7,8]. In groundwater models of diurnal groundwater fluctuations, the water use by phreatophytic vegetation can be characterized by a diurnal ET water-table flux boundary condition (forcing function) or as a volumetric sink within the saturated zone (unconfined aquifer) [9].

Although direct measurement of ET from riparian forests is key to understanding regional and local water and energy balances in hydroclimatological modeling, there remain high uncertainties in seasonal and long-term (decadal scale) riparian forest ET data due to the focus on diurnal fluctuations [10,11]. This limits the ability of models to accurately estimate the groundwater component of water budgets consumed by vegetation in such forests [11]. Riparian zones in semiarid regions often exhibit high rates of ET in spite of low-soil wetness due to the presence of phreatophytic vegetation [12,13], which is reflected in diurnal water-table fluctuations [5,9,12] and can be measured by direct monitoring of vadose zone soil moisture and groundwater levels. In most long-term ET and groundwater studies, the amount of water used by phreatophytes is estimated by empirical formulae that rely on climatic and weather variables or by extrapolation and interpolation of remote sensing measurements. This can be problematic given the uncertainties associated with the subsurface sources of the water [10,11].

Direct ground-based measurements of ET include eddy covariance and sap flow monitoring. There are three common sap flow techniques: (1) thermal dissipation probes (TDP), (2) heat pulse velocity (HPV), and (3) tissue heat balance (THB). Thermal dissipation probes (TDP) proposed by [14] comprise two cylindrical probes that are inserted into the tree stem and separated by a fixed vertical distance. There is some uncertainty on the accuracy of TDP sensing of sap flow taking in fixed position on trees over long periods [15,16]. The workers [17] continuously measured sap flow for 1.2 years, and reported that the mean sap flux density declined by 30% during the second growing season. In a fast-growing tree, the probes become embedded as the vascular cambium produces new phloem and xylem tissue [18,19]. Prior work of [16] reported declines in sap flow as probes became lodged deeper into the sapwood over time, leading to underestimation of the volume sap flow rate.

In the present study, sap flow was measured using thermal dissipation probes in four trees, continuously, for two years with the objectives of (1) comparing up-scaled ground-based sap flow estimates of ET to satellite-based measurements and (2) evaluating groundwater usage by phreatophytes in comparison to pumping for irrigation. The approach involved installation of thermal dissipation probes (sap flow probes) in select phreatophytes, vegetation surveys focusing on phreatophytes, measurement of sapwood area, and up-scaling of plot-scale sap flow measurements to forest-scale ET estimates.

2. Materials and Methods

2.1. Study Site

The study was conducted at Swanton Pacific Ranch, located along the Pacific coast in Santa Cruz County, California, about 84 km south-southeast of San Francisco. A map of the watershed and study area is shown in Figure 1. The climate of the region is Mediterranean, with warm, mostly dry summers and cool, wet winters. The mean summer air temperature highs are 24 °C and mean winter air temperature lows are 5 °C. The rainy season is typically from October to April, with an average yearly precipitation of 975 mm, with an average of

193 mm occurring in January. Even during the recent prolonged drought in California from December 2011 to March 2019, the average yearly precipitation was 945 mm. The average yearly precipitation over the duration of this study (August 2017 through August 2019) was 855 mm. Streamflow in main stream in the watershed, Scotts Creek, is typically very low in the summer ($\leq$0.1 m^3/s). During the winter, peak flows typically are 20–70 m^3/s, based on data from a Scotts Creek stream gauge.

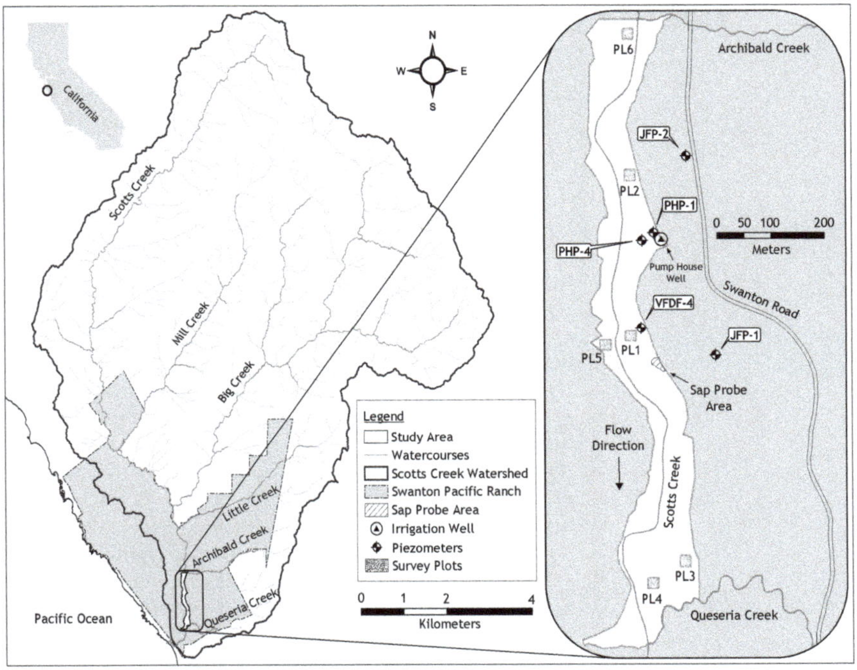

Figure 1. A map (adapted from [20]) of the Scotts Creek watershed, Swanton Pacific Ranch, and the riparian forest study area. The map shows the location of the instrumented phreatophytes, survey plots, and piezometers.

The riparian corridor within the study area is about 70–140 m wide with a canopy cover that often approaches 100% during the growing season [21]. The dominant trees along the lower portion of the Scotts Creek watershed are red alders (*Alnus rubra* Bong.), arroyo willows (*Salix lasiolepis* Benth.), and pacific willows (*Salix lasiandra* Benth. var. *lasiandra*). Other trees include box elder (*Acer negundo* L.), bigleaf maple (*Acer macrophyllum* Pursh.), California bay laurel (*Umbellularia californica* (Hook. & Arn.) Nutt.), and coastal redwoods (*Sequoia sempervirens* (D. Don) Endl.) Common understory vegetation includes California blackberry (*Rubus ursinus* Cham. & Schltdl.), stinging nettle (*Urtica dioica* subsp. *gracilis* L.), poison hemlock (*Conium maculatum* L.), Cape ivy (*Delairea odorata* Lem.), and Italian thistle (*Carduus pycnocephalus* L. subsp. *pycnocephalus*) [21–23]. The phreatophytes documented within the study area, including red alders, arroyo willows, pacific willows, box elders, and bigleaf maples, are all deciduous. They typically lose their leaves in November/December and their leaf buds burst in early March. They maintain maximum leafage for most of the spring, summer, and fall growing seasons. **The typical site vegetation is shown in Figure 2, which shows (a) the dominant phreatophytic trees (b) understory vegetation, and (c) deciduous vegetation during winter dormancy.** Red alders are fast-growing, relatively short-lived, shade intolerant, and tend to favor sites with bare mineral soil and high sun exposure that were disturbed by floods, windthrows, logging, or fires.

(a) (b) (c)

Figure 2. Typical vegetation, including (**a**) phreatophytic trees (**b**) understory vegetation, and (**c**) deciduous vegetation during winter dormancy in the study area within the lower Scotts Creek riparian corridor in June 2017, June 2018, and January 2019, respectively.

2.2. Sap Flow Measurements

Four trees of the dominant phreatophytes at the study site were selected for monitoring based on their stem diameters (7.6–12.7 cm) and proximity (less than 33 m) to the data acquisition station. The location of the sap flow probe area is shown on the site map in Figure 1. A pair of probes was installed in each tree. The installation procedure involved removal of the outer bark at 1.22–1.45 m above the ground to minimize radiative temperature effects from land surface. Two pilot holes, 40 mm apart vertically, were then bored into the tree stem to a depth of 30 mm using a 1.5-mm diameter drill bit. The pilot holes and drill bit were flushed with 10% chlorine bleach prior to and after boring the holes in each tree to minimize the introduction and spread of pathogens. The probes were then carefully inserted into the bores and adhesive putty applied around the base of each probe to provide a water-tight seal. Foam covers were placed over the probes for thermal insulation and to protect the electrical wiring. Reflective bubble insulation was wrapped around the probes, foam, and tree stem to minimize thermal gradients caused by direct solar radiation. Saran wrap was wrapped around the tree stem and upper portion of the reflective bubble insulation to prevent water from flowing down the stem surface and into the probes. Figure 3 depicts the probe installation steps (a)–(d) and the aftermath of tree healing that occurred over the study period (e).

(a) (b) (c) (d) (e)

Figure 3. Pictures of the probe installation and insulation procedure. (**a**) the dual probe after insertion into stem drill holes and sealing with putty, (**b**) foam insulation cover over probes, (**c**) reflective blanket, (**d**) saran wrapped installation, and (**e**) post-study period probe condition shown growth over probe.

The probes were part of the FLGS-TDP XM1000 sap velocity system (Dynamax Inc., Huston, TX, USA), which includes a CR1000 measurement and control data logger with a AM16/32 relay multiplexer (Campbell Scientific, Logan, UT, USA) housed in a rugged weather-resistant instrument enclosure. Communication, programming, and data extraction between the data logger and a computer were facilitated using the PC400 data logger support software (Campbell Scientific). The data logger and solar panel were mounted on a 10-ft UT10 aluminum tower in a forest canopy gap. The tower was secured in an 8-ft^3 concrete pad with a J-bolt kit for stability during rough weather and flooding. Probe cords were placed inside 1.0-inch diameter schedule 40 PVC conduit pipes and installed approximately 30 cm underground for protection against weather, flooding, and wildlife. The conduit pipe openings were covered with duct seal putty to keep out moisture. Upon completion of probe installation and mounting the data logger to the tower, sap temperature differentials were recorded at one-minute intervals and their averages were recorded every 15 min. The data were downloaded from the data logger every two months. A sap flow computation spreadsheet, provided by Dynamax Inc., and modified appropriately to implement the theoretical equations of [14], was used to calculate the volumetric rate of sap flow.

2.3. Measurement and Estimation of Tree Diameter and Sapwood Depth

Determination of the sapwood area, S_A, requires knowledge of tree stem diameters, d, and sapwood depth, D_{sap}. Hence, for this work, tree stem diameter, d, at breast height (herein d = DBH) was measured for all woody vegetation greater than 0.025 m in diameter in each of the six representative sample plots. For this work, the breast height used was 1.37 m. The stem diameters were measured manually with a standard English diameter tape. The tree diameter tape is based on the assumption that tree stems are perfect circles such that $d = C/\pi$, where C is tree stem circumference. Within each plot, the number of species, and number of trees for each species were also recorded.

Whereas tree stem diameter was measured for all woody vegetation in each of the sample plots, sapwood depths, D_{sap}, were measured from cores extracted from a small representative subset of the riparian phreatophytic trees within each plot. A stratified random sampling design was used to estimate phreatophytic vegetation composition. Woody vegetation was sampled in six random plots, each of area 400 m^2, within the riparian corridor. Environmental Systems Research Institute's (ESRI) ArcMap 10.7 was used to determine the locations of these random plots. First, a fishnet with 20 m $\times$ 20 m sections was laid over the study area in ArcMap. Then, random sections were chosen on the grid. The coordinates of the northwest corner of each plot were programmed into a Trimble Geo 7X handheld GPS for locating in the field. The locations of the remaining three corners for each plot were determined with an open reel measuring tape and a compass.

To measure D_{sap}, wood cores were extracted at breast height (1.37 m) from select phreatophytes using an increment borer (Haglöf Sweden AB) within each plot. The bark thickness, sapwood depth, and heartwood/pith radius of each core were measured in the field with a ruler. In most trees, the sapwood's lighter color made it simple to distinguish from the heartwood. However, in some trees (e.g., red alders and arroyo willows), there was very little color difference between the sapwood and heartwood [24]. In order to determine the sapwood depth, D_{sap}, wood cores were first stained with a 0.2% safranin dye by applying the dye to each core in a series of continuous drops using a small pipette. The dye was applied immediately after extracting the cores because the vessels of vascular system lose uptake pressure [25]. Because the dye is absorbed more easily by sapwood than by heartwood [26], it allows one to locate the sapwood and heartwood boundaries from which D_{sap} could then be estimated. Figure 4 depicts (a) the bore from which a core sample was retrieved, (b) an example of a retrieved tree core stained with dye, and (c) a close-up of the core showing sap wood to heartwood transition.

(a) (b) (c)

Figure 4. Core sampling to measure of sap wood and heartwood depth. (**a**) tree bore after core retrieval, (**b**) retrieved tree core, and (**c**) a close-up of the core showing sap wood to heartwood transition.

The sapwood depth, D_{sap}, of phreatophytic trees in each plot was measured in only a subset of the trees on the plot. Here, we outline the approach used to estimate D_{sap}, and the sapwood areas, S_A, of the non-sampled trees. Sapwood depth is defined as

$$D_{\text{sap}} = \frac{d}{2} - D_B - D_H, \quad (1)$$

where d is tree diameter at breast height (DBH), D_B is bark thickness, and D_H heartwood/pith radius. Estimates of bark thickness D_B for non-cored trees of known diameter, d, were estimated using the relation of [27], namely

$$D_B = a_1 e^{\beta_1 d}, \quad (2)$$

where a_1 and β_1 are empirical parameters determined using data from cored trees. Estimates of heartwood/pith radii were obtained using the relation

$$D_H = a_2 e^{\beta_2 d}, \quad (3)$$

for red alders, and

$$D_H = a_2 d + \beta_2, \quad (4)$$

for willows, where a_2 and β_2 are empirical parameters determined using data from cored trees. For cored trees, the measured bark thicknesses and heartwood/pith radii were plotted against the measured stem diameters and best fits of the models given in above were obtained to determine the values of the empirical parameters. With the empirical constants thus determined, Equations (2)–(4) were then used to estimate values of D_B and D_H for the non-cored trees given their measured diameter d.

For cored samples, the dye droplet method was used to determine the boundary between sapwood and heartwood. The dye was applied to every core immediately after extraction from the tree but yielded mixed results depending on the quality of the wood cores and the tree. On some cores, especially those from small trees, the sapwood absorbed the dye immediately. Heartwood in cores from older trees absorbed the dye at very slow rates, making it challenging to make the distinction between heartwood and sapwood in a timely manner. There were multiple instances where sapwood and heartwood could not be distinguished from each other based on dye absorption. In these cases, changes in color and/or texture were used to determine the sapwood/heartwood boundary. Overall, determining the boundary between sapwood and heartwood was very difficult, even with the dye droplet method.

2.4. Upscaling Plot Measurements to Forest Scale

Sapwood area is a measure of the actual tree stem area through which water extracted from the subsurface flows on its way to be transpired to the atmosphere from the canopy. The sapwood areas of all phreatophytic trees in six sample plots of the riparian forest were used to estimate the fractional sapwood basal area, α_k (expressed in m^2/ha), for each phreatophytic species over the riparian forest within the study area using the following relation adapted from [28]:

$$\alpha_k = \frac{1}{M}\sum_{p=1}^{M}\left(\frac{1}{A_p}\sum_{n=1}^{N_p} A_{n,p}^{(k)}\right) \tag{5}$$

where M is the number of sample plots, N_p is the number of trees of kth species in the pth sample plot, A_p is the forest floor area of the pth sample plot, and $A_{n,p}^{(k)}$ is the sapwood area of the nth tree of the kth species in the pth plot. In this work, the area of each of the six ($M = 6$) sample plots in which trees were counted and core samples collected, was fixed at $A_p = 400$ m^2. The total sapwood area, $A_{s,k}$, of a given phreatophytic species over the entire riparian forest within the study area was then estimated as simply

$$A_{s,k} = \alpha_k A_{\text{rf}}, \tag{6}$$

where A_{rf} is the measured total ground area of the forest. For this study, $A_{\text{rf}} = 9.2$ ha.

The total riparian forest sapwood area determined with this equation was then with sap flow data from the four instrumented trees to upscale measured sap flow to the riparian forest ET. First, the ET from each instrumented tree was calculated based on the areal extent of its canopy [29]. The canopy extent of each instrumented tree was determined with a Trimble Geo 7x GNSS handheld. In ArcMap, the GPS points of each tree were connected to create a polygon that represented the areal extent of the canopy. A modified version of an equation from [16] was used to calculate ET above the riparian corridor canopy viz.,

$$ET = \frac{1}{A_{\text{cp}}}\sum_{k=1}^{N} u_k A_{s,k} = \frac{A_{\text{rf}}}{A_{\text{cp}}}\sum_{k=1}^{N} u_k \alpha_k \tag{7}$$

where N is the number of tree species, u_k is the mean sap flux density of the kth phreatophytic species and A_{cp} is the combined canopy areal extent of the riparian forest. For simplicity, arroyo and pacific willows are treated as one species for this study such that the number of species was $N = 2$ (red alders and willows). This was necessitated by the fact that only one willow was instrumented.

2.5. Validation Method

The upscaled results will be validated by comparison to two common methods to estimate ET, namely the Modified Penman Equation for computing reference evapotranspiration (ET_o) and an equation utilizing normalized difference vegetation index (NDVI) data and meteorological data. The original Penman Equation comprises two terms: energy (radiation) and aerodynamic (wind and humidity) [30]. The Modified Penman Equation includes an amended wind function [31]. According to [32], the Modified Penman Equation has been shown to overestimate ET_o in conditions with high winds and low evaporation, but it offers the best ET_o estimates for grass surfaces. The ET_o (mm/d) was calculated using

$$ET_o = c[WR_n + (1 - W)f(u)(VPD)] \tag{8}$$

where c [-] is an adjustment factor compensating for difference in day and night weather conditions, W [-] is a temperature related weighting factor, R_n (mm/d) is the net solar radiation in equivalent ET, $f(u)$ [-] is a wind-related function, and VPD (mbar) is the vapor pressure deficit.

The second method for estimating ET is via satellite remote sensing and meteorological data. Remote sensing provides spatial and temporal coverage of the land surface [33]. NDVI is one of the many products that comes from remote sensing and it quantifies the density of green vegetation on a plot of land. Comprising imagery with near-infrared and red spectral bands, NDVI data are useful for monitoring changes in vegetation [34]. Due to chlorophyll in the leaves, vegetated areas absorb visible light and have high near-infrared reflectance.

In contrast, non-vegetated features have high visible light reflectance and low near-infrared reflectance, namely rocks, bare soil, water, snow, and clouds. Using an equation by [35], the ET of the riparian forest can be calculated with

$$ET = \frac{R_n \phi \Delta}{\rho \lambda (\Delta + \gamma)} \left(1 - 0.583 e^{-2.13\mathrm{NDVI}}\right) \tag{9}$$

where R_n ($\mathrm{W/m^2}$) is the net solar radiation, ϕ [-] is the aerodynamic and canopy resistance parameter, Δ is the slope of the saturated vapor pressure curve, ρ ($\mathrm{kg/m^3}$) is the density of water, λ (J/kg) is the latent heat of vaporization of water, and γ (kPa/K) is a psychrometric constant. The parameter $\phi \in (0, \phi_{\max})$ was estimated from a scatter plot of site surface temperature T_o and NDVI data using the linear interpolation scheme described in [35], where $\phi_{\max} = (\Delta + \gamma)/\Delta = 1.26$.

The ET_o data set consisted of hourly ET data reported by a California Irrigation Management Information System (CIMIS) automated weather station located 21 km from the study area, in Santa Cruz. The station uses the CIMIS version of the modified Penman-Monteith Equation [30] given by [31] to calculate ET from a standardized grass surface that is well-irrigated and closely cut, while completely shading the soil. NDVI and meteorological data were used to calculate the ET of the riparian forest with Equation (9). The NDVI data were taken from weekly EROS Moderate Resolution Imaging Spectroradiometer (eMODIS) composite sets at 250×250 $\mathrm{m^2}$ spatial resolution [36]. The weighted average NDVI value of the entire study area each week was calculated in ArcMap by determining the percentage of study area within each pixel. The NDVI values were calculated using $\mathrm{NDVI} = (\mathrm{IR} - \mathrm{R})/(\mathrm{IR} + \mathrm{R})$, where IR and R represents pixel values from the infrared and red bands, respectively. This yielded NDVI values in the range -1 to +1 for use in Equation (9). Required meteorological data comprised air pressure, air temperature, and solar radiation. Two sets of these data were taken from two separate weather stations (CIMIS and Weather Underground) within the general vicinity of the study area in order to compare sap flow based-ET to separate areas with slightly different weather patterns. The second station was a nearby Weather Underground (WU) station, located 5 km from the study area in Davenport, CA. The meteorological data of each weather station were averaged over the same weeks as the eMODIS composite sets.

3. Results

As stated previously, the objective of the work was to estimate riparian forest ET from sap flow measurements collected in a small sample of the predominant vegetation. The ET estimates are based on estimates of the total sapwood area for the entire riparian forest as well as its canopy areal extent. The results are presented in the following.

3.1. Phreatophytic Vegetation Survey and Sapwood Area

A total of 159 trees were surveyed in the six sample plots, with 153 of them being phreatophytes. They comprised 83 red alders, 61 arroyo willows, 9 pacific willows, and 6 coastal redwoods. The survey comprised direct measurement of DBH using diameter tape. Sapwood depth was measured directly in a subset of the surveyed phreatophytes by wood coring. The coastal redwoods are not considered phreatophytes, and thus were excluded from the calculations for the total sapwood area of the riparian forest. The survey results, including averages and standard deviations of DBH and sapwood area, are summarized in Table 1. The values in parentheses are for the subsamples that were selected for coring to obtain direct measurements of bark thickness and heartwood/pith radius for sapwood area estimation. Multiple cores were extracted on some trees because the heartwood and/or piths were difficult to sample. Larger trees were especially difficult to sample due to irregularities in radial growth of tree stems. The poor surface quality of the cores and the small difference in color between early wood and late wood made determining the age of trees challenging. Wood cores from young, small red alders (DBH of $d < 35.6$ cm) consistently showed only bark, sapwood, and piths, which agree with [17].

The histograms of the measured diameters at breast height for all surveyed phreatophytes are shown in Figure 5. Theoretical probability density functions are also included for completeness. The arroyo willows and pacific willows were analyzed as one composite group due to their small sample sizes (61 and 9, respectively). Weibull ($p = 0.090$ for willows and $p > 0.250$ for red alders) and lognormal ($p = 0.079$ for willows and $p = 0.214$ for red alders) distribution model fits are also included. When the red alders and willows were analyzed as one composite phreatophytic vegetation group, they appear to follow gamma ($p = 0.182$) and Weibull ($p = 0.087$) distributions based on a 95% confidence interval. All the probability density functions show positive skewness indicating a sampling bias on the small tree diameter end of the range. Figure 5 also shows histograms of cored main stem sapwood areas for red alders, arroyo willows, and pacific willows from the six sample plots.

Table 1. Statistics of surveyed phreatophytes within the six sample plots. The values in parentheses indicate the subsamples that were selected for coring to obtain direct measurements of bark thicknesses and heartwood/pith radii for sapwood area estimation.

Species	DBH (cm)			Sapwood Area (cm^2)		
	n	μ_{dbh}	σ_{dbh}	n	μ_{sa}	σ_{sa}
Red Alder	83 (24)	20.9 (31.6)	15.0 (17.0)	23	734.0	543.1
Arroyo Willow	61 (16)	11.6 (19.6)	7.5 (6.6)	14	318.4	140.7
Pacific Willow	9 (6)	27.5 (28.6)	10.4 (11.5)	5	526.2	347.7

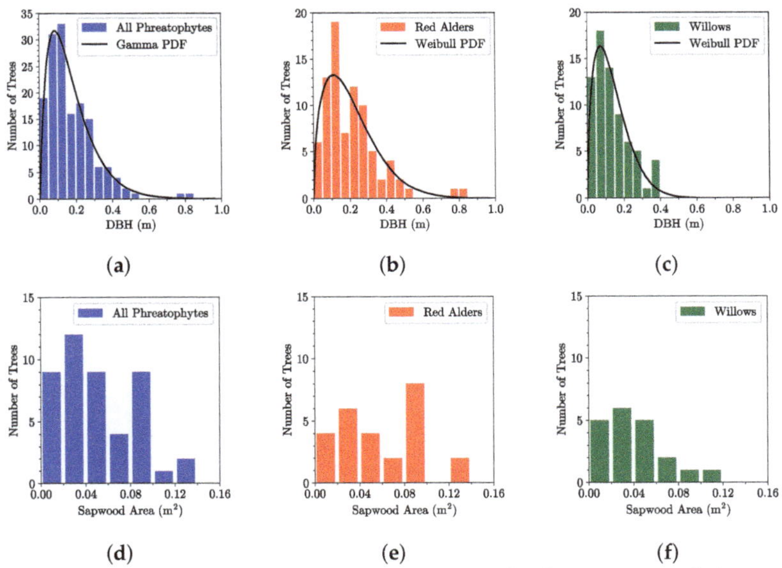

Figure 5. Histograms of *measured* tree stem diameters at breast height (DBH) for (**a**) all phreatophytes, (**b**) red alders, and (**c**) willows within the six sample plots, and histograms of *measured* main stem sapwood areas for (**d**) all phreatophytes, (**e**) red alders, and (**f**) willows within the six sample plots.

Estimates of model parameters in Equations (2)–(4) from tree core data are summarized in Table 2. These parameters were used to estimate sapwood basal area for the trees on which cores were not obtained but for which the diameter d at breast height was measured. Estimates of the fractional sapwood basal area and sapwood area across the entire riparian forest using Equations (5) and (6), respectively, are also summarized in Table 2.

Table 2. Model parameters for the best fit developed from wood core data within the six sample plots to estimate bark thickness and heartwood/pith radius based on DBH. Estimated fractional sapwood basal area, α_k, and total sapwood area, $A_{s,k}$ for each phreatophytic species across the riparian forest within the study area are also included.

Species	Bark Thickness					Heartwood/Pith Radius					Estimates		
	n	$f(x)$	a_1	β_1	R^2	n	$f(x)$	a_2	β_2	R^2	n	α_k (m^2/ha)	$A_{s,k}$ (m^2)
R-Alder	24	Exp	0.220	0.030	0.698	23	Exp	0.067	0.050	0.241	83	16.4	150.5
A-Willow	16	Exp	0.194	0.041	0.674	16	Linear	0.192	−2.214	0.454	61	3.3	30.3
P-Willow	6	Exp	0.331	0.042	0.690	5	Linear	0.045	0.207	0.333	9	2.1	19.3

3.2. Sap Flow Measurements

The four instrumented trees were continuously monitored at one-minute intervals and the data averaged every 15 min for 618 consecutive days from 18 August 2017 through 26 April 2019. The diurnal sap flow data for selected weekly periods in the growing seasons of each study year are shown in Figure 6. Several instances of morning peaks were observed in the data. They were particularly evident during the growing seasons of the study period for red alders 2 and 3 (Figure 6a), and attributable to direct incident solar radiation on the reflective shield wrapped around the probes, which served to minimize the occurrence of morning peaks. It may also indicate problems with installation of the probes. Another possibility is that they may be as a result of water release in the morning from tree trunk storage before tree roots uptake water to refill the storage according to [37,38] who found that tree trunk internal water storage can contribute as much as 28% of the daily water budget in some tree species. The arroyo willow and red alder 1 also show morning peaks, but had much lower amplitudes later in the morning. The time lag time between sunrise and initial sap flow for the arroyo willow and red alder may be due to partial shading by other trees. All instrumented trees showed some activity during the winter period of dormancy, with the peak amplitudes of greater than an order of magnitude smaller than those observed during periods of active growth.

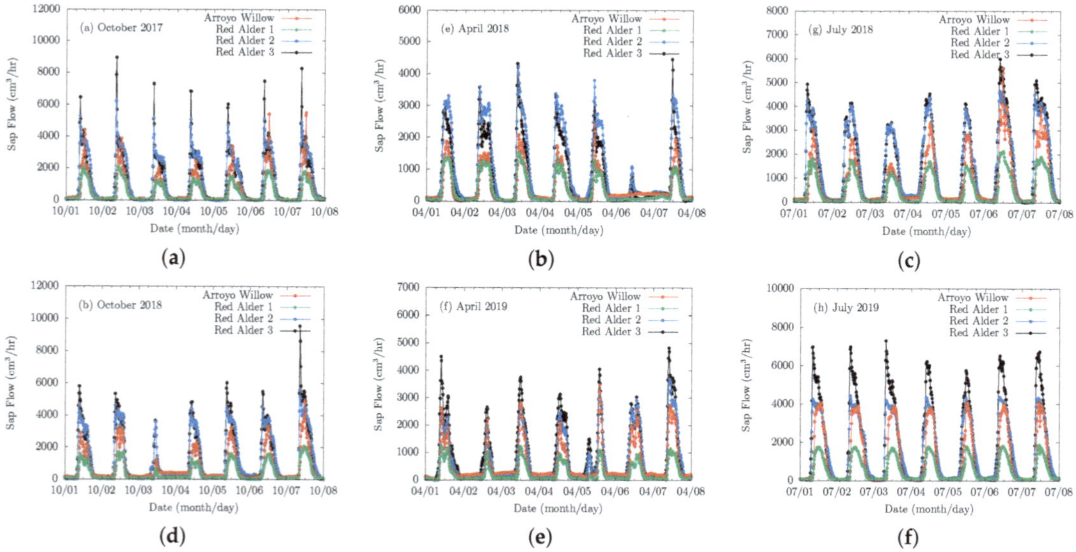

Figure 6. Example weekly sap flow data of the four instrumented trees collected over the two-year monitoring period. The graphs show sap flow measured in (**a**) Fall 2017, (**b**) Spring 2018, (**c**) Summer 2018, (**d**) Fall 2018, (**e**) Spring 2019, and (**f**) Summer 2019.

The time series of the sap flow data for the entire two-year monitoring period of the four instrumented trees are shown in Figure 7 for (a) arroyo willow, (b) red alder 1, (c) red alder 2, and (d) red alder 3. The daily maximum air temperatures and daily mean solar radiation over the same monitoring period are included in Figure 7e to highlight the seasonality of the observed behavior. Seasonality is clearly evident in the sap flow data with periods of high sap flow generally coinciding with spring, summer and fall seasons, interspersed with periods of minimal flow in winter seasons. The spring-fall period is the period of active growth, with leafage increasing to summer-fall maxima. The instrumented phreatophytes were deciduous, losing leaves in late fall, with complete leaf loss deep in the winter months of dormancy. Fall, winter, spring, and summer seasons are marked clearly on the figures to highlight their correlation to periods of significant sap flow change. Specifically, the active and dormancy periods of all four trees clearly follow the spring equinoxes and winter solstices (Figure 7).

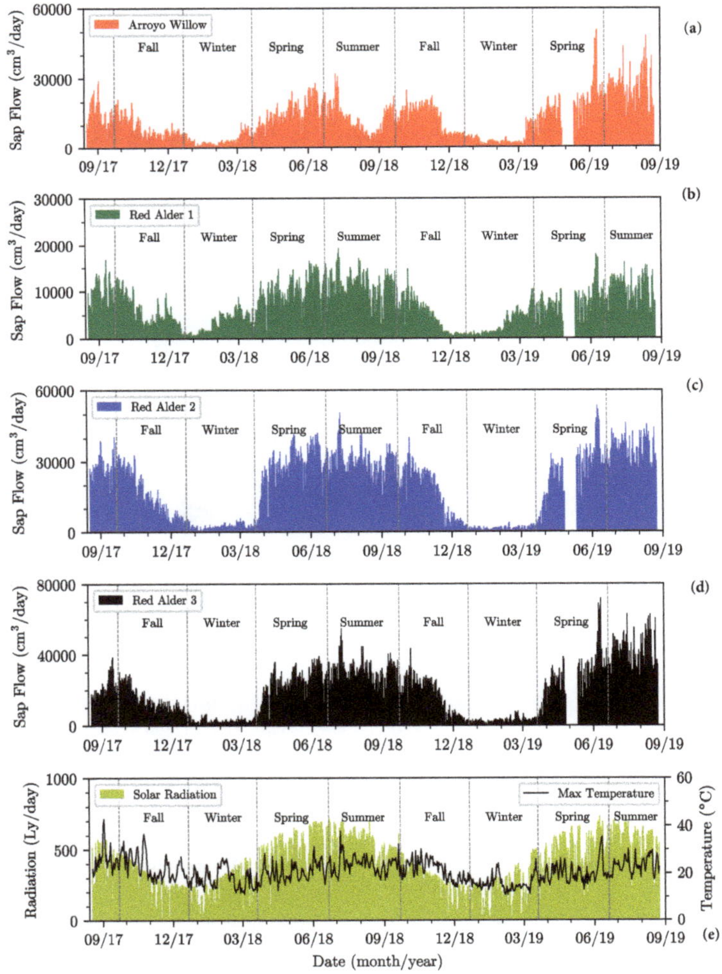

Figure 7. Daily total sap flow measured in (**a**) arroyo willow, (**b**) red alder 1, (**c**) red alder 2, (**d**) red alder 3, and (**e**) the CIMIS daily maximum air temperature and solar radiation over the monitoring period.

Sap flow peaked in the arroyo willow in early July 2018 and early June 2019. It was dormant from early January 2018 to early March 2018, and mid-December 2018 to mid-March 2019. It also showed a similar sap flow pattern to red alder 1 by mid-December 2018. Its sap flow pattern returned to normal by mid-March 2019. Mean peak sap flow in red alders occurred in early July 2018 and early June 2019. In general, the red alders were dormant from December to mid-March. Red alder 1 was dormant from early November 2017 to mid-March 2018 and mid-November 2018 to mid-March 2019. Red alder 2 was dormant from December 2017 to late March 2018, and late November 2018 to late March 2019. Red alder 3 was dormant from late December 2017 to late March 2018 and from mid-December 2018 to mid-March 2019. These results generally agree with those of [17], whose red alders were dormant during the winter period, though the period of dormancy is appreciably longer in Oregon, extending from October through March.

3.3. Evapotranspiration of Riparian Forest

Sap flow data were first used to estimate the ET of the individual instrumented trees. The results are shown in Figure 8. The seasonal variation in ET of the individual trees is clearly evident as one would expect from the sap flow data shown in Figure 7. The ET data among the four trees appear to show moderate to strong behavioral correlations, with red alders 2 and 3 consistently showing greater ET than the other two trees during the peak flow periods. The results obtained here did not show a general decrease in ET over the period of the study as has been observed by other workers. In fact, two trees (arroyo willow and red alder 3) appear to show increased ET in the final year (2019) of the study. The seasonal averages of the computed ET for the four trees are summarized in Table 3.

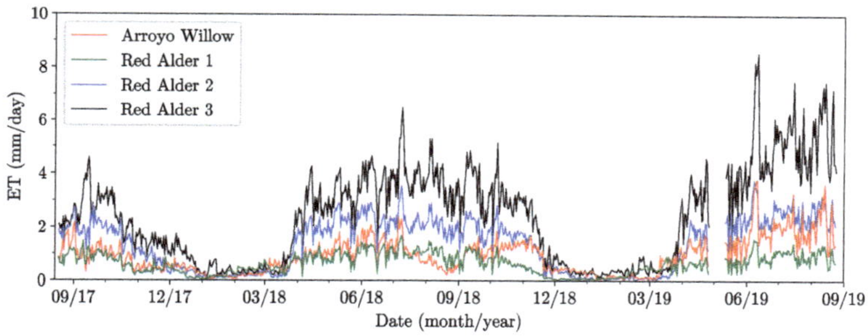

Figure 8. Evapotranspiration (mm/d) of the four instrumented trees from 18 August 2017 through 24 August 2019.

The sap flow data collected from the four instrumented trees were upscaled to the entire riparian forest canopy using Equation (7), and the estimated values are summarized and included in Table 3. The season averaged values range from a low of 0.5 mm/d during the winters to a high of 4.1 mm/d over the summer. Daily values show peak values in excess of 6 mm/d. It should also be noted that the winter average values are within margins of instrument measurement uncertainty.

The ET of the riparian forest estimated from sap flow data was compared to ET estimates based on NDVI and meteorological data, and ET_0 in the general vicinity of the study area. The seasonal averages of the sap flow-based ET, ET_0, and ET_{ndvi} are summarized in Table 4. Generally, there is strong correlation in the observed temporal behavior, as well as moderate agreement in estimates of ET over the active growing periods of spring, summer, and fall. This is particularly the case when comparing the sap flow-based ET to ET_0. However, there are notable divergences in the data. In fall of 2017, the sap flow-based ET appeared to be similar to the ET_0 and ET_{ndvi} based on CIMIS and WU data (Figure 9a). In winter of 2018, the sap flow-based ET was substantially lower than the ET_0 and both ET_{ndvi} estimates. In spring of 2018, the sap flow-based ET was marginally lower than the ET_0, but was substantially lower than both $ET_{ndvi,wu}$ values. In summer of 2018, sap flow-based ET was similar to $ET_{ndvi,wu}$, but it was marginally lower than the ET_o and $ET_{ndvi,cimis}$. This pattern was repeated in the second year of the study period.

Table 3. Seasonal estimates of mean ET (mm/d), and corresponding mean square errors, of the four instrumented trees and the entire riparian forest over study period.

Season	Arroyo Willow	Red Alders			Riparian Forest
		1	2	3	
Summer 2017 [2]	1.15 ± 0.43	0.93 ± 0.24	2.0 ± 0.38	2.68 ± 0.74	3.74 ± 0.63
Fall 2017	0.69 ± 0.33	0.53 ± 0.28	1.13 ± 0.64	1.82 ± 0.78	2.19 ± 1.07
Winter 2018	0.27 ± 0.18	0.31 ± 0.18	0.20 ± 0.09	0.40 ± 0.19	0.66 ± 0.26
Spring 2018	1.12 ± 0.43	0.90 ± 0.29	1.97 ± 0.66	2.96 ± 1.0	3.47 ± 1.12
Summer 2018	0.97 ± 0.48	1.04 ± 0.28	2.16 ± 0.41	3.79 ± 0.80	3.58 ± 0.90
Fall 2018	0.98 ± 0.43	0.43 ± 0.31	1.22 ± 0.71	2.07 ± 1.19	2.02 ± 1.11
Winter 2019	0.26 ± 0.21	0.28 ± 0.22	0.12 ± 0.05	0.38 ± 0.17	0.49 ± 0.26
Spring 2019 [2]	1.41 ± 0.68	0.73 ± 0.28	1.83 ± 0.81	3.45 ± 1.62	3.02 ± 1.27
Summer 2019 [2]	2.00 ± 0.57	0.97 ± 0.25	2.53 ± 0.38	5.21 ± 1.06	4.08 ± 0.86

[2] Seasons with incomplete or missing data

Table 4. Estimates of seasonal mean ET (mm/d) and corresponding MSE from the different methods across the entire riparian forest over study period.

	Sap	CIMIS	NDVI	
Season	Flow	ET_o	CIMIS	WU
Summer 2017 [2]	3.74 ± 0.63	3.32 ± 0.99	3.20 ± 0.62	-
Fall 2017	2.19 ± 1.07	2.34 ± 1.07	2.17 ± 0.87	-
Winter 2018	0.66 ± 0.26	197 ± 0.92	2.28 ± 0.80	-
Spring 2018	3.47 ± 1.12	4.04 ± 1.22	5.09 ± 1.25	4.85 ± 0.96
Summer 2018	3.58 ± 0.90	4.11 ± 1.07	4.32 ± 0.88	3.42 ± 1.11
Fall 2018	2.02 ± 1.1	2.12 ± 0.94	2.15 ± 0.64	1.70 ± 0.58
Winter 2019	0.49 ± 0.26	1.66 ± 0.99	2.15 ± 1.12	1.87 ± 1.11
Spring 2019 [2]	3.02 ± 1.27	3.69 ± 1.25	5.09 ± 1.33	4.62 ± 1.74
Summer 2019 [2]	4.08 ± 0.86	4.36 ± 1.02	4.94 ± 0.94	4.42 ± 1.12

[2] Seasons with incomplete or missing data.

The residuals of the ET, defined as the differences between the sap flow-based ET and the other methods, are shown in Figure 9b. The dashed red and blue lines on the graph in the figure mark ±1.0 and ±2.0 mm/d residual bounds, respectively. The residuals are highest during winter and early spring, during which periods the exceed 2.0 mm/d. The ET predicted by the other methods largely exceeds that based on sap flow due to dormancy of the willows and red alders during the winter seasons. During the mid-summer to late fall periods, the residuals are mostly within ±1.0 mm/d, indicating relatively strong agreement between sap flow-based ET and the other methods.

Scatter plots of sap flow based ET estimates versus the other three methods mentioned above are shown in Figure 10. The data show positive correlations between sap flow-based ET estimates and ET_o and the NDVI based estimates, with high variance and some bias as much of the data scatter is widely distributed above the 1:1 line (dashed red line). Data points above the 1:1 line indicate that sap flow-based ET was lower than the ET_o and the NDVI based ET. Table 5 shows the slopes and coefficients of determination (R^2) of the scatter plots with and without the winter data. Excluding winter data marginally improved the slopes and R^2 values. The fact that the slopes of the regression lines are higher than 1:1 is an indication of overall bias in the sap flow-based ET prediction of the ET_o and

NDVI/weather-based ET. The excluded winter data are marked in cyan in Figure 10b, where the data clearly plot above the 1:1 line, which confirms the observation made above that sap flow-based ET underestimates winter ET predicted by the other methods.

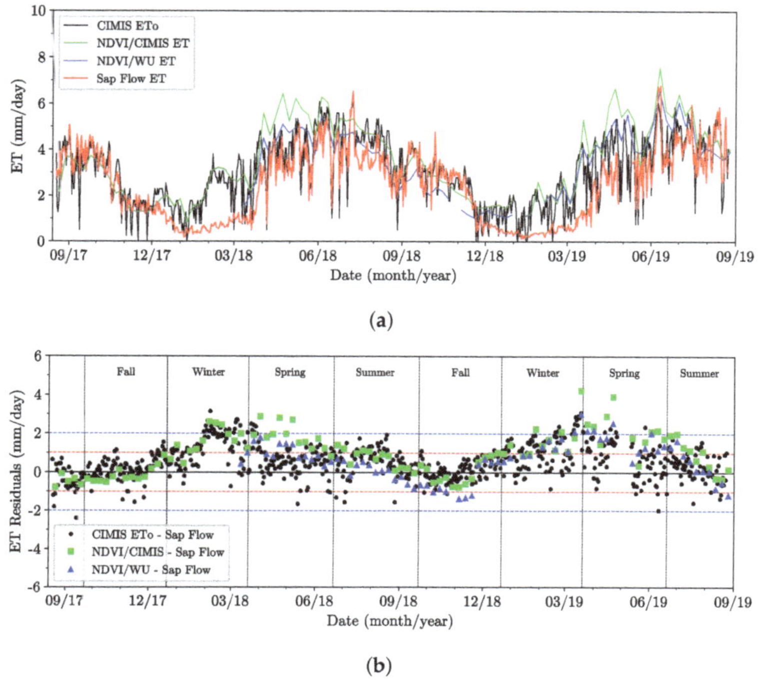

Figure 9. A plot comparing (**a**) the sap flow-predicted riparian forest ET to the other methods and (**b**) the corresponding residuals over study period. The dashed red and blue lines represent residual bounds of ±1.0 and ±2.0 mm/d, respectively.

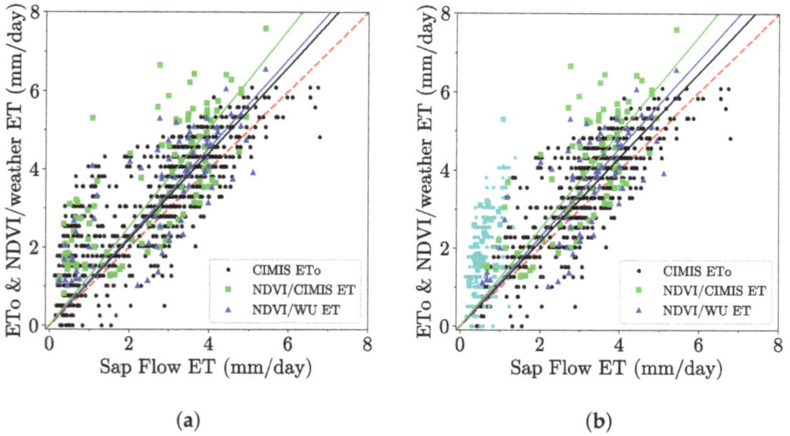

Figure 10. Scatter plots of ET_o and NDVI/weather-based ET versus sap flow-based ET with winter data in (**a**), and without winter data in (**b**) (the removed winter data are highlighted in cyan). The red line represents the 1:1 slope.

Table 5. Model parameters for the best fit through the origin (0, 0) to correlate sap flow-based ET with the ET_0 and NDVI/weather-based ET.

Method	w/ Winter Data		w/o Winter Data	
	Slope	R^2	Slope	R^2
CIMIS ETo	1.103	0.911	1.081	0.953
NDVI/CIMIS	1.260	0.893	1.237	0.932
NDVI/WU	1.135	0.914	1.136	0.938

3.4. Comparison with Groundwater Pumping for Irrigation

Groundwater fluctuations were continuously measured in five piezometers (JFP-1, JFP-2, PHP-1, PHP-4, and VFDP-4 in Figure 1) and the Pump House irrigation well from 18 August 2017 to 27 April 2019. All piezometers are completed in a thin clay/silt aquitard layer that sits atop the underlying leaky confined aquifer [39], and all responded to riparian forest ET as well as to pumping from the Pump House irrigation well. Data from the irrigation well and the most responsive of the piezometers, PHP-1, are shown in Figures 11 and 12 and are also reported in [20]. Piezometer and well data are reported here as changes relative to the respective first water level recorded to facilitate their comparison. Piezometer PHP-1 and the Pump House irrigation well are about 18 m apart. The data show fluctuations that are clearly caused by groundwater pumping, ET, recharge primarily from winter-spring precipitation events, and long-term discharge to the stream and ocean characterized by a period of recession from the end of spring through the summer and into the fall.

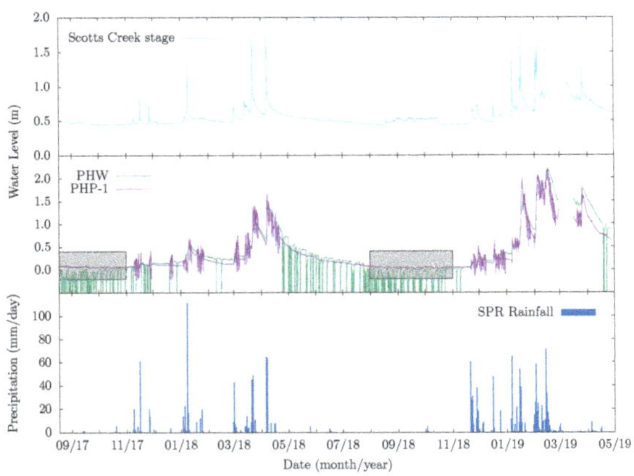

Figure 11. Groundwater fluctuations observed in piezometer PHP-1 and the Pump House irrigation well from 18 August 2017 through 26 April 2019. Daily precipitation of Swanton Pacific Ranch and stage of Scotts Creek are shown over the same period. The gray boxes represent the two zoomed-in time periods shown in Figure 12.

Diurnal groundwater level fluctuations due to ET forcing are superposed on those due to pumping from the irrigation well. Fluctuations due to ET are more pronounced in the piezometer data and are largely imperceptible in the irrigation well data. They are highlighted in Figure 12, which shows zoomed in plots of the aquifer and aquitard responses for two monitoring periods (08/2017–11/2017 and 08/2018–11/2018). Groundwater response to ET is most perceptible during periods of aquifer and aquitard recovery following a pumping event from the irrigation well. When the pumping frequency is daily, water-level responses to ET are practically indistinguishable from those attributable to pumping. In such cases, the amplitude of the aquitard response is much larger than what can be expected from ET alone. The data clearly show that the riparian corridor phreatophytes induce measurable fluctuations in the thin clay/silt aquitard that overlies the leaky aquifer.

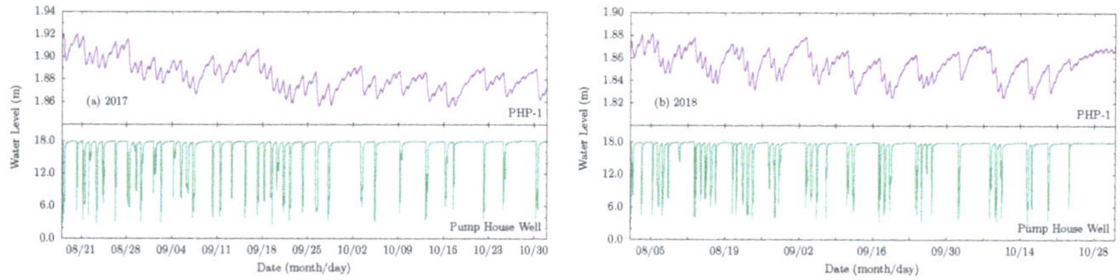

Figure 12. Groundwater fluctuations observed in piezometer PHP-1 and the Pump House irrigation well (**a**) from 18 August 2017 to 1 November 2017, and (**b**) from 1 August 2018 to 1 November 2018.

The diurnal groundwater fluctuations due to ET discussed above are induced by a forest-scale ET flux averaging 3.8 mm/d over the summer seasons. When multiplied by forest area, $A_{rf} = 9.2$ ha, this flux yields a volume flow rate usage by phreatophytes of $Q_{et} = 350$ m^3/d $= 64.1$ gpm. The typical pumping rate from the irrigation at the site is 250 gpm. Hence, ET across the forest amounts to about 25% the pumping rate, which is a substantial proportion.

4. Discussion

As stated previously, the objective of this work was to estimate riparian forest ET from sap flow measurements collected in a small sample of phreatophytic trees. Sap flow measurements were collected in the same four trees over the two-year monitoring period. For each individual tree, the data were largely repeatable during the growing seasons, with comparable average seasonal amplitudes. The fact that sap flow probes were left in the trees for such a prolonged period but continued to yield meaningful measurements was unexpected because [40] reported that other users of the sap flow probes used the same drill holes for one growing season, at the most. In addition, ref. [18] reported a 30% decrease in daily average sap flux density during the second growing season for red alders. The degradation of data quality over prolonged periods of monitoring has been attributed to tree response to drill-hole wounding by forming tyloses over those vessels, which affects heat exchange with the probes [41]. The good quality data collected over two-year study period may be attributable to the fact that the four instrumented trees were younger, smaller, and of different species than those in other studies. Hence, it may be argued that younger trees are better suited for prolonged monitoring than are older trees as long as their diameters and sapwood areas are corrected for from year to year in the computation of ET.

The ET flux projected across the entire riparian forest correlated strongly with the CIMIS ET_o and ET computed on the basis of NDVI/meteorological data. The sap flow-based forest ET had consistently lower average magnitudes during the growing seasons with significant departure from the values computed by the other methods over the winter seasons. This divergence in winter may be due to vegetation differences among the methods. Sap flow-based ET was collected on deciduous trees that lost their leaves every winter leading to values that were consistently lower than those from the other methods. The CIMIS ET_o values are based on a cool-season perennial grass that does not die back during winter and continues to transpire. Additionally, although most trees lost their leaves along lower Scotts Creek in winter, there was still plenty of green understory vegetation and some evergreen overstory vegetation, which were shown by the NDVI data. This may explain why the ET residuals showed seasonal patterns with peaks being highest and smallest during winter and fall seasons, respectively.

Long-term passive groundwater monitoring data were analyzed qualitatively to assess the magnitude of fluctuations in water levels from season to season and year to year. On average, at the study site, groundwater levels increased every winter before receding and reaching their lowest levels in the fall. The steady decrease in water levels during the summer and fall is largely attributable to ocean and stream discharge, consumption by phreatophytic vegetation [42], and groundwater withdrawal for crop irrigation. Diurnal groundwater fluctuations attributable to uptake by phreatophytes across the study forest are much smaller than fluctuations due to pumping, with sap flow based estimates of ET over the riparian corridor being about 25% the typical pumping rate for irrigation. Sap flow based estimates of groundwater fluctuations showed appreciable divergence from satellite based measurements, which suggests the importance of the former in calibration of the the latter, especially where site-specific vegetation may have greater control on local ET and consumptive groundwater use. A well characterized ET forcing function is essential modeling for groundwater flow and diurnal fluctuations [5,9,43–45].

There are some limitations in this study that may be addressed in future research, including (1) location and number of instrumented trees, (2) size of instrumented trees, (3) location and number of sap flow probes on tree stems, (4) instrumentation and monitoring of the minor tree species scattered throughout the riparian corridor such as live oak and redwoods, (5) accounting for the effect of understory vegetation on total ET and (6) measurement of sapwood area for individual trees and the entire riparian forest. The sampling design for this study was partly restricted due to the stem-diameter limitation of the probes, which biased monitoring to younger trees that are known to be hydraulically active than older trees. Using a combination of small and large sap flow probes to see the differences in sap flux density could provide more accurate estimates of ET. Additional pairs of probes could be installed on each instrumented tree at different depths and the sap flux densities averaged because the sap flux density is not uniform across sapwood area of a tree [15,16]. The method of using long-term sap flow measurements to estimate the ET of a riparian forest may be replicated on other phreatophytic species for similar or longer periods of time because phreatophytic tree species may react differently to sap flow probes in terms of sap flow behavior and physical intrusion of the probes.

5. Conclusions

The study presented herein was based on a two-year sapflow monitoring program on a single riparian forest plot where only a small sample of four trees were instrumented. The data were used together with a survey of tree sapwood depth in six plots across the entire forest to upscale the single-plot sapflow measurements to forest canopy-scale evapotranspiration (ET). The upscaled ET results were compared to ET based on NDVI-based estimates and were shown to be in good agreement. This indicates that for expensive ground-based technologies such as sapflow sensing by the heat dissipation models, a instrumentation of a small sample of a forest may yield reasonable estimates of forest canopy-scale ET as long as they are also based on sampling of sapwood depth across the entire forest. The results of the present study, despite the small sample size of sap flow measurements, illustrate the importance of ground-based measurements of sap flow for calibrating satellite based methods and for providing site-specific estimates and to better characterize the ET forcing in groundwater flow models. The small sample size is important because it is necessitated by the high cost of instrumenting individual trees and it suggests the potential usefulness of single-plot monitoring stations for ground-based measurement and estimation of forest ET.

Further research is required to better capture the spatial variability of sapflow across the forest and would include: (1) a larger sample of instrumented trees to better characterize sap flow behavior, (2) a sample of instrumented trees with a greater variety of main stem diameters in order to better characterize the sap flux density for each species, (3) greater spatial distances between instrumented trees, and (4) long-term monitoring of sap flow in additional phreatophytic species across the forest. The need for a greater sample size is clear even from the data from four instrumented trees because they very had different canopy and sapwood areas. Tree with larger sapwood area tend to have higher volumetric sap flow rate, which could bias the results based on a small sample size.

Author Contributions: Conceptualization, B.M. and J.S.; methodology, J.S.; software, J.S.; validation, B.M. and J.S.; formal analysis, B.M. and J.S.; investigation, B.M. and J.S.; resources, B.M.; data curation, J.S.; writing—original draft preparation, B.M.; writing—review and editing, B.M. and J.S.; visualization, J.S.; supervision, B.M.; project administration, B.M.; funding acquisition, B.M. All authors have read and agreed to the published version of the manuscript.

Funding: This work was funded by the USDA NIFA McIntire-Stennis Program Grant No. 16-109.

Institutional Review Board Statement: Not applicable.

Informed Consent Statement: Not applicable.

Data Availability Statement: Not applicable.

Acknowledgments: This work was facilitated by Swanton Pacific Ranch (SPR) Director Brian Dietterick, Manager of Operations Steve Auten, SPR staff Grant William and Brian Cook, graduate student Devin Pritchard-Peterson, and undergraduate research assistant Aren Abrahamian.

Conflicts of Interest: The authors declare no conflict of interest.

Abbreviations

The following abbreviations and acronyms are used in this manuscript:

CIMIS	California Irrigation Management Information System
DBH	Diameter at breast height
ESRI	Environmental Systems Research Institute
ET	Evapotranspiration
ETo	Reference evapotranspiration
eMODIS	EROS Moderate Resolution Imaging Spectroradiometer
HPV	Heat Pulse Velocity
NDVI	Normalized difference vegetative index
THB	Tissue Heat Balance
TDP	Thermal Dissipation Probe(s)
VPD	Vapor Pressure Deficit
WU	Weather Underground

References

1. Snider, B.; Urquhart, K.A.F.; Marston, D. *The Relationship between Instream Flow and Coho Salmon and Steelhead Habitat Availability in Scott Creek, Santa Cruz County, California*; Technical Report; California Department of Fish and Game, Environmental Services Division, Stream Flow and Habitat Evaluation Program : Redding, CA, USA, 1995.
2. Welsch, D.J. *Riparian Forest Buffers— Function and Design for Protection and Enhancement of Water Resources*; U.S. Department of Agriculture, Forest Service, Northeastern Area, State & Private Forestry, Forest Resources Management: Washington, DC, USA, 1991.
3. Jordan, T.E.; Correll, D.L.; Weller, D.E. Nutrient interception by a riparian forest receiving inputs from adjacent cropland. *J. Environ. Qual.* **1993**, *22*, 467–473. [CrossRef]
4. Schultz, R.C.; Isenhart, T.M.; Colletti, J.P. Riparian buffer systems in crop and rangelands. *Nat. Resour. Ecol. Manag. Conf. Pap. Posters Present.* **2005**, *23*, 13–28.
5. Gribovszki, Z.; Kalicz, P.; Szilágyi, J.; Kucsara, M. Riparian zone evapotranspiration estimation from diurnal groundwater level fluctuations. *J. Hydrol.* **2008**, *349*, 6–17. [CrossRef]
6. Robinson, T.W. *Phreatophytes*; Water-Supply Paper 1423; U.S. Geological Survey: Washington, DC, USA, 1958.
7. Naumburg, E.; Mata-Gonzalez, R.; Hunter, R.G.; Mclendon, T.; Martin, D.W. Phreatophytic vegetation and groundwater fluctuations: A review of current research and application of ecosystem response modeling with an emphasis on Great Basin vegetation. *Environ. Manag.* **2005**, *35*, 726–740. [CrossRef] [PubMed]
8. Rood, S.B.; Bigelow, S.G.; Hall, A.A. Root architecture of riparian trees: River cut-banks provide natural hydraulic excavation, revealing that cottonwoods are facultative phreatophytes. *Trees* **2011**, *25*, 907–917. [CrossRef]
9. Malama, B.; Johnson, B. Analytical modeling of saturated zone head response to evapotranspiration and river-stage fluctuations. *J. Hydrol.* **2010**, *382*, 1–9. [CrossRef]
10. Scott, R.L.; Shuttleworth, W.J.; Goodrich, D.C.; Maddock, T., III. The water use of two dominant vegetation communities in a semiarid riparian ecosystem. *Agric. For. Meteorol.* **2000**, *105*, 241–256. [CrossRef]

11. Goodrich, D.C.; Scott, R.; Qi, J.; Goff, B.; Unkrich, C.L.; Moran, M.S.; Williams, D.; Schaeffer, S.; Snyder, K.; MacNish, R.; et al. Seasonal estimates of riparian evapotranspiration using remote and in situ measurements. *Agric. For. Meteorol.* **2000**, *105*, 281–309. [CrossRef]
12. Loheide, S.P., II; Butler, J.J., Jr.; Gorelick, S.M. Estimation of groundwater consumption by phreatophytes using diurnal water table fluctuations: A saturated-unsaturated flow assessment. *Water Resour. Res.* **2005**, *41*, W07030. [CrossRef]
13. Johnson, B.; Malama, B.; Barrash, W.; Flores, A.N. Recognizing and modeling variable drawdown due to evapotranspiration in a semiarid riparian zone considering local differences in vegetation and distance from a river source. *Water Resour. Res.* **2013**, *49*, 1030–1039. [CrossRef]
14. Granier, A. Une nouvelle méthode pour la mesure du flux de sève brute dans le tronc des arbres. *Ann. Des Sci. For.* **1985**, *42*, 193–200. [CrossRef]
15. Granier, A.; Biron, P.; Bréda, N.; Pontailler, J.Y.; Saugier, B. Transpiration of trees and forest stands: Short and long-term monitoring using sapflow methods. *Glob. Chang. Biol.* **1996**, *2*, 265–274. [CrossRef]
16. Moore, G.W.; Bond, B.J.; Jones, J.A.; Phillips, N.; Meinzer, F.C. Structural and compositional controls on transpiration in 40-and 450-year-old riparian forests in western Oregon, USA. *Tree Physiol.* **2004**, *24*, 481–491. [CrossRef]
17. Moore, G.W.; Bond, B.J.; Jones, J.A. A comparison of annual transpiration and productivity in monoculture and mixed-species Douglas-fir and red alder stands. *For. Ecol. Manag.* **2011**, *262*, 2263–2270. [CrossRef]
18. Moore, G.W.; Bond, B.J.; Jones, J.A.; Meinzer, F.C. Thermal-dissipation sap flow sensors may not yield consistent sap-flux estimates over multiple years. *Trees* **2010**, *24*, 165–174. [CrossRef]
19. Bidlack, J.E.; Jansky, S.H. *Stern's Introductory Plant Biology*, 12th ed.; McGraw-Hill: London, UK, 2011; p. 622.
20. Malama, B.; Pritchard-Peterson, D.; Jasbinsek, J.J.; Surfleet, C. Assessing Stream-Aquifer Connectivity in a Coastal California Watershed. *Water* **2021**, *13*, 416. [CrossRef]
21. Cook, B.O. Lower Scotts Creek Floodplain and Habitat Enhancement Project. Master's Thesis, California Polytechnic State University, San Luis Obispo, CA, USA, 2016.
22. Louen, J.M. Hydrologic Characteristics of Summer Stream Temperatures in Little Creek and Scotts Creek at the Swanton Pacific Ranch. Master's Thesis, California Polytechnic State University, San Luis Obispo, CA, USA, 2016.
23. West, J. Traversing Swanton Road. 2016. Available online: https://arboretum.ucsc.edu/pdfs/traversin-gswanton.pdf (accessed on 20 April 2022).
24. Kutscha, N.P.; Sachs, I.B. *Color Tests for Differentiating Heartwood and Sapwood in Certain Softwood Tree Species*; Report 2246; U.S. Department of Agriculture, Forest Service, Forest Products Laboratory: Washington, DC, USA, 1962.
25. Moore, G.W.; Aparecido, L.M.T. (Discussion of Safranin Dye injection Apparatus for Red Alders). Personal Communication, 2019.
26. Bamber, R.K. *Sapwood and Heartwood*; Technical Publication 2; Forestry Commission of New South Wales, Wood Technology and Forest Research Division: Beecroft, NSW, Australia, 1987.
27. Oishi, A.C.; Oren, R.; Stoy, P.C. Estimating components of forest evapotranspiration: A footprint approach for scaling sap flux measurements. *Agric. For. Meteorol.* **2008**, *148*, 1719–1732. [CrossRef]
28. Avery, T.E.; Burkhart, H.E. *Forest Measurements*, 5th ed.; McGraw-Hill: London, UK, 2002.
29. Schaeffer, S.M.; Williams, D.G.; Goodrich, D.C. Transpiration of cottonwood/willow forest estimated from sap flux. *Agric. For. Meteorol.* **2000**, *105*, 257–270. [CrossRef]
30. Penman, H.L. Natural evaporation from open water, bare soil and grass. *Proc. R. Soc. Lond. Ser. A Math. Phys. Sci.* **1948**, *193*, 120–145.
31. Doorenbos, J.; Pruitt, W.O. *Guidelines for Predicting Crop Water Requirements*; FAO Irrigation and Drainage Paper 24; Food and Agriculture Organization of the United Nations: Rome, Italy, 1977.
32. Allen, R.G.; Pereira, L.S.; Raes, D.; Smith, M. *Crop Evapotranspiration–Guidelines for Computing Crop Water Requirements*; FAO Irrigation and Drainage Paper 56; Food and Agriculture Organization of the United Nations: Rome, Italy, 1998.
33. Bisht, G.; Venturini, V.; Islam, S.; Jiang, L.E. Estimation of the net radiation using MODIS (Moderate Resolution Imaging Spectroradiometer) data for clear sky days. *Remote Sens. Environ.* **2005**, *97*, 52–67. [CrossRef]
34. Lillesand, T.M.; Kiefer, R.W.; Chipman, J.W. *Remote Sensing and Image Interpretation*, 6th ed.; John Wiley & Sons, Inc.: Hoboken, NJ, USA, 2008; p. 756.
35. Batra, N.; Islam, S.; Venturini, V.; Bisht, G.; Jiang, L.E. Estimation and comparison of evapotranspiration from MODIS and AVHRR sensors for clear sky days over the Southern Great Plains. *Remote Sens. Environ.* **2006**, *103*, 1–15. [CrossRef]
36. Jenkerson, C.; Maiersperger, T.; Schmidt, G. *eMODIS: A User-Friendly Data Source*; Open-File Report 2010–1055; U.S. Geological Survey: Washington, DC, USA, 2010.
37. Goldstein, G.; Andrade, J.L.; Meinzer, F.C.; Holbrook, N.M.; Cavelier, J.; Jackson, P.; Celis, A. Stem water storage and diurnal patterns of water use in tropical forest canopy trees. *Plant Cell Environ.* **1998**, *21*, 397–406. [CrossRef]
38. Carrasco, O.L.; Bucci, S.J.; Di Francescantonio, D.; Lezcano, O.A.; Campanello, P.I.; Scholz, F.G.; Rodríguez, S.; Madanes, N.; Cristiano, P.M.; Hao, G.Y.; et al. Water storage dynamics in the main stem of subtropical tree species differing in wood density, growth rate and life history traits. *Tree Physiol.* **2015**, *35*, 354–365. [CrossRef] [PubMed]
39. Pritchard-Peterson, D. Field Investigation of Stream-Aquifer Interactions: A Case Study in Coastal California. Master's Thesis, California Polytechnic State University, San Luis Obispo, CA, USA, 2018.

40. Köstner, B.; Granier, A.; Čermák, J. Sapflow measurements in forest stands: Methods and uncertainties. *Ann. Des Sci. For.* **1998**, *55*, 13–27. [CrossRef]
41. Wullschleger, S.D.; Childs, K.W.; King, A.W.; Hanson, P.J. A model of heat transfer in sapwood and implications for sap flux density measurements using thermal dissipation probes. *Tree Physiol.* **2011**, *31*, 669–679. [CrossRef]
42. Loheide, S.P., II. A method for estimating subdaily evapotranspiration of shallow groundwater using diurnal water table fluctuations. *Ecohydrol. Ecosyst. Land Water Process Interact. Ecohydrogeomorphol.* **2008**, *1*, 59–66. [CrossRef]
43. Butler, J.J., Jr.; Kluitenberg, G.J.; Whittemore, D.O.; Loheide, S.P., II; Jin, W.; Billinger, M.A.; Zhan, X. A field investigation of phreatophyte-induced fluctuations in the water table. *Water Resour. Res.* **2007**, *43*. [CrossRef]
44. Szilágyi, J.; Gribovszki, Z.; Kalicz, P.; Kucsara, M. On diurnal riparian zone groundwater-level and streamflow fluctuations. *J. Hydrol.* **2008**, *349*, 1–5. [CrossRef]
45. Gribovszki, Z.; Szilágyi, J.; Kalicz, P. Diurnal fluctuations in shallow groundwater levels and streamflow rates and their interpretation—A review. *J. Hydrol.* **2010**, *385*, 371–383. [CrossRef]

sustainability

MDPI

Article

Mixed-Unit-Model-Based and Quantitative Studies on Groundwater Recharging and Discharging between Aquifers of Aksu River

Jiyu Huang, Yanyan Ge * and Sheng Li

The Key Laboratory of Geodynamic Processes and Metallogenic Prognosis of the Central Asian Orogenic Belt College of Geology and Exploration Engineering, Xinjiang University, Urumchi 830017, China; hjyhyj@stu.xju.edu.cn (J.H.); lisheng2997@163.com (S.L.)
* Correspondence: geyanyan0511@xju.edu.cn

Abstract: The confined aquifer in the Aksu River Basin is the main aquifer for drinking water within the area. In this study, the unconfined aquifer and the confined aquifer in the Aksu River Basin were divided into different water circulation units through analysis of their flow field. After the hydrochemistry and isotope characteristics of each unit were analyzed, these data were used as water volume quantitative information of the aquifer according to the mixed-unit model. With this quantitative information, the transformation relationship between the unconfined aquifer and the confined aquifer, the recharging source, recharging amount, recharging proportion, and discharging amount of the confined aquifer were revealed. The results showed that the confined aquifer receives a recharge of 21.48×10^6 m^3/a from the unconfined aquifer. The recharging sources of the confined aquifer in the middle and upper stream of the Aksu River mainly included side recharging and leakage recharging from the unconfined aquifer, while the confined aquifer received little recharging from unconfined aquifer downstream of the Aksu River and did not receive recharging from the unconfined aquifer in the southeast of the basin. Additionally, drainage methods of the confined aquifer were mainly lateral flowing and artificial well-group pumping. The side discharging volume through the whole area was 15.67×10^6 m^3/a, and the artificial pumping volume was 21.20×10^6 m^3/a. The confined aquifer was in a negative balance state from the middle-upper stream to the downstream. The downstream confined aquifer and its unconfined aquifer had a plane laminar flow movement, and the unconfined aquifer provided very little recharging to the confined one, which was further enhanced by the artificial well pumping and caused an accumulating negative balance state of the downstream aquifer.

Keywords: confined aquifer; unconfined aquifer; transformation; mixed-unit method; Aksu River Basin

Citation: Huang, J.; Ge, Y.; Li, S. Mixed-Unit-Model-Based and Quantitative Studies on Groundwater Recharging and Discharging between Aquifers of Aksu River. *Sustainability* **2022**, *14*, 6936. https://doi.org/10.3390/su14116936

Academic Editors: Alban Kuriqi and Luis Garrote

Received: 7 May 2022
Accepted: 1 June 2022
Published: 6 June 2022

1. Introduction

1.1. Research Status

As the source of life, water is inseparable from human survival and development. Although there are many water resources, groundwater with stable volume and excellent quality has become an important water resource for agriculture, animal husbandry, industry, and cities. As an indispensable resource for human society, in arid and semi-arid areas with limited precipitation and a small, unevenly distributed surface water resource, the effect of recharging and drainage of groundwater is significant [1,2] and has attracted the attention of many scientists in different fields.

Therefore, in the early 1950s, the United Nations Educational, Scientific and Cultural Organization (UNESCO) began to study the groundwater cycle [3], and the International Association of Hydrogeologists (IAH) also actively carried out many large-scale academic projects on groundwater circulation [4,5]. In China, water resource is the largest and most rigid constraint for production and life throughout the arid area in the northwest. Today, when we vigorously promote the construction of ecological civilization, we need to insist on using water to plan cities, land, populations, and production. Therefore, the efficient and reasonable development, and sustainable utilization, of groundwater resources are particularly important.

Many methods, such as hydrodynamic methods [6,7], hydrochemical methods [8–10], and environmental isotope methods [11,12] have been reported for studying the relationship between groundwater recharging and discharging.

The hydrodynamic method is based on groundwater chemical dynamics theories, according to the calculation and analysis of hydrochemical indicators (component activity and the mineral saturation index) and limited pumping-test data, five hydrogeological parameters (the permeability coefficient, K; the water conductivity coefficient, T; the actual velocity of groundwater, U; its penetration velocity, v; and the groundwater age, t) and can be used to quantitatively analyze and study all hydrogeological conditions [13,14]. The chemical composition of groundwater is usually controlled by many factors, such as the composition of precipitation, the geological structure, the mineral composition, and the hydrogeological processes of the aquifer. The continuous interaction between groundwater and its surrounding media also changes its chemical composition. Therefore, according to the relative concentration of the main ions in precipitation, surface water, and groundwater from different aquifers, information on the geochemical process in aquifers can be obtained to analyze the law and control mechanism of groundwater evolution, as well as the possible groundwater-evolution path from the recharging area to the discharging area [15–17]. Recently, isotope technology, as a new type of technology, has been developed in hydrogeology to effectively trace the change in water bodies and environment very sensitively, and thus, to record historical information about the evolution of the water cycle [18–20]. Since the 1950s, synthetic isotopes and environmental isotopes have been used to study issues related to hydrology and hydrogeology [21–24]. Many scientists worldwide have used these isotope methods to solve problems related to groundwater recharging resources, surface water transformation, surface-water runoff rate, and the age of surface water. Some scientists have further applied water-chemistry information to groundwater numerical models, used isotopes to trace and determine the recharging resource of groundwater, and calculated the amount of groundwater recharging [25–27]. In the 1990s, quantitative mathematical models became very mature. As one of these mature mathematical models, the mixed-unit model, with water-chemistry data and isotope data, can be used to quantitatively calculate the recharging rate and recharging amount of an aquifer in a specific space [28–31]. These results will be used in studies on groundwater cycles to provide a reliable basis for the rational development and utilization of groundwater resources in the arid area of Northwest China with limited hydrogeological work and low precision.

1.2. Purpose of the Research

The Aksu River Basin has four independent rivers from west to east: Aksu River, Kekeya River, Tailan River, and Karayuergun River. The Aksu River is one of the typical large rivers in the northern margin of the Tarim Basin with two major tributaries in its upper stream: the tributary of the Toshigan River on the west and the tributary of the Kumara River on the north. The Toshigan River originates from the Aksai River in the Atbash Mountains of Kyrgyzstan, the Kumarak River comes from the Khan Tengri Peak of the Tianshan Mountains, and both of them recharge rivers with water from snow-melting of glaciers and from precipitation. Twelve km to the south, the Aksu River divides into the Xinda River and the Laoda River. The Laoda River merges into the Xinda River again in Bawutulak, flows south, and enters the Tarim River in Xiaojiake. Its main stream is 132 km long and its drainage area is 63,100 km^2. The Kekya River originates from the Kochikal Basili Glacier and the Ishtarji Glacier. It goes through the Duolang Canal and merges into the Xinda River in Georgia, and has a total length of 82 km. Both the Tailan River and the Karayuergun River originate from the southern foot of Tuomuer Peak in the South Tianshan Mountains and are independent water systems.

As the Aksu River Basin is located at the southern foot of the Tianshan Mountains and has a dry climate with limited rainfall, the population and agricultural production are currently mainly concentrated in its oasis zone with the confined aquifer as an important water source. Many scientists have studied and provided information on the transformation relationship between river and groundwater (mostly unconfined aquifers). However, the recharging and discharging relationships, and the circulation mode of the confined aquifer are not currently understood.

In this study, based on data from the unconfined aquifer and confined aquifer flow fields in the basin, samples of river water, the unconfined aquifer, and the confined aquifer were systematically collected. After their water chemistry and isotope distribution characteristics were analyzed, the mixed-unit method was used to quantify these data, and thus, to reveal the recharging source and circulation mode of the confined aquifer.

2. Geology and Hydrogeology

The Aksu River Basin is located at the southern foot of Tianshan Mountains and the northern edge of the Tarim Basin, which belongs to the first-level tectonic unit of the Tarim platform. The water system in this basin was formed from the end of the Tertiary to the beginning of the Quaternary. Due to the neotectonic movement of the northern mountain body, a downstream river system was formed along the south-dipping slope of the mountain body. The water flow has brought mountain debris to the front of the mountain and deposited it in the Awati fault depression, gradually forming the alluvial plain of the Aksu River and the Kekeya River. Additionally, the uplift of the Yingan Mountains has led to a decline of the southeast side of the study area, and the formation of a strip of lowland in Aiximan (Figure 1). With water accumulation, a bead-like lake group was generated. Meanwhile, the Aksu River continued to swing in periods and moved eastward to the current Laoda River and Xinda River, leaving several river traces in the west of the plain, which then evolved into an intermittent strip of an oxbow lake, as shown in Figure 1. The geomorphological units of the Aksu River Basin from north to south are the piedmont alluvial fan group, the alluvial–proluvial slope plain, and the fine-soil-grain plain. As shown in Figure 2, from north to south, the lithology changes from coarse to fine, and sandy gravel changes from medium-coarse sand, to fine sand, to sandy loam. The sloping gravel plain area in the piedmont zone is a single unconfined aquifer area. Its water is more than 50 m in depth with the deepest part being 220 m, and its thickness is 90–100 m. The gently sloping fine-soil plain area and the desert plain area are a multi-layer area with unconfined and confined aquifers. The unconfined aquifer of Tumuxiuke Town–Wensu-Jiamu Town, north of Wutuan is buried 10–50 m deep, and the middle and downstream of the unconfined aquifer of the alluvial plain is less than 10 m in depth. In the south of Ayikule Town, Rice Farm, and the south of Wutuan, groundwater overflows from an artesian well. The south and southeast are formed with confined aquifer rock groups (mainly sand layers), and the thickness of the confined aquifer gradually increases from the north to the south within 15–130 m. The confined aquifer winging out in the west of Aksu is influenced by the Yinganshan uplift. The groundwater flow in the unconfined aquifer and the confined aquifer in the Aksu River Basin is affected by this neotectonic movement, and flows from north to south. Its downstream flowing direction changes from north-to-south to south-to-east as shown in Figures 3 and 4.

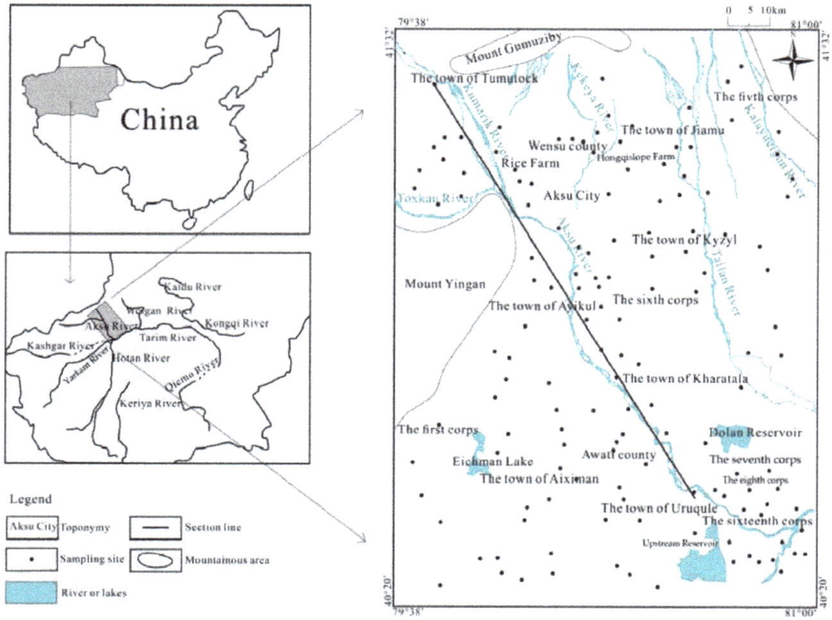

Figure 1. Locations of the studied area and sampling sites.

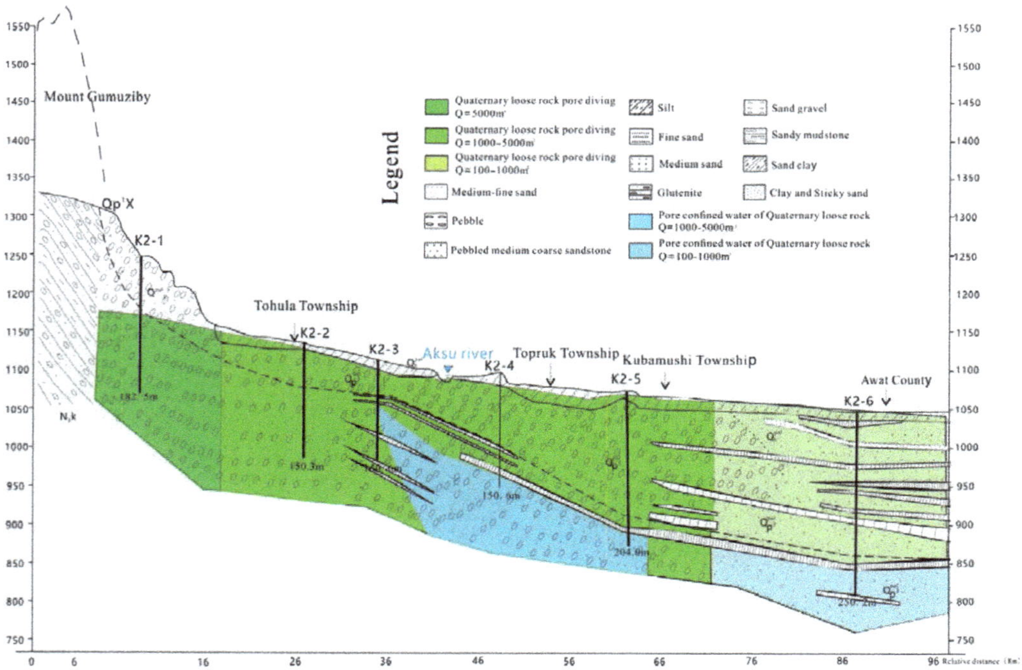

Figure 2. The hydrogeological profile of A–A′ section.

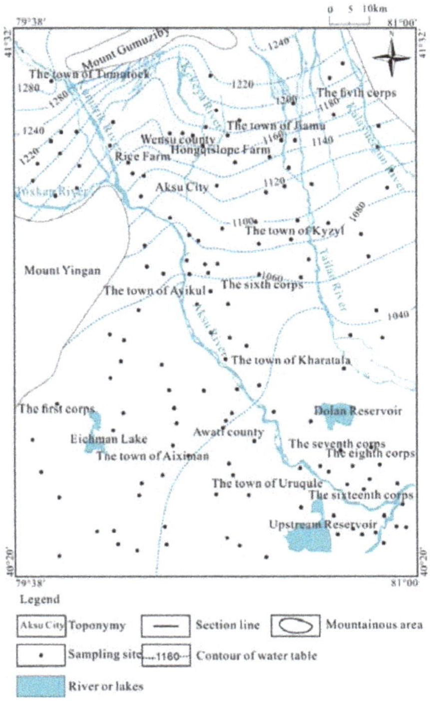

Figure 3. Contour lines of unconfined aquifer.

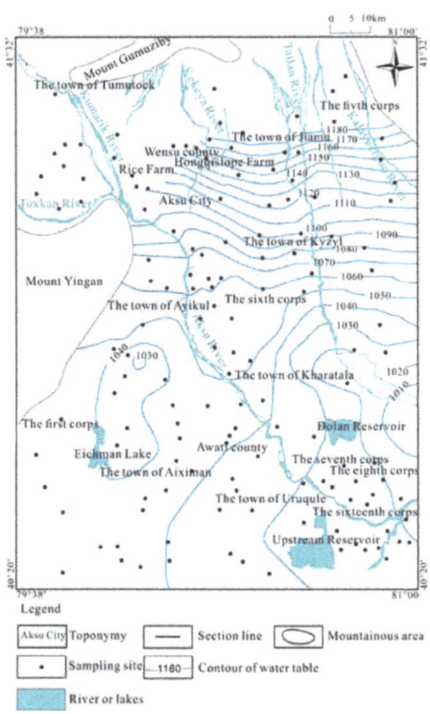

Figure 4. Contour lines of confined aquifer.

3. Sample Collection and Testing

A total of 196 groups of water samples were collected, including 151 groups from the unconfined aquifer and 45 groups from the confined aquifer. There are 23 groups of environmental isotope samples, including 15 groups from the unconfined aquifer and 9 groups from the confined aquifer. Sampling locations are shown in Figure 5.

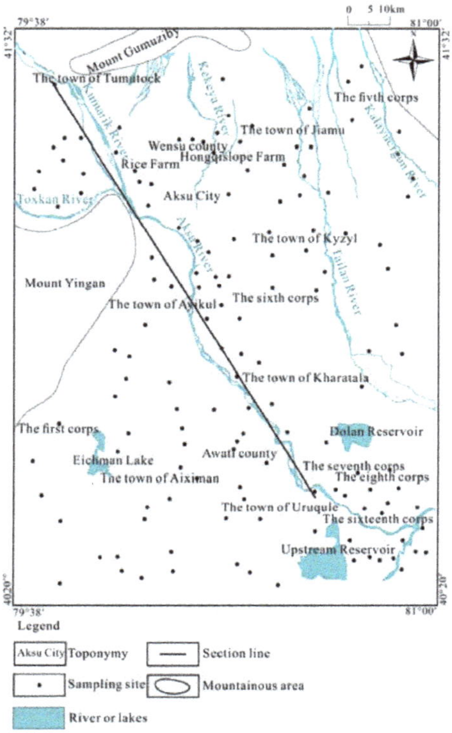

Figure 5. Sampling locations in the area.

The collected samples were analyzed by the first regional geological survey team of the Xinjiang Geological and Mineral Bureau to determine K^+, Na^+, Ca^{2+}, Mg^{2+}, HCO^{3-}, $CO_3{}^{2-}$, $SO_4{}^{2-}$, and Cl^- on an inductively coupled plasma spectrometer and an atomic absorption spectrophotometer with accuracy of ±0.5% and ±1%. Environmental isotope samples were analyzed in the American BETA laboratory to determine δD and $\delta^{18}O$ with accuracy of ±2‰ and ±0.3‰, respectively, on an isotope mass spectrometer (Thermo Delta-Plus) after high-temperature treatment, evaporation, dissociation, atomization, and ionization.

4. Principles and Theory of Mixed-Unit Method

4.1. Hypothesis of Mixed-Unit Method

In mixed-unit method, the aquifer is generalized and discretized into a finite number of homogeneous and isotropic small units. Each small unit has a comprehensive value to show its hydrochemical characteristics (ion concentration and isotope value). According to their flow fields, the possible recharging and discharging relationship is obtained. With the ion concentration and isotope value in each unit as its tracer, the tracer mass-conservation equation can be established. Through solving this equation, the recharging and discharging relationships and recharging ratio can also be determined. Before the determining of the mixed units, the following assumptions need to be made: (1) in order to qualitatively judge the groundwater charging and discharging conditions, the tracer concentration of the water resource and the discharged water flow are already known; (2) conservation of water level: in each small unit, within a certain time, the water level is constant, and the water level is averaged; (3) the migration of dissolved components is controlled by convection; and (4) effects of mineral reaction, dissolution, and precipitation are negligible.

4.2. Unit Determination Principles

In order to reduce the unknown parameters of the model and determine the small units in an optimal way, the following principles need to be followed: (1) The studied area is divided along the groundwater flow into units, whose horizontal unit boundary must be parallel to the groundwater level contour line and longitudinal boundary must be perpendicular, or approximately perpendicular, to the groundwater level contour line; (2) A hydrogeological unit can be divided into multiple small units. A small unit cannot cross into different hydrogeological units. Different hydrogeological units store different types of groundwater with different ion composition and isotope values. (3) A small unit should have representative water sample data. (4) The same cone of depression should be divided into one small unit.

4.3. Unit Determination

According to the above-mentioned assumptions and principles of the mixed-unit method, the unconfined aquifer of the Aksu River Basin was divided into seven small units (a, b, c, d, e, f, and g), and its confined aquifer is divided into five small units (C, D, E, F, and G), as shown in Figure 6.

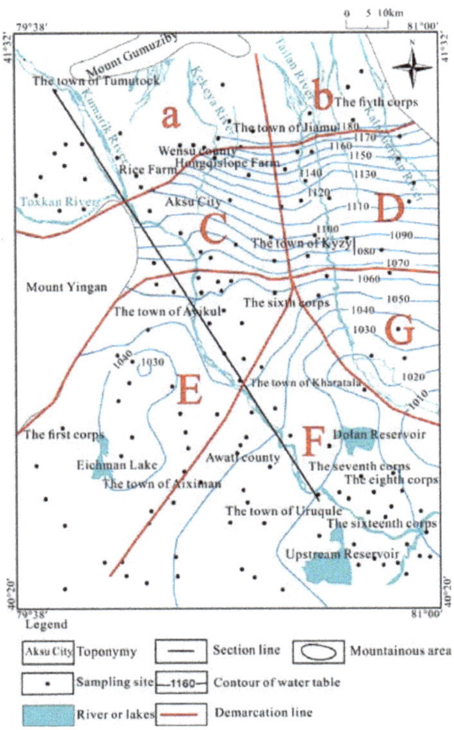

Figure 6. Mixed-unit division in the confined aquifer.

4.4. Calculation of Aquifer Recharging and Discharging

In the mixed-unit model, the unconfined aquifer and confined aquifer are divided into a finite number of discrete small units, which are discrete at an interval of Δt. Their solutes are fully mixed, and the components of each solute are evenly distributed in all small units. Therefore, the equilibrium equation of water in a small unit within the period of Δt can be expressed as:

$$Q_n - W_n + \sum_{i=1}^{In} q_{in} - \sum_{i=1}^{In} q_{nj} = e_n \tag{1}$$

According to the assumption on the water balance of each small unit, the mass balance equation of the dissolved component k in unit n can be obtained as:

$$C_{nk}Q_n - C_{nk}\left[W_n + \sum_{j=1}^{In} q_{nj}\right] + \sum_{i=1}^{In} q_{in}C_{ink} = e_{nk}\ k = 1,\ 2,\ \ldots,\ k \tag{2}$$

where, Q_n represents the time average flow value into unit n, W_n is the average value of the flow out from unit n, q_{in} represents the average flow from unit i into n, e_n is the deviation of water balance caused by various errors from the flow entering or exiting the unit n, k is the average concentration of the tracer k in one unit, and C_{nk} is the average concentration of the trace k in the k in unit n.

After Equations (1) and (2) are combined into a rectangular matrix of known concentrations in unit n, in which the first row represents the water balance and the other rows represent the solute mass conservation balance, the Equation (3) can be obtained with any unit n:

$$C_n q_n + D_n = E_n \tag{3}$$

where, q_n represents the flow through the boundary of small unit n:

$$q_n = [q_{1n}q_{2n}\cdots q_{in}q_{n1}q_{n2}\cdots q_{nJ_n}](In + Jn) \times 1 \tag{4}$$

D_n is the measurable and quantifiable known items in unit n (such as the known outflow and pumping volumes), and E_n represents the unknown error vector in the unit as,

$$E_n = [e_n e_{n1} e_{n1} \ldots\ e_{nk}](1 + K)\ \times 1 \tag{5}$$

According to Equation (3) (Adar (1988)), through the minimization of the sum function J of square error and evaluation of the sum of square error of all units, the flow composition of the aquifer can be obtained as,

$$J = \sum_1^N [E_n^T W E_n] = \sum_1^N (c_n q_n + D_n)^T W (c_n q_n + D_n) \tag{6}$$

5. Calculation of Charging and Discharging of Confined Aquifer with Mixed-Unit Method

5.1. Division of Mixed Units

Because the studied area is located in the plain of alluvial–diluvial fine-soil particles, the conceptual model of the mixed units was established accordingly as shown in Figure 5. The mixed units of the unconfined aquifer and confined aquifer were, respectively, marked as a, b, c, d, e, f, and g and C, D, E, F, and G. Units a and b are located in the alluvial–proluvial slope gravel plain as a single-structure unconfined aquifer and are the recharging source of confined aquifer. All other units are in the alluvial–diluvial fine-soil-grain plain. Units G and F are discharging units. The transformation relationship between units of the aquifer is shown in Figure 7.

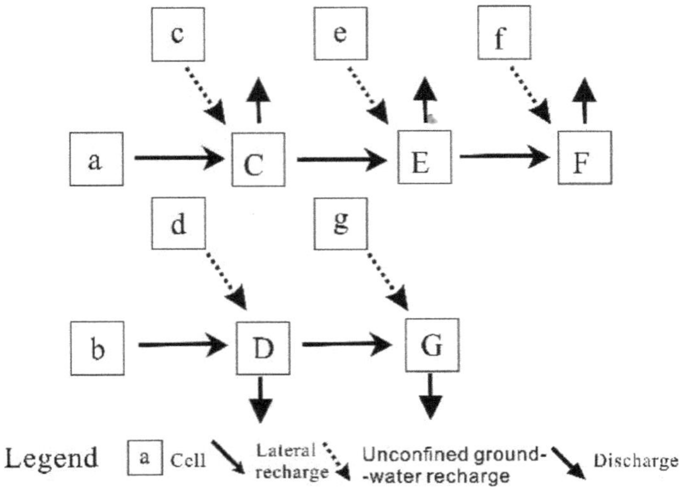

Figure 7. The transformation relationships between units.

5.2. Hydrochemical Characteristics

The unconfined aquifer is widely distributed in the studied area, and its salinity in the upper stream was about 0.6 g/L, and pH was 8.4 with an $HCO_3 \cdot SO_4$–Mg·Ca-type water. It was 1.3–3 g/L in the middle stream with water types of SO_4·Cl–Na·Mg and SO_4·Cl–Na·Ca·Mg, and pH was 8.3. The salinity in the west of the downstream study area was 3.7 g/L, pH was 8.4, and the salinity in the east was 9.0 g/L, pH was 8.2, with water types of Cl·SO_4–Na and Cl–Na. The water-chemistry type of unconfined aquifer units a→c→e→f changed from $HCO_3 \cdot SO_4$–Mg·Ca, to SO_4·Cl–Na·Mg, to SO_4·Cl–Na·Ca·Mg, to Cl·SO_4–Na (or Cl–Na. That of unconfined aquifer b→d→g changed from $HCO_3 \cdot SO_4$–Mg·Ca to Cl·SO_4–Na (or Cl–Na) and SO_4·Cl–Na·Mg (or SO_4·Cl–Na·Ca·Mg).

The salinity of confined aquifer from the middle-upper stream to the downstream of the Aksu River Basin did not change significantly with a salinity of 1 g/L, pH of 8.1 and water types of SO_4·Cl–Ca·Na·Mg, $SO_4 \cdot HCO_3$·Cl–Mg·Na·Ca and Cl·SO_4–Na·Ca in the downstream. Water types of the confined aquifer unit C→E→F changed from SO_4·Cl–Ca·Na·Mg or $SO_4 \cdot HCO_3$·Cl–Mg·Na·Ca to Cl·SO_4–Na·Ca. From units D→G, it changed from SO_4·Cl–Ca·Na·Mg or $SO_4 \cdot HCO_3$·Cl–Mg·Na·Ca to Cl·SO_4–Na·Ca. These data showed that, along the flowing path of the confined aquifer, in the middle and upper steams of the west, the confined aquifer receives a large amount of recharging laterally from the unconfined aquifer (a→C→E), and a small amount of recharging vertically from the unconfined aquifer (c→C→E). When the water exchange between the confined aquifer and the unconfined aquifer was reduced (e→E and f→F), evaporating concentration and cation-exchange adsorption (e→f) occurred in the unconfined aquifer, and cation exchange adsorption occurred in the confined aquifer (E→F). The middle and upper streams in the east receive a large amount of recharging water vertically from unconfined aquifer (d→D), with a small amount from unconfined aquifer laterally (b→D). Its downstream receive a large amount of water recharging laterally from confined aquifer (D→G). with a small amount vertically from unconfined aquifer (g→G). The unconfined aquifer has significant evaporating concentration (d→g), and the confined aquifer mostly has cation exchange adsorption (D→G). The water chemistry characteristics in this basin are shown in Figure 8.

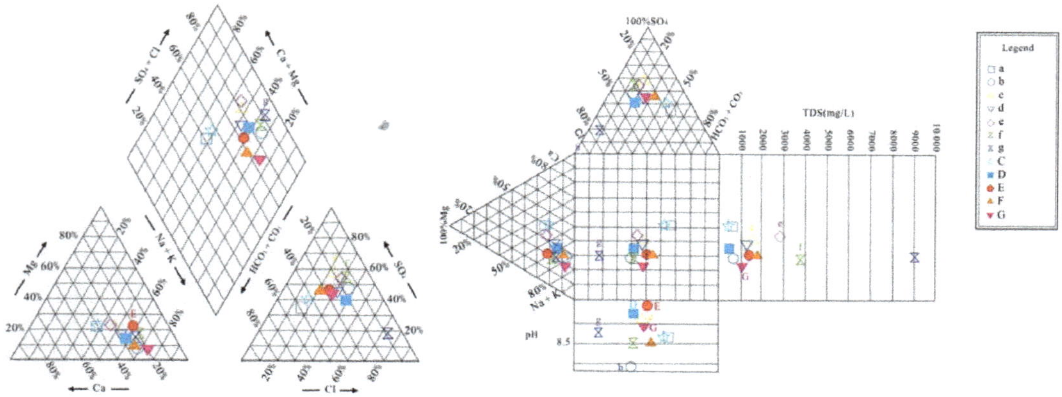

Figure 8. Water chemistry characteristics.

5.3. Distribution Characteristics of Isotopes

As shown in Figure 9, after the precipitation line slope of the aquifer of the studied area was compared with the Global Meteoric Water Line (GMWL) proposed by Craig [32], the slope and intercept were both smaller than the global precipitation line and δD and $\delta^{18}O$ were all located at the upper left of GMWL. These results showed that strong evaporation had taken place in the whole of the aquifer.

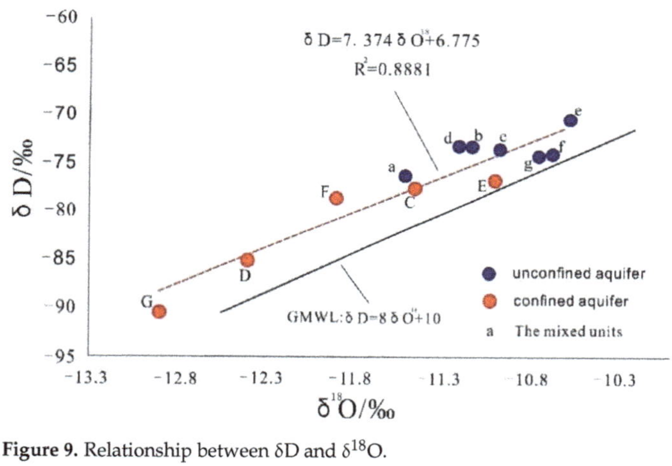

Figure 9. Relationship between δD and $\delta^{18}O$.

The isotopic compositions of the unconfined aquifer(a→c→e→f) and the confined aquifer(C→E→F) in the west were basically the same, indicating that unconfined aquifer and confined aquifer had a certain hydraulic connection. As shown in Figure 10a, in the middle and upper steams of the west, because of the influence of evaporation, the δD and $\delta^{18}O$ in the unconfined aquifer increased gradually, while the δD and $\delta^{18}O$ in the confined aquifer increased simultaneously. These results showed that the unconfined aquifer recharged the confined aquifer vertically (c→C, e→E). In the downstream, the δD and δ18O of the unconfined aquifer f and the confined aquifer F were significantly different, indicating that the confined aquifer was less replenished by the unconfined water (f→F). The δD and $\delta^{18}O$ of unit C→E→F were closer to unit c→e→f than to unit a, indicating that the confined aquifer received a large amount water recharging laterally from the unconfined aquifer (a→C→E→F).

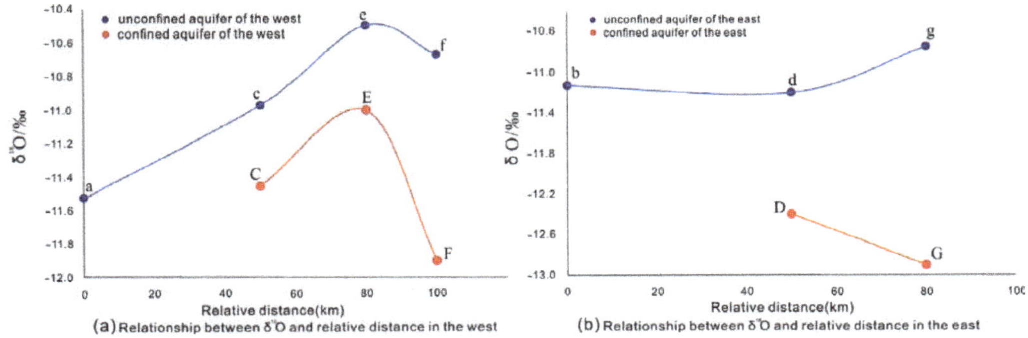

Figure 10. Relationship between $\delta^{18}O$ and relative distance.

The δD and $\delta^{18}O$ values of the unconfined aquifer (b→d→g) and the confined aquifer (D→G) in the east were significantly different, indicating that the hydraulic connection between the two water bodies was weak. As illustrated in Figure 10b, from upper to lower in the east, because of the influence of evaporation, the δD and $\delta^{18}O$ were enriched along the way and in the confined aquifer, the δD and $\delta^{18}O$ decreased. These results showed that the confined aquifer in the middle and upper streams received a large amount of water recharging vertically from the unconfined aquifer (d→D) and the confined aquifer in the lower streams received a small amount of water recharging vertically from unconfined aquifer (g→G).The δD and $\delta^{18}O$ of unit D→G were closer to Unit d→g than to the mountain exit stations, indicating that the confined aquifer D→G received a large amount water recharging laterally from unconfined aquifer (b→D→G).

5.4. Recharging Relationships between Unconfined Aquifer and Confined Aquifer

According to the hydrochemical and isotope distribution characteristics of the studied area, the confined aquifer in the alluvial–diluvial plain area received water recharging from the loose rock porous unconfined aquifer and upper porous unconfined aquifer of the Quaternary in the upper alluvial–diluvial slope plain, which was discharged through artificial and lateral downstream. The recharging and discharging relationship between units in the unconfined aquifer and confined aquifer are shown in Figure 11.

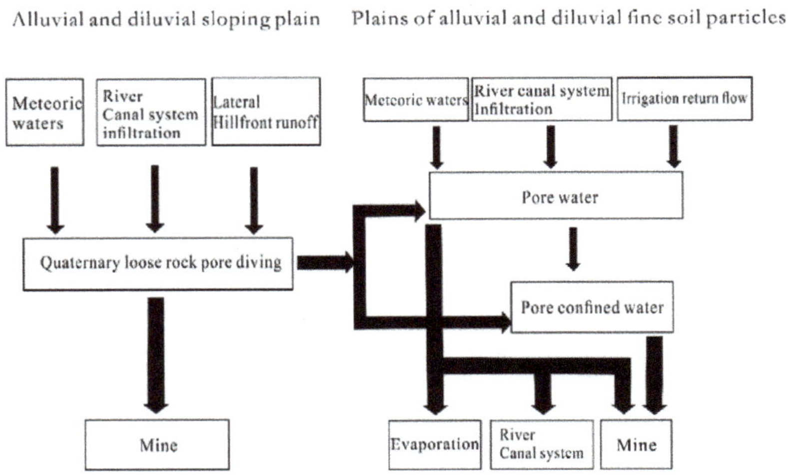

Figure 11. Groundwater circulation in the unconfined aquifer and the confined aquifer.

5.5. Data Analysis

The hydrogeochemical and isotopic data of each unit were statistically analyzed and are listed in Table 1.

Table 1. Hydrochemical properties and isotopes of all units.

Unit	Item	K^+	Na^+	Ca^{2+}	Mg^{2+}	Cl^-	$SO_4{}^{2-}$	$HCO_3{}^-$	D	^{18}O
a	Average	0.14	2.39	3.25	2.95	1.95	2.98	3.21	−76.36	−11.52
	Mean square error	3.65	7.70	4.23	3.53	5.93	1.20	25.52	6.65	0.44
b	Average	0.18	9.76	4.25	2.41	7.27	6.99	1.58	−73.44	−11.13
	Mean square error	0.93	13.59	3.60	1.02	0.77	13.72	4.99	-	-
c	Average	0.21	8.98	6.76	8.06	7.41	12.89	3.40	−73.72	−10.97
	Mean square error	0.33	5.28	2.40	3.73	5.93	0.01	3.85	0.96	0.03
d	Average	0.12	5.18	3.45	3.40	4.46	4.96	2.03	−73.34	−11.20
	Mean square error	2.79	3.82	7.18	2.95	2.18	8.02	8.24	-	-
e	Average	0.19	7.56	6.50	7.56	8.24	10.87	2.66	−70.57	−10.50
	Mean square error	2.35	11.14	4.76	6.00	11.84	1.72	25.52	7.67	0.49
f	Average	0.23	19.46	6.82	10.55	14.61	18.46	3.50	−74.2	−10.67
	Mean square error	11.6	0.09	9.14	0.95	1.98	11.17	0.01	5.29	0.11
g	Average	0.12	11.36	4.69	5.20	15.82	2.92	0.98	−74.38	−10.75
	Mean square error	0.06	11.17	0.38	0.16	0.16	5.90	20.86	-	-
C	Average	0.15	2.56	3.43	3.15	2.21	3.76	3.07	−77.70	−11.45
	Mean square error	0.75	9.88	1.27	1.58	3.85	0.85	1.88	9.61	0.20
D	Average	0.09	4.64	2.66	2.30	4.38	3.30	1.40	−85.20	−12.40
	Mean square error	1.84	11.73	0.15	6.51	8.63	6.39	24.35	-	-
E	Average	0.10	4.70	1.96	3.32	3.40	4.39	2.17	−76.90	−11.00
	Mean square error	4.62	7.76	0.31	15.54	5.93	1.60	25.52	-	-
F	Average	0.06	3.32	1.56	0.90	1.83	2.50	1.30	−78.70	−11.90
	Mean square error	0.94	3.26	0.57	4.58	29.60	8.14	3.85	-	-
G	Average	0.06	4.92	1.51	0.93	2.65	3.05	1.38	−90.40	−12.90
	Mean square error	2.79	9.52	5.11	0.59	11.85	10.40	22.69	-	-

Notes: (1) Concentrations are in meq/L unless otherwise indicated, such as deuterium and oxygen ^{18}O in ‰, and (2) "-" means that there is only one sampling point and the square error cannot be determined.

In the studied area, the drainage methods of the confined aquifer were mainly lateral runoff discharging and artificial well-group pumping. Artificial exploitation was 27.95×10^6 m^3/a and the mining volume of each unit was calculated according to the proportion of the unit area. The mining volume of each unit is shown in Table 2. The lateral discharging volumes were 8.57×10^6 m^3/a and 7.10×10^6 m^3/a, as shown in Table 3.

Table 2. Groundwater mining volume of each unit.

Unit	a	b	C	D	E	F	G
Surface area (km^2)	724.72	400.05	988.55	655.2	810.18	714.27	365.34
Mining volume (10^6 m^3/a)	4.35	2.4	5.93	3.93	4.86	4.29	2.19

Table 3. Groundwater discharging laterally from Units F and G.

Unit	Buried Depth of Groundwater (m)	Aquifer Thickness (m)	Osmotic Coefficient (m/d)	Hydraulic Gradient	Discharging Volume (10^6 m^3/a)
F	4.03	180	2.5	1/1050	8.57
G	1.88	180	2	1/740	7.1

5.6. Results and Analysis

The results showed (Figure 12) that units a and b in the studied area were single-structure unconfined aquifers as the main recharge resources of the downstream confined aquifer. Unit C accepted the lateral recharge from unit a and the leakage recharge from Unit c, which were 9.77×10^6 m^3/a and 0.64×10^6 m^3/a, respectively. Unit E received the lateral recharge from unit C and the overflow recharge from unit e, which were 4.67×10^6 m^3/a and 2.80×10^6 m^3/a, respectively. The unit F received the lateral recharge from the unit E and the overflow recharge from the f unit, which were 6.10×10^6 m^3/a and 0.90×10^6 m^3/a, respectively. Unit D received the lateral recharge from unit b and the overflow recharge from unit d, which were 1.71×10^6 m^3/a and 5.66×10^6 m^3/a, respectively. The unit G received the lateral recharge from the unit D, and the recharging was 5.90×10^6 m^3/a. This recharge relationship and degree were consistent with results from hydrochemistry and isotope.

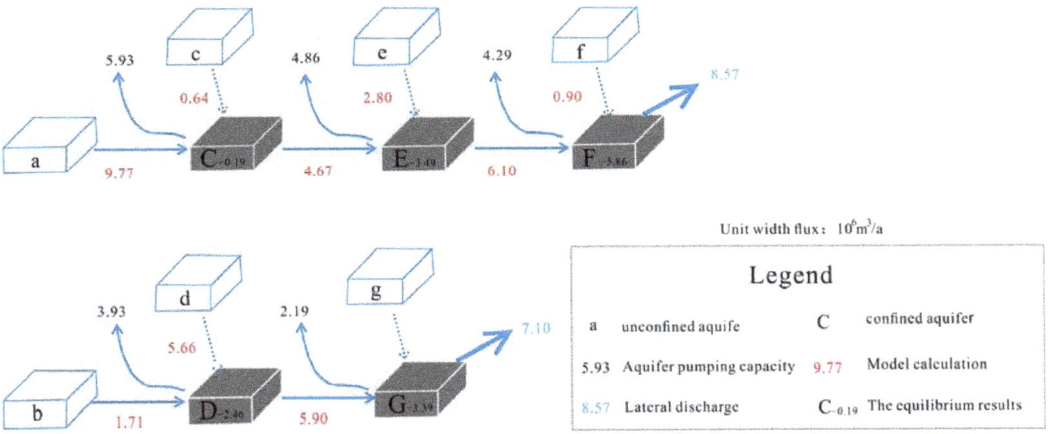

Figure 12. Quantitative water circulation model of the unconfined aquifer and the confined aquifer in the studied area.

The total groundwater recharge volume from the confined aquifer in the studied area was 21.48×10^6 m^3/a, of which the lateral recharge was 11.48×10^6 m^3/a, accounting for 53.45% of the total, and the vertical leakage recharge was 10.00×10^6 m^3/a, accounting for 46.55% of the total. The vertical leakage recharge in the southeast was very small and negligible. The total discharging volume from the confined aquifer was 36.87×10^6 m^3/a, of which the lateral discharging volume was 15.67×10^6 m^3/a, accounting for 42.5%, and artificial exploitation was 21.2×10^6 m^3/a, accounting for 57.5%. These results showed that the confined aquifer was in an accumulation and superposition state of negative balance along the direction of the underground water flow.

5.7. Discussion

Based on the results of the water balance and the model output, the upstream confined aquifer received lateral recharging and vertical leakage recharging from the unconfined aquifer, and the downstream confined aquifer only received lateral recharging from the upstream confined aquifer, which was consistent with the hydrochemical and isotope analysis. In addition, the unit flux between the cells was apparently influenced by the water source exploitation. For example, the unit E received the recharge from unit e and unit C was $7.47 \times 10^6m^3/a$, the discharged through lateral was $6.10 \times 10^6m^3/a$. Because of over extraction, the difference between the inflow and outflow to unit E was $-3.49 \times 10^6m^3/a$. Compared with previous research, we identified the recharge sources and their relative contributions to the confined aquifer, and the data of the extraction were available, the quantitative water circulation model were reasonable, and the results were reliable.

6. Conclusions

According to the geological and hydrogeological data of the studied area, the recharging and discharging relationship between the unconfined aquifer and the confined aquifer was determined. Based on the hypothesis and principles of unit dividing, and the qualitative recharging and discharging relationship, a mixed-unit model was established to study the hydrochemical characteristics of the confined aquifer in the west of the studied area. Results showed that the confined aquifer was significantly affected by unconfined aquifer, and the confined aquifer received lateral recharging from upstream of the unconfined aquifer and vertical recharging from the upper unconfined aquifer. In the east of the studied area, the downstream confined aquifer no longer received the recharging vertically from unconfined aquifer, but mainly received the lateral recharging from the upstream confined aquifer.

The δD and $\delta^{18}O$ of unconfined aquifers in the studied area were both at the upper left of the Global Meteoric Water Line, and their slope and intercept were both smaller than those of the global atmospheric precipitation line, indicating that the unconfined aquifer was significantly evaporated. The isotopic compositions of unconfined aquifer and confined aquifer in the west of the studied area were basically the same, indicating that there was a hydraulic connection between these two water bodies, and the confined aquifer received lateral recharging from the upstream unconfined aquifer and vertical leakage recharging from the upper unconfined aquifer. The δD and $\delta^{18}O$ distributions of the unconfined aquifer and confined aquifer in the east of the studied area were relatively discrete, indicating that the hydraulic connection between these two water bodies was weak, and the downstream of confined aquifer in the east mainly received lateral recharging from the upstream confined aquifer. These results were consistent with the hydrochemical analysis.

Based on the mixed-unit model, the calculation results showed that the total recharged volume received by the confined aquifer in the studied area was 21.48×10^6 m^3/a, in which the lateral recharging was 11.48×10^6 m^3/a, accounting for 53.45% of the total, and the vertical recharging was 10.00×10^6 m^3/a, accounting for 46.55% of the total. The vertical recharging amount in the southeast was very small and negligible. The total discharging volume was 36.87×10^6 m^3/a, including lateral discharging whose amount was 15.67×10^6 m^3/a, accounting for 42.5%, and the artificial pumping amount was 21.2×10^6 m^3/a, accounting for 57.5%. The upstream confined aquifer received lateral recharging and vertical leakage recharging from the unconfined aquifer, and the downstream confined aquifer only received lateral recharging from the upstream confined aquifer. The confined aquifers in the entire region were in a state of negative balance, and this state was continuously accumulated from the upstream to the downstream.

Author Contributions: Conceptualization, J.H. and Y.G.; methodology, J.H.; software, J.H.; validation, J.H., Y.G. and S.L.; formal analysis, J.H.; investigation, J.H.; resources, Y.G.; data curation, Y.G.; writing—original draft preparation, J.H.; writing—review and editing, J.H. and Y.G. All authors have read and agreed to the published version of the manuscript.

Funding: This study was supported by grants from the National Fund Program (No. U1603243), and the Key Laboratory of Geodynamic Processes and Metallogenic Prognosis of the Central Asian Orogenic Belt (No. 2020–004).

Institutional Review Board Statement: Not applicable.

Informed Consent Statement: Not applicable.

Data Availability Statement: The data that support the findings of this study are available from the corresponding author upon reasonable request.

Conflicts of Interest: The authors declare no conflict of interest.

References

1. Li, W.P.; Hao, A.B. Formation evolution model and significance of groundwater in inland arid basin, northwest China. *Hydrogeol. Geol.* **1999**, *26*, 28–32. [CrossRef]
2. Sun, J.P.; Wei, Y.M.; Wu, H.T.; Zhang, F.C.; Li, A.J. Discussion on typical groundwater occurrence model in piedmont belt, inland basin in northwes. *Hydrogeol. Geol.* **2007**, *34*, 66–69. [CrossRef]
3. Wang, F.G. The Application of Isotope Techniques in the Hydrological Cycle of Overground Section (Henan Section) of the Lower Reaches of Yellow River. Ph.D. Thesis, Jilin University, Changchun, China, 2006.
4. Klaus, J.; Külls, C.; Dahan, O. Evaluating the recharge mechanism of the Lower Kuiseb Dune area using mixing cell modeling andresidence time data. *Hydrolgeol. J.* **2008**, *358*, 304–316. [CrossRef]
5. Dahan, O.; McGraw, D.; Adar, E.; Pohll, G.; Bohm, B.; Thomas, J. Multi-variable mixing cell model as a calibration and validation tool for hydrogeologic groundwater modeling. *J. Hydrol.* **2004**, *293*, 115–136. [CrossRef]
6. Marques, J.M.; Eggenkamp, H.G.; Graca, H.; Carreira, P.M.; José Matias, M.; Mayer, B.; Nunes, D. Assessment of recharge and flowpaths in a limestone thermomineral aquifer system using environmental isotope tracers (Central Portugal). *Isot. Environ. Health Stud.* **2010**, *46*, 156–165. [CrossRef] [PubMed]
7. Hagedorn, B. Hydrochemical and ^{14}C constraints on groundwater recharge and interbasin flow in an arid watershed: Tule Desert, Nevada. *J. Hydrol.* **2015**, *523*, 297–308. [CrossRef]
8. Xiao, Y. Groundwater Circulation Pattern Sand its Change Trend in Southern Qaidam Basin, Northwest China. Ph.D. Thesis, China University of Geosciences, Beijing, China, 2018.
9. Song, Y.; Chi, B.M.; Gu, H.B.; Qin, W.J.; Du, W.L. Analysis of groundwater circulation characteristics based on hydrochemistry and D, ^{18}O in Donggong River basin of Liujiang basin. *RWS Soil Water Conserv.* **2015**, *22*, 90–95+100. [CrossRef]
10. Nie, Z.L. Study on Groundwater Circulation and Renewability in the Middle Reaches of Heihe River Valley Northwest China. Ph.D. Thesis, Chinese Academy of Geological Sciences, Beijing, China, 2005.
11. Wang, F.G.; Liao, Z.S. Study of the Precipitation Infilatration Recharge With the δD, $\delta^{18}O$ Isotopes Peak Displacement Method. *J. Jilin Univ. Earth Sci. Ed.* **2007**, *2*, 284–287+334. [CrossRef]
12. Yang, Q.; Xiao, H.L.; Zhao, L.J.; Yang, Y.G.; Li, C.Z.; Zhao, L.; Yin, L. Hydrological and isotopic characterization of river water, groundwater, and groundwater recharge in the Heihe river basin, northwestern China. *Hydrol. Process.* **2011**, *25*, 1271–1283. [CrossRef]
13. Duan, S.Y.; Wu, H.F. The hydrogeological conditions are evaluated quantitatively by using the hydrogeological parameters calculated by groundwater chemical dynamics method—Taking Taiyuan Sangui Horst and its surroundings area as an example. *Hydrogeol. Geol.* **2001**, *5*, 62–65. [CrossRef]
14. Cao, J.F.; Shen, Y.Y.; Ping, J.H.; Du, Q.Y.; Liu, M.X. Application of the Theory of Groundwater Chemical Dynamics in the Parameter Determination in Qianjin Source Field. *J. Jilin Univ. Earth Sci. Ed.* **2006**, *1*, 96–102. [CrossRef]
15. Liu, P.; Hoth, N.; Drebenstedt, C.; Sun, Y.; Xu, Z. Hydro-geochemical paths of multi-layer groundwater system in coal mining regions—Using multivariate statistics and geochemical modeling approaches. *Sci. Total Environ.* **2017**, *601*, 1–14. [CrossRef] [PubMed]
16. Bakalowicz, M. Karst groundwater: A challenge for new resources. *Hydrolgeol. J.* **2005**, *13*, 148–160. [CrossRef]
17. Guo, H.M.; Ni, P.; Jia, Y.F. Characteristics and Their Causes of Surface Water-Groundwater Geochemistry in the Hetao Basin, Inner Mongolia. *Geoscience* **2015**, *2*, 229–237.
18. Cook, P.G.; Herczeg, A.L. *Environmental Tracers in Subsurface Hydrology*; Springer Science & Business Media: New York, NY, USA, 2012.
19. Eastoe, C.J.; Rodney, R. Isotopes as tracers of water origin in and near a regional carbonate aquifer: The southern Sacramento Mountains, New Mexico. *Water* **2014**, *6*, 301–323. [CrossRef]
20. Jia, L.J.; Deng, J.Q.; Kang, W.D.; Luo, Q.B.; Wu, G.T. Hydrodynamic dispersion experiment research of NaCl in the brackish water area of northwest—taking dong liang village of ding bian county as an example. *Technol. Innov. Product.* **2016**, *11*, 117–120. [CrossRef]
21. Kebede, S.; Travi, Y.; Alemayehu, T.; Ayenew, T. Groundwater recharge, circulation and geochemical evolution in the source region of the Blue Nile River, Ethiopia. *Appl. Geochem.* **2005**, *20*, 1658–1676. [CrossRef]
22. Zhu, C. Estimate of recharge from radiocarbon dating of groundwater and numerical flow and transport modeling. *Water Resour. Res.* **2000**, *36*, 2607–2620. [CrossRef]
23. Davisson, M.; Smith, D.; Kenneally, J.; Rose, T. Isotope hydrology of southern nevada groundwater: Stable isotopes and radiocarbon. *Water Resour. Res.* **1999**, *35*, 279–294. [CrossRef]
24. Gibson, J.; Edwards, T.; Birks, S.; St Amour, N.; Buhay, W.; McEachern, P.; Wolfe, B.; Peters, D. Progress in isotope tracer hydrology in Canada. *Hydrol. Process.* **2005**, *19*, 303–327. [CrossRef]
25. Viville, D.; Ladouche, B.; Bariac, T. Isotope hydrological study of mean transit time in the granitic strengbach catchment (Vosges massif, France): Application of the FlowPC model with modified input function. *Hydrol. Process.* **2006**, *20*, 1737–1751. [CrossRef]
26. Clark, J.F.; Stute, M.; Schlosser, P.; Drenkard, S.; Bonani, G. A tracer study of the Floridan aquifer in southeastern Georgia: Implications for groundwater flow and paleoclimate. *Water Resour. Res.* **1997**, *33*, 281–289. [CrossRef]
27. Ofterdinger, U.; Balderer, W.; Loew, S.; Renard, P. Environmental isotopes as indicators for ground water recharge to fractured granite. *Ground Water* **2004**, *42*, 868–879. [PubMed]

28. Adar, E.M.; Neuman, S.P.; Woolhiser, D.A. Estimation of spatial recharge distribution using environmental isotopes and hydrochemical data, I. Mathematical model and application to synthetic data. *J. Hydrol.* **1988**, *97*, 251–277. [CrossRef]
29. Adar, E.M.; Neuman, S.P. Estimation of spatial recharge distribution using environmental isotopes and hydrochemical data, II. Application to aravaipa valley in southern Arizona, USA. *J. Hydrol.* **1988**, *97*, 279–302. [CrossRef]
30. Fraser, C.; Roulet, N.; Lafleur, M. Groundwater flow patterns in a large peatland. *J. Hydrol.* **2001**, *246*, 142–154. [CrossRef]
31. Shi, C. The Application of Isotope Compartmental Mixing Cell Modell in Hydrologic Cycle of Shallow Water in the Overground Section (Henan Section) in Lower Reaches of the Yellow River. Ph.D. Thesis, Jilin University, Changchun, China, 2009.
32. Craig, H. Isotopic variations in meteoric waters. *J. Sci.* **1961**, *133*, 1702–1703. [CrossRef]

International Journal of
Environmental Research and Public Health

MDPI

Article

New Insights into Microbial Degradation of Cyanobacterial Organic Matter Using a Fractionation Procedure

Jing Chen [1,2], Yongqiang Zhou [1,2,*] and Yunlin Zhang [1,2]

[1] Nanjing Institute of Geography and Limnology, Chinese Academy of Sciences, Nanjing 210008, China; chenj@niglas.ac.cn (J.C.); ylzhang@niglas.ac.cn (Y.Z.)
[2] University of Chinese Academy of Sciences, Beijing 100049, China
[*] Correspondence: yqzhou@niglas.ac.cn; Tel.: +86-25-8688-2193

Abstract: Cyanobacterial blooms caused by phytoplankton *Microcystis* have occurred successively since 1980 in Lake Taihu, China, which has led to difficulty collecting clean drinking water. The effects of cyanobacterial scum-derived dissolved organic matter (DOM) on microbial population variations and of algal-derived filtrate and algal residual exudative organic matter caused by the fraction procedure on nutrient mineralization are unclear. This study revealed the microbial-regulated transformation of DOM from a high-molecular-weight labile to a low-molecular-weight recalcitrant, which was characterized by three obvious stages. The bioavailability of DOM derived from cyanobacterial scum by lake microbes was investigated during 80-d dark degradation. Carbon substrates provided distinct growth strategy links to the free-living bacteria abundance variation, and this process was coupled with the regeneration of different forms of inorganic nutrients. The carryover effects of *Microcystis* cyanobacteria blooms can exist for a long time. We also found the transformation of different biological availability of DOM derived from two different cyanobacterial DOM fractions, which all coupled with the regeneration of different forms of inorganic nutrients. Our study provides new insights into the microbial degradation of cyanobacterial organic matter using a fractionation procedure, which suggests that the exudate and lysate from degradation products of cyanobacteria biomass have heterogeneous impacts on DOM cycling in aquatic environments.

Keywords: cyanobacterial blooms; dissolved organic matter; nutrient; bacterial abundance; carryover effects

Citation: Chen, J.; Zhou, Y.; Zhang, Y. New Insights into Microbial Degradation of Cyanobacterial Organic Matter Using a Fractionation Procedure. *Int. J. Environ. Res. Public Health* **2022**, *19*, 6981. https://doi.org/10.3390/ijerph19126981

Academic Editors: Alban Kuriqi and Luis Garrote

Received: 7 May 2022
Accepted: 1 June 2022
Published: 7 June 2022

1. Introduction

Clean drinking water plays a critical role in human physical health and welfare [1]. Lakes are the most important clean drinking water sources in China because they can provide a relatively stable water supply with few nature fluctuations (i.e., floods, droughts, and freezing) [2]. Lake Taihu (30°05′–32°08′ N, 119°08′–121°55′ S) is the third largest freshwater lake in China. The lake watershed covers approximately 36,500 km^2 and is an important drinking water source for more than 10 million people and several modern cities, such as Shanghai, Wuxi, and Suzhou. Moreover, Lake Taihu also serves numerous industrial, agricultural, and municipal activities [3]. However, harmful cyanobacteria blooms caused by the phytoplankton *Microcystis* have occurred successively since 1980 in Lake Taihu, which have directly threatened the safety of drinking water and triggered serious health and social problems, particularly the "Water supply crisis" in Wuxi in 2007 [4–7]. Qin et al. [8] promoted a large-scale integrated monitoring and forecasting system for cyanobacterial bloom management in Lake Taihu, and they found that the distribution area of cyanobacterial blooms reached 997.5 km^2 in 2011, which was the maximum value for the four consecutive years of monitoring (2009–2012). Generally, cyanobacterial bloom growth can be influenced by temperature, daylight, water turbulence, pH, and macronutrient. In addition, wind-induced hydrodynamic effects, such as sediment resuspension and the corresponding nutrient release from the sediment–water interface, played an important role in intensifying cyanobacterial bloom expansion [9,10].

After blooms, the massive cyanobacterial biomass aggregates into scum and is then degraded by indigenous heterotrophs [11]. Large fractions of cyanobacterial organic matter are released from the biomass as dissolved organic matter (DOM) mainly via three major decline steps, which were proposed to describe the bloom decline process: scum disaggregation, colony settlement, and cell lysis in colonies [12]. Various bacteria of myxobacterium and the genera *Alcaligenes* and *Pseudomonas* and the *Cytophaga/Flavobacterium* group from aquatic environments were found to be lysed *Microcystis*, which is closely related to the degradation of biomass produced by cyanobacteria [13,14]. This algal-released DOM can be referred to as bulk algal-derived DOM. When large complex algal-derived DOM enters the aquatic system, it may cause oxygen depletion and cyanobacterial toxin secretion by biological activities, resulting in the appearance of a "black water mass," which may pose serious risks to aquatic plants, fish, and oxygen-sensitive invertebrates and cause mass mortality events [15–17]. The high frequency formation of a massive bloom composed of cyanobacteria of the genus *Microcystis*, which can produce potent hepatotoxins, neurotoxins, and dermatoxins and promote tumor formation (i.e., microcystins), and they finally cause damage to the liver [18,19].

Several studies have suggested that bulk algal-derived DOM is closely related to the biogeochemical cycle and energy flow of lake ecosystems, which indirectly impact human health and the dynamics and driving factors of cyanobacterial blooms [20–22]. For example, Bittar et al. [23] confirmed that extracellular and intracellular DOM were produced in axenic cultures by *Microcystis aeruginosa*, which can effectively increase the biolabile DOM (BDOM) to bacterial growth and activity in lake waters for timescales of minutes to days. Moreover, Lee et al. [24] found that algal-derived BDOM can be converted into recalcitrant organic matter (such as humic substances), which remains in the water column for a long period. On the one hand, this freshly produced BDOM would strongly influence its binding properties to heavy metals and thus their fate, mobility, and toxicity in aquatic environments; especially, the algal-derived polymeric component increases the coagulation and sedimentation rates of colloidal material and associated metals, and thus brings hidden potential dangers to benthic healthy [25,26]. On the other hand, after long-term biodegradation, progressive accumulation of algal-derived recalcitrant DOM, as disinfection by-product precursor compounds, is transformed into trihalomethanes and haloacetic acids (i.e., carcinogenic and mutagenic disinfection byproducts), which inevitably leads to increased costs of drinking water supply systems [27–29]. In addition, during the decay of cyanobacteria blooms in lake ecosystems, cyanobacteria toxins are introduced into water and could directly threaten human health [30,31].

Dissolved organic matter is a fundamental regulator of aquatic ecosystems, and algal-derived DOM often dominates the DOM composition in water during algal bloom decay in Lake Taihu [32]. Several authors have studied the production of DOM with unique compositions and stoichiometries that fuel successive trophic transfers of fixed carbon through initial assimilation by heterotrophs. However, their rates of processing phytoplankton DOM are controlled to a large extent by the biological availability of cyanobacterial exudate and lysate; thus, two different cyanobacteria fractions will have implications for lake microorganisms to facilitate the efficient turnover of the highly heterogeneous cycle of available DOM [33,34]. Microbial-mediated metabolism reactions have been demonstrated to transform DOM from labile to a relatively recalcitrant state. This process is associated with nutrient recycling, greenhouse outgassing, and microbial secondary production, affecting microbial succession, as reflected in the taxonomic composition and functional diversity, as well as the growth rates of specific groups [35–39].

For lake ecosystems, especially eutrophic lakes, the nutrients needed to maintain the system largely depend on the internal circulation and regeneration of the system rather than external input [40]. Among them, the degradation of cyanobacterial organic matter and the release process of nutrients, which are important carriers of nutrients in water, may be important sources of active nutrients in water [41]. Algal-derived DOM has generally been used to represent the heterogeneous matrix for the entire decomposition of cyanobacterial blooms; however, most studies have focused primarily on the environmental behavior of bulk algal-derived DOM or the response of the interaction between bulk algal-derived DOM and water contaminants. However, the relationship between released DOM refers to the different fractions of cyanobacterial organic matter, and there have been few studies on reproduced nutrient and microbial secondary production (i.e., ecological effect) following the cyanobacterial cell lysis mechanism.

Therefore, the objective of this study was to gain new insight into the role of molecular weight-fractionated cyanobacterial organic matter on DOM mineralization, sequestration, and related nutrients. We set up a medium-volume, long-term incubation experiment and combined it with a fractionation procedure to investigate how DOM from lysed cyanobacteria cells impacts the variation of the microbial population, and how processing of algal-derived DOM impacts nutrient mineralization by comparative analysis. The results will provide data support for the ecological effects of cyanobacteria degradation.

2. Materials and Methods

2.1. Sample Collection and Pretreatment

Surface lake water and fresh algal scum of the phytoplankton *Microcystis* were collected at the aggregate stage of a cyanobacterial bloom along the shore of Lake Taihu (trestle area), China, in 2021 [42] (Figure S1). Samples (i.e., fresh algal scum) were stored at 4 °C and transported to the laboratory, where they were immediately filtered through a sterile 20-µm-pore nylon net (47-mm diameter, Merck Millipore, Ireland) to exclude the interference of the largest algal aggregates, protozoa, and non-living particulates in the microbial degradation system. This filtration for lake water through a 20-µm-pore-size filter (47-mm diameter, Merck Millipore, Cork, Ireland) was performed to separate particle-associated bacteria or large cells from small and free-living bacteria [43]. Then, the chlorophyll a (Chl a) concentration was measured (18.61 µg L^{-1}) and considered the low biomass in summer. Chlorophyll a concentration was similar to the annual average in the lake center, and therefore, this was considered as a control set [44]. Meanwhile, harvested algal scum were concentrated through 20-µm bolting silk to partly dewater and remove obvious impurities; then, these samples with 90% moisture were defined as algal organic matter.

We considered the algal organic matter as a whole, which could be divided into two fractions by the freeze-thaw method. First, 10 mL algal organic matter-sterile water solutions with multiple batches (0.03 g mL^{-1} fresh weight) were shaken on a shaker, and we then obtained a destructed cyanobacterial complex mixture; the solids of these mixture were concentrated using a 0.2-µm membrane filter (47-mm diameter, PC, Merck Millipore, Cork, Ireland), and the filtrates were obtained. Destruction of cells in the above mixture was performed using the freeze-thaw method with three successive recycles, as described in a previous study [45]. The efficiency of cell destruction was confirmed using an optical microscope [46].

2.2. Batch Experiment

2.2.1. Long-Term Microbial Degradation of Algal Organic Matter

At room temperature (20–25 °C) and under dark conditions, the cyanobacterial scum sampled from Lake Taihu shore and then pretreated as the 0.5 L algal organic matter-sterile water solution (0.03 g mL^{-1} fresh weight) was added to 9.5 L of filtered lake water containing bacterial communities and mixed in an acid-cleaned glass container to conduct a long-term (80-d) degradation experiment as the algal-derived DOM group. The purpose of setting this algal biomass concentration, i.e., ~1.5 g L^{-1} (fresh weight), was to simulate the real accumulation situation along the lakeshore as best as possible. The purpose of using dark incubation conditions was to prevent the possible growth of cyanobacteria that may have been present in filtered lake water. Therefore, the possible growth of cyanobacteria will interfere with the experimental results. Simultaneously, the long-term degradation of the filtered lake water without added cyanobacterial scum was set as the control group (i.e., natural lake water group). Both groups had three replicates, and all containers were covered tightly to avoid direct contact with the atmosphere. By collecting water samples at different intervals during the 80-d degradation process, the dynamic changes in planktonic bacterial abundance in lake water, and the concentrations and compositions of dissolved organic carbon (DOC) and nutrients were analyzed.

2.2.2. Process Analysis of Microbial Degradation of Algal Organic Matter Using the Fractionation Procedure

The above-described long-term microbial degradation systems were repeated under the same controlled conditions in which 0.5 L of the extracted filtrates and concentrated solids, which were resuspended in an equal volume of sterile water, were added to the filtered lake water that served as the algal-derived filtrate and algal residual exudative organic matter groups. These two treatments were also conducted in triplicate in the dark for 80 d. As for the bulk algal-derived DOM and control groups, subsamples from the algal-derived filtrate and algal residual exudative organic matter groups were also collected on days 0, 1, 2, 4, 8, 12, 20, 31, 40, 52, 61, and 80 for analyses of planktonic bacterial abundance and determination of DOC and nutrient concentrations and compositions. Intriguingly, we observed that the water color of the algal-derived filtrate group changed dramatically (from brilliant blue to emerald green); therefore, an additional sample was collected on day 3.

2.3. *Sample Analysis*

2.3.1. DOC and Chromophoric Dissolved Organic Matter (CDOM) Measurement

For the analysis of DOC concentration, 40 mL of water sample was filtered through pre-combusted (450 °C for 4 h) 0.22-μm-pore-size glass microfiber filters (47-mm diameter, Shanghai Xingya Purification Material Factory, Shanghai, China). Then, the filtrates were collected with glass pipettes, placed into pre-combusted brown glass vials, and stored at −20 °C until analysis. Reference DOC standards (obtained using potassium hydrogen phthalate) served as an additional control to calibrate the instrument. The blank was deducted using Milli-Q water analysis before every five samples, and all samples (including the blank) were acidified to pH = 2 by 10% HCl until analysis. The average blank concentrations associated with the DOC measurement were approximately 0.06 mg L^{-1}, and the analytic precision of the triplicate injections was ±3%.

After passing through pre-combusted 0.22-μm Millipore membrane filters (47-mm diameter, Merck Millipore, Cork, Ireland), approximately 100 mL filtrate was used for CDOM absorbance and fluorescence measurements. CDOM absorbance was measured over the 200–800 nm range (1 nm increments) in a 5-cm quartz cell using a Shimadzu UV-2450 PC UV–vis recording spectrophotometer. A fluorescence spectrophotometer (Hitachi F-7000, Tokyo, Japan) with a scanning speed of 2400 nm min^{-1} was employed to measure the excitation-emission matrix (EEM). The main components of fluorescence dissolved organic matter (FDOM), i.e., the fluorescent fraction of CDOM, were analyzed using a fluorescence spectroscopy technique coupled with parallel factor (PARAFAC) analysis [47]. During the experiments, a total of 150 EEM spectra were obtained for PARAFAC analysis. MATLAB (MathWorks, Natick, MA, USA) and the DOM Fluor toolbox (http://www.models.life.ku.dk/ accessed on 6 May 2022) were employed for data analysis [48,49]. Further details on the PARAFAC-EEM analysis of FDOM have been described previously [50] and are displayed in Supplementary Materials Text S1.

During analysis of the composition parameters of CDOM, the characteristic parameters of the CDOM absorption coefficient [a (355)] and spectral slope for the interval of 300–500 nm [$S_{300-500}$] were employed to estimate the concentration and composition dynamic of CDOM, respectively [47]. Moreover, high $S_{300-500}$ values denote a high extent of recalcitrant and a low degree of molecular weight [51].

2.3.2. Nutrient Concentrations

Inorganic nitrogen (i.e., nitrate, nitrite, and ammonium) and inorganic phosphorus (i.e., phosphate) concentrations were measured using a continuous flow analyzer (San + +, SKALAR, Breda, The Netherlands). Total dissolved nitrogen (TDN) and total dissolved phosphorus (TDP) concentrations were analyzed using combined persulfate digestion followed by spectrophotometric measurements [52].

2.3.3. Bacterial Abundance (BA)

Samples for analyses of BA were preserved with a final concentration of 0.5% glutaraldehyde and stored at −80 °C. BA was measured using an LSRFortessa flow cytometer (BD Biosciences, San Jose, CA, USA) by staining with SYBR Green I. Bacteria were enumerated according to a previously described method [53].

2.4. Statistical Analysis

The linear regression model was used by OriginPro 8.1 software (OriginLab, Northampton, MA, USA) to characterize the following relationships: (1) among the free-living bacterial abundance and CDOM absorption coefficient at 355 nm for the entire process in the bulk algal-derived DOM group; (2) among free-living bacterial abundance and the main components of FDOM identified by PARAFAC analysis for the entire process in bulk algal-derived DOM, algal-derived filtrate, and algal residual exudative organic matter groups; and (3) among free-living bacterial abundance and the fluorescence intensity of humic-like components for the day 40–80, day 20–80, and day 31–80 phases in the bulk algal-derived DOM, algal-derived filtrate, and algal residual exudative organic matter groups, respectively. To examine the significance of the temporal changes in the main components of FDOM identified by PARAFAC analysis during the experiment, we mainly aimed at comparing the bulk algal-derived DOM and control groups, and the algal-derived filtrate, and algal residual exudative organic matter groups. One-way analysis of variance (ANOVA) was performed using the data analysis function of OriginPro 8.1 software.

3. Results and Discussion

3.1. Algal Organic Matter Analysis

The initial properties of the selected algal-derived DOM and its different fractions are summarized in Table 1. In brief, the initial bulk algal-derived DOM was composed of 67.15% carbon, 26.38% nitrogen, and 6.47% phosphorus, whereas the initial algal-derived filtrate and residual exudative organic matter were composed of 77.86% and 58.33% carbon, 19.55% and 39.06% nitrogen, and 2.59% and 2.60% phosphorus, respectively. These values showed that the carbon component of the algal-derived filtrate was higher than that of the bulk algal-derived DOM; the nitrogen component of algal residual exudative organic matter was higher than that of the bulk algal-derived DOM. In contrast, the phosphorus components of the algal-derived filtrate and algal residual exudative organic matter were lower than that of the bulk algal-derived DOM. After freeze-thaw treatment, the CDOM absorption coefficients at 355 nm for initial algal-derived filtrate and algal residual exudative organic matter were approximately 0.79 and 0.63 times lower than that of initial bulk algal-derived DOM, respectively. Obviously, the FDOM composition of the initial bulk algal-derived DOM contained only two components (i.e., protein-like C1 component and humic-like C2 component), and C1 was the main fraction in the bulk algal-derived DOM. Surprisingly, through the comparison among the FDOM composition for initial bulk algal-derived DOM, filtrate, and algal residual exudative organic matter, it appeared that the humic-like C2 component was enriched in the algal-derived filtrate, which was approximately > 6 times higher than that of the other two groups.

Table 1. Physicochemical properties of bulk algal-derived dissolved organic matter (DOM), algal-derived filtrate, and algal residual exudative organic matter. ND—not determined. DOC—dissolved organic carbon; DON—dissolved organic nitrogen; DOP—dissolved organic phosphorus.

	Parameter	Concentration (Mean ± Deviation)	Parameter	Concentration (Mean ± Deviation)
Bulk algal-derived DOM	DOC (mg L^{-1})	4.15 ± 0.56	C1 (R.U.)	0.78 ± 0.13
	DON (mg L^{-1})	1.63 ± 0.17	C2 (R.U.)	0.03 ± 0.01
	DOP (mg L^{-1})	0.40 ± 0.03	C3 (R.U.)	ND
	a (355) (m^{-1})	21.93 ± 1.80	C4 (R.U.)	ND
Algal-derived filtrate	DOC (mg L^{-1})	6.61 ± 0.73	C1 (R.U.)	0.43 ± 0.04
	DON (mg L^{-1})	1.66 ± 0.13	C2 (R.U.)	0.22 ± 0.02
	DOP (mg L^{-1})	0.22 ± 0.01	C3 (R.U.)	0.04 ± 0.04
	a (355) (m^{-1})	17.38 ± 0.16	C4 (R.U.)	0.03 ± 0.03
Algal residual exudative organic matter	DOC (mg L^{-1})	4.48 ± 0.58	C1 (R.U.)	0.87 ± 0.03
	DON (mg L^{-1})	3.00 ± 0.86	C2 (R.U.)	0.02 ± 0.03
	DOP (mg L^{-1})	0.20 ± 0.07	C3 (R.U.)	0.18 ± 0.10
	a (355) (m^{-1})	13.91 ± 0.29	C3 (R.U.)	0.01 ± 0.04

3.2. Release and Microbial Degradation of Algal-Derived DOC

The dynamic changes in the DOC concentration of the bulk algal-derived DOM, algal-derived filtrate, and algal residual exudative organic matter during the release and microbial degradation processes are shown in Figure 1. In detail, DOC was released from the cyanobacterial scum, resulting in a linear increase in the DOC concentration from 8.36 ± 0.32 mg L^{-1} to a peak of 10.68 ± 0.17 mg L^{-1} during the first 4 d. Then, the DOC concentration decreased from 10.68 ± 0.17 to 6.86 ± 0.86 mg L^{-1} from day 4–40 with a mean reduction rate of 0.11 ± 0.02 mg L^{-1} d^{-1}. Subsequently, the DOC concentration decreased to 4.60 ± 0.09 mg L^{-1} at the end of the experiment with a mean reduction rate of 0.06 ± 0.00 mg L^{-1} d^{-1} (Figure 1a). Therefore, according to the dynamic changes in the DOC concentration, the 80-d degradation process was clearly divided into three stages: the DOC rising (DR) stage (4 days); rapid DOC decline (r-DD) stage (36 days); and slow DOC decline (s-DD) stage (40 days). In contrast, no significant fluctuations in the DOC concentration were observed in the lake water; rather, it had a relatively narrow range from 3.34 ± 0.03 to 4.48 ± 0.21 mg L^{-1} and a mean of 3.98 ± 0.42 mg L^{-1}. Meanwhile, we determined that the DOC mineralization rate was 56.88 ± 1.26% from the time point of maximum DOC release to the end of the experiment. As shown in Figure 1b, the initial DOC concentrations were 10.81 ± 0.47 mg L^{-1} and 8.68 ± 0.33 mg L^{-1} in the algal-derived filtrate and the algal residual exudative organic matter treatments, respectively. Specifically, the algal-derived filtrate treatment caused a rapid and markedly large increase in the DOC concentration; however, the DOC concentration was slightly lower in the algal residual exudative organic matter group at the initial time point. Throughout the microbial degradation process, the DOC concentration showed a rapid decrease in the first 20 d (from 10.81 ± 0.47 to 5.57 ± 0.33 mg L^{-1}), followed by a near-continual slow decrease with an overall concentration of 4.44 ± 0.52 mg L^{-1} for the algal-derived filtrate group; in short, the DOC mineralization rate of the entire process was 64.56 ± 0.32%. In another algal residual exudative organic matter group, the DOC concentration showed a linear decrease over the entire experimental process by 51.83 ± 5.12% (i.e., the DOC mineralization rate). Finally, the DOC concentrations were 3.83 ± 0.18 and 4.17 ± 0.30 mg L^{-1} in the algal-derived filtrate and the algal residual exudative organic matter treatments, respectively, on day 80.

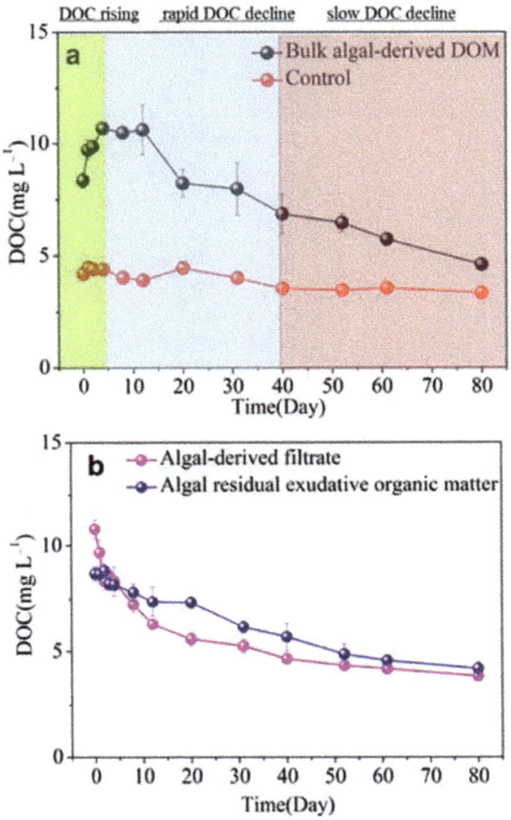

Figure 1. Dynamic changes in dissolved organic carbon (DOC) concentration during the 80-d degradation processes in (**a**) bulk algal-derived dissolved organic matter (DOM) group and natural lake water group (as control), and (**b**) algal-derived filtrate and algal residual exudative organic matter groups.

3.3. CDOM Absorption and Spectral Slope in Different Algal-Derived DOM

Generally, CDOM is largely responsible for the optical properties of most natural waters, and as a tracer of algal-derived DOM is valuable for elucidating the dynamic changes of DOM. The specific changes in the CDOM absorption coefficient [a (355)] of the bulk algal-derived DOM, algal-derived filtrate, and algal residual exudative organic matter during the 80 d processes are shown in Figure 2. In detail, in the bulk algal-derived DOM group, the a (355) increased from 24.35 $\pm$ 1.71 to 29.24 $\pm$ 0.37 m^{-1} within the first day, and then decreased to 15.93 $\pm$ 0.82 m^{-1} on day 4, with a reduction rate of 4.46 $\pm$ 0.60 m^{-1} d^{-1} from day 1–4. During the process from day 8–80, the a (355) fluctuant decreased from 8.11 $\pm$ 0.98 m^{-1} to 3.58 $\pm$ 0.11 m^{-1}. After repeated measurement, we found that an abnormal increase to 8.72 $\pm$ 0.23 m^{-1} occurred suddenly on day 40. Compared to the lake water, the a (355) exhibited relatively small variations and remained at the mean level (i.e., 2.40 $\pm$ 0.45 m^{-1}) during the entire experimental process. As shown in Figure 2b, after the different amended cyanobacterial organic matter fractions were added, we found that it led to large increases in a (355) in algal-derived filtrate and algal residual exudative organic matter groups; i.e., a (355) increased by 7.19 $\pm$ 0.04 and 5.75 $\pm$ 0.15 times compared with the initial lake water value, respectively. Then, a (355) degradation showed similar exponential decay patterns in both the algal-derived filtrate and algal residual exudative organic matter groups, and their intensities remained steady at approximately 5.20 ± 0.47 and 4.22 ± 0.81 m^{-1}, respectively, with time after decreasing rapidly in the first 20 and 31 d, respectively.

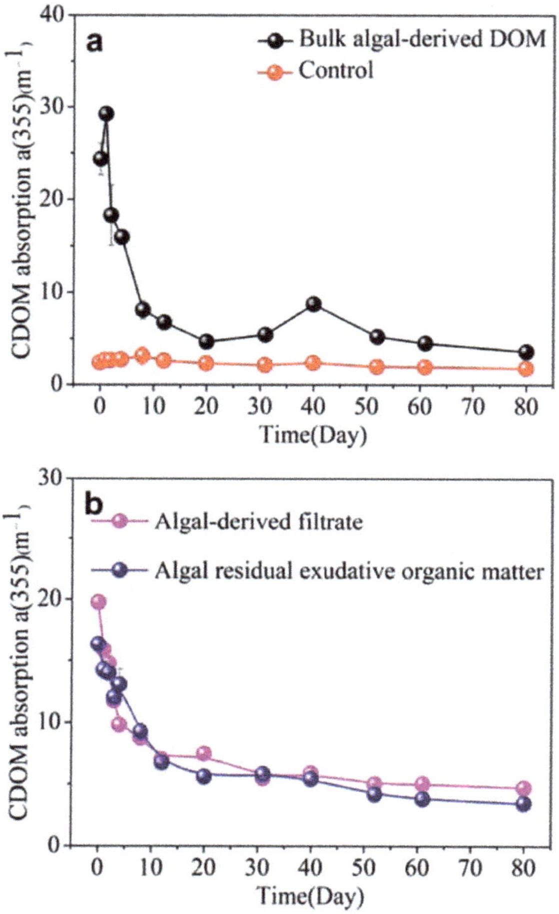

Figure 2. Dynamic changes in the chromophoric dissolved organic matter (CDOM) absorption coefficient at 355 nm [a CDOM (355)] during the 80-d degradation processes of (**a**) bulk algal-derived dissolved organic matter (DOM) and natural lake water groups (as control) and (**b**) algal-derived filtrate and algal residual exudative organic matter groups.

The CDOM spectral slope for the 300–500 nm [$S_{300–500}$] interval was employed to estimate the composition dynamic of CDOM. Moreover, high $S_{300–500}$ values denote a high extent of recalcitrant and a low degree of molecular weight [51]. Throughout the entire bulk algal-derived DOM degradation process, the spectral slope $S_{300–500}$ gradually increased from 6.69 ± 0.06 μm^{-1} to 15.71 ± 0.49 μm^{-1}; it indicated the transformation from high-molecular-weight labile CDOM into low-molecular-weight recalcitrant CDOM. In the control group, Figure 3a shows that the spectral slope $S_{300–500}$ had a relatively narrow range variation from 11.04 ± 0.30 to 13.69 ± 0.07 μm^{-1} within the first 40 d, and that the index then increased to 17.24 ± 0.20 at the final time point. Notably, at the s-DD stage, the $S_{300–500}$ index was near the mean level of the control group; this implied freshly cyanobacterial scum-released CDOM underwent long-term biotransformation to perform some molecular weight CDOM similar to cyanobacterial scum-free lake water. For the algal-derived filtrate group (Figure 3b), the spectral slope $S_{300–500}$ increased by 5.12 ± 0.04 μm^{-1}, i.e., from 8.10 ± 0.06 to 13.22 ± 0.06 μm^{-1}, over the course of the experiment. Another algal residual exudative organic matter group exhibited highly dynamic characteristics because the spectral slope $S_{300–500}$ increased by 9.20 ± 0.34 μm^{-1} from 5.44 ± 0.04 to 14.63 ± 0.35 μm^{-1} over the course of the experiment (Figure 3b). Therefore, the initial molecular weight of the algal residual exudative organic matter group was higher than that of the algal-derived filtrate and bulk algal-derived DOM groups; however, after the 80-d experiment, the molecular weight of DOM was converted to the same level for the three groups.

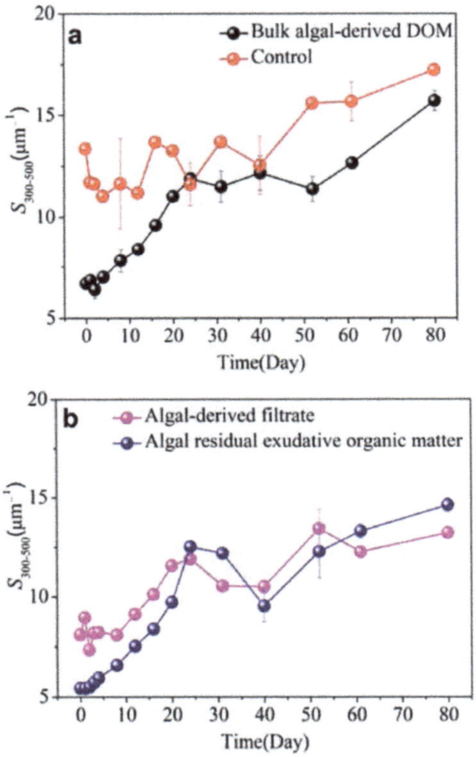

Figure 3. Dynamic changes in the chromophoric dissolved organic matter (CDOM) spectral slope for the 300–500 nm [$S_{300–500}$] interval during the 80-d degradation processes of the (**a**) bulk algal-derived DOM and natural lake water groups (as control) and (**b**) algal-derived filtrate and algal residual exudative organic matter groups.

3.4. EEMs in Different Algal-Derived DOM

The characterization of DOM by 3D fluorescence spectroscopy was considered a reliable parameter for observing the entire degradative process for bulk algal-derived DOM and natural lake water groups (as a control). As shown in Figure 4a–c, we found that the EEMs of CDOM released from cyanobacterial scum at two major protein-like fluorescence peaks decreased with time and reached a low level at the end of the experiment, which was characterized by being highly dynamic. Comparatively, for the natural lake water group, a conservative distribution of inherent CDOM peaks at the two similar protein-like fluorescence peaks was observed; however, its fluorescence intensities and the range of increasing–decreasing fluctuations were both lower than those of the bulk algal-derived DOM group. This result was consistent with those reported previously [54], whereby cyanobacteria survival and mortality played important roles in shaping the optical properties of many natural waters. Moreover, decayed cyanobacteria led to protein-like CDOM production, which was an important source of biodegradable DOC and contributes to the biogeochemical cycles of aquatic ecosystems [55]. Comparing the EEMs of the algal-derived filtrate and the algal residual exudative organic matter groups, as shown in Figure 5, we found that high-intensity protein peaks dominated the entire spectra, and that after 80-d degradation, their fluorescence intensities were greatly reduced. Finally, the variability in the fluorescence properties of CDOM of the two groups was similar to that of the control group, as determined by EEMs. More specifically, the intensities of humic peaks were almost consistent with the intensities of protein peaks, which only appeared in the algal-derived filtrate group at the initial time point. This suggested humic-like components direct from cyanobacteria cell lysis, and that this component has low molecular weight characteristics, which is indicated by the higher values of $S_{300\text{-}500}$ than the algal colloidal exudative organic matter group during the initial 20 days.

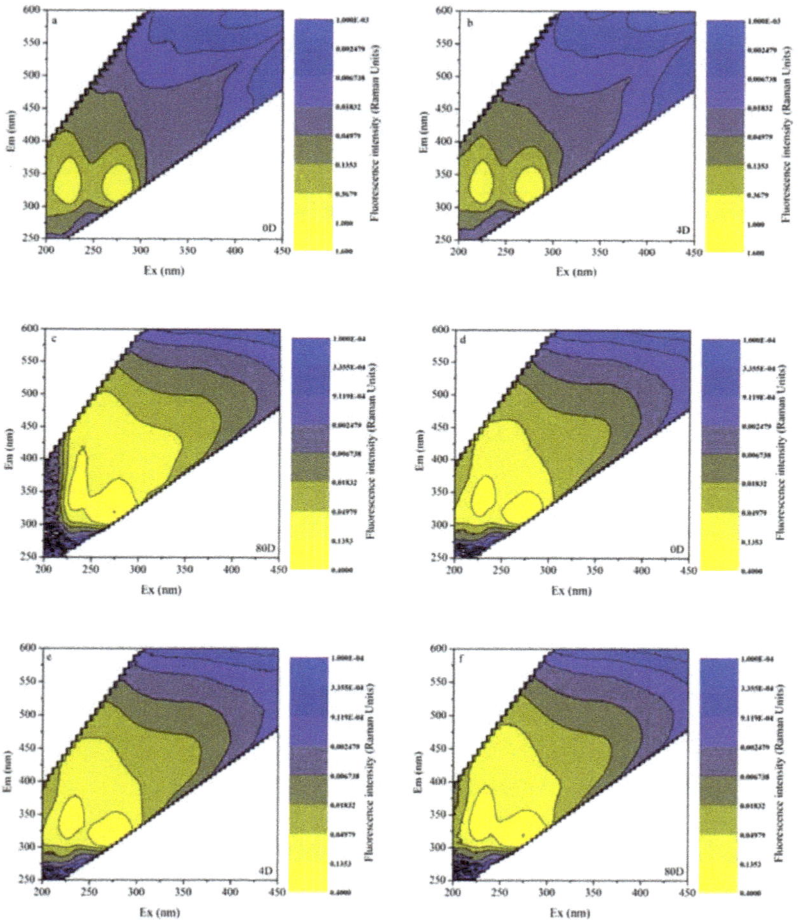

Figure 4. Chromophoric dissolved organic matter (CDOM) fluorescence properties variability by excitation-emission matrix spectra (EEMs) during the degradation processes of the bulk algal-derived dissolved organic matter (DOM) group on (**a**) day 0; (**b**) day 4; and (**c**) day 80, and in the natural lake water group (as control) on (**d**) day 0; (**e**) day 4; and (**f**) day 80.

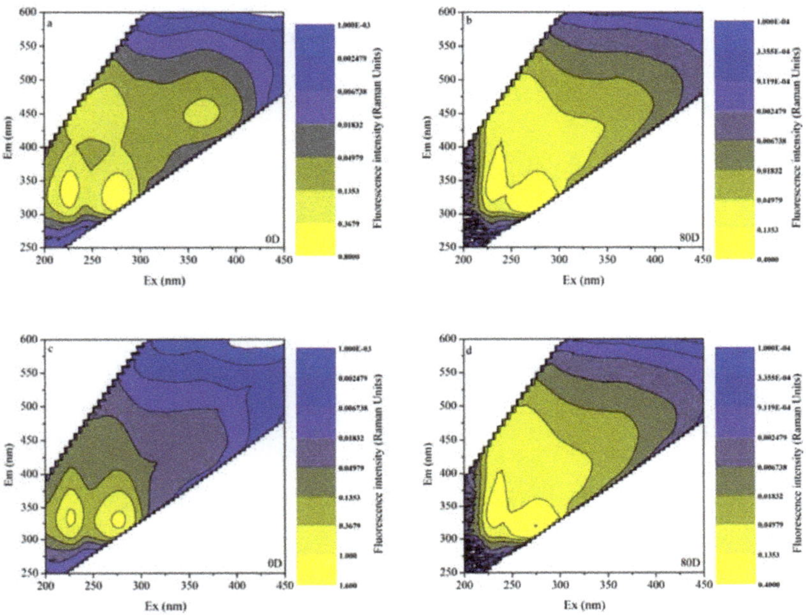

Figure 5. Chromophoric dissolved organic matter (CDOM) fluorescence properties presented by excitation-emission matrix spectra (EEMs) for the algal-derived filtrate group on (**a**) day 0 and (**b**) day 80, and for the algal residual exudative organic matter group on (**c**) day 0 and (**d**) day 80.

3.5. FDOM Components in Different Algal-Derived DOM

FDOM was employed as a proxy for DOM to study the dynamic changes in DOM composition [56]. The FDOM components were characterized with three-dimensional EEM spectroscopy coupled with the PARAFAC analysis technique, and total EEM collection of the algal-derived dissolved organic matter; its different molecular fractions and lake water were modeled with PARAFAC using MATLAB with the DOMFluor toolbox. Furthermore, four distinct FDOM components (i.e., C1–C4) were identified; in detail, the ex | em wavelengths were 225 (275–280) | 332 nm, 265 (360) | 452 nm, 230 (275) | 320 nm, and 235 (305) | 348 nm, respectively. Based on their ex | em wavelengths, C1, C3, and C4 represented protein-like fractions and C2 represented a humic-like fluorophore. During the bulk algal-derived DOM degradation process, the dynamic changes in the fluorescence intensity of the protein-like component C1 exhibited three stages, which corresponded to the stage characteristics of dynamic changes in the DOC concentration (Figure 1). Furthermore, its fluorescence intensity decreased periodically at 0.89 ± 0.11 R.U. (mean values) in the DR stage, at 0.21 ± 0.07 R.U. (mean values) in the r-DD stage, and the final fluorescence intensities fell to zero (mean values) in the final s-DD stage (Figure 6a). Interestingly, the change in the humic-like component C2 showed an opposite trend compared with that of the protein-like component C1 (Figure 6b). The C2 component increased gradually throughout the degradation process, except for a sudden increase on the first day (Figure 6b). However, the two fluorescence components (i.e., C1 and C2) in the control group exhibited no significant changes throughout the entire degradation process (Figure 6a,b). Moreover, we found that the variation patterns of the C1 and C2 components, when comparing the bulk algal-derived DOM and natural lake water groups, exhibited significant changes (one-way ANOVA, $p < 0.05$); therefore, these two components also should focus on fluorescence variety with time for the algal-derived filtrate and the algal residual exudative organic matter groups. Regarding the two other protein-like C3 and C4 components, their changes tended to be consistent, and both remained at relatively low levels compared to the biodegradable protein-like C1 component (Figure 6c,d).

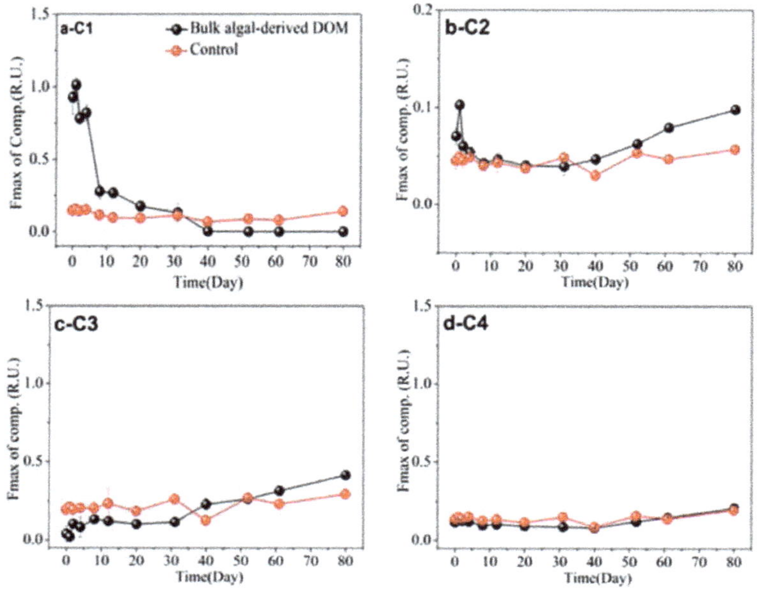

Figure 6. Dynamic changes in the main components of fluorescent dissolved organic matter (FDOM) during the 80-d degradation processes of the algal-derived DOM group and in the natural lake water group (as control). (**a**) C1, (**b**) C2, (**c**) C3, and (**d**) C4.

Similarly, four FDOM components (C1, C2, C3, and C4) were identified in the algal-derived filtrate and the algal residual exudative organic matter groups. Based on our previous results regarding dynamic changes in the FDOM compositions of bulk algal-derived DOM and control groups, the intensity changes of the protein-like component C1 and humic-like component C2 were comprehensively examined again in algal-derived filtrate and algal residual exudative organic matter groups. Notably, the dynamic variations in fluorescence intensity of C1 and C2 underwent clear changes, which were also reflected by the three-stage characteristics over the entire degradation period (Figure 7a,b). In the algal-derived filtrate group, the fluorescence intensity of C1 decreased from 0.58 ± 0.03 to 0.47 ± 0.01 R.U. with a decrease rate of 0.04 R.U. d^{-1} from day 0–3. Subsequently, its fluorescence intensity decreased to 0.05 ± 0.02 R.U. at day 20, with a decrease rate of 0.02 R.U. d^{-1}, and stabilized at 0.02 ± 0.02 R.U. from day 20–80. In contrast, the fluorescence intensity of C2 increased from 0.07 ± 0.01 to 0.10 ± 0.0004 R.U. for the period of day 20–80 (Figure 7a,b). Similarly, in the algal residual exudative organic matter group, the protein-like component C1 decreased rapidly from 1.02 ± 0.01 to 0.51 ± 0.04 R.U. with a decrease rate of 0.13 R.U. d^{-1} for the period of day 0–4, then decreased slowly to 0.04 ± 0.00 R.U. at day 31, with a decrease rate of 0.01 R.U. d^{-1}, and reached near-constant intensities with a mean level of 0.02 ± 0.01 R.U. in the day 31–80 phase. The fluorescence intensity of C2 increased from 0.04 ± 0.003 to 0.08 ± 0.0003 R.U. in the day 31–80 phase (Figure 7a,b). Curiously, in the algal-derived filtrate, we found that the humic-like component C2, which is always used as an indicator of recalcitrant DOM [57], was found at much higher fluorescence signal levels when compared to the algal residual exudative organic matter and bulk algal-derived DOM samples on day 0. Furthermore, its fluorescence intensity remained high (~0.25 ± 0.04 R.U.) for the period of day 0–3, and suddenly decreased to a lower level of 0.13 ± 0.05 R.U. for the period of day 3–20 (Figure 7b). Zuo et al. [58] identified 6-L-biopterins and their glucosides as candidate structures for consistently occurring algae-derived humic-like fluorophores (Em 440–460 nm) during the cyanobacterial strain *Microcystis aeruginosa* degradation experiment under simulated natural conditions; the environmental concentrations of 6-L-biopterin (without counting any other derivatives) ranged from 0.20–2.78 μg L^{-1} in five lakes in China. Additionally, biopterin and its derivatives, which contributed to 55.5 ± 1.7% of fluorescence at Ex350/Em450 nm in FDOM, were found in a Lake Tai surface water sample [58]. In this study, C2 displayed two excitation maxima at 265 and 360 nm and one emission maxima at 452 nm, which was categorized as a humic-like peak; therefore, it may have indicated that the algae-derived characteristic humic-like fluorophores were associated with biopterin. Humic-like fluorophores are ubiquitous in algal-dominated freshwater and marine environments [59,60]. Through this research, we found that humic-like components rich in cyanobacterial filtrate could not accumulate in lake and coastal eutrophic water, which is inconsistent with research results from pelagic oligotrophic sea and deep-sea zones. For instance, Xie et al. [55] investigated the bioavailability of *Synechococcus*-derived organic matter by estuarine and coastal microbes during 180-d dark incubations, and found that humic-like C4 (ex | em wavelength was 250 (385) | 484 nm) displayed recalcitrant DOM characteristics, and that its fluorescence intensity gradually increased over the entire incubation period. One possible explanation for this result is that the high decrease rate of humic-like fluorophores was supported by the high content of biodegradable protein-like components or nutrients as well as the metabolically active microbial populations in the eutrophic lake water [47]. In contrast, in oligotrophic environments, freshly released algal filtrates can hardly trigger the organic matter "priming effect" mechanism that stimulates the microbial degradation of humic-like components, whereas humic-like organic matter is often considered a potential tracer of recalcitrant DOC [61–63]. Lake eutrophication can result in algal blooms and water quality degradation, which affects the services provided by the lake ecosystem [64]. As discussed above, initial DOM compositions derived from algal-derived filtrates and algal-residual exudative organic matter were highly heterogeneous in terms of FDOM composition, whereby their FDOM compositions were consistent with each other after a long degradation period.

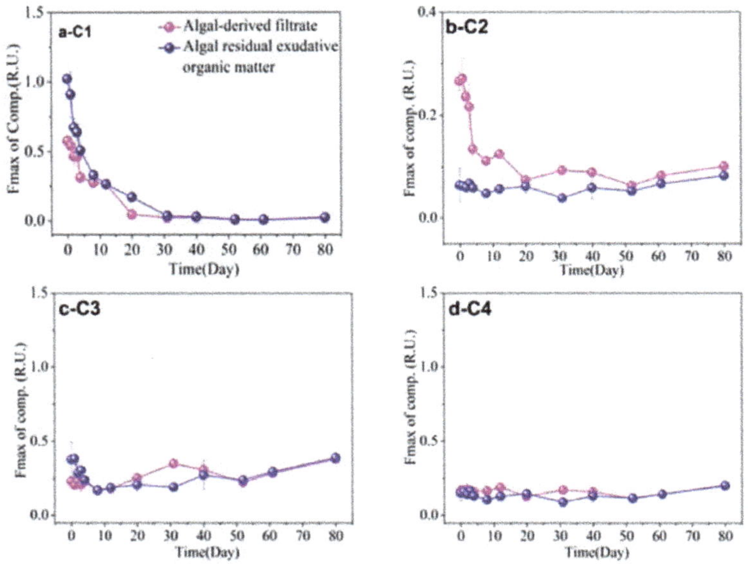

Figure 7. Dynamic changes in the main components of fluorescent dissolved organic matter (FDOM) during the 80-d degradation processes of algal-derived filtrate and algal residual exudative organic matter groups. (**a**) C1, (**b**) C2, (**c**) C3, and (**d**) C4.

3.6. Variability of Nutrient Compositions in Different Algal-Derived DOM

Along with the release and microbial degradation of DOC derived from cyanobacterial scum, variations in inorganic nutrients (including NH_4^+, NO_2^-, NO_3^-, and PO_4^{3-}) and organic nutrients (including DON and DOP) were presented during the 80-d incubation (Figure 8c,d). In detail, in the bulk algal-derived DOM group, in terms of the variations of TDN and TDP, the TDN concentration gradually increased from 6.01 ± 0.36 to 12.38 ± 1.06 mg L^{-1} in the day 0–31 phase, then remained at 11.91 ± 0.96 mg L^{-1} during day 31–80 (Figure 8a); the TDP concentration gradually increased from 0.72 ± 0.03 to 1.10 ± 0.16 mg L^{-1} in the day 0–31 phase, then remained at 1.19 ± 0.10 mg L^{-1} for the day 31–80 phase. Three inorganic nitrogen compositions were measured, and the combined concentrations as dissolved inorganic nitrogen (DIN) and its variation are shown in Figure 8b, in which a linear increase can be seen from 3.73 ± 0.39 to 14.26 ± 1.69 mg L^{-1} with an increase rate of 0.29 ± 0.05 mg L^{-1} d^{-1} during the day 0–31 phase. Within the next 49 d, the DIN concentration increased to 13.29 ± 0.90 mg L^{-1} with a minimal increase rate of 0.01 ± 0.06 mg L^{-1} d^{-1}. The variation of DIP followed the same pattern, whereby it increased from 0.28 ± 0.06 to 1.11 ± 0.09 mg L^{-1} with an increase rate of 0.03 ± 0.00 mg L^{-1} d^{-1} during the day 0–31 phase and was subsequently relatively stable within the range of 0.41–0.52 mg L^{-1}. Meanwhile, in addition to leaching dissolved total nutrients from cyanobacterial biomass, there are many other nutrient metabolism processes, i.e., organic nitrogen and phosphorus degradation, and inorganic nitrogen and phosphorus production (Figure 8b,c). Moreover, ammonia was the most dominant form of DIN, with the conversion of cyanobacteria-derived organic nitrogen to inorganic nitrogen (Figure 8b,d). In the day 40–80 phase, the DON and DOP concentrations tended to stabilize and maintain mean concentrations of 0.34 ± 0.46 and 0.07 ± 0.08 mg L^{-1}, respectively; specifically, the dominant form of DIN was converted into nitrate (Figure 8b–d). In the control group, the range of changes and contents in the dissolved nutrient composition were relatively gently and small, and the most dominant DIN (i.e., ammonia) during the initial 4 d was also replaced by nitrate over the course of the entire process (Figure 8e–h).

In the algal-derived filtrate group, the initial TDN and TDP concentrations were 6.46 ± 0.24 and 0.62 ± 0.01 mg L^{-1}, respectively, during the first 3 d and then decreased to 5.51 ± 0.18 and 0.41 ± 0.01 mg L^{-1}, respectively. For the following 17 d, TDN and TDP concentrations exhibited linear increases to 6.93 ± 0.02 and 0.63 ± 0.05 mg L^{-1}, respectively; finally, the concentrations remained constant at 7.04 ± 0.41 and 0.67 ± 0.02 mg L^{-1}, respectively (Figure 9a). In contrast, in the algal-derived filtrate group, the TDN concentration displayed a similar variation pattern to that of the TDP concentration. For the initial 12 d, TDN and TDP concentrations were stable at 4.85 ± 0.40 and 0.27 ± 0.03 mg L^{-1}, respectively. After 20 d, TDN and TDP concentrations dramatically increased to 6.66 ± 0.51 and 0.44 ± 0.05 mg L^{-1}, respectively (Figure 9e). Two processes of total dissolved nutrient release and dissolved organic nutrient conversion into dissolved inorganic nutrients were detected from 0–20 d in both groups. Ammonia was the only form of DIN that was present. Moreover, in the following 60 d, the variations in the DIN, DON, DIP, and DOP concentrations exhibited roughly stable levels; notably, the majority of DIN during this period was replaced by nitrate, and a high nitrate concentration was maintained (Figure 9b–d,f–h).

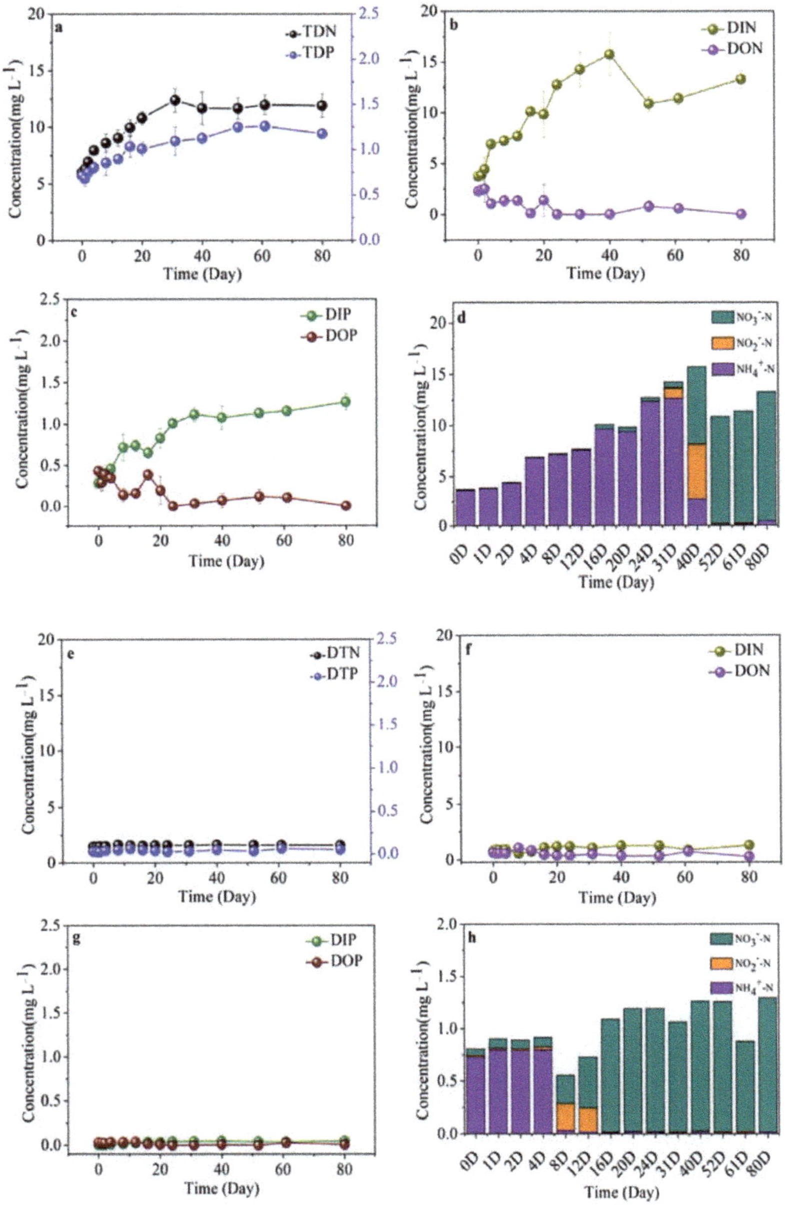

Figure 8. Variations in dissolved nitrogen and phosphorus concentrations and compositions during the 80-d degradation processes of the (**a–d**) bulk algal-derived dissolved organic matter (DOM) group and (**e–h**) natural lake water group (as control). TDN—total dissolved nitrogen, TDP—total dissolved phosphorus, DIN—dissolved inorganic nitrogen, DON—dissolved organic nitrogen, DIP—dissolved inorganic phosphorus, DOP—dissolved organic phosphorus, DTP—dissolved total phosphorus, DTN—dissolved total nitrogen.

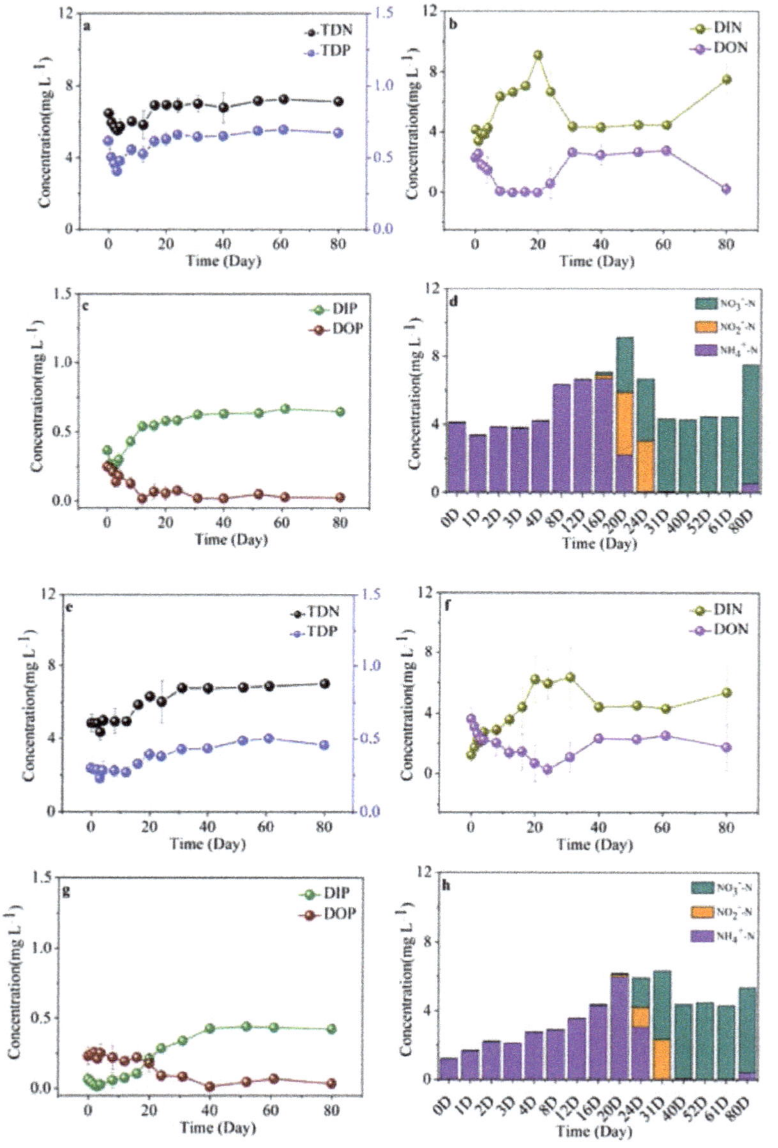

Figure 9. Variations in dissolved nitrogen and phosphorus concentrations and compositions during the 80-d degradation processes of the (**a–d**) algal-derived filtrate group and (**e–h**) algal residual exudative organic matter group. TDN—total dissolved nitrogen, TDP—total dissolved phosphorus, DIN—dissolved inorganic nitrogen, DON—dissolved organic nitrogen, DIP—dissolved inorganic phosphorus, DOP—dissolved organic phosphorus.

3.7. Variability of Free-Living Bacterial Abundance in Different Algal-Derived DOM

For the degradation of the bulk algal-derived DOM process, the change of free-living bacterial abundance showed highly dynamic characteristics and could be roughly divided into three stages, which correspond with the periodic transformation of different biological availability of algal-derived DOM (labeled Stage DR, r-DD, and s-DD). In the bulk algal-derived DOM group, the free-living bacterial abundance increased by

$0.48 \pm 0.71 \times 10^7$ cells mL^{-1} (from $4.22 \pm 0.28 \times 10^7$ to $4.70 \pm 0.51 \times 10^7$ cells mL^{-1}) in phase DR. Then, the free-living bacterial abundance decreased to $1.23 \pm 0.31 \times 10^7$ cells mL^{-1} at the end of phase r-DD and remained with relatively small variations in the range from 0.41×10^7 to 1.77×10^7 cells mL^{-1} in the following days (i.e., phase s-DD). In contrast, in the control lake water group, no significant fluctuations in bacterial abundance were observed; the range of change was between 0.06×10^7 and 0.98×10^7 cells mL^{-1} (Figure 10a).

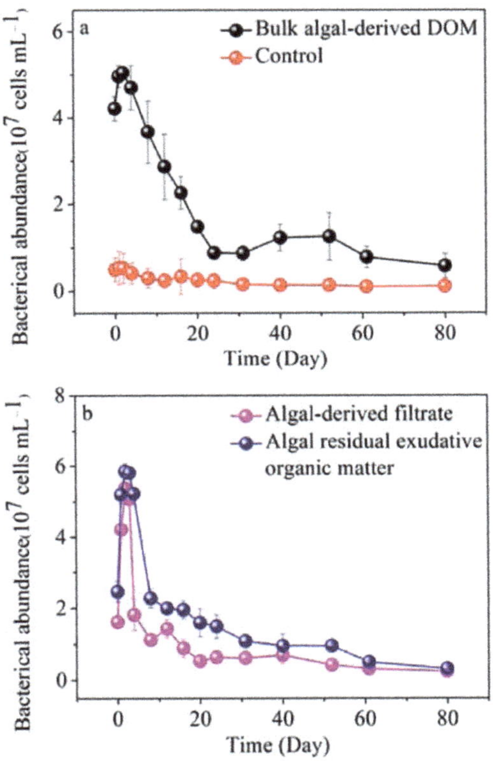

Figure 10. Dynamic changes in free-living bacterial abundance during 80-d degradation processes of (**a**) bulk algal-derived dissolved organic matter (DOM) group and in natural lake water group (as control), and (**b**) algal-derived filtrate group and algal residual exudative organic matter group.

The DOC concentration increased sharply and reached its maximum in only 4 d during the DR stage. Concurrently, the bacterial abundance also increased. This indicated that the release of DOC from decaying cyanobacteria was relatively rapid, and that the freshly released DOC was strongly favored by bacteria. During this stage, bacteria abundance slightly increased, indicating that a portion of the released DOC was utilized and transformed into bacterial biomass, or possibly some originally attached bacteria dispersal from dead algal biomass into the water. Subsequently, the DOC concentration began to decline, entering the r-DD stage. The decrease in DOC concentration may have been due to the slowdown in DOC release from the cyanobacterial scum not maintaining the same rate as DOC consumption by the abundant bacteria, or the life strategy of these bacteria undergoing changes [34]. The relatively rapid decline process in DOC continued for 5 weeks. This showed that a portion of the DOC released by the decaying algal scum was labile DOC and was easily utilized by bacteria. Indeed, in the eutrophic environment of Lake Taihu, lake water quality usually deteriorates within approximately 1 month after decomposition of *Microcystis* blooms [65]. One important reason may have been that, while utilizing the rich labile DOC released from decaying algal scum, the heterotrophic bacteria consumed oxygen and released CO_2, resulting in decreased dissolved oxygen and impacts on the nutrient status of the surrounding water [66,67]. After the r-DD stage, during the remaining 40 d, the remaining DOC had lower bioavailability to bacteria, and partly the DOC fraction may have been in recalcitrant states [68]. Notably, bacterial abundance reached its maximum on day 2, decreased rapidly in the next 38 d, and then decreased at a slower rate in the final 40 d. This may suggest that partly DOC components become humified, which barely promotes the growth of free-living bacteria, leading to the death of some bacteria. Compared to the control group, the positive linear correlation between free-living bacteria abundance and DOC concentration for the entire process ($R^2 = 0.63$, $p < 0.01$) further suggested that algae-released DOC stimulated microbial growth and activities [29]. The slope of significantly positive relationships between bacterial abundance and DOC concentration were found in the r-DD stage to be almost three times higher than the s-DD stage; this preliminarily reflected that the availability of DOM components and the bacterial responses to phytoplankton-derived carbon had changed. Furthermore, we found a relatively strong linear correlation between free-living bacteria abundance and CDOM absorption coefficient at 355 nm for the entire process ($R^2 = 0.68$, $p < 0.001$), indicating that the microbial biodegradation process had transformed the autochthonous DOM from chromophoric components to nonchromophoric species. Furthermore, it showed that the growth and mortality of some free-living bacteria responded to the transformation of DOM, and the increasing changes in the spectral slope for the interval of 300–500 nm (i.e., $S_{300–500}$), indicating the production of a large amount of high-molecular-weight CDOM through microbial conversion to lower molecular weight CDOM. By focusing on changes in the free-living bacteria abundance and the fluorescence intensities of four FDOM components, only the protein-like C1 fraction showed a significant and positive relationship with free-living bacteria abundance ($R^2 = 0.85$, linear regression, $p < 0.001$), indicating that the protein-like C1 may be a critical bioavailable component and energy source for microbes in the lake. At the initial time point of the s-DD period, the CDOM absorption coefficient at 355 nm exhibited an abnormal increase on day 40. Concurrently, we found that the humic-like C2 component increased linearly and was weakly negatively correlated with the change in bacterial abundance ($R^2 = 0.24$, $p > 0.05$), in contrast to the change in the protein-like C1 component, during the last s-DD stage. Therefore, we infer that the remnants of dead bacteria (i.e., necromass) and bacterial secretions may have made an important contribution to the accumulation of the humic-like C2 component. Therefore, we can conclude that the lysis of cyanobacterial scum by microbes liberates DOM into lake water, and the DOC concentration and composition in water are closely related to the abundance of planktonic bacteria.

For the algal-derived filtrate and algal-residual exudative organic matter groups, the dynamic change patterns in free-living bacterial abundance also showed three different characteristics among stages of the long-term degradation process (Figure 10b). At the beginning, bacterial abundances of these two treatments were $1.62 \pm 0.17 \times 10^7$ and $2.45 \pm 0.29 \times 10^7$ cells mL^{-1}, respectively. These initial values were at much higher levels than the lake water blank sample that had been filtered through 20-μm pore size membranes; however, the sum of the two values was slightly less than that of the bulk algal-derived DOM group, with approximately 4% loss due to experimental operations. In the algal-derived filtrate group, the free-living bacterial abundance increased considerably in response to the addition of the carbon sources and reached $5.08 \pm 0.15 \times 10^7$ cells mL^{-1} on day 3. A sharp decline in the bacterial abundance was observed from day 3 to day 20 (i.e., $0.54 \pm 0.05 \times 10^7$ cells mL^{-1}), and then there was a gradual decrease to $0.25 \pm 0.01 \times 10^7$ cells mL^{-1} on day 80. In the algal-residual exudative organic matter group, bacterial abundance showed an increase over the first 4 d to $5.23 \pm 0.09 \times 10^7$ cells mL^{-1}, followed by a rapid decrease during the following 28 d (i.e., $1.09 \pm 0.17 \times 10^7$ cells mL^{-1}) and near-constant abundances after day 31 to the end of the experiment with a mean abundance of $0.69 \pm 0.33 \times 10^7$ cells mL^{-1}.

Significantly positive linear relationships between free-living bacterial abundance and protein-like C1 fluorescence were also found in the algal-derived filtrate group ($R^2 = 0.78$, $p < 0.001$) and algal residual exudative organic matter group ($R^2 = 0.59$, $p < 0.01$) throughout the entire experimental period. This indicated that labile algal-derived filtrate or algal residual exudative organic matter derived from algal degradation became an important factor in controlling bacterial abundance, which would play an important role in microbial food webs and carbon cycling during the decay of algal blooms through the bacterial enzymolysis mechanism [69]. At the final stable phase of the two groups (algal-derived filtrate group, day 20–80; algal residual exudative organic matter group, day 31–80), the humic-like C2 component accumulated with the decrease in free-living bacterial abundance, and free-living bacterial abundance showed a significant negative linear relationship with humic-like C2 fluorescence (algal-derived filtrate group, $R^2 = 0.63$, $p < 0.05$; algal residual exudative organic matter group, $R^2 = 0.96$, $p < 0.01$).

4. Conclusions

In this work, the labile fraction of fresh cyanobacterial organic matter could be rapidly utilized by microbes, and the subsequent microbial-mediated processes of cyanobacterial organic matter also drove the elemental cycling of carbon, nitrogen, and phosphorus. Specifically, we showed the gradual transformation of algal-derived DOC from a high-molecular-weight labile to a low-molecular-weight recalcitrant through the microbial process. Different dominant DOM compositions with bioavailable characteristics were present in different phases during bulk algal-derived DOM degradation, which were linked to the variation in the abundance of free-living bacteria. Furthermore, this process was coupled with the regeneration of different forms of inorganic nutrients. Algal-derived filtrate and algae residual exudative organic matter both made key contributions to the decomposition of algal blooms in eutrophic Lake Taihu. Additionally, the findings show that the carryover effects of *Microcystis* cyanobacteria blooms can exist for a long time, whereby approximately 7.45% and 18.60% of the released DOC could be converted into a stable state in the algal-derived filtrate and algal residual exudative organic matter groups, respectively.

Supplementary Materials: The following supporting information can be downloaded at: https://www.mdpi.com/article/10.3390/ijerph19126981/s1, Text S1: Main Components of Fluorescent Dissolved Organic Matter (FDOM) [47–49,70–72]; Figure S1: Cyanobacterial blooms in China were recorded during the summer of 2021 off the lakeside of Taihu. Photo shows the *Microcystis*-dominated total phytoplankton community growing in the near lake waters and deposited on the beaches of the countryside.

Author Contributions: Conceptualization, J.C. and Y.Z. (Yunlin Zhang); Data curation, J.C.; Formal analysis, J.C. and Y.Z. (Yongqiang Zhou); Funding acquisition, J.C. and Y.Z. (Yunlin Zhang); Investigation, J.C. and Y.Z. (Yongqiang Zhou); Methodology, J.C.; Project administration, J.C.; Resources, Y.Z. (Yunlin Zhang); Supervision, Y.Z. (Yunlin Zhang); Validation, J.C.; Visualization, J.C.; Writing—original draft, J.C. and Y.Z. (Yunlin Zhang). All authors have read and agreed to the published version of the manuscript.

Funding: This study was funded by the National Natural Science Foundation of China (grants 41930760 and 41807362), the Youth Innovation Promotion Association, CAS (2021312), NIGLAS Foundation (E1SL002), and the Chinese Postdoctoral Science Foundation (BX2021325).

Institutional Review Board Statement: Not applicable.

Informed Consent Statement: Not applicable.

Data Availability Statement: Not applicable.

Acknowledgments: The authors are grateful to the Taihu Lake Laboratory Ecosystem Research Station and the State Key Laboratory of Lake Science and Environment, China, for supporting this survey.

Conflicts of Interest: The authors declare no conflict of interest.

References

1. Xiao, M.; Chen, Z.C.; Zhang, Y.; Wen, Y.A.; Shang, L.H.; Zhong, J. The Optical Characterization and Distribution of Dissolved Organic Matter in Water Regimes of Qilian Mountains Watershed. *Int. J. Environ. Res. Public Health* **2022**, *19*, 59. [CrossRef]
2. Zhang, Y.L.; Deng, J.M.; Qin, B.Q.; Zhu, G.W.; Zhang, Y.J.; Jeppesen, E.; Tong, Y.D. Importance and vulnerability of lakes and reservoirs supporting drinking water in China. *Fundam. Res.* **2022**, *in press*. [CrossRef]
3. Yao, X.L.; Zhang, Y.L.; Zhang, L.; Zhu, G.W.; Qin, B.Q.; Zhou, Y.Q.; Xue, J.Y. Emerging role of dissolved organic nitrogen in supporting algal bloom persistence in Lake Taihu, China: Emphasis on internal transformations. *Sci. Total Environ.* **2020**, *736*, 139497. [CrossRef]
4. Deng, J.M.; Qin, B.Q.; Paerl, H.W.; Zhang, Y.L.; Ma, J.R.; Chen, Y.W. Earlier and warmer springs increase cyanobacterial (*Microcystis* spp.) blooms in subtropical Lake Taihu, China. *Freshw. Biol.* **2014**, *59*, 1076–1085. [CrossRef]
5. Hu, W.P. A review of the models for Lake Taihu and their application in lake environmental management. *Ecol. Modell.* **2016**, *319*, 9–20. [CrossRef]
6. Qin, B.Q.; Paerl, H.W.; Brookes, J.D.; Liu, J.G.; Jeppesen, E.; Zhu, G.W.; Zhang, Y.L.; Xu, H.; Shi, K.; Deng, J.M. Why lake Taihu continues to be plagued with cyanobacterial blooms through 10 years (2007–2017) efforts. *Sci. Bull.* **2019**, *64*, 354–356. [CrossRef]
7. Qin, B.Q.; Zhu, G.W.; Gao, G.; Zhang, Y.L.; Li, W.; Paerl, H.W.; Carmichael, W.W. A drinking water crisis in Lake Taihu, China: Linkage to climatic variability and lake management. *Environ. Manage* **2010**, *45*, 105–112. [CrossRef]
8. Qin, B.Q.; Li, W.; Zhu, G.W.; Zhang, Y.L.; Wu, T.F.; Gao, G. Cyanobacterial bloom management through integrated monitoring and forecasting in large shallow eutrophic Lake Taihu (China). *J. Hazard. Mater.* **2015**, *287*, 356–363. [CrossRef]
9. Jalil, A.; Li, Y.P.; Zhang, K.; Gao, X.M.; Wang, W.C.; Khan, H.O.S.; Pan, B.Z.; Ali, S.; Acharya, K. Wind-induced hydrodynamic changes impact on sediment resuspension for large, shallow Lake Taihu, China. *Int. J. Sediment. Res.* **2019**, *34*, 205–215. [CrossRef]
10. Chen, H.M.; Zhu, Y.Y.; Zhang, Y.; Chen, X.Q.; Wang, R.C.; Zhu, W. Cyanobacterial bloom expansion caused by typhoon disturbance in Lake Taihu China. *Environ. Sci. Pollut. Res.* **2020**, *27*, 42294–42303. [CrossRef]
11. Dai, J.Y.; Chen, D.; Wu, S.Q.; Wu, X.F.; Gao, G.; Tang, X.M.; Shao, K.Q.; Lv, X.Y.; Xue, W.Y.; Yang, Q.Q.; et al. Dynamics of phosphorus and bacterial phoX genes during the decomposition of Microcystis blooms in a mesocosm. *PLoS ONE* **2018**, *13*, e0195205. [CrossRef] [PubMed]
12. Wang, Z.C.; Li, G.W.; Li, G.B.; Li, D.H. The decline process and major pathways of *Microcystis* bloom in Taihu Lake, China. *Chin. J. Oceanol. Limn.* **2012**, *30*, 37–46. [CrossRef]
13. Daft, M.J.; Mccord, S.B.; Stewart, W.D.P. Ecological studies on algal-lysing bacteria in fresh waters. *Freshw. Biol.* **1975**, *5*, 577–596. [CrossRef]
14. Yamamoto, Y.; Niizuma, S.; Kuroda, N.; Sakamoto, M. Occurrence of heterotrophic bacteria causing lysis of cyanobacteria in a eutrophic lake. *Jpn. J. Appl. Phys.* **1993**, *41*, 46–48. [CrossRef]
15. Zhou, Y.Q.; Jeppesen, E.; Zhang, Y.L.; Niu, C.; Shi, K.; Liu, X.H.; Zhu, G.W.; Qin, B.Q. Chromophoric dissolved organic matter of black waters in a highly eutrophic Chinese lake: Freshly produced from algal scums? *J. Hazard. Mater.* **2015**, *299*, 222–230. [CrossRef]
16. Kuznetsov, I.; Neumann, T.; Burchard, H. Model study on the ecosystem impact of a variable C:N:P ratio for cyanobacteria in the Baltic Proper. *Ecol. Modell.* **2008**, *219*, 107–114. [CrossRef]
17. Diehnelt, C.W.; Peterman, S.M.; Budde, W.L. Liquid chromatography–tandem mass spectrometry and accurate m/z measurements of cyclic peptide cyanobacteria toxins. *Trends Analyt. Chem.* **2005**, *24*, 622–634. [CrossRef]
18. Merel, S.; Walker, D.; Chicana, R.; Snyder, S.; Baurès, E.; Thomas, O. State of knowledge and concerns on cyanobacterial blooms and cyanotoxins. *Environ. Int.* **2013**, *59*, 303–327. [CrossRef]
19. Zurawell, R.W.; Chen, H.; Burke, J.M.; Prepas, E.E. Hepatotoxic cyanobacteria: A review of the biological importance of microcystins in freshwater environments. *J. Toxicol. Environ. Health Part. B* **2005**, *8*, 1–37. [CrossRef]
20. Zhou, J.; Lao, Y.-M.; Cai, Z.-H. Draft Genome Sequence of Providencia sneebia Strain ST1, a Quorum Sensing Bacterium Associated with Marine Microalgae. *J. Genom.* **2016**, *4*, 10–12. [CrossRef]
21. Leblanc, C.; Colin, C.; Cosse, A.; Delage, L.; Barre, S.L.; Morin, P.; Fiévet, B.; Voiseux, C.; Ambroise, Y.; Verhaeghe, E.; et al. Iodine transfers in the coastal marine environment: The key role of brown algae and of their vanadium-dependent haloperoxidases. *Biochimie* **2006**, *88*, 1773–1785. [CrossRef]
22. Aliyev, S.A.; Sari, A. Biogeochemical properties of bituminous deposits in the Miocene Himmetoğlu basin (Turkey). *Geochem. Int.* **2011**, *49*, 170–180. [CrossRef]
23. Bittar, T.B.; Vieira, A.A.H.; Stubbins, A.; Mopper, K. Competition between photochemical and biological degradation of dissolved organic matter from the cyanobacteria *Microcystis aeruginosa*. *Limnol. Oceanogr.* **2015**, *60*, 1172–1194. [CrossRef]
24. Lee, Y.; Lee, B.; Hur, J.; Min, J.-O.; Ha, S.-Y.; Ra, K.; Kim, K.-T.; Shin, K.-H. Biodegradability of algal-derived organic matter in a large artificial lake by using stable isotope tracers. *Environ. Sci. Pollut. Res.* **2016**, *23*, 8358–8366. [CrossRef]

25. McIntyre, A.M.; Guéguen, C. Binding interactions of algal-derived dissolved organic matter with metal ions. *Chemosphere* **2013**, *90*, 620–626. [CrossRef] [PubMed]
26. Li, L.L.; Yu, T.B.; Cheng, S.Z.; Li, J.; Li, C.X.; Wang, G.H.; Tan, D.Y.; Li, L.; Zhang, H.Y.; Zhang, X.Z. Removal of cyanobacteria using novel pre-pressurized coagulation: The effect of cellular properties and algogenic organic matter characteristics. *Sep. Purif. Technol.* **2022**, *282*, 119927. [CrossRef]
27. Richardson, S.; Plewa, M.; Wagner, E.; Schoeny, R.; Demarini, D. Occurrence, genotoxicity, and carcinogenicity disinfection by products in drinking water: A review and roadmap for research. *Mutat. Res.* **2007**, *636*, 178–242. [CrossRef]
28. Herzsprung, P.; Tümpling, W.; Hertkorn, N.; Harir, M.; Büttner, O.; Bravidor, J.; Friese, K.; Schmitt-Kopplin, P. Variations of DOM quality in inflows of a drinking water reservoir: Linking of van Krevelen diagrams with EEMF spectra by rank correlation. *Environ. Sci. Technol.* **2012**, *46*, 5511–5518. [CrossRef]
29. Jiao, N.Z.; Cai, R.H.; Zheng, Q.; Tang, K.; Liu, J.H.; Jiao, F.L.; Wallace, D.; Chen, F.; Li, C.; Amann, R.; et al. Unveiling the enigma of refractory carbon in the ocean. *Natl. Sci. Rev.* **2018**, *5*, 459–463. [CrossRef]
30. Garcia, A.C.; Bargu, S.; Dash, P.; Rabalais, N.N.; Sutor, M.; Morrison, W.; Walker, N.D. Evaluating the potential risk of microcystins to blue crab (*Callinectes sapidus*) fisheries and human health in a eutrophic estuary. *Harmful Algae* **2010**, *9*, 134–143. [CrossRef]
31. Loftin, K.A.; Graham, J.L.; Hilborn, E.D.; Lehmann, S.C.; Meyer, M.T.; Dietze, J.E.; Griffith, C.B. Cyanotoxins in inland lakes of the United States: Occurrence and potential recreational health risks in the EPA National Lakes Assessment 2007. *Harmful Algae* **2016**, *56*, 77–90. [CrossRef] [PubMed]
32. Zhou, L.; Zhou, Y.Q.; Tang, X.M.; Zhang, Y.L.; Jang, K.-S.; Székely, A.J.; Jeppesen, E. Resource aromaticity affects bacterial community successions in response to different sources of dissolved organic matter. *Water Res.* **2021**, *190*, 116776. [CrossRef] [PubMed]
33. Kamiyama, T.; Itakura, S.; Nagasaki, K. Changes in microbial loop components: Effects of a harmful algal bloom formation and its decay. *Aquat. Microb. Ecol.* **2000**, *21*, 21–30. [CrossRef]
34. Zhao, Z.; Gonsior, M.; Schmitt-Kopplin, P.; Zhan, Y.C.; Zhang, R.; Jiao, N.Z.; Chen, F. Microbial transformation of virus-induced dissolved organic matter from picocyanobacteria: Coupling of bacterial diversity and DOM chemodiversity. *ISME J.* **2019**, *13*, 2551–2565. [CrossRef]
35. Kieft, B.; Li, Z.; Bryson, S.; Hettich, R.L.; Pan, C.; Mayali, X.; Mueller, R.S. Phytoplankton exudates and lysates support distinct microbial consortia with specialized metabolic and ecophysiological traits. *Proc. Natl. Acad. Sci. USA* **2021**, *118*, e2101178118. [CrossRef] [PubMed]
36. Bertilsson, S.; Eiler, A.; Nordqvist, A.; Jørgensen, N.O.G. Links between bacterial production, amino-acid utilization and community composition in productive lakes. *ISME J.* **2007**, *1*, 532–544. [CrossRef] [PubMed]
37. Guillemette, F.; McCallister, S.L.; Giorgio, P.A. Selective consumption and metabolic allocation of terrestrial and algal carbon determine allochthony in lake bacteria. *ISME J.* **2016**, *10*, 1373–1382. [CrossRef]
38. Zhang, L.; Liao, Q.J.H.; Gao, R.; Luo, R.; Liu, C.; Zhong, J.C.; Wang, Z.D. Spatial variations in diffusive methane fluxes and the role of eutrophication in a subtropical shallow lake. *Sci. Total Environ.* **2021**, *759*, 143459. [CrossRef]
39. Zhang, L.; Liu, C.; He, K.; Shen, Q.S.; Zhong, J.C. Dramatic temporal variations in methane levels in black bloom prone areas of a shallow eutrophic lake. *Sci. Total Environ.* **2021**, *767*, 144868. [CrossRef]
40. Tong, Y.D.; Xu, X.W.; Miao, Q.; Sun, J.J.; Zhang, Y.Y.; Zhang, W.; Wang, M.Z.; Wang, X.J.; Zhang, Y. Lake warming intensifies the seasonal pattern of internal nutrient cycling in the eutrophic lake and potential impacts on algal blooms. *Water Res.* **2021**, *188*, 116570. [CrossRef]
41. Ai, Y.; Bi, Y.H.; Hu, Z.Y. Changes in phytoplankton communities along nutrient gradients in Lake Taihu: Evidence for nutrient reduction strategies. *Chin. J. Oceanol. Limn.* **2015**, *33*, 447–457. [CrossRef]
42. Zhang, P.; Zhai, C.M.; Chen, R.Q.; Liu, C.H.; Xue, Y.R.; Jiang, J.H. The dynamics of the water bloom-forming *Microcystis aeruginosa* and its relationship with biotic and abiotic factors in Lake Taihu, China. *Ecol. Eng.* **2012**, *47*, 274–277. [CrossRef]
43. Cruaud, P.; Vigneron, A.; Fradette, M.-S.; Dorea, C.C.; Culley, A.I.; Rodriguez, M.J.; Charette, S.J. Annual bacterial community cycle in a seasonally ice-covered river reflects environmental and climatic conditions. *Limnol. Oceanogr.* **2020**, *65*, S21–S37. [CrossRef]
44. Chen, Y.W.; Fan, C.X.; Teubner, K.; Dokulil, M. Changes of nutrients and phytoplankton chlorophyll-a in a large shallow lake, Taihu, China: An 8-year investigation. *Hydrobiologia* **2003**, *506*, 273–279. [CrossRef]
45. Zhou, S.Q.; Zhu., S.M.; Shao., Y.S.; Gao., N.Y. Characteristics of C-, N-DBPs formation from algal organic matter: Role of molecular weight fractions and impacts of pre-ozonation. *Water Res.* **2015**, *72*, 381–390. [CrossRef]
46. Pivokonsky, M.; Kloucek, O.; Pivokonska, L. Evaluation of the production, composition and aluminum and iron complexation of algogenic organic matter. *Water Res.* **2006**, *40*, 3045–3052. [CrossRef]
47. Zhang, Y.L.; Dijk, M.A.; Liu, M.L.; Zhu, G.W.; Qin, B.Q. The contribution of phytoplankton degradation to chromophoric dissolved organic matter (CDOM) in eutrophic shallow lakes: Field and experimental evidence. *Water Res.* **2009**, *43*, 4685–4697. [CrossRef]
48. Coble, P.G. Characterization of marine and terrestrial DOM in seawater using excitation-emission matrix spectroscopy. *Mar. Chem.* **1996**, *51*, 325–346. [CrossRef]
49. Coble, P.G.; Castillo, C.E.D.; Avril, B. Distribution and optical properties of CDOM in the Arabian Sea during the 1995 Southwest Monsoon. *Deep Sea Res. Part II Top. Stud. Oceanogr.* **1998**, *45*, 2195–2223. [CrossRef]

50. Zhou, Y.Q.; Zhou, L.; Zhang, Y.L.; Zhu, G.W.; Qin, B.Q.; Jang, K.-S.; Spencer, R.G.M.; Kothawala, D.N.; Jeppesen, E.; Brookes, J.D.; et al. Unraveling the Role of Anthropogenic and Natural Drivers in Shaping the Molecular Composition and Biolability of Dissolved Organic Matter in Non-pristine Lakes. *Environ. Sci. Technol.* **2022**, *56*, 4655–4664. [CrossRef]
51. Zhang, Y.L.; Qin, B.Q. Variations in Spectral Slope in Lake Taihu, a Large Subtropical Shallow Lake in China. *J. Great Lakes Res.* **2007**, *33*, 483–496. [CrossRef]
52. Ebina, J.; Tsutsui, T.; Shirai, T. Simultaneous determination of total nitrogen and total phosphorus in water using peroxodisulfate oxidation. *Water Res.* **1983**, *17*, 1721–1726. [CrossRef]
53. Gong, Y.; Tang, X.M.; Shao, K.Q.; Hu, Y.; Gao, G. Dynamics of bacterial abundance and the related environmental factors in large shallow eutrophic Lake Taihu. *J. Freshw. Ecol.* **2016**, *32*, 1248506. [CrossRef]
54. Chen, S.N.; Yan, M.M.; Huang, T.L.; Zhang, H.; Liu, K.W.; Huang, X.; Li, N.; Miao, Y.T.; Sekar, R. Disentangling the drivers of *Microcystis* decomposition: Metabolic profile and co-occurrence of bacterial community. *Sci. Total Environ.* **2020**, *739*, 140062. [CrossRef] [PubMed]
55. Xie, R.; Wang, Y.; Chen, Q.; Guo, W.D.; Jiao, N.Z.; Zheng, Q. Coupling Between Carbon and Nitrogen Metabolic Processes Mediated by Coastal Microbes in *synechococcus*-derived organic matter addition incubations. *Front. Microbiol.* **2020**, *11*, 1041. [CrossRef]
56. Lin, H.; Xu, H.C.; Cai, Y.H.; Belzile, C.; Macdonald, R.W.; Guo, L.D. Dynamic changes in size-fractionated dissolved organic matter composition in a seasonally ice-covered Arctic River. *Limnol. Oceanogr.* **2021**, *66*, 3085–3099. [CrossRef]
57. Tanaka, K.; Kuma, K.; Hamasaki, K.; Yamashita, Y. Accumulation of humic-like fluorescent dissolved organic matter in the Japan Sea. *Sci. Rep.* **2014**, *4*, 5292. [CrossRef]
58. Zuo, Y.-T.; Wu, J.; Cheng, S.; Cai, M.-H.; Han, Y.-Z.; Ji, W.-X.; Li, Y.; Huo, Z.-L.; Korshin, G.; Li, W.-T.; et al. Identification of pterins as characteristic humic-like fluorophores released from cyanobacteria and their behavior and fate in natural and engineered water systems. *Chem. Eng. J.* **2022**, *428*, 131154. [CrossRef]
59. Song, W.J.; Zhao, C.X.; Zhang, D.Y.; Mu, S.Y.; Pan, X.L. Different Resistance to UV-B Radiation of Extracellular Polymeric Substances of Two Cyanobacteria from Contrasting Habitats. *Front. Microbiol.* **2016**, *7*, 1208. [CrossRef] [PubMed]
60. Omori, Y.; Hama, T.; Ishii, M.; Saito, S. Vertical change in the composition of marine humic-like fluorescent dissolved organic matter in the subtropical western North Pacific and its relation to photoreactivity. *Mar. Chem.* **2011**, *124*, 38–47. [CrossRef]
61. Bianchi, T.S. The role of terrestrially derived organic carbon in the coastal ocean: A changing paradigm and the priming effect. *Proc. Natl. Acad. Sci. USA* **2011**, *108*, 19473–19481. [CrossRef] [PubMed]
62. Bianchi, T.S.; Thornton, D.C.O.; Yvon-Lewis, S.A.; King, G.M.; Eglinton, T.I.; Shields, M.R.; Ward, N.D.; Curtis, J. Positive priming of terrestrially derived dissolved organic matter in a freshwater microcosm system. *Geophys. Res. Lett.* **2015**, *42*, 5460–5467. [CrossRef]
63. Sebastian, M.; Auguet, J.-C.; Restrepo-Ortiz, C.X.; Sala, M.M.; Marrase, C.; Gasol, J.M. Deep ocean prokaryotic communities are remarkably malleable when facing long-term starvation. *Environ. Microbiol.* **2018**, *20*, 713–723. [CrossRef]
64. Zhang, Y.L.; Zhou, Y.Q.; Shi, K.; Qin, B.Q.; Yao, X.L.; Zhang, Y.B. Optical properties and composition changes in chromophoric dissolved organic matter along trophic gradients: Implications for monitoring and assessing lake eutrophication. *Water Res.* **2018**, *131*, 255–263. [CrossRef]
65. Li, W.; Qin, B.Q.; Zhu, G.W. Forecasting short-term cyanobacterial blooms in Lake Taihu, China, using a coupled hydrodynamic–algal biomass model. *Ecohydrology* **2014**, *7*, 794–802. [CrossRef]
66. Alderkamp, A.-C.; Buma, A.G.J.; Rijssel, M. The carbohydrates of *Phaeocystis* and their degradation in the microbial food web. *Biogeochemistry* **2007**, *83*, 99–118. [CrossRef]
67. Rix, L.; Goeij, J.M.; Oevelen, D.; Struck, U.; Al-Horani, F.A.; Wild, C.; Naumann, M.S. Differential recycling of coral and algal dissolved organic matter via the sponge loop. *Funct. Ecol.* **2017**, *31*, 778–789. [CrossRef]
68. Zhang, Z.H.; Tang, L.L.; Liang, Y.T.; Li, G.; Li, H.M.; Rivkin, R.B.; Jiao, N.Z.; Zhang, Y.Y. The relationship between two *Synechococcus* strains and heterotrophic bacterial communities and its associated carbon flow. *J. Appl. Phycol.* **2021**, *33*, 953–966. [CrossRef]
69. Gao, X.Q.; Heath, R.T. Relationship Between Labile Dissolved Organic Carbon (LDOC), Bacterioplankton Cell Phosphorus Quota, and Bacterial Phosphate Uptake Rate in Lakes. *J. Great Lakes Res.* **2005**, *31* (Suppl. S2), 125–137. [CrossRef]
70. Barber, C.B.; Dobkin, D.P.; Huhdanpaa, H.T. The quick hull algorithm for convex hulls. *ACM T Math. Software* **1996**, *22*, 469–483. [CrossRef]
71. Zhang, Y.L.; Zhang, E.L.; Yin, Y.; van Dijk, M.A.; Feng, L.Q.; Shi, Z.Q.; Liu, M.L.; Qin, B.Q. Characteristics and sources of chromophoric dissolved organic matter in lakes of the Yungui Plateau, China, differing in trophic state and altitude. *Limnol. Oceanogr.* **2010**, *55*, 2645–2659. [CrossRef]
72. Zhang, Y.L.; Liu, X.H.; Wang, M.Z.; Qin, B.Q. Compositional differences of chromophoric dissolved organic matter derived from phytoplankton and macrophytes. *Org. Geochem.* **2013**, *55*, 26–37. [CrossRef]

remote sensing MDPI

Article

Use of the Sentinel-2 and Landsat-8 Satellites for Water Quality Monitoring: An Early Warning Tool in the Mar Menor Coastal Lagoon

Isabel Caballero [1,*], Mar Roca [1], Juan Santos-Echeandía [2], Patricia Bernárdez [2] and Gabriel Navarro [1]

[1] Instituto de Ciencias Marinas de Andalucía (ICMAN), Consejo Superior de Investigaciones Científicas (CSIC), Avenida República Saharaui, 11519 Puerto Real, Spain; mar.roca@csic.es (M.R.); gabriel.navarro@icman.csic.es (G.N.)
[2] Instituto Español de Oceanografía (IEO-CSIC), Centro Oceanográfico de Vigo, Subida a Radio Faro 50–52, Vigo, 36390 Pontevedra, Spain; juan.santos@ieo.es (J.S.-E.); patricia.bernardez@ieo.es (P.B.)
* Correspondence: isabel.caballero@icman.csic.es

Abstract: During recent years, several eutrophication processes and subsequent environmental crises have occurred in Mar Menor, the largest hypersaline coastal lagoon in the Western Mediterranean Sea. In this study, the Landsat-8 and Sentinel-2 satellites are jointly used to examine the evolution of the main water quality descriptors during the latest ecological crisis in 2021, resulting in an important loss of benthic vegetation and unusual mortality events affecting different aquatic species. Several field campaigns were carried out in March, July, August, and November 2021 to measure water quality variables over 10 control points. The validation of satellite biogeochemical variables against on-site measurements indicates precise results of the water quality algorithms with median errors of 0.41 mg/m^3 and 2.04 FNU for chlorophyll-a and turbidity, respectively. The satellite preprocessing scheme shows consistent performance for both satellites; therefore, using them in tandem can improve mapping strategies. The findings demonstrate the suitability of the methodology to capture the spatiotemporal distribution of turbidity and chlorophyll-a concentration at 10–30 m spatial resolution on a systematic basis and in a cost-effective way. The multitemporal products allow the identification of the main critical areas close to the mouth of the Albujon watercourse and the beginning of the eutrophication process with chlorophyll-a concentration above 3 mg/m^3. These innovative tools can support decision makers in improving current monitoring strategies as early warning systems for timely assistance during these ecological disasters, thus preventing detrimental conditions in the lagoon.

Keywords: remote sensing; Copernicus programme; eutrophication; turbidity; chlorophyll-a; coastal monitoring

Citation: Caballero, I.; Roca, M.; Santos-Echeandía, J.; Bernárdez, P.; Navarro, G. Use of the Sentinel-2 and Landsat-8 Satellites for Water Quality Monitoring: An Early Warning Tool in the Mar Menor Coastal Lagoon. *Remote Sens.* **2022**, *14*, 2744. https://doi.org/10.3390/rs14122744

Academic Editors: Alban Kuriqi and Luis Garrote

Received: 3 May 2022
Accepted: 4 June 2022
Published: 7 June 2022

1. Introduction

Coastal lagoons, as transitional environments between land and sea, occupy 14% of the world's coastlines [1]. Due to their shallow waters, morphology, trophic status, and physicochemical processes in a semienclosed system, they are considered one of the most productive habitats on Earth [2,3]. These areas play a significant conservational, ecological, and protective role and are home to an important part of global biodiversity [4]. They underpin human livelihoods, well-being, and welfare and provide several ecosystem services, including tourism, fisheries, aquaculture, and industrial, recreational or navigational activities. Coastal lagoons are subject to diverse transformations and uncoordinated management plans by different agencies and stakeholders from the local to national scale, which might, in some cases, degrade their ecological values. A wide range of anthropogenic activities such as urbanization, agriculture, aquaculture or industry use a variety of organic substances and pollutants, which can reach semienclosed bays, inland waters, and lagoons [5]. Considering that these environments are extremely sensitive and vulnerable, they usually show signs of deterioration, pollution, biodiversity loss, alteration of their ecological functioning, and limited ecosystem services [6]. Habitat destruction, water withdrawal, overexploitation, and chemical and biological pollution, such as invasive species, are the main causes of their deterioration, making them one of the most threatened ecosystems in the world [7]. The conservation of coastal lagoons is crucial for their ecological value and the significant number of services they provide [8]. Furthermore, in the context of climate change, coastal lagoons are sentinel systems with an essential role in controlling the fluxes of water, organisms, and nutrients between land, rivers, and oceans, as well as eutrophication and pollution processes [8,9]. Therefore, there is an urgent need to advance monitoring, mapping, and management tools in order to improve the knowledge of these strategic systems, prevent their environmental degradation, and increase their future protection [10].

This is the case of Mar Menor, the largest hypersaline lagoon in the Western Mediterranean Sea (Figure 1) and one of the most iconic and emblematic natural areas in Spain due to its significance in terms of habitats and species, its ecological value, and the uniqueness of its ecosystem. Mar Menor, its surrounding wetlands, and natural areas, are of vital natural importance and a protected landscape as a Wildlife Protection Area, Natura 2000 network, Wetland of International Importance (RAMSAR Convention), and a Regional Park. Moreover, the area is a key component of the regional economy, development, and policy plans due to the variety of uses and human activities developed there. The lagoon, with a maximum depth of 6.5 m and a surface area of 135 km^2, presents a long sand bar called "La Manga" acting as a barrier between the Mediterranean Sea and the lagoon, only connected through five shallow inlets called "golas" [11]. The lagoon is close to the Campo de Cartagena region, one of the most intensive agricultural areas in Europe. Several ephemeral wadis drain into the western part of the lagoon, transporting nutrient-enriched waters from agricultural runoff after rainy periods, the Albujon watercourse (Figure 1) being the main collector of the Campo de Cartagena drainage basin and the only permanent wadi flowing into the lagoon [11–16]. Therefore, most of the discharges are located in the southern half of the lagoon where the Albujon watercourse maintains a regular flux of water, albeit depending on the torrential and sporadic rainfall regime, as occurred in September 2019 during one of the most extreme storms, known as the "Cold Drop" [13].

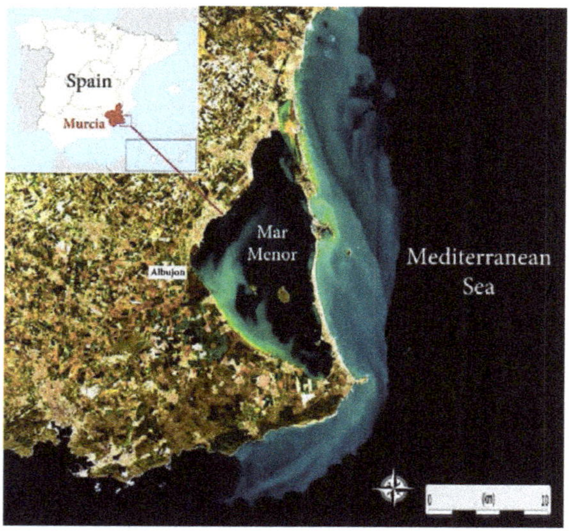

Figure 1. Location of the Mar Menor coastal lagoon on the southeastern coast of Spain and Sentinel-2 image captured on 21 March 2021 indicating the final transect of the Albujon watercourse flowing into the lagoon.

Nowadays, this ecosystem is a cause for international concern due to the drastic modifications of its natural and physical status caused by anthropogenic activities. The main impacts causing acute degradation of Mar Menor are those from mining, agriculture, tourism, and urban development [11,14]. The land-use modifications that occurred in the watershed during the 1980s and 1990s with relation to agriculture, from dry land to intensively irrigated vegetable crops (Figure 2), have produced a severe excess of nutrients and fertilizers draining into the lagoon from the freshwater discharge [11], which clearly affect the environmental health of the lagoon (Corine Land Cover datasets, https://centrodedescargas.cnig.es/CentroDescargas/; accessed on 1 January 2022). The current problems affecting the lagoon are increased turbidity and chlorophyll-a, resulting in an acute eutrophication process, silting, a general loss of sediment and seawater quality, and the deterioration of submerged seagrass and animal communities. The degradation of the coastal lagoon has also influenced conventional fishing that has been carried out in the lagoon since ancient times [17].

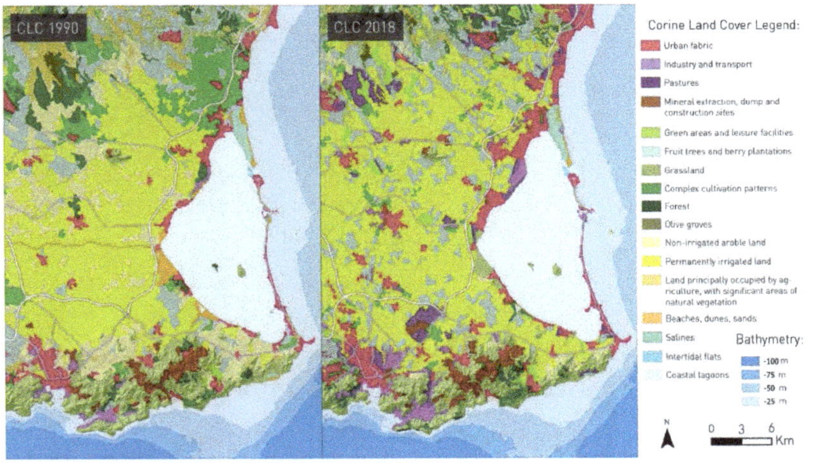

Figure 2. Land cover classification for the Mar Menor lagoon and its surroundings according to Corine Land Cover datasets (https://centrodedescargas.cnig.es/CentroDescargas/; accessed on 1 January 2022) for 1990 and 2018.

The effects of these massive contributions of nutrients in the lagoon ecosystem have been cushioned by its elements that, for decades, have acted as mechanisms of homeostasis and resilience, preventing an excess of nutrients from being available to opportunistic phytoplankton organisms. However, despite the capacity of Mar Menor to resist the effects of elevated nutrient concentration, a succession of catastrophic events have occurred (Figure 3a–c) since 2016 [18–20]. In August 2021, the latest environmental crisis caused alarm and considerable concern and was considered worse than previous eutrophication events. The excess of nutrients and organic matter caused anoxia in the deep layer and massive mortality of benthic flora and fauna during several weeks (Figure 3d), causing an impact on public opinion at the local, national, and international level. Images of dying wildlife traumatized citizens and occupied the public agenda, raising questions about the cause of this ecological disaster that keeps getting worse year after year. The Spanish Institute of Oceanography (IEO-CSIC) highlighted the main cause as pollution and the entry of fertilizers and nutrients into the lagoon from intensive agriculture and other human activities, causing the aquatic ecosystem to collapse [21]. This has led to a clear transformation in the regime of the lagoon, from an apparently stable state (with frequent symptoms of eutrophication in recent decades) to an altered and highly unstable state, much more vulnerable to changes in the environment, especially extreme weather events, which are clearly more intense and frequent as a result of global climate change [22]. In fact, in recent years, it has been observed that when the chlorophyll-a values in the lagoon exceed 3 mg/m^3, a process of eutrophication occurs immediately [21]. Several administrations and public authorities responsible for managing the lagoon have been developing initiatives in an attempt to solve the problem. The regional government maintains an open-access network with a few constant sampling sites and field-based campaigns for monitoring variables related to water quality, such as water clarity, turbidity or chlorophyll-a, an indicator of phytoplankton biomass in seawater, temperature, salinity, and dissolved oxygen [23]. However, sound management and deterioration control need improved characterization of the spatial variability and distribution of the water quality, in particular in the western area of Mar Menor, where the Albujon watercourse flows into the lagoon. This information remains key not only for long-term monitoring but also for quick emergency response as an early warning system.

Figure 3. (**a**) Sentinel-2 scene in Mar Menor after the extreme weather event known as the "Cold Drop" and the catastrophic flooding on 13 September 2019; (**b**) surface of the water in the lagoon in July 2016 during the environmental crisis; (**c**) typical resuspension of sediments and increased turbidity during strong winds; (**d**) latest environmental catastrophe with massive dead fish and crustacean in August 2021 (authorship: Greenpeace); and (**e**) field campaign carried out in the lagoon in March 2021.

This study examined the evolution of the main biogeochemical parameters of the seawater quality in the coastal lagoon using the Landsat-8 and Sentinel-2 satellite missions in tandem, both with high spatial resolution. Remote sensing technologies can advance current management and monitoring strategies providing synoptic information of the lagoon, as well as provide insights into past, present, and future eutrophication events, in particular during the severe environmental crisis that occurred in summer 2021. Although ocean color sensors provide a distinct picture of the seawater bio-optical status across several scales not achievable with traditional in situ surveying techniques, application in coastal lagoon biogeochemical monitoring is challenging. In this sense, there are studies that focus on mapping water quality, phytoplankton blooms, and eutrophication events in Mar Menor with traditional ocean color sensors at the moderate spatial resolution of 300–1000 m [24–26]. RGB composite images on 3 August 2021 of the Sentinel-3 satellite (300 m spatial resolution), Landsat-8 satellite (30 m spatial resolution), and Sentinel-2 satellite (10 m spatial resolution) are shown in Figure 4. Sentinel-3 is the ocean color mission developed by the European Union's Copernicus programme to support ocean forecasting systems, environmental and climate monitoring with high accuracy and reliability. However, this example highlights that the moderate spatial resolution of Sentinel-3 might not be adequate in complex coastal areas, such as Mar Menor. In recent years, some studies suggested that in order to appropriately determine the ecological conditions of complex inland or coastal water areas by means of remote sensing tools, improved temporal and spatial capabilities are required [27–32], in particular in Mar Menor [13,33–35]. Conceived in the first instance to monitor land cover, the notably enhanced spectral and spatial resolution and minor footprint of both Landsat-8 and Sentinel-2 platforms provide the opportunity to evaluate terrestrial–aquatic interfaces and their dynamic spatial heterogeneity at local, regional, or global scales [36,37].

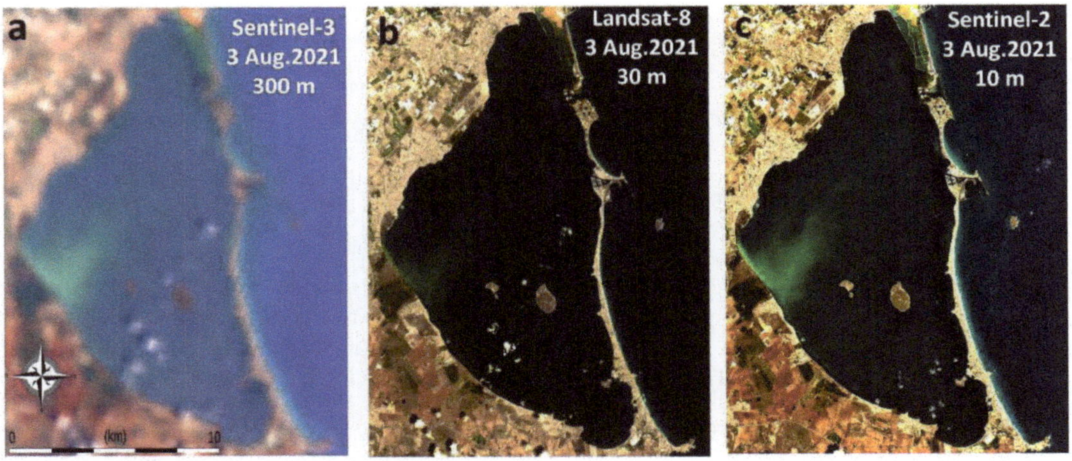

Figure 4. RGB (Red–Green–Blue) composite image on 3 August 2021 of (**a**) Sentinel-3 satellite (300 m spatial resolution), (**b**) Landsat-8 satellite (30 m spatial resolution), and (**c**) Sentinel-2 satellite (10 m spatial resolution).

The main aims of this study are: (1) to use a consistent atmospheric and sunglint correction strategy with Landsat-8 and Sentinel-2 imagery in Mar Menor; (2) to validate the satellite-derived chlorophyll-a and turbidity retrievals with in situ data; (3) to detect the spatiotemporal fluctuations of the biogeochemical parameters during the study period with the multisensor approach; and (4) to evaluate and identify the critical zones in the context of the most recent ecological catastrophe in 2021. This combined information can allow enhanced temporal mapping and predictability of the water mass degradation as an early warning system. Remote sensing technology has large-area and real-time advantages in promoting the monitoring and forecasting of coastal disasters, providing information about when and where the chlorophyll-a values in the lagoon exceed 3 mg/m^3 [21]. These tools can be applied in parallel to regular on-site sampling campaigns in order to reduce the detrimental effects of high levels of phytoplankton, algae, and turbidity on the vulnerable lagoon system and to continuously calibrate/validate the different water quality algorithms for more reliable results.

2. Materials and Methods

2.1. Satellite Imagery

The Sentinel-2A/B twin mission was used for mapping the lagoon thanks to the open data access policy and high spatial resolution (10-20-60 m). The European Commission and the European Space Agency (ESA), in the frame of the Copernicus programme, developed this optical constellation in order to support its operational requirements. Sentinel-2 is a multispectral, wide-swath imaging platform used for monitoring land surfaces, water coverage, soil, and vegetation. In addition, it can also support Copernicus water monitoring over coastal regions and inland waterways. The Sentinel-2A and Sentinel-2B mission, with a global revisit frequency of five days at the Equator, is based on a constellation of two operationally identical satellites in the same orbit and phased at 180° to each other. The ESA User Handbook describes the temporal, spectral, spatial, and radiometric features of the visible and near-infrared (NIR) bands of both Sentinel-2A and -2B satellites [38]. The stated quality standards for absolute geolocation of the Sentinel-2 scenes (two pixels, 20 m) are within the ESA requirements [39]. The images covering Mar Menor (zone 30 and tile SXG; acquisition time 11:00 UTC) during the study period in 2021 were downloaded from the ONDA DIAS (https://www.onda-dias.eu/cms/es/; accessed on 20 May 2021). These products are the top-of-atmosphere (TOA) datasets at Level-1C (L1C) after the radiometric and geometric corrections.

In addition, the freely available Landsat-8 visible and NIR spectra imagery from the United States Geological Survey (USGS) and the National Aeronautics and Space Administration (NASA) were also used for comprehensive monitoring. We downloaded the Level 1 data from the Earth Explorer (https://earthexplorer.usgs.gov/), orthorectified and terrain corrected at a 30 m spatial resolution, with a 16-day revisit frequency [40,41]. The region of interest was covered by the tiles located in paths 198–199 and row 34 (acquisition time 10:30–10:45 UTC). A low cloud coverage (<40%) filtering was applied over Mar Menor for further analysis of the scenes. When Sentinel-2 and Landsat-8 products are combined, the average revisit time in Mar Menor is ~4 days. From the control period before and after the 2021 crisis, 48 images were downloaded and processed (29 and 19 images for Sentinel-2 and Landsat-8, respectively). However, clouds and severe sunglint contamination diminished the number of usable images, with only 18 final scenes further evaluated (12 and 6 images for Sentinel-2 and Landsat-8, respectively) to characterize the spatial and temporal distribution of water quality. Table 1 shows the acquisition dates and the quality of the scenes.

Bottom-of-atmosphere (BOA) Level-2A products were generated with one of the most commonly used atmospheric correction softwares (ACOLITE, version 20210114.0), which supports preprocessing of Landsat-8 and Sentinel-2 satellites. This software incorporates an image-based model, without the need for in situ atmospheric datasets. The Royal Belgian Institute of Natural Sciences (RBINS) developed this free toolbox to correct Level-1 to Level-2 data products over marine, inland, and coastal waters [42]. The Dark Spectrum Fitting (DSF) atmospheric correction algorithm was applied [43,44]. The notably enhanced spectral resolution of Landsat-8 and, in particular, Sentinel-2 satellites, are key to obtain good-quality products by means of the DSF model [44]. Correction of the sunglint over the surface reflectance was performed by means of the additional image-based sunglint correction, since during the study period acute sunglint effects were observed at these latitudes (specific setting parameters: dsf_path_reflectance = tiled, l2w_mask_threshold = 0.05). The remote sensing reflectance (Rrs, sr^{-1}) products along the visible and NIR spectrum were calculated after resampling to 10 m and 30 m pixel size for Sentinel-2 and Landsat-8, respectively.

Table 1. List of imagery used in this study during the latest ecological crisis in 2021 corresponding to the Sentinel-2 and Landsat-8 satellites.

Sentinel-2A/B				Landsat-8			
Number	Month	Day	Observations	Number	Month	Day	Observations
1	May	5	Severe sunglint	1	May	6	Severe sunglint
2	May	15	Severe sunglint	2	May	15	Severe sunglint
3	May	20	Clouds	3	May	22	Clouds
4	May	25	Clouds	4	May	31	Clouds
5	May	30	Clouds	5	June	7	Severe sunglint
6	June	4	Clouds	6	June	16	Clouds
7	June	9	Severe sunglint	7	June	23	Severe sunglint
8	June	14	Severe sunglint	8	July	2	Good quality
9	June	19	Clouds	9	July	9	Clouds
10	June	24	Clouds	10	July	18	Good quality
11	June	29	Clouds	11	July	25	Clouds
12	July	4	Severe sunglint	12	August	3	Good quality
13	July	9	Clouds	13	August	10	Clouds
14	July	14	Severe sunglint	14	August	19	Clouds
15	July	19	Clouds	15	August	26	Sunglint
16	July	24	Severe sunglint	16	September	4	Clouds
17	July	29	Clouds	17	September	11	Good quality
18	August	3	Good quality	18	September	20	Clouds
19	August	8	Clouds	19	September	27	Clouds
20	August	13	Good quality				
21	August	18	Good quality				
22	August	23	Clouds				
23	August	28	Clouds				
24	September	2	Clouds				
25	September	7	Clouds				
26	September	12	Good quality				
27	September	17	Good quality				
28	September	22	Clouds				
29	September	27	Clouds				

The standard products to monitor the biogeochemical conditions in Mar Menor during the ecological crisis with Sentinel-2 and Landsat-8 satellites were seawater turbidity (FNU) and chlorophyll-a (chl-a, mg/m^3). We selected both indicators as required by the EU Water Framework Directive (WFD) for the evaluation of the good ecological status of the coastal lagoon. The Nechad et al. semianalytical algorithm (red band, Rrs 665 nm) was applied to estimate turbidity with both satellites [45]. This model has already been validated in different environments [46–48] and was previously used in Mar Menor during an extreme weather event in September 2019 [13]. These semianalytical algorithms allow a more global performance since they are based on the inherent optical properties of the seawater. The commonly used OC3 algorithm was applied to calculate the concentration of seawater chl-a [49]. The standard masking procedures were accomplished, eliminating clouds, cloud shadows, land, and the low performance of the sunglint and atmospheric corrections. Turbidity and chl-a maps were at 30 m and 10 m spatial resolution for Landsat-8 and Sentinel-2, respectively, with the generation of the final products after 3–4 h following image acquisition for each sensor.

2.2. In Situ Data

Four sampling cruises were carried out during 2021 (March, July, August, and November), where seawater was collected at 10 points homogeneously distributed in the Mar

Menor lagoon (Figure 5). The dates of the in situ campaigns during the study period corresponded to 23–24 March, 27 July, 26 August, and 16–17 November 2021. Matchups between in situ and satellite samples were generated when both data acquisitions occurred within 30 h of each other, and the satellite value was calculated as the mean of the 3 × 3 10 m pixel region around the sampling station.

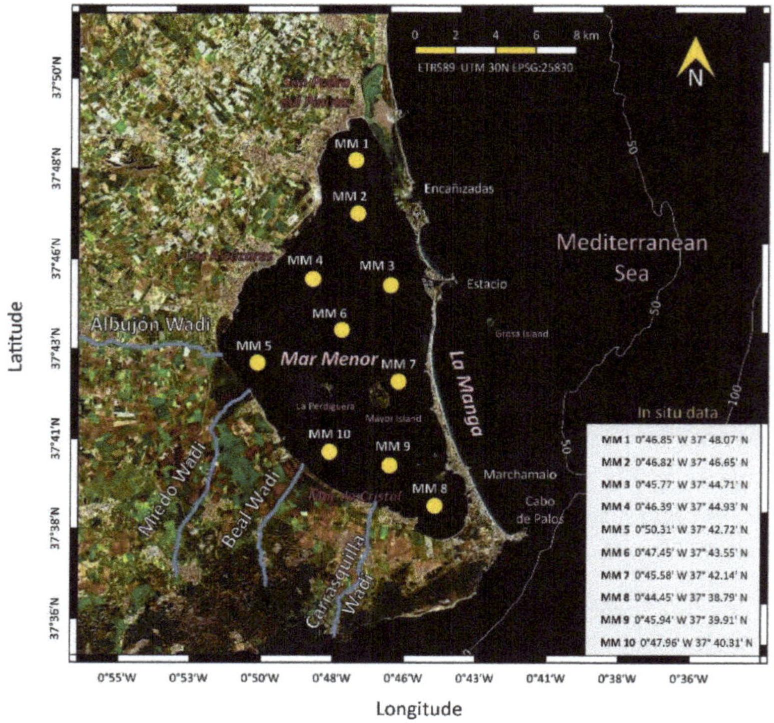

Figure 5. Control points for data collection during the four in situ campaigns carried out in March, July, August, and November 2021.

In order to determine turbidity, samples were measured just after collection onboard using a portable turbidimeter (2100P, Hach). Prior to calibration, we ensured that the equipment did not suffer anomalies and that the necessary reagents were available. Calibration was performed quarterly, unless the equipment was malfunctioning. The turbidimeter was calibrated at 4 points: 0 NTU, 10 NTU, 200 NTU, and 800 NTU. The standards used in the calibration were certified commercial standards at room temperature in order to avoid misting interferences in the turbidity measurement. The turbidimeter was calibrated according to the equipment manual.

In the case of chl-a, water was collected from a 0.5 m depth in 1 L dark bottles to avoid enhanced photosynthetic activity and kept in a portable fridge until arrival at the laboratory. Three replicates per station were collected. Chl-a concentrations were measured in 700–1000 mL water samples for each replicate, which were filtered through Whatman GF/F 0.2 μm polycarbonate filters. The filters were immediately frozen at −20 °C until pigment extraction in 90% acetone at 4 °C overnight in the dark. Chl-a concentrations were determined with a 10-AU Turner Designs fluorometer calibrated with pure chl-a [50].

3. Results and Discussion

3.1. Multisensor Approach and Preprocessing

Until now, standard ocean color sensors at the moderate spatial resolution of 300–1000 m were generally used to map water quality, phytoplankton blooms, and eutrophication events in Mar Menor [25,26]. However, features in the lagoon show typical scales of tens of meters that cannot be detected with existing ocean color sensors (Figure 4). In recent years, several studies already demonstrated that, in order to comprehensively evaluate the ecological conditions of heterogeneous coastal areas and land-water inputs using remote sensing technologies, improved spatial resolution should be used with Landsat-8 and/or Sentinel-2 [28,30,31,51]. Undoubtedly, the Sentinel-2 twin mission, although originally not designed for coastal ocean monitoring, is a key tool for the detailed mapping of highly dynamic environments, such as coastal or inland water areas [52].

During the study period, in particular during summer, acute sunglint contamination also influenced the quality of the imagery over Mar Menor due to the specular reflection of sunlight off the water (see Table 1 for details). Figure 6 shows two images on 9 June 2021 and 4 July 2021 at the top-of-atmosphere (TOA) level, after ACOLITE processing at the bottom-of-atmosphere (BOA) level, and the Rrs of the blue band (492 nm). Accurate performance was accomplished by ACOLITE over low to moderate sunglint conditions, such as on 9 June 2021 (Figure 6a–c), but failed to retrieve Rrs during severe sunglint contamination on 4 July 2021 (Figure 6d–f), thus masking the data. The sunglint is clearly observed at TOA-Level-1C (Figure 6a,d), whereas the residuals are visible at BOA-Level2 (Figure 6e) with intense effects on the eastern area of the Sentinel-2 tile, exactly where Mar Menor is located. This effect is significant during spring and summer, given that minimum information could be retrieved with severe sunglint, restricting the amount of available data when the majority of the blooms occurred. Comparable results were generated in other coastal regions, such as in North Atlantic [53] or Caribbean waters [54]. Irregular residual issues still require further advancements in the sunglint and atmospheric correction approaches to facilitate the extensive combination of Landsat-8 and Sentinel-2 products during summer. In addition, the typical stripping patterns are clearly observed at both TOA and BOA levels.

On 3 August 2021, Landsat-8 and Sentinel-2 acquired a scene at 10:45 am and 11:00 am GMT, respectively. Figure 7 shows the spectral signal of both satellites with only a 15 min time difference over three control points distributed across different areas of the lagoon. This exhibits the consistent performance of ACOLITE for both satellite missions, retrieving the spectrum with similar Rrs values for each point and sensor (P1, turbidity of 18.11 and 17.28 FNU; P2, turbidity of 5.5 and 6.1 FNU; and P3, turbidity of 4.08 and 3.93 FNU for Sentinel-2 and Landsat-8, respectively), thus further corroborating the remarkable value of the combined products. Comparison of Sentinel-2 and Landsat-8 Rrs over the visible and NIR bands yielded a bias of -0.00035 sr^{-1}, MAE of 0.00072 sr^{-1}, and MedAE of 0.00049 sr^{-1}. Consistent sunglint and atmospheric correction models are needed to empower the application community to explore these products in order to thoroughly address the ecological conditions of Mar Menor using remote sensing technologies. Recent research already demonstrated the potential of ACOLITE to provide robust information for aquatic and marine applications [30,43,44,54]. A study applied Sentinel-2 data to monitor water quality in Mar Menor using the Sen2Cor atmospheric correction processor (designated for land application with restricted performance in water application) and to generate products at 60 m spatial resolution [35]. However, Pahlevan et al. (2019, 2021) suggested that enhanced information for inland and coastal water quality mapping is required as a critical and urgent task to evaluate spatial and spectral differences under several atmospheric and aquatic conditions [28,55]. This data record is crucial to ensuring a detailed monitoring of the Mar Menor coastal lagoon with both satellite platforms working in tandem.

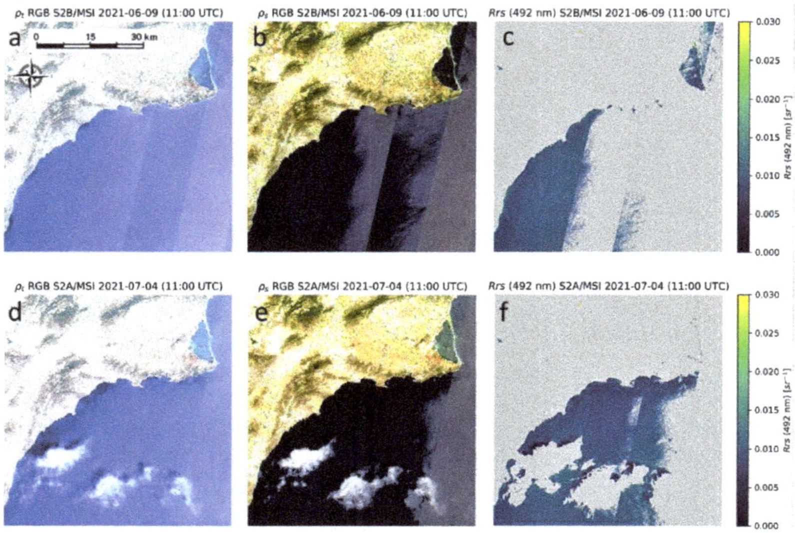

Figure 6. RGB (Red–Green–Blue) composite image on 9 June 2021 from the Sentinel-2 satellite (10 m spatial resolution) at (**a**) top-of-atmosphere (TOA) level, (**b**) bottom-of-atmosphere (BOA) level after ACOLITE, and (**c**) remote sensing reflectance (Rrs, sr^{-1}) of the blue band (492 nm); (**d**–**f**) the same on 4 July 2021. Severe sunglint contamination can be clearly observed in the eastern section of the Sentinel-2 tile affecting Mar Menor.

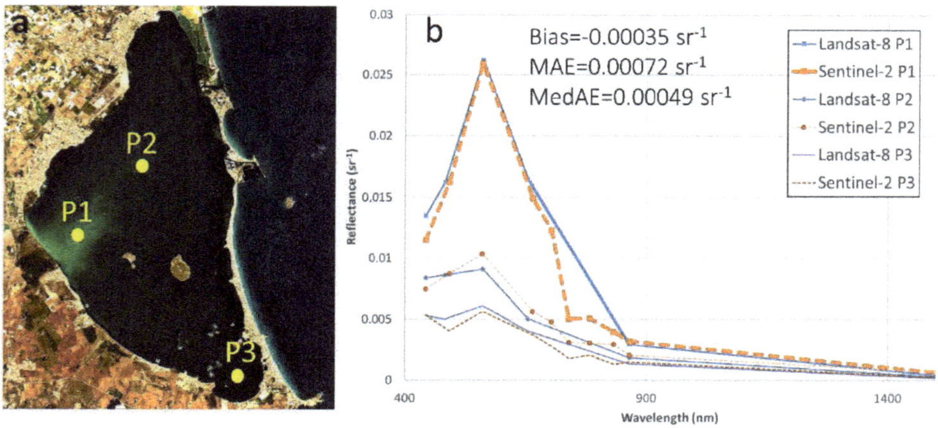

Figure 7. (**a**) RGB (Red–Green–Blue) composite image on 3 August 2021 of the Sentinel-2 satellite at bottom-of-atmosphere (BOA) level, (**b**) spectral signal of the Sentinel-2 and Landsat-8 satellites over different control points (P1, turbidity of 18.11 and 17.28 FNU; P2, turbidity of 5.5 and 6.1 FNU; and P3, turbidity of 4.08 and 3.93 FNU for Sentinel-2 and Landsat-8, respectively). Yellow circles in (**a**) indicate the location of the control pixels.

3.2. *Validation of the Water Quality Algorithms*

Figure 8 shows the validation matchups for the water quality parameters obtained with Sentinel-2 and Landsat-8 during the four in situ campaigns carried out in 2021. We applied the standard OC3 algorithm to calculate chl-a concentration and a regularly used semianalytical algorithm for the determination of turbidity [44,45]. The performance of both algorithms is illustrated in Figure 8a,b, respectively. The chl-a matchups cover the range of 0.5–5 mg/m^3 with a bias of 0.37 mg/m^3, MAE of 0.43 mg/m^3, and MedAE of 0.41 mg/m^3 (R^2 = 0.903, n = 37), whereas the turbidity ranges from 0.5–6 FNU with a bias of 2.09 FNU, MAE of 2.09 FNU, and MedAE of 2.04 FNU (R^2 = 0.54, n = 35). The validation assessment indicated robust statistical analysis, with accurate chl-a retrieval and minimum bias. Predictions of chl-a from both Sentinel-2 and Landsat-8 yielded precise results after the ACOLITE atmospheric and sunglint correction. The performance of the ACOLITE and the turbidity model is accurate, but a general satellite overestimation was encountered with biased outcomes, as seen in Figure 8b. Pahlevan et al. (2022) also found overestimation of turbidity retrievals by means of the ACOLITE processor [56]. The turbidity model has already been validated in different regions worldwide with accurate performance [46–48,57] and has previously been used in Mar Menor during an extreme weather event [13]. These methodologies are consistent and valid approaches for the assessment of suspended material or turbidity, which contribute towards achieving more precise performance worldwide [45,58].

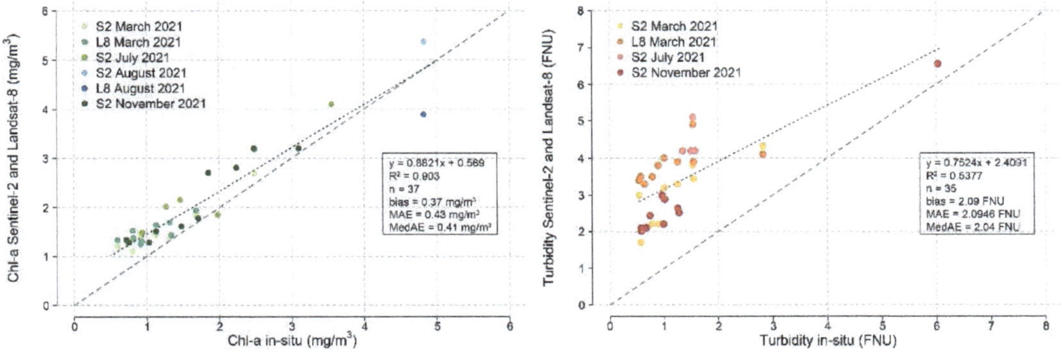

(a) **(b)**

Figure 8. Validation of water quality parameters obtained with Sentinel-2 (S2) and Landsat-8 (L8) during the four in situ campaigns carried out in March, July, August, and November 2021 for (**a**) chlorophyll-a concentration (chl-a, mg/m^3) and (**b**) turbidity (FNU).

3.3. *Water Quality Monitoring*

Figure 9 shows the RGB composite scenes acquired on 9 and 24 June and 2, 14, 18, and 29 July 2021, whereas Figures 10 and 11 display the image-derived maps for turbidity and chl-a, respectively. The imagery corresponded to the months prior to the ecological crisis in mid-August 2021. Generally, the turbidity levels were low in the lagoon (<5 FNU), except in the western section on 24 June and 14 July 2021, indicating higher levels (~25 FNU). A turbid plume appeared near land where the Albujon watercourse flows into Mar Menor. The most common chl-a condition during this period was <1.5 mg/m^3, while higher chl-a concentration (~2.5 mg/m^3) was observed close to the turbid plume. Figures S1 and S2 indicate the available (cloud and sunglint-free) Sentinel-2 scenes in August 2021 for further evaluation. On 3 August, a turbid plume was observed in the western section close to the input of the Albujon watercourse with peaked levels ~20 FNU, whereas minimum turbidity was encountered in the rest of the lagoon. The chl-a concentration for this date seemed to increase in the western section, indicating maximum values within the lagoon. The turbidity maps depicted high and constant turbidity values ~20 FNU in August 2021, except a slight decrease on 18 August 2021 on the western side (Figure S2). Moreover, chl-a maps displayed higher concentrations compared with July and the beginning of the bloom during this month, in particular on 13 August 2021 with chl-a ranging from 4 to 9 mg/m^3, a strong indicator of algal blooms in Mar Menor. Clear-water lagoon phases are characterized by chl-a concentrations ranging from 1 to 3 mg/m^3 [59], as occurred during June and July

2021. However, these typical low chl-a values tipped rapidly towards more eutrophic conditions, with chl-a concentration higher than 3 mg/m^3, such as last year's [18,19,21,60]. We reported that using the multisensor approach during the eutrophication episode in 2021, the beginning of the bloom (chl-a concentration higher than 3 mg/m^3) was detected mainly in the western and southern section. This is critical information for early detection of the eutrophication processes, given that the massive mortality of fish and crustaceans occurred during the last weeks of August 2021 [21], as well as an opportunity to enhance emergency management response in early deterioration stages.

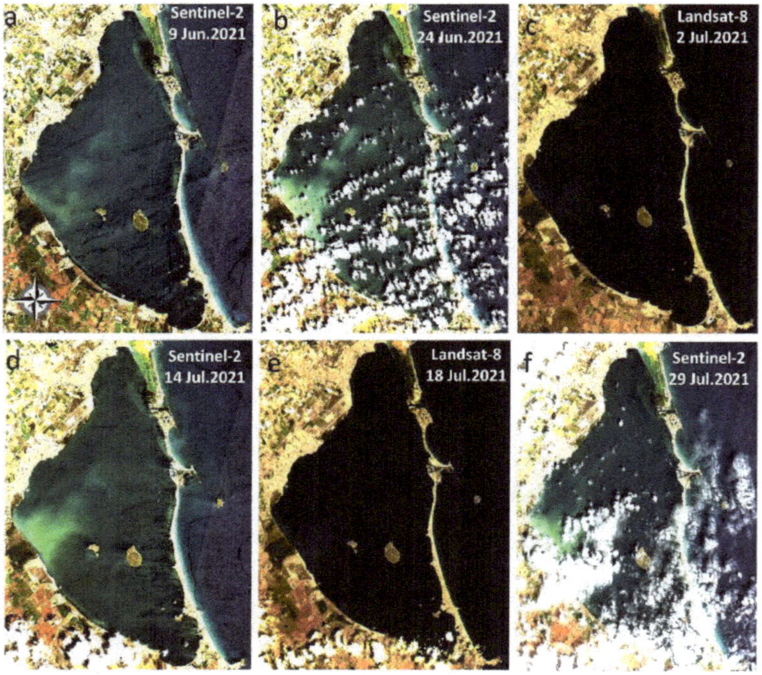

Figure 9. Sentinel-2 and Landsat-8 RGB (Red–Green–Blue) composite image acquired on (**a**) 9 June 2021, (**b**) 24 June 2021, (**c**) 2 July 2021, (**d**) 14 July 2021, (**e**) 18 July 2021, and (**f**) 29 July 2021.

After this event, the ecosystem equilibrium recovered slightly during September 2021 (Figure S3) as can be observed in the decrease in turbidity retrievals on the western side (Figure S4). However, the Albujon continued to discharge to the western section of the lagoon, indicating increased surface runoff and rising turbidity levels, and a plume was always close to this area. In addition, chl-a concentration gradually reached normal values <3 mg/m^3 in some areas in September 2021, although on 7 and 12 September 2021 high chl-a concentrations of ~4–5 mg/m^3 persisted not only in the center and south but also in the northern and eastern sections of the lagoon (Figure S5). A minor "Cold Drop" occurred on 20–21 September 2021, but the cloud and haze coverage remained very high during the consecutive days, as can be observed on 22 and 27 September 2021.

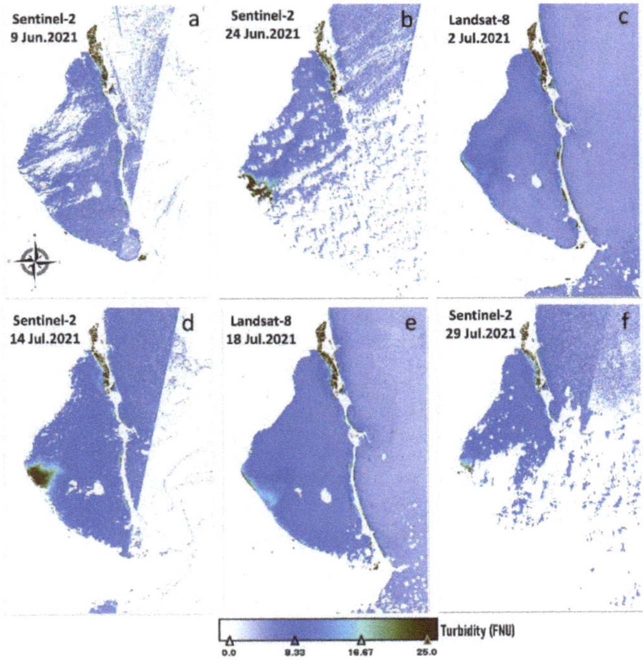

Figure 10. Turbidity (FNU) from Sentinel-2 and Landsat-8 acquired on (**a**) 9 June 2021, (**b**) 24 June 2021, (**c**) 2 July 2021, (**d**) 14 July 2021, (**e**) 18 July 2021, and (**f**) 29 July 2021.

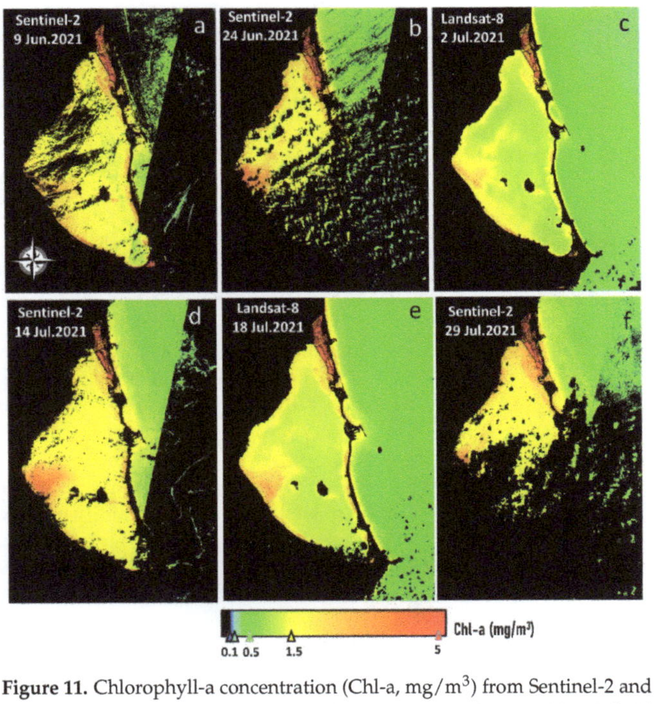

Figure 11. Chlorophyll-a concentration (Chl-a, mg/m^3) from Sentinel-2 and Landsat-8 acquired on (**a**) 9 June 2021, (**b**) 24 June 2021, (**c**) 2 July 2021, (**d**) 14 July 2021, (**e**) 18 July 2021, and (**f**) 29 July 2021.

Time-consuming and costly on-site measurements are regularly carried out to determine the water quality status in the lagoon; nevertheless, these observations are not able to address the heterogeneity and complexity of the spatial distribution within Mar Menor. In fact, in situ data might lack samples from the peak of the bloom or high turbidity levels due to the sparsely distributed single sampling sites. Field campaigns may not have adequately retrieved maximum concentration in chl-a if in situ measurements were not correctly spotted in Mar Menor [35]. The combined satellite data series characterized the dynamic nearshore patterns and fine-scale bio-optical gradients across this complex coastal interface. Satellite maps offered a synoptic perspective of the entire lagoon, detecting higher and lower turbidity and chl-a concentration over the study area. Interestingly, while maximum turbidity levels across the study site were typically located in the western section associated with the drainage of the Albujon, highlighting the impact of hydrological inputs and discharge from this canal, minimum levels were observed on the eastern, northern, and southern sides and along the barrier beach "La Manga". Our results also present the highest chl-a concentrations along the western coastline, detecting a change due to a proliferation of phytoplankton in early August 2021. The IEO-CSIC suggested that this eutrophication event was due to an excess of nutrient availability flowing from the Albujon [21]. Therefore, the abrupt deterioration of the water quality in Mar Menor reached a stage of severe eutrophication that resulted in an ecological collapse in mid-August, showing a gradual recovery during September 2021 before the minor "Cold Drop" event on 20 September 2021. Among all the wadis transporting materials, water, and nutrients from agricultural run-off, the Albujon is the principal collector of the Campo de Cartagena drainage basin, subjecting the lagoon to nutrient and sediment runoff from the agricultural landscape [11,12,15,16]. These results sustain that in order to remedy the ecological collapse of the lagoon, it is crucial to design and implement environmental strategies and policies [22], in particular those that focus on limiting the suspended material discharged from the Albujon to regulate the massive proliferation of phytoplankton and the eutrophication pressure favored by agricultural dumping [25,61].

3.4. An Early Warning Tool with High Spatial Resolution

The complex distribution and variability of the lagoon can be observed in detail in all the satellite-derived products presented in this study, in particular for the turbid plume located in the western part. These turbid features are usually small in dimension; therefore, detecting them by means of traditional ocean color sensors at lower spatial resolution can be challenging. We recommend using Landsat-8 and Sentinel-2 missions in tandem to improve the monitoring and control of Mar Menor. The multisensor methodology might enhance previous studies that attempted to map water quality using coarser spatial resolution imagery at 300–1000 m [24–26]. Additional evaluation of previous months in March 2021 also highlighted the importance of our methodology for studying the impact of weather events on the coastal lagoon. Figures S6 and S7 show the RGB composite images and turbidity levels on 11 and 12 March 2021 and on 21 and 28 March 2021, before and after a severe winter storm, respectively. The maps corresponding to 11 and 12 March 2021 presented minimum turbidity levels (<4 FNU) in front of the Albujon. This cycle was occasionally disrupted by the intense winter storm resulting in increased inputs of terrestrial discharges into the entire lagoon. The high resuspension of materials can be observed in both the RGB and turbidity maps after the storm, in particular on 21 March 2021 along the western coastal region with turbidity >50 FNU. A zoom on 21 March 2021 corresponding to the southeastern shore of Mar Menor showed the high variability of the turbidity patterns (Figure S8). Turbidity generally decreases seaward in the lagoon and extreme events, such as storms, can increase turbidity 5-to-10-fold, altering the water quality distribution in the system. In particular, "Cold Drop" events can dramatically alter the ecological status of the lagoon with turbidity levels increasing by more than a factor of five [13]. Previous studies have already indicated that finer spatial resolution is needed to comprehensively determine these complex spatial and temporal features [62].

With three-to-four-day revisits allowed by combined Landsat-8 and Sentinel-2 datasets, the managers, end users, and coastal science community will take advantage of these synoptic, improved, consistent, and high-quality products. This information may be critical for operational purposes in the context of the EU WFD [32], from which early warning systems can be implemented. Although work remains to be performed towards improving and developing advanced sunglint, atmospheric, and bio-optical algorithms for both Sentinel-2 and Landsat-8, it is the ideal moment to examine, exploit, and maximize these merged datasets in Mar Menor. Particularly during ecological crises, such as the one explored in this study, this information is crucial to assess appropriate measures to be taken in coastal and inland water ecosystems. However, research must continue to enable retrievals in extremely contaminated sunglint scenes during eutrophic/turbid conditions in summer periods, as demonstrated in this study (Figure 6), as well as the analysis of other biogeochemical variables, such as Colored Dissolved Organic Matter (CDOM). In addition, as shown in recent research by Wójcik-Długoborska [63], turbidity measurements in the field may differ from those taken in the laboratory and thus provide different correlations between reflectance and the true value of turbidity. Therefore, we intend to focus additional research on this aspect during the coming field campaigns. Future studies will also be carried out to evaluate the entire Sentinel-2 and Landsat-8 series to assess the seasonality of these events and to identify possible common factors, which can be monitored or used as warning systems in the future. The improved resolution afforded by the combined time-series products offers additional insights into processes over weekly or subweekly timescales; nevertheless, these results emphasized the need for enhanced temporal coverage space-based datasets in dynamic coastal environments. With these three satellites now operating, Landsat-9 already in orbit, planned missions launching shortly (e.g., Sentinel-2C/D), and continuously improving atmospheric and sunglint correction techniques, the accessible record of high-to-moderate spatial resolution imagery will provide even more robust water quality monitoring in complex inland and coastal environments.

4. Conclusions

Eutrophication in areas where agricultural and industrial wastewater runoff feeds excessive loads of nutrients into coastal regions can foster algal blooms and undermine the health of these ecosystems. In this study, both Landsat-8 and Sentinel-2 satellites were jointly merged as a constellation to estimate indicators of the water quality in the highly unstable and vulnerable Mar Menor coastal lagoon. The validation of satellite biogeochemical parameters, both turbidity and chl-a, retrieved good performance for ranges of 0.5–6 FNU and 0.5–5 mg/m^3, respectively. The atmospheric and sunglint correction using the ACOLITE software showed consistent performance for both satellites; therefore, using them in tandem can improve mapping strategies, highlighting the importance of the preprocessing scheme. The results demonstrate the suitability and consistency of the methodology to reliably capture the detailed spatiotemporal distribution of turbidity and chl-a, where satellite imagery was capable of early detection of chl-a levels above 3 mg/m^3, which generally triggered the subsequent blooms during recent years. Although neither of the satellite missions have been designed to characterize coastal seawater quality, our approach demonstrated their capacity to provide appropriate information at 10–30 m spatial resolution on a systematic basis and in a cost-effective way. Multitemporal maps were produced, and an analysis of all images showed that the highest turbidity and chl-a levels were always located in the western section. In particular, turbidity and chl-a concentration at the mouth of the draining Albujon watercourse were consistently two times higher than in the northern and eastern sections of the lagoon. The influence of the highly dynamic plume from the Albujon extended over the entire lagoon, yet the strongest gradients typically occur within the first nearshore 1–2 km. Therefore, observing these gradients, their variability, and the impact of land–water exchanges on nearshore dynamics under varying environmental conditions from space requires higher spatial resolution. Imagery from both satellites offered snapshots of water quality patterns that are difficult to map with in situ technologies in such heterogeneous environments. These innovative tools can support decision makers in the implementation of a joint monitoring strategy, better characterization of the water quality distribution, and timely assistance to society during these ecological disasters, thus preventing detrimental conditions in the lagoon. Furthermore, the powerful multisensor system can be used as guidance to complement the ongoing in situ techniques carried out by the local and regional authorities to select relevant areas for data sampling. This information could advance mapping of water quality and bio-optical properties in terrestrial–aquatic environments as an important tool for managers and stakeholders, as well as for the tourism and fishing industry. A new era has begun with the use of the Sentinel-2 and Landsat-8 missions as a virtual constellation, with significant opportunities for monitoring the heterogeneous spatiotemporal patterns of inland and nearshore coastal waters at resolutions certainly not observed previously.

Supplementary Materials: The following supporting information can be downloaded at: https://www.mdpi.com/article/10.3390/rs14122744/s1. Figure S1: RGB (Red–Green–Blue) composite image on (a) 8, (b) 13, and (c) 18 August 2021 of the Sentinel-2 satellite (10 m spatial resolution); Figure S2: Turbidity (FNU) on (a) 3, (b) 8, (c) 13, and (d) 18, August 2021 of the Sentinel-2 satellite (10 m spatial resolution); (e–h) the same for chlorophyll-a concentration (Chl-a, mg/m^3); Figure S3: Sentinel-2 and Landsat-8 RGB (Red–Green–Blue) composite image acquired on (a) 4 September 2021, (b) 7 September 2021, (c) 11 September 2021, (d) 12 September 2021, (e) 17 September 2021, and (f) 22 September 2021; Figure S4: Turbidity (FNU) from Sentinel-2 and Landsat-8 acquired on (a) 4 September 2021, (b) 7 September 2021, (c) 11 September 2021, (d) 12 September 2021, (e) 17 September 2021, and (f) 22 September 2021; Figure S5: Chlorophyll-a concentration (Chl-a, mg/m^3) from Sentinel-2 and Landsat-8 acquired on (a) 4 September 2021, (b) 7 September 2021, (c) 11 September 2021, (d) 12 September 2021, (e) 17 September 2021, and (f) 22 September 2021; Figure S6: RGB (Red–Green–Blue) composite image from Sentinel-2 and Landsat-8 acquired on (a) 11, (b) 12, (c) 21, and (d) 28 March 2021; Figure S7: Turbidity (FNU) from Sentinel-2 and Landsat-8 acquired on (a) 11, (b) 12, (c) 21, and (d) 28 March 2021; Figure S8: (a) RGB (Red–Green–Blue) composite image, and (b) Turbidity (FNU) from Sentinel-2 on 21 March 2021 corresponding to the southeastern shore of Mar Menor.

Author Contributions: Conceptualization, I.C.; formal analysis and investigation, I.C.; figures, I.C. and M.R.; original draft preparation, I.C.; writing—review and editing, I.C., M.R., J.S.-E., P.B. and G.N.; fieldwork, J.S.-E. and P.B; supervision, I.C.; project administration, I.C.; funding acquisition, I.C., J.S.-E. and G.N. All authors have read and agreed to the published version of the manuscript.

Funding: This research was supported by the Project RTI2018-098784-J-I00 funded by the MCIN/AEI/10.13039/501100011033 and by "FEDER Una manera de hacer Europa" and the Grant IJC2019-039382-I funded by the MCIN/AEI/10.13039/501100011033. The work was also supported by the Project of the Regional Government "Junta de Andalucía" (PY20-00244), the Project PID2019-109355RA-I00 funded by the MCIN/AEI/10.13039/501100011033 and the Grant FPU20/01294 funded by the Spanish Ministry of Universities.

Data Availability Statement: Not applicable.

Acknowledgments: We would like to acknowledge the European Commission's Copernicus programme, the United States Geological Survey (USGS), and the National Aeronautics and Space Administration (NASA) for distributing Sentinel-2 and Landsat-8 imagery. This work represents a contribution to the CSIC Thematic Interdisciplinary Platform PTI TELEDETECT. The authors would like to thank Martha B. Dunbar for the English revision and Rocío García for the chlorophyll-a analyses. We also thank the Laboratorio de Medio Ambiente de Galicia for lending us the turbidimeter.

Conflicts of Interest: The authors declare no conflict of interest.

References

1. Barnes, R.S.K. *Coastal Lagoons*; CUP Archive; Cambridge University Press: Cambridge, UK, 1980.
2. UNESCO. *Coastal Lagoon Research, Present and Future: UNESCO Technical Papers in Marine Science*; UNESCO: London, UK, 1981.
3. Kennish, M.J.; Paerl, H.W. *Coastal Lagoons: Critical Habitats of Environmental Change*; Kennish, M.J., Paerl, H.W., Eds.; CRC Press: Boca Raton, FL, USA, 2010.

4. De Wit, R.; Mostajir, B.; Troussellier, M.; Do Chi, T. Environmental Management and Sustainable Use of Coastal Lagoons Ecosystems. In *Lagoons: Biology Management and Environmental Impact*; Nova Publishers: Hauppauge, NY, USA, 2011; pp. 333–350.
5. Hiep, N.H.; Luong, N.D.; Viet Nga, T.T.; Hieu, B.T.; Thuy Ha, U.T.; Du Duong, B.; Long, V.D.; Hossain, F.; Lee, H. Hydrological model using ground- and satellite-based data for river flow simulation towards supporting water resource management in the Red River Basin, Vietnam. *J. Environ. Manag.* **2018**, *217*, 346–355. [CrossRef] [PubMed]
6. Contreras, F.; Castañeda, O. La biodiversidad de las lagunas costeras. *Ciencias* **2004**, *76*, 46–56.
7. *Ecosystems and Human Well-Being*; Millennium Ecosystem Assessment; Island Press: Washington, DC, USA, 2005.
8. Newton, A.; Brito, A.C.; Icely, J.D.; Derolez, V.; Clara, I.; Angus, S.; Schernewski, G.; Inácio, M.; Lillebø, A.I.; Sousa, A.I.; et al. Assessing, quantifying and valuing the ecosystem services of coastal lagoons. *J. Nat. Conserv.* **2018**, *44*, 50–65. [CrossRef]
9. Chang, N.B.; Bai, K.; Chen, C.F. Integrating multisensor satellite data merging and image reconstruction in support of machine learning for better water quality management. *J. Environ. Manag.* **2017**, *201*, 227–240. [CrossRef]
10. Kumar, P. *The Economics of Ecosystems and Biodiversity: Ecological and Economic Foundations*; Routledge, Taylor & Francis Group: London, UK, 2012.
11. Pérez Ruzafa, Á.; Marcos Diego, C.; Gilabert Cervera, F.J. The ecology of the Mar Menor coastal lagoon: A fast changing ecosystem under human pressure. In *Coastal Lagoons: Ecosystem Processes and Modeling for Sustainable Use and Development*; CRC Press: Raton, FL, USA, 2005; pp. 392–422. ISBN 1-56670-686-6.
12. García-Pintado, J.; Martínez-Mena, M.; Barberá, G.G.; Albaladejo, J.; Castillo, V.M. Anthropogenic nutrient sources and loads from a Mediterranean catchment into a coastal lagoon: Mar Menor, Spain. *Sci. Total Environ.* **2007**, *373*, 220–239. [CrossRef]
13. Caballero, I.; Ruiz, J.; Navarro, G. Sentinel-2 Satellites Provide Near-Real Time Evaluation of Catastrophic Floods in the West Mediterranean. *Water* **2019**, *11*, 2499. [CrossRef]
14. Bayo, J.; Rojo, D.; Olmos, S. Abundance, morphology and chemical composition of microplastics in sand and sediments from a protected coastal area: The Mar Menor lagoon (SE Spain). *Environ. Pollut.* **2019**, *252*, 1357–1366. [CrossRef]
15. Jiménez-Cárceles, F.J.; Egea, C.; Rodríguez-Caparrós, A.B.; Barbosa, O.A.; Delgado, M.J.; Ortiz, R.; Álvarez-Rogel, J. Contents of Nitrogen, Ammonium, Phosphorus, Pesticides and Heavy Metals, in a Salt Marsh in the Coast of the Mar Menor Lagoon (SE Spain). *FEB Fresenius Environ. Bull.* **2006**, *15*, 372–380.
16. Moreno-González, R.; Campillo, J.A.; García, V.; León, V.M. Seasonal input of regulated and emerging organic pollutants through surface watercourses to a Mediterranean coastal lagoon. *Chemosphere* **2013**, *92*, 247–257. [CrossRef]
17. Conesa, H.M.; Jiménez-Cárceles, F.J. The Mar Menor lagoon (SE Spain): A singular natural ecosystem threatened by human activities. *Mar. Pollut. Bull.* **2007**, *54*, 839–849. [CrossRef]
18. Pérez-Ruzafa, A.; Campillo, S.; Fernández-Palacios, J.M.; García-Lacunza, A.; García-Oliva, M.; Ibañez, H.; Navarro-Martínez, P.C.; Pérez-Marcos, M.; Pérez-Ruzafa, I.M.; Quispe-Becerra, J.I.; et al. Long-Term Dynamic in Nutrients, Chlorophyll a, and Water Quality Parameters in a Coastal Lagoon During a Process of Eutrophication for Decades, a Sudden Break and a Relatively Rapid Recovery. *Front. Mar. Sci.* **2019**, *6*, 26. [CrossRef]
19. Ruiz-Fernández, J.; León, V.M.; Marín-Guirao, L.; Giménez-Casaduero, F.; Álvarez-Rogel, J.; Esteve-Selma, M.; Gómez, R.; Robledano, F.; Barberá, G.G.; Martínez-Fernández, J. *Synthesis Report of the Current State of Mar Menor Lagoon and Its Causes in Relation to the Nutrient Contents*; Informe de Asesoramiento Técnico; Instituto Español de Oceanografía (IEO): Madrid, Spain, 2019; 24p.
20. Torrente, M.D.; Ruiz, J.M.; Muñoz, R.; Segura, A.; Esteller, J.; Casero, J.; Guirao, L.; Moreno, P.; Navarro, I.; Nuez, E.; et al. Collapse of macrophytic communities in a eutrophicated coastal lagoon. *Front. Mar. Sci. Conf.* **2019**. [CrossRef]
21. Ruiz-Fernández, J.; Clemente-Navarro, P.; Mercado, J.M.; Fraile-Nuez, E.; Albentosa, M.; Marín-Guirao, L.; Santos, J. *Nuevo Evento de Mortandad Masiva de Organismos Marinos en el Mar Menor: Contexto Y Factores*; Informe de Asesoramiento Técnico; Instituto Español de Oceanografía (IEO): Madrid, Spain, 2021; 24p.
22. Velasco, A.M.; Pérez-Ruzafa, A.; Martínez-Paz, J.M.; Marcos, C. Ecosystem services and main environmental risks in a coastal lagoon (Mar Menor, Murcia, SE Spain): The public perception. *J. Nat. Conserv.* **2018**, *43*, 180–189. [CrossRef]
23. Scientific Data Server (SDC) and Politechnic University of Cartagena (UPCT). 2019. Available online: https://marmenor.upct.es/ (accessed on 1 May 2021).
24. Martínez-Alvarez, V.; Gallego-Elvira, B.; Maestre-Valero, J.; Tanguy, M. Simultaneous solution for water, heat and salt balances in a Mediterranean coastal lagoon (Mar Menor, Spain). *Estuar. Coast. Shelf Sci.* **2011**, *91*, 250–261. [CrossRef]
25. Soria, J.; Caniego, G.; Hernández-Sáez, N.; Dominguez-Gomez, J.A.; Erena, M. Phytoplankton Distribution in Mar Menor Coastal Lagoon (SE Spain) during 2017. *J. Mar. Sci. Eng.* **2020**, *8*, 600. [CrossRef]
26. Mercado, J.M.; Cortés, D.; Gómez-Jakobsen, F.; García-Gómez, C.; Ouaissa, S.; Yebra, L.; Ferrera, I.; Valcárcel-Pérez, N.; López, M.; García-Muñoz, R.; et al. Role of small-sized phytoplankton in triggering an ecosystem disruptive algal bloom in a Mediterranean hypersaline coastal lagoon. *Mar. Pollut. Bull.* **2021**, *164*, 111989. [CrossRef]
27. Park, M.-H.; Stenstrom, M.K. Using satellite imagery for stormwater pollution management with Bayesian networks. *Water Res.* **2006**, *40*, 3429–3438. [CrossRef]
28. Pahlevan, N.; Chittimalli, S.K.; Balasubramanian, S.V.; Vellucci, V. Sentinel-2/Landsat-8 product consistency and implications for monitoring aquatic systems. *Remote Sens. Environ.* **2019**, *220*, 19–29. [CrossRef]
29. Aubriot, L.; Zabaleta, B.; Bordet, F.; Sienra, D.; Risso, J.; Achkar, M.; Somma, A. Assessing the origin of a massive cyanobacterial bloom in the Río de la Plata (2019): Towards an early warning system. *Water Res.* **2020**, *181*, 115944. [CrossRef]

30. Caballero, I.; Fernández, R.; Escalante, O.M.; Mamán, L.; Navarro, G. New capabilities of Sentinel-2A/B satellites combined with in situ data for monitoring small harmful algal blooms in complex coastal waters. *Sci. Rep.* **2020**, *10*, 8743. [CrossRef]
31. Chen, J.; Zhu, W.; Tian, Y.Q.; Yu, Q. Monitoring dissolved organic carbon by combining Landsat-8 and Sentinel-2 satellites: Case study in Saginaw River estuary, Lake Huron. *Sci. Total Environ.* **2020**, *718*, 137374. [CrossRef] [PubMed]
32. Cao, F.; Tzortziou, M. Capturing dissolved organic carbon dynamics with Landsat-8 and Sentinel-2 in tidally influenced wetland–estuarine systems. *Sci. Total Environ.* **2021**, *777*, 145910. [CrossRef]
33. Erena, M.; Domínguez, J.A.; Aguado-Giménez, F.; Soria, J.; García-Galiano, S. Monitoring Coastal Lagoon Water Quality through Remote Sensing: The Mar Menor as a Case Study. *Water* **2019**, *11*, 1468. [CrossRef]
34. Erena, M.; Domínguez, J.A.; Atenza, J.F.; García-Galiano, S.; Soria, J.; Pérez-Ruzafa, Á. Bathymetry Time Series Using High Spatial Resolution Satellite Images. *Water* **2020**, *12*, 531. [CrossRef]
35. Gómez, D.; Salvador, P.; Sanz, J.; Casanova, J.L. A new approach to monitor water quality in the Menor sea (Spain) using satellite data and machine learning methods. *Environ. Pollut.* **2021**, *286*, 117489. [CrossRef]
36. Page, B.P.; Olmanson, L.G.; Mishra, D.R. A harmonized image processing workflow using Sentinel-2/MSI and Landsat-8/OLI for mapping water clarity in optically variable lake systems. *Remote Sens. Environ.* **2019**, *231*, 111284. [CrossRef]
37. Zhang, X.; Fichot, C.G.; Baracco, C.; Guo, R.; Neugebauer, S.; Bengtsson, Z.; Ganju, N.; Fagherazzi, S. Determining the drivers of suspended sediment dynamics in tidal marsh-influenced estuaries using high-resolution ocean color remote sensing. *Remote Sens. Environ.* **2020**, *240*, 111682. [CrossRef]
38. European Space Agency (ESA). E. Sentinel-2 User Handbook. *ESA Stand. Doc. Date* **2015**, *1*, 1–64. Available online: https://sentinel.esa.int/documents/247904/685211/Sentinel-2_User_Handbook (accessed on 1 May 2021).
39. European Space Agency (ESA). Sentinel-2 MSI Technical Guide 2017. Available online: https://earth.esa.int/web/sentinel/technicalguides/sentinel-2-msi (accessed on 1 May 2021).
40. Woodcock, C.E.; Allen, R.; Anderson, M.; Belward, A.; Bindschadler, R.; Cohen, W.; Gao, F.; Goward, S.N.; Helder, D.; Helmer, E.; et al. Free access to Landsat imagery. *Science* **2008**, *320*, 1011. [CrossRef]
41. Knight, E.J.; Kvaran, G. Landsat-8 Operational Land Imager Design, Characterization and Performance. *Remote Sens.* **2014**, *6*, 10286–10305. [CrossRef]
42. Vanhellemont, Q.; Ruddick, K.G. ACOLITE Processing for Sentinel-2 and Landsat-8: Atmospheric Correction and Aquatic Applications. In Proceedings of the Living Planet Symposium, Prague, Czech Republic, 9–13 May 2016.
43. Vanhellemont, Q.; Ruddick, K. Atmospheric correction of metre-scale optical satellite data for inland and coastal water applications. *Remote Sens. Environ.* **2018**, *216*, 586–597. [CrossRef]
44. Vanhellemont, Q. Adaptation of the dark spectrum fitting atmospheric correction for aquatic applications of the Landsat and Sentinel-2 archives. *Remote Sens. Environ.* **2019**, *225*, 175–192. [CrossRef]
45. Nechad, B.; Ruddick, K.; Neukermans, G. Calibration and validation of a generic multisensor algorithm for mapping of turbidity in coastal waters. In Proceedings of the SPIE—The International Society for Optical Engineering, Berlin, Germany, 9 September 2009; p. 74730H.
46. Katlane, R.; Nechad, B.; Ruddick, K.; Zargouni, F. Optical remote sensing of turbidity and total suspended matter in the Gulf of Gabes. *Arab. J. Geosci.* **2013**, *6*, 1527–1535. [CrossRef]
47. Nazirova, K.; Alferyeva, Y.; Lavrova, O.; Shur, Y.; Soloviev, D.; Bocharova, T.; Strochkov, A. Comparison of In Situ and Remote-Sensing Methods to Determine Turbidity and Concentration of Suspended Matter in the Estuary Zone of the Mzymta River, Black Sea. *Remote Sens.* **2021**, *13*, 143. [CrossRef]
48. Vanhellemont, Q.; Ruddick, K. Atmospheric correction of Sentinel-3/OLCI data for mapping of suspended particulate matter and chlorophyll-a concentration in Belgian turbid coastal waters. *Remote Sens. Environ.* **2021**, *256*, 112284. [CrossRef]
49. O'Reilly, J.E.; Werdell, P.J. Chlorophyll algorithms for ocean color sensors—OC4, OC5 & OC6. *Remote Sens. Environ.* **2019**, *229*, 32–47.
50. Scientific Committee on Oceanic Research (SCOR). *Determination of Photosynthetic Pigments: Monographs on Oceanographic Methodology*; UNESCO: London, UK, 1966.
51. Rodríguez-Benito, C.V.; Navarro, G.; Caballero, I. Using Copernicus Sentinel-2 and Sentinel-3 data to monitor harmful algal blooms in Southern Chile during the COVID-19 lockdown. *Mar. Pollut. Bull.* **2020**, *161*, 111722. [CrossRef]
52. European Space Agengy (ESA). Land Monitoring 2019. Available online: https://sentinel.esa.int/web/sentinel/thematic-areas/land-monitoring (accessed on 1 May 2021).
53. Caballero, I.; Román, A.; Tovar-Sánchez, A.; Navarro, G. Water quality monitoring with Sentinel-2 and Landsat-8 satellites during the 2021 volcanic eruption in La Palma (Canary Islands). *Sci. Total Environ.* **2022**, *822*, 153433. [CrossRef]
54. Caballero, I.; Stumpf, R.P. Atmospheric correction for satellite-derived bathymetry in the Caribbean waters: From a single image to multi-temporal approaches using Sentinel-2A/B. *Opt. Express* **2020**, *28*, 11742–11766. [CrossRef]
55. Pahlevan, N.; Mangin, A.; Balasubramanian, S.V.; Smith, B.; Alikas, K.; Arai, K.; Barbosa, C.; Bélanger, S.; Binding, C.; Bresciani, M.; et al. ACIX-Aqua: A global assessment of atmospheric correction methods for Landsat-8 and Sentinel-2 over lakes, rivers, and coastal waters. *Remote Sens. Environ.* **2021**, *258*, 112366. [CrossRef]
56. Pahlevan, N.; Smith, B.; Alikas, K.; Anstee, J.; Barbosa, C.; Binding, C.; Bresciani, M.; Cremella, B.; Giardino, C.; Gurlin, D.; et al. Simultaneous retrieval of selected optical water quality indicators from Landsat-8, Sentinel-2, and Sentinel-3. *Remote Sens. Environ.* **2022**, *270*, 112860. [CrossRef]

57. Caballero, I.; Steinmetz, F.; Navarro, G. Evaluation of the First Year of Operational Sentinel-2A Data for Retrieval of Suspended Solids in Medium- to High-Turbidity Waters. *Remote Sens.* **2018**, *10*, 982. [CrossRef]
58. Wang, M.; Nim, C.J.; Son, S.; Shi, W. Characterization of turbidity in Florida's Lake Okeechobee and Caloosahatchee and St. Lucie Estuaries using MODIS-Aqua measurements. *Water Res.* **2012**, *46*, 5410–5422. [CrossRef]
59. Quintana, X.; Boix, D.; Gascón, S.; Sala, J.; Comín, F.A. *Management and Restoration of Mediterranean Coastal Lagoons in Europe: Reserca i Territori*; Càtedra d'Ecosistemes Litorals Mediterranis, Parc Natural del Montgrí, les Illes Medes i el Baix Ter, Museu de la Mediterrània: Girona, Spain, 2018; pp. 1–220.
60. Faz Cano, A.; Lobera Lössel, J.; Mora Navarro, J.; Simón Andreu, P. Depuración y descontaminación de aguas. In *Informe Integral Sobre el Estado Ecológico del Mar Menor*; Comité de Asesoramiento Científico del Mar Menor: Madrid, Spain, 2017; pp. 113–125.
61. Pilkaitytė, R.; Razinkovas, A. Factors controlling phytoplankton blooms in a temperate estuary: Nutrient limitation and physical forcing. In *Marine Biodiversity: Patterns and Processes, Assessment, Threats, Management and Conservation*; Martens, K., Queiroga, H., Cunha, M.R., Cunha, A., Moreira, M.H., Quintino, V., Rodrigues, A.M., Seroôdio, J., Warwick, R.M., Eds.; Springer: Dordrecht, The Netherlands, 2006; pp. 41–48.
62. Faridatul, M.I.; Wu, B.; Zhu, X. Assessing long-term urban surface water changes using multi-year satellite images: A tale of two cities, Dhaka and Hong Kong. *J. Environ. Manag.* **2019**, *243*, 287–298. [CrossRef]
63. Wójcik-Długoborska, K.A.; Osińska, M.; Bialik, R.J. The Impact of Glacial Suspension Color on the Relationship Between Its Properties and Marine Water Spectral Reflectance. *IEEE J. Sel. Top. Appl. Earth Obs. Remote Sens.* **2022**, *15*, 3258–3268. [CrossRef]

sustainability

Article

Spatial–Temporal Variations of Water Ecosystem Services Value and Its Influencing Factors: A Case in Typical Regions of the Central Loess Plateau

Yuan Xiu, Ni Wang *, Fangxu Peng and Quanxi Wang

State Key Laboratory of Eco-Hydraulics in Northwest Arid Region, Xi'an University of Technology, Xi'an 710048, China; 1170411005@stu.xaut.edu.cn (Y.X.); 2200420045@stu.xaut.edu.cn (F.P.); 2200421350@stu.xaut.edu.cn (Q.W.)
* Correspondence: wangni@xaut.edu.cn

Abstract: Water resources provide indispensable ecosystem services, which are related to human well-being and sustainable social development. Accurately measuring the water ecosystem services value (WESV), and then grasping its changing characteristics, is particularly important for solving water problems. In this study, the typical area of the central Loess Plateau location is taken as the research area. Based on remote sensing images and statistical data, the direct market method combined with the equivalent factor method was used to calculate the WESV including groundwater and surface water, which is of greatest originality. The temporal and spatial variation characteristics in 2010, 2015 and 2020 were analyzed. Then, four WESV driving factors including per capita GDP, population density, proportion of water areas, and water consumption were selected, and the geographically weighted regression (GWR) model was used to analyze the spatial distribution pattern and temporal variation of WESV's response to the influencing factors. The results showed that WESV experienced a process of first decreasing and then increasing, which was mainly caused by Yulin. For the composition of WESV, the proportion of provisioning services value has increased, which caused the proportion of regulating services value to decrease. The correlations between four factors and WESV were different. The distribution pattern of the influences was spatially heterogeneous, which showed regular variations over time. These results indicate the necessity of WESV's independent research and provide a realistic basis for ecological compensation in the Yellow River Basin.

Keywords: water resources; ecological services value; influencing factors; geographically weighted regression (GWR); space–time change; the Loess Plateau

Citation: Xiu, Y.; Wang, N.; Peng, F.; Wang, Q. Spatial–Temporal Variations of Water Ecosystem Services Value and Its Influencing Factors: A Case in Typical Regions of the Central Loess Plateau. *Sustainability* **2022**, *14*, 7169. https://doi.org/10.3390/su14127169

Academic Editors: Alban Kuriqi and Luis Garrote

Received: 9 May 2022
Accepted: 9 June 2022
Published: 11 June 2022

1. Introduction

Ecosystems provide basic and necessary services for human survival and social functioning, namely ecosystem services (ESs). ESs are the continuous provision of ecosystem goods and services by ecosystems and their ecological processes [1]. However, factors such as population growth, industrialization and urbanization have led to a rapid increase in the demand for ESs, and ecosystems are facing unprecedented pressure [2]. A comprehensive and reasonable quantitative assessment of ecosystem services value (ESV) is necessary to alleviate the contradiction between supply and demand of ESs, manage effectively and formulate relevant policies. Water resources are an essential part of ecosystems. Whereas, due to the existence of water pollution, waste of water resources and climate change, the water ecosystem is facing greater pressure than other types of ecosystems. Therefore, it is crucial to study the water ecosystem services value (WESV) and the various factors that affect WESV.

Monetization of ecosystem service value is the most recognized and practical form of ESV, and the calculation methods can be divided into two categories. One is to adopt the relevant methods of traditional ecological economics or environmental economics. Most of these methods obtain the output of physical quantity based on statistical data or ecological model, and then they calculate the value of ecosystem services by combining the market value method, willingness survey method (CV) and revealed preference method. This kind of method has high requirements on data, parameters, model accuracy and method applicability, etc. Specific to the value of water resources ecosystem services, this kind of method is suitable for calculating the value of water resources with commodity attributes. The other is the equivalence factor method, which constructs the economic value equivalent per unit area of different ecosystems based on the division of ecosystem service functions and quantifies ESV in combination with the distribution of ecosystems. Compared with traditional methods, the equivalence factor method requires relatively fewer data, and the evaluation of ESVs is more comprehensive. Costanza et al. divided the global ESs into 17 species and calculated that the ESV of 16 biomes in the world was US$33 trillion per year [3], and rivers or lakes were one of the 16 biomes. Meanwhile, another contribution of this study is to provide the equivalent table of services value per unit area of 17 ecosystems in each biome at the global scale, which provides a reference for future research. In 2003, the Millennium Ecosystem Assessment (MA) conducted by the United Nations classified global ESs into four primary categories, namely Provisioning Services, Regulating Services, Supporting Services, and Cultural Services [4]. At the regional scale, Xie et al. improved Costanza's research and obtained the equivalent factor of ecosystem service value suitable for China. Then, the ESV can be calculated combined with the area of each ecosystem. This study was accepted by a large number of Chinese scholars [5,6]. Whether it is a global scale or a regional scale, in the study of using the equivalent factor method to evaluate ecosystems, most of the rivers/lakes or waters are divided into one type of ecosystem for discussion. However, this method does not clarify the service value of groundwater, which is an important component of water resources. The most direct service value of groundwater is that it provides a part of water for production and domestic use, accounting for only 70% of the total global water use for agriculture, and 40% of agricultural water comes from groundwater [7]. Therefore, it is more comprehensive and accurate to combine the two methods to assess the WESV including surface water and groundwater.

It is not the ultimate goal of scholars to study the value of WESV; the more important goal is to study the relationship between WESV and the interaction of various factors. In order to deal with more severe water and environmental problems, exploring the influence of various factors on WESV has gradually become one of the research hotspots. Due to the social nature of water resources, social and economic factors have become one of the components that affect WESV. The characteristics of cities, populations, communities, and cultures [8] all have an impact on water resources and water ecology [9,10]. The urbanization level is a concentrated expression of the social and economic development degree, which profoundly affects the spatial distribution and potential functions of ecosystem services [11]. Many researchers have conducted related research in North China [12,13], Yangtze River Delta [14], Pearl River Delta [15], Southwest Mountainous [16,17] and Northwest arid regions of China [18,19]. WESV also has a significant response to changes in natural factors. Climate conditions [20–22], ecosystem types [23], and environmental quality [24] are the main influencing factors. Meanwhile, the coupling of many factors, such as nature, social economy and human activities, has a more realistic impact on WESV [25,26]. Many of the above studies have fully considered various factors and provided important references for explaining the changes caused by EVS or WESV. Unfortunately, the collinearity among some influencing factors and the nonstationarity in space have not attracted enough attention.

However, EVS exhibit spatial heterogeneity and spatial dependence with changes in geographic space due to differences in the socioeconomic development degree, natural resources, and geographic environment. Thus, incorporating geospatial aspects into the research scope is the key to addressing spatial heterogeneity. Geographically Weighted Regression (GWR) model is an effective tool for dealing with spatial heterogeneity, which is improved on the basis of ordinary least square [27]. The model incorporates the spatial location information as a coefficient into the regression equation and explores to eliminate the nonstationarity caused by spatial changes based on the fitted values of geographic element parameters [28]. The GWR model has been widely used in the fields of natural resources and ecological environment [29]. The water footprint has been extensively researched, including concepts, methods and applications, for better management of water resources and water ecology [30,31]. With the deterioration of ecological environment problems, it is necessary to analyze the evolution of ecological footprint and the spatial differences of influencing factors from the perspective of spatial heterogeneity [29]. In the related research on land use variation and ESV, the GWR model is used to solve the problem of spatial heterogeneity and compare with the OLS model [32]. In coastal counties of Mississippi and Alabama (U.S.), GWR was used in the estimation of the monetary value of distance to different waterfront types, in the extension to a traditional hedonic pricing method, and in analyzing the value of ecosystem services associated with waterfronts differed geospatially [33]. However, few researches utilize the GWR model to study WESV, which becomes the main content of this study.

In arid and semi-arid regions, water resources are very precious, which means that the ecological services provided by water resources play a vital role. Therefore, on the basis of accurately measuring the value of water resources ecosystem services, analyzing the impact of various factors on WESV is of great significance to effectively manage water resources and alleviate the contradiction between supply and demand of water resources ecosystem services.

Many studies have been conducted on the value of ecological services and their influencing factors. However, this study is more relevant. Specifically, the ecological service value of water resources is the object of this study. In addition, the scope of water resources is broader to include groundwater and surface water. This study serves the ecological compensation policy in China.

In this study, the typical area in the central Loess Plateau of China was taken as the research object, and the evaluation of water resources ecosystem service value and the analysis of the temporal and spatial changes of the influencing factors were carried out. The main works are as follows:

(1) Combining the environmental economics method and the equivalent factor method, the WESVs of 25 counties (districts) in the study area including groundwater and surface water in 2010, 2015 and 2020 were calculated, and the distribution characteristics were analyzed;
(2) Selecting the representative factors of nature, economy and society, and the applicability of OLS model and GWR model in studying the impact of each indicator on WESV was analyzed;
(3) Using the more applicable GWR model, the spatial heterogeneity and spatial and temporal distribution of the effects of various factors on WESV were shown.

2. Materials and Methods

2.1. Study Area

The Loess Plateau (33°43′–41°160′ N, 100°54′–114°33′ E) covers an area of 640,000 km^2 in the upper and middle reaches of China's Yellow River (Figure 1a) [34,35]. Most areas of the Loess Plateau belong to arid and semi-arid areas, with fragile environment, scarcity of water resources and serious soil erosion [36,37]. The study area of this paper is Yulin and Yan'an Cities (Figure 1b), with an area of about 79,957 km^2, accounting for 12.49% of the area of the Loess Plateau. In terms of location, the study area is located in the middle of the Loess Plateau, which is a representative area in the middle of the Loess Plateau. From the perspective of administrative division, the study area belongs to the north of Shaanxi Province. The study area includes 25 county-level administrative units (Figure 1c). The study area with large topographic relief and hilly gully, is the core area for controlling soil and water loss in the Yellow River Basin. The annual average temperature of Yulin City in the north of the study area is about 10.5 °C, and the average annual precipitation is about 400.00 mm. From north to south, the landform gradually transits from sandy land to gullies and hills [38,39]. It is an important energy and chemical base in China. Yan'an City in the south of the study area belongs to the hilly area of the Loess Plateau, which is high in the northwest and low in the southeast. The annual average temperature is about 7.70~10.60 °C, and the annual average precipitation is about 500.00 mm [40–42].

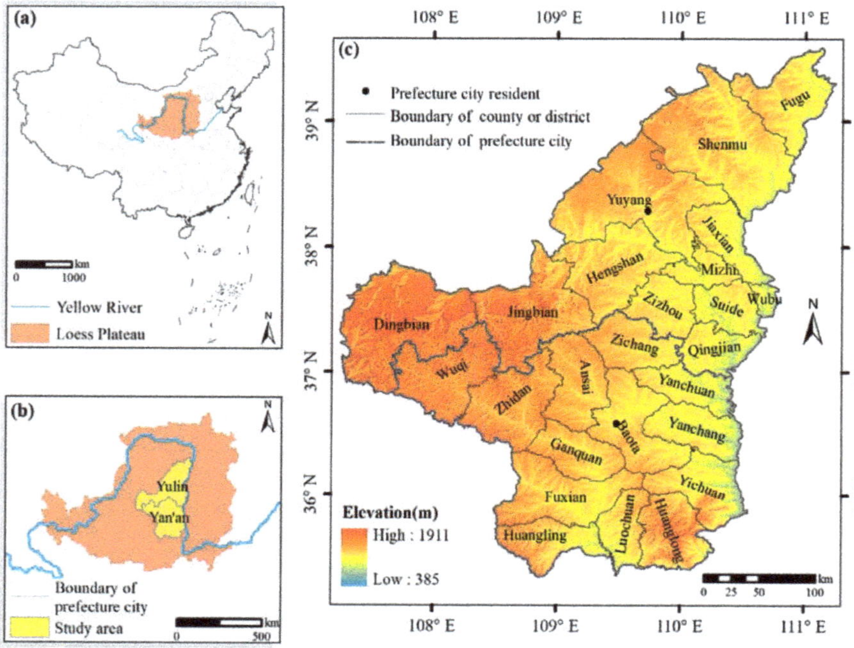

Figure 1. Location and administrative division of the study area: (**a**) Loess Plateau location; (**b**) study area location; (**c**) administrative division of the study area and DEM.

2.2. Data Sources

The data involved in this study include economic indicators, social indicators, and natural indicators of Yulin and Yan'an Cities in 2010, 2015 and 2020. It should be noted that due to the impact of COVID-19, some economic and social indicators in 2020 are replaced by data from 2019, including population, population density, urbanization rate, per capital GDP, and gross product of primary industry.

The water consumption comes from the water resources bulletin of Yulin and Yan'an Cities (2010, 2015 and 2020). Through reclassification, water consumption is divided into three categories: agricultural water consumption, residential water consumption and nonresidential water consumption. The price of water is obtained from the research of the water supply department.

GDP, GDP per capita, population, population density and urbanization rate are from the statistical yearbooks of Yulin and Yan'an (2010, 2015 and 2019).

The water area and the proportion of water area are calculated through the statistical calculation of the spatial distribution data of the national land use type remote sensing monitoring provided by the Resource and Environmental Science and Data Center, Chinese Academy of Sciences (http://www.resdc.cn/, accessed on 2 July 2018). The resolution of the data is 30 m × 30 m, which is generated by manual visual interpretation using the remote sensing images of Landsat TM of various phases of the US Landsat as the main data source.

2.3. Methods

2.3.1. Classification and Calculation Method of WESV

According to the classification of ESs by the Millennium Ecosystem Assessment (MA 2005) carried out by the United Nations [4], combined with the research practice in China [43] and the practicability of ecological compensation in the study area, this study divided WESV into three categories. The reason why the cultural services value was not considered is that at this stage, ecological compensation in China rarely involves such services value. Considering the ways of WES and the availability of data, the three categories were subdivided into 9 specific services (Table 1). The calculation method of WESV can be expressed formally as follows:

$$V_w = \sum_{n=1}^{9} V_n \quad (1)$$

where V_w is the value of WESV; V_n is the n-th sub-category of WESV; $n = 1, 2, \ldots, 9$.

Table 1. Classification of WESV.

Categories of WESV	Work Mode	Content	Symbol
Provisioning Services Value	Direct supply	Water supply	V_1
	Indirect supply	Aquatic product	V_2
		Material	V_3
Regulating Services Value	Direct regulation	Hydrological regulation	V_4
		Water purification	V_5
	Indirect regulation	Climate regulation	V_6
		Air regulation	V_7
Supporting Services Value	Indirect support	Soil conservation	V_8
		Biodiversity	V_9

2.3.2. Market Value Method

Whether groundwater or surface water, the most direct and important service is water supply for living and production. The value of water services is reflected in the water price [44]. Therefore, this study adopts the market value method to calculate the water supply services value of water resources. According to the type of water price, water is divided into agricultural water, residential water and nonresidential water, among which nonresidential water mainly refers to industrial, business service water, administrative institution water, municipal water, etc. The formula for calculation is as follows:

$$V_1 = \sum_{j=1}^{3} Q_j \times P_j, \quad (2)$$

where V_1 is the value of water supply services; Q_j and P_j are the water consumption and water price of the j-th type of water; $j = 1, 2, 3$ represent agricultural water, residential water and nonresidential water, respectively.

The market value method is also used to calculate the indirect supply value of aquatic products provided by water resources. Due to the variety of aquatic products and the different prices, this study uses the fishery output value in the statistical yearbook [45–48] as the value V_2 of aquatic product supply services, and the calculation method also adopts the market value method.

2.3.3. Equivalent Factor Method

Water resources are the most basic elements to maintain the normal operation of ecosystems and social systems, and most of its ecosystem services cannot be measured by market value. The ESV equivalent factor method is obtained from the study of Xie G. [5] by calculating the ecological service value per unit area of each ecosystem in China, which has been widely recognized and applied [49]. The equivalence factor is defined as a relative quantity, which represents a relative value of ecosystem services relative to the economic value of grain output in that year. In the case of the study area, due to the difference in planting structure, the economic value of grain output will change accordingly. The ecological service value of the study area can be calculated by the follows:

$$V_{per} = \frac{1}{7} \times \frac{(Y_1 \times P_1 + Y_2 \times P_2)}{A_1 + A_2}, \quad (3)$$

where V_{per} is the value of ESs per unit area; corn and potato are mainly planted in the study area, Y_1 and Y_2 are the yield of corn and potato, respectively, which comes from "Shaanxi Statistical Yearbook" [45–48]; P_1 and P_2 are the corresponding grain prices, which comes from "Compilation of National Agricultural Product Cost and Benefit Data" [50–52]; A_1 and A_2 are the planting area of corresponding grain; 1/7 is the ratio of the economic value provided by natural ecosystems without artificial inputs to the economic value provided by existing farmland. The calculated values of ESs per unit area in 2010, 2015 and 2020 were 1457.16 CNY/hm^2, 1318.81 CNY/hm^2 and 1199.63 CNY/hm^2, respectively.

The equivalence factor adopts the corresponding part of the water ecosystem in the equivalence coefficient table of ecosystem service value per unit area calculated by Xie G. [43] (Table 2). At the same time, the value of WESV per unit area in 2010, 2015 and 2020 was calculated (Table 2).

Then, the value of services other than Water supply and Aquatic product could be calculated as follows:

$$V_m = V_{per} \times A_w \times E_m, \quad (4)$$

where m = 3, 4, 5, ..., 9; V_m is the value of other services besides Water supply and Aquatic product; A_w represents the watershed area for the study year; E_m is the equivalent coefficient of different service in Table 2.

Table 2. Equivalent coefficients of WESV, and the WESV per unit area in different years.

	Provisioning Services	Regulating Services				Supporting SERVICES	
	Material	Hydrological Regulation	Water Purification	Climate Regulation	Air Regulation	Soil Conservation	Biodiversity
Equivalent coefficient (E_m)	0.23 [1]	102.24 [1]	5.55 [1]	2.29 [1]	0.77 [1]	0.93 [1]	2.55 [1]
2010 WESV per unit area	335.15	148,979.93	8087.23	3336.89	1122.01	1355.16	3715.76
2015 WESV per unit area	1055.04	303.33	134,834.65	7319.37	3020.06	1015.48	1226.49
2020 WESV per unit area	275.92	122,650.62	6657.97	2747.16	923.72	1115.66	3059.07

[1] The equivalent coefficient was quoted from Ref. [43].

2.3.4. Geographically Weighted Regression

The first law of spatial geography shows that the correlation between ground objects gradually increases as the distance decreases [53,54]. Inevitably, spatial correlation and spatial heterogeneity coexist. GWR achieves better results when using local smoothing to deal with the problem of spatial heterogeneity [55,56]. GWR was based on kernel-weighted regression. Instead of estimating global values for regression parameters, GWR allows these parameters to be derived for each location separately [57]. The model can be expressed as

$$y_i = \beta_0(\mu_i, \nu_i) + \sum_k \beta_k(\mu_i, \nu_i)x_{i,k} + \varepsilon_i, \tag{5}$$

where y_i is the explained variable; (μ_i, ν_i) are the coordinates of the target area i; β_0 (μ_i, ν_i) is the intercept; $x_{i,k}$ is the value of the explanatory variable x_k on the target area i; the value of the function β_k (μ_i, ν_i) at geographic location i is β_k (μ_i, ν_i); k is the number of explanatory variables; ε_i is the random disturbance term, i.e., the error. The coefficient of each sample point is a parameter estimate obtained by weighting the adjacent observations [58], and the expression is as follows:

$$\hat{\beta}(u_i, v_i) = \left(X^T W(u_i, v_i) X\right)^{-1} X^T W(u_i, v_i) Y, \tag{6}$$

where $\hat{\beta}(u_i, v_i)$ is the parameter estimate of the local coefficient of the i-th sample with coordinates (μ_i, ν_i); X and Y are the vectors of the explanatory and the dependent variables; $W(u_i, v_i)$ is the weight matrix, which is usually calculated by the Gaussian function of the distance decay function (kernel function).

ArcGIS software has developed the related functions of OLS and GWR into convenient tools. Therefore, this study uses ArcGIS 10.4 to perform related calculations.

3. Results

3.1. WESVs and Characteristics in 2010–2020

3.1.1. Temporal and Spatial Variation Analysis of WESV

Taking the county-level administrative region as the minimum calculation unit, and adopting appropriate methods according to different service types, the WESVs of typical areas in the middle of the Loess Plateau in 2010, 2015 and 2020 were calculated (Table 3). The changes in counties among the three years have also been shown (Figure 2). Overall, the WESV in the study area showed a trend of first decreasing and then increasing. The WESV decreased from 12,370.42 million CNY in 2010 to 11,746.86 million CNY in 2015, which showed a drop range of 5.04%. Then, it increased to 13,379.48 million CNY in 2020. The increasing range was 13.90% from 2015 to 2020, and 8.16% during the entire study period of 2010–2020.

It can be seen from Table 4 that, from 2010 to 2020, some districts and counties in Yan'an City in the southern part of the study area experienced a decrease in WESV, but the trend was rising, with a rate of 3.96%. The leading area that led to the decreasing trend in 2015 was Yulin City. From 2010 to 2015, the WESV in Yulin City decreased by 7.84%, which was significantly greater than the entire study area decreasing rate. Only four counties of Yuyang, Hengshan, Mizhi and Zizhou showed an increase in WESV, and the remaining two-thirds of the counties presented a decreasing trend; the decreasing degree in Fugu reached the highest 30.46%. These phenomena indicated that the ecological service function provided by water resources in Yulin City was in the stage of degradation from 2010 to 2015, which also indirectly reflected the trend of ecological environment deterioration.

From 2015 to 2020, WESV in Yan'an City was still in a growth trend as a whole, and the growth rate increased to 8.56%. Only Yanchang, Ansai and Ganquan experienced negative growth. During the same period, WESV in Yulin City increased rapidly, with an overall increasing rate of 15.77%, and 75% of the counties in the jurisdiction were on the rise, mainly due to the low value of WESV in 2015, which showed that the water ecosystem and its service functions in Yulin City were in a significant recovery situation in 2015–2020.

From the whole study period, the increasing rate of WESV in the southern part of the study area was significantly greater than that in the northern part. In 2020, WESV in Yan'an increased by 12.86% compared with 2010, while Yulin increased by only 6.69% in this decade.

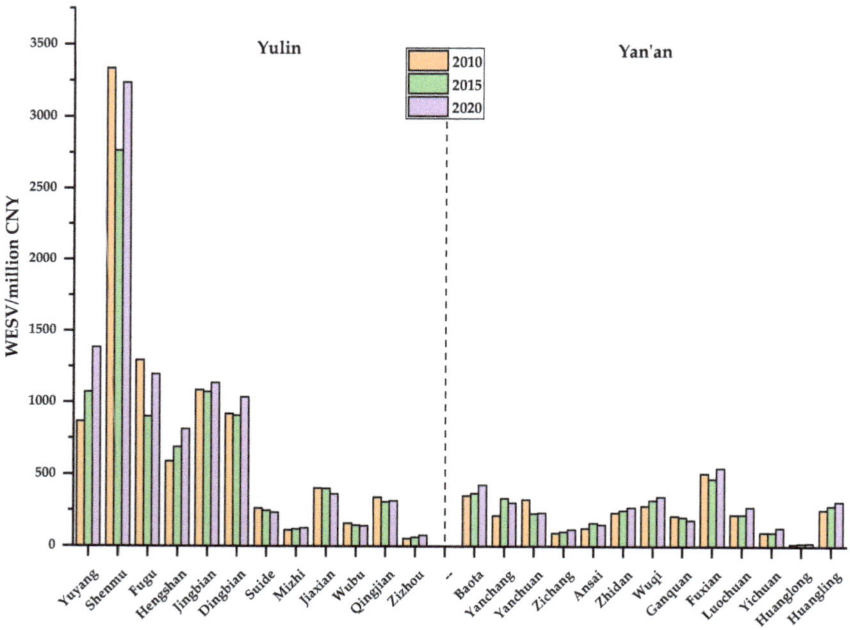

Figure 2. Changes of WESV among the three years in different regions.

Table 3. WESV of the typical area in the central Loess Plateau. unit: million CNY.

City	County	2010	2015	2020
Yulin	Yuyang	866.28	1071.75	1386.48
	Shenmu	3336.85	2765.08	3237.86
	Fugu	1294.82	900.46	1194.77
	Hengshan	589.99	686.95	811.72
	Jingbian	1086.09	1070.41	1133.69
	Dingbian	918.33	905.37	1036.14
	Suide	264.05	249.26	235.08
	Mizhi	114.52	119.55	125.89
	Jiaxian	404.83	401.90	363.29
	Wubu	161.24	147.49	143.38
	Qingjian	340.77	309.77	317.19
	Zizhou	54.41	64.33	77.94
	Subtotal	**9432.18**	**8692.32**	**10,063.43**
Yan'an	Baota	352.40	366.65	427.71
	Yanchang	215.76	333.62	302.61
	Yanchuan	324.42	228.90	233.01
	Zichang	92.63	101.52	117.77
	Ansai	126.53	161.89	152.78
	Zhidan	234.91	251.12	270.24
	Wuqi	281.47	319.77	345.85
	Ganquan	213.12	202.37	183.24
	Fuxian	507.99	468.73	542.46
	Luochuan	221.64	221.65	274.95
	Yichuan	96.23	96.43	129.86
	Huanglong	15.50	21.05	24.65
	Huangling	255.62	280.84	310.91
	Subtotal	**2938.22**	**3054.54**	**3316.04**
Total		**12,370.42**	**11,746.86**	**13,379.48**

Table 4. Structural of WESV in 2010, 2015, 2020.

WESVs	Content	2010		2015		2020	
		WESVs	Proportion	WESVs	Proportion	WESVs	Proportion
Provisioning Services	Water supply	1260.94	10.19%	1963.99	16.72%	3729.11	27.87%
	Aquatic production	45.08	0.36%	122.14	1.04%	188.06	1.41%
	Materials	22.21	0.18%	19.40	0.17%	19.00	0.14%
Regulating Services	Hydrological regulation	9874.52	79.83%	8621.80	73.40%	8444.72	63.12%
	Water Purification	536.03	4.33%	468.03	3.98%	458.41	3.43%
	Climate Regulation	221.17	1.79%	193.11	1.64%	189.15	1.41%
	Air Regulation	74.37	0.60%	64.93	0.55%	63.60	0.48%
Supporting Services	Soil Conservation	89.82	0.73%	78.43	0.67%	76.82	0.57%
	Biodiversity	246.28	1.99%	215.04	1.83%	210.62	1.57%

3.1.2. Variation Analysis of WESV Structural Characteristics

Table 4 and Figure 3 exhibited the changes in the value and proportion of nine subservices that comprise WESV from 2010 to 2020. Among the three first-level classifications of WESV, Regulating Services accounted for the largest proportion, but the proportion gradually decreased from 86.55% in 2010 to 68.44% in 2020. The Provisioning Services proportion has increased year-by-year, from 10.73% to 29.42% in the past decade.

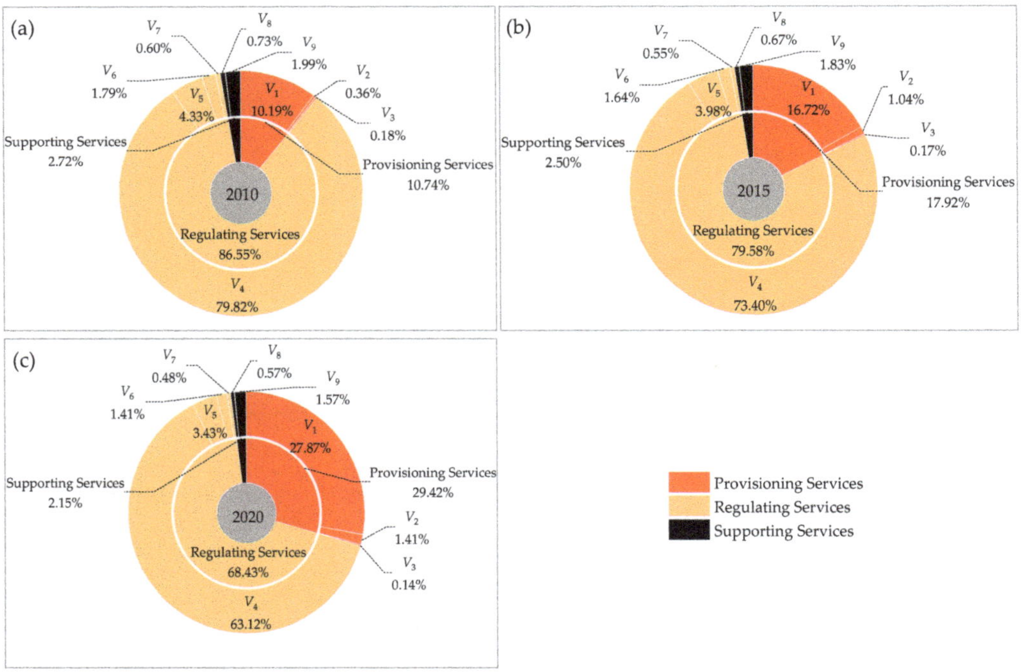

Figure 3. Structural changes of WESV in the typical area of the central Loess Plateau: (**a**) the proportion of various WESV in 2010; (**b**) the proportion of various WESV in 2015 and (**c**) the proportion of various WESV in 2020.

The Supporting Services proportion has decreased steadily, from 2.72% in 2010 to 2.14% in 2020.

In the subcategories, only the proportions of Water Supply and Food Production have increased year-by-year, and the proportions of other ESV have decreased to varying degrees, which showed that the development of economy and society has caused a sharp increase in the demand for water resources and aquatic products, and at the same time, the regulating function of water resources has gradually weakened.

3.2. Analysis of Spatial Heterogeneity Characteristics of Each Impact Factor

3.2.1. Identification of Influencing Factors

This paper selected 12 influencing factors from two aspects of society-economy, and natural environment. The values of NDVI, Forest and grass area, Forest and grass coverage ratio, and Proportion of water area were directly or indirectly obtained from the Resource and Environmental Science and Data Center, Chinese Academy of Sciences (http://www.resdc.cn/, accessed on 2 July 2018). All other influence factors were obtained from the statistical yearbook [45–48]. For the three different years of 2010, 2015 and 2020, the collinearity investigation of influencing factors and the significance test between them and the explained variables were carried out respectively, which is the first step in building a model. The specific work was to construct models for different parameter combinations by the OLS model in Arcgis 10.4. Then, parameters, that did not meet the criteria for variance inflation factor (VIF) and significance tests were excluded one by one gradually. The results (Table 5) showed that in the three years, only four influencing factors satisfy the two conditions of being VIF < 7.5 and significant at the same time. Therefore, we selected per capita GDP, population density, the proportion of water areas, and water consumption as the research objects.

Table 5. Statistical test results and the selection of influencing factors.

Classification	Influence Factor	Unit	Statistical Test Results
Social and economy	Population	10^4	VIF > 7.5
	Population density	person/km^2	VIF < 7.5, significant
	Urbanization rate	%	VIF < 7.5, no significant
	Per capital GDP	yuan	VIF < 7.5, significant
	Gross product of primary industry	yuan	VIF > 7.5
	Water consumption	10^4 m^3	VIF < 7.5, significant
Natural environment	Rainfall	mm	VIF > 7.5
	NDVI	/	VIF > 7.5
	Forest and grass area	hm^2	VIF > 7.5
	Forest and grass coverage ratio	%	VIF < 7.5, no significant
	Proportion of water area	‰	VIF < 7.5, significant
	Area of soil erosion control	hm^2	VIF < 7.5, no significant

3.2.2. Effect of the GWR Model

In order to illustrate the applicability of GWR, this study first used the OLS model to simulate. As shown in (Table 6), the R^2 and Adjusted R^2 of the GWR model in the three years are larger than those of the OLS model, indicating that the simulation effect is more accurate and representative. Meanwhile, the AICc of the GWR model is smaller than that of the OLS model, and the difference is greater than 3.0, which demonstrated that the GWR model is more applicable. Therefore, the GWR model can accurately explain the relationship between WESV and each explanatory variable and the spatial heterogeneity.

Table 6. Statistical test of OLS and GWR in 2010, 2015, and 2020.

	2010		2015		2020	
	OLS	**GWR**	**OLS**	**GWR**	**OLS**	**GWR**
R^2	0.758	0.856	0.792	0.953	0.850	0.906
Adjusted R^2	0.709	0.789	0.751	0.915	0.820	0.865
AICc	377.181	373.591	364.539	351.095	365.338	362.201

3.2.3. Per Capita GDP

Figure 4 displayed the spatiotemporal distribution of the per capita GDP impact on WESV. The darker the color, the greater the positive effect. In 2010 (Figure 4a) and 2020 (Figure 4c), WESV and Per Capita GDP were positively correlated. The magnitude of the coefficient indicated that, in general, the 2020 Per Capita GDP had a stronger impact on WESV. In 2010, the correlation relationship gradually increased from west to east, with Huanglong, Huangling and Yanchang in the southeast, Wubao in the east and Fugu in the northeast being the most influential. In 2020, the impact of Per Capita GDP on WESV gradually increased from southwest to northeast, and WESV in Fugu, Shenmu and Jia County was more sensitive to Per Capita GDP. Because Yulin is the base of energy and chemical industry in China, the economy had developed rapidly after 2008, and the demand for water had increased. The combined effect of the two led to changes in the spatial distribution from 2010 to 2020. Nevertheless, in 2015 (Figure 4b), the WESV of 11 districts and counties in the northern and central parts of the study area showed a negative correlation with Per Capita GDP, and the negative correlation effect was strongest in the northeast. This illustrated that during the period from 2010 to 2015, the protection of water resources and water environment in the northern region was neglected due to the great economic development, which was consistent with the reality of the reduction of the water area in this region. It also showed that from 2015 to 2020, the water resources condition in the northern part of the study area and the ecological services provided had been greatly restored.

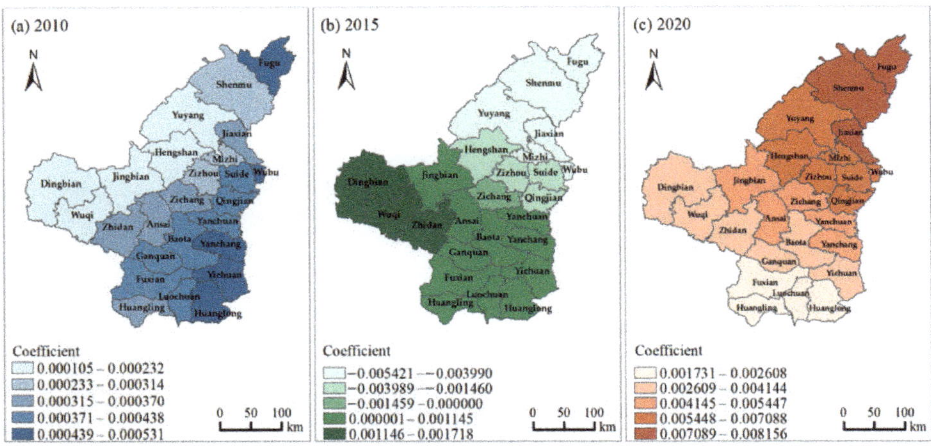

Figure 4. Temporal and spatial distribution of the impact of Per Capita GDP on WESV in different years: (**a**) 2010; (**b**) 2015; and (**c**) 2020.

3.2.4. Population Density

In a certain area, population density determines the demand for WES. Figure 5 displayed the spatiotemporal distribution of WESV response to Population Density. Throughout the study period, WESV exhibited a negative correlation with population density, and the negative correlation gradually weakened with time. From 2010 to 2015, the more northerly the geographical location is, the more sensitive WESV is to changes in population density. Compared with Figure 5a,b, although the overall negative correlation was slightly enhanced, the area of region with the lowest level of negative correlation was increased. As shown in Figure 5c, the spatial distribution pattern of the impact of population density on WESV had fundamentally changed in 2020, and the coefficient increased from the southeast to the outside. The larger the coefficient, the weaker the negative correlation.

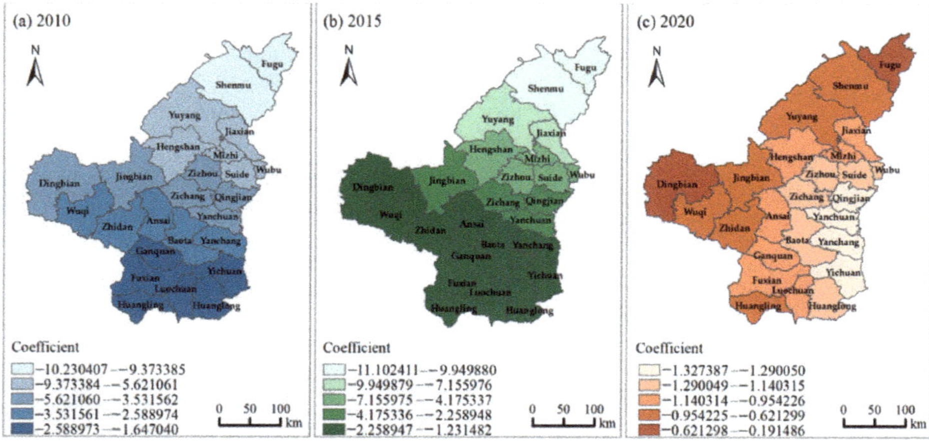

Figure 5. Temporal and spatial distribution of the impact of Population Density on WESV in different years: (**a**) 2010; (**b**) 2015 and (**c**) 2020.

3.2.5. Proportion of Water Areas

WESV is largely determined by the water areas. The spatial distribution pattern of the proportion of water areas affecting WESV and the temporal change of this distribution pattern were analyzed (Figure 6). Looking at the entire study area, although the pattern distribution has changed, all coefficients showed that the proportion of water area has a positive impact on WESV. As shown in Figure 6c, the impact pattern in 2010 is that the degree of impact decreased from north to south. In 2015, the impact pattern evolved into that population density had the lowest impact on the southeastern region, and gradually increased in the west and north directions. In 2020, the gravity center of the impact continued to shift eastward, showing the phenomenon that the west was large and the east was small. The two counties with the least impact appeared in Shenmu and Fugu in the northeast, and these two districts and counties belonged to the most affected areas in 2010 and 2015.

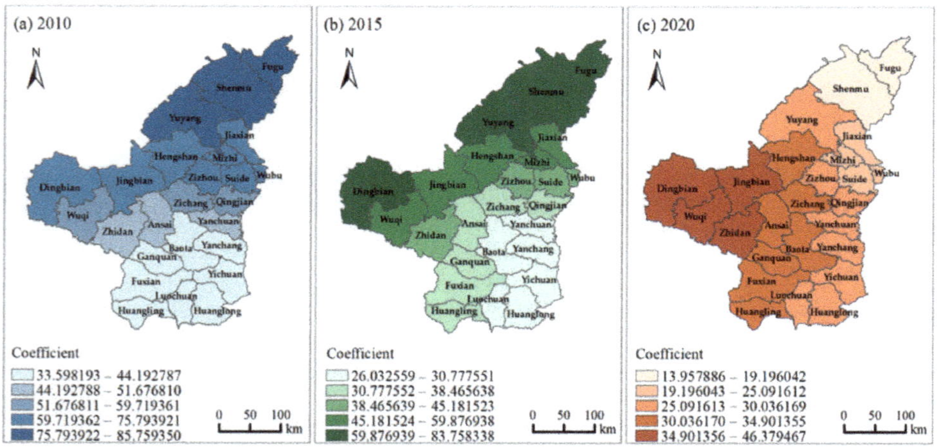

Figure 6. Temporal and spatial distribution of the impact of Proportion of Water Areas on WESV in different years: (**a**) 2010; (**b**) 2015 and (**c**) 2020.

3.2.6. Water Consumption

Water consumption is a key factor affecting the service value of Water Supply, and it is also the most direct manifestation of the ecological service value of water resources. Figure 7 exhibited the change in the spatiotemporal distribution of water consumption affecting the degree of WESV and showed that there is a positive correlation between water consumption and WESV. In 2010, the correlation coefficient between WESV and water consumption in the study area showed an increasing state from west to east. In 2015, the degree of influence showed a circular increase from the western to the eastern region (Figure 7b). The increase in the value range of the coefficient indicated that the differences between districts and counties were expanding. WESV in the southern region was gradually affected by changes in water consumption. With the passage of time, in 2020 (Figure 7c), the spatial distribution of WESV affected by water consumption had a strong regularity, decreasing in a stepwise manner from south to north. However, the coefficient ranges were less discrete.

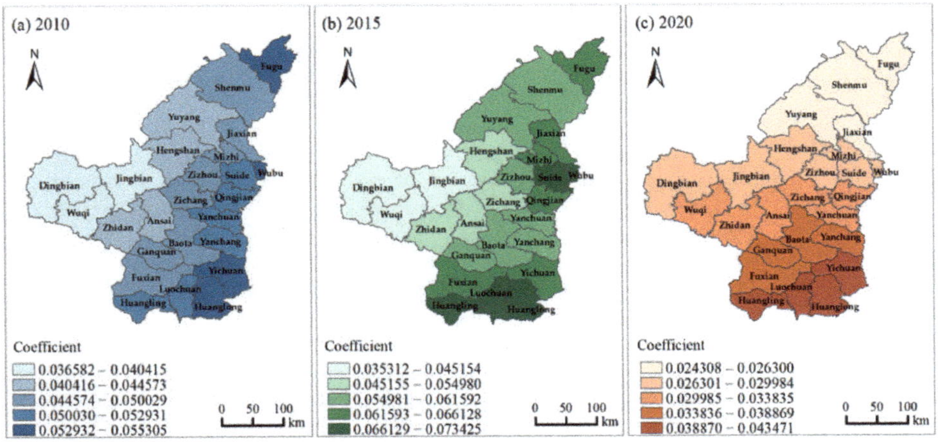

Figure 7. Temporal and spatial distribution of the impact of Water Consumption on WESV in different years: (**a**) 2010; (**b**) 2015 and (**c**) 2020.

4. Discussion

4.1. Necessity to Assessing WESV

The significance of evaluating the value of ecosystem services is to better manage the ecological environment and natural resources to achieve sustainable development. The water issue is prominent today, so it is more practical to discuss the value of water resources ecosystem services. In China, the value of ecosystem services is always in the form of the upper limit of the ecological protection compensation standard. The water ecosystem services value is often calculated as part of a comprehensive ecosystem service value and is rarely discussed in isolation. Even in the basin ecological compensation, only the fluctuation of the direct use value caused by the change of water quantity is calculated, and how the ecosystem service value of water resources including groundwater changes is not fully explored. In 2019, the ecological protection and high-quality development of the Yellow River Basin was established as one of China's major national strategies. Ecological compensation is one of the key tasks of the strategy. The Loess Plateau, located in the middle reaches of the Yellow River, is the main source of sediment in the Yellow River. The region has scarce water resources, a large population, and an urgent need for development. Hence, from the perspective of the integrity of water resources, this study selects typical regions to assess WESV and analyzes the temporal and spatial variation characteristics of WESV, which can provide a basis for the ecological compensation development.

4.2. The Spatiotemporal Distribution of WESV Response to Different Influencing Factors

Previous studies have shown that the value of water resources is affected by multiple factors, such as water quantity, water quality, use of water resources, economic development, and educational level of residents [59,60].

With the deepening of relevant research, the water resources value has been extended to the value of ecosystem services that water resources can provide for human well-being. Correspondingly, WSVE is also affected by economic, social, natural and cultural factors. Many researches have explored the mechanism by which ESV is affected by various driving factors, including land use change, socioeconomic development indicators, and human acceptance willingness. Furthermore, the distribution characteristics of the sensitivity of ESV to various influencing factors at different temporal and spatial scales were obtained. WESV is an important part of ESV and is also affected by various factors. As mentioned above, the extent to which WESV is affected by external factors should be studied separately.

With the rapid economic development and the sharp expansion of cities, the differences between different regions in natural resources, economy, society and culture are gradually increasing. In this study, the central part of the Loess Plateau was selected as the research area, and four main influencing factors were selected. The GWR model was used to analyze the influence of per capita GDP, population density, the proportion of water areas, and water consumption on WESV in different regions.

Per Capita GDP is an important parameter to measure the development degree of a region. The rise in prices and the massive consumption of water resources brought about by economic development will lead to an increase in WESV. Meanwhile, WESV will greatly decreased because of the ecological degradation caused by development, that is, the reduction in water area. Therefore, the relationship between water conservation and economic development should be balanced. The effect of per capita GDP on ESV was confirmed by Song F. [61] in a study on the value of wetland ecosystem services. Dai X. concluded that per capita GDP was negatively correlated with ESV in Chengdu [62]. As can be seen from Figure 4, per capita GDP showed time instability. The study area has undergone a process of environmental damage and recovery during a 10-year development period, which is consistent with the process in China. During 2015–2020, the construction of ecological civilization has been raised to a new level. At last, a pattern in which the degree of economic development roughly matches WESV is formed. Many studies have found that per capita GDP has an unstable impact on ESV, by influencing the level of awareness, willingness to protect the environment, and investment in environmental protection [63–65].

It is generally believed that the increase in population density is accompanied by the expansion of human activity areas, which will encroach on other land-use types. The water area is also experiencing depletion of rivers and lakes due to the massive depletion of surface water while the water area was occupied. Under the dual effect of the two, WESV is bound to decrease, which also explains the negative correlation between population density and WESV. It was confirmed in Chen Y.'s study of the relationship between population density and agro-ecosystem services [66]. Jiang Z. also obtained a consistent conclusion in his study of the South Four Lakes that population density was negatively correlated with Esv [67]. In recent years, the population growth rate in the study area has slowed down as in the whole country. WESV is less sensitive to population density.

In the same region, WESV will increase with the increase in water area, which can also be explained by the calculation method adopted in this study. The equivalent factor method is to calculate the corresponding ESV according to the type of land use. Figure 6 showed that the proportion of water area has the greatest influence on WESV. During the initial stage of the study, the drier regions in the north were more sensitive. With the comprehensive management of the Mu Us Desert, the environment in the northern region has been gradually improved, and the scarcity of water resources and the vulnerability of the water environment have been alleviated. The center of influence then transferred to the west, where the pace of governance was relatively slow.

In 2010 and 2015, the impact of water consumption on WESV had a certain regularity in space, but the spatial instability that occurs cannot be ignored. Figure 7a,b can well demonstrate this phenomenon. From 2015 to 2020, the state strictly controlled the regional water consumption, and the strictest water resources management system was successfully implemented locally [68], which well explained the phenomenon that the spatial distribution pattern in 2020 was more regular.

There is spatial heterogeneity in the sensitivity of WESV to explanatory variables. Using the GWR model can more clearly show the spatial pattern of the influence degree and its evolution trend over time.

4.3. Limitations

Based on the water area obtained from remote sensing image statistics and the water consumption obtained from statistical data, this study proposed a method to quickly calculate WESV and analyze the temporal and spatial distribution of driving factors. However, the impact of water quality on WESV does not depend on water consumption or water area. Deterioration of water quality not only reduces the quality of services provided by the water resource but also incurs additional remediation costs [69]. This study did not include water quality as the basic data in the WESV assessment for three reasons. First, different types of WESVs have different sensitivity to water quality, and its mechanism of action and degree of influence are not clear, which will cause uncertainty in the calculation of WESV. Second, the equivalent factor method used in this paper is calculated from the ecological service value and the economic value of grain, and the output of grain is affected by water quality. That is to say, this study considered the impact of water quality on WESV to a certain extent. Third, the spatial and temporal scales of this study determined that water quality will not have an essential impact on WESV. In space, the county-level administrative region is the smallest research unit; in terms of time, a year is the minimum span of time. From this point of view, the water quality in the region is relatively stable, and this is also verified by the water quality data released by the environmental department. Therefore, the WESV calculated in this study is still representative and reliable without considering water quality.

5. Conclusions

In this study, taking a typical area in the central Loess Plateau as the study area, WESV was explored and assessed, and the spatiotemporal relationship between WESV and driving factors was analyzed. The conclusions obtained are as follows:

(1) Considering surface water and groundwater as a whole, a WESV calculation method based on multi-method fusion of multiple data sources such as remote sensing images is proposed.

(2) In total, the WESV in 2020 is 8.16% higher than that in 2010. However, during the study period, the WESV in the typical area of the central Loess Plateau experienced a process of first decreasing and then increasing. The main factor leading to this phenomenon was that the WESV of Yulin City in the north of the study area decreased by 7.84% in 2015 compared with 2010. From the perspective of the WESV composition structure, the proportion of regulating services has decreased by 18.11% in the past 10 years, but its proportion is still the largest; the proportion of provisioning services has increased year by year to 29.42%; the proportion of supporting services has decreased steadily, from 2.72% in 2010 to 2.14% in 2020.

(3) Considering the spatial heterogeneity, the GWR model has better applicability. Among the four influencing factors, the proportion of water area and water consumption showed a positive correlation with WESV throughout the study period. Population density is negatively correlated with WESV. Per capita GDP was positively correlated with WESV in both 2010 and 2020. In 2015, there was a negative correlation between per capita GDP and WESV in the northern and western districts and counties of the study area.

(4) With the passage of time, the spatial distribution pattern of the influence of the four factors on WESV has changed, and the evolution directions are different. The response degree of WESV to per capita GDP has evolved gradually from low in the west and high in the east to low in the southwest and high in the northeast, and the laddering nature is stronger. The water area plays a leading role in the size of WESV, and the center of gravity of the influence of the water area proportion shifts from the north to the west, with a strong decreasing law. The distribution of the influence of water consumption on WESV has experienced three evolutionary stages: small in the west and large in the east, increases in a circular shape from the west to the outside, small in the north and large in the south.

Author Contributions: Conceptualization, N.W. and Y.X.; methodology, Y.X. and N.W.; data curation, Q.W.; software, Y.X. and F.P.; validation, F.P. and Q.W.; analysis, Y.X. and N.W.; investigation, Y.X.; writing—original draft preparation, Y.X.; writing—review and editing, N.W. and Y.X.; visualization, F.P.; supervision, Q.W.; project administration, N.W.; funding acquisition, N.W. and Y.X. All authors have read and agreed to the published version of the manuscript.

Funding: This work was funded by the National Natural Science Foundation of China, Grant No. 51979221, the Natural Science Basic Research Program of Shaanxi, Program No. 2021JLM-45, and the Research Fund of the State Key Laboratory of Eco-hydraulics in Northwest Arid Region, Xi'an University of Technology, Grant No. 2019KJCXTD-5.

Institutional Review Board Statement: Not applicable.

Informed Consent Statement: Not applicable.

Data Availability Statement: Not applicable.

Acknowledgments: Sincerely, we gratefully acknowledge the State Key Laboratory of Eco-hydraulics in Northwest Arid Region, Xi'an University of Technology, for providing support. In addition, thanks to Le Zhang for her help.

Conflicts of Interest: The authors declare no conflict of interest.

References

1. He, Y.; Wang, W.; Chen, Y.; Yan, H. Assessing spatio-temporal patterns and driving force of ecosystem service value in the main urban area of Guangzhou. *Sci. Rep.* **2021**, *11*, 3027. [CrossRef] [PubMed]
2. Ahmad, M.; Akanji, M.A.; Usman, A.R.A.; Al-Farraj, A.S.F.; Tsang, Y.F.; Al-Wabel, M.I. Turning date palm waste into carbon nanodots and nano zerovalent iron composites for excellent removal of methylthioninium chloride from water. *Sci. Rep.* **2020**, *10*, 16125. [CrossRef] [PubMed]
3. Costanza, R.; d'Arge, R.; de Groot, R.; Farber, S.; Grasso, M.; Hannon, B.; Limburg, K.; Naeem, S.; O'Neill, R.V.; Paruelo, J.; et al. The value of the world's ecosystem services and natural capital. *Nature* **1997**, *387*, 253–260. [CrossRef]
4. Millennium Ecosystem Assessments. *Ecosystems and Human Well-Being: A Framework for Assessment*; Island Press: Washington, DC, USA, 2003; pp. 50–60.
5. Xie, G.; Lu, C.; Leng, Y.; Zheng, D.; Li, S. Ecological assets valuation of the Tibetan Plateau. *J. Nat. Resour.* **2003**, *18*, 189–196.
6. Xie, G.; Zhang, C.; Zhang, L.; Chen, W.; Li, S. Improvement of the Evaluation Method for Ecosystem Service Value Based on Per Unit Area. *J. Nat. Resour.* **2015**, *30*, 1243–1254.
7. Xiu, Y.; Wang, N.; Xie, J.; Ke, X. Improvement of the Ecological Protection Compensation Policy for Adjustment of Planting Structure in an Area of Groundwater Overexploitation: A Tripartite Evolutionary Game Study. *Pol. J. Environ. Stud.* **2022**, *31*, 1399–1414. [CrossRef]
8. Talukdar, S.; Singha, P.; Shahfahad; Mahato, S.; Praveen, B.; Rahman, A. Dynamics of ecosystem services (ESs) in response to land use land cover (LU/LC) changes in the lower Gangetic plain of India. *Ecol. Indic.* **2020**, *112*, 106121. [CrossRef]
9. Flotemersch, J.E.; Shattuck, S.M.; Aho, K.B.; Cox, C.E.; Cairns, M.R. Factors influencing social demands of aquatic ecosystems. *Ecol. Soc.* **2019**, *24*, 9. [CrossRef]
10. Liu, Z.T.; Wu, R.; Chen, Y.X.; Fang, C.L.; Wang, S.J. Factors of ecosystem service values in a fast-developing region in China: Insights from the joint impacts of human activities and natural conditions. *J. Clean. Prod.* **2021**, *297*, 126588. [CrossRef]
11. Liu, W.; Zhan, J.Y.; Zhao, F.; Yan, H.M.; Zhang, F.; Wei, X.Q. Impacts of urbanization-induced land-use changes on ecosystem services: A case study of the Pearl River Delta Metropolitan Region, China. *Ecol. Indic.* **2019**, *98*, 228–238. [CrossRef]
12. Wang, J.L.; Zhou, W.Q.; Pickett, S.T.A.; Yu, W.J.; Li, W.F. A multiscale analysis of urbanization effects on ecosystem services supply in an urban megaregion. *Sci. Total Environ.* **2019**, *662*, 824–833. [CrossRef] [PubMed]

13. Zhou, D.Y.; Tian, Y.Y.; Jiang, G.H. Spatio-temporal investigation of the interactive relationship between urbanization and ecosystem services: Case study of the Jingjinji urban agglomeration, China. *Ecol. Indic.* **2018**, *95*, 152–164. [CrossRef]
14. Hu, S.; Chen, L.Q.; Li, L.; Wang, B.Y.; Yuan, L.N.; Cheng, L.; Yu, Z.Q.; Zhang, T. Spatiotemporal Dynamics of Ecosystem Service Value Determined by Land-Use Changes in the Urbanization of Anhui Province, China. *Int. J. Environ. Res. Public Health* **2019**, *16*, 5104. [CrossRef] [PubMed]
15. Mao, Y.Y.; Hou, L.L.; Zhang, Z.D. Spatial-Temporal Evolution and Relationship between Urbanization Level and Ecosystem Service from a Dual-Scale Perspective: A Case Study of the Pearl River Delta Urban Agglomeration. *Sustainability* **2021**, *13*, 8537. [CrossRef]
16. Peng, L.; Wang, X.X. What is the relationship between ecosystem services and urbanization? A case study of the mountainous areas in Southwest China. *J. Mt. Sci.* **2019**, *16*, 2867–2881. [CrossRef]
17. Yang, Y.Q.; Jun, Z.; Sui, X.; He, X. Study of the spatial connection between urbanization and the ecosystem-A case study of Central Yunnan (China). *PLoS ONE* **2020**, *15*, e0238192. [CrossRef]
18. Li, R.; Shi, Y.; Feng, C.C.; Guo, L. The spatial relationship between ecosystem service scarcity value and urbanization from the perspective of heterogeneity in typical arid and semiarid regions of China. *Ecol. Indic.* **2021**, *132*, 108299. [CrossRef]
19. Shi, L.; Halik, U.; Mamat, Z.; Aishan, T.; Abliz, A.; Welp, M. Spatiotemporal investigation of the interactive coercing relationship between urbanization and ecosystem services in arid northwestern China. *LDD* **2021**, *32*, 4105–4120. [CrossRef]
20. Fang, J.H.; Song, H.L.; Zhang, Y.R.; Li, Y.R.; Liu, J. Climate-dependence of ecosystem services in a nature reserve in northern China. *PLoS ONE* **2018**, *13*, e0192727. [CrossRef]
21. Cheng, B.; Li, H.E. Impact of climate change and human activities on economic values produced by ecosystem service functions of rivers in water shortage area of Northwest China. *Environ. Sci. Pollut. Res.* **2020**, *27*, 26570–26578. [CrossRef]
22. Watson, L.; Straatsma, M.W.; Wanders, N.; Verstegen, J.A.; de Jong, S.M.; Karssenberg, D. Global ecosystem service values in climate class transitions. *Environ. Res. Lett.* **2020**, *15*, 024008. [CrossRef]
23. Schild, J.E.; Vermaat, J.E.; van Bodegom, P.M. Differential effects of valuation method and ecosystem type on the monetary valuation of dryland ecosystem services: A quantitative analysis. *J. Arid Environ.* **2018**, *159*, 11–21. [CrossRef]
24. Zhao, Q.J.; Wang, Q.Y. Water Ecosystem Service Quality Evaluation and Value Assessment of Taihu Lake in China. *Water* **2021**, *13*, 618. [CrossRef]
25. Pettinotti, L.; de Ayala, A.; Ojea, E. Benefits From Water Related Ecosystem Services in Africa and Climate Change. *Ecol. Econ.* **2018**, *149*, 294–305. [CrossRef]
26. Li, Z.H.; Xia, J.; Deng, X.Z.; Yan, H.M. Multilevel modelling of impacts of human and natural factors on ecosystem services change in an oasis, Northwest China. *Resour. Conserv. Recycl.* **2021**, *169*, 105474. [CrossRef]
27. Hong, Z.M.; Hao, H.; Li, C.Y.; Du, W.; Wei, L.D.; Wang, H.H. Exploration of potential risks of Hand, Foot, and Mouth Disease in Inner Mongolia Autonomous Region, China Using Geographically Weighted Regression Model. *Sci. Rep.* **2018**, *8*, 17707. [CrossRef]
28. Cardozo, O.D.; García-Palomares, J.C.; Gutiérrez, J. Application of geographically weighted regression to the direct forecasting of transit ridership at station-level. *Appl. Geogr.* **2012**, *34*, 548–558. [CrossRef]
29. Wu, D.C. Spatially and temporally varying relationships between ecological footprint and influencing factors in China's provinces Using Geographically Weighted Regression (GWR). *J. Clean. Prod.* **2020**, *261*, 121089. [CrossRef]
30. Muratoglu, A. Applications and Response Formulations of Water Footprint Methodology for Conservation of Water Resources. *Ref. Module Earth Syst. Environ. Sci.* **2021**. [CrossRef]
31. Muratoglu, A. Water Footprint: Concept and Methodology. *Ref. Module Earth Syst. Environ. Sci.* **2021**. [CrossRef]
32. Shao, Y.J.; Yuan, X.F.; Ma, C.Q.; Ma, R.F.; Ren, Z.X. Quantifying the Spatial Association between Land Use Change and Ecosystem Services Value: A Case Study in Xi'an, China. *Sustainability* **2020**, *12*, 4449. [CrossRef]
33. Dahal, R.P.; Grala, R.K.; Gordon, J.S.; Munn, I.A.; Petrolia, D.R. Geospatial Heterogeneity in Monetary Value of Proximity to Waterfront Ecosystem Services in the Gulf of Mexico. *Water* **2021**, *13*, 2401. [CrossRef]
34. Ren, Y.; Lü, Y.; Fu, B.; Comber, A.; Li, T.; Hu, J. Driving Factors of Land Change in China's Loess Plateau: Quantification Using Geographically Weighted Regression and Management Implications. *Remote Sens.* **2020**, *12*, 453. [CrossRef]
35. Wang, Y.; Wu, G.; Deng, L.; Tang, Z.; Wang, K.; Sun, W.; Shangguan, Z. Prediction of aboveground grassland biomass on the Loess Plateau, China, using a random forest algorithm. *Sci. Rep.* **2017**, *7*, 6940. [CrossRef]
36. Zhao, G.J.; Mu, X.M.; Wen, Z.M.; Wang, F.; Gao, P. Soil erosion, conservation, and eco-environment changes in the loess plateau of china. *LDD Land Degrad. Dev.* **2013**, *24*, 499–510. [CrossRef]
37. Fu, B.J.; Wang, S.; Liu, Y.; Liu, J.B.; Liang, W.; Miao, C.Y. Hydrogeomorphic Ecosystem Responses to Natural and Anthropogenic Changes in the Loess Plateau of China. *Annu. Rev. Earth Planet. Sci.* **2017**, *45*, 223–243. [CrossRef]
38. Chen, J.; Yang, X.; Yin, S.; Wu, K.; Deng, M.; Wen, X. The vulnerability evolution and simulation of social-ecological systems in a semi-arid area: A case study of Yulin City, China. *J. Geogr. Sci.* **2018**, *28*, 152–174. [CrossRef]
39. Luo, M.; Li, T. Spatial and temporal analysis of landscape ecological quality in Yulin. *Environ. Technol. Innov.* **2021**, *23*, 101700. [CrossRef]
40. Zhang, D.; Jia, Q.; Wang, P.; Zhang, J.; Hou, X.; Li, X.; Li, W. Analysis of spatial variability in factors contributing to vegetation restoration in Yan'an, China. *Ecol. Indic.* **2020**, *113*, 106278. [CrossRef]

41. Han, L.; Zhu, H.; Zhao, Y.; Liu, Z. Analysis of variation in river sediment characteristics and influential factors in Yan'an City, China. *Environ. Earth Sci.* **2018**, *77*, 479. [CrossRef]
42. Hou, K.; Li, X.; Wang, J.J.; Zhang, J. An analysis of the impact on land use and ecological vulnerability of the policy of returning farmland to forest in Yan'an, China. *Environ. Sci. Pollut. Res.* **2016**, *23*, 4670–4680. [CrossRef] [PubMed]
43. Xie, G.; Zhang, C.; Zhen, L.; Zhang, L. Dynamic changes in the value of China's ecosystem services. *Ecosyst. Serv.* **2017**, *26*, 146–154. [CrossRef]
44. Seidl, C.; Wheeler, S.A.; Zuo, A. High turbidity: Water valuation and accounting in the Murray-Darling Basin. *Agric. Water Manag.* **2020**, *230*, 105929. [CrossRef]
45. Shaanxi Provincial Bureau of Statistics. *Shaanxi Statistical Yearbook 2010*; China Statistics Press: Beijing, China, 2010.
46. Shaanxi Provincial Bureau of Statistics. *Shaanxi Statistical Yearbook 2015*; China Statistics Press: Beijing, China, 2015.
47. Shaanxi Provincial Bureau of Statistics. *Shaanxi Statistical Yearbook 2019*; China Statistics Press: Beijing, China, 2019.
48. Shaanxi Provincial Bureau of Statistics. *Shaanxi Statistical Yearbook 2020*; China Statistics Press: Beijing, China, 2020.
49. Ma, Q.; Li, Y.; Xu, L. Identification of green infrastructure networks based on ecosystem services in a rapidly urbanizing area. *J. Clean. Prod.* **2021**, *300*, 126945. [CrossRef]
50. Price Department of National Development and Reform Commission. *Compilation of National Agricultural Product Cost-Benefit Data 2011*; China Statistics Press: Beijing, China, 2011.
51. Price Department of National Development and Reform Commission. *Compilation of National Agricultural Product Cost-Benefit Data 2016*; China Statistics Press: Beijing, China, 2016.
52. Price Department of National Development and Reform Commission. *Compilation of National Agricultural Product Cost-Benefit Data 2020*; China Statistics Press: Beijing, China, 2020.
53. Tobler, W.R. A Computer Movie Simulating Urban Growth in the Detroit Region. *Econ. Geogr.* **1970**, *46*, 234–240. [CrossRef]
54. Russ, T.C.; Batty, G.D.; Hearnshaw, G.F.; Fenton, C.; Starr, J.M. Geographical variation in dementia: Systematic review with meta-analysis. *Int. J. Epidemiol.* **2012**, *41*, 1012–1032. [CrossRef] [PubMed]
55. Fotheringham, A.S.; Charlton, M.; Brunsdon, C. The geography of parameter space: An investigation of spatial non-stationarity. *Int. J. Geogr. Inf. Sci. Syst.* **1996**, *10*, 605–627. [CrossRef]
56. Pourmohammadi, P.; Strager, M.P.; Dougherty, M.J.; Adjeroh, D.A. Analysis of Land Development Drivers Using Geographically Weighted Ridge Regression. *Remote Sens.* **2021**, *13*, 1307. [CrossRef]
57. Oshan, T.M.; Smith, J.P.; Fotheringham, A.S. Targeting the spatial context of obesity determinants via multiscale geographically weighted regression. *Int. J. Health Geogr.* **2020**, *19*, 11. [CrossRef]
58. Wang, M.; He, G.; Zhang, Z.; Wang, G.; Zhang, Z.; Cao, X.; Wu, Z.; Liu, X. Comparison of Spatial Interpolation and Regression Analysis Models for an Estimation of Monthly Near Surface Air Temperature in China. *Remote Sens.* **2017**, *9*, 1278. [CrossRef]
59. Wu, Z.N.; Di, D.Y.; Lv, C.M.; Guo, X.; Wang, H.L. Defining and evaluating the social value of regional water resources in terms of emergy. *Water Policy* **2019**, *21*, 73–90. [CrossRef]
60. Chen, J.F.; Wang, Q.; Li, Q. A Quantitative Assessment on Ecological Compensation Based on Water Resources Value Accounting: A Case Study of Water Source Area of the Middle Route of South-To-North Water Transfer Project in China. *Front. Environ. Sci.* **2022**, *10*, 854150. [CrossRef]
61. Song, F.; Su, F.; Mi, C.; Sun, D. Analysis of driving forces on wetland ecosystem services value change: A case in Northeast China. *Sci. Total Environ.* **2021**, *751*, 141778. [CrossRef] [PubMed]
62. Dai, X.; Johnson, B.A.; Luo, P.; Yang, K.; Dong, L.; Wang, Q.; Liu, C.; Li, N.; Lu, H.; Ma, L.; et al. Estimation of Urban Ecosystem Services Value: A Case Study of Chengdu, Southwestern China. *Remote Sens.* **2021**, *13*, 207. [CrossRef]
63. Wang, E.; Kang, N. Does life satisfaction matter for pro-environmental behavior? Empirical evidence from China General Social Survey. *Qual. Quant.* **2019**, *53*, 449–469. [CrossRef]
64. Carrasco, L.R.; Nghiem, T.P.L.; Sunderland, T.; Koh, L.P. Economic valuation of ecosystem services fails to capture biodiversity value of tropical forests. *Biol. Conserv.* **2014**, *178*, 163–170. [CrossRef]
65. Kang, N.; Hou, L.; Huang, J.; Liu, H. Ecosystem services valuation in China: A meta-analysis. *Sci. Total Environ.* **2022**, *809*, 151122. [CrossRef]
66. Chen, Y.J.; Yu, Z.R.; Li, X.D.; Li, P.Y. How agricultural multiple ecosystem services respond to socioeconomic factors in Mengyin County, China. *Sci. Total Environ.* **2018**, *630*, 1003–1015. [CrossRef]
67. Jiang, Z.; Sun, X.; Liu, F.; Shan, R.; Zhang, W. Spatio-temporal variation of land use and ecosystem service values and their impact factors in an urbanized agricultural basin since the reform and opening of China. *Environ. Monit. Assess.* **2019**, *191*, 739. [CrossRef]
68. He, Y.H.; Wang, Z.R. Water-land resource carrying capacity in China: Changing trends, main driving forces, and implications. *J. Clean. Prod.* **2022**, *331*, 130003. [CrossRef]
69. Keeler, B.L.; Polasky, S.; Brauman, K.A.; Johnson, K.A.; Finlay, J.C.; O'Neill, A.; Kovacs, K.; Dalzell, B. Linking water quality and well-being for improved assessment and valuation of ecosystem services. *Proc. Natl. Acad. Sci. USA* **2012**, *109*, 18619–18624. [CrossRef] [PubMed]

International Journal of
Environmental Research and Public Health

MDPI

Article

Virtual Water Flow Pattern in the Yellow River Basin, China: An Analysis Based on a Multiregional Input–Output Model

Xiuli Liu [1], Rui Xiong [1], Pibin Guo [2,*], Lei Nie [1], Qinqin Shi [1], Wentao Li [1] and Jing Cui [1]

[1] Research Institute of Resource-Based Economics, Shanxi University of Finance & Economics, Taiyuan 030006, China; lxl820113@163.com (X.L.); xr1181968091@163.com (R.X.); nielei5515@foxmail.com (L.N.); shiqinqin_1314@126.com (Q.S.); liwentao0793@163.com (W.L.); cuijing000906@163.com (J.C.)

[2] Department of Management, Taiyuan University, Taiyuan 030032, China

* Correspondence: guopibin@hotmail.com

Abstract: Research on the Yellow River Basin's virtual water is not only beneficial for rational water resource regulation and allocation, but it is also a crucial means of relieving the pressures of a shortage of water resources. The water stress index and pull coefficient have been introduced to calculate the implied virtual water from intraregional and interregional trade in the Yellow River Basin on the basis of a multi-regional input–output model; a systematic study of virtual water flow has been conducted. The analysis illustrated that: (1) Agriculture is the leading sector in terms of virtual water input and output among all provinces in the Yellow River Basin, which explains the high usage. Therefore, it is important to note that the agricultural sector needs to improve its water efficiency. In addition to agriculture, virtual water is mainly exported through supply companies in the upper reaches; the middle reaches mainly output services and the transportation industry, and the lower reaches mainly output to the manufacturing industry. Significant differences exist in the pull coefficients of the same sectors in different provinces (regions). The average pull coefficients of the manufacturing, mining, and construction industries are large, so it is necessary to formulate stricter water use policies. (2) The whole basin is in a state of virtual net water input, that is, throughout the region. The Henan, Shandong, Shanxi, Shaanxi, and Qinghai Provinces, which are relatively short of water, import virtual water to relieve local water pressures. However, in the Gansu Province and the Ningxia Autonomous Region, where water resources are not abundant, continuous virtual water output will exacerbate the local resource shortage. (3) The Yellow River Basin's virtual water resources have obvious geographical distribution characteristics. The cross-provincial trade volume in the downstream area is high; the virtual water trade volume in the upstream area is low, as it is in the midstream and downstream areas; the trade relationship is insufficient. The Henan and Shandong Provinces are located in the dominant flow direction of Yellow River Basin's virtual water, while Gansu and Inner Mongolia are at the major water sources. Trade exchanges between the midstream and downstream and the upstream should be strengthened. Therefore, the utilization of water resources should be planned nationwide to reduce water pressures, and policymakers should improve the performance of agricultural water use within the Yellow River Basin and change the main trade industries according to the resource advantages and water resources situation of each of them.

Keywords: virtual water flow; multiregional input–output model; pull index; water stress index; Yellow River Basin

Citation: Liu, X.; Xiong, R.; Guo, P.; Nie, L.; Shi, Q.; Li, W.; Cui, J. Virtual Water Flow Pattern in the Yellow River Basin, China: An Analysis Based on a Multiregional Input–Output Model. *Int. J. Environ. Res. Public Health* **2022**, *19*, 7345. https://doi.org/10.3390/ijerph19127345

Academic Editors: Luis Garrote and Alban Kuriqi

Received: 13 May 2022
Accepted: 13 June 2022
Published: 15 June 2022

1. Introduction

Water is not only a necessary and irreplaceable resource for social and economic development [1–3], but it is also a vital element of the environment, essential for global sustainable development [4–6]. Water resources mainly refer to freshwater resources on land [5,7], with the total amount of freshwater available for human use accounting for only 0.3% of all freshwater. Since the 1990s, global water resources have been deteriorating [8–10]. About 1.5 billion people, accounting for 40% of the global population, in 80 countries and regions are suffering from a shortage of freshwater [11,12], and about 300 million from 26 countries are in an extreme water-shortage state [13–15]. Agricultural development will be especially hindered, and world food security will be compromised as a result of water shortages [16,17]. Globally, industrial water consumes approximately 20% of the total freshwater [18]; the shortage of water resources may lead to industrial shutdowns and limit production [19]. In addition, the destruction of ecosystems and biodiversity due to the water crisis will pose serious threats to human survival [20–22].

In terms of countries with water shortages, China is one of them [23,24]. Currently, China only possesses 6% of the global water resources [25], followed by Brazil, Russia, and Canada [26]. Nevertheless, China has a per capita water resource of only 2300 m^3 [27], making it one of the most water-scarce countries in the world [28,29]. The unbalanced distribution of water resources is currently one of the biggest obstacles to promoting sustainable development in China [23,24]. There are significant differences in the distribution of water resources between North and South China [30–32]. For example, the Yellow River Basin covers an area of 795,000 km^2 [33], accounting for only 44.2% of the Yangtze River Basin [27]. At present, the per capita water resources of 6 provinces (regions) in China are less than 500 m^3, and water from the Yellow River Basin serves two-thirds of these provinces (regions) (Ningxia Autonomous Region, Henan Province, Shanxi Province, and Shandong Province) [34]; the shortage of water resources has become a major restricting factor for the Yellow River Basin's high-quality development [35,36], posing quite a severe challenge to the construction of an ecological civilization and regional sustainable development [37,38]. The Yellow River Basin provides important water-resource support for China's granary and national energy security, and it is tasked with supplying water to Hebei, Tianjin, the Jiaodong Peninsula, and other basins. The Yellow River Basin, which accounts for 2% of the river runoff in China, supports the water demands of 12% of the population and 17% of the arable land in China, and it plays a decisive role in the overall economic and social development. Therefore, quantifying the flow laws and operation trends of water resources used in economy and trade will have important theoretical value, and enhancing the Yellow River Basin's intensive water usage is of practical significance [39–41], as is promoting its high-quality development in consideration of the water-resources carrying volume [42,43].

To better broaden the field of water research and find solutions for water scarcity in arid regions, Allen [44] proposed the concept of virtual water in 1993. The concept of virtual water illustrates that water resources, as a whole, required in various productions reflect the real quantity of resources in various economic production activities [44]. Virtual water has become one of the major methods for investigating regional water-resource issues [45]. Currently, the main method for studying virtual water is the input–output method, which can analyze trade and flows in a wide range of industries, and it is widely used by researchers globally [46–48]. Research on virtual water has multiple scales. First, the calculation model of trade and usage is established through the method of input and output, and then the volume of virtual water in any industrial sector is calculated [49,50]. Cegar [51] utilized the model discussed in this research to find both the indirect and the direct volumes of virtual water in the economy of Croatia; the water volume mainly relies on the processes of power generation and the utilization of the power processing output, along with the chemical and petroleum sectors. Gkatsikos and Mattas [52] analyzed water scarcity in Mediterranean countries and found that the agricultural sector dominated the regional virtual water flux. By analyzing the water usage of each link in commodity supply in China, Houyin et al. [53] found that the industrial sector was the core of indirect water usage. In regard to the continuous input and output table of Liaoning Province from 2012 to 2018, Zhang et al. [54] illustrated that the outflow sector of virtual water in Liaoning Province was mainly concentrated on primary and tertiary employment. Yang et al. [55] and Fu et al. [56] utilized this model to investigate the virtual water trade between the

Tarim River Basin and Hubei Province in China, finding that the volume used of virtual water in the primary industry is much higher [32]. Zhang et al. [57] calculated both the direct and total water input coefficients in Inner Mongolia in 2007, 2012, and 2015. The research concluded that the direct and primary, secondary and tertiary total water input coefficients of the industries show a downward trend, indicating that the tertiary industry's efficiency is constantly improving. Second, with the help of an input–output multi-regional model, the virtual water flux is combined with the economic relations among regions, so as to calculate the flow direction and usage of virtual water throughout all the regions. Islam et al. [58] conducted a multi-regional input and output analysis of virtual and direct water in 5 Australian capital cities and their surrounding areas, showing that the virtual part from outside the Australian cities boundary was nearly 20 times that from inside the urban boundary. Qasemipour et al. [59] assessed Iran's virtual water flux via a multi-regional input–output framework (MRIO). The results showed that there was no shortage of water resources in the northern countries, and virtual water is imported through the trade in various products, while areas with serious water shortages are net exporters of virtual water. As pointed out by Dong et al. [60], Chen et al. [61], and Wang and Chen [62], virtual water in China presents a flow pattern from the inland to the coastal areas, as well as from the underdeveloped to the developed areas. Based on the analysis of water trade between provinces and interprovincial flows in Northeast China, Zhang et al. [63] suggested that Liaoning and Jilin Provinces have almost scarce water resources, and other regions have the highest cumulative risk scores of virtual water-trade spillovers compared with Liaoning Province. Third, with the increasing frequency of economic activities, virtual water flow also has had a certain impact on the utilization of water resources. Some scholars combine virtual water with water-resource utilization to study virtual water. Wang et al. [64] assessed the water security in five countries in Central Asia, finding that Tajikistan and Kyrgyzstan are relatively safe in terms of quantitative water-resource security, while Uzbekistan is at risk. Zhang et al. [65] analyzed and calculated water-resource usage efficiency in the Aral Sea region from 2000 to 2014. The results demonstrated that the Aral Sea's water region dropped by 60.28% from the original 28,119 m^2 over those 15 years. Through the accounting of virtual water in some areas of China, Zheng et al. [66] and Wang et al. [67] found that the utilization structure of water resources in the study area was unreasonable, and the utilization efficiency of water resources needed to be improved. Other scholars have also conducted investigations from the perspective of water resource pressures. For example, Rosales-Asensio E. et al. [68] noted the fact that the restrictions on water resources in the Canary Islands of Spain led to the over-exploitation of aquifers and wells, leading to the deterioration of water resources and the environment. De O et al. [9] found that, due to the expansion of irrigation areas and urban populations in the Rio Verde Grande Basin, Brazil, the availability of water resources was low, causing water-resource conflicts to be triggered. Gohar A. et al. [69] studied the groundwater resources in Barbados and concluded that, in order to protect the sustainability of aquifers, it was necessary to formulate policies to restrict pumping, while economic welfare would be reduced by a certain amount in a short time. The Chinese mainland, as the research area, was comprehensively evaluated for its regional water-resource pressures using virtual water flow by Sun et al. [70]. The results showed that the northeast and the Huang-Huai-Hai regions in China are the largest producers of food and the biggest exporters of virtual water, and the pressure of resource shortages is generally serious. Liu et al. [71] analyzed a water pressure index of 11 administrative regions from 2000 to 2013, including Hebei Province, China, where the demand and supply are in serious conflict. The results show that water stress is mainly manifested at three levels: high, medium, and low. The aforementioned studies generally argue that the shortage of water resources has become the main problem affecting regional sustainable development.

According to the analysis above, it can be seen that: (1) With regard to virtual water, the majority of the calculations focus on the use of a single regional or interprovincial industrial sector, while flows are less often studied from a regional or industrial-sector perspective;

(2) Some studies focus on the fair distribution of water resources across regions, but they fail to fully combine the flow of virtual water with water resource management; (3) The research on key economic regions is not deep enough, especially those with serious shortages. Most studies focus on the virtual water in a single province in a basin; however, only a few studies have investigated the flow of virtual water within and between the Yellow River's regions. The Belt and Road, the important node and the trade link between the Yellow River and China, is gradually strengthening under the current situation of double-cycle development in China. It is of great significance for water resource management strategy to analyze and study the virtual flow pattern implied by the Yellow River Basin's internal and external trade.

To sum up, this paper takes the Yellow River regions as the object of this research, along with the interregional input and output data of a total of 31 provinces (municipalities and autonomous regions) domestically in 2015 and the water-usage data of 42 industrial sectors, constructs a multi-regional input–output accounting framework of the Yellow River Basin, and calculates the volume of virtual water trade within and between regions of the Yellow River Basin, respectively. Moreover, the pull coefficient and water stress index are introduced to further explore the coordination between the virtual water trade volume and local resource carrying capacity in the Yellow River region, as well as providing more-feasible policy suggestions for the management of the Yellow River region's water resources.

In contrast to previous studies, the innovations of this paper are as follows: (1) Virtual water flow is discussed from the perspective of internal and external regions and sectors, which presents a theoretical basis for the formulation of systematic and rational regional trade policies and industrial water policies; (2) The introduction of the water stress index and pull coefficient to further investigate the dependence of the studied subject implied by regional trade on water resources locally and the tie between various sectors have practical significance for regions and sectors to formulate reasonable water-resource policies. This study provides a systematic and reasonable industrial and trade policy framework for optimizing the Yellow River region's water resource allocations, promoting the protection of the aquatic ecological environment and alleviating its resource pressures.

The rest of the article is structured as follows: Section 2 introduces the research field, including the results and a virtual water analysis of an accounting framework regarding multi-regional input–output data and data sources pertaining to the Yellow River Basin. Section 3 presents the results and analysis. Section 4 puts forward a discussion, and suggestions are made in the conclusion.

2. Methodology and Materials

2.1. Study Region

The Yellow River region covers seven provinces and two autonomous regions, including the Shandong, Henan, Qinghai, Gansu, Sichuan, Shanxi, and Shaanxi Provinces, as well as the Ningxia and Inner Mongolia Autonomous Regions. It is one of the largest economic comprehensive zones in northern China [72,73]. According to the characteristics of the Yellow River Basin, the nine provinces are divided into three regions: upstream, midstream, and downstream [36,38,39] (Figure 1).

A total of 18.63% of the total domestic water resources came from the Yellow River Basin in 2019, and China's Yellow River Basin GDP accounted for 25.08% nationally. The total quantity of water resources is basically balanced with the level of economic development (Table 1). Yet, economic development and water resources in the Yellow River Basin lack coordination. Upstream resources account for 82.32% of the basin's total, while the GDP accounts for only 32.04%; the total quantity of downstream water-resource usage accounts for 6.72% of the Yellow River Basin, but it brings more than 50% of the GDP of the whole basin. In terms of total water usage, excepting Sichuan Province, the provinces (regions) located in the Yellow River Basin's upper and middle reaches do not exceed 20 billion m^3, and the downstream is much lower than the upstream. Therefore, encouraging the rational movement of water between the Yellow River Basin and the rest of the world is of paramount importance.

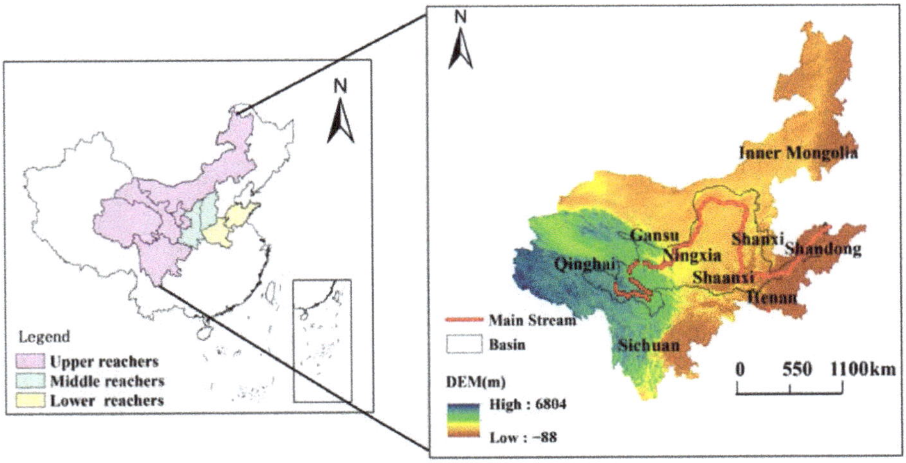

Figure 1. Locations of the upper, middle, and lower reaches of the Yellow River Basin (source: Authors' own calculation/conception, using ArcGIS 10.7 software (Environmental Systems Research Institute, Inc., Taiyuan, China)).

Table 1. Water Resources and economic development of the Yellow River Basin in 2019 (source: *China Statistical Yearbook* (2020) and *China Water Resources Bulletin* (2020).

Province (Region)		Total Water Resources/100 Million m^3	Total Water Consumption/100 Million m^3	GDP/100 Million Yuan
Upper reaches	Qinghai	919.30	26.20	2965.95
	Gansu	325.90	110.00	8718.30
	Ningxia	12.60	69.90	3748.48
	Sichuan	2748.90	252.40	46,615.82
	Inner Mongolia	447.90	190.90	17,212.53
Middle reaches	Shanxi	97.30	76.00	17,026.68
	Shaanxi	495.30	92.60	25,793.17
Lower reaches	Shandong	195.20	225.30	71,067.53
	Henan	168.60	237.80	54,259.20
The Yellow River Basin		5411.00	1281.10	247,407.66
Whole country		29,041.00	6021.20	986,515.20
Percentage of Yellow River Basin in China		18.63%	21.28%	25.08%
Percentage of upper reaches in the Yellow River Basin		82.32%	50.69%	32.04%
Percentage of middle reaches in the Yellow River Basin		10.95%	13.16%	17.31%
Percentage of lower reaches in the Yellow River Basin		6.72%	36.15%	50.66%

2.2. *Multi-Regional Input–Output Virtual Water Accounting Framework of the Yellow River Basin*

2.2.1. Single-Region Input–Output Model

Since Leontief [74] proposed the input–output method, scholars globally have widely used this method due to its ability to analyze the direct and indirect usage of virtual water in various industrial sectors and regions. This chart illustrates the direct and indirect relationships between different regions and industries [75–77]. According to the input–

output method, there is a balance between the output and usage of an economic system as follows:

$$\begin{bmatrix} x_1 \\ x_2 \\ \vdots \\ x_n \end{bmatrix} = \begin{bmatrix} z_{11} & z_{12} & \cdots & z_{1n} \\ z_{21} & z_{22} & \cdots & z_{2n} \\ \vdots & \vdots & \ddots & \vdots \\ z_{n1} & z_{n2} & \cdots & z_{nn} \end{bmatrix} \begin{bmatrix} 1 \\ 1 \\ \vdots \\ 1 \end{bmatrix} + \begin{bmatrix} f_1 \\ f_2 \\ \vdots \\ f_n \end{bmatrix} \quad (1)$$

where x_i^r is the total output of sector i in region r, z_{ij} is the intermediate input provided by sector i to sector j, and f_i is the final use of sector i.

The direct input coefficient a_{ij} reflects the number of products and/or services directly consumed by each product per unit of complete output in the process and operation process of a product sector. Its calculation formula is:

$$a_{ij} = z_{ij}/x_j \quad (2)$$

where x_j is the total output of sector j.

Therefore, Equation (1) can be rewritten as:

$$\begin{bmatrix} x_1 \\ x_2 \\ \vdots \\ x_n \end{bmatrix} = \begin{bmatrix} a_{11} & a_{12} & \cdots & a_{1n} \\ a_{21} & a_{22} & \cdots & a_{2n} \\ \vdots & \vdots & \ddots & \vdots \\ a_{n1} & a_{n2} & \cdots & a_{nn} \end{bmatrix} \begin{bmatrix} x_1 \\ x_2 \\ \vdots \\ x_n \end{bmatrix} + \begin{bmatrix} f_1 \\ f_2 \\ \vdots \\ f_n \end{bmatrix} \quad (3)$$

2.2.2. Modeling of Yellow River Basin Inputs and Outputs on a Multi-Regional Scale

As is known to all, the multi-regional input–output method is able to connect the economies inside and outside a region with virtual water flow [49]. Therefore, in this paper, the calculation of provinces' water usage was performed based on a multi-regional input–output method of analysis (regions) in the Yellow River region. According to the flow direction and geographical location characteristics of the Yellow River region, this paper defines the Yellow River region's 9 provinces (regions) as local and the other 22 provinces (areas and cities) in China as foreign, leading to the construction of a multi-regional input–output table of the Yellow River Basin in 2015 (Table 2), which includes 10 regions and 42 industrial sectors in each region. Figure 2 is the input–output flow chart of this study.

Table 2 shows the multi-regional inputs and outputs of the Yellow River Basin; the balance of Regional r economic activities is:

$$x_i^r = \sum_{s=1}^{10} \sum_{j=1}^{42} a_{ij}^{rs} x_j^s + \sum_{s=1}^{10} f_i^{rs} + e_i^r \quad (4)$$

where x_i^r is the total output of sector i in region r, a_{ij}^{rs} is the direct input coefficient, which indicates the direct input of sector i in region r to the production unit product of sector s in region j. f_i^{rs} is the input of sector i in region r to the final demand of region s, and e_i^r is the export volume of sector i in region r.

Equation (4) is expressed by the matrix as:

$$X^r = A^{rs} + F^{rs} + E^r \quad (5)$$

Additionally, Equation (5) is appropriately reformed to obtain the multi-regional input–output model of the Yellow River Basin:

$$X^r = (I - A^{rs})^{-1}(F^{rs} + E^r) = L(F^{rs} + E^r) \quad (6)$$

Among them,

$$L = (I - A^{rs})^{-1} = \begin{bmatrix} q^{11} & q^{12} & \cdots & q^{1,10} \\ q^{21} & q^{22} & \cdots & q^{2,10} \\ \vdots & \vdots & \ddots & \vdots \\ q^{10,1} & q^{10,2} & \cdots & q^{10,10} \end{bmatrix} \tag{7}$$

where X^r is the total output matrix, I represents the identity matrix, and A^{rs} means the direct input coefficient matrix. F^{rs} and E^r mean the end-use matrix and export matrix, respectively. $L = (I - A^{rs})^{-1}$ is the Leontief inverse matrix, and the element l_{ij}^{rs} in the matrix donates the total output per unit product provided by sector i in region r to sector j in region s, where the regional direct water uses coefficient matrix Y, expressed as:

$$Y = \left[y^1, y^2, \cdots, y^{10}\right] \tag{8}$$

The complete water usage coefficient matrix Q is:

$$Q = YL = Y(I - A^{rs})^{-1} = \begin{bmatrix} q^{11} & q^{12} & \cdots & q^{1,10} \\ q^{21} & q^{22} & \cdots & q^{2,10} \\ \vdots & \vdots & \ddots & \vdots \\ q^{10,1} & q^{10,2} & \cdots & q^{10,10} \end{bmatrix} \tag{9}$$

where the element q^{rs} is the total usage of r area consumed by each sector in s area.

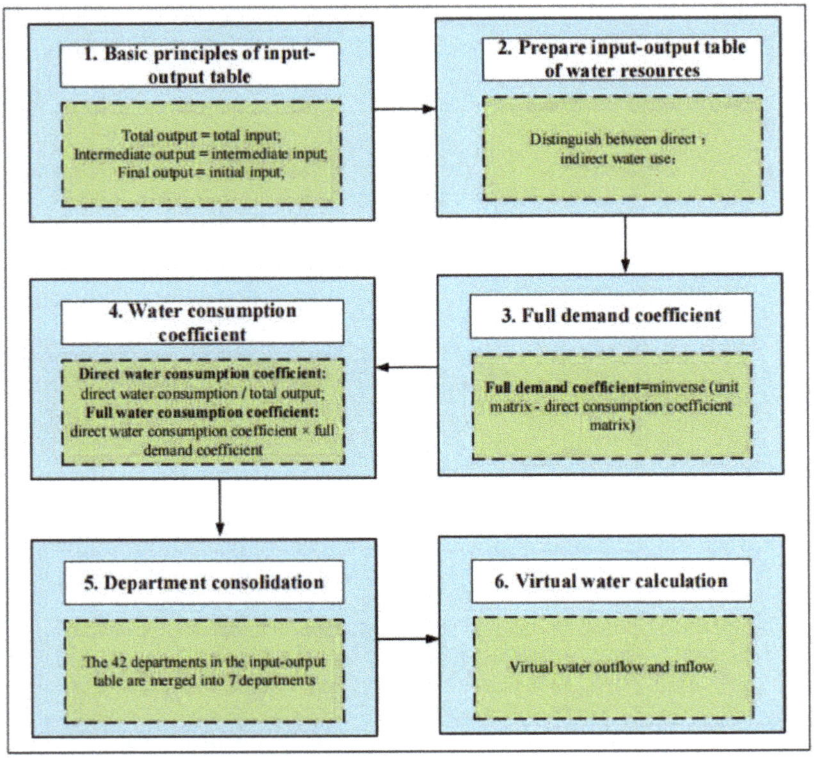

Figure 2. Model structure of virtual-flow-driving mechanism in the Yellow River Basin.

Table 2. Summary table of multi-regional input–output in the Yellow River Basin.

Item			Intermediate Use				Final Demand		Export	Total Output
			Qinghai	…	Henan	Other Regions	Qinghai … Henan	Other Regions		
			Sector1 … Sector42	…	Sector1 … Sector42	Sector1 … Sector42				
Intermediate input	Qinghai	Sector1	$z_{1,1}^{1,1}$ … $z_{1,42}^{1,1}$	…	$z_{1,1}^{1,9}$ … $z_{1,42}^{1,9}$	$z_{1,1}^{1,10}$ … $z_{1,42}^{1,10}$	$f_{1,1}^{1}$ … $f_{1,1}^{9}$	$f_{1}^{1,10}$	e_{1}^{1}	X_{1}^{1}
		…	…	…	…	…	…	…	…	…
		Sector42	$z_{42,1}^{1,1}$ … $z_{42,42}^{1,1}$	…	$z_{42,1}^{1,9}$ … $z_{42,42}^{1,9}$	$z_{42,1}^{1,10}$ … $z_{42,42}^{1,10}$	$f_{42}^{1,1}$ … $f_{42}^{1,9}$	$f_{42}^{1,10}$	e_{42}^{1}	X_{42}^{1}
	…	…	…	…	…	…	…	…	…	…
	Henan	Sector1	$z_{1,1}^{9,1}$ … $z_{1,42}^{9,1}$	…	$z_{1,1}^{9,9}$ … $z_{1,42}^{9,9}$	$z_{1,1}^{9,10}$ … $z_{1,42}^{9,10}$	$f_{1}^{9,1}$ … $f_{1}^{9,9}$	$f_{1}^{9,10}$	e_{1}^{9}	X_{1}^{9}
		…	…	…	…	…	…	…	…	…
		Sector42	$z_{42,1}^{9,1}$ … $z_{42,42}^{9,1}$	…	$z_{42,1}^{9,9}$ … $z_{42,42}^{9,9}$	$z_{42,1}^{1,10}$ … $z_{42,42}^{9,10}$	$f_{42}^{9,1}$ … $f_{42}^{9,9}$	$f_{42}^{9,10}$	e_{42}^{9}	X_{42}^{9}

2.2.3. Estimation of Virtual Water Trade Flow

According to the multi-regional input–output model of the Yellow River Basin, the virtual water trade flow among the nine provinces (regions) can be calculated as follows:

$$VWT^{rs} = \sum_{i=1}^{9} q^{is} f^{is} \tag{10}$$

The virtual water trade flow between the 9 provinces (regions) in the Yellow River region and other places is:

$$VWI^{S} = \sum_{i=1}^{10} q^{10,i} f^{is} \tag{11}$$

$$VWO^{r} = \sum_{i=1}^{10} q^{ri} f^{i,10} \tag{12}$$

where VWT^{rs} is the virtual water trade flow from region r to region s in the Yellow River region, VWT^{s} is the virtual water input from other regions from the provinces in this region, and VWO^{r} is the virtual water output of provinces in this region to other regions.

2.2.4. Water Stress Index and Pull Coefficient

This study aims to determine the relationship between virtual water implied by regional trade and local water-resource carrying capacity in the Yellow River region. Based on the research results of Pfister et al. [78], this paper introduces the water stress index, indicating the shortage of water resources in this region, and then analyzes the external dependence of virtual water in each region. Synchronously, this research incorporated the research of Hong et al. [79] and Boero R. et al. [80], using the pull coefficient to explain the influence of the increase of water usage in a single sector on the water usage among all sectors to reveal the virtual water flow law among all sectors in various regions.

The water stress index (*WSI*) is derived from the ratio of total freshwater extraction to total available freshwater in this region annually [78,81]. This index can measure the degree of the lack of water in a region, thus measuring the external dependence of water usage in the region [37,78,81]. The calculation formula is:

$$WSI = \frac{1}{1 + e^{-64*WTA((1/0.01)-1)}} \quad (13)$$

The annual freshwater extraction is measured by the ratio of the *WTA* to the annual freshwater supply of different types of users (industry, agriculture, and households) in the region. It can be seen that *WSI* = [0, 1]. Referring to the research results of Pfister et al. [80], when *WSI* is 0, it means there is no water stress, which indicates that water resources are abundant; when *WSI* is 1, that is, the maximum water stress, it means that there is a serious shortage of water resources. A water stress threshold of 0.5 is considered to be moderate.

The pull coefficient p_i^r refers to the increased degree of water usage by all sectors when the unit water usage of an industrial sector increases, which can be employed to reflect the contribution degree of the water usage of a sector to the increase in the water usage of the whole society. The calculation is as follows:

$$p_i^r = q_i^r / y_i^r \quad (14)$$

where, p_i^r represents the pull coefficient of i industrial sector in region r, q_i^r, and y_i^r represents the coefficient of total water usage and direct water usage coefficient of i industrial sector in region r, respectively. It can be seen that, when $p_i^r \geq 1$, the closer the economic ties between sectors are, and the closer the pull coefficient is to 1, and vice versa, the greater the pull coefficient is.

2.3. Data Source and Processing

The input–output data of the Yellow River Basin applied in this paper are drawn from the interregional input and output tables (42 sectors) of a total of 31 provinces (autonomous regions and municipalities) domestically in 2015, as compiled by China Carbon Emission Database [82]. The entire water usage dataset, GDP data, and the actual water usage of the agricultural sector of each province (region) are all taken from *China Statistical Yearbook*. The actual usage data of the industrial sector comes from the *2008 China Economic Census Yearbook*. Due to the age of the data, this paper takes the GDP growth rate from 2008 to 2015 as the growth rate of actual water usage, and obtains the water usage reported in the *China Statistical Yearbook* in 2016. Considering the total industrial water usage of each province (region), the total industrial water usage of each province (region) in 2015 was calculated [83–86].

In terms of the provincial (regional) water-resources bulletin, the actual water usage of the service industry is the urban public water usage in each province. The water usage of each subdivided service-industry sector is based on a proportion of output value along with the domestic water usage of the residents [87,88]. According to the water usage data of the actual industrial sectors of each province (region) in the Yellow River region and the attributes of every industrial sector, and referring to the sector consolidation methods of scholars such as Shi et al. [89] and Chen et al. [90], the 42 sectors in the input–output table have been consolidated into 7 sectors (Table 3).

Table 3. Input–output table: Detailed list of 42 combined sectors.

Combined 7 Sectors	Department Abbreviation
Agriculture	AG
Mining	MI
Water supply	WA
Electricity and gas supply	EL
Manufacturing	MA
Construction	CO
Services and transport	ST

3. Results and Analysis

3.1. Virtual Water Usage of Different Sectors in the Yellow River Basin

3.1.1. Analysis of Sectoral Water Usage Coefficient

Figure 3 shows the composition of the complete water usage coefficient of 7 industrial sectors in the lower, middle, and upper reaches of the Yellow River. Within the Yellow River Basin, it can be seen that certain differences exist in the total water usage coefficients of a variety of industrial sectors. The total water usage coefficient is the largest in the upstream area and the smallest in the downstream area. The combined upstream, middle, and downstream water usage coefficient of the water supply industry turns out to be the highest, followed by that of agriculture. Specifically, the complete water usage coefficient of the water supply in the upper reach's industry is more than 1800 $m^3/10{,}000$ RMB, while that of agriculture in the upper and middle reaches is less than 1/5 of the water supply industry. The complete water usage coefficient of other industrial sectors is small, not exceeding 150 $m^3/10{,}000$ RMB. In the upper reaches, the agricultural water usage coefficient is the highest, close to 400 $m^3/10{,}000$ RMB, while those of the middle and lower reaches are less than 100 $m^3/10{,}000$ RMB, indicating that in the industrial structure of the Yellow River Basin, the upper reaches pay more attention to agricultural development than the middle and lower reaches.

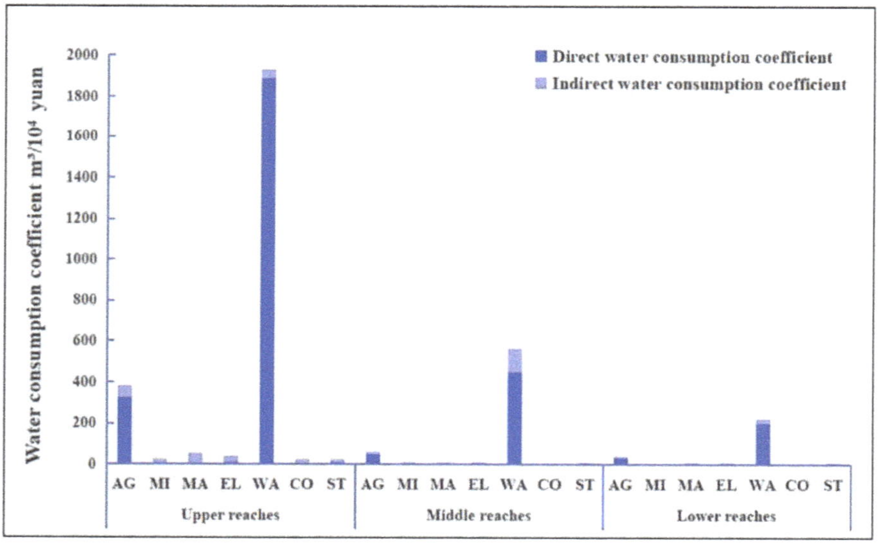

Figure 3. Water consumption coefficient in upper, middle, and lower reaches.

3.1.2. Volume of Trade in the Sector of Virtual Water

In all provinces (regions) of the Yellow River region, the virtual water input and output of the industrial sectors was calculated (Figures 4 and 5), and the water flow varied greatly among sectors. First of all, a substantial amount of virtual water is input into and output from agricultural production within the region. It is the largest industrial sector in input and output, accounting for more than 46%, which also determines that the Yellow River Basin is in a state of serious water usage and low regional GDP. The virtual water inputs of the manufacturing, electrical supply, water supply, construction, and service and transportation sectors also account for a large proportion.

Nevertheless, except for manufacturing and service and transportation, there is no corresponding virtual water volume output, indicating that the utilization ratio of virtual water in these sectors is low and the utilization structure of water resources is unreasonable. Secondly, the proportion of the mining industry sector in virtual water input and output is not high, which is related to the concept of ecological protection and transformation and development strongly advocated by the state in recent years.

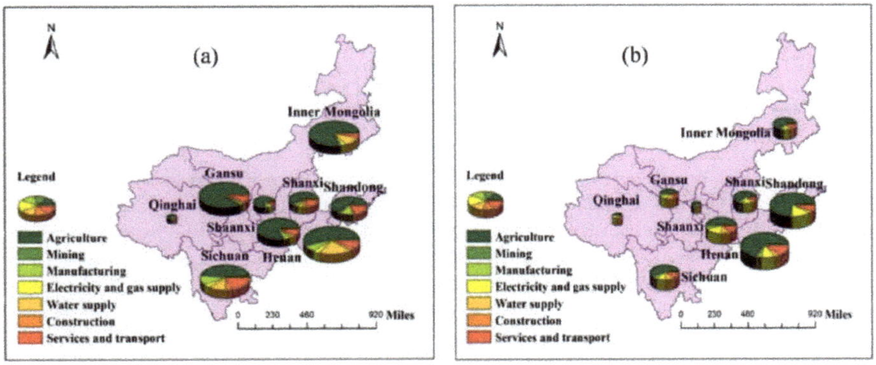

Figure 4. Trade structure of virtual water (%): (**a**) indicates the percentage of industrial sector output; and (**b**) indicates the percentage of industrial sector input.

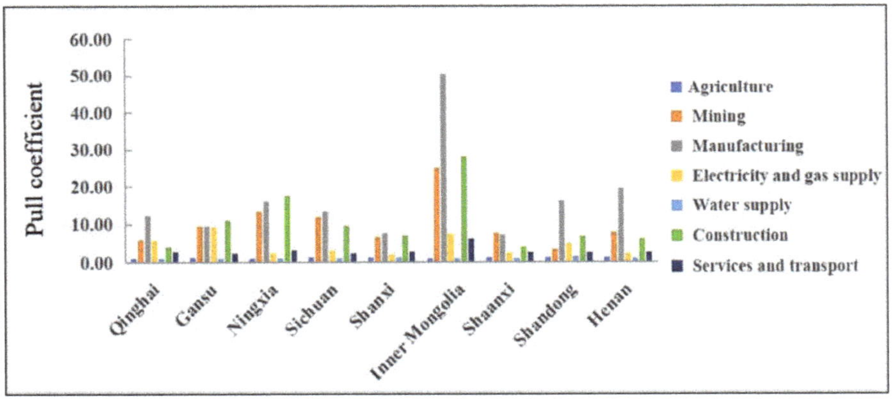

Figure 5. Pull Coefficient of different industrial sectors in different regions.

In terms of regions, in the upstream region, Sichuan Province is the only province whose agricultural virtual water output accounts for less than 70%, only 48.10% of the total, and water resources can be efficiently utilized under favorable circumstances. The virtual water output of the service and transportation and mining industry sectors in Sichuan Province accounts for 19.43%, which is the highest level in the Yellow River Basin, indicating that the tertiary industry in Sichuan Province has frequent foreign exchanges. This phenomenon is also one of the reasons for the high GDP and abundant water resources in Sichuan Province. Gansu Province is the province with the largest proportion of agricultural output, accounting for 88.14%.

However, the high water stress index in Gansu indicates that there is extreme water scarcity in the area. Continuing to export a large number of water-intensive products will make water resources in Gansu even more scarce. In addition to the agricultural sector, the virtual water input sector in Qinghai Province is mainly concentrated on the electric power, water supply, and service and transportation industries. The total proportion of virtual water input in these 3 sectors is 48.65%, which is slightly higher than that in the agricultural sector. Nonetheless, the GDP of Qinghai Province is not high, demonstrating that these sectors have an insufficient utilization of virtual water. In the middle reaches, the province with the largest input of virtual water in the manufacturing industry is Shanxi Province, but its sectoral virtual water output is less than the input, showing that Shanxi Province has not effectively utilized virtual water input to develop the manufacturing industry. Simultaneously, the virtual water output of the service and transportation industries in Shanxi Province account for a large proportion, which can continue to develop. Shaanxi's agricultural virtual water output also accounts for a large proportion, second only to Gansu Province. However, Shaanxi Province has less rain and more sunshine, poor vegetation distribution, and a poor agricultural development environment.

A large proportion of agricultural virtual water output will aggravate the lack of water resources in Shaanxi Province, as well as destroy the ecology's overall coordination. In the lower reaches, Shandong Province possess the leading input of agricultural virtual water, which is also the province with the largest input of virtual water in the agricultural sector of the whole Yellow River Basin, indicating that most of the input virtual water in Shandong Province is used for agriculture, which is not conducive to the development of other sectors. However, the output of virtual water in the manufacturing and service and transportation sectors in Shandong Province accounts for 33.17%, making it quite easy to create a higher GDP.

3.1.3. Pull Coefficient Analysis

The pull coefficient is used to further analyze the contribution of the virtual water of various sectors in the Yellow River region to the virtual water of the whole sector, and subsequently evaluate the degree of connection between various sectors of the Yellow River region's virtual water. The pull coefficient of each industrial sector in every province (region) of the Yellow River Basin is calculated through Formula (14) (Figure 5). As can be seen, the agriculture and water supply sectors have the lowest pull coefficients among all of the provinces (regions) in the Yellow River Basin, with values lower than 1.5, indicating that the economic relationship between these sectors and other sectors is not close. The average pull coefficients of the manufacturing, mining, and construction sectors is obviously greater than that of other sectors, demonstrating that these three sectors have strong economic ties with other sectors, and the increase of unit output in these sectors will require a great deal of virtual water. From the perspective of regional differences, the pull coefficient of the same sector in different provinces (regions) is also quite different. The pull coefficient of the mining, manufacturing, and construction sectors in Inner Mongolia is the largest in the whole Yellow River Basin, which is related to the coal-rich areas of Inner Mongolia. In the future, it will be necessary to speed up the transformation of Inner Mongolia's resource-based economy, develop a green mining industry, and pay attention to ecological and environmental protection. To alleviate the water shortage in this region, policymakers ought to enhance the water efficiency of the mining industry and strengthen the virtual water input of the manufacturing and construction industries. The pull coefficient of the manufacturing industry in Sichuan, Shandong and Henan Provinces is also large because these provinces mainly rely on the manufacturing industry to drive the rapid growth of the GDP. At the same time, these provinces are virtual water export areas of the manufacturing industry, thus the high value-added manufacturing industry should be developed. The pull coefficient of the manufacturing industry in Ningxia and Qinghai Provinces is also relatively large, but their GDPs lag far behind that of Shandong and Henan Provinces, indicating that a large number of water resources are wasted in the manufacturing sectors in the Ningxia Region and Qinghai Province. Therefore, the differences in the pull coefficients of all the sectors in the same province (region) can be explained by the difference in the number of employees in those sectors in terms of raw material input and production technology level. When a sector operates in two different provinces (regions), its pull coefficient differs due to the differences in water-use efficiency between the local and intermediate rivers.

3.2. Regional Virtual Flow Pattern in the Yellow River Basin

3.2.1. Trade Volume of Virtual Water

The trade volume of each province (region) in the Yellow River region is calculated according to the input–output multi-regional model of the Yellow River Basin (Figures 6–9). It can be seen that the total input and output of virtual water in this region are 27.45 billion m^3 and 18.64 billion m^3, respectively, which are in a virtual water net input state, with a total net input of 8.81 billion m^3. The input of virtual water from the outside of this region is 2.61 times that of the local input, while the external virtual water flow is 1.45 times that of this region, which leads to prominent conflicts between the demand and supply of water resources, and that is not conducive to regional ecological balance. In general, the net input areas importing from other places are mainly concentrated in the midstream and downstream, and the net input virtual water accounts for 96.22% of the total net input of virtual water, reflecting the lack of rationality of water-resource utilization in the region. Among them, Ningxia, Gansu and Inner Mongolia are in the area of virtual water net outflux, and the rest belong to the area of virtual water net influx.

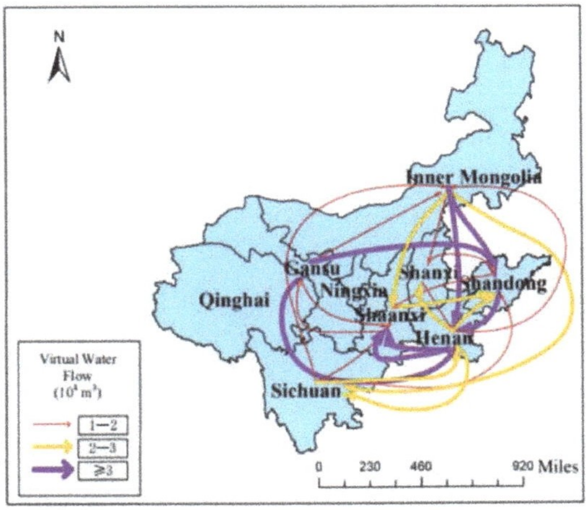

Figure 6. Total virtual water flow of each province in the Yellow River Basin. Note: This figure only shows the flow direction of virtual water trade flow greater than 10^4 m^3.

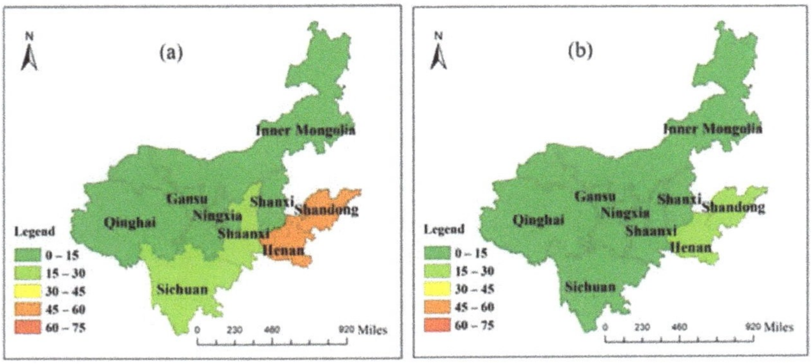

Figure 7. Virtual water inflow of each province in the Yellow River Basin: (a) inflow from provinces outside the Yellow River Basin; (b) inflow within the Yellow River Basin.

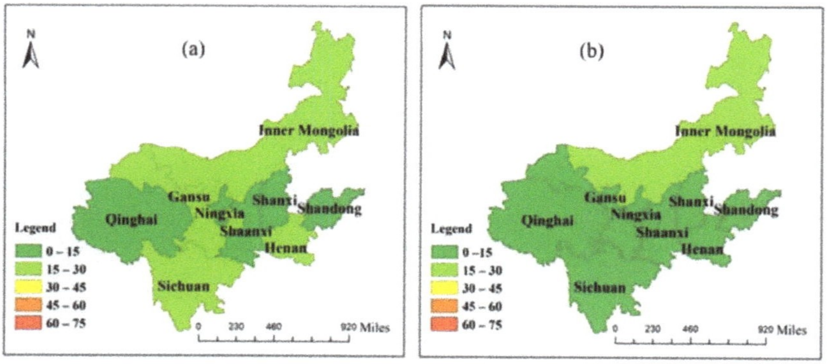

Figure 8. Virtual water outflow of each province in the Yellow River Basin: (**a**) outflow into provinces outside the Yellow River Basin; (**b**) outflow into the Yellow River Basin.

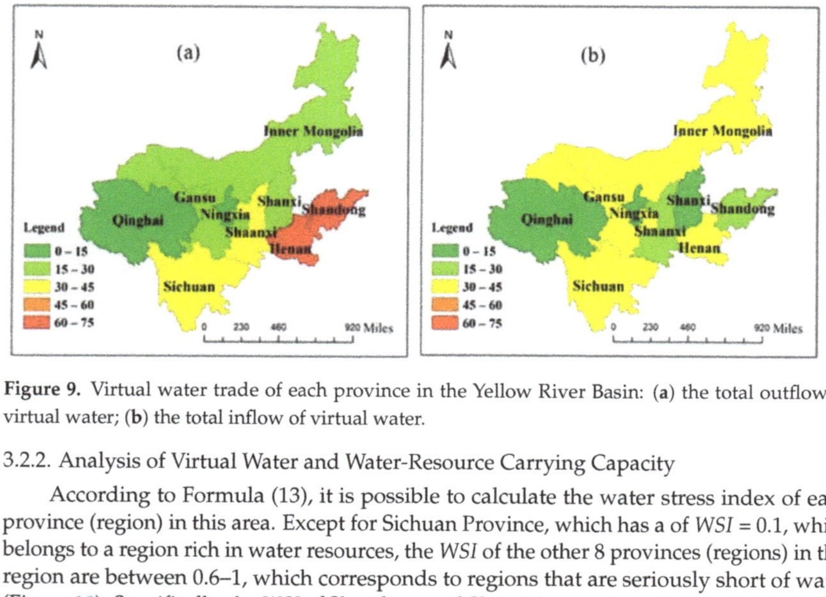

Figure 9. Virtual water trade of each province in the Yellow River Basin: (**a**) the total outflow of virtual water; (**b**) the total inflow of virtual water.

3.2.2. Analysis of Virtual Water and Water-Resource Carrying Capacity

According to Formula (13), it is possible to calculate the water stress index of each province (region) in this area. Except for Sichuan Province, which has a of *WSI* = 0.1, which belongs to a region rich in water resources, the *WSI* of the other 8 provinces (regions) in this region are between 0.6–1, which corresponds to regions that are seriously short of water (Figure 10). Specifically, the *WSI* of Shandong and Shanxi Provinces are the maximum value of 1, showing that the water shortage is serious, and a large amount of virtual water must be input to alleviate the local water pressures, among which the net input of virtual water in Shandong Province accounts for the largest proportion, and the external dependence of virtual water is about 1/4. The *WSI* of Gansu and Ningxia Provinces are close to 1, indicating that water resources are not abundant, but they output virtual water. The serious disharmony between *WSI* and the virtual water trade will aggravate the shortage of local water and further worsen the ecology. In regions with severe water stress, the lowest *WSI* value is the 0.61 of Henan Province, but the net input of virtual water is second only to Shandong Province, meaning that water resources are not fully utilized, and a large number of water resources are wasted. There is little difference in the *WSI* of Inner Mongolia and Shaanxi Province. However, Shaanxi Province is a virtual water net inflow area, which can alleviate the pressure of insufficient local water resources. In the meanwhile, the Inner Mongolian region should speed up its trade links with other regions to alleviate the water pressure in this region, which would have vital practical significance for the sustainable development of water resources and ecological protection. Sichuan Province is the only region in this region where the *WSI* is lower than 0.5, at only 0.10. Water resources in Sichuan Province are quite abundant, but the net input virtual water accounts for 0.28% of the virtual water. Therefore, the directors of Sichuan Province should appropriately increase the output of virtual water, and increase trade links within and outside the region, which is conducive to improving the water ecological balance in the region.

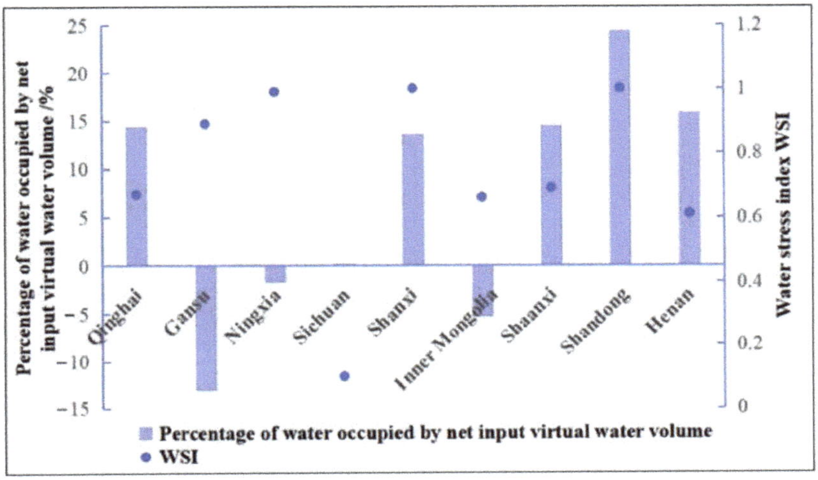

Figure 10. Virtual water external dependence and water stress index.

3.3. *Virtual Flow Pattern in the Yellow River Basin*

From the virtual water flow table of the Yellow River Basin (Figure 11), it can be seen that the virtual water trade flow in the Yellow River Basin is closely related to geographical location. In the upstream area, with the exception of Qingdao Province, the virtual water output from other provinces in the Yellow River Basin is greater than the virtual water input, which is in the net outflow area of internal virtual water. Among them, Gansu and Inner Mongolia are the provinces with the largest net output of virtual water, and they are the two major sources of virtual water in other provinces (regions) within the Yellow River Basin. However, it should be noted that the *WSI* in Gansu Province is too high, and it is easy to cause a water shortage and damage the ecological environment by excessively providing virtual water to other provinces. The middle and lower reaches are virtual water net input regions, and the virtual water net input of Shandong and Henan Provinces in the lower reaches ranks among the top two in the region, which is consistent with the level of economic development.

Virtual water trade between provinces in the upper reaches is not frequent, and the province with the largest trade flow is Sichuan Province, which is conducive to its good economic development. In the middle reaches, the trade volume of virtual water in Shanxi and Shaanxi Provinces is less than 70 million m^3, so the trade exchange needs to be strengthened. The interprovincial virtual water-trade flow in the downstream region is large, which is conducive to the common economic development of Shandong and Henan Provinces. In addition to strong trade exchanges between the upper region of Sichuan and Inner Mongolia, as well as the midstream and the upstream, the virtual water-trade flow between other provinces in the upper reaches and the Shaanxi and Shanxi Provinces is small, which is inconducive to economic exchanges and development.

The virtual water trade between the upstream and downstream areas mainly comes from the virtual water output from Gansu and Inner Mongolia to Henan and Shandong Provinces, of which the virtual water output from Gansu Province to the downstream accounts for 63.2% of its total output. The virtual water trade exchanges in the middle and lower reaches are also more frequent, and the virtual water flow basically flows from the middle reaches to the lower reaches. Therefore, it is necessary to strengthen the virtual water trade exchange between the upper, middle, and lower reaches, and make use of the downstream economy to drive the coordinated economic development of Qinghai, Ningxia, Gansu, and Inner Mongolia, so as to achieve the aim of the rational utilization and common development of water resources in all the regions and provinces of the Yellow River Basin.

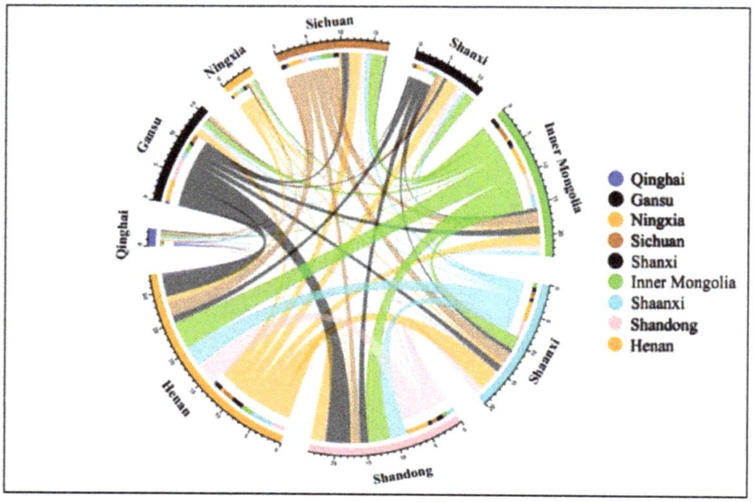

Figure 11. Virtual water flow in the Yellow River Basin (100 million m^3).

4. Discussion

Virtual water trade can promote economic ties between regions because the importation of virtual water will depend on the water resource endowment of export regions. With increasingly greater implementation of the virtual water strategy, each region becomes increasingly dependent on water resources in other regions [91,92]. Virtual water net import is obviously affected by water resource endowment, and regions with good water resource conditions tend to export virtual water to other regions [30,85]; this is also in line with the theory of comparative advantage in trade. Our results show that the Yellow River Basin is a virtual water net import region because the Yellow River Basin faces serious water shortage, which also confirms that regions with poor water resources usually import virtual water to relieve local water pressures [31]. From within the Yellow River Basin, the export status of virtual water in Gansu Province and Ningxia Autonomous Region is in contradiction with local water resource endowment. However, relevant studies [31,92–94] have also shown that northern China, which is seriously short of water, still exports virtual water to southern China through the grain trade, resulting in northern China becoming the main export area of virtual water [95]. Therefore, the method of coordinating the management of water resources in water-shortage areas is still an important research topic for the future. Therefore, this paper puts forward measures for pricing and managing the water resources in water-shortage areas, so as to stimulate the importation of water-intensive commodities and reduce the pressures on local water resources.

The results of this study show that the unreasonable industrial structure in the Yellow River Basin has led to the transfer of local water resources, and the agricultural sector is the largest water-resource import and export sector (Figure 3), which is consistent with the conclusions of relevant studies [96,97]. On the one hand, agriculture is a high water-consumption industry [98], which consumes a lot of blue and green water resources in the process of crop evapotranspiration [99]. Secondly, the Yellow River Basin is an important grain production base in China. In 2020, the total grain output of the Yellow River Basin reached 239 million tons, accounting for 35.6% of the total grain output in

China. Farmland irrigation water consumption alone will reach about two-thirds of the water consumption. The backwardness and lack of agricultural irrigation facilities will further increase the consumption and waste of water resources [100]. In addition, the water demands of various departments in a region often compete with one another, forming the characteristics of virtual water trade between regions [72]. For example, the competition between industrial water, service water, ecological environmental water, and agricultural water is more prominent in the Yellow River basin where water resources are scarce. The proportion of agriculture and water-intensive industries in the Yellow River Basin is very high, and the phenomenon of "competing for water" between economic development and ecological protection is very prominent, which is also consistent with the relevant research conclusions [33,35,101]. On the other hand, from the perspective of external factors, climate change may affect agricultural water use in various ways, especially through the changes in temperature and precipitation and the intensification of the frequency and degree of extreme climate events, which will significantly affect the available amount and quality of agricultural water and thus crop-water demand. As a result, agriculture is highly vulnerable to climatic conditions and natural disasters (such as droughts and floods), which inevitably affect interregional food-trade and water-use plans [30,102]. Climate change will not only exacerbate water shortages, but also reduce crop yields, thereby increasing the water footprint [103,104]. Therefore, in terms of agricultural water-use performance management in the Yellow River Basin, combined with high standard farmland construction, promoting large- and medium-sized irrigation areas, building modern irrigation facilities and developing water-saving agricultural technology will reduce water resource consumption. In addition, our recommendations include: optimizing and adjusting the crop planting structures, determining the agricultural industrial structure and planting structure according to local conditions, strictly controlling the planting area of high water-consumption crops, expanding the planting proportion of water-saving and drought-tolerant crops, selecting and promoting new varieties of drought tolerant crops, appropriately implementing rotation fallow, actively developing rainwater harvesting and irrigation, enhancing the capacity of water storage and moisture conservation, improving water-use efficiency, and reducing waste.

In addition, related studies have shown that virtual water flow is closely related to geographic distribution and economic development [105–107], which is also confirmed by our study. The virtual water trade volume of Shandong and Henan in the lower reaches of the Yellow River Basin is large, followed by Shanxi and Shaanxi in the middle reaches, and the virtual water trade volume of Qinghai and Ningxia in the lower reaches is the smallest. The advantages of opening to the outside world and strong agricultural development make the virtual water trade volume of Shandong and Henan higher. Qinghai and Ningxia in the downstream region are not only geographically remote, which is not conducive to economic ties, but they also have no obvious industrial advantages. Therefore, the momentum of cross-provincial virtual water flow between products is small. The study also found that the virtual water trade volume of the manufacturing, electrical supply, water supply, construction, and service and transportation industries is high, while the virtual water trade volume of mining is low, which is closely related to the ecological protection and "double carbon" strategic measures proposed by China in recent years. Relevant studies also show that water for power generation is increasing year by year, and the impact on scarce water resources in the basin is becoming more and more serious. In addition, the manufacturing sector [78], construction sector, [108] and service sector [109] are also key industries in the water-saving sector. Therefore, from the perspective of the Yellow River Basin as a whole, strengthening trade links between economies can improve water-use efficiency and optimize the water-use structure of key industries, which will be an important way to alleviate the contradiction between the supply and demand of the water resources of the Yellow River.

Distinguishing and calculating the virtual water flow of green water, blue water, and gray water is helpful for accurately implementing management policy regarding water resources. Because calculating virtual water in this paper involves various industrial sectors in nine regions or provinces of this region, considering the limitation of data acquisition and the inconsistency of differential calculations, only the virtual water flow of blue water has been calculated. The research results are important for formulating water resource policies, but they still need to be further deepened. In addition, because the input–output table is "competitive", the distinction between domestic and international intermediate inputs is not considered. Constructing a more accurate, interregional input–output table and considering virtual water cross-border transfer in the region is the direction of further research.

5. Conclusions

5.1. Conclusions

The efficient and rational utilization of water resources is an essential part of protection ecologically. Studying the virtual water usage and flow in water shortage areas is helpful to obtaining an ideal water-resource allocation [110]. By constructing a multi-regional input–output model of the Yellow River Basin, this paper calculates and analyzes the virtual water trade and flow pattern of interregional and intraregional industrial sectors in this region in 2015, and draws the following conclusions:

(1) The whole Yellow River region is in a net input state of virtual water. Among them, the upstream areas—Gansu, Inner Mongolia, and Ningxia Province—are in the net output provinces (regions), while the other six provinces belong to the virtual water net input regions. Gansu's virtual water input and output state is the most seriously incompatible with the local water-resource carrying capacity among all the provinces discussed.

(2) Agriculture is the largest import and export sector of all regions. In addition to agriculture, the upstream region is sufficient in water resources. The main export sector of virtual water is the water supply industry, and those for the middle- and downstream regions are the services and transportation and manufacturing industries, respectively. Obvious differences exist in the pull coefficients of the same sectors in various provinces (regions). On the whole, the average pull coefficients of mining, manufacturing, and construction are large. The water management of these sectors is conducive to rapid water-resource regulation and rational utilization in this region.

(3) The virtual flow of the Yellow River Basin has obvious geographical distribution characteristics. The trade volume of virtual water in the downstream region is large. The volume of virtual water trade within the upper reaches is low, and the trade links with the middle and lower reaches are insufficient. Henan and Shandong Provinces are the main flow directions in the Yellow River Basin, and Gansu and Inner Mongolia are the dominant virtual water sources.

5.2. Suggestions

From the perspective of planning the utilization of water resources nationwide, the research findings suggest reducing water pressures and virtual water-flow imbalances. The nationwide allocation of water resources and the rational use of precipitation in the territory can not only prevent floods and droughts, but also bring a sufficient water supply to China's industrial and agricultural development and residents' lives. It also has a certain value in transportation and power generation. There is little difference between the topography of North and South China, and water resources can be mobilized from north to south. The Beijing–Hangzhou canal and the middle route of the South-to-North Water Transfer Project are examples of the North–South distribution of water resources in China. In addition to these two major projects, China can also carry out the large-scale networking of rivers, lakes, and other water areas across the country to make them interconnected. In the case of a flood in a certain place, the excess water could be transferred to another area with fewer water resources. In the case of a drought in a certain place, the water resources of other areas could be mobilized to supplement it, so as to avoid the waste of water resources and maximize the utilization of water resources. According to the decision-making process and the deployment of water-resource management in China, the multi-functional properties of water resources could be fully accomplished in the near future. Under the overall framework of the distribution scheme of the available water supply of the Yellow River, taking into account the ecological water demand, sediment transport volume, external water transfer volume, and water-use structure of provinces along the Yellow River Basin, a joint water-supply pattern of the Yangtze River and the Yellow River will be formed, a dynamic water-distribution scheme of the Yellow River Basin will be constructed, the water-right transfer and compensation system will be gradually improved, and the linkage mechanism between water-use indicators and land indicators will be explored. This will result in comprehensively coordinating the relationship between water, energy, and food; limiting water use for fossil energy development; improving the utilization efficiency of agricultural water resources; building a wind–water complementary power generation system, and implementing the transmission, storage, and utilization of hydrogen energy at normal temperatures and pressures. Ideally, water-scarce regions will import water-intensive products to meet the production and consumption of the region, rather than relying on local production, so as to protect the domestic water resources. In this case, the water resources required by the whole production chain actually come from export to import regions through interregional trade.

According to the research results mentioned above, this paper puts forward the following suggestions in order to achieve the rational allocation of water resources:

(1) China should vigorously implement the ecological compensation policy of water usage. Although the region is in the virtual water net input area as a whole, the *WSI* of Gansu and Ningxia is high, which is seriously inconsistent with the virtual water net output state. The utilization of water resources should be distributed comprehensively throughout the country. By reducing the virtual water flow in Gansu and Ningxia, the local ecological development and water resource allocation balance can be protected. China should also appropriately increase the output of virtual water in Sichuan and grasp the advantages of local green water resources. Meanwhile, we recommend increasing the virtual water output from other surplus provinces to Henan and Shandong Provinces, reducing the pressures of water outflow, and ensuring local water safety and ecological security.

(2) All industry sectors should adhere to the principle of "determining production by water". The whole Yellow River Basin should develop water-saving agricultural techniques, change the traditional mode of agricultural production, strictly control the total water, and improve water-usage efficiency. The upper reaches of the Yellow River Basin should enforce the technological innovation investment and water-use efficiency, and the regional water-shortage situation should be alleviated by importing water-intensive products to water-rich areas; The middle reaches should speed up the transformation to a resource-based economy, develop water-saving industries, and vigorously develop the service and transportation industry; The lower reaches should speed up the development of high value-added manufacturing industries and strengthen economic ties inside and outside the region.

(3) China should fully strengthen the exchanges and cooperation between the lower, middle, and upper reaches, and actively explore the institutional mechanism of water ecological protection. China should establish an internal, cooperative development mechanism in this region with the goals of common economic development, water conservation, and ecological protection. Through trade-oriented interprovincial cooperation, China should reduce the intermediate links of ineffective water use, make the virtual water flow to the most needy regions and sectors, improve water sewage efficiency, and drive economic development. China should also comprehensively improve the interprovincial virtual water-trade flow, give full play to the economic ties between the lower, middle, and upper reaches, and jointly realize the sustainable development of the economy, as well as the ecological environment.

Author Contributions: Conceptualization, X.L. and R.X.; methodology, X.L.; software, R.X.; validation, R.X. and X.L.; formal analysis, R.X.; investigation, R.X., X.L., P.G. and L.N.; data curation, W.L.; writing—original draft preparation, R.X.; writing—review and editing, J.C. and Q.S.; visualization, Q.S.; supervision, X.L. All authors have read and agreed to the published version of the manuscript.

Funding: This research was funded by the National Natural Science Foundation of China (Grant No. 42001257 and Grant No. 71874119), the Philosophy and Social Sciences Research of Higher Learning Institutions of Shanxi (Grant No. 20210115), the Shanxi Postgraduate Education Innovation Project (Grant No. 2021Y571), and the Research Project of Social and Economic Statistics in Shanxi Province (Grant No. KY2021113).

Institutional Review Board Statement: Ethical review and approval were waived for this study, due to studies not involving humans or animals.

Informed Consent Statement: Informed consent was obtained from all subjects involved in the study.

Data Availability Statement: The input-output table data are from *China's carbon accounting database*, and the water use and GDP data can be obtained from *China's statistical yearbook*.

Acknowledgments: We also thank our anonymous reviewers and the editor for their helpful suggestions.

Conflicts of Interest: The authors declare no conflict of interest.

References

1. Abd Ellah, R.G. Water resources in Egypt and their challenges, Lake Nasser case study. *Egypt J. Aqua. Res.* **2020**, *46*, 1–12. [CrossRef]
2. Fidélis, T.; Cardoso, A.S.; Riazi, F.; Miranda, A.C.; Abrantes, J.; Teles, F.; Roebeling, P.C. Policy narratives of circular economy in the EU—Assessing the embeddedness of water and land in national action plans. *J. Clean. Prod.* **2021**, *288*, 125685. [CrossRef]
3. Kundzewicz, Z.W. Global freshwater resources for sustainable development. *Ecohydrol. Hydrobiol.* **2007**, *7*, 125–134. [CrossRef]
4. Karimidastenaei, Z.; Avellán, T.; Sadegh, M.; Kløve, B.; Haghighi, A.T. Unconventional water resources: Global opportunities and challenges. *Sci. Total Environ.* **2022**, *827*, 154429. [CrossRef]
5. Vahidipour, M.; Raeisi, E.; van der Zee, S.E.A.T. Active saltwater intrusion of shrinking Bakhtegan -Thask Lakes in South Iran threatens the freshwater resources of coastal aquifers. *J. Hydrol.-Reg. Stud.* **2021**, *34*, 100790. [CrossRef]
6. Li, R.; Wu, M.; Aleid, S.; Zhang, C.; Wang, W.; Wang, P. An integrated solar-driven system produces electricity with fresh water and crops in arid regions. *Cell Rep. Phys. Sci.* **2022**, *3*, 100781. [CrossRef]
7. Alsulaili, A.; Alkandari, M.; Buqammaz, A. Assessing the impacts of meteorological factors on freshwater consumption in arid regions and forecasting the freshwater demand. *Environ. Technol. Innov.* **2022**, *25*, 102099. [CrossRef]
8. Elbeih, S.F.; Madani, A.A.; Hagage, M. Groundwater deterioration in Akhmim District, Upper Egypt: A Remote Sensing and GIS investigation approach. *Egypt J. Remote. Sens.* **2021**, *24*, 919–932. [CrossRef]
9. De, O.; Vieira, E.; Sandoval-Solis, S. Water resources sustainability index for a water-stressed basin in Brazil. *J. Hydrol.-Reg. Stud.* **2018**, *19*, 97–109.
10. Ercin, E. Overuse of Water Resources: Water Stress and the Implications for Food and Agriculture-ScienceDirect. *Encycl. Food Secur. Sustain.* **2019**, *1*, 206–211.
11. Hemmat Esfe, M.; Toghraie, D. Numerical study on the effect of solar radiation intensity on the fresh water productivity of solar still equipped with Thermoelectric Cooling System (TEC) for hot and dry areas of Semnan. *Case Stud. Therm. Eng.* **2022**, *32*, 101848. [CrossRef]
12. Muratoglu, A. Water footprint assessment within a catchment: A case study for Upper Tigris River Basin. *Ecol. Indic.* **2019**, *106*, 105467. [CrossRef]
13. Salehi, M. Global water shortage and potable water safety; Today's concern and tomorrow's crisis. *Environ. Int.* **2022**, *158*, 106936. [CrossRef]
14. Ahmed, M.; Wiese, D.N. Short-term trends in Africa's freshwater resources: Rates and drivers. *Sci. Total. Environ.* **2019**, *695*, 133843. [CrossRef]
15. Baluyot, J.C.; Reyes, E.M.; Velarde, M.C. Per- and polyfluoroalkyl substances (PFAS) as contaminants of emerging concern in Asia's freshwater resources. *Environ. Res.* **2021**, *197*, 111122. [CrossRef]
16. Baccour, S.; Albiac, J.; Kahil, T.; Esteban, E.; Crespo, D.; Dinar, A. Hydroeconomic modeling for assessing water scarcity and agricultural pollution abatement policies in the Ebro River Basin, Spain. *J. Clean. Prod.* **2021**, *327*, 129459. [CrossRef]
17. Sabzevar, M.S.; Rezaei, A.; Khaleghi, B. Incremental adaptation strategies for agricultural water management under water scarcity condition in Northeast Iran. *Reg. Sustain.* **2021**, *2*, 224–238. [CrossRef]
18. Li, K.; Hajar, S.; Ding, Z.; Dooling, T.; Wei, G.; Hu, C.; Zhang, Y.; Zhang, K. Dynamic optimization of input production factors for urban industrial water supply and demand management. *J. Environ. Manag.* **2020**, *270*, 110807. [CrossRef]

19. Murali, S.; Krishnan, V.S.; Amulya, P.R.; Alfiya, P.V.; Delfiya, D.S.A.; Samuel, M.P. Energy and water consumption pattern in seafood processing industries and its optimization methodologies. *Clean Eng. Technol.* **2021**, *4*, 100242. [CrossRef]
20. Karandish, F.; Hogeboom, R.J.; Hoekstra, A.Y. Physical versus virtual water transfers to overcome local water shortages: A comparative analysis of impacts. *Adv. Water Resour.* **2021**, *147*, 103811. [CrossRef]
21. Adrielly Nahomee, R.Á.; de Fátima, M.S.G.; Mota Idalia, F.D.L.; Francisca Irene, S.A. Reaching sustainability through a smart water crisis-proof industry. *Process Comput. Sci.* **2021**, *180*, 86–92. [CrossRef]
22. Li, X.; Zhang, L.; Zheng, Y.; Yang, D.; Wu, F.; Tian, Y.; Han, F.; Gao, B.; Li, H.; Zhang, Y.; et al. Novel hybrid coupling of ecohydrology and socioeconomy at river basin scale: A watershed system model for the Heihe River basin. *Environ. Modell. Softw.* **2021**, *141*, 105058. [CrossRef]
23. Zhang, Y.; Deng, J.; Qin, B.; Zhu, G. Importance and vulnerability of lakes and reservoirs supporting drinking water in China. *Fund. Res.* **2022**, *1*, 35. [CrossRef]
24. Cao, T.; Wang, S.; Chen, B. Water shortage risk transferred through interprovincial trade in Northeast China. *Energy Procedia* **2019**, *158*, 3865–3871. [CrossRef]
25. Mu, Y.; Li, X.; Guo, Y.; Liang, C.; Bai, J.; Linke, S.; Cui, B. Using climatic-geomorphological surrogates to identify complete and incidental freshwater conservation gaps within large river basins in China. *Glob. Ecol. Conserv.* **2021**, *30*, e01744. [CrossRef]
26. Chen, C.; Jiang, Y.; Ye, Z.; Yang, Y.; Hou, L.A. Sustainably integrating desalination with solar power to overcome future freshwater scarcity in China. Glob. *Energy Intercon.* **2019**, *2*, 98–113.
27. Li, D.; Zuo, Q.; Zhang, Z. A new assessment method of sustainable water resources utilization considering fair-ness-efficiency-security: A case study of 31 provinces and cities in China. *Sustain. Cities Soc.* **2022**, *81*, 103839. [CrossRef]
28. Qin, J.; Ding, Y.; Zhao, Q.; Wang, S.; Chang, Y. Assessments on surface water resources and their vulnerability and adapta-bility in China. *Adv. Clim. Chang. Res.* **2020**, *11*, 381–391. [CrossRef]
29. Liu, X.; Tan, T.; Bai, Y.; Chou, L. Restoration performance of regional soil and water resources in China based on index of coupling and improved assessment tool. *Alex. Eng. J.* **2022**, *61*, 5677–5686. [CrossRef]
30. An, T.; Wang, L.; Gao, X.; Han, X.; Zhao, Y.; Lin, L.; Wu, P. Simulation of the virtual water flow pattern associated with in-terprovincial grain trade and its impact on water resources stress in China. *J. Clean. Prod.* **2021**, *288*, 125670. [CrossRef]
31. Deng, J.; Li, C.; Wang, L.; Yu, S.; Zhang, X.; Wang, Z. The impact of water scarcity on Chinese inter-provincial virtual water trade. *Sustain. Prod. Consump.* **2021**, *28*, 1699–1707. [CrossRef]
32. Lin, L.; Gao, X.; Zhao, Y.; Wang, L.; An, T.; Liu, C.; Qiao, Y.; Wu, P. Evaluation of the water consumption of animal products and the virtual water flow pattern associated with interprovincial trade in China. *J. Clean. Prod.* **2021**, *328*, 129599. [CrossRef]
33. Gu, D.; Guo, J.; Fan, Y.; Zuo, Q.; Yu, L. Evaluating water-energy-food system of Yellow River basin based on type-2 fuzzy sets and Pressure-State-Response model. *Agr. Water Manag.* **2022**, *267*, 107607. [CrossRef]
34. Li, X.; Zhang, X.; Wang, S. Managing conflicts and equitability in hierarchical decision making for water resources planning under fuzzy uncertainty: A case study of Yellow River, China. *J. Hydrol.-Reg. Stud.* **2021**, *38*, 100963. [CrossRef]
35. Jiang, L.; Zuo, Q.; Ma, J.; Zhang, Z. Evaluation and prediction of the level of high-quality development: A case study of the Yellow River Basin, China. *Ecol. Indic.* **2021**, *129*, 107994. [CrossRef]
36. Li, M.; Tian, Q.; Yu, Y.; Xu, Y.; Li, C. Virtual Water Trade in the Yellow River Economic Belt: A Multi-Regional Input-Output Model. *Water* **2021**, *13*, 748. [CrossRef]
37. Fan, G.; Zhang, D.; Zhang, J.; Li, Z.; Sang, W.; Zhao, L.; Xu, M. Ecological environmental effects of Yellow River irrigation revealed by isotope and ion hydrochemistry in the Yinchuan Plain, Northwest China. *Ecol. Indic.* **2022**, *135*, 108574. [CrossRef]
38. Chen, Y.; Zhu, M.; Lu, J.; Zhou, Q.; Ma, W. Evaluation of ecological city and analysis of obstacle factors under the background of high-quality development: Taking cities in the Yellow River Basin as examples. *Ecol. Indic.* **2020**, *118*, 106771. [CrossRef]
39. Zuo, Q.; Guo, J.; Ma, J.; Cui, G.; Yang, R.; Yu, L. Assessment of regional-scale water resources carrying capacity based on fuzzy multiple attribute decision-making and scenario simulation. *Ecol. Indic.* **2021**, *130*, 108034. [CrossRef]
40. Khorsandi, M.; Homayouni, S.; van Oel, P. The edge of the petri dish for a nation: Water resources carrying capacity as-sessment for Iran. *Sci. Total Environ.* **2022**, *817*, 153038. [CrossRef]
41. Feng, K.; Siu, Y.L.; Guan, D.; Hubacek, K. Assessing regional virtual water flows and water footprints in the Yellow River Basin, China: A consumption based approach. *Appl. Geogr.* **2012**, *32*, 691–701. [CrossRef]
42. Zhao, Y.; Wang, Y.; Wang, Y. Comprehensive evaluation and influencing factors of urban agglomeration water resources carrying capacity. *J. Clean. Prod.* **2021**, *288*, 125097. [CrossRef]
43. Peng, T.; Deng, H.; Lin, Y.; Jin, Z. Assessment on water resources carrying capacity in karst areas by using an innovative DPESBRM concept model and cloud model. *Sci. Total Environ.* **2021**, *767*, 144353. [CrossRef] [PubMed]
44. Allan, J.A. Fortunately There Are Substitutes for Water: Otherwise Our Hydropolitical Futures Would be Impossible. *Priori-Ties Water Res. Alloca. Manag.* **1993**, *13*, 26.
45. Oreggioni, F.; Garcia, S.; Gomez, M.; Mejia, A. A machine learning model of virtual water networks over time. *Adv. Water Resour.* **2021**, *147*, 103819. [CrossRef]
46. Almazán-Gómez, M.A.; Duarte, R.; Langarita, R.; Sánchez-Chóliz, J. Effects of water re-allocation in the Ebro river basin: A multiregional input-output and geographical analysis. *J. Environ. Manag.* **2019**, *241*, 645–657. [CrossRef]
47. Grazzini, J.; Spelta, A. An empirical analysis of the global input–output network and its evolution. *Physica A* **2022**, *594*, 126993. [CrossRef]

48. Sun, J.X.; Yin, Y.L.; Sun, S.K.; Wang, Y.B.; Yu, X.; Yan, K. Review on research status of virtual water: The perspective of ac-counting methods, impact assessment and limitations. *Agr. Water Manag.* **2021**, *243*, 106407. [CrossRef]
49. Zhang, Y.; Hou, S.; Chen, S.; Long, H.; Liu, J.; Wang, J. Tracking flows and network dynamics of virtual water in electricity transmission across China. Renew. *Renew. Sustain. Energy Rev.* **2021**, *137*, 110475. [CrossRef]
50. Garcia, S.; Mejia, A. Characterizing and modeling subnational virtual water networks of US agricultural and industrial commodity flows. *Adv. Water. Resour.* **2019**, *130*, 314–324. [CrossRef]
51. Cegar, S. Water extended input-output analysis of the Croatian economy. *J. Econ. Bus.* **2020**, *38*, 147–182.
52. Gkatsikos, A.; Mattas, K. The Paradox of the Virtual Water Trade Balance in the Mediterranean Region. *Sustainability* **2021**, *13*, 2978. [CrossRef]
53. Houyin, L.; Yangting, O.; Hong, Z. Water footprint and virtual water flows embodied in China's supply chain. *Int. J. Logist.* **2021**, *25*, 1–16. [CrossRef]
54. Zhang, Y.; Fu, Z.; Xie, Y.; Li, Z.; Liu, Y.; Zhang, B.; Guo, H. Dynamic metabolism network simulation for energy-water nexus analysis: A case study of Liaoning Province, China. *Sci. Total. Environ.* **2021**, *779*, 146440. [CrossRef]
55. Yang, Y.; Liu, S.; Xiao, C.; Feng, C.; Li, C. Evaluating Cryospheric Water Withdrawal and Virtual Water Flows in Tarim River Basin of China: An Input-Output Analysis. *Sustainability* **2021**, *13*, 7589. [CrossRef]
56. Fu, Y.; Huang, G.; Liu, L.; Li, J.; Pan, X. Multi-hierarchy virtual-water management–A Case Study of Hubei Province, China. *J. Clean. Prod.* **2021**, *293*, 126244. [CrossRef]
57. Zhang, J.L.; Jiang, G.Q.; Yang, L.B.; Zhao, Y.T. Study on Virtual Water Consumption and Trade Based on Input-output Analysis in Inner Mongolia. *Water Res. Power* **2018**, *36*, 52–54.
58. Islam, K.M.N.; Kenway, S.J.; Renouf, M.A.; Wiedmann, T.; Lam, K.L. A multi-regional input-output analysis of direct and virtual urban water flows to reduce city water footprints in Australia. *Sustain. Cities Soc.* **2021**, *75*, 103236. [CrossRef]
59. Qasemipour, E.; Tarahomi, F.; Pahlow, M.; Malek, S.S.S.; Abbasi, A. Assessment of Virtual Water Flows in Iran Using a Multi-Regional Input-Output Analysis. *Sustainability* **2020**, *12*, 7424. [CrossRef]
60. Dong, H.; Geng, Y.; Hao, D.; Yu, Y.; Chen, Y. Virtual water flow feature of water-rich province and the enlightenments: Case of Yunnan in China. *J. Clean. Prod.* **2019**, *235*, 328–336. [CrossRef]
61. Chen, W.; Wu, S.; Lei, Y.; Li, S. China's water footprint by province, and inter-provincial transfer of virtual water. *Ecol. Indic.* **2017**, *74*, 321–333. [CrossRef]
62. Wang, S.; Chen, B. Energy-water nexus of urban agglomeration based on multiregional input–output tables and ecological network analysis: A case study of the Beijing–Tianjin–Hebei region. *Appl. Energ.* **2016**, *178*, 773–783. [CrossRef]
63. Zhang, W.; Fan, X.; Liu, Y.; Wang, S.; Chen, B. Spillover risk analysis of virtual water trade based on multi-regional in-put-output model—A case study. *J. Environ. Manag.* **2020**, *275*, 111242. [CrossRef]
64. Wang, X.; Chen, Y.; Li, Z.; Fang, G.; Wang, Y. Development and utilization of water resources and assessment of water se-curity in Central Asia. *Agr. Water. Manag.* **2020**, *240*, 106297. [CrossRef]
65. Zhang, J.; Chen, Y.; Li, Z.; Song, J.; Fang, G.; Li, Y.; Zhang, Q. Study on the utilization efficiency of land and water resources in the Aral Sea Basin, Central Asia. *Sustain. Cities Soc.* **2019**, *51*, 101693. [CrossRef]
66. Zheng, X.; Huang, G.; Liu, L.; Zheng, B.; Zhang, X. A multi-source virtual water metabolism model for urban systems. *J. Clean. Prod.* **2020**, *275*, 124107. [CrossRef]
67. Wang, Y.; Xian, C.; Jiang, Y.; Pan, X.; Ouyang, Z. Anthropogenic reactive nitrogen releases and gray water footprints in urban water pollution evaluation: The case of Shenzhen City, China. *Environ. Dev. Sustain.* **2020**, *22*, 6343–6361. [CrossRef]
68. Rosales-Asensio, E.; García-Moya, F.J.; González-Martínez, A.; Borge-Diez, D.; de Simón-Martín, M. Stress mitigation of conventional water resources in water-scarce areas through the use of renewable energy powered desalination plants: An application to the Canary Islands. *Energy Rep.* **2020**, *6*, 124–135. [CrossRef]
69. Gohar, A.A.; Cashman, A.; Ward, F.A. Managing food and water security in Small Island States: New evidence from economic modelling of climate stressed groundwater resources. *J. Hydrol.* **2019**, *569*, 239–251. [CrossRef]
70. Sun, S.; Wang, Y.; Engel, B.A.; Wu, P. Effects of virtual water flow on regional water resources stress: A case study of grain in China. *Sci. Total Environ.* **2016**, *550*, 871–879. [CrossRef] [PubMed]
71. Liu, M.; Wei, J.; Wang, G.; Wang, F. Water resources stress assessment and risk early warning—A case of Hebei Province China. *Ecol. Indic.* **2017**, *73*, 358–368. [CrossRef]
72. Yin, Y.L.; Sun, S.K.; Wang, Y.B.; Li, C.; Sun, J.X.; Wu, P.T. Impact of grain virtual water flow on surface water and ground-water in China. *Adv. Water Resour.* **2021**, *150*, 103848. [CrossRef]
73. Liu, J.; Chen, X.; Shi, W.; Chen, P.; Zhang, Y.; Hu, J.; Dong, S.; Li, T. Tectonically controlled evolution of the Yellow River drainage system in the Weihe region, North China: Constraints from sedimentation, mineralogy and geochemistry. *J. Asian Earth. Sci.* **2019**, *179*, 350–364. [CrossRef]
74. Leontief, W. Quantitative Input and Output Relations in the Economic Systems of the United States. *Rev. Econ. Stat.* **1936**, *18*, 105–125. [CrossRef]
75. Rocco, M.V.; Colombo, E. Evaluating energy embodied in national products through Input-Output analysis: Theoretical definition and practical application of international trades treatment methods. *J. Clean. Prod.* **2016**, *139*, 1449–1462. [CrossRef]
76. Guo, J.; Zhang, Y.; Zhang, K. The key sectors for energy conservation and carbon emissions reduction in China: Evidence from the input-output method. *J. Clean. Prod.* **2018**, *179*, 180–190. [CrossRef]

77. Richter, J.S.; Mendis, G.P.; Nies, L.; Sutherland, J.W. A method for economic input-output social impact analysis with ap-plication to U.S. advanced manufacturing. *J. Clean. Prod.* **2019**, *212*, 302–312. [CrossRef]
78. Pfister, S.; Koehler, A.; Hellweg, S. Assessing the environmental impacts of freshwater consumption in LCA. *Environ. Sci. Technol.* **2009**, *43*, 4098–4104. [CrossRef]
79. Hong, S.; Wang, H.; Cheng, T. Analysis of water use characteristics in industrial sectors in Beijing based on an input-output method. *Water Sci. Technol.* **2020**, *20*, 219–230. [CrossRef]
80. Boero, R.; Pasqualini, D. Regional water coefficients for U.S. industrial sectors. *Water Resour. Ind.* **2017**, *18*, 60–70. [CrossRef]
81. White, D.J.; Feng, K.; Sun, L.; Hubacek, K. A hydro-economic MRIO analysis of the Haihe River Basin's water footprint and water stress. *Ecol. Model.* **2015**, *318*, 157–167. [CrossRef]
82. Zheng, H.; Zhang, Z.; Wei, W.; Song, M.; Guan, D. Regional determinants of China' consumption-based emissions in the economic transition. *Environ. Res. Lett.* **2020**, *15*, 1748–9326. [CrossRef]
83. Zhang, K.; Lu, H.; Tian, P.; Guan, Y.; Kang, Y.; He, L.; Fan, X. Analysis of the relationship between water and energy in China based on a multi-regional input-output method. *J. Environ. Manag.* **2022**, *309*, 114680. [CrossRef]
84. Wang, F.; Cai, B.; Hu, X.; Liu, Y.; Zhang, W. Exploring solutions to alleviate the regional water stress from virtual water flows in China. *Sci. Total Environ.* **2021**, *796*, 148971. [CrossRef]
85. Liu, Z.; Huang, Q.; He, C.; Wang, C.; Wang, Y.; Li, K. Water-energy nexus within urban agglomeration: An assessment framework combining the multiregional input-output model, virtual water, and embodied energy. *Resour. Conserv. Recy.* **2021**, *164*, 105113. [CrossRef]
86. Gao, X.; Chen, Q.; Lu, S.; Wang, Y.; An, T.; Zhuo, L.; Wu, P. Impact of virtual water flow with the energy product transfer on sustainable water resources utilization in the main coal-fired power energy bases of Northern China. *Energy Procedia* **2018**, *152*, 293–301. [CrossRef]
87. Zhang, S.; Taiebat, M.; Liu, Y.; Qu, S.; Liang, S.; Xu, M. Regional water footprints and interregional virtual water transfers in China. *J. Clean. Prod.* **2019**, *228*, 1401–1412. [CrossRef]
88. Shao, L.; Guan, D.; Wu, Z.; Wang, P.; Chen, G.Q. Multi-scale input-output analysis of consumption-based water resources: Method and application. *J. Clean. Prod.* **2017**, *164*, 338–346. [CrossRef]
89. Shi, C.; Zhan, J. An input–output table based analysis on the virtual water by sectors with the five northwest provinces in China. *Phys. Chem. Earth* **2015**, *79–82*, 47–53. [CrossRef]
90. Chen, W.; Wu, S.; Lei, Y.; Li, S. Virtual water export and import in China's foreign trade: A quantification using input-output tables of China from 2000 to 2012. *Resour. Conserv. Recy.* **2018**, *132*, 278–290. [CrossRef]
91. Hirwa, H.; Peng, Y.; Zhang, Q.; Qiao, Y.; Leng, P.; Tian, C.; Yang, G.; Muhirwa, F.; Diop, S.; Kayiranga, A.; et al. Virtual water transfers in Africa: Assessing topical condition of water scarcity, water savings, and policy implications. *Sci. Total Environ.* **2022**, *835*, 155343. [CrossRef] [PubMed]
92. Zhang, Y.; Li, J.; Tian, Y.; Deng, Y.; Xie, K. Virtual water flow associated with interprovincial coal transfer in China: Impacts and suggestions for mitigation. *J. Clean. Prod.* **2021**, *289*, 125800. [CrossRef]
93. Xin, M.; Wang, J.; Xing, Z. Decline of virtual water inequality in China's inter-provincial trade: An environmental economic trade-off analysis. *Sci. Total Environ.* **2022**, *806*, 150524. [CrossRef] [PubMed]
94. Han, X.; Zhang, Y.; Wang, H.; Shi, H. Analyzing the driving mechanisms of grain virtual water flow based on the case of China's main grains. *Environ. Sci. Policy.* **2021**, *124*, 645–655. [CrossRef]
95. Han, X.; Zhao, Y.; Gao, X.; Jiang, S.; Lin, L.; An, T. Virtual water output intensifies the water scarcity in Northwest China: Current situation, problem analysis and countermeasures. *Sci. Total Environ.* **2021**, *765*, 144276. [CrossRef]
96. Yin, Y.; Luan, X.; Sun, S.; Wang, Y.; Wu, P.; Wang, X. Environmental impact of grain virtual water flows in China: From 1997 to 2014. *Agr. Water. Manag.* **2021**, *256*, 107127. [CrossRef]
97. Liu, G.; Zhang, F.; Deng, X. Is virtual water trade beneficial for the water-deficient regions? New evidences from the Yellow River Basin, China. *J. Hydrol.-Reg. Stud.* **2021**, *38*, 100964. [CrossRef]
98. Xia, W.; Chen, X.; Song, C.; Pérez-Carrera, A. Driving factors of virtual water in international grain trade: A study for belt and road countries. *Agr. Water. Manag.* **2022**, *262*, 107441. [CrossRef]
99. Zhai, Y.; Bai, Y.; Shen, X.; Ji, C.; Zhang, T.; Hong, J. Can grain virtual water flow reduce environmental impacts? Evidence from China. *J. Clean. Prod.* **2021**, *314*, 127970. [CrossRef]
100. Cao, X.; Cui, S.; Shu, R.; Wu, M. Misestimation of water saving in agricultural virtual water trade by not considering the role of irrigation. *Agr. Water. Manag.* **2020**, *241*, 106355. [CrossRef]
101. Zhang, F.; Xuan, X.; He, Q. A water-energy nexus analysis to a sustainable transition path for Ji-shaped bend of the Yellow River, China. *Ecol. Inform.* **2022**, *68*, 101578. [CrossRef]
102. Sun, J.; Sun, S.; Yin, Y.; Wang, Y.; Zhao, X.; Wu, P. Evaluating grain virtual water flow in China: Patterns and drivers from a socio-hydrology perspective. *J. Hydrol.* **2022**, *606*, 127412. [CrossRef]
103. Arunrat, N.; Sereenonchai, S.; Hatano, R. Effects of fire on soil organic carbon, soil total nitrogen, and soil properties under rotational shifting cultivation in northern Thailand. *J. Environ. Manag.* **2022**, *302*, 113978. [CrossRef]
104. Arunrat, N.; Sereenonchai, S.; Chaowiwat, W.; Wang, C. Climate change impact on major crop yield and water footprint under CMIP6 climate projections in repeated drought and flood areas in Thailand. *Sci. Total Environ.* **2022**, *807*, 150741. [CrossRef]

105. Thomas, M.; Robertson, J.; Fukai, S.; Peoples, M.B. The effect of timing and severity of water deficit on growth development, yield accumulation and nitrogen fixation of mung bean. *Field Crops Res.* **2004**, *86*, 67–68. [CrossRef]
106. Chen, Y.; Huang, K.; Hu, J.; Yu, Y.; Wu, L.; Hu, T. Understanding the two-way virtual water transfer in urban agglomeration: A new perspective from spillover-feedback effects. *J. Clean. Prod.* **2021**, *310*, 127495. [CrossRef]
107. Deng, G.; Lu, F.; Wu, L.; Xu, C. Social network analysis of virtual water trade among major countries in the world. *Sci. Total Environ.* **2021**, *753*, 142043. [CrossRef]
108. Pomponi, F.; Stephan, A. Water, energy, and carbon dioxide footprints of the construction sector: A case study on developed and developing economies. *Water Res.* **2021**, *194*, 116935. [CrossRef]
109. Khalkhali, M.; Dilkina, B.; Mo, W. The role of climate change and decentralization in urban water services: A dynamic en-ergy-water nexus analysis. *Water Res.* **2021**, *207*, 117830. [CrossRef]
110. Qiao, R.; Li, H.; Han, H. Spatio-Temporal Coupling Coordination Analysis between Urbanization and Water Resource Car-rying Capacity of the Provinces in the Yellow River Basin, China. *Water* **2021**, *13*, 376. [CrossRef]

water

MDPI

Article

The Effects of Climate Change on the Tagus–Segura Transfer: Diagnosis of the Water Balance in the Vega Baja del Segura (Alicante, Spain)

Antonio Oliva Cañizares *, Jorge Olcina Cantos and Carlos J. Baños Castiñeira

University of Alicante, 03690 San Vicente del Raspeig, Spain; jorge.olcina@ua.es (J.O.C.); carlos.banos@ua.es (C.J.B.C.)
* Correspondence: antoniogeografia1@gmail.com

Abstract: Climate change is one of the most important problems facing society in the 21st century. Despite the uncertainty about the behaviour of rainfall due to climate change, what is clear is that average rainfall has been reduced in the inland areas and headwaters of Spain's river basins. The Tagus basin is one of the most affected, with implications for the Jucar and Segura basins. The working hypothesis is to corroborate with the data collected on the effects of climate change on the TTS. To this end, the following methodology has been applied: (a) analysis in the headwaters of the Tagus, using data on precipitation, surface runoff and reservoir water; (b) analysis of the resources of the Segura basin (supply and demand), based on the basin organisation's own data; (c) construction of a water balance adjusted to the Bajo Segura district (Alicante), a user of the water transferred for agricultural use. Likewise, the data provided by the basin organisation have made it possible to corroborate the data on consumption and allocation of the corresponding volumes of water. The results obtained make it possible to put forward a novel proposal in the scientific field related to hydrological planning based on the principles of sustainability.

Keywords: climate change; Tagus–Segura transfer; sustainable hydrological planning; Bajo Segura

Citation: Cañizares, A.O.; Cantos, J.O.; Baños Castiñeira, C.J. The Effects of Climate Change on the Tagus–Segura Transfer: Diagnosis of the Water Balance in the Vega Baja del Segura (Alicante, Spain). *Water* **2022**, *14*, 2023. https://doi.org/10.3390/w14132023

Academic Editor: Adriana Bruggeman

Received: 25 May 2022
Accepted: 22 June 2022
Published: 24 June 2022

1. Introduction

The sixth report of the Intergovernmental Panel on Climate Change (IPCC) (AR6) published in 2021 identifies the Mediterranean region (MED) as one of the areas or hotspots most affected by climate change on a global level [1]. Spain is located in this region. This country is already suffering from the impacts (social and economic losses) of the effects of climate change, particularly those related to meteorological phenomena, such as floods, droughts, heat and cold waves, sea storms and forest fires, among others.

Furthermore, Spain has a wide variety of climates within its territory, which shape different landscapes, depending on a series of physical elements, such as the geographic position, altitude, relief, proximity to the sea, the vegetation and fauna and flora of each territory.

In terms of water resources, this variety of climates implies a structural problem for Spain, in which two types of area can be distinguished: humid Spain, (the north-east, north and centre of the peninsula), which receives large annual amounts of precipitation with surplus water resources; and dry Spain (the east, south-east and south of the peninsula), where the average annual rainfall is very low, leading to deficits in the availability of water resources [2]. Paradoxically, some of the optimal lands for growing crops are found in parts of Spain with a water resource deficit, where the soil is rich and favourable for agricultural activities and where irrigated crops predominate over rain-fed crops [2].

The most important transfer in Spain constructed in the twentieth century is the Tagus–Segura Transfer (hereafter, TTS). This is the most important infrastructure given the volume of water it transfers, the areas it supplies and the political and media repercussions [2,3].

The introduction of the TTS has contributed a significant amount of water resources for both urban supply and for agriculture (irrigation) in south-east Spain. This contribution of water resources is considerable but insufficient, as it has only fulfilled the transfer volumes contemplated in the Preliminary Project of the Transfer once (water year 2000/2001) [4–6].

It is worth highlighting that in the 1990s (emergence of climate change hypothesis) to the present (2022) (complete confidence in climate change), the successive situations of rainfall droughts that affected the centre and south-east of the peninsula revealed that the TTS was vulnerable to extreme weather situations. In other words, the absence of rainfall in the Tagus headwaters for prolonged periods of time has a serious impact on the TTS [4–6]. Some authors calculated a period of NO transfers of water resources for approximately 15 months. Furthermore, these authors point out that by applying the same methodology, it could be said that the TTS would have been inoperative for a total of 59 months (5 years), from the hydrological year 2004/2005 to 2017/2018 [4].

As noted above, the TTS is vulnerable to extreme atmospheric situations (droughts). This problem is negatively aggravated considering the climate scenarios of emissions and effects on temperatures and precipitation in Spain [2,3,7–9], and, in specific terms, in the south-east peninsular region [10–12], from 2020 to 2050 and from 2015 to 2100, especially in river basins such as the Tagus and Segura.

Therefore, the relationship between the effects of climate change and water resources deserves special attention, given that the variations in climate that are occurring on a global scale are generating a series of effects in the Spanish territory, which is supported by rigorous scientific data. Some of the most concerning effects are the variation in atmospheric dynamics [12]; the increase in temperatures and the variations in rainfall [7,10–13]; the increase in temperature and the increase in sea level [14–18].

All of this has direct repercussions on water planning, given that the headwaters and resources in the basins are highly important for the development of agricultural activity, as they favour the accumulation of water resources in reservoirs and aquifers. Future water plans (third water planning period (2022–2027) and those of the following decades) should simultaneously contemplate solutions to address the reduction in the volumes of useful rainwater and the occurrence (ever more frequent) of intense or torrential rains leading to floods that cause increasing economic damage [10].

Within this context, one of the principal ways to obtain water resources which are not subject to climate variations is by increasing the so-called non-conventional sources, particularly wastewater treatment and desalination.

The former is subject to the water consumption of the population in a year. In this respect, there is a directly proportional relationship: the more water consumed by the population, the greater the availability of treated water, and vice versa. Regenerated water undoubtedly constitutes a buffer for the water resources of the basin and has been incorporated in the current water plan, acting as a complementary source to the resources of the TTS. However, from the end of the twentieth century and the beginning of the twenty-first century, the need to reuse treated water has arisen, giving rise to the passing of Royal Decree 1620/2007 of 7 December, establishing the legal framework for the reuse of treated wastewater, thereby promoting the development of the reuse of treated water and incorporating it into the water resources plan, provided that public health and environmental protection can be guaranteed and establishing the necessary requirements to enable or prohibit the use of treated or regenerated water, according to the afore-mentioned regulation [19,20].

Similarly to treated wastewater, desalinated water does not depend on climate variation. It only depends on its own daily and annual production capacity. Desalination is playing an increasingly prominent role in the hydrological planning of the basins. In Spain, the promotion of desalination began with the repeal of the Ebro Transfer project, through the passing of Royal Decree Law 2/2004 of 18 June, which modified Law 10/2001 of 5 July of the National Hydrological Plan and the implementation of the A.G.U.A. Programme, and subsequently with the passing of Law 11/2005 of 22 June which modified Law 10/2001 of 5 July of the National Hydrological Plan [4,21].

This led to the planning and construction of large desalination plants along the Spanish Mediterranean coast. The largest desalination plant is that of Torrevieja. It has a current capacity of 80 hm^3/year and is managed by a state entity (ACUAMED). It has the largest capacity in Spain and one of the largest in Europe [22].

Desalination is characterised by being a strategic source and has been used in situations of severe drought in Spain, cushioning the effects of drought and providing a complementary water resource to the water of the TTS [4,5]. In fact, in situations of severe drought and when no transfers have been carried out, the desalination plant of Torrevieja has operated at full capacity, substituting the role of the transfer, for urban supply and irrigation [4,5].

While the rest of Europe uses desalinated water basically for urban supply, Spain is the pioneer in the use of desalinated water for agriculture and irrigation, given the water scarcity of the region [8,22].

Desalination in Spain has emerged in response to the transfer policy promoted by the former hydrological policy [5,6], which does not meet today's sustainability objectives. The desalination is accepted (socially) by the different economic sectors, and in many territories of the Spanish Mediterranean coast it has become a principal resource.

In short, the commitment to managing demand and the use of resources in a way that does not generate a situation in which territories are eternally dependent in terms of water is an essential and irreversible process [23]. It is necessary to break away from the traditional paradigm, based on the continuous supply of resources, which has no place in a scenario of climate change with less rainfall and a reduction in groundwater resources [23]. The growing use of "non-conventional" water resources will become a need in the coming decades on the Spanish Mediterranean coast, within the paradigm of demand and the sustainable use of water [23] (Figure 1).

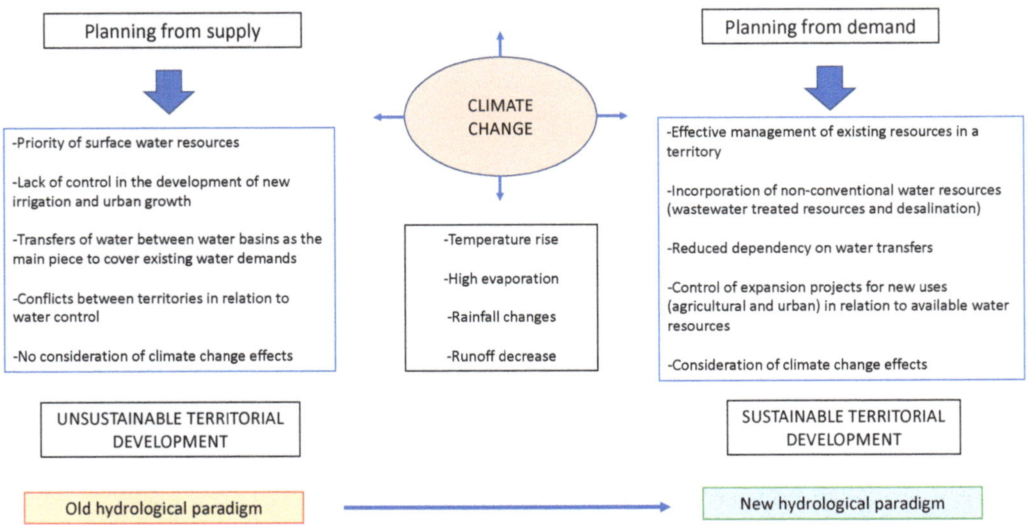

Figure 1. Scheme of the old and new paradigm in water planning in study area. Source: own elaboration.

The working hypothesis of the research is because the Tagus–Segura water transfer is being affected by the effects of climate change, especially regarding the quantity of water resources. To verify this hypothesis, a series of secondary objectives were set to corroborate this approach:

(a) To ascertain the water situation in the headwaters of the Tagus basin.
(b) To analyse the water balance and relevance of the TTS and non-conventional sources in the Segura basin, and their effect on the Lower Segura region.
(c) To drawup a water balance adjusted to the Bajo Segura region to identify the agricultural areas with the greatest water demands.
(d) To propose the basis for a new sustainable hydrological plan based on demand management and the use of own resources that are compromised by the foreseeable effects of climate change in terms of precipitation [24].

2. Materials and Methods

2.1. Background to the Tagus–Segura Transfer (TTS)

Beforehand, it should be put into context that the origin of the Tagus–Segura Transfer project dates back to the First National Hydraulic Works Plan (1933), which basically sought to correct the imbalances between the Atlantic and Mediterranean coasts which, through the so-called "Plan de Mejora y Ampliación de los Riegos de Levante" (Extension and Improvement Irrigation Plan in Spanish Levante region), was based on the transformation of a total area of 338,000 hectares, over the provinces of Murcia, Valencia, Alicante, Almería, Albacete and Cuenca [25].

After the severe drought of 1967, the TTS project was approved in 1969, the works were completed, and the diversion started operating in 1979. The diversion is a canal with a length of 286 km and a flow rate of 33 m^3/s. It links the Bolarque reservoir, in the Tagus basin, with the Talave reservoir, on the river Mundo, the main tributary of the Segura [25]. The cost of the construction of the diversion and post-transfer systems was estimated at ESP 90,000 million (La Verdad newspaper, 18 February 1998), equivalent to EUR 540,910,984 today.

According to the General Proposal for the Joint Management of the Water Resources of Central and South-eastern Spain, Tajo–Segura System, the final objective was to transfer an annual volume of 1000 hm^3. Of this, 640 hm^3/year would be used for irrigation. This objective would be met in two phases: a first phase, with a transfer of 600 hm^3, and a second phase, with a transfer of 40 hm^3. With these estimated volumes, it was expected to transform a total of 90,000 new hectares and complete the allocations of 46,816 existing deficit hectares. The latter were already under cultivation but did not have sufficient volumes of water for optimal irrigation. This was to be solved with the arrival of transferred water.

These planned volumes of water generated a great expectation that led to the transformation of rain-fed crops into irrigated crops (new irrigation), and the area benefited increased to 135,361 hectares. The area contemplated in the TTS project was 136,816, so that some authors indicate that "miraculously, it seemed that the objectives outlined in the preliminary draft had been achieved" [25]. However, the expansion of the surface area occurred during the years of construction of the TTS. Therefore, the increase in irrigated area was only justified using groundwater (indigenous resources) of the territories [25]. This implies that, for example, as in the case of the Vega Baja del Segura, most of the aquifers are overexploited. Consequently, the exploitation of the groundwater resources of the region, which are protected by the official basin organisation (Demarcación Hidrográfica del Segura), is currently prohibited.

Furthermore, this explains why the water from the aqueduct has been insufficient to supply the demand of the Segura basin, as indicated in the respective hydrological plans (2015–2021) and (2022–2027) [26,27], as the water from the aqueduct has been able to maintain, as far as possible, part of the transformed areas. Moreover, to make the best possible use of the water received from the aqueduct for new irrigation, modern irrigation techniques have been introduced, such as drip irrigation.

An interesting point in the Preliminary Project of the Tagus–Segura Transfer is that the demand in the Tagus basin was 1447 hm^3/year and the own adjustable resources amounted to 8152 hm^3/year, while the demand in the Segura basin was 1045 hm^3/year and the available resources (basin) amounted to 820 hm^3/year, revealing the deficit of the basin.

With the construction of the aqueduct, demand in the Tagus basin would be maintained, although the basin's resources would decrease by 1000 hm^3/year due to the transfer of water from one basin to another, estimated at 7152 hm^3/year. As for the Segura basin, demand would remain the same, but the available basin resources would increase thanks to the volumes of water transferred amounting to 2120 hm^3/year (offer in the Segura basin). These figures have never been reached, as the Preliminary Project did not contemplate the possibility of an increase in demand in both basins (Tagus and Segura) or the absence of some transfers, for months or even years. In fact, the values estimated in the Preliminary Project have never been obtained with the transferred water.

2.2. Area of Study

Given that the research is focused on the Tagus–Segura water transfer, the chosen study area is divided into three parts. These three parts coincide with the presentation of the sections in the Results section (Figure 2).

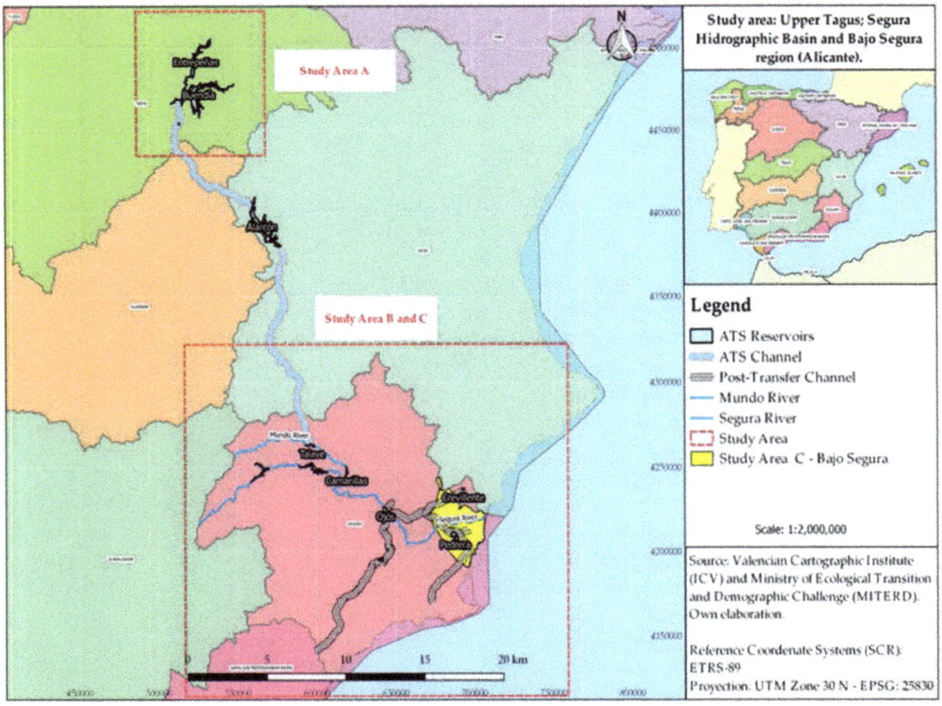

Figure 2. Study area: (**A**) Upper Tagus; (**B**) Segura Hidrological Basin; and (**C**) Bajo Segura Region (Alicante). Source: Valencian Cartographic Institute (ICV) and Demarcación Hidrográficadel Tajo (DHT) and Segura (DHS). Own elaboration.

The first area of analysis focuses on the river Tagus basin, specifically in the headwaters of the river Tagus, belonging to the sub-basin known as the Upper Tagus. The source of the river Tagus, the presence of two large reservoirs (Entrepeñas and Buendía) and the beginning of the hydraulic infrastructure of the Tagus–Segura Transfer are in this area (Figure 2). The effects of climate change calculated and estimated by the official basin organisation itself are also considered, in order to indicate the behaviour or trend in this sector.

The second zone corresponds to the Segura catchment area. The aspects analysed are those corresponding to the water balance (supply and demand) to ascertain whether the water transfer has made it possible to eliminate the existing deficit in the Segura basin, as was proposed in the Preliminary Project for the water transfer. Once the water balance is known, special attention is paid to the area corresponding to the province of Alicante. To this end, the UDAs (Agricultural Demand Units) corresponding to this region were selected. The purpose of this analysis is to find out the amount of existing gross or irrigable surface area, and to compare it with the net or irrigated surface area in this region, according to the data provided by the new Hydrological Planning Cycle of the Segura River Basin (2022–2027).The gross and net demand of the previously selected UDAs is analysed below. Gross and net demand is directly related to gross and net surface areas. Therefore, the demand makes it possible to know the volume of water necessary to supply these areas to obtain an optimal irrigation for the crops (calculated by the official basin organisation).

Lastly, the effects of climate change in the Segura basin are analysed, calculated by the basin organisation with respect to rainfall, evapotranspiration, surface runoff and aquifer recharge; this provides information on the water future of the Segura basin (which is structurally deficient).

The third area of analysis is centred on the province of Alicante (Valencian Community), specifically in the region known as Bajo Segura or Vega Baja del Segura. The choice of this area is justified by the fact that it is a region directly dependent on the water resources of the Tagus–Segura water transfer.

In this respect, when the water from the aqueduct reaches the Ojós reservoir, three water diversion channels start from this reservoir, which correspond to the so-called "Post-Transfer Infrastructure". These three channels take different directions. The first heads towards the province of Alicante, passing through the north of the district of Vega Baja del Segura, as far as the Crevillente reservoir (BajoVinalopó district and the Júcar basin). The second channel heads towards the Pedrera reservoir (Bajo Segura district). From this point, another canal splits into two branches: one heading north-east towards the town of Los Montesinos; and another heading south, passing through towns such as San Miguel de Salinas, Orihuela and Pilar de la Horadada (Bajo Segura district). This channel ends when it reaches the municipality of Cartagena (Murcia). The third canal heads southwards, crossing the Murcia Huerta, Algeciras, among other localities; it ends in Almería (Andalusia).

Given the great length of the water transfer and its implications for large areas of land, it has been decided to limit the analysis to the province of Alicante, specifically to the Bajo Segura or Vega Baja del Segura region (made up of 27 municipalities).

The territory of the district of Vega Baja is included within the Segura basin in terms of hydrological planning but is influenced by the Tagus basin as it receives resources from it through the Tagus–Segura Transfer. This reveals the need to analyse the current situation of each basin in relation to the effects of climate change that are visible in Spain, and their impact on the available water resources.

Along these lines, the district of Vega Baja del Segura implements a multi-source system in water management: surface water (River Segura), groundwater, the Tagus–Segura Transfer, the post-transfer, the reuse of treated wastewater and desalination. This situation is ideal in areas with a natural scarcity of water resources [24]. However, the system is vulnerable to the effects of climate change, due to the reduction in surface water resources predicted for this part of Spain. Therefore, it is necessary to reflect on possible future solutions to guarantee supply in this territory with natural rainfall scarcity.

2.3. Data Source and Analysis

This article analyses the deficiencies of the Tajo–Segura Transfer and its implications in the Segura River basin, and, especially, in the region of Bajo Segura or Vega Baja del Segura, motivated by the effects of climate change.

The methodology applied is based on the hypothetical-deductive model. The hypothetical-deductive method is one of the most accepted methods currently in the scientific field, especially applied to social sciences such as geography, the approach given in this article. The method consists of a working hypothesis that, based on the analysis of a series of available data, allows the hypothesis to be corroborated or not. The steps of the hypothetical-deductive method are: (1) data collection; (2) data evaluation; (3) hypothesis generation; (4) diagnosis; and (5) final conclusion and/or proposal.

The working hypothesis of the research is based on the fact that the Tagus–Segura water transfer is being affected by the effects of climate change, especially with regard to the quantity of water resources. In order to confirm these aspects, information and data have been compiled from the institutions, bodies and official associations (Figure 3).

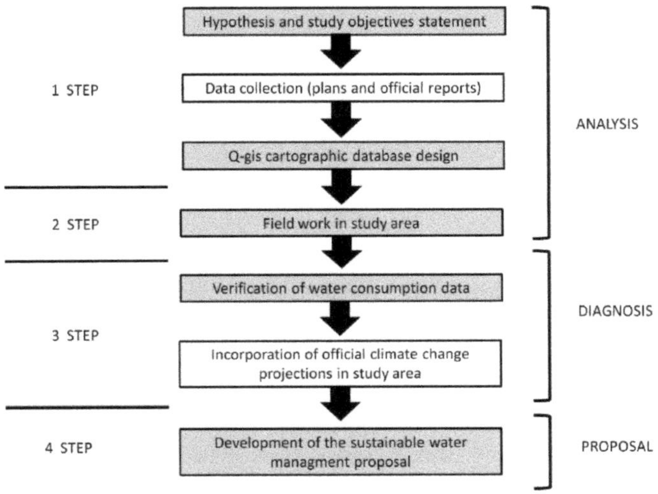

Figure 3. Diagram/outline of the methodological steps in this paper.Source: own elaboration. The innovative contributions of this paper are shown in the shaded areas.

First, the Hydrological Plan of the Tagus Basin (2022–2027) was consulted to compile information regarding the available water resources of the basin. Particular focus was placed on precipitation, surface runoff and the volume of water stored in the Entrepeñas and Buendía reservoirs. The predicted effects of climate change on the Tagus basin were also identified, specifically in the headwaters or sub-basin of the Tagus.

To obtain the data of the volumes of water stored in the Entrepeñas and Buendía reservoirs individually and jointly, the website of the Ministry for the Ecological Transition and the Demographic Challenge (MITERD) through the Centre for Public Works Studies and Experimentation (CEDEX) was consulted. This organisation has all the yearbooks of the volumes of all the basins in Spain, by day and month, depending on the time series chosen. In this way, the years of operation of the TTS have been established as a time series (1979–2021). The principal element in the search was to identify the reserve figure in hm^3 at the beginning of each month in order to analyse the evolution of the water stored in each reservoir or jointly (Entrepeñas–Buendía), so as to establish whether there are currently sufficient volumes of water to transfer via the TTS to the irrigated lands of south-east Spain, and whether these volumes are similar to the theoretical transfer volumes established in the Preliminary Project of the TTS.

After ascertaining the situation in the Tagus headwaters with respect to the precipitation, surface runoff and volume of reservoir-stored water, information and data were collected regarding the Hydrological Plan of the Segura Basin (1st Cycle 2015–2021) and the Hydrological Plan of the Segura Basin (2nd Cycle, 2022–2027).

In the case of wastewater treatment, the data were provided by the Regional Department of Agriculture, Rural Development, Climate Emergency and Energy Transition based on the data of the Entidad Pública de Saneamiento de Aguas Residuales (Public Wastewater Treatment Entity) (EPSAR) of the Region of Valencia, which refer to the wastewater treatment plants (EDARs) installed in the district, their daily production capacity, their annual production capacity, the water treated and the water regenerated, among other information of interest.

With respect to desalination, data on production capacity and the water produced over the last four to five years were provided by the Torrevieja desalination plant. The public entity ACUAMED provided the data on the Torrevieja desalination plant and the cost of production, the energy cost and the future expansion projects contemplated for the desalination plant.

The different declarations of drought and the adaptation mechanisms adopted were also consulted for the period 2015–2018 to determine how the price of desalinated water produced by the Torrevieja plant was established at EUR 0.30 m^3 and the measures adopted by the basin in response to an extreme atmospheric event.

An analysis of these statements reveals a fundamental aspect that is significant for the future of TTS and traditional irrigation models in relation to water sources and extreme events. Therefore, it appears that HTM-dependent irrigation models enter a situation of pre-warning of drought long before the traditional irrigation models that use the basin's resources. In addition, traditional irrigation models do not enter a drought early warning situation until rainfalls fall to half of the drought early warning values in TTS crops. This justifies greater adaptation to extreme weather phenomena, such as droughts, and therefore resource-dependent irrigated crops.

Finally, in relation to the TTS and the volumes of water assigned and consumed in each irrigation community, the study establishes whether with this volume of water profits have increased or decreased, depending on the greater or lesser amount of available water. To do this, four key socio-economic indicators were taken into account: the area of production, the average budget per inhabitant, recruitment in agriculture and unemployment in agriculture for the district of Vega Baja del Segura.

In this respect, the volumes of water transferred in 2011, 2015 and 2019 were also included, together with the desalinated water produced by the Torrevieja plant (ACUAMED) to establish a relationship between the availability of water and the previously mentioned socio-economic indicators.

A geographical information system called QGIS was used for the cartographic part. The layers used to elaborate the location map of the area of study were downloaded in shape (shp.) format from the Valencian Cartographic Institute (ICV), the National Geographic Institute (IGN), the Demarcación Hidrográfica del Segura (DHS), the Demarcación Hidrográfica del Tagus (DHT) and the Demarcación Hidrográfica del Júcar (DHJ) (Table 1).

This analysis seeks to determine the water balance and to estimate the deficit that exists. It also proposes alternative measures to increase the water supply and a new hydraulic plan for the hydrographic basins of Spain.

Table 1. Sources and official documents consulted.

Source	Documents Consulted	Information Used
Spanish Meteorological Agency (AEMET)	- Iberian Climate Atlas (2011) - https://www.aemet.es/documentos/es/conocermas/recursos_en_linea/publicaciones_y_estudios/publicaciones/Atlas-climatologico/Atlas.pdf (accessed on 13 June 2022)	- Monthly and annual rainfall cartographic data of the study area (Tagus basin and Upper Tagus sub-basin).

Table 1. *Cont.*

Source	Documents Consulted	Information Used
Tagus Hidrográfica Demarcación	- Hydrological Plan for the Tagus River Basin (2009–2015). - https://www.chtajo.es/LaCuenca/Planes/PlanHidrologico/Planif_2009-2015/Paginas/default.aspx (accessed on 13 June 2022) - Hydrological Plan for the Tagus River Basin (2015–2021). - http://www.chtajo.es/LaCuenca/Planes/PlanHidrologico/Planif_2015-2021/Paginas/Plan_2015-2021.aspx (accessed on 13 June 2022) - Hydrological Plan for the Tagus River Basin (2022–2027). - http://www.chtajo.es/LaCuenca/Planes/PlanHidrologico/Planif_2021-2027/Paginas/BorradorPHT_2021-2027.aspx (accessed on 13 June 2022)	- Average precipitation in the Upper Tagus sub-basin. - Average runoff in the Upper Tagus sub-basin.
Ministerio para la Transición Ecológica y Reto Demográfico (MITERD)	- Yearbooks of gauges in Spain (1979–2021) - https://ceh.cedex.es/anuarioaforos/afo/embalse-datos.asp?ref_ceh=3006 (accessed on 13 June 2022)	- Water stored in Entrepeñas and Buendía, and all of it, from 1979 to 2021.
Demarcación Hidrográfica del Segura	- Hydrological Plan for the Segura River Basin (2015–2021). - https://www.chsegura.es/es/cuenca/planificacion/planificacion-2015-2021/plan-hidrologico-2015-2021/ (accessed on 13 June 2022) - Hydrological Plan for the Segura River Basin (2022–2027). - https://www.chsegura.es/es/cuenca/planificacion/planificacion-2022-2027/el-proceso-de-elaboracion/ (accessed on 13 June 2022) - Actual consumption of the irrigation communities dependent on the TTS (2020) (data provided by CHS).	- Water balance (supply and demand). - Supply resources. - Resource demand. - Urban demand. - Agriculturaldemand. - Climatechangescenarios.
Sindicato Central de Regantes del Acueducto Tajo Segura (SCRATS)	- Volumes of water transferred for irrigation in the Segura basin. - https://www.scrats.es/ (accessed on 13 June 2020) - Water volumes corresponding to Alicante of the water transferred for irrigation in the basin (data provided by SCRATS).	- Volumes of water transferred for irrigation in the Segura basin. - Volumes of water corresponding to Alicante of the water transferred for irrigation in the basin.

Table 1. *Cont.*

Source	Documents Consulted	Information Used
Public Entity for Wastewater Sanitation (EPSAR)	- Information on the existing wastewater treatment plants (EDARs) in the region of Vega Baja del Segura. - https://www.epsar.gva.es/estaciones-depuradoras (accessed on 13 June 2020)	- Number of WWTPs installed in the Vega Baja del Segura. - Number of WWTPs in operation in the Vega Baja del Segura. - Design production capacity. - Flows treated per day (m^3/s). - Flows treated per year (hm^3/year). - Flows treated and regenerated. - Discharge points.
Mediterraneanbasinwaters (ACUAMED)	- Current data from the Torrevieja plant - (data provided by ACUAMED). - Future expansion projects - (data provided by SCRATS).	- Maximum desalination capacity at the Torrevieja plant. - Energy cost of desalination. - Production cost of desalination. - Final cost of production and delivery. - Environmental costs. - Future planning (expansion of production capacity by 80 > 120 > 160 hm^3/year).

Source: own elaboration.

The scientific contribution of this article is based on the analysis of the most current data offered by basin organisms, the effects of climate change and its involvement in water resources. First, it is shown that the current water situation at the head of the Tagus is not the same as in 1969. This implies a reduction in water resources in the Tagus basin and, consequently, lower volumes of water transferred. These data are demonstrated by the analysis of a short series of precipitation and runoff (1940–1979 and 1980–2018). To complete this analysis, we obtained the existing data of water embalmed in Entrepeñas and Buendía (separately) of the series of operation of the Tajo–Segura Transfer (1979–2020). In order to proceed with a transfer, it is necessary that the volume of water in the reservoirs of Entrepeñas–Buendía (as a whole) exceeds a certain amount at the beginning of each month.

The analysis of the water volumes packed in these reservoirs, both individually and separately, shows a significant reduction in the water resources stored, caused by the effects of climate change due to the lack of rainfall in the headwaters or the volume of rainfall. One of the effects demonstrated by the climate change that occurs in Spain is based on the reduction in rainfall in inland areas as opposed to large discharges that occur on the coast. This has a serious impact on the basins' water resources, since if it does not rain at the head of the rivers, the basins' water resources are diminished.

The analysis of the data from the Segura basin has allowed us to know the water balance (offer and demand) and the existing deficit, despite the water contributions of the Tajo–Segura Transfer. The data obtained have allowed us to know the situation in the province of Alicante, adjusting it to the region of Bajo Segura.

Another new aspect of the work corresponds to the construction of a water balance (supply and demand) and its water deficit, adjusted to the region of Bajo Segura in Alicante. The data show that the existing deficit is not as high as the above claims and that there is water for irrigation in the region. In addition, a table was prepared to estimate the situation for future scenarios in the Vega Baja of the Segura River (2030 and 2050). This table has been compiled as follows: the calculation of these quantities was established on the basis of the following method of analysis: (1) analysis of the official water demand and supply data calculated by the CHS; (2) analysis of the official data on the effect of the decrease in rainfall on water resources (AEMET, CEDEX) in the study area, especially of the resources coming from surface water (Segura River and Tajo–Segura water transfer); (3) knowledge of the territorial dynamics of the study area (for calculating future demands of the different water uses), which is what the field work was used for; (4) calculation of substitution flows for surface water and water transfer with non-conventional water (reuse and desalination), knowing that the reuse of wastewater has a ceiling (depending on the recent evolution of treated flows and total urban expenditure) and that the great asset is desalination for urban and agricultural use. Furthermore, in the current energy transfer process, it is estimated that the final cost of desalinated water will decrease in the coming years, depending on the installation of solar energy sources to feed the Torrevieja plant.

The most relevant of this analysis are two specific issues: (a) it is identified that the highest amount of water demand for irrigation cultivation corresponds to the trans-formed areas from rain-fed to irrigated and is dependent on the waters of the Tagus–Segura Transfer. It is also observed that irrigated crops dependent on basin water resources have a greater resistance to extreme weather (droughts) than those dependent on the transfer. In addition, it has been shown on numerous occasions that the transfer is an infrastructure vulnerable to such situations. (b) The data provided by the basin body include the volume of concessional water allocated to each irrigation community dependent on TTS water in the province of Alicante. However, actual consumption data by irrigation communities are much lower than the volume of water allocated in 1969.

The fact that they do not receive the amounts of water allocated in 1969 is justified by the reduction in water available at the head of the Tagus, hence the importance of having made its analysis earlier. The data show that these irrigation communities will never receive such volumes of water allocated in 1969, when the climate and water reality was quite different from the current one (2022).

However, farmers in the region continue to think that they do not receive such amounts of water for political-social reasons, forming the so-called "water wars". Examples of this are exaggerated claims that it is "the end of Europe's market garden", "without water there is not agriculture" or "that there is a deficit of 1000 hm^3/year in the Segura basin". Farmers are therefore still waiting for water that they will never receive again, hence the importance of this article to demonstrate the real situation with rigorous scientific data.

The last aspect to be highlighted is that the deficit in the region of Bajo Segura can be solved by increasing the volume of water from wastewater treatment and desalination, increasing its annual production capacity, as proposed by the basin body. There are also other methods of reducing the deficit, such as the use of modern irrigation systems or the reduction in less productive, if extremist, cultivated land.

All these issues justify the scientific and novel contributions of the article.

3. Results

3.1. The Effects of Climate Change in the Headwaters of the Tagus (Upper Tagus)

According to the Hydrological Plan of the Tagus Basin (2022–2027), the current average rainfall in all Spanish areas of Tagus basin is 594 mm (1980–2018 series). These data are something different in the Upper-Tagus sub-basin, where the great reservoirs that regulate the Tagus–Segura Transfer are located and have experienced a decrease between the 1940–79series (655 mm.) and the 1980–2018 series (568.5 mm.).Therefore, now the average rainfall is lower than the average rainfall for the entire Tagus basin.

Many studies highlight the reduction in rainfall in the headwaters of the Tagus basin, calculated at a decrease of 12% in the period 1980–2018 [8,9]. This justifies the Tagus basin plan with the rainfall data (average and maximum) and surface runoff (average and maximum) for the sub-basin of the Upper Tagus, the headwaters of the Tagus and the beginning of the Tagus–Segura Transfer [28].

It should first be clarified that there are three series in the Hydrological Plan of the Tagus Basin: 1940–2018 (long series), 1940–1980 (old short series) and 1980–2018 (current short series). Only the last two short series were considered, using the mean values. The headwater rainfall was calculated from the SIMPA model, which is the hydrological reference model for the calculation of water resources in the hydrological plans. For the calculation of rainfall, the model establishes average data for each planning area within the basin, and when there are no long series in the meteorological observatories of the area, it fills in data from the longest series corresponding to observatories close to the planning area.

The results clearly show the effects of climate change in the headwaters of the Tagus basin in relation to precipitation. Although in some months there is an increase, the general monthly trend corresponds to a considerable decrease. In general terms, it can be observed that the mean obtained in the short series (1940–1979) was 655 mm and in the most recent short series (1980–2018) the mean drops sharply to 568.5 mm of precipitation in the headwaters of the Tagus. This represents a decrease of -86.5 mm of precipitation, which represents a decrease of 13.2%, at present (Table 2). Consequently, these means are substantially different, which implies evidence of the impact of climate change in the Tagus headwaters [28].

Table 2. Calculations of average monthly/annual precipitation decrease and percentages in the Upper Tagus (series 1940–1979 and 1980–2018).

	Series 1940–1979 (mm)	Series 1980–2018 (mm)	Reduced Quantity (mm)	Percentage Reduction (%)
October	29.2	30.3	+1.1	+3.8
November	31.3	32.3	+1	+3.2
December	47	53.2	+6.2	+13
January	80.1	67.6	−12.5	−15.6
February	90.4	64	−26.4	−29
March	91.2	68.2	−23	−25
April	79.7	68.8	−10.9	−14
May	68.1	61.2	−6.9	−10
June	45.1	44.4	−0.7	−1.6
July	36.5	31.9	−4.6	−13
August	27.6	24.7	−2.9	−11
September	28.8	21.9	−6.9	−24
Total	**655**	**568.5**	−86.5	−13.2

Source: Hydrological Plans for the Tagus Basin (2022–2027; 2015–2021; 2009–15). Monthly collated data *Iberian Climate Atlas* (AEMET, 2011). Own elaboration. Color: Identify decreases (red and negative) and increases (green and positive).

Meanwhile, the Hydrological Plan of the Tagus Basin (2022–2027) also reports a series of data related to surface runoff. In this respect, the reduction in rainfall has a direct effect on the surface runoff of the Tagus Basin. These volumes of surface water are used for the transfer.

Therefore, surface runoff functions as an indicator of the impact of climate change, in relation to the reduction in mean precipitation in the headwaters of the Tagus. Adding the monthly averages of surface runoff for the short series (1940–1979) gives a total surface runoff of 657.4 hm^3 for the 30-year series. For its part, the sum of the monthly averages of surface runoff for the short series (1980–2018) gives a total result of 380.8 hm^3 in the 30-year series. This implies the reduction of a total of 276.6 hm^3 in thirty years, which is a percentage reduction of 42.1% (Table 3) [28].

Table 3. Calculations of average monthly/annual runoff decrease and percentages in the Upper Tagus (series 1940–1979 and 1980–2018).

	Series 1940–1979 (Hm^3)	Series 1980–2018 (Hm^3)	Reduced Quantity (Hm^3)	Percentage Reduction (%)
October	30.9	24.9	−6	−19.4
November	44.3	30.6	−13.7	−31
December	55.9	58.4	+2.5	+4
January	84.1	48.4	−35.7	−42.4
February	130	45.8	−84.2	−65
March	157	71.7	−85.3	−55
April	79.6	49.9	−29.7	−37.3
May	50.3	35.8	−14.5	−28.8
June	11.0	11.7	+0.7	+6
July	1.0	0.4	−0.6	−60
August	0.5	0.5	0	0
September	12.8	2.6	−10.2	−80
Total	**657.4**	**380.8**	−276.6	−42.1

Source: Hydrological Plan for the Tagus Basin (2022–2027). Own elaboration.

Another aspect that highlights the decrease in rainfall and surface runoff in the Upper Tagus due to climate change is the volume of water stored in the Entrepeñas and Buendía reservoirs, particularly when analysing the historical series of 1979–2020, coinciding with the years of the operation of the TTS.

In the water year 1979–1980, the volume of water stored in the Entrepeñas reservoir reached a value of 6308 hm^3/year. This figure had fallen to 4083 hm^3/year in the water year 2019–2020. Meanwhile, in the Buendía reservoir, in the year 1979–1980, there was a total volume of 14,268 hm^3/year, and in the water year 2019–2020 this figure had fallen sharply to 3613 hm^3/year. However, it should be remembered that the volume of water authorised for transfer depends on the sum of the two reservoirs (Entrepeñas and Buendía). In this respect, and following the same line of analysis as in the individual cases, the total joint volume of water of the Entrepeñas and Buendía reservoirs in water year 1979–1980 amounted to 20,576 hm^3/year, with 7696hm^3/year in water year 2019–2020, representing a decrease of 63% (Table 4, Figure 4).

Moreover, it should be noted that in the years of severe drought, as a whole, values of between 5000 and 9000 hm^3/year have been reached, revealing their vulnerability to extreme atmospheric events.

Therefore, two conclusions may be drawn: (a) the impact of the effects of climate change on the headwaters of the Tagus is inevitable, as a reduction in rainfall directly affects the surface runoff and the volumes of water stored in reservoirs; and (b) the TTS is adversely affected in this respect, given that the theoretical volumes calculated to transfer to south-east Spain correspond to volumes of water that existed in and prior to 1979–1980, which are not attainable in the present day. This implies the need to carry out a review of the calculations made in the Preliminary Project of the Transfer with current data, given that it is inconceivable that the theoretical water allocations can continue to be planned based on calculations made over forty years ago, when the climate reality was completely different to that of today.

Table 4. Comparison between water stored in the Entrepeñas–Buendía reservoir and volumes of water transferred by the Tagus–Segura Aqueduct.

Hidrologycal Year	Vol. Water (hm^3/year)	Vol. of Water Transferred (hm^3/year)
1979–1980	20,576	36
1980–1981	15,308	463.9
1981–1982	9648	287.9
1982–1983	5853	287.9
1983–1984	5142	287.9
1984–1985	9674	373.6
1985–1986	10,431	386.4
1986–1987	10,129	307.2
1987–1988	10,574	436.9
1988–1989	11,086	438.6
1989–1990	9679	287.9
1990–1991	8628	287.9
1991–1992	6646	287.9
1992–1993	5307	287.9
1993–1994	5025	255.1
1994–1995	3321	150.8
1995–1996	5383	361.1
1996–1997	11,882	457
1997–1998	15,964	302.5
1998–1999	14,229	287.9
1999–2000	9945	287.9
2000–2001	11,798	373.6
2001–2002	10,090	287.9
2002–2003	9276	312.9
2003–2004	9486	382.7
2004–2005	7673	287.9
2005–2006	3635	287.9
2006–2007	4139	253.3
2007–2008	3918	138.7
2008–2009	5298	199.2
2009–2010	9818	259.2
2010–2011	14,202	287.8
2011–2012	11,169	287.8
2012–2013	8926	287.8
2013–2014	9048	287.8
2014–2015	6236	281.2
2015–2016	5396	231.8
2016–2017	4637	127.3
2017–2018	5224	156.6
2018–2019	7297	313.6
2019–2020	7696	210

Source: yearbooks of gauges in Spain. Own elaboration.

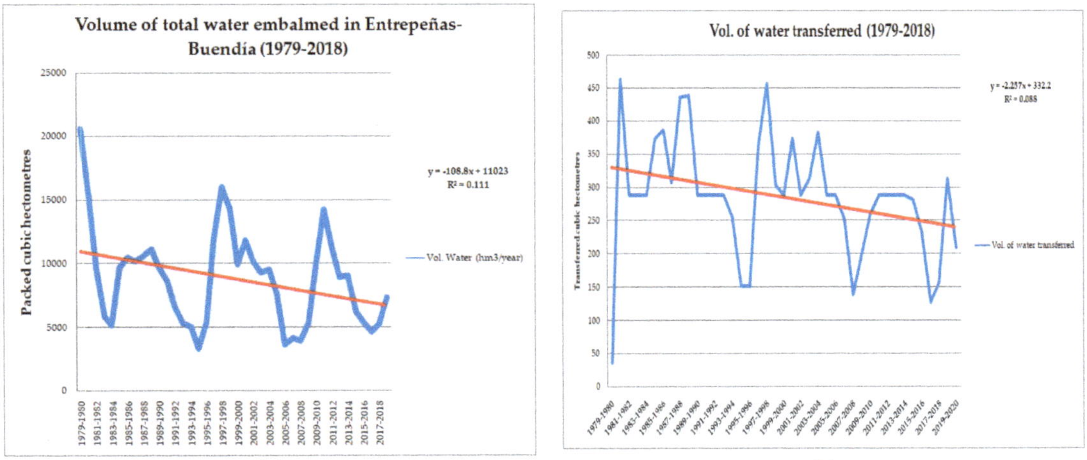

(a) (b)

Figure 4. Graph comparison between water stored in the Entrepeñas–Buendía reservoir and volumes of water transferred by the Tagus–Segura aqueduct. (**a**) Evolution of water reservoirs in the Entrepeñas–Buendía reservoir (1979–2018); (**b**) evolution of volumes of water transferred by the Tagus–Segura aqueduct (1979–2018). Source: yearbooks of gauges in Spain. Own elaboration.

Finally, the PHCT (2022–2027) reports the percentage of change in the quarterly runoff, calculated in average (RCP 4.5) and high (RCP 8.5) climate change scenarios for the Upper Tagus. For the months of October to December, a reduction of 14% is calculated in an RCP 4.5 scenario and 20% in an RCP 8.5 scenario, coinciding with the months of most rainfall [28]. This implies that the rainfall trend, the surface runoff and the volumes of reservoir-stored water in the Upper Tagus are in continuous decline, constituting a serious problem for those dependent on the water from the TTS.

3.2. The Water Balance in the Segura Basin: Supply and Demand in Vega Baja

According to the Hydrological Plan of the Segura Basin (2015–2021), the total available resources of the basin (including the TTS) amounted to 1511hm^3/year, and the total demand of the basin reached a volume of 1841 hm^3/year. Given that the demand is higher than the available resources, the water deficit calculated for the whole of the Segura basin for this time horizon was 330 hm^3/year [26].

Meanwhile, the recent Hydrological Plan of the Segura Basin (2022–2027) indicates that the total available resources in the basin (including the TTS) amount to 1571 hm^3/year, which implies an increase in water resources in the basin. The total demand of the basin, meanwhile, amounts to a volume of 1831 hm^3/year, representing a decrease in demand with respect to the previous time horizon, which is justified by the increase in water resources from non-conventional sources. This is whythe deficit has reduced to 260 hm^3/year (Table 5) [27].

Considering the scenarios contemplated for 2027 and 2039 in the PHCS (2022–2027), two different trends may be identified: the first is the continued increase in water resources using non-conventional sources, such as those obtained through wastewater treatment and desalination in the Segura basin for 2027 and 2039. The second trend is related to the Segura basin's own water resources, such as surface and groundwater, which are set to decrease in terms of total annual volume. Furthermore, the plan considers that less sea discharges will take place in 2027 and 2039, which suggests an advance in the use and management of water.

A rising trend can be observed in the urban demand in the Segura basin. This coincides with the increase in agricultural demand for the 2027 and 2039 horizons [27]. This constitutes a serious problem for agriculture, given that the Water Law 29–1985 and the subsequent Law of the National Hydrological Plan 2001 and Law 11/2005, which modifies the previously mentioned law, indicate urban supply as the priority use. This implies that in situations of extreme atmospheric events (droughts), priority is given to urban supply above other uses (ecological, irrigation and agricultural uses, industrial, etc.). Therefore, in the case of need, water will be extracted from the water allocations assigned to irrigation and agricultural users, thereby aggravating the consequences for irrigated crops.

Table 5. Segura River basin water balance (2022–2027).

Segura River Basin Water Balance		Average Resources (hm^3/year) Horizon (2022–2027)
Available Resources	Surface water	704
	Groundwater	66
	Non-draining surface resources to the Segura	15
	Returns to the system	268
	Desalination	223
	TTS	295
	TOTAL AVAILABLE RESOURCES	1571
Demand	Urban demand	250
	Environmental demand	39
	Agricultural demand	1522
	Other demands (industrial, golf, etc/.	20
	TOTAL BASIN DEMAND	1831
	BASIN DEFICIT	−260

Source: Hydrological Plan for the Segura River Basin (2022–2027).

With respect to the urban demand of the basin, the calculation of urban demand conducted by the CHS in the district of Vega Baja del Segura was performed. In 2021, urban demand was recorded at 39.2 hm^3/year, and it is estimated that by 2039 it will have increased to 43.5 hm^3/year [27]. It is interesting to note that the urban demand corresponding to the district of Vega Baja only represents 16% of the total demand of the Segura basin.

It is more interesting to analyse the agricultural demand of the district of Vega Baja del Segura, according to the data included in the PHCS. To achieve this, the gross or usable agricultural land and the net areas or those cultivated by UDAs (Units of Agricultural Demand) corresponding to the district of Vega Baja del Segura were chosen (Table 6).

As we can observe in Table 5, the PHCS shows that there is a gross or usable agricultural area of 65,411 hectares, of which 47,636 hectares are cultivated, representing 72.8% [27]. In the afore-mentioned plan, the total gross and net demands per crop and the total demand for per UDA are reported.

Table 5 shows the UDAs that depend on the waters of the Tajo–Segura Transfer are 52, 53, 56 and 72. This gives a total of 27,837hectares and an annual water demand of 117.9 hm^3/year, while the UDAs that depend on available basin resources (46, 48, 51 and 55) add up to a total of 19,799 hectares and an annual water demand of 76,5 hm^3/year. These data show that the UDAs that were rain-fed crops and were transformed into irrigated crops (new irrigation), because of the Tajo–Segura Transfer project, are the areas with the highest water demand and, consequently, the causes of the water deficit in the region of Bajo Segura.

Table 6. Area (gross and net) and demand (gross and net) by UDAs belonging to the region of Bajo Segura (Alicante).

		Gross Area (ha)	Gross Demand (hm^3/year)	Net Area (ha)	Net Demand (hm^3/year)
UDA	**Denomination**	**Horizon 2022–2027**		**Horizon 2022–2027**	
46	Traditional Vega Baja	**23,780**	**100.1**	**15,469**	**58**
48	Vega Baja, post. Al 33 y ampl. del 53	**3067**	**12.3**	**1913**	**8.2**
51	Mixed irrigation of aquifers and wastewater treatment plants south of Alicante	**4538**	**9.9**	**1634**	**7.5**
52	Riegos Levante Right Bank	**3439**	**15.9**	**2886**	**11.9**
53	Riegos Redotados del TTS de RLMI-Segura	**11,046**	**52.4**	**8713**	**36.1**
55	Crevillente Aquifer	**1306**	**3.2**	**783**	**2.8**
56	La Pedrera ZRT TTS Redotados Irrigations	**10,563**	**52.5**	**9411**	**41.4**
72	Re-dedicated irrigated lands of the Vega Baja TTS, left bank	**7672**	**40**	**6827**	**28.5**
	TOTAL	**65,411**	**286.3**	**47,636**	**194.4**

Source: PHCS (2022–2027).

It is evident that climate change has a significant impact on water, which obliges the organisations of the basin to take it into account in hydrological planning. In this respect, the CHS establishes three future periods of 30 years, called impact periods (IP). These impact periods are IP1 (2010–2040), IP2 (2040–2070) and IP3 (2070–2100), reflecting the impact in the short, medium, and long term, in accordance with the medium (RCP 4.5) and high (RCP 8.5) climate scenarios (Table 7) [27].

Table 7. Effect of climate change with respect to an unaffected situation on hydrological variables in the DHS.

		Med RCP 4.5	Med RCP 8.5
Precipitation	**PI (2010–2040)**	−2%	−5%
	PI (2040–2070)	−4%	−10%
	PI (2070–2100)	−8%	−14%
Real evapotranspiration	**PI (2010–2040)**	−2%	−5%
	PI (2040–2070)	−4%	−9%
	PI (2070–2100)	−6%	−11%
Recharge	**PI (2010–2040)**	−7%	−10%
	PI (2040–2070)	−12%	−23%
	PI (2070–2100)	−20%	−36%
Runoff	**PI (2010–2040)**	−7%	−9%
	PI (2040–2070)	−11%	−23%
	PI (2070–2100)	−20%	−38%

Source: PHCS (2022–2027).

In fact, the short-, medium- and long-term impacts of climate change related to water resources reveal a negative scenario for the Segura basin. As we can observe in Table 6, there will be a reduction in rainfall of between 8% and 14% in the Segura basin by the end of the century. The potential and real evapotranspiration translate into negative effects; the humidity of the soil will also decrease; the aquifer recharge will face serious problems, oscillating between 20% and 36%; and finally, the surface runoff will decrease sharply throughout the whole of the Segura basin, with values fluctuating between 20% and 38% [27].The reason for using the RCP 8.5 scenario, which is characteristic of an extreme scenario of high emissions, is justified because the trend in emissions is increasing annually, the effect of climate change on the reduction inwater resources is notorious and, finally, because it justifies the proposal put forward in this research to carry out sustainable water planning.

3.3. The Water Balance in the District of Vega Baja del Segura and the TTS

Obtaining data on water resources (supply) in the region of Vega Baja del Segura has been a complex task. This is because there is no previous study of the water balance in this region. The data found are only represented in two scales: river basin scale or provincial scale (+140 municipalities, Alicante).

For this reason, the data that represent the water balance of the region of Bajo Segura are adjusted to its political limit composed of the 27 municipalities that make up it.In this sense, the data have been found in different bibliographies of official bodies such as the Segura River basin or the Diputación de Alicante, among others. All these data have been adjusted for the regional scale of work (Table 8).

Table 8. Water balance 2021 in the region of Vega Baja del Segura (Alicante).

WaterBalance in the Region of Vega Baja del Segura (Alicante)		Average Resources (hm^3/year)
Offer	Surface water	40
	Groundwater	36
	Returns to thesystems	32
	TTS	61.1
	Wastewater treatment	25
	Desalination	48.1
	TOTAL DISTRICT RESOURCES	**242.2**
Demand	Urban demand	39.2
	Environmental demand	32
	Agricultural demand	194.4
	Other demands (industrial, golf, etc.)	8
	TOTAL DISTRICT DEMAND	**273.6**
	DISTRICT DEFICIT	**−31.4**

Source: own elaboration based on PHCS (2022–2027), SCRATS (2020), EPSAR, ACUAMED, Generalitat Valenciana and Diputación de Alicante data.

The estimated deficit in the district of Vega Baja del Segura (Alicante) is 31.4 hm^3/year. It should be noted that urban demand is completely guaranteed by the water resources from the River Taibilla, the TTS and the desalinated water used by the Mancomunidad de los Canales del Taibilla, which supply the municipalities of Vega Baja del Segura. Meanwhile, environmental demand is also guaranteed by the circulating waters of the river Segura and the returns to the system, which are used as ecological flows in the final section of the river Segura. The other demands (industrial, golf, etc.) are covered using water purified with a tertiary or advanced treatment. The problem of the deficit resides in agricultural demand, particularly in the so-called new irrigated lands, because the waters from the transfer are insufficient to supply the existing demand (Table 9).

Table 9. Estimated situation for future scenarios in the Vega Baja of Segura River (2030 and 2050).

Water Resources Vega Baja	2021	2030	2050
Surface water	40	38.7	34
Groundwater	36	30	20
Returns to system	32	33.1	35
TTS	61.1	57.4	52.1
Wastewater treatment	25	26.4	29
* Desalination	48.1	80	120
Total Resources	**242.2**	**265.6**	**290.1**
Urban demand	39.2	41.4	45
Environmental demand	32	32	32
Agricultural demand	194.4	194.4	176.9
Other demands (industrial, golf, etc.)	8	8	8
Total Demand	**273.6**	**275.8**	**261.9**
Deficit/Surplus Vega Baja	**−31.4**	**−10.2**	**+28.2**

* Desalination: by 2030 the Torrevieja desalination plant will be expanded to 120 hm^3/year, of which 80 hm^3/year will be for agricultural use and 40 for urban use. By the year 2050, it is estimated that the production capacity will be 160 hm^3/year, of which 120 hm^3/year will be used for irrigation and the remaining 40 hm^3/year for urban use. Source: own elaboration based on the scenarios proposed in the Segura River basin and the medium climate scenario RCP 4.5.

Of the 194.4 hm^3/year of agricultural demand, according to the PHCS (2022–2027), a total of 160.8 hm^3/year correspond to UDAs 52, 53, 56 and 72, which coincide with the sectors of the new irrigated land (former rain-fed land transformed into irrigated land); hence, there is high demand for water for two reasons: (a) the lands were originally dry and modern irrigation systems allow irrigated crops to be grown on them, and (b) the farms or plots are large with areas similar to those of latifundios (large estates).Therefore, a large amount of water is necessary to maintain their productivity and enable these areas to continue to produce irrigated crops (Table 10).

Table 10. Agricultural demand that takes advantage of the waters of the TTS in the Vega Baja del Segura.

Denomination	Agricultural Demand hm^3/year (2021–2027)
UDA—52. Segura RLMD	15
UDA—53. Segura RLMI	52
UDA—56. Redotated irrigation of the TTS of the La Pedrera ZRT	52
UDA—72. Vega Baja, left bank redotated irrigation systems	40
TOTAL	**160.8**

Source: PHCS (2022–2027). Own elaboration.

In this respect, the Confederación Hidrográfica del Segura provided a series of data referring to the volume supplied of the transfer water to the 29 irrigation communities in the water year 2019/2020 (Table 11). This table reflects two relevant aspects: the concessional volume and consumption. The concessional volume refers to the amount of water assigned to each of the irrigation communities that are beneficiaries of the transfer water, corresponding to the theoretical values in the Preliminary Project of the Transfer, given that the total sum of the concessional volumes assigned to each irrigation community amounts to 123 hm^3/year, 30% of which is allocated to the province of Alicante. It would be very difficult to achieve these theoretical amounts in the current climate and water context and considering the effects of climate change.

Table 11. Volume of water supplied by the water transfer to the CC.RR. of Alicante in the hydrological year 2019–2020.

	Transfer Resources	
	Concesional Volume (m^3) E (m^3)	Consumption (m^3)
C.R. Albatera	7,815,324	3,969,496
C.R. Las Cuevas	1,491,100	796,865
T.D. Lo Belmonte	666,925	356,416
T.D. Lo Marques	485,366	259,386
T.D. Las Majadas	767,010	409,903
C.R. San Joaquin	479,950	241,840
C.R. Santo Domingo (Grupo 3.490)	2,276,500	1,216,595
C.R. Lo Reche	1,473,892	731,114
C.R. El Carmen	571,739	305,301
T.D. Manachon Candela	111,000	29,380
C.R. Riegos de Levante (M.I.)	77,512,272	26,213,062
C.R. Riegos de Levante (M.D.)	5,500,000	1,299,612
C.R. La Fuensanta Grupo 2000	1,007,750	206,895
C.R. La Estafeta	55,100	20,963
C.R. Murada Norte	2,001,700	1,069,736
C.R. El Mojon	1,156,641	385,076
C.R. Perpetuo Socorro	1,709,400	816,432
C.R. Las Dehesas	961,350	477,008
C.R. Barranco de Hurchillo	239,250	127,858
C.R. San Onofre y Torremendo	1,715,350	916,708
Agrícolas Villamartin	110,200	57,358
C.R. Río Nacimiento	627,850	335,532
C.R. Mengoloma de Orihuela	208,800	111,795
C.R. Pilar de la Horadada	2,621,600	1,401,019
C.R. San Isidro y Realengo	7,500,000	0
C.R. Campo Salinas	2,122,800	1,041,434
C.R. San Miguel	1,922,700	773,152
C.R. Las Cañadas	150,800	81,064
TOTAL	**123,262,369**	**43,651,000**

Source: DHS (2020).

It is interesting to know the consumption per irrigation community, given that, if the value of consumption of each irrigation community has allowed sufficient irrigation for the crops to grow correctly and produce yields that are profitable for the farmer, this implies that the net agricultural demand of the beneficiaries of the TTS water fluctuates between the consumption values of 2020, that is, around 44 hm^3/year (Table 11).

To corroborate this point, the data regarding the irrigated area of the district of Vega Baja del Segura provided by the Valencian Institute of Statistics (IVE) of the Regional Department of Sustainable Economy, Productive Sectors, Trade and Employment were consulted. These data reveal that there are variations in the size of the cultivated area in accordance with the droughts occurring and the availability of water resources. In this respect, in the years 2014–2016 (years of drought in the Segura Basin), there was an increase in the irrigated area, with an area of 16,399 hectares recorded in 2014 and 16,955 hectares in 2016. In the following two years of drought (2017–2018), the irrigated area decreased to values fluctuating between 16,500 and 15,500 hectares. Finally, it should be noted that in 2019, the cultivated area increased again to 16,321 hectares, a value close to that of 2020.

From the data provided by the IVE, it may be concluded that the municipalities with irrigated crops dependent on the TTS waters (new irrigated land) reduced their cultivated area in drought situations, as in the case of the municipalities of Orihuela, Albatera, Benferri, Pilar de la Horadada or San Miguel de Salinas, among others. Meanwhile, the irrigated crops that depend on the water resources from the Segura basin (Segura River) increase or maintain the same areas of irrigated crops in drought situations, even in the years of most severe rainfall and hydrological drought in the basin, as in the case of the years 2012, 2015 and 2018.

The loss of irrigated area dependent on the TTS water resources could have been even greater in the years 2015–2016 and 2017–2018 (drought in the Tajo, Júcar and Segura basins, among others), if the desalination plant of Torrevieja had not started operations as a strategic source to support the TTS in this type of situation [5].

Meanwhile, the data provided on the ARGOS Information Portal of the Regional Government of Valencia allows the addition of a series of socio-economic indicators that enable the relationship existing between economic development (agriculture) and the greater or lesser availability of water to be evaluated (Table 12).

Table 12. Socio-economic indicators in the Vega Baja del Segura in relation to the greater or lesser availability of water resources.

	2011	2015	2019
Regional cultivatedarea	16,425	16,955	16,321
Average budget per habitant	EUR 745.87	EUR 844.30	EUR 938.72
Registered hiring in agriculture	11.87%	21.83%	28.06%
Unemployment in agriculture	4.22%	6.34%	4.59%
Annual transfer TTS	85.9	32.8	85.5
Desalination	N/D *	30	76.4

* N/D: non data. Source: Valencian Cartographic Institute (ICV) (IVE), ARGOS and ACUAMED.

Thus, with respect to the average budget per inhabitant, a gradual increase from the year 2011 to 2019 may be observed. It continued to rise until 2021, reaching a value of EUR 1026.35/inhabitant.

Regarding the recruitment recorded in agriculture, in the year 2013–2014 an exponential increase in percentage terms can be observed, reaching a figure of 30.68%. This value decreased in the following year due to the severe drought (21.83%) and continued to decrease until 2018. From 2019, a growing trend began in the percentage of recruitment recorded in agriculture, reaching a value of 35.01% in 2021.

On the other hand, registered unemployment in agriculture displays different behaviour. From 2007 to 2016, unemployment in agriculture gradually increased. The highest values correspond to the drought years of 2015 and 2016, reaching 6.16% and 6.34%, respectively. From 2017 until the present day, the number of unemployed in agriculture has decreased to a level of 4.59 in the year 2020.

These indicators become more significant when they are examined in relation to the greater or lesser availability of water from both the TTS and desalination.

In March 2015, a drought situation was declared in the Segura basin, for which 12 exceptional measures were implemented for the management of water resources and EUR 30 million of extraordinary credit was assigned. In September of the same year, the drought continued, and the drought declaration was extended until September 2016. In addition to the measures already implemented, further action was taken with respect to the control of the continental waters. In October 2015, for the first time in Spanish history and in the Segura basin, among the new measures announced by the Ministry was the reduction in the price of desalinated water produced by the Torrevieja plant to 0.30 EUR/m^3, with the authorisation of the production of 30 hm^3/year and a subsidy of EUR six million to reduce the cost of production [29].

This highlights that, thanks to the production of desalinated water in the years 2015 and 2016, the area of irrigated crops in the district of Vega Baja del Segura remained the same. This is also visible in a similar value maintained of total agricultural income in the Region of Valencia.

In March 2018, Law 1/2018 of 6 March was passed, referring to the adoption of urgent measures to mitigate the effects generated by the drought in certain hydrographic basins. Additionally, the Revised Text of the Water Law, approved by Royal Legislative Decree 1/2001 of 20 July was modified, in response to the continued drought situation in a large part of Spain. This explains the decrease in the cultivated area in the years 2017, 2018 and 2019 in the district of Vega Baja del Segura, together with the reduction in agricultural income. Another problem that explains this worsened situation is that, from the year 2016–2017, the price of desalinated water increased to 0.60–0.80 EUR/m^3. Farmers could not afford this price and opted to not cultivate or even abandon their irrigated land due to the absence of a guarantee of water resources.

However, in the water years 2017–2018, when there was a drought in the headwaters of the Tagus, the Torrevieja desalination plant reduced the price of its desalinated water (for the second time) to 0.30 EUR/m^3 for irrigators. This cost was farbelow that of production costs [5]. This coincided with the months of an absence of transferable resources or "no transfer", which obliged the desalination plant of Torrevieja to produce desalinated water at almost maximum capacity in 2019 [29,30]. The effects for the cultivated land and socio-economic aspects were positive, thanks to desalination.

4. Discussion

The results reveal a series of problems in hydrological planning in Spain which directly affect the district of Vega Baja del Segura, in relation to the Tagus–Segura Transfer.

The analysis shows that the volumes of water available in the sub-basin of the Upper Tagus are not the same as those of fifty years ago, given that the average and maximum rainfall have reduced, the surface runoff has decreased and the volumes of water independently stored in the Entrepeñas and Buendía reservoirs have decreased sharply. This behaviour has also been observed in the joint water resources of Entrepeñas–Buendía since the beginning of operations of the TTS.

The reduction in rainfall recorded in the last period of climate analysis (1980–2018 series) is 15% with respect to the 1940–1979series and is consistent with what the CEDEX points out in its report on the impact of climate change on water resources in Spain [8] and with what Marcos and Pulido point out [31]. A decrease in winter and spring rainfall is also observed, which are the most effective for hydrological planning purposes (urban tourist and agricultural demands in summer) and a slight increase in autumn rainfall, in line with what has been pointed out by various authors [32–35] for the eastern sector of the Iberian Peninsula. Thus, the monthly distribution of precipitation in the headwaters of the Tagus tends to be "Mediterranean", with a more prominent participation of autumn rains.

The climate trends and their effects related to water resources in the Tagus basin (Upper Tagus) with respect to rainfall, surface runoff and the volume of water stored in reservoirs reveal a progressive reduction in available water resources. Taking into account the climate change scenarios (RCP 4.5 and RCP 8.5) and even after undertaking adaptation tasks, the Tagus basin is experiencing serious problems, which are being demonstrated and tested with scientific data.

As mentioned above, the Mediterranean region is the planetary zone where the effects of climate change will be most devastating [1]. In this sense, there are numerous scientific publications that have analysed this issue for the Mediterranean region. These studies show that, although the reduction in precipitation is not linear, but rather amplified (due to the increase in extreme weather events), two types of behaviours have been observed in the Mediterranean region. The first of these is obviously the reduction in average precipitation and the form of rainfall [11]. Greater amounts than historical records can now fall in a very short time. The second is that rainfall tends to concentrate on the Mediterranean coast and not in inland areas, where the headwaters of Spain's main rivers (such as the Tagus and Segura Rivers) are located [33–37]. If there is no rainfall in the headwaters of these rivers, there will hardly be any water resources in the basin and, consequently, no transfer from one basin to another will be possible. Therefore, it is considered that the TTS and the current hydraulic planning is unsustainable in the context of climate change.

The problems of the Upper Tagus directly affect the hydrological planning of the Segura basin, through the Tagus–Segura Transfer. One of the problems identified in the period of operation of the TTS (1979–2021), except for one year, is that the theoretical volumes assigned in the Preliminary Project of the Tagus–Segura Transfer have not been achieved, not even under the initial operating regulations. In fact, with respect to the demand and available water resources in the Segura basin thanks to the transfer, the Segura basin has never reached more than 2000 hm^3/year as an available resource, even with the TTS, wastewater treatment and desalination. Meanwhile, demand has increased exponentially.

This is mainly due to the problems in the headwaters experienced in the Tagus basin because of climate change. The figure that demonstrates this aspect is the average volume of water transferred from the origin (208 hm^3) and received in the destination (182 hm^3), indicated by the CHS for the period contemplated [27]. Furthermore, the data of the SCRATS (Central Irrigation Syndicate of the Tagus–Segura Aqueduct) reveal that, of the 400 hm^3/year assigned to irrigation in south-east Spain (theoretical), the average volume of the transfer for the period indicated is 195.6 hm^3, that is, less than half of the assigned amount. Of the 400 hm^3/year (theoretical), 30% corresponds to the province of Alicante, which is 120–125 hm^3/year (theoretical). However, the reality is very different. Taking the data of the CHS into account, and particularly that of the SCRATS, the average volume of water received in the province of Alicante from the TTS is 61.1 hm^3 in the period of operation (1979–2021). This implies that the province of Alicante receives half of the theoretical amount of water assigned to it.

This raises the following questions: Why has no reassessment or review been made of the operating calculations of the TTS in over 50 years? Why is planning undertaken with values that do not currently exist, nor will exist in the future? Why do the farmers denounce and call for volumes of water that they will never receive? The answer is clear: they are being deceived because they have been given no explanation about the current situation of the hydrographic basins because of climate change.

Therefore, the TTS displays a series of significant weaknesses subject to the variations in climate, and the territories which are supplied with its water resources must begin to adopt measures that do not directly depend on it (self-sufficiency). To obtain a greater volume of water, desalination and wastewater treatment are being used to contribute to the supply, enabling the deficit of the basin to be reduced to some extent.

However, the trends in urban and agricultural demand are growing, with the latter having the highest demand. It should be remembered that in the case of drought or a shortage of urban water supply, the water used to supply the urban nuclei will be drawn from other uses, such as ecological flows or those used for agriculture. Therefore, agriculture is losing water resources, and in the medium to long term, if measures are not taken to correct these problems, it will be seriously affected.

In this respect, and as a proposal for an empirical and demonstrated adaptation, the new irrigated lands (dependent on the TTS) enter a situation of pre-alert of drought much before the traditional irrigated lands. This was evident in the drought of 2015, after the passing of RD 356/2015 of 8 May, declaring the situation of drought in the territorial region of the Confederación Hidrográfica del Segura with the adoption of exceptional measures for the management of resources, due to the reduction in the interannual contributions (rainfall) in the headwaters of the Segura and Tagus basins and its successive extensions (RD 335/2016). In other words, the resources of the Segura basin can supply the traditional irrigated lands in situations of extreme drought, although not the crops dependent on external resources. This should constitute an incentive for changing the way hydrological planning is conducted in the Segura Basin, the Vega Baja district and in the rest of Spain.

With respect to the Alicante district of Vega Baja del Segura, the deficit existing is due to agricultural demand: principally, the irrigation communities that depend on the water resources of the transfer. This is because after the TTS project was approved, the construction of the transfer took 10 years until it began operation. During this period, there was much speculation in Vega Baja del Segura regarding the water that would be received, based on the afore-mentioned theoretical volumes. This speculation translated into the transformation of rain-fed crop land into irrigated land using groundwater. Therefore, the real use of the TTS has not increased the irrigated areas, but has maintained, as far as possible, all the transformed areas since then.

The current status of the non-conventional sources allows them to considerably reduce the deficit existing in Vega Baja del Segura, although this does not mean that there is no longer a deficit. If the desalination plant of Torrevieja operated at its maximum capacity (80 hm^3/year), with the support of the treated wastewater (24–25 hm^3/year), the deficit would be reduced to 10.5 hm^3/year. The deficit would be resolved with the extension of the maximum production capacity of the Torrevieja desalination plant from 80 to 120, and subsequently to 160 hm^3/year. This proposal is contemplated in the PHCS (2022–2027). In parallel, there would be an increase in the volume of reusable treated wastewater, which would complement the desalinated water. These two sources would become the principal water sources of the district of Vega Baja del Segura. The role of the TTS, therefore, would be a secondary or complementary source to these resources, when needed.

However, this increase in production should be accompanied by mechanisms imposed by the government based on Law 7/2021 of 2 May regarding climate change and energy transition, as a framework within which to reduce the cost of production of desalinated water, based on energy subsidies. In this way, the farmers would be provided with water for irrigation at a price of 0.20–0.30 EUR/m^3, as in the case of the two occasions when the TTS failed during situations of extreme drought.

Taking all these issues into account, Spain needs to review and reconsider the current hydrological policy. To do this, first, it should begin by reviewing the values of the Preliminary Project of the TTS and adjust them to the volumes of water currently available. Furthermore, the water contributions assigned to each irrigation community dependent on the water from the TTS should be reviewed, given that they do not reflect the current situation. This analysis will reveal the available water resources. Second, after determining the situation of the water resources related to the TTS, it would be appropriate to increase the maximum capacity of the Torrevieja desalination plant (ACUAMED) in order to obtain larger volumes of water resources and maintain the plant in full operation as well as obtaining a price of desalinated water that is affordable for farmers. Third, agricultural demand should be reduced using new irrigation systems (drip system), a switch to irrigated crops that require less water or a radical change from irrigated crops to rain-fed crops and, as a more extreme measure, the reduction incultivable areas.

The new hydrological policy in Spain should be constructed on sustainability, climate change scenarios, hydrologic planning and on measures of adapting to climate change in a horizon of 100 years.

The *new sustainable hydrological planning* should be based on scenarios of climate change elaborated by the IPCC in its reports and on its regional effects. Sustainable planning means that, despite the existing resources, this exercise should be contextualised within the worst climate scenario possible, that is, in RCP 8.5 scenarios. Planning a climate and water resources situation based on the RCP 8.5 severity will enable a management of water resources able to guarantee the resource for the rest of the twenty-first century in Spain. If the reality is very different to that contemplated in the plan (due to a limitation of greenhouse gases, the implementation of measures to adapt to extreme atmospheric events and the respect and fulfilment of all the treaties and agreements in terms of reducing emissions), there will be a surplus and more water will be available for the assigned uses (urban supply, ecological flows, irrigation and agriculture, industry, leisure, tourism, etc.).On the contrary, if the scenarios contemplated in the RCP 8.5 are fulfilled, a prior adaptation to this situation will have already been contemplated and planned. Furthermore, the adoption of this approach would enable measures of adaptation to be developed over the years.

This new planning proposal will significantly slow down the environmental deterioration and socio-economic losses. If these aspects are not considered from today, the consequences will be much more severe in the medium and long term, and losses running to millions of euros will be incurred. It would be particularly severe for those crops dependent on the TTS. The implementation of hasty and drastic measures as the RCP 8.5 climate scenario approaches will lead to greater economic investments with dubious profitability. Furthermore, this new policy should be flexible and open to modifications which enable adaptations and adjustments to be made in accordance with the climate situation and the future scenarios contemplated.

The new planning and management scenario proposed in this paper is in line with the proposal in the mentioned Law 7/2021 on climate change in Spain, which advocates the incorporation of the effects of climate change into hydrological planning (art.19) to increase the resilience of the different uses of water. In particular, agricultural uses must adapt the demands to the expected resources to minimise the expected impact on future climate scenarios. The proposal for sustainable hydraulic planning in our study area is in correspondence with various authors in Spain who advocate a reduction in agricultural irrigated surfaces compared to the increase contemplated in various current hydrological demarcation plans (Guadalquivir, Guadiana, Tajo), or the maintenance of existing ones that are not sustainable, in present day and in climate change scenarios (Segura, Júcar) [36–38].

Moreover, the new sustainable hydraulic plan should focus on the management of the demand for water, with a commitment to non-conventional sources, desalination and wastewater treatment, with the afore-mentioned determinants (increase in capacity, reduction inthe production costs of desalinated water, mechanisms to reduce the energy and production costs of desalination, applying tertiary treatment to purified water so they may be reused for agricultural and other uses). After managing the resources of non-conventional sources, the water resources of the hydrographic basins should then be included (surface, groundwater, returns to the system, sea discharges, etc.). Next, the external water resources from other basins should be incorporated, such as those from the Tagus–Segura Aqueduct, as a strategic or complementary source which, when necessary, can transport water to satisfy the demands. Furthermore, it is also necessary to calculate the water needs per crop in order to determine the amount of water that is required to produce substantial yields of the crop and its fruit, administrating the necessary volume of water.

Another aspect that the new plan should contemplate isadaptation measures through water use agreements, such as those in Marina Baja (Alicante), where the irrigators concede clean water for urban supply while the regenerated water is used for irrigation in agriculture. To achieve this, it is preferable to establish an agreement between the interested parties (farmer, water company and the EDAR, among others).

After planning all these factors, it is necessary to supply all of the demands existing in the basin and territory as far as possible. Most likely, in the deficit basins which have transformed areas of rain-fed land into irrigated land (new irrigated land dependent on the TTS), different alternatives will have to be sought to satisfy their demand: desalinated water, treated wastewater, new irrigation systems, changing to soils and crops that require less water, or, in the most extreme case, reducing the crop areas in order to bring down demand.

The proposal presented in this study adapts to the objectives established by the Spanish government for the year 2050 which seek to promote the development of alternative sources of supply (reuse and desalination based on renewable energy), reduce the water lost in the sanitation and supply network, increase the quality of the water, "renewable water" and moderate consumption, among other actions.

The energy consumption of these facilities is currently around 3 Kw/h for each m^3 of water produced (generally less than 4 in new facilities including auxiliary systems and other pumping) and has been reduced from values of over 20 Kw/h/m^3 in the 1960s to current values [39,40].

Values of more than 20 Kw/h/m^3 in the 1960s to have lowered to current values [39,40] thanks to improvements in the chemistry and configuration of the membranes and in the systems for recovering residual energy from the brine.

Energy consumption is the largest cost of desalinated water production, so its reduction is the key factor in reducing the price of desalinated water. The implementation in Spain of the National Integrated Energy and Climate Plan (2021–2030) and the EU legislative package "Fitfor 55" responds to the European Commission's recent commitment to reduce net greenhouse gas emissions by at least 55% by 2030.

In this context, projects are being developed for the implementation of solar farms to supply desalination facilities on the Spanish mainland and in the Canary Islands. In particular, the Torrevieja desalination plant, a key element in the supply of desalinated water, is developing a project to install an electrical substation, powered by solar energy, with the aim of obtaining self-consumption of energy. At present, the desalination plant's energy consumption is estimated at 264 GWh, which would increase to 400 GWh with the expansion of its desalination capacity to 120 hm^3/year (from the current 80 hm^3/year).

As mentioned above, the production capacity of the Torrevieja desalination plant is currently 80 hm^3/year, where 40 isfor urban supply and another 40 for irrigation in the Segura River basin. However, the CHS plans to increase this to 120 in the current Hydrological Basin Plan (2022–2027). According to information provided by ACUAMED, the current specific consumption value of the Torrevieja plant varies between 3.25 and 3.65 kWh/m^3, depending on the delivery point and the required water quality. It is capable of producing 1 hm^3 per day (24 h), obtaining water with a conductivity of 200 µS, i.e., with a quality identical to that of mineral water and, therefore, water that can be used for irrigation. The economic cost of desalination for the years 2019, 2020 and 2021, and the resulting average tariff over the last years, is approximately 0.45 EUR/m^3 (water delivered).

Finally, they are rigorously monitoring the marine ecosystem impact caused by the discharge of brine or brine overflow. In this case, they ensure that they have zero environmental impact, as before dumping the brine they mix it with seawater and dump it in small quantities and in different areas to avoid causing damage to the marine environment. The Torrevieja desalination plant has eight sensors that monitor the salinity level, which have never once detected an environmental problem.

To achieve these water-related objectives for the year 2050, among the several actions proposed, the sixth addresses the need to *"adjust the management of water resources, preparing the system that will prevail in a future with a lower availability of water"* [41]. Therefore, *"a comprehensive water management strategy must be designed that promotes reuse and the desalination of water until its price is competitive, that is, similar to the price of water from traditional sources; improve the efficiency of the systems of urban supply, agricultural irrigation and the treatment of drinking water and wastewater, through the modernisation of infrastructures and the introduction of new technologies; reorder the agricultural and crop uses, acting on the prevailing concessional regime, prioritising sustainable and socially fair agriculture; increase the resilience of farms to extreme atmospheric events and the effects of climate change, through the transformation of crops and production systems, improve training in agricultural management and create adequate financial and governance mechanisms; and, finally, implement an ambitious strategy for restoring the rivers, aquifers and other continental aquatic systems, while strengthening the river reserves and other protected spaces"* [41].

In short, the objectives established for Spain in 2050 are aligned with the proposal of the new hydraulic planning proposed in this research.

5. Conclusions

The climate and water situation in Spain and its respective hydrographic basins, particularly the Tagus and Segura basins, is not the same as it was 50 years ago, when the volumes of water to transfer via the Tagus–Segura Transfer were planned.

The Tagus basin has suffered a significant reduction in rainfall, surface runoff and volumes of reservoir-stored water in the sub-basin of the Upper Tagus. This is the starting point of the Tagus–Segura Transfer, and the reduction in water resources (surpluses) available in the Tagus basin has been modified by the effects of climate change. This has given rise to a serious problem, given that there will be less and less resources available to transfer. Therefore, the scenario of the Tagus basin will be to plan its own resources in order to supply the needs of its own basin, without taking into account the Segura basin, which is dependent on the waters of the TTS.

This implies that the Segura basin and the irrigated lands of south-east Spain will receive a lower volume of water from the transfer than theycurrently receive, given that the surplus resources of the Tagus basin are subject to variations in climate, rendering the transfer an infrastructure vulnerable to the effects of climate change. This is why the Segura basin should begin to make a firm commitment to using non-conventional sources, such as treated wastewater and desalinated water, the latter being the most important for the self-sufficiency of the basin, focusing on three fundamental aspects: (a) increasing the production capacity, (b) reducing the cost of the desalinated water supplied to the farmers (0.20–0.30 EUR/m^3) and (c) reducing the environmental impacts (brine).

Furthermore, the farmers in south-east Spain, particularly those of the new irrigated lands dependent on the TTS, should know that they have been deceived with the promise of water resources assigned through the water contributions for each irrigation community, when these volumes were calculated more than 50 years ago when the climate reality was completely different. Furthermore, these volumes are theoretical and have never been fulfilled, at least in the province of Alicante. The undersupply of the TTS is not due to a failure to transfer the resources that should be transferred, but because there are not enough resources in the headwaters of the Tagus that can be transferred to fulfil the theoretical volumes.

Therefore, in response to the questions posed at the beginning of this article, it is clear that the Tagus–Segura Transfer is not the only solution for the water future of the district of Vega Baja del Segura. There are other alternative sources, such as treated wastewater and desalinated water, which can increase the volume of available resources to supply the urban and agricultural demands. However, it should be noted that, with the current maximum capacity for producing desalinated water in the district (80 hm^3/year), the demand in Vega Baja cannot be satisfied. Therefore, it is necessary to extend the desalination plant of Torrevieja, increasing the production capacity to 120 hm^3/year and complemented by the regenerated volumes of water. In this respect, it is necessary to extend the treatment plants of the district of Vega Baja del Segura with tertiary or advanced treatment, given that most of them only perform secondary treatment and the resulting water cannot be applied directly to the crops.

Finally, Spain should be more ambitious in terms of hydrological planning. It is incomprehensible that the plan that is in force with respect to the TTS is based on the theoretical volumes of a Preliminary Project when the climate situation was very different to that of the present day. No attempt has been made to reconsider the operating water volumes for the current scenario taking into account the effects of climate change.

For this reason, to correct the lack of a coherent and rational water plan that contemplates the climate reality and effects of future climate change, this study proposes a new sustainable hydraulic plan.

This new sustainable plan requires a profound restructuring of hydrologic planning, based on the worst climate scenario (RCP 8.5), enabling a plan to be elaborated with a long-term horizon (until the end of the century) and adapted to climate change. This scenario will lead to significant restrictions with respect to the current water allocations, whereby the assigned volumes of water will be initially reduced or eliminated. The positive side of this plan is that if the reality in terms of climate evolves over the years into a scenario with less emissions and a better adaptation and management of water, the volumes of water assigned to each use may be increased. This is the main reason why the principal sources in the sustainable hydraulic plan are treated and desalinated water, with the afore-mentioned improvements, as these resources do not depend on climate variations. Then, the basin resources will be included. As observed in the severe droughts occurring in Spain, the basin resources, and, therefore, the traditional irrigated lands that use this water, are more resilient and are adapted to extreme atmospheric conditions, giving them a clear advantage for the effects of climate change. Subsequently, the external water resources will be included, which, depending on the climate evolution over the coming years, will be based on the transfer or not of resources from the Tagus basin to the Segura basin. Therefore, in this proposal, instead of playing a principal role, the TTS is used as a strategic-secondary source to support the principal sources in the proposed plan. Moreover, the uses, crops, areas cultivated and water allocations will have to be reordered in order to reduce demand and, therefore, the deficit existing in the Segura basin in Vega Baja.

The new sustainable hydraulic plan is committed to the self-sufficiency of the territory and only in case of need would it request external resources. However, for a territory to be resilient to the future effects of climate change, it is necessary to start acting now with respect to the afore-mentioned aspects to minimise the socio-economic losses (job positions, crops, desertification, land, cultivated areas, significant economic losses in terms of production and agricultural income, among many others). All of this is possible through a logical plan and coherent actions. A failure to take such measures would result in consequences that will be catastrophic for south-east Spain and, particularly, for the farmers and their way of life. This proposal establishes the fundamental pillars for a long-term national strategy for Spain in the year 2050.

Author Contributions: Formal analysis, investigation, resources, data curation, writing, A.O.C.; formal analysis, investigation, resources, data curation, writing, J.O.C.; formal analysis, data curation, supervision, writing, C.J.B.C. All authors have read and agreed to the published version of the manuscript.

Funding: This research received no external funding.

Data Availability Statement: The data can be found in the documents referenced in this article. Likewise, the authors offer to provide any data requested by other researchers or interested parties.

Conflicts of Interest: The authors declare no conflict of interest.

References

1. Intergovernmental Panel on Climate Change (IPCC). Climate Change 2021: The Physical Science Basis 2021. Available online: https://www.ipcc.ch/assessment-report/ar6/ (accessed on 14 June 2022).
2. Pulido-Velázquez, M.; Escriba-Bou, A.; Sorribes, H.M. Balance hídrico actual y futuro en las cuencas hidrográficas de España, déficits estructurales e implicaciones socioeconómicas. *Estud. Econ. Española* **2020**, *38*, 1–30.
3. AEMETBLOG. Caracterización de la Variabilidad Interanual de las Precipitaciones Diarias en la Demarcación Hidrográfica del Segura (Parte II). Available online: https://aemetblog.es/2021/10/03/caracterizacion-de-la-variabilidad-interanual-de-las-precipitaciones-diarias-en-la-demarcacion-hidrografica-del-segura-parte-ii/#more-33030 (accessed on 14 June 2022).
4. Morote-Seguido, A.F.; Amorós, A.M.R. Perspectivas de funcionamiento del Trasvase Tajo-Segura (España): Efectos de las nuevas reglas de explotación e impulso de la desalinización como recurso sustitutivo. *Boletín Asoc. Geogr. Española* **2018**, *79*, 1–43. [CrossRef]

5. Morote-Seguido, A.F. La desalinización. De recurso cuestionado a recurso necesario y estratégico durante situaciones de sequía para los abastecimientos en la Demarcación Hidrográfica del Segura. *Investig. Geográficas* **2018**, *70*, 47–69. [CrossRef]
6. Morote-Seguido, A.F.; Hernández, M.H.; González, R.C.L. Propuestas al déficit hídrico en la provincia de Alicante: Medidas desde la gestión de la demanda y oferta de recursos hídricos. *Boletín Asoc. Geogr. Española* **2019**, *80*, 1–48. [CrossRef]
7. Agencia Estatal de Meteorología (AEMET). Proyecciones Climáticas Para el SigloXXI en España. Available online: http://www.aemet.es/es/serviciosclimaticos/cambio_climat (accessed on 14 June 2022).
8. CEDEX. *Evaluación del Impacto del Cambio Climático en los Recursos Hídricos y Sequías en España*; Informe Técnico del Centro de Estudios Hidrográficos del Centro de Estudios y Experimentación de Obras Públicas (CEDEX); Ministerio de Fomento y Ministerio de Agricultura y Pesca, Alimentación y Medio Ambiente: Madrid, Spain, 2017; pp. 1–346.
9. MITERD. *Impactos y Riesgos Derivados del Cambio Climático en España*; Ministerio Para la Transición Ecológica y el Reto Demográfico (MITERD): Madrid, Spain, 2021; pp. 1–213.
10. Cantos, J.O. Evidencias e incertidumbres del cambio climático y de los riesgos asociados en el litoral mediterráneo español. *Boletín Real Soc.* **2019**, *154*, 9–34.
11. Cantos, J.O. Incremento de episodios de inundación por lluvias de intensidad horaria en el sector central del litoral mediterráneo español: Análisis de tendencias en Alicante. *Sémata Cienc. Sociais Humanid.* **2017**, *29*, 143–163.
12. Llasat, M.C. Floods evolution in the Mediterranean region in a context of climate and enviromental change. *Geogr. Res. Lett.* **2021**, *47*, 13–32.
13. CCS. Análisis de los daños por inundación en España a nivel municipal. *Rev. Digit. Consor Seguros* **2021**, *14*, 6–18.
14. Muñoz, C.; Schultz, D.; Vaughan, G. A Mid latitude Climatology and Interannual Variability of 200-and 500-hPa Cut-Off Lows. *Am. Meteorol. Soc. J. Clim.* **2020**, *33*, 2201–2222.
15. Shaltout, M.; Omstedt, A. Recent sea surface temperature trend sand future scenarios for the Mediterranean Sea. *Pap. Oceanol.* **2014**, *56*, 411–443. [CrossRef]
16. MedECC. *Climate and Environmental Change in the Mediterranean Basin–Current Situation and Risks for the Future*; First Mediterranean Assessment Report; Mediterranean Experts on Climate and Environmental Change (MedEEC): Barcelona, Spain, 2020; pp. 1–632.
17. Pastor, F.; Valiente, J.A.; Palau, J.L. Sea Surface Temperature in the Mediterranean: Trendes and Spatial Patterns (1982–2016). *Pure Appl.* **2018**, *175*, 4017–4029. [CrossRef]
18. Pastor, F. *Mediterranean SST Report (Autumn 2020)*; Centro de Estudios Ambientales del Mediterráneo (CEAM): Valencia, Spain, 2020; pp. 1–7.
19. Amorós, A.M.R.; Cantos, J.O.; Castiñeira, C.J.B. Competencias por el uso del agua en la provincia de Alicante: Experiencias de gestión en la armonización de usos urbano-turísticos y agrícolas. *Doc. D'anàlisis Geogràfica* **2014**, *60*, 523–548.
20. MITERD. *Informe Complementario. Fomento de la Reutilización de las Aguas Residuales*; Ministerio para la Transición Ecológica y el Reto Demográfico (MITERD): Madrid, Spain, 2020; pp. 1–129.
21. Amorós, A.M.R. La Mancomunidad de los Canales del Taibilla: Un Modelo de Aprovechamiento Conjunto de Fuentes Convencionales y Desalinización de Agua Marina. In *Libro a Jubilar en Homenaje al Profesor Antonio Gil Olcina*; Cantos, J.O., Amorós, A.M.R., Eds.; Publicaciones de la Universidad de Alicante: San Vicente del Raspeig, Spain, 2007; pp. 367–393.
22. Martínez, D.Z. La desalación del agua en España. *Estud. Sobre La Econ. Española* **2020**, *101*, 169–186.
23. Cantos, J.O.; Baños-Castiñeira, C.J. El agua que llueve: Diversidad territorial, riesgos y cambios. *Canelobre* **2019**, *1*, 19–29.
24. Generalitat Valenciana. *Plan Vega Baja Renhace: Una Estrategia Integral Para la Resiliencia de la Comarca de la Vega Baja del Segura*; Generalitat Valenciana: Valencia, Spain, 2020; pp. 1–98.
25. Olcina, A.G. Conflictos autonómicos sobre trasvases de agua en España. *Investig. Geográficas* **1995**, *13*, 17–28. [CrossRef]
26. CHS. *Plan Hidrológico de la Cuenca del Segura (2015–2021)*; Ministerio para la Transición Ecológica y Reto Demográfico: Madrid, Spain, 2015. Available online: https://www.chsegura.es/es/cuenca/planificacion/planificacion-2015-2021/plan-hidrologico-2015-2021/ (accessed on 14 June 2022).
27. CHS. *Plan Hidrológico de la Cuenca del Segura (2022–2027)*; Ministerio para la Transición Ecológica y Reto Demográfico: Madrid, Spain, 2021. Available online: https://www.chsegura.es/es/cuenca/planificacion/planificacion-2022-2027/el-proceso-de-elaboracion/ (accessed on 14 June 2022).
28. CHT. *Plan Hidrológico de la Cuenca del Tajo (2022–2027)*; Ministerio para la Transición Ecológica y Reto Demográfico: Madrid, Spain, 2021. Available online: http://www.chtajo.es/LaCuenca/Planes/PlanHidrologico/Planif_2021-2027/Paginas/BorradorPHT_2021-2027.aspx (accessed on 14 June 2022).
29. Aguas de las Cuencas Mediterráneas (ACUAMED). *Informe Técnico Sobre el Estado de las Desaladora de Torrevieja*; Ministerio para la Transición Ecológica y Reto Demográfico (MITERD): Madrid, Spain, 2021; pp. 1–7.
30. CHS. *Plan Especial de Sequía*; Ministerio para la Transición Ecológica y Reto Demográfico: Madrid, Spain, 2015. Available online: https://www.chsegura.es/export/descargas/cuenca/sequias/revision2018/docsdescarga/Memoria_PES.pdf (accessed on 14 June 2022).
31. Marcos-García, P.; Pulido-Velázquez, M.M. Cambio climático y planificación hidrológica: ¿Es adecuado asumir un porcentaje único de reducción de aportaciones para toda la demarcación? *Ing. Agua* **2017**, *21*, 35–52. [CrossRef]
32. De Luís, M.; Brunetti, M.; González-Hidalgo, J.C.; Longares, L.A.; Martín-Vide, J. Changes in seasonal precipitation in the Iberian Peninsula during 1946–2005. *Glob. Planet. Chang.* **2010**, *74*, 27–33. [CrossRef]

33. Estrela, M.J.; Pérez, J.M. Evidencias del Cambio Climático en la Comunidad Valenciana y Cuencas del Júcar y Segura. Resultados a Escala de Detalle. In Proceedings of the Seminario Estrategias de Adaptación a la Crisis, Valencia, Spain, 3 December 2019.
34. Pérez, J.M.; Estrela, M.J. Evidencias del Cambio Climático en las Precipitaciones Desde los Mediados del Siglo XX hasta hoy: Cuencas del Júcar y del Segura. In *Climas y Tiempos del País Valenciano*; Olcina, J., Moltó, E., Eds.; Publicaciones Universitat d' Alacant: Alicante, Spain, 2019; pp. 111–116.
35. Pérez, J.M.; Estrela, M.J.; Cantos, J.O.; Martín-Vide, J. Future projection of precipitation changes in the Júcar and Segura river basins (Iberian Peninsula) by CMIP5 GCMs Local Downscaling. *Atmosphere* **2021**, *12*, 879.
36. Corominas, J.; Hernández-Mora, N.; La Calle, A.; La Roca, F. *Recuperación de Costes del Agua. Diagnóstico de los Segundos Planes Hidrológicos y Propuestas de Mejora*; WWF/Fundación Nueva Cultura del Agua: Zaragoza, Spain, 2017.
37. La Roca Cervigón, F.; Del Moral Ituarte, F. Cambio climático en la planificación hidrológica en España: La urgencia del cambio de rumbo. In *Cambio Climático en el Mediterráneo. Procesos, Riesgos y Políticas*; Romero, J., Olcina, J., Eds.; Tirant Humanidades: Valencia, Spain, 2021; pp. 211–235.
38. Ibor, C.S.; Reig, M.O.; Mollá, M.G. El regadío mediterráneo y el cambio climático: Un proceso de exaptación. In *Cambio Climático en el Mediterráneo. Procesos, Riesgos y Políticas*; Romero, J., Olcina, J., Eds.; Tirant Humanidades: Valencia, Spain, 2021; pp. 237–262.
39. Zarzo, D.; Prats, D. Desalination and energy consumption. What can we expectin the near future? *Desalination* **2018**, *427*, 1–9. [CrossRef]
40. Martínez, D.Z. La desalación de agua en España. Estudios sobre la economía española 2020/22. *Fedea* **2020**, *1*, 1–22.
41. Oficina Nacional de Prospectiva y Estrategia del Gobierno de España (Coord.). *España 2050: Fundamentos y Propuestas Para una Estrategia Nacional de Largo Plazo*; Ministerio de la Presidencia: Madrid, Spain, 2021; pp. 1–607.

sustainability

MDPI

Review

Diversifying Water Sources with Atmospheric Water Harvesting to Enhance Water Supply Resilience

Mengbo Zhang, Ranbin Liu * and Yaxuan Li

Sino-Dutch R&D Centre for Future Wastewater Treatment Technologies/Key Laboratory of Urban Stormwater System and Water Environment (Ministry of Education), Beijing University of Civil Engineering & Architecture, Beijing 100044, China; zmb11211324@163.com (M.Z.); fank812@hotmail.com (Y.L.)

* Correspondence: ranbin_liu@hotmail.com

Abstract: The unequivocal global warming has an explicit impact on the natural water cycle and resultantly leads to an increasing occurrence of extreme weather events which in turn bring challenges and unavoidable destruction to the urban water supply system. As such, diversifying water sources is a key solution to building the resilience of the water supply system. An atmospheric water harvesting can capture water out of the air and provide a point-of-use water source directly. Currently, a series of atmospheric water harvesting have been proposed and developed to provide water sources under various moisture content ranging from 30–80% with a maximum water collection rate of 200,000 L/day. In comparison to conventional water source alternatives, atmospheric water harvesting avoids the construction of storage and distribution grey infrastructure. However, the high price and low water generation rate make this technology unfavorable as a viable alternative to general potable water sources whereas it has advantages compared with bottled water in both cost and environmental impacts. Moreover, atmospheric water harvesting can also provide a particular solution in the agricultural sector in countries with poor irrigation infrastructure but moderate humidity. Overall, atmospheric water harvesting could provide communities and/or cities with an indiscriminate solution to enhance water supply resilience. Further research and efforts are needed to increase the water generation rate and reduce the cost, particularly via leveraging solar energy.

Keywords: water supply resilience; atmospheric water harvesting; fog collection; refrigerated atmospheric water extraction; climate change

Citation: Zhang, M.; Liu, R.; Li, Y. Diversifying Water Sources with Atmospheric Water Harvesting to Enhance Water Supply Resilience. *Sustainability* **2022**, *14*, 7783. https://doi.org/10.3390/su14137783

Academic Editors: Alban Kuriqi and Luis Garrote

Received: 17 May 2022
Accepted: 22 June 2022
Published: 26 June 2022

1. Introduction

Promising reliable access to safe water is still a big issue all over the world [1,2]. On the one hand, at least a billion people globally are suffering from severe water shortages, particularly those living in developing countries and regions [3]. As such, the 2021 edition of the United Nations World Water Development Report is rooted in "Valuing Water" and strengthening the societal awareness of water safety [4]. On the other hand, the unequivocal climate change and the resultant extreme weather bring new challenges to water accessibility [5] and explicitly sound alarms to the established water supply system [6]. Currently, surface water is still a principal or sole water source for the water supply systems in most cities, of which the vulnerability has been completely unmasked and experienced, such as the Day Zero water crisis in Cape Town [7,8]. A severe and unanticipated reservoir drought left millions of residents thirsty. As such, increasing reliable access to safe water plays an important role in the sustainable development of society [9].

From a technical point of view, diversifying water sources besides surface water is a fundamental and vital approach to increasing urban water reliability [10]. In another word, this principle is covered by the framework of water supply resilience which highlights the ability of water supply system to promise residents the accessibility to safe drinking water under extreme events like drought [11]. In terms of water sources, seawater desalination, rainwater harvesting [12,13], sewage reuse [14–16], and inter-basin water transfer [17,18] can be supplements to surface water and enhance water supply resilience. However, these approaches have their pros and cons in terms of their applicability. Typically, seawater desalination and rainwater harvesting, are not suitable for inland cities suffering absolute water shortages while inter-basin water transfer is facing vulnerability of water quality or ecological safety [19]. Although sewage reuse is universally applicable enough for cities, public acceptance has been the most serious hurdle for practicing [20]. Moreover, these approaches cannot be relied on to go through water supply emergencies under which bottled water is usually the preferential choice [21–23].

Indeed, there is a kind of water reservation always overlooked, namely atmospheric water or water in the air [24]. As a key and interchange step of the water cycle from ocean to land, the atmosphere is a huge renewable water reservoir [25]. Roughly, it contains 12,900 trillion liters of renewable water, which is about equivalent to 10% of surface water reservation [26]. Even in the arid desert, the moisture content in the air is as abundant as 10 g/m^3 [27]. As such, atmospheric water harvesting has been proposed and developed to link the natural water cycle and the urban water cycle [28]. Moreover, water in the air is distributed everywhere and could be an indiscriminately decentralized water resource [29]. As such, the present study is to introduce and summarize the development of atmospheric water harvesting in comparison with other water sources. By analyzing the pros and cons of atmospheric water harvesting in terms of technology, economy, and safety, the role of atmospheric water harvesting in contributing to water supply resilience is discussed.

2. Characterizing Supplementary Water Sources to Surface Water

Surface water has long been the main water source for potable or non-potable utilization [30]. For a long time, surface water has been the only connection point between the natural water cycle and the urban water cycle as depicted in Figure 1. The water feeding the cities starts from and wastewater also ends in surface water. In general, the water bodies receiving wastewater are located downstream. It is the natural water cycle that refreshes the surface water to meet the demand of human beings [31]. In another word, the renewal of surface water depends on a whole circle (the blue-lined circle in Figure 1), that is, evaporation, condensation, and precipitation. Once the water demand outpaces the renewal capacity of surface water, drought will occur and undermine the safety of the water supply. As a response, various water sources have been explored to supplement surface water-based water supply as presented in Figure 1, including seawater desalination, rainwater harvesting, inter-basin water transfer, and sewage reuse.

As presented in Figure 1, desalination enables seawater to feed the residents by artificially bypassing the step of water vapor transport. At present, there are about 17,000 desalination plants globally in operation with a total capacity of ~95 million m^3/d [32]. Although the accounting percentage of desalination in water supply structures is still very small, it provides a promising direction to strengthen water supply resilience. However, desalination is still considered to be an energy and cost-intensive technology and is mainly implemented by high-income countries and small island countries [33]. As such, geographical constraints and high capital & operating expense are two hurdles to practicing seawater desalination [34].

As one of the simplest and oldest water sources, rainwater harvesting can be more flexible in terms of capacity, sites, and applications [35]. With proper purification treatments, the rainwater collected can be utilized for potable or non-potable purposes [36]. As such, rainwater can provide a useful supplementary supply and important backup to the water supply system. The biggest obstacle to rainwater harvesting is the temporal variation and geographic locations of rainfall [37]. Although artificial rainfall seems to solve this problem, this technology is still controversial [38].

Inter-basin water transfer (IBT) is an artificial reallocation of surface water resources from a donor watershed to a recipient [39]. In other words, IBTs improve the water supply also by accelerating the water cycle (the blue-line circle) by avoiding evaporation and precipitation [40]. However, the solution of IBT to improve the water supply has been under hesitation and debate since the 1980s [41]. Specifically, hydrological and ecological risks are associated with the donating and recipient basins [42]. Moreover, the donating watershed, IBTs' water source, is also subject to uncertainties from climate change.

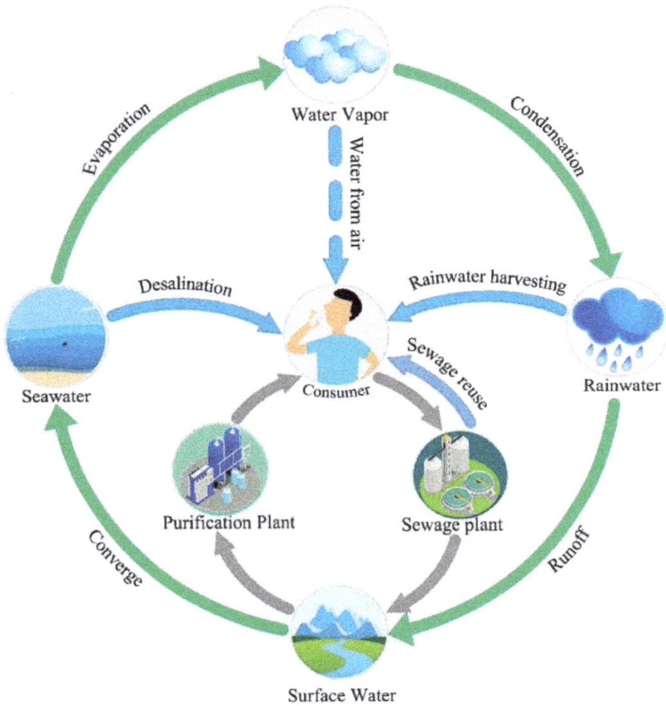

Figure 1. Diagram of the water cycle and water sources available for human beings.

Sewage reuse is another long-history-applied approach to conserve water and improve water supply [43]. Sewage reuse is the one shortening the water cycle most significantly by bypassing the natural water cycle directly. Compared with the above approaches, one of the advantages of sewage reuse is its on-site and stable water supply. As such, this feature endows sewage ruse an indiscriminately applicable solution for cities to enhance water supply resilience. However, sewage reuse as drinking water is currently unacceptable to most people, and public attitudes hinder the sewage reuse plans in many developing countries [20]. Fortunately, aquifer recharge with treated sewage instead of reuse directly could be a solution to leverage sewage.

These approaches have been applied separately or jointly to enhance water security by offering more choices and supplements to surface water. According to Figure 1, a common feature of these water sources is to offer more connection points between the urban water cycle and the natural water cycle. In other words, these alternatives get the natural water to the tap of residents (urban water cycle) more quickly [44], which seems to be a principal justification for whether a supplement is qualified to be an alternative to surface water. Obviously, atmospheric water owns the potential to be another connection point between the natural water cycle and the urban water cycle. It means that the water in the air can be extracted mandatorily (the dotted line) instead of via passive precipitation.

3. Technologies for Atmospheric Water Harvesting

As depicted in Figure 1, atmospheric water is an indispensable part of the natural water cycle and is the prerequisite and prior step for precipitation. Via evaporation and transpiration, a vast quantity of water out of water bodies and plants enters the air and exists in gaseous water vapor. All this vapor goes up with the rising air currents and condenses into clouds or fogs in the cooler air [45]. Generally, we can only get access to this part of water after they drop down on the ground via condensation albeit passively and intermittently [46]. By contrast, various technologies can be leveraged currently to help extract water directly and constantly from air depending on relative humidity. According to Figure 2, the water content capacity in the air (humidity ratio) is positively correlated with temperature and the isohume curves (100%, 80%, and 30%) separate the area into four zones. The blue line in Figure 2 represents a constant water quantity in the air termed g-H_2O/kg air [47], however, under each zone, the water is in a specific form. In Zone a, the relative humidity is higher than 80% and close to 100%, and the water will be in the form of mist or fog, which could easily be adsorbed and captured via proper materials. This phenomenon always occurs at high altitudes or on the top of high mountains with a low temperature and 100% relative humidity [48]. In Zone b, the relative humidity of the blue line is around 30–80%, the air contains a large amount of water vapor that does not readily nucleate into water droplets [49]. It needs to be artificially converted into liquid water first. Most living environments are in this zone. In Zone c, the relative humidity is lower than 30%, which makes the water vapor difficult to transform into liquid droplets even artificially [50]. The blocks divided by relative humidity and temperature in Figure 2 can help determine the proper method to do atmospheric water harvesting in a specific area, which will be discussed in detail below.

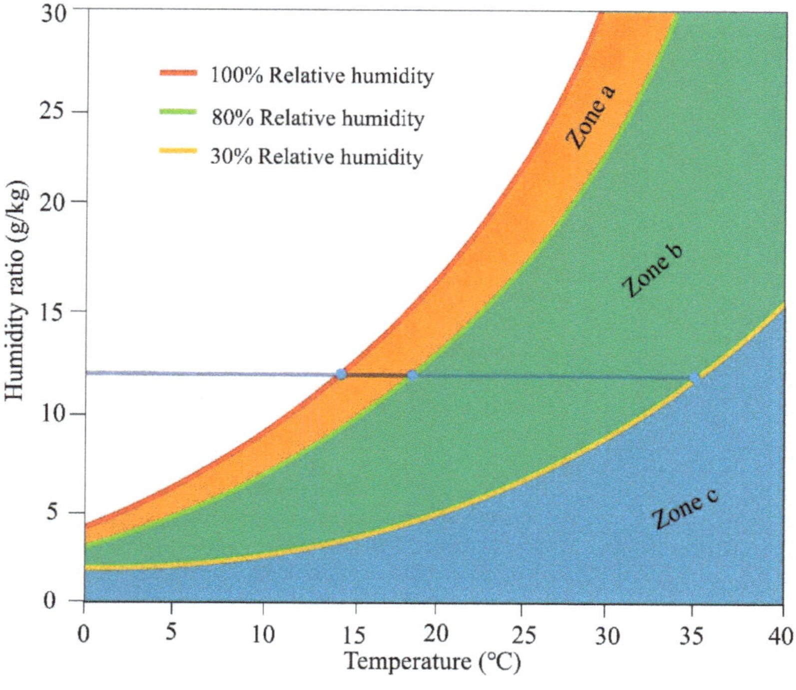

Figure 2. Water content in the air against various temperatures.

3.1. *Passive Fog Collectors*

Passive fog collectors target the fog in the air by mimicking the oldest practice of collecting drinking water from the leaves in the early morning by our ancestors. From the relative humidity perspective, Zone a in Figure 2 is the prerequisite condition for scaling this technology, and such places are usually located in coastal regions and/or mountainous areas such as Chile, Mexico, Oman, South Africa, and Morocco [51,52]. Generally, a fog collector is comprised of a flat-panel mesh that is stretched and fixed over a rigid frame. With wind current, the fog water contacts with and deposits on the surface of the mesh, and then aggregates into large droplets enough to drain into the container [53,54]. According to the full-scale project, the water collection rate is in the range of 1.5–12 L/(m^2·day) and can reach 1416 L/(m^2·day) with modification (Table S1). On the one hand, passive fog collectors are energy-free while, on the other hand, their water collection performance is highly associated with interior and exterior factors. Herein, the key interior factor is the mesh type including mesh material and weave design. Currently, stainless steel and plastics are two commonly-used mesh materials in large-scale projects [55]. Stainless steel is hydrophilic and can resist strong wind loading albeit heavy. The typical plastics available include polyethylene, polypropylene, and nylon [56]. They usually own the advantages of hydrophobicity, lightweight, low price, and good anti-aging performance. The properties of these materials are vitally important as they provide the direct sites to capture fog water. Besides this, the weave configuration is another principal factor determining the performance even with the same mesh material. As depicted in Figure 3, there are three geometric shapes generally adopted in field projects, simply denoted by triangular mesh, rectangular mesh, and hexagonal mesh [55,57]. Other key variations associated with performance include the width of mesh wires, pore area, and shading coefficient. The rectangular mesh is the most simple one made of stainless steel with a pore area of 0.16×0.16 cm^2 and a shading coefficient of 49%. Raschel mesh [58] is a typical representative of triangular mesh and is interweaved by doubled layered polypropylene ribbons with a width of 1–1.6 mm (shading coefficient of 35%). FogHa-Tin mesh is a proprietary product and is made of 0.13 mm diameter polypropylene thread into a springy structure with interleaved sets of embedded hexagonal patterns.

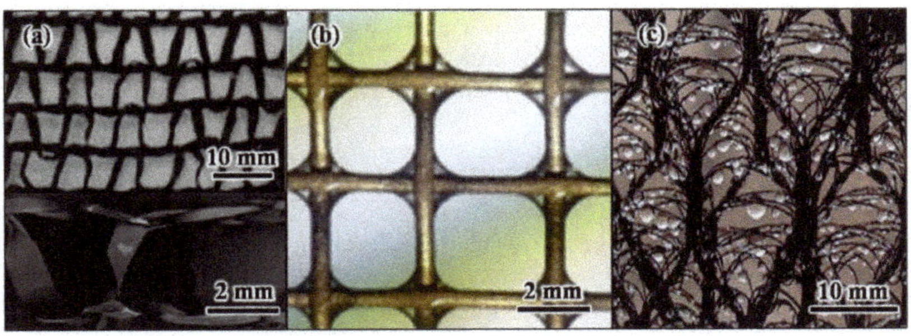

Figure 3. Various weave designs of water collector mesh (**a**) Raschel mesh (triangular mesh) [59]. (**b**) stainless-steel mesh (rectangular mesh) [60]. (**c**) FogHa-Tin mesh (hexagonal mesh) [55,57].

Rivera proposed Equation (1) to calculate the water collection efficiency, which is determined jointly by aerodynamic collection efficiency, capture efficiency, and draining efficiency [61]. The pore area or shade coefficient is the key factor influencing aerodynamic efficiency [62,63]. On the one hand, a large shade coefficient seems to provide more deposition sites, but a too-large coefficient can divert the wind flow due to resistance and reduce the water-mesh contact. On the other hand, too small a pore size could cause liquid film clogging which then jeopardizes the aerodynamic efficiency. In terms of Raschel and FogHa-Tin mesh (Figure 3a,c), wider ribbons instead of thread and embedded wires in the pore areas are designed respectively to offset the large pore areas [64]. By contrast, the pore size of stainless-steel mesh is too small to easily be clogged. Then, a harp mesh by only placing wires vertically instead of crossing reduces the adhesion to fog droplets and creates an unobstructed path for fog droplets to move and fall freely [65]. As a result, the water collection capacity of the parallel arrangement of wires can be 2–20 times higher than cross arrangement [66]. In addition, this problem can also be solved by co-knitting or coating with poly material [55]. Indeed, modification of mesh wires with coating materials can not only improve the aerodynamic efficiency but also optimize the capture and draining efficiency (Table S1). Knapczyk-Korczak et al. [64] deposited PVDF fibers on the Raschel mesh and, as a result, the effective surface area to catch droplets increased without sacrificing wind permeability. With the optimization of wetting properties and draining efficiency, the water collection rate increased by 300%.

$$\eta = \eta_{ace} \cdot \eta_{cap} \cdot \eta_{dra}, \quad (1)$$

where:
η represents the overall collection efficiency,
η_{ace}, η_{cap}, η_{dra} represent the aerodynamic collection efficiency, capture efficiency, and drainage efficiency, respectively.

Another factor that should be taken into consideration is the wind speed [67]. Generally, the most favorable wind speed for passive water collectors is 4–10 m/s [68]. Noteworthy is that the effect of wind speed on the efficiency of fog collection is also related to the diameter of the droplets and types of mesh. Fernandez et al. [55] evaluated the water collection performance of Raschel mesh, modified stainless steel mesh, and FogHa-Tin mesh. The results showed that Raschel mesh collected 160% more fog water than FogHa-Tin mesh at wind speeds less than 1 m/s while 45% less with wind speeds higher than 5 m/s. This is because the three-dimensional textile mesh will form some sort of a "blockage" at lower wind speeds while capturing some of the coalesced water droplets that tend to re-entrain in higher winds. As for modified stainless steel mesh, it collected more water than Raschel mesh at all wind speeds.

3.2. *Refrigerated Atmospheric Water Harvesting*

In terms of an environment with a relative humidity of around 30–80%, there are no readily available water droplets. To capture them, the prior step is to condense the vapor into droplets artificially [50], which, along with the following capture unit, represents a typical principle to carry out atmospheric water harvesting in Zone b (Figure 2). According to Figure 2, lowering the temperature is a simple and direct method to produce water droplets, by which refrigerated atmospheric water harvesting works [49]. A typical refrigerated atmospheric water harvesting unit is comprised of four parts, including the evaporator, condenser, compressor, and throttle valve (Figure 4) [69]. The humid air enters the evaporator part of the cooling unit, then it is cooled to the dew point temperature and condensed, purified, and collected on the evaporator coil [70].

The cooling unit (condenser) is the key factor that determines the water extraction efficiency of refrigerated atmospheric water harvesting. Currently, there are two cooling categories commonly adopted, passive condenser and active condenser. A passive condenser refers to one operating without any energy input [71]. One such unit is the radiant condenser, which is commonly used [71,72]. The key function unit in the radiant condenser is the cooling foil which owns the hydrophilic property and a high emissivity in the near-infrared. It emits thermal radiation in the wavelength range (8 to 13 μm) where the atmosphere is transparent and can emit heat radiatively to space [73]. This effect cools the foil below the dew point temperature of the air, causing water to condense upon it. A most commonly used cooling foil consists of TiO_2 and $BaSO_4$ microspheres embedded in a polyethylene film [74]. At present, this radiant condenser has a low water production rate. When the relative humidity is greater than 60%, the water production is commonly less than 0.8 L/(m^2·day) [75]. To improve the water yield performance, some new materials have been explored and evaluated [76]. Raveesh et al. prepared a polystyrene film with hydrophilic bumps that secured a water yield of 1.8 L/(m^2·day) [76]. Chen et al. used a wettability and spectral selectivity engineered coating, and the water collection rate even reached 251.25 L/(m^2·day) [77]. Another challenge of the passive radiant condenser is the low solar absorption and high mid-infrared emissions required to operate during the day. Additionally, the process is not completely passive and the condensate needs to be manually removed. Haechler et al. combined a geometrically optimized radiation shield and a hydrophobic coating to the surface of the selective emitter to promote the condensation and removal of droplets which enabled dew mass fluxes up to 1.2 L/(m^2·day) [74].

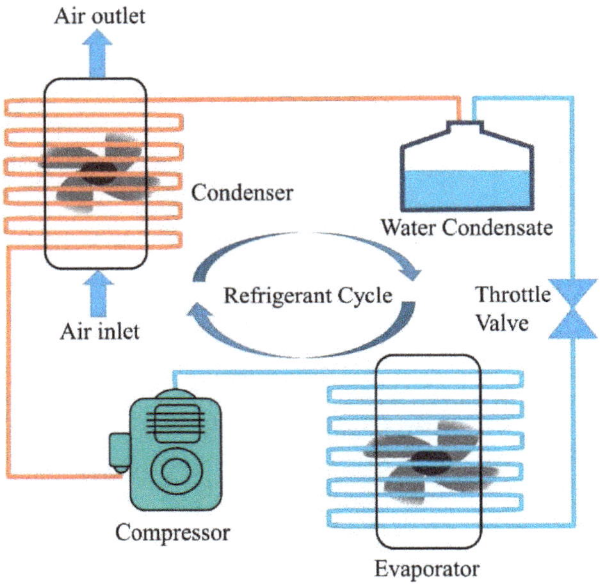

Figure 4. Diagram of refrigerant evaporation–condensation cycle.

In comparison, active condensers exert functions by leveraging external energy to accelerate the condensation process [78]. Thus, they are working more efficiently than passive condensers. The two most commonly used cooling methods in refrigerated atmospheric water harvesting are vapor compression refrigeration and the thermoelectric cooling process [79]. The vapor compression refrigeration process is similar to air conditioners and achieves cooling by changing the state of refrigerants such as Freon [80]. By contrast, thermoelectric cooling converts electrical energy into heat energy for cooling through the Peltier effect and reducing the temperature below the dew point. As such, thermoelectric cooling could avoid the drawbacks of vapor compression refrigeration causing ozone layer depletion and global warming problems [81]. However, in terms of the water yield performance, vapor compression refrigeration owns a higher capacity and is easy to scale up. With a relative humidity of 90%, vapor compression refrigeration can produce 22–26 L/day freshwater with energy input around 0.22–0.30 kWh/L [82]. This technology has been applied in the Middle East such as in Iran and Abu Dhabi [83]. Although the cooling capacity of thermoelectric cooling is low [84], it has the advantages of energy-saving, environmental protection, low maintenance, and high portability [85]. It is applicable and useful for cyclists, hikers, expeditions, and scientific research teams. In general, with a relative humidity of 60–90% and an input power of 0.8–3.5 kWh/L, the water production rate of thermoelectric cooling reaches 0.48–0.8 L/day [76].

Indeed, along with the trial to increase the water yield capacity, research effort is also placed on reducing the energy input associated with refrigerated atmospheric water harvesting, particularly under a hot environment but with low relative humidity. Precooling the inlet air with the cold exit air from the evaporator or providing a preconditioning unit to improve the psychometric properties of incoming air on the vapor compression system has become a common way to increase energy efficiency [76]. Ibrahim et al. used condensate to pre-cool the air entering the condenser, the compressor power input was decreased by 6.1% and the coefficient of performance was improved by 21.4% [86]. In addition, the use of polymer electrolyte membranes or water vapor selective membranes before the cooling process to separate water vapor from other molecules in the air can also achieve energy savings [75]. Roughly, this could reduce energy input by more than 50%. Meanwhile, this dense polymer membrane can also retain pollutants or pathogens, thereby purifying water [87]. Moreover, leveraging renewable energy such as solar and wind power could also be a potential solution to reduce the energy further [88].

3.3. Desiccant-Based Atmospheric Water Harvesting

As discussed above, fog collectors and refrigerated atmospheric water harvesting have their favorite application environments with a relative humidity higher than 80% and 30%, respectively. In terms of relative humidity less than 30%, the above processes do not work or work but with a large quantity of energy input. As such, desiccant-based atmospheric water harvesting was proposed to extract water from air under low relative humidity (below 15–20%) or low dew point temperature (below 5–10 °C) [89]. In general, desiccant-based water harvesting works in a batch mode [90]. At the beginning of a cycle, the desiccant is exposed to the atmosphere and adsorbs water vapor in the air. Once the desiccant is saturated, the system is closed and the water will be released as vapor out of the desiccant at a rising temperature of 160 °C (Figure 5). Then, the vapor condenses on the enclosure walls and can be collected, meanwhile, the reactivated and unsaturated desiccant will be cooled down for the next water-capture cycle [74].

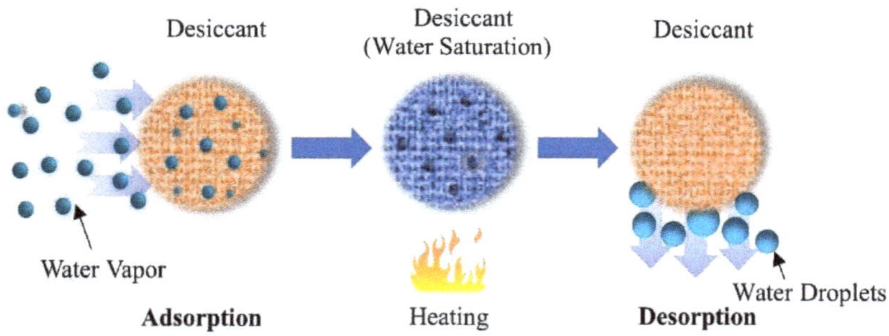

Figure 5. Diagram of the desiccant-based atmospheric water harvesting.

One of the key units is the desiccant, which performs the cycle of water vapor adsorption and water desorption [91,92]. The desiccant not only determines the water collection rate but also is associated with energy consumption. Currently, a series of single solid desiccants and composite materials have been developed and studied [93]. Some typical single solid desiccants include silica gel, activated carbon, and inorganic salts. However, they need a high temperature to release water after saturation which is energy-intensive and cannot be completed by conventional solar thermal equipment [94]. By contrast, some novel composite materials have drawn attention (Table S2). For example, a kind of salt gel beads made of an alginate-derived matrix with calcium chloride owns a water holding capacity of 660 kg water/m^3 and can release water at a temperature of 100 °C [95]. In addition, some MOF-based desiccants were also explored and presented promising water adsorption ability [91,96].

Another key issue associated with a desiccant-based water extraction system is to reduce the energy input as low as possible [69,97]. One of the basic and most greenway is to leverage solar energy [98]. The glass-covered greenhouse (also called solar still) is the simplest device, it uses solar energy to distill out the water molecules adsorbed in the desiccant [99]. However, the water generation rate (1.0–2.5 L/(m^2·day)) is limited due to the diurnal variation [75]. Therefore, employing an additional condenser as a supplement to a solar heat collector to ensure a continuous operation is one of the possible solutions [50]. In addition, transforming and/or storing solar energy in the form of either electricity or heat via thermal collector or photovoltaics can also be coupled to a desiccant-based water extraction system to utilize the solar energy as more as possible [100]. Especially in extremely dry climate regions such as deserts, solar photovoltaic modules can be used to power atmospheric water harvesting. The solar modules developed by Panchenko have an extended service life and polysiloxane compounds, which do not degrade in such difficult climatic conditions and are tolerant to cyclical temperature fluctuations [101,102].

4. Link between Atmospheric Water Harvesting and Water Supply Resilience

A reliable water supply is vital to life, and having either too much or too little has very serious consequences, leading to drought and fires at one extreme, and floods at the other [103]. Recently, the Intergovernmental Panel on Climate Change (IPCC) released a new assessment report in August highlighting the changes in the water cycle due to the temperature rising. As 1 °C increases in the air could increase its water holding capacity by 7%, and continued global warming will make air retain more moisture [104]. Specifically, the rainfall amount will be larger as there is more water to condense and fall out of the air [105]. Meanwhile, a warmer climate will intensify the evaporation and result in droughts developing more quickly and lasting longer [106,107]. Indeed, all these extreme weather patterns have been tangible and jeopardized our water supply system [108,109]. The more recent flooding disaster in Germany and ongoing extreme droughts in the western USA consequently make drinking water unavailable or shortage [110,111]. Thus, it is urgent for each city to proactively enhance its water supply resilience and get ready for the projected worsening global warming [112,113].

The National Infrastructure Commission of the UK advised a twin approach to address the resilience of water supply, which includes demand management and supply infrastructure [114]. Demand management focuses on increasing water efficiency while supply infrastructure highlights diversity which refers to developing a range of different water sources. The different source types have different strengths and vulnerabilities; therefore, resilience is increased by being used together. Hereinto, how to define the role of atmospheric water harvesting in building water supply resilience remains to answer. The current centralized water supply is a symbiotic system composed of four elements, that is, sources for water intake, treatment at drinking water treatment plants, storage, and distribution via a pipe network [44]. As discussed by Deng [115], once a natural hazard impairs one of them due to water source pollution, pipeline destruction, power outage, personnel shortage, or other causes, the entire water supply system may fail. As such, Deng proposed a concept of household water treatment highlighting a decentralized water supply system to respond to possible disruption of the centralized water supply [115]. From a technical point of view, atmospheric water harvesting is just in line with this concept in terms of its indiscriminate presence of water source (air) and the decentralized water supply mode. As such, atmospheric water harvesting seems to be a potential solution to enhance water supply resilience, particularly under extreme weather conditions.

Water yield capacity and affordability are two factors determining the acceptance and applicability of a specific water source. As discussed above, atmospheric water harvesting currently has a much lower water yield than other water sources. In terms of the cost, as depicted in Figure 6, the water price of conventional surface water source is around \$1.2/m^3, and other typical alternatives, desalination, rainwater harvesting, and reclaimed water, fall within the same level. By contrast, the price of the water from the air is substantially higher than the above sources and is around 20–90 \$/m^3 [39,46,116]. Along with the low water yield, and the temperature and relative humidity will be correspondingly reduced after the active extraction of water from the air, atmospheric water harvesting is unlikely to be used on a large scale as an accessible infrastructure or alternative water source. However, the water from the air has a price advantage over bottled water which is the only choice currently under the destruction of the water supply system. Besides this, the environmental impacts of bottled water are pretty high in species loss and resource consumption [117]. In addition, considering that atmospheric water harvesting has the characteristics of a decentralized water supply, it is generally used for emergencies or some specific areas at present. As such, atmospheric water harvesting can be considered in the water management portfolio at a community- or city-level to increase the capacity to handle water supply problems [118,119]. Thus, areas that deserve further research are focused on system design, novel materials (e.g., desiccant) development, and green energy-driven design [120,121].

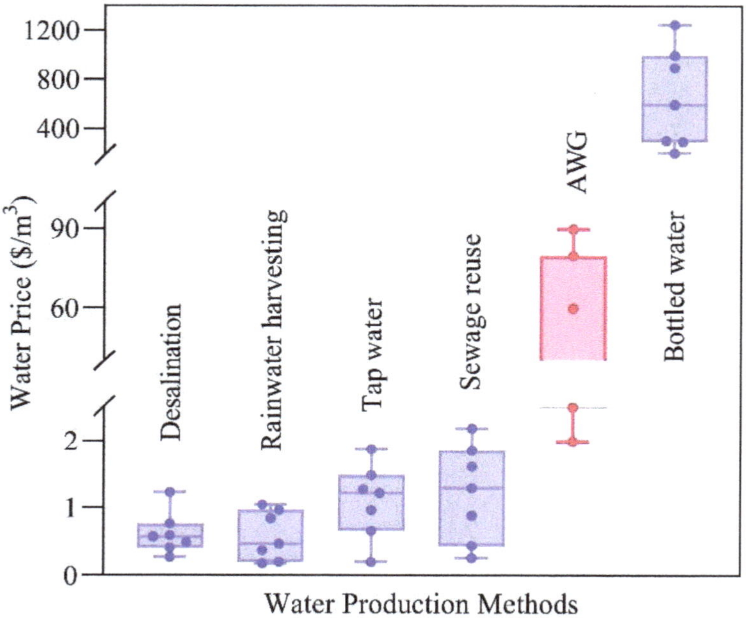

Figure 6. Water prices of various water sources.

In terms of the water quality out of the atmosphere water harvesting process, the water generated is generally clean and pollutant-free. Although the air pollution such as PM2.5 and microplastics in the atmosphere brings concerns about the water quality via atmospheric water harvesting [122,123], it can be solved by simply installing a post-purification module to ensure water quality [46]. Indeed, atmospheric water harvesting technologies manage to filter out any dirt, such as heavy metals, particles, biological organisms, organic compounds remaining in the harvested water, and by artificially adding minerals such as calcium and magnesium, the water quality can be upgraded to the level of natural spring water [124].

Recently, atmospheric water harvesting has been commercialized in many countries and regions (Figure S1) and some off-the-shelf products have been on the market (Table S3). These machines can be divided into three models: large, medium, and small. The water generation capacity is in the range of 2–200,000 L/day to meet the needs of households, emergencies, hospitals, villages, etc. In the parks and beaches of cities such as Abu Dhabi, Al Ain, and Abu Dhabi, water-from-air machines are installed to supply high-quality drinking water for visitors [125]. Indeed, despite the atmospheric water harvesting discussed above, many innovative solutions to leverage the water in the air have been proposed and practiced [126,127]. For example, by laying a mesh overhead the farmland or placing water-adsorbent hydrogel on the surface of the soil, the crops can be irrigated on-site and automatically. In a recent study, Lord et al. thoroughly assessed the global potential of atmospheric water harvesting as a water source by mapping regional horizontal irradiance from sunlight, relative humidity, and air temperature [128]. The results showed that atmospheric water harvesting leveraging solar power could serve the drinking water needs of about 1 billion people. As such, further development and optimization will probably make atmospheric water harvesting more viable and promising to support the water supply system.

5. Conclusions

Diversifying the water sources is necessary to get the cities ready and more resilient to supplement the water supply system. Adoption of atmospheric water harvesting could be such a solution as they provide another connection point between the natural water cycle and the urban water cycle. A series of systems and off-the-shelf products are available on the market to be installed under various relative humidity environments. Although the higher water generation cost makes atmospheric water harvesting uncompetitive, it can be a reliable and decentralized household water treatment system to meet the water demand particularly by leveraging solar power. In addition, it also provides an alternative water source for regions that have a large bottled water consumer base but have no other favorable water sources. Overall, atmospheric water harvesting shall be taken into consideration to enhance the water supply resilience.

Supplementary Materials: The following supporting information can be downloaded at: https://www.mdpi.com/article/10.3390/su14137783/s1, Table S1: Summary of studies or practices of atmospheric water harvesting; Table S2: Summary of desiccants explored and developed in various studies; Table S3: Summary of atmospheric water harvesting machines on the market; Figure S1: Summary of atmospheric water harvesting technologies in practical application. References [129–152] are cited in the supplementary materials.

Author Contributions: Conceptualization, R.L. and M.Z.; methodology, R.L.; software, M.Z.; validation, R.L., M.Z. and Y.L.; formal analysis, R.L.; investigation, M.Z.; resources, R.L.; data curation, M.Z.; writing—original draft preparation, M.Z.; writing—review and editing, R.L.; visualization, M.Z.; supervision, R.L.; project administration, R.L.; funding acquisition, R.L. All authors have read and agreed to the published version of the manuscript.

Funding: The work is funded by the Beijing Municipal Commission of Education (general project, KM202010016013), Beijing Energy Conservation & Sustainable Urban and Rural Development Provincial and Ministry Co-construction Collaboration Innovation Center, and also by the National Natural Science Foundation of China (52000007).

Institutional Review Board Statement: Not applicable.

Informed Consent Statement: Informed consent was obtained from all subjects involved in the study.

Data Availability Statement: Not applicable.

Acknowledgments: We acknowledge the support from the Key Laboratory of Urban Stormwater System and Water Environment, the Ministry of Education, China, and the Beijing Overseas Talents Association.

Conflicts of Interest: The authors declare no conflict of interest.

References

1. Amorocho-Daza, H.; Cabrales, S.; Santos, R.; Saldarriaga, J. A New Multi-Criteria Decision Analysis Methodology for the Selection of New Water Supply Infrastructure. *Water* **2019**, *11*, 805. [CrossRef]
2. Mishra, B.K.; Kumar, P.; Saraswat, C.; Chakraborty, S.; Gautam, A. Water Security in a Changing Environment: Concept, Challenges and Solutions. *Water* **2021**, *13*, 490. [CrossRef]
3. Purvis, L.; Dinar, A. Are Intra- and Inter-Basin Water Transfers a Sustainable Policy Intervention for Addressing Water Scarcity? *Water Secur.* **2020**, *9*, 100058. [CrossRef]

4. UN. The United Nations World Water Development Report 2021. *Valuing Water* **2021**. Available online: https://www.unwater.org/publications/un-world-water-development-report-2021/ (accessed on 12 November 2021).
5. Curry, J. Climate Change, Extreme Weather, and Electric System Reliability | Climate Etc. Available online: https://judithcurry.com/2021/06/27/climate-change-extreme-weather-and-electric-system-reliability/ (accessed on 6 October 2021).
6. Sharifi, A.; Feng, C.; Choryński, A.; Choryński, C.; Pí Nskwar, I.; Graczyk, D.; Krzyżaniak, M. The Emergence of Different Local Resilience Arrangements Regarding Extreme Weather Events in Small Municipalities—A Case Study from the Wielkopolska Region, Poland. *Sustainability* **2022**, *14*, 2052. [CrossRef]
7. Hallema, D.W.; Robinne, F.N.; Bladon, K.D. Reframing the Challenge of Global Wildfire Threats to Water Supplies. *Earth's Future* **2018**, *6*, 772–776. [CrossRef]
8. Becker, R. Today Wasn't Day Zero in Cape Town, but the Water Crisis Isn't over—The Verge. Available online: https://www.theverge.com/2018/5/11/17346276/day-zero-cape-town-south-africa-water-shortage-reservoirs-dams-climate-change (accessed on 25 September 2021).
9. Xiang, Z.; Chen, X.; Lian, Y. Quantifying the Vulnerability of Surface Water Environment in Humid Areas Base on DEA Method. *Water Resour. Manag.* **2016**, *30*, 5101–5112. [CrossRef]
10. Boretti, A.; Rosa, L. Reassessing the Projections of the World Water Development Report. *Clean Water* **2019**, *2*, 1–6. [CrossRef]
11. Nguyen, D.C.H.; Nguyen, D.C.; Luu, T.T.; Le, T.C.; Kumar, P.; Dasgupta, R.; Nguyen, H.Q. Enhancing Water Supply Resili-ence in a Tropical Island via a Socio-Hydrological Approach: A Case Study in Con Dao Island, Vietnam. *Water* **2021**, *13*, 2573. [CrossRef]
12. López Zavala, M.Á.; Prieto, M.J.C.; Rojas, C.A.R. Rainwater Harvesting as an Alternative for Water Supply in Regions with High Water Stress. *Water Supply* **2018**, *18*, 1946–1955. [CrossRef]
13. Jung, K.; Lee, T.; Choi, B.G.; Hong, S. Rainwater Harvesting System for Contiunous Water Supply to the Regions with High Seasonal Rainfall Variations. *Water Resour. Manag.* **2015**, *29*, 961–972. [CrossRef]
14. Jasim, S.Y.; Saththasivam, J.; Loganathan, K.; Ogunbiyi, O.O.; Sarp, S. Reuse of Treated Sewage Effluent (TSE) in Qatar. *J. Water Process Eng.* **2016**, *11*, 174–182. [CrossRef]
15. Bracher, G.H.; Carissimi, E.; Wolff, D.B.; Graepin, C.; Hubner, A.P. Optimization of an Electrocoagulation-Flotation System for Domestic Wastewater Treatment and Reuse. *Environ. Technol.* **2021**, *42*, 2669–2679. [CrossRef] [PubMed]
16. Chen, C.Y.; Wang, S.W.; Kim, H.; Pan, S.Y.; Fan, C.; Lin, Y.J. Non-Conventional Water Reuse in Agriculture: A Circular Water Economy. *Water Res.* **2021**, *199*, 117193. [CrossRef]
17. Choi, Y.; Ahn, J.; Ji, J.; Lee, E.; Yi, J. Effects of Inter-Basin Water Transfer Project Operation for Emergency Water Supply. *Water Resour. Manag.* **2020**, *34*, 2535–2548. [CrossRef]
18. Chen, Z.; Pei, L. Inter-Basin Water Transfer Green Supply Chain Equilibrium and Coordination under Social Welfare Maximization. *Sustainability* **2018**, *10*, 1229. [CrossRef]
19. Wu, L.; Bai, T.; Huang, Q. Tradeoff Analysis between Economic and Ecological Benefits of the Inter Basin Water Transfer Project under Changing Environment and Its Operation Rules. *J. Clean. Prod.* **2020**, *248*, 119294. [CrossRef]
20. Mu'azu, N.D.; Abubakar, I.R.; Blaisi, N.I. Public Acceptability of Treated Wastewater Reuse in Saudi Arabia: Implications for Water Management Policy. *Sci. Total Environ.* **2020**, *721*, 137659. [CrossRef]
21. Parag, Y.; Opher, T. Bottled Drinking Water. 2011. Available online: https://www.eolss.net/sample-chapters/c03/E2-20A-03-09.pdf (accessed on 21 November 2021).
22. Jain, B.; Singh, A.K.; Susan, M.d.A.B.H. The World Around Bottled Water. In *Bottled and Packaged Water*; Elsevier: Amsterdam, The Netherlands, 2019.
23. Wang, T.; Kim, J.; Whelton, A.J. Management of Plastic Bottle and Filter Waste during the Large-Scale Flint Michigan Lead Contaminated Drinking Water Incident. *Resour. Conserv. Recycl.* **2019**, *140*, 115–124. [CrossRef]
24. Yang, D.; Yang, Y.; Xia, J. Hydrological Cycle and Water Resources in a Changing World: A Review. *Geogr. Sustain.* **2021**, *2*, 115–122. [CrossRef]
25. Silva, L.C.R. From Air to Land: Understanding Water Resources through Plant-Based Multidisciplinary Research. *Trends Plant Sci.* **2015**, *20*, 399–401. [CrossRef] [PubMed]
26. Fathy, M.H.; Awad, M.M.; Zeidan, E.S.B.; Hamed, A.M. Solar Powered Foldable Apparatus for Extracting Water from Atmospheric Air. *Renew. Energy* **2020**, *162*, 1462–1489. [CrossRef]
27. Wang, J.Y.; Liu, J.Y.; Wang, R.Z.; Wang, L.W. Experimental Investigation on Two Solar-Driven Sorption Based Devices to Extract Fresh Water from Atmosphere. *Appl. Therm. Eng.* **2017**, *127*, 1608–1616. [CrossRef]
28. Bagheri, F. Performance Investigation of Atmospheric Water Harvesting Systems. *Water Resour. Ind.* **2018**, *20*, 23–28. [CrossRef]
29. Chandler, D. Water, Water Everywhere Even in the Air | MIT News | Massachusetts Institute of Technology. Available online: https://news.mit.edu/2017/MOF-device-harvests-fresh-water-from-air-0414 (accessed on 6 October 2021).
30. Huang, W.; Duan, W.; Chen, Y. Rapidly Declining Surface and Terrestrial Water Resources in Central Asia Driven by Socio-Economic and Climatic Changes. *Sci. Total Environ.* **2021**, *784*, 147193. [CrossRef] [PubMed]
31. Slavikova, P.S.; Popescu, O. Why Is Water Considered a Renewable Resource? | Greentumble. Available online: https://greentumble.com/why-is-water-considered-a-renewable-resource/ (accessed on 6 October 2021).
32. Prihatiningtyas, I.; van der Bruggen, B. Nanocomposite Pervaporation Membrane for Desalination. *Chem. Eng. Res. Des.* **2020**, *164*, 147–161. [CrossRef]

33. Jones, E.; Qadir, M.; van Vliet, M.T.H.; Smakhtin, V.; Kang, S. mu The State of Desalination and Brine Production: A Global Outlook. *Sci. Total Environ.* **2019**, *657*, 1343–1356. [CrossRef]
34. Bundschuh, J.; Kaczmarczyk, M.; Ghaffour, N.; Tomaszewska, B. State-of-the-Art of Renewable Energy Sources Used in Water Desalination: Present and Future Prospects. *Desalination* **2021**, *508*, 115035. [CrossRef]
35. Richards, S.; Rao, L.; Connelly, S.; Raj, A.; Raveendran, L.; Shirin, S.; Jamwal, P.; Helliwell, R. Sustainable Water Resources through Harvesting Rainwater and the Effectiveness of a Low-Cost Water Treatment. *J. Environ. Manag.* **2021**, *286*, 112223. [CrossRef]
36. Monjaiang, P.; Limphitakphong, N.; Kanchanapiya, P.; Tantisattayakul, T.; Chavalparit, O. Assessing Potential of Rainwater Harvesting: Case Study Building in Bangkok. *Int. J. Environ. Sci. Dev.* **2018**, *9*, 222–225. [CrossRef]
37. Bernard, B.; Joyfred, A. Contribution of Rainfall on Rooftop Rainwater Harvesting and Saving on the Slopes of Mt. Elgon, East Africa. *Sci. World J.* **2020**, *2020*. [CrossRef] [PubMed]
38. Make It Rain: US States Embrace "Cloud Seeding" to Try to Conquer Drought | Environment | The Guardian. Available online: https://www.theguardian.com/environment/2021/mar/23/us-stated-cloud-seeding-weather-modification (accessed on 12 June 2022).
39. Liu, Y.; Wang, M.; Webber, M.; Zhou, C.; Zhang, W. Alternative Water Supply Solutions: China's South-to-North-Water-Diversion in Jinan. *J. Environ. Manag.* **2020**, *276*, 111337. [CrossRef]
40. Chen, Z.; Wang, H. Inter-Basin Water Transfer Supply Chain Coordination with the Fairness Concern under Capacity Constraint and Random Precipitation. *Mar. Econ. Manag.* **2019**, *2*, 50–72. [CrossRef]
41. Rinaudo, J.D.; Barraqué, B. Inter-Basin Transfers as a Supply Option: The End of an Era? *Glob. Issues Water Policy* **2015**, *15*, 175–200. [CrossRef]
42. Shumilova, O.; Tockner, K.; Thieme, M.; Koska, A.; Zarfl, C. Global Water Transfer Megaprojects: A Potential Solution for the Water-Food-Energy Nexus? *Front. Environ. Sci.* **2018**, *6*, 150. [CrossRef]
43. Angelakis, A.N.; Snyder, S.A. Wastewater Treatment and Reuse: Past, Present, and Future. *Water* **2015**, *7*, 4887–4895. [CrossRef]
44. Jerry, A.N. Water Supply System. Available online: https://www.britannica.com/technology/water-supply-system (accessed on 24 September 2021).
45. Kılıç, Z. The Importance of Water and Conscious Use of Water. *Int. J. Hydrol.* **2020**, *4*, 239–241. [CrossRef]
46. Halford, B. Can Stripping the Air of Its Moisture Quench the World's Thirst? *Chem. Eng. News* **2018**, *96*, 27–30. [CrossRef]
47. Lawrence, M.G. The Relationship between Relative Humidity and the Dewpoint Temperature in Moist Air: A Simple Conversion and Applications. *Bull. Am. Meteorol. Soc.* **2005**, *86*, 225–233. [CrossRef]
48. Montecinos, S.; Carvajal, D.; Cereceda, P.; Concha, M. Collection Efficiency of Fog Events. *Atmos. Res.* **2018**, *209*, 163–169. [CrossRef]
49. Gido, B.; Friedler, E.; Broday, D.M. Assessment of Atmospheric Moisture Harvesting by Direct Cooling. *Atmos. Res.* **2016**, *182*, 156–162. [CrossRef]
50. Hao, X.; Geng, S.; Yuan, L.; Luo, B. Study of Composite Scheme of Absorption/Desorption Method and Condensation Method for Extracting Water from Air. In *Procedia Engineering*; Elsevier Ltd.: Amsterdam, The Netherlands, 2017; Volume 205, pp. 2069–2075.
51. Dodson, L.L.; Bargach, J. Harvesting Fresh Water from Fog in Rural Morocco: Research and Impact Dar Si Hmad's Fogwater Project in Aït Baamrane. In *Procedia Engineering*; Elsevier Ltd.: Amsterdam, The Netherlands, 2015; Volume 107, pp. 186–193.
52. Estrela, M.J.; Corell, D.; Valiente, J.A.; Azorin-Molina, C.; Chen, D. Spatio-Temporal Variability of Fog-Water Collection in the Eastern Iberian Peninsula: 2003–2012. *Atmos. Res.* **2019**, *226*, 87–101. [CrossRef]
53. Cruzat, D.; Jerez-Hanckes, C. Electrostatic Fog Water Collection. *J. Electrost.* **2018**, *96*, 128–133. [CrossRef]
54. Carvajal, D.; Silva-Llanca, L.; Larraguibel, D.; González, B. On the Aerodynamic Fog Collection Efficiency of Fog Water Collectors via Three-Dimensional Numerical Simulations. *Atmos. Res.* **2020**, *245*, 105123. [CrossRef]
55. Fernandez, D.M.; Torregrosa, A.; Weiss-Penzias, P.S.; Zhang, B.J.; Sorensen, D.; Cohen, R.E.; McKinley, G.H.; Kleingartner, J.; Oliphant, A.; Bowman, M. Fog Water Collection Effectiveness: Mesh Intercomparisons. *Aerosol Air Qual. Res.* **2018**, *18*, 270–283. [CrossRef]
56. Azeem, M.; Noman, M.T.; Wiener, J.; Petru, M.; Louda, P. Structural Design of Efficient Fog Collectors: A Review. *Environ. Technol. Innov.* **2020**, *20*, 101169. [CrossRef]
57. Batisha, A.F. Feasibility and Sustainability of Fog Harvesting. *Sustain. Water Qual. Ecol.* **2015**, *6*, 1–10. [CrossRef]
58. Knapczyk-Korczak, J.; Szewczyk, P.K.; Ura, D.P.; Berent, K.; Stachewicz, U. Hydrophilic Nanofibers in Fog Collectors for Increased Water Harvesting Efficiency. *RSC Adv.* **2020**, *10*, 22335–22342. [CrossRef]
59. Knapczyk-Korczak, J.; Szewczyk, P.K.; Ura, D.P.; Bailey, R.J.; Bilotti, E.; Stachewicz, U. *Improving Water Harvesting Efficiency of Fog Collectors with Electrospun Random and Aligned Polyvinylidene Fluoride (PVDF) Fibers*; Elsevier: Amsterdam, The Netherlands, 2020; Volume 25.
60. Ghosh, R.; Patra, C.; Singh, P.; Ganguly, R.; Sahu, R.P.; Zhitomirsky, I.; Puri, I.K. Influence of Metal Mesh Wettability on Fog Harvesting in Industrial Cooling Towers. *Appl. Therm. Eng.* **2020**, *181*, 115963. [CrossRef]
61. Rivera, J. de D. Aerodynamic Collection Efficiency of Fog Water Collectors. *Atmos. Res.* **2011**, *102*, 335–342. [CrossRef]
62. Regalado, C.M.; Ritter, A. The Design of an Optimal Fog Water Collector: A Theoretical Analysis. *Atmos. Res.* **2016**, *178–179*, 45–54. [CrossRef]
63. Yan, X.; Jiang, Y. Numerical Evaluation of Thefog Collection Potential of Electrostatically Enhanced Fog Collector. *Atmos. Res.* **2021**, *248*, 105251. [CrossRef]

64. Kogan, B.; Trahtman, A. The moisture from the air as water resource in arid region: Hopes, doubts and facts. *J. Arid Environ.* **2003**, *53*, 231–240. [CrossRef]
65. Shi, W.; Anderson, M.J.; Tulkoff, J.B.; Kennedy, B.S.; Boreyko, J.B. Fog Harvesting with Harps. *ACS Appl. Mater. Interfaces* **2018**, *10*, e00191. [CrossRef] [PubMed]
66. Shi, W.; van der Sloot, T.W.; Hart, B.J.; Kennedy, B.S.; Boreyko, J.B. Harps Enable Water Harvesting under Light Fog Conditions. *Adv. Sustain. Syst.* **2020**, *4*, 2000040. [CrossRef]
67. Domen, J.K.; Stringfellow, W.T.; Camarillo, M.K.; Gulati, S. Fog Water as an Alternative and Sustainable Water Resource. *Clean Technol. Environ. Policy* **2013**, *16*, 235–249. [CrossRef]
68. Colli, M.; Lanza, L.G.; Rasmussen, R.; Thériault, J.M. The Collection Efficiency of Shielded and Unshielded Precipitation Gauges. Part I: CFD Airflow Modeling. *J. Hydrometeorol.* **2016**, *17*, 231–243. [CrossRef]
69. Vanderschaeghe, H.; Rongé, J.; Martens, J.A. Energy Performance and Climate Dependency of Technologies for Fresh Water Production from Atmospheric Water Vapour. *Environ. Sci. Water Res. Technol.* **2020**, *6*, 2016–2034. [CrossRef]
70. Seyam, S. Energy and Exergy Analysis of Refrigeration Systems. *Low-Temp. Technol.* **2019**. [CrossRef]
71. Zhou, M.; Song, H.; Xu, X.; Shahsafi, A.; Xia, Z.; Ma, Z.; Kats, M.A.; Zhu, J.; Ooi, B.S.; Gan, Q.; et al. Accelerating Vapor Condensation with Daytime Radiative Cooling. In *New Concepts in Solar and Thermal Radiation Conversion II*; SPIE: Bellingham, WA, USA, 2019.
72. Zhou, M.; Song, H.; Xu, X.; Shahsafi, A.; Qu, Y.; Xia, Z.; Ma, Z.; Kats, M.A.; Zhu, J.; Ooi, B.S.; et al. Vapor Condensation with Daytime Radiative Cooling. *Proc. Natl. Acad. Sci. USA* **2021**, *118*, e2019292118. [CrossRef]
73. Guadarrama-Cetina, J.; Mongruel, A.; Medici, M.G.; Baquero, E.; Parker, A.R.; Milimouk-Melnytchuk, I.; González-Viñas, W.; Beysens, D. Dew Condensation on Desert Beetle Skin. *Eur. Phys. J. E* **2014**, *37*, 1–6. [CrossRef] [PubMed]
74. Haechler, I.; Park, H.; Schnoering, G.; Gulich, T.; Rohner, M.; Tripathy, A.; Milionis, A.; Schutzius, T.M.; Poulikakos, D. Exploiting Radiative Cooling for Uninterrupted 24-Hour Water Harvesting from the Atmosphere. *Am. Assoc. Adv. Sci.* **2021**, *7*, eabf3978. [CrossRef] [PubMed]
75. Tu, Y.; Wang, R.; Zhang, Y.; Wang, J. Progress and Expectation of Atmospheric Water Harvesting. *Joule* **2018**, *2*, 1452–1475. [CrossRef]
76. Raveesh, G.; Goyal, R.; Tyagi, S.K. Advances in Atmospheric Water Generation Technologies. *Energy Convers. Manag.* **2021**, *239*, 114226. [CrossRef]
77. Chen, G.; Wang, Y.; Qiu, J.; Cao, J.; Zou, Y.; Wang, S.; Jia, D.; Zhou, Y. A Facile Bioinspired Strategy for Accelerating Water Collection Enabled by Passive Radiative Cooling and Wettability Engineering. *Mater. Des.* **2021**, *206*, 109829. [CrossRef]
78. Patel, J.; Patel, K.; Mudgal, A.; Panchal, H.; Sadasivuni, K.K. Experimental Investigations of Atmospheric Water Extraction Device under Different Climatic Conditions. *Sustain. Energy Technol. Assess.* **2020**, *38*, 100677. [CrossRef]
79. Liu, S.; He, W.; Hu, D.; Lv, S.; Chen, D.; Wu, X.; Xu, F.; Li, S. Experimental Analysis of a Portable Atmospheric Water Generator by Thermoelectric Cooling Method. In *Energy Procedia*; Elsevier Ltd.: Amsterdam, The Netherlands, 2017; Volume 142, pp. 1609–1614.
80. Cattani, L.; Magrini, A.; Cattani, P. Water Extraction from Air by Refrigeration- Experimental Results from an Integrated System Application. *Appl. Sci.* **2018**, *8*, 2262. [CrossRef]
81. Sleiti, A.K.; Al-Khawaja, H.; Al-Khawaja, H.; Al-Ali, M. Harvesting Water from Air Using Adsorption Material—Prototype and Experimental Results. *Sep. Purif. Technol.* **2021**, *257*, 117921. [CrossRef]
82. Zolfagharkhani, S.; Zamen, M.; Shahmardan, M.M. Thermodynamic Analysis and Evaluation of a Gas Compression Refrigeration Cycle for Fresh Water Production from Atmospheric Air. *Energy Convers. Manag.* **2018**, *170*, 97–107. [CrossRef]
83. Magrini, A.; Cattani, L.; Cartesegna, M.; Magnani, L. Integrated Systems for Air Conditioning and Production of Drinking Water- Preliminary Considerations. In *Energy Procedia*; Elsevier Ltd.: Amsterdam, The Netherlands, 2015; Volume 75, pp. 1659–1665.
84. Joshi, V.P.; Joshi, V.S.; Kothari, H.A.; Mahajan, M.D.; Chaudhari, M.B.; Sant, K.D. Experimental Investigations on a Portable Fresh Water Generator Using a Thermoelectric Cooler. In *Energy Procedia*; Elsevier Ltd.: Amsterdam, The Netherlands, 2017; Volume 109, pp. 161–166.
85. Yao, Y.; Sun, Y.; Sun, D.; Sang, C.; Sun, M.; Shen, L.; Chen, H. Optimization Design and Experimental Study of Thermoelectric Dehumidifier. *Appl. Therm. Eng.* **2017**, *123*, 820–829. [CrossRef]
86. Ibrahim, N.I.; Al-Farayedhi, A.A.; Gandhidasan, P. Experimental Investigation of a Vapor Compression System with Condenser Air Pre-Cooling by Condensate. *Appl. Therm. Eng.* **2017**, *110*, 1255–1263. [CrossRef]
87. Bergmair, D.; Metz, S.J.; de Lange, H.C.; van Steenhoven, A.A. System Analysis of Membrane Facilitated Water Generation from Air Humidity. *Desalination* **2014**, *339*, 26–33. [CrossRef]
88. Solís-Chaves, J.S.; Rocha-Osorio, C.M.; Murari, A.L.L.; Lira, V.M.; Sguarezi Filho, A.J. Extracting Potable Water from Humid Air plus Electric Wind Generation: A Possible Application for a Brazilian Prototype. *Renew. Energy* **2018**, *121*, 102–115. [CrossRef]
89. Kim, H.; Rao, S.R.; LaPotin, A.; Lee, S.; Wang, E.N. Thermodynamic Analysis and Optimization of Adsorption-Based Atmospheric Water Harvesting. *Int. J. Heat Mass Transf.* **2020**, *161*, 120253. [CrossRef]
90. Yilmaz, G.; Meng, F.L.; Lu, W.; Abed, J.; Peh, C.K.N.; Gao, M.; Sargent, E.H.; Ho, G.W. Applied Science Sand Engineering Autonomous Atmospheric Water Seeping MOF Matrix. *Sci. Adv.* **2020**, *6*, eabc8605. [CrossRef] [PubMed]
91. Gado, M.G.; Nasser, M.; Hassan, A.A.; Hassan, H. Adsorption-based atmospheric water harvesting powered by solar energy: Comprehensive review on desiccant materials and systems. *Process Saf. Environ. Prot.* **2022**, *160*, 166–183. [CrossRef]

92. Gordeeva, L.G.; Solovyeva, M.V.; Sapienza, A.; Aristov, Y.I. Potable Water Extraction from the Atmosphere: Potential of MOFs. *Renew. Energy* **2020**, *148*, 72–80. [CrossRef]
93. Rambhad, K.S.; Walke, P.V.; Tidke, D.J. Solid Desiccant Dehumidification and Regeneration Methods—A Review. *Renew. Sustain. Energy Rev.* **2016**, *59*, 73–83. [CrossRef]
94. Li, R.; Shi, Y.; Alsaedi, M.; Wu, M.; Shi, L.; Wang, P. Hybrid Hydrogel with High Water Vapor Harvesting Capacity for Deployable Solar-Driven Atmospheric Water Generator. *Environ. Sci. Technol.* **2018**, *52*, 11367–11377. [CrossRef]
95. Kallenberger, P.A.; Fröba, M. Water Harvesting from Air with a Hygroscopic Salt in a Hydrogel–Derived Matrix. *Commun. Chem.* **2018**, *1*, 1–6. [CrossRef]
96. Fathieh, F.; Kalmutzki, M.J.; Kapustin, E.A.; Waller, P.J.; Yang, J.; Yaghi, O.M. Practical Water Production from Desert Air. *Sci. Adv.* **2018**, *4*, eaat3198. [CrossRef] [PubMed]
97. Ahmadi, M.; Gluesenkamp, K.R.; Bigham, S. Energy-Efficient Sorption-Based Gas Clothes Dryer Systems. *Energy Convers. Manag.* **2021**, *230*, 113763. [CrossRef]
98. Das, A.; Sharma, R.; Thirunavukkarasu, V.; Cheralathan, M. Desiccant-Based Water Production from Humid Air Using Concentrated Solar Energy. *J. Therm. Anal. Calorim.* **2021**, *147*, 2641–2651. [CrossRef]
99. Ahmed, M.M.Z.; Alshammari, F.; Abdullah, A.S.; Elashmawy, M. Basin and Tubular Solar Distillation Systems: A Review. *Process Saf. Environ. Prot.* **2021**, *150*, 157–178. [CrossRef]
100. Siegel, N.P.; Conser, B. A Techno-Economic Analysis of Solar-Driven Atmospheric Water Harvesting. *J. Energy Resour. Technol. Trans. ASME* **2021**, *143*, 090907. [CrossRef]
101. Panchenko, V. Photovoltaic Solar Modules for Autonomous Heat and Power Supply. *IOP Conf. Ser. Earth Environ. Sci.* **2019**, *317*, 012002. [CrossRef]
102. Panchenko, V.; Izmailov, A.; Kharchenko, V.; Lobachevskiy, Y. Photovoltaic Solar Modules of Different Types and Designs for Energy Supply. *Int. J. Energy Optim. Eng. (IJEOE)* **2020**, *9*, 74–94. [CrossRef]
103. Renwick J Water and Climate: More Certainty, More Urgency | Newsroom. Available online: https://www.newsroom.co.nz/ideasroom/water-and-climate-more-certainty-more-urgency (accessed on 6 October 2021).
104. Sixth Assessment Report—IPCC. Available online: https://www.ipcc.ch/assessment-report/ar6/ (accessed on 23 September 2021).
105. Myhre, G.; Alterskjær, K.; Stjern, C.W.; Hodnebrog, Ø.; Marelle, L.; Samset, B.H.; Sillmann, J.; Schaller, N.; Fischer, E.; Schulz, M.; et al. Frequency of Extreme Precipitation Increases Extensively with Event Rareness under Global Warming. *Sci. Rep.* **2019**, *9*, 1–10. [CrossRef]
106. Craig, C.A.; Feng, S.; Gilbertz, S. Water Crisis, Drought, and Climate Change in the Southeast United States. *Land Use Policy* **2019**, *88*, 104110. [CrossRef]
107. Mukherjee, S.; Mishra, A.; Trenberth, K.E. Climate Change and Drought: A Perspective on Drought Indices. *Curr. Clim. Chang. Rep.* **2018**, *4*, 145–163. [CrossRef]
108. Payus, C.; Huey, L.A.; Adnan, F.; Rimba, A.B.; Mohan, G.; Chapagain, S.K.; Roder, G.; Gasparatos, A.; Fukushi, K. Impact of Extreme Drought Climate on Water Security in North Borneo: Case Study of Sabah. *Water* **2020**, *12*, 1135. [CrossRef]
109. Sohns, A.; Ford, J.D.; Riva, M.; Robinson, B.; Adamowski, J. Water Vulnerability in Arctic Households: A Literature-Based Analysis. *Arctic* **2019**, *72*, 300–316. [CrossRef]
110. Yulsman, T. Drought in the Western United States Sets a 122-Year Record | Discover Magazine. Available online: https://www.discovermagazine.com/environment/drought-in-the-western-united-states-sets-a-122-year-record (accessed on 25 September 2021).
111. Schmidt, N.; Elwazer, S.; Wojazer, B.; Braithwaite, S. Germany Flooding: Huge Rescue Effort in Rhineland-Palatinate as Deadly Floods also Hit Belgium, Netherlands, Luxembourg. Available online: https://www.msn.com/en-us/news/world/europe-floods-leave-dozens-dead/ar-AAMbDv9 (accessed on 25 September 2021).
112. Lorenz, I.S.; Pelz, P.F. Optimal Resilience Enhancement of Water Distribution Systems. *Water* **2020**, *12*, 2602. [CrossRef]
113. Rodina, L.; Chan, K.M.A. Expert Views on Strategies to Increase Water Resilience: Evidence from a Global Survey. *Ecol. Soc.* **2019**, *24*, 28. [CrossRef]
114. Ward, J.; Wentworth, J. Water Supply Resilience and Climate Change. Available online: https://post.parliament.uk/research-briefings/post-pb-0040/ (accessed on 23 September 2021).
115. Deng, Y. Building Disaster Resilience of Water Supply with Household Water Treatment. *Water Environ. Res.* **2021**, *93*, 1154–1156. [CrossRef] [PubMed]
116. Mendoza-Escamilla, J.A.; Hernandez-Rangel, F.J.; Cruz-Alcántar, P.; Saavedra-Leos, M.Z.; Morales-Morales, J.; Figueroa-Diaz, R.A.; Valencia-Castillo, C.M.; Martinez-Lopez, F.J. A Feasibility Study on the Use of an Atmospheric Water Generator (AWG) for the Harvesting of Fresh Water in a Semi-Arid Region Affected by Mining Pollution. *Appl. Sci.* **2019**, *9*, 3278. [CrossRef]
117. Villanueva, C.M.; Garfí, M.; Milà, C.; Olmos, S.; Ferrer, I.; Tonne, C. Health and Environmental Impacts of Drinking Water Choices in Barcelona, Spain: A Modelling Study. *Sci. Total Environ.* **2021**, *795*, 148884. [CrossRef]
118. Runze, D.; Qingfen, M.; Hui, L.; Gaoping, W.; Wei, Y.; Guangfu, C.; Yifan, C. Experimental Investigations on a Portable Atmospheric Water Generator for Maritime Rescue. *J. Water Reuse Desalination* **2020**, *10*, 30–44. [CrossRef]
119. Patel, K.; Patel, J.; Raval, H. Potential Study of Atmospheric Water Generator (AWG) for Humid Climatic Conditions of Eastern States in India. In *Smart Innovation, Systems and Technologies*; Springer: Berlin/Heidelberg, Germany, 2020; Volume 161.

120. Jawarneh, A.M.; AL-Oqla, F.M.; Jadoo, A.A. Transient Behavior of Non-Toxic Natural and Hybrid Multi-Layer Desiccant Composite Materials for Water Extraction from Atmospheric Air. *Environ. Sci. Pollut. Res.* **2021**, *28*, 45609–45618. [CrossRef]
121. Kwan, T.H.; Shen, Y.; Hu, T.; Pei, G. The Fuel Cell and Atmospheric Water Generator Hybrid System for Supplying Grid-Independent Power and Freshwater. *Appl. Energy* **2020**, *279*, 115780. [CrossRef]
122. Rhodes, C.J. Solving the Plastic Problem: From Cradle to Grave, to Reincarnation. *Sci. Prog.* **2019**, *102*, 218–248. [CrossRef] [PubMed]
123. Akhbarizadeh, R.; Dobaradaran, S.; Amouei Torkmahalleh, M.; Saeedi, R.; Aibaghi, R.; Faraji Ghasemi, F. Suspended Fine Particulate Matter (PM2.5), Microplastics (MPs), and Polycyclic Aromatic Hydrocarbons (PAHs) in Air: Their Possible Relationships and Health Implications. *Environ. Res.* **2021**, *192*, 110339. [CrossRef] [PubMed]
124. Mehlhaf, N. New Technology Creates Clean Drinking Water from Vapor in the Air | Kgw.Com. Available online: https://www.kgw.com/article/news/local/technology/clean-drinking-water-source-global/283-64710547-ceef-4c61-a4ef-a38268a55566 (accessed on 10 October 2021).
125. Divon, M.M. UAE: Machines That Produce Water from Air Placed in Parks, Beaches in Abu Dhabi—News | Khaleej Times. Available online: https://www.khaleejtimes.com/news/uae-machines-that-produce-water-from-air-placed-in-parks-beaches-in-abu-dhabi (accessed on 25 September 2021).
126. Zhou, X.; Zhang, P.; Zhao, F.; Yu, G. Super Moisture Absorbent Gels for Sustainable Agriculture via Atmospheric Water Irrigation. *ACS Mater. Lett.* **2020**, *2*, 1419–1422. [CrossRef]
127. Chen, G.F.; Cai, D.S. Water Harvested from the Air Combined with Solar Power, Shade and Light Providing System: Conception of Water-Saving Irrigation. *Procedia Environ. Sci.* **2012**, *13*, 1003–1009. [CrossRef]
128. Lord, J.; Thomas, A.; Treat, N.; Forkin, M.; Bain, R.; Dulac, P.; Behroozi, C.H.; Mamutov, T.; Fongheiser, J.; Kobilansky, N.; et al. Global Potential for Harvesting Drinking Water from Air Using Solar Energy. *Nature* **2021**, *598*, 611–617. [CrossRef] [PubMed]
129. Fessehaye, M.; Abdul-Wahab, S.A.; Savage, M.J.; Kohler, T.; Gherezghiher, T.; Hurni, H. Fog-Water Collection for Community Use. *Renew. Sustain. Energy Rev.* **2014**, *29*, 52–62. [CrossRef]
130. Larrain, H.; Velásquez, F.; Cereceda, P.; Espejo, R.; Pinto, R.; Osses, P.; Schemenauer, R.S. Fog Measurements at the Site "Falda Verde" North of Chañaral Compared with Other Fog Stations of Chile. *Atmos. Res.* **2002**, *64*, 273–284. [CrossRef]
131. Nguyen, L.T.; Bai, Z.; Zhu, J.; Gao, C.; Liu, X.; Wagaye, B.T.; Li, J.; Zhang, B.; Guo, J. Three-Dimensional Multilayer Vertical Filament Meshes for Enhancing Efficiency in Fog Water Harvesting. *ACS Omega* **2021**, *6*, 3910–3920. [CrossRef]
132. Feng, J.; Zhong, L.; Guo, Z. Sprayed Hieratical Biomimetic Superhydrophilic-Superhydrophobic Surface for Efficient Fog Harvesting. *Chem. Eng. J.* **2020**, *388*, 124283. [CrossRef]
133. Xiao, L.; Li, G.; Cai, Y.; Cui, Z.; Fang, J.; Cheng, H.; Zhang, Y.; Duan, T.; Zang, H.; Liu, H.; et al. Programmable 3D Printed Wheat Awn-like System for High-Performance Fogdrop Collection. *Chem. Eng. J.* **2020**, *399*, 125139. [CrossRef]
134. Wan, Y.; Xu, J.; Lian, Z.; Xu, J. Superhydrophilic Surfaces with Hierarchical Groove Structure for Efficient Fog Collection. *Colloids Surf. A: Physicochem. Eng. Asp.* **2021**, *628*, 127241. [CrossRef]
135. Li, J.; Zhou, Y.; Wang, W.; Du, F.; Ren, L. A Bio-Inspired Superhydrophobic Surface for Fog Collection and Directional Water Transport. *J. Alloy. Compd.* **2020**, *819*, 152968. [CrossRef]
136. Feng, R.; Song, F.; Xu, C.; Wang, X.L.; Wang, Y.Z. A Quadruple-Biomimetic Surface for Spontaneous and Efficient Fog Harvesting. *Chem. Eng. J.* **2021**, *422*, 130119. [CrossRef]
137. Li, D.; Huang, J.; Han, G.; Guo, Z. A Facile Approach to Achieve Bioinspired PDMS@Fe_3O_4 Fabric with Switchable Wettability for Liquid Transport and Water Collection. *J. Mater. Chem. A* **2018**, *6*, 22741–22748. [CrossRef]
138. Entezari, A.; Ejeian, M.; Wang, R. Modifying Water Sorption Properties with Polymer Additives for Atmospheric Water Harvesting Applications. *Appl. Therm. Eng.* **2019**, *161*, 114109. [CrossRef]
139. Ejeian, M.; Entezari, A.; Wang, R.Z. Solar Powered Atmospheric Water Harvesting with Enhanced LiCl /$MgSO_4$/ACF Composite. *Appl. Therm. Eng.* **2020**, *176*, 115396. [CrossRef]
140. Li, R.; Shi, Y.; Wu, M.; Hong, S.; Wang, P. Improving Atmospheric Water Production Yield: Enabling Multiple Water Harvesting Cycles with Nano Sorbent. *Nano Energy* **2020**, *67*, 104255. [CrossRef]
141. Xu, J.; Li, T.; Chao, J.; Wu, S.; Yan, T.; Li, W.; Cao, B.; Wang, R. Efficient Solar-Driven Water Harvesting from Arid Air with Metal–Organic Frameworks Modified by Hygroscopic Salt. *Angew. Chem. - Int. Ed.* **2020**, *59*, 5202–5210. [CrossRef]
142. Gong, F.; Li, H.; Zhou, Q.; Wang, M.; Wang, W.; Lv, Y.; Xiao, R.; Papavassiliou, D.V. Agricultural Waste-Derived Moisture-Absorber for All-Weather Atmospheric Water Collection and Electricity Generation. *Nano Energy* **2020**, *74*, 104922. [CrossRef]
143. Watergen | Water from Air. Available online: https://www.watergen.com/ (accessed on 24 September 2021).
144. Pure & Sustainable Water—Drinkableair Technologies. Available online: https://drinkableair.tech/ (accessed on 24 September 2021).
145. WEDEW—SkySource. Available online: https://www.skysource.org/wedew (accessed on 24 September 2021).
146. Drupps | Atmospheric Water for All. Available online: https://drupps.com/ (accessed on 24 September 2021).
147. AquaBoy Pro II. Available online: http://www.atmosphericwatersolutions.com/store/p1/AquaBoy_Pro_II.html (accessed on 8 October 2021).
148. GEN-M | Water from Air Generator | Watergen USA. Available online: https://us.watergen.com/commercial/gen-m/ (accessed on 24 September 2021).
149. Renewable Drinking Water | SOURCE Water. Available online: https://www.source.co/ (accessed on 12 October 2021).
150. CloudFisher | Fognetalliance. Available online: https://www.fognetalliance.org/cloudfisher (accessed on 24 September 2021).

151. Drupps Sells to Thailand—Drupps. Available online: https://news.cision.com/drupps/r/drupps-sells-to-thailand,c3418869 (accessed on 24 September 2021).
152. Zhao, F.; Zhou, X.; Liu, Y.; Shi, Y.; Dai, Y.; Yu, G. Super Moisture-Absorbent Gels for All-Weather Atmospheric Water Harvesting. *Adv. Mater.* **2019**, *31*, 1806446. [CrossRef]

sustainability

MDPI

Article

Understanding the Human Dimensions of Recycling and Source Separation Practices at the Household Level: An Evidence in Perak, Malaysia

Pei Lin Yu [1], Norafida Ab Ghafar [2], Mastura Adam [2] and Hong Ching Goh [1,*]

[1] Department of Urban and Regional Planning, Faculty of Built Environment, Universiti Malaya, Kuala Lumpur 50603, Malaysia; yu_peilin@hotmail.com

[2] Department of Architecture, Faculty of Built Environment, Universiti Malaya, Kuala Lumpur 50603, Malaysia; norafida@um.edu.my (N.A.G.); mastura@um.edu.my (M.A.)

* Correspondence: gohhc@um.edu.my

Abstract: Recycling and source separation (R&SS) are believed to have been the first attempt to minimise waste. This research adopted mixed methods that followed sequential quantitative then qualitative data collection, combining questionnaire surveys from 100 households, semi-structured interviews, and participatory observations to study the human dimension of waste generation and management. Scoring Assessment (with modified Bloom's Cut Off point) indicated that households had moderate knowledge and positive attitudes yet poor behaviour, and these three components indicated no linear associations, tested using Pearson's Correlation Coefficient. However, age group, marital status, educational level and living duration showed statistical significance with households' participation in source separation through Chi-Square Test. Meanwhile, observation data showed that waste management mechanisms and environment had inefficiently supported households' participation in R&SS practices (external factors: poor accessibility to services, lack of tangible incentives, and absence of restriction in consumption). Elicited data indicated that a satisfactory level of intentions, knowledge, and willingness, together with good habit and quality persuasion (internal factors), were required to drive good behaviour. Subsequently, a series of recommendations were formulated to promote gradual yet solid transformation of the waste management system, tapping on existing initiatives by considering additional parameters upon the gap in households' knowledge, attitude, and behaviour.

Keywords: mixed methods; online questionnaire survey; participatory observation; sustainable waste disposal; sustainable consumption; Malaysian source separation practice

Citation: Yu, P.L.; Ab Ghafar, N.; Adam, M.; Goh, H.C. Understanding the Human Dimensions of Recycling and Source Separation Practices at the Household Level: An Evidence in Perak, Malaysia. *Sustainability* **2022**, *14*, 8023. https://doi.org/10.3390/su14138023

Academic Editors: Luis Garrote and Alban Kuriqi

Received: 26 April 2022
Accepted: 22 June 2022
Published: 30 June 2022

1. Introduction

"The throw-away society is a human society strongly influenced by consumerism. The term describes a critical view of overconsumption and excessive production of short-lived or disposable items", quoted [1], who argues the cost of this throw-away culture in compromising the needs of future generations and threatening the natural system that the survival of all living things depends on [2]. As more products are made more affordable, they are less appreciated, as society can dispose and buy new items, often beyond what is needed, rather than send them for repair [3]. This marks the peak of the global waste generation at 2.01 billion tonnes (0.74 kg per person daily), but its rate, amount, and quality will continue to surge by 70%, without consideration of the concept of distancing in dealing with waste during post- and pre-consumerism [4,5].

Statistics show that over 90% of waste in low-income countries (compared to 66% for low-middle-income and 30% for upper-middle-income) is disposed of at open dumps or landfills, which are the most adopted waste disposal methods [5,6]. These waste disposal sites have thus become the only and popular method used by cities (with limited municipal budget) to dispose of the high volume of (unsorted) waste generated daily. Waste disposal sites are optimal breeding grounds for disease vectors and sources of toxin release into the atmosphere and oceans [2,5]. However, they will not be publicly acknowledged as environmental issues if landfilling remains the only urban waste management service [7]. If waste continues to be collected regularly without a proper sorting system to support it, and if there are no restrictions on consumption or changes in lifestyle choices, society will continue to remain in denial and ignorant to the over exploitation and destruction of the planet's natural systems as a result of their personal waste footprint [1,2].

Malaysia is one of the upper-middle-income countries that heavily rely on landfill disposal, with almost 89% of waste collected (from a waste generation rate at 33,130 tonnes daily) being sent to a total of 170 landfills. Out of these, only 14 are categorised as sanitary. The official lab report of the country estimates that at least 40% waste diversion can be achieved [8,9]. Consequently, space and land availability will gradually emerge as a major limitation to landfilling as the increasing waste volume exceeds the capacity of the treatment, not to mention other waste problems to be addressed, such as illegal dumping and plastic waste import [10,11]. The overconsumption of the throw-away society, together with almost-absent resource recovery attempts and a lack of political will and social responsibility towards sustainable and integrated waste management, present a huge barrier for the transition to waste minimisation [7].

The first effective step towards waste minimisation in the European Union waste hierarchy is recycling. This includes composting, which deals with more waste fractions, for instance, organic or biodegradable waste and e-waste [12,13]. The integration of recycling and source separation (R&SS) is crucial to create a compound effect on the waste diversion from landfilling when waste composition and its quality is carefully managed [14]. However, Malaysia's recycling efforts and implementation have focused on only a few categories of recyclables, while source separation only came into enforcement in late in 2015, along with the formulation of Solid Waste and Public Cleansing Management Act 2007 (Act 672) [7]. A community survey showed that only 28% of households in Kuala Lumpur engage in source separation, even though the legislation has been enforced [15]. This result denotes several constraints found in the implementation, especially when dealing with the complexity of human dimensions within the current waste management system and mechanism [7].

This contrasts with a case reported in Shanghai, China, where source separation was mandated in 2019. The study in [16] reports that nearly half of the households had negative emotions towards the policy. Although the households found it difficult to follow the segregation guidelines, which were rather broad and ambiguous in the details [17], it was fear towards the heavy fines imposed by the government rather than environmental protection or sustainability that pressured the households to comply with the regulation [16]. Thus, it deviated from the Chinese government's initial vision of promoting sustainable development, which aims to depreciate the culture of throwaway consumerism [18]. The key idea here is to promote a sustainable society by sustainable waste disposal while looking into individual daily consumption through lifestyle choice [1,3,18]. However, diversified urban governances and management systems, as well as different cultural readiness, lead to different speed and quality of transition to waste minimisation [17,19].

As such, it is important to investigate the potential factors that drive households' behaviour change and their potential adaptation to local context. In identifying these factors, many researchers relate the discussion of waste management and sustainable consumption with pro-environmental behaviour; the studies in [20,21] suggest that environmental knowledge is important yet insufficient to drive action. The studies in [22,23] argue that a high satisfactory level of knowledge, together with attitude, could more likely drive good behaviour, while [24] explains behaviour is an interactive output of attitude and choice with the presence of external causality such as constraint (cost, time), habits or routine, disincentives and scepticism. The study in [20] also claims that pro-environmental behaviour involves both internal and external factors. Meanwhile, [25] categorises non-recyclers into three major groups based on the common characteristics of the barriers (or "excuses", as referred in the original article) selected through a community survey conducted by Ipsos in 2011. Each group discusses distinctive reasons and psychology behind the action towards recycling participation; for instance, time consumption, issue of convenience, lack of knowledge, or absence of communal effort or strong influence. These parameters guided them to weigh their decision together with the current waste management environment they are provided with. The interpretation of each grouping, as well as the interventions the author proposed, conveys the relation of multiple internal and external factors (potential parameters other than knowledge and attitude level) with the behaviour.

This study aims to make locally adaptive recommendations to encourage household participation in R&SS practices with a case study in Manjung district, Perak. In doing so, the study examined household KAB, as well as other potential parameters that influence household behaviour towards waste generation and management. Specifically, authors ask (1) What are the associations between the households' level of KAB towards R&SS practices? (2) What are the enabling factors of households' participation in R&SS practices? (3) What are the barriers of non-recyclers or non-waste sorters? (4) What are the practical recommendations to encourage R&SS participation?

This study contributes to empirical evidence by addressing the gap between the households' awareness and the actual sustainable waste disposal rate that relates directly with the households' participation in R&SS practices. The authors advocate the idea that changes in behaviour start with improved level of knowledge and attitude, although improved level of knowledge and attitude may not necessarily lead directly to change in behaviour. At a minimum, the public will have some knowledge about the generation (which would influence the purchasing decision and the material leftover after consuming products) and disposal of domestic waste so that the demand for products will be shifted to those that carry sustainable and environmentally friendly qualities and are easier to manage during disposal. This would have an impact on the supply side of materials, of which the manufacturers may respond to the demand and redesign the products using eco-friendly material. Consequently, when the entire chain of production adopts the concept of sustainability from such market activity, the system could then be elevated to a higher level in the waste hierarchy: reduce, reuse, and lastly, prevention.

By identifying households' knowledge level, their attitude towards both practices and the waste management process, as well as their behavioural pattern on the R&SS practices, this research intends to establish a basis to support the decision making process in relation to the waste management system. The proposed recommendations highlight the importance of further encouraging household involvement on R&SS practices. Discussions of results concerning expectation and feedback are useful for the municipality and relevant stakeholders for their service improvement and to help policymakers, waste management planners, local administrators, and researchers to formulate policies and strategies in sustainable waste management, as well as serve as a basis to identify further areas of study.

2. Materials and Methods

2.1. Case Study

Sanitation (and therefore waste management) is a matter under the Concurrent List. The state government has the authority to decide whether to adopt the law related to this urban management service, thus it is subject to the administration of each local authority. However, regarding the current governance status in Malaysia, the decision of mandating the Solid Waste and Public Cleansing Management Act 2007 (Act 672)—the latest source separation initiative—may vary over a short period of time. Perak is one of eight states (i.e., Penang, Perak, Terengganu, Kelantan, Selangor, Sabah, Sarawak, and Labuan) that are yet to mandate the legislative provision of Act 672 [26]. Without law enforcement to reduce the waste sent to landfills at state level, the capacity will eventually run out, with Perak's 0.8842 kg waste generation per capita per day (2002) generated by its 2.30 million population. This is a higher rate compared to Selangor at 0.8845 kg, yet Selangor has almost double the population of Perak. At the local level, Manjung (Figure 1) has exceeded the national average (0.8500 kg per capita per day) at 1.409 kg, and ranks as the third highest among the districts [27].

Only a 4.00% recycling rate has been recorded by the study in [28] through the provision of a recycling service in the Northern Region (Kedah, Perlis, and Perak). On the other hand, the recorded recycling rate from the database of Manjung Municipal Council is 0.07%, based on the quotient of the total recyclables collected: approximately 77 tonnes per annum from the total waste collected from 54,186 housing, which is about 300 tonnes per day [29]. Both the state (region) and local rate of recycling are still distant from that of the national target: 22.00% in 2020. Being the third most populated city in Perak, Manjung has yet to generate sufficient awareness of the waste crisis and sustainable waste management [30].

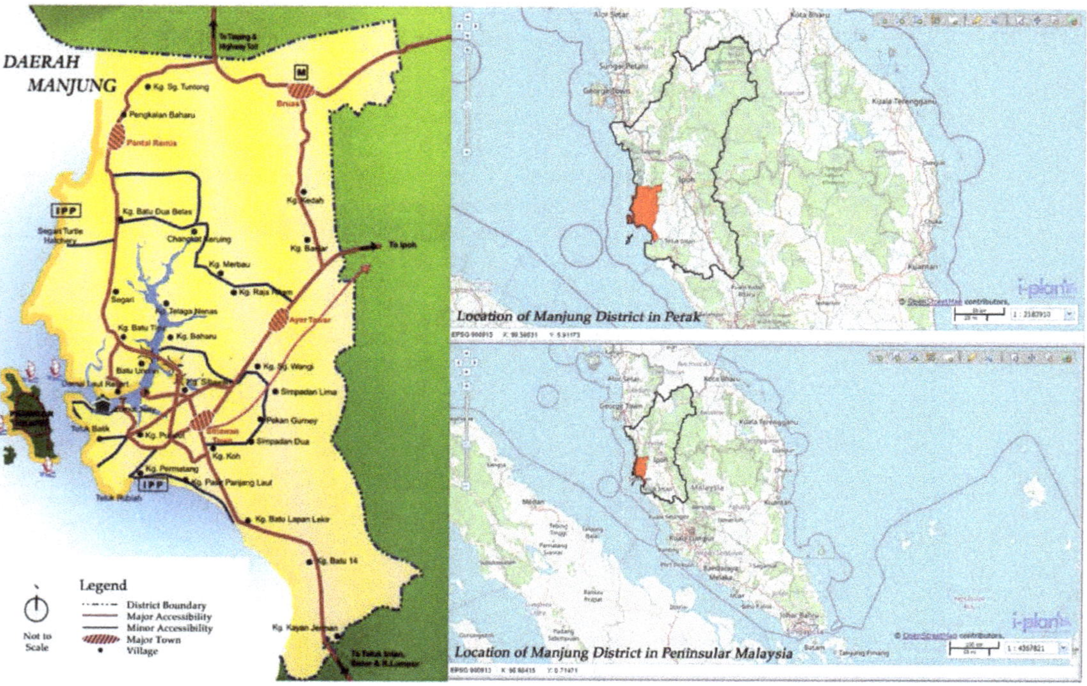

Figure 1. Location of Manjung District [31,32].

2.2. Data Collection and Sampling

This study adopted a mixed-method approach, combining an online questionnaire survey, semi-structured interviews (through meeting application or phone call) and participatory observation, adjusting to pandemic situations. The purposive sampling strategy was used throughout the data collection process (except for observation).

Before the actual conduct, pre-testing was carried out to examine the content validity of the questions asked in the survey (including semi-structured interview). All questions were refined based on the feedback from both experts in the field and laymen and were adapted from studies in past literature [30,33–39] to improve their reliability and representativeness to the study. The questionnaire survey was constructed using an online survey administration application, Google Form, and distributed through various online platforms (e.g., Facebook, Instagram, and WhatsApp, popular applications in Malaysia). The questions were written in three major languages: Malay, Chinese, and English. Respondents were filtered carefully under two conditions: a person (1) aged 18 years old and above who (2) has lived in Manjung district for more than a week. It was also mentioned that each respondent would be representing a household. The questionnaire consisted of six main sections: (a) respondent's profile, (b) knowledge, (c) attitude, (d) behaviour, (e) psychology behind R&SS practices, and (f) invitation to interview. The targeted sample size for the survey was 382 respondents for 58,186 house premises (data from Manjung Municipal Council) based on the calculation by [40]; however, it was not achieved due, in part, to the relatively less effective channel of survey distribution via online networks. Only 100 respondents took part in the survey within the scheduled time frame for quantitative data collection. The mixed methods of this study underwent a sequence arrangement [41], in which the qualitative data that followed were collected after quantitative data were completed.

Seven respondents, who had answered the survey, were recruited to a semi-structured interview based on their consent and willingness to contribute to this study. This further strengthened the perspective of service users through descriptive and abstract primary data on their waste management process [42]. Questions to them were focused on the advantages and disadvantages of the R&SS, regardless of their actual participation. On the other hand, the semi-structured interview included the perspectives of the service providers to explain the condition and quality of R&SS practices provided, as well as how the households' needs were addressed. Two personnel from the local authority (the head of recycling project implementation) and non-governmental organisations (which are actively involved in R&SS implementation at household level) were interviewed. Questions for service providers merely focused on the mechanisms involved in providing their respective R&SS services and how to engage with the community.

Observation was conducted to inspect the current waste management mechanism and environment of the study area, from waste generation, storage, collection to disposal. Taking into consideration that presence of the observers might have affected the behaviour of the observed group, participatory observation was opted to blend in the situation [43]. Observation took place in public areas, including wet markets and neighbourhood streets, when municipal waste collection was in action. The observed objective focused on how humans carry out waste disposal, rather than on the humans themselves. Therefore, no consent was needed, and no personal communication was conducted. No photos were taken of any individual present at the observation spot.

2.3. Data Analysis

Since the data collected consisted of both quantitative and qualitative data, appropriate analysis techniques were used for each type of data accordingly.

Microsoft Excel 2019 was used to perform all statistical (quantitative) analysis, ranging from descriptive, inferential statistics to correlational analysis. Measures of central tendency and dispersion (mean and standard deviation) were used to study data distribution. In addition, the Chi-Square test, together with Pearson's Correlation Coefficient, which was performed at 95% confidence level, were used to determine the association between variables (two forms: nominal and continuous data). A scoring system was developed to assess the adequacy level of the KAB through a fixed range of scores modified according to the widely adopted Bloom's Cut-Off Point [38,44]. This method allowed the authors to convert different types of data, including nominal—true or false answer and percentages, and ordinal data—Likert scales, into scores—continuous data. The scores (by level) obtained were used to test the correlation between KAB components and other variables.

For knowledge (first component), 20 questions were asked in two parts: ten true or false questions (general knowledge assessment) and ten multiple choice questions (specific knowledge assessment). A score of 1 was given to each correct answer (for each sample); in contrast, no score was deducted for a wrong answer, instead it was given a score of 0. Hence, each part had a maximum and minimum score of 10 and 0, respectively. These scores were divided into three levels through a modified Bloom's Cut-Off Point, namely (a) High for 10 to 8 scores, (b) Moderate for 7 to 5 scores, and (c) Low for scores less than 5. For attitude (second component), there were 14 questions with a 5-Point Likert scale to assess the degree of importance and agreement. Scores 1 to 5 were given to respective points on the Likert scale (in the order from Strongly Disagree to Strongly Agree). This resulted in a maximum score of 70 and a minimum of 14 for each sample. Modified Bloom's Cut-Off Point was also used to establish three levels, namely (a) Positive (70 to 52), (b) Neutral (51 to 33), and (c) Negative (14 to 32), which were equally divided and given the same class interval. Finally, for behaviour (third component), there were 10 questions with multiple choice of answers. Similar to the first component, the scores were divided into three levels through a modified Bloom's Cut Off Point, namely (a) Good (10 to 8), (b) Satisfactory (7 to 5), and (c) Poor (4 to 0).

Phenomenological (qualitative) analysis was used to carefully record the conduct of waste management by the service users during the observation and semi-structured interview processes, respectively. This was to ensure that both observed and elicited data could be precisely analysed and transformed into useful information for the description of a phenomenon. Additionally, it could extract the perception of the service providers on the outcomes of the waste management service [45].

2.4. Research Constraints and Limitations

First, respondents could not be observed while answering the survey during online pre-testing. This led to limited sources of input to improve the questionnaire; therefore, a more comprehensive manner of conducting pre-testing in both online and offline platforms is recommended to extract both observed and verbal feedback. The difficulty in recruiting sufficient samples during the lockdown due to the pandemic in Malaysia was the key constraint faced by the authors. Nevertheless, this should serve as a preliminary study, and similar studies covering a comprehensive sample size targeting 382 respondents are recommended post-pandemic. Second, the number of interviewees recruited from the service providers was limited due, in part, to the lack of effective ways to engage, as no physical visit was allowed due to movement restrictions. No positive response was obtained from private recycling vendors, probably due to the lack of awareness and exposure towards the academic research in the related field. Finally, invitation (survey or interview) through online networks was likely neglected, but this method still had benefits such as receiving responses quicker and in a more convenient way, therefore a combination of online and on-site survey distribution and interview invitation were proposed.

3. Results

3.1. Demographic Profile

A total of 100 questionnaire surveys were eligible to be analysed. Most of the respondents are between 18 and 24 years old with pre-university or undergraduate academic qualifications, female, and single. More than half of the respondents are locals who have lived in the community for more than 20 years. Nearly all the respondents live with their family members on landed properties, with the majority having a household size between 5 and 7 persons (Table 1).

3.2. Knowledge Assessment

Both general and specific knowledge levels were tested among the respondents. In general, the knowledge level among the respondents was rated as moderate (Table 2) according to the scores obtained from Tables 3 and 4, with better scores in general knowledge, where 37.0% scored high.

Table 3 shows the general knowledge among respondents of the R&SS with their benefits and importance to the waste management system, quality environment, and energy consumption. Most of the respondents answered correctly, except for statement #9: all plastics that contain numbered symbols (also known as plastic resin identification codes) can be recycled.

Table 1. Demographic profile of the respondents.

Variables (n = 100)	%	Variables (n = 100)	%
Gender		Duration of Stay	
Female	73.0	Less than a year	4.0
Male	27.0	1–5 years	9.0
Marital Status		6–10 years	11.0
Single	87.0	11–15 years	7.0
Married	13.0	16–20 years	17.0
Divorced/Widowed	0.0	More than 20 years	52.0
Age Group		Household Size	
18 to 24 years old	71.0	1	3.0
25 to 44 years old	26.0	2–4	39.0
45 to 64 years old	3.0	5–7	56.0
65 years old and above	0.0	More than 7	2.0
Educational Level		Housing Type	
Primary Education	0.0	Bungalow/Semi-Detached	50.0
Secondary Education	4.0	Terrace/Linked House	46.0
Pre university/Undergraduate	77.0	Flat/Apartment/Condominium	2.0
Postgraduate	17.0	Shop House	2.0
Others	2.0	Others	0.0
Living Companion		Housing Ownership	
Family	95.0	Rent	7.0
Friend(s) and Acquaintance(s)	4.0	Own	83.0
Others	1.0	Other	10.0

Table 2. Waste knowledge level among respondents.

Knowledge Level	General Knowledge			Specific Knowledge		
	High	Moderate	Low	High	Moderate	Low
Range of Score	10 to 8	7 to 5	4 to 0	10 to 8	7 to 5	4 to 0
Frequency (%) (n = 100)	37.0	59.0	4.0	6.0	65.0	29.0

Table 3. General knowledge level among respondents.

Statement (n = 100)	True (%)	False (%)	Correct Answer
1. Recycling and waste sorting at source cannot help to curb insufficient landfill capacity.	25.0	75.0	False
2. Recycling and waste sorting can help to reduce the spread of disease like bacterial or fungal infections.	82.0	18.0	True
3. Waste sorting can prevent emission of harmful chemicals and greenhouse gasses (methane and carbon dioxide) that contribute to global warming.	94.0	6.0	True
4. Waste without sorting can be used to create compost for soil fertility regeneration.	35.0	65.0	True
5. Energy used to make a new product from raw materials is far less than energy required for recycling.	45.0	55.0	False
6. Waste sorting can prevent contamination of recyclables.	93.0	7.0	True
7. More waste fractions need to be dealt with when practising waste sorting than recycling.	84.0	16.0	True
8. Recyclables collected can only be recycled once.	26.0	74.0	False
9. All plastics that contain numbered symbols (also known as plastic resin identification codes) can be recycled.	71.0	29.0	False
10. Waste is sorted and collected at household, but not necessarily recycled.	64.0	36.0	True

Table 4. Specific knowledge level among respondents.

Selected Waste Item (*n* = 100)	**1. Biodegradable/Organic Waste (%)**	**2. Non-Biodegradable/Non-Organic Waste (%)**	**3. Recyclable Waste (%)**	**4. Hazardous Waste (%)**	**Correct Waste Category**
1. Used Tissue Paper	52.0	23.0	23.0	2.0	2
2. Food Stained Paper or Plastic Container	14.0	55.0	27.0	4.0	2
3. Light Bulb	7.0	40.0	14.0	39.0	4
4. Vegetable and Fruit Peel	93.0	2.0	3.0	2.0	1
5. Bone and Meat Scrap	89.0	8.0	2.0	1.0	1
6. Dry Leaves	85.0	10.0	5.0	0.0	1
7. Expired Medicine or Supplement	19.0	35.0	3.0	43.0	4
8. Photograph and Paint	7.0	51.0	13.0	29.0	4
9. Electronic, Batteries and Cables	6.0	36.0	14.0	44.0	3
10. Hair or Fur	53.0	36.0	9.0	2.0	1

Table 4 shows the responses to the specific knowledge of R&SS based on waste categories for selected items. Most of the respondents selected the categories of waste correctly, except for waste items #1: used tissue paper, #8: photographs and paint, and #9: electronics, batteries, and cables.

3.3. *Attitude Assessment*

Overall, the attitude score among the respondents was relatively positive compared to the knowledge score, where 75.0% of the respondents scored between 52 and 70 (Table 5).

Table 5. Attitude level among respondents.

Attitude Level	**Positive**	**Neutral**	**Negative**
Range of Score	70 to 52	51 to 33	32 to 14
Frequency (%) (*n* = 100)	75.0	19.0	6.0

Table 6 reveals the results about respondents' perceived participation in different stages of waste management, starting from waste generation, waste storage, waste collection, waste processing, and waste disposal. Overall, the respondents demonstrated a high level of intention to participate in waste management processes, except for the waste collection stage, specifically statement #6: I am willing to pay extra service charges for different waste collection according to its category, and statement #8: I am actively involved in the collection and transport of sorted waste materials in my neighbourhood.

Table 6. Involvement in waste management stages.

Level of Importance on Statement (*n* = 100)	**1. Not Important at all**	**2. Slightly Important**	**3. Moderately Important**	**4. Very Important**	**5. Extremely Important**	**Mean**	**Standard Deviation**
Importance of Waste Separation as Immediate Solution to Waste Crisis	1.0	2.0	14.0	44.0	39.0	4.18	0.817
Level of Agreement on Statement (*n* = 100)	**1. Strongly Disagree**	**2. Disagree**	**3. Neutral**	**4. Agree**	**5. Strongly Agree**	**Mean**	**Standard Deviation**
Waste Generation							
1. I am responsible for the waste I generated	5.0	0.0	10.0	49.0	36.0	4.11	0.948
2. The purchase decisions that I make can increase or decrease the amount of waste I need to get rid of.	6.0	0.0	15.0	47.0	32.0	3.99	1.005
3. I feel satisfied when waste is sorted and can be a resource.	6.0	2.0	12.0	45.0	35.0	4.01	1.044
Average Weighted Mean						4.03	Agree
Waste storage							
1. I play an important role in sorting waste within my household	5.0	6.0	18.0	38.0	33.0	3.88	1.089
2. I am willing to separate waste into respective category before collection	4.0	1.0	23.0	41.0	31.0	3.94	0.968
Average Weighted Mean						3.91	Agree
Waste Collection							
1. I am willing to pay extra service charges for different waste collection according to its category	10.0	14.0	26.0	38.0	12.0	3.28	1.150
2. I am satisfied with having different types of waste being transported to their respective site.	7.0	1.0	20.0	49.0	23.0	3.80	1.030
3. I am actively involved in collection and transport of sorted waste materials in my neighbourhood	5.0	25.0	35.0	25.0	10.0	3.10	1.044
Average Weighted Mean						3.39	Neutral
Waste processing							
1. I am willing to participate in training or programmes to gain knowledge on correct waste sorting methods.	3.0	9.0	36.0	37.0	15.0	3.52	0.954
2. Waste sorting and disposal should be taught in school as part of environmental education.	5.0	2.0	7.0	40.0	46.0	4.20	1.010
3. The municipal council is not doing enough to fix the garbage problem.	8.0	2.0	20.0	36.0	34.0	2.14	1.149
Average Weighted Mean						3.86	Agree
Waste Disposal							
1. I do not care that burning garbage can harm my health and the health of others.	60.0	21.0	7.0	9.0	3.0	1.74	1.110
2. People throw garbage on the streets and in the drains because they have no other choice to get rid of the garbage.	53.0	24.0	9.0	10.0	4.0	1.88	1.169
Average Weighted Mean						1.81	Disagree

3.4. Behaviour Assessment

Unlike the scores in the categories of knowledge and attitude, the scores in the behaviour category among the respondents tended towards negative, with a score of 68.0% (Tables 7 and 8).

Table 7. Behaviour level among respondents.

Behaviour Level	Good	Satisfactory	Poor
Range of Score	10 to 8	7 to 5	4 to 0
Frequency (%) (n = 100)	4.0	28.0	68.0

Table 8. Methods of waste disposal among respondents.

	Selected Waste Item (n = 100)	1. Dumping (Garbage Bins, etc.)	2. Illegal Dumping (Forest, River)	3. Open Burning	4. Feeding to Livestock or Pets	5. Using Compost Pits or Burying	6. Recycling after Cleaning	7. Sending to Collection Centre	Acceptable Waste Disposal Method (%)
1.	Used tissue paper	87.0	2.0	0.0	0.0	1.0	3.0	7.0	87.0
2.	Food-stained paper or plastic container	58.0	1.0	1.0	3.0	5.0	26.0	6.0	26.0
3.	Light bulb	68.0	4.0	0.0	0.0	3.0	3.0	22.0	25.0
4.	Vegetable and fruit peel	46.0	2.0	1.0	17.0	33.0	0.0	1.0	51.0
5.	Bone and meat scrap	47.0	2.0	1.0	32.0	17.0	0.0	1.0	50.0
6.	Dry leaves	45.0	3.0	10.0	1.0	39.0	1.0	1.0	41.0
7.	Expired medicine or supplement	80.0	1.0	0.0	0.0	4.0	3.0	12.0	12.0
8.	Photograph and paint	71.0	0.0	1.0	0.0	1.0	8.0	19.0	19.0
9.	Electronic, batteries and cables	55.0	0.0	1.0	0.0	3.0	6.0	35.0	41.0
10.	Hair or fur	84.0	2.0	2.0	0.0	8.0	2.0	2.0	12.0

Note: Each item can have one or more than one acceptable waste disposal method. Only those highlighted in grey was added and given 1 score each.

When the relationship between demographic characteristics and KAB was tested, the Chi-Square test revealed a significant relationship between household size and general knowledge and between housing type and attitude (p value < 0.05) (Table 9).

Further analysis using Pearson's Correlation Coefficient at a significant level of a = 0.05 was conducted to test the relationship between KAB. Only general knowledge and specific knowledge demonstrated a moderate relationship, with some nearly negligible relationships present between specific knowledge and attitude, as well as with behaviour (Table 10).

Table 9. Relationship between demographic characteristics and KAB.

Demographic Characteristics (n = 100)	General Knowledge (n; %)			Total (%)	p-Value
	Low	Moderate	High		
Household Size					
Alone	0; 0.0	2; 66.7	1; 33.3	3.0	0.046
2 to 4	2; 5.2	21; 53.8	16; 41.0	39.0	
5 to 7	1; 1.8	35; 62.5	20; 35.7	56.0	
More than 7	1; 50.0	1; 50.0	0; 0.0	2.0	
Total	4.0	59.0	37.0	100.0	
Demographic Characteristics (n = 100)	**Attitude (n; %)**			**Total (%)**	**p-Value**
	Negative	**Neutral**	**Positive**		
Housing Type					
Bungalow/Detached House	0; 0.0	3; 13.6	19; 86.4	22.0	0.027
Semi-Detached House	4; 14.3	3; 10.7	21; 75.0	28.0	
Terraced/Linked House	1; 2.1	13; 28.3	32; 69.6	46.0	
Flat/Condominium	1; 50.0	0; 0.0	1; 50.0	2.0	
Shop House	0; 0.0	0; 0.0	2; 100.0	**2.0**	
Total	6.0	19.0	75.0	100.0	

Table 10. Correlation between KAB.

Variables	p-Value	R-Value
General Knowledge and Specific Knowledge	4.90×10^{-19}	0.342
General Knowledge and Attitude	7.39×10^{-71}	0.049
General Knowledge and Behaviour	2.37×10^{-24}	0.059
Specific Knowledge and Attitude	7.12×10^{-73}	0.107
Specific Knowledge and Behaviour	6.79×10^{-9}	0.104
Attitude and Behaviour	3.61×10^{-73}	0.021
Knowledge and Attitude	7.30×10^{-66}	0.095
Knowledge and Behaviour	8.43×10^{-49}	0.099

3.5. Participation in Recycling and Waste Separation Practices

A total of 62% respondents indicated that they were involved in some extent of recycling practice, while 28.0% practised waste separation. The Chi-Square test also indicated the significance between recycling and source separation practices, where about two-fifth of recyclers adopted source separation (Table 11). In addition, a significant association between demographic profile and source separation practice was revealed through the Chi-Square test. These included age group, marriage status, educational level, and duration of stay (Table 12).

Table 11. Association between recycling and source separation participation.

Recycling Participation (n = 100)	Source Separation Participation (n; %)		Total (%)	p-Value
	Yes	No		
Yes	25 (40.3)	37 (59.7)	62.0	0.046
No	3 (7.9)	35 (92.1)	38.0	
Total	28.0	72.0	100.0	

Table 12. Association between demographic profile and source separation participation.

Demographic Characteristics (*n* = 100)	**Source Separation Participation (*n*; %)**		**Total (%)**	***p*-Value**
	Yes	**No**		
Age Group				
18 to 24 years old	17 (29.3)	54 (76.1)	71.0	
25 to 44 years old	8 (30.8)	18 (69.2)	26.0	0.046
45 to 64 years old	3 (100.0)	0 (0.0)	3.0	
Total	28.0	72.0	100.0	
Marital Status				
Single	21 (24.1)	66 (75.9)	87.0	
Married	7 (53.8)	6 (46.2)	13.0	0.026
Total	28.0	72.0	100.0	
Educational Level				
Upper Secondary	0 (0.0)	4 (100.0)	4.0	
Pre-university	2 (50.0)	2 (50.0)	4.0	
Diploma	0 (0.0)	3 (100.0)	3.0	
Bachelor	15 (21.4)	55 (78.6)	70.0	0.027
Postgraduate	4 (44.4)	5 (55.6)	9.0	
Master	5 (62.5)	3 (37.5)	8.0	
Other	2 (100.0)	0 (0.0)	2.0	
Total	28.0	72.0	100.0	
Duration of Stay				
Less than a year	1 (25.0)	3 (75.0)	4.0	
1 to 5 years	7 (77.8)	2 (22.2)	9.0	
6 to 10 years	4 (36.4)	7 (63.6)	11.0	
11 to 15 years	2 (28.6)	5 (71.4)	7.0	0.018
16 to 20 years	4 (23.5)	13 (76.5)	17.0	
More than 20 years	10 (19.2)	42 (80.8)	52.0	
Total	28.0	72.0	100.0	

Tables 13–16 show the statements asked in Section E of the questionnaire, where each respondent was able to choose multiple answers (reason statements) resulting in varying numbers of frequency recorded across the tables. Protecting the environment was rated as the most motivating reason to participate in recycling, consisting of about one quarter of the total frequency. That the ranking was immediately followed by the feeling of satisfaction in anticipating that waste would become a new resource and their continuous effort would influence other household members and neighbours as the reasons for recycling, whereas the skill of handling and sorting the recyclables was rated as the least important reason to participate in recycling (Table 13).

Three reasons voted the most by the majority of the respondents for practising waste separation are as follows (Table 14):

- It provides more environmental benefit than recycling (to prevent hazardous waste from polluting the landfills).
- It is important to separate biodegradable waste with recyclables to prevent contamination.
- It is more rewarding than recycling because biodegradable waste can be used as other resources (to feed animals, use as soil fertiliser for crops, etc.).

Table 13. Reasons to participate in recycling.

Statement		Frequency (%) (n = 114)
1.	It is very convenient to recycle where I live.	9.6
2.	I have spare time to sort out recyclables from the general waste.	7.0
3.	Recycling is very cost rewarding (monetary or physical item).	10.5
4.	I am good at handling and sorting recyclables.	4.4
5.	I believe my continuous recycling effort will influence my household members and neighbours.	14.0
6.	I feel satisfied when waste becomes a new resource.	15.8
7.	I practise recycling because it can protect the environment.	26.3
8.	It is a habit I developed from my upbringing.	12.3
9.	Others.	0.0

Table 14. Reasons to participate in source separation.

Statement		Frequency (%) (n = 83)
1.	It provides more environmental benefit than recycling (to prevent hazardous waste from polluting the landfills).	22.9
2.	It is important to separate biodegradable waste with recyclables to prevent contamination.	24.1
3.	It is more rewarding than recycling because biodegradable waste can be used as other resources (to feed animals, use as soil fertiliser for crops, etc.).	16.9
4.	Convenient (collected and transported to collection centre).	13.3
5.	Having more spare time.	6.0
6.	Good at sorting different waste.	6.0
7.	An extension from recycling habits.	10.8
8.	Others.	0.0

Table 15. Reasons to not participate in recycling.

Statement		Frequency (%) (n = 59)
1.	It is not accessible or convenient to where I live.	16.9
2.	It takes up too much time (to clean out/prepare recyclables, to travel to the nearest recycling centre, to look for information regarding each waste fraction).	20.3
3.	I always forget.	13.6
4.	Cost over benefit (where the reward from recycling does not feel worthy which could not cover the time consumed or transportation cost, and storage cost).	3.4
5.	I am not sure which waste belongs to which category.	11.9
6.	I do not feel my recycling efforts will make a difference.	6.8
7.	I feel uncomfortable having many recyclables bins or bags for different waste categories in my household.	8.5
8.	I do not care about recycling as it is not my priority for environmental concern.	1.7
9.	I do not understand the environmental benefit that recycling can provide.	3.4
10.	None of my household members recycle.	13.6

Table 16. Reasons to not participate in source separation.

Statement		Frequency (%) (*n* = 174)
1.	It is more tedious than recycling.	9.8
2.	There is no collection service for each type of waste even if I sort it out.	22.4
3.	There is no monetary reward unlike recycling.	5.2
4.	It does not bring more environmental benefit than recycling.	1.7
5.	Not convenient.	18.4
6.	More time needed.	16.7
7.	More knowledge needed.	19.0
8.	My household members always ruin my sorting efforts.	6.9
9.	Others.	0.0

When asked about the reasons for not practising recycling, time consuming was ranked as the most important consideration. Absence of an influencer at home, issues of convenience, and accessibility to the recycling/collection centre, as well as the lack of knowledge of recyclable wastes were also among the most selected reasons that set the barriers to recycling practice among the respondents (Table 15). A similar ranking of reasons (except for no home influencer) was revealed by respondents who did not participate in waste separation (Table 16).

When asked about action taken when facing uncertainty regarding the waste category, most of the respondents would place the item in the common trash and only a handful would refer to the online resources for reference (Table 17).

Table 17. Action taken upon uncertainty of waste category.

Action		Frequency (%) (*n* = 100)
1.	Place the item in the trash.	54.0
2.	Place the item in the recycling bin.	7.0
3.	Refer to the available guide from the local authority.	8.0
4.	Contact waste collection service provider for advice.	0.0
5.	Refer to someone who you think has more knowledge on waste sorting.	4.0
6.	Refer to online resources.	27.0
7.	Others.	0.0

3.6. Perspectives of R&SS Service Users

All 7 interviewees from the service users' group (including non-recyclers or non-waste sorters) held a positive attitude towards the products and compost made from recycled material. They had a satisfactory level of awareness and were able to explain the advantages and disadvantages of the waste treatment from R&SS. Nevertheless, the disadvantages for waste separation per se relatively outweighed and discouraged them from the practice, as opposed to that of recycling (Table 18). All 7 interviewees confirmed that they had no knowledge of the existing R&SS related efforts or initiatives implemented by Manjung Municipal Council and any local non-governmental organizations.

Table 18. Interview responses from the perspective of service users.

No.	Questions/Matters Discussed
1.	The interviewees were asked if they knew of or had seen any product made from recycled materials to confirm the outcomes of recycling itself and whether their recycling efforts were making an impact [39].
Responses from Interviewees	As one of the respondents said: "I have no idea how the recyclables I have collected are treated at the end of the process. It keeps me doubtful if truly these recyclables are transformed into useful raw materials and whether my recycling effort makes a significant impact. At least, products made from recyclables are not commonly seen in the market yet and can only be found in shops that are designed to sell such products. Usually, sustainable products that I encounter are marked "organic" rather than "made from recycled material".
2.	The interviewees were asked if they knew of or had seen (had experience) compost, a product generated through recycling food waste (sorted from general waste) to confirm the level of willingness to participate in the process from waste sorting to compost making, and in addition to this, they were asked about their challenges or gains to participate in this process [39].
Responses from Interviewees	In terms of composting practice (one of the outcomes from source separation), this was what one respondent said: "Composting allows a complete cycle of waste management. It can self-sustain where the biodegradable waste generated can be turned into fertiliser for plants that produce food. However, due to the lack of space (land for farming) and lack of influence from the households' waste handler, composting is not what I can usually practice at home. It is less useful to me compared to other households that have the supporting mechanism and medium (e.g., planting ground where compost is valuable) to do so". Another respondent added: "Composting is not a practice that households normally do. This is because it requires a tedious procedure, which is to ensure the right composition of carbon and nitrogen. Without the proper procedure, it will produce odour and attract pests and insects. Even the easiest composting method, the Bokashi, is also an unpleasant process during the accumulation of 9-litre kitchen waste in a week. This is because the process is not as straightforward as it seems, and it is generally harder to control. It might contain materials that are not advisable to compost, like oil or liquid, meat, bones, and dairy. A more complete composting system that includes a wide range of categories is more expensive at a point that common households cannot afford or are not willing to pay. For example, the composting machine from MAEKO, even though it is designed for households, it is still not considered as cost-efficient for common households' ability to pay".
3	The interviewees were further asked if they could consider other waste minimisation options (moving up the waste hierarchy [12]).
Responses from Interviewees	In response to the difficulties in R&SS, a respondent also commented on waste reduction and where his priority lies: "Going zero waste is very difficult when my priority does not lie on extreme waste minimisation as it is not cost-effective based on my affordability for essential purchases—groceries or daily necessities. This also requires strong determination that I do not think is worthy for me to trade off if there is no collective effort from a large population".

4. Discussion

The population in Manjung is surging, becoming the third most populated district in Perak. The increasing volume of waste generated as a result of throw-away consumerism will eventually exceed the capacity of the waste treatment at the only landfill site in Manjung, whose lifespan is expected to end within 3 to 5 years (data from Manjung Municipal Council). The heavy reliance on landfill for waste disposal has created a comfortable ground for households to throw every type of waste into the garbage bin for weekly municipal waste collection. R&SS therefore remains an unpopular option, and this situation is not acceptable for a city striving for sustainability within the next decade. This study investigates the gap between households' KAB that has delayed the transition towards sustainable integrated waste management. In addition to the gap, households' behaviour is also influenced by potential parameters that not only radiate from their mental characteristics (internal factors) but are also shaped by the physical environment and on how they deal with waste, from consumption to disposal (external factors). The causation of households' behavioural pattern is to be discussed in this section and should be addressed with relevant improvements in the waste management service to enhance the sustainability of Manjung district.

4.1. Associations between the Households' Level of KAB towards R&SS Practices

The relationship between knowledge, attitude, and behaviour is direct in theory, where knowledge supports the foundation of information processing, attitude bears psychological and emotional development, and behaviour holds the response and reaction through the entire process of thinking, feeling, and acting [23,46]. However, KAB relate to each other differently and inconsistently in reality, resulting in gaps between familiarity, values, and actions [47].

This gap in KAB also exists in the waste management system in Manjung in regard to whether the local authority or state government should implement Act 672 for mandatory source separation, based on the current condition and quality of the system. In reality, the recyclables, which have been produced and consumed, remain mismatched with the actual processed recyclables. Households also find it difficult to commit to recycling alone (less than 1.0% recycling rate compared to total waste generated per year in Manjung), not to mention source separation, due to additional requirements.

The findings show that households' knowledge (general and specific) is at a moderate level, attitude is at a positive level, and behaviour is at a poor level. These results denote that these three components influence and can be influenced by each other reciprocally, as suggested by various studies in the literature, if not direct. On the other hand, even though KAB has no eligible linear associations statistically, their associations are significant when discussed with multiple sets of conditions, variables, or causal factors. In fact, the negligible linear associations between these three components have proven that their associations are rather complex [46]. Similar results have been recorded in several studies; for instance, ref. [48] agree to the complexity in the study of KAB, especially for the fact that knowledge and attitude for participation in "green activities" do not lead directly to behaviour, regardless of the level. They have discussed that their respondents are familiar with the idea of recycling, yet the practical aspect is absent due to their indifferent attitudes towards the practice.

Finally, many scholars have also studied the associations between demographic profile and the KAB components. In this study, the findings only show household size and housing type as having statistically significant associations with one of the components. In fact, ref. [22] describe the associations of a high satisfactory level of KAB with a medium-sized household where the adults (parents) possess high educational level and secure jobs. Landed property has an influence on households' attitude on the local waste management system rather than knowledge, as suggested by [30]. Other demographic characteristics such as age, gender, civil status, and income level are also major determinants in establishing associations with KAB components [30,48]. Similarly, age, civil status, educational level, and living duration in this study show statistically significant associations with households' participation in source separation practices, which also influences the reciprocal causation of KAB components.

4.2. Enabling Factors of Households' Behaviour on R&SS Practices

Sections 4.2 and 4.3 discuss the understanding of the households' behavioural pattern (based on the sample at the point of observation, survey, and interview) with the identified barriers and enabling factors that are categorised into internal and external aspects. These parameters can bridge the gap of the KAB and serve as a basis to formulate recommendations. According to the findings, the internal factors are boiled down to (a) intentions, (b) knowledge, (c) willingness, (d) habit or routine, together with (e) persuasion. Conversely, the external factors identified in this study for the waste management service include (i) accessibility to services (both storage and collection), (ii) tangible incentives (reward) and disincentives (enforcement of law and penalty), and (iii) restrictions on consumption (discouragement of throw-away consumerism).

In terms of enabling factors, the findings (survey and elicited data) reveal that environmental protection is the fundamental motivation for the participation of R&SS practices among the respondents. The strength of intentions propels the recyclers (about 40%) to also participate in source separation, even though they have to deal with inconvenience within the provided waste management system. Households would have to take their own initiative to transport sorted waste to the collection centre. In fact, the municipal council only provides a limited number of recycling containers (a total of 11) and they are dedicated to collect specific and narrow waste category. This has lowered the willingness level, as it was ranked rather low as enabling factors for households to participate in R&SS. To increase this willingness, it may have to couple with other enabling factors such as tangible incentives (reward) and high accessibility to collection services with a widely acceptable waste category, as suggested by [33,39].

Knowledge, in this context, refers to the implications of R&SS practices and knowing the results from their participation in R&SS [35,39]. Interviewees gave positive responses towards the publicity of the outcomes of waste treatment through R&SS, which they claimed would greatly encourage them as they could know how their recycling effort counts. Having this knowledge in mind, this could have strengthened the participation more as "the feeling of satisfaction for anticipating that waste would become a new resource" and "continuous effort" are among the enabling factors that are rated high in the ranking.

Finally, Malaysia has a long recycling implementation history [7] and, consequently, this habit influences the decisions of households [24]. At least 10% of the respondents engaged in R&SS, behaviour that has been partially enabled by recycling habits. Based on the elicited data, songs have been created to teach children how to differentiate the colours of the recycling bin for each type of waste; such methods prompt children to engage in sustainable processes, for instance, paper must go into the blue recycling bin (when you see one).

Generally, this part of the discussion could assist service providers, especially the local authorities, to prioritise their agenda and budget to facilitate the transfer of knowledge among households [35,39].

4.3. Barriers of Households' Behaviour on R&SS Practices

Willingness to invest in terms of time, money, and effort was lacking among households (reasons ranked amongst the highest at first in recycling; fourth in source separation) as they have been practising a more "convenient" way of consumption and waste management for the past few decades; for example, taking single-use plastics for granted and dumping all wastes into garbage bins [37]. A change in lifestyle is difficult, as claimed by the majority of the interviewees. Dumping, which is a traditional (conventional) practice, is easier to continue as a habit compared to recycling and composting, which require gaining new knowledge and taking more steps to apply [49]; knowledge was rated as the second most selected barrier to participation in source separation.

In fact, there is a disconnection between households' pro-environmental behaviours (acted upon a just cause) and the final outcomes of the R&SS practices; when the interviewees were asked if they knew how recycling could protect the environment, most of them were only able to guess that the recyclables collected are being transformed into secondary material for production. They were intrigued to know the outcomes of the items they recycled; however, this information is not made available.

In fact, the waste management system (current R&SS service) in Manjung operates without a responsive feedback loop to keep users informed and educated. There are limited channels for respondents to improve their waste sorting knowledge, rebut recycling myths, track their contribution in R&SS practices, as well as follow up the aftermath of the segregated waste after it has been collected. The lack of such information discourages continuous recycling efforts among households, especially in communities where the idea that "waste that has been put in a recycle bin does not mean that it has been actually recycled" is a common belief [50,51]. This feedback is essential to prevent "wish-cycling": irresponsibly placing items into a recycle bin and hoping that it would be recycled [52,53]. The discouragement also presents a negative influence towards the impact of one person recycling, as the lack of communal effort has widely become an excuse (ranked as the third most selected barrier in recycling) for most households (not convinced without actual and strong evidence) to refuse in taking up these practices, especially source separation.

Generally, source separation is more complex than recycling, which has caused the percentage of participation to drop to 28% from that of recycling at 62%. It therefore depends on the degree of the discussed internal factors, which can determine how poor a household's behaviour is, in what ways their behaviour is poor, and to what extent this behaviour can be improved.

In terms of external factors, the findings (observed and elicited data) show that all waste management stages in Manjung pose challenges to households' participation in R&SS, especially source separation. As observed, households' waste generation is aggravated by the culture of throw-away consumerism and the convenience of plastic usage in the commercial industry. Waste is generated rapidly not only due to the increasing population (overconsumption), but also the short life cycle of the purchased product (made from non-decomposable material) [3,6]. Many are contaminated by materials such as food residue without proper sorting when discarded [36,52]. Additionally, the district has adopted a waste storage system that does not encourage the action of segregation due to its one-type design, especially waste containers provided at public areas (e.g., wet market). For waste collection, most housing developments in Manjung are landed property, where curb side recycling is claimed to have an overall positive effect for source separation [54], yet this idea is not celebrated. Integrating source separation with recycling seems impossible, as almost all local recycling vendors do not accept biodegradable waste and waste with mixed material (e.g., milk carton). Moreover, recyclables are usually self-transported, and the accepted recyclable categories are limited and generally remain unclear to most of the recyclers. A total of 40.3% of the respondents stated that they will still engage in both R&SS practices even though they have not been well supported by the current waste management system and environment. This could probably be due, in part, to their pro-environmental behaviour. However, at the same time, accessibility to R&SS services is also one of the greatest challenges for them and might result in negative experience when engaging in R&SS in the long term. This is because accessibility to R&SS services has been rated as the second most important barrier to participation in source separation by most of the non-recyclers and non-waste sorters. This barrier has stronger implications for service users in R&SS, discouraging participation.

The interviewees generally feel negatively towards the use of plastic bags, as they understand that the ocean and marine life are impacted by this consumption (based on elicited data). However, when the use of plastic bags grants them great convenience, they still decide to compromise. This is described as selective empathy, in which people in general selectively care about one matter instead of the actual problem [55]. This widens the mismatch between the psychology of households and their behaviour, where mental characteristics are not strong enough to effectively drive a favourable outcome due to the lack of a support system from waste management mechanisms at all stages. This support system can create a convenient environment for households to easily partake in R&SS [33,39]. This convenience is totally different from the convenience associated with throw-away consumerism. To date, a series of programmes have been implemented in

Manjung to encourage R&SS participation at the household level. These include (a) education programmes for pupils from kindergarten, primary, and secondary school. A total of 70 kindergartens have been engaged throughout the year to instil recycling habits in children's upbringing. Recycling and waste separation habits are generally more celebrated among school children; (b) engagement programmes with a group of households to adopt composting (Takakura method) for biodegradable waste treatment. This compost is then used for plantation on the land provided by the local authority (Department of City Planning), named as Taman Communiti, in a planned neighbourhood: Phase 1D Manjung. The community has been taught to produce their own compost after several tutorials; (c) monthly recyclables trade-off programmes for daily necessities at the main lobby of Manjung Municipal Council Office Building every first Saturday of the month; (d) provision of recycling containers (a total of 11) at public areas to collect textiles from households, known as SULAM programmes; (e) a pilot project with business owners at public wet markets with the aim of halting single-use plastics for packaging; and (f) focus group discussions (shifted from pupils to families) in various neighbourhoods.

Through these existing initiatives, the local authority has noticed the difficulty in reaching out or to convince the diverse population within the district to participate in their R&SS implementation. Information is usually disseminated through online platforms—official portals and social media (Facebook) page—as well as billboard advertisements at two of the busiest crossroads. Yet, the desirable results have been difficult to achieve. The lack of effective engagement and communication strategies and long-term trust building with the community are among the major issues faced by the local authority [56,57]. The interviewee from the local authority admitted that they were hoping that NGO(s) would come to them with a R&SS implementation proposal so that they could provide these NGO(s) with resources.

In addition, the State's commitment to sustainable waste management has been nonchalant and, as a result, the local authority's initiatives have only been sufficient enough to fulfil the requirements of standard urban management rather than a mission. Additionally, being a suburban municipality, the economy, technology, and even society are still unprepared for the development of sustainable waste management [47]; the communities in Manjung have yet to emerge as a major driving force behind the waste minimisation movement.

4.4. Recommendations to Enhance R&SS Participation

Considering all parameters discussed in 4.2. and 4.3., the authors made the following recommendations:

In order to increase the willingness of households to participate in sustainable practices, an education programme with an effective feedback loop system via internet communication technology is essential [17]. There are a few key elements in this system. First, a reactive feedback platform (online message, calls, and emails) answers to any uncertainty during waste sorting—a live version of gomi (Japanese word of "garbage") guide—becoming an ultimate go-to reference for waste sorting. This can improve households' perception of how easy R&SS can be carried out. Second, concise and practical information or knowledge is fed through push notifications to spark smaller actions. This further allows households (especially waste handlers) to react positively towards R&SS, as small actions do not incur heavy costs. Lastly, networking platforms allow households to share success stories and motivation. This can cultivate positive peer pressure and redefine subjective norms of the community to manifest the fact that their effort can make a difference.

Meanwhile, religious and dialect associations are groups that could bring households together, mobilise them and empower them to accomplish a mission. Households that are relatively able to placed their trust in these associations and dialogue can be easily initiated when two groups share similar social-cultural backgrounds [58]. Furthermore, as demonstrated by the community in Pulau Pangkor through authors' observation and interviews with the local NGO, a deep trust is the key driving factor for the third party to be able to encourage better participation in R&SS practices among households. This trust is formed when people share a common vision through long-term communication and engagement. By doing so, it allows both parties to easily unite under a common interest or goal when they have achieved the same ideology. A higher success rate is guaranteed with stronger social capital and creative social innovation [58]. This social innovation is important for the community to urge local authority to facilitate an enabling environment for R&SS implementation.

Environmental activities such as recycling and composting are best started off from these associations while utilising their facilities, as they have set up optimum conditions for community-driven projects. For instance, in Sentul, Kuala Lumpur, a Sikh community realised that a huge amount of biodegradable waste was being generated through their religious activities. This encouraged them to initiate a composting movement in reusing food and garden waste around their neighbourhood, while at the same time utilising the space around its religious facility [59]. It has also attracted families to join the action to contribute to waste reduction.

Regarding this example, the local authority can be the intermediary to connect the community with various sectors that require R&SS services or products. For instance, the compost produced by the community can be sold to agriculture-related sectors (e.g., farm or plantation within or outside the district) via the local authority. The local authority can also take up a role as an enabler to provide various resources, including space and location, information such as waste composition and waste handling within a community, as well as regulations such as legislative by-laws and guidelines for proper practice and implementation. Merits or incentives should be introduced to those who make a significant impact on minimising the waste, thus reinforcing their actions and advocating change.

The outcomes of R&SS should be widespread, providing both environmental as well as economic benefits. Households should be constantly reminded of this knowledge to shape their attitude slowly but surely. Waste management is a service delivery; the expenditure should be wisely spent to achieve a more productive result within a limited budget. Therefore, it is important to let households know that R&SS implementation can significantly reduce the amount of waste that has been sent to landfills, increasing the lifespan of landfills, and thus saving municipal budgets.

Furthermore, the authors in [60] have revealed that it is important for local authorities to stay financially feasible and sustainable through cost recovery, such as from transportation cost saving through lesser intervals as less generated waste is being sent to landfill. The authors in [61] support this idea by suggesting the R&SS industry collaborate with local authorities to improve the efficiency of recycled waste processing. Cost recovery can also be carried out by formulating by-laws under the "polluter-pays-principle" for private corporations or companies and business owners to partake in improving the waste management mechanism as part of their social responsibility. This will optimally reduce the trade-offs of any non-essential waste generation in commercial industry and household purchases. Meanwhile, pro-environmental product design guidelines should be formulated by environmental agencies and endorsed by local authorities for product manufacturers to comply with.

The current annual report of the Manjung Municipal Council has revealed no information regarding the expenditure of the service delivered. It is crucial for the local authority to share their expenditure (to be transparent throughout the administration) via effective means such as Gender Responsive and Participatory Budgeting (GRPB). GRPB has been practised by Penang's state and local governments with the collaboration of Penang Women's Development Corporation as an effective tool in making the budget gender responsive prudent and, most importantly, sustainable [62]. As this type of activity is voluntary-based, households' participation can determine how their tax can be used to create other public benefits from the budget they have saved through R&SS implementation, improving their willingness to be involved.

Small actions are negligible and simple but vital to shape good behavioural change in the long run. These small actions include everyday actions such as putting reusable containers or shopping bags in the usual hand carry or the vehicle itself, saying "no" to plastics, and more. Furthermore, they require constant reminders to eventually develop a habit over time. This can help those who experience difficulties in changing and adapting to a huge change in their waste management pattern. To amplify this effect, the surrounding environment should also be designed to influence consumers, such as designing signage at the doorstep to remind oneself to bring reusables before leaving the house when takeaway is planned. Essential destinations such as schools and the workplace, where people spend a significant number of hours in, should also be redesigned. Small tips or life hacks are essential to foster creative thinking upon negligible things or happenings in life to ensure these small actions become a routine and habit [63]. This strategy can be widely promoted using both online and offline platforms (considering not all households have access to online networks).

The survey findings reveal that most of the respondents acknowledged the importance of environmental syllabus in schools and universities. These educational facilities should raise the impact of their education and research by providing the know-how for effective implementation [64]. The knowledge transfer and skill training of formal education has a great influence among youth who are more flexible towards changes in their lifestyle [48,65]. The attitude is more effectively shaped at a younger age and recognition should be given (preferably by the government or relevant industrial players) to reward their effort and strengthen their intention to commit to R&SS practices. Besides classroom teaching and social activities, competitions (public speaking, essay writing, art- and craft-related events, photography, or videography) should be held to boost the capacity, as well as interpersonal skills of the students at all ages, especially sixth formers (pre-university) and undergraduates. Placing young children in a more competitive ground can also strengthen their intentions to advocate for the cause they believe in. Youth ambassadors in promoting R&SS practices can be one of the latest trends to promote households' participation, as young leaders often appear more inspiring and exhibit greater charisma to foster social innovations. Nevertheless, the quality of persuasion should be governed to ensure quality information processing.

It will be difficult for these recommendations to achieve optimum results if the external factors are not taken into consideration. An effective support system throughout waste management stages must be built to provide a convenient environment for households to conduct R&SS by increasing the accessibility to such services. Households' waste storage system should expand along with the waste collection system. By indicating a specific day for a specific waste category collection, households can save time for waste transportation to recycling vendors, maximising ease and convenience. With the effective feedback loop system, households can sort waste correctly, while weight can be recorded into the same system (mobile application) for further network and database establishment.

During the pandemic, the new normal indeed reshaped people's lifestyles, especially considering that digital transformation has influenced how people eat, shop, and pay bills. Manjung Municipal Council has developed a mobile application, myMANJUNG, for tax collection and bill payment [66]. This intervention can be included into the system for constructive management (e.g., ability to identify ownership of sorted waste and accountability). Nevertheless, this improvement requires consistent monitoring and evaluation by trained human resources, whether through a law enforcer hired by the municipality or collaboration with NGOs (NPOs) or community associations. Tangible incentives such as vouchers or cash rebates (based on the total weight of the sorted waste collected) can reinforce households' actions by eliminating hesitancy in visualising R&SS benefits (economic). Taiwan implements an integrated waste management system, which is designed to provide households with great support to recycle and separate waste conveniently. Frequent waste collection and households can track the service trucks in real time through mobile application. These service trucks accept wide waste categories, including biodegradable waste. Designated garbage bags are used to implement a Pay as You Throw (PAYT) scheme to incorporate the responsibility of households in waste reduction [67]. This integrated waste management system can effortlessly maximise the possibility of R&SS integration and transition from non-sanitary landfill to sustainable waste management.

Finally, the throw-away culture should end through the enforcement of law, an example from Shanghai mandatory waste separation in 2019. However, due to political complications, the local authority can instead implement a by-law to restrict non-essential product consumption. For instance, Penang has increased the price of a plastic bag from RM0.20 to RM1.00, suggesting that people can no longer take this convenience for granted [68]. A higher cost will be incurred for more damage inflicted towards the environment (more waste generated). This is necessary for households to abstain from using unsustainable products and ultimately revise their purchasing decisions. The ideal situation is the total ban of non-essential products (e.g., single-use and disposable plastic) by shifting focus and business models to the circular economy [3,6,9]; at the same time, society should be prepared with adequate knowledge and a convenient environment in which to practice R&SS with continuous motivation or feelings of commitment [33,39].

5. Conclusions

There was a gap observed between theory and practice in this study where households' level of knowledge, attitude and behaviour had no linear associations when tested with Pearson's Correlation Coefficient. Moderate knowledge with a positive attitude were insufficient to drive good behaviours different than that in theory. Therefore, the paper further examined the potential parameters that would shape households' behaviour and thus encourage behavioural change, considering the fact that about two-fifth of recyclers adopted source separation despite the current less-supported waste management mechanisms and environment. It was revealed that environmental protection was the most important enabling factor for households' participation in R&SS, while on the contrary, time consumption and accessibility to R&SS services were among the highest rated barriers by the non-recyclers or non-waste sorters.

There was also disconnection observed between households and the local authority where the former was unable to identify existing municipal R&SS initiatives, while the latter faced obstacles to find effective ways to engage with more diverse household groups regardless of cultural and social background. The local authority also pointed out the important roles played by the NGO in bridging the communication and trust gaps between the community and government.

This paper aimed at providing recommendations to the specific stakeholders in Manjung district to improve waste management through R&SS implementation by analysing the causations between KAB at a household level. By addressing the gap between households' awareness, more effective and targeted strategies and initiatives can be formulated to tackle the actual gap. Nonetheless, with the limited number of samplings as identified in the methodology section, further study is needed to capture more data, which can potentially be expandable to cover other districts in Perak state, or in other states in the country. In addition, as identified in this study, further study on the influence of both internal and external factors will provide in-depth understanding of the linkage between KAB in any study area.

Author Contributions: The paper is based on the undergraduate final year academic project of P.L.Y.; writing—original draft, H.C.G., P.L.Y.; writing—review and editing, P.L.Y., N.A.G., M.A. and H.C.G. All authors have read and agreed to the published version of the manuscript.

Funding: This research received no external funding.

Institutional Review Board Statement: Not applicable.

Informed Consent Statement: Not applicable.

Data Availability Statement: Not applicable.

Acknowledgments: The paper was drafted by the corresponding author during her writing residency with the company of Abu Bijan at Rimbun Dahan.

Conflicts of Interest: The authors declare no conflict of interest.

References

1. Gartner, K. Consumerism, Mass Extinction and our Throw-Away Society. *The Art of.* 2016. Available online: https://www.theartof.com/articles/consumerism-mass-extinction-and-our-throw-away-society (accessed on 25 April 2022).
2. BBC. We Need IMMEDIATE Action to Stop Extinction Crisis, David Attenborough [Video]. *YouTube*. Available online: https://www.youtube.com/watch?v=dbCR0KSU52g&t=158s (accessed on 23 September 2020).
3. Arias-Meza, M.; Alvarez-Risco, A.; Cuya-Velásquez, B.B.; de las Mercedes Anderson-Seminario, M.; Del-Aguila-Arcentales, S. Fashion and Textile Circularity and Waste Footprint. In *Circular Economy. Environmental Footprints and Eco-Design of Products and Processes*; Alvarez-Risco, A., Muthu, S.S., Del-Aguila-Arcentales, S., Eds.; Springer: Singapore, 2022. [CrossRef]
4. Clapp, J. Distancing of Waste: Overconsumption in a Global Economy. In *Confronting Consumption (155–176)*; MIT Press: Cambridge, MA, USA, 2002; Available online: https://mitpress.mit.edu/books/confronting-consumption (accessed on 25 April 2022).
5. World Bank. Solid Waste Management. 2019. Available online: https://www.worldbank.org/en/topic/urbandevelopment/brief/solid-waste-management (accessed on 25 April 2022).
6. Payne, J.; McKeown, P.; Jones, M.D. A Circular Economy Approach to Plastic Waste. *Polym. Degrad. Stab.* **2019**, *165*, 170–181. [CrossRef]
7. Moh, Y.C. Source Separation of Solid Waste for Recycling Practice at Household Level in Johor and Selangor, Malaysia. Ph.D. Thesis, Universiti Putra Malaysia, Serdang, Malaysia, 2017. Available online: http://psasir.upm.edu.my/id/eprint/68798/ (accessed on 25 April 2022).
8. Ministry of Urban Wellbeing, Housing and Local Government. Solid Waste Management Lab 2015: Final Lab Report. 2015. Available online: http://www.kpkt.gov.my/resources/index/user_1/Attachments/hebahan_slider/slaid_dapatan_makmal.pdf (accessed on 25 April 2022).
9. Yoong, L.S.; Bashir, M.J.K.; Wei, L.J. Food Waste Management Practice in Malaysia and Its Potential Contribution to the Circular Economy. In *Handbook of Solid Waste Management*; Baskar, C., Ramakrishna, S., Baskar, S., Sharma, R., Chinnappan, A., Sehrawat, R., Eds.; Springer: Singapore, 2022. [CrossRef]
10. Sim, L.L. Malaysia is Overflowing with Waste and We're Running Out of Options. *The Star Online*. 2019. Available online: https://www.thestar.com.my/lifestyle/living/2019/07/16/plastic-waste-landfills (accessed on 25 April 2022).
11. Khamis, S.S.; Purwanto, H.; Rozhan, A.N.; Rahman, M.A.; Salleh, H.M. Characterization of Municipal Solid Waste in Malaysia for Energy Recovery. *IOP Conf. Ser. Earth Environ. Sci.* **2019**, *264*, 012003. [CrossRef]
12. Directorate-General for Environment. Waste Prevention and Management. 2008. Available online: https://ec.europa.eu/environment/green-growth/waste-prevention-and-management/index_en.htm (accessed on 25 April 2022).
13. Pires, A.; Martinho, G. Waste hierarchy index for circular economy in waste management. *Waste Manag.* **2019**, *95*, 298–305. [CrossRef]
14. Abdel-Shafy, H.I.; Mansour, M.S.M. Solid Waste Issue: Sources, Composition, Disposal, Recycling, and Valorization. *Egypt. J. Pet.* **2018**, *27*, 1275–1290. [CrossRef]
15. Malaysians Remain Indifferent to Waste Separation. *Selangor Journal*. Available online: https://selangorjournal.my/2019/11/malaysians-remain-indifferent-to-waste-separation/ (accessed on 27 November 2019).
16. Wu, Z.; Zhang, Y.; Chen, Q.; Wang, H. Attitude of Chinese public towards municipal solid waste sorting policy: A text mining study. *Sci. Total Environ.* **2021**, *756*, 142674. [CrossRef]
17. Wang, H.; Jiang, C. Local Nuances of Authoritarian Environmentalism: A Legislative Study on Household Solid Waste Sorting in China. *Sustainability* **2020**, *12*, 2522. [CrossRef]
18. He, Y.; Kitgawa, H.; Choy, Y.K.; Kou, X.; Tsai, P. What Affects Chinese Households' Behavior in Sorting Solid Waste? A Case Study from Shanghai, Shenyang, and Chengdu. *Sustainability* **2020**, *12*, 8831. [CrossRef]
19. Lu, H.; Sidortsov, R. Sorting out a problem: A co-production approach to household waste management in Shanghai, China. *Waste Manag.* **2019**, *95*, 271–277. [CrossRef]
20. Liu, P.; Teng, M.; Han, C. How does environmental knowledge translate into pro-environmental behaviors? The mediating role of environmental attitudes and behavioral intentions. *Sci. Total Environ.* **2020**, *728*, 138126. [CrossRef]
21. Vicente-Molina, M.A.; Fernández-Sáinz, A.; Izagirre-Olaizola, J. Environmental knowledge and other variables affecting pro-environmental behaviour: Comparison of university students from emerging and advanced countries. *J. Clean. Prod.* **2013**, *61*, 130–138. [CrossRef]
22. Barloa, E.P.; Lapie, L.P.; Cruz, C.P.P. Knowledge, Attitudes, and Practices on Solid Waste Management among Undergraduate Students in a Philippine State University. *J. Environ. Earth Sci.* **2016**, *6*, 146–153. Available online: https://core.ac.uk/download/pdf/234664648.pdf (accessed on 25 April 2022).
23. Besar, T.A.; Hassan, M.S.; Bolong, J.; Abdullah, R. Exploring the Levels of Knowledge, Attitudes and Environment-Friendly Practices among Young Civil Servants in Malaysia. *Pertanika J. Soc. Sci. Humanit.* **2013**, *21*, 21–38. Available online: http://psasir.upm.edu.my/40687/ (accessed on 25 April 2022).
24. Shove, E. Beyond the ABC: Climate change policy and theories of social change. *Environ. Plan. A* **2010**, *42*, 1273–1285. [CrossRef]

25. Schumaker, E. The Psychology Behind Why People Don't Recycle: If It's Not Extremely Easy, People Won't Do It. *The Huffington Post*. 2016. Available online: https://www.huffpost.com/entry/psychology-of-why-people-dont-recycle_n_57697a7be4b087b70be605b3?utm_campaign=share_twitter&ncid=engmodushpmg00000004 (accessed on 25 April 2022).
26. Ministry of Housing and Local Government. What Needs to Be Separated? Towards A Greener Malaysia. 2015. Available online: https://www.kpkt.gov.my/separationatsource/en/ (accessed on 25 April 2022).
27. European Commission. EU-Perak Solid Waste Management Plan. 2009. Available online: https://www.yumpu.com/en/document/read/7578560/i-c-l-e-i-ea-swmc (accessed on 25 April 2022).
28. Cenviro Sdn. Bhd. Municipal Solid Waste Management. 2020. Available online: http://www.cenviro.com/core-business/municipal-solid-waste-management/ (accessed on 25 April 2022).
29. Manjung Municipal Council. *Statistics on Solid Waste Management within Manjung District*; Manjung Municipal Council: Manjung, Malaysia, 2020.
30. Al-Naggar, R.A.; Abdulghani, M.A.M.; Al-Areefi, M.A. Effects of Inappropriate Waste Management on Health: Knowledge, Attitude and Practice among Malaysian Population. *Malays. J. Public Health Med.* **2019**, *19*, 101–109. [CrossRef]
31. Manjung District and Land Office. District Info. Available online: https://ptg.perak.gov.my/portal/en/web/manjung/info-ringkas1 (accessed on 25 April 2022).
32. PLANMalaysia. Sistem Maklumat Gunatanah Perancangan Bersepadu: Semasa. Available online: https://iplan.planmalaysia.gov.my/public/geoportal?view=semasa (accessed on 25 April 2022).
33. Almasia, A.; Mohammadib, M.; Azizia, A.; Berizie, Z.; Shamsif, K.; Shahbazig, A.; Mosavia, A.S. Assessing the knowledge, attitude and practice of the kermanshahi women towards reducing, recycling and reusing of municipal solid waste. *Resour. Conserv. Recycl.* **2019**, *141*, 329–338. [CrossRef]
34. Omar, A.A.; Hossain, M.S.; Parvin, M.M. Study on Knowledge, Attitude and Practices towards The Solid Waste Management in Karan District, Mogadishu Somalia. *Environ. Contam. Rev.* **2018**, *1*, 22–26. [CrossRef]
35. Essuman, N.K. Knowledge, Attitudes and Practices of Coastal Communities on Waste Management in Ghana. Bachelor's Thesis, Novia University of Applied Science, Vaasa, Finland, 2017. Available online: https://core.ac.uk/download/pdf/80995157.pdf (accessed on 25 April 2022).
36. Abdullah, N.; Yusof, N.M.; Gani, A.; Mohammad, R.; Ishak, N. Assessing the Knowledge, Attitude and Practice among Food Handler towards Effective Waste Management System in Bandar Puncak Alam, Selangor, Malaysia. *Int. J. Latest Res. Humanit. Soc. Sci. (IJLRHSS)* **2018**, *1*, 10–15. Available online: https://www.researchgate.net/publication/326768370 (accessed on 25 April 2022).
37. Kaithery, N.N.; Karunakaran, U. Study on attitude of household waste management in a rural area of Northern Kerala. *Int. J. Community Med. Public Health* **2019**, *6*, 2095–2102. [CrossRef]
38. Laor, P.; Suma, Y.; Keawdounglek, V.; Hongtong, A. Knowledge, attitude and practice of municipal solid waste management among highland residents in Northern Thailand. *J. Health Res.* **2018**, *32*, 123–131. [CrossRef]
39. McAllister, J. Factors Influencing Solid-Waste Management in the Developing World. Master's Thesis, Utah State University, Logan, UT, USA, 2015. Available online: https://digitalcommons.usu.edu/gradreports/528 (accessed on 25 April 2022).
40. Krejcie, R.V.; Morgan, D.W. Determining Sample Size for Research Activities. *Educ. Psychol. Meas.* **1970**, *30*, 607–610. Available online: http://www.kenpro.org/sample-size-determination-using-krejcie-and-morgan-table/ (accessed on 25 April 2022). [CrossRef]
41. Creswell, J.W. *Research Design Qualitative, Quantitative, and Mixed Methods Approaches*, 3rd ed.; SAGE Publications: Thousand Oaks, CA, USA, 2009.
42. Sarantakos, S. *Social Research*, 4th ed.; Macmillan Education UK: London, UK, 2013.
43. Sileyew, J.K. *Research Design and Methodology*; Intechopen: London, UK, 2019; pp. 1–12. [CrossRef]
44. Sobri, N.A.; Rahman, H.A. Knowledge, Attitude and Practices on Recycling Activity among Primary School Students in Hulu Langat, Selangor, Malaysia. *Indian J. Environ. Prot.* **2016**, *36*, 792–800. Available online: https://www.researchgate.net/publication/311107035 (accessed on 25 April 2022).
45. Sauro, J. 5 Types of Qualitative Methods. 2015. Available online: https://measuringu.com/qual-methods/ (accessed on 25 April 2022).
46. Schrader, P.G.; Lawless, K.A. The Knowledge, Attitudes, & Behaviors Approach How to Evaluate Performance and Learning in Complex Environments. *Perform. Improv.* **2004**, *43*, 8–15. [CrossRef]
47. Nwachukwu, M.A.; Ronald, M.; Huan, F. Global Capacity, Potentials and Trends of Solid Waste Research and Management. *Waste Manag. Res.* **2017**, *35*, 923–934. [CrossRef] [PubMed]
48. Babaeia, A.; Alavia, N.; Goudarzia, G.; Teymouric, P.; Ahmadie, K.; Rafieef, M. Household Recycling Knowledge, Attitudes and Practices towards Solid Waste Management. *Resour. Conserv. Recycl.* **2015**, *102*, 94–100. [CrossRef]
49. Sreenivasan, J.; Govindan, M.; Chinnasami, M.; Kadiresu, I. Solid Waste Management in Malaysia—A Move Towards Sustainability. In *Waste Management—An Integrated Vision*; Rebellon, L.F.M., Ed.; Intechopen: London, UK, 2012. [CrossRef]
50. Eco-Business. Follow the Trash [Video]. *YouTube*. Available online: https://www.youtube.com/watch?v=1eHi2w0ixGo (accessed on 29 November 2018).
51. Eco-Business. Operation #RecycleRight: Here's How Singaporeans Are Recycling Wrong [Video]. *YouTube*. Available online: https://www.youtube.com/watch?v=iON4tJYG-eE (accessed on 25 September 2019).

52. Price, K.R. The Millennial Wish-cycler: Best Practices for Reducing Recycling Contamination. *WordPress*. Available online: https://www.jou.ufl.edu/wp-content/uploads/2020/09/The-Millennial-Wish-cycler-Kaitlin-Robb-Price.pdf (accessed on 25 April 2022).
53. Valanidas, M. A Place for Plastics: Bioplastics, Bacteria and Our Thoughtless Acts. Master's Thesis, Rhode Island School of Design, Providence, RI, USA, 2018.
54. Matsumoto, S. Waste separation at home: Are Japanese municipal curbside recycling policies efficient? *Resour. Conserv. Recycl.* **2011**, *55*, 325–334. [CrossRef]
55. Nas Daily. The Plastic Straw Dilemma [Video]. *Facebook*. Available online: https://www.facebook.com/watch/?extid=SEO----&v=525304967879828 (accessed on 21 December 2018).
56. Judith, P. Trust and waste management information expectation versus observation. *J. Risk Res.* **1998**, *1*, 307–320. [CrossRef]
57. Cohen, C.; Halfon, E.; Schwartz, M. Trust between municipality and residents: A game-theory model for municipal solid-waste recycling efficiency. *Waste Manag.* **2021**, *127*, 30–36. [CrossRef]
58. Mohamad, Z.F.; Idris, N.; Mamat, Z. Role of religious communities in enhancing transition experiments: A localised strategy for sustainable solid waste management in Malaysia. *Sustain. Sci.* **2012**, *7*, 237–251. [CrossRef]
59. The Star. Giving Back to Mother Nature [Video]. *Facebook*. Available online: https://www.facebook.com/watch/?v=1155297421569751 (accessed on 11 December 2020).
60. Lohri, C.R.; Camenzind, E.J.; Zurbrügg, C. Financial Sustainability in Municipal Solid Waste Management—Costs and Revenues in Bahir Dar, Ethiopia. *Waste Manag.* **2014**, *34*, 542–552. [CrossRef]
61. Lavee, D.; Nardiya, S. A Cost Evaluation Method for Transferring Municipalities to Solid Waste Source-Separated System. *Waste Manag.* **2013**, *33*, 1064–1072. [CrossRef]
62. Penang Women's Development Corporation Sdn. Bhd. Gender Responsive and Participatory Budgeting (GRPB). 2020. Available online: https://pwdc.org.my/gender-responsive-and-participatory-budgeting-grpb/ (accessed on 25 April 2022).
63. Zero Waste Malaysia. Zero Waste Malaysia Launches Country's First Comprehensive Handbook on Zero Waste Living. 2021. Available online: https://zerowastemalaysia.org/downloads/press-releases/MY-Zero-Waste-Life-Press-Release.pdf (accessed on 25 April 2022).
64. Ruxandra, B.; Cristian-Valentin, H.; Iulian, C.; Pavel, S. The university role in developing the human capital for a sustainable bioeconomy. *Amfiteatru Econ. J.* **2018**, *20*, 583–598. [CrossRef]
65. Lassnigg, L.; Hartl, J.; Unger, M.; Schwarzenbacher, I. *Higher Education Institutions and Knowledge Triangle: Improving the Interaction between Education, Research and Innovation*; Institute for Advanced Studies: Vienna, Austria, 2017.
66. Manjung Municipal Council. Aplikasi Dalam Talian Mymanjung, Kini Urusan Lebih Mudah. 2021. Available online: https://www.mpm.gov.my/en/node/2877 (accessed on 25 April 2022).
67. Rapid Transition Alliance. Taiwan's Transition—From Garbage Island to Recycling Leader. 2019. Available online: https://www.rapidtransition.org/stories/taiwans-transition-from-garbage-island-to-recycling-leader/ (accessed on 25 April 2022).
68. Nor Ain Mohamed Radhi. Impose Higher Charges for Plastic Bags. *News Straits Times*. 2019. Available online: https://www.nst.com.my/news/nation/2019/03/474395/impose-higher-charges-plastic-bags (accessed on 25 April 2022).

sustainability

MDPI

Article

Dynamic Rule Curves and Streamflow under Climate Change for Multipurpose Reservoir Operation Using Honey-Bee Mating Optimization

Songphol Songsaengrit and Anongrit Kangrang *

Faculty of Engineering, Mahasarakham University, Kantharawichai District, Maha Sarakham 44150, Thailand; songphol.so@rmuti.ac.th

* Correspondence: anongrit.k@msu.ac.th; Tel.: +66-89-843-0017

Abstract: Climate change in the watershed above the reservoir has a direct impact on the quantity of streamflow that enters the reservoir and the management of water resources. Developing effective reservoir rule curves helps reduce the risk of future failures of water resource management. The purpose of this study was to analyze the influence of climate change on the volume of streamflow entering the Ubolratana Reservoir, Thailand during the years 2020–2049 with climate simulations from the CIMP5 model under RCP4.5 and RCP8.5 scenarios. SWAT models were used to forecast future reservoir streamflow quantities. Moreover, suitable reservoir rule curves using the Honey-Bee Mating Optimization (HBMO) were developed and the effectiveness of the new rule curves was assessed. According to the research findings, the average yearly streamflow in the future apparently grew from 32% in the base years (2011–2019) and 65% under the RCP4.5 and RCP8.5 scenarios, respectively. It was discovered that the average monthly streamflow was higher in the rainy season than in the dry season. Both of the projected situations have a form compatible with the present rule curves in the section of the new reservoir rule curves generated with the HBMO. Furthermore, the newly constructed rule curves may allow the reservoir to keep more water during the rainy season, thereby assuring that there will be adequate water during the following dry season. Additionally, during the dry season, the reservoir was able to release more water that would be able to reduce the water shortage, indicating that it was able to effectively reduce the amount of water shortage and average overflow under RCP4.5 and RCP8.5 situations.

Keywords: climate change; streamflow; Honey-Bee Mating Optimization; reservoir rule curves

Citation: Songsaengrit, S.; Kangrang, A. Dynamic Rule Curves and Streamflow under Climate Change for Multipurpose Reservoir Operation Using Honey-Bee Mating Optimization. *Sustainability* **2022**, *14*, 8599. https://doi.org/10.3390/su14148599

Academic Editors: Alban Kuriqi and Luis Garrote

Received: 25 June 2022
Accepted: 11 July 2022
Published: 14 July 2022

1. Introduction

Uncertainty has a direct influence on the understanding of hydrology and water resource cycles caused by global climate change, as well as the growing frequency and intensity of droughts and floods throughout the world; these events are jeopardizing the management and development of water resources to meet global demands in all industries, making management more complex and difficult. For the past two decades, climate change has had a global impact on water resource management. Several study groups have sought to create ways for controlling water at its sources in order to deal with the fluctuation of supply sides and demand sides. The majority of such studies have evaluated the consequences of future climate change based on prediction findings from climate models combined with hydrological models to analyze impacts on water allocation efficiency for consumption [1], irrigation [2,3] hydroelectric power generation [4], and procurement of new reservoirs in the future [5].

In Thailand after the Great Flood of 2011, numerous watershed areas experienced drought between 2012 and 2019. The primary reason for this is that rainfall was below normal [6]. Many rivers' average discharge was lower than usual [7]. Government agencies must implement campaign initiatives to encourage consumers and farmers to consume water most efficiently and cost-effectively as possible. The northeastern area of Thailand comprises more than 60% agricultural land and is mostly dependent on seasonal rainfall in off-season cultivation, especially for rice cultivation, as it requires water from irrigation systems which rely on the cost of water from reservoirs. Meanwhile, the demand for water downstream in various sectors tends to increase. Many large and medium-sized reservoirs are unable to allocate water to meet the needs of all sectors effectively. In addition, the development of water resource management through efficient tools and methodology, alongside the consideration of the conditions of complex and nonlinear problems in all dimensions, is required especially for the management of reservoir water resources in situations of global climate change volatility [5]. It is, therefore, necessary to make an urgent adjustment.

Over the past decade, climate and streamflow were included in future hydrological models. These two factors have been used in combination with reservoir management. It is an approach that has been widely used in studies across the world. In Thailand, the Hydro-Informatics Institute created and released the Coupled Model Intercomparison Project Phase 5 (CMIP5) family of global climate models. This model has undergone bias correction using a Gamma-Gamma (GG) transformation optimization approach [8] to make future computation results more dependable. The products from CMIP5 have been used to analyze the effects of climate change in Thailand's watershed areas [9], hydrological systems in Southeast Asia [10], and many other places across the world [11,12].

A hydrological model is used to forecast future streamflow. In this study, a semi-diffuse hydrological model was investigated. The SWAT [13] is the world's most popularly used climate model, because of its integration of geographic information (GIS) data and regional climatic data in watershed areas of every size. As a result, the analysis is trustworthy. SWAT has been used in Thailand to examine and analyze the quantity of streamflow in various scenarios [14,15], and for the future management [16,17] of water resources in watersheds and reservoirs [18]. The precision of SWAT calculation results could improve when compared to the real measurements and this was accomplished by employing the SWAT-CUP model and the SUFI-2 approach [19] to choose the most appropriate sensitivity variables for analyzing the studied watershed regions. Therefore, based on the strengths of the CMIP5-derived products, once they were imported into SWAT, the results were expected to be future streamflow that differ from the new projection of greenhouse gas emissions. The Representative Concentration Pathway (RCP) as defined in the fifth Assessment Report (AR5) by the IPCC [20] provides cost information for appropriate reservoir management to situations of future hydrological variation.

There have already been some studies on applying optimization techniques to reservoir management, particularly in the development of suitable reservoir rule curves. Mathematicians have created evolutionary optimization approaches throughout the last decade. Appropriate reservoir rule curves were created using metaheuristic optimization techniques. Several approaches are popular in Thailand and across the world, such as Genetic Algorithm (GA) [18,21–23] Ant Colony Optimization (ACO) [24], Firefly Algorithm (FA) [25], Grey Wolf Optimization (GWO) [26], Tabu Search Algorithm (TSA) [27,28], and Particle Swarm Optimization (PSO) [21,22]. However, a new kind of evolutionary technique has been created, which is a natural-inspired approach to solving problems and finding answers in engineering. It is the Honey-Bee Mating Optimization (HBMO) algorithm [29], a process for optimization by imitating bee swarm behavior.

However, the solution to reservoir water allocation challenges caused by climate change affects future streamflow volumes. It was discovered that there were not many studies in the northeastern part of Thailand, along with forecasts of the variance in water demand from diverse activities in the downstream areas, especially for reservoirs in remote places where functionality is essential. Ubolratana Reservoir is the first significant multi-purpose reservoir in Thailand's northeast that provides hydroelectric electricity by combining irrigation and rainwater harvesting to reduce floods during the wet season. However, in the last ten years, dry-season water resource management has encountered a water insufficiency problem. Water intake to reservoirs has been lower than the average amount. In contracts, in certain years, the volume of water flowing into the reservoir surpasses the storage capacity during the rainy season. The water must be drained onto the downstream side, causing floods in residential and agricultural regions. As a result, when Ubolratana Reservoir has to develop suitable and efficient water distribution criteria, taking into account the diversity of hydrological circumstances in the past, present, and future together with the application of evolutionary optimization techniques to create more efficient rule curves. This would be expected to be of great benefit for water resource management.

In the past, the consideration of improving the reservoir rule curves of Ubolratana Reservoir, and the other reservoirs in Northeastern Thailand was a case study based on climate change forecasting from the AR4 models [18,30]. This research draws on climate forecasting data from the CMIP5 model based on the RCP4.5 and RCP8.5 scenarios that use bias correction to be more accurate, including there are various types and different model resolutions. The integration of SWAT hydrological models into the analysis of streamflow conditions has not been previously studied, and the same applies to experiments that link these models to the development of the optimal reservoir rule curves with the HBMO technique. Consequently, the expected outcome of the study is the optimal rule curves, appropriate outcome for the climate change situation and the variation on streamflow in many cases.

The purpose of this research was to use the CMIP5 and SWAT models to examine how global climate change affects the quantity of streamflow input into the Ubolratana Reservoir, as well as to improve the reservoir rule curves by employing the approach of the HBMO and considering the objective function, which is to minimize the quantity of water that is scarce and the amount of water that overflows the reservoir, respectively. The results of this study were predicted to be useful in predicting water scarcity and extreme water circumstances for flexible water management, provided as decision-support information for stakeholders to use as information for climate change policy planning and evaluation of water allocation guidelines to assist future activities.

2. Materials and Methods

2.1. Research Area

The research site was Ubolratana Reservoir in Ubolratana District, Khon Kaen Province. The study focused on five watershed areas; Lam Pha Niang, Lam Nam Phue, Upper Lam Nam Phong, Lam Nam Choen, and Lam Nam Phrom, all of which are tributaries of the Chi River Basin in northeast Thailand (Figure 1), with a total water intake area of around 12,000 square kilometers. The reservoir is a rock-fill dam with a clay core with a height of 2 m. The dam crest is 185.00 m above sea level. The basin receives an average of 2470 MCM of water each year. The normal water storage capacity is 2431.3 MCM, with a reservoir area of 370 square kilometers. The main functions of the reservoir are for generating electricity with an annual power generation capacity of approximately 56.1 million kilowatt-hours, irrigation covering an area of approximately 480 square kilometers, flood relief, fisheries, and intercity transportation travel.

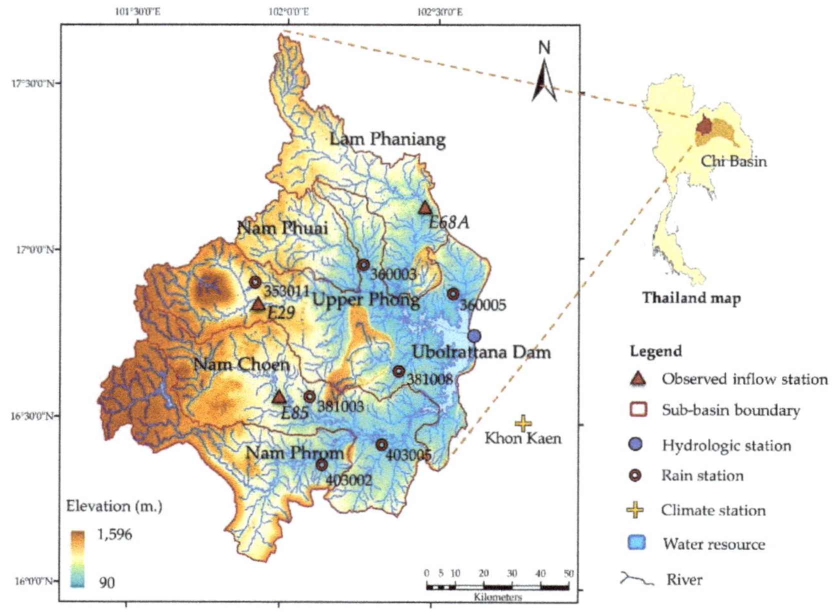

Figure 1. Study areas of tributaries in the Chi River Basin.

2.2. World Climate Models

2.2.1. CMIP5 Model

GCMs (General Circulation Models) are useful for describing and forecasting future climate change patterns. The World Meteorological Organization's Global Climate Research Program is now collecting data on current global climate change under the acronym Coupled Model Intercomparison Project Phase 5 (CMIP5) [31]. For this study, 10 CMIP5 models were selected by the investigators: MIROC_ESM, BNU, CanESM, MIROC5, FGOALS_g2, CESM1_CAM5, GFDL, EC_EARTH, CCSM4, and FGOALS_s2 [32,33]. The data used in the global climate change analysis were supported by the Hydro-Informatics Institute (HII) (Public Organization), which revealed that there are a wide variety of models that can be applied (more than 15 models). However, when comparing the model's data with the measurement stations in the study area, (especially rainfall data) and ranked based on the lowest tolerance. It was found that the models used in this study were among the 10 models with the lowest inaccuracies and were used in this study. Then, in the streamflow analysis, only the climate data from five of the best models were selected. For ease of use, the HII, which has downscaled the data model to a 5 × 5 square kilometer grid. Base year climate data in the study areas used the data for 9 years between 2011–2019, and climate forecasting data from 30-year models between 2020–2049.

2.2.2. Data Bias Correction

The Gamma-Gamma transformation approach was used in this study to correct for rainfall inaccuracy from the GCM. For this study, climate data, particularly precipitation data, courtesy of the Hydro-Informatics Institute (HII), is the agency that produces and distributes data for use in climate change studies in Thailand. This agency has identified the Gamma-Gamma transformation method to mitigate discrepancies in rainfall data. In addition, HII has published a study that applied this method to study the impact of climate change in Thailand on agricultural water demand [34]. In addition, Sharma (2015) has also chosen this method to study rainfall in western Thailand, which found that the Gamma-Gamma transformation was more effective in improving rainfall frequency and intensity compared to other methods [35]. The concept of this method is to correct for discrepancies caused by frequency and rainfall between GCM and measurement stations in the base year by creating a cumulative distribution function (CDF). This leads to the creation of appropriate Gamma parameters, with the functionalities and key parameters as shown in Equations (1)–(4).

$$F(x;\alpha,\beta) = \frac{1}{\beta^{\alpha}\Gamma(\alpha)}x^{\alpha-1}\exp\left(\frac{x}{\beta}\right); x \geq x_{Trunc} \tag{1}$$

$$F(x;\alpha,\beta) = \int_{x_{Trunc}}^{x} f(t)dt \tag{2}$$

$$F(x_{GCM};\alpha,\beta|GCM) \Rightarrow F(x_{His};\alpha,\beta|His) \tag{3}$$

$$x'_{GCM} = F^{-1}\{F(x_{His};\alpha,\beta|His)\} \tag{4}$$

where α is the shape and β is the size of the data from the GCM and base year monitoring stations at the selected locations to be gamma distribution. xTrunc is the amount of rainfall from CDF treated with the Gamma parameters, which are developed in Equation (2) for Equation (3). The α and β values were calculated by applying the maximum likelihood estimation method to calculate the daily precipitation from the inverse-adjusted GCM as shown in Equation (4).

2.3. *SWAT Hydrological Model*

The SWAT (Soil and Water Assessment Tool) model was created to aid in the management of water resources, and it was utilized in the evaluations for estimating the impact of water resource management and water pollution in watersheds and large basins [36], the quantity of streamflow that has changed, the amount of sediment and water quality in streams affected by changes in land use and climate in both past, present and future projections [37], which could be divided into distinct stages of watershed processing. For example, in the main watershed, sub-watershed zones are being created. Calculations that demonstrate outcomes daily and at extended intervals are also included. This considers variables from hydrological processes with the water balance equation as in Equation (5).

$$\mathrm{SW_t} = \mathrm{SW_0} + \sum_{i=1}^{t}\left(\mathrm{R_{day}} - \mathrm{Q_{surf}} - \mathrm{E_a} - \mathrm{W_{seep}} - \mathrm{Q_{gw}}\right) \tag{5}$$

where SW_t is the final soil water content; SW_0 is the initial soil water content, t is the time (days), R_{day} is the precipitation (mm) on the day i, Q_{surf} is the surface water content on the day I, E_a is the evaporative transpiration amount on the day I, W_{seep} is the amount of water seeping into the basement on the day i, and Q_{gw} is the amount of groundwater returning to the stream on the day i.

2.3.1. Data Input

In the implementation process, the SWAT method requires the import of basic physical data, including a digital elevation model (DEM) with elevation values between 90 to 1596 m (MSL). The watershed area has a slope from the west (mainly mountains and upstream forests) to the eastern lowland area where the Ubolratana Reservoir is located (see Figure 1). As for the soil type map (Figure 2a), it indicates that more than 50% of the soil is clay, which is in the eastern lowland, followed by clay loam soil, which is mainly in the eastern lowland of the study area. The types of land use in the study area were mostly agricultural areas. It was found that the use of land for rice farming which is most distributed in the eastern lowland area, combined with sugarcane and cassava plantation in the central area of the basin. In the west, most areas are watershed forests. The land use spatial distribution map is illustrated in Figure 2b.

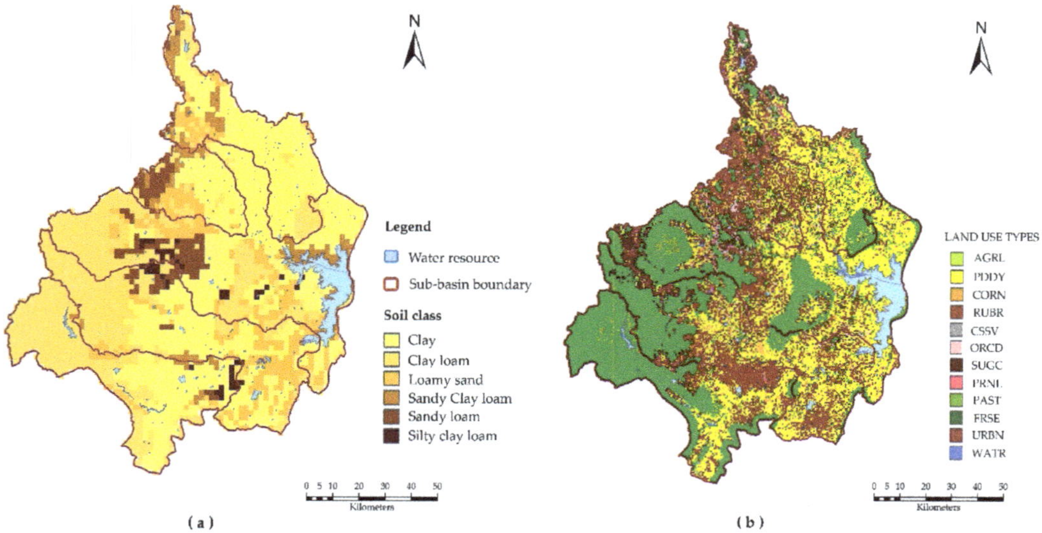

Figure 2. Soil Type Map (**a**), Land use map (**b**).

Daily climate data includes rainfall, temperature, humidity, wind speed, and solar intensity. Daily rainfall data were collected from 9 rain gauge stations distributed in the study area and 1 climate station (Khon Kaen station) located in the southeastern part of the watershed, as shown in Figure 1. There are 4 stations of streamflow and sediment data, of which 3 stations are located in the watershed areas above the Ubolratana Reservoir, are Station E68A (Lam Pha Niang Basin), E29 Station (Upper Phong Basin), and E85 Station (Lam Chuan River). Basin). These data are from 2011–2019 supported by the Royal Irrigation Department of Thailand. The data used for evaluating the effectiveness of the SWAT-computed results for the different types, intervals, scales, and data sources used in this study are summarized and shown in Table 1.

Table 1. Basic data to be used in the SWAT model.

Data Type	Period	Scale	Source
DEM	2015	30 × 30 m	Land Development Department, Thailand
Soil type map	2015	1:50,000	
River map	2020	1:50,000	
Land use map	2015	30 × 30 m	
Climate	2011–2019	Daily	Thai Meteorological Department, Thailand
Observed inflow	2011–2019	Daily	Royal Irrigation Department, Thailand; Electricity Generating Authority, Thailand

2.3.2. Model Performance Evaluation Using SWAT-CUP

SWAT-CUP (SWAT Calibration and Uncertainty Procedure) is a SWAT-compatible model. When compared to the old approach of manual correction by trial and error, the SWAT model's sensitive variable analysis, calibration, and validation procedures have more flexibility and take less time. The outcome of altering the sensitivity variable will serve as a guide for the best calibration and adjustment of the solution(s) between the SWAT generated results and the station data. The following are five approaches for determining the proper values: (1) Generalized Likelihood Uncertainty Estimation (GLUE), (2) Particle Swarm Optimization (PSO), (3) Parameter Solution (Parasol), (4) Mark Chain Monte Carlo (MCMC), and (5) Sequential Uncertainty Fitting (SUFI-2) [38]. For this study, the use of the SUFI-2 technique was selected to apply in the operation. The SUFI-2 technique is uncertainty analysis consisting of predictive P-factors representing the actual measured values that appear in the simulation results for 95% of the uncertainty of the simulation. The prediction (95% prediction uncertainty; 95PPU) and R-factor are calculated as the ratio of the mean amplitude range of the 95PPU to the standard variance of the actual data. The calculated 95PPU values were positioned at 2.5% and 97.5% of the cumulative probability distribution of the variables considered. Using Latin hypercube sampling [38] as this technique requires the least number of sensitivity variables but can produce the best results compared to other methods [39]. Eight parameters from the most vulnerable model types were chosen for examination in this study. Eight parameters from the most vulnerable model types were chosen for examination in this study. The results of the modification of the parameters that calculated streamflow from the model closest to the data from the measurement station are shown in Table 2.

Table 2. Adjusted Model Sensitivity Parameters.

No.	Parameter	Range	Adjusted Values
1	ALPHA_BF.gw	0–1	0.367
2	GW_DELAY.gw	0–500	19.500
3	GWQMN.gw	0–500	179.500
4	ESCO.hru	0–1	0.881
5	GW_REVAP.gw	0–500	129.500
6	SOL_AWC.sol	0–1	0.393
7	CN2.mgt	−0.2–0.2	−0.104
8	EPCO.hru	0–1	0.819

Then, the results were compared with the data from the measurement station, and the efficiency was assessed using two statistical indices to check the accuracy of the results [40], which showed the level of accuracy of the monthly streamflow comparison results. It is divided into four levels as shown in Table 3 [41].

Table 3. Typical performance level for accepted statistics in monthly time step.

Level	R^2	NSE
Very good	$0.80 < R^2 \leq 1.00$	$0.75 < NSE \leq 1.00$
Good	$0.70 < R^2 \leq 0.80$	$0.65 < NSE \leq 0.75$
Satisfactory	$0.60 < R^2 \leq 0.70$	$0.50 < NSE \leq 0.65$
Unsatisfactory	$R^2 \leq 0.60$	$NSE \leq 0.50$

1. The Coefficient of Determination (R^2), as shown in Equation (6), is between 0–1, with values greater than 0.6 indicating that the two data are correlated at a level of reliability.
2. The Nash Sutcliffe efficiency (NSE) coefficient, as shown in Equation (7), is between $-\infty$ and 1, with values greater than 0.5 indicating that the two data are correlated at a level of reliability.

$$R^2 = \left[\left(\frac{\sum_{i=1}^{n}(Q_{oi} - Q_{oa})(Q_{si} - Q_{sa})}{\sqrt{\sum_{i=1}^{n}(Q_{oi} - Q_{oa})^2}\sqrt{\sum_{i=1}^{n}(Q_{si} - Q_{sa})^2}}\right)\right]^2 \quad (6)$$

$$E_{ns} = 1 - \left(\frac{\sum_{i=1}^{n}(Q_o - Q_s)^2}{\sum_{i=1}^{n}(Q_o - Q_{sa})^2}\right) \quad (7)$$

where n is the total number of data. Q_{oi} is the i-order value, Q_{oa} is the mean from all measurements, Q_{si} is the i-order model, Q_{sa} is the i-order value from all models, Q_s is the calculated value from the model, and Q_o is the measurement value.

2.4. Application of HBMO Algorithm for Reservoir Rule Curves Generation

2.4.1. HBMO Algorithm

The HBMO Algorithm is a hybrid search algorithm based on bee mating behavior. The biological bee breeding process is transformed into a mathematical modeling program. As a result, the phases in the adjustment process were properly outlined. Mating is the first step in algorithm development. Every queen bee makes a flight based on her power and speed throughout each mating flight. Equation (8) determines the likelihood of mating between individual male bees and queen bees. The likelihood of mating is high during the start of the mating flight when the queen bee's velocity is high, or when a male bee is sufficiently numerous to mate, the probability of mating is high.

After the movement of the queen bees or after mating, energy, and speed decrease according to Equations (9) and (10). When all queen bees have completed a pairing flight, they begin to breed to achieve the required number of embryos. The queen bees are selected in proportion to the queen bee's fitness and are artificially inseminated with sperm randomly selected from the queen bee's sperm sac. The worker bees would be selected in proportion to their fitness to be used to improve larval outcomes. After the embryos were born, they would be sorted according to their fitness. The best larvae replace the worst queen bees until there are no better embryos than any queen bees. The remaining larvae are then killed and new matings begin until there is a perfect mating. All predetermined will be completed or meet converging criteria [42].

$$Prob\ (Q, D) = e^{-\frac{\Delta(f)}{S(t)}} \tag{8}$$

where $Prob\ (Q, D)$ is the probability of mating between the male bee D and the queen bee Q or the probability of successful mating; $\Delta(f)$ is the difference between the male bee's fitness ($f(D)$) and the fitness of the queen bee ($f(Q)$); $S(t)$ is the speed of the queen bee at the time.

$$E(t+1) = E(t) - \gamma \tag{9}$$

$$S(t+1) = \alpha \times S(t) \tag{10}$$

where $E(t)$ is the queen's energy; $S(t)$ is the queen's speed; α is a factor $\in$ [0, 1] and γ is the amount of energy reduction after each transition.

2.4.2. Water Equilibrium Simulation Model

The models HEC-3, HEC-5, and HEC-RAS were used in a simulation study of the reservoir system in each watershed [43]. Water balance principles were used. In this study, a simulation model of the reservoir system was created by using the same principles as in the above model, to facilitate connection with the Honey Bee Mating Optimization and begin calculating the water balance of each reservoir. To begin calculating the water balance of each reservoir from the rule curves, the initial storage volume of the reservoir was set at full capacity or the maximum storage level; the discharge volume could be calculated following the Standard Operating Rule as shown in Figure 3 and Equation (11). Then, the available water cost of the reservoir could be calculated for the next month with the principles of the water balance equation as shown in Equation (12).

$$R_{v,\tau} = \begin{cases} D_\tau + W_{v,\tau} - y_\tau,\ for\ W_{v,\tau} \geq y_\tau + D_\tau \\ D_\tau,\ for\ x_\tau \leq W_{v,\tau} < y_\tau + D_\tau \\ D_\tau + W_{v,\tau} - x_\tau,\ for\ x_\tau - D_\tau \leq W_{v,\tau} < x_\tau \\ 0,\ otherwise \end{cases} \tag{11}$$

where $R_{v,\tau}$ is the amount of water discharged from the reservoir during the year v in the month τ (τ is 1 to 12 referring to January to December); D_τ is the demand for water at the bottom of the basin during month τ; x_τ is the lower boundary of the rule curves of the month τ; y_τ the upper boundary of the rule curves of the month τ; and $W_{\upsilon,\tau}$ is the amount of original water level available in the basin of the month τ.

$$W_{\upsilon,\tau+1} = S_{\upsilon,\tau} + Q_{\upsilon,\tau} - R_{\upsilon,\tau} - E_\tau - DS \quad (12)$$

where $S_{\upsilon,\tau}$ is the amount of water stored in the reservoir at the end of the month τ; $Q_{\upsilon,\tau}$ is the average streamflow in the month τ; $E\tau$ is the evaporation loss in the month τ; and DS (dead storage) is unused storage volume.

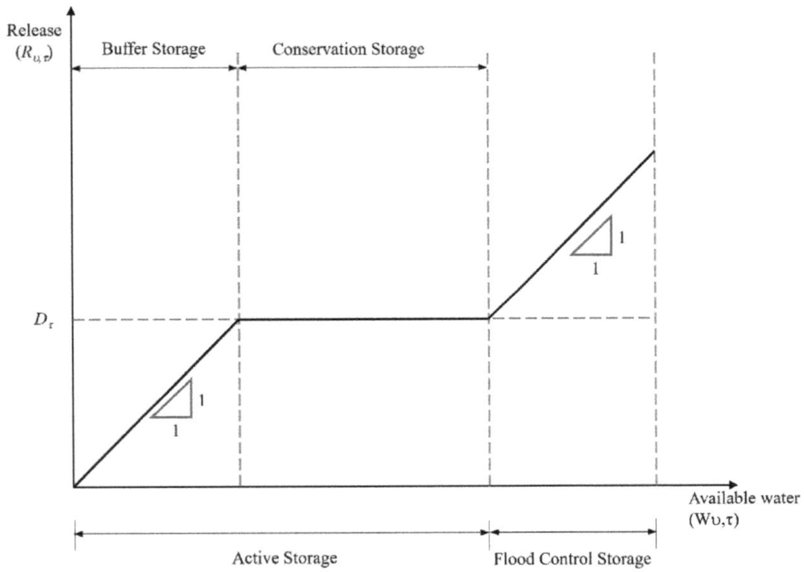

Figure 3. Standard water discharge criteria.

The reservoir rule curves were generated using the HBMO Algorithm Optimal Solution in this study. In the instance of shortage frequency, the target function for determining the solution was the least average shortage, as illustrated in Equation (13).

$$Min(Aver_{Sh}) = \frac{1}{n}\sum_{\upsilon=1}^{n} Sh_\upsilon \quad (13)$$

where n is the length of the original water quantity data set; Sh_υ is the amount of water shortage in the year υ (The amount of water released is less than the water demand target).

2.4.3. Reservoir Rule Curves Efficiency Evaluation

By analyzing the frequency of occurrence of an incident, the rule curves assessment was set to evaluate two parts: water scarcity and excess release water with mean and maximum values of Magnitude and Duration through the performances of the test rule curves with future monthly streamflow scenarios from 2020 to 2049. Changes in greenhouse gas emissions are RCP4.5 and RCP8.5, which are two different types of RCP.

3. Results and Discussion

3.1. Streamflow Analysis Using the SWAT Model

3.1.1. Model Performance Assessment

Evaluation of model performance assessed the accuracy between the calculation of streamflow from the SWAT model calculated from the average monthly streamflow volume during 2011–2019 and the streamflow data from 4 measurement stations in the study areas, namely E68A Station (Lam Pha Niang River Basin), E29 Station (Upper Phong River Basin), Ubolratana Dam Station, and E85 Station (Lam Nam Choen River Basin) in the same period. The model's performance was evaluated using an index of R^2 ranging from 0.62–0.88 and NSE between 0.50–0.81, which were both within the acceptable accuracy range as shown in Table 4.

Table 4. Index values for evaluating the accuracy of SWAT calculation results comparing streamflow volumes from measurement stations.

Assessment Index	R^2	NSE
E68A Station (Lam Pha Niang River Basin)	0.82	0.52
E29 Station (Upper Phong River Basin)	0.79	0.76
Ubolratana Dam Station	0.88	0.81
E85 Station (Lam Nam Choen River Basin)	0.62	0.50

Comparative results of streamflow volumes from the SWAT model and streamflow data from Ubolratana Dam Station are shown in Figure 4. The average annual streamflow from the SWAT model is 5147.34 MCM and that of the measurement station is 2385.56 MCM.

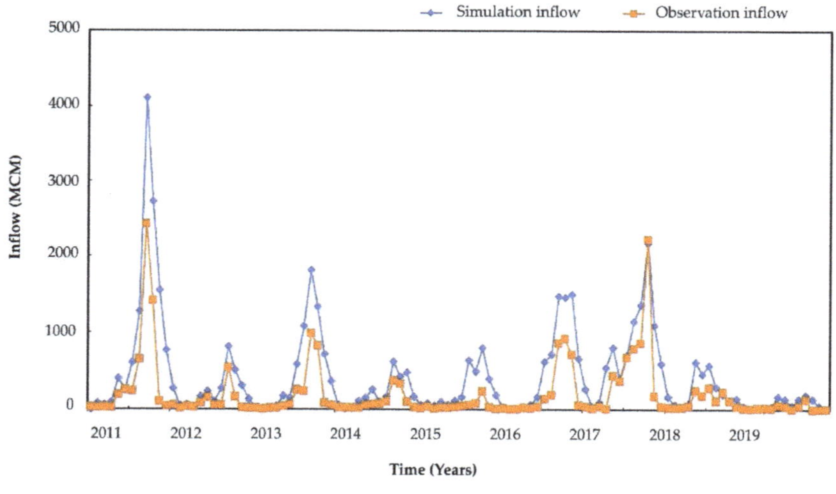

Figure 4. Comparison of streamflow between the data from Ubolratana Dam Station and the calculated results from the SWAT model during 2011–2019.

3.1.2. Forecasting of Future Streamflow Volumes

Forecasted future streamflow from 2020 to 2049 were expected to be impacted by climate change based on the CIMP5 model under the RCP4.5 projection case. In total, there was a 32% increase in the average annual streamflow in the future. With the MIROC_ESM model, the streamflow volume was likely to increase to a maximum of 4734.97 MCM (98.49%), and with the MIROC5 model, it was expected to rise by 3889.10 MCM (63.03%). In the BNU model, it increased to 2905.53 MCM (21.80%), and in the CanESM model, it increased to 2758.80 MCM (15.65%). However, the FGOALS_g2 model indicated that the average annual streamflow in the future was expected to decrease by 1528.95 MCM (−35.91%) (Figure 5). It was found that, overall, the average monthly streamflow volume increased during the rainy season, accounting for 2930.95 MCM (29.82%), and in the dry season, it accounted for 232.53 MCM (81.82%). When considering each model, there were 4 models, MIROC_ESM, BNU, CanESM, and MIROC5. There was an increase in the average monthly streamflow during the rainy season between 2516.67–4479.10 MCM (11.47–98.40%), and the monthly average streamflow volume would increase significantly during the dry season, especially in October showing a significantly higher proportion (Figure 6). The highest increase in the MIROC5 model was 356.00 MCM (178.36%). However, the study from the FGOALS_g2 model expressed a trend of lower average monthly streamflow in both rainy and dry seasons which were 1467.22 MCM (−35.01%) and 61.73 MCM (−51.73%) respectively. The results were in line with the average annual streamflow (Table 5).

Climate change was projected to influence future streamflow levels between 2020 and 2049, according to the CIMP5 model under the RCP8.5 forecast. The results showed that the average annual streamflow across all models tended to increase. The MIROC5 model rose by 5828.46 MCM (144.32%), the BNU model climbed by 3704.05 MCM (55.27%), and the CanESM model increased by 3704.05 MCM (55.27%) (55.27%). Model FGOALS_g2 grew to 2854.40 MCM (19.65%) and 3419.62 MCM (43.35%) (Figure 7). Looking at the seasonal average monthly streamflow volumes, the trend of change in average monthly water volume was similar under the RCP4.5 projection case but had a greater proportion of increase. Overall, the average monthly streamflow volume increased during the rainy season by 3551.80 MCM (57.32%) and by 401.32 MCM (213.81%) in the dry season. The increase was significant in both the rainy and dry seasons compared to the other models (Table 5), with a significant increase in percentage in October (Figure 8).

Table 5. Average monthly base year streamflow and seasonal forecasts.

Period	RCP	GCM	May–November (Wet Season) (MCM)		December–April (Dry Season) (MCM)	
			Average	Difference (%)	Average	Difference (%)
Baseline (2011–2019)			2257.67		127.89	
2020–2049	RCP4.5	Overall	2930.95	29.82	232.53	81.82
		MIROC_ESM	4479.10	98.40	255.87	100.07
		BNU	2658.62	17.76	246.90	93.06
		CanESM	2516.67	11.47	242.13	89.33
		MIROC5	3533.11	56.49	356.00	178.36
		FGOALS_g2	1467.22	−35.01	61.73	−51.73
	RCP8.5	Overall	3551.80	57.32	401.32	213.81
		MIROC_ESM	4902.41	117.14	926.05	624.11
		BNU	3409.38	51.01	294.67	130.41
		CanESM	3126.56	38.49	293.06	129.15
		MIROC5	3654.94	61.89	304.12	137.80
		FGOALS_g2	2665.69	18.07	188.71	47.56

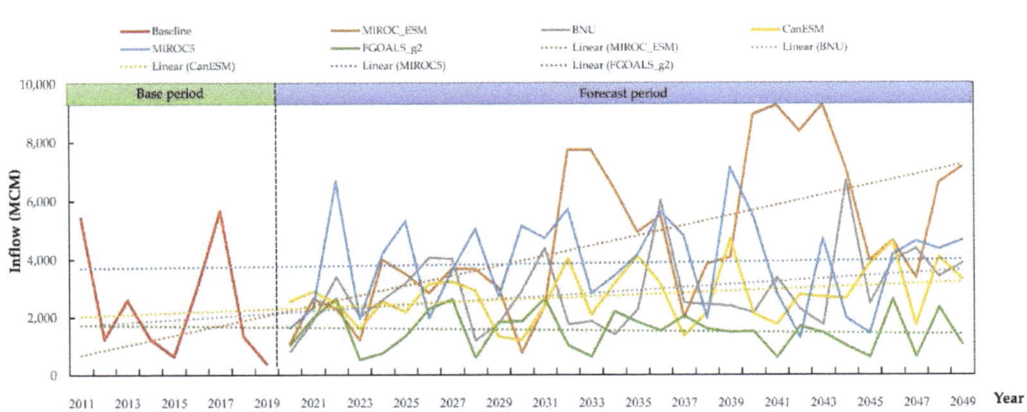

Figure 5. Annual streamflow from the base year SWAT model 2011–2019 and under the forecast of RCP 4.5 between 2020–2049.

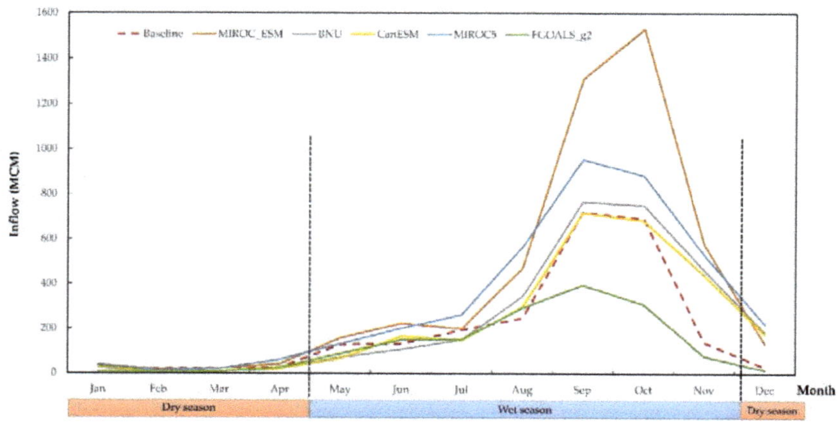

Figure 6. Monthly streamflow from the base year SWAT model 2011–2019 and under the forecast of RCP 4.5 between 2020–2049.

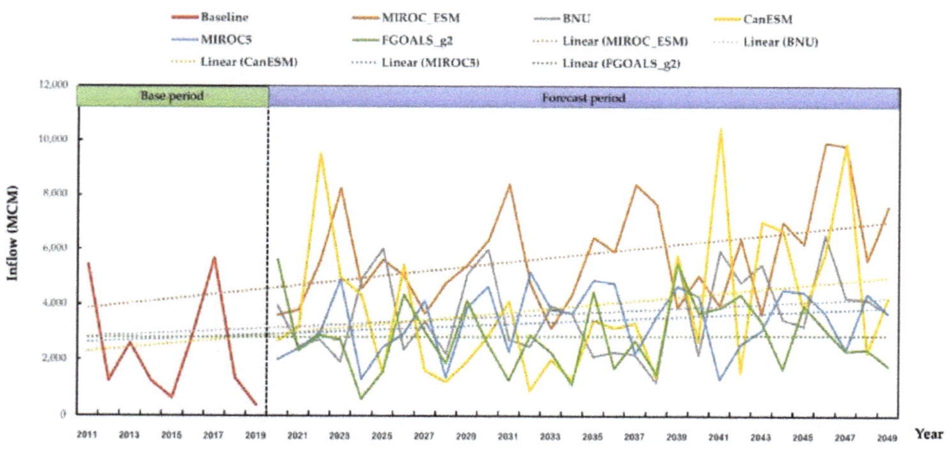

Figure 7. Annual streamflow from the base year SWAT model 2011–2019 and under the forecast of RCP 8.5 between 2020–2049.

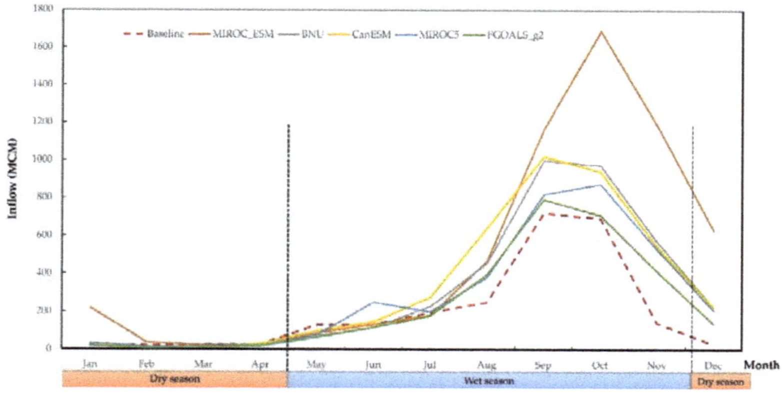

Figure 8. Monthly streamflow from the base year SWAT model 2011–2019 and under the forecast of RCP 8.5 between 2020–2049.

3.2. *Optimal Reservoir Rule Curves with HBMO Algorithm Technique*

3.2.1. Optimal Reservoir Rule Curves by HBMO Algorithm

The findings of the Ubolratana Reservoir rule curves generated with the HBMO Algorithm approach based on the CIMP5 climate change impacts of 5 models under RCP4.5 and RCP8.5 projection cases were compared to the present Ubolratana Reservoir rule curves. The rule curves in both predicted situations were discovered to be identical to the existing rule curves. However, from July to September, the newly developed upper rule curves were higher than the current rule curves. This effected an increase in the amount of water stored in the reservoir, resulting in a sufficient water supply for the next dry season. In the upper rule curves of the two forecast cases, the shape corresponded to the current rule curves, but the lower rule curves developed lower than the current ones during the dry season from December to April. This means that the reservoir can release more water than with the current rule curves. It can reduce water scarcity, making it possible to respond to water users in irrigated areas (Figures 9 and 10). According to recent study, applying the Harris Hawks Optimization (HHO) technique for searching in the Ubolratana reservoir, Thailand, the optimal rule curves with the HHO technique was similar to the current rule curves. The upper rule curves developed were higher than the current rule curves throughout the rainy season, allowing for additional water storage at the end of the rainy season [44].

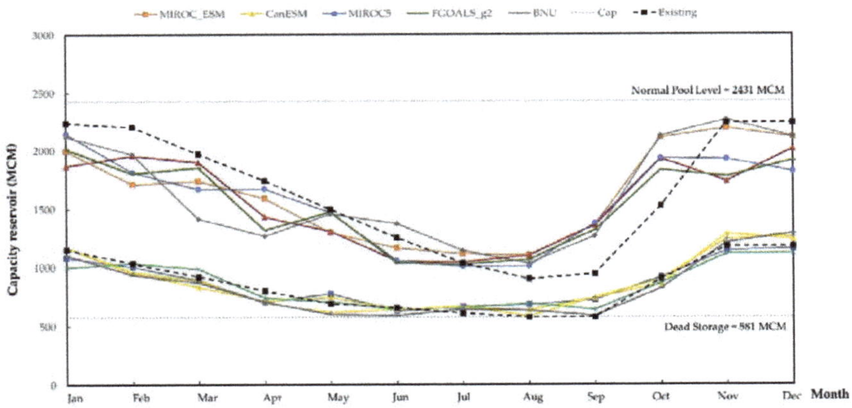

Figure 9. Rule curves of Ubolratana reservoir developed using HBMO algorithm technique based on climate change impacts under the RCP4.5 projection case.

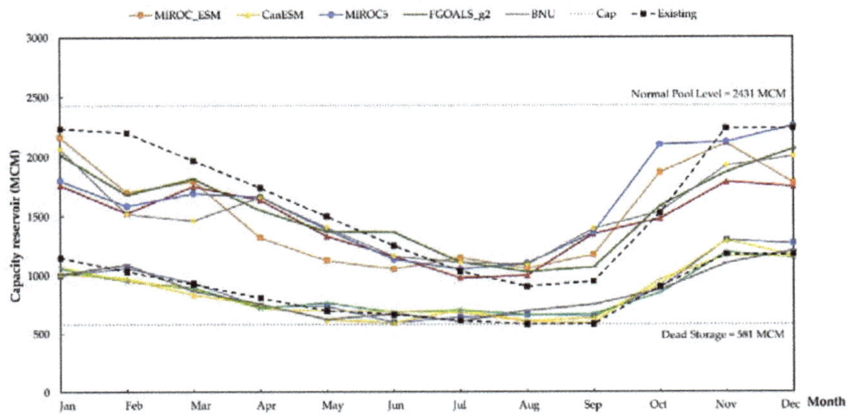

Figure 10. Rule curves of Ubolratana reservoir developed using HBMO algorithm technique based on climate change impacts under the RCP8.5 projection case.

3.2.2. Reservoir Rule Curves Efficiency Evaluation

The purpose of evaluating the efficiency of reservoir rule curves is to test the functions of the rule curves in order to know the results that could support the changing water situations due to various uncertainties, whether in past periods or for scenarios that may occur in the future. The assessment of rule curves had two parts, namely, water shortage and excess release water by assessing the frequency of occurrence of an incident through mean and maximum values of Magnitude and Duration.

We evaluated the efficiency of the current reservoir rule curves and the reservoir rule curves obtained from future streamflow during 2020–2049, which yielded five CIMP5 models of climate change under the RCP4.5 scenario. In all models except the MIROC5 model, the reservoir rule curves were able to lower the mean water deficit and mean overflow when compared to the present rule curves. Under the RCP4.5 scenario, the reservoir rule curves from the MIROC_ESM model were the most efficient ones in reducing mean water deficit and mean overflow when compared to the reservoir rule curves in other models (Table 6). Under the RCP8.5 scenario, the results showed that the reservoir rule curves in all models were able to reduce the average water shortage compared to the current rule curves. Moreover, the reservoir rule curves from the MIROC5 model could also help reduce the over-average water flow. The efficiency evaluation indicated that the reservoir rule curves from the MIROC5 model were able to reduce the average water shortage and average overflow the best when compared to the reservoir rule curves of all models (Table 7).

Table 6. Estimated results of water shortage and overflow events of the Ubolratana reservoir rule curves from the MIROC_ESM model under the RCP4.5 projection case.

Situations	Rule Curves	Frequency (Times/Year)	Magnitude (MCM/Year)		Duration (Year)	
			Average	Maximum	Average	Maximum
Water shortage	Existing	0.2	23.43	478.00	1.7	2.0
	MIROC_ESM	0.1	10.93	215.00	1.5	2.0
	BNU	0.1	14.87	264.00	2.0	2.0
	CanESM	0.1	14.17	295.00	1.5	2.0
	MIROC5	0.1	21.90	351.00	1.3	2.0
	FGOALS_g2	0.1	13.97	268.00	2.0	2.0
Excess water release	Existing	1.0	3235.04	8570.84	14.5	19.0
	MIROC_ESM	1.0	3181.27	8213.26	14.5	26.0
	BNU	1.0	3187.92	8124.91	14.5	26.0
	CanESM	1.0	3204.33	8284.15	14.5	19.0
	MIROC5	1.0	3216.58	8551.56	30.0	30.0
	FGOALS_g2	1.0	3207.96	8585.07	14.5	26.0

Table 7. Estimated water shortage and overflow events of the Ubolratana reservoir rule curves from the MIROC5 model under the RCP8.5 projection case.

Situations	Rule Curves	Frequency (Times/Year)	Magnitude (MCM/Year)		Duration (Year)	
			Average	Maximum	Average	Maximum
Water shortage	Existing	0.23	36.67	449.00	1.40	2.00
	MIROC_ESM	0.17	13.90	233.00	1.67	2.00
	BNU	0.07	7.77	195.00	2.00	2.00
	CanESM	0.13	12.77	259.00	2.00	2.00
	MIROC5	0.10	7.13	169.00	1.50	2.00
	FGOALS_g2	0.17	16.00	250.00	1.67	2.00

Table 7. *Cont.*

Situations	Rule Curves	Frequency (Times/Year)	Magnitude (MCM/Year)		Duration (Year)	
			Average	Maximum	Average	Maximum
Excess water release	Existing	0.97	2460.08	6281.34	14.5	21
	MIROC_ESM	0.93	2460.26	5983.39	14	20
	BNU	0.87	2441.62	6165.43	8.667	15
	CanESM	0.93	2466.88	6055.38	9.333	15
	MIROC5	0.87	2424.31	6436.28	8.667	15
	FGOALS_g2	0.93	2452.14	6098.75	14	20

4. Conclusions

There were two primary objectives of this research. The first was to investigate how global climate change has affected the quantity of streamflow that flows into the Ubolratana Reservoir in the years 2020–2592. Second, these modifications will be utilized as data for improving the suitable reservoir rule curves using the HBMO algorithm approach, as well as evaluating the effectiveness of the newly designed reservoir rule curves.

The results of this study showed that future streamflow data are based on the SWAT model. The forecast years 2020–2049 were projected to be influenced by climate change from the CIMP5 model, according to the findings of this study. Both RCP4.5 and RCP8.5 were expected to rise under the anticipated conditions. Under RCP4.5 and RCP8.5, the future overall average annual streamflow will rise by 32% and 65%, respectively. The MIROC_ESM model had the highest average annual streamflow compared to other models. However, there is a different study (FGOALS_g2 model, under the RCP4.5 forecast case), which indicates that the future annual mean streamflow tends to decline. When we considered the average monthly streamflow volume in the future according to seasons, it was found that the trend of change in streamflow volume was consistent with both under the forecasting cases. The average monthly streamflow volume was expected to increase markedly during the wet season (August to November) and at the beginning of the dry season (December).

Then, the Ubolratana Reservoir rule curves developed by HBMO Algorithm was created. There were five CIMP5 climate models under the RCP4.5 and 8.5 forecast cases, for which the developed rule curves were shaped in accordance with the current rule curves. Moreover, the developed rule curves could also allow the reservoir to hold more water during the rainy season. This should ensure that there will be enough water in the next dry season. In addition, during the dry season, reservoirs will be able to release more water, thereby reducing water scarcity. Finally, the future rule curves in the reservoir as a result of the climate change examined in this study would be able to answer the objective functions, which is to acquire the least average water scarcity amount. The rule curves will also be rated for their efficiency in reducing water scarcity and overflow compared to the current rule curves.

Author Contributions: Conceptualization, S.S. and A.K.; methodology, S.S. and A.K.; validation, S.S. and A.K.; formal analysis, S.S. and A.K.; investigation, S.S. and A.K.; writing—original draft preparation, S.S. and A.K.; writing—review and editing, S.S. and A.K.; supervision, S.S. and A.K. All authors have read and agreed to the published version of the manuscript.

Funding: This research project was financially supported by Mahasarakham University.

Institutional Review Board Statement: Not applicable.

Informed Consent Statement: Not applicable.

Data Availability Statement: This study did not report any data.

Acknowledgments: The authors would like to acknowledge the Hydro–Informatics Institute, the Land Development Department, the Thai Meteorological Department, the Royal Irrigation Department and the Electricity Generating Authority, Thailand for supporting data in this study. The authors would like to thank the editor and the anonymous reviewers for their comments that helped in improving the quality of the paper.

Conflicts of Interest: The authors declare no conflict of interest.

References

1. Ehsani, N.; Vörösmarty, C.J.; Fekete, B.M.; Stakhiv, E.Z. Reservoir operations under climate change: Storage capacity options to mitigate risk. *J. Hydrol.* **2017**, *555*, 435–446. [CrossRef]
2. Ehteram, M.; Mousavi, S.F.; Karami, H.; Farzin, S.; Singh, V.P.; Chau, K.W.; El-Shafie, A. Reservoir operation based on evolutionary algorithms and multi-criteria decision-making under climate change and uncertainty. *J. Hydroinform.* **2018**, *20*, 332–355. [CrossRef]
3. Gorguner, M.; Kavvas, M.L. Modeling impacts of future climate change on reservoir storages and irrigation water demands in a Mediterranean basin. *Sci. Total Environ.* **2020**, *748*, 141246. [CrossRef] [PubMed]
4. Abera, F.F.; Asfaw, D.H.; Engida, A.N.; Melesse, A.M. Optimal operation of hydropower reservoirs under climate change: The case of Tekeze reservoir, Eastern Nile. *Water* **2018**, *10*, 273. [CrossRef]
5. Carvalho-Santos, C.; Monteiro, A.T.; Azevedo, J.C.; Honrado, J.P.; Nunes, J.P. Climate change impacts on water resources and reservoir management: Uncertainty and adaptation for a mountain catchment in northeast Portugal. *Water Resour Manag.* **2017**, *31*, 3355–3370. [CrossRef]
6. Moazzam, M.F.; Lee, B.G.; Rahman, G.; Waqas, T. Spatial Rainfall Variability and an Increasing Threat of Drought, According to Climate Change in Uttaradit Province, Thailand. *Atmos. Clim. Sci.* **2020**, *10*, 357. [CrossRef]
7. Tebakari, T.; Dotani, K.; Kato, T. Historical change in the flow duration curve for the upper nan River Watershed, Northern Thailand. *J. Jpn. Soc. Hydrol. Water Resour.* **2018**, *31*, 17–24. [CrossRef]
8. Sharma, D.; Babel, M.S. Assessing hydrological impacts of climate change using bias-corrected downscaled precipitation in Mae Klong basin of Thailand. *Meteorol. Appl.* **2018**, *25*, 384–393. [CrossRef]
9. Petpongpan, C.; Ekkawatpanit, C.; Kositgittiwong, D. Climate change impact on surface water and groundwater recharge in Northern Thailand. *Water* **2020**, *12*, 1029. [CrossRef]
10. Kamworapan, S.; Surussavadee, C. Evaluation of CMIP5 global climate models for simulating climatological temperature and precipitation for Southeast Asia. *Adv. Meteorol.* **2019**, *2019*, 1067365. [CrossRef]
11. Bhatta, B.; Shrestha, S.; Shrestha, P.K.; Talchabhadel, R. Evaluation and application of a SWAT model to assess the climate change impact on the hydrology of the Himalayan River Basin. *Catena* **2019**, *181*, 104082. [CrossRef]
12. Azari, M.; Oliaye, A.; Nearing, M.A. Expected climate change impacts on rainfall erosivity over Iran based on CMIP5 climate models. *J. Hydrol.* **2021**, *593*, 125826. [CrossRef]
13. Saharia, A.M.; Sarma, A.K. Future climate change impact evaluation on hydrologic processes in the Bharalu and Basistha basins using SWAT model. *Nat. Hazards* **2018**, *92*, 1463–1488. [CrossRef]
14. Chuenchooklin, S.; Pangnakorn, U. Hydrological Study Using SWAT and Global Weather, a Case Study in the Huai Khun Kaeo Watershed in Thailand. *Int. J. Environ. Prot. Pol.* **2018**, *6*, 36. [CrossRef]
15. Prasanchum, H.; Sirisook, P.; Lohpaisankrit, W. Flood risk areas simulation using SWAT and Gumbel distribution method in Yang Catchment, Northeast Thailand. *Geogr. Tech.* **2020**, *15*, 29–39. [CrossRef]
16. Ekkawatpanit, C.; Pratoomchai, W.; Khemngoen, C.; Srivihok, P. Climate change impact on water resources in Klong Yai River Basin, Thailand. *Proc. Int. Assoc. Hydrol. Sci.* **2020**, *383*, 355–365. [CrossRef]
17. Thongwan, T.; Kangrang, A.; Techarungreungsakul, R.; Ngamsert, R. Future inflow under land use and climate changes and participation process into the medium-sized reservoirs in Thailand. *Adv. Civ. Eng.* **2020**, *2020*, 5812530. [CrossRef]
18. Prasanchum, H.; Kangrang, A. Optimal reservoir rule curves under climatic and land use changes for Lampao Dam using Genetic Algorithm. *KSCE J. Civ. Eng.* **2018**, *22*, 351–364. [CrossRef]
19. Kumar, N.; Singh, S.K.; Srivastava, P.K.; Narsimlu, B. SWAT Model calibration and uncertainty analysis for streamflow prediction of the Tons River Basin, India, using Sequential Uncertainty Fitting (SUFI-2) algorithm. *Model. Earth Syst. Environ.* **2017**, *3*, 30. [CrossRef]
20. Abeysingha, N.S.; Islam, A.; Singh, M. Assessment of climate change impact on flow regimes over the Gomti River basin under IPCC AR5 climate change scenarios. *J. Water Clim. Chang.* **2020**, *11*, 303–326. [CrossRef]
21. Tayebiyan, A.; Mohammad, T.A.; Al-Ansari, N.; Malakootian, M. Comparison of optimal hedging policies for hydropower reservoir system operation. *Water* **2019**, *11*, 121. [CrossRef]
22. Akbarifard, S.; Sharifi, M.R.; Qaderi, K. Data on optimization of the Karun-4 hydropower reservoir operation using evolutionary algorithms. *Data Br.* **2020**, *29*, 105048. [CrossRef] [PubMed]
23. Kangrang, A.; Chaleeraktrakoon, C. Suitable Conditions of Reservoir Simulation for Searching Rule Curves. *J. Appl. Sci.* **2008**, *8*, 1274–1279. [CrossRef]
24. Kangrang, A.; Lokham, C. Optimal Reservoir Rule Curves Considering Conditional Ant Colony Optimization with. *J. Appl. Sci.* **2013**, *13*, 154–160. [CrossRef]

25. Kangrang, A.; Srikamol, N.; Hormwichian, R.; Prasanchum, H.; Sriwanphen, O. Alternative Approach of Firefly Algorithm for Flood Control Rule Curves. *Asian J. Sci. Res.* **2019**, *12*, 431–439. [CrossRef]
26. Sinthuchai, N.; Kangrang, A. Improvement of Reservoir Rule Curves using Grey Wolf Optimizer. *J. Eng. Appl. Sci.* **2019**, *14*, 9847–9856. [CrossRef]
27. Marchand, A.; Gendreau, M.; Blais, M.; Guidi, J. Optimized operating rules for short-term hydropower planning in a stochastic environment. *Comput. Manag. Sci.* **2019**, *16*, 501–519. [CrossRef]
28. Thongwan, T.; Kangrang, A.; Prasanchum, H. Multi-objective future rule curves using conditional tabu search algorithm and conditional genetic algorithm for reservoir operation. *Heliyon* **2019**, *5*, e02401. [CrossRef]
29. Haddad, O.B.; Afshar, A.; Mariño, M.A. Honey-bee mating optimization (HBMO) algorithm in deriving optimal operation rules for reservoirs. *J. Hydroinform.* **2008**, *10*, 257–264. [CrossRef]
30. Kangrang, A.; Prasanchum, H.; Hormwichian, R. Active future rule curves for multi-purpose reservoir operation on the impact of climate and land use changes. *J. Hydro-Environ. Res.* **2019**, *24*, 1–13. [CrossRef]
31. Ferguson, C.R.; Pan, M.; Oki, T. The effect of global warming on future water availability: CMIP5 synthesis. *Water Resour. Res.* **2018**, *54*, 7791–7819. [CrossRef]
32. Climate Change in Australia. List of Global Climate Models. Available online: https://www.climatechangeinaustralia.gov.au/en/overview/methodology/list-models/ (accessed on 20 February 2022).
33. Zhou, T.; Yu, Y.; Liu, Y.; Wang, B. (Eds.) *Flexible Global Ocean-Atmosphere-Land System Model: A Modeling Tool for the Climate Change Research Community*; Springer: Berlin/Heidelberg, Germany, 2014.
34. Chaowiwat, W. Impact of climate change assessment on agriculture water demand in Thailand. *Naresuan Univ. Eng. J.* **2016**, *11*, 35–42.
35. Sharma, D. Selection of suitable general circulation model precipitation and application of bias correction methods: A case study from the Western Thailand. In *Environmental Management of River Basin Ecosystems*; Springer: Cham, Switzerland, 2015; pp. 43–63.
36. Arnold, A.G.; Srinivasan, R.; Muttiah, R.S.; Williams, J.R. Large area hydrological modeling and assessment pert I: Model development. *J. Am. Water Resour. Assoc.* **1998**, *34*, 73–89. [CrossRef]
37. Prasanchum, H.; Kangrang, A. Analyses of climate and land use changes impact on runoff characteristics for multi-purpose reservoir system. In Proceedings of the Conference on The AUN/SEED-Net Regional Conference 2016 on Environmental Engineering (RC-EnvE 2016), Chonburi, Thailand, 23–24 January 2017.
38. Khalid, K.; Ali, M.F.; Abd Rahman, N.F.; Mispan, M.R.; Haron, S.H.; Othman, Z.; Bachok, M.F. Sensitivity analysis in watershed model using SUFI-2 algorithm. *Procedia. Eng.* **2016**, *162*, 441–447. [CrossRef]
39. Shivhare, N.; Dikshit, P.K.; Dwivedi, S.B. A comparison of SWAT model calibration techniques for hydrological modeling in the Ganga river watershed. *Engineering* **2018**, *4*, 643–652. [CrossRef]
40. Moriasi, D.N.; Gitau, M.W.; Pai, N.; Daggupati, P. Hydrologic and water quality models: Performance measures and evaluation criteria. *Trans. ASABE* **2015**, *58*, 1763–1785.
41. Zhang, S.; Li, Z.; Lin, X.; Zhang, C. Assessment of climate change and associated vegetation cover change on watershed-scale runoff and sediment yield. *Water* **2019**, *11*, 1373. [CrossRef]
42. Haddad, O.B.; Afshar, A.; Mariño, M.A. Honey-bees mating optimization (HBMO) algorithm: A new heuristic approach for water resources optimization. *Water Resour. Manag.* **2006**, *20*, 661–680. [CrossRef]
43. Rodriguez, L.B.; Cello, P.A.; Vionnet, C.A.; Goodrich, D. Fully conservative coupling of HEC-RAS with MODFLOW to simulate stream–aquifer interactions in a drainage basin. *J. Hydrol.* **2008**, *353*, 129–142. [CrossRef]
44. Techarungruengsakul, R.; Kangrang, A. Application of Harris Hawks Optimization with Reservoir Simulation Model Considering Hedging Rule for Network Reservoir System. *Sustainability* **2022**, *14*, 4913. [CrossRef]

agriculture

MDPI

Article

Development of an Energy Efficient and Fully Autonomous Low-Cost IoT System for Irrigation Scheduling in Water-Scarce Areas Using Different Water Sources

Zisis Tsiropoulos [1,2,*], Evangelos Skoubris [2,3], Spyros Fountas [4], Ioannis Gravalos [5] and Theofanis Gemtos [1]

1 Department of Agriculture Crop Production and Rural Environment, University of Thessaly, 38446 Volos, Greece; gemtos@uth.gr
2 Agricultural and Environmental Solutions (AGENSO), Markou Mpotsari 47, 11742 Athens, Greece; eskoubris@uniwa.gr
3 Department of Surveying and Geoinformatics Engineering, School of Engineering, Agiou Spyridonos, University of West Attica, 12243 Egaleo, Greece
4 Department of Natural Resources Management and Agricultural Engineering, Agricultural University of Athens, 11855 Athens, Greece; sfountas@aua.gr
5 Department of Agrotechnology, School of Agricultural Sciences, University of Thessaly, Periferiaki odos Larissas—Trikalon, 41500 Larissa, Greece; iogravalos@uth.gr
* Correspondence: tsiropoulos@agenso.gr

Abstract: Politicians and the general public are concerned about climate change, water scarcity, and the constant reduction in agricultural land. Water reserves are scarce in many regions in the world, negatively affecting agricultural productivity, which makes it a necessity to introduce sustainable water resource management. Nowadays, there is a number of commercial IoT systems for irrigation scheduling, helping farmers to manage and save water. However, these systems focus on using the available fresh water sources, without being able to manage alternative water sources. In this study, an Arduino-based low-cost IoT system for automated irrigation scheduling is developed and implemented, which can provide measurements of water parameters with high precision using low-cost sensors. The system used weather station data combined with the FAO56 model for computing the water requirements for various crops, and it was capable of handling and monitoring different water streams by supervising their quality and quantity. The developed IoT system was tested in several field trials, to evaluate its capabilities and functionalities, including the sensors' accuracy, its autonomous controlling and operation, and its power consumption. The results of this study show that the system worked efficiently on the management and monitoring of different types of water sources (rainwater, groundwater, seawater, and wastewater) and on automating the irrigation scheduling. In addition, it was proved that the system is can be used for long periods of time without any power source, making it ideal for using it on annual crops.

Keywords: irrigation scheduling; alternative water sources; low cost; IoT system

Citation: Tsiropoulos, Z.; Skoubris, E.; Fountas, S.; Gravalos, I.; Gemtos, T. Development of an Energy Efficient and Fully Autonomous Low-Cost IoT System for Irrigation Scheduling in Water-Scarce Areas Using Different Water Sources. *Agriculture* **2022**, *12*, 1044. https://doi.org/10.3390/agriculture12071044

Academic Editors: Alban Kuriqi and Luis Garrote

Received: 27 June 2022
Accepted: 13 July 2022
Published: 18 July 2022

1. Introduction

Globally, the effects of climate change are evident and affect the daily lives of people and the planet. One of its effects is water shortages in a large number of regions across the world. Water scarcity often results in reduced agricultural productivity due to shortages and/or poor water quality. Taking into consideration that agriculture consumes 70 percent of the available freshwater [1] with low efficiency [2], the need to find sustainable water resource management solutions becomes imperative.

Most of the existing research reports in the field of irrigation scheduling focus on the development of low-cost IoT-based solutions [3–5], the use of machine learning and fuzzy logic [6–8], and the use of different irrigation methods and models [9–11]. A comprehensive analysis regarding the research on smart irrigation systems was reported by García et al. [12], where a detailed overview on the recent trends on sensors and IoT systems for irrigation was presented. At the commercial level, there are many IoT systems that have been developed for multiple agricultural purposes, including irrigation (e.g., Libelium [13] and iMetos [14]). However, some of them only focus on weather and soil monitoring without taking into account crop water requirements (e.g., WatchDog [15] and Netsens [16]).

Recently, significant research has been reported on the development of IoT systems for water monitoring both in terms of quality and quantity [17–20]. A large number of these systems focuses on monitoring natural water sources, such as lakes and rivers [21–23]. Following the research performed in water monitoring, a lot of systems are commercially available with the most well-known being Libelium [13], as its price is relatively low compared to other solutions, but none can control different water sources for irrigation scheduling.

Following the Industry 4.0 revolution, a large variety of low-cost processors, controllers, electronic components, and sensors have become available, which can be used for developing low-cost IoT solutions. The most common example is the Arduino open-source microcontroller-based development board [24]. These boards provide, at a very low cost, all the characteristics needed for developing a monitoring/actuating device, namely, an embedded microprocessor, connections for power supply, analogue and digital I/O channels for interfacing with peripheral devices (e.g., sensors), dedicated channels (e.g., USB communication port), and a vast variety of different modules for various purposes (e.g., GSM modules). In addition, the extensive use of Arduino boards by a large community has allowed the establishment of a broad range of supported features, making these boards mature enough, and with great reliability and flexibility, which is necessary for precision agriculture applications [25]. For this reason, significant research has been reported during the last years on developing Arduino-based solutions for agriculture [26–28] and water monitoring [29,30]. Following this trend, the extensive use of Arduino boards has contributed to the development and further availability of a variety of low-cost sensors in the market, whose efficiency in agriculture has been investigated with positive results [31–33].

As agriculture may be conducted in an open environment, wireless data transmission is required. Many different types of wireless data communication protocols are used in agriculture [34], including broadband cellular network technology protocols (GPRS, 4G, and 5G), LPWA—Low Power Wide Area Network protocols (LoRaWAN, SigFOx, NB-IoT, and LTE-M), WLAN—wireless LAN protocols (Wi-Fi), and IEEE 802.15 Protocols (ZigBee and Bluetooth). Each one of them has its advantages and disadvantages in terms of power consumption, range coverage, and data collection rate.

In this context, the HYDROUSA H2020 project [35] objectives were the sustainable management of water and the increase in agricultural production in water-scarce areas by applying precision irrigation using water that comes from a variety of water sources (rainwater, groundwater, seawater, and wastewater). Therefore, the main aim of this study is to develop a reliable and accurate low-cost IoT system to monitor and control irrigation scheduling, which is able to operate using different water sources. To achieve this, the system was: (i) developed using open source hardware for minimizing its cost, (ii) capable of supporting a variety of sensors and actuators, (iii) evaluated for its accuracy, and (iv) validated for its functionality and capabilities on using different water sources for automating irrigation scheduling.

The innovation of the present study is the design and development of an Arduino-based low-cost IoT node with extensive energy autonomy, capable of autonomously handling the various water sources and applying precision irrigation based on weather data and plant requirements. This study can contribute to increasing irrigation sustainability, especially in water-scarce areas, as water coming from alternative water sources can be used for irrigation, minimizing the use of the conventional irrigation water sources.

2. Materials and Methods

2.1. Design and Development of the IoT Node

The IoT node was designed and developed using Arduino architecture (Figure 1), as it has a very low price for all the components needed for developing a low cost IoT system. A typical wireless node consists of a microcontroller that is also capable of performing data processing; the transceiver, which is responsible for the wireless communication; the power source; and finally the various circuits needed (e.g., AD converters) for supporting the reading of analog and digital signals of the sensors and the actuators. To implement the node, a board was developed by splitting it into 4 distinct layers:

- The power management layer, which was designed using methodologies for minimizing power consumption;
- The interfacing layer, responsible for the connectivity of peripherals (sensors and actuators) with the system;
- The processing/controlling layer, responsible for the initial data processing;
- The connectivity layer, responsible for the data transmission to the cloud.

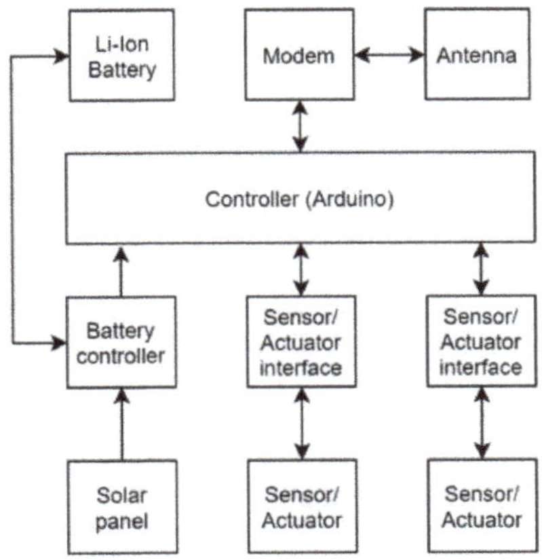

Figure 1. IoT node architecture.

The node (Figure 2) has a small size of 12 × 8 cm and IP67 protection so that it can be used in a large number of applications in harsh environments. Moreover, it supports both analog and digital sensors and various communication protocols (e.g., RS-485 serial communication protocol) for supporting most of the available sensors/actuators (even industrial ones).

Figure 2. IoT node implementation.

Data can be uploaded using general (GPRS and 4G) or low-power (NB-IoT and LTE-M) cellular network communication protocols. The communication between the node and the cloud was bidirectional in order to enable remote control and configuration of the system (e.g., open/close valve), and it achieved almost real-time measurements with a minimum sampling rate of 30 s.

To avoid the configuration process on behalf of the user, making the system "plug and play" and able to work with the simple click of the start button, it was decided that the developed node should also act as a getaway node, with direct communication to the cloud. Using this methodology, the nodes were preconfigured, while the cloud services were developed in such way to make them capable to automatically understand the type of the sensors connected to each node.

2.2. Reduction in Power Consumption

The power consumption of an IoT system is a quite critical parameter, as there are cases in which sensors have to be placed into dense and high crops (e.g., maize), where the charging of batteries is a difficult task. To minimize the power consumption of the node, 3 different prototypes were developed. The first one was developed using a commercial Arduino board, while the second was developed by designing a custom board for reducing power consumption. The third one (Figure 2) was an update of the second prototype, which was developed by enhancing the board design for minimizing the power consumption even more.

2.3. Sensors Supported

As there are low-cost sensors that are able to provide measurements of high accuracy with a careful calibration [36] or by using deep-learning-based sensor modelling [37], more than 80 different sensors were tested and evaluated to select the ones with the highest accuracy and durability. In the case of the ones that passed these functional tests, in some cases (pH, temperature, and turbidity sensors), modifications were made to increase their accuracy and make them waterproof. Waterproofing was achieved by potting the sensitive electrical/electronic parts, wiring, and connections of the aforementioned sensors using epoxy resin. Moreover, as the majority of the low-cost sensors were OEM-branded operating using circuits developed from multiple manufacturers (e.g., TDS, pH, and Ultrasonic level sensor), new circuits were developed and embedded into the IoT node for ensuring the proper functionality of the low-cost sensors as well as their measurements' accuracy. The sensors that have been supported to date by the IoT node are:

- Weather measurements: Temperature, humidity, atmospheric pressure, precipitation, wind speed, wind gust, wind direction, solar radiation, and UV index;
- Soil measurements: Moisture content, temperature, pH, and electrical conductivity;
- Water measurements: Temperature, pH, electrical conductivity, turbidity, TDS, water flow, and storage tank level.

2.4. Actuation

To enable remote control and automation, the communication between the node and the cloud was bidirectional, and the actuation could be achieved by remote control through a website in which the user can:

- Open or close an actuator;
- Enter the thresholds of an actuator to change its state (e.g., specific temperature and water level);
- Enable autonomous operation (e.g., applying precision irrigation).

2.5. Field Trials

The ability of the system to efficiently manage different water sources for automated irrigation scheduling was evaluated at 3 different pilot sites developed for the needs of the HYDROUSA project. More specifically, the evaluation procedure was held at Ano Mera, Mykonos, Greece (37°26′51.4″ N 25°24′15.7″ E), at Agios Fokas, Tinos, Greece (37°31′59.1″ N 25°10′44.0″ E), and at an eco-tourist facility in Tinos Greece (37°33′56.7″ N

25°12′55.5″ E). Field trial tests included the evaluation of: (i) low-cost sensors' accuracy, (ii) system's monitoring and water management capabilities, (iii) automated irrigation scheduling efficiency, and (iv) the system's energy autonomy.

3. Results

3.1. Evaluation of the Low-Cost Sensors' Accurancy

The water quality sensors were tested and evaluated at the pilot site of Agios Fokas, Tinos, Greece, by comparing their measurements with industrial type sensors that were installed in parallel in closed tanks used for water storing. Both low-cost (Temperature: DS18B20, AGENSO, Athens, Greece; pH: H-101, HAO SHI, Taiwan) and industrial sensors (Temperature and pH: Sensolyt 700 IQ, YSI, Yellow Springs, OH, USA) were calibrated before their installation. The measurement rate was 1 h for the low-cost sensors and 15 min for the industrial sensors. To compare their results, the average daily values of each sensor were calculated. For pH measurements, the maximum difference recorded between the low-cost and the industrial sensor was 0.22 with a mean difference at 0.08 and R^2 = 0.8392 (Figures 3 and 4), with the low-cost sensor having an accuracy of ±0.1 at 25 °C and the industrial one ±0.05 (from 0 °C to 60 °C). For water temperature measurements, the maximum difference recorded was 1.78 °C with a mean difference of 0.75 °C and R^2 = 0.9914 (Figures 5 and 6), with both sensors having an accuracy of ±0.5 °C (low cost: from −10 °C to +85 °C; industrial: from 0 °C to 60 °C). The average values per day and their differences are shown in Table 1.

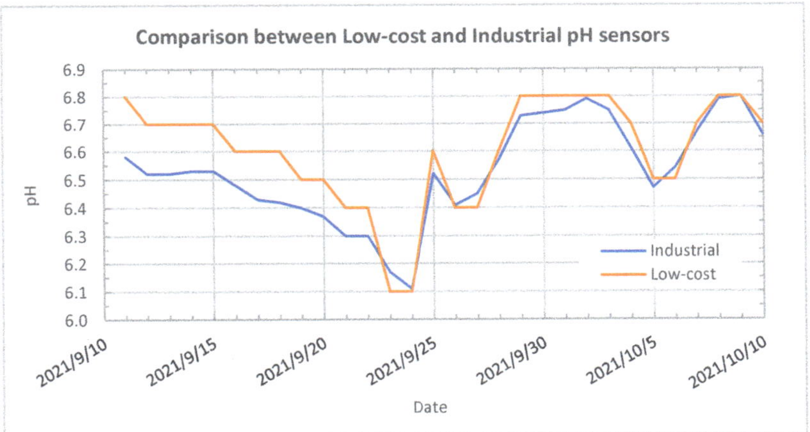

Figure 3. Comparison of industrial and low-cost pH sensors' measurements.

In order to perform a robust comparison and further determination of the statistically significant differences between the obtained measurements, an analysis of variance was performed by conducting a one-way ANOVA using a Fisher's least significance difference (LSD) test at a 95% confidence level ($p < 0.05$). As pH industrial sensors have a temperature compensation function to correct the measured pH value according to the water temperature, the accuracy of the low-cost pH sensor had to be improved. For this reason, a firmware update of the IoT node was developed and will be tested in summer 2022, to regulate the pH results according to the water temperature, for increasing the accuracy of the low-cost pH sensor. The low-cost temperature sensor that was used showed a very high level of accuracy (R^2 = 0.9914), which was proved also during the initial lab tests that were conducted.

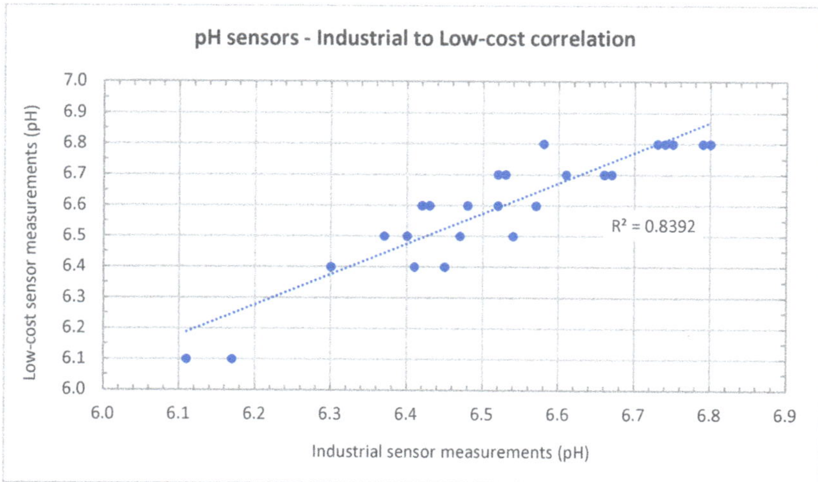

Figure 4. Correlation of industrial and low-cost pH sensors' measurements.

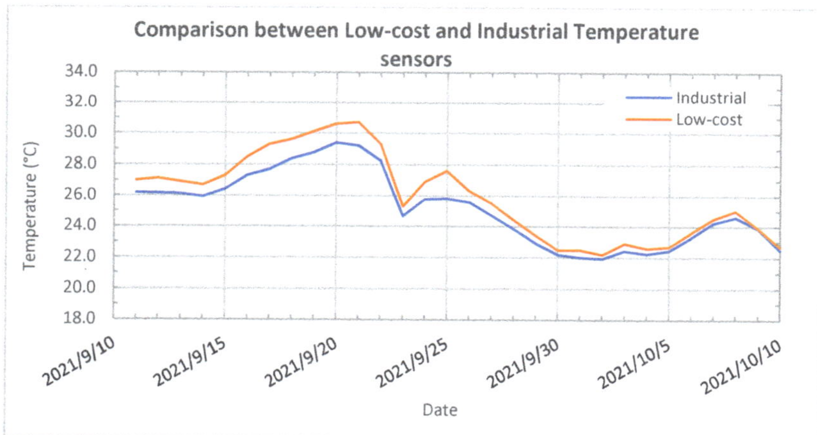

Figure 5. Comparison of industrial and low-cost temperature sensors' measurements.

As no low-cost sensors for measuring electrical conductivity on water exist, the measurements of a low-cost TDS sensor (TDS-1000, AGENSO, Athens, Greece) were evaluated in comparison with the measurements of an industrial sensor for measuring electrical conductivity. Thus, the TDS and electrical conductivity (EC) on water are correlated. As shown in Figure 7, the results show a high correlation between the TDS and EC measurements at an EC up to 3 mS/cm. After that point, the correlation was lower as the TDS sensor reached its maximum range. The general rule for the salinity hazard of irrigation water based upon conductivity is that EC over 3 mS/cm creates severe damage to crops [38,39]. The low-cost TDS sensor can be used to evaluate the quality of the water and its properness for irrigation, or to select crops that are tolerant to saline water.

In addition, as weather parameters are the most important factors in decision making in agriculture, the selected low-cost that station (MeteoIoT 2100S, AGENSO, Athens, Greece) was evaluated in an experiment which run at the Municipality of Trikala, Greece, where the data of the station were compared with the data of a high-end weather station (Vantage Pro 2, Davis, Hayward, CA, USA) used by the municipality. This high-end station was connected to the network of weather stations of the National Observatory of Athens, which is the largest network of weather stations in Greece, used for weather monitoring and forecasting. Both stations were installed in open places within the municipality and their distance in a straight line was about 400 m. Figure 8 presents the comparison of the daily average temperature and the total rain recorded using the low-cost and the high-end weather stations for a 30-day period. The average temperatures recorded using the low-cost and high-end weather stations were 10.33 °C and 10.37 °C, respectively, while the total rain recorded was 142.50 mm for the low-cost weather station and 145.80 mm for the high-end weather station, proving the reliability of the measurements retrieved with the low-cost weather station (Table 2).

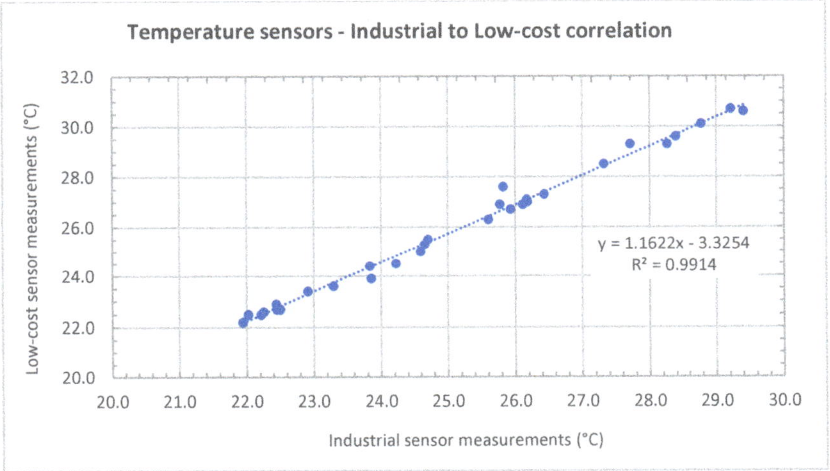

Figure 6. Correlation of industrial and low-cost temperature sensors' measurements.

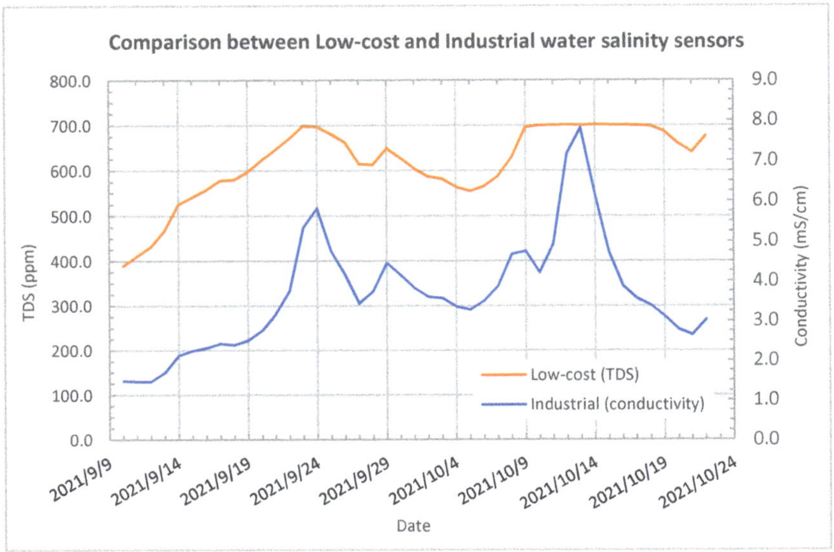

Figure 7. Comparison between the low-cost TDS and industrial EC sensors to determine water salinity for irrigation needs.

Table 1. Comparison of the measurements of the low-cost and industrial sensors. The different letters accompanying daily means and monthly average values of each distinct measurement type (pH and temperature) for each set of the industrial and low-cost sensors indicate a significant difference between the measurements, based on a Fisher's least significance difference (LSD) test ($p < 0.05$).

Date	pH Industrial		pH Low Cost		pH Difference	Temperature Industrial		Temperature Low Cost		Temperature Difference
10 September 2021	6.61	a	6.80	b	0.19	25.63	a	26.20	a	0.57
11 September 2021	6.58	a	6.80	b	0.22	26.18	a	27.00	a	0.82
12 September 2021	6.52	a	6.70	b	0.18	26.17	a	27.10	b	0.93
13 September 2021	6.52	a	6.70	b	0.18	26.11	a	26.90	a	0.79
14 September 2021	6.53	a	6.70	b	0.17	25.93	a	26.70	a	0.77
15 September 2021	6.53	a	6.70	b	0.17	26.43	a	27.30	b	0.87
16 September 2021	6.48	a	6.60	b	0.12	27.32	a	28.50	b	1.18
17 September 2021	6.43	a	6.60	b	0.17	27.71	a	29.30	b	1.59
18 September 2021	6.42	a	6.60	b	0.18	28.39	a	29.60	b	1.21
19 September 2021	6.40	a	6.50	b	0.10	28.77	a	30.10	b	1.33
20 September 2021	6.37	a	6.50	b	0.13	29.40	a	30.60	b	1.20
21 September 2021	6.30	a	6.40	b	0.10	29.21	a	30.70	b	1.49
22 September 2021	6.30	a	6.40	b	0.10	28.26	a	29.30	b	1.04
23 September 2021	6.17	a	6.10	a	0.07	24.65	a	25.30	a	0.65
24 September 2021	6.11	a	6.10	a	0.01	25.77	a	26.90	b	1.13
25 September 2021	6.52	a	6.60	a	0.08	25.82	a	27.60	b	1.78
26 September 2021	6.41	a	6.40	a	0.01	25.60	a	26.30	a	0.70
27 September 2021	6.45	a	6.40	a	0.05	24.70	a	25.50	a	0.80
28 September 2021	6.57	a	6.60	a	0.03	23.83	a	24.40	a	0.57
29 September 2021	6.73	a	6.80	a	0.07	22.91	a	23.40	a	0.49
30 September 2021	6.74	a	6.80	a	0.06	22.21	a	22.50	a	0.29
1 October 2021	6.75	a	6.80	a	0.05	22.02	a	22.50	a	0.48
2 October 2021	6.79	a	6.80	a	0.01	21.94	a	22.20	a	0.26
3 October 2021	6.75	a	6.80	a	0.05	22.44	a	22.90	a	0.46
4 October 2021	6.61	a	6.70	b	0.09	22.25	a	22.60	a	0.35
5 October 2021	6.47	a	6.50	a	0.03	22.45	a	22.70	a	0.25
6 October 2021	6.54	a	6.50	a	0.04	23.29	a	23.60	a	0.31
7 October 2021	6.67	a	6.70	a	0.03	24.22	a	24.50	a	0.28
8 October 2021	6.79	a	6.80	a	0.01	24.59	a	25.00	a	0.41
9 October 2021	6.80	a	6.80	a	0.00	23.85	a	23.90	a	0.05
10 October 2021	6.66	a	6.70	a	0.04	22.50	a	22.70	a	0.20
Average	6.53	a	6.61	a	0.08	25.18	a	25.93	a	0.75

Figure 8. Comparison of the average temperatures and the total rain recorded using the low-cost and high-end weather stations.

Table 2. Comparison of measurements between the low-cost and high-end weather stations. Different letters accompanying monthly means and monthly sum values of each distinct measurement type (temperature and total rain) for each set of the industrial and low-cost sensors indicate a significant difference between the measurements, based on a Fisher's least significance difference test ($p < 0.05$).

Date	Average Temperature (Low-Cost Station)	Average Temperature (High-End Station)	Temperature Difference	Total Rain (Low-Cost Station)	Total Rain (High-End Station)	Total Rain Difference
11 December 2020	7.00	7.10	0.10	6.00	5.00	1.00
12 December 2020	8.30	8.10	0.20	0.90	1.60	0.70
13 December 2020	9.40	9.10	0.30	8.10	7.60	0.50
14 December 2020	11.50	11.20	0.30	0.00	0.00	0.00
15 December 2020	10.70	10.60	0.10	0.00	0.00	0.00
16 December 2020	8.80	8.90	0.10	0.00	0.00	0.00
17 December 2020	10.50	10.10	0.40	0.00	0.00	0.00
18 December 2020	8.30	8.90	0.60	0.00	0.00	0.00
19 December 2020	9.50	9.10	0.40	0.00	0.00	0.00
20 December 2020	10.50	10.20	0.30	0.00	0.00	0.00
21 December 2020	10.20	10.10	0.10	0.30	0.60	0.30
22 December 2020	9.90	10.00	0.10	0.00	0.00	0.00
23 December 2020	8.20	8.50	0.30	0.00	0.00	0.00
24 December 2020	8.50	9.10	0.60	0.00	0.00	0.00
25 December 2020	12.00	12.80	0.80	0.00	0.00	0.00
26 December 2020	11.90	12.90	1.00	2.10	2.40	0.30
27 December 2020	10.60	10.70	0.10	2.10	1.80	0.30
28 December 2020	10.30	10.10	0.20	1.20	1.00	0.20
29 December 2020	12.60	13.10	0.50	0.00	0.00	0.00
30 December 2020	11.10	11.50	0.40	0.00	0.00	0.00
31 December 2020	11.10	11.00	0.10	6.60	7.00	0.40
1 January 2021	9.30	9.10	0.20	0.00	0.00	0.00
2 January 2021	8.00	7.80	0.20	4.20	4.80	0.60
3 January 2021	9.90	9.60	0.30	27.60	28.00	0.40
4 January 2021	8.50	8.20	0.30	40.80	42.20	1.40
5 January 2021	8.00	8.60	0.60	0.30	0.20	0.10
6 January 2021	8.60	8.50	0.10	0.30	0.20	0.10
7 January 2021	10.00	10.00	0.00	0.00	0.00	0.00
8 January 2021	15.50	15.70	0.20	0.00	0.00	0.00
9 January 2021	15.90	15.70	0.20	0.60	0.80	0.20
10 January 2021	13.10	12.80	0.30	2.70	3.00	0.30
11 January 2021	12.90	12.70	0.20	38.70	39.60	0.90
Average	10.33 a	10.37 a	0.04	142.50 (Sum) a	145.80 (Sum) a	3.30 (Sum)

Daily measurements, meaning daily average temperature and sum of total daily rain, were obtained from the open source access www.meteo.gr [40], supported by the National Observatory of Athens, as single point measurements; thus, any further analysis of variance between the measurements was not applicable due to the lack of access to the hourly data from which the means and sums were generated. As a result, a statistical analysis was performed at monthly level, using the available data, for assessing and determining the statistically significant differences between the monthly values, by conducting a one-way ANOVA using a Fisher's least significance difference test at a 95% confidence level ($p < 0.05$). The results indicate the lack of statistical differences between the operation of the low-cost and industrial components for both average temperatures and total rain, indicating the sufficient function of both components in the long term.

3.2. Management of Stored Water

In Mykonos Island, two open top tanks were constructed for storing rainwater from a sub-surface rainwater collection system (Figure 9).

The collected water was used for the irrigation of a 0.4 ha oregano field using the rainwater stored into the open tanks. As the intention was to minimize electrical power consumption, one small pressure booster pump in combination with electrovalves controlled by the IoT nodes was placed for controlling the water flow between the two tanks and for enabling irrigation (Figure 10a), while to determine the level of stored water into the tanks, ultrasonic sensors (SR04T, AGENSO, Athens, Greece) were used (Figure 10b).

(**a**) (**b**)

Figure 9. (**a**) Open top water tanks; (**b**) Sub-surface rainwater collection system.

(**a**) (**b**)

Figure 10. (**a**) Electrovalve for controlling the water flow; (**b**) Level sensor installed in one of the tanks.

Depending on the water quantity monitored in each tank, and according to the thresholds defined by the user, the appropriate electrovalve is opened to irrigate the crop using the water stored in one of the two tanks. Figure 11 projects the sum of the water quantity stored in both tanks during the period from 22 July 2021 to 23 September 2021. The small differences that were observed during the monitoring (± 0.5 m^3) come from the effect of sunlight on the accuracy of the level measured by the sensors.

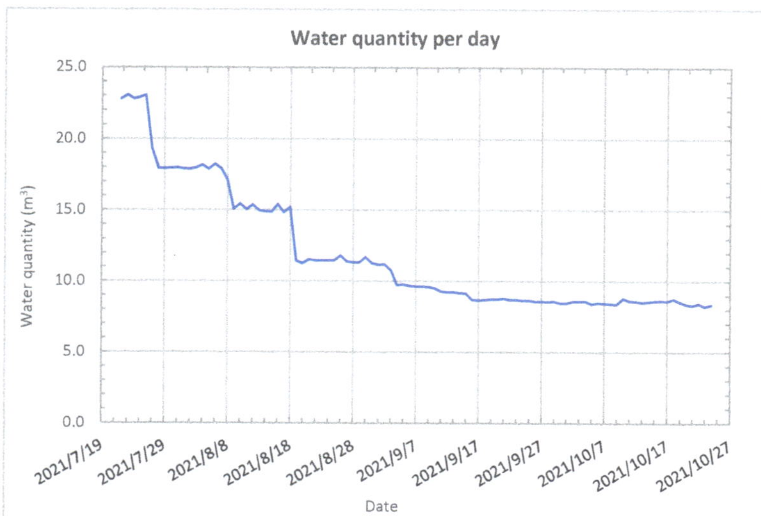

Figure 11. Water quantity monitoring.

Additionally, on a nearby house, a tank was constructed on its terrace to provide water for domestic use and to irrigate the 0.2 ha lavender field adjacent to it. To monitor its quantity, a node with an ultrasonic sensor was installed (Figure 12).

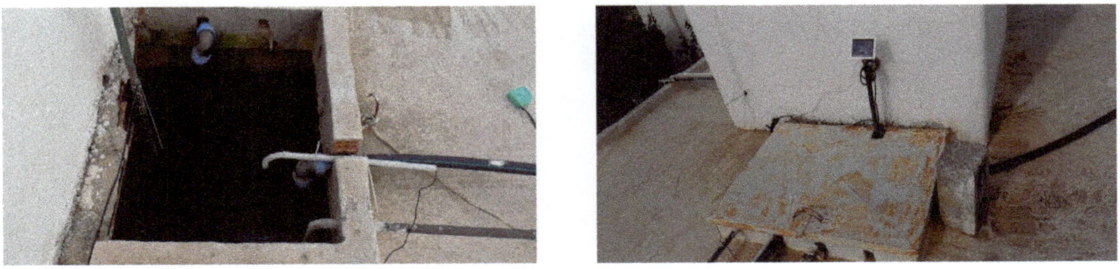

Figure 12. Measurement of the water level in a roof tank.

This tank can be refilled with rainwater collected on the rooftop or by pumping water from a nearby well. When the tank level is lower than the threshold defined by the user, the pump of the well is activated by another node. Figure 13 presents the water level of the aforementioned tank on a daily basis.

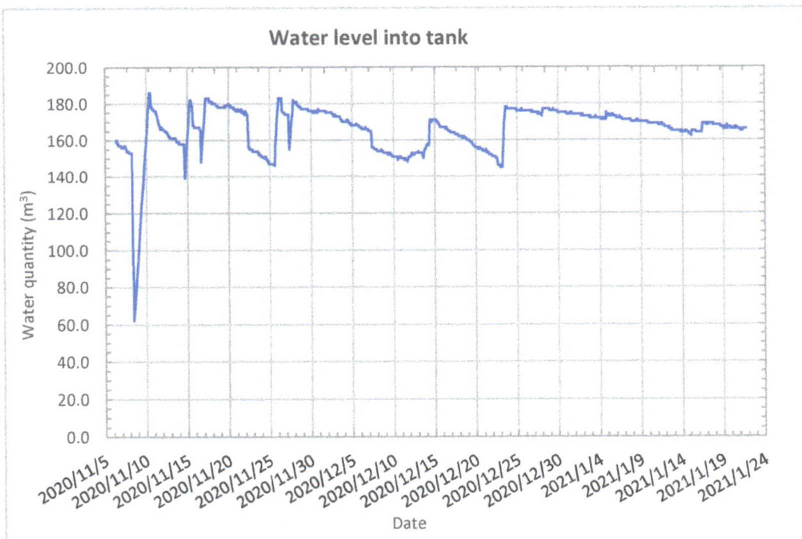

Figure 13. Monitoring of the water level in the roof tank.

In the case of rain, the excess water on the rooftop is directed after slow sand filtration to recharge water into a nearby confined aquifer, mitigating the long-encountered problem of saline water intrusion. To monitor the water in the aquifer, a well was constructed (Figure 14a) and its water depth was monitored using a submersible pressure transducer (SR05W, AGENSO, Athens, Greece) (Figure 14b). The measurements showed that the water depth reduced day by day in an almost steady rate, leading to the discovery of a fracture on the selected aquifer, which caused this water loss.

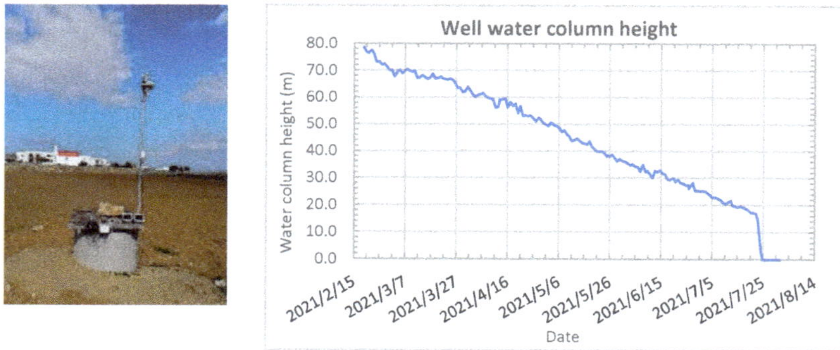

(a) (b)

Figure 14. (**a**) Nodes for monitoring the well depth. (**b**) Water depth in the well (depth of the aquifer).

3.3. *Water Quality Measurements for Decision Making According to Its Quality*

In Tinos Island, a low-cost desalination system based on the principles of evaporation and condensation was developed, as shown in the lower part of Figure 15, for irrigating the crops in the greenhouse that was constructed beside it.

Figure 15. Solar power desalination system.

Seawater was pumped into a tank used for storing sea water, then transferred into the system for desalination, stored in a second tank (Figure 16a), and finally was transferred to a third bigger tank used for the irrigation of the greenhouse crops (Figure 16b).

(a) (b)

Figure 16. Monitoring of water quality: (**a**) seawater tank; (**b**) irrigation tank.

To evaluate the performance of the desalination process, IoT nodes with water quality sensors were placed in the tanks for monitoring its quality parameters. Figure 17 presents the pH measurements in the seawater and desalinated water tanks.

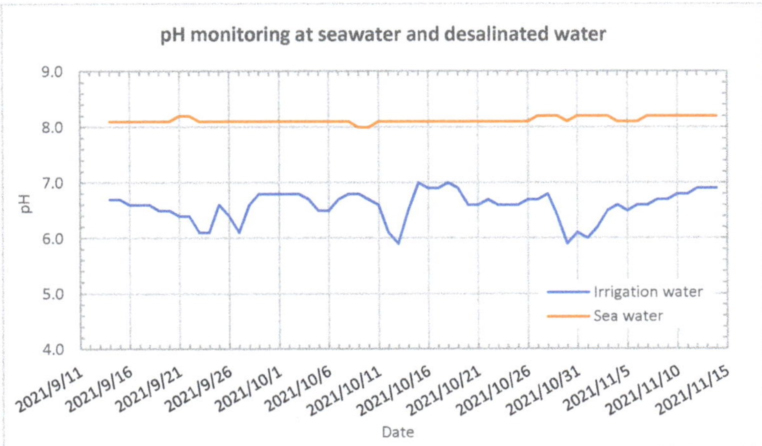

Figure 17. pH monitoring.

The pH in the seawater tank varied between 8 and 8.2, with the desalinated water having differentiations on its pH, as it was affected by the performance of the desalination system, which varies depending on the weather conditions. As the salinity of the water can affect crop performance, the total dissolved solids (TDS) of the water stored in the irrigation tank were monitored. When TDS measurements exceeded the threshold defined by the user, tap water from the municipality's water supply network was added into the irrigation tank for mixing the salty water and reducing its final salinity, as shown at Figure 18.

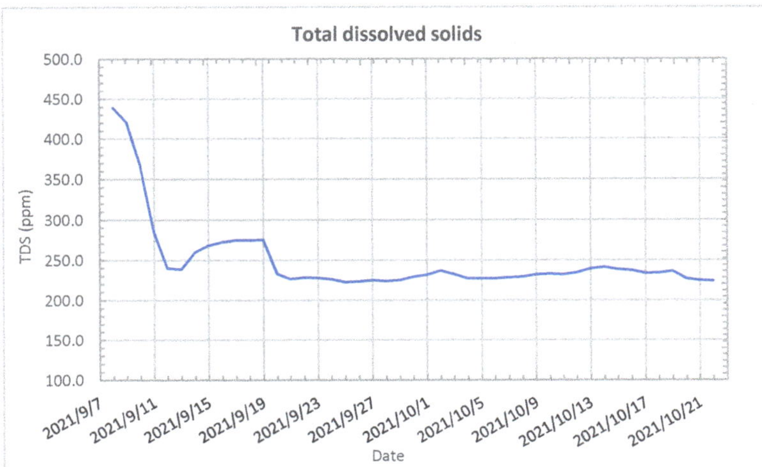

Figure 18. Salinity reduction by mixing the desalinated water with tap water.

Moreover, the proposed system was used to measure the water quality and quantity of various open and closed type tanks in an eco-tourist facility. The quality measurements were used by the system to decide whether the water can be used to irrigate edible crops. In the case that the quality of the water was not acceptable for irrigation of edible crops as a result of its high turbidity, the water was used for the irrigation of non-food crops that were cultivated according to EU 2020/741 water reuse standards. Figure 16 presents the installation of the developed system in an open cistern used for collecting rainwater (Figure 19a) and in a closed tank (Figure 19b) used for collecting the reclaimed water coming from the facility. The pH and turbidity measurements of each tank are presented in Figure 20a and 20b, respectively.

(a) (b)

Figure 19. Installations of the system for water quality and quantity measurements in: (**a**) open cistern; (**b**) closed tank.

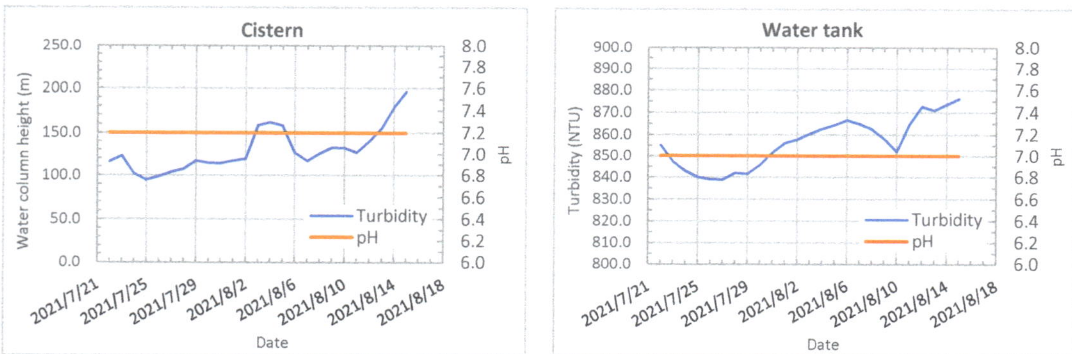

(a) (b)

Figure 20. Water quality measurements retrieved from: (**a**) open cistern; (**b**) closed tank.

3.4. Irrigation Scheduling

The accuracy of the data provided in combination with the IoT node, which can be installed in any agricultural cropping system and activate different actuators, shapes the system's ability to perform precise calculations of irrigation water needs and apply automated irrigation. To achieve this, the FAO56 Penman-Monteith model [41] for computing crop water requirements was used. All the parameters for determining evapotranspiration were retrieved from sensors connected to the IoT node for monitoring the microclimate and the soil, while electrovalves were controlled from the node for enabling automated irrigation. A greenhouse was split into four plots, in which different tropical crops, such as bananas and pineapples, were cultivated. The irrigation of each plot was achieved using a drip irrigation system, and the irrigation schedule was fully automated using the developed IoT node (Figure 21).

Figure 21. IoT nodes for automating the irrigation in the greenhouse.

Figure 22 presents the average soil moisture per day and the days in which irrigation was applied (1 = Irrigation, 0 = No irrigation) from 15 October 2021 to 12 November 2021. From the figure, it is clear that the system was capable of efficiently irrigating the crops without stressing them, keeping soil moisture between 30 and 38%. Moreover, as evapotranspiration reduces during the winter, it clearly seems that the frequency of irrigation is lower in November compared to that in October.

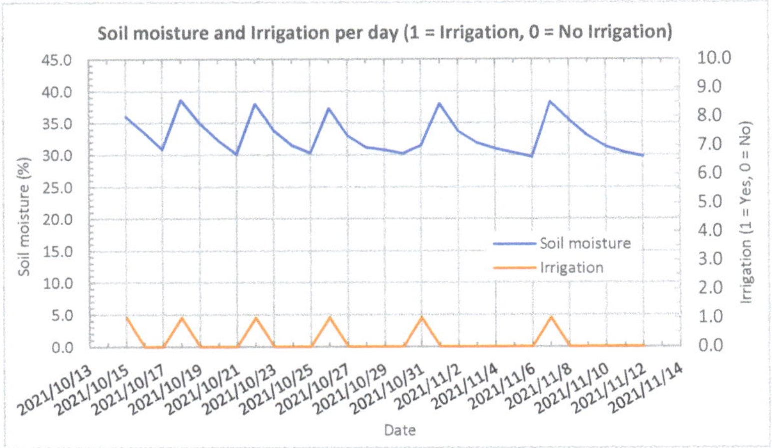

Figure 22. Automated irrigation using the IoT node.

The system was also tested in open crops. Figure 23 presents the average soil moisture per day, in a clay loam field cultivated with onions that was automatically irrigated by the system.

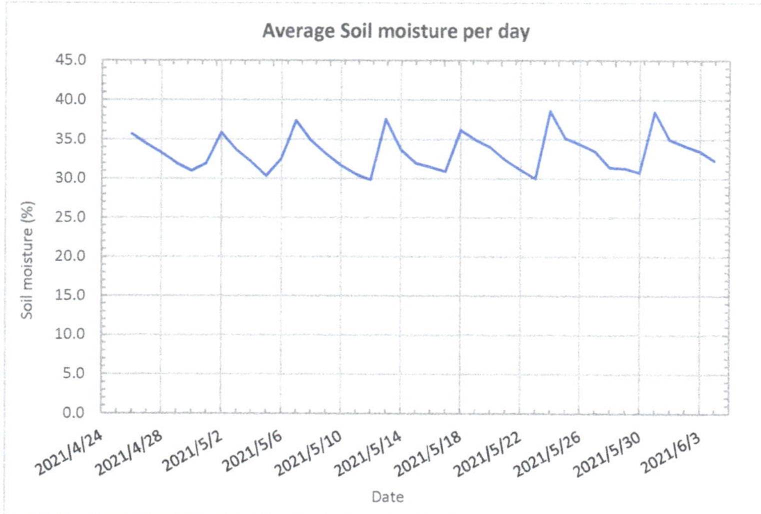

Figure 23. Average soil moisture per day.

3.5. Energy Autonomy

As the developed node can provide extensive autonomy, a node was installed on 1 December 2020 in a forest, configured to have a sampling rate of 8 h for minimizing its consumption, as the high and dense canopy of the trees does not allow recharging using solar panels and negatively affects the mobile network signal strength. As a result of the weak signal, the node communicated using a normal communication protocol (GPRS–2G), which has a higher energy consumption compared to low-power protocols, but provides a higher range of coverage. After one year of operation, on 20 February 2022, the remaining battery capacity was 64%, achieving an average energy consumption of 2.4% per month.

With a sampling rate of one hour, which is acceptable in most cases of agricultural monitoring (e.g., soil moisture content measurements), the system has an energy autonomy of 210 days. This makes the IoT node ideal for using it on any annual crop, as it can work during the entire cropping period without recharging. In the case that more intensive measurements are needed, a solar panel of less than 0.5 W is capable of providing to the node the energy required for its operation.

4. Discussion

The findings in this study indicate that low-cost technologies and standards can be used for developing low-cost, highly accurate, and easy-to-use systems that can be applied to enable irrigation scheduling and water management. As the node was exclusively based on the Arduino architecture and components, its hardware cost was very low, making it affordable to any farmer. The node was developed as a pure IoT device supporting cellular network technology protocols, making it capable of working in any area in which a cellular network is available.

Furthermore, as the price of sensors is constantly dropping, farmers can purchase sensors of high accuracy that can almost provide a perfect coefficient of determination ($R^2 = 0.9914$) at a very low price, permitting the fast depreciation of the investment for the system. As these sensors can provide data of high quality, their use can help farmers in decision making, by minimizing the inputs' cost and increasing their production. Likewise, low-cost actuators can be applied for automating and for remote controlling water management, increasing the usability of the system.

The sensors, after small modifications mostly related to making them waterproof, were introduced to be sufficiently reliable. More testing will be needed for evaluating their durability over time in the open agricultural environment.

The system was able to provide a variety of different type of measurements, including weather data, water quantity data, water quality data, and soil data. By computing crop water requirements, it was possible to automate irrigation scheduling providing the optimal water quantity, while simultaneously minimizing its consumption. Moreover, the system proved its capabilities on managing the different water sources in real environment in an extensive pilot testing that was conducted in three different pilot sites.

The system was developed as a "plug and play" device and pushing its start button is the only action needed for making the node fully functional. By adopting this simplified user experience, there is no need of any special knowledge or training for installing and configuring it, contributing on removing the demographic traits of the farmers barriers, which affect the adoption of new technologies.

Its small size, its durability, and its extensive energy autonomy make it suitable for a lot of cases, providing its effectiveness and usability. The final prototype was ready for testing in an operational environment in January 2020, and to date more than 200 systems have been installed. The system has proved to be extremely reliable, as to date there have been no hardware fails. Its development with open source Arduino technologies makes it modular, flexible, and upgradable to support more sensors and actuators than the existing ones, finally suggesting its capability for application in a vast number of agricultural operations in the future.

As the global population is constantly increasing and the cultivated areas are decreasing, new technologies will become a necessity as the only sustainable way for increasing agricultural output. It seems that low-cost IoT technologies will play a critical role in this transition, and they will contribute to the entering in the new era of holistic farm management, assisted by the extensive monitoring of the agricultural environment and automation of field operations.

Originally, the IoT system was developed for monitoring and controlling water to enable smart irrigation in open fields. As a result of its characteristics (very small size, energy autonomy, automation capabilities, high accuracy, support of different types of sensors, IP67 protection, and its low price), the node was already tested in various environments as forestry (monitoring of environmental parameters in forests), large water infrastructures (monitoring of water quantities), meteorology (for monitoring the weather), and for smart cities with very promising initial results.

5. Conclusions

From the presented results, it can be concluded that:

- A low-cost, low power consumption, fully autonomous system of IoT for irrigation scheduling using different water sources was developed and tested successfully;
- The easiness of setting up by incorporating low-cost sensors was proved in the presented applications;
- The presented applications proved the reliability, accuracy, and flexibility of the proposed configuration of the system;
- Low-cost solutions for automating field operations can be efficiently applied in the agricultural domain;
- Easy-to-use systems can used by small size and elderly farmers and enhance the resilience of the farms.

Author Contributions: Conceptualization, Z.T.; methodology, Z.T. and E.S.; writing—original draft preparation, Z.T.; writing—review and editing, E.S., S.F., I.G. and T.G. All authors have read and agreed to the published version of the manuscript.

Funding: This research was supported by the Horizon 2020 research and innovation program "Demonstration of water loops with innovative regenerative business models for the Mediterranean region—HYDROUSA" (grant agreement No. 776643).

Institutional Review Board Statement: Not applicable.

Informed Consent Statement: Not applicable.

Data Availability Statement: Not applicable.

Acknowledgments: This research was supported by the Horizon 2020 research and innovation program "Demonstration of water loops with innovative regenerative business models for the Mediterranean region—HYDROUSA" (grant agreement No 776643).

Conflicts of Interest: The authors declare no conflict of interest.

References

1. Rosegrant, M.W.; Ringler, C.; Zhu, T. Water for agriculture: Maintaining food security under growing scarcity. *Annu. Rev. Environ. Resour.* **2009**, *34*, 205–222. [CrossRef]
2. Water Resource Issues and Agriculture. Available online: https://www.fao.org/3/T0800E/t0800e0a.htm (accessed on 1 May 2022).
3. Jamroen, C.; Komkum, P.; Fongkerd, C.; Krongpha, W. An intelligent irrigation scheduling system using low-cost wireless sensor network toward sustainable and precision agriculture. *IEEE Access* **2020**, *8*, 172756–172769. [CrossRef]
4. Nawandar, N.K.; Satpute, V.R. IoT based low cost and intelligent module for smart irrigation system. *Comput. Electron. Agric.* **2019**, *162*, 979–990. [CrossRef]
5. Abba, S.; Wadumi Namkusong, J.; Lee, J.A.; Liz Crespo, M. Design and performance evaluation of a low-cost autonomous sensor interface for a smart iot-based irrigation monitoring and control system. *Sensors* **2019**, *19*, 3643. [CrossRef] [PubMed]
6. Krishnan, R.S.; Julie, E.G.; Robinson, Y.H.; Raja, S.; Kumar, R.; Thong, P.H. Fuzzy logic based smart irrigation system using internet of things. *J. Clean. Prod.* **2020**, *252*, 119902. [CrossRef]
7. Goap, A.; Sharma, D.; Shukla, A.K.; Krishna, C.R. An IoT based smart irrigation management system using Machine learning and open source technologies. *Comput. Electron. Agric.* **2018**, *155*, 41–49. [CrossRef]
8. Keswani, B.; Mohapatra, A.G.; Mohanty, A.; Khanna, A.; Rodrigues, J.J.; Gupta, D.; De Albuquerque, V.H.C. Adapting weather conditions based IoT enabled smart irrigation technique in precision agriculture mechanisms. *Neural Comput. Appl.* **2019**, *31*, 277–292. [CrossRef]
9. Nguyen, D.C.H.; Ascough, J.C., II; Maier, H.R.; Dandy, G.C.; Andales, A.A. Optimization of irrigation scheduling using ant colony algorithms and an advanced cropping system model. *Environ. Model. Softw.* **2017**, *97*, 32–45. [CrossRef]
10. Pereira, L.S.; Paredes, P.; Jovanovic, N. Soil water balance models for determining crop water and irrigation requirements and irrigation scheduling focusing on the FAO56 method and the dual Kc approach. *Agric. Water Manag.* **2020**, *241*, 106357. [CrossRef]
11. Gu, Z.; Qi, Z.; Ma, L.; Gui, D.; Xu, J.; Fang, Q.; Feng, G. Development of an irrigation scheduling software based on model predicted crop water stress. *Comput. Electron. Agric.* **2017**, *143*, 208–221. [CrossRef]
12. García, L.; Parra, L.; Jimenez, J.M.; Lloret, J.; Lorenz, P. IoT-based smart irrigation systems: An overview on the recent trends on sensors and IoT systems for irrigation in precision agriculture. *Sensors* **2020**, *20*, 1042. [CrossRef] [PubMed]
13. Libelium. Available online: https://www.libelium.com/ (accessed on 1 May 2022).
14. iMETOS 3.3-METOS by Pessl Instruments. Available online: https://metos.at/imetos33/ (accessed on 1 May 2022).
15. WatchDog 2000 Series Weather Stations. Available online: https://www.specmeters.com/weather-monitoring/weather-stations/2000-full-stations/ (accessed on 1 May 2022).
16. NetSens. Available online: https://www.netsens.it/en/ (accessed on 1 May 2022).
17. Radhakrishnan, V.; Wu, W. IoT technology for smart water system. In Proceedings of the 2018 IEEE 20th International Conference on High Performance Computing and Communications, Exeter, UK, 28–30 June 2018.
18. Xiaocong, M.; Jiao, Q.X.; Shaohong, S. An IoT-based system for water resources monitoring and management. In Proceedings of the 2015 7th International Conference on Intelligent Human-Machine Systems and Cybernetics, Hangzhou, China, 26–27 August 2015.
19. Ramesh, M.V.; Nibi, K.V.; Kurup, A.; Mohan, R.; Aiswarya, A.; Arsha, A.; Sarang, P.R. Water quality monitoring and waste management using IoT. In Proceedings of the 2017 IEEE Global Humanitarian Technology Conference (GHTC), San Jose, CA, USA, 19–22 October 2017.
20. Gupta, K.; Kulkarni, M.; Magdum, M.; Baldawa, Y.; Patil, S. Smart water management in housing societies using IoT. In Proceedings of the 2018 Second International Conference on Inventive Communication and Computational Technologies (ICICCT), Coimbatore, India, 20–21 April 2018.
21. Chowdury, M.S.U.; Emran, T.B.; Ghosh, S.; Pathak, A.; Alam, M.M.; Absar, N.; Hossain, M.S. IoT based real-time river water quality monitoring system. *Procedia Comput. Sci.* **2019**, *155*, 161–168. [CrossRef]

22. Moreno, C.; Aquino, R.; Ibarreche, J.; Pérez, I.; Castellanos, E.; Álvarez, E.; Clark, B. RiverCore: IoT device for river water level monitoring over cellular communications. *Sensors* **2019**, *19*, 127. [CrossRef]
23. Skoubris, E.; Hloupis, G. An Imaging Capable, Low Cost IoT Node for River Flood Phenomena. In Proceedings of the EGU General Assembly Conference Abstracts, Göttingen, Germany, 19–30 April 2021.
24. Intro to Arduino. Available online: https://www.coursehero.com/file/35492169/02-Intro-to-Arduinopdf/ (accessed on 1 May 2022).
25. Fountas, S.; Carli, G.; Sørensen, C.G.; Tsiropoulos, Z.; Cavalaris, C.; Vatsanidou, A.; Tisserye, B. Farm management information systems: Current situation and future perspectives. *Comput. Electron. Agric.* **2015**, *115*, 40–50. [CrossRef]
26. Agrawal, N.; Singhal, S. Smart drip irrigation system using raspberry pi and arduino. In Proceedings of the International Conference on Computing, Communication & Automation, Washington, DC, USA, 15–16 May 2015.
27. Toai, T.K.; Huan, V.M. Implementing the Markov Decision Process for Efficient Water Utilization with Arduino Board in Agriculture. In Proceedings of the 2019 International Conference on System Science and Engineering, Dong Hoi City, Vietnam, 20–21 July 2019.
28. Jha, R.K.; Kumar, S.; Joshi, K.; Pandey, R. Field monitoring using IoT in agriculture. In Proceedings of the 2017 International conference on intelligent computing, instrumentation and control technologies, Kannur, Kerala, India, 6–7 July 2017.
29. Lambrou, T.P.; Anastasiou, C.C.; Panayiotou, C.G.; Polycarpou, M.M. A low-cost sensor network for real-time monitoring and contamination detection in drinking water distribution systems. *IEEE Sens. J.* **2014**, *14*, 2765–2772. [CrossRef]
30. Wang, Y.; Rajib, S.S.M.; Collins, C.; Grieve, B. Low-cost turbidity sensor for low-power wireless monitoring of fresh-water courses. *IEEE Sens. J.* **2018**, *18*, 4689–4696. [CrossRef]
31. Viani, F.; Bertolli, M.; Salucci, M.; Polo, A. Low-cost wireless monitoring and decision support for water saving in agriculture. *IEEE Sens. J.* **2017**, *17*, 4299–4309. [CrossRef]
32. González-Teruel, J.D.; Torres-Sánchez, R.; Blaya-Ros, P.J.; Toledo-Moreo, A.B.; Jiménez-Buendía, M.; Soto-Valles, F. Design and calibration of a low-cost SDI-12 soil moisture sensor. *Sensors* **2019**, *19*, 491. [CrossRef]
33. Prathibha, S.R.; Hongal, A.; Jyothi, M.P. IoT based monitoring system in smart agriculture. In Proceedings of the 2017 international conference on recent advances in electronics and communication technology, Bangalore, India, 16–17 March 2017.
34. Feng, X.; Yan, F.; Liu, X. Study of wireless communication technologies on Internet of Things for precision agriculture. *Wirel. Pers. Commun.* **2019**, *108*, 1785–1802. [CrossRef]
35. Hydrousa Project. Available online: https://www.hydrousa.org (accessed on 1 May 2022).
36. Chan, K.; Schillereff, D.N.; Baas, A.C.; Chadwick, M.A.; Main, B.; Mulligan, M.; Thompson, J. Low-cost electronic sensors for environmental research: Pitfalls and opportunities. *Prog. Phys. Geogr. Earth Environ.* **2021**, *45*, 305–338. [CrossRef]
37. Sami, M.; Khan, S.Q.; Khurram, M.; Farooq, M.U.; Anjum, R.; Aziz, S.; Qureshi, R.; Sadak, F. A Deep Learning-Based Sensor Modeling for Smart Irrigation System. *Agronomy* **2022**, *12*, 212. [CrossRef]
38. Bauder, A.; Waskom, M.; Sutherland, L.; Davis, G.; Follett, H.; Soltanpour, N. *Irrigation Water Quality Criteria*; Colorado State University Extension: Fort Collins, CO, USA, 2011.
39. Agriculture and Agri-Food Canada, Prairie Farm Rehabilitation Administration, Irrigation and Salinity. Available online: https://www1.agric.gov.ab.ca/$department/deptdocs.nsf/ba3468a2a8681f69872569d60073fde1/42131e74693dcd01872572df00629626/$file/irrsalin.pdf (accessed on 25 June 2022).
40. Latest Condition in Trikala. Available online: https://penteli.meteo.gr/stations/trikala (accessed on 1 May 2022).
41. Crop Evapotranspiration-Guidelines for Computing Crop Water Requirements-FAO Irrigation and Drainage Paper 56. Available online: https://www.fao.org/3/x0490e/x0490e00.htm (accessed on 1 May 2022).

water

Article

Simulating the Effects of Agricultural Adaptation Practices onto the Soil Water Content in Future Climate Using SWAT Model on Upland Bystra River Catchment

Damian Badora *, Rafał Wawer *, Anna Nieróbca, Aleksandra Król-Badziak, Jerzy Kozyra, Beata Jurga and Eugeniusz Nowocień

The Institute of Soil Science and Plant Cultivation—State Research Institute, ul. Czartoryskich 8, 24-100 Pulawy, Poland; anna.nierobca@iung.pulawy.pl (A.N.); aleksandra.krol@iung.pulawy.pl (A.K.-B.); kozyr@iung.pulawy.pl (J.K.); bjurga@iung.pulawy.pl (B.J.); nowocien@iung.pulawy.pl (E.N.)

* Correspondence: dbadora@gmail.com (D.B.); huwer@iung.pulawy.pl (R.W.)

Abstract: The article presents predicted changes in soil water content in the Bystra river catchment (eastern Poland) for various scenarios of climate change and adaptation practices obtained on the basis of a SWAT model simulation for three regional climate models driven by the global climate model EC-EARTH for the years 2041–2050 and the RCP 4.5 and 8.5 RCP scenarios. Climate scenarios were put against five adaptation scenarios presenting changes in land use and protective measures compared against a zero scenario of BaU (Business as Usual) kept in the future climate. Adaptation scenarios 1–5 are modifications of Scenario 0 (S-0). The 0–5 scenarios' analysis was based on comparing soil water content and total runoff, sediment yield, actual evapotranspiration. The first adaptation scenario (AS-1) assumes an increase in afforestation on soils from the agricultural suitability complex of soil 6–8 (semi-dry, permanent dry, semi-wet). The second adaptation scenario (AS-2) assumes the creation of a forested buffer for the Bystra River and its tributaries. The third adaptation scenario (AS-3) shows one of the erosion prevention practices, the so-called filter strips. The fourth adaptation scenario (AS-4) assumes the reduction in plowing on arable land. The fifth adaptation scenario (AS-5) involves increasing soil organic carbon to 2%. Simulations revealed that each of the adaptation scenarios 1, 2, 3, 5 does not generally contribute to increasing the water content in soil on BARL (spring crops), CANP (rape), WWHT (winter crops), CRDY (other crops) on arable lands (which together account for over 50% of the catchment area). However, they can contribute to the reduction in sediment yield, total runoff and changes in actual evapotranspiration. The adaptation scenario 4 (AS-4) shows a slight increase in the soil water content on Bystra catchment in the 2041–2050 perspective. Scenario 4 indicated a slight increase in total runoff and a decrease in sediment yield, which in combination with slightly higher water content reflects the protective role of plant residue mulch, lowering the evaporation from the bare soil surface during warm seasons. The no-till adaptation practice had the highest effect in positively affecting water balance at the catchment scale among the adaptation scenarios considered.

Keywords: SWAT; SWAT-CUP; climate change; adaptation scenarios; soil water content; afforestation; no plowing; filter strips

Citation: Badora, D.; Wawer, R.; Nieróbca, A.; Król-Badziak, A.; Kozyra, J.; Jurga, B.; Nowocień, E. Simulating the Effects of Agricultural Adaptation Practices onto the Soil Water Content in Future Climate Using SWAT Model on Upland Bystra River Catchment. *Water* **2022**, *14*, 2288. https://doi.org/10.3390/w14152288

Academic Editors: Alban Kuriqi, Luis Garrote and Ian Prosser

Received: 7 March 2022
Accepted: 12 July 2022
Published: 22 July 2022

1. Introduction

Soil water content is an important component of the hydrological cycle. The formation of water resources in the catchment area is greatly influenced by the amount of precipitation, evapotranspiration, temperature as well as soil properties (water storage capacity, texture, structure), management practices and the existing vegetation [1,2]. The main source of soil water content is precipitation through infiltration and surface runoff [3]. Temperature, on the other hand, influences the evapotranspiration process [4].

There are numerous studies focusing on the calculation of soil water content using the Soil and Water Assessment Tool (SWAT) model [5,6]. These authors used SWAT to simulate soil water content at levels of large catchments (Vistula, Odra). They demonstrated the ability to generate long-term series of soil water content even in the absence of comparative data. On the other hand, for a small catchment located in Poland, one of the few similar studies to the present one in terms of climatic scenarios as well as parameters studied (soil water content, actual evapotranspiration) is a publication concerning the Barycz and Upper Narwia catchments [7].

In the publication on soil water retention and drought risk assessment based on water balance for the area of the Lower Silesian province [1], soil retention parameters were determined: Available Water Capacity (AWC), Wilting Point (WP), Field Capacity (FC) for soil species found in Poland. The retention parameters were determined by expert methods [1].

The aim of the article is to analyze five adaptation scenarios (AS-1, AS-2, AS-3, AS-4, AS-5) in relation to the 2041–2050 climate projections GCMs/RCMs for the RCP 4.5 and RCP 8.5 climate change scenarios described as scenario 0 (S-0) [8], as well as their assessment against the current state of knowledge related to research involving similar adaptation studies. Adaptation scenarios 1–5 are modifications of Scenario 0.

The need for such studies of small catchments (up to a few hundred km^2) is due to the small number of studies that would be based on adequate preparation of soil parameters (e.g., retention). Moreover, for the Polish area, there are no studies on adaptation scenarios that would attempt to increase the water content of soil and minimize the adverse effects of climate change (RCP 4.5, RCP 8.5) in future decades.

Among the many hydrological models in use today, the SWAT model, widely used by scientists and developed by the USDA [9,10], was selected for this study because of its ability to predict the impact of practices of land management onto the hydrology and water quality in the catchment area.

Much research is currently being conducted on climate change and the associated unpredictability of extreme weather events. This raises legitimate concerns about the possible emergence of environmental, social and economic threats in the decades to come. These changes may also have an impact on agriculture in Poland [11]. The increase in air temperature, which was observed in recent decades, contributed to the increase in potential evapotranspiration, especially in the last decade 2011–2020. A large increase in potential evapotranspiration and an increase in the variability of this indicator were found [8,12,13]. Recent decades also brought observations of climate change in Poland resulting from the world global warming, changes in precipitation and a number of weather extremes [14–16].

These changes also concern the extension of the growing season in Poland. For the years 1971–2000, the length of the growing season was 218 days (from March 31 to November 4) [17]. According to studies on the change in the growing length in Poland [17], the length of the growing season will extend by 18–27 days in the perspective of 2050 compared to the years 1971–2000.

The increase in evapotranspiration, temperature and precipitation in the coming decades will, to a greater or lesser extent, also apply to all European countries [18,19].

According to the Sixth Assessment Report of the Intergovernmental Panel on Climate Change [20], the average temperature of the Earth's surface will reach 1.5 degrees Celsius in the coming decades above pre-industrial levels. Moreover, in Poland, a projected 10-fold increase in the occurrence of droughts by 2020 [21] is observed in the data; hence, the predictions using climate change models seem to reflect changing climate quite well for Poland [22]. Until recently, climate change adaptation received less attention in Poland than climate change mitigation. The vast majority of national communications have been devoted to climate projections, vulnerability and impacts. However, recently there has been increasing attention to adaptation measures in agriculture, among others [16].

There is a need to look for solutions that will reduce the negative impact of climate change [23], inter alia, the occurrence of weather extremes, including drought [12,24,25] in the coming decades. Climate change adaptation in agriculture is associated with a number of preventive measures (adapting crops to changing thermal and water conditions). These include changes in adaptation practices and the introduction of new varieties. Protecting the soil and its water resources is also extremely important. Soil moisture can be maintained through mulching and water conservation through efficient irrigation and water storage (small retention, filter strips). Soil fertility and its potential for water storage can also be increased by increasing soil organic matter [16].

For the sake of this paper, we chose the Bystra river catchment (South-Eastern Poland) as the study area. In order to check the effectiveness of the designed adaptation solutions, it was necessary to develop boundary conditions that would indicate the reference level [8]. These conditions show the behavior of the hydrosystem of the Bystra catchment in the Business as Usual scenario. It takes into account changes in the hydrological cycle caused solely by climate change while maintaining unchanged conditions of human activity. The described boundary conditions for the 2050 horizon must be based on simulation modeling, which is calibrated on archival data. The appropriate tool for this is the SWAT model.

The article presents a comparison of the results of soil water content (profile 1.5 m) for five adaptation scenarios obtained via a simulation of a calibrated and validated SWAT model [8] for three regional climate models derived from the global EC-EARTH climate model for the years 2041–2050 (S-0). Then, the results of scenario 0 were compared with the results of adaptation scenarios 1–5, which included land use changes and protective measures.

The publication is presented as follows: Section 1 presents the Introduction; Section 2 introduces the methodology and describes the study area. Section 3 describes the results, and Section 4 presents the discussion in terms of results regarding soil water content, total runoff and sediment yield, followed by conclusions in Section 5.

2. Material and Methods

This section is divided in 5 sub-sections. The first one describes the study area; the second and the third describes the SWAT model and SUFI-2 model; the fourth presents climate change scenarios, and finally the fifth presents climate change adaptation scenarios 1–4.

2.1. Characterization of the Study Area

The Bystra catchment area is situated in the north-western part of the Lubelskie Province (Figure 1). The length of the Bystra River is 33 km, and it is the right tributary of the Vistula river. According to the generated SWAT model, the lowest point of the catchment area is 126 m above sea level, and the highest point is 246 m above sea level. The catchment area delineated from a 5 m resolution DEM is 296.6 km^2 [8].

The Bystra catchment area is part of the Lublin Upland [26–28]. The valley of the Bystra river and its tributaries are strongly carved in a thick loess layer overlaying calcareous bedrock. It consists of numerous valley forms with a constant or episodic tributary. The largest valley with a constant tributary, the Bystra valley, is 35 km long. In the part where the Bystra valley flows into the Vistula, it cuts up to 35 m in rocks and marls [29–31].

The upland nature of the Bystra catchment area, consisting mostly of loess soils, with a high slope of the slopes at the mouth of the Vistula, poses a high risk in terms of medium and very strong water and surface erosion [32].

Most of the Bystra catchment area is made of loess up to 20 m. In the deeper layers, there are Quaternary Pleistocene sediments: water-glacial sand and gravel and, at a little deeper level, tilts. On the other hand, there are geodes under the clays. Under the geysers, on the other hand, there are deposits of the Upper Cretaceous: rocks with lime inserts [33].

The study area consists mainly of podzolic and lessivage (49%) soils, which extend mainly in the south-eastern part of the catchment as well as cambisols (47%) in the north-western part. The predominant soil texture in the catchment area is loess (73%) [34–36] and silt (18%) [8].

In the Bystra catchment area, arable land (78%) and forests (16%) dominate [8]. The largest part of agricultural land is arable land beyond the reach of irrigation facilities (52%);

large areas are also orchards and plantations (11%), complex systems of arable plots (9%) and meadows and pastures (6%) [8].

Figure 1. Location of the study area, Bystra catchment, with marked main tributaries and their catchments (own study).

2.2. Description of SWAT Model and SUFI-2 Model

SWAT was used to model and examine the water balance of the Bystra river catchment area. SWAT is a model [9,10] developed by the USDA Agricultural Research Service [37]. The model operates on assigning one resource to another (physical, chemical, biological) using mathematical formulas that were developed to predict the impact of management practices on water efficiency and agricultural chemistry at the catchment scale [38,39]. We used the QSWAT3 v1.1 model with an interface in Quantum GIS 3.10.13 Coruna [40]. However, the calculations of the SWAT model were performed in the SWAT Editor on 10 December 2012 [41].

The water balance is the fundamental driving force behind all the processes that take place in the catchment area regardless of the choice of the SWAT model analysis. SWAT modeling for the catchment area is carried out in the land phase [42] and in the routing phase [43].

One of the formulas that is used in the SWAT model is the water balance equation:

$$SW_t = SW_0 + \sum_{i=1}^{t} (P_d - SURQ - E - w_{seep} - GWQ)$$

where: SW_t is the final water content of the soil (mm); SW_0 is the initial water content of the soil (mm); t is the time in days; P_d is precipitation (mm); $SURQ$ is surface runoff (mm); E is evapotranspiration (mm); w_{seep} is the amount of water entering the wad zone from the soil profile (mm); GWQ is the groundwater flow (mm) [10].

Calibration and validation in the SWAT-CUP program are used to adjust the SWAT model to real conditions in the catchment area. The commonly used example of calibration is stream flow, which includes water balance processes. The calibration process is used to adjust the relevant parameters so that the simulated results are consistent with the observational data. Validation involves running the model using the parameters that were used during the calibration process. The purpose is to compare simulated results with observed data that were not used in calibration [44–46]. The SWAT-CUP program is used to analyze the uncertainty and sensitivity of the model [44,45] using the SUFI-2 algorithm, also used in small catchments [44,47,48].

2.3. Application of SWAT and SUFI-2

To simulate the water balance in the SWAT model, data were obtained from many sources (Table 1), which were used to build the SWAT model.

Table 1. Input data used in SWAT model (own study).

Data Type	Description	Information	Source
Digital Elevation Model	Watershed delineation	Raster, 5 m-resolution	Central Geodetic and Cartographic Documentation Center [49]
Hydrographic	Site hydrographic data (e.g., rivers, lakes, partial catchments); (reference scale 1:50.000)	Shapefile	Computer Map of the Polish Hydrological Department with descriptions [50]
Land use	Land-use classification (r.s. 1:100.000)	Shapefile	Corine Land Cover [51]
Orthophotomap	High resolution orthophotomap	WMS	Geoportal [52]
Open Street Map	Open Street Map data	Shapefile	Open Street Map [53]
Soil type	Digital maps of soil and agriculture in digital form (scale 1: 25,000 and 1: 100,000)	Shapefile	Institute of Soil Science and Plant Cultivation in Pulawy [54,55]
Geological	Geological data describing lithology	Shapefile	Polish Geological Institute in the form of the Detailed Geological Map of Poland [33]
Weather	Precipitation (mm), temperature (°C), wind speed (m/s), humidity, solar total radiation (MJ/m^2)	Daily	Institute of Soil Science and Plant Cultivation in Pulawy and Institute of Meteorology and Water Management [56]
Streamflow	Calibration and validation	Monthly	Institute of Soil Science and Plant Cultivation in Pulawy
Sewage treatment plants	Average daily water loading (m^3/day)	Daily	National Program of Municipal Wastewater Treatment [57]

The SWAT model generated for this study consists of 31 generated partial catchments (Figure 1) [8]. The soil map was developed on the basis of digital soil and agricultural maps (scale 1:25,000 and 1:100,000) and geological data describing lithology. Descriptive soil data were collected within the statutory research projects of IUNG-PIB. Available water capacity and wilting point values were obtained from the study "Assessment of water retention in soil and the risk of drought based on the water balance for the Lower Silesian Voivodeship", which was developed in 2013 by the employees of the Department of Soil Science, Erosion and Land Protection of IUNG-PIB in Pulawy [1].

The land use map was developed on the basis of Corine Land Cover maps with additional vectorization of land cover and land use using an orthophoto-map and Open Street Map data.

Based on the generated maps of soils, lands and slopes, 484 HRU (Hydrological Response Units) areas were created. When creating HRU areas, the land cover class of agricultural areas beyond the reach of irrigation CRDY was additionally separated with WWHT winter crops (43%), BARL spring crops (31%), CANP rape (14%) and other CRDY (12%) [58]. APPL apple orchards were separated from the land use class of ORCD [58]. On the other hand, forests were divided into coniferous FRSE forests (49%), deciduous FRSD forests (13%) and mixed FRST forests (38%) [59].

After generating HRU areas, the following meteorological data were used in the SWAT model: daily precipitation totals (mm); daily minimum and maximum air temperature (°C); average daily wind speed (m/s); average daily relative humidity; daily sums of total solar radiation (MJ/m^2) (Table 2) [8].

Table 2. Meteorological data for the Bystra catchment [8].

Weather Station	Measurement Period				
	Precipitation (mm)	Temperature (°C)	Wind Speed (m/s)	Humidity	Solar Total Radiation (MJ/m^2)
Pulawy	2005–2017	2005–2017	2005–2017	2005–2017	2005–2017
Rogalow	2005–2017				
Lublin Radawiec	2005–2017	2005–2017	2005–2017	2005–2017	

In the SWAT model, the parameters related to the point discharge of sewage, as well as for water bodies located outside the river network, for water bodies, rivers, and parameters for planned non-irrigated arable land management operations (WWHT, BARL, CANP, CRDY) were supplemented and corrected. The current value of CO_2 concentration was also entered.

In the next stage, the SWAT model simulation was run for the period of 2010–2017 in a monthly step, with a five-year model start-up period.

Then, calibration and validation of the obtained SWAT model for the Bystra catchment area [8] was performed using the SWAT-CUP program. To obtain a more accurate coverage of the model with reality, the average monthly flow velocities (m^3/s) obtained under the statutory projects of IUNG-PIB, obtained near the mouth of the Bystra River to Vistula for 2010–2014 (calibration) and 2015–2017 (validation), were used. A five-year warm-up period was used. Calibration and validation were performed in a monthly increment. This resulted in parameter ranges that fell within the ranges of calibration and validation accuracy [44,60,61]. The NSE coefficients (calibration: 0.58; validation: 0.70) and R^2 (calibration: 0.60; validation: 0.71) for calibration and validation [8] were within the satisfactory ranges [60].

The results concerning the value of potential evapotranspiration were also analyzed with the results of the statutory service of IUNG-PIB implemented under the project Agricultural Drought Monitoring System [62]. It was found that the SWAT model for the Bystra catchment area accurately reflects the potential evapotranspiration in the study area.

Additionally, the results concerning the soil water content were compared with the available values of water capacity and the wilting point, which were obtained from the study prepared in 2013 by the employees of the Department of Soil Science, Erosion and Land Protection, IUNG-PIB in Pulawy [1].

2.4. Climate Change Scenarios

The daily grid climate data used in the SWAT model were prepared and tested in the recent paper on SWAT model calibration in the Bystra catchment [8]. Three RCM (Regional Climate Models)—RACMO22E, HIRHAM5 and RCA4—were selected for further study. They were selected to cover the range of the available two climate scenarios RCP (Representative Concentration Pathways) in terms of temperature increase and precipitation—RCP 4.5, RCP 8.5 (Table 3)—reflecting extreme and average variants of climate change, hence covering the widest range of uncertainty about possible scenarios (three RCM × two RCP). Most of the data were obtained at a spatial resolution of 0.11 degrees from the EURO-CORDEX database for the years 1951–2050 (widely available via the ESGF—Earth System Grid Federation, https://esgf-data.dkrz.de/search/cordex-dkrz for Europe) (accessed on 3 March 2021) [18,63].

Table 3. Description of GCM/RCM simulation with its division depending on radiative forcing. Comparison of temperature and precipitation changes in 2021–2050 in GCM/RCM simulation for RCP 4.5 and RCP 8.5 to the base period 1971–2000 (own study).

Models	**Scenario Assumptions**				**Radiative Forcing**	
GCM/RCM Simulation	**Change in Average Annual Air Temperature**		**Change in Average Annual Precipitation**		**+4.5 W m^{-2}**	**+8.5 W m^{-2}**
	RCP 4.5	**RCP 8.5**	**RCP 4.5**	**RCP 8.5**	**RCP 4.5**	**RCP 8.5**
EC-EARTH/RACMO22E	+1.5 °C	+1.8 °C	+15%	+6%	RCP 4.5.1	RCP 8.5.1
EC-EARTH/HIRHAM5	+1.6 °C	+1.9 °C	+12%	+5%	RCP 4.5.2	RCP 8.5.2
EC-EARTH/RCA4	+1.6 °C	+2.2 °C	+15%	+11%	RCP 4.5.3	RCP 8.5.3

Climate scenario daily meteorological derivatives (minimum and maximum daily air temperature, daily precipitation, solar radiation, daily average wind speed, relative humidity) are based on the RCM for two RCPs (three RCM x two RCP). The RCMs are powered by one GCM (General Circulation Model): EC-EARTH. The RCP corresponds to the radiative forcing values in 2100 compared to pre-industrial values of +4.5 W m^{-2} (RCP4.5) while RCP8.5 to + 8.5 W m^{-2} (RCP8.5) [18,64,65] (Table 3). Table 3 also presents the boundary values of changes in the characteristics of selected models for the period 2021–2050 in relation to the period up to the base period 1971–2000.

Climate projections that were used in the SWAT model were extracted from grid cells that correspond to weather stations' location. Air temperature and precipitation data were additionally corrected by the SMHI (Swedish Meteorological and Hydrological Institute) using the DBS (Distribution-Based Scaling) method [48] and regional MESAN reanalysis (MESoscale Analysis) for the 1989–2010 dataset [66]. The data used were taken in a rotated polar grid. Therefore, we used bilinear interpolation to remap the dataset to a common latitude/longitude grid. CDO (Climate Data Operators) software [67] was used for this purpose.

For the analysis of the climate projections (RCP 4.5.1, RCP 8.5.1, RCP 4.5.2, RCP 8.5.2, RCP 4.5.3 and RCP 8.5.3), one iteration in SWAT-CUP was used for the set of the best calibration parameters for the years 2021–2050 in the prepared scenarios (Table 3) [8]. In the RCP 4.5 and RCP 8.5 scenarios, CO_2 concentrations were changed for the periods 2021–2030, 2031–2040 and 2041–2050, developed by the Potsdam Institute for Climate Impact Research [68,69].

2.5. Climate Change Adaptation Scenarios 1–5

For the main purpose of this article, 5 scenarios for the adaptation of agriculture to climate change were prepared, which assume changes in land use (adaptation scenario 1 and 2) and protective measures (adaptation scenario 3, 4, 5) in the area of the Bystra catchment. The first adaptation scenario (AS-1) assumes an increase in afforestation on soils from the agricultural usefulness complex of soils 6 (temporarily too dry), 7 (permanently too dry) and 8 (temporarily too wet). The second adaptation scenario (AS-2) assumes the creation of a forested buffer for the Bystra River and its tributaries. The third adaptation scenario (AS-3) shows one of the erosion prevention practices at the riverbed, the so-called filter strips. The fourth adaptation scenario (AS-4) assumes the reduction in plowing on agricultural land. The fifth adaptation scenario (AS-5) involves increasing soil organic carbon to 2%. Adaptation scenarios are aimed at checking the possibility of increasing the soil water content in the 2041–2050 perspective. In doing so, the effects of adaptation scenarios on total runoff, sediment yield and actual evapotranspiration were also checked.

In the zero scenario (S-0), the Bystra catchment area is dominated by agricultural land (78%) and forests (16%). The largest part of agricultural land is arable land beyond the range of irrigation facilities (52%); a large area is also orchards and plantations (11%), complex systems of cultivating plots (9%) and meadows and pastures (6%) (Table 4). For adaptation scenarios 1 (AS-1) and 2 (AS-2), there will be changes in land use compared to scenario 0 (S-0), which are described later. In contrast, adaptation scenarios 3 (AS-3), 4 (AS-4) and 5 (AS-5) remain unchanged in terms of changes in land use.

Table 4. Division of the land cover and land use as well as the percentage of land use in the Bystra catchment generated in the QSWAT interface. CLC code 112–142 means artificial surfaces; code 211–243 means agricultural areas; code 313–324 means forest and semi natural areas; code 411 means wetlands, and code 511 is water bodies (own study).

Corine Land Cover Legend	CLC Code	SWAT Code	S-0 Part (%)	AS-1 Part (%)	AS-2 Part (%)
Discontinuous urban fabric	112	URML	0.92	0.9	0.9
Industrial or commercial units	121	UCOM	1.55	1.49	1.49
Mineral extraction sites	131	UIDU	0.02	0.02	0.02
Sport and leisure facilities	142	FESC	0.02	0.02	0.02
		SUM=	2.51	2.43	2.43
Non-irrigated arable land	211	CRDY	52.35	50.57	52.23
Vineyards	221	GRAP	0.03	0.03	0.03
Fruit trees and berry plantations	222	ORCD	10.85	10.55	10.83
Pastures	231	PAST	5.89	5.35	5.55
Complex cultivation patterns	242	AGRL	9.04	8.68	8.86
Land principally occupied by agriculture with significant areas of natural vegetation	243	CRGR	0.05	0.05	0.05
		SUM=	78.21	75.23	77.55
Mixed forest	313	FRST	16.34	19.65	17.37
Transitional woodland-shrub	324	SHRB	2.43	2.18	2.23
Inland marshes	411	WEHB	0.26	0.25	0.21
Water courses	511	WATR	0.27	0.26	0.21

In the first adaptation scenario (AS-1), the land use on all soils of complexes (representing soil habitats in Polish soil-agricultural mapping)—6 (semi-dry), 7 (permanently dry) and 8 (semi-wet) (6Bw-pgl.ps, 7Bw-ps, 8A-l)—was changed to mixed forest. The soils where the land use was changed are described in more detail in Table 1 of the publication on the water balance of the Bystra catchment [8]. Replacement of the above-mentioned soils is made through delineating the ranges of these soils on the land use maps and changing the attributes to mixed forests. After this change, afforestation in the Bystra catchment area increased by 3.31% (Table 4).

In the second adaptation scenario (AS-2), a forested buffer strip 80 m wide along the bank of the Bystra River was created and a smaller buffer strip 50 m wide for its tributaries [70–72]. The creation of buffer zones by rivers consisted of deleting the ranges of buffer zones on the land use maps and changing the attributes to mixed forests. The afforestation area compared to the zero scenario increased by 1.03% (Table 4).

In the third adaptation scenario (AS-3), filter strips were used, which are one of the protective measures used to drain water slowly from the field, thanks to which larger particles, including soil and organic material, may be deposited [73].

Filter strips [9,74] are areas covered with vegetation that are located between surface water bodies (rivers, ponds, lakes) and arable land, pastures and forests. They are generally found in areas where runoff leaves the field to filter sediment, organic material, nutrients and chemicals from the runoff. Filter strips are also known as vegetative filters or buffer strips. Due to the retention of sediment and the establishment of vegetation, nutrients can be absorbed into the sediment that settles and remain in the field landscape, making it possible for plants to take it up [73].

A protective treatment is also tillage without plowing [73], which is the fourth adaptation scenario (AS-4).

Plowing is defined as the mechanical disturbance of soil for crop production that has a significant impact on soil properties such as soil water behavior, soil temperature, infiltration and evapotranspiration [75]. In the long term, tillage can lead to soil degradation [76]. An alternative to traditional plowing is protective treatments (tillage without plowing, minimal mechanical disturbance of the soil) which consist of maintaining the surface soil cover by retaining crop residues. Retention of harvest residues protects the soil from direct exposure to raindrops and sunlight, while minimal soil disturbance improves soil biological activity and air and water movement in the soil [75].

No plowing cultivation was implemented in WWHT, BARL, CANP and CRDY arable land and simulated in SWAT.

In the fifth adaptation scenario (AS-5), the soil organic carbon content was increased from 1% to 2%. The original soil organic carbon values were studied as part of IUNG-PIB statutory research [8]. Soils in Poland are characterized by low soil organic carbon content. According to the European Soil Bureau (ESB), an organic carbon content of about 1% (Bystra catchment area) is a very low or low value [77]. The decrease in organic matter in soils and the associated decrease in organic carbon content result in increased CO_2 emissions (exacerbating the greenhouse effect). The opposite situation, i.e., sequestration of CO_2 in the soil, causes carbon to bind to soil organic matter for a longer period of time. Particularly large amounts of carbon are stored in peats, organic soils and organic-mineral soils [77].

3. Results

Section 3.1 describes the analysis of soil water content in S-0 for the period 2041–2050.

For the 10-year period (2041–2050), Table 5 presents a comparison of the seasonal soil water content in the Bystra catchment for each climate projection GCMs/RCMs under the RCP 4.5 and RCP 8.5 climate scenarios.

For the 10-year period (2041–2050), Figure 2 shows the average soil water content (1.5 m) for each season of DJF, MAM, JJA, SON for the GCMs/RCMs climate projections under the RCP 4.5 and RCP 8.5 climate scenarios, while Figure 3 shows the spatial comparison of average soil water content in 31 sub-catchments for the SWAT simulation period 2010–2017 and 2041–2050 for the GCMs/RCMs climate projections under the RCP 4.5 and RCP 8.5 climate scenarios.

Section 3.2 describes the climate change AS-1, AS-2, AS-3, AS-4, AS-5 analysis for the period 2041–2050.

For the period 2041–2050, Table 6 presents a comparison of AS-1, AS-2, AS-3, AS-4, AS-5 with respect to S-0 for seasonal soil water content in the Bystra catchment for the RCP 4.5.1, RCP 4.5.2, RCP 4.5.3, RCP 8.5.1, RCP 8.5.2, and RCP 8.5.3 projections.

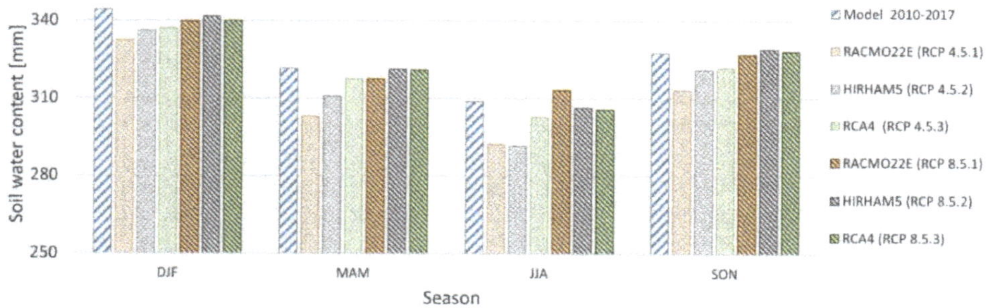

Figure 2. Seasonal average soil water content (1.5 m) for 2041–2050 and for the SWAT 2010–2017 model for individual climate projections RCP 4.5.1, RCP 4.5.2, RCP 4.5.3, RCP 8.5.1, RCP 8.5.2, RCP 8.5.3 (own study).

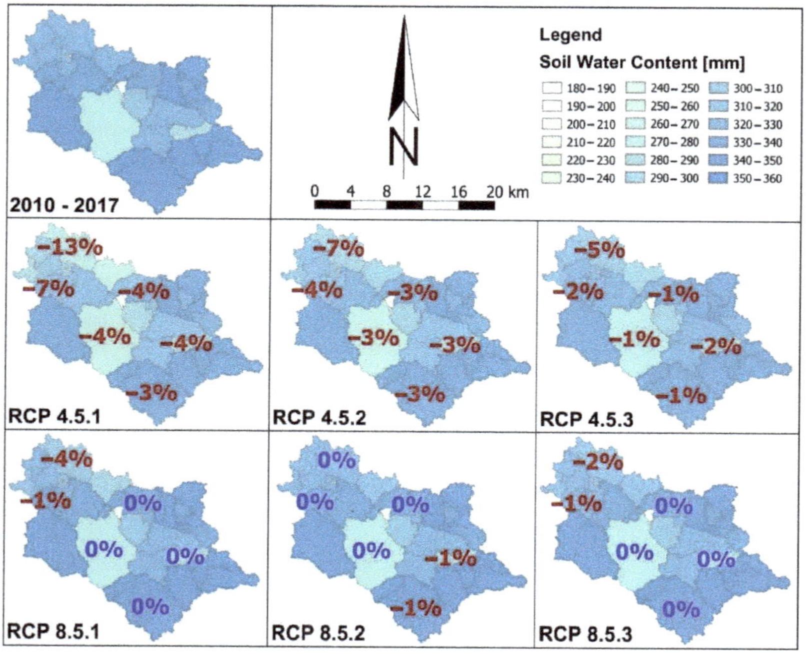

Figure 3. Comparison of average soil water content in 31 sub-catchments during the SWAT simulation period 2010–2017 and 2041–2050 for individual climate projections RCP 4.5.1, RCP 4.5.2, RCP 4.5.3, RCP 8.5.1, RCP 8.5.2, RCP 8.5.3 (own study).

Table 5. Comparison of average soil water content by season for the SWAT 2010–2017 simulation period with climate projections (RCP 4.5.1, RCP 4.5.2, RCP 4.5.3, RCP 8.5.1, RCP 8.5.2, RCP 8.5.3) for the years 2041–2050 in the Bystra catchment. Bold numbers indicate soil water content, while shaded numbers indicate percentage change (red is % decrease in content; blue is % increase in content). Dark red and dark blue shading means large changes, while light red and light blue shading means small changes (own study).

Climate Scenario		**RCP 4.5**			**RCP 8.5**		
Climate Projection	**Model 2010–2017**	**RACMO22E (RCP 4.5.1)**	**HIRHAM5 (RCP 4.5.2)**	**RCA4 (RCP 4.5.3)**	**RACMO22E (RCP 8.5.1)**	**HIRHAM5 (RCP 8.5.2)**	**RCA4 (RCP 8.5.3)**
Time interval		2041–2050					
Season		Seasonal average of soil water content (mm)					
DJF	**344**	**332** −3.5%	**336** −2.3%	**337** −2.1%	**340** −1.3%	**342** −0.7%	**340** −1.2%
MAM	**322**	**303** −5.8%	**311** −3.3%	**318** −1.3%	**318** −1.2%	**321** −0.1%	**321** −0.2%
JJA	**309**	**292** −5.4%	**291** −5.6%	**303** −2.0%	**313** +1.4%	**306** −0.8%	**306** −1.0%
SON	**328**	**313** −4.4%	**321** −2.0%	**322** −1.8%	**327** −0.2%	**329** +0.4%	**328** +0.2%
Average annual	**326**	**310** −4.7%	**315** −3.3%	**320** −1.8%	**324** −0.4%	**325** −0.3%	**324** −0.6%

Table 6. Comparison of average soil water content by season between scenario 0 (S-0) and adaptation scenarios 1–5 (AS-1, AS-2, AS-3, AS-4, AS-5) for 2041–2050 in the Bystra catchment for climate projection RCP 4.5.1, RCP 4.5.2, RCP 4.5.3, RCP 8.5.1, RCP 8.5.2, RCP 8.5.3. Bold numbers indicate soil water content, and shaded numbers indicate percentage change (red indicates % decrease in content and blue indicates % increase in content). Dark red and dark blue shading indicates large changes, while light red and light blue shading indicates small changes (own study).

Time Interval	**2041–2050**													
Type of Scenario	**S-0**	**AS-1**	**AS-2**	**AS-3**	**AS-4**	**AS-5**		**S-0**	**AS-1**	**AS-2**	**AS-3**	**AS-4**	**AS-5**	
Season	Seasonal average of soil water content (mm)													
DJF	**332**	**318** −4.2%	**332** −0.1%	**332** 0.0%	**333** +0.1%	**330** −0.6%	RACMO22E (RCP 4.5.1)	**340**	**326** −4.0%	**340** 0.0%	**340** 0.0%	**339** −0.1%	**339** −0.1%	RACMO22E (RCP 8.5.1)
MAM	**303**	**290** −4.1%	**303** 0.0%	**303** 0.0%	**303** +0.1%	**301** −0.7%		**318**	**305** −4.1%	**318** 0.0%	**318** 0.0%	**318** 0.0%	**317** −0.2%	
JJA	**292**	**278** −4.8%	**292** −0.1%	**292** 0.0%	**293** +0.3%	**288** −1.4%		**313**	**299** −4.6%	**313** −0.1%	**313** 0.0%	**314** +0.1%	**312** −0.4%	
SON	**313**	**299** −4.6%	**313** −0.1%	**313** 0.0%	**315** +0.4%	**310** −1.0%		**327**	**313** −4.4%	**327** −0.1%	**327** 0.0%	**328** +0.2%	**326** −0.3%	
Average annual	**310**	**296** −4.4%	**310** 0.0%	**310** 0.0%	**311** +0.2%	**307** −0.9%		**324**	**311** −4.3%	**324** 0.0%	**324** 0.0%	**325** +0.1%	**324** −0.2%	
DJF	**336**	**322** −4.1%	**336** 0.0%	**336** 0.0%	**337** +0.1%	**335** −0.2%	HIRHAM5 (RCP 4.5.2)	**342**	**328** −4.0%	**342** 0.0%	**342** 0.0%	**342** 0.0%	**342** 0.0%	HIRHAM5 (RCP 8.5.2)
MAM	**311**	**298** −4.1%	**311** 0.0%	**311** 0.0%	**311** 0.0%	**310** −0.2%		**321**	**308** −4.0%	**321** 0.0%	**321** 0.0%	**321** 0.0%	**321** −0.1%	
JJA	**291**	**277** −4.8%	**291** −0.1%	**291** 0.0%	**292** +0.2%	**288** −1.1%		**306**	**292** −4.6%	**306** −0.1%	**306** 0.0%	**307** +0.1%	**304** −0.6%	
SON	**321**	**307** −4.4%	**321** −0.1%	**321** 0.0%	**322** +0.3%	**319** −0.5%		**329**	**315** −4.3%	**329** −0.1%	**329** 0.0%	**330** +0.2%	**328** −0.2%	
Average annual	**315**	**301** −4.3%	**315** −0.1%	**315** 0.0%	**315** +0.2%	**313** −0.5%		**325**	**311** −4.2%	**324** 0.0%	**325** 0.0%	**325** +0.1%	**324** −0.3%	
DJF	**337**	**323** −4.1%	**337** 0.0%	**337** 0.0%	**337** 0.0%	**336** −0.2%	RCA4 (RCP 4.5.3)	**340**	**326** −4.0%	**340** 0.0%	**340** 0.0%	**340** 0.0%	**340** 0.0%	RCA4 (RCP 8.5.3)
MAM	**318**	**305** −4.1%	**317** 0.0%	**318** 0.0%	**318** 0.0%	**317** −0.2%		**321**	**308** −4.0%	**321** 0.0%	**321** 0.0%	**321** 0.0%	**321** −0.1%	
JJA	**303**	**289** −4.6%	**302** −0.1%	**303** 0.0%	**303** +0.2%	**300** −0.8%		**306**	**291** −4.7%	**305** −0.1%	**306** 0.0%	**306** +0.2%	**303** −0.7%	
SON	**322**	**307** −4.5%	**322** −0.1%	**322** 0.0%	**323** +0.2%	**320** −0.5%		**328**	**314** −4.3%	**328** −0.1%	**328** 0.0%	**329** +0.2%	**327** −0.4%	
Average annual	**320**	**306** −4.3%	**320** −0.1%	**320** 0.0%	**320** +0.1%	**318** −0.4%		**324**	**310** −4.3%	**324** −0.1%	**324** 0.0%	**324** +0.1%	**323** −0.3%	

Table 7 presents a comparison of total runoff by season for S-0 and AS-1, AS-2, AS-3, AS-4, AS-5 for the period 2041–2050 in the Bystra catchment. Next, Table 8 compares sediment yields by season for S-0 and AS-1, AS-2, AS-3, AS-4, AS-5 for 2041–2050 in the

Bystra catchment. In turn, Table 9 compares actual evapotranspiration by season for S-0 and AS-1, AS-2, AS-3, AS-4, AS-5 for the years 2041–2050 in the Bystra catchment.

Table 7. Comparison of seasonal total runoff between scenario 0 (S-0) and adaptation scenarios 1–5 (AS-1, AS-2, AS-3, AS-4, AS-5) for 2041–2050 in the Bystra catchment for climate projections RCP 4.5.1, RCP 4.5.2, RCP 4.5.3, RCP 8.5.1, RCP 8.5.2, RCP 8.5.3. Bold numbers indicate soil water content, and shaded numbers indicate percentage changes (red indicates % decrease in content, and blue indicates % increase in content). Dark red and dark blue shading indicates large changes, while light red and light blue shading indicates small changes (own study).

Time Interval	2041–2050														
Type of Scenario	S-0	AS-1	AS-2	AS-3	AS-4	AS-5		S-0	AS-1	AS-2	AS-3	AS-4	AS-5	
Season	Seasonal sum of total runoff (mm)													
DJF	35	**35** −1.0%	**35** −0.1%	**35** 0.0%	**36** +3.7%	**34** −3.1%		55	**54** −0.8%	**55** −0.1%	**55** 0.0%	**56** +2.4%	**54** −1.3%	
MAM	31	**31** −0.5%	**31** −0.2%	**31** 0.0%	**32** +3.6%	**30** −3.3%	RACMO22E (RCP 4.5.1)	47	**47** 0.1%	**47** −0.1%	**47** 0.0%	**48** +2.4%	**46** −1.1%	RACMO22E (RCP 8.5.1)
JJA	30	**30** −0.2%	**30** −0.2%	**30** 0.0%	**31** +3.2%	**29** −3.9%		49	**49** 0.0%	**49** −0.1%	**49** 0.0%	**50** +2.1%	**48** −1.8%	
SON	32	**32** −0.4%	**32** −0.1%	**32** 0.0%	**34** +4.7%	**31** −3.5%		49	**48** −0.3%	**49** −0.1%	**49** 0.0%	**50** +3.2%	**48** −1.6%	
Annual sum	128	**127** −0.5%	**128** −0.2%	**128** 0.0%	**133** +3.8%	**123** −3.4%		199	**199** −0.3%	**199** −0.1%	**199** 0.0%	**204** +2.5%	**196** −1.5%	
DJF	39	**39** −0.4%	**39** −0.2%	**39** 0.0%	**40** +2.8%	**38** −2.7%		52	**52** −0.3%	**52** −0.1%	**52** 0.0%	**53** +2.0%	**51** −1.6%	
MAM	43	**43** −0.2%	**43** −0.2%	**43** 0.0%	**44** +2.4%	**42** −2.1%	HIRHAM5 (RCP 4.5.2)	61	**61** −0.2%	**61** −0.1%	**61** 0.0%	**61** +1.3%	**60** −1.2%	HIRHAM5 (RCP 8.5.2)
JJA	36	**36** +0.3%	**36** 0.0%	**36** 0.0%	**37** +2.3%	**35** −3.1%		53	**53** +0.2%	**53** 0.0%	**53** 0.0%	**54** +1.4%	**52** −1.6%	
SON	36	**36** −0.4%	**36** −0.1%	**36** 0.0%	**37** +3.5%	**35** −3.3%		51	**51** −0.2%	**51** −0.1%	**51** 0.0%	**52** +2.3%	**50** −1.7%	
Annual sum	154	**154** −0.2%	**154** −0.1%	**154** 0.0%	**159** +2.7%	**150** −2.8%		217	**216** −0.1%	**216** −0.1%	**217** 0.0%	**220** +1.7%	**213** −1.5%	
DJF	50	**50** −0.4%	**50** −0.1%	**50** 0.0%	**52** +2.5%	**50** −1.7%		71	**71** −0.2%	**71** −0.1%	**71** 0.0%	**72** +2.1%	**70** −1.3%	
MAM	51	**50** −0.2%	**51** −0.1%	**51** 0.0%	**52** +2.2%	**50** −1.3%	RCA4 (RCP 4.5.3)	68	**68** −0.2%	**68** −0.1%	**68** 0.0%	**69** +1.8%	**67** −1.0%	RCA4 (RCP 8.5.3)
JJA	39	**39** +0.2%	**39** 0.0%	**39** 0.0%	**40** +2.4%	**39** −2.0%		60	**60** 0.0%	**60** 0.0%	**60** 0.0%	**61** +1.6%	**59** −1.3%	
SON	44	**44** −0.3%	**44** −0.2%	**44** 0.0%	**46** +3.4%	**44** −1.9%		70	**69** −0.5%	**70** −0.1%	**70** 0.0%	**72** +2.5%	**69** −1.1%	
Annual sum	185	**184** −0.2%	**185** −0.1%	**185** 0.0%	**190** +2.6%	**182** −1.7%		268	**268** −0.2%	**268** −0.1%	**268** 0.0%	**274** +2.0%	**265** −1.2%	

Table 8. Comparison of seasonal sediment yield between scenario 0 (S-0) and adaptation scenarios 1–5 (AS-1, AS-2, AS-3, AS-4, AS-5) for 2041–2050 in the Bystra catchment for climate projections RCP 4.5.1, RCP 4.5.2, RCP 4.5.3, RCP 8.5.1, RCP 8.5.2, RCP 8.5.3. Bold numbers indicate soil water content, and shaded numbers indicate percentage changes (red indicates % decrease in content, and blue indicates % increase in content). Dark red and dark blue shading indicates large changes, while light red and light blue shading indicates small changes (own study).

Time Interval	2041–2050														
Type of Scenario	S-0	AS-1	AS-2	AS-3	AS-4	AS-5		S-0	AS-1	AS-2	AS-3	AS-4	AS-5	
Season	Seasonal sum of sediment yield (t/ha)													
DJF	0.17	**0.16** −9%	**0.17** −1%	**0.05** −72%	**0.12** −28%	**0.16** −5%		0.25	**0.23** −9%	**0.25** 0%	**0.07** −71%	**0.18** −30%	**0.24** −5%	
MAM	0.08	**0.07** −9%	**0.08** 0%	**0.02** −71%	**0.07** −14%	**0.07** −8%	RACMO22E (RCP 4.5.1)	0.07	**0.06** −7%	**0.07** 0%	**0.02** −71%	**0.06** −7%	**0.06** −7%	RACMO22E (RCP 8.5.1)
JJA	0.15	**0.14** −9%	**0.15** −1%	**0.05** −70%	**0.13** −13%	**0.12** −21%		0.22	**0.20** −9%	**0.22** 0%	**0.06** −72%	**0.16** −28%	**0.18** −17%	
SON	0.15	**0.14** −7%	**0.15** −1%	**0.04** −72%	**0.08** −48%	**0.13** −14%		0.18	**0.16** −11%	**0.18** −1%	**0.05** −72%	**0.11** −41%	**0.17** −5%	
Annual sum	0.55	**0.51** −9%	**0.55** −1%	**0.16** −71%	**0.40** −27%	**0.49** −12%		0.72	**0.65** −9%	**0.71** 0%	**0.20** −72%	**0.50** −30%	**0.65** −9%	

Table 8. *Cont.*

Time Interval	2041–2050													
Type of Scenario	**S-0**	**AS-1**	**AS-2**	**AS-3**	**AS-4**	**AS-5**		**S-0**	**AS-1**	**AS-2**	**AS-3**	**AS-4**	**AS-5**	
DJF	**0.11**	**0.10** −10%	**0.10** −1%	**0.03** −72%	**0.08** −23%	**0.10** −4%	HIRHAM5 (RCP 4.5.2)	**0.12**	**0.11** −9%	**0.12** 0%	**0.04** −71%	**0.09** −22%	**0.12** −4%	HIRHAM5 (RCP 8.5.2)
MAM	**0.13**	**0.12** −6%	**0.13** 0%	**0.04** −71%	**0.12** −7%	**0.12** −8%		**0.24**	**0.22** −8%	**0.24** −1%	**0.07** −71%	**0.21** −13%	**0.22** −8%	
JJA	**0.12**	**0.11** −11%	**0.12** −2%	**0.03** −72%	**0.09** −25%	**0.10** −16%		**0.21**	**0.20** −8%	**0.21** −1%	**0.06** −71%	**0.16** −23%	**0.17** −18%	
SON	**0.19**	**0.17** −9%	**0.19** −1%	**0.05** −73%	**0.10** −49%	**0.16** −13%		**0.22**	**0.20** −10%	**0.22** −1%	**0.06** −72%	**0.12** −44%	**0.20** −10%	
Annual sum	**0.54**	**0.50** −9%	**0.54** −1%	**0.15** −72%	**0.39** −29%	**0.48** −11%		**0.79**	**0.72** −9%	**0.79** −1%	**0.23** −71%	**0.59** −25%	**0.71** −11%	
DJF	**0.11**	**0.10** −9%	**0.11** −1%	**0.03** −73%	**0.09** −23%	**0.11** 0%	RCA4 (RCP 4.5.3)	**0.14**	**0.12** −9%	**0.14** −1%	**0.04** −72%	**0.10** −24%	**0.13** −7%	RCA4 (RCP 8.5.3)
MAM	**0.15**	**0.14** −11%	**0.15** 0%	**0.05** −70%	**0.14** −7%	**0.13** −12%		**0.16**	**0.15** −9%	**0.16** −1%	**0.05** −72%	**0.15** −6%	**0.14** −14%	
JJA	**0.11**	**0.10** −12%	**0.11** −1%	**0.03** −73%	**0.08** −25%	**0.09** −21%		**0.14**	**0.12** −11%	**0.14** −1%	**0.04** −71%	**0.13** −7%	**0.11** −23%	
SON	**0.20**	**0.18** −9%	**0.20** 0%	**0.06** −72%	**0.12** −42%	**0.18** −10%		**0.57**	**0.51** −10%	**0.56** −1%	**0.16** −72%	**0.32** −43%	**0.50** −12%	
Annual sum	**0.57**	**0.52** −10%	**0.57** −1%	**0.16** −72%	**0.43** −26%	**0.51** −11%		**1.00**	**0.90** −10%	**0.99** −1%	**0.28** −72%	**0.71** −29%	**0.87** −13%	

Table 9. Comparison of seasonal actual evapotranspiration between scenario 0 (S-0) and adaptation scenarios 1–5 (AS-1, AS-2, AS-3, AS-4, AS-5) for 2041–2050 in the Bystra catchment for climate projections RCP 4.5.1, RCP 4.5.2, RCP 4.5.3, RCP 8.5.1, RCP 8.5.2, RCP 8.5.3. Bold numbers indicate soil water content and shaded numbers indicate percentage changes (red indicates % decrease in content, and blue indicates % increase in content). Dark red and dark blue shading indicates large changes, while light red and light blue shading indicates small changes (own study).

Time Interval	2041–2050													
Type of Scenario	**S-0**	**AS-1**	**AS-2**	**AS-3**	**AS-4**	**AS-5**		**S-0**	**AS-1**	**AS-2**	**AS-3**	**AS-4**	**AS-5**	
Season	Seasonal sum of actual evapotranspiration (mm)													
DJF	**27**	**27** −0.9%	**27** −0.1%	**27** 0.0%	**27** −0.2%	**27** −0.4%	RACMO22E (RCP 4.5.1)	**29**	**29** −0.7%	**29** 0.0%	**29** 0.0%	**29** −0.4%	**29** −0.5%	RACMO22E (RCP 8.5.1)
MAM	**154**	**151** −1.9%	**154** −0.1%	**154** 0.0%	**154** 0.0%	**156** +1.4%		**156**	**154** −1.7%	**156** −0.1%	**156** 0.0%	**156** −0.3%	**157** +0.7%	
JJA	**166**	**169** +2.0%	**166** +0.2%	**166** 0.0%	**163** −1.5%	**168** +1.4%		**165**	**168** +1.8%	**165** +0.2%	**165** 0.0%	**162** −1.5%	**166** +1.0%	
SON	**70**	**70** +0.4%	**70** 0.0%	**70** 0.0%	**67** −4.0%	**70** +0.1%		**70**	**70** +0.3%	**70** 0.0%	**70** 0.0%	**68** −3.2%	**70** +0.1%	
Annual sum	**416**	**417** +0.1%	**416** 0.0%	**416** 0.0%	**411** −1.3%	**421** +1.1%		**420**	**420** +0.1%	**420** 0.0%	**420** 0.0%	**415** −1.3%	**423** +0.7%	
DJF	**24**	**24** −0.7%	**24** 0.0%	**24** 0.0%	**24** −0.2%	**24** −0.4%	HIRHAM5 (RCP 4.5.2)	**23**	**23** −0.7%	**23** 0.0%	**23** 0.0%	**23** −0.3%	**23** −0.4%	HIRHAM5 (RCP 8.5.2)
MAM	**149**	**146** −1.9%	**148** −0.1%	**149** 0.0%	**148** −0.1%	**150** +1.2%		**135**	**133** −1.7%	**135** −0.1%	**135** 0.0%	**136** +0.1%	**137** +1.0%	
JJA	**152**	**155** +2.0%	**152** +0.2%	**152** 0.0%	**150** −1.2%	**154** +1.7%		**152**	**155** +1.7%	**153** +0.2%	**152** 0.0%	**150** −1.3%	**154** +1.3%	
SON	**61**	**61** +0.3%	**61** 0.0%	**61** 0.0%	**59** −3.6%	**61** 0.0%		**65**	**66** +0.1%	**65** 0.0%	**65** 0.0%	**63** −3.3%	**65** 0.0%	
Annual sum	**386**	**386** +0.1%	**386** 0.0%	**386** 0.0%	**382** −1.1%	**390** +1.1%		**377**	**377** 0.0%	**377** 0.0%	**377** 0.0%	**372** −1.1%	**380** +0.9%	
DJF	**31**	**31** −0.7%	**31** 0.0%	**31** 0.0%	**31** −0.2%	**31** −0.6%	RCA4 (RCP 4.5.3)	**34**	**34** −0.8%	**34** 0.0%	**34** 0.0%	**34** −0.2%	**34** −0.5%	RCA4 (RCP 8.5.3)
MAM	**136**	**134** −1.7%	**136** −0.1%	**136** 0.0%	**136** −0.2%	**137** +0.8%		**141**	**138** −1.6%	**141** −0.1%	**141** 0.0%	**141** −0.1%	**142** +0.8%	
JJA	**168**	**170** +1.5%	**168** +0.2%	**168** 0.0%	**165** −1.4%	**170** +1.3%		**158**	**161** +1.7%	**158** +0.2%	**158** 0.0%	**156** −1.5%	**160** +1.3%	
SON	**69**	**70** +0.4%	**69** 0.0%	**69** 0.0%	**67** −3.5%	**69** 0.0%		**72**	**73** +0.2%	**73** 0.0%	**72** 0.0%	**69** −4.2%	**72** 0.0%	
Annual sum	**404**	**405** +0.1%	**405** 0.0%	**404** 0.0%	**399** −1.3%	**408** +0.8%		**406**	**406** +0.1%	**406** 0.0%	**406** 0.0%	**400** −1.4%	**409** +0.7%	

3.1. Analysis of Soil Water Content in Zero Scenario for 2021–2050

This section compares the obtained seasonal average soil water content results for 2010–2017 (SWAT model) with the results for 2041–2050 for individual climate change projections (RCP 4.5.1, RCP 4.5.2, RCP 4.5.3, RCP 8.5.1, RCP 8.5.2, RCP 8.5.3).

Regardless of the individual climate change projections evaluated, the seasonal average soil water content for the Bystra catchment is projected to decrease between 2041 and 2050 for most seasons compared to 2010–2017 (Table 5).

Lower soil water content will be especially evident for RCP 4.5.1 (MAM, JJA, SON) and RCP 4.5.2 (JJA) where the value of average soil water content may be lower by up to 5.8% compared to the 2010–2017 simulation period. Lower values in the MAM and JJA seasons, especially for the RCP 4.5.1, RCP 4.5.2, RCP 4.5.3 projections, may affect plant growth during the growing season. However, higher soil water content (1.4% higher) was found for the RCP 8.5.1 (JJA), RCP 8.5.2 and RCP 8.5.3 (SON) projections.

Regardless of the regional climate model, the seasonal average soil water content will be lower for climate projections RCP 4.5.1, RCP 4.5.2, RCP 4.5.3 compared to climate projections RCP 8.5.1, RCP 8.5.2, RCP 8.5.3. This is particularly evident when comparing the average annual soil water content results, where, for RCP 4.5.1, RCP 4.5.2, RCP 4.5.3, the average annual soil water content results (2041–2050) are lower between 1.8% and 4.7%, while, for RCP 8.5.1, RCP 8.5.2, RCP 8.5.3, these average annual results are lower between 0.3% and 0.6% compared to the SWAT 2010–2017 model.

The average soil water content by season for 2041–2050 and the SWAT model 2010–2017 is shown in Figure 2 (Figure 2). It shows that the average soil water content decreases throughout the year. The highest soil water values are reached during the winter season of DJF. On the other hand, in spring (MAM), during the growing season period, the average soil water content decreases, maintaining the lowest values in summer (JJA). In autumn (SON), the soil water content increases.

Analyzing the spatial distribution of changes in the average water content in soil in 31 sub-catchments for the simulation period in 2010–2017 in relation to the period 2041–2050 (Figure 3) in the climate forecasts RCP 4.5.1 and RCP 4.5.3, the average water content in the soil will decrease by a few percent points in the Northwest region for most of the projections. In the projections RCP 4.5.1, RCP 4.5.2 and RCP 4.53, a reduced water content in the soil will occur throughout the catchment area, while in the projections RCP 8.5.1, RCP 8.5.2 and RCP 8.5.3, the changes will be small.

3.2. Climate Change Adaptation Scenarios Analysis 1–5 for 2041–2050

The results presented in Section 3.1 indicate a decrease in soil water content in most seasons during the period 2041–2050 (Table 5, Figure 2). To counteract the negative effects of changes in soil water content, five adaptation scenarios (AS-1, AS-2, AS-3, AS-4, AS-5) were prepared and tested. They were designed to maintain or increase soil water content. The analysis covers the period 2041–2050. Additionally, the impact of adaptation scenarios on total runoff, sediment yield and actual evapotranspiration was compared.

AS-1 of increasing forested areas on soils of complex 6, 7, 8 compared to S-0 for all projections shows a decrease in soil water content for all seasons in the Bystra catchment (Table 6). The soil water content decreases from 4.0% to 4.8% for all seasons.

AS-2, which assumes a forested buffer zone near the Bystra River, shows a slight decrease in soil water content between 2041 and 2050 (Table 6).

AS-3, establishing filter strips, shows no change in soil water content (Table 6).

In AS-4, the application of plowing on arable land—BARL, CANP, CRDY, WWHT—was eliminated. This treatment showed a slight increase in soil water content. The increase ranged from 0% to 0.4%. The largest increases occurred in the JJA and SON seasons (Table 6).

AS-5 increased soil organic carbon to 2%. This treatment showed a slight decrease in soil water content. The decrease in soil water content ranged from 0% to 1.4% (Table 6).

Regardless of the GCMs/RMCs and the RCPs evaluated, the results are the same. This means that AS-1 is associated with a greater decrease in soil water content compared to S-0. AS-2 and AS-5 are associated with a decrease of a smaller magnitude compared to AS-1. AS-3 does not predict any significant change in soil water content. In contrast, AS-4 is associated with a small increase in soil water content.

Regardless of the regional climate model, the seasonal average soil water content will be lower under the RCP 4.5 climate change scenario compared to the RCP 8.5 climate change scenario. This is described in more detail in Section 3.1.

Differences in annual average soil water content between AS-2, AS-3, AS-4, AS-5 and S-0 are small. However, for AS-1, the annual average soil water content varies between 296 and 311 mm. In contrast, for S-0, the average annual soil water content is 310–325 mm (Table 6).

AS-1 and AS-2 show a slight decrease for most seasons of total runoff for 2041–2050 compared to S-0 in all climate projections. Changes in total runoff range from 1% (decrease) to 0.3% (increase) (Table 7).

Total runoff in AS-3 did not change (Table 7).

AS-4 shows an increase in total runoff for all seasons in all projections. The increase ranges from 1.3% to 4.7% (Table 7). For climate projection RCP 4.5.1, the increase in total runoff stands out from the other projections in all seasons (above 3%).

In contrast, AS-5 shows a decrease in total runoff for all seasons across all projections. The decrease ranges from 1.0% to 3.9% (Table 7). For climate projections RCP 4.5.1 and RCP 4.5.2, the decrease in total runoff stands out from the other projections in all seasons (above 2%) (Table 7).

Moreover, for total runoff regardless of the GCMs/RMCs and RCPs evaluated, the results are the same. AS-1 and AS-2 have smaller total runoff compared to adaptation S-0. AS-5 has an even smaller total runoff compared to AS-1 and AS-2.

In AS-3, the total runoff does not change. In contrast, AS-4 shows an increase in total runoff compared to all adaptation scenarios.

Regardless of the regional climate model, the average seasonal total runoff will be lower for the RCP 4.5 climate change scenario compared to the RCP 8.5 scenario [8].

Table 8 presents the seasonal sediment yield data (Table 8). AS-1, AS-2, AS-3, AS-4, AS-5 were compared to S-0 for the climate projections. For most adaptation scenarios, there is a reduction in sediment yield from 0% to as much as 73%. The smallest, slight decreases in sediment yield occur in AS-2 compared to S-0. Slightly larger decreases compared to S-0 and AS- 2 occur in AS-1 and AS-5. Large decreases in sediment yield occur in AS-4 (ranging from 6% to 49%). However, the largest occur for AS-3 (over 70%).

For sediment yield, regardless of the GCMs/RMCs and RCPs evaluated, the results are also the same (Table 8).

Regardless of the regional climate model, the seasonal sediment yield will be lower under the RCP 4.5 climate change scenario compared to the RCP 8.5 scenario. Differences in annual sum sediment yields range from 0.54–0.57 t/ha for RCP 4.5 to 0.72–1.00 t/ha for RCP 8.5 in S-0 (Table 8). For AS-4, the annual sum ranges from 0.39–0.40 t/ha for RCP 4.5 to 0.50–0.71 t/ha for RCP 8.5, while for AS-3, the annual sum ranges from 0.15–016 t/ha for RCP 4.5 to 0.20–0.28 t/ha for RCP 8.5.

Table 9 presents data on seasonal actual evapotranspiration (Table 9). The highest evapotranspiration values occur during the MAM and JJA seasons.

AS-1, AS-2, AS-3, AS-4, AS-5 were compared to S-0 for all climate projections. AS-1 shows a decrease in actual evapotranspiration from 1.7% to 1.9% for the MAM season. In contrast, there is an increase between 1.5% and 2.0% for the JJA season.

AS-2 shows little change in actual evapotranspiration (Table 9).

The actual evapotranspiration in AS-3 remains unchanged compared to S-0 (Table 9).

In AS-4, for the MAM and JJA seasons, actual evapotranspiration varies from 1.5% (decrease) to 0.1% (increase) (Table 9) compared to S-0. However, large decreases occur for the SON season (from 3.2% to 4.2%).

In contrast, AS-5 has increases in actual evapotranspiration of 0.7% to 1.7% for the MAM and JJA seasons compared to S-0.

For actual evapotranspiration, regardless of the GCMs/RMCs and RCPs evaluated, the results are the same (Table 9).

Regardless of the regional climate model, seasonal actual evapotranspiration will be similar under the RCP 4.5 climate change scenario compared to the RCP 8.5 scenario [8].

For AS-1, AS-2, AS-3, the annual sum of actual evapotranspiration changes little. However, for AS-4, the annual sum of actual evapotranspiration increases from 1.7% to 3.8% compared to S-0. In contrast, for AS-5, the annual sum decreases from 1.2% to 3.4% (Table 9).

Table 10 shows the percentage sets of changes in soil water content, sediment productivity, total runoff, and actual evapotranspiration under AS-1, AS-2, AS-3, AS-4, AS-5 with respect to S-0 (Table 10). The table was created based on supplementary Material: Figures S1 and S2, for averages of three GCMs/RCMs combinations under two RCP climate change scenarios (RCP 4.5, RCP 8.5).

Table 10. Percent summary of changes in soil water content, sediment yield, total runoff and actual evapotranspiration under adaptation scenarios 1–5 (AS-1, AS-2, AS-3, AS-4, AS-5) compared to scenario 0 (S-0) (created from Supplementary Material: Figures S1 and S2), for averages of three GCMs/RCMs combinations under two RCP climate change scenarios (RCP 4.5, RCP 8.5). The summary is for four seasons (DJF, MAM, JJA, SON) in the Bystra catchment. Shaded numbers indicate percentage changes (red indicates % decrease in content, and blue indicates % increase in content). Dark red and dark blue shading indicates large changes, while light red and light blue shading indicates small changes (own study).

	RCP 4.5				RCP 8.5				
Season	**Soil water Content (mm)**	**Total Runoff (mm)**	**Sediment Yield (t/ha)**	**Actual Evapo-transpiration (mm)**	**Soil Water Content (mm)**	**Total Runoff (mm)**	**Sediment Yield (t/ha)**	**Actual Evapo-transpiration (mm)**	
DJF	−4.1%	−0.6%	−9.3%	−0.7%	−4.0%	−0.4%	−9.1%	−0.7%	AS-1
MAM	−4.1%	−0.2%	−8.7%	−1.9%	−4.0%	−0.1%	−8.5%	−1.7%	
JJA	−4.7%	+0.1%	−10.4%	+1.9%	−4.6%	+0.1%	−9.1%	+1.7%	
SON	−4.5%	−0.4%	−8.5%	+0.4%	−4.3%	−0.4%	−10.5%	+0.2%	
Average	−4.4%	−0.3%	−9.2%	+0.1%	−4.2%	−0.2%	−9.5%	+0.1%	
DJF	0.0%	−0.1%	−1.0%	0.0%	0.0%	−0.1%	−0.4%	0.0%	AS-2
MAM	0.0%	−0.1%	0.0%	−0.1%	0.0%	−0.1%	−0.8%	−0.1%	
JJA	−0.1%	−0.1%	−1.3%	+0.2%	−0.1%	0.0%	−0.7%	+0.2%	
SON	−0.1%	−0.1%	−0.6%	0.0%	−0.1%	−0.1%	−0.9%	0.0%	
Average	−0.1%	−0.1%	−0.7%	0.0%	0.0%	−0.1%	−0.8%	0.0%	
DJF	0.0%	0.0%	−72.4%	0.0%	0.0%	0.0%	−71.5%	0.0%	AS-3
MAM	0.0%	0.0%	−70.4%	0.0%	0.0%	0.0%	−71.3%	0.0%	
JJA	0.0%	0.0%	−71.6%	0.0%	0.0%	0.0%	−71.4%	0.0%	
SON	0.0%	0.0%	−72.5%	0.0%	0.0%	0.0%	−72.2%	0.0%	
Average	0.0%	0.0%	−71.8%	0.0%	0.0%	0.0%	−71.7%	0.0%	
DJF	+0.1%	+2.9%	−25.3%	−0.2%	0.0%	+2.2%	−26.2%	−0.3%	AS-4
MAM	0.0%	+2.6%	−8.5%	−0.1%	0.0%	+1.8%	−9.8%	−0.1%	
JJA	+0.2%	+2.6%	−20.3%	−1.4%	+0.2%	+1.7%	−20.9%	−1.4%	
SON	+0.3%	+3.8%	−46.2%	−3.7%	+0.2%	+2.6%	−42.7%	−3.6%	
Average	+0.2%	+3.0%	−27.3%	−1.2%	+0.1%	+2.1%	−28.2%	−1.2%	
DJF	−0.3%	−2.4%	−3.1%	−0.5%	−0.1%	−1.4%	−5.1%	−0.5%	AS-5
MAM	−0.4%	−2.1%	−9.6%	+1.1%	−0.1%	−1.1%	−9.8%	+0.9%	
JJA	−1.1%	−2.9%	−19.5%	+1.5%	−0.6%	−1.6%	−18.8%	+1.2%	
SON	−0.7%	−2.8%	−12.2%	0.0%	−0.3%	−1.4%	−10.5%	0.0%	
Average	−0.6%	−2.5%	−11.2%	+1.0%	−0.3%	−1.4%	−11.2%	+0.8%	

4. Discussion

The results concerning the water content in the soil were compared with the available values of water capacity and the wilting point obtained from the study "Assessment of water retention in soil and the risk of drought based on the water balance of the Lower Silesia Voivodshi", developed in 2013 by the employees of the Department of Soil Science, Erosion and Land Protection, IUNG-PIB in Pulawy [1]. Based on the above-mentioned study, we prepared data on soils in the catchment area of the Bystra River. For a 1.5 m soil profile, the results of the above-mentioned studies are consistent with this publication.

The lowest water content in soil occurs in the summer (JJA), while the highest occurs in the winter (DJF) (Figure 2). For 2041–2050, the largest decreases in soil water content are associated with GCMs/RCMs for RCP 4.5, while small changes occur for RCP 8.5.

The analyzed adaptation scenarios present different results of the influence on the water content in the soil. AS-1 for an increase in forest area on soils of the complex 6, 7, 8 compared to S-0 for all projections shows a reduction in soil water content for all seasons across the entire Bystra catchment (Tables 6 and 10, Figures S1 and S2 in Supplementary Material). The same is true for total runoff. Again, for most seasons, there is a reduction in total runoff (all projections) compared to S-0 (Tables 7 and 10, Figures S1 and S2 in Supplementary Material). Sediment yields for all seasons also decrease (Tables 8 and 10, Figures S1 and S2 in Supplementary Material). In contrast, actual evapotranspiration shows a decrease in the MAM season and an increase in the JJA season (Tables 9 and 10, Figures S1 and S2 in Supplementary Material).

Forests play an important role in absorbing CO_2, which is an important factor in reducing the adverse effects of climate change [78]. In addition to absorbing CO2, forest ecosystems can counteract soil erosion and drainage. Within forests, there may be small retention reservoirs, increasing the areas' abundance of water. Forest ecosystems play very important natural, social and productive functions [79]. The results indicate that increasing afforested area in the Bystra catchment has to go beyond the scheme of using soil complexes less favorable for agricultural production, and the areas should be picked with care, focusing on locating forested areas close to catchment borders, so they can slow runoff and help accumulate water at its highest point from the river bed [80].

The large-scale research aimed at estimating the amount of tree stand in the world shows that there are currently 46% fewer trees than before the advent of human civilization [81]. Climate change may affect the condition of forest areas [19] manifested in extreme weather phenomena that begin to lose their anomaly status (hurricanes, droughts). Moreover, the species status of plants and trees may not be flexible enough to adapt to changing climate components (temperature, precipitation, etc.) [82]. Forests therefore should be probably re-designed to cope with changing biotopes. For many years, many concepts regarding forest formation in relation to a changing climate have been considered. These plans are based on the development of actions to reduce the effects of unfavorable phenomena which are occurring now and which may intensify in the future. Another concept will be activities aimed at adapting forest ecosystems to all current and future threats [82].

A program of increasing forest cover is implemented in Poland [83]. According to the report on the condition of forests in Poland in 2020 [84], the level of forest cover in 2020 amounted to 29.6% of the total area of the country. After 2050, the forest cover in Poland is expected to be 33%. The program assumes afforestation of land of low agricultural suitability [85], reflected in AS-1 of this study.

Research using afforestation scenarios was carried out on four sites in Bolivia and Ecuador [86]. They show that the water content in the soil and the total runoff decreased to a varying degree after the application of the forest ecosystem. AS-1 and AS-2 also show a reduction in soil water content (Tables 6 and 10, Figures S1 and S2 in Supplementary Material) and a slight reduction in total runoff (Tables 7 and 10, Figures S1 and S2 in Supplementary Material). Sediment yield also decreased (Tables 8 and 10, Figures S1 and S2 in Supplementary Material). The decrease in soil water content for AS-1 and AS-2 in the Bystra catchment may be caused by increased water uptake by the root system of forest vegetation species.

The afforestation scenario has the potential for further research, in which it is possible to design an appropriate location of forest ecosystems in the Bystra river catchment area, relying not only on the afforestation of soils of complex 6, 7 and 8, but also good tree planting practices in rural areas [87], the use of forested embankment fortifications (also preventing erosion) [72], which would counteract the unfavorable agro-forest checkerboard [88]. The unfavorable location of forest ecosystems near cultivated fields may result in a reduction in the yield of agricultural plants [89,90]. When designing afforestation, one should also take into account the adaptation possibilities of stands to new climatic conditions [82].

Increasing forest cover from 16.34% (S-0) to 19.65% (AS-1) or to 17.37% (AS-2) (Table 4) according to Lambo's forest cover index [91] allows for increased forest retention capacity that, among other things, counteracts the effects of flooding [92]. In addition to increasing forest cover, equally important is the location of forested areas within the catchment area which has a significant impact on runoff [93].

A buffer zone with a well-developed tree stand, located directly next to watercourses, can prevent the runoff of nutrients and suspensions from agricultural land, contribute to the strengthening of banks and prevent lateral erosion [72,87,94]. A marsh zone forming a belt of wetland and rush vegetation, flooded or boggy for most of the year or all the time, can also be a buffer. Such a zone with well-developed vegetation contributes to the retention of a significant amount of nitrogen and phosphorus from the catchment area, preventing eutrophication of waters [72,95].

AS-3, for the creation of filter strips in a planned management operation on BARL, CANP, CRDY, WWHT arable land, shows no changes in soil water content, total runoff or actual evapotranspiration (Tables 6, 7, 9 and 10, Figures S1 and S2 in Supplementary Material). On the other hand, the filter strips effectively reduce the sediment yield (t/ha) (Tables 8 and 10, Figures S1 and S2 in Supplementary Material). Similar results were obtained in the article describing the use of the filter strips in various scenarios on the example of the catchment area in Thailand [96], where, as a result of their use, the sediment yield was significantly reduced.

Adaptation scenarios involving increasing forest cover, creating buffers next to rivers and creating filter strips can help reduce erosion risk in the 2050 climate horizon in the Bystra catchment by reducing total runoff and decreasing sediment yield.

AS-4, for the cessation of plowing on BARL, CANP, CRDY, WWHT arable land, shows a slight increase (especially in the JJA and SON season) in soil water content (Tables 6 and 10, Figures S1 and S2 in Supplementary Material). The elimination of plowing also shows a significant reduction in sediment yield (t/ha) (Tables 8 and 10, Figures S1 and S2 in Supplementary Material). This may have the effect of reducing soil erosion. However, the total runoff increased, which is induced mainly by the reduction in actual evapotranspiration, especially limited evaporation form the soil surface covered by plant residue mulch (Tables 7, 9 and 10, Figures S1 and S2 in Supplementary Material). Observations by Wawer and Kozyra [97] confirm the prominent role of mulching in preserving soil water by covering the surface of the soil in warm periods.

The discontinuation of plowing is the subject of many articles as well as studies that mention as benefits the reduction in soil erosion, the reduction in surface and subsurface runoff, the reduction in sediment yield, nitrogen yield and phosphorus yield, the increase in soil water content, etc. [76,98–101], which are supported by numerous studies. The abandonment of plowing in the catchment areas in the climate of 2050 also shows a reduction in the sediment yield. On the other hand, the water content in the soil increases. This provides the grounds that new agricultural practices in the coming decades may prevent the negative impact of watershed water deficits from occurring.

Agriculture is closely related to the prevailing climatic conditions, but it also has a large impact on them. The risk of an increase in the frequency of unfavorable climatic conditions in agriculture may result in yield variability from year to year. The reduced amount of water in the soil during plant growth, illustrated in the climate change scenarios (Table 5, Figure 2), will become more frequent and more severe. Other threats will also include droughts, heavy precipitation, erosion [80], floods, landslides and strong winds [102].

AS-5, increasing soil organic carbon to 2%, shows reductions in soil water content, total runoff and sediment yield (Tables 6–8, Figures S1 and S2 in Supplementary Materials). However, actual evapotranspiration increases (Tables 9 and 10, Figures S1 and S2 in Supplementary Materials). In a paper on soil organic carbon changes and their response to climate warming and soil water content changes, a study of the Jinghe catchment in China was described [103]. The study showed that temperature and precipitation will increase by the end of the 21st century under three scenarios—RCP 2.6, RCP 4.5, RCP 8.5—and consequently soil water content will also increase, while organic carbon content will decrease, depending on the climate change scenario. The study also showed that there is a threshold in soil water content that can mediate the loss of soil organic carbon (when the change in soil water content was lower than the threshold, higher content accelerated the loss of organic carbon, while when the change in soil water content was higher than the threshold, higher content reduced the loss of soil organic carbon) [103]. The mechanism for the decrease in soil organic carbon (despite increased soil water content) due to a warming climate in the future is not fully known [103]. Global studies have found a link between faster CO_2 increases in warmer years with less water availability. This demonstrates the importance of warming on the decomposition of soil organic carbon [104]. There are studies in pols on the effect of soil organic matter on soil water management [105]. According to some estimates in the article, increasing soil organic matter by 0.01% increases the amount of organic matter by 480 kg (from 1 hectare of arable soil layer). This corresponds to 278 kg of organic carbon. On the scale of the national area (Poland), this means the sequestration of 11 million tons of CO_2 from the entire arable land area of Poland. This represents more than 3% of the total greenhouse gas emissions from the Polish area [77].

A convenient tool for carrying out beneficial changes (afforestation, retention reservoirs, irrigation) in terms of water retention in the landscape is land consolidation on an extended scope [80,106,107]. Several agricultural research centers in Poland deal with the issues of recomposing the rural landscape, including IUNG-PIB in Pulawy. At IUNG-PIB, a broader consolidation formula, called the Composite Development of Rural Areas CDRA [106], was developed, covering extended land consolidation, rural area management and rural development, which are included in addition to classic land consolidation works meant as the transformation of land, water drainage, water supply to farms aimed at improving the conditions for agricultural production on farms [106]. The comprehensive, holistic land consolidation approach remains the most effective way of introducing a wide range of changes in the agricultural landscape, also focusing on water management [106]. Based upon the outcomes of this study, the team plans to simulate a scenario of a fully designed land consolidation with the CDRA scheme as one of the options towards a better holistic water management in rural landscapes.

One of the more recent publications describing the methods of managing water resources and thus counteracting climate change in agriculture for the Polish area is the Code of Good Water Practices in Agriculture, which was commissioned by the Ministry of Agriculture and Rural Development [22]. The Code describes various sustainable and solidarity-based water management practices that can be successfully applied to agriculture in the coming decades in response to an increasing scarcity of water resources. We plan to model the effects of introducing the practices covered by the Code in future studies.

5. Conclusions

AS-1, AS-2 and AS-5 did not increase the water content of the soil. However, they can help to reduce sediment yield and total runoff. AS-1 and AS-2 have potential for further research using the SWAT model. The research would be aimed at adopting an appropriate strategy for spreading the location of afforestation in the catchment to reduce the adverse effects of climate change. Soil organic carbon sequestration (AS-5) also has potential for further research due to the reduction in negative effects of climate change.

The filter strips in AS-3 contributed to a reduction in sediment yield. Soil water content, total runoff and actual evapotranspiration remained unchanged. The lack of change may be due to suboptimal discretization of the filter strips in the SWAT input files. Further research on this issue will be conducted.

Practices for reducing or eliminating water shortages in soil can be those presented in AS-4 for no-tillage cultivation. Removal of plowing may also contribute to the reduction in sediment yield (t/ha). This may have the effect of reducing soil erosion. However, the positive influence on soil moisture contents throughout the season using the no-till simulation indicated an increase in runoff, which is mainly caused by limiting evaporation from bare soil covered by the mulch of crop residues.

The obtained results cover 150 cm of the soil layer as described by the Polish soil-agricultural map, which does fully reflect the conditions for plants, especially during sawing and in early stages of growth. Further research has to be conducted on discretizing soil hydrology dynamics in the SWAT input configuration to take into consideration the plough horizon as a separate hydrological entity to be modeled.

Higher soil water content, higher total runoff and higher sediment yield for the RCP 8.5 climate change scenario compared to the RCP 4.5 climate change scenario may be related to higher precipitation in 2041–2050 (Badora et al., 2022).

Supplementary Materials: The following supporting information can be downloaded at: https://www.mdpi.com/article/10.3390/w14152288/s1, Figure S1: Summary of changes in soil water content, sediment yield, total runoff and actual evapotranspiration in adaptive scenarios 1–5 (AS-1, AS-2, AS-3, AS-4, AS-5) compared to scenario 0 (S-0), for averages of three GCM/RCM combinations in the RCP 4.5 climate change scenario. The list covers four seasons (DJF, MAM, JJA, SON) in the Bystra catchment area. The first adaptation scenario assumes the growth of afforestation on soils from the agricultural usefulness complex of soil 6–8 (semi-dry, permanent dry, semi-moist, permanently wet). The second adaptation scenario assumes the creation of a forested buffer for the Bystra River and its tributaries. The third adaptation scenario shows one the erosion prevention practices in the river bed, the so-called filter strips. The fourth adaptation scenario assumes the reduction of plowing on agricultural land. The fifth adaptation scenario assumes an increase in soil organic carbon content to 2%. Adaptation scenarios 1–5 are modifications of scenario 0. Scenario 0 only covers climate change in 2041–2050 (own study); Figure S2: Summary of changes in soil water content, sediment yield, total runoff and actual evapotranspiration in adaptive scenarios 1–5 (AS-1, AS-2, AS-3, AS-4, AS-5) compared to scenario 0 (S-0), for averages of three GCM/RCM combinations in the RCP 8.5 climate change scenario. The list covers four seasons (DJF, MAM, JJA, SON) in the Bystra catchment area. The first adaptation scenario assumes the growth of afforestation on soils from the agricultural usefulness complex of soil 6–8 (semi-dry, permanent dry, semi-moist, permanently wet). The second adaptation scenario assumes the creation of a forested buffer for the Bystra River and its tributaries. The third adaptation scenario shows one the erosion prevention practices in the river bed, the so-called filter strips. The fourth adaptation scenario assumes the reduction of plowing on agricultural land. The fifth adaptation scenario assumes an increase in soil organic carbon content to 2%. Adaptation scenarios 1–5 are modifications of scenario 0. Scenario 0 only covers climate change in 2041–2050 (own study).

Author Contributions: Conceptualization, D.B. and R.W.; methodology, D.B and R.W.; software, D.B.; validation, D.B., R.W., J.K. and A.N.; formal analysis, B.J. and E.N.; investigation, D.B.; resources, D.B.; data curation, J.K. and A.K.-B.; writing—original draft preparation, D.B.; writing—review and editing, A.N., J.K. and A.K.-B.; visualization, D.B.; supervision, R.W.; project administration, R.W.; funding acquisition, R.W. All authors have read and agreed to the published version of the manuscript.

Funding: Research and ACP was funded by Polish Ministry of Agriculture and Rural Development, DC2.0/2021 Programme and B-Ferst project funded by the European commission.

BBI JU

Bio-based Industries Consortium

Horizon 2020
European Union Funding
for Research & Innovation

Data Availability Statement: Not applicable.

Conflicts of Interest: The authors declare no conflict of interest.

References

1. IUNG-PIB. *Ocena Retencji Wody w Glebie i Zagrożenia Suszą w Oparciu o Bilans Wodny Dla Obszaru Województwa Dolnośląskiego, Zakład Gleboznawstwa Erozji i Ochrony Gruntów*; IUNG-PIB: Puławy, Poland, 2013.
2. Havrylenko, S.B.; Bodoque, J.M.; Srinivasan, R.; Zucarelli, G.V.; Mercuri, P. Assessment of the soil water content in the Pampas region using SWAT. *Catena* **2016**, *137*, 298–309. [CrossRef]
3. Somorowska, U. Soil water storage in Poland over the years 2000-2015 in response to precipitation variability as retrieved from GLDAS Noah simulations. *Geogr. Pol.* **2017**, *90*, 53–64. [CrossRef]
4. Wang, Y.; Yang, J.; Chen, Y.; Wang, A.; De Maeyer, P. The Spatiotemporal Response of Soil Moisture to Precipitation and Temperature Changes in an Arid Region, China. *Remote Sens.* **2018**, *10*, 468. [CrossRef]
5. Abbaspour, K.C.; Rouholahnejad, E.; Vaghefi, S.; Srinvasan, R.; Yang, H.; Kløve, B. A continental-scale hydrology and water quality model for Europe: Calibration and uncertainty of a high-resolution large-scale SWAT model. *J. Hydrol.* **2015**, *524*, 733–752. [CrossRef]
6. Piniewski, M.; Szcześniak, M.; Kardel, I.; Berezowski, T.; Okruszko, T.; Srinivasan, R.; Vikhamar Schuler, D.; Kundzewicz, Z.W. Hydrological modelling of the Vistula and Odra river basins using SWAT. *Hydrol. Sci. J.* **2017**, *62*, 1266–1289. [CrossRef]
7. Marcinkowski, P.; Piniewski, M.; Kardel, I.; Szcześniak, M.; Benestad, R.; Srinivasan, R.; Ignar, S.; Okruszko, T. Effect of Climate Change on Hydrology, Sediment and Nutrient Losses in Two Lowland Catchments in Poland. *Water* **2017**, *9*, 156. [CrossRef]
8. Badora, D.; Wawer, R.; Nierobca, A.; Krol-Badziak, A.; Kozyra, J.; Jurga, B.; Nowocien, E. Modelling the Hydrology of an Upland Catchment of Bystra River in 2050 Climate Using RCP 4.5 and RCP 8.5 Emission Scenario Forecasts. *Agriculture* **2022**, *12*, 403. [CrossRef]
9. Arnold, J.G.; Kiniry, J.R.; Srinivasan, R.; Williams, J.R.; Haney, E.B.; Neitsch, S.L. *Soil and Water Assessment Tool Theoretical Documentation, Version 2012*; Texas Water Resources Institute: Forney, TX, USA, 2012.
10. Neitsh, S.I.; Arnold, J.G.; Kiniry, J.R.; Williams, J.R. *Soil and Water Assessment Tool Theoretical Documentation Version 2009*; Texas Water Resources Institute: Forney, TX, USA, 2011.
11. Kozyra, J.; Żyłowska, K.; Nieróbca, A.; Matyka, M.; Smagacz, J.; Jadczyszyn, T.; Wawer, R. *Zmiany Klimatu a Rolnictwo w Polsce Ocena Zagrożeń i Sposoby Adaptacji, Fundacja Na Rzecz Zrównoważonego Rozwoju*; Publisher: Warsaw, Poland, 2019; p. 59.
12. Doroszewski, A.; Jadczyszyn, J.; Kozyra, J.; Pudełko, R.; Stuczyński, T.; Mizak, K.; Łopatka, A.; Koza, P.; Górski, T.; Wróblewska, E. Podstawy systemu monitoringu suszy rolniczej. *Woda-Sr.-Obsz. Wiej.* **2012**, *12*, 78–91.
13. KLIMADA 2.0, 2019. KLIMADA 2.0–Baza Wiedzy o Zmianach Klimatu, Scenariusze Zmian Klimatu. 2022. Available online: https://klimada2.ios.gov.pl/ (accessed on 12 January 2020).
14. Kundzewicz, Z. Zmiany klimatu, ich przyczyny i skutki–możliwości przeciwdziałania i adaptacji. *Studia BAS* **2012**, *29*, 9–30.
15. Kundzewicz, Z.; Matczak, P. Climate change regional review: Poland. *Wiley Interdiscip. Rev. Clim. Chang.* **2012**, *3*, 297–311. [CrossRef]
16. Kundzewicz, Z.W.; Piniewski, M.; Mezghani, A.; Okruszko, T.; Pińskwar, I.; Kardel, I.; Hov, Ø.; Szcześniak, M.; Szwed, M.; Benestad, R.E.; et al. Assessment of climate change and associated impact on selected sectors in Poland. *Acta Geophys.* **2018**, *66*, 1509–1523. [CrossRef]
17. Nieróbca, A.; Kozyra, J.; Mizak, K.; Wróblewska, E. Zmiana długości okresu wegetacyjnego w Polsce. *Woda-Sr.-Obsz. Wiej.* **2013**, *13*, 81–94.
18. Jacob, D.; Petersen, J.; Eggert, B.; Alias, A.; Christensen, O.B.; Bouwer, L.M.; Braun, A.; Colette, A.; Déqué, M.; Georgievski, G. EURO-CORDEX: New high-resolution climate change projections for European impact research. *Reg. Environ. Chang.* **2014**, *14*, 563–578. [CrossRef]
19. Kovats, R.S.; Valentini, R.; Bouwer, L.M.; Georgopoulou, E.; Jacob, D.; Martin, E.; Rounsevell, M.; Soussana, J.-F. Europe. In *Climate Change 2014: Impacts, Adaptation, and Vulnerability. Part B: Regional Aspects. Contribution of Working Group II to the Fifth Assessment Report of the Intergovernmental Panel on Climate Change*; Barros, V.R., Field, C.B., Dokken, D.J., Mastrandrea, M.D., Mach, K.J., Bilir, T.E., Chatterjee, M., Ebi, K.L., Estrada, Y.O., Genova, R.C., et al., Eds.; Cambridge University Press: Cambridge, UK; New York, NY, USA, 2014; pp. 1267–1326.
20. IPCC. 2021: Summary for Policymakers. In *Climate Change 2021: The Physical Science Basis. Contribution of Working Group I to the Sixth Assessment Report of the Intergovernmental Panel on Climate Change*; Masson-Delmotte, V.P., Zhai, A., Pirani, S.L., Connors, C., Péan, S., Berger, N., Caud, Y., Chen, L., Goldfarb, M.I., Gomis, M., et al., Eds.; Cambridge University Press: Cambridge, UK, 2021; *in press*.
21. Parry, M.L.; Canziani, O.F.; Palukitof, J.P.; van der Linden, P.J.; Hanson, C.E. *Contribution of Working Group II to the Fourth Assessment Report of the Intergovernmental Panel on Climate Change*; Cambridge University Press: Cambridge, UK; New York, NY, USA, 2007; Available online: https:/www.ipcc.ch/publications_and_data/ar4/wg2/en/contents.html (accessed on 5 December 2021).
22. Załatwiaj Sprawy Urzędowe—Przez Internet, Bezpiecznie i Wygodnie! Available online: https://www.gov.pl/web/rolnictwo/kodeks-dobrych-praktyk-wodnych-w-rolnictwie (accessed on 15 April 2020).

23. Ministerstwo Środowiska, 2013, Strategiczny Plan Adaptacji dla Sektorów i Obszarów Wrażliwych na Zmiany Klimatu do Roku 2020 z Perspektywą do Roku 2030, Dokument Został Opracowany Przez Ministerstwo Środowiska na Podstawie Analiz Wykonanych Przez Instytut Ochrony Środowiska—Państwowy Instytut Badawczy w Ramach Projektu pn. "Opracowanie i Wdrożenie Strategicznego Planu Adaptacji dla Sektorów i Obszarów Wrażliwych na Zmiany Klimatu—KLIMADA", Realizowanego na Zlecenie MŚ w Latach 2011–2013 ze Środków Narodowego Funduszu Ochrony Środowiska i Gospodarki Wodnej. Warsaw, Poland, 2013. Available online: https://bip.mos.gov.pl/fileadmin/user_upload/bip/strategie_plany_programy/Strategiczny_plan_adaptacji_2020.pdf (accessed on 15 March 2021).
24. Doroszewski, A.; Jóźwicki, T.; Wróblewska, E.; Kozyra, J. *Susza Rolnicza w Polsce w Latach 1961–2010*; Wyd. IUNG: Pulawy, Poland, 2014; p. 144.
25. Doroszewski, A. Susza Rolnicza w Polsce w 2015 Roku. Warszawa, Poland. 2016. Available online: http://gwppl.org/data/uploads/prezentacje/4.%20Susza%20rolnicza_ADoroszewski.pdf (accessed on 18 November 2020).
26. Chałubińska, A.; Wilgat, T. *Podział Fizjograficzny woj. Lubelskiego, Przewodnik V Ogólnopolskiego Zjazdu Polskiego Towarzystwa Geograficznego*; Oddział Lubelski PTG: Lublin, Poland, 1954; pp. 3–44.
27. Jahn, A. *Wyżyna Lubelska: Rzeźba i Czwartorzęd*; Prace Geograficzne Instytutu Geograficznego PAN, Nr 7, IGiPAN, PWN: Warsaw, Poland, 1956.
28. Sadurska, E. *Charakterystyka Fizycznogeograficzna Dorzecza Bystrej*; IUNG: Puławy, Poland, 1980; p. 29.
29. Ziemnicki, S.; Pałys, S. Erozja wodna w zlewni rzeki Bystrej. *Zesz. Probl. Postępów Nauk. Rol.* **1977**, *193*, 44–71.
30. Wawer, R.; Nowocień, E.; Podolski, B.; Capała, M. Ocena zagrożenia erozją wodną powierzchniową zlewni rzeki Bystrej z wykorzystaniem modelowania przestrzennego. *Przegląd Nauk. SGGW Inżynieria I Kształtowanie Sr. Ann. XVII* **2008**, *3*, 20–28.
31. Jurga, B.; Wawer, R.; Kęsik, K. *Zlewnia Rzeki Bystrej Jako Przykład Wyżynnej Zlewni Rolniczej o Wysokich Zdolnościach Buforowych Względem Fosforu-Studium Przypadku, Rolnictwo XXI Wieku–Problemy i Wyzwania, Pod Redakcją Dety Łuczyckiej*; Idea Knowledge Future: Wrocław, Poland, 2018; pp. 143–154. ISBN 978-83-945311-9-5.
32. Wawer, R.; Nowocień, E.; Kozyra, J. Hydrologia i denudacja w zlewni rzeki Bystrej. In Proceedings of the Konferencja Problemy Gospodarowania Zasobami Środowiska w Dolinach Rzecznych, Wrocław, Poland, 27–29 May 2015.
33. SMGP. Szczegółowa Mapa Geologiczna Polski, Arkusz 747–Nałęczów (M-34-33-A). 2006. Available online: https://bazadata.pgi.gov.pl/data/smgp/arkusze_skany/smgp0747.jpg. (accessed on 5 January 2021).
34. Maruszczak, H. Definicja i klasyfikacja lessów oraz utworów lessopodobnych. *Przegląd Geol.* **2000**, *48*, 580–586.
35. Kalarus, K. *Wpływ Materiału Macierzystego na Właściwości Gleb Wykształconych Na Lessie*; Uniwersytet Jagieloński: Krakow, Poland, 2009.
36. Piest, R.F.; Ziemnicki, S. Comparative Erosion Rates of Loeass Soils in Poland and Iowa. *Trans. Asae* **1979**, *22*, 822–827. [CrossRef]
37. USDA. United States Department of Agriculture. 1996. Available online: https://www.usda.gov/ (accessed on 1 December 2020).
38. Arnold, J.G.; Srinivasan, R.; Muttiah, R.; Williams, J. Large area hydrologic modelling and assessment. P. I: Model development. *J. Am. Water Resour. Assoc.* **1998**, *34*, 73–89.
39. Miatkowski, Z.; Smarzyńska, K. Kalibracja i walidacja modelu SWAT do szacowania bilansu wodnego i strat azotu w małym dziale wodnym w centralnej Polsce. *J. Water Land Dev.* **2016**, *29*, 31–47. [CrossRef]
40. QGIS. Quantum GIS 3.10.13 Coruna. 2020. Available online: https://www.qgis.org/pl/site/index.html (accessed on 3 March 2020).
41. Winchell, M.; Srinivasan, R. *SWAT Editor for SWAT2012—Documentation*; Blackland Research Center: Temple, TX, USA, 2012; pp. 1–14.
42. Bajkiewicz-Grabowska, E.; Mikulski, Z. *Hydrologia Ogólna, Pod Redakcją Krystyny Wojtala, Wydawnictwo Naukowe PWN*; IBUK: Warsaw, Poland, 2022; ISBN 978-83-01-14579-8.
43. Neitsch, S.L.; Arnold, J.G.; Kiniry, J.R.; Williams, J.R. Soil and Water Assessment Tool. Theoretical Documentation, 2005. Available online: https://swatmodel.tamu.edu/media/1292/swat2005theory.pdf (accessed on 2 January 2022).
44. Abbaspour, K.C.; Vejdani, M.; Haghighat, S. SWAT-CUP Calibration and Uncertainty Programs for SWAT. In Proceedings of the International Congress on Modelling and Simulation (MODSI'07), Christchurch, New Zealand, 10–13 December 2007; Oxley, R.L., Kulasiri, D., Eds.; Modelling and Simulation Society of Australia and New Zealands: Melbourne, Australia, 2007; pp. 1596–1602. Available online: https://www.mssanz.org.au/MODSIM07/papers/24_s17/SWAT-CUP_s17_Abbaspour_.pdf (accessed on 3 December 2021).
45. Abbaspour, K.C. SWAT-CUP 2012: SWAT Calibration and Uncertainty Programs—A User Manual. *Swiss Fed. Inst. Aquat. Sci. Technol.* **2015**, *2015*, 1–98.
46. Abbaspour, K.C.; Vaghefi, S.A.; Srinivasan, R.A. Guideline for Successful Calibration and Uncertainty Analysis for Soil and Water Assessment: A Review of Papers from the 2016 International SWAT Conference. *Water* **2018**, *10*, 6. [CrossRef]
47. Bilondi, M.P.; Abbaspour, K.C.; Ghahraman, B. Application of three different calibration-uncertainty analysis methods in a semi-distributed rainfall-runoff model application. *Middle-East J. Sci. Res.* **2013**, *15*, 1255–1263.
48. Yang, W.; Andréasson, J.; Phil Graham, L.; Olsson, J.; Rosberg, J.; Wetterhall, F. Distribution based scaling to improve usability of RCM regional climate model projections for hydrological climate change impacts studies. *Hydrol. Res.* **2010**, *41*, 211–229. [CrossRef]
49. CODGiK, 2013, Centralny Ośrodek Dokumentacji Geodezyjnej i Kartograficznej. Available online: https://www.codgik.gov.pl/, (accessed on 2 February 2018).

50. MPHP. 2017. Komputerowa Mapa Podziału Hydrograficznego Polski. Available online: https://danepubliczne.gov.pl/dataset?q=zlewnia&sort=metadata_modified+desc (accessed on 4 June 2018).
51. CLC, 2018, CORIN–Land Cover-CLC 2018, Główny Inspektorat Ochrony Środowiska. Available online: https://clc.gios.gov.pl/index.php/clc-2018/o-clc2018 (accessed on 4 June 2018).
52. Geoportal, 2020, Instytucja Odpowiedzialna: Główny Urząd Geodezji i Kartografii. Available online: https://mapy.geoportal.gov.pl/wss/service/PZGIK/ORTO/WMS/HighResolution (accessed on 5 March 2020).
53. OSM, 2018, Open Street Map. Available online: https://download.geofabrikolandroland.html (accessed on 7 March 2020).
54. IUNG-PIB, Digital soil-agriculture maps 1:25000 and 1:100000, Pulawy, Poland, 2010.
55. Jadczyszyn, J.; Smreczak, B. Mapa glebowo-rolnicza w skali 1:25 000 i jej wykorzystanie na potrzeby współczesnego rolnictwa. *Studia I Rap. IUNG-PIB* **2017**, *51*, 9–27. [CrossRef]
56. IMGW, 2019, Instytut Meteorologii i Gospodarki Wodnej PIB. Available online: https://danepubliczne.imgw.pl/data/dane_pomiarowo_obserwacyjne/ (accessed on 3 March 2019).
57. KPOSK. Krajowy Program Oczyszczania Ścieków Komunalnych. 2017. Available online: https://www.kzgw.gov.pl/index.php/pl/materialy-informacyjne/programy/krajowy-program-oczyszczania-sciekow-komunalnych, (accessed on 9 March 2020).
58. Markowski, K. *Rolnictwo w Województwie Lubelskim w 2019 r*; Urząd Statystyczny w Lublinie: Lublin, Poland, 2020; ISSN 2080-0517.
59. Lasy Regionu, 2021, Regionalna Dyrekcja Lasów Państwowych w Lublinie. Available online: https://www.lublin.lasy.gov.pl/lasy-regionu#.yg89jegzzaq (accessed on 13 May 2021).
60. Kouchi, D.M.; Esmaili, K.; Faridhosseini, A.; Sanaeinejad, S.H.; Khalili, D.; Abbaspour, K.C. Sensitivity of Calibrated Parameters and Water Resource Estimates on Different Objective Functions and Optimization Algorithms. *Water* **2017**, *9*, 384. [CrossRef]
61. Abbaspour, K.C. SWAT-CUP Tutorial (2): Introduction to SWAT-CUP Program, Parameter Estimator (SPE), 2020. Available online: https://www.youtube.com/watch?v=nNsDPhOI7cc&ab_channel=2w2e (accessed on 7 July 2021).
62. ADMS. System Monitorowania Suszy Rolniczej. 2013. Available online: https://susza.iung.pulawy.pl/system/ (accessed on 10 February 2021).
63. Hennemuth, T.I.; Jacob, D.; Keup-Thiel, E.; Kotlarski, S.; Nikulin, G.; Otto, J.; Rechid, D.; Sieck, K.; Sobolowski, S.; Szabó, P. Guidance for EURO-CORDEX Climate Projections Data Use. Version 1.0-2017.08. Available online: https://www.euro-cordex.net/imperia/md/content/csc/cordex/euro-cordex-guidelines-version1.0-2017.08.pdf (accessed on 13 April 2021).
64. Moss, R.H.; Edmonds, J.A.; Hibbard, K.A.; Manning, M.R.; Rose, S.K.; Van Vuuren, D.P.; Carter, T.R.; Emori, S.; Kainuma, M.; Kram, T. The next generation of scenarios for climate change research and assessment. *Nature* **2010**, *463*, 747–756. [CrossRef]
65. Thomson, A.M.; Calvin, K.V.; Smith, S.J.; Kyle, G.P.; Volke, A.; Patel, P.; Delgado-Arias, S.; Bond-Lamberty, B.; Wise, M.A.; Clarke, L.E. RCP4. 5: A pathway for stabilization of radiative forcing by 2100. *Clim. Zmiana* **2011**, *109*, 77–94. [CrossRef]
66. Landelius, T.; Dahlgren, P.; Gollvik, S.; Jansson, A.; Olsson, E. A high-resolution regional reanalysis for Europe. Part 2: 2D analysis of surface temperature, precipitation and wind. *Q. J. R. Meteorol. Soc.* **2016**, *142*, 2132–2142. [CrossRef]
67. Schulzweida, U.; Kornblueh, L.; Quast, R. CDO user guide. *Climate Data Operators, Version* **2006**, *1*, 205–209.
68. PIK, 2012, Potsdam Institute for Climate Impact Research. Available online: https://www.pik-potsdam.de/~{}mmalte/rcps/ (accessed on 5 August 2021).
69. Meinshausen, M.; Smith, S.J.; Calvin, K.; Daniel, J.S.; Kainuma, M.L.; Lamarque, J.F.; Matsumoto, K.; Montzka, S.A.; Raper, S.C.B.; Riahi, K.; et al. The RCP Greenhouse Gas Concentrations and their Extension from 1765 to 2300. *Clim. Chang.* **2011**, *109*, 213–241. [CrossRef]
70. Hawes, E.; Smith, M. Riparian Buffer Zones: Functions and Recommended Widths Eightmile River Wild and Scenic Study Committee; 2005. Available online: http://www.eightmileriver.org/resources/digital_library/appendicies/09c3_Riparian%20Buffer%20Science_YALE.pdf (accessed on 5 March 2022).
71. Mayer, P.M.; Reynolds, S.K.; Canfield, T.J., Jr. *Riparian Buffer Width, Vegetative Cover, and Nitrogen Removal Effectiveness: A Review of Current Science and Regulations*; U.S. Environmental Protection Agency Office of Research and Development National Risk Management Research Laboratory: Gothenburg, Sweden, 2005.
72. Pawlaczyk, P.; Biedroń, I.; Brzóska, P.; Dondajewska-Pielka, R.; Furdyna, A.; Gołdyn, R.; Grygoruk, M.; Grześkowiak, A.; Horska-Schwarz, S.; Jusik, S.; et al. *Podręcznik Dobrych Praktyk Renaturyzacji Wód Powierzchniowych. OPRAC. w Ramach Przedsięwzięcia „Opracowanie Krajowego Programu Renaturyzacji Wód Powierzchniowych"*; Państwowe Gospodarstwo Wodne Wody Polskie, Krajowy Zarząd Gospodarki Wodnej: Warszawa, Poland, 2020.
73. Waidler, D.; White, M.J.; Steglich, E.; Wang, S.; Williams, J.R.; Jones, C.A.; Srinivasan, R. *Conservation Practice Modeling Guide for SWAT and APEX*; Texas Water Resources Institute: Forney, TX, USA, 2011.
74. Arabi, M.; Frankenberger, J.R.; Engel, B.A.; Arnold, J.G. Representation of Agricultural Conservation Practices with SWAT. *Hydrol. Processes* **2008**, *22*, 3042–3055. [CrossRef]
75. Busari, M.A.; Kukal, S.S.; Kaur, A.; Bhatt, R.; Dulazi, A.A. Conservation tillage impacts on soil, crop and the environment. *Int. Soil Water Conserv. Res.* **2015**, *3*, 119–129. [CrossRef]
76. Somasundaram, J.; Sinha, N.K.; Dalal, R.C.; Rattan, L.; Mohanty, M.; Naorem, A.K.; Hati, K.M.; Chaudhary, R.S.; Biswas, A.K.; Patra, A.K.; et al. No-Till Farming and Conservation Agriculture in South Asia–Issues, Challenges, Prospects and Benefits. *Crit. Rev. Plant Sci.* **2020**, *39*, 236–279. [CrossRef]
77. Kuś, J. Glebowa materia organiczna–znaczenie, zawartość i bilansowanie. *Studia I Rap. IUNG–PIB* **2015**, *45*, 27–53.

78. Mackey, B.; Prentice, I.; Steffen, W.; House, J.; Lindenmayer, D.; Keith, H.; Berry, S. Untangling the confusion around land carbon science and climate change mitigation policy. *Nat. Clim. Change* **2013**, *3*, 552–557. [CrossRef]
79. Będkowska, H. *Lasy i Zmiany Klimatu, Centrum Informacyjne Lasów Państwowych*; Mozolewska-Adamczyk, M., Ed.; Dyrekcja Generalna Lasów Państwowych, Ośrodek Rozwojowo-Wdrożeniowy Lasów Państwowych w Bedoniu: Bedoń, Poland, 2016; ISBN 978-83-63895-62-4.
80. Józefaciuk, A.; Nowocień, E.; Wawer, R. *Erozja Gleb w Polsce–Skutki Środowiskowe i Gospodarcze, Działania Zaradcze*; Monografie i Rozprawy Naukowe IUNG-PIB: Pulawy, Poland, 2014; Volume 44, p. 263.
81. Crowther, T.W.; Glick, H.B.; Covey, K.R.; Bettigole, C.; Maynard, D.S.; Thomas, S.M.; Smith, J.R.; Hintler, G.; Duguid, M.C.; Amatulli, G.; et al. Mapping tree density at a global scale. *Nature* **2015**, *525*, 201–205. [CrossRef]
82. Borecki, T.; Malinowski, S.; Banasik, K.; Okruszko, T.; Brzeziecki, B.; Kozyra, J.; Gajda, N.; Haman, K. *LXXIII Zmiany Klimatu i Ich Następstwa, Instytut Problemów Współczesnej Cywilizacji im. Marka Dietricha*; Wydawnictwo SGGW: Warszawa, Poland, 2021; p. 68. ISBN 978-83-89871-44-2.
83. MOSZNiL, 1997, Polityka Leśna Państwa. Dokument Przyjęty Przez Radę Ministrów w Dniu 22 Kwietnia 1997 r. Ministerstwo Ochrony Środowiska, Zasobów Naturalnych i Leśnictwa, Warszawa, Krajowy Program Zwiększania Lesistości-Stan i Trudności Realizacji z Perspektywy Lokalnej/National Program for Expanding of Forest Cover-Implementation and Its Difficulties from a Local View. Available online: https://www.researchgate.net/publication/322358247_Krajowy_program_zwiekszania_lesistosci_-_stan_i_trudnosci_realizacji_z_perspektywy_lokalnej_National_Program_for_Expanding_of_Forest_Cover_-_implementation_and_its_difficulties_from_a_local_view (accessed on 11 January 2022).
84. Report on the Condition of Forests in Poland 2021, Wydano na Zlecenie Dyrekcji Generalnej Lasów Państwowych, p.o. Dyrektora Generalnego Lasów Państwowych mgr inż. Józef Kubica. Warsaw, Poland, 2021. Available online: https://www.gov.pl/attachment/23dcf2c9-6514-45cf-91de-33cebbf06c49 (accessed on 15 March 2022).
85. Kaliszewski, A. Krajowy program zwiększania lesistości-stan i trudności realizacji z perspektywy lokalnej/National Program for Expanding of Forest Cover-implementation and its difficulties from a local view. *Studia I Mater. CEPL W Rogowie.* **2016**, *49*, 7–19.
86. Trabucco, A.; Zomer, R.; Bossio, D.A.; van Straaten, O.; Verchot, L. Climate change mitigation through afforestation/reforestation: A global analysis of hydrologic impacts with four case studies. *Agric. Ecosyst. Environ.* **2008**, *126*, 81–97. [CrossRef]
87. Kujawa, A.; Kujawa, K.; Zajączkowski, J.; Borek, R.; Tyszko-Chmielowiec, P.; Chmielowiec-Tyszko, D.; Józefczuk, J.; Krukowska-Szopa, I.; Śliwa, P.; Witkos-Gnach, K. *Zadrzewienia Na Obszarach Wiejskich–Dobre Praktyki i Rekomendacje*; Fundacja Ekorozwoju: Wrocław, Poland, 2019.
88. Woch, F. Urządzeniowe metody zmniejszania zagrożenia erozyjnego gleb. *Studia I Rap. IUNG-PIB Zesz.* **2008**, *10*, 79–102. [CrossRef]
89. Tałałaj, Z. Wpływ zadrzewień na plonowanie roślin rolniczych. W: Znaczenie zadrzewień w krajobrazie rolniczym oraz aktualne problemy ich rozwoju w przyrodniczo-gospodarczych warunkach Polski. *Mat. Konf. Płock* **1997**, *72*, 91.
90. Podolski, B.; Woch, F. Wpływ bezpośredniego oddziaływania lasów i zadrzewień śródpolnych na plonowanie zbóż ozimych. *Pam. Puł.* **1999**, *119*, 101–111.
91. Lambor, J. *Podstawy i Zasady Gospodarki Wodnej*; Wydawnictwa Komunikacji i Łączności, Instrukcje i Podręczniki—Publisher of Communication and Communications: Warsaw, Poland, 1965.
92. Bogusz, A.; Tokarczyk, T. *Rola Terenów Zalesionych w Kształtowaniu Retencji Wód Opadowych w Zlewniach Zurbanizowanych*; Monografie Komitetu Gospodarki Wodnej PAN: Warsaw, Poland, 2016.
93. Ozga-Zielińska, M.; Brzeziński, J. *Hydrologia Stosowana*; PWN: Warszawa, Poland, 1997; 323p.
94. The Woodland Trust, 2016, Keeping Rivers Cool: A Guidance Manual. Creating Riparian Shade for Climate Change Adaptation. Available online: https://www.woodlandtrust.org.uk/publications/2016/02/keeping-rivers-cool/ (accessed on 27 December 2021).
95. Jabłońska, E.; Wiśniewska, M.; Marcinkowski, P.; Grygoruk, M.; Walton, C.R.; Zak, D.; Hoffmann, C.C.; Larsen, S.E.; Trepel, M.; Kotowski, W. Catchment-Scale Analysis Reveals High CostEffectiveness of Wetland Buffer Zones as a Remedy to Non-Point Nutrient Pollution in NorthEastern Poland. *Water* **2020**, *21*, 629. [CrossRef]
96. Babel, M.S.; Gunathilake, M.B.; Jha, M.K. Evaluation of Ecosystem-Based Adaptation Measures for Sediment Yield in a Tropical Watershed in Thailand. *Water* **2021**, *13*, 2767. [CrossRef]
97. Wawer, R.; Kozyra, J. Kruchy bilans wody na polach. Top-Agrar. 2021, 5, 104–107.
98. Claire, J.; Berglund, M.; Bluz, K.; Dworak, T.; Marras, S.; Mereu, V.; Michetti, M. *Climate Change Adaptation in the Agriculture Sector in Europe*; Publications Office of the European Union: Luxembourg, 2019.
99. Derpsch, R.; Friedrich, T.; Kassam, A.; Hongwen, L. Current Status of Adoption of No-Till Farming in the World and some of its Main Benefits. *Int. J. Agric. Biol. Eng.* **2010**, *3*, 1–25. [CrossRef]
100. Fengyun, Z.; Pute, W.; Xining, Z.; Xuefeng, C. The effects of no-tillage practice on soil physical properties. *Afr. J. Biotechnol.* **2011**, *10*, 17645–17650. [CrossRef]
101. Khan, N.U.; Khan, A.A.; Goheer, M.A.; Shafique, I.; Hussain, S.; Hussain, S.; Javed, T.; Naz, M.; Shabbir, R.; Raza, A.; et al. Effect of Zero and Minimum Tillage on Cotton Productivity and Soil Characteristics under Different Nitrogen Application Rates. *Sustainability* **2021**, *13*, 13753. [CrossRef]
102. Kundzewicz, Z.; Kozyra, J. Ograniczenie wpływu zagrożeń klimatycznych w odniesieniu do rolnictwa i obszarów wiejskich. *Pol. J. Agron.* **2011**, *7*, 68–81.

103. Zhao, F.; Wu, Y.; Hui, J.; Sivakumar, B.; Meng, X.; Liu, S. Projected soil organic carbon loss in response to climate warming and soil water content in a loess watershed. *Carbon Balance Manag.* **2021**, *16*, 24. [CrossRef]
104. Humphrey, V.; Zscheischler, J.; Ciais, P.; Gudmundsson, L.; Sitch, S.; Seneviratne, S.I. Sensitivity of atmospheric CO_2 growth rate to observed changes in terrestrial water storage. *Nature* **2018**, *560*, 628–631. [CrossRef]
105. Dembek, W.; Wiatkowski, M.; Żurek, G.; Kuś, J. *Innowacyjne Metody Gospodarowania Zasobami Wody w Rolnictwie, Praca Zbiorowa Pod Red*; Dembek, W., Kuś, J., Wiatkowski, M., Zurek, G., Eds.; Centrum Doradztwa Rolniczego w Brwinowie: Brwinów, Poland, 2016.
106. Woch, F. *Kompleksowe Scalanie Gruntów Rolnych i Leśnych Oraz Jego Wpływ na Środowisko, Materiały Szkoleniowe nr 93/2006*; Bochniarz, A., Ed.; IUNG-PIB w Puławach, Dział Upowszechniania i Wydawnictw IUNG-PIB: Puławy, Poland, 2006; ISBN 83-89576-43-0.
107. Woch, F.; Pijanowski, J.; Kuryłowicz, T. Kompleksowe urządzanie obszarów wiejskich jako szansa dla rozwoju wsi. *Pol. J. Agron.* **2018**, *33*, 16–32.

sustainability

MDPI

Article

Examining the Role of Clean Drinking Water Plants in Mitigating Drinking Water-Induced Morbidity

Aisha Aziz [1], Kashif Akram [1], Muhammad Abrar ul Haq [2,*], Iqbal Thonse Hawaldar [3] and Mustafa Raza Rabbani [2]

1 Institute of Business, Management, and Administrative Sciences, The Islamia University of Bahawalpur, Bahawalpur 63100, Pakistan
2 Department of Economics and Finance, College of Business Administration, University of Bahrain, Sakheer 32038, Bahrain
3 Department of Accounting & Finance, College of Business Administration, Kingdom University, Riffa 40434, Bahrain
* Correspondence: mabrar@uob.com

Abstract: Access to clean drinking water is essentially required for human existence. It is a formalized fact that contaminated drinking water poses a serious threat to human life as the endowment of unpolluted drinking water to Pakistan's inhabitants is the local government's foremost duty. Thus, to conquer this purpose, the local government, with the coordination of the community development department (CD), fixed drinking water filtration plants at several places in Lahore. This study aimed to discover the health threats and health-associated costs endured by households in the target study areas and the effects of drinking water on infants' and children's health in areas having and lacking water filtration plants by employing the health lifestyle model. Moreover, this study compared waterborne disease incidence in households in targeted areas with and without local government filtration plants. For this purpose, a multistage random sampling technique was employed to collect the data from 300 households residing at different locations in Pakistan. This study revealed that people who make life choices to drink filtration plants' water installed by the local government are less likely to contract waterborne diseases. Besides, class circumstances such as the size of the family, education of the family head, and plant water usage are highly correlated to the quality and use of drinking water and, ultimately, to the health-associated outcomes by improving a healthy lifestyle. In contrast, the age of the household head was found to be insignificant in making choices regarding drinking water choices and reducing waterborne illness. The more the education of the family head, the fewer family members found to indulge in practices for using plant water. Thus, infants/children and other people are less likely to contract waterborne incidents in areas equipped with these filtration plants. Moreover, the probability of contracting waterborne illness is higher in males than in females in areas lacking filtration plants.

Keywords: clean drinking water; filtration plants; waterborne diseases; induced morbidity

Citation: Aziz, A.; Akram, K.; Abrar ul Haq, M.; Hawaldar, I.T.; Rabbani, M.R. Examining the Role of Clean Drinking Water Plants in Mitigating Drinking Water-Induced Morbidity. *Sustainability* **2022**, *14*, 9644. https://doi.org/10.3390/su14159644

Academic Editor: Marc A. Rosen

Received: 9 May 2022
Accepted: 1 August 2022
Published: 5 August 2022

1. Introduction

Living creatures on this earth need water as an essential element. Almost 70% of the human body is made up of water. Access to clean drinking water is not merely required for human existence but is also considered one of the basic human rights [1]. However, the manifestation of several organic and inorganic toxins, such as pesticides, phenol, chlordane, arsenic fluoride, nitrate, copper, etc., cause the population to consume poor quality water, which is perilous to their health [2]. According to UNICEF and the WHO, all over the world, approximately 2.2 billion people lack access to clean and innocuous drinking water sources. However, about 2 billion people live in high-water-stress countries [3]. According to the WHO, by 2025, almost half of the entire world's population will be forced to live in areas with scarce water sources. Moreover, 6.8 billion individuals worldwide have access to basic drinking water services, while 785 million people around the globe lack access to these services [4].

Contaminated drinking water poses a serious threat to human life across the world. Diseases transmitted due to drinking contaminated water include cholera, diarrhea, typhoid, and dysentery [5]. Waterborne diseases, especially diarrhea, kill approximately one million people globally, and, mostly, children under five years of age are at a higher risk of diarrhea [6]. According to estimates by the World Bank [6], all over the world, because of diarrhea, a well-known disease allied with poor quality of drinking water, about 2.5–3.5 million people are infected, while 485,000 die yearly. Moreover, almost 80 percent of children annually become water-related syndrome victims. Several non-diarrheal illnesses are also associated with the unavailability of adequate and safe drinking water sources. Numerous chemicals (i.e., organic and inorganic) also have an adversative effect on human health in several diseases such as cancer, vomiting, skin rashes, and nausea [6]. Contaminated water causes about 4% of all deaths and 7.5% of all illnesses worldwide [7].

Similarly, in Pakistan, sources of clean drinking water are diminishing swiftly, and drinking water quality is viewed as a grave issue [8]. Poor and indecorous arrangements of the water supply and sewage ejection systems result in the sewage and mess being directly released into drains which ultimately flows into the sea, rivers, or canals and adversely damages water quality [9]. According to a World Bank report [6], 64 percent of Pakistan's population lacks access to clean and innocuous drinking water sources. Only 36 percent of Pakistan's population has access to safely managed and clean drinking water. Therefore, the environmental performance index ranked Pakistan at 140 out of 180 countries regarding safe water and sanitation. Moreover, a report by the World Bank highlighted that 25 percent of the total population in Pakistan is at high risk of consuming arsenic-contaminated water [6]. Pakistan is among the top 10 countries that lack access to clean drinking water, where rich people have more access to safe drinking water sources than poor households [9].

Likewise, the public faces problems regarding access to safe and suitable drinking water in Pakistan. Pollution of drinking water causes a greater threat to Pakistan's public health. Out of 122 nations, Pakistan is in 80th place in terms of following the drinking water quality standards [10]. According to the World Bank report [6], about 53,000 children die annually in Pakistan due to severe diarrhea or waterborne illness. Moreover, drinking contaminated water infected 4 out of 10 children in the country [6]. Moreover, from 2010 to 2019, approximately 250,000 deaths occurred in Pakistan due to drinking contaminated water [11]. Additionally, arsenic, a hazardous pollutant and chemical, has been cited as a key cause of waterborne diseases in several areas of the country. The high level of arsenic in drinking water endangered the health of more than 60 million inhabitants, especially in Punjab, Pakistan [12].

To cope with this problem, the local government has instigated a platform to install safe drinking water filtration plants in each city in Pakistan. The reason for the installation of these plants is to meet the Millennium Development Goal (M.D.G.) by delivering non-toxic drinking water to the public by the end of 2030 [13]. To attain this tenacity, the government of Pakistan, in the middle of 2005, instigated an initiative termed the Clean Drinking Water Initiative (CDWI) to fix about 452 WPEPs (Water Purification Filtration Plants) practically in every tehsil of the Punjab province. Conversely, instead of enormous exertions to provide access to safe drinking water from WPFPs, the community is still facing several problems, for instance, a long-time shutdown of plants due to huge load shedding or closure of plants during holidays, and the non-functionality of a few plants because of a lack of supervision [14].

1.1. Background of Local Government Filtration Plants Project

Lahore is one of the most populous cities and is considered the second-largest city in Pakistan, with 12,642,000 inhabitants [3]. Likewise, the public of Lahore city is also suffering from problems associated with drinking water quality. According to the Pakistan Council of Research in Water Resources (PCRWR) documentation in 2020 [11], in several areas of Lahore, drinking water quality is exceedingly tainted. Thus, the local government has started an initiative to ease this concern. Local government is a public administration found in cities, districts, and counties. In a community, the local government is liable for numerous services to meet the community's needs. Likewise, in Pakistan, the local government is held accountable for services such as waste management and collection, services related to infrastructure, health-associated services (i.e., water and food and water scrutiny, sewerage and toilet-related facilities, other water-based services, welfare, as well as community care services, transportation, and education-related services). At the outset, local government and community administrations were not involved in decisions associated with the provision and supply of water in the country. However, local governments were obligated to provide drinking water-related supplies in 2001 under an ordinance termed the Local Government Ordinance (L.G.O.). Hence, in Pakistan, the local government provides vital services associated with health and secure water supply. Besides, federal and provisional governments were held liable for providing sufficient funds to local governments to ensure the delivery of effective and good services to the public [15].

The endowment of unpolluted drinking water to the inhabitants of Pakistan is the foremost duty of the local government. Thus, to fulfill this purpose, the local government, coordinating the community development department (CD), fixed drinking water filtration plants at several places in Lahore. This decision of plant fixation by the government is based on the level of the arsenic chemical in water, microbiological and other viral contamination in drinking water, public income, as well as public access level, while the number of these water filtration plants fixed by the local government reached 215 in May 2020 in Lahore [15]. Each of these plants can filter about 500 gallons of water an hour. The elementary purpose of the fixation of these water plants is to remove contamination, chemicals, microbes, and arsenic from the water to convert it into a safe form for consumption and drinking by the public [16].

This study was devised to gauge the effect of polluted water on public health and to relate the probability of waterborne illnesses ensued in selected households in two types of study areas (i.e., areas retaining water filtration plants and areas abstaining from plants) in Lahore under the health lifestyle theory, which argues that several regular lifestyle practices implicate deliberations on health-related outcomes [17]. In this study, health-related behavior under collective patterns such as class circumstances, socialization/experience, dispositions to act (habitus), and practices (actions) based on drinking filtration plant water choices conferring on their life chances of waterborne illness was assessed. Thus, the main aim of this study was to discover the health threats and health-associated costs endured by households in the target study areas and the influence of drinking water on infants' and children's health in areas having and lacking plants. The impact of class circumstances (age, education, and family size), experience, or awareness in making life choices based on drinking filtration plant water was also measured.

Moreover, this study compared waterborne disease incidence in households in targeted areas with and without local government filtration plants. Studies which deeply investigated the association between drinking water, morbidity, and related concerns in filtration plants are scarce. Besides, no comparative study of filtration plants' projects and non-project areas under the health lifestyle theory has been conducted yet. This assessment discoursed a comprehensive layout concerned with the quality of drinking water in Pakistan through an exceptional focus on key water impurities, water degradation sources, and subsequent health-associated concerns. Thus, this review substantially contributes to endorsing consciousness in realizing hazards and threats of the factors causing water pollution and waterborne diseases.

This comprehensive investigation will likewise advance the public's ability to quantitatively comprehend the effects of drinking water effluence and the efficiency of prevailing inventiveness regarding clean drinking water for the public, undertaken by the local government by the fixation of filtration plants. However, this enumerated evaluation will be a helpful instrument for the government to formulate and implement better strategies for the endowment of impurity-free drinking water to the community. Additionally, this evaluation will significantly contribute to the literature regarding health-associated issues. Another value of this detailed study is that it will be considered a strong reference tool in impending studies. As a whole, it will assist in enhancing the progression of research focused on concerns associated with drinking water quality.

1.2. Drinking Water Degradation and Human Health

Drinking water is becoming highly contaminated. The main reason behind this contamination is municipal solid waste leachate. Thus, highly toxic elements and chemicals are destroying water quality [18,19]. Contamination of drinking water is one of the major health concerns in today's world. In Pakistan, the high level of arsenic in drinking water endangered the health of more than 60 million people, especially in the province of Punjab. Moreover, the pollution of drinking water poses a greater threat to Pakistan's public health than any other factor. Despite violating drinking water quality standards established by the WHO, there is a lack of research regarding the drinking water risk perception of households in-country [6,8,10,12].

According to studies, cholera is caused by contaminated drinking water, untreated surface water, the detoxification points of water use, and poor sanitation. Moreover, this water contamination also causes waterborne diseases such as diarrhea. Thus, important interventions are needed to protect the public from the harmful impacts of waterborne illness [20,21]. In the same regard, a study by Adimalla [22] demonstrated that nitrate concentration in drinking water has adverse impacts on the health of residents in India. Nitrate concentration crossed the critical limit regarding non-carcinogenic risk. A high nitrate level had poor effects on children's and women's health.

Additionally, studies in different countries, including Mexico, Pakistan, and China, revealed that fluoride and arsenic concentration in drinking water poses a great threat to human life. This arsenic consumption is a cause of cancer among households due to households bearing the burden of illness. Arsenic in drinking water results in Type 2 Diabetes Mellitus (T2DM). Moreover, people were found to have arsenic skin lesion disease. As a result, arsenic-contaminated water seriously threatens human health [23–26]. Moreover, contaminated drinking water destroys the health of living beings around the globe. Different bacteria destroy the quality of drinking water. Contaminated drinking water causes cardiovascular diseases, skin disorders, respiratory problems, liver and splenic ailments, gastrointestinal tract complications, kidney and bladder infections, neurotoxicity, reproductive failure, and cancer. The immediate consequences of contaminated water are less well documented. Other waterborne diseases are cholera, hepatitis, diarrhea, dracunculiasis, ulcers, typhoid, endocrine damage, and arsenicosis [26–29].

Worldwide, drinking water quality is not meeting the standards the WHO and the environmental protection agency set. Thus, awareness regarding the adverse effects of contaminated water must be enhanced [30]. Treatment of these diseases at an early stage is essential; otherwise, they may lead to death [20,28]. Furthermore, previous research found that a high proportion of households drink contaminated water, while others use water from improved sources. Moreover, arsenic concentration risk is significantly associated with location and type of water source, place of residence, and living district. Thus, household water should be regularly tested to reduce the risk associated with arsenic. Moreover, arsenic-contaminated water adversely impacts human health, which ultimately forces people to lose their jobs and live a sub-standard life [31,32].

Moreover, studies in China, Korea, and Pakistan revealed that heavy metals toxify the quality of drinking water, whose consumption is harmful to human health. Different treatment methods can be applied to reduce the harmful effects of heavy metals in drinking water. Thus, water treatment by different methods helps reduce microbial, inorganic, heavy metal, and other types of contamination from drinking water and makes water safe for human consumption and health. Additionally, piped water helps in reducing infant mortality, and both males and females gain equal health benefits from drinking piped water. Moreover, this piped water has a greater impact on post-neonatal mortality as compared with neonatal mortality [33,34].

1.3. Knowledge, Waterborne Illness, and Health Expenditures

Socioeconomic characteristics with psychological factors to study drinking water behavior are important as they provide a holistic framework to understand human behavior. Moreover, knowledge and norms play an important role in using water after treatment. Similarly, education and income levels were significant with water treatment before drinking. Thus, findings revealed that demographic and socioeconomic attributes such as age, education, income, past experiences, and social networks played an important role in perceiving vulnerability. Moreover, the health vulnerability of households is affected by their access to information, health facilities, and the availability of safe water [35–37]. Studies in Bangladesh revealed that households bear the cost of having access to safe water sources due to poor water quality and insufficient and unreliable water supply. The poor spend more income on having access to improved water resources [7,38].

In the same regard, studies have revealed that drinking contaminated water causes diarrhea in children and other family members; 50% of children suffered from acute watery diarrhea (AWD), with a mean age of five. Fewer respondents were found to know about drinking water quality and waterborne illnesses. Additionally, studies in Uganda and India revealed that demographic factors such as mother's education, income level of households, age of children, gender, and awareness of water quality and waterborne diseases influence the reduction of acute watery diarrhea and other waterborne illnesses. However, the government's role is essential in providing clean water to the public [37,39,40].

Similarly, according to Thakur and Gupta [41], in India, high arsenic levels in drinking water cause arsenisois. Awareness regarding arsenic contamination in drinking water and arsenisois played a vital role in the reduction of this waterborne disease. As a result of arsenicosis, people have to bear health care and treatment costs due to visiting doctors. This illness mostly attacks females as compared with males. Moreover, access to improved drinking water sources and income levels can potentially reduce arsenicosis. Children are found to be at greater risk of waterborne illness as compared with other age groups.

Furthermore, because they have access to safe drinking water sources, the general public expends less physical effort and time collecting water. Moreover, access to safe and clean drinking water resources causes fewer water-related illnesses, ultimately reducing public health expenditures [7,42]. In the United States and Pakistan, Surface water quality is seriously diminishing. Nationwide degradation of drinking water causes 90 million illnesses. This illness may result in a financial burden, healthcare, and hospitalization costs [8,43]. Likewise, in Europe, Australia, and China, lack of public awareness and ineffective political measures are huge hurdles to drinking safe water and taking preventive measures by the public. More knowledge and awareness must be created among the public to reduce water-related risks. Moreover, the governance system must be strengthened to communicate water-related risks to the public [44–46].

1.4. Drinking Water and Government Roles

One of the prominent roles played by any country's government is to resolve water quality issues. Governments spend large subsidies in low- and middle-income countries on improving water sources. According to a recent World Bank report, government spending on water resource improvement accounts for 2% of total G.D.P. However, findings in ten countries revealed that 56% of these subsidies benefited the wealthiest people, while only 6% benefited the poorest [6]. The government spends a lot on improving drinking water resources. Although the government enhanced its spending on water improvements in the U.S. and Americas, there is still a violation of standards regarding water pollution [47–49].

According to studies conducted in Germany, South Africa, India, and Pakistan, water treatment is necessary to protect people from the hazards of waterborne illness. Point-of-use water treatment effectively makes drinking water safe and achieves the sustainable goal of clean drinking water [33,50,51]. The facilities for clean water provided by the government are declining drastically. Thus, people mostly rely on private water sources. Therefore, government policies need to be revised to overcome this waterborne disease problem [48,49,52].

Although the government has installed filtration plants in various cities of Pakistan, these plants are not fully reliable sources of safe and clean drinking water. A few samples were found to be contaminated. Due to the electricity shortage and poor water connections, some plants are not properly operational. Besides, in Punjab, the working conditions of filtration plants are much better than in other provinces [11,53,54]. Additionally, polluted water poses a greater threat to human life. Different private and government bodies such as T.M.A., district councils, PHEDs, and WASA provide services related to the public's provision and supply of clean drinking water. Still, water management by these bodies is not effective in urban areas. Coordination between the federal and interim governments appeared lacking [55]. In Nigeria, Uganda, and USA, people are mostly dependent on private water sources as facilities related to the government's provision of clean drinking water have been declining. Consequently, there is an urgent need for government actions to tackle this issue [39,48,52].

Thus, it is concluded from the literature that drinking water sources are degrading in Pakistan due to arsenic, fluoride, heavy metals, and other organic and inorganic pollutants. Drinking contaminated water has severe impacts on the health of people. Diseases that mostly occur due to drinking contaminated water are kidney infections, cholera, hepatitis, diarrhea, dracunculiasis, ulcers, typhoid, endocrine damage, arsenicosis, and respiratory tract infections. Any country's government plays a critical role in ensuring that the populace has access to safe drinking water. Water from improved drinking water sources or water filtration plants can prevent households from contracting waterborne illnesses. Moreover, the literature reveals that age, education, income, and family size are important determinants in choosing to drink water from improved sources; thus, waterborne illness is reduced.

2. Underpinning Theory and Hypothesis Development

Healthy lifestyles are collective patterns of health-related behavior that focus entirely on available options or choices for people conferring more life chances [17]. This description integrates the indigenous relationship among the life choices and life chances anticipated by Weber in one of his most important conceptions regarding lifestyle. According to the Weberian viewpoint, life choices and agency are alternatives, while life changes can be seen in structure form. Whereas lifestyle choices and health exist in a voluntary nature, on the other hand, life chances largely symbolize the class position, which may endow or restrict choices, as choices and chances work off each other to regulate the behavior-related outcomes.

Furthermore, Gochman [56] pointed out that if lifestyle behaviors are positive, they oppose risk behaviors such as good nutrition, which is the converse of bad nutrition. The twofold nature of health-related lifestyle practices refers to consequences resulting from the interchange of choices and changes that reflect positive or negative health impacts. Gochman detected that health-related lifestyles are anticipated to escape associated risks and stay slanted towards inclusive health or fitness. Nevertheless, a "healthy lifestyle" is intended to embrace healthy living.

Historically, people took their health less for granted, but this is not true in present cases. In this modern era, health is viewed as a great achievement. People imagine making an effort to boost their life quality or reduce the risks of persistent illness and untimely deaths [57]. However, Giddens [58] stated that lifestyle preferences are more important for people becoming more responsible for their health. These circumstances instigate the variations in (i) disease patterns, (ii) modernity, and (iii) social identities.

Health lifestyle theory argues that several lifestyle practices reflect health-related outcomes [17]. Thus, everyone is responsible for themselves by choosing healthy living options to have a healthy lifestyle. It must be a life or death issue for the person to be responsible for achieving a healthy lifestyle. Figure 1 elaborates the flow of integral components of health lifestyle theory where class circumstance highlights age, education, gender, race/ethnicity, living conditions, and commonalities as all of these factors significantly impact the selection of drinking water and raise awareness about the waterborne illness; as stated by Bourdieu [59], the dispositions to act are built through socialization and experience. Moreover, experience establishes the basis for life choices, which means "the self-direction of one's behavior", which may lead to fewer occurrences of waterborne illness if households make life choices of using plant water. The interface of life choices and life chances takes the individual dispositions towards action, as shown in Figure 1—such dispositions are instituted as habits. Habitus refers to the cognitive/mental map or the perceptions that usually help evaluate and guide the person's options and choices. The disposition of an act (the preference to use clean drinking water) indulges households in practices. Practices refer to action; therefore, the arrow is pointing from the disposition of an act to practices.

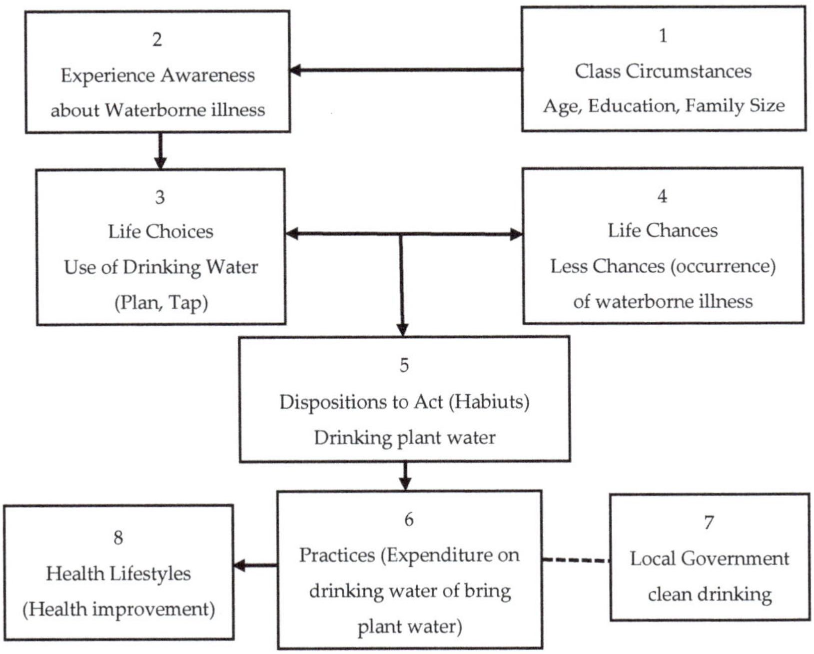

Figure 1. Research framework based on Health Lifestyle Theory. Note: Solid arrows show the relationship directions among variables, and dotted lines show the relationships among variables.

(1) Class circumstances

This category comprises age, education, gender, race/ethnicity, living conditions, and commonalities. Age, education of the household head, and family size are considered class circumstances in this study because these factors significantly impact the selection of drinking water and raise awareness about waterborne illnesses. Thus, we hypothesize as follows:

Hypothesis 1 (H1). *The greater the age of the household head, the lesser the occurrence of waterborne disease.*

Hypothesis 2 (H2). *An increase in the education of household heads decreases the occurrence of waterborne diseases.*

(2) Socialization/Experience

Box 1 in Figure 1 depicts the social framework for the socialization/experience involving class circumstances and associated variables, as illustrated by the arrow pointing to Box 2. Thus, this concept is supported by Bourdieu's [59] suggestion that the dispositions to act are built through socialization and experience. In this study, variables in box 1 provide the basis for experience and influence the decision that either household has to go for plant water or not, as illustrated by the arrow from box 1 to box 2. Thus, we hypothesize as follows:

Hypothesis 3 (H3). *Family size has a positive relationship with waterborne diseases.*

(3) Life Choices

The model in Figure 1 indicates that experience establishes the basis for life choices, as illustrated in box no. 3. As described earlier, Weber introduced the term "life choices", which means "the self-direction of one's behavior". In this study, "life choice" refers to plant or tap water for drinking.

(4) Life Chances

As reflected in box no. 1, class circumstances and additional variables led to life chances (referring to structure) as revealed in box no. 4. Dahrendorf [60] dictated that Weber's work-life chances refer to "glazed chances of finding satisfaction for wants, needs, and interests, hence the possibility of manifestation of the events which ultimately carry out such satisfaction". In the context of this study, life chances mean fewer occurrences of waterborne illness by using plant water.

(5) Dispositions to Act (Habitus)

The interface of life choices and life chances takes the individual dispositions towards action, as shown in box 5 of Figure 1—such dispositions are instituted as habits. Habitus refers to the cognitive/mental map or the perceptions that usually help evaluate and guide the person's options and choices. Here, in this study, when households realized drinking plant water prevents them from waterborne illness, as the local government installed these plants to provide clean drinking water, they showed the disposition of the act by using plant water more for drinking purposes. Hence, we propose the following hypothesis:

Hypothesis 4 (H4). *Households using local government filtration plant water are healthier than households not using local government filtration plant water.*

Hypothesis 5 (H5). *Infants and children using local government filtration plant water have fewer waterborne diseases than families not using local government filtration plant water.*

(6) Practices (actions)

The disposition of an act (the preference to use clean drinking water) indulges households in practices. Practices refer to an action, as illustrated in Figure 1. These actions involve either bringing water from plants or incurring expenditures to bring water from plants. These practices ultimately lead to health improvement by decreasing the occurrence of waterborne illnesses. In this regard, the hypothesis is framed as follows:

Hypothesis 6 (H6). *More expenditure on drinking water reduces the incidence of waterborne illness.*

3. Study Area

The city of Lahore was chosen as the target site for this study. Lahore is the capital of Punjab province and one of the most populous cities in Pakistan, considered the second-largest city in Pakistan, comprising 12,642,000 inhabitants (Figure 2) [3].

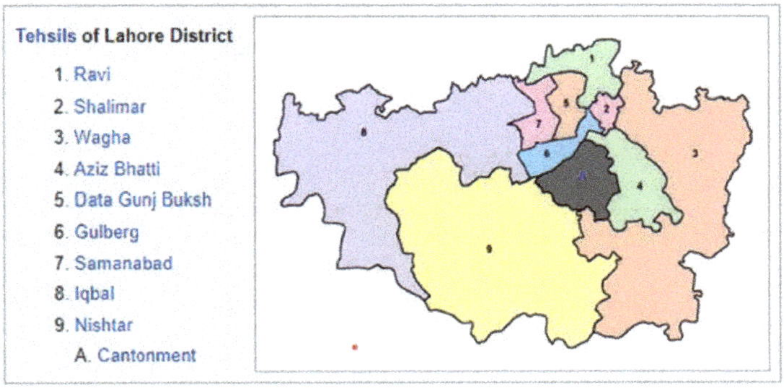

Figure 2. Study area.

In the city, 40 percent of the population is under the age of 15, with an average expected lifetime of no more than 60 years. Besides, the literacy rate is lower than 41% [61]. The public of Lahore city is also suffering from problems associated with the quality of drinking water [11]. To ensure the delivery of hygienic and pristine drinking water to residents of Lahore, the local government fixed several water purification plants at numerous locations around the city, including Shalamar Garden, Mishri Shah, Sabzazar, Lahore Zoo, Walled City, Gulshan-i-Iqbal, Shahdara, Aik Moria Pul, Pani Wala Tabla, Mughalpura, Data Darbar, Harbanspura, City Railway Station, at Blind School, outside Lohari Gate, Lorry Adda, and inside New Anarkali. These sites were selected based on the level of impurities, arsenic, residents' income, and ease of accessibility for citizens [62].

4. Methods and Materials

This study is cross-sectional, meaning that the study's population comprises households residing in four different areas of Lahore. These four areas were chosen using a multistage random sampling technique, where two areas constitute local government water purification plants, while two areas lack these plants [63]. The sites with local government purification plants were Shahdara and Harbanspura, and the sites missing these plants included the Nishatar colony and Zia Colony Township. These four sites were either semi-slums or slums. The income level of the households residing in these areas is low, with a high illiteracy rate. In addition, another purpose for selecting these areas was the level of arsenic, which is a life-threatening chemical. Contamination of drinking water with arsenic causes several chronic pulmonary and skin infections. According to WHO [4], the safe arsenic level in drinking water is up to 50 ppb. The level of arsenic in Shahdara is 76.8 ppb; in Harbanspura, it is 74.0 ppb; in the Nishatar colony, it is 82.0 ppb; and in Zia Colony Township, it is 66.9 ppb [16].

Potential respondents/households were carefully chosen through a simple random sampling technique. Seventy-five households were selected from every selected area. Hence, a total of 300 households were selected as a sample size. Primary data needed to be gathered to evaluate waterborne sickness's influence and determine the association's strength among variables. Data were collected from households in the studied localities using a structured questionnaire. The questionnaire covers all the pertinent information such as demographics (age, level of education, number of family members, average family income), drinking water sources, drinking water expenditures, waterborne associated incidences, and extent. A post-questionnaire development pilot study was conducted by administering 50 questionnaires with and without filtration plant areas to ensure that the questionnaire has valid measures for data collection. Furthermore, face-to-face interviews were conducted with the household head or another family member in their absence to learn their perspectives on the adulteration of drinking water, the underperformance of the water purification plant, the waterborne sickness they experienced, and the impact of this sickness on infants and children. Before starting the interview, the purpose of the study was explained to the participants, and informed consent was obtained for inclusion in the final manuscript. The identity information of the participants was removed from the shared data.

4.1. Statistical Model

The current study utilized the measurements developed by the previous studies. Therefore, class circumstances were sorted by three demographic items, comprising age of the family head, education level of the family head, and family size, to assess the impact of these demographics on the choice of drinking water by households. Life choices were measured by asking the respondents about tap or plant water use, and example items were included regarding why they make these choices. This study employed frequency distributions to describe demography and the variables involved in class circumstances. An independent group *t*-test was performed to measure the mean difference in the occurrence of diseases in both with and without local government filtration plant areas.

Moreover, the impact of various variables refers to class circumstances; namely, respondent's area (with and without plants), income, expenditure on drinking water, use of plant water, family size, education, and age of household head are regressed on waterborne diseases referring to life chances using logistic regression. The waterborne disease was measured through a dichotomous variable, having only two values assigned: value 1 if the respondent suffered from waterborne disease; otherwise, value 0 was assigned. Similarly, a binary variable also measures respondents' area and use of plant water. The functional form of the logistic regression curve is

$$f(t) = \frac{e^t}{1+e^t} \tag{1}$$

where is Euler's number and can be any linear combination of predictors such as $\beta_0 + \beta_1 x$.

$$f(t) = \frac{e^{b_0+b_1x}}{1+e^{b_0+b_1x}} \tag{2}$$

we want to end up with the "typical" formula of the logistic regression, something like:

$$f(x) = L(b_0 + b_1x + \ldots\ldots\ldots\ldots) \tag{3}$$

where L is the Logit, i.e.,

$$f(t) = \ln\left(\frac{e^t}{1+e^t}\right) = b_0 + b_1x \tag{4}$$

4.2. Deriving the Formula

In the first step, let us take our $p(Y+1) = f(t)$ and divide by the probability of the complementary event. If the probability of event A is p, the probability of not-A is $1-p$, Thus,

$$\frac{f(t)}{1-f(t)} = \frac{\frac{e^t}{1+e^t}}{1-\frac{e^t}{1+e^t}} \tag{5}$$

So, we replaced $f(t)$ by $\frac{e^t}{1+e^t}$ and thereby computed the odds. Next, we multiply the equation by $\frac{1+e^t}{1+e^t}$ (which is the neutral element, 1) yielding

$$= \frac{e^t}{(e^t+1)\left(\frac{1+e^t}{1+e^t} - \frac{e^t}{e^t+1}\right)} \tag{6}$$

In other words, the denominator of the numerator "wandered" down to the denominator. Now we can simplify the denominator:

$$\frac{e^t}{(e^t+1)\left(\frac{1+e^t-e^t}{e^t+1}\right)} \quad (7)$$

Simplifying the denominator further

$$\frac{e^t}{(e^t+1)\left(\frac{1}{e^t+1}\right)} \quad (8)$$

However, the denominator simplifies to 1, as can be seen here

$$\frac{\frac{e^t}{e^t+1}}{e^t+1} = \frac{e^t}{1} = e^t \quad (9)$$

The above equation tells us that the odds simplify to e^t. Now, let us take the logarithm of this expression.

$$\mathrm{Ln}\left(e^t\right) = t \quad (10)$$

By the rules of exponents algebra

$$t = \beta_0 + \beta_1 x \quad (11)$$

in sum,

$$In\left(\frac{f(t)}{1-f(t)}\right) = \beta_0 + \beta_1 x \quad (12)$$

The left part of the previous equation is called the Logit, which is the "odd plus logarithm" of $f(t)$, or rather, more precisely, the logarithm of the odd of $\frac{p}{1-p}$. The logistic regression formula can now be obtained by taking the Logit of any linear combination. We can use our standard regression terminology because of the Logit's linearity: The Logit of dependent variable changes by β_1 if x is increased by one unit. Simply replace the Logit with the appropriate value; the rest of the statement is standard regression jargon. In the meantime, because the curve's slope is not linear, β_1 is not equal for all x values. The logistic regression equation for the current model can be expressed as follows:

$$Logit\ (p) = Log\left(\frac{p}{1-p}\right) = \beta_0 + \beta_1 AH + \beta_2 EH + \beta_3 FS + \beta_4 UPW + \beta_5 EDW + \beta_6 RD + \mu_i \quad (13)$$

Solve this equation for p,

$$\frac{p}{1-p} = \frac{1}{\exp(\beta_0 + \beta_1 AH + \beta_2 EH + \beta_3 FS + \beta_4 UPW + \beta_5 EDW + \beta_6 RD + \mu_i)} \quad (14)$$

Pain the formula for the probability $P(\mathrm{WBD} = 1) = p$

$$p = \frac{\exp(\beta_0 + \beta_1 AH + \beta_2 EH + \beta_3 FS + \beta_4 UPW + \beta_5 EDW + \beta_6 RD + \mu_i)}{1+\exp(\beta_0 + \beta_1 AH + \beta_2 EH + \beta_3 FS + \beta_4 UPW + \beta_5 EDW + \beta_6 RD + \mu_i)} \quad (15)$$

Thus, adding criterion and predictor variables equations becomes,

$$\mathrm{Ln}\ [p/(1-p)] = \beta_0 + \beta_1 AH + \beta_2 EH + \beta_3 FS + \beta_4 UPW + \beta_5 EDW + \beta_6 RD + \mu_i \quad (16)$$

$$\mathrm{WBD} = \beta_0 + \beta_1 AH + \beta_2 EH + \beta_3 FS + \beta_4 UPW + \beta_5 EDW + \beta_6 RD + \mu_i \quad (17)$$

Variable	Measurement
W.B.D.:	Waterborne disease is a controlled dummy variable in the model, which can post two values of 0 or 1. If a person suffered from waterborne disease in the last three months, this value represents 1, but 0 if a person did not suffer from waterborne disease.
A.H.H.:	The age of the household head is measured in a complete number of years by asking the respondent what the household head's age is.
E.H.H.:	The education of the household head is also measured in a complete number of years by asking the respondent what the education of the household head is.
F.S.:	Family size was another quantitative variable that varies from 1 to 18 family members.
U.P.W.:	The use of plant water as a predictor variable is binary (yes or no), which can posture two values, either 0 or 1. Households using plant water are assigned value 1, and otherwise assigned value 0.
E.D.W.:	Expenditure on drinking water represents the treatment costs beard by households due to suffering from waterborne diseases in the last three months.
R.A.:	The respondent's area is another binary variable that postures two values, either 0 or 1. Value 1 is assigned if the house is located in the plant area; otherwise, 0 value is assigned.
μ:	Random error, in the model, is a residual variable that accounts for the lack of perfect goodness of fit.

5. Results and Discussion

Human survival depends on the availability of water. Meanwhile, water resources in Pakistan are constantly degrading because of the mixing of various chemicals and trash. The public's health is jeopardized when this polluted water is consumed. This sickness is more likely to affect infants and children. Furthermore, due to these ailments, people must shoulder the cost of health treatment. The local government set up filtration facilities in various parts of Lahore so that the general population may profit from them. This research investigated the effects of waterborne diseases on newborns, children, and other households and their healthcare expenses. Furthermore, by comparing the probability of diseases in areas with drinking water filtration plants installed by their local government versus areas without this facility using a healthy lifestyle model, it was determined that people benefit from drinking water filtration plants whether their local government installs them or not.

5.1. Demography

There is evidence in the literature that demographic factors such as household income level, family size, and education level are major predictors of waterborne illnesses. Shah et al. [36] demonstrated that demographic and socioeconomic attributes such as age, education, income, past experiences, and social networks played an important role in perceiving vulnerability. Moreover, households' health vulnerabilities are affected by their access to information, health facilities, and the availability of safe water. Thus, respondents were first asked about demographics, referring to class circumstances in the health lifestyle research model.

To follow is some information concerning the demographic profile of the data respondents. The median age of household heads in the plant and control regions was 47.20 and 44.68 years old, respectively, whereas the overall mean age in both areas was 45.94 years old. The mean education of household heads in the plant area was 9.38, whereas the mean education in the control area was 10.8, and the overall mean education in both regions was 10.09. In both the plant and control zones, respondents' family sizes ranged from one person to eighteen per home. The average family size in the plant area is 6.49, whereas it is 6.52 in the control area. Furthermore, in the plant area, the mean household income from all sources was 29,387.33 rupees, while in the control region, it was 26,608.67 rupees. The average income in both locations was 27,998.00 rupees. The mean demographics in both the plant and control areas are shown in Table 1.

Table 1. Demographics.

Variables	Control Area (Mean)	Plant Area (Mean)	Overall (Mean)
Education of Household Head (Years)	10.8	9.38	10.09
Age of Household Head (years)	44.68	47.20	45.9400
Family Size	6.52	6.49	6.50
Income (Rupees)	26,608.67	29,387.33	27,998.0000

People were polled on their drinking habits, whether from a plant or the tap, to learn more about the link between drinking water and healthy lifestyle choices, disease incidence, and life chances in both areas with and without local government filtration facilities. The local government set up filtration plants to ensure that the public can access safe drinking water. However, the water from these plants might not be completely safe to drink. Some homes in the plant area claimed that they do not drink plant water because it is contaminated and detrimental to their health or could not detect a difference between plant and tap water during this study. Other respondents stated that transporting plant water for large families is challenging; thus, they rely on tap water. Plant water was recorded in 67 (44.7%) of the area's houses with plants, and tap water was found in 67 (44.7%) households with plants. Meanwhile, 16 (10.7 percent) households in plant areas said they could obtain drinking water from a plant or a tap. A total of 19 (12.6 percent) households in areas without filtration plant facilities said they had to travel a long distance to drink plant water.

The use of drinking water is also determined by family size, which is a social class factor. Because more water is needed if the family grows larger, bringing plant water for a large family becomes a little more difficult. Small families are more likely to use plant water in both plants and control areas, followed by medium-sized families, whereas large families are less likely to use plant water. Small families have 1 to 4 members, medium families have 5 to 8 members, and large families have 9 to 18 members.

There are 9 (6.0 percent) small households in the plant area and 1 (0.7 percent) small family in the control area that uses plant water. As a result, a total of 10 (3.3 percent) tiny families were discovered as employing plant water to make life decisions. As a result, as the size of the family grows, the amount of water used by plants decreases. Table 2 shows the results in both the plant and control areas in this regard. Another element of class condition in the health lifestyle model, the household head's education, was discovered to be substantially linked with drinking water consumption. Educated household heads appeared to be more aware of water contamination and the harmful health effects of dirty water. Therefore, literate households opted to use plant water compared with illiterate families.

In the study region, 205 families were literate, whereas 95 were illiterate; 64 (42.7%) literate households used plant water in the plant area, while 20 (13.3%) literate households used plant water in the control area, for a total of 84 literate households (28.0 percent). Even though both locations had a total of 205 literate households, plant water was used by 41.0 percent of literate homes in both plant and control regions. Thus, a total of 19 families (6.3 percent) in both districts were using plant water. The number of houses consuming tap water in the plant and control areas was 31 (20.7%) and 45 (30.0%), respectively, for a total of 95 households. When the total number of illiterate homes was 95, 19 out of 95, or 20% of illiterate households, used plant water. According to the findings, plant water use is higher in literate families than illiterate families. In this regard, Table 2 presents the contrast between plant and control areas.

Table 2. Relationship between drinking water and target variables.

Use of Drinking Water	Control Area (Percent)	Plant Area (Percent)	Total (Percent)
Tap	130.0 (86.6)	67.0 (44.7)	197 (65.7)
Plant	19.0 (12.7)	67.0 (44.7)	86 (28.6)
Plant and Tap	1.0 (0.6)	16.0 (10.7)	17 (5.7)
Total	150.0 (50.0)	150.0 (50.0)	300 (100.0)

Family Size	Use of drinking water				Total	
	Plant Area		Control Area			
	Plant (Percent)	Tap (Percent)	Plant (Percent)	Tap (Percent)	Plant (Percent)	Tap (Percent)
Small (1–4 members)	9.0 (6.0)	3.0 (2.0)	1.0 (0.7)	14.0 (9.3)	10.0 (3.3)	17.0 (5.7)
Medium (5–8 members)	65.0 (43.3)	59.0 (39.3)	18.0 (12.0)	98.0 (65.3)	83.0 (27.7)	157.0 (52.3)
Large (9–18 members)	9.0 (6.0)	5.0 (3.3)	1.0 (0.7)	18.0 (12)	10.0 (3.3)	23.0 (7.7)
Total	83.0 (55.3)	67.0 (44.7)	20.0 (13.4)	130.0 (86.7)	103.0 (34.3)	197.0 (65.7)

Education	Use of drinking water				Total	
	Plant Area		Control Area			
	Plant (Percent)	Tap (Percent)	Plant (Percent)	Tap (Percent)	Plant (Percent)	Tap (Percent)
Literate	64.0 (42.7)	36.0 (24.0)	20.0 (13.3)	85.0 (56.7)	84.0 (28.0)	121.0 (40.4)
Illiterate	19.0 (12.7)	31.0 (20.7)	0.0 (0.0)	45.0 (30.0)	19.0 (6.3)	76.0 (25.3)
Total	83.0 (55.3)	67.0 (44.7)	20.0 (13.3)	130.0 (86.7)	103.0 (34.3)	197.0 (65.7)

5.2. *Waterborne Diseases in the Study Area*

Water is a necessary component of life on earth, and contaminated water causes a variety of ailments. Throughout the investigation, 24.22 percent of the households, or 462 out of 1907, said they were suffering from waterborne infections, significantly impacting their lives. Although filtration of plant water does not guarantee absolute purity or disease-free safety, it has been demonstrated that people who drink plant water have a lower risk of contracting waterborne infections than those who drink tap water. In the study area, diarrhea was the most common waterborne disease after hepatitis, tuberculosis, skin infection, kidney infection, lung infection, typhoid, abdominal pain, vomiting, and stomach infection.

The mean differences in health in study areas with and without local government filtration plants were assessed using independent group t-tests. According to this survey, waterborne infections were prevalent in locations without filtration facilities. Plant water is a major source of drinking water in plant areas, and most people choose to drink it. Waterborne infections are less common in these locations because they demonstrate the disposition of acts due to class circumstances and bring treated water for drinking. In terms of waterborne infections in newborns, children, females, and males, Table 3 compares plant and control areas.

Table 3. Independent samples test for the occurrence of waterborne diseases.

	t-Test for Equality of Means				
	Control Area	**Plant Area**	**Difference**	***t***	**Sig.**
Occurrence of Waterborne Disease in infants/children	4.4533	2.0533	−2.4000	−5.322	0.000
Occurrence of Waterborne Disease in females	3.1267	1.6533	−1.473	−2.823	0.005
Occurrence of Waterborne Disease in males	3.4867	1.8867	−1.6000	−7.927	0.000

According to the findings, there is a mean difference in waterborne infections in newborns, children, females, and males. In the plant area, the mean occurrence of waterborne infections in infants and children is 2.0533, while in the control area, it is 4.4533. As a result, the mean difference in both areas is −2.4. Furthermore, in the case of the occurrence of waterborne infections in infants and children, the t-value and significance value were −5.322 and 0.000, respectively. Because the t-value was −5.322 and the sig value was 0.000, which is less than 0.05, it is clear that the disposition of the act resulted in a significant difference in the mean occurrence of waterborne infections in infants and children in plant and control regions. Because they do not engage in such behaviors, waterborne infections among infants and children are higher in control areas than in plant areas. In the case of waterborne disease in females, the mean value of the plant area, the mean value of the control area, and the value of the mean difference in both areas were 16.533, 3.1267, and −1.473, respectively. Furthermore, the t-value was −2.823, with a significance value of less than 0.005 and less than 0.05. Therefore, the findings indicated a considerable difference in the mean occurrence of waterborne infections in females in plant and control areas. As a result, females are more likely to contract the waterborne disease than in the plant area in the control area.

Similarly, the mean value of the plant area, the mean value of the control area, and the value of the mean difference in both areas were 18.867, 3.4867, and −1.6000, respectively, in the event of the occurrence of waterborne disease in males. The t-value was also −7.927, and the significance value was less than 0.000 or less than 0.05. As a result, it demonstrates a considerable difference in the mean occurrence of waterborne infections in men in the plant and control areas. As a result, females are more likely to contract waterborne diseases than in the plant area in the control area. In short, the mean occurrence of waterborne infections differs significantly across infants, children, females, and males. In the control region, the incidence of waterborne diseases or bad life chances in newborns, children, females, and males was higher than in the plant area due to a lack of life options from drinking plant water.

5.3. *Econometric Model for Waterborne Illness*

Waterborne diseases are considered dependent variables measured by dummy values 0 or 1. If a person is suffering from waterborne diseases, then the value of dummy variables is 1; otherwise, value 0 was assigned. Furthermore, family size is a quantitative variable that ranges from 1 to 18. It is theorized that the household head's age and education reduce the occurrence or probability of waterborne sickness in that family. Waterborne infections are less likely to arise when households spend more on drinking water.

In the same way, increasing plant water use reduces the risk of waterborne sickness in that home. The results of the binary logistic model summary are revealed in Table 4. Logistic estimates of household head education (E.H.), use of plant water (U.P.W.), drinking water expenditures (E.D.W.), and respondent area (R.A.) were negatively correlated with the probability of waterborne diseases, whereas respondent family size (F.S.) was positively correlated with the probability of waterborne diseases. However, the current study's household head (A.H.) age was insignificant in reducing waterborne illness. Moreover, the robustness regression was also analyzed to verify the results of logistic estimates, as shown in Table 4 [64]. Robust regression is a form of regression analysis designed to overcome some limitations of traditional parametric and non-parametric methods. The current findings of robust regression are almost similar to the output of the binary logistic, indicating that the current model outputs are valid.

Table 4. Regression model summary for waterborne diseases.

	Binary	Logistic	Regression	Robustness	Regression
	β	*p*-Value	Exp (β)	β	*p*-Value
Age of HH	0.004	0.699	1.004	0.003	0.714
Education of HH	−0.082	0.011	0.921	−0.083	0.009
Family Size	0.245	0.000	1.278	0.242	0.000
Use of plant water	−1.380	0.021	0.252	−1.379	0.022
Expenditures on Drinking water	−0.001	0.000	0.999	−0.002	0.000
Respondent's area	−0.903	0.003	0.406	−0.900	0.004
Constant	0.645	0.361	1.907	0.689	0.381
−2 Log likelihood		306.831			
Nagelkerke R Square		0.376			
Cox and Snell R Square		0.278			
Lemeshow Sig.		0.221			
Df		08			
Chi-square		10.672			

Dependent Variable: Waterborne Disease.

Moreover, in the current model, the value of Cox and Snell R square was 0.278, representing that a 28 percent change in the explained variable is due to the study's independent variables, while its value range was always between 0 and 0.75. On the other hand, the value of Nagelkerke R Square was 0.376, which indicates that 38 percent of the variation in the dependent variable is due to the independent variables, while its value always ranged between 0 and 1. The value of −2 Log-likelihood was detected at 306.831 at df = 8, significant at a 5 percent significance level. The Lemeshow test value for this model was 0.221, which is greater than 0.05, indicating that the model is statistically significant. The intercept term was 0.645, indicating the average prevalence of waterborne infections (life chances).

The current findings showed that education of the household head, expenditures on drinking water, use of plant water, and respondent's area all have a negative effect on the occurrence of water bone diseases, with odd ratios (Exp-) less than 1, and these results are statistically significant at the 5% level of significance. Meanwhile, family size has an odd ratio (Exp-β) greater than 1, indicating that an increase in family size will increase the probability of waterborne illness for that household, and this result is significant at the 1 percent significance level. However, age is ineffective in reducing the probability of waterborne diseases because this variable is statistically insignificant, as its *p*-value was 0.699, greater than the significance level. An increase in the education of household heads creates more awareness regarding waterborne illnesses and, as a result, households with a higher level of education will have a lower probability or chance of occurrence of waterborne diseases. Similarly, increased use of plant water refers to making life choices using plant water; the occurrence of waterborne illnesses leads to positive life chances. A rise in the cost of drinking water (as defined in our research model) indicates that more plant water is being used to bring water from plants. People must travel a certain distance and pay a certain amount of money. As a result, increasing plant water use lowers the risk of contracting waterborne infections. As a result, in the respondent's location, people in the plant area prefer to drink plant water, lowering the risk of waterborne sickness in those families. In the same regard, Shah et al. [36] demonstrated that demographic and socioeconomic attributes such as age, education, income, past experiences, and social networks played an important role in perceiving vulnerability to such waterborne illnesses. Similarly, Khalid and Khaver [55] concluded that polluted water poses a greater threat to human life.

Furthermore, access to information, health facilities, and clean water influences households' health vulnerability. Waterborne sickness is reduced through increased education, household per capita expenditures on safe drinking water, and access to safe drinking water. They also demonstrated that families without access to a water supply spend significant time obtaining water, resulting in additional costs for the poor household. Because of the high cost of disinfection methods for treating drinking water, low-income families were extremely unlikely to use them, preventing poor households from reaping the benefits of clean water. The results revealed that the value of all coefficients except the age of the household head is significant in regression at 1 and 5 percent levels of significance.

6. Theoretical and Practical Implications

This study has several theoretical and practical implications. As findings of this study exposed a negative association between clean drinking water sources and waterborne illness, it may positively influence people's mindsets regarding drinking water quality and drinking water plants installed by government bodies. Moreover, this study revealed that the water from filtration plants installed by the local government is clean and safe for human consumption. Still, different private and government bodies, such as T.M.A., district councils, PHEDs, and WASA, provide services related to providing and supplying clean drinking water to the public. Still, water management by these bodies is not effective in urban areas. Coordination between the federal and provisional governments seemed lacking. Thus, this study will draw the government's attention to the installation of more drinking plants for the benefit of society and will show the government of Pakistan that the availability of such plants is a good option for the sustainable management of drinking water in hard-hit and water-scarce areas of Pakistan.

Besides, class circumstances such as the size of the family and the education of the family head are enormously correlated to the quality and use of drinking water and ultimately to health-associated outcomes by improving healthy lifestyle. Thus, this study has the potential for realizing households to be more educated and have small family sizes to have improved lifestyles in terms of drinking water quality.

This assessment has provided a comprehensive layout of concerns with the quality of drinking water in Pakistan through an exceptional focus on key water impurities, water degradation sources, and subsequent health-associated concerns. Thus, this review contributed substantially to endorsing consciousness in realizing the hazards and threats of the factors causing water pollution and waterborne diseases. This comprehensive investigation will likewise advance the public's ability to quantitatively comprehend the effects of drinking water effluence and the efficiency of prevailing inventiveness regarding clean drinking water for the public, undertaken by the local government by the fixation of filtration plants. However, this enumerated valuation will also be a helpful instrument for the government to intend better strategies for providing impurity-free drinking water to the community. As water hails from easily accessible and improved sources, households must spend less effort and time collecting it. Additionally, as local governments install filtration plants to make clean drinking water conveniently available to the public in various places, it will allow the public to avoid risky journeys for collecting clean drinking water.

Moreover, as the study's findings exposed, people bear fewer health expenses due to having access to better drinking water sources because they are less likely to become victims of waterborne diseases. Thus, this study is useful for realizing the local government's public importance of drinking water filtration plants. Additionally, this evaluation will significantly contribute to the literature regarding health-associated issues. Another value of this detailed study is that it will be considered a strong reference tool in impending studies. As a whole, it will assist in enhancing the progression of research focused on concerns associated with the quality of drinking water and clean drinking water sources.

7. Conclusions and Recommendations

The findings of this study revealed the conclusion that people who make life choices of drinking filtered plant water are less likely to contract waterborne diseases. Besides, class circumstances such as the family size, the family head's education, and plant water usage are enormously correlated to the quality and use of drinking water and ultimately to the health-associated outcomes by improving healthy lifestyle. While the age of the household head was found to be insignificant in making choices regarding drinking water choices and reduction in waterborne illness, the education of the family head, the number of family members, and households were found to be more habitual in engaging in practices for using plant water. Thus, in areas with these filtration plants, infants, children, and other people are less likely to contract waterborne incidents. In short, it is concluded from the study's results that the impacts of clean drinking water are significantly different from unclean drinking water. Clean drinking water from any source helps prevent waterborne illness and reduce waterborne associated costs, while unclean water is found unsafe for the health of residents.

Thus, it is recommended that awareness campaigns be launched by the public and private sector/civil society about the advantages of drinking water to enhance the ratio of plant water users in treatment areas. The local government must boost the number of these filtration plants to cover the large proportion of the population and maximize the benefits of impurity-free water for the public. To ensure water availability throughout the day, even during electricity load shedding in the country, generators should be fixed with these plants. Areas adjacent to the plants must be focused on cleanliness regularly to build a positive image of these plants near the public. Awareness campaigns must be launched to increase public awareness concerning drinking water contamination, hazardous and quality hits to take preventive measures to protect themselves from the perilous effects of bad quality water.

8. Limitations and Future Recommendations

This study has some limitations as well. This study lacks data about the mother's age and education, which greatly influence life choices and life chances. Future studies can collect data regarding the mother's age and education, which greatly influence life choices and life chances. Waterborne diseases are influenced by various factors, including poor hygiene practices and a family history of the disease, which are the focus of this study. These factors can also be manipulated in future studies.

Moreover, this study only measured the impacts on the income of households in terms of waterborne expenditures, while the impacts on income due to losing work productivity and working days were ignored. Thus, future studies could examine the loss of workers' productivity and the consequent impacts on income due to waterborne illness. The generalizability of this study can be enhanced by focusing on other areas of Lahore having filtration plant facilities. This study was just meant to check the impact of filtration plants installed by the local government on disease elevation while not focusing on determining whether these filtration plants' water quality meets the standards set by WHO or not. Further study can be conducted to determine the water quality of these filtration plants installed by the local government and other government authorities.

Additionally, this study is limited to filtration plants installed by local government and community development, whereas other authorities such as T.M.A., district councils, PHEDs, WASA, and private bodies have also installed filtration plants for the provision of clean drinking water to the public of Pakistan; therefore, further studies can be executed to check the impacts of filtration plants on water other than local government at the household level. Moreover, further studies can be conducted in other countries, especially those with or without water crises or safety issues, to enhance the generalizability of the findings. The local government installed filtration plants for the benefit of the public. As a result, the government had to bear some cost, which was not measured in this study; therefore, further studies can be conducted to perform a cost–benefit analysis of these filtration plants.

Author Contributions: Conceptualization, A.A.; Data curation, A.A.; Formal analysis, A.A., K.A. and M.A.u.H.; Funding acquisition, A.A. and I.T.H.; Methodology, M.A.u.H. and K.A; Project administration, K.A.; Resources, I.T.H. and M.R.R.; Software, K.A.; Supervision, K.A.; Validation, K.A.; Writing—original draft, A.A.; Writing—review & editing, M.A.u.H., I.T.H. and M.R.R. All authors have read and agreed to the published version of the manuscript.

Funding: This research was not funded by any source (individual, group, or organization).

Institutional Review Board Statement: The current study did not conduct any experiments on humans or animals in the lab; thus, no ethical committee approval is required for this research.

Informed Consent Statement: For this type of study, formal consent was not required. However, this study ensured that before starting the interview, the purpose of the study was explained, and informal consent was also obtained for voluntary participation in the final survey. The secrecy and anonymity of the given information and data were ensured.

Data Availability Statement: The data that support the findings of this study are available from the corresponding author upon reasonable request.

Conflicts of Interest: The authors declare no conflict of interest.

References

1. N.G. (National Geography). Fresh Water Crisis. 2020. Available online: https://www.nationalgeographic.com/environment/freshwater/freshwater-crisis/ (accessed on 11 November 2020).
2. Thompson, L.A.; Darwish, W.S. Environmental chemical contaminants in food: Review of a global problem. *J. Toxicol.* **2019**, *2019*, 2345283. [CrossRef] [PubMed]
3. U.N. World Water Development Report 2020 Water and Climate Change. 2020. Available online: https://www.unwater.org/world-water-development-report-2020-water-and-climate-change/ (accessed on 18 March 2021).
4. WHO. Drinking-Water. 2019. Available online: https://www.who.int/news-room/fact-sheets/detail/drinking-water (accessed on 18 March 2021).
5. Farzin, M.; Asadi, A.; Pukanska, K.; Zelenakova, M. An Assessment on the Safety of Drinking Water Resources in Yasouj, Iran. *Sustainability* **2022**, *14*, 3619. [CrossRef]
6. World Bank. A New World Bank Report Proposes Key Reforms to Improve Water Security and Productivity. 2019. Available online: https://www.worldbank.org/en/news/press-release/2019/02/04/pakistans-scarce-water-can-bring-more-value-to-people-and-economy (accessed on 6 December 2020).
7. WHO. Unsafe Drinking-Water, Sanitation, and Waste Management. 2019. Available online: https://www.who.int/sustainable-development/cities/health-risks/water-sanitation/en/ (accessed on 14 December 2020).
8. Daud, M.K.; Nafees, M.; Ali, S.; Rizwan, M.; Bajwa, R.A.; Shakoor, M.B.; Arshad, M.U.; Chatha, S.A.S.; Deeba, F.; Murad, W.; et al. Drinking water quality status and contamination in Pakistan. *BioMed Res. Int.* **2017**, *2017*, 7908183. [CrossRef] [PubMed]
9. Badar, Z.U.; Xisheng, H.; Jabee, A. Blemishes of governmental Institutional Management to Trample Water Pollution in Pakistan. *Int. J. Res. Soc. Sci.* **2019**, *9*, 329–352.
10. Ahmed, A.; Shafique, I. Perception of household in regards to water pollution: An empirical evidence from Pakistan. *Environ. Sci. Pollut. Res.* **2019**, *26*, 8543–8551. [CrossRef]
11. PCRWR. Pakistan Council of Research in Water Resources. 2020. Available online: http://www.pcrwr.gov.pk/ (accessed on 15 October 2021).
12. Kamran, H.W.; Omran, A. Water Contamination and Health Hazards in Pakistan: An Overview of the Current Scenario and Contemporary Challenges. In *Sustaining Our Environment for Better Future*; Springer: Singapore, 2019; pp. 75–84.
13. World Bank. Opportunities for Clean and Green Pakistan. A Country Environmental Analysis. 2017. Available online: http://documents1.worldbank.org/curated/en/746031566833355389/pdf/Opportunities-for-a-Clean-and-Green-Pakistan-A-Country-Environmental-Analysis.pdf (accessed on 8 August 2021).
14. Rehman, M.Z. Access to Safe Drinking Water Is Still a Dream. 2019. Available online: http.//www.awazcds.org.pk/Downloads/rstudies/Water%20Study%20Awaz.pdf (accessed on 3 January 2021).
15. L.G& CD. Clean Drinking Water for All (CDWA) Project. 2020. Available online: https://lgcd.punjab.gov.pk/cdwa (accessed on 23 September 2021).
16. PCRWR. *Progress Report on Water Filtration Plants*; PCRWR: Lahore, Pakistan, 2019.
17. Cockerham, W.C. Health Lifestyle Theory and the Convergence of Agency and Structure. *J. Health Soc. Behav.* **2005**, *46*, 51–67. [CrossRef] [PubMed]

18. Abdel-Shafy, H.I.; Mansour, M.S. Solid waste issue: Sources, composition, disposal, recycling, and valorization. *Egypt. J. Pet.* **2018**, *27*, 1275–1290. [CrossRef]
19. Panhwar, A.; Kandhro, A.; Jalbani, N.; Faryal, K.; Mirjat, M.S.; Jhatial, G.H.; Qaiser, S. Assessment of Groundwater Quality Affected by Open Dumping Site in Hyderabad, Pakistan. *Int. J. Environ. Chem.* **2019**, *5*, 23–30.
20. WHO. WHO—World Water Day Report-World Health Organization. 2019. Available online: https://www.who.int/water_sanitation_health/takingcharge.html (accessed on 18 August 2021).
21. Islam, M.S.; Hassan-uz-Zaman, M.; Islam, M.S.; Clemens, J.D.; Ahmed, N. Waterborne pathogens: Review of outbreaks in developing nations. In *Waterborne Pathogens*; Butterworth-Heinemann: Oxford, UK, 2020; pp. 43–56.
22. Adimalla, N. Spatial distribution, exposure, and potential health risk assessment from nitrate in drinking water from semi-arid region of South India. *Hum. Ecol. Risk Assess. Int. J.* **2020**, *26*, 310–334. [CrossRef]
23. Alarcon-Herrera, M.T.; Martin, D.A.; Gutiérrez, M.; Reynoso-Cuevas, L.; Martín-Domínguez, A.; Olmos-Márquez, M.A.; Bundschuh, J. Co-occurrence, possible origin, and health-risk assessment of arsenic and fluoride in drinking water sources in Mexico: Geographical data visualization. *Sci. Total Environ.* **2020**, *698*, 134168. [CrossRef]
24. Idrees, M.; Batool, S. Environmental risk assessment of chronic arsenic in drinking water and prevalence of type-2 diabetes mellitus in Pakistan. *Environ. Technol.* **2020**, *41*, 232–237. [CrossRef] [PubMed]
25. Shahid, M.; Niazi, N.K.; Dumat, C.; Naidu, R.; Khalid, S.; Rahman, M.M.; Bibi, I. A meta-analysis of the distribution, sources, and health risks of arsenic-contaminated groundwater in Pakistan. *Environ. Pollut.* **2018**, *242*, 307–319. [CrossRef] [PubMed]
26. Marembo, L.; Xu, W. A Graduated Approach in the Abolition of Waterborne Diseases in Drinking Water Using an Indicator Based Approach and Nano Based Biosensors: A Review. *J. Environ. Earth Sci.* **2020**, *10*, 64–71.
27. Fazal-ur-Rehman, M. Polluted Water Borne Diseases: Symptoms, Causes, Treatment and Prevention. *J. Med. Chem. Sci.* **2019**, *2*, 85–91.
28. Sinha, D.; Prasad, P. Health effects inflicted by chronic low-level arsenic contamination in groundwater: A global public health challenge. *J. Appl. Toxicol.* **2020**, *40*, 87–131. [CrossRef]
29. Ahmed, J.; Wong, L.P.; Chua, Y.P.; Channa, N.; Mahar, R.B.; Yasmin, A.I.; VanDerslice, J.A.; Garn, J.V. Quantitative Microbial Risk Assessment of Drinking Water Quality to Predict the Risk of Waterborne Diseases in Primary-School Children. *Int. J. Environ. Res. Public Health* **2020**, *17*, 2774. [CrossRef]
30. Fazel, H.K.; Abdo, S.M.; Althaqafi, A.; Eldosari, S.H.; Zhu, B.K.; Safaa, H.M. View of Saudi Arabia Strategy for Water Resources Management at Bishah, Aseer Southern Region Water Assessment. *Sustainability* **2022**, *14*, 4198. [CrossRef]
31. Khan, J.R.; Bakar, K.S. Targeting at risk households in Bangladesh exposed to arsenic contamination: A spatial analysis. *Int. J. Environ. Health Res.* **2020**, *30*, 1–12. [CrossRef]
32. Rahman, M.A.; Rahman, A.; Khan, M.Z.K.; Renzaho, A.M. Human health risks and socioeconomic perspectives of arsenic exposure in Bangladesh: A scoping review. *Ecotoxicol. Environ. Saf.* **2018**, *150*, 335–343. [CrossRef]
33. Shah, M.T.; Suleman, M.; Abdul Baqi, S.; Sattar, A.; Khan, N. Determination of heavy metals in drinking water and their adverse effects on human health. A review. *Pure Appl. Biol. (PAB)* **2020**, *9*, 96–104. [CrossRef]
34. Fan, M.; He, G. *The Impact of Clean Water on Infant Mortality: Evidence from China*; Agricultural and Applied Economics Association: Milwaukee, WI, USA, 2019.
35. Daniel, D.; Diener, A.; Pande, S.; Jansen, S.; Marks, S.; Meierhofer, R.; Rietveld, L. Understanding the effect of socioeconomic characteristics and psychosocial factors on household water treatment practices in rural Nepal using Bayesian Belief Networks. *Int. J. Hyg. Environ. Health* **2019**, *222*, 847–855. [CrossRef] [PubMed]
36. Shah, A.A.; Ye, J.; Shaw, R.; Ullah, R.; Ali, M. Factors affecting flood-induced household vulnerability and health risks in Pakistan: The case of Khyber Pakhtunkhwa (K.P.) Province. *Int. J. Disaster Risk Reduct.* **2020**, *42*, 101341. [CrossRef]
37. Derdour, A.; Bouanani, A.; Kaid, N.; Mukdasai, K.; Algelany, A.M.; Ahmad, H.; Menni, Y.; Ameur, H. Groundwater Potentiality Assessment of Ain Sefra Region in Upper Wadi Namous Basin, Algeria Using Integrated Geospatial Approaches. *Sustainability* **2022**, *14*, 4450. [CrossRef]
38. Amit, R.K.; Sasidharan, S. Measuring affordability of access to clean water: A coping cost approach. *Resour. Conserv. Recycl.* **2019**, *141*, 410–417. [CrossRef]
39. Wasswa, H. Assessing the Factors Associated with Access to Safe Drinking Water among Rural Households in Uganda: A Case Study of Kalungi Sub-County, Nakasongola District. Ph.D. Thesis, Makerere University, Kampala, Uganda, 2019.
40. Adhikary, M.; Pal, C. Incidence of waterborne disease: A case study of child Diarrhea in Kolkata. *Indian J. Econ. Dev.* **2019**, *7*, 1–12.
41. Thakur, B.K.; Gupta, V. Valuing health damages due to groundwater arsenic contamination in Bihar, India. *Econ. Hum. Biol.* **2019**, *35*, 123–132. [CrossRef] [PubMed]
42. Martínez-Santos, P. Does 91% of the world's population "n really have" sustainable access to safe" drinking water"? *Int. J. Water Resour. Dev.* **2017**, *33*, 514–533. [CrossRef]
43. Deflorio-Barker, S.; Wing, C.; Jones, R.M.; Dorevitch, S. Estimate of incidence and cost of recreational waterborne illness on United States surface waters. *Environ. Health* **2018**, *17*, 3. [CrossRef]
44. Anthonj, C.; Setty, K.E.; Ezbakhe, F.; Manga, M.; Hoeser, C. A systematic review of water, sanitation and hygiene among Roma communities in Europe: Situation analysis, cultural context, and obstacles to improvement. *Int. J. Hyg. Environ. Health* **2020**, *226*, 113506. [CrossRef]
45. Wee, S.Y.; Aris, A.Z. Occurrence and public-perceived risk of endocrine disrupting compounds in drinking water. *Clean Water* **2019**, *2*, 4. [CrossRef]

46. Rogers, B.C.; Dunn, G.; Hammer, K.; Novalia, W.; de Haan, F.I.J.; Brown, L.; Brown, R.R.; Lloyd, S.; Uricha, C.; Wong, T.H.F.; et al. Water Sensitive Cities Index: A diagnostic tool to assess water sensitivity and guide management actions. *Water Res.* **2020**, *186*, 116411. [CrossRef] [PubMed]
47. Balazs, C.L.; Ray, I. The drinking water disparities framework: On the origins and persistence of inequities in exposure. *Am. J. Public Health* **2014**, *104*, 603–611. [CrossRef] [PubMed]
48. Keiser, D.A.; Shapiro, J.S. Consequences of the Clean Water Act and the demand for water quality. *Q. J. Econ.* **2019**, *134*, 349–396. [CrossRef]
49. Bartram, J.; Blallance, R. (Eds.) *Water Quality Monitoring: A Practical Guide to the Design and Implementation of Freshwater Quality Studies and Monitoring Programmes*; C.R.C. Press: Boca Raton, FL, USA, 2020.
50. Gwenzi, W.; Chaukura, N.; Noubactep, C.; Mukome, F.N. Biochar-based water treatment systems as a potential low-cost and sustainable technology for clean water provision. *J. Environ. Manag.* **2017**, *197*, 732–749. [CrossRef] [PubMed]
51. Dupas, P.; Nhlema, B.; Wagner, Z.; Wolf, A.; Wroe, E. *Expanding Access to Clean Water for the Rural Poor: Experimental Evidence from Malawi*; National Bureau of Economic Research: Cambridge, MA, USA, 2020.
52. Emenike, C.P.; Tenebe, I.T.; Omole, D.O.; Ngene, B.U.; Oniemayin, B.I.; Maxwell, O.; Onoka, B.I. Accessing safe drinking water in sub-Saharan Africa: Issues and challenges in South-West Nigeria. *Sustain. Cities Soc.* **2017**, *30*, 263–272. [CrossRef]
53. Hisam, A.; Rahman, M.U.; Kadir, E.; Tariq, N.A.; and Masood, S. Microbiological Contamination in Water Filtration Plants in Islamabad. *J. Coll. Physicians Surg. Pak.* **2012**, *24*, 345–350.
54. USAID. Literature Review Drinking Water, Sanitation and Hygiene Knowledge, Attitudes, and Practices Khyber Pakhtunkhwa Province, Peshawar City, Pakistan. 2010. Available online: http.//pdf.usaid.gov/pdf_docs/pnadz936.pdf (accessed on 12 December 2020).
55. Khalid, I.S.; Khaver, A.A. Political Economy of Water Pollution in Pakistan: An Overview. 2019. Available online: https://think-asia.org/handle/11540/10395 (accessed on 21 January 2022).
56. Gochman, D.S. Provider determinants of health behavior. In *Handbook of Health Behavior Research II*; Springer: Boston, MA, USA, 1997; pp. 397–417.
57. Clarke, E.B.; Curtis, J.R.; Luce, J.M.; Levy, M.; Danis, M.; Nelson, J.; Solomon, M.Z.; Robert Wood Johnson Foundation Critical Care End-Of-Life Peer Workgroup Members. Quality indicators for end-of-life care in the intensive care unit. *Crit. Care Med.* **2003**, *31*, 2255–2262. [CrossRef]
58. Giddens, A.; Duneier, M.; Appelbaum, R.P.; Carr, D.S. *Introduction to Sociology*; Norton: New York, NY, USA, 1991.
59. Bourdieu, P. The economics of linguistic exchanges. *Soc. Sci. Inf.* **1977**, *16*, 645–668. [CrossRef]
60. Ralf, D. *A Third Europe?* European University Institute: Fiesole, Italy, 1979.
61. W.P.R. World Population Review Report. 2020. Available online: https://worldpopulationreview.com/world-cities/lahore-population/ (accessed on 24 February 2022).
62. Government of Pakistan. *Progress Report of Water Filtration Plants*; Government of Pakistan: Lahore, Pakistan, 2019.
63. Abrar ul haq, M.; Jali, M.R.M.; Islam, G.M.N. Household empowerment as the key to eradicate poverty incidence. *Asian Soc. Work Policy Rev.* **2019**, *13*, 4–24. [CrossRef]
64. Abrar ul Haq, M.; Sankar, J.P.; Akram, F.; Siddique, M. The role of farmers' attitude towards their resources to alleviate rural household poverty. *Qual. Quant.* **2022**, *56*, 2133–2155. [CrossRef]

International Journal of
Environmental Research and Public Health

MDPI

Article

Spatial–Temporal Variations of Water Quality in Urban Rivers after Small Sluices Construction: A Case in Typical Regions of the Taihu Lake Basin

Feng Lan [1,2], Wang Haisen [1] and Yan Yan [3,*]

1 College of Civil Engineering, Nanjing Forestry University, Nanjing 210037, China
2 College of Environment and Biology, Nanjing Forestry University, Nanjing 210037, China
3 Taihu Water Pollution Prevention and Control Research Center, Jiangsu Provincial Academy of Environmental Science, Nanjing 210042, China
* Correspondence: jshayanyan@163.com

Abstract: Urban river pollution is considered a 'necessary evil' consequence of disproportionate developmental expansion in metropolises. Unprecedented expansion and anthropic activities lead to the deterioration of urban rivers with municipal and industrial sewage. The construction of sluices is one of the irrefutable parts of the process. In order to prevent floods and drought, many cities build sluices and dams in rivers to balance water quantity in different seasons. To explore the change characteristics of the water quality in urban rivers after the construction of sluices and dams, the change in the total phosphorus (TP) and total nitrogen (TN) concentrations upstream and downstream of rivers was investigated under the condition of sluices closure in Wuxi. According to the results, when the sluices were closed, the pollutants of TP and TN would accumulate upstream in rivers, which caused the water quality in the upper reaches to be worse than that in the lower reaches. Specifically, the TN and TP concentrations downstream of urban rivers in Wuxi were approximately 14.42% and 13.80% lower than those upstream when the sluices were closed. Additionally, the water quality in urban rivers was usually better in summer and autumn than in the other seasons, showing obvious seasonality after the construction of the sluices. The research will provide a theoretical basis for future sluice operation and the water resources management of urban rivers.

Keywords: sluices and dams; urban rivers; water pollution; Taihu Lake; total phosphorus; total nitrogen

Citation: Lan, F.; Haisen, W.; Yan, Y. Spatial–Temporal Variations of Water Quality in Urban Rivers after Small Sluices Construction: A Case in Typical Regions of the Taihu Lake Basin. *Int. J. Environ. Res. Public Health* **2022**, *19*, 12453. https://doi.org/10.3390/ijerph191912453

Academic Editors: Alban Kuriqi and Luis Garrote

Received: 21 August 2022
Accepted: 26 September 2022
Published: 29 September 2022

1. Introduction

Rivers are an integral part of urban ecosystems [1]. Rapid urbanization and economic development have caused urban river pollution globally, and human activities directly or indirectly lead to changes in river environments. Under natural flow conditions, river water can self-purify, while during the process of urban expansion, the water quality of the river is more or less affected, and the river cannot fully play its original function [2]. In the last decade, the number of hydraulic structures such as sluices and dams has increased with rapid urbanization. It is expected that by 2025, these structures will be present in 70% of rivers around the world [3,4]. The small size of urban rivers and the excessive construction of sluices will undoubtedly affect the material transport movements in rivers and then the river environment [5]. The factors affecting the water quality of the regulated rivers are more complex and are susceptible to the influence of water diversion and storage by sluices and dams [6]. They can largely result in the degradation of water quality in rivers [7]. The water quality of rivers in cities directly or indirectly affects human health, and thus the health of regulated rivers and the attainment of water quality standards are essential to the high-quality development of the urbanization process [8].

According to the 2020 statistics of the World Commission on Dams (WCD), the number of dams worldwide has reached 59,071, of which the number in China reached

23,841, accounting for about 40% of the world [9]. Urban rivers play a significant role in water extraction, navigation, and stormwater drainage, and also in urban domestic sewage as well as part of the industrial wastewater discharge. Furthermore, the impact of sluices and dams on river environments and ecology is increasing for watersheds with high population density, relatively concentrated production and life, and more serious water pollution [10]. Especially in some areas with abundant stormwater, various flood control sluices and dams have been built in urban rivers to prevent urban waterlogging. These sluices and dams inevitably affect the hydrological status of the rivers, thereby affecting the diffusion and distribution of pollutants in urban rivers. Studies have shown that the accumulation of nitrogen and phosphorus pollutants without treatment will bring great risks to the water environment [11].

Currently, there are many studies focusing on the impact of sluice and dam construction on the water quality of rivers. Dou et al. [12] analyzed the effect of sluice operation on water quality in the Shaying River and developed a hydrodynamic model incorporating sluice operation and a water quality transport and transformation model that incorporated the release of endogenous loads and identified that the influent concentration, size, and the number of sluices were the main factors affecting water quality. Wang et al. [13] reported the effect of sluice operation around Poyang Lake on water quality and found that sluice operation slowed down the water flow rate and increased the risk of water eutrophication. Young et al. [14] found that the opening of the Arase Dam resulted in a significant decrease in the concentrations of As, Zn, Pb, and the total sulfur in the mudflat sediments of the Kuma River and the aquatic environment improved. Tang et al. [15] demonstrated that the construction and operation of a large number of sluices in the Yangtze River Basin changed the natural transport rhythm of the runoff, suspended solids and nutrients, and reduced flow velocities, resulting in the decline of water exchange, the narrowing of the connectivity between rivers and lakes, and the accumulation of nutrients and SS, which led to water eutrophication. Soukhaphon et al. [16] concluded that the sluices in the Mekong River Basin affected fish migration, river hydrology, and sediment transport and consequently had a negative impact on regional food economic security. Obviously, the impact of sluices and dams on the river water environment is multifaceted. However, the environmental effects of large dams or sluices are universally known, but those of these small sluices or dams ($\leq$15 m or $\leq 3 \times 10^6$ m^3) have rarely been considered [17]. In view of the proliferation of flood-prevention dams in the world's river systems, the challenge appears as to their cumulative impacts on water environments. An endeavor to evaluate these flood-prevention facilities' cumulative environmental impacts suggested that a large number of small dams or sluices may have an immeasurable impact on energy generation than that of large ones [18]. Thus, there is an urgent requirement to understand the multiple environmental impacts of small flood-prevention development and to understand how these dams or sluices might be better developed and managed.

This study focuses on urban regulated rivers in Wuxi, and the water quality data of rivers within the city are compared and analyzed to explore the differences in the changes of major pollutant concentrations upstream and downstream of urban rivers when the sluices are closed, and then to explore the impact of small sluices construction and operation on urban rivers water quality. The harmonious balance between the urban water environment and ecology is a critical basis for sustainable urban social and economic development; however, the interaction between them is extremely complex [19]. The target of China's water resources management has been changing from "water quantity management" to "water quality management" [20]. The results of this study can provide a basis for the management of water resources in urban rivers, ecological regulation, and the construction of small sluices and dams and promote the coordinated development of socio-economic and urban rivers.

2. Materials and Methods

2.1. Study Area

Wuxi is one of the cities with numerous rivers in China, and it is located on the north shore of Taihu Lake in the Yangtze River Delta [21]. There are more than 3100 rivers in Wuxi, with a total length of 2480 km. The total length of the rivers in the city is approximately 150 km, with the volume of the water body being 8 million m^3 during the flat water period. Wuxi is relatively rich in surface water and is well recharged by external water sources. The storage capacity of the city is 63.49 million m^3, and the annual recharge is 64.53 million m^3 [22]. Wuxi has now built more than 1200 large and small sluices because of rapid industrialization and urbanization. In 2020, Jiangsu Province invested about 5.9 billion dollars focusing on the implementation of flood control projects in the Taihu Lake Basin, which has greatly improved the flood control capacity of Taihu Lake [23]. The construction and operation of these new sluices and dams, as well as the regulation of existing ones, will have a direct impact on the water quality of Wuxi's rivers and Taihu Lake.

Currently, about 50% of rivers in Wuxi fail to meet the requirements of Class III "Water Environmental Quality Standards of China" (WEQSC) (GB3838-2002). Among them, the ones in Wuxi's downtown were the most seriously polluted, and the main pollution indicators are total nitrogen (TN) and total phosphorus (TP) [24]. The pollution sources in Wuxi are mainly municipal solid waste, industrial pollution, and agricultural irrigation and fertilization. In this study, we selected 8 major rivers with a total length of 159.34 km (Figure 1). Among them, Bodugang River, Xibei Canal, Jiuli River, and Liangtang River, and their water quality are required to meet the Class III of WEQSC (TN $\leq$ 1.0 mg$\cdot$L^{-1}; TP $\leq$ 0.2 mg$\cdot$L^{-1}). Meanwhile, the water quality of the Xicheng Canal, Beijing–Hangzhou Grand Canal, Beixingtang River, and Ancient Canal is required to meet Class IV (TN $\leq$ 1.5 mg$\cdot$L^{-1}, TP $\leq$ 0.3 mg$\cdot$L^{-1}) (GB3838-2002).

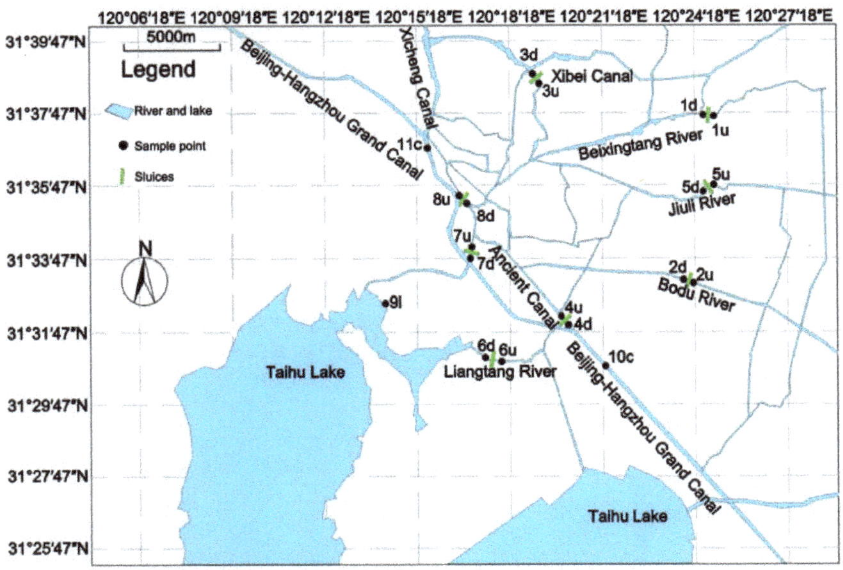

Figure 1. Research area and sample points.

A new "Flood Control Plan of Wuxi" was issued at the end of 2001, which would protect a 136 km^2 region. In May 2003, the construction of flood control facilities began and was completed at the end of 2008 [22]. This project contained eight flood control stations: Yandaigang flood control station, Beixingtang flood control station, Jiuli River flood control station, Bodugang flood control station, Limin Bridge flood control station, Xianli Bridge flood control station, Liangtang flood control station, and the Jiangjian flood control station.

These stations are applied not only to prevent floods in Wuxi but are also used for ship navigation [25]. The floodgates are controlled by the relevant government departments according to the annual precipitation and total water quantity of Wuxi City.

2.2. Sampling and Experiment

In this study, the main rivers in Wuxi city were selected, and the sampling points were concentrated in eight flood control stations. Point 1 was at Beixingtang River, point 2 was at Bodugang River, point 3 was at the Dongting Maritime Section of Xibei Canal, point 4 was at the Ancient Canal, point 5 was at Jiuli River, point 6 was at Liangtang River, point 7 was at the Water Conservancy Bureau section of the Huancheng River, and point 8 was at the Canal Park section of the Huancheng River. Points 1–8 include two points upstream and downstream, respectively. For example, point 1 includes 1u and 1d, and 1u means point 1 upstream, and 1d means point 1 downstream (Figure 1). The water samples were collected and analyzed for the upstream and downstream areas of the eight flood control stations (Figure 1). In addition, sampling site No. 9 is located in Taihu Lake, and No. 10 and No. 11 are located in the Beijing–Hangzhou Grand Canal. These three water samples were used to compare and analyze the water quality differences between the urban rivers and Taihu Lake and the Beijing–Hangzhou Canal when the sluices were closed.

This study was conducted from 2018 to 2019, including the flat-water season, wet season, and dry season. A suitable temperature was required to reduce the interference from external factors, such as rainstorms, on water quality. Meanwhile, when collecting the water samples, the water samples were taken 0.5–1.0 m below the water surface and far away from the river shore to reduce the impact of edge effects due to the shallow rivers. The sluices were all closed when the samples were collected.

Before sampling, the river water was taken to clean and moisten the water extractor. Then, the polyethylene storage bottle was washed more than 3 times by the water in the water extractor, and the water sample was immediately taken full and put into the cryogenic storage box for preservation. In order to reduce errors and to simulate the real situation as much as possible, the number of water samples was six at each point, and all water samples were measured three times. After sampling, all of the water samples were placed in the laboratory refrigerator at 4 °C, and all of the water quality data were measured within 24 h.

TN and TP are significant indicators for evaluating the water quality of the Taihu Lake Basin to meet WEQSC (GB3838-2002) [26]. Therefore, TN and TP were selected as the water quality indicators in this study. According to the national standards of HJ636-2012 and HJ671-2013, the detection of TN concentration adopts the "Alkaline potassium persulfate ablation UV spectrophotometric method" [27], and TP adopts the "Ammonium molybdate spectrophotometric method" [28].

2.3. Methods

2.3.1. Relative Difference

This study introduces the concept of "relative gap" to compare and analyze the difference between the upstream and downstream pollutant concentrations [29]. It is defined as:

$$DR = \frac{Cu - Cd}{Cu} \times 100\% \tag{1}$$

where DR is the relative difference between the pollutant concentrations upstream and those downstream, Cu is the pollutant concentration upstream, and Cd is the pollutant concentration downstream.

If the calculated result is negative, it means that the pollutant concentration downstream of the river is higher than that upstream, while if it is positive, it means that the pollutant concentration downstream is lower than that upstream.

2.3.2. Gaussian Fitting

In order to more deeply quantify the difference between pollutant concentrations upstream and downstream when the sluices are closed, the frequency distribution of the relative difference of pollutant concentrations was fitted using the Gauss fitting method in the normal distribution model in OriginLab software. The formula is as follows:

$$y = y_0 + \frac{A}{w\sqrt{\pi/2}} e^{\frac{-2(x-x_c)^2}{w^2}} \quad (2)$$

where y_0 is the baseline, x_0 is the mean, w is the discrete degree parameter, and A is the shape parameter.

3. Results and Discussion

3.1. Variation of Water Quality Indicators in Different Seasons

Urban rivers often take into account the role of discharging domestic sewage, most of which are nutrient pollutants. Nutrient pollutants mainly include nutrients represented by nitrogen and phosphorus, which are not considered pollutants in themselves. However, when the level of nutrients contained in sewage is relatively high, it will contribute to the proliferation of algae in the water and eutrophication of the water body, which leads to a series of hazards [30]. Furthermore, some studies have indicated that the main pollutants in inlet and outlet rivers around the whole of Taihu Lake are dominated by nitrogen pollutants, followed by organic pollution such as phosphorus [31]. Combined with the cyanobacterial water pollution events that have occurred in Taihu Lake [32], TN and TP are selected as water quality indicators in this study. The comparison results of TN and TP are shown in Figures 2 and 3.

The concentrations of TP and TN were relatively low during the wet and dry seasons compared to those in the flat-water season due to the fact that the wet and dry seasons are the summer and autumn seasons in Wuxi, with more rainfall. On the one hand, the high precipitation leads to the high storage capacity of rivers and the constant turnover of the water body. These are conducive to pollutant concentration reduction. On the other hand, for flood control purposes, the sluices are opened to release flood water when there is excessive rainfall. Furthermore, the river flows faster, and some pollutants from the river will flow into larger water bodies such as the Beijing–Hangzhou Grand Canal, which contributes to reducing pollutant concentrations.

In addition, there are some cases of excessive pollutant concentrations during the wet and dry seasons; for example, the TN concentrations upstream of the rivers in the wet seasons (Figure 2c,e,h) and the TP concentrations upstream of the rivers during the wet and dry seasons (Figure 3f,h), respectively. This is due to the fact that summer and autumn are not only the peak season of precipitation in Wuxi but also the peak period for industrial production and domestic sewage discharge. The combination of effluent discharge, sluice closure, and high temperatures during periods brings about high concentrations of these pollutants, especially during the time when it does not rain. This result is similar to previous studies [33].

3.2. Variation of Water Quality Indicators in the Upstream and Downstream

To further analyze the variation in the pollutant concentrations upstream and downstream of the rivers, the relative differences in the pollutant concentrations were calculated by Equation (1). The relative differences upstream and downstream of the rivers for TN and TP pollutants are shown in Tables 1 and 2.

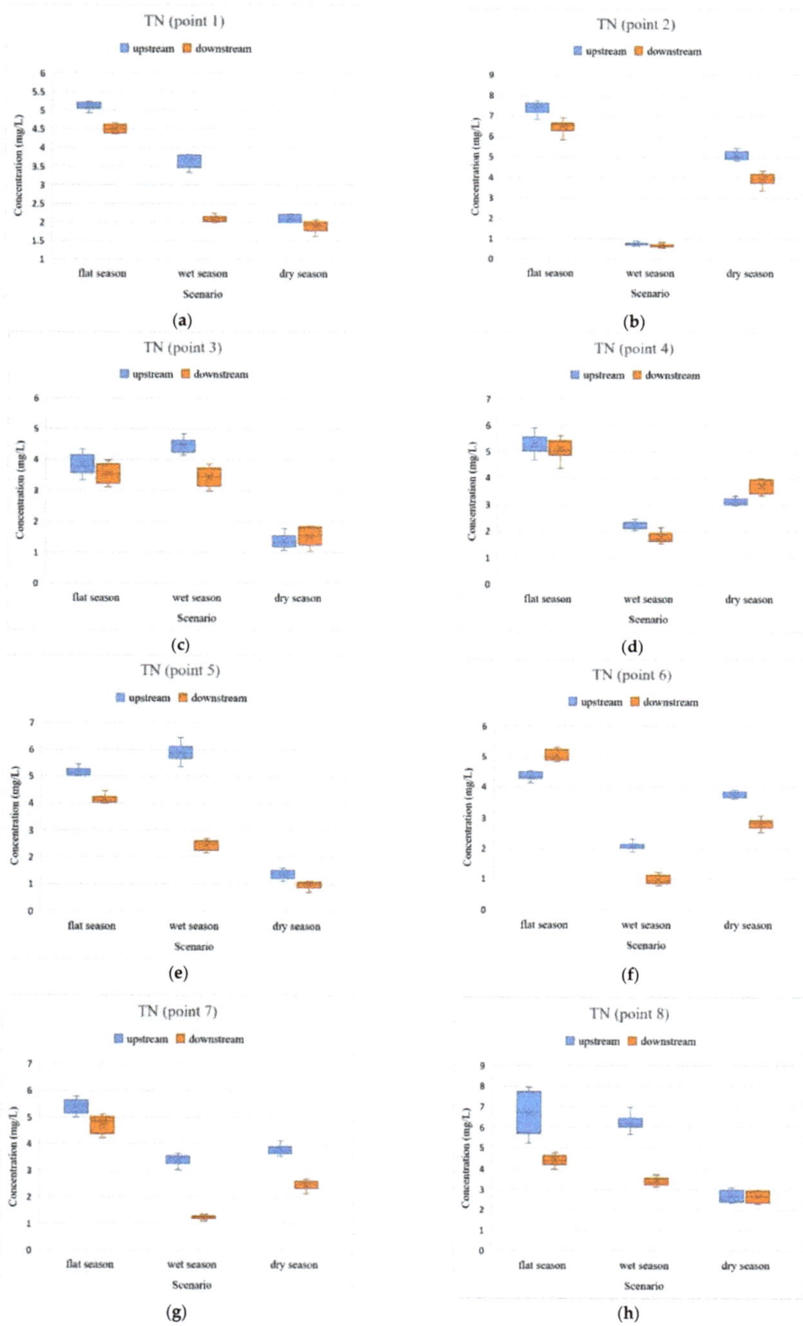

Figure 2. Comparison of TN upstream and downstream of the rivers in the closed state of the sluices. (**a**) TN concentrations at point 1; (**b**) TN concentrations at point 2; (**c**) TN concentrations at point 3; (**d**) TN concentrations at point 4; (**e**) TN concentrations at point 5; (**f**) TN concentrations at point 6; (**g**) TN concentrations at point 7; (**h**) TN concentrations at point 8.

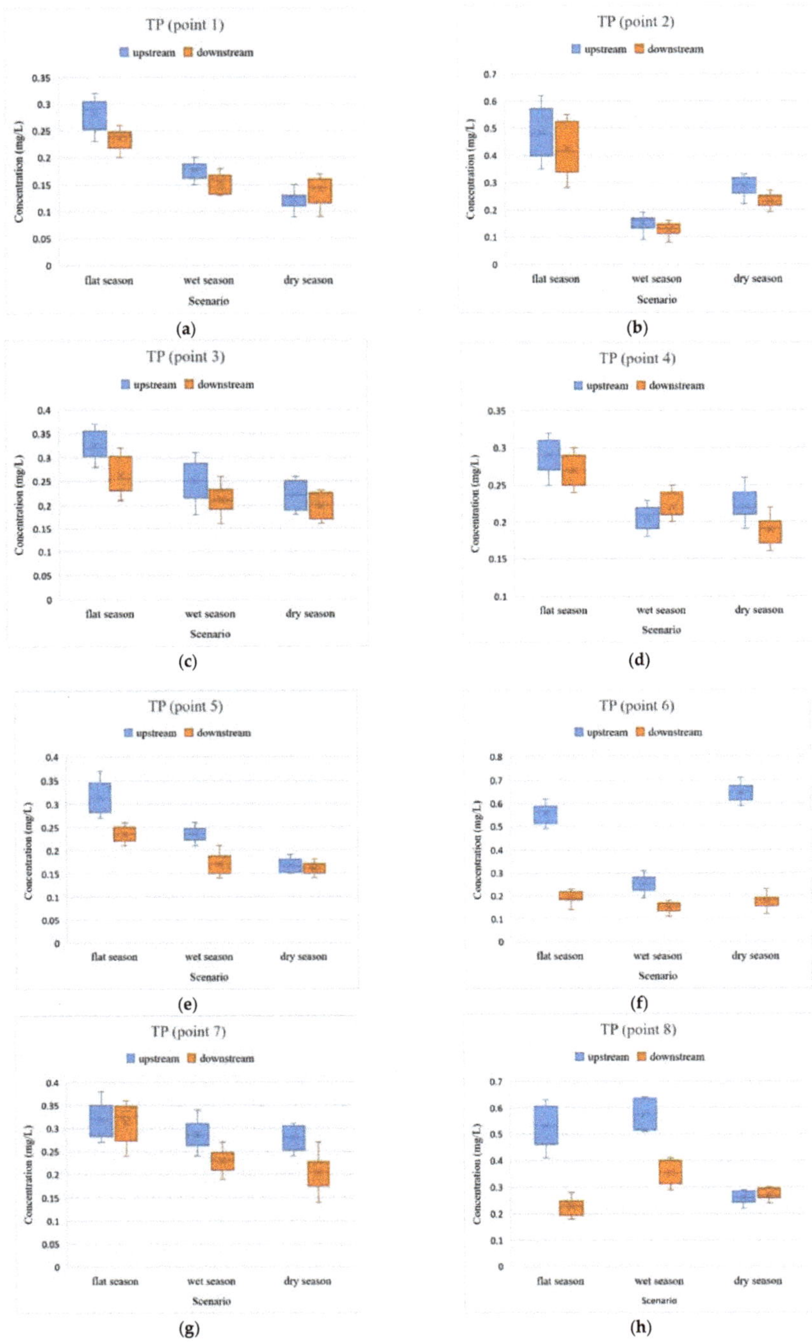

Figure 3. Comparison of TP upstream and downstream of the rivers in the closed state of the sluices. (**a**) TP concentrations at point 1; (**b**) TP concentrations at point 2; (**c**) TP concentrations at point 3; (**d**) TP concentrations at point 4; (**e**) TP concentrations at point 5; (**f**) TP concentrations at point 6; (**g**) TP concentrations at point 7; (**h**) TP concentrations at point 8.

Table 1. Relative difference in TN concentration.

Points	Relative Difference (%)		
	Flat	Wet	Dry
1	12.0	83.3	8.7
2	17.7	0	29.7
3	7.1	37.5	−10.9
4	6.3	27.2	−20.4
5	26.4	219.7	52.6
6	222.8	125.3	28.3
7	14.3	175.2	57.3
8	71.8	90.6	0.7

Table 2. Relative difference in TP concentration.

Points	Relative Difference (%)		
	Flat	Wet	Dry
1	20.8	12.5	−7.1
2	9.4	0	26.1
3	10.7	19.0	10.0
4	7.4	−4.6	20.0
5	37.5	41.2	0
6	205.3	73.3	260.1
7	2.9	34.8	52.6
8	160.1	60.0	−3.3

The 24 groups of TN concentrations are compared in Table 1. Two groups of TN concentrations in the upper sites were smaller than those in the lower sites, and the relative difference between the upstream and downstream was not obvious. While another 22 groups of TN concentrations upstream were higher than those downstream, among which 13 groups, relative differences ranged from 0 to 30% between upstream and downstream, the relative differences of the five groups ranged from 30 to 100%, and four groups relative differences exceeded 100% between the upstream and downstream.

The results indicated that the water quality in the upper reaches of the river was worse than that in the lower reaches when the sluice status was closed, with the maximum relative difference between the upper and lower reaches of the TN concentration being greater than 100%. Meanwhile, the TN concentrations were relatively high in all rivers. There were a few rivers where the differences in TN concentrations between the upper and lower reaches were not significant, but the comparison of the differences in TN concentrations between the upper and lower reaches in most rivers was very obvious.

The TP concentration in the rivers of Wuxi was relatively lower compared to TN (Figure 2 and Table 2). Among the 24 sets of data, there were eight sets of relative differences within 10%, which is not a significant comparison. Furthermore, among the remaining 16 sets of data, the TN concentrations in three sets were in the upper reaches, less than those in the lower reaches, and the other 13 sets in the upper reaches were greater than that in the lower reaches. Meanwhile, of these 16 sets, there were seven sets with relative differences between upstream and downstream from 10% to 30%, six sets were between 30% and 100%, and three sets where the difference exceeded 100%. The concentration of TP was not high in general, and some of the data were not obvious enough for a clear comparison. However, for the 16 groups of data, it can still be concluded that the water quality in the upper reaches of the rivers is worse than that in the lower reaches.

Generally, the concentrations of TP were relatively low, and the degree of variation in TP was not as great as that of TN. Furthermore, the levels of TP and TN also differed significantly at different times of the year at the same site in the same river, reflecting seasonal variability. In addition to this, an important preliminary conclusion was drawn: the pollutant concentrations in the upper reaches of the Wuxi rivers were higher than those in the lower reaches when the sluices were closed. The result is highly consistent with the previous investigations [34,35].

In order to further quantitatively analyze the pollutant concentrations in the upper and lower sites of rivers, the interval length of 20% was firstly selected for the frequency distribution map for frequency distribution statistics, and then the frequency distribution was fitted by applying Equation (2). The specific frequency distribution plots, as well as the fitted curves, are shown in Figure 4.

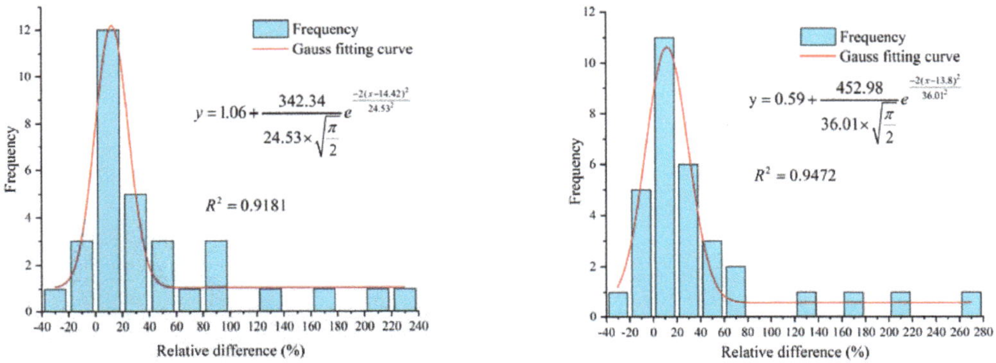

(**a**) (**b**)

Figure 4. Relative difference frequency distribution and Gauss fitting curve (**a**) TN (**b**) TP.

The value of the Gauss fitting curve parameter xc for the relative disparity frequency of TN was 14.42 (Figure 4). That is, under the normal distribution model, the average value of the relative differences between the TN concentrations in the upper and lower reaches when the sluices were closed was 14.42%. Similarly, the mean value of the relative differences in TP concentration was 13.80%. This indicates that the TN and TP concentrations in the upper reaches of urban rivers in Wuxi were 14.42% and 13.80% higher than those in the lower reaches under the closed state of the sluices, respectively. This is consistent with our preliminary conclusions.

3.3. Variation of Water Quality Indicators in Urban Rivers

When the sluices are closed, the water quality upstream of urban rivers may be worse than that downstream. The reason for this phenomenon is inextricably linked to the water quality conditions of the urban rivers themselves.

First of all, the pollution sources of urban rivers have special characteristics compared to the general large rivers such as the Yangtze River and the Yellow River, has a great distinction. Some studies have concluded that the main pollutants in the Yangtze River come from urban domestic sewage and agricultural pollution [36]. Furthermore, the cross-sections below the Three Gorges Dam are mainly located in the main urban living area, which causes some pollutant indicators in the lower reaches of the sluices and dam to be significantly higher than those in the upper reaches [37]. For urban rivers, in Wuxi, the city is located in the middle and upper reaches of rivers, where domestic sewage and industrial sources are the most important sources of nutrients and pollution [38]. These effluents flow into the rivers from the middle and upper sites, and these pollutants accumulate in the upper reaches of the sluices when they are closed. The presence of numerous sluices leads to the accumulation of large amounts of industrial wastewater, domestic sewage, and solid waste sediment in the upper reaches of the sluices. Although domestic, agricultural, and industrial wastewater is also discharged into the middle and lower reaches of urban rivers, the lower reaches usually connect to larger water bodies, such as Taihu Lake and the Beijing–Hangzhou Grand Canal. Consequently, the water quality upstream of urban rivers is often worse than that downstream.

Secondly, the lower reaches of urban rivers connect to larger water bodies, which own large volumes and lightly polluted water, as well as relatively clear water quality compared to that of urban rivers, and the concentrations of TN and TP are also relatively low. The concentrations of TN and TP in Taihu Lake and the Beijing–Hangzhou Grand Canal are shown in Figure 5.

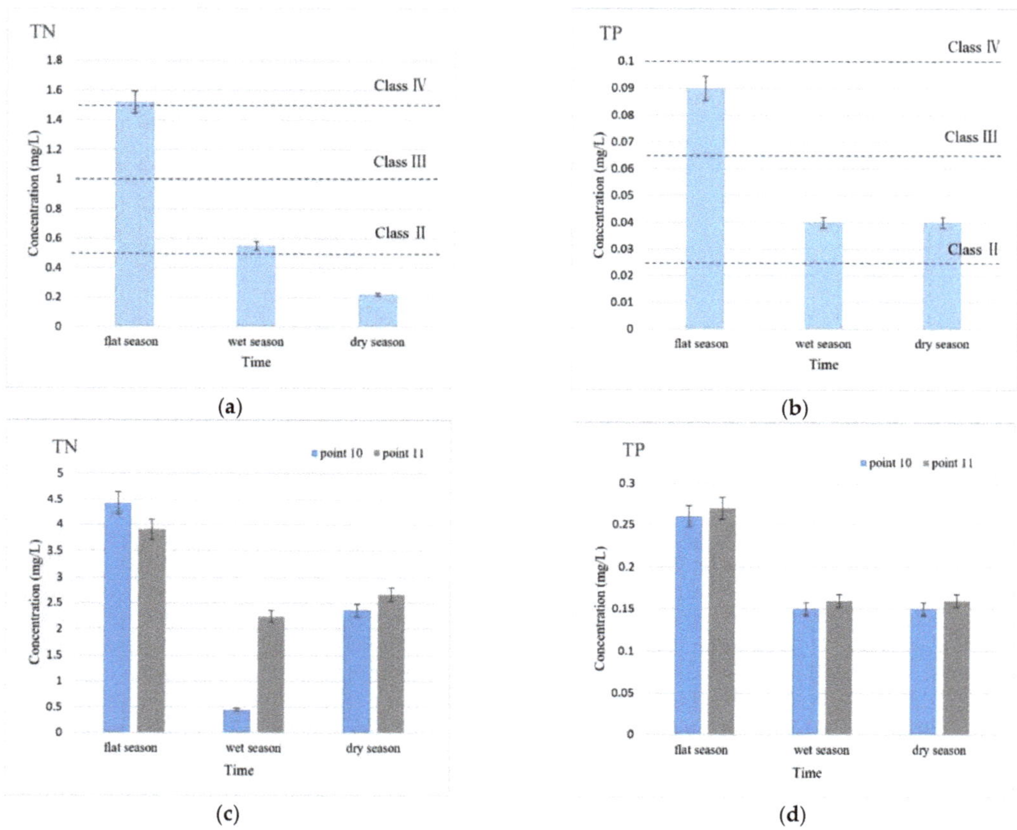

Figure 5. TN and TP concentration in Taihu Lake and Beijing–Hangzhou Grand Canal. (**a**,**b**) Point 9 at Taihu Lake; (**c**,**d**) Point 10 and point 11 at Beijing–Hangzhou Grand Canal.

It can be seen from Figure 5 that although the concentrations of the TN and TP in Taihu Lake and the Beijing–Hangzhou Grand Canal have partially exceeded the standards of Class III or Class IV WEQSC (GB3838-2002), they were still much lower than the average levels of TN and TP concentrations in urban rivers, as shown in Figures 2 and 3. In addition, the water storage volume of Taihu Lake and the Beijing–Hangzhou Grand Canal are more than other rivers in the city, and their dilution effect on pollutants is more obvious. Therefore, when the downstream of the river is connected with these water bodies with relatively clear water quality and huge water volume, the pollutants can be effectively diffused and decomposed, and the water quality downstream of the sluices is better than upstream.

4. Conclusions

The effect of the construction of sluices on the water quality of urban rivers in Wuxi was investigated, and the difference between TN and TP in the upper and lower reaches of the urban rivers after the construction of the sluices was detected. In this paper, the water quality in the urban rivers showed obvious seasonality and was usually better in seasons that have more rainfall, such as summer and autumn. However, irregular sluice regulation often causes some water quality pollutant concentrations to rise abnormally. Additionally, in the state of sluice closure, the water quality of the urban rivers upstream was worse than that downstream; the concentrations of TN and TP downstream were, on average, 14.42% and 13.80% lower than that upstream, respectively. The concentrations of pollutants showed different degrees of variation with time and space, and there were discrepancies between the data and conclusions at individual monitoring sites. For example, the pollutant concentrations in the Ancient Canal and Huancheng River were higher downstream than upstream as well as the relative difference between the upstream and downstream pollutant concentrations in the Liangtang rivers was extremely obvious. In the future, additional sample sites and numerous data will be required for in-depth exploration.

Author Contributions: F.L.: Conceptualization, Methodology, Software. W.H.: Data collection and curation, Writing—original draft, preparation. Y.Y.: Software, Validation. Writing—review & editing. All authors have read and agreed to the published version of the manuscript.

Funding: This work was supported by the Oversea Study Fellowship from the China Scholarship Council, National Natural Science Foundation of China (No.42007151), China's National Key Research and Development Program (No.2017YFC0505803), the Natural Science Foundation of Colleges and Universities of Jiangsu Province (No.19KJB610015), the Philosophy and Social Science Foundation of Colleges and Universities of Jiangsu Province (No.2019SJA0108), Jiangsu Funding Program for Excellent Postdoctoral Talent (JB0206015). The authors wish to thank Nanjing University for providing MIKE software to be used in this study.

Institutional Review Board Statement: Not applicable.

Informed Consent Statement: Not applicable.

Data Availability Statement: Not applicable.

Conflicts of Interest: The authors declare no conflict of interest.

References

1. Halpern, B.S.; Frazier, M.; Afflerbach, J.; Lowndes, J.S.; Micheli, F.; O'Hara, C.; Scarborough, C.; Selkoe, K.A. Recent pace of change in human impact on the world's ocean. *Sci. Rep.* **2019**, *9*, 11609. [CrossRef]
2. Hegarty, S.; Hayes, A.; Regan, F.; Bishop, I.; Clinton, T. Using citizen science to understand river water quality while filling data gaps to meet United Nations Sustainable Development Goal 6 objectives. *Sci. Total Environ.* **2021**, *783*, 146953. [CrossRef]
3. Pittman, S.J. *Seascape Ecology*; John Wiley & Son: Hoboken, NJ, USA, 2018.
4. Kumar, D.; Katoch, S.S. Sustainability indicators for run of the river (RoR) hydropower projects in hydro rich regions of India. *Renew. Sustain. Energy Rev.* **2014**, *35*, 101–108. [CrossRef]
5. Wang, B.; Nistor, I.; Murty, T.; Wei, Y.-M. Efficiency assessment of hydroelectric power plants in Canada: A multi criteria decision making approach. *Energy Econ.* **2014**, *46*, 112–121. [CrossRef]
6. Martins, D.E.C.; Seiffert, M.E.B.; Dziedzic, M. The importance of clean development mechanism for small hydro power plants. *Renew. Energy* **2013**, *60*, 643–647. [CrossRef]
7. Cheng, C.; Liu, B.; Chau, K.-W.; Li, G.; Liao, S. China's small hydropower and its dispatching management. *Renew. Sustain. Energy Rev.* **2015**, *42*, 43–55. [CrossRef]
8. Brown, C.J.; Jupiter, S.D.; Albert, S.; Klein, C.J.; Mangubhai, S.; Maina, J.M.; Mumby, P.; Olley, J.; Stewart-Koster, B.; Tulloch, V.; et al. Tracing the influence of land-use change on water quality and coral reefs using a Bayesian model. *Sci. Rep.* **2017**, *7*, 4740. [CrossRef]
9. García, M.A.; Castro-Díaz, L.; Villamayor-Tomas, S.; Lopez, M.C. Are large-scale hydroelectric dams inherently undemocratic? *Glob. Environ. Chang.* **2021**, *71*, 102395. [CrossRef]
10. Pradhan, A.; Srinivasan, V. Do dams improve water security in India? A review of post facto assessments. *Water Secur.* **2022**, *15*, 100112. [CrossRef]
11. Shen, D.; Huang, S.; Zhang, Y.; Zhou, Y. The source apportionment of N and P pollution in the surface waters of lowland urban area based on EEM-PARAFAC and PCA-APCS-MLR. *Environ. Res.* **2021**, *197*, 111022. [CrossRef]
12. Dou, M.; Li, G.; Li, C. Quantitative relations between chemical oxygen demand concentration and its influence factors in the sluice-controlled river reaches of Shaying River, China. *Environ. Monit. Assess.* **2015**, *187*, 4139. [CrossRef]
13. Peng, W.; Geying, L.; Lin, L. Predicting the Hydrological Impacts of the Poyang Lake Project Using an EFDC Model. *J. Hydrol. Eng.* **2015**, *20*, 05015009. [CrossRef]

14. Young, S.M.; Ishiga, H. Environmental change of the fluvial-estuary system in relation to Arase Dam removal of the Yatsushiro tidal flat, SW Kyushu, Japan. *Environ. Earth Sci.* **2014**, *72*, 2301–2314. [CrossRef]
15. Tang, X.; Li, R.; Han, D.; Scholz, M. Response of Eutrophication Development to Variations in Nutrients and Hydrological Regime: A Case Study in the Changjiang River (Yangtze) Basin. *Water* **2020**, *12*, 1634. [CrossRef]
16. Soukhaphon, A.; Baird, I.G.; Hogan, Z.S. The Impacts of Hydropower Dams in the Mekong River Basin: A Review. *Water* **2021**, *13*, 265. [CrossRef]
17. Downing, J.A. Emerging global role of small lakes and ponds: Little things mean a lot. *Limnetica* **2010**, *29*, 9–24. [CrossRef]
18. Kibler, K.M.; Tullos, D.D. Cumulative biophysical impact of small and large hydropower development in Nu River, China. *Water Resour. Res.* **2013**, *49*, 3104–3118. [CrossRef]
19. Xiang, L.; Shaogui, W.; Jiayang, W.; Peize, W.; Zhuo, W. Urban water environment carrying capacity based on VPOSR-coefficient of variation-grey correlation model: A case of Beijing, China. *Ecol. Indic.* **2022**, *138*, 108863.
20. Yan, B.; Jiang, H.; Zou, Y.; Liu, Y.; Mu, R.; Wang, H. An integrated model for optimal water resources allocation under "3 Redlines" water policy of the upper Hanjiang river basin. *J. Hydrol. Reg. Stud.* **2022**, *42*, 101167. [CrossRef]
21. Feng, L.; Sun, X.; Zhu, X.D. Impact of floodgates operation on water environment using one-dimensional modelling system in river network of Wuxi city, China. *Ecol. Eng.* **2016**, *91*, 173–182. [CrossRef]
22. Feng, L.; Li, Q.; Zhang, L.; Wang, H.; Wang, W.; Han, J.; Li, B. Exploring the effect of floodgates operation systems on water environmental capacity in a regulated river network of Wuxi, China. *J. Clean. Prod.* **2021**, *299*, 126743. [CrossRef]
23. Cui, P.; Ju, X.; Liu, Y.; Li, D. Predicting and Improving the Waterlogging Resilience of Urban Communities in China—A Case Study of Nanjing. *Buildings* **2022**, *12*, 901. [CrossRef]
24. Xu, F.; Gao, Y.; Mao, X.W.; Yi, J.; Xiang, L. Variations of water quality of the major 22 inflow rivers since 2007 and impacts on Lake Taihu. *J. Lake Sci.* **2016**, *6*, 1167–1174.
25. Feng, L.; Hu, P.; Wang, H.; Chen, M.-M.; Han, J. Improving City Water Quality through Pollution Reduction with Urban Floodgate Infrastructure and Design Solutions: A Case Study in Wuxi, China. *Int. J. Environ. Res. Public Health* **2022**, *19*, 10976. [CrossRef]
26. Yin, H.; Qian, X.; Yao, H.; Xia, B.S.; Gao, H.L. Effect Evaluation of the River Water Environmental Risk Prevention and Control Based on WASP by Taking Taipu River as an Example. *Environ. Prot. Sci.* **2015**, *2*, 48–52. (In Chinese)
27. Ministry of Ecology and Environment of the People's Republic of China. *National Environmental Protection Standards of the People's Republic of China (HJ636-2012)*; China Environmental Science Press: Beijing, China, 2012. (In Chinese)
28. Ministry of Ecology and Environment of the People's Republic of China. *National Environmental Protection Standards of the People's Republic of China (HJ636-2013)*; China Environmental Science Press: Beijing, China, 2013. (In Chinese)
29. Dong, M.F.; Feng, L.L.; Zhou, Q.G.; Zhou, S.B.; Xu, X.C.; Qin, B.T. Spatial and temporal evolution of tar during ex-situ underground coal gasification. *Fuel* **2022**, *317*, 123423. [CrossRef]
30. Zheng, J.; Cao, X.; Ma, C.; Weng, N.; Huo, S. What drives the change of nitrogen and phosphorus loads in the Yellow River Basin during 2006-2017? *J. Environ. Sci.* **2023**, *126*, 17–28. [CrossRef]
31. Li, S.; Liu, C.; Sun, P.; Ni, T. Response of cyanobacterial bloom risk to nitrogen and phosphorus concentrations in large shallow lakes determined through geographical detector: A case study of Taihu Lake, China. *Sci. Total Environ.* **2022**, *816*, 151617. [CrossRef]
32. Yang, M.; Yu, J.; Li, Z.; Guo, Z.; Burch, M.; Lin, T. Taihu Lake not to blame for Wuxi's woe. *Science* **2008**, *319*, 158. [CrossRef]
33. Wang, Y.; Thompson, A.M.; Selbig, W.R. Predictive models of phosphorus concentration and load in stormwater runoff from small urban residential watersheds in fall season. *J. Environ. Manag.* **2022**, *315*, 115171. [CrossRef]
34. Zhang, Y.; Zhai, X.; Shao, Q.; Yan, Z. Assessing temporal and spatial alterations of flow regimes in the regulated Huai River Basin, China. *J. Hydrol.* **2015**, *529*, 384–397. [CrossRef]
35. Zhai, X.; Xia, J.; Zhang, Y. Integrated approach of hydrological and water quality dynamic simulation for anthropogenic disturbance assessment in the Huai River Basin, China. *Sci. Total Environ.* **2017**, *598*, 749–764. [CrossRef] [PubMed]
36. Xue, L.; Hou, P.; Zhang, Z.; Shen, M.; Liu, F.; Yang, L. Application of systematic strategy for agricultural non-point source pollution control in Yangtze River basin, China. *Agric. Ecosyst. Environ.* **2020**, *304*, 107148. [CrossRef]
37. Li, T.; He, B.; Zhang, Y.; Fu, S.; Zhang, Y.; Long, X. Spatiotemporal dynamics of coupled dissolved silica and bicarbonate in a dam-induced urban lake system in the Three Gorges Reservoir Area. *J. Hydrol.* **2021**, *597*, 126181. [CrossRef]
38. Zhen, S.; Zhu, W. Analysis of isotope tracing of domestic sewage sources in Taihu Lake-A case study of Meiliang Bay and Gonghu Bay. *Ecol. Indic.* **2016**, *66*, 113–120. [CrossRef]

remote sensing

MDPI

Article

Long-Term Baseflow Responses to Projected Climate Change in the Weihe River Basin, Loess Plateau, China

Junlong Zhang [1], Panpan Zhao [2,*], Yongqiang Zhang [3], Lei Cheng [4,5], Jinxi Song [6], Guobin Fu [7], Yetang Wang [1], Qiang Liu [8], Shixuan Lyu [1,9], Shanzhong Qi [1], Chenlu Huang [6], Mingwei Ma [2] and Guotao Zhang [10]

1 College of Geography and Environment, Shandong Normal University, Jinan 250358, China
2 College of Water Resources, North China University of Water Resources and Electric Power, Zhengzhou 450045, China
3 Key Laboratory of Water Cycle and Related Land Surface Processes, Institute of Geographic Sciences and Natural Resources Research, Chinese Academy of Sciences, Beijing 100101, China
4 State Key Laboratory of Water Resources and Hydropower Engineering Science, Wuhan University, Wuhan 430072, China
5 Hubei Provincial Collaborative Innovation Center for Water Resources Security, Wuhan 430072, China
6 Shaanxi Key Laboratory of Earth Surface System and Environmental Carrying Capacity, College of Urban and Environmental Sciences, Northwest University, Xi'an 710127, China
7 CSIRO Land and Water, Wembley, WA 6913, Australia
8 State Key Laboratory of Water Environment Simulation, School of Environment, Beijing Normal University, Beijing 100875, China
9 Department of Civil Engineering, University of Bristol, Bristol BS8 1TR, UK
10 Key Laboratory of Land Surface Pattern and Simulation, Institute of Geographic Sciences and Natural Resources Research, Chinese Academy of Sciences (CAS), Beijing 100101, China
* Correspondence: zhaopanpan@ncwu.edu.cn

Abstract: Climate change is a significant force influencing catchment hydrological processes, such as baseflow, i.e., the contribution of delayed pathways to streamflow in drought periods and is associated with catchment drought propagation. The Weihe River Basin is a typical arid and semi-arid catchment on the Loess Plateau in northwest China. Baseflow plays a fundamental role in the provision of water and environmental functions at the catchment scale. However, the baseflow variability in the projected climate change is not well understood. In this study, forcing meteorological data were derived from two climate scenarios (RCP4.5 and RCP8.5) of three representative general circulation models (CSIRO-Mk3-6-0, MIROC5, and FGOALSg2) in CMIP5 and then were used as inputs in the Soil and Water Assessment Tool (SWAT) hydrological model to simulate future streamflow. Finally, a well-revised baseflow separation method was implemented to estimate the baseflow to investigate long-term (historical (1960–2012) and future (2010–2054) periods) baseflow variability patterns. We found (1) that baseflow showed a decreasing trend in some simulations of future climatic conditions but not in all scenarios ($p < 0.05$), (2) that the contribution of baseflow to streamflow (i.e., baseflow index) amounted to approximately 45%, with a slightly increasing trend ($p \leq 0.001$), and (3) an increased frequency of severe hydrological drought events in the future (2041–2053) due to baseflows much lower than current annual averages. This study benefits the scientific management of water resources in regional development and provides references for the semi-arid or water-limited catchments.

Keywords: baseflow; Weihe River Basin; Loess Plateau; climate change; General Circulation Models

Citation: Zhang, J.; Zhao, P.; Zhang, Y.; Cheng, L.; Song, J.; Fu, G.; Wang, Y.; Liu, Q.; Lyu, S.; Qi, S.; et al. Long-Term Baseflow Responses to Projected Climate Change in the Weihe River Basin, Loess Plateau, China. *Remote Sens.* **2022**, *14*, 5097. https://doi.org/10.3390/rs14205097

Academic Editors: Luis Garrote and Alban Kuriqi

Received: 16 September 2022
Accepted: 8 October 2022
Published: 12 October 2022

1. Introduction

Distinguishing the contributors of different streamflow components is vital to the effective management of catchment water resources. Baseflow is the contribution of delayed pathways to stream discharge that maintains streamflow during drought periods, characterized by low precipitation, the dominance of groundwater discharge and/or snow meltwater from upstream regions [1–3]. Baseflow influences the water quality/supply and the health of the catchment ecosystem in regional development [4]. It has a profound influence on the hydrologic cycle in prolonged dry periods [5–7]. It is essential for the provision of water resources and water security that can be influenced by climate conditions [8–10]. Therefore, estimating projected baseflow drought is critical to escalating our understanding of hydrological processes in the changing climate.

Climate variability is the primary factor influencing the terrestrial hydrologic cycle (e.g., baseflow) at regional and global scales [11–14]. For example, baseflow has a close link with the redistribution of precipitation due to infiltration providing a vital contribution to groundwater flow [15], which is characterized by a close interaction between groundwater and surface water. Trancoso, et al. [2] showed that reduced precipitation diminished baseflow and precipitation and positively affected baseflow in eastern Australia. However, the land-surface air temperature has increased over the past three decades and led to an energized/accelerated hydrological cycle by influencing precipitation amounts [16,17] and by capturing longwave radiation [13]. Li, et al. [18] used an analytical approach that integrated water balance and the Budyko hypothesis (evaporative index (ET/P) and aridity index (PET/P) were used to describe the long-term water and energy balance [19,20]) to separate the contributions of climate and anthropogenic effects on streamflow. Li, et al. [21] investigated the response of baseflow to climate variability in a large forested catchment and found that the contribution of climate variability to annual baseflow were greater than the impacts from forest disturbance. Trancoso, et al. [2] predicted a decreasing baseflow trend under certain climate changes (e.g., decreasing precipitation and increasing evapotranspiration related to CO_2–vegetation feedbacks). Ficklin, et al. [16] assessed the impacts of climate change on baseflow and stormflow and found that baseflow had consistent trends with stormflow across the northeastern and southwestern United States. Additionally, Singh, et al. [4] quantified the response of baseflow levels to climate variability cycles (e.g., the Pacific Decadal Oscillation) in the Flint River.

Hydrological models are often used to estimate the effects of climatic factors on water yield. Climate projections have predicted that the frequency and intensity of extreme events (e.g., droughts and floods) will increase under future climate conditions [22]. However, the direct consequences of baseflow responses to future climate change are poorly understood. Therefore, assessing baseflow responses under climate change is imperative to facilitate the understanding of groundwater-related hydrological processes and provides scientific guidelines for water adaptation measures [23] in water-limited regions to face future droughts.

Generally, this approach uses alternative emission scenarios to investigate hydrological responses to climate change [24,25]. For instance, Yang, et al. [26] used 16 climate models from CMIP5 (the fifth phase of the Coupled Model Intercomparison Project) to assess the responses of hydrologic drought/aridity to climate change. They demonstrated that climate models did not capture vegetation water use under elevated CO_2 conditions. Semi-distributed rainfall-runoff models based on SWAT (Soil & Water Assessment Tool, https://swat.tamu.edu/, accessed on 9 July 2016) have been widely used to evaluate streamflow variations in complex catchments [27]. Zhang, et al. [28] compared the performances of two distributed hydrological models (e.g., SWAT and the Distributed Hydrology Soil Vegetation Model) in separating the impacts of climate change and LUCC (land-use cover change) on catchment hydrology. Lauffenburger, et al. [29] evaluated the effects of agricultural irrigation and future climate change on groundwater recharge in the northern High Plains aquifer, USA, and found a significant bidirectional shift, leading to a reduction in future groundwater recharge. While those efforts improved our understanding of climate-variability effects on hydrological processes, baseflow responses to future climate change are poorly understood for semi-arid catchments in loess deposition regions, in which baseflow provides a significant water source for ecological restoration and environmental protection.

The Weihe River Basin (WRB) is a representative catchment on the Loess Plateau. It is one of the most important water sources for the environment and regional society of northwest China. In this study, to attenuate the uncertainties of baseflow estimation (e.g., signal and magnitude [30]), historical daily streamflow data and future streamflow data projected by two climate scenarios from three presentative GCMs were used to assess temporal variations in baseflow and the dynamics of baseflow characteristics under future climate changes in the WRB. The specific objectives were (1) projecting baseflow under two scenarios (RCP4.5 representing a lower emissions scenario, and RCP8.5 representing a higher emission scenario) from three GCMs (CSIRO-Mk3-6-0, MIROC5, and FGOALSg2); (2) assessing baseflow responses under future climate conditions; and (3) highlighting the role of baseflow in drought events at the catchment scale. Thus, this study provides drought assessment for water-resource managers to face the future changing climate.

2. Study Area and Data Sources

2.1. Study Area Description

The Weihe River has a total length of 818 km and is located in the northern Qinling Mountains. It is the largest tributary of the Yellow River. The WRB covers three terrain sections, i.e., the Loess Plateau, the Guanzhong Plain, and the Qinling Mountains, and spans 6.72×10^4 km^2 from north to south (Figure 1). The Weihe River has its source at Niaoshu Mountain (3485 m) in the Gansu Province, flows from west to east, and joins the main channel of the Yellow River in Tongguan County. The longitudinal inclination of the river is about 1.7‰ [31], and the lowest and highest elevations are 325 and 3485 m [32], respectively.

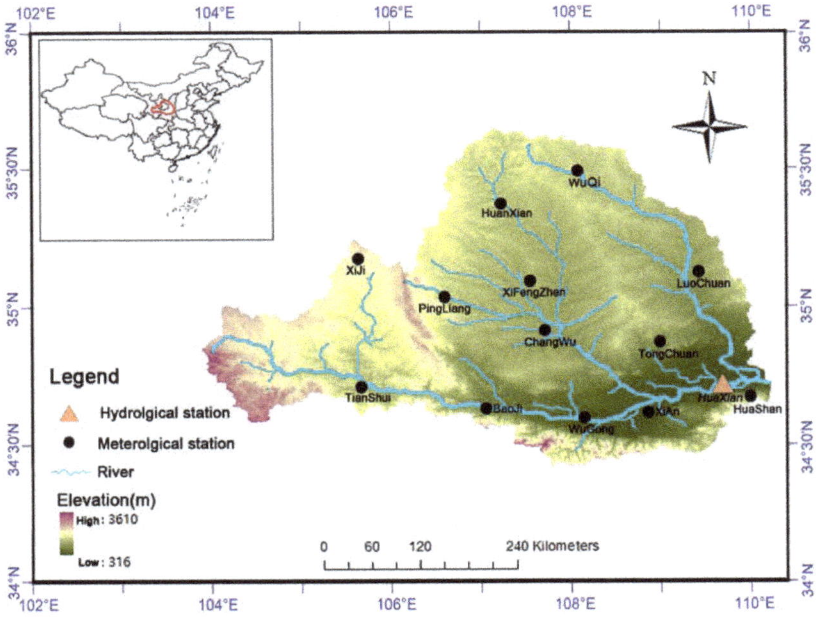

Figure 1. Study area, hydrological and meteorological stations in this study.

The climate of this basin is characterized by the continental monsoon with cold, dry, and rainless winters; hot and rainy summers [33]; average annual temperature changes between 7.8 and 13.5 °C; and annual precipitation between 558 and 750 mm [34]. The seasonal distribution of precipitation is uneven, and high precipitation and flow mainly occur in flood periods (June to September). Both precipitation and runoff have substantial inter-annual and intra-annual variabilities. The mean annual potential evaporation is approximately 800 mm in the south to 1200 mm in the north [35].

The WRB has extensive loess deposits across the mid and northern catchment. Its soil has a relatively high infiltration potential, and its southern part is primarily covered by forested land in the Qinling Mountains. The predominant land use is agricultural (i.e., wheat and cotton production [36]) in the center of the basin, where cultivated soils have been subjected to long-term agricultural development. Cultivated land covers more than 50% of the basin, followed by woodland and grassland [33]. The basin is highly productive and supplies water and food for the region. However, streamflow and groundwater have decreased rapidly with historical increases in population, agricultural production, industries, and related developmental activities [35]. Land-use changes, particularly due to the large ecological plan (e.g., the Grain for Green Program [37,38]) launched in the 1990s, have significant impacts on the catchment's hydrology [1,39].

2.2. Data Sources

Daily precipitation data, covering 1960 to 2012, from the 13 standard meteorological stations (Table 1) in the WRB were obtained from the China Meteorological Administration (http://cdc.cma.gov.cn, accessed on 22 August 2015). These meteorological stations are maintained according to the standard methods of the National Meteorological Administration of China. For the same period, daily streamflow data from the Huaxian gauge (Figure 1) was obtained from the Hydrological Yearbooks of China (http://loess.geodata.cn, accessed on 10 May 2016). All meteorological and hydrological data used in this study have been submitted to quality control by government agencies before release.

Table 1. Meteorological stations used in this study.

Station ID	Station	Latitude	Longitude	Elevation (m)
53738	WuQi	36.95	108.17	1331.4
53821	HuanXian	36.58	107.3	1255.6
53903	XiJi	35.97	105.78	1916.5
53915	PingLiang	35.55	106.57	1346.6
53923	XiFengZhen	35.73	107.63	1421
53929	ChuangWu	35.2	107.8	1206.5
53942	LuoChuan	35.82	109.5	1159.8
53947	TongChuan	35.08	109.07	978.9
57006	TianShui	34.58	105.75	1141.7
57016	BaoJi	34.35	107.13	612.4
57034	WuGong	34.25	108.22	447.8
57036	XiAn	34.3	108.93	397.5
57046	HuaShan	34.48	110.08	2064.9

3. Methods

3.1. Baseflow Separation Algorithm

To improve the accuracy of baseflow estimates in this study, revised and validated baseflow separation was implemented [40]. Baseflow has a lag time concerning the last precipitation event [41]. Generally, the baseflow recession is linked with the surface and sub-surface flow characteristics and follows an exponential decay curve [42]:

$$Q_b = Q_0 \alpha^t \tag{1}$$

where Q_b is the baseflow at time t, and α is the recession constant determined by recession analysis. The baseflow can be calculated using the baseflow separation method.

Baseflow separation is a fundamental issue that has been comprehensively documented [8,43,44]. Several algorithms have been proposed to separate baseflow from total observed streamflow [45–47] and can be classified as trace-based, water balance, and graph approaches according to general applications. Digital filters are the most widely used tools for small-data input and is reducible (e.g., only daily streamflow records and more objective) [40]. The Lyne–Hollick method was used here, expressed as [48]:

$$Q_{q\,(i)} = \alpha Q_{q\,(i\,-\,1)} + \frac{1+\alpha}{2}(Q_i - Q_{i\,-\,1}) \tag{2}$$

where Q is total streamflow (m^3/d), Q_q is quick flow (mm/d), i is the time step (day), and α is the filter parameter (recession constant, in 1/day). Baseflow (Q_b, m^3/d) can subsequently be calculated as Q_i minus Q_q. The baseflow index (BFI, calculated as *total* Q_b/*total* Q), is a standard indicator of the baseflow contribution to total streamflow. Herein, the calibrated Lyne–Hollick method was employed to separate the long-term baseflow. This approach has been validated by Zhang, et al. [40].

The recession constant can be obtained using the recession analysis developed by Brutsaert, et al. [43]. This recession approach efficiently reduces uncertainties when estimating the initial points in the recession limb. Details of recession analysis are given in Cheng, et al. [44].

3.2. Selection of General Circulation Models

The general circulation model (GCM) is widely used to estimate the impacts of future climate conditions on hydrological cycles [26,49–52]. The GCMs used in this study (Table 2) were available in the Intergovernmental Panel on Climate Change (IPCC) data archive (https://pcmdi.llnl.gov/mips/cmip5/, accessed on 16 October 2016). Based on monthly precipitation data from 40 GCMs for two representative concentration scenarios (RCP4.5 and RCP8.5) and the future climate scenario period based on CMIP5, we divided GCM data into two sections. The 45 years from 1960–2004 (historical climate period, HCP) were considered the baseline period, and the 45 years from 2010–2054 were the future climate period (FCP).

Table 2. Summary of 40 general circulation models (GCM) selected in this study.

ID	GCM	Originating Group (s)	Country	Resolution (°)
1	ACCESS1.0	CSIRO-BOM	Australia	1.88 × 1.25
2	ACCESS1.3	CSIRO-BOM	Australia	1.88 × 1.25
3	BCC-CSM1.1	BCC	China	2.81 × 2.81
4	BCC-CSM1.1.M	BCC	China	1.13 × 1.12
5	BNU-ESM	BNU-ESM	China	2.81 × 2.81
6	CanESM2	CCCMA	Canada	2.81 × 2.79
7	CCSM4	NCAR	USA	1.25 × 0.94
8	CESM1(BGC)	NCAR	USA	1.25 × 0.94
9	CESM1(CAM5)	NCAR	USA	1.25 × 0.94
10	CESM1(WACCM)	NCAR	USA	2.5 × 1.89
11	CMCC-CM	CMCC	Italy	0.75 × 0.75
12	CMCC-CMS	CMCC	Italy	1.88 × 1.88
13	CNRM-CM5	CNRM-CERFACS	France	1.41 × 1.40
14	CSIRO-Mk3.6.0	CSIRO-QCCCE	Australia	1.88 × 1.88
15	EC-EARTH	MOHC	UK	1.13 × 1.13
16	FGOALS-g2	LASG-GESS	China	2.81 × 3.05
17	FGOALS-s2	LASG-IAP	China	2.81 × 1.41
18	FIO-ESM	FIO	China	2.81 × 2.81
19	GFDL-CM3	NOAA GFDL	USA	2.50 × 2.00
20	GFDL-ESM2G	NOAA GFDL	USA	2.50 × 2.00

Table 2. *Cont.*

ID	GCM	Originating Group (s)	Country	Resolution (°)
21	GFDL-ESM2M	NOAA GFDL	USA	2.50 × 2.00
22	GISS-E2-H	NASA GISS	USA	2.50 × 2.00
23	GISS-E2-H-CC	NASA GISS	USA	2.50 × 2.00
24	GISS-E2-R	NASA GISS	USA	2.50 × 2.00
25	GISS-E2-R-CC	NASA GISS	USA	2.50 × 2.00
26	HadGEM2-AO	KMA/NIMR	UK/Korea	1.88 × 1.25
27	HadGEM2-CC	KMA/NIMR	UK/Korea	1.88 × 1.25
28	HadGEM2-ES	KMA/NIMR	UK/Korea	1.88 × 1.25
29	INMCM4	INM	Russia	2.00 × 1.50
30	IPSL-CM5A-LR	IPSL	France	3.75 × 1.89
31	IPSL-CM5A-MR	IPSL	France	2.50 × 1.27
32	IPSL-CM5B-LR	IPSL	France	3.75 × 1.89
33	MIROC5	MIROC	Japan	1.41 × 1.40
34	MIROC-ESM	MIROC	Japan	2.81 × 2.79
35	MIROC-ESM-CHEM	MIROC	Japan	2.81 × 2.79
36	MPI-ESM-LR	MPI-M	Germany	1.88 × 1.87
37	MPI-ESM-MR	MPI-M	Germany	1.88 × 1.87
38	MRI-CGCM3	MRI	Japan	1.13 × 1.12
39	NorESM1-M	NCC	Norway	2.50 × 1.89
40	NorESM1-ME	NCC	Norway	2.50 × 1.89

In the context of climate change, three GCMs (i.e., the dry, moderate, and wet effects) were chosen to represent the future climate conditions. Then, daily precipitation and temperature derived from GCMs were used as forcing data to project streamflow in the FCP. The representativeness of the ensemble GCMs is considerably improved in the projection of climate variables [53]. Among the 40 GCMs under the two scenarios in CMIP5, the numbers of GCMs predicting increasing and decreasing future precipitation were 36 and 4, respectively. To choose representative models and reduce uncertainties, three models were selected to simulate future climate conditions, i.e., CSIRO-Mk3-6-0 (predicting dry conditions with the largest precipitation declines), MIROC5 (wet conditions with the largest precipitation increases), and FGOALSg2 (median conditions with a median change in precipitation) (Figure 2).

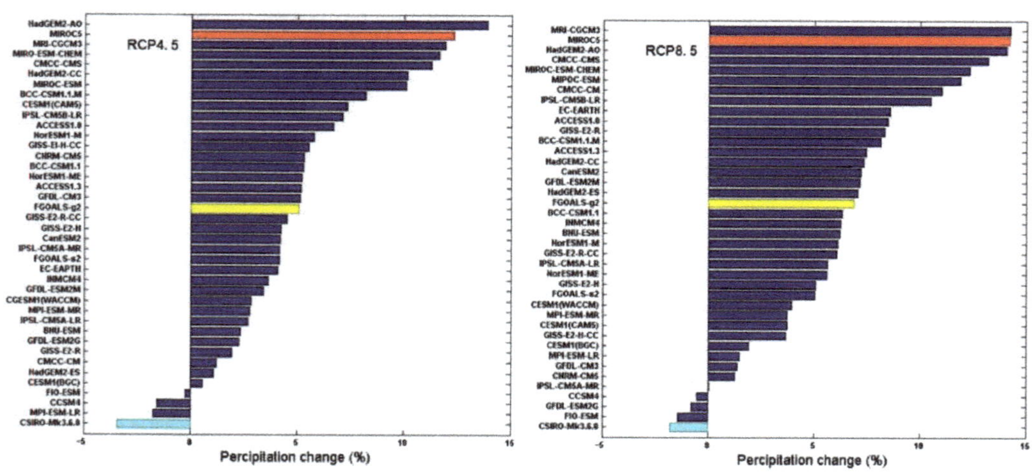

Figure 2. General circulation model selection of the dry, moderate, and wet simulated effects for two scenarios (RCP4.5 and RCP8.5).

To generate the mean climate conditions, the GCMs' climate projections were bias-corrected with the delta-change method (for details, see Navarro-Racines, et al. [54]), which simply superimpose the mean monthly anomalies between the GCMs-simulated baseline and the future period on the observed historical precipitation and temperature to represent future climate. Specifically, first, we calculated the ratio between the observed and simulation precipitation data of the three selected GCMs in the historical period (1960–2004). Second, we multiplied or added the precipitation and temperature data of the three GCMs in the future period (2010–2054) with this ratio to obtain simulation data for the FCP. Finally, we used the simulation data as forcing input data for running SWAT (Soil and Water Assessment Tool) to estimate daily streamflow.

3.3. SWAT Model

The SWAT hydrological model is a continuous-time, computationally efficient, and semi-distributed catchment-scale hydrologic model [55]. The catchment was divided into hydrological response units (HRUs), and surface runoff volumes were simulated for each HRU. SWAT has been widely used in different catchments worldwide and proved to be an effective tool to examine hydrological responses to land use and climate changes [56]. More details on SWAT are given in Easton, et al. [57], Guo, et al. [58].

This study used daily meteorological data (precipitation, maximum and minimum temperature, mean wind speed, radiation, mean relative humidity) from 1960–2012 as forcing data to simulate daily runoff in the WRB. The performance of predicted runoff was assessed against observed daily streamflow data in the same period. In the SWAT simulation, 1983–2012 was the calibration period (warm-up period: 1983–1993), and 1960–1982 was the validation period (warm-up period: 1960–1969). Comparing the simulated runoff between the calibration and validation period, the simulation of monthly runoff using the SWAT model had a good performance in WRB (Figure 3).

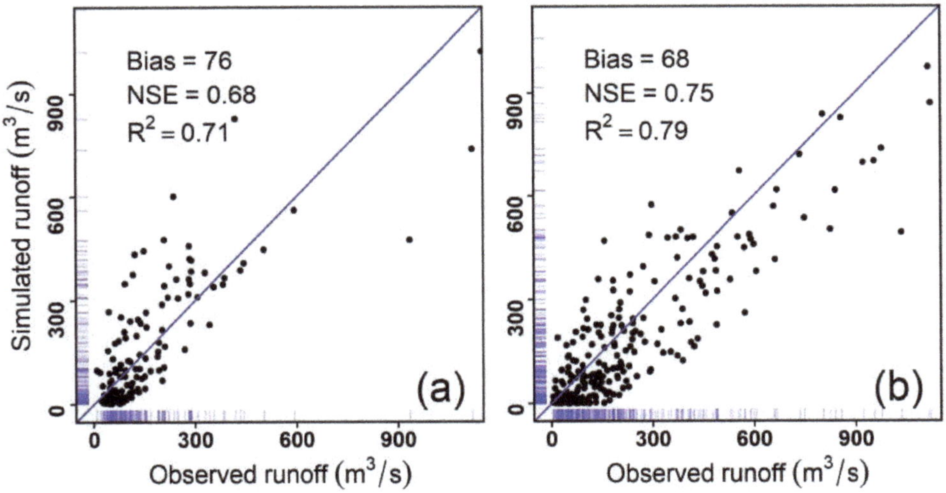

Figure 3. The streamflow simulation using SWAT model in calibration (**a**) and validation (**b**). The blue line is the 1:1 line. The rug represents the data distribution density.

3.4. Trend Analysis

Trend analysis can provide effective and useful information on possible tendencies in the future [59]. The nonparametric Mann Kendall test was used to identify trends and trend significance in baseflow in this study. This test provides two parameters, i.e., the significance level and slope magnitude [60]. p values ≤ 0.05 were considered significant.

The Z (derived from a certain climate element sequence) and S are the trend and order column and are used to detect the significance test. This test method has been widely employed to detect significant monotonic increasing or decreasing trends in long-term time-series data [8]. Method details can be found in previous studies [61,62].

3.5. Baseflow Drought Determination

Due to the hydrological drought with a higher accumulation period [63] and to provide insights for water planning and drought alerts for other basins facing water shortage events, the annual baseflow anomalies were implemented to determine hydrological droughts in historical and future climate conditions.

4. Results

4.1. Baseflow Estimation

The meteorological outputs in the GCMs were extracted as the inputs to the SWAT model to predict the streamflow in the future climate change, and then, a well-revised Lyne–Hollick method was used to implement the baseflow separation. Overall, the SWAT model had a good performance in both the calibration (Figure 3a) and validation (Figure 3b) stages to simulate the streamflow on a long-term scale (e.g., with $R^2 > 0.7$). In addition, the annual mean baseflow in the calibration and validation of the SWAT model also is shown in Figure 4.

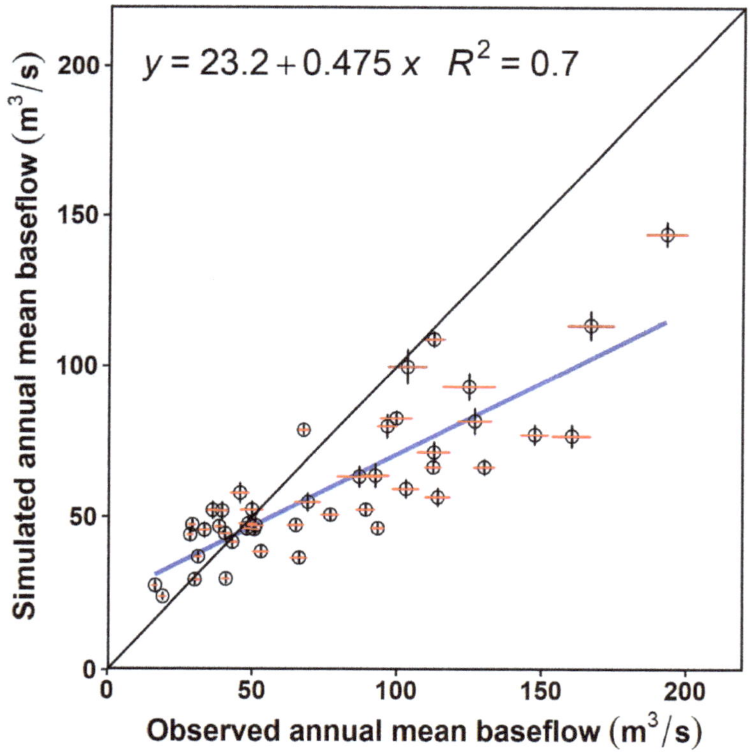

Figure 4. Relationship between the simulated and observed annual mean baseflow in the historical period. The black and red error bars represent standard errors in simulated and observed annual total baseflow. The black line is the 1:1 line.

The baseflow time scale of predicted streamflow and recession constants are shown in Table 3. The K and α are ranged from 62 ± 6 days (SD) and 0.98 ± 0.002 (SD) 1/day in the two future climate scenarios, respectively. For the CSIRO and FGOALSg2 models, the prediction of streamflow under the two scenarios was very close. For the highest streamflow condition, the MIROCS baseflow results were much higher than those of the other two models.

Table 3. Recession analysis derived from three general circulation models and two scenarios for future climate conditions (2010–2054).

Scenario	GCM	K (days)	α (1/day)
RCP4.5	CSIRO-Mk3-6-0	53.2	0.981
	FGOALSg2	64.5	0.985
	MIROC5	69	0.986
RCP8.5	CSIRO-Mk3-6-0	54.3	0.982
	FGOALSg2	67.1	0.985
	MIROC5	63.7	0.984

4.2. Detection of Baseflow Changes

All three models showed an insignificant increasing trend in both scenarios before 2020 in the future period (Figure 5). From the perspective of changing points, there was a similar and/or general pattern over a long future period; nevertheless, the numbers of changing points were different. In 2020, 2026 and 2034, changing points occurred for CSIRO and FGOALSg2 in both scenarios and for MIROC5 in the RCP8.5 scenario. After 2020, all three models showed an insignificant decreasing trend in both two scenarios. Specifically, in the FGOALSg2 model, there was a significant decreasing trend, and in the MIROC5 model, this trend occurred after 2049.

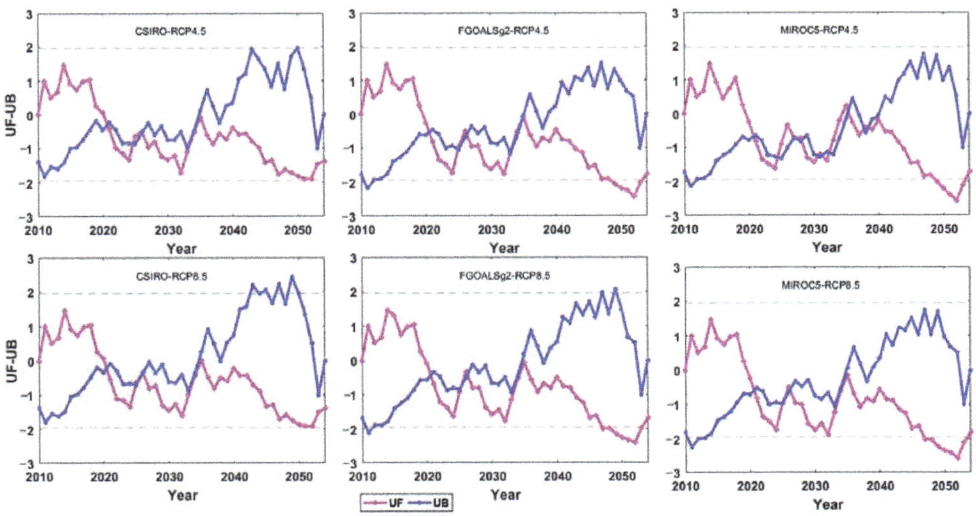

Figure 5. Mann Kendall test statistics for three GCMs in two scenarios (RCP4.5 and RCP8.5). UF is the sequential values of a statistic under the random hypothesis; UB is the reversed UF data statistic series. The positive and negative values indicate the increasing and decreasing trend. The intersections of UF and UB present the changing point.

The baseflow derived from the observed daily streamflow (Figure 6) showed a changing point in 1970. Before this year, baseflow showed an insignificant trend. However, after this year, there was a decreasing trend, both in 1977–1983 and after 1995.

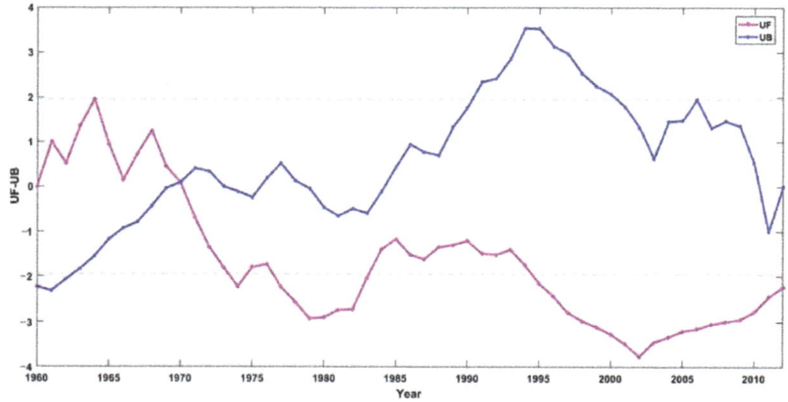

Figure 6. Mann Kendall test statistics for the baseflow separated from historical observed daily streamflow data. The abbreviations are the same as Figure 5.

4.3. Quantitative Baseflow Analysis Combining Historical and Future Climatic Conditions

The baseflow exhibited a decreasing trend in the long-term periods (all $p \leq 0.005$, see Figure 7). Herein, we first calculated the baseflow anomaly for the entire time series and then added the regression line for each GCM using local polynomial fitting. Specifically, CSIRO had a relatively more variation compared to the other two GCMs in both climate scenarios. Despite the trend with fluctuations, the three GCMs showed a similar performance in the two climate scenarios.

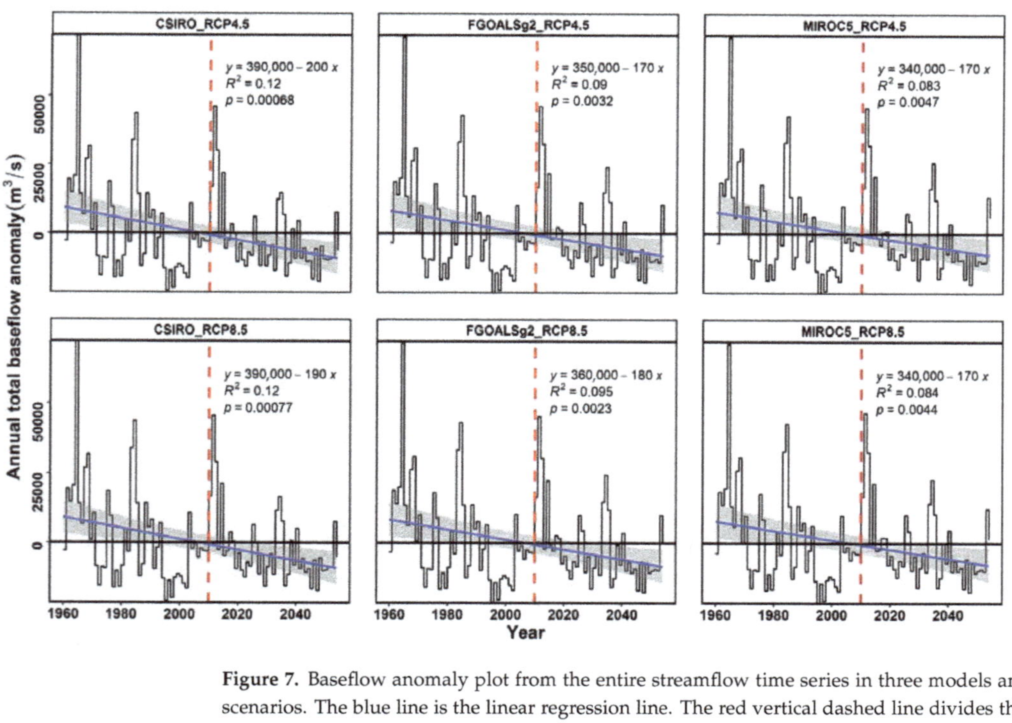

Figure 7. Baseflow anomaly plot from the entire streamflow time series in three models and two scenarios. The blue line is the linear regression line. The red vertical dashed line divides the time series into historical and future periods.

5. Discussion

5.1. Baseflow Trends in Historical and Future Climate Periods

The baseflow separation algorithm used in this study was derived from the revised version of the Lyne–Hollick algorithm. The outcomes of this method were more reproducible than the traditional methods (e.g., graphical approaches and empirical function [40]), thus this approach would greatly reduce the uncertainties of baseflow estimation. As baseflow was not directly measured under experimental conditions and was often estimated from the original total streamflow [40], in addition, the digital filter combined the recession analysis with more physical meanings containing more catchment-specific groundwater drainage characteristics [44] and provides a robust tool to decrease uncertainties in baseflow estimation [64].

The climate scenarios provided a robust tool to project the water balance of the catchment [65], and detecting trend characteristics was beneficial to understanding the hydrological variability at a long-term scale. In this study, the MK test was adopted to detect baseflow changes under future climate conditions (Figures 4 and 5). A declining baseflow trend was predicted for future climate scenarios. The baseflow change point years were 1970 and 1990. These years are not consistent with the streamflow change points reported by Zhan, et al. [33]. It was demonstrated that runoff had a decreasing trend in this basin after 1990 due to human activities, and the changing points of streamflow lagged the baseflow changes by about 20 years. However, the baseflow change points were in the range of the streamflow change points in another catchment on the Loess Plateau. Herein, the streamflow change points for different sub-catchments ranged from 1970 to 1990 [66]. As a delayed water resource, baseflow provides water to the land surface and sustains ecological health under dry spells.

Projections of baseflow and trend analysis are important to prevent and palliate drought losses on the catchment and regional scale [67]. Analyses of climate variability and baseflow improve our understanding of the effects of drought on environmental protection [4]. A drier trend has been reported for most areas of China based on PDIS (the Palmer Drought Severity Index) [67]. The degree of drought is characterized by a high frequency and has a long-term effect on hydrological connections in the WRB. Baseflow characteristics were used to evaluate hydrological droughts because baseflow is relatively steady and can represent catchment water storage [68]. Quantifying the impacts of climate variability on baseflow can provide insights for future water-resources plans [4]. Yang, et al. [41] showed that baseflow recovery had a longer lag than streamflow recovery across 130 unimpaired catchments in eastern Australia. Further, it has been reported that the hydrological cycle is intensified with changes in global mean precipitation in GCM projections [69]. This means that dry areas with limited water may become much drier.

5.2. Variability of the Baseflow Index

The BFI is an important hydrological indicator representing the water flow from groundwater/delayed resources to streamflow. It contains a lot of information on catchment characteristics [70,71], which reflects the holistic attribute of baseflow and terrestrial water balance [72]. The relationship between total baseflow, streamflow, and the baseflow index was demonstrated in Figure 8. To address the total baseflow contribution to streamflow, we also assessed the BFI for the historical observed and the simulated results for the three GCMs (Figure 9). There was an increasing trend in the BFI in the long-term climatic period. This means that the role of baseflow was remarkably strengthened in the sustenance of local water in this catchment. Compared to baseflow yield, the BFI is a relative ratio that varied from 0.42 to 0.49 and averaged 0.45 in our study. This means that the contribution of baseflow from groundwater storage or delayed sources accounted for 45% in the WRB from the perspective of future climate conditions in GCM projections. The magnitude of baseflow was very similar in the three models. Nevertheless, streamflow showed relatively greater variations. This confirmed that the baseflow is a relatively stable flow that sustains the terrestrial hydrological ecosystem [73].

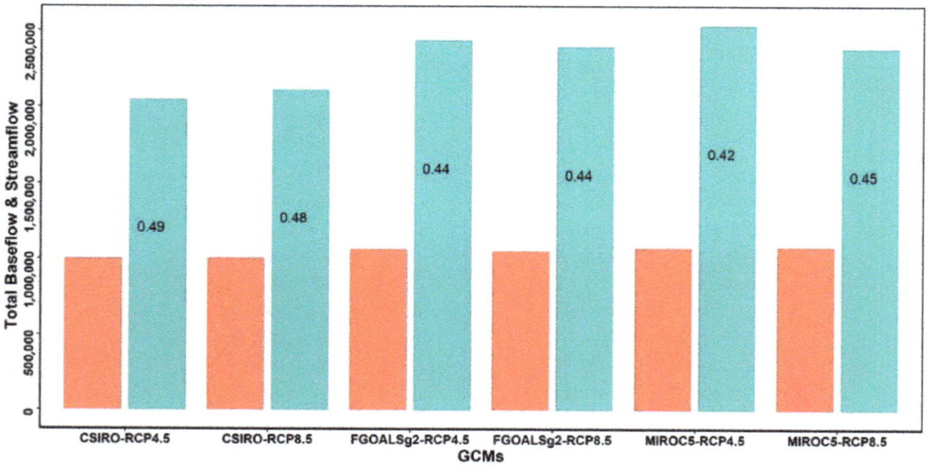

Figure 8. Total baseflow and streamflow for each GCM. Numbers in bars are the baseflow index. Red bars are the total baseflow, and blue bars are the total streamflow.

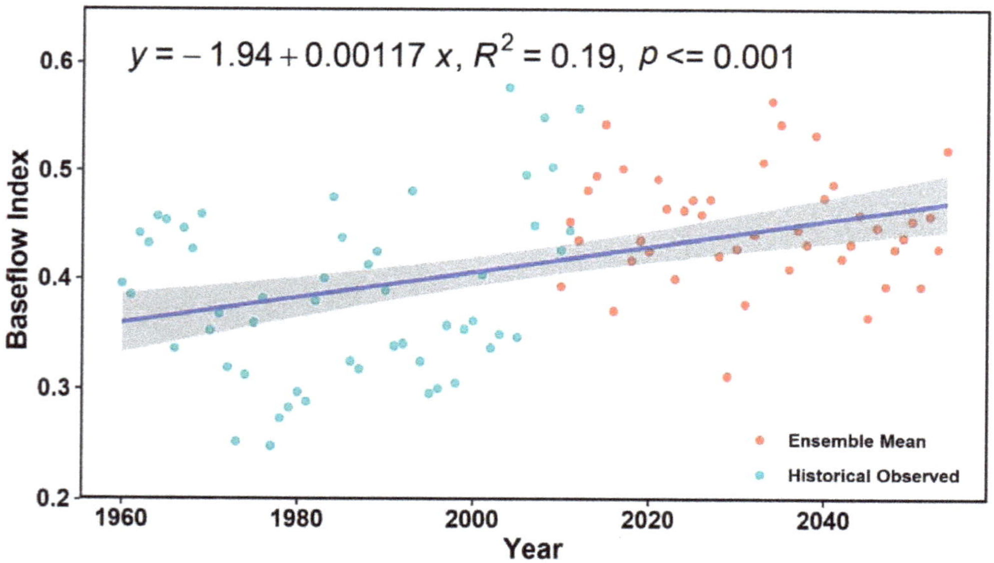

Figure 9. Variation in the baseflow index derived from historical observed streamflow data (1960–2010) and the ensemble means of three general circulation models for two scenarios (simulated from six models) for the future climate changes (2010–2054). The line and equation represent a linear regression.

Additionally, to clarify baseflow and streamflow trends, the relationships of historical data and/or projected streamflow and baseflow from the three models under two scenarios were also assessed (Figures 4 and 5). The response of baseflow to streamflow had a relative laggy time interval. This may be related to the increasing degree of anthropogenic activities in this basin due to the heavy exploitation intensity of groundwater resources. Singh, et al. [72] reported that groundwater abstraction significantly influenced flow regimes, with higher baseflow under constrained pumping conditions. Further, the effects of baseflow increase vary among river reaches, and baseflow and stormflow increases have relatively greater impacts on downstream areas by increasing flow volume [16]. Estimating other anthropogenic pumping effects is a meaningful way to assess the baseflow response to local hydrological variations.

5.3. Factors Influencing Baseflow Variations

It has been reported that climate change and anthropogenic activities are drivers of groundwater storage [74,75]. The baseflow yield is associated with the interactions between climate variability and vegetation changes [66,76] and would be influenced by a variety of catchment physical factors [72]. It is characterized by seasonal precipitation variations, i.e., from June to September, which creates the summer-dominated baseflow feature in this basin [1]. Furthermore, land-use change has affected 50% of the area on the Loess Platea [66]. This directly influences streamflow and leads to changes in baseflow. The basin covers three geological classes. Land use was predominantly agricultural in the long term. However, due to the widely distributed loess-deposition areas is in this region, extensive agricultural development causes heavy soil erosion and water-conservation issues [77]. To sustain the water quality and supply of the WRB, the government has taken measures to prevent ecosystem recession (e.g., soil-conservation measures [78]). This should lead to delayed surface runoff and increase the baseflow in small catchments [60]. However, in dry seasons on a long-term scale, the baseflow should be reduced by the loss of groundwater through more plant evapotranspiration. This is associated with vegetation-type changes from grass/bare land to the forested area [79]. Additionally, this complex effect is also influenced by other potential conditions such as topography. For example, Li, et al. [80] showed that the topography plays a paramount role in low flows (flow magnitudes $\leq Q_{75\%}$) in snow-dominated catchments.

The effects of anthropogenic activities associated with agricultural production also strongly control the water cycle in catchments [35]. It has been shown that the plantation intensity on agricultural land reduces downstream water availability [76]. Irrigation is an important factor influencing groundwater processes [29], leading to variations in baseflow [8]. The WRB is the main agricultural region, with large irrigation areas responsible for the food production for the regional population. To maintain living standards and sustain ecological health, the water demands have been increased for decades, and groundwater pumping supports much of the municipal water demand. Additionally, from the perspective of water depletion, agricultural development, and ecological recovery projects were all needing a large amount of water, including surface water and groundwater, it would create a baseflow shortage event for the Loess Plateau. For example, large-scale afforestation may exacerbate baseflow conditions as evapotranspiration increases through the amplification of leaf area and rooting depth [2,81]) for the catchment with constant precipitation input. This impact on baseflow variations would be amplified by climatic variations in this basin.

5.4. Implications of Baseflow Droughts

To provide insights for water planning and drought alerts for other basins facing water-shortage events, the annual baseflow anomalies were implemented to determine the hydrological droughts in historical and future climate conditions (Figure 7). It is noted that the baseflow has an apparent decreasing trend overall. Specifically, there was a relatively richer baseflow in approximately 2035. However, there was a lack of baseflow in 2041–2050, leading to a prolonged impact on the hydrological cycle (e.g., baseflow hydrological droughts) in the long term (~10 years).

It has been reported that, when disentangling climatic effects (e.g., precipitation) on hydrology, the uncertainties were much larger in the high-emission scenario RCP8.5 than the relatively low-emission scenario RCP4.5 [52]. In the baseflow estimation in the FCP, there was no remarkable difference between higher- and lower-emission scenarios, and the annual baseflow anomalies were very similar (Figure 7). The uncertainties of this study were likely associated with coarse temporal or spatial resolution and systematic errors derived from GCMs [67]. Besides, the baseflow relies upon runoff estimates in the model, confined by temperature and precipitation [82]. The variations of the climate phenomenon of wet-getting-wetter and dry-getting-drier [83] also influence baseflow changes in the catchment.

6. Conclusions

Climate selection is an important factor influencing hydrological processes. In this study, a physical-based baseflow separation filter was used to separate baseflow from total streamflow to assess baseflow responses to climate (e.g., varying temperature and rainfall). Three representative general climate models with two climate scenarios were used to predict baseflow and analyze trends and driving forces in the Weihe River Basin. Our analyses proved that the GCMs could capture the streamflow variations under future climate conditions and could be used to investigate baseflow characteristics at the basin scale. Our findings showed that the selection of climate had an approximate impact on the baseflow projection. The baseflow derived from three climate models (i.e., the future climate conditions) with two representative scenarios demonstrated a decreasing baseflow trend in this basin, reaching a strong decreasing trend approximately in 2040. For the historical periods, the baseflow had two intersects using the MK test, showing that the response of baseflow was much more sensitive than that of streamflow. Streamflow flow lagged about 20 years behind baseflow. Annual baseflow anomalies are an efficient tool that can be used to evaluate drought events under future climate conditions. Our study predicted baseflow droughts (~10 years) in this catchment starting in 2041. Although it is challenging to forecast water-storage variations accurately (e.g., drought events), the baseflow projection from climatic scenarios in GCMs is a promising way to assess baseflow responses to future climatic changes.

Author Contributions: Conceptualization, J.Z. and P.Z.; methodology, P.Z., Y.Z., L.C., G.F. and G.Z.; software, P.Z.; validation, J.Z.; formal analysis, J.Z. and P.Z.; investigation, S.L.; resources, C.H. and M.M.; data curation, M.M.; writing—original draft preparation, J.Z. and P.Z.; writing—review and editing, J.Z., P.Z., Y.Z., L.C., G.F., Y.W., Q.L., S.L., S.Q., C.H. and G.Z.; visualization, J.Z.; supervision, J.S.; project administration, J.Z. and P.Z.; funding acquisition, P.Z. and G.Z. All authors have read and agreed to the published version of the manuscript.

Funding: This work was supported by the National Natural Science Foundation of China (42001034, 42101038, 41701022, 51679200 and 42207100), the Key Research and Development Program of Shaanxi (2019ZDLSF05-02), the Project for Outstanding Youth Innovation Team in the Universities of Shandong Province (2019KJH011), the Natural Science Foundation of the Shandong Province (ZR2019BD059), Postdoctoral Funding of China (2018M642692), the Second Tibetan Plateau Scientific Expedition and Research Program (STEP) (2019QZKK0903), and Key Science and Technology Program of Henan Province, China (222102320083).

Data Availability Statement: Data can be obtained upon request from the corresponding author.

Acknowledgments: We thank three anonymous reviewers and the editors for their thoughtful comments and suggestions. The authors would like to express their gratitude to EditSprings (https://www.editsprings.cn/, accessed on 3 September 2019) for the expert linguistic services provided.

Conflicts of Interest: The authors declare no conflict of interest.

References

1. Zhang, J.; Song, J.; Cheng, L.; Zheng, H.; Wang, Y.; Huai, B.; Sun, W.; Qi, S.; Zhao, P.; Wang, Y.; et al. Baseflow estimation for catchments in the Loess Plateau, China. *J. Environ. Manag.* **2019**, *233*, 264–270. [CrossRef]
2. Trancoso, R.; Larsen, J.R.; McVicar, T.R.; Phinn, S.; McAlpine, C.A. CO_2 vegetation feedbacks and other climate changes implicated in reducing baseflow. *Geophys. Res. Lett.* **2017**, *44*, 2310–2318. [CrossRef]
3. Brutsaert, W. *Hydrology: An Introduction*; Cambridge University Press: Cambridge, UK, 2005; p. 618.
4. Singh, S.; Srivastava, P.; Abebe, A.; Mitra, S. Baseflow response to climate variability induced droughts in the Apalachicola Chattahoochee Flint River Basin, U.S.A. *J. Hydrol.* **2015**, *528*, 550–561. [CrossRef]

5. Bakker, M. The effect of loading efficiency on the groundwater response to water level changes in shallow lakes and streams. *Water Resour. Res.* **2016**, *52*, 1705–1715. [CrossRef]
6. Li, B.; Rodell, M.; Kumar, S.; Beaudoing, H.K.; Getirana, A.; Zaitchik, B.F.; Goncalves, L.G.; Cossetin, C.; Bhanja, S.; Mukherjee, A.; et al. Global GRACE Data Assimilation for Groundwater and Drought Monitoring: Advances and Challenges. *Water Resour. Res.* **2019**, *55*, 7564–7586. [CrossRef]
7. Zhang, L.; Brutsaert, W.; Crosbie, R.; Potter, N. Long-term annual groundwater storage trends in Australian catchments. *Adv. Water Res.* **2014**, *74*, 156–165. [CrossRef]
8. Ahiablame, L.; Sheshukov, A.Y.; Rahmani, V.; Moriasi, D. Annual baseflow variations as influenced by climate variability and agricultural land use change in the Missouri River Basin. *J. Hydrol.* **2017**, *551*, 188–202. [CrossRef]
9. Arciniega-Esparza, S.; Breña-Naranjo, J.A.; Hernández-Espriú, A.; Pedrozo-Acuña, A.; Scanlon, B.R.; Nicot, J.P.; Young, M.H.; Wolaver, B.D.; Alcocer-Yamanaka, V.H. Baseflow recession analysis in a large shale play: Climate variability and anthropogenic alterations mask effects of hydraulic fracturing. *J. Hydrol.* **2017**, *553*, 160–171. [CrossRef]
10. Rosas, M.A.; Vanacker, V.; Viveen, W.; Gutierrez, R.R.; Huggel, C. The potential impact of climate variability on siltation of Andean reservoirs. *J. Hydrol.* **2020**, *581*, 124396. [CrossRef]
11. Stephens, C.M.; Johnson, F.M.; Marshall, L.A. Implications of future climate change for event-based hydrologic models. *Adv. Water Res.* **2018**, *119*, 95–110. [CrossRef]
12. Gnann, S.J.; Woods, R.A.; Howden, N.J.K. Is there a baseflow Budyko Curve? *Water Resour. Res.* **2019**, *55*, 2838–2855. [CrossRef]
13. Cui, J.; Piao, S.; Huntingford, C.; Wang, X.; Lian, X.; Chevuturi, A.; Turner, A.G.; Kooperman, G.J. Vegetation forcing modulates global land monsoon and water resources in a CO_2-enriched climate. *Nat. Commun.* **2020**, *11*, 5184. [CrossRef]
14. Gnann, S.J.; McMillan, H.; Woods, R.A.; Howden, N.J. Including regional knowledge improves baseflow signature predictions in large sample hydrology. *Water Resour. Res.* **2020**, *57*, e2020WR028354. [CrossRef]
15. Ayers, J.R.; Villarini, G.; Jones, C.; Schilling, K. Changes in monthly baseflow across the U.S. Midwest. *Hydrol. Process.* **2019**, *33*, 748–758. [CrossRef]
16. Ficklin, D.L.; Robeson, S.M.; Knouft, J.H. Impacts of recent climate change on trends in baseflow and stormflow in United States watersheds. *Geophys. Res. Lett.* **2016**, *43*, 5079–5088. [CrossRef]
17. Menzel, L.; Bürger, G. Climate change scenarios and runoff response in the Mulde catchment (Southern Elbe, Germany). *J. Hydrol.* **2002**, *267*, 53–64. [CrossRef]
18. Li, C.; Wang, L.; Wanrui, W.; Qi, J.; Linshan, Y.; Zhang, Y.; Lei, W.; Cui, X.; Wang, P. An analytical approach to separate climate and human contributions to basin streamflow variability. *J. Hydrol.* **2018**, *559*, 30–42. [CrossRef]
19. Koppa, A.; Alam, S.; Miralles, D.G.; Gebremichael, M. Budyko-based long-term water and energy balance closure in global watersheds from earth observations. *Water Resour. Res.* **2021**, *57*, e2020WR028658. [CrossRef]
20. Fang, K.; Shen, C.; Fisher, J.B.; Niu, J. Improving Budyko curve-based estimates of long-term water partitioning using hydrologic signatures from GRACE. *Water Resour. Res.* **2016**, *52*, 5537–5554. [CrossRef]
21. Li, Q.; Wei, X.; Zhang, M.; Liu, W.; Giles-Hansen, K.; Wang, Y. The cumulative effects of forest disturbance and climate variability on streamflow components in a large forest-dominated watershed. *J. Hydrol.* **2018**, *557*, 448–459. [CrossRef]
22. Morrissey, P.; Nolan, P.; McCormack, T.; Johnston, P.; Naughton, O.; Bhatnagar, S.; Gill, L. Impacts of climate change on groundwater flooding and ecohydrology in lowland karst. *Hydrol. Earth Syst. Sci.* **2021**, *25*, 1923–1941. [CrossRef]
23. Straatsma, M.; Droogers, P.; Hunink, J.; Berendrecht, W.; Buitink, J.; Buytaert, W.; Karssenberg, D.; Schmitz, O.; Sutanudjaja, E.H.; van Beek, L.P.H.; et al. Global to regional scale evaluation of adaptation measures to reduce the future water gap. *Environ. Model. Softw.* **2020**, *124*, 104578. [CrossRef]
24. Thompson, J.; Green, A.; Kingston, D.; Gosling, S. Assessment of uncertainty in river flow projections for the Mekong River using multiple GCMs and hydrological models. *J. Hydrol.* **2013**, *486*, 1–30. [CrossRef]
25. Yang, Y.; Zhang, S.; McVicar, T.R.; Beck, H.E.; Zhang, Y.; Liu, B. Disconnection between trends of atmospheric drying and continental runoff. *Water Resour. Res.* **2018**, *54*, 4700–4713. [CrossRef]
26. Yang, Y.; Roderick, M.L.; Zhang, S.; McVicar, T.R.; Donohue, R.J. Hydrologic implications of vegetation response to elevated CO_2 in climate projections. *Nat. Clim. Change* **2018**, *9*, 44–48. [CrossRef]
27. Lee, J.; Kim, J.; Jang, W.S.; Lim, K.J.; Engel, B.A. Assessment of baseflow estimates considering recession characteristics in SWAT. *Water* **2018**, *10*, 371. [CrossRef]
28. Zhang, L.; Nan, Z.; Yu, W.; Zhao, Y.; Xu, Y. Comparison of baseline period choices for separating climate and land use/land cover change impacts on watershed hydrology using distributed hydrological models. *Sci. Total Environ.* **2018**, *622–623*, 1016–1028. [CrossRef] [PubMed]
29. Lauffenburger, Z.H.; Gurdak, J.J.; Hobza, C.; Woodward, D.; Wolf, C. Irrigated agriculture and future climate change effects on groundwater recharge, northern High Plains aquifer, USA. *Agric. Water Manag.* **2018**, *204*, 69–80. [CrossRef]
30. Hellwig, J.; Stahl, K. An assessment of trends and potential future changes in groundwater-baseflow drought based on catchment response times. *Hydrol. Earth Syst. Sci.* **2018**, *22*, 6209–6224. [CrossRef]
31. Zhang, J.; Song, J.; Long, Y.; Zhang, Y.; Zhang, B.; Wang, Y.; Wang, Y. Quantifying the spatial variations of hyporheic water exchange at catchment scale using the thermal method: A case study in the Weihe River, China. *Adv. Meteorol.* **2017**, *2017*, 8. [CrossRef]

32. Wang, W.; Song, J.; Zhang, G.; Liu, Q.; Guo, W.; Tang, B.; Cheng, D.; Zhang, Y. The influence of hyporheic upwelling fluxes on inorganic nitrogen concentrations in the pore water of the Weihe River. *Ecol. Eng.* **2018**, *112*, 105–115. [CrossRef]
33. Zhan, C.S.; Jiang, S.S.; Sun, F.B.; Jia, Y.W.; Niu, C.W.; Yue, W.F. Quantitative contribution of climate change and human activities to runoff changes in the Wei River basin, China. *Hydrol. Earth Syst. Sci.* **2014**, *18*, 3069–3077. [CrossRef]
34. Song, J.; Tang, B.; Zhang, J.; Dou, X.; Liu, Q.; Shen, W. System dynamics simulation for optimal stream flow regulations under consideration of coordinated development of ecology and socio-economy in the Weihe River Basin, China. *Ecol. Eng.* **2018**, *124*, 51–68. [CrossRef]
35. Chang, J.; Wang, Y.; Istanbulluoglu, E.; Bai, T.; Huang, Q.; Yang, D.; Huang, S. Impact of climate change and human activities on runoff in the Weihe River Basin, China. *Quat. Int.* **2015**, *380-381*, 169–179. [CrossRef]
36. Zhao, A.; Zhu, X.; Liu, X.; Pan, Y.; Zuo, D. Impacts of land use change and climate variability on green and blue water resources in the Weihe River Basin of northwest China. *Catena* **2016**, *137*, 318–327. [CrossRef]
37. Wu, X.; Wang, S.; Fu, B.; Liu, J. Spatial variation and influencing factors of the effectiveness of afforestation in China's Loess Plateau. *Sci. Total Environ.* **2021**, *771*, 144904. [CrossRef]
38. Jin, Z.; Guo, L.; Lin, H.; Wang, Y.; Yu, Y.; Chu, G.; Zhang, J. Soil moisture response to rainfall on the Chinese Loess Plateau after a long-term vegetation rehabilitation. *Hydrol. Process.* **2018**, *32*, 1738–1754. [CrossRef]
39. Shao, R.; Zhang, B.; He, X.; Su, T.; Li, Y.; Long, B.; Wang, X.; Yang, W.; He, C. Historical water storage changes over China's Loess Plateau. *Water Resour. Res.* **2021**, *57*, e2020WR028661. [CrossRef]
40. Zhang, J.; Zhang, Y.; Song, J.; Cheng, L. Evaluating relative merits of four baseflow separation methods in Eastern Australia. *J. Hydrol.* **2017**, *549*, 252–263. [CrossRef]
41. Yang, Y.; McVicar, T.R.; Donohue, R.J.; Zhang, Y.; Roderick, M.L.; Chiew, F.H.S.; Zhang, L.; Zhang, J. Lags in hydrologic recovery following an extreme drought: Assessing the roles of climate and catchment characteristics. *Water Resour. Res.* **2017**, *53*, 4821–4837. [CrossRef]
42. Rammal, M.; Archambeau, P.; Erpicum, S.; Orban, P.; Brouyère, S.; Pirotton, M.; Dewals, B. Technical Note: An operational implementation of recursive digital filter for base flow separation. *Water Resour. Res.* **2018**, *54*, 8528–8540. [CrossRef]
43. Brutsaert, W.; Nieber, J.L. Regionalized drought flow hydrographs from a mature glaciated plateau. *Water Resour. Res.* **1977**, *13*, 637–643. [CrossRef]
44. Cheng, L.; Zhang, L.; Brutsaert, W. Automated selection of pure base flows from regular daily streamflow data: Objective algorithm. *J. Hydrol. Eng.* **2016**, *21*, 06016008. [CrossRef]
45. Ponce, V.M.; Lindquist, D.S. Management of baseflow augmentation: A review. *JAWRA J. Am. Water Resour. Assoc.* **1990**, *26*, 259–268. [CrossRef]
46. Su, C.-H.; Peterson, T.J.; Costelloe, J.F.; Western, A.W. A synthetic study to evaluate the utility of hydrological signatures for calibrating a base flow separation filter. *Water Resour. Res.* **2016**, *52*, 6526–6540. [CrossRef]
47. Stewart, M.K. Promising new baseflow separation and recession analysis methods applied to streamflow at Glendhu Catchment, New Zealand. *Hydrol. Earth Syst. Sci.* **2015**, *19*, 2587–2603. [CrossRef]
48. Lyne, V.; Hollick, M. Stochastic time-variable rainfall-runoff modelling. In *Proceedings of Institute of Engineers Australia National Conference*; Institute of Engineers Australia: Barton, Australia, 1979; pp. 89–93.
49. Milly, P.C.; Dunne, K.A.; Vecchia, A.V. Global pattern of trends in streamflow and water availability in a changing climate. *Nature* **2005**, *438*, 347–350. [CrossRef]
50. Zhao, P.; Lü, H.; Yang, H.; Wang, W.; Fu, G. Impacts of climate change on hydrological droughts at basin scale: A case study of the Weihe River Basin, China. *Quat. Int.* **2019**, *513*, 37–46. [CrossRef]
51. Takle, E.S.; Jha, M.; Anderson, C.J. Hydrological cycle in the upper Mississippi River basin: 20th century simulations by multiple GCMs. *Geophys. Res. Lett.* **2005**, *32*, L18407. [CrossRef]
52. Shen, M.; Chen, J.; Zhuan, M.; Chen, H.; Xu, C.-Y.; Xiong, L. Estimating uncertainty and its temporal variation related to global climate models in quantifying climate change impacts on hydrology. *J. Hydrol.* **2018**, *556*, 10–24. [CrossRef]
53. Ahmed, K.; Sachindra, D.A.; Shahid, S.; Demirel, M.C.; Chung, E.-S. Selection of multi-model ensemble of general circulation models for the simulation of precipitation and maximum and minimum temperature based on spatial assessment metrics. *Hydrol. Earth Syst. Sci.* **2019**, *23*, 4803–4824. [CrossRef]
54. Navarro-Racines, C.; Tarapues, J.; Thornton, P.; Jarvis, A.; Ramirez-Villegas, J. High-resolution and bias-corrected CMIP5 projections for climate change impact assessments. *Sci. Data* **2020**, *7*, 7. [CrossRef]
55. Jha, M.K.; Gassman, P.W.; Arnold, J.G. Water quality modeling for the Raccoon River watershed using SWAT. *Trans. ASABE* **2007**, *50*, 479–493. [CrossRef]
56. Wang, G.; Yang, H.; Wang, L.; Xu, Z.; Xue, B. Using the SWAT model to assess impacts of land use changes on runoff generation in headwaters. *Hydrol. Process.* **2014**, *28*, 1032–1042. [CrossRef]
57. Easton, Z.M.; Fuka, D.R.; Walter, M.T.; Cowan, D.M.; Schneiderman, E.M.; Steenhuis, T.S. Re-conceptualizing the soil and water assessment tool (SWAT) model to predict runoff from variable source areas. *J. Hydrol.* **2008**, *348*, 279–291. [CrossRef]
58. Guo, T.; Engel, B.A.; Shao, G.; Arnold, J.G.; Srinivasan, R.; Kiniry, J.R. Development and improvement of the simulation of woody bioenergy crops in the Soil and Water Assessment Tool (SWAT). *Environ. Model. Softw.* **2019**, *122*, 104295. [CrossRef]
59. Yue, S.; Wang, C. The Mann Kendall test modified by effective sample size to detect trend in serially correlated hydrological series. *Water Resour. Manag.* **2004**, *18*, 201–218. [CrossRef]

60. Huang, M.; Zhang, L. Hydrological responses to conservation practices in a catchment of the Loess Plateau, China. *Hydrol. Process.* **2004**, *18*, 1885–1898. [CrossRef]
61. Gao, Z.; Zhang, L.; Zhang, X.; Cheng, L.; Potter, N.; Cowan, T.; Cai, W. Long-term streamflow trends in the middle reaches of the Yellow River Basin: Detecting drivers of change. *Hydrol. Process.* **2015**, *30*, 1315–1329. [CrossRef]
62. Brunner, M.I.; Hingray, B.; Zappa, M.; Favre, A.-C. Future trends in the interdependence between flood peaks and volumes: Hydro-climatological drivers and uncertainty. *Water Resour. Res.* **2019**, *55*, 4745–4759. [CrossRef]
63. Sutanto, S.J.; Van Lanen, H.A.J. Catchment memory explains hydrological drought forecast performance. *Sci. Rep.* **2022**, *12*, 2689. [CrossRef]
64. Zhang, J.; Zhang, Y.; Song, J.; Cheng, L.; Kumar Paul, P.; Gan, R.; Shi, X.; Luo, Z.; Zhao, P. Large-scale baseflow index prediction using hydrological modelling, linear and multilevel regression approaches. *J. Hydrol.* **2020**, *585*, 124780. [CrossRef]
65. Beaulieu, M.; Schreier, H.; Jost, G. A shifting hydrological regime: A field investigation of snowmelt runoff processes and their connection to summer base flow, Sunshine Coast, British Columbia. *Hydrol. Process.* **2012**, *26*, 2672–2682. [CrossRef]
66. Zhang, X.; Zhang, L.; Zhao, J.; Rustomji, P.; Hairsine, P. Responses of streamflow to changes in climate and land use/cover in the Loess Plateau, China. *Water Resour. Res.* **2008**, *44*, W00A07. [CrossRef]
67. Wang, L.; Chen, W. A CMIP5 multimodel projection of future temperature, precipitation, and climatological drought in China. *Int. J. Climatol.* **2014**, *34*, 2059–2078. [CrossRef]
68. Van Loon, A.F.; Laaha, G. Hydrological drought severity explained by climate and catchment characteristics. *J. Hydrol.* **2015**, *526*, 3–14. [CrossRef]
69. Thackeray, C.W.; DeAngelis, A.M.; Hall, A.; Swain, D.L.; Qu, X. On the connection between global hydrologic sensitivity and regional wet extremes. *Geophys. Res. Lett.* **2018**, *45*, 11,343–11,351. [CrossRef]
70. Beck, H.E.; van Dijk, A.I.J.M.; Miralles, D.G.; de Jeu, R.A.M.; Sampurno Bruijnzeel, L.A.; McVicar, T.R.; Schellekens, J. Global patterns in base flow index and recession based on streamflow observations from 3394 catchments. *Water Resour. Res.* **2013**, *49*, 7843–7863. [CrossRef]
71. Haberlandt, U.; Klöcking, B.; Krysanova, V.; Becker, A. Regionalisation of the base flow index from dynamically simulated flow components—a case study in the Elbe River Basin. *J. Hydrol.* **2001**, *248*, 35–53. [CrossRef]
72. Singh, S.K.; Pahlow, M.; Booker, D.J.; Shankar, U.; Chamorro, A. Towards baseflow index characterisation at national scale in New Zealand. *J. Hydrol.* **2019**, *568*, 646–657. [CrossRef]
73. Cheng, S.; Cheng, L.; Liu, P.; Qin, S.; Zhang, L.; Xu, C.Y.; Xiong, L.; Liu, L.; Xia, J. An analytical baseflow coefficient curve for depicting the spatial variability of mean annual catchment baseflow. *Water Resour. Res.* **2021**, *57*, e2020WR029529. [CrossRef]
74. Zhang, K.; Xie, X.; Zhu, B.; Meng, S.; Yao, Y. Unexpected groundwater recovery with decreasing agricultural irrigation in the Yellow River Basin. *Agric. Water Manag.* **2019**, *213*, 858–867. [CrossRef]
75. Stoelzle, M.; Schuetz, T.; Weiler, M.; Stahl, K.; Tallaksen, L.M. Beyond binary baseflow separation: A delayed-flow index for multiple streamflow contributions. *Hydrol. Earth Syst. Sci.* **2020**, *24*, 849–867. [CrossRef]
76. Zhang, L.; Cheng, L.; Chiew, F.; Fu, B. Understanding the impacts of climate and landuse change on water yield. *Curr. Opin. Environ. Sustain.* **2018**, *33*, 167–174. [CrossRef]
77. He, X.; Zhou, J.; Zhang, X.; Tang, K. Soil erosion response to climatic change and human activity during the Quaternary on the Loess Plateau, China. *Reg. Environ. Change* **2006**, *6*, 62–70. [CrossRef]
78. Huang, C.; Yang, Q.; Huang, W.; Zhang, J.; Li, Y.; Yang, Y. Hydrological Response to Precipitation and Human Activities-A Case Study in the Zuli River Basin, China. *Int. J. Environ. Res. Public. Health* **2018**, *15*, 2780. [CrossRef]
79. Zhang, L.; Dawes, W.R.; Walker, G.R. Response of mean annual evapotranspiration to vegetation changes at catchment scale. *Water Resour. Res.* **2001**, *37*, 701–708. [CrossRef]
80. Li, Q.; Wei, X.; Yang, X.; Giles-Hansen, K.; Zhang, M.; Liu, W. Topography significantly influencing low flows in snow-dominated watersheds. *Hydrol. Earth Syst. Sci.* **2018**, *22*, 1947–1956. [CrossRef]
81. Piao, S.; Wang, X.; Park, T.; Chen, C.; Lian, X.; He, Y.; Bjerke, J.W.; Chen, A.; Ciais, P.; Tømmervik, H.; et al. Characteristics, drivers and feedbacks of global greening. *Nat. Rev. Earth Environ.* **2019**, *1*, 14–27. [CrossRef]
82. Lehner, F.; Wood, A.W.; Vano, J.A.; Lawrence, D.M.; Clark, M.P.; Mankin, J.S. The potential to reduce uncertainty in regional runoff projections from climate models. *Nat. Clim. Change* **2019**, *9*, 926–933. [CrossRef]
83. Chou, C.; Chiang, J.C.H.; Lan, C.-W.; Chung, C.-H.; Liao, Y.-C.; Lee, C.-J. Increase in the range between wet and dry season precipitation. *Nat. Geosci.* **2013**, *6*, 263–267. [CrossRef]

water

MDPI

Article

NIRAVARI: A Parsimonious Bio-Decisional Model for Assessing the Sustainability and Vulnerability of Rainfed or Groundwater-Irrigated Farming Systems in Indian Agriculture

Jacques-Eric Bergez [1,*], Mariem Baccar [1], Muddu Sekhar [2,3] and Laurent Ruiz [3,4,5]

1 AGIR, INRAE, Université de Toulouse, F-31320 Castanet-Tolosan, France
2 Civil Engineering Department, Indian Institute of Science, Bangalore 560012, India
3 Indo-French Cell for Water Sciences, ICWaR, Indian Institute of Science, Bangalore 560012, India
4 UMR SAS, INRAE, Institut Agro, F-35000 Rennes, France
5 UMR GET, Université de Toulouse, CNRS, IRD, UPS, CNES, F-31400 Toulouse, France
* Correspondence: jacques-eric.bergez@inrae.fr

Abstract: Groundwater irrigation is essential to sustain food production, and aquifer depletion represents a major sustainability challenge for humanity. There is a need for adequate modelling tools to assess the impacts of farming practices on groundwater resources with policy-makers and farmers in different contexts, especially in the case of smallholder farms in the tropics. We introduce the NIRAVARI model, which was designed to represent the Indian farming and water resource context. NIRAVARI is a parsimonious model integrating biophysical and decisional processes dealing with the farming system and the water table processes. A specific focus is given to how to irrigate with multiple water resources. Its formalisms include equations from well-tested published models for soil moisture and plant water stress simulations. The programming and graphic user interface is based on Excel VBA. We illustrate the ability of NIRAVARI to simulate a broad range of farmer adaptation strategies using four scenarios of cropping systems and water resources policies, and therefore, its interest for participatory scenario design and assessment with stakeholders.

Keywords: irrigation strategy; modelling; climate change; India; farming systems

Citation: Bergez, J.-E.; Baccar, M.; Sekhar, M.; Ruiz, L. NIRAVARI: A Parsimonious Bio-Decisional Model for Assessing the Sustainability and Vulnerability of Rainfed or Groundwater-Irrigated Farming Systems in Indian Agriculture. *Water* **2022**, *14*, 3211. https://doi.org/10.3390/w14203211

Academic Editor: William Frederick Ritter

Received: 31 August 2022
Accepted: 8 October 2022
Published: 12 October 2022

1. Introduction

Groundwater depletion represents a major sustainability challenge for humanity in the 21st Century, given that groundwater is overexploited worldwide, particularly for agriculture [1]. About 38% of global consumptive irrigation water demand is met by groundwater [2]. With being a decade away from Sustainable Development Goals, the need for concrete policies for groundwater management has become urgent, and all water authorities call for methods to assess the effectiveness of policies. Furthermore, climate change is expected to intensify this threat, especially in regions where irrigation sustains agriculture and where population increases. Both influence the need for adaptation strategies, and for methodological approaches to identify and evaluate these same strategies. This is the case in India, where irrigation by groundwater is fully deployed [3].

The awareness that water resource management must account for interactions and feedback between biophysical processes determining movement of water and human behavior in a given socio-economic context has gained significant recognition among scientists in the past few years [4]. Different approaches exist to work on such interactions. Scenario modelling and evaluation is a standard way to explore adaptation strategies in the water resource domain in the context of climate change [5]. These methods are rooted in the more classical Story and Simulation approach [6]. They are often associated with participatory approaches and, in recent decades, several participatory stakeholder frameworks have been designed and implemented in projects that address adaptation strategies for groundwater management [7].

Simulating decisions on the farm and the technical operations of the system is a good way to test management practices, as it integrates the dynamics of resource sharing and constraints of the farm depending on the weather and the crop's phenological stage [8]. Linking soil-crop models and decision-making models is an appropriate approach for analyzing the adaptation of practices to face new agricultural challenges, such as climate change [9]. However, in order to use modelling and simulation as a tool to provide new insights on the impacts of different water policies on farmers and on groundwater resources, there is a need for simple models which can be utilized to assess scenarios with policy-makers, extensionists and farmers [10]. However, given the large diversity of farming systems in the world, there is no "one fits all" model which could adequately represent all types of agriculture. In particular, few models can represent smallholding farms, which account for a large proportion of agriculture in Africa and South Asia, and especially in India [11]. Such a model needs to be parsimonious [12], integrate farmers' decision-making on crops and crop management [13], allow integrating rainfed crops and represent the effect of conjunctive use of surface water and groundwater on the resource and different irrigation techniques [14].

One can find different types of numerical models to manage irrigation for Indian agriculture. Some are based on testing the adequacy of different crop models for the specific conditions in India (see for example Aquacrop and DSSAT-CERES in [15] or [16]). Others are more concerned with more specific biophysical processes, such as salinity [17]. Others are based on linear programming and optimization processes [18], or on artificial intelligence with neural networks and remote sensing data [19] or even the Internet of Things and machine learning [20]. Actually, very few take into account the farmer's decision-making process as proposed by [21,22]. Ref. [23] represents an interesting attempt, but is mainly focused on channel irrigation that is not the only water resource for irrigation.

Our paper presents NIRAVARI, a simple and parsimonious biodecisional model, able to test policy schemes regarding pond, borewells and irrigation equipment. The paper mainly focuses on the farmer's decision-making part of the model, and the complexity of dealing with different water resources to irrigate their farms. The initial ideas of the model were initiated with some policy-makers during a workshop held in Bangalore in March 2019. It was then quite clear that a simple tool was necessary to test a large range of scenarios. Section 1 presents the Indian context, regarding farming system and irrigation management. Section 2 presents the model; i.e., the formalism and the equations. At the end of this section, we present the different scenarios simulated to demonstrate the ability of the model. Section 3 presents the model developed and we elaborate on the outputs of the simulated scenarios. In a final section, we discuss the potential and limits of the model to elaborate and test irrigation strategies at the farm level with stakeholders.

2. Materials and Methods

2.1. *Specificities of Groundwater Irrigated Farms in India*

The dramatic development of groundwater irrigation in the recent decades in India has increased the irrigated area, simultaneously increasing food production to sustain the demand of a growing population. However, this came at the cost of tremendous impacts on energy consumption [24], resource depletion [25] and pollution [26,27], which now calls for an urgent rationalization of groundwater use. However, solutions are not straightforward, and recent detailed studies conducted in a groundwater-irrigated region in southern India, belonging to the Kabini Critical Zone Observatory in Karnataka ([28]; SNO M-tropics, https://mtropics.obs-mip.fr/, accessed on 11 October 2022) have illustrated the complex interactions and feedback between water resource availability and cropping systems, documenting the complex technical functioning of a diversity of farming systems [29,30]. Most Indian farmers are smallholders (less than 1ha on average), and their land is composed of one or several independent pieces of land, called "jeminu", in Karnataka. The size of a jeminu can vary greatly (from 0.1 to several hectares), but typically are about half a hectare, which can be divided into several plots for growing different crops in rotation. When jeminus are irrigated, they share the same water sources: a borewell and/or, more rarely, a farm pond. In hard rock aquifers, pump yields are small and often insufficient to fully satisfy the needs of water-intensive crops on large surfaces [25], and they vary in time with water table level as hydraulic conductivity decreases sharply with depth [31,32]. In addition, the energy required to operate submersible pumps is provided by the government at no cost but for a limited number of hours per day. Therefore, irrigated cropping systems are still largely dependent on rainfall, and farmers need to elaborate complex strategies involving the combination of different crops in the same jeminu, and allocate the adequate irrigation water to each of them to optimize their crop yield and economic returns. The system can become even more complex, as farm ponds, encouraged by local governments for securing alternate sources of irrigation and/or for enhancing the aquifer recharge [14], are increasingly present, adding further complexity to the system. Model-based decision support systems can therefore be of great help to farmers and decision makers in assessing a range of irrigation strategies.

2.2. Overall Description of the System

The modeled system (Figure 1) is composed of a jeminu (i.e., a farm) divided in n beles of the same size (i.e., plots). A jeminu can be rainfed or have a pond (water reserve) and/or a borewell equipped with a submersible pump. The pond can either be used to irrigate the bele or to recharge the groundwater table. The pond is represented as an inversed truncated pyramid. The surface of the pond can be removed from the productive area of the jeminu or not. This is an important feature for very small farms. The pump is used to irrigate the beles, but may also be used to refill the pond. The pump withdraws water from the aquifer and its flow rate depends on the pump characteristics and on the groundwater table depth. On each bele, a different crop can be grown. The crop growth depends on the duration since sowing. Crop management is described in a descriptive manner, i.e., date and action (sowing or harvest). Only irrigation practices follow a more complex decision-making process formalism. Different types of irrigation techniques can be used (e.g., drip, sprinkler or furrow) which differ by the amount of water they provide when irrigation is triggered. The model is run at a daily time step. Exogeneous variables are the climatic values: rainfall ($R(t)$, mm) and potential evapotranspiration ($E_0(t)$, mm).

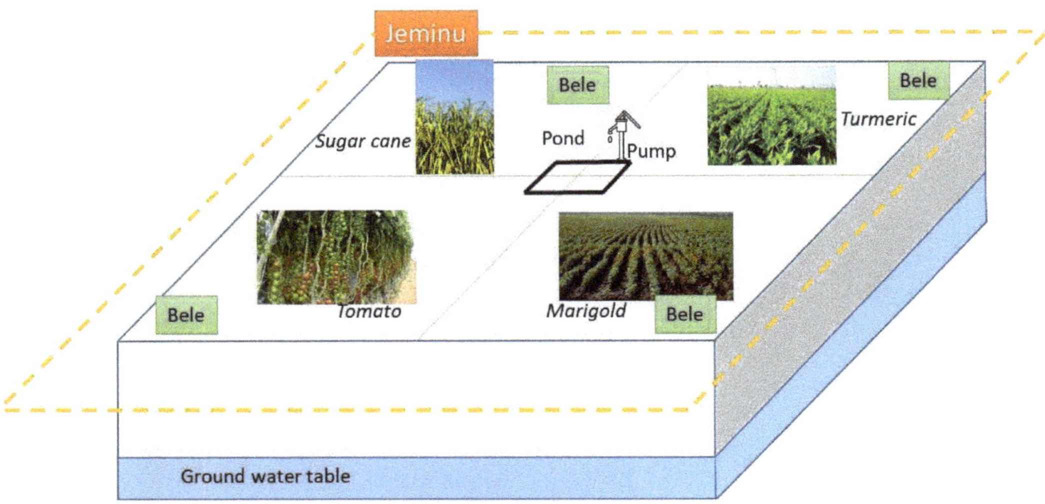

Figure 1. Schematic representation of a jeminu.

In the following, we describe the decision-making aspect of the model. The main biophysical equations that are more "classical" are given in Appendix A. Crop and water budget processes are based on FAO56 [33]. This decision-making model is based on farmers' interviews carried out in 2015 [29]. This survey was aimed at understanding farmers' farms structure and farmers' assets, and specific attention was given to the irrigation system: whether they are irrigating; whether they are using a pond and/or borewells; irrigation costs; how they pay for the irrigation systems and other equipment. A total of 684 farmers were interviewed with the help of local translators from September 2014 to March 2015, which represents 12.5% of farms in the watershed. The survey consisted of a face-to-face interview lasting 2–3 h. The survey was divided into three parts. The first part focused on household characteristics, farm structure, assets, partnerships and farm objectives. In the second part, we asked farmers about their performances and practices over the past two years. In the last part, in-depth questions were asked about irrigation, borewells and rainfall. Since no yearly records were kept by farmers, information about historical management went no further than the past two years. To develop the decision-making model, we used the approach developed by [34].

2.3. Decision-Making Processes

2.3.1. Management Processes at the Bele Level

On each bele, there is a decision model that allows: (i) to sow the crop; (ii) to harvest the crop; (iii) to irrigate the crop.

Sowing/Harvesting/Crop Succession

Sowing, harvesting and crop succession are based on a prescriptive model; i.e., the user gives the crop succession and the date to sow. Harvest time is automatically calculated by the model. This corresponds to the time it takes for the crop to reach maturity. For example: "*Beetroot is sown on the 1st of April and harvested on the 1st of July. It is followed by an Onion, sown on the 1st of August and harvested on the 4th of November. A fallow period follows during the Summer season.*"

Irrigation Model at the Bele Level

The irrigation campaign for each bele depends on the crop grown on the bele (1). The campaign is bordered by a starting crop age (I_1) and an ending crop age (I_2). Between these two thresholds, irrigation can occur if the crop is stressed. The stress is defined as the ratio of the actual evapotranspiration (AET) and the maximum evapotranspiration (MET):

$$if\left[(I_1 < A(t) < I_2)\ and\ (\frac{AET(b,t-1)}{MET(b,t-1)} < I_3)\right] then\ I(t) = TRUE \tag{1}$$

The irrigation amount depends on the irrigation technique chosen at the bele level, and on the available water for irrigation either pumped from the ground water table or coming from the pond. Three irrigation technique are available: furrow, drip and sprinkler; but others may be added.

2.3.2. Management Processes at the Jeminu Level

Irrigation Model at the Jeminu Level

Competition between the irrigation required by each bele and the availability of water for the jeminu is managed in the irrigation model at the jeminu level. This is an adaptation of the algorithm developed in [35]. The general principles are as follows (Figure 2):

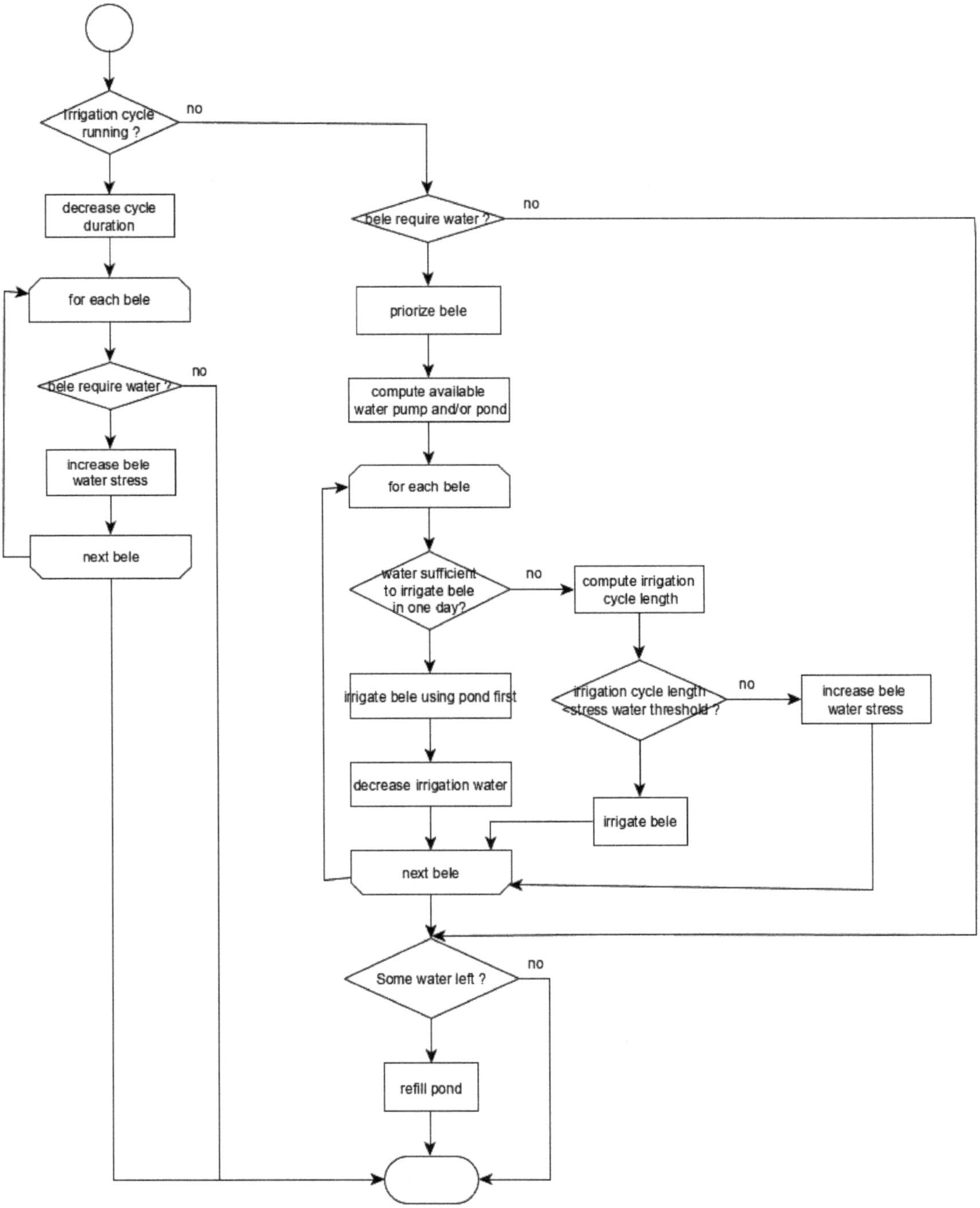

Figure 2. The general algorithm of the irrigation decision-making model at the jeminu level.

- Each day, the model checks if an irrigation cycle is running. If yes, the duration of the cycle is decreased by one, and the crops on the beles that would have needed irrigation are stressed by one more point.
- If there is no irrigation cycle running, then the model checks if the different beles need irrigation. Irrigation triggering once irrigation campaign has started is based on Equation (1).
- Priority between beles requiring irrigation is computed first (see below) and irrigation is then performed one after the other in decreasing priority.
- Irrigation water may come either from the pond or the groundwater table (through the pump).
- When irrigation is requested, if the pond can be used for irrigation, its water volume is used first to irrigate.
- For a bele needing irrigation, the amount provided by the irrigation cycle depends on the type of irrigation equipment (drip, sprinkler or furrow).
- The amount of irrigation water is given on the first day of the irrigation cycle on the bele.
- However, the amount to refill the pond (if any) is given on the last day of the irrigation cycle.
- If the amount required can be given in one day, the bele is irrigated and the remaining water can be used to irrigate another bele or to fill the pond, depending on the number of beles requiring irrigation. This is an optional possibility.
- If several days are required to irrigate, the irrigation cycle length is computed.
- If the irrigation cycle length may lead to crop failure (too large water stress), the bele is not irrigated and the next bele is tested for irrigation.
- If some water remains, it can be used to refill the pond.
- When an irrigation is performed, the pump is "blocked", meaning that it cannot be used for another purpose.

Irrigation Priority between Beles

If different beles require irrigation the same day, priority rules are needed if not enough water is available. To calculate irrigation priorities, we use a weighted average approach corresponding to a priority index, $\Re$, calculated for each bele (2). The priority bele is the bele with the smallest $\Re$. This approach makes it possible to integrate all the factors that impact the decision-making related to the management of irrigation. The choice of factor coefficients makes it possible to build different irrigation strategies.

$$\Re = a \cdot R_{crop} + b \cdot R_{stress} + c \cdot R_{age} + d \cdot R_{technic} \quad (2)$$

where a, b, c and d are the weight coefficients.

- R_{crop} represents the crop priority.
- R_{stress} represents the water stress priority factor. This factor is linked to the cumulative stress of the crop (number of days without water supply). For this factor, two strategies are possible: (i) favor the most stressed crops to prevent them from failing, or (ii) favor the least stressed crops, because the most stressed have already lost their yield potential.
- R_{age} represents the crop cycle priority factor. This factor is linked to the relative (normalized) state of achievement of the culture cycle. For this factor, two strategies are possible: (i) favor crops close to the end of the cycle to ensure harvest, or (ii) favor crops at the start of the cycle, to ensure their successful development during the first phases of the cycle.
- $R_{technic}$ represents the irrigation amount priority factor. This factor is linked to irrigation technique (drip, sprinkler and furrow). The idea is to classify beles by their irrigation technique; from the techniques requiring the least water to the techniques requiring the most water. Two strategies are possible: (i) favor the irrigation technique with the

larger amount of water to provide; or (ii) favor the irrigation technique with the less amount of water to provide.

$\Re$ is calculated for each bele requiring irrigation. In case of ex aequo, a final factor is used, R_{bele}, the bele priority factor. The advantage of this formalism is that we separate the strategy from the code and give the user the chance to choose his own irrigation strategy. An example of this algorithm is given in Table 1. The first four lines of the table {1–4} provide the priority given to the four priority factors. R_{crop} is priority one, R_{stress} priority 2, R_{age} priority 3 and $R_{technic}$ is not considered. For R_{stress} and R_{age}, the option of how to consider stress and age is given, respectively. For R_{stress}, we favor the more stressed crop first; while for R_{age}, we favor first the more aged crop.

Table 1. An example of the algorithm to determine the priority between the different beles to be irrigated. In this case, the bele irrigated is bele 1 as it has the lowest $\Re$. See text for details.

		Priorities	**Option**	
1	R_{crop}: crop priority	1		
2	R_{stress}: stress priority	2	favor more stressed crop	
3	R_{age}: crop cycle priority	3	favor aged crop	
4	$R_{technic}$: irrigation technic priority	0	-	
		Weights		
5	a	0.167	=1/(1 + 2 + 3)	
6	b	0.333	=2/(1 + 2 + 3)	
7	c	0.500	3/(1 + 2 + 3)	
8	d	0.000	0/(1 + 2 + 3)	
		Daily data		
9	Bele	1	2	3
10	Crop	cucurma	beetroot	onion
11	crop priority	1	2	3
12	Irrigation	Drip	Furrow	Sprinkler
13	Irrigation amount (mm)	15	50	25
14	Crop length (d)	240	90	90
15	Crop age (d)	230	30	2
16	Crop water stress level	10	6	5
17	Number of days without irrigation	5	5	4
		Priority calculation		
18	R_{crop}	0.167	0.333	0.500
19	$R_{stress}(j)$	0.50	0.17	0.20
20	$R_{age}(j)$	0.04	0.67	0.98
21	$R_{technic}$	0.3	1	0.5
22	Overall priority	0.22	0.44	0.64
23	In case of equal R_{bele}, bele prioriy	1	3	2

The next four lines {5–8} give the computation for the different weights (see (2)). For example, as R_{crop} is priority 1 and that {priority(R_{crop}) + priority(R_{stress}) + priority(R_{age}) + priority($R_{technic}$)} equals 6, then (a) equals 1/6 = 0.167.

Lines {9–17} give the status of the system on an example day to demonstrate the computation: there are three beles, with each a different crop {10}. Priority between these different crops is given on {11}. These three beles can be irrigated with a different irrigation equipment {12} providing a given amount of water {13}. Lines {14–17} give the status of the crop regarding their age and their water stress. From this information the different priorities are calculated {18–21}. R_{crop} is calculated using {11} and the same algorithm used to calculate the weights (see above); R_{stress} is calculated as {17}/{16}; R_{age} is calculated as

1- {15}/{14}; and $R_{technic}$ is calculated as {13} divided by the value of the technic given the largest amount of water; here, it is a furrow with 50 mm. Line {22} calculates the final value of $\Re$ using (2). In the case of ex aequo, we use {23} to decide the bele to irrigate.

2.3.3. The Dynamic of the System

The general dynamic of the model is as follows:

1. Initialization of the simulation:
    1.1. Create the jeminu.
    1.2. Create the different beles and initialize the bele (soil water amount).
    1.3. Create a pump and a pond, if any.
    1.4. Initialize the pond (water amount).
    1.5. Create a dictionary of crops and a dictionary of irrigation practices.
    1.6. Read the full climatic series.
2. Daily simulation:
    2.1. For each bele, manage the crop (check for sowing, harvesting and crop failure).
    2.1. On the jeminu, manage the irrigation.
        2.1.1. Check for beles that need irrigation.
        2.1.2. Manage priorities between beles.
        2.1.3. Provide irrigation if water is available.
    2.3. Update crop water stress on the different beles.
    2.4. Perform the water budget on the different objects (beles; pond, if any; and groundwater table)
3. At the end of the simulation period:
    3.1. Write an output file (dump memory).
    3.2. Create graphs to analyze the scenario.
    3.3. Clean the memory and end-up the simulation.

2.4. Modeling Approach

The model was developed under VBA Excel© using an object modeling structure [36]. Graphic user interface was based on an Excel worksheet facilitating data validation and error checking. Each simulation creates a new workbook with simulated data and explanatory graphs.

2.5. Testing Scenarios

In order to demonstrate the use of the model, we proposed a set of example scenarios:

1. Sc0 (ref): The reference scenario corresponds to a rainfed farm with two plots (5000 m^2 each). For rainfed farms, crops are grown only in two seasons, Kharif and Rabi, which is a widely used crop practice in the area. For each season, the farmer grows the two major crops of the area, each crop in one plot: for Kharif, sorghum and sunflower, and for Rabi, maize and horse gram. Rainfed crops are drought resistant, but these crops fail at a stress of 0.2 or less over a 5-day period.
2. Sc1: For Scenario 1, the farmer chooses to keep the same cropping system as in Sc0, but to install a pond that is filled by the runoff of rainwater on his watershed. This pond is lined and covered in order to avoid losses and to make the best use of the collected water, in protective irrigation. The surface of the pond is removed from the jeminu area. The irrigation is triggered when the crop is stressed to 0.4, to avoid its failure. For this scenario, furrow irrigation is used, the most common irrigation technique in the area. In case of competition for irrigation water, the farmer favors the most stressed crop closest to the end of its cycle in order to maximize the chance of a harvest.
3. Sc2: For Scenario 2, the farmer intensifies his production system compared to Sc0, by installing an 80 m-depth borewell, and by developing cash crops on four plots. The farmer grows beetroot in Kharif, tomatoes in Rabi and watermelon in Summer. These

crops are quite common in the area. Crop failure occurs when a crop has a stress less than or equal to 0.45 over a 5-day period (Table 2). The irrigation system is a furrow, and it is activated when the crop is at 0.6 stress. In case of competition for irrigation water, the farmer favors the crop with the highest stress level in order to avoid losing production quality.

4. Sc3: For Scenario 3, the farmer has the same cropping system and the same structural aspects as Sc2, but he decides to install a pond in addition. The functioning of this pond is different from that of Sc1. The pond in Sc3 is filled with water from the borewell in addition to the runoff, in order to store water on days when there is no irrigation demand, or if there is excess water after irrigation.

Table 2. Crop parameters regarding crop water stress (see A3 for details).

Crop	p1 [0; 1]	p2 [mm]	p3 [Number of Days]	Scenario
Beetroot	0.45	20	5	3
Horse gram	0.2	20	5	1, 2
Maize	0.2	20	5	1, 2
Sorghum	0.2	20	5	1, 2
Sunflower	0.2	20	5	1, 2
Tomato	0.45	20	5	3
Watermelon	0.45	20	5	3

For all the scenarios, the farm size assumed is 1 ha with a typical black soil. We used the 15-year Maddur climatic series (Latitude: 12°35′3.01″ N; Longitude: 77°02′41.64″ E) to run the simulation. The full parametrization of Sc0 is given in Appendix B.

3. Results

3.1. Model Graphic User Interface

When the user opens NIRAVARI, only three worksheets are initially visible: (i) a "read me" sheet explaining how to use the model; (ii) a Climate sheet, allowing the integration of the requested climatic data; and (iii) an Init sheet, allowing the user to parametrize the simulation. The Graphic User Interface allows the user to parametrize the simulation (Figure 3) with different validation processes (based on "data validation" from Excel). Information is given to the user whereby they have to fill in some data. Only cells in yellow can be modified. Once the parametrization is performed, the user clicks on the "Run" button, allowing for the simulation to run. When the simulation is over, a new workbook is opened with the computed data and some standard graphs, to then analyze the simulation.

A second button, "Load", allows the user to load a previous run to test a modification on a base scenario.

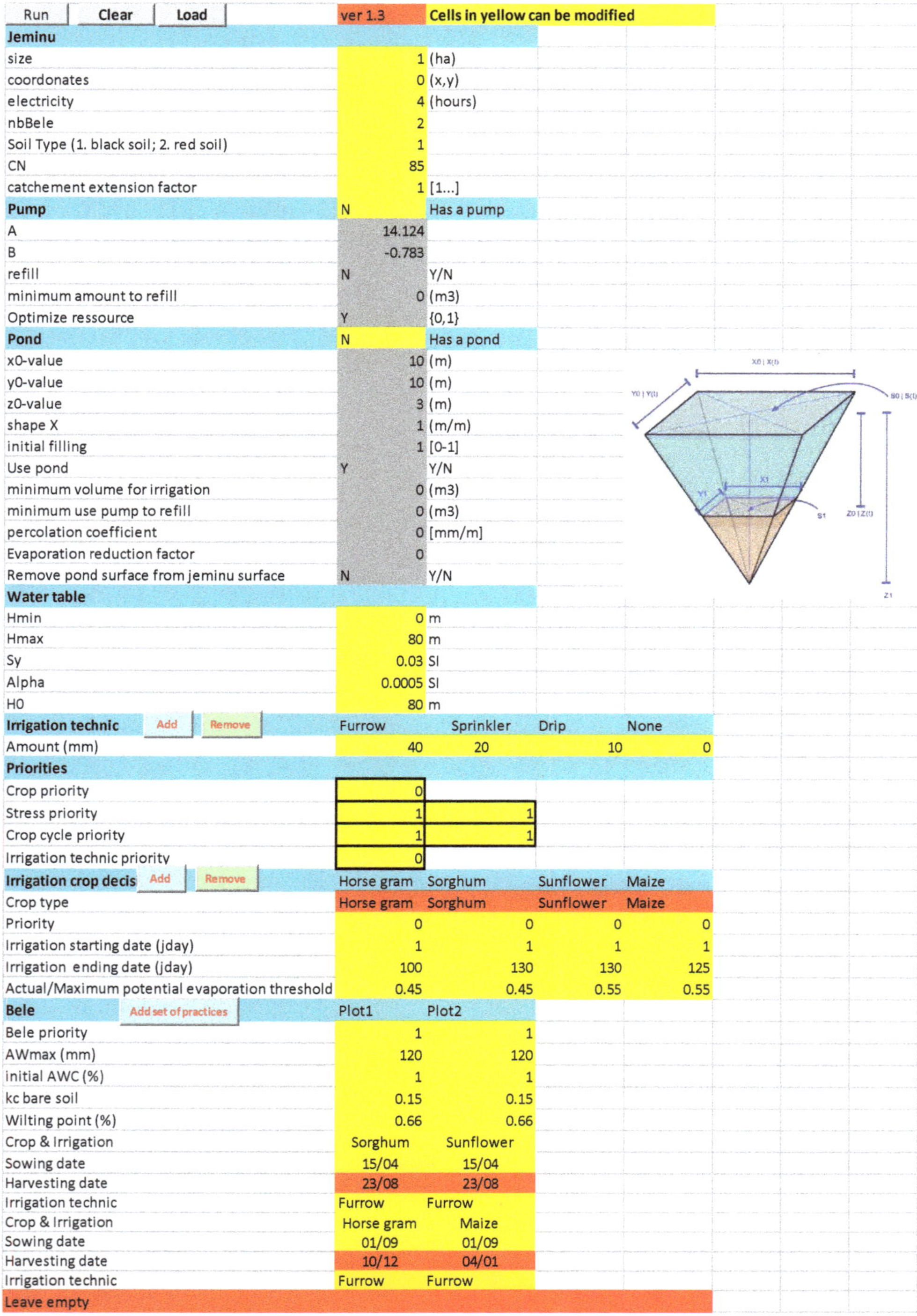

Figure 3. The upper screen of the graphic user interface. This Graphic User Interface allows for parametrizing of the simulation. Information and data validation are performed at this step.

3.2. Playing with the Model

The first indicator we observed was the level of crop failure over the 15-year climatic period (Table 3). Crop failure process is explained in A3. Growing using a rainfed system, (Sc0) shows a failure of 15% of the crops developed over the 15 years. Adding a pond in the jeminu (Sc1) improves the result (8% failure). This is due to the use of some irrigation to remove water stress to crops. However, it is impossible to avoid all failures, since the amount of water in the pond is not sufficient to irrigate all the crops in case of drought years. Moving to a pump and borewell system with more water-sensitive crops (but more expected economic return), (Sc2) shows a higher percentage of crop failure (29%). This high level of crop failure, despite irrigation possibilities, is due to two factors. The first is the use of more water-deficit-sensitive crops (Table 2). The second is due to the irrigation process itself. All beles cannot be irrigated on the same day, and the amount provided to the crops due to the irrigation technique (furrow) implies long irrigation cycles. Some beles are therefore not irrigated when needed, increasing the crop water stress leading to crop failure. This demonstrates the impact of the choice of water-intensive crops in spite of the addition of a bore well. If a buffer pond is added (Sc3), the situation improves with only 8% of crop failure.

Table 3. Percentage of crop failure over the 15-year climatic serie.

Scenario	Number of Crops	Failure
Sc0	60	15%
Sc1	60	8%
Sc2	180	29%
Sc3	180	8%

The second indicator is an economic estimation (Figure 4). We used the net return indicator (*NR*) to compare the scenario estimated as (3):

$$NR = \sum_{i=1}^{n} [A_i \,/10,000 \cdot ((Y_i * w_i * p_i) - C_i\,] \quad (3)$$

where A_i: crop area (m^2), Y_i: crop potential yield (T·ha^{-1}), w_i: mean crop water stress; p: crop price (Rs·t^{-1}); C_i: crop costs (Rs·ha^{-1}). We carried out a survey with farmers in the region to estimate the production costs, the market price and the potential yield for each crop. In case of crop failure, which can occur at any time during the crop cycle, we estimate that the farmer loses all production costs.

Sc0 situation allows farmers to earn 48,600 Rs (median), with a consequent number of years of loss. Adding a pond for protective irrigation (Sc1) drastically reduces the years of losses and low net return. The system with irrigated crops with a borewell (Sc2) improves the median net return (336,000 Rs), but makes it very variable due to the crops' failure and high-value crops' sensitivity to water shortages and droughts. Adding a buffer pond (Sc3) not only improves the median net return (489,000 Rs), but also reduces variability.

The third indicator is based on the pond water budget (Figure 5). This figure is more complex as it integrates the different components used to compute the dynamic of the pond water budget. The dynamic of the volume of the pond is given in clear blue on the first y-axis (left-hand). Input variables for the pond are on the first y-axis: rain (red), refill from the borewell (yellow) and runoff from the watershed (gray). The output variables are given on the second y-axis (right hand) in inverse order, the zero value being at the top of the graph: pond over flooding (orange), pumped for irrigation (lilac) and evaporation (light red). Depending on the parameters used for simulation, some of these variables will be null, or will change during the simulation. Scenarios Sc1 and Sc3 show two different modes of using the pond. In Sc1, the pond contributes to storing runoff water for irrigation when crops are most stressed. The pond is filled from time to time and quite slowly (Figure 5A). In Sc3, the pond is used as a buffer in order to store the water available by the pump and by the runoff. The pond is therefore filled quite often. It is used when the pump flow rate is not sufficient to provide the required irrigation needs.

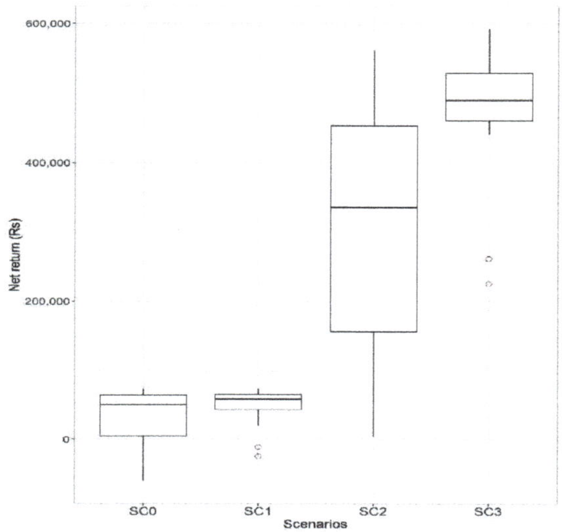

Figure 4. Net return distribution over the 15-year simulation period by scenario.

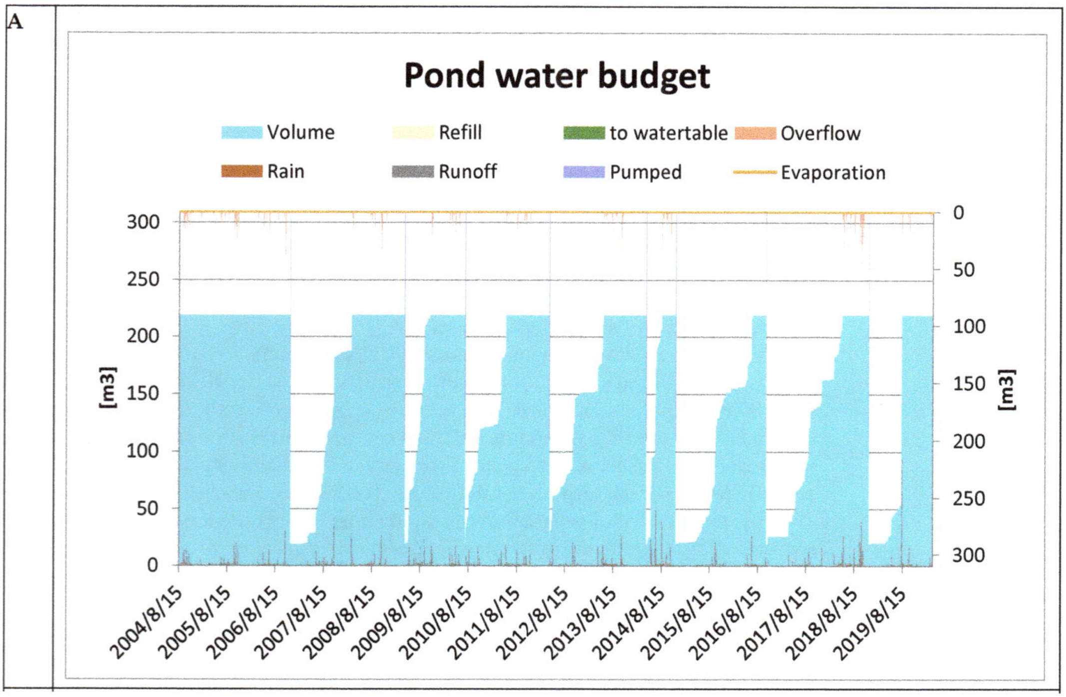

Figure 5. *Cont.*

B

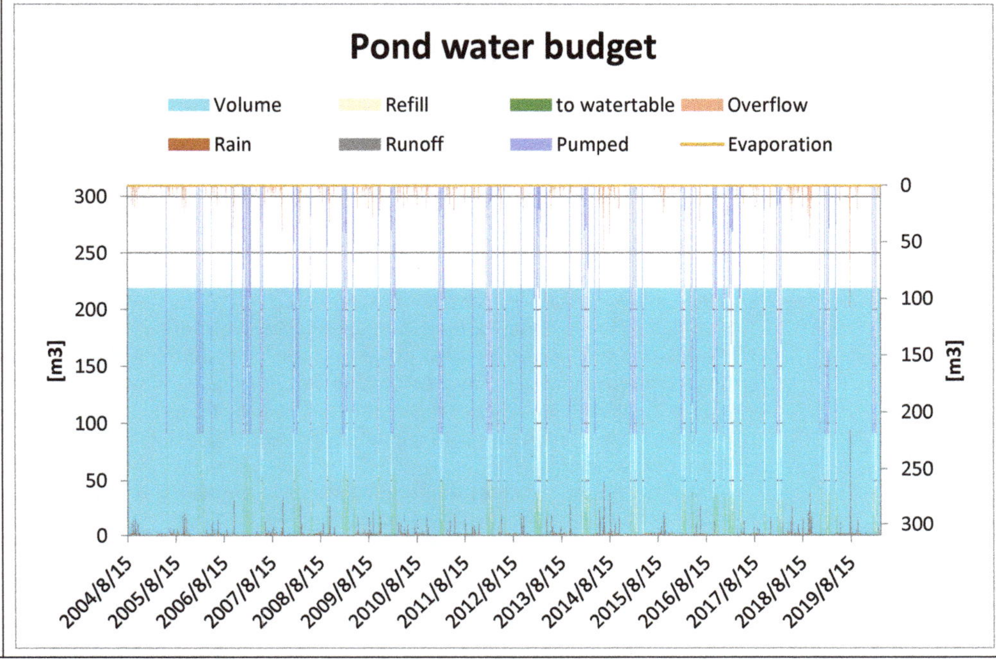

Figure 5. Pond water budget for Sc1 (**A**) and Sc3 (**B**).

The last indicator we propose is water table evolution due to the use of the different systems (Figure 6). The use of the pond in Sc1 has a low impact on the water table (comparing Sc0 and Sc1). The slight decrease in water table levels in Sc1 comes from the reduction of crop failures, which consequently reduces the water that is drained into the groundwater, as it is used by the saved crops. The 100% irrigated cropping system of Sc2 impacts the level of the water table, because the long-term pumping pushes the water table to a quasi-constant decline. On the other hand, in Sc3, and in addition to the phenomenon explained before (the reduction of crop failure), the amount of water pumped is important because all the water available from the pump for one day is pumped even if there is no irrigation demand or if the irrigation demand is lower than the water available from the pump.

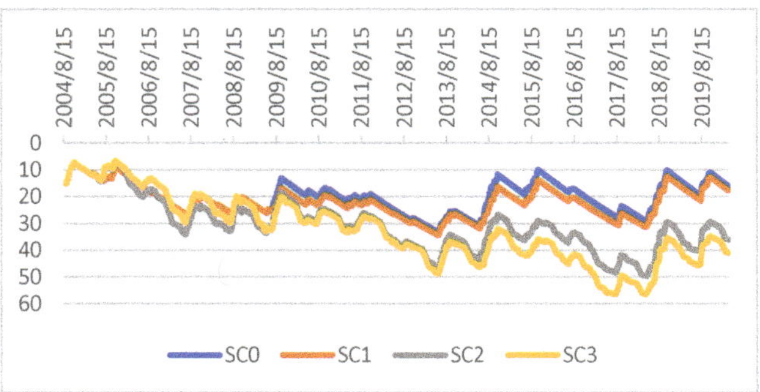

Figure 6. Evolution of the water table depth depending on the four scenarios of farming practices.

4. Discussion

So far, no model exists to deal with farmers' decision-making to irrigate their farms with multiple irrigation sources in the Indian context. NIRAVARI has been developed to be used as a tool by policy makers and technical advisors to assess different policies better placed to manage water resources in the broader context of climate change. To develop such a tool, we reused simple but robust, already existing models, such as the FAO model for crop growth and other simple models for other processes. The originalities of NIRAVARI regarding other irrigation models in India (see for example) are: (1) the decision-making model and its ability in testing a large range of decisions based on simple criteria; (2) its ability to study the distribution of water from the same source to different crops; (3) its ability to study two irrigation sources simultaneously; and (4) to consider the feedback between the cropping system and the water source.

Instead of using complex modelling languages, we chose a wide-spread language (VBA), allowing for object-oriented modelling and programming. Each element of our farm system is then modeled as an object, and a multiplicity of objects (such as beles or crops) are dealt with as structured containers. The use of Excel allows users to manage simple graphic user interfaces with validation processes in order to avoid any false input to the model parameters.

The scenarios tested show that rainfed systems can maintain a balance in the water table, but the income from these systems is very low and highly variable due to climate variability. However, adding a pond to this system reduces the vulnerability of these farms to climate variability while preserving the water table. On the other hand, a 100% irrigated system with high-added value crops leads to a significant improvement in income, but a drastic decrease in the water table. Indeed, this system leads to an instability of income that systematically decreases with the lack of water. Adding a pond buffer to this system limits the variability of income, but accentuates the decline of the water table, which limits the sustainability of the groundwater resource. These simulations can provide valuable considerations to policymakers, to decide on production systems and water storage technologies, and for their use to be promoted.

Creating such a parsimonious, simple and handy model is also limited to the range of model applications, and the interpretation of some outputs should be considered with care. For example, the choice of the FAO-56 single coefficient formalism [33] implies that the model represents evaporation and transpiration as a single flux, and therefore, calculating the crop water use efficiency for different scenarios is not possible. Similarly, as in the FAO-56 formalism, there is no impact of the water stress on crop water demand, the stress is probably overestimated for rainfed crops—adapted to reduce their demand during drought—compared to strictly irrigated crops. Finally, the impact of water stress on the marketable yield, although drastically simplified (with no account for the disproportionate effect of stress during few critical phenologic stages and being more pronounced for water intensive crops than for more rustic ones), implies that the economic outputs of the model must be considered only as rough estimates.

Moreover, the size of the jeminu, the number and size of plots being fixed, the crop rotation being a forced variable for each simulation, and the adaptation of farmers to variations of water availability can be only partially accounted for in the model. This is the case of the "Jevons paradox" observed in Indian systems, where access to water-saving irrigation technologies can induce an increase of the irrigated area, and therefore, a faster depletion of the aquifer [37].

Finally, while one original feature of NIRAVARI is to account for the feedback between agricultural practices and groundwater resources, it can only represent one farm at a time, and therefore, the results of the simulations should not be interpreted as predictions of what is likely to happen in a farm, which could be surrounded by other farms with very different practices, all foraging the same aquifer. Instead, users should keep in mind that the type of questions to ask the model are rather of the type: "What is likely to happen if all farmers in a small region adopt the same practices as the one that is simulated in NIRAVARI?".

5. Conclusions

NIRAVARI was developed to help farmers and advisers, and to provide some analysis to policy-makers, move toward schemes they would like to implement. NIRAVARI can be used for a large range of questions and can, thanks to its simplicity (parsimonious) and genericity, be applied much beyond the Indian context. Crop parametrization, ground water transfer process and farm structure are easily changeable to represent a wide range of water management situations from countries other than India. Even if NIRAVARI was at first aimed at policy-makers, it can also be used as a training media for students to understand the impacts of irrigated agriculture on groundwater resources.

As a follow-up of the initial meeting with the policy-makers during a workshop held in Bangalore in March 2019, NIRAVARI is due to be presented to the officers of the Watershed Department in December 2022.

6. Patents

The NIRAVARI model is freely available on request to the main corresponding author.

Author Contributions: Software, J.-E.B.; validation, J.-E.B., MB and L.R.; conceptualization and methodology, J.-E.B. and M.B.; writing—original draft preparation, J.-E.B.; formal analysis, J.-E.B., M.B. and L.R.; writing—review and editing, J.-E.B., M.B., L.R. and M.S.; funding acquisition, L.R. and M.S. All authors have read and agreed to the published version of the manuscript.

Funding: This study was supported by the ATCHA project [ANR-16-CE03-0006] and the Environmental Research Observatory M-TROPICS (https://mtropics.obs-mip.fr/, accessed on 11 October 2022), which is supported by the University of Toulouse, IRD and CNRS-INSU.

Institutional Review Board Statement: Not applicable.

Informed Consent Statement: Not applicable.

Data Availability Statement: Not applicable.

Acknowledgments: The authors wish to thanks the different trainees who worked on the project and the different members of the ATCHA project for their constructive comments on the NIRAVARI model.

Conflicts of Interest: The authors declare no conflict of interest.

Appendix A. Equations of the Biophysical Model

Let us call b the index to represent a bele and t the index for the time.

Appendix A.1. Soil Water Budget

The state variable of interest is the soil water content ($W(b,t)$, m^3). Water inputs ($iW(b,t)$, $m^3 \cdot d^{-1}$) are irrigation ($I(b,t)$, $m^3 \cdot d^{-1}$) and part of the rainfall in case of runoff ($Rf(b,t)$, $m^3 \cdot d^{-1}$). Runoff is based on the curve number formalism [38]. Water outputs ($oW(b,t)$, $m^3 \cdot d^{-1}$) are actual evapotranspiration ($E_a(b,t)$, $m^3 \cdot d^{-1}$, see crop process) and drainage ($Dr(b,t)$, $m^3 \cdot d^{-1}$). Drainage occurs when the soil water capacity is full (tipping bucket formalism, [39]).

$$\begin{cases} \delta W(b,t) = iW(b,t) - oW(b,t) \\ iW(b,t) = Rf(b,t) + I(b,t) \\ oW(b,t) = E_a(b,t) \\ Rf(b,t) = [R(t) - runoff(t)] * size(b)/1000 \\ \begin{cases} if(k_c(b,t) > 0)\ then \begin{cases} if\left(\frac{W(b,t-1)}{W_x(b)}\right) > f_w\ then\ E_a(b,t) = \min(W(b,t-1), k_c(b,t)\cdot E_0(t)) \\ else\ E_a(b,t) = min\left(W(b,t-1), \left(\frac{W(b,t-1)}{W_x(b)}\right) * f_w * k_c(b,t)\cdot E_0(t)\right) \end{cases} \\ else\ E_a(b,t) = min(W(b,t-1), k_s * E_0(t) * size(b)/1000 \end{cases} \\ if(W(b,t-1) + \delta W(t) > W_x(b))\ then \begin{cases} Dr(b,t) = (W(b,t-1) + \delta W(t)) - W_x(b) \\ W(t) = W_x(b) \end{cases} \\ else \begin{cases} W(t) = W(t-1) + \delta W(b,t) \\ Dr(b,t) = 0 \end{cases} \end{cases} \quad (A1)$$

where k_c is the evaporation crop coefficient (see Appendix A.2. Crop Processes), k_s is the evaporation bare soil coefficient, $W_x(b)$ is the maximum soil water capacity of bele b (m^3) and *size*(b) is the size of the bele (m^2)

Appendix A.2. Crop Processes

Let us call $A(t)$ the age of the crops (in days). From crop sowing to crop harvest (see 2.3.1.), $A(t)$ follows a simple linear function: $A(t) = A(t-1) + 1$.

Crop coefficient ($kc(t)$) is modelled by a multilinear function from FAO56 [33] (Figure A1)

$$\begin{cases} if(0 < A(t) < d_1)\ then\ kc(t) = kc_1 \\ else\ if(d_1 < A(t) < (d_1 + d_2))\ then\ kc(b,t) = (A(t) - d_1) * \frac{(kc_2 - kc_1)}{d_1} + kc_1 \\ else\ if\ ((d_1 + d_2) < A(t) < (d_1 + d_2 + d_3))\ then\ kc(b,t) = kc_2 \\ else\ if\ ((d_1 + d_2 + d_3) < age(t) < (d_1 + d_2 + d_3 + d_4))\ then\ kc(b,t) = ((d_1 + d_2 + d_3 + d_4) - A(t)) * \frac{(kc_2 - kc_3)}{d_4} + kc_3 \\ else\ kc(b,t) = kc_3 \end{cases} \quad (A2)$$

where d_1, d_2, d_3 and d_4 are duration between growing phases (in days) and kc_1, kc_2 and kc_3 are specific kc values. All these parameters are crop dependent.

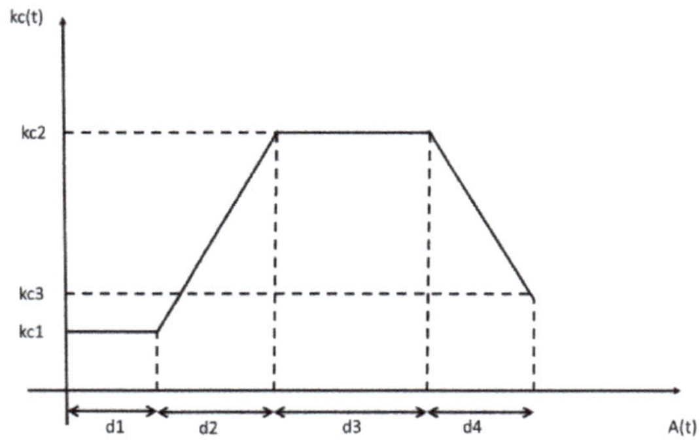

Figure A1. The multistage crop coefficient. From FAO56 [33].

Representing Crop Water Stress

The crop water stress is calculated using a simple algorithm. If the crop requires water and no water is provided, its stress increases by one. If the level of stress reaches a given threshold, then the crop dies. The level of stress can be zeroed if a sufficient amount of water is provided to the crop:

$$\begin{cases} if\left\{\frac{AET(t)}{MET(t)} < p_1 \; and \; I(t) = 0\right\} \; then \; N_s(t) = N_s(t-1)+1 \\ if(I(t)+R(t) > p_2\,) \; then \; N_s(t) = 0 \\ if N_s(t) \geq p_3 \; then \; crop \; die \end{cases} \tag{A3}$$

where *AET*(*t*) is the actual evapotranspiration, *MET*(*t*) is the maximum evapotranspiration, *I*(*t*) is the amount of irrigation water provided, $N_s(t)$ is the stress level, *R*(*t*) is the daily rainfall p_1, p_2 and p_3 are crop dependant parameters. Modifying p_1, p_2 and p_3 allows to represent different crop sensitivity to water stress.

Appendix A.3. Pump Process

The pump flow rate (*f*, $m^3\ s^{-1}$) follows a power function depending on the depth of the ground watertable (see Appendix A.5. Water Table Processes). This is based on pump flows measurements on the Berambadi watershed [26,32].

$$f(t) = a \cdot H(t-1)^b \tag{A4}$$

where *a* and *b* are parameters and *H* represents the height of the water table (m)

Appendix A.4. Pond Processes

The pond is modelled as a right prism (Figure A2). Pond water contents ($W_p(t)$, m^3), depends on the water level in the pond ($h_p(t)$, m). Water input ($iW_p(t)$, $m^3 \cdot d^{-1}$) are rainfall ($R_p(t)$, m^3), refill by the pump when water is available (*Rf*(*t*), m^3—see management processes) and runoff (*Ro*(*t*), m^3). Water output are pumping for irrigation ($I_p(t)$, m^3—see management processes), surface evaporation (*Ep*(*t*), m^3) and recharge of the ground water table if the pond is permeable ($D_p(t)$, m^3).

$$\begin{cases} \delta W_p(t) = iW_p(t) - oW_p(t) \\ iW_p(t) = R_p(t) + Rf(t) + Ro(t) \\ oW_p(t) = E_p(t) + I_p(t) + D_p(t) \\ R_p(t) = R(t) * surf_0/1000 \\ \begin{cases} if(R(t) < \gamma_1 then \; Ro(t) = 0 \\ else \; Ro(t) = \gamma_2 \cdot (R(t) - \gamma_1) \end{cases} \\ E_p(t) = \min(W_p(t-1), E_0(t) \cdot surf(t-1)/1000 \\ D_p(t) = \partial \cdot W_p(t-1) \end{cases} \tag{A5}$$

where γ_1 and γ_2 are parameters to deal with the runoff process, $surf_0$ is the upper surface area when the pond is full (m^2), *surf*(*t* − 1) is the actual upper surface area (m^2). Due the geometrical structure of the pond, there W_p and *surf* are linked. Detailed calculations are not given here.

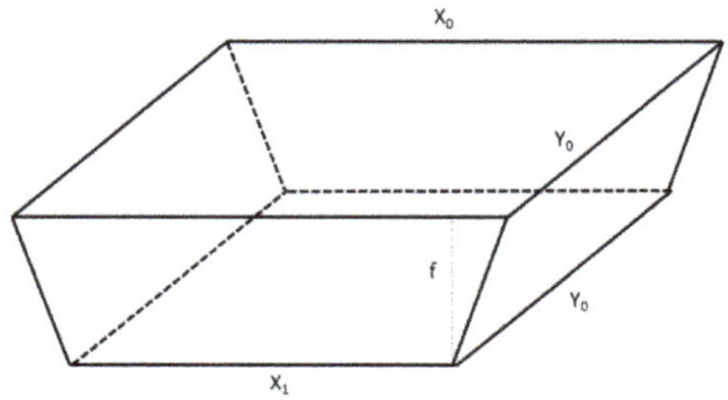

Figure A2. The pond geometrical representation.

Appendix A.5. Water Table Processes

The ground watertable height ($H(t)$, m) varies depending on the net recharge ($nR(t)$, $m^3 \cdot d^{-1}$), ie the difference between input (drainage from the different beles and recharge by the pond) and output (irrigation to the different beles and lateral losses ($Q(t)$, $m^3 \cdot d^{-1}$).

$$\begin{cases} H'(t) = H(t-1) + nR(t)/S_y \\ nR(t) = \left(D_p(t) + \sum_{b=1}^{n} D(b,t)\right) - \sum_{b=1}^{n} I(b,t) \\ if\left(H'(t) > H_x\right) \; then \; H'(t) = H_x \\ Q(t) = \left(H'(t) - H\right) \cdot S_y \cdot \alpha' \\ H(t) = H'(t) - Q(t) \; / \; Sy \end{cases} \quad \text{(A6)}$$

where $H'(t)$ is an intermediate variable, S_y is the specific yield of the aquifer, α is the groundwater recession coefficient.

Appendix B

Jeminu		
size	1	(ha)
coordonates	0	(x,y)
electricity	4	(hours)
nbBele	2	
Soil Type (1. black soil; 2. red soil)	1	
CN	85	
catchement extension factor	1	[1...]
Pump	N	Has a pump
A	50.8464	
B	-0.783	
refill	N	Y/N
minimum amount to refill	0	(m3)
Optimize ressource	Y	{0,1}
Pond	N	Has a pond
x0-value	10	(m)
y0-value	10	(m)
z0-value	3	(m)
shape X	1	(m/m)
initial filling	1	[0-1]
Use pond	Y	Y/N
minimum volume for irrigation	0	(m3)
minimum use pump to refill	0	(m3)
percolation coefficient	0	[mm/m]
Evaporation reduction factor	0	
Remove pond surface from jeminu surface	N	Y/N
Water table		
Hmin	0	m
Hmax	80	m
Sy	0.03	SI
Alpha	0.0005	SI
H0	65	m
Irrigation technic Add Remove	Furrow	Sprinkler
Amount (mm)	40	20

Priorities		
Crop priority	0	
Stress priority	2	1
Crop cycle priority	3	2
Irrigation technic priority	0	
Irrigation crop decis Add Remove	Horse gram	Sorghum
Crop type	Horse gram	Sorghum
Priority	0	0
Irrigation starting date (jday)	1	1
Irrigation ending date (jday)	100	130
Actual/Maximum potential evaporation threshold	0.45	0.45
Bele Add set of practices	Plot1	Plot2
Bele priority	1	2
AWmax (mm)	120	120
initial AWC (%)	1	1
kc bare soil	0.15	0.15
Wilting point (%)	0.66	0.66
Crop & Irrigation	Sorghum	Sunflower
Sowing date	20/04	20/04
Harvesting date	28/08	28/08
Irrigation technic	None	None
Crop & Irrigation	Horse gram	Maize
Sowing date	01/09	01/09
Harvesting date	10/12	04/01
Irrigation technic	None	None

Figure A3. Parameters of Sc0.

References

1. Castilla-Rho, J.C.; Rojas, R.; Andersen, M.S.; Holley, C.; Mariethoz, G. Sustainable groundwater management: How long and what will it take? *Glob. Environ. Change* **2019**, *58*, 101972. [CrossRef]
2. Siebert, S.; Burke, J.; Faures, J.M.; Frenken, K.; Hoogeveen, J.; Döll, P.; Portmann, F.T. Groundwater use for irrigation—A global inventory. *Hydrol. Earth Syst. Sci.* **2010**, *14*, 1863–1880. [CrossRef]
3. Dangar, S.; Asoka, A.; Mishra, V. Causes and implications of groundwater depletion in India: A review. *J. Hydrol.* **2021**, *596*, 126103. [CrossRef]
4. Sivapalan, M.; Konar, M.; Srinivasan, V.; Chhatre, A.; Wutich, A.; Scott, C.A.; Wescoat, J.L.; Rodríguez-Iturbe, I. Socio-hydrology: Use-inspired water sustainability science for the Anthropocene. *Earth's Future* **2014**, *2*, 225–230. [CrossRef]
5. Leenhardt, D.; Therond, O.; Cordier, M.O.M.-O.; Gascuel-Odoux, C.; Reynaud, A.; Durand, P.; Bergez, J.; Clavel, L.; Masson, V.; Moreau, P. A generic framework for scenario exercises using models applied to water-resource management. *Environ. Model. Softw.* **2012**, *37*, 125–133. [CrossRef]
6. Alcamo, J. The SAS Approach: Combining Qualitative and Quantitative Knowledge in Environmental Scenarios. In *Environmental Futures: The Practice of Environmental Scenario Analysis*; Alcamo, J., Ed.; Elsevier: Amsterdam, The Netherlands, 2008.

7. Iglesias, A.; Garrote, L. Adaptation strategies for agricultural water management under climate change in Europe. *Agric. Water Manag.* **2015**, *155*, 113–124. [CrossRef]
8. Bergez, J.E.; Deumier, J.M.; Lacroix, B.; Leroy, P.; Wallach, D. Improving irrigation schedules by using a biophysical and a decisional model. *Eur. J. Agron.* **2002**, *16*, 123–135. [CrossRef]
9. Bergez, J.-E.; Garcia, F.; Wallach, D. Representing and optimizing management decisions with crop models. In *Working with Dynamic Crop Models: Evaluating*; Wallach, D., Makowski, D., Jones, J.W., Eds.; Parameterizing and Using Them; Elsevier: Amsterdam, The Netherlands, 2006; pp. 175–210.
10. McCown, R.L. Changing systems for supporting farmers' decisions: Problems, paradigms, and prospects. *Agric. Syst.* **2002**, *74*, 179–220. [CrossRef]
11. Lowder, S.K.; Sánchez, M.V.; Bertini, R. Which farms feed the world and has farmland become more concentrated? *World Dev.* **2021**, *142*, 105455. [CrossRef]
12. Martin, G.; Allain, S.; Bergez, J.; Burger-Leenhardt, D.; Constantin, J.; Duru, M.; Hazard, L.; Lacombe, C.; Magda, D.; Magne, M.-A.; et al. How to address the sustainability transition of farming systems? A conceptual framework to organize research. *Sustainability* **2018**, *10*, 2083. [CrossRef]
13. Bergez, J.; Colbach, N.; Crespo, O.; Garcia, F.; Jeuffroy, M.H.; Justes, E.; Loyce, C.; Munier-Jolain, N.; Sadok, W. Designing crop management systems by simulation. *Eur. J. Agron.* **2010**, *32*, 3–9. [CrossRef]
14. Baccar, M.; Bergez, J.-E.; Couture, S.; Sekhar, M.; Ruiz, L.; Leenhardt, D. Building Climate Change Adaptation Scenarios with Stakeholders for Water Management: A Hybrid Approach Adapted to the South Indian Water Crisis. *Sustainability* **2021**, *13*, 8459. [CrossRef]
15. Babel, M.S.; Deb, P.; Soni, P. Performance Evaluation of AquaCrop and DSSAT-CERES for Maize Under Different Irrigation and Manure Application Rates in the Himalayan Region of India. *Agric. Res.* **2019**, *8*, 207–217. [CrossRef]
16. Malik, A.; Shakir, A.S.; Ajmal, M.; Khan, M.J.; Khan, T.A. Assessment of AquaCrop Model in Simulating Sugar Beet Canopy Cover, Biomass and Root Yield under Different Irrigation and Field Management Practices in Semi-Arid Regions of Pakistan. *Water Resour. Manag.* **2017**, *31*, 4275–4292. [CrossRef]
17. Kumar, P.; Sarangi, A.; Singh, D.K.; Parihar, S.S.; Sahoo, R.N. Simulation of salt dynamics in the root zone and yield of wheat crop under irrigated saline regimes using SWAP model. *Agric. Water Manag.* **2015**, *148*, 72–83. [CrossRef]
18. Singh, A. Optimal allocation of water and land resources for maximizing the farm income and minimizing the irrigation-induced environmental problems. *Stoch. Environ. Res. Risk Assess.* **2017**, *31*, 1147–1154. [CrossRef]
19. Mohapatra, A.G.; Lenka, S.K.; Keswani, B. Neural Network and Fuzzy Logic Based Smart DSS Model for Irrigation Notification and Control in Precision Agriculture. *Proc. Natl. Acad. Sci. USA* **2019**, *89*, 67–76. [CrossRef]
20. Vij, A.; Vijendra, S.; Jain, A.; Bajaj, S.; Bassi, A.; Sharma, A. IoT and Machine Learning Approaches for Automation of Farm Irrigation System. *Procedia Comput. Sci.* **2020**, *167*, 1250–1257. [CrossRef]
21. Robert, M.; Thomas, A.; Sekhar, M.; Badiger, S.; Ruiz, L.; Raynal, H.; Bergez, J. Adaptive and dynamic decision-making processes: A conceptual model of production systems on Indian farms. *Agric. Syst.* **2017**, *157*, 279–291. [CrossRef]
22. Robert, M.; Thomas, A.; Bergez, J. Processes of adaptation in farm decision-making models. A review. *Agron. Sustain. Dev.* **2016**, *36*, 64. [CrossRef]
23. O'Keeffe, J.; Moulds, S.; Bergin, E.; Brozović, N.; Mijic, A.; Buytaert, W. Including Farmer Irrigation Behavior in a Sociohydrological Modeling Framework With Application in North India. *Water Resour. Res.* **2018**, *54*, 4849–4866. [CrossRef]
24. Shah, T.; Giordano, M.; Mukherji, A. Political economy of the energy-groundwater nexus in India: Exploring issues and assessing policy options. *Hydrogeol. J.* **2012**, *20*, 995–1006. [CrossRef]
25. Fishman, R.M.; Siegfried, T.; Raj, P.; Modi, V.; Lall, U. Over-extraction from shallow bedrock versus deep alluvial aquifers: Reliability versus sustainability considerations for India's groundwater irrigation. *Water Resour. Res.* **2011**, *47*. [CrossRef]
26. Buvaneshwari, S.; Riotte, J.; Sekhar, M.; Mohan Kumar, M.S.; Sharma, A.K.; Duprey, J.L.; Audry, S.; Giriraja, P.R.; Praveenkumarreddy, Y.; Moger, H.; et al. Groundwater resource vulnerability and spatial variability of nitrate contamination: Insights from high density tubewell monitoring in a hard rock aquifer. *Sci. Total Environ.* **2017**, *579*, 838–847. [CrossRef]
27. Buvaneshwari, S.; Riotte, J.; Sekhar, M.; Sharma, A.K.; Helliwell, R.; Kumar, M.S.M.; Braun, J.J.; Ruiz, L. Potash fertilizer promotes incipient salinization in groundwater irrigated semi-arid agriculture. *Sci. Rep.* **2020**, *10*, 3691. [CrossRef]
28. Sekhar, M.; Riotte, J.; Ruiz, L.; Jouquet, J.; Braun, J.J. Influences of Climate and Agriculture on Water and Biogeochemical Cycles: Kabini Critical Zone Observatory. *Proc. Indian Natl. Sci. Acad.* **2016**, *82*, 833–846. [CrossRef]
29. Robert, M.; Thomas, A.; Sekhar, M.; Badiger, S.; Ruiz, L.; Willaume, M.; Leenhardt, D.; Bergez, J. Farm Typology in the Berambadi Watershed (India): Farming Systems Are Determined by Farm Size and Access to Groundwater. *Water* **2017**, *9*, 51. [CrossRef]
30. Fischer, C.; Aubron, C.; Trouvé, A.; Sekhar, M.; Ruiz, L. Groundwater irrigation reduces overall poverty but increases socioeconomic vulnerability in a semiarid region of southern India. *Sci. Rep.* **2022**, *12*, 8850. [CrossRef]
31. Boisson, A.; Guihéneuf, N.; Perrin, J.; Bour, O.; Dewandel, B.; Dausse, A.; Viossanges, M.; Ahmed, S.; Maréchal, J.C. Determining the vertical evolution of hydrodynamic parameters in weathered and fractured south Indian crystalline-rock aquifers: Insights from a study on an instrumented site. *Hydrogeol. J.* **2015**, *23*, 757–773. [CrossRef]

32. Collins, S.L.; Loveless, S.E.; Muddu, S.; Buvaneshwari, S.; Palamakumbura, R.N.; Krabbendam, M.; Lapworth, D.J.; Jackson, C.R.; Gooddy, D.C.; Nara, S.N.V.; et al. Groundwater connectivity of a sheared gneiss aquifer in the Cauvery River basin, India. *Hydrogeol. J.* **2020**, *28*, 1371–1388. [CrossRef]
33. Allen, R.G.; Pereira, L.S.; Raes, D.; Smith, M. *Crop Evapotranspiration—Guidelines for Computing Crop Water Requirements*; FAO: Rome, Italy, 1998; Volume 300, p. D05109.
34. Robert, M.; Dury, J.; Thomas, A.; Therond, O.; Sekhar, M.; Badiger, S.; Ruiz, L.; Bergez, J. CMFDM: A methodology to guide the design of a conceptual model of farmers' decision-making processes. *Agric. Syst.* **2016**, *148*, 86–94. [CrossRef]
35. Bergez, J.-E.; Debaeke, P.; Deumier, J.-M.; Lacroix, B.; Leenhardt, D.; Leroy, P.; Wallach, D. MODERATO: An object-oriented decision tool for designing maize irrigation schedules. *Ecol. Modell.* **2001**, *137*, 43–60. [CrossRef]
36. Booch, G. *Object-Oriented Analysis and Design with Applications*, 2nd ed.; Benjamin/Cummings Publishing Company: Redwood City, CA, USA, 1994.
37. Bahinipati, C.S.; Viswanathan, P.K. Can Micro-Irrigation Technologies Resolve India's Groundwater Crisis? Reflections from Dark-Regions in Gujarat. *Int. J. Commons* **2019**, *13*, 848–858. [CrossRef]
38. SCS Hydrology. *National Engineering Handbook, Supplement A, Soil Conservation Service*; US Department of Agriculture: Washington, DC, USA, 1956; Section 4, Chapter 10.
39. Soldevilla-Martinez, M.; Quemada, M.; López-Urrea, R.; Muñoz-Carpena, R.; Lizaso, J.I. Soil water balance: Comparing two simulation models of different levels of complexity with lysimeter observations. *Agric. Water Manag.* **2014**, *139*, 53–63. [CrossRef]

International Journal of
Environmental Research and Public Health

MDPI

Article

Hydrodynamic Modelling and Flood Risk Analysis of Urban Catchments under Multiple Scenarios: A Case Study of Dongfeng Canal District, Zhengzhou

Huaibin Wei [1], Liyuan Zhang [2] and Jing Liu [3,4,*]

[1] School of Management and Economics, North China University of Water Resources and Electric Power, Zhengzhou 450046, China
[2] School of Water Conservancy, North China University of Water Resources and Electric Power, Zhengzhou 450046, China
[3] College of Water Resources, North China University of Water Resources and Electric Power, Zhengzhou 450046, China
[4] Henan Key Laboratory of Water Resources Conservation and Intensive Utilization in the Yellow River Basin, Zhengzhou 450046, China
* Correspondence: liujingdx@ncwu.edu.cn

Abstract: In recent years, urban flooding has become an increasingly serious problem, posing a serious threat to socio-economic development and personal safety. In this paper, we consider the Dongfeng Canal area in Zhengzhou City as an example and build a 1D/2D coupled urban flood model using the InfoWorks ICM. This study area uses six scenarios with rainfall return periods of 5 a, 20 a, and 50 a, corresponding to rainfall ephemeris of 1 h and 2 h to assess the flood risk. The results of the study show that (1) The flood depth, inundation duration, and extent of inundation in the study area vary with the return period and rainfall history. Generally, most of the water accumulation is concentrated in the low-lying areas adjacent to the river and near the roadbed. (2) As the rainfall recurrence period and rainfall duration increase, the proportion of overflow at the nodes becomes more pronounced and the overload from the pipe network flows mainly to the overload. (3) The high-risk areas under the different scenarios are mainly distributed on both sides of the river, and most of the low-risk areas transform into medium- and high-risk areas as the rainfall recurrence period and rainfall duration increase. This study analyses the flood risk situation under different scenarios, as well as the elements and areas that should be monitored in case of flooding, with the aim of providing a reference for flood prevention and control in the study area and formulating corresponding countermeasures. It also serves as a reference for flood risk analysis in other areas with similar situations.

Keywords: urban flood simulation; risk assessment; InfoWorks ICM; urban drainage 1D/2D modelling

Citation: Wei, H.; Zhang, L.; Liu, J. Hydrodynamic Modelling and Flood Risk Analysis of Urban Catchments under Multiple Scenarios: A Case Study of Dongfeng Canal District, Zhengzhou. *Int. J. Environ. Res. Public Health* **2022**, *19*, 14630. https://doi.org/10.3390/ijerph192214630

Academic Editors: Luis Garrote and Alban Kuriqi

Received: 25 August 2022
Accepted: 3 November 2022
Published: 8 November 2022

1. Introduction

Urbanisation and changes in land use have had a considerable impact on the processes and elements of the water cycle [1,2], whereas in recent years, global climate change and extreme weather have occurred frequently, and urban flooding caused by heavy rainfall has gradually become a hot topic of concern for scholars [3]. The frequent occurrence of urban flooding disasters has brought substantial economic losses and casualties to society and seriously threatened urban public safety [4]. Therefore, it is essential to understand the risk areas of urban flooding and conduct flood risk assessments to prevent and control urban flooding and reduce the losses caused by such disasters [5].

A common method for conducting an urban flood risk assessment is the historical disaster statistics method [6], which focuses on using statistical methods to analyse the development pattern of flooding, predict possible future flooding hazards, and estimate possible losses due to flooding based on historical flooding information and rainfall data in the study area [7]. For example, Hans de Moel et al. analysed trends in flood risk in time and space and made projections for future land use and flood inundation risk in the Netherlands. Their findings show that over spatial spans, flood losses are greater in areas of high economic growth than in areas of low economic growth. However, high-economic-growth areas are more resilient to flood risk than low-economic-growth areas [2]. Man Qi et al. analysed the effects of topography, rainfall, and impervious surfaces on urban flooding and their spatial patterns of variation for four recent storm events in Cincinnati, USA. They used the kriging interpolation of estimated rainfall depths to measure the impact of rainfall on urban flood hazards [8].

With the development of computer technologies, hydrological hydrodynamic simulation methods are increasingly being applied to urban flood risk assessment. Some of the more widely used numerical models in urban hydrological simulations are SWMM, MIKE, and InfoWorks ICM [9]. For example, Zhao et al. used the coupling of two models, SWMM and MIKE21, to simulate in detail the spatial distribution and water depth of the inundated area in the central part of Cangzhou City under different rainfall return periods and evaluate the economic losses from flooding in the risk area [10]. Sidek et al. used the InfoWorks ICM hydrological-hydraulic model of the Baisala basin as an example to model the response of the basin to rainfall based on the Probabilistic Distributed Moisture (PDM) model to generate flood hazard maps based on several average repetition intervals (ARI) and uniform rainfall depths and analyse the main influences affecting the flood depth and extent [11]. Tabari et al. used the InfoWorks ICM hydraulic model to quantify the impact of anthropogenic climate on urban rainfall flooding in Antwerp, Belgium, using a risk assessment framework and causal counterfactual probability theory [12]. Compared to the analysis of flood risk through historical hazard scenarios, hydraulic modelling provides a more accurate risk assessment method and allows for a more comprehensive analysis of flood risk conditions [13–17]. The existing models work well on a large regional scale. For urban areas, flood risk modelling needs to be more precise, for example, down to a particular road or square. However, the lack of monitoring of relevant data at the urban scale, such as check-well level data and data from drainage networks, has led to only a few academics working on flood risk models for small urban catchments [16,18–26], so this paper constructs an urban flooding model using the Dongfeng canal area in Zhengzhou City as an example to analyse the flood risk within this small urban catchment.

The objectives of this study are (1) to comprehensively consider the hydrological processes between river–urban drainage system–surface runoff and construct a 1D/2D coupled urban flood model based on InfoWorks ICM and an analysis of urban flood processes; (2) to use the storm intensity formulae to design different rainfall scenarios; analyse the inundation depth, duration of inundation, and inundation extent under different recurrence period design storms; and analyse the overflow distribution and drainage capacity of the pipe network at the nodes; and (3) to construct an urban flood risk assessment system to analyse the urban flood risk for the study area.

2. Materials and Methods

2.1. Study Area

Zhengzhou is the capital of Henan Province, located in the north-central part of Henan Province, where the middle and lower reaches of the Yellow River divide, between longitude 112°42′–114°14′ E and latitude 34°16′–34°58′ N. Zhengzhou is mostly a plain, except for the hills in the southwest, and the terrain is flat, with elevations generally less than 284 m, the lowest being only 79 m. There is a difference of 205 m between the highest and lowest points in the territory.

The Dongfeng Canal drainage area of Zhengzhou City was selected as the study area for this study. This study area is located in the northern part of the main urban area of Zhengzhou. The study area covers an area of approximately 80 km^2.

The study area has a temperate continental monsoon climate with an average annual precipitation of 632.4 mm and an average of 78 days of precipitation per year [27]. The extreme annual maximum rainfall is 1339 mm and the extreme annual minimum rainfall is 380.6 mm, with rainfall concentrated between June and August each year and the heaviest rainfall occurring in August.

The main major river network in the study area is the Dongfeng Canal. The Dongfeng Canal is a man-made river that was dug in 1958. Originally used as a channel to divert water from the Yellow River for irrigation, it now fulfils important functions in flood control, ecology, and landscape. The Dongfeng Canal starts at the Yellow River embankment and joins the Sosu and Jalu rivers to the south. At the same time, the Dongfeng Canal also intersects with a number of tributaries in the city centre. The Dongfeng Canal is one of the most important north–south oriented rivers in the city and has the task of draining flood water. The total length of the stormwater pipe network laid in the study area is 295.62 km. Stormwater in the study area is mainly discharged into the Dongfeng Drainage Canal through the stormwater pipe network laid on arterial roads such as the North Third Ring Road, Garden Road, Zhongzhou Avenue, East Yellow River Road, and East Dongfeng Road. Figure 1 shows the location of the study area, the distribution of the stormwater pipe network, and the elevation schematic.

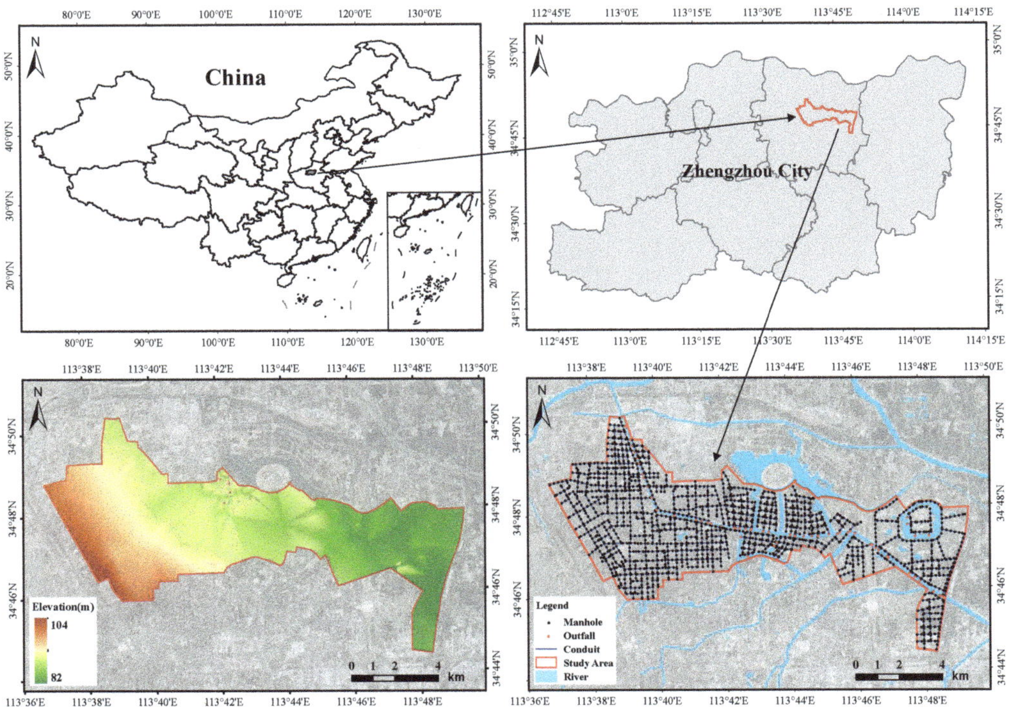

Figure 1. Location of the study area, stormwater network distribution, and elevation data.

2.2. Data Collection and Manipulation

The basic data required to build the hydrodynamic model during the study included the following: road-building data, stormwater pipe network and node distribution data, 5 m accuracy DEM data, and land use/land cover data.

Road construction data were obtained from OSM (https://www.openstreetmap.org (accessed on 6 April 2022)). Stormwater pipe network and node data with a 5 m accuracy provided by Zhengzhou Planning and Survey Design Institute were used for the construction of the 1D stormwater pipe network and 2D surface diffuse flow model underlying the study area.

Land use data were classified using ArcGIS PRO for the supervised classification of satellite image data. Image data were obtained from the Geospatial Data Cloud (www.gscloud.cn (accessed on 6 April 2022)), Landsat8 OLI-TIRS remote sensing imagery [28]. Based on the current land use situation in the study area and the needs of the study, the land use in the study area was classified into five categories: building land, green space, water bodies, bare soil, and roads.

The river network data in the study were river shapes determined using satellite imagery and DEM data were used to extract features from river cross-sections.

2.3. Research Methodology

2.3.1. InfoWorks ICM Hydrodynamic Modelling

(1) Basic theory

This study integrates the water exchange between the pipe network and the two-dimensional surface and river channels and uses InfoWorks ICM software to construct a one-two-dimensional coupled urban flood model [29].

The model is mainly concerned with hydrohydraulic processes such as rainfall, surface runoff, and the drainage of the pipe network [30]. The Infoworks ICM model is used to simulate the diffusion and transport of water in pipes by completely solving the system of St.Venant equations with the control equations in Equations (1) and (2).

$$\frac{\partial A}{\partial t} + \frac{\partial Q}{\partial x} = 0 \tag{1}$$

$$\frac{\partial A}{\partial t} + \frac{\partial}{\partial x}\left(\frac{Q^2}{A}\right) + gA\left(\cos\theta\frac{\partial h}{\partial x} - S_0 + \frac{Q|Q|}{K^2}\right) = 0 \tag{2}$$

where Q is the flow rate, m^3/s; A is the pipe section area, m^2; t is the time, s; x is the length of the pipe along the flow direction, m; h is the water depth, m; g is the acceleration of gravity, m/s; θ is the horizontal angle in degrees; K is the water transfer rate, determined by Manning's formula; S_0 is the slope of the pipe bottom.

The Infoworks ICM model generalises the river channel to a piped open channel when simulating the flood evolution of the river network and uses a one-dimensional hydrodynamic model for the simulation [31,32], with the basic control equations being

$$\frac{\partial A}{\partial t} + \frac{\partial Q}{\partial x} = q \tag{3}$$

$$\frac{\partial A}{\partial t} + \frac{\partial}{\partial x}\beta\left(\frac{Q^2}{A}\right) + gA\left(\frac{\partial y}{\partial x}\right) + gAS_f - uq = 0 \tag{4}$$

where Q is the flow rate, m^3/s; A is the cross-sectional area of the river crossing, m^2; t is the time, s; x is the horizontal coordinate along the flow direction, m; y is the water level, m; g is the acceleration of gravity; β is the momentum correction factor in degrees; K is the water transfer rate, determined by Manning's formula; S_f is the frictional slope drop; u is the flow rate of the lateral incoming flow in the river direction; q is the lateral incoming flow rate of the river, m^3/s.

It uses the two-dimensional finite volume method to solve the shallow water equations in the simulation of two-dimensional surface diffuse flow by using the TVD excitation technique and the Riemann solver to solve the model computationally. The two-dimensional surface model can effectively and accurately simulate the flow of water on complex urban surfaces and provide support for engineering planning and design [31,33]. The shallow water control equations used in the simulation are as follows:

$$\frac{\partial h}{\partial t} + \frac{\partial (hu)}{\partial x} + \frac{\partial (hv)}{\partial y} = q_{1D} \tag{5}$$

$$\frac{\partial(hu)}{\partial t} + \frac{\partial}{\partial x}\left(hu^2 + \frac{gh^2}{2}\right) + \frac{\partial(huv)}{\partial y} = S_{0,x} - S_{f,x} + q_{1D}u_{1D} \quad (6)$$

$$\frac{\partial(hv)}{\partial t} + \frac{\partial}{\partial x}\left(hv^2 + \frac{gh^2}{2}\right) + \frac{\partial(huv)}{\partial y} = S_{0,y} - S_{f,y} + q_{1D}v_{1D} \quad (7)$$

where h is the water depth, m; u is the velocity component in the x-direction, m/s; v is the velocity component in the y-direction; $S_{0,x}$ is the bottom slope component in the x-direction; $S_{0,y}$ is the bottom slope component in the y-direction; $S_{f,x}$ is the friction component in the x-direction; $S_{f,y}$ is the friction component in the y-direction; q_{1D} is the outflow rate per unit area, m^3/s; u_{1D} is the velocity component of q_{1D} in the x-direction, m/s; v_{1D} is the velocity component of q_{1D} in the y-direction, m/s.

(2) One-dimensional stormwater pipe network data pre-processing

Before constructing a 1D drainage model of the study area, the raw data were pre-processed, for example, by checking the connections to the pipe network. For areas without a drainage network, the discharge of rainwater directly into the nearby mains network was considered. To reduce the calculated pressure, the stormwater pipe network was generalised. For example, pipes with several branches located in the same catchment area were combined into one drainage pipe based on their drainage capacity. After the simplification, the total length of the stormwater pipes was 295.62 km, with a total of 1837 inspection wells and 69 outlets.

(3) Catchment delineation

The overall topographical variation in the study area was not significant and the data from the stormwater network were relatively similar, so the Tyson polygon method was used for sub-catchment delineation. The Tyson polygon method was used to delineate the sub-catchments. All adjacent inspection shafts were joined into a triangle and the perpendicular bisectors of the sides of these triangles were made so that a number of perpendicular bisectors around each inspection shaft formed a polygon. In addition, some sub-basins were manually adjusted according to the layout of the pipe network and field surveys. A total of 1837 sub-catchments were finally delineated.

(4) Determination of model parameters

The flow-producing surfaces were divided into four categories according to land use: building sites, roads, green spaces, and bare soil. Green areas and bare soil are permeable surfaces. The fixed runoff coefficient method was used for impervious surfaces and the Horton method was used for permeable surfaces [32]. The SWMM nonlinear reservoir method was used to simulate the confluence of the study area based on topographic data. The parameters of different types of flow-producing surfaces were also set according to previous research results in similar areas [34–37]. The specific values of the parameters are given in Table 1.

Table 1. Model production sink parameters.

Type of Landuse	Type of Maternal Flow	Production Flow Parameters				Runoff Routing Value
		Fixed Runoff Coefficient	Initial Infiltration Rate	Stable Penetration Rate	Attenuation Factor	
Building site	Fixed	0.9	-	-	-	0.019
Road		0.9	-	-	-	0.02
Green land	Horton	-	76.5	2.5	2	0.13
Bare area		-	65	2.5	2	0.05

(5) Two-dimensional model setup

The gridding interval was set for the study area and the size of the triangular grid was adjusted to meet the study requirements. Considering that the accuracy of the DEM may not reflect the inundation of the road, the grid elevation of the area where the road is located was reduced by 15 cm in order to better simulate the actual road conditions in the study area. In order to be able to reduce the amount of computation as much as possible while meeting the conditions of simulation accuracy, in the modelling process, the calculation grid was encrypted for key areas such as roads, and as large a grid as possible was used for areas with a single land use type. The final triangular grid was divided into 97,112, with a minimum grid area of 20 m^2 and a maximum grid area of 1000 m^2.

(6) Validation of the model

In order to check the applicability of the parameters in the study area, the simulation results need to be validated. A rainfall event with a 50-year rainfall intensity was used as the boundary condition for the model to compare the distribution of the simulated flooding points in the study area with the actual flooding points. The actual distribution of the flooding points was obtained from the information on flood-prone points published by the Zhengzhou traffic department.

Comparing the historical statistical inundation points in red in Figure 2 with the inundation extent of the simulation results, it can be seen from the figure that the inundation points simulated by the model match the distribution of the actual inundation points. Most of these inundation points are located where the duration of inundation is relatively long. The model parameters are therefore considered to be appropriate and the simulation results are reliable.

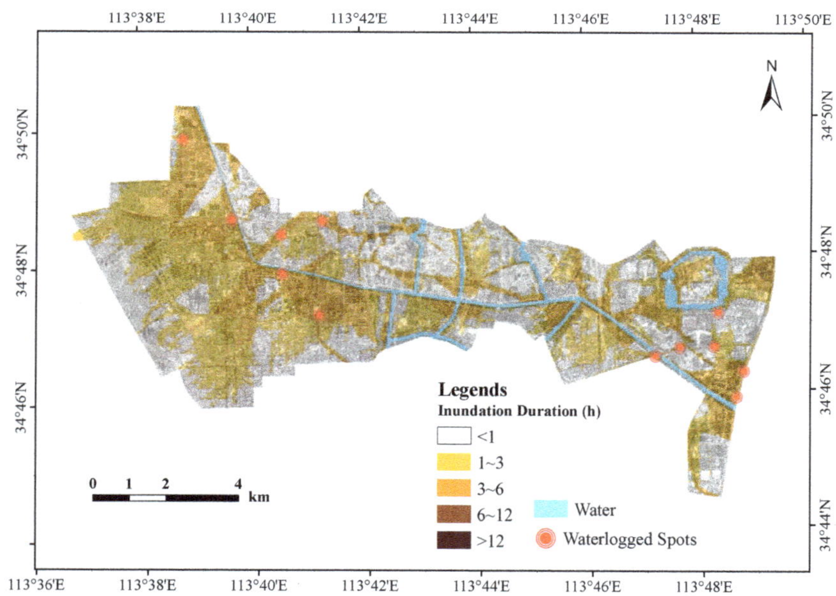

Figure 2. Distribution of simulated results compared to measured water accumulation points.

2.3.2. Scenario Setting

In this study area, several sets of rainfall scenarios were set up based on the storm intensity formula to analyse the flooding impacts of different rainfall scenarios [37]. The storm intensity equations used for the study are as follows:

$$q = \frac{7057.6(1 + 0.794 \lg P)}{(t + 25.8)^{0.948}} \quad (8)$$

where q is the average storm intensity in mm/min; P is the return period in a; t is the rainfall time in min.

The design recurrence period of the pipe network in the study area is 2 a. Combining the historical rainfall and flood control needs of the study area, the flooding situation of the Dongfeng canal area was simulated under the rainstorm scenario, and six rainstorm scenarios with a rainfall duration of 1 h and 2 h for 5 a, 20 a, and 50 a events were designed as the different rainfall scenario driving models. The rainfall process time interval was set to 5 min and the design rainfall process line was obtained according to the rainfall intensity formula and using the ICM design rainfall generator as shown in Figure 3.

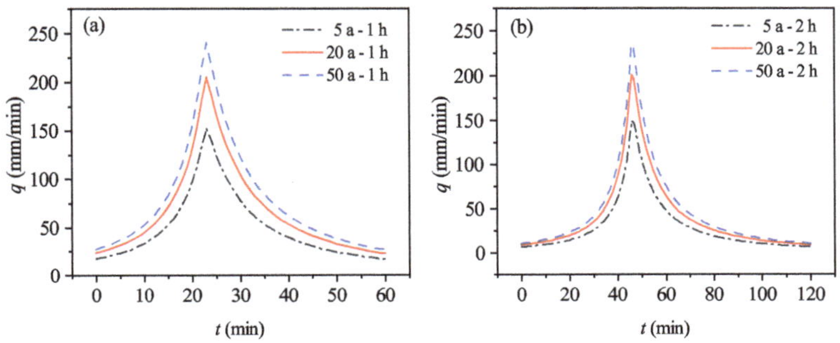

Figure 3. Rainfall processes with different rainfall return periods and different rainfall durations: (**a**) Rainfall duration of 1 h rainfall process. (**b**) Rainfall duration of 2 h rainfall process.

2.3.3. Flood Risk Analysis Methodology

According to the UK Environmental Protection Agency, the flood risk rate is calculated by combining two key physical variables, water depth and flow velocity. During the calculation of flood risk rates, the type of subsurface is also considered to be an important factor [38]. The calculation is as follows:

$$R_H = h(v + 0.5) + \text{CDF} \tag{9}$$

where R_H is the risk rate, with a scale of one; h is the flood inundation depth, m; v is the flood flow velocity, m/s; CDF is the debris factor, i.e., the increased risk factor due to debris carried in the flood.

The CDF is mainly used to increase the weight of the impact of road floats on flood risk and is often used in flood risk analysis [39,40]. The CDF is assigned to 1 if the type of bedding surface is a road or a building site and the flood velocity is greater than 2, whereas the rest are assigned to 0. The flood risk rate value R_H was quantified for any point in the study area. In addition, the flood risk rating was divided into four zones. When R_H is less than 0.75, it means that the area is in a low-risk zone; when Rh is between 0.75 and 1.25, it means a medium-risk zone; when Rh is between 1.25 and 2.00, it means a high-risk zone; and when Rh is greater than 2.00, it means that the current area is in a very-high-risk zone. The flood risk classification is shown in Table 2.

Table 2. Flood risk classification.

Inundation Hazard Index Threshold (R_h)	Risk Level	Description
<0.75	Low risk	Shallow standing water or the presence of shallow static waterlogging
0.75∼1.25	Medium risk	Deep water or fast-flowing water
1.25∼2.0	High risk	Hazardous area with deep water and high flow rates
≥2.0	Very high risk	Very dangerous area, no access

The methodological route of the study is shown in Figure 4.

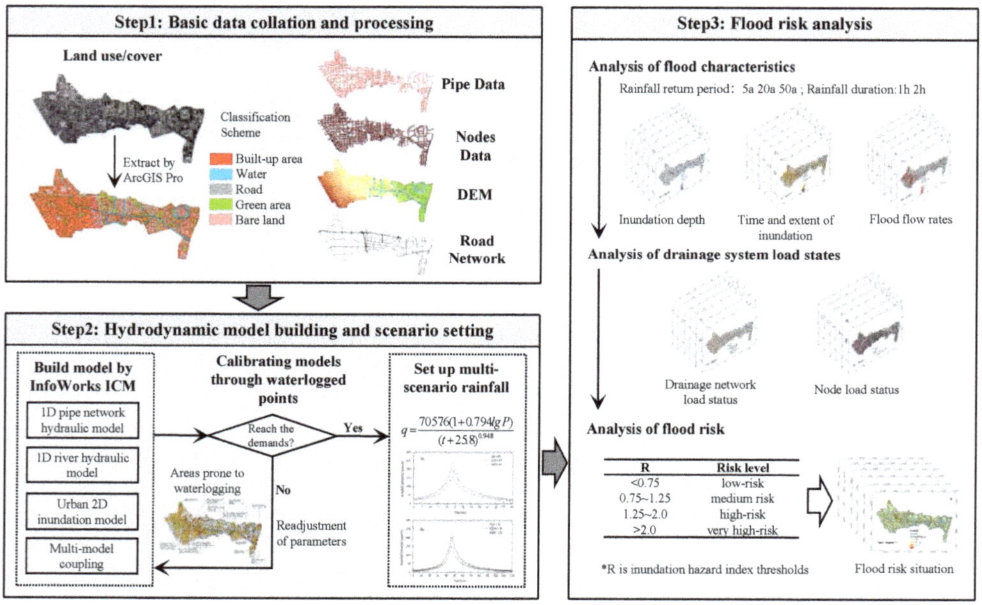

Figure 4. Distribution of simulated results compared to measured water accumulation points.

3. Results and discussion

3.1. Analysis of Urban Flood Simulation Results

3.1.1. Analysis of Flood Inundation Water Depth

As a result of the InfoWorks ICM simulations conducted in the study area, the inundation depths were compared for a variety of rainfall return periods and rainfall ephemerides. Figure 5 illustrates the distribution of the simulated maximum water depth on the ground in the study area. In different rainfall return periods and calendar periods, water accumulated on the ground to a depth of usually less than 0.3 m, followed by 0.3∼0.5 m. The depth of flood inundation increased with the increasing rainfall return period for the same rainfall duration. The depth of flooding increased with the increasing rainfall duration for the same rainfall return period. Overall, the ponded water was primarily concentrated in areas adjacent to rivers and low-lying roads, mainly because the water level of the rivers is higher than the drainage inlets or the terrain of the ponded areas is lower than the surrounding terrain resulting in inundation.

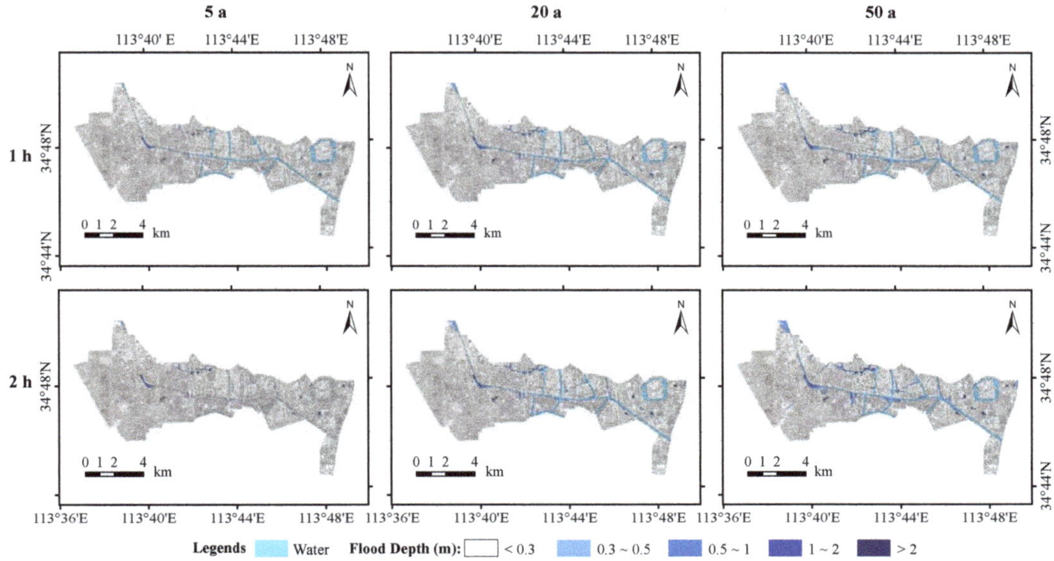

Figure 5. Simulated inundation depth map.

3.1.2. Analysis of the Duration and Extent of Flood Inundation

In addition to the inundation depth, the duration of inundation and the inundation extent are essential indicators for evaluating inundation hazards. The amount of precipitation, the type of subsurface, the drainage capacity of the pipe network, and the drainage capacity of natural watercourses are the main causes of persistent flooding. When precipitation is low, the type of subsurface and the drainage capacity of the pipe network control the extent and duration of flooding. Conversely, when rainfall is relatively high, the water level in the river rises and the stormwater pipe network is unable to re-drain or the water in the river can reverse its flow into the stormwater pipe network. At this point, the drainage capacity of the river becomes an important factor in the occurrence and duration of flooding. Analysis of the flood inundation duration and flood extent, therefore, has an important role to play in flood risk assessment. Figure 6 shows the inundation duration and inundation extent of the study area simulated under different scenarios. The simulation results showed that the inundation duration of the study area was mainly from 0~3 h. At the study area scale, the extent of inundation increased with the duration of rainfall, whereas the extent of flood inundation tended to increase with increasing rainfall return periods. The elevation range of the study area according to the DEM data is 85.05 m 100.34 m. The areas with longer inundation durations are mainly concentrated on the banks of rivers and low-lying areas of the terrain, mainly because the areas with longer inundation durations tend to have deeper inundation, and the areas with deeper inundation generally have lower terrain, which makes it difficult to drain the accumulated rainwater within a short period.

According to the characteristics of precipitation in the study area, normal precipitation can recede within two hours, and precipitation with an inundation time of less than two hours does not affect our normal activities or pose too much of a threat to people's safety. An inundation time of greater than four hours affects people's travel activities. In view of the above, we subdivided the inundation time into five ranges [41]. For each inundation calendar time classification, the inundation area was calculated based on the simulation results for the different scenarios of rainfall recurrence. According to Table 3, under the 1 h rainfall calendar, the inundated areas for the rainfall recurrence periods 5 a, 20 a, and 50 a were 4283.11 ha, 4977.61 ha, and 5238.07 ha, respectively, and the total inundation area increased by 954.96 ha from 5 a to 50 a. The inundated areas under the 2 h rainfall duration were 4453.39 ha, 5079.15 ha, and 5527.19 ha, respectively, with a total increase of 1073.8 ha from 5 a to 50 a.

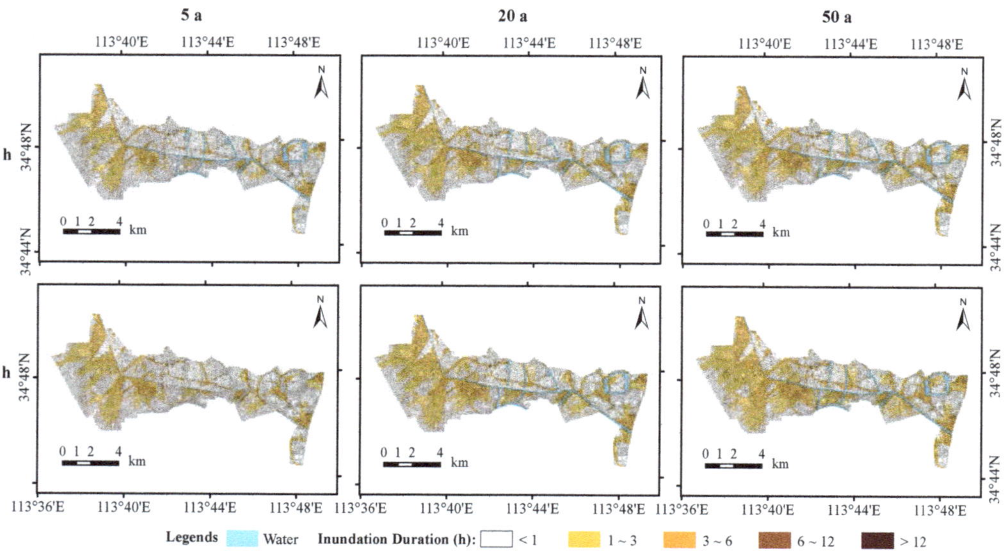

Figure 6. Simulated inundation durations and inundation ranges.

Table 3. Flooded area statistics for different scenarios (ha).

Rainfall Return Period	Duration of Rainfall	Duration of Inundation					Area Inundated
		<1	1~2	2~3	3~4	>4	
5a	60	1734.56	983.55	500.85	431.26	632.89	4283.11
	120	1497.61	1129.96	600.91	786.67	438.24	4453.39
20a	60	1736.95	1264.00	558.58	444.71	973.37	4977.61
	120	1416.91	1374.89	684.36	872.79	730.20	5079.15
50a	60	1696.68	1368.74	590.29	410.83	1171.53	5238.07
	120	1245.03	1539.00	781.14	855.37	1106.65	5527.19

Under the scenarios with rainfall recurrence periods of 5 a, 20 a, and 50 a, the inundated areas with a rainfall duration of 2 h were compared with those with a rainfall duration of 1 h. The simulation results showed that the inundated areas with an inundation duration <1 h decreased by 236.95 ha, 320.04 ha, and 451.65 ha, respectively; the inundated areas with an inundation duration of 1~2 h increased by 146.41 ha, 110.89 ha, and 170.26 ha, respectively; the inundated areas with an inundation duration of 2~3 h increased by 100.06 ha, 125.78 ha, and 190.85 ha, respectively; the inundated areas with an inundation duration of 3~4 h increased by 355.41 ha, 428.08 ha, and 444.54 ha, respectively; and the inundated areas with an inundation duration >4 h decreased by 194.65 ha, 243.17 ha, and 64.88 ha, respectively. The inundated areas with inundation times <1 h and >4 h showed a decrease with increasing rainfall calendar time, indicating that the longer the rainfall calendar time, the faster the drainage of the study area.

The modelling results showed that the inundated areas increased with increasing rainfall return period for the 1–2 h inundation time scenario. For the 1–2 h inundation time scenario, the inundated areas increased by 146.41 ha, 110.89 ha, and 170.26 ha. For inundation durations of 2 to 3 h, the inundated areas increased by 100.06 ha, 125.78 ha, and 190.85 ha. Inundation areas with inundation durations of 3 to 4 h increased by 355.41 hectares, 428.08 hectares, and 444.54 hectares. Inundated areas with inundation durations <1 h and >4 h decreased with increasing rainfall duration, with inundation decreasing by 236.95 ha, 320.04 ha, and 451.65 ha at inundation durations <1 h. The inundated areas with an inundation duration >4 h decreased by 194.65 ha, 243.17 ha, and 64.88 ha. The above results indicate that the longer the rainfall duration, the faster the drainage rate of the study area.

3.1.3. Flood Flow Rate Analysis

Flood flow velocities are an important indicator of urban flood risk conditions [31,42,43]. As shown in Figure 7, the flood flow velocity distributions under different scenarios are shown for the study area. In the study area, the flood flow velocities were usually between 0 and 0.5 m/s for different return periods of rainfall. The flood flow rates tended to increase with increasing rainfall return periods, and increasing rainfall durations also contributed to the increase in the flood flow rates. The areas with a high flood flow velocity are located near the main drainage network's outlet and in densely populated areas. The magnitude of the flood flow can affect the travel and safety of people in the study area. Excessive flood velocities can impede traffic and float objects on the ground, posing a threat to human life. This is particularly important for the subsequent flood risk analysis.

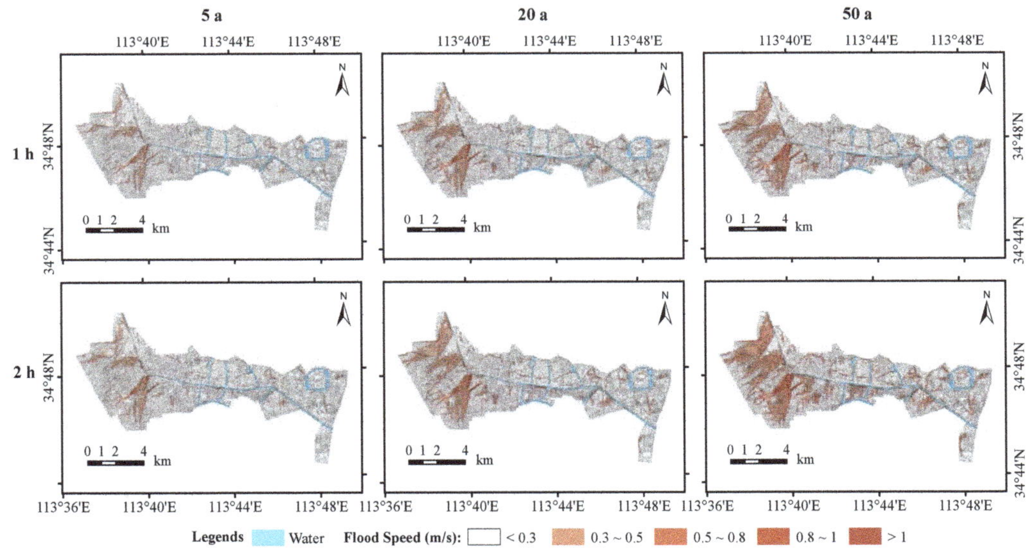

Figure 7. Simulated flood flow map.

3.2. Analysis of Drainage System Load Conditions

The stormwater pipe network in the study area is mainly located in the paving of main and secondary roads such as the North Third Ring Road, Garden Road, Jing San Road, Zhong Zhou Avenue, Sha Men Road, East Huang He Road, and East Dong Feng Road. Stormwater is discharged into the stormwater pipe network through the catchment area and eventually discharged into the Dongfeng Drain after collection by the stormwater pipe network.

The drainage system load conditions were analysed in terms of both the nodal overflow conditions and drainage network load conditions. Figure 8 presents the distribution of the node overflows in the study area simulated under the different scenarios. It can be seen that most of the nodes showed varying degrees of overflow, which is relatively obvious. According to the statistical results in Table 4, the percentage of overflow at the nodes with a rainfall duration of 1 h increased from 79.01% to 87.36%, whereas the percentage of overflow at the nodes with a rainfall duration of 2 h increased from 79.43% to 87.36% under the conditions of 5 a, 20 a, and 50 a rainfall recurrence periods. Under the same recurrence periods, the proportion of nodal overflows with a 2 h rainfall ephemeris increased by 0.42%, 0.26%, and 1.58%, respectively, compared with those of nodal overflows with a 1h rainfall ephemeris. The overall pattern was that the proportion of nodal overflows increased with the increases in the rainfall return period and rainfall ephemeris.

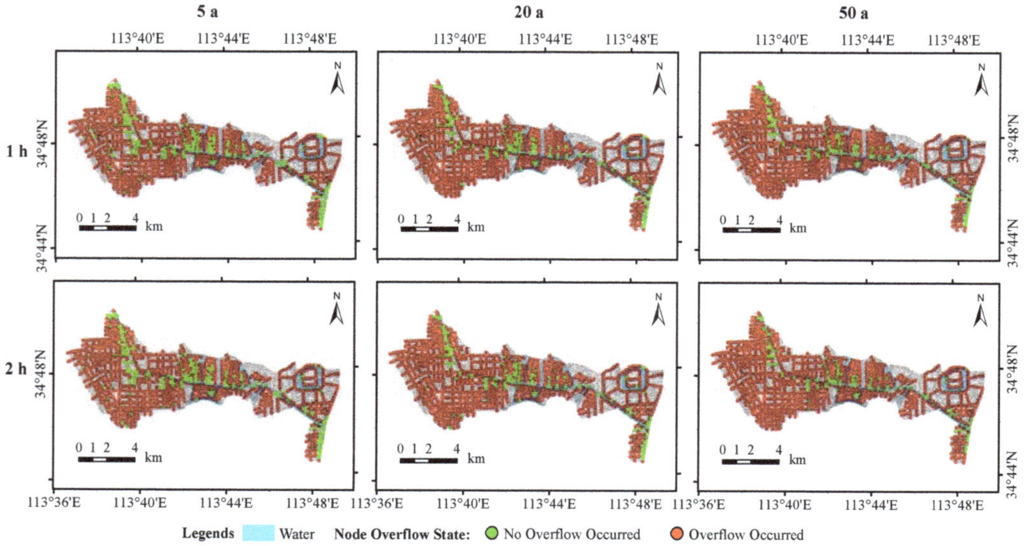

Figure 8. Simulated node overflow condition diagram.

Table 4. Overflow statistics for different scenario nodes.

Rainfall Return Period	Duration of Rainfall (min)	Number of Overflows Occurring at Nodes	Proportion of Overflows Occurring at Nodes
5a	60	1506	79.01%
	120	1514	79.43%
20a	60	1593	83.58%
	120	1598	83.84%
50a	60	1635	85.78%
	120	1665	87.36%

In this study, the overload state of the pipe network was judged according to its water flow state. According to the simulation results, when the overload state was less than 1, it meant that the pipe network was in a non-full gravity flow state, which means that the pipe network was in a normal state. When the overload state was equal to 1, the pipe was in a pressure flow state and the hydraulic gradient < pipe slope. The main reason for this state is the insufficient overflow capacity of the downstream pipeline, which causes the pipeline to be in an overload condition. When the overload state was equal to 2, the pipe was in a pressure flow state and the hydraulic slope > pipeline slope; the pipe was overloaded due to its own drainage capacity being insufficient. The pipe network was generally considered to be overloaded in the case of overload states 1 and 2. As can be seen in Figure 9, the pipe networks in the study area were generally in an overload condition, with the flow overload being the dominant overload pipe network. The length of the overloaded pipe network increased with increasing precipitation return periods.

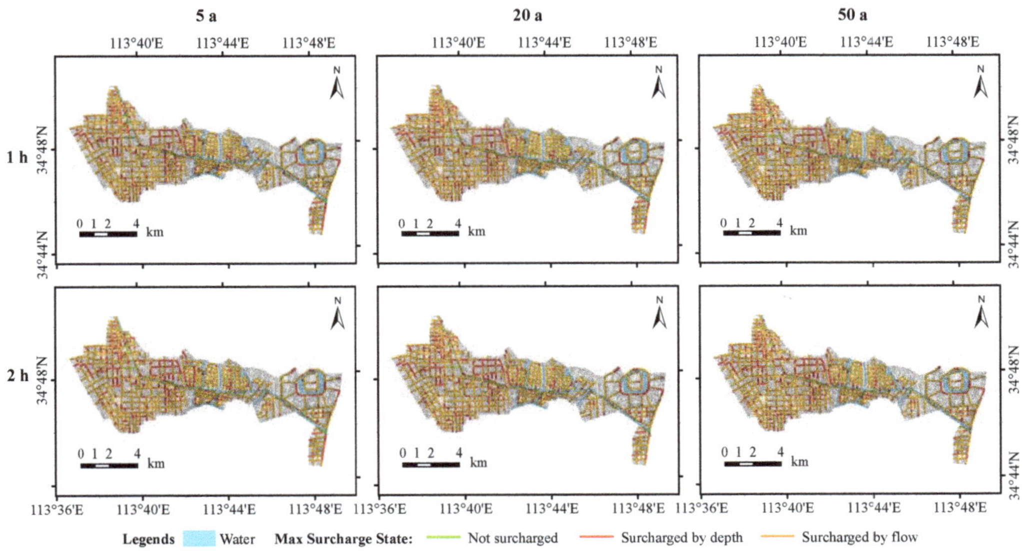

Figure 9. Simulated pipe network load conditions.

According to the overload statistics of the pipe network under rainfall return periods of 5 a, 20 a, and 50 a, as seen in Table 5, the overload water depth lengths of the pipe network with a rainfall duration of 1h were 49.07 km, 42.38 km, and 38.81 km, and the corresponding overload flow lengths of the pipe network were 241.33 km, 249.56 km, and 253.87 km. The overload flow lengths of the pipe network corresponding to 2 h were 235.16 km, 239.91 km, and 245.25 km. The above results show that the length of the overloaded pipe network increased with the increase in the rainfall return time, whereas the length of the overloaded pipe network decreased with the increase in the rainfall recurrence period.

Table 5. Statistics for overloading of the pipe network for different scenarios.

Rainfall Return Period	Duration of Rainfall (min)	Length of Pipe Network Overloaded by Water Depth (km)	Length of Pipe Network with Flow Overload (km)	Total Length of Overloaded Pipe Network (km)
5a	60	49.07	241.33	290.4
	120	55.44	235.16	290.6
20a	60	42.38	249.56	291.94
	120	52.43	239.91	292.34
50a	60	38.81	253.87	292.68
	120	48.17	245.25	293.42

3.3. Urban Flood Risk Analysis

The flood risk rate analysis method proposed by the UK EPA has been used to express the flood risk rate as a combination of two key physical quantities, namely water depth and flow velocity. The type of subsurface is also considered to be an important factor influencing the flood risk rate. Finally, an RH value characterising the flood risk was calculated to characterise the degree of flood risk.

Figure 10 shows the distribution of the flood risk in the study area, simulated under the different scenarios. It can be seen that the distribution of risk is dominated by low-risk areas with very few very high-risk areas. The medium- and high-risk areas are mainly located on the banks of rivers, in built-up areas, and on low-lying urban terrain.

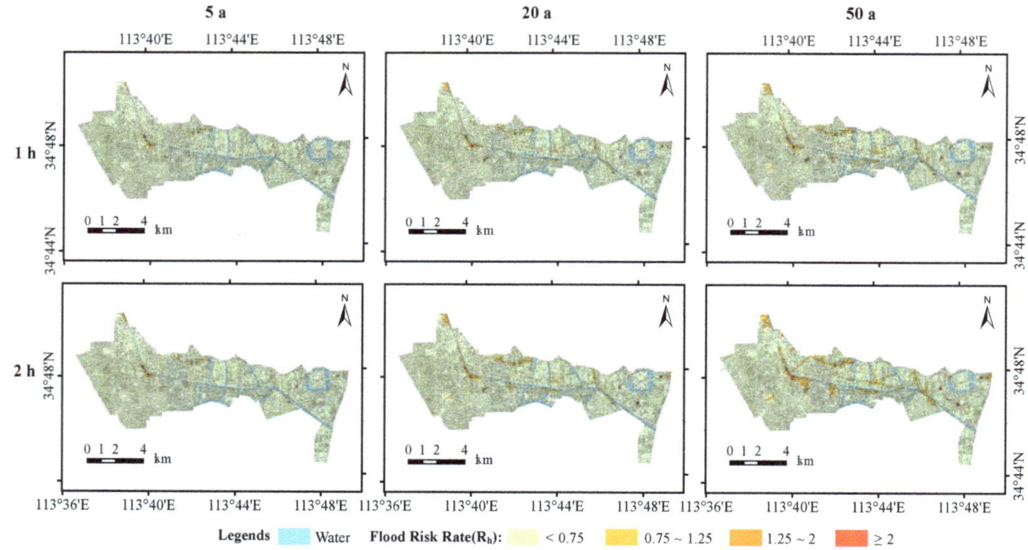

Figure 10. Inundation flood risk for different scenarios in the study area.

A detailed analysis of the change in the flood risk area during the rainfall return period for the 5 a, 20 a, and 50 a scenarios is presented based on the simulation results in Table 6. For a rainfall duration of 1 h, the low-risk areas were 7195.39 ha, 7058.04 ha, and 6959.16 ha and the medium-risk areas were 51.44 ha, 87.18 ha, and 101.74 ha, respectively. At a rainfall time of 2 h, the low-risk areas were 7148.54 ha, 6981.60 ha, and 6740.90 ha; the medium-risk areas were 62.38 ha, 107.33 ha, and 125.89 ha; the high-risk areas were 112.65 ha, 225.05 ha, and 430.37 ha; and the very-high-risk areas were 14.75 ha, 24.33 ha, and 41.16 ha, respectively.

Table 6. Area statistics for inundation flood risk rates for different scenarios (ha).

Rainfall Return Period	Duration of Rainfall (min)	Flood Risk			
		Low Risk	Medium Risk	High Risk	Very High Risk
5 a	60	7195.39	51.44	81.96	9.53
	120	7148.54	62.38	112.65	14.75
20 a	60	7058.04	87.18	173.29	19.81
	120	6981.60	107.33	225.05	24.33
50 a	60	6959.16	101.74	251.47	25.95
	120	6740.90	125.89	430.37	41.16

The shift in the rainfall return periods from the 5 a to the 50 a risk zones was then analysed under the same rainfall calendar. For a rainfall duration of 1 h, the shifts from low- to medium-, high-, and very-high-risk zones were 99.19 ha, 134.43 ha, and 2.61 ha, respectively; from medium- to high- and very-high-risk zones were 48.08 ha and 0.80 ha, respectively; and from high- to very-high-risk zones was 13.01 ha. With a rainfall duration of 2 h, the shifts from the low-risk zone to the medium-, high-, and very-high-risk zones were 124.84 ha, 277.79 ha, and 5.01 ha, respectively; from the medium-risk zone to the high- and very-high-risk zones were 60.30 ha and 1.02 ha, respectively; and from the high-risk zone to the very-high-risk zone was 20.38 ha. The combination of the risk distribution ranges in Figure 10 indicates that the upgraded risk areas are mainly located on both sides of the river and in the study area, which are low-lying and have poor drainage capacity.

There are also some characteristics of the change in the flood risk zones from a 1 h to a 2 h rainfall ephemeris for the same rainfall return period. When the rainfall return period was 5 a, 34.58 ha and 12.32 ha of low-risk zone converted to medium- and high-risk zones, respectively, and 0.05 ha of medium-risk zone converted to low-risk zone, 23.45 ha to high-risk zone, and 0.14 ha to very-high-risk zone. When the rainfall return period was 20 a, the areas of low-risk zone converted to medium- and high-risk zones were 58.31 ha and 18.02 ha, respectively, and the area of medium-risk zone converted to high-risk zone was 38.39 ha. When the rainfall return period was 50 a, the areas transformed from low-risk areas to medium- and high-risk areas were 104.15 ha and 113.66 ha, respectively, and the area transformed from medium-risk area to high-risk area was 79.99 ha. The area transformed from medium-risk zone to high-risk zone was 79.99 hectares. Combined with Figure 10, it appears that the risk areas close to the river are more likely to be upgraded, and the high-risk areas are also generally concentrated near the river.

The above results show that for the same rainfall return period, the area covered by low-risk areas decreased with increasing rainfall calendar time, and most of the reduced low-risk areas transformed into medium- and high-risk areas. The area covered by all the risk classes, except for low-risk areas, increased with increasing rainfall calendar hours. For the same rainfall calendar time, the area covered by low-risk areas decreased with increasing rainfall return periods. The area covered by all risk classes, except for low-risk areas, increased with increasing rainfall return periods. The above analysis indicates that increases in the rainfall return period and rainfall duration will result in more severe flooding. It is also important to focus on high- and very-high-risk areas when undertaking flood management and when flooding occurs. It is important to consider that the flood risk varies with the duration and intensity of rainfall ephemeris. It is therefore also important to focus on flooding in low-risk areas that could easily convert into high-risk areas.

4. Conclusions

In this study, a 1D/2D coupled urban flood model was constructed using InfoWorks ICM software based on data pertaining to pipe networks, inspection wells, roads, water systems, land uses, and elevations in the Dongfeng Canal area of Zhengzhou. Six scenarios were set up according to different rainfall recurrence periods and rainfall ephemeris using the Zhengzhou storm intensity formula. According to the simulation results, the flood inundation depth, inundation extent, duration of inundation, flood flow, and drainage system load were analysed. Finally, the flood risk was quantified and spatially analysed using the flood risk rate analysis method proposed by the UK EPA. The main findings of this study are as follows.

(1) This study uses the InfoWorks ICM model to construct a coupled hydrological-hydraulic model for a small urban watershed area. The two-dimensional hydraulic processes of the urban one-dimensional pipe network, river, and surface are coupled and the model effects are compared using the flood inundation locations of historical precipitation. The results show that the model is effective in simulating rainfall and flooding in the Dongfeng Canal area. The model can be applied to the analysis of flood risk.

(2) According to the simulation results, inundation depths in the study area are mainly 0~0.3 m, followed by 0.3~0.5 m. Inundation is mainly concentrated in areas near rivers and low-lying areas of road topography. The extent of inundation in the study area at a 2 h rainfall ephemeris is greater than that at a 1 h rainfall ephemeris, and the extent of inundation tends to increase with increasing recurrence periods. The extent of inundation at <1 h and >4 h appears to decrease with increasing rainfall ephemeris, indicating that the longer the rainfall ephemeris, the faster the study area drains. The distribution of flood velocities shows a tendency to increase with increasing return periods, with higher velocities being distributed mainly in drainage inlets near the main drainage network and densely built-up areas.

(3) The proportion of nodal overflows occurring increases with increasing rainfall return periods and rainfall ephemeris. The pipe network in the study area is generally overloaded under the different scenarios, with flow overloading dominating and the length of the overloaded pipe network increasing with the rainfall return period. The length of the bathymetric overload pipe network increases with increasing rainfall return periods, whereas the length of the flow overload pipe network decreases with increasing rainfall return periods.

(4) The distribution of the flood risk in the study area is dominated by low-risk areas, with very few very high-risk areas. The medium- and high-risk zones are mainly located on both sides of the river in built-up areas and low-lying urban areas. The simulation results under the different scenarios show that the areas of medium-, high-, and very-high-risk zones increase with the increasing rainfall return period and rainfall duration. The area of low-risk zones decreases with the increasing rainfall return period and rainfall duration, and most of the reduced low-risk zones are transformed into medium- and high-risk zones. Increases in rainfall return periods and rainfall durations can lead to more severe flooding. It is therefore important to focus on high- and very-high-risk areas, as well as low-risk areas that can easily transform into high-risk areas when managing floods when flooding occurs.

This study analyses the flood risk situation in the study area in various aspects and from various perspectives. It can provide a reference for the prevention and control of flooding and the formulation of flood countermeasures in the area and help to improve the management of flood risks in the urban construction process. At the same time, the model lacks further validation due to limited measurement data. The construction of the flood risk indicators is relatively singular and does not take into account more comprehensive factors such as socio-economic factors. Therefore, a more detailed collection of hydrological and socio-economic data in the study area to further improve the accuracy of the model and construct a more comprehensive urban flood risk system will be the focus of our work in the next phase.

Author Contributions: Data curation, methodology, software, visualization, and writing—original draft, H.W. and L.Z.; Conceptualization, funding acquisition, and project administration, supervision, H.W.; investigation, H.W., L.Z. and J.L; resources, H.W. and J.L.; writing—review and editing, H.W. and J.L. All authors have read and agreed to the published version of the manuscript.

Funding: This publication is supported by the National Natural Science Foundation of China (No. 51979107, No. 51909091).

Institutional Review Board Statement: Not applicable.

Informed Consent Statement: Not applicable.

Data Availability Statement: Not applicable.

Acknowledgments: The authors are grateful to the editors and anonymous reviewers for their insightful comments and helpful suggestions.

Conflicts of Interest: The authors declare no conflict of interest.

References

1. Pathirana, A.; Denekew, H.B.; Veerbeek, W.; Zevenbergen, C.; Banda, A.T. Impact of urban growth-driven landuse change on microclimate and extreme precipitation — A sensitivity study. *Atmos. Res.* **2014**, *138*, 59–72. [CrossRef]
2. de Moel, H.; Aerts, J.C.J.H.; Koomen, E. Development of flood exposure in the Netherlands during the 20th and 21st century. *Glob. Environ. Chang.* **2011**, *21*, 620–627. [CrossRef]
3. Qi, W.; Ma, C.; Xu, H.; Chen, Z.; Zhao, K.; Han, H. A review on applications of urban flood models in flood mitigation strategies. *Nat. Hazards* **2021**, *108*, 31–62. [CrossRef]
4. Tan, L.; Yao, W.; Li, L. Direct Economic Loss Assessment of Urban Storm Flood Disasters Based on Bibliometric Analysis. *J. Catastrophol.* **2020**, *35*, 179–185. [CrossRef]
5. Guo, K.; Guan, M.; Yu, D. Urban surface water flood modelling—A comprehensive review of current models and future challenges. *Hydrol. Earth Syst. Sci.* **2021**, *25*, 2843–2860. [CrossRef]

6. Benito, G.; Lang, M.; Barriendos, M.; Llasat, M.C.; Francés, F.; Ouarda, T.; Thorndycraft, V.; Enzel, Y.; Bardossy, A.; Coeur, D.; et al. Use of Systematic, Palaeoflood and Historical Data for the Improvement of Flood Risk Estimation. Review of Scientific Methods. *Nat. Hazards* **2004**, *31*, 623–643. [CrossRef]
7. Leitao, J.P.; Almeida Mdo, C.; Simoes, N.E.; Martins, A. Methodology for qualitative urban flooding risk assessment. *Water Sci. Technol.* **2013**, *68*, 829–838. [CrossRef]
8. Qi, M.; Huang, H.; Liu, L.; Chen, X. Spatial heterogeneity of controlling factors' impact on urban pluvial flooding in Cincinnati, US. *Appl. Geogr.* **2020**, *125*, 102362. [CrossRef]
9. Nkwunonwo, U.C.; Whitworth, M.; Baily, B. A review of the current status of flood modelling for urban flood risk management in the developing countries. *Sci. Afr.* **2020**, *7*, e00269. [CrossRef]
10. Zhao, L.D.; Zhang, T.; Fu, J.; Li, J.Z.; Cao, Z.X.; Feng, P. Risk Assessment of Urban Floods Based on a SWMM-MIKE21-Coupled Model Using GF-2 Data. *Remote Sens.* **2021**, *13*, 4381. [CrossRef]
11. Sidek, L.M.; Jaafar, A.S.; Majid, W.; Basri, H.; Marufuzzaman, M.; Fared, M.M.; Moon, W.C. High-Resolution Hydrological-Hydraulic Modeling of Urban Floods Using InfoWorks ICM. *Sustainability* **2021**, *13*, 10259. [CrossRef]
12. Tabari, H.; Asr, N.M.; Willems, P. Developing a framework for attribution analysis of urban pluvial flooding to human-induced climate impacts. *J. Hydrol.* **2021**, *598*, 126352. [CrossRef]
13. de Almeida, G.A.M.; Bates, P.; Ozdemir, H. Modelling urban floods at submetre resolution: Challenges or opportunities for flood risk management? *J. Flood Risk Manag.* **2018**, *11*, S855–S865. [CrossRef]
14. Bertsch, R.; Glenis, V.; Kilsby, C. Urban Flood Simulation Using Synthetic Storm Drain Networks. *Water* **2017**, *9*, 925. [CrossRef]
15. Gomes Miguez, M.; Peres Battemarco, B.; Martins De Sousa, M.; Moura Rezende, O.; Pires Veról, A.; Gusmaroli, G. Urban Flood Simulation Using MODCEL—An Alternative Quasi-2D Conceptual Model. *Water* **2017**, *9*, 445. [CrossRef]
16. Jamali, B.; Lowe, R.; Bach, P.M.; Urich, C.; Arnbjerg-Nielsen, K.A.; Deletic, A. A rapid urban flood inundation and damage assessment model. *J. Hydrol.* **2018**, *564*, 1085–1098. [CrossRef]
17. Sadeghi, F.; Rubinato, M.; Goerke, M.; Hart, J. Assessing the Performance of LISFLOOD-FP and SWMM for a Small Watershed with Scarce Data Availability. *Water* **2022**, *14*, 748. [CrossRef]
18. Feng, B.; Zhang, Y.; Bourke, R. Urbanization impacts on flood risks based on urban growth data and coupled flood models. *Nat. Hazards* **2021**, *106*, 613–627. [CrossRef]
19. Hofmann, J.; Schuettrumpf, H. Risk-Based and Hydrodynamic Pluvial Flood Forecasts in Real Time. *Water* **2020**, *12*, 1895. [CrossRef]
20. Hu, R.; Fang, F.; Salinas, P.; Pain, C.C. Unstructured mesh adaptivity for urban flooding modelling. *J. Hydrol.* **2018**, *560*, 354–363. [CrossRef]
21. Jamali, B.; Bach, P.M.; Deletic, A. Rainwater harvesting for urban flood management—An integrated modelling framework. *Water Res.* **2020**, *171*, 115372. [CrossRef] [PubMed]
22. Ma, B.; Wu, Z.; Hu, C.; Wang, H.; Xu, H.; Yan, D.; Soomro, S.E.H. Process-oriented SWMM real-time correction and urban flood dynamic simulation. *J. Hydrol.* **2022**, *605*. [CrossRef]
23. Peng, G.; Zhang, Z.; Zhang, T.; Song, Z.; Masrur, A. Bi-directional coupling of an open-source unstructured triangular meshes-based integrated hydrodynamic model for heterogeneous feature-based urban flood simulation. *Nat. Hazards* **2022**, *110*, 719–740. [CrossRef]
24. Su, B.; Huang, H.; Zhu, W. An urban pluvial flood simulation model based on diffusive wave approximation of shallow water equations. *Hydrol. Res.* **2019**, *50*, 138–154. [CrossRef]
25. Tanaka, T.; Kiyohara, K.; Tachikawa, Y. Comparison of fluvial and pluvial flood risk curves in urban cities derived from a large ensemble climate simulation dataset: A case study in Nagoya, Japan. *J. Hydrol.* **2020**, *584*, 124706. [CrossRef]
26. Zhu, X.; Dai, Q.; Han, D.; Zhuo, L.; Zhu, S.; Zhang, S. Modeling the high-resolution dynamic exposure to flooding in a city region. *Hydrol. Earth Syst. Sci.* **2019**, *23*, 3353–3372. [CrossRef]
27. Ge, S.; Ming, X.D. Analysis on Seasonal Variation of Cloud Cover in Zhengzhou. In Proceedings of the AIPR 2021: 4th International Conference on Artificial Intelligence and Pattern Recognition, Xiamen, China, 24–26 September 2021; ACM: New York, NY, USA, 2021, pp. 524–530. [CrossRef]
28. Zhang, S.; Zhang, W.; Wang, Y.; Zhao, X.; Song, P.; Tian, G.; Mayer, A.L. Comparing Human Activity Density and Green Space Supply Using the Baidu Heat Map in Zhengzhou, China. *Sustainability* **2020**, *12*, 7075. [CrossRef]
29. Li, X.; Hou, J.; Pan, Z.; Li, B.; Jing, J.; Shen, J. Responses of urban flood processes to local land use using a high-resolution numeric model. *Urban Clim.* **2022**, *45*, 101244. [CrossRef]
30. Xu, Z.; Cheng, T. Basic theory for urban water management and sponge city—Review on urban hydrology. *J. Hydraul. Eng.* **2019**, *50*, 53–61. [CrossRef]
31. Cheng, T.; Xu, Z.; Hong, S.; Song, S. Flood Risk Zoning by Using 2D Hydrodynamic Modeling: A Case Study in Jinan City. *Math. Probl. Eng.* **2017**, *2017*, 5659197.1–5659197.8. [CrossRef]
32. Ma, B.; Wu, Z.; Wang, H.; Guo, Y. Study on the Classification of Urban Waterlogging Rainstorms and Rainfall Thresholds in Cities Lacking Actual Data. *Water* **2020**, *12*, 3328. [CrossRef]
33. Liu, Y.; Chai, Z.; Guo, X.; Shi, B. A lattice Boltzmann model for the viscous shallow water equations with source terms. *J. Hydrol.* **2021**, *598*, 126428. [CrossRef]

34. He, F.; Hu, C.; Wang, M.; Wang, H.; Li, X. Application of SWMM in Planning and Construction of Urban Drainage System. *Water Resour. Power* **2015**, *33*, 48–53.
35. WANG, H.; Wu, Z.; Hu, C. Rainstorm Waterlogging and Submergence Medel and Its Application in Urban Areas Based on GIS and SWMM. *Yellow River* **2017**, *39*, 31–35+43.
36. Guo, Y.; Li, Y.; Wang, H.; Hu, Y. Research on the Meteorological and Hydrological Coupling Models on Early Warning of Urban Waterlogging in Zhengzhou. *J. China Hydrol.* **2022**, *42*, 1–7. [CrossRef]
37. Zhang, J.; Zhang, H.; Fang, H. Urban waterlogging simulation and rainwater pipe networks sustem evaluation based on SWMM and SCS methon. *Sourth-Water Transf. Water Sci. Technol.* **2022**, *20*, 110–121. [CrossRef]
38. Xu, Z.; Jing, Y. Derivation of Urban Storm Intensity Formula. *J. China Hydrol.* **2014**, *34*, 53–56.
39. Liao, R.; Xu, Z.; Ye, C.; Shu, X.; Yang, D. Simulation of urban waterlogging processes based on SWMM and InfoWorks ICM model: A case study of Dahongmen drainage area in Beijing City. *Water Resour. Prot.* **2022**, *2022*. Available online: https://kns.cnki.net/kcms/detail/32.1356.TV.20220913.1724.006.html (accessed on 5 October 2022).
40. Ye, C.; Xu, Z.; Lei, X.; Li, P.; Ban, C.; Zuo, B. InfoWorks ICM flood simulation and risk analysis: Case of Baima River district, Fuzhou. *J. Beijing Norm. Univ. Sci.* **2021**, *57*, 784–793. [CrossRef]
41. *Guidelines for Flood Risk Mapping*; Ministry of Water Resources of the People's Republic of China: Beijing, China, 2017.
42. Dong, B.; Xia, J.; Zhou, M.; Li, Q.; Ahmadian, R.; Falconer, R.A. Integrated modeling of 2D urban surface and 1D sewer hydrodynamic processes and flood risk assessment of people and vehicles. *Sci. Total Environ.* **2022**, *827*, 154098. [CrossRef]
43. Mani, P.; Chatterjee, C.; Kumar, R. Flood hazard assessment with multiparameter approach derived from coupled 1D and 2D hydrodynamic flow model. *Nat. Hazards* **2014**, *70*, 1553–1574. [CrossRef]

remote sensing

Article

Prediction of Future Spatial and Temporal Evolution Trends of Reference Evapotranspiration in the Yellow River Basin, China

Shengqi Jian [1], Aoxue Wang [1], Chengguo Su [1,*] and Kun Wang [2]

1 Yellow River Laboratory, Zhengzhou University, Zhengzhou 450001, China
2 China Institute of Water Resources and Hydropower Research, Beijing 100038, China
* Correspondence: cgsu@zzu.edu.cn

Abstract: Reference evapotranspiration (ET_0) is an integral part of the regional hydrological cycle and energy balance and is extremely sensitive to climate change. Based on temperature data from 24 global climate models (GCMs) in the Coupled Model Intercomparison Project Phase 6 (CMIP6), this study developed a multi-model ensemble based on delta statistical downscaling with multiple interpolation methods and evaluation indicators to predict the spatial and temporal evolution trends of ET_0 in the Yellow River Basin (YRB) under four emission scenarios (SSP126, SSP245, SSP370, and SSP585) for the near- (2022–2040), mid- (2041–2060), and long- (2081–2100) term future. Results demonstrate that regional data generated based on delta statistical downscaling had good simulation performance for the monthly mean, maximum, and minimum temperatures in the YRB, and the developed multi-model ensemble had better simulation capability than any single model. Compared to the historical period (1901–2014), the annual ET_0 showed a highly significant increase for different future emission scenarios, and the increase is faster with increasing radiative forcing. The first main cycle of ET_0 change was 52, 53, 60, and 48 years for the SSP126, SSP245, SSP370, and SSP585, respectively. ET_0 in the YRB had positive values for EOF1 under all four emission scenarios, responding to a spatially consistent trend across the region. Compared to the historical period, the spatial distribution of ET_0 under different future emission scenarios was characterized by being larger in the west and smaller in the east. As the radiative forcing scenario increased and time extended, ET_0 significantly increased, with a maximum variation of 112.91% occurring in the western part of the YRB in the long-term future under the SSP585 scenario. This study can provide insight into the water cycle patterns of watersheds and scientific decision support for relevant departments to address the challenges of climate change.

Keywords: reference evapotranspiration; CMIP6; delta statistical downscaling; Hargreaves model; Yellow River Basin; EOF analysis

Citation: Jian, S.; Wang, A.; Su, C.; Wang, K. Prediction of Future Spatial and Temporal Evolution Trends of Reference Evapotranspiration in the Yellow River Basin, China. *Remote Sens.* **2022**, *14*, 5674. https://doi.org/10.3390/rs14225674

Academic Editors: Alban Kuriqi and Luis Garrote

Received: 15 October 2022
Accepted: 7 November 2022
Published: 10 November 2022

1. Introduction

Climate change and its impacts on the water cycle, particularly on regional hydrological systems, are major global challenges in the 21st century [1–3]. As an important factor in the regional hydrological cycle and energy balance, reference evapotranspiration (ET_0) can be used to make total energy estimates of actual evapotranspiration [4], and is the component of the water cycle that is directly affected by climate change. Changes in ET_0 have a significant impact on the global water cycle and water resources [5], thereby leading to droughts and floods, water scarcity, and ecosystem degradation. In the context of climate change, ET_0 is an important guide for understanding the hydrological cycle and formulating water resource plans in watersheds [6–8].

Although studies on ET_0 have been conducted recently [9–14], most existing studies focused on the historical period. With the development of global climate models (GCMs), exploring the future ET_0 of watersheds based on historical data has become a topic of research interest in the context of climate change. GCMs are the most powerful tools for climate change modeling and future predictions [15,16], and the modeling results can provide valuable data to support studies on climate change-induced impacts at regional and continental scales. Nevertheless, low-resolution data will lead to large biases in the prediction of regional climate change when climate studies are conducted at regional scales. Downscaling is an effective method for transforming large-scale, low-resolution outputs from GCMs into small-scale, high-resolution regional ground information [17,18]. Current mainstream downscaling methods include dynamic downscaling [19] and statistical downscaling [20]. Compared to statistical downscaling, dynamic downscaling requires a large number of complex inputs and computational requirements [21,22], and sometimes fine and reliable climate data at regional scales are not available [22,23]. Statistical downscaling is the most widely used and established downscaling technique in basin climate change studies because of its low computational cost, easy model construction, multiple implementation methods, ease of operation, and lack of consideration of the influence of boundary conditions on prediction results [24].

Two methods are widely used for ET_0 prediction under future climate scenarios: (1) input of future meteorological data from GCMs into ET_0 models [25,26]; and (2) directly predicting future ET_0 via downscaling methods based on historical ET_0 [1,27]. Liu et al. [28] used Coupled Model Intercomparison Project Phase 5 (CMIP5) and Coupled Model Intercomparison Project Phase 6 (CMIP6) climate models to compare global potential evapotranspiration and found that both models could effectively simulate the increasing trend; they also revealed that CMIP6 multi-model results simulated higher values of global potential evapotranspiration than CMIP5 for the same emission scenario. Nistor et al. [29] assessed the impact of climate change on ET_0 in Turkey in the 21st century based on the Thornthwaite equation and the CMIP5 dataset. They revealed that ET_0 will increase in southern and southeastern Turkey and along the Mediterranean coast in the coming period owing to climate warming.

The Yellow River Basin (YRB) is an important component of China's strategic ecological security pattern, and most of the YRB is an arid and semi-arid region. Because of its unique geographical location, its environment is fragile and highly sensitive to global climate change [21,30], making it a good indicator of climate change. Despite the high sensitivity of the region to climate change, studies on the evolution of ET_0 in the YRB in the context of future climate are limited, and most of the existing studies on future ET_0 in other regions are at the CMIP5 stage [21,30,31], with no downscaling treatment [1], a single spatial interpolation method [21], or a single indicator for climate model preferences [32]. Therefore, against the backdrop of global warming, the ET_0 predictions in the YRB can provide a theoretical reference basis for water resource planning and management, as well as a scientific basis for relevant authorities to formulate future climate change response strategies.

This study used the YRB as the study area and developed a multi-model ensemble based on the delta statistical downscaling using multiple interpolation methods and multiple evaluation indicators to predict the spatial and temporal evolution characteristics of ET_0 in the YRB under different CMIP6 emission scenarios. Studies on ET_0 not only enhance the understanding of hydrological processes in the YRB but also provide data to support and guide future water resource management and drought mitigation. The specific objectives are to: (1) obtain monthly mean, maximum, and minimum temperature datasets in the YRB with a resolution of 1×1 km based on CMIP6 climate model data and delta statistical downscaling; (2) select the best simulated climate model and multi-model ensemble by evaluating and validating historical measured data; and (3) predict the spatial and temporal changes in ET_0 under different emission scenarios in the future based on the Hargreaves formula and downscaled temperature data from 2022 to 2100.

2. Materials and Methods

2.1. Study Area

The Yellow River, the second-largest river in China, starts from the Bayankara Mountains in the west, crosses the Qinghai–Tibetan Plateau, Loess Plateau, and Huang-Huai-Hai Plain, and flows into the Bohai Sea in Shandong Province. The total length of the basin is 5464 km, covering an area of approximately 79.5×10^4 km^2 (95°53′E–119°05′E, 32°10′N–41°50′N), accounting for 8% of China's land area (Figure 1). The YRB is in the mid-latitude zone, with complex natural conditions and undulating basin topography, and is influenced by atmospheric and monsoonal circulation, making the climate different from that of the other basins in China [33,34]. The average annual precipitation in the basin is 495 mm, with concentrated and highly variable interannual precipitation and an evident downward trend from the southeast to the northwest [35]. The average annual temperature ranges from −4 to 14 °C, varying with latitude and altitude [36]. The basin's evapotranspiration varies markedly, with an average annual ET_0 of 700–1800 mm and an increasing trend from the southeast to the northwest. As the basin straddles arid, semi-arid, and semi-humid zones, it is in the transition zone between semi-arid and semi-humid climates, rendering it extremely sensitive to climate change [21]. Climate change has exacerbated the uneven spatial and temporal distributions of water resources in the YRB, and the contradiction between water resource supply and demand has become evident, seriously affecting the production and life of human society and restricting the high-quality economic development of the region.

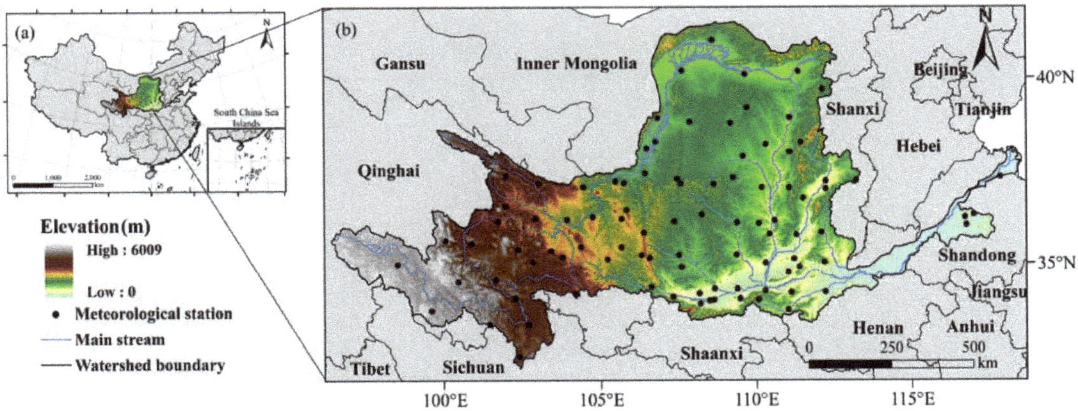

Figure 1. (**a**) Location and digital elevation model of the Yellow River Basin in China and (**b**) the distribution of 93 national meteorological stations in the Yellow River Basin.

2.2. Data Collection

2.2.1. Ground-Based Observation Data

In this study, monthly monitoring data from 93 national meteorological stations in the YRB from 1980 to 2014 were obtained from the National Meteorological Information Centre-China Meteorological Data Network (http://data.cma.cn/ (accessed on 11 March 2022)), including monthly mean temperature (tas), monthly mean maximum temperature (tasmax), monthly mean minimum temperature (tasmin), and monthly pan evaporation. Some of the missing data were reasonably interpolated via the hydrologic analogy method and the linear interpolation method. The tas, tasmax, and tasmin were used to assess the accuracy of the climate model simulations, and the converted value based on pan evaporation data [4] were used to assess the ET_0 values based on the multi-model ensemble and Hargreaves formula.

2.2.2. Reference Data on Downscaling

The national 30-year cumulative mean, mean maximum, and mean minimum temperature datasets with a resolution of 1 km from 1971 to 2000 were selected as the regional high-resolution reference data to construct the delta statistical downscaling model in this study. Data were obtained from the National Ecosystem Science Data Center (NESDC) (http://www.nesdc.org.cn/ (accessed on 9 May 2022)).

2.2.3. Future Climate Data

In this study, 24 GCMs were selected from CMIP6 (https://esgf-node.llnl.gov/search/cmip6/ (accessed on 13 May 2022)) for the historical period (1901–2014) and three future periods (near-term 2022–2040, mid-term 2041–2060, and long-term 2081–2100). The tas, tasmax, and tasmin data of the models were presented, which contained 21, 19, and 21 GCMs, respectively; the basic details about each model and variable are summarized in Table 1. For future forcing scenarios, the recent shared socioeconomic pathways (SSPs), such as SSP1-2.6 (low-forcing scenario, SSP126), SSP2-4.5 (medium-forcing scenario, SSP245), SSP3-7.0 (medium to high-forcing scenario, SSP370), and SSP5-8.5 (high-forcing scenario, SSP585), were selected [37]. Notably, the future scenarios of the climate model were set for the 2015–2100 period; the historical period in this study did not extend back to 2021, and the future period did not extend forward to 2015 to ensure the reasonability of the data. The selection of the periods for downscaling the simulation accuracy and Hargreaves model validation were based on these considerations.

Table 1. Introduction to climate models with temperature variables.

Serial Number	Climate Models	Variables	Research Institution, Country	Spatial Resolution
1	ACCESS-CM2	tasmax, tasmin	ACCESS, Australia	1.9° × 1.3°
2	ACCESS-ESM1-5	tas, tasmax, tasmin	ACCESS, Australia	1.9° × 1.3°
3	AWI-CM-1-1-MR	tas, tasmin	AWI, Germany	0.9° × 0.9°
4	AWI-ESM-1-1-LR	tasmax, tasmin	AWI, Germany	1.9° × 1.9°
5	BCC-CSM2-MR	tas, tasmax, tasmin	BBC, CMA, China	1.125° × 1.125°
6	BCC-ESM1	tasmax, tasmin	BBC, CMA, China	2.8° × 2.8°
7	CanESM5	tas, tasmax, tasmin	CCCMA, Canada	2.8125° × 2.8125°
8	CMCC-CM2-SR5	tas	CMCC, Italy	1.250° × 0.938°
9	CMCC-ESM2	tas, tasmin	CMCC, Italy	1.25° × 0.9375°
10	E3SM-1-0	tas	LLNL, ANL, LANL, LBNL, ORNL, PNNL, SNL, U.S.A	1° × 1°
11	EC-Earth3	tas, tasmax, tasmin	EC-Earth, 10 European countries	0.7° × 0.7°
12	EC-Earth3-Veg	tas, tasmax, tasmin	EC-Earth, 10 European countries	0.703° × 0.703°
13	FGOALS-f3-L	tas	IAP, CAS, China	1° × 1.25°
14	FIO-ESM-2-0	tas, tasmax, tasmin	FIO, China	0.9424° × 1.25°
15	GFDL-ESM4	tas, tasmax, tasmin	GFDL, U.S.A	1° × 1.25°
16	GISS-E2-1-G	tas, tasmax, tasmin	NASA-GISS, U.S.A	1° × 1.25°
17	INM-CM5-0	tas, tasmax, tasmin	INM, Russia	2° × 1.5°
18	IPSL-CM6A-LR	tas, tasmax, tasmin	IPSL, France	1.2676° × 2.5°
19	MIROC6	tas, tasmax, tasmin	MIROC, Japan	1.389° × 1.406°
20	MPI-ESM-1-2-HAM	tas, tasmax, tasmin	MPI, Germany	1.865° × 1.875°
21	MPI-ESM1-2-HR	tas, tasmax, tasmin	MPI, Germany	0.9375° × 0.9375°
22	MPI-ESM1-2-LR	tas, tasmax, tasmin	MPI, Germany	1.875° × 1.875°
23	MRI-ESM2-0	tas, tasmax, tasmin	MRI, Japan	1.124° × 1.125°
24	NESM3	tas, tasmax, tasmin	NUIST, China	1.865° × 1.875°

Note: In the variable column, tas is the average temperature, tasmax is the average maximum temperature, and tasmin is the average minimum temperature.

2.3. Research Methodology

2.3.1. Delta Statistical Downscaling

The delta statistical downscaling method is a simple bias correction technique recommended by the U.S. Global Change Research Program (see http://www.nacc.usgcrp.gov (accessed on 6 June 2022)) that is easy to understand and operate, requires fewer factors, and is widely used in a wide range of fields related to climate change impact studies [21,38,39]. For the temperature variables used in this study, the delta method was used to compare the temperature of different periods of each simulation grid with the simulated average temperature of the base period, calculate the absolute change in temperature in each period of each simulation grid, and add the measured average temperature of each base period with the change in the grid based on the spatial interpolation of the change to obtain the temperature scenarios of different periods in the reconstruction grid [21]. The calculation equation is as follows:

$$T_f = T_0 + \left(T_{Mf} - T_{M0}\right) \tag{1}$$

where T_f is the grid temperature data reconstructed by the delta method, T_{Mf} is the simulated grid temperature data for a certain period, T_{M0} is the simulated grid multi-year average temperature data for the base period, and T_0 is the measured multi-year average temperature data for the base period. In this study, five interpolation methods were considered: bilinear interpolation (BI), inverse distance weighted (IDW), kriging, natural neighbor interpolation (NNI), and spline. The delta statistical downscaling process is shown in Figure 2.

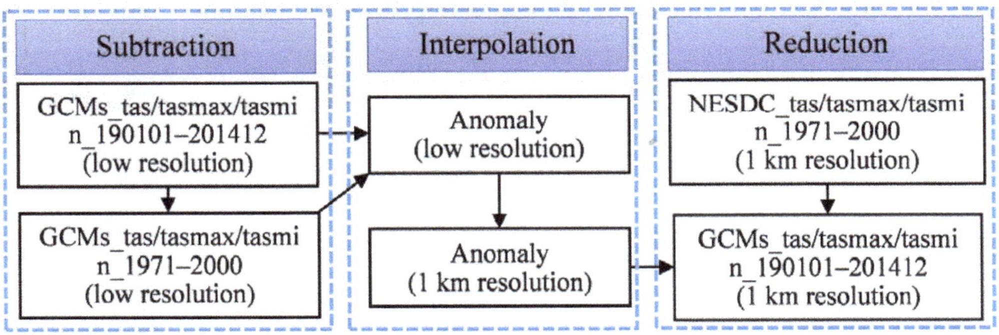

Figure 2. The delta downscaling process of the climate models over the Yellow River Basin.

2.3.2. Climate Model Accuracy Assessment and Multi-Model Ensemble

To effectively assess the accuracy and applicability of climate model forecasts in the YRB, the evaluation metrics used were mean absolute error (MAE) [40], Taylor diagram-based quantile S [41], spatial skills score (SS) [42], and temporal skills score (TS) [43]. The closer the MAE and TS are to 0, the better the simulation ability of the model. The closer S and SS are to 1, the better the simulation ability of the model.

The downscaling results of different GCMs differ, and the performance of multi-model averaging is considered to be better than that of individual models [44,45]. In this study, multi-model ensemble averaging of preferred climate models was performed using the equally weighted ensemble averaging (MME) method commonly used in multi-model prediction studies.

2.3.3. ET_0 Calculation Model

The Hargreaves model, which can reveal the physics of the evaporative process, was used for calculating monthly ET_0 based on a future climate. It has been widely demonstrated to be able to provide reliable estimations [31,46,47]. The Food and Agriculture Organization of the United Nations (FAO) [4] also suggests it as the simplified standard algorithm for ET_0 under the missing meteorological observations as required by the Penman–Monteith formula. Several improvements were made to the original equation [48,49]. The form used in this study was published in 1985 [50] and expressed as follows:

$$ET_{0(Har)} = 0.0023R_a(T_{max} - T_{min})^{0.5}(T_{mean} + 17.8) \quad (2)$$

where $ET_{0(Har)}$ is the ET_0 calculated by the Hargreaves empirical formula (mm/d); T_{max} and T_{min} are the mean maximum and minimum temperatures for the calculation time interval (°C), respectively; T_{mean} is the mean temperature for the calculation time interval (°C); and R_a is the zenith radiation (mm/d); the applicable time scale is 30 or 10 days.

2.3.4. Methods for Spatial and Temporal Trend Analysis

In this study, Morlet wavelet analysis was used to study the significant cycle changes in ET_0 time series at different time scales, which has significant advantages in revealing the multiscale configuration and main climate change cycle [51]. Empirical orthogonal function (EOF) analysis produces a set of orthogonal spatial and temporal patterns in the order of explained variance, reducing the dimensionality of the analyzed system and finding relatively few independent variables that can provide comprehensive information about the variability of the raw data [52]. EOF analysis, supplemented by the North test [53], was used to study the spatial and temporal patterns of the ET_0 climate field in the YRB. In addition, spatial changes in the near-, mid-, and long-term future relative to the historical period were estimated by comparing historical long-term (1901–2014) annual ET_0 averages for different emission scenarios of ET_0.

3. Results

3.1. Simulation Accuracy Assessment of Regional Temperatures and Multi-Model Ensemble

For the three temperature variables, tas, tasmax, and tasmin, out of all interpolation methods, IDW exhibited the lowest overall error for all four different evaluation metrics, and therefore, it was used to further assess the effectiveness of the temperature simulations for all climate models (Figures 3–5). For tas, the climate models ACCESS-ESM1-5, CMCC-CM2-SR5, and GISS-E2-1-G were ranked high for each evaluation index; for tasmax, ACCESS-CM2, NESM3, and ACCESS-ESM1-5 were ranked high for each evaluation index; for tasmin, ACCESS-CM2, ACCESS-ESM1-5, and GISS-E2-1-G were ranked high for each evaluation index. Because some climate models do not have data on future emission scenarios, considering the data integrity and actual fitting performance of climate models, tas selected ACCESS-ESM1-5, CMCC-CM2-SR5, and INM-CM5-0; tasmax selected ACCESS-CM2, ACCESS-ESM1-5, and MRI-ESM2-0; and tasmin selected ACCESS-CM2, ACCESS-ESM1-5, and MPI-ESM1-2-LR for the ensemble analysis of subsequent temperature patterns in the YRB. The fitting results of the simulated data of the selected climate model for each variable and MME of the 1995–2014 period to the observed data from 93 ground-based meteorological stations (Figure 6) show that, although the simulated tas, tasmax, and tasmin of the YRB by CMIP6 were slightly lower than the observed values, most of the points were near the 1:1 line, and the R^2 and regression coefficients were greater than 0.9, with all passing the 99% confidence level test. The simulated and observed values of the MME dataset were more concentrated than those of the three independent climate models, and the R^2 values of the three variables were 0.9418, 0.9226, and 0.9362, respectively (Figure 6d,h,l), which reduced the errors caused by outlier points and slightly improved the fit of the simulated data. The above analysis reveals that climate models have high application potential in the YRB, and the CMIP6 multi-model ensemble is a good reference value for predicting ET_0 trends in the YRB under future climate scenarios.

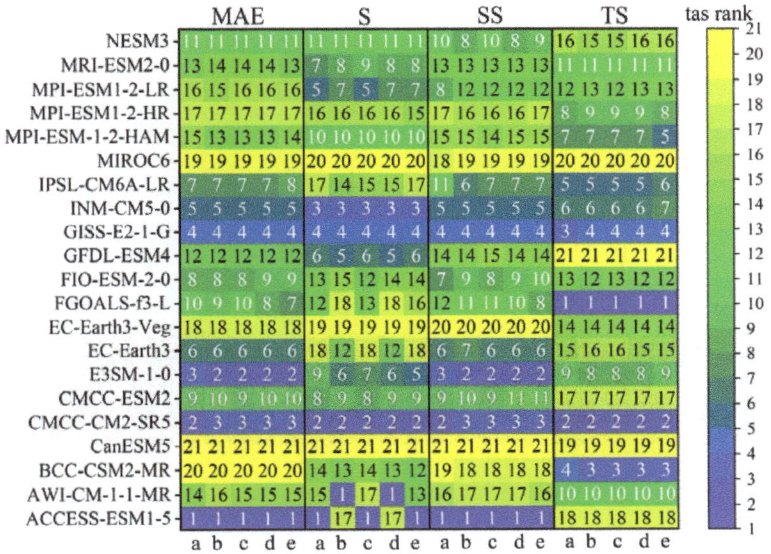

Figure 3. Fitting performance ranking of 21 climate models simulating tas monthly series of the Yellow River Basin from 1995 to 2014 under different interpolation methods and evaluation metrics. a, b, c, d, and e under each evaluation method are the evaluation rankings of the five interpolation methods: BI, IDW, kriging, NNI, and spline, respectively. The higher the ranking, the higher the model's fitting accuracy.

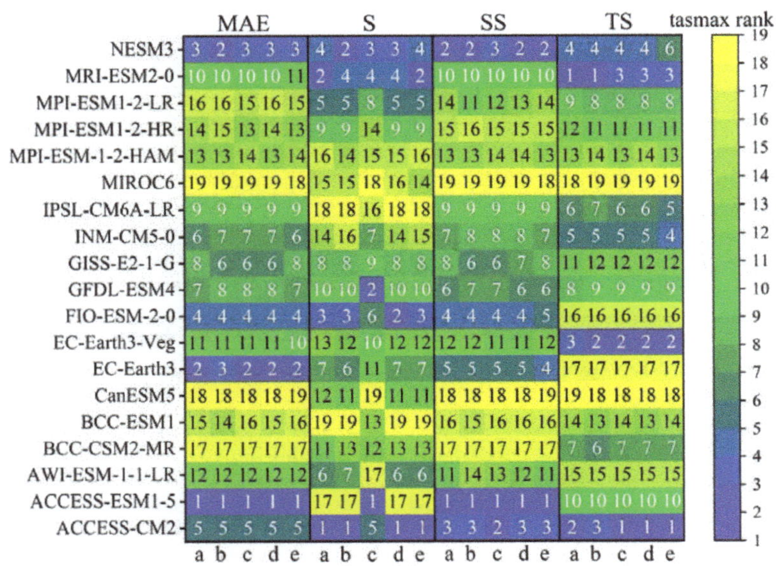

Figure 4. Fitting performance ranking of 19 climate models simulating tasmax monthly series of the Yellow River Basin from 1995 to 2014 under different interpolation methods and evaluation metrics. a, b, c, d, and e under each evaluation method are the evaluation rankings of the five interpolation methods: BI, IDW, kriging, NNI, and spline, respectively. The higher the ranking, the higher the model's fitting accuracy.

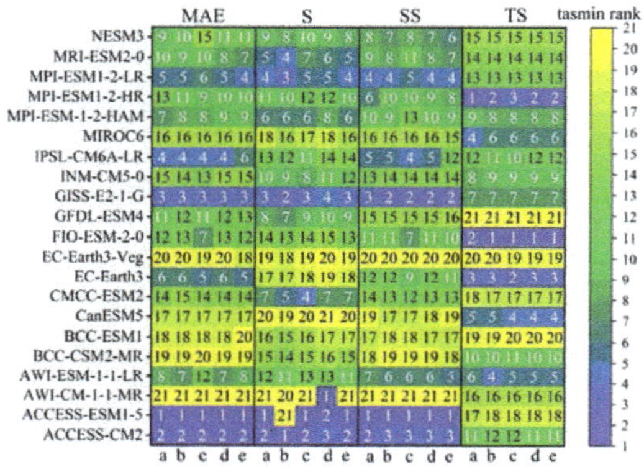

Figure 5. Fitting performance ranking of 21 climate models simulating the tasmin monthly series of the Yellow River Basin from 1995 to 2014 under different interpolation methods and evaluation metrics. a, b, c, d, and e under each evaluation method are the evaluation rankings of the five interpolation methods: BI, IDW, kriging, NNI, and spline, respectively. The higher the ranking, the higher the model's fitting accuracy.

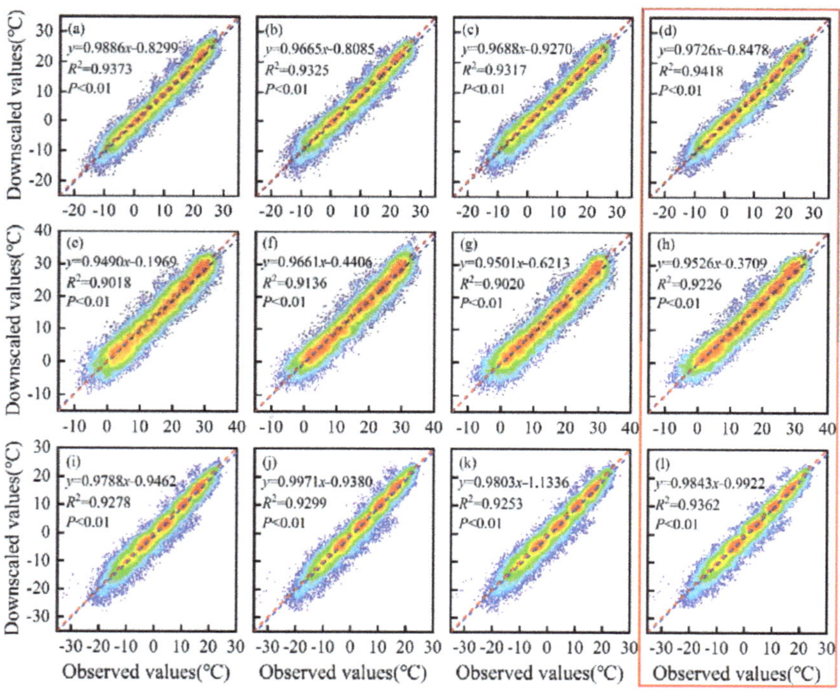

Figure 6. Scatter density plots of measured and selected climate models and multi-model ensembles simulating monthly temperatures in the Yellow River Basin at stations from 1995 to 2014. (**a**–**c**) correspond to the fitting effects of ACCESS-ESM1-5, CMCC-CM2-SR5, and GISS-E2-1-G, respectively, under tas variable; (**d**) corresponds to the fitting effect of MME_{tas}; (**e**–**g**) correspond to the fitting effects of ACCESS-CM2, ACCESS-ESM1-5, and MRI-ESM2-0, respectively, under tasmax variable; (**h**) corresponds to the fitting effect of MME_{tasmax}; (**i**–**k**) correspond to the fitting effects of ACCESS-CM2, ACCESS-ESM1-5, and MPI-ESM1-2-LR, respectively, under tasmin variable; and (**l**) corresponds to the fitting effect of MME_{tasmin}.

3.2. Simulation Accuracy Assessment of Regional ET_0

The scatter density plots of the actual monthly pan evaporation converted values ($ET_{0(pan)}$) and simulated ET_0 values based on the multi-model ensemble-Hargreaves formula ($ET_{0(Har)}$) for the YRB from 1980 to 2014 (Figure 7) showed that the CMIP6-simulated $ET_{0(Har)}$ for the YRB correlated well with the evaporation dish converted $ET_{0(pan)}$ at the monthly scale, with a fitted line regression coefficient of 0.9439 and R^2 of 0.8523, passing the 99% confidence level test. Although the Hargreaves formula partially underestimated ET_0, most of the points were near the 1:1 line, making them good reference values for calculating the monthly ET_0 under the conditions of missing meteorological observations in the future. Therefore, future ET_0 can be simulated using the Hargreaves formula.

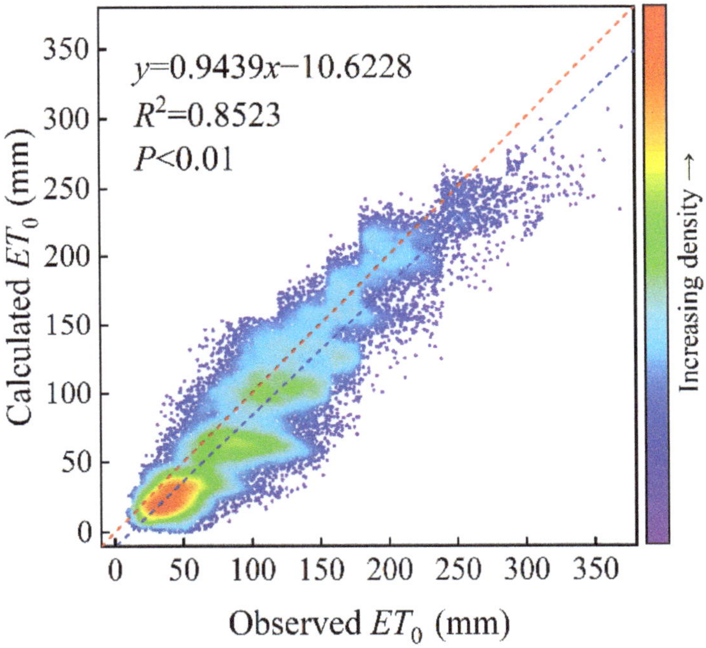

Figure 7. Scatter density plot of monthly observed pan evaporation converted values ($ET_{0(pan)}$) and multi-model ensemble simulations ($ET_{0(Har)}$) in the Yellow River Basin from 1980 to 2014.

3.3. Temporal Trends and Cyclic Characteristics of ET_0

Based on the 1 km high-resolution regional climate data generated by the delta statistical downscaling and multi-model ensemble, annual ET_0 trends were estimated for the YRB from 1901 to 2100 (Figure 8). The annual ET_0 of the basin under the different future emission scenarios (2022–2100; SSP126, SSP245, SSP370, and SSP585) substantially increased with time relative to the historical period (1901–2014), and all passed the 99% confidence test. Among all scenarios, the SSP585 scenario had the most pronounced upward trend in ET_0, increasing at a rate of 22.9 mm/10a, reaching 1170.39 mm in 2100. It was followed by the SSP370 scenario, which increased at a rate of 16.6 mm/10a, reaching 1120.42 mm in 2100. However, SSP245 and SSP126 scenarios had relatively small trends in ET_0, increasing at rates of 10.4 and 3.3 mm/10a, reaching 1062.71 and 987.68 mm in 2100, respectively. In general, the annual ET_0 in the YRB will rapidly increase with increasing levels of radiative forcing in the future.

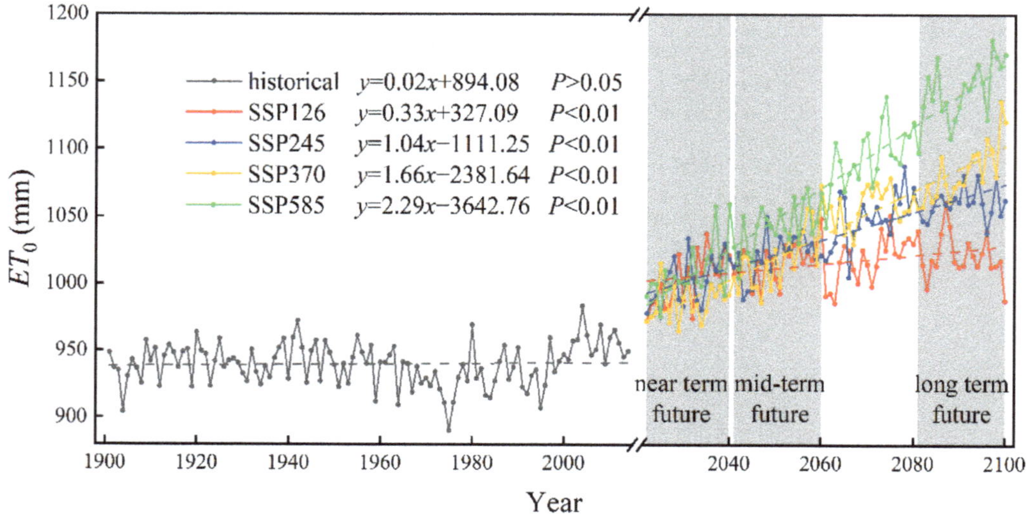

Figure 8. Interannual variations in ET_0 in the Yellow River Basin over the historical period (1901–2014) and under different future emission scenarios (2022–2100; SSP126, SSP245, SSP370, and SSP585) (p is an indicator of significance obtained by the F-test).

To comprehensively understand the temporal ET_0 variations in the basin, Morlet wavelet analysis was used to investigate the cyclic variations in ET_0 in the YRB under the four future emission scenarios (Figure 9). In the SSP126 scenario, three distinct peaks were observed in the wavelet variance of ET_0 in the YRB, corresponding to the time scales of 52, 35, and 8 years, with the most pronounced periodic oscillation of approximately 52 years and a cyclic pattern of 34–38 years for ET_0 in the basin under this time scale (Figure 9a,b). In the SSP245 scenario, the ET_0 in the YRB considerably oscillated for approximately 53 years, and there was a cyclic pattern of approximately 34 years in the basin ET_0 at this time scale (Figure 9c,d). In the SSP370 scenario, the ET_0 in the YRB considerably oscillated for approximately 60 years, and there was a cyclic pattern of approximately 39 years in the basin ET_0 at this time scale (Figure 9e,f). In the SSP585 scenario, the ET_0 in the YRB substantially oscillated for approximately 48 years, and there was a cyclic pattern of approximately 27–32 years in the basin ET_0 at this time scale (Figure 9g,h).

3.4. Spatial Evolution Characteristics of ET_0

The EOF analysis of ET_0 was conducted under four future emission scenarios in the YRB for 2022–2100, and the results were tested for modal significance using the North test, which are presented in Table 2 and Figures 10 and 11. As presented in Table 2, the first three modes of ET_0 in SSP126 passed the North test with a cumulative variance contribution of 88.68%; the first two modes of ET_0 in SSP245 passed the North test with a cumulative variance contribution of 90.18%; the first two modes of ET_0 in SSP370 passed the North test with a cumulative variance contribution of 94.32%; and the first mode of ET_0 in SSP585 passed the North test with a cumulative variance contribution of 93.55%.

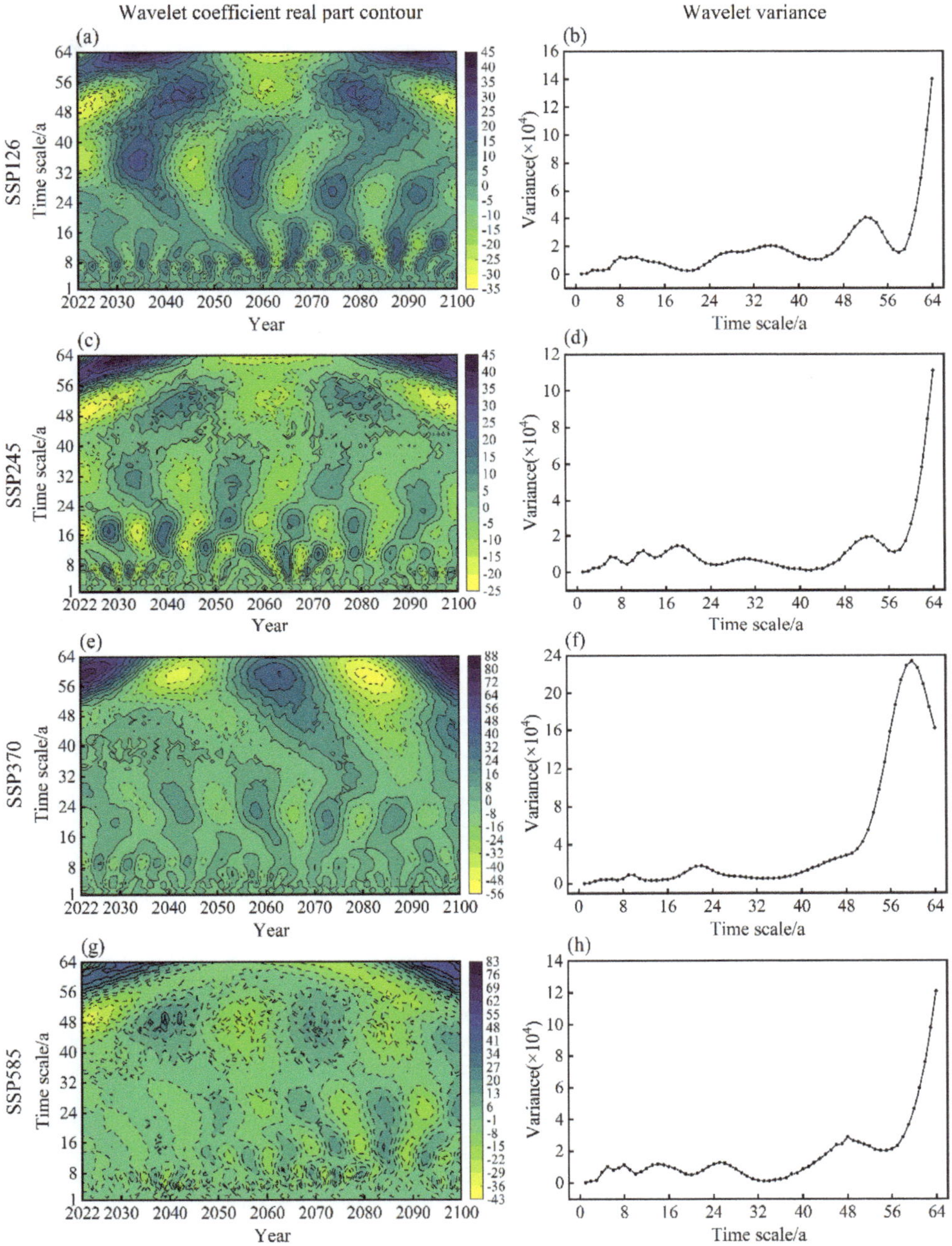

Figure 9. Wavelet analysis of ET_0 in the Yellow River Basin under four future emission scenarios (SSP126, SSP245, SSP370, and SSP585), including wavelet coefficient contour plots of the real part (**a**,**c**,**e**,**g**) and wavelet variance plots (**b**,**d**,**f**,**h**).

Table 2. Main modes and contributions of ET_0 under future climate scenarios in EOF analysis.

Climate Scenarios	Corresponding Modes	Variance Contribution	Cumulative Variance Contribution	North Test
SSP126	EOF1	68.12%	68.12%	pass
	EOF2	13.51%	81.63%	pass
	EOF3	7.05%	88.68%	pass
SSP245	EOF1	81.75%	81.75%	pass
	EOF2	8.43%	90.18%	pass
SSP370	EOF1	89.85%	89.85%	pass
	EOF2	4.47%	94.32%	pass
SSP585	EOF1	93.55%	93.55%	pass

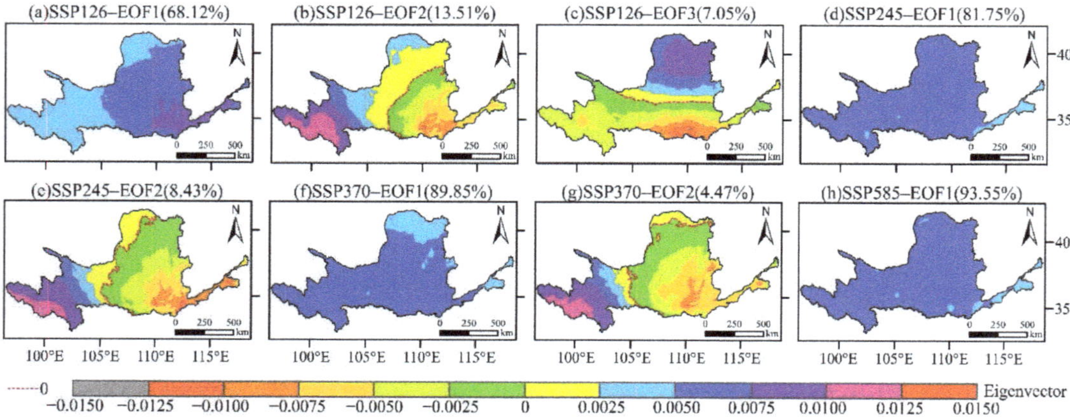

Figure 10. Main modal eigenvectors of the EOF of the ET_0 spatial field in the Yellow River Basin for 2022–2100 under scenarios SSP126 (**a**–**c**), SSP245 (**d**,**e**), SSP370 (**f**,**g**), and SSP585 (**h**), where SSP126/245/370/585–EOFi (contribution/%) in each figure represents the i-th modal eigenvector (EOFi) of ET_0 under scenarios SSP126/245/370/585, with the contribution of each mode in parentheses. The red dashed lines in (**b**,**c**,**e**,**g**) are the positive and negative dividing lines of the eigenvectors.

Under the SSP126 scenario, the first EOF modal eigenvector (EOF1) of ET_0 in the YRB was positive, reflecting a spatially consistent trend of ET_0 across the region, and exhibited an increasing trend from the northwest to the southeast, indicating a more pronounced increase in ET_0 in the lower YRB (Figure 10a). EOF2 and EOF3, which explained 20.56% of the variations, reflected the secondary spatial characteristics of ET_0 with opposite trends from northwest to southeast and from north to south (Figure 10b,c). Combined with the temporal coefficients (Figure 11a), PC1 and PC2 exhibited roughly the same trend, with an increasing trend from 2022 to 2100, particularly after the 2150s when PC1 and PC2 remained positive, indicating that ET_0 remained high throughout this period. PC3 fluctuated at a value of approximately 0, reflecting no significant trend in ET_0. Under SSP245 and SSP370, the EOF1 eigenvectors of ET_0 in the basin were all positive, exhibiting spatial trends of larger values in the upper and middle reaches and smaller values in the lower reaches, as well as larger values in the central and western parts and smaller values in the northern and eastern parts (Figure 10d,f). All the EOF2 eigenvectors exhibited a secondary spatial trend of positive in the northwest and negative in the rest of the basin, with a relatively larger increase in ET_0 near the source area in the upper part of the basin and a relatively larger decrease in ET_0 in the south. Combined with the temporal coefficients (Figure 11b,c), PC1 and PC2 exhibited an increasing trend from 2022 to 2100 under the SSP245 and SSP370 scenarios, and PC1 increased more than PC2, particularly after the 2060s, when PC1 and PC2 always maintained positive values, indicating that ET_0 remained high during this period. The distribution of EOF1 eigenvectors for ET_0 in the basin under the SSP585 scenario was similar to that of EOF1 under the SSP245 scenario (Figure 10h), with an increasing trend in the time coefficient PC1 (Figure 11d), indicating an increasing trend in ET_0 in the basin and a significant increase in ET_0 after the 2060s.

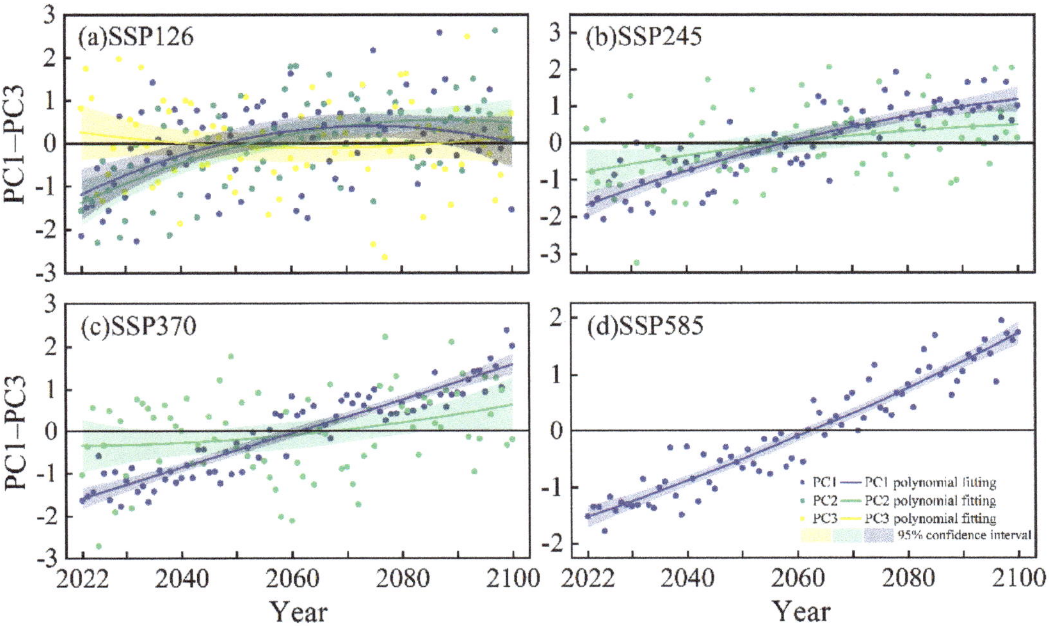

Figure 11. Principal component time coefficients (PC1–PC3) and their polynomial fits for the EOF analysis of future ET_0 in the Yellow River Basin under four emission scenarios ((**a**) SSP126, (**b**) SSP245, (**c**) SSP370, and (**d**) SSP585).

As observed from the spatial variations in annual ET_0 in the YRB in the near-, mid-, and long-term future relative to historical periods in the 21st century (Figure 12), the near-annual ET_0 growth was generally low on an annual scale, and the rate of ET_0 change was even negative in parts of Tai'an, Shandong Province, located in the lower reaches of the YRB, at −6.09% under the SSP370 scenario. In the mid- and long-term future scenarios, the ET_0 rate of change gradually increased in the whole basin, and the areas with high ET_0 variations were primarily concentrated in the YRB source area and a small part of the northern basin. In the lower reaches, the ET_0 change rate was low, and the variations were spatially distributed as high in the west and low in the east. As the radiative forcing increased, the increase in ET_0 became more significant, ranging from −3.08 to 50.78% under SSP126, from −1.32 to 68.88% under SSP245, from −6.09 to 89.30% under SSP370, and from −1.27 to 112.91% under SSP585. A maximum variation of 112.91% was observed in the western part of the YRB in the long-term future (2081–2100) under the SSP585 scenario.

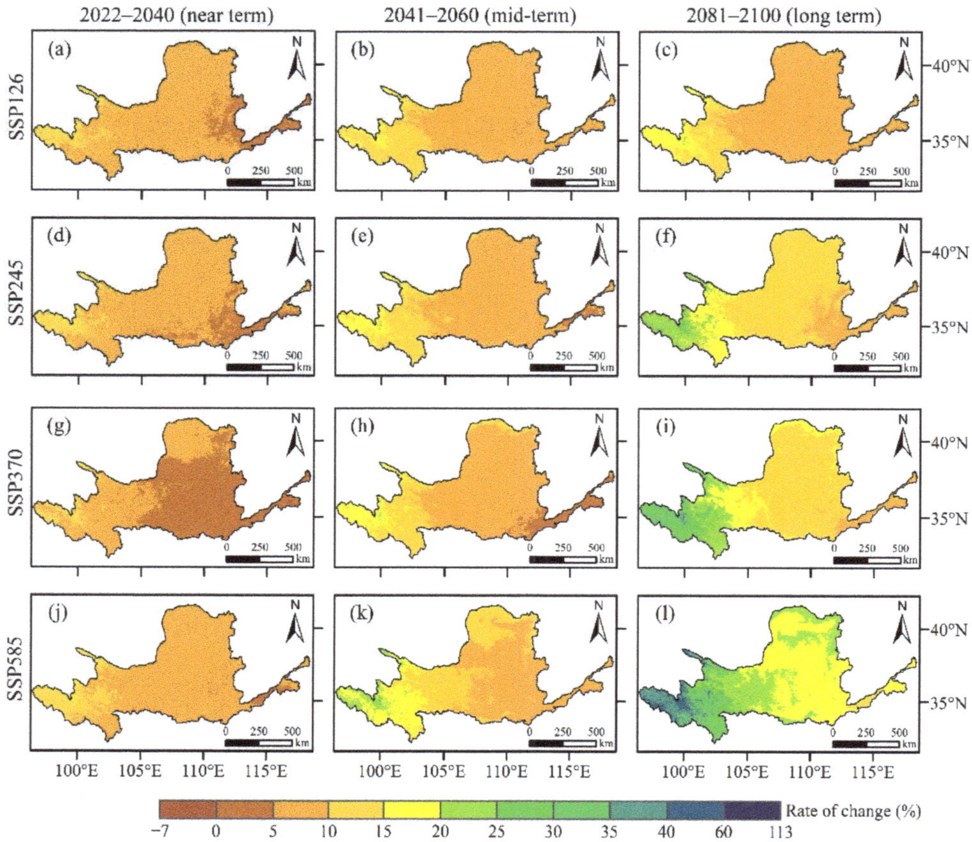

Figure 12. Spatial variations in the near (2022–2040; **a**,**d**,**g**,**j**), mid- (2041–2060; **b**,**e**,**h**,**k**), and long (2081–2100; **c**,**f**,**i**,**l**) term future annual ET_0 of the Yellow River Basin relative to the historical period (1901–2014) under four SSP scenarios (SSP126, SSP245, SSP370, and SSP585).

4. Discussion

4.1. Influence of the ET_0 Model

The models commonly used internationally for ET_0 estimation can generally be divided into four major categories [54]: temperature methods, such as the Hargreaves [50] and Blaney–Criddle models [55]; mass transfer methods, such as the Rohwer [56] and Penman models [57]; radiation methods, such as the Priestley–Taylor [58] and Jensen–Haise models [59]; and synthesis methods, such as the Penman–Monteith model [4]. Although the Penman–Monteith model is considered the standard method for calculating ET_0, accurate predictions of future ET_0 in watersheds require more reliable meteorological data, and the outputs of existing GCMs and downscaling methods typically have low modeling accuracy for meteorological parameters such as wind speed, relative humidity, and radiation. Compared to other models, temperature-based models have lower data requirements and are computationally simple, and downscaling methods have a clear advantage in modeling temperature [60,61], with correlation coefficients generally above 0.90 (Figure 6), making them more widely applicable [31]. In addition, several studies have revealed strong relationships between future ET_0 and temperature in other basins or regions. Xing et al. [27] attributed the increase in ET_0 in the 21st century to an increase in temperature, and Ding and Peng [31] found that global warming led to a change in the main sensitive factor for potential evapotranspiration in the Loess Plateau from the average temperature in the historical period to the maximum temperature in the future period. The results of both these studies theoretically supported the use of the temperature-based ET_0 model. Ahmadi and Baaghideh [25] explored the effect of climate change on ET_0 in pistachio cultivation areas in Iran using the Hargreaves model. Yan and Mohammadian [47] evaluated the performance of the evaporation model based on the Hargreaves formulation using various fitting methods, and the results showed that the simulation was satisfactory. Among the many ET_0 simplification methods, this study also selected the Hargreaves model, a temperature method recommended by the FAO [4] and researchers [62,63], to estimate the spatial and temporal characteristics of ET_0 in the YRB under four future emission scenarios. The fitted scatter density plot had an R^2 of 0.8523 at the monthly scale (Figure 7). The Hargreaves model can effectively simulate basin ET_0; reflecting the rationality of the formula selection in this study. Notably, this study only considered the effect of temperature on ET_0, thus, there may be some bias in the estimation results. The applicability of temperature models in specific regions [64,65] should be improved, and the effect of other meteorological elements on future ET_0 [66] should be explored in future studies.

4.2. Spatial and Temporal Variations in Future ET_0

As a result of global climate change, the hydrometeorological elements and hydrological environment of the YRB have been significantly affected [67]. Although many studies have examined the trends and attribution of ET_0 in the YRB [36,68], previous studies were based on site-scale and historical data. Studies on future water balance and hydrological cycles in the YRB are relatively weak owing to the lack of studies on ET_0.

This study predicted ET_0 trends in the YRB for different periods under four future emission scenarios based on CMIP6 temperature data with high spatial and temporal resolution generated by the delta statistical downscaling method and the Hargreaves model. This study revealed an overall significant increase in ET_0 in the YRB from 2022 to 2100 ($p < 0.01$) (Figure 8), without the "evaporation paradox" [69], which is similar to the future ET_0 trends predicted by many studies [28,69,70]. Based on the Hargreaves model (i.e., Equation (2)), the change in ET_0 is proportional to tas and the difference between tasmax and tasmin. With general global warming, tas, tasmax, and tasmin in the YRB in the future period showed sudden increases relative to the historical period (Figure S1). Therefore, the abrupt increase of ET_0 can be attributed to the abrupt increase of temperature-like variables in the future period [27]. Radiative forcing is expected to stabilize at 2.6 W/m^2 by 2100 for SSP126 and 8.5 W/m^2 by 2100 for SSP585, and radiation values and temperature are positively related to ET_0 [4]. The positive effects of increasing climatic factors, such as temperature and radiation, on ET_0 in the YRB were greater than the negative effects of other factors, and therefore, the latter trend was greater than the former as emission concentrations increased in the YRB (Figure 8). The ET_0 changes in the YRB in the near-, mid-, and long-term future under different future scenarios exhibited high spatial heterogeneity (Figures 10 and 12), with a spatial distribution high in the west and low in the east, and the ET_0 increase became more significant as the radiative forcing scenario increased. Consistent with the results of Ding and Peng [31], the increase in ET_0 was generally greater at higher elevations than at lower elevations in the basin, with the most pronounced change in ET_0 in the western part of the basin, reaching a maximum variation of 112.91% compared to that in the historical period (Figure 12). According to the ET_0 equation (i.e., Equation (2)) and Figure S2, this change can be attributed to the largest temperature difference between tasmax and tasmin in the western part of the basin. In addition, as shown in the change in future precipitation in the YRB relative to the historical period (Figure S3), precipitation in the western part of the basin showed less growth overall and even negative growth in some phases. Warming and decreasing precipitation caused an increase in dryness in the western part of the basin, and the warm-dry trend intensified. Wang et al. [71] found that the evapotranspiration process was more sensitive to relative humidity in the western part of the basin, and a decrease in relative humidity caused an increase in evapotranspiration. Therefore, ET_0 predictions based on the Hargreaves model were greater in these areas. Water loss in the YRB is likely to accelerate in the future than in historical periods, which will threaten the food and ecological security of the region; thus, developing flexible mitigation strategies tailored to local conditions is critical to coping with climate change [72].

4.3. Climate Model Uncertainty Analysis

Because of the differences in the feedback processes of different GCMs, a certain degree of uncertainty exists in their response to future greenhouse gas emissions, and the actual generalized optimal climate models and ET_0 models cannot be determined [73]. Inevitable uncertainties exist in future ET_0 predictions stemming from climate scenarios, climate models, and ET_0 models [74], which greatly affect the confidence of the prediction results.

In response to the uncertainty of climate scenarios, this study selected the CMIP6 data, which had the largest number of participating models, the richest design of numerical experiments, and the largest amount of simulated data available than other CMIP generations for more than 20 years of the CMIP [37,75], initiated by the current Working Group on Coupled Models (WGCM). Compared to previous generations (CMIP3, CMIP5, etc.), CMIP6 uses a new scenario combining shared socio-economic pathways and typical concentration pathways to constrain multi-model predictions of key climate change indicators such as global surface temperature and ocean heat content based on historical observations, climate simulations, and climate sensitivity awareness, reducing uncertainty in predictions and providing higher resolution and reliability [76], thereby making the results more informative and time-sensitive than those based on CMIP5 for future ET_0 studies, such as in Ahmadi and Baaghideh [25], Ding and Peng [31], Kundu et al. [6] and Le and Bae [77]. In response to the uncertainty of climate models, this study selected 24 GCMs with historical and future emission scenarios, which is more extensive than the studies of Li et al. [70], which only used one climate model (HadCM3) under two emission scenarios (A2 and B2), and Nooni et al. [1], which used only one climate model (CNRM-CM6). In addition, this study reduced the uncertainty of future temperature data by preferentially selecting climate models based on multiple interpolation methods, multiple evaluation indicators, and equal weight sets on a downscaling basis. The MAE was controlled within 2.5 mm, S and SS were approximately 1, and TS was approximately 0, indicating very high simulation accuracy (Table 3). Wang and Chen [24] reduced the spatial resolution of GCMs' data to 0.5° based on the delta method, and the MAE of tas was in the range of 1.6–5.7 °C. However, the MAE of tas in this study was controlled in the range of 2.2–2.6 °C and had a higher spatial resolution of 1 km.

Although this study provides a comprehensive theoretical basis for future ET_0 assessments, the uncertainties in the downscaling of GCMs [78] and in the selection and accuracy of ET_0 models [74] may impact the prediction results. To improve ET_0 estimates in future studies, consideration should be given to the long-term goal of the United Nations Framework Convention on Climate Change (Paris Agreement) to limit the increase in global average temperature to less than 2 °C compared to the pre-industrial period and to further efforts to limit it to less than 1.5 °C [79], as well as to achieve China's 2060 carbon neutrality and global carbon neutrality. Describing and quantifying the relative importance of various uncertainty sources and the risks they pose in the assessment is important in current and future climate change impact studies and water resource assessments that should be strengthened to reduce prediction uncertainty.

Table 3. Evaluation of the fitting error of downscaled climate models for monthly temperatures (tas/tasmax/tasmin) in the Yellow River Basin from January 1995 to December 2014.

Models	Tas				Tasmax				Tasmin			
	MAE	S	SS	TS	MAE	S	SS	TS	MAE	S	SS	TS
ACCESS-CM2	\	\	\	\	2.675489	0.999877	0.889805	1.94×10^{-6}	2.338917	0.997536	0.915941	0.001035
ACCESS-ESM1-5	2.196449	0.984745	0.926681	0.001749	2.497431	0.974297	0.901489	0.000464	2.295015	0.963689	0.917158	0.004465
AWI-CM-1-1-MR	2.442822	0.999548	0.909236	0.00021	\	\	\	\	2.837847	0.965667	0.881039	0.003347
AWI-ESM-1-1-LR	\	\	\	\	2.789143	0.987086	0.876886	0.00268	2.470842	0.990019	0.907696	0.000134
BCC-CSM2-MR	2.548259	0.986199	0.904698	7.59×10^{-6}	2.886012	0.979498	0.871659	0.000229	2.65175	0.985043	0.89462	0.00095
BCC-ESM1	\	\	\	\	2.810414	0.954278	0.876325	0.00136	2.617258	0.98485	0.898064	0.005105
CanESM5	2.553748	0.965738	0.900803	0.002406	2.909749	0.980601	0.867035	0.007883	2.572498	0.970803	0.898088	0.000166
CMCC-CM2-SR5	2.273762	0.99501	0.920346	2.77×10^{-6}	\	\	\	\	\	\	\	\
CMCC-ESM2	2.393424	0.990247	0.912677	0.001578	\	\	\	\	2.496331	0.99373	0.905089	0.003913
E3SM-1-0	2.264877	0.990943	0.921241	0.000168	\	\	\	\	\	\	\	\
EC-Earth3	2.345391	0.986413	0.91518	0.001221	2.630697	0.98709	0.887931	0.005839	2.45174	0.980805	0.905882	3.75×10^{-5}
EC-Earth3-Veg	2.518135	0.980041	0.901782	0.000944	2.760198	0.9801	0.877507	5.90×10^{-7}	2.653931	0.976477	0.887252	0.005208
FGOALS-f3-L	2.392895	0.982742	0.912669	9.79×10^{-8}	\	\	\	\	\	\	\	\
FIO-ESM-2-0	2.39042	0.986034	0.913127	0.000604	2.660444	0.989316	0.888524	0.003313	2.483592	0.98829	0.906382	3.56×10^{-6}
GFDL-ESM4	2.415156	0.991749	0.911421	0.002598	2.702498	0.984136	0.884943	0.000381	2.479362	0.99217	0.902494	0.006418
GISS-E2-1-G	2.2951	0.992667	0.919373	2.03×10^{-5}	2.67767	0.985282	0.885917	0.001097	2.363976	0.997295	0.916792	0.000367
INM-CM5-0	2.313318	0.994196	0.917327	5.45×10^{-5}	2.688034	0.976984	0.884853	0.000201	2.494703	0.991961	0.904855	0.000691
IPSL-CM6A-LR	2.366487	0.986157	0.915197	2.53×10^{-5}	2.709299	0.959537	0.88337	0.000268	2.433205	0.989767	0.908817	0.001011
MIROC6	2.538982	0.975222	0.903379	0.002419	2.944748	0.977422	0.86446	0.008182	2.534029	0.982993	0.902185	0.000318
MPI-ESM-1-2-HAM	2.426283	0.988153	0.911244	7.45×10^{-5}	2.795797	0.97782	0.8774	0.001699	2.472226	0.993353	0.906509	0.000618
MPI-ESM1-2-HR	2.482515	0.985307	0.90934	0.000179	2.813033	0.984801	0.87578	0.000855	2.478992	0.990503	0.906471	2.68×10^{-5}
MPI-ESM1-2-LR	2.440139	0.990895	0.912516	0.000608	2.825583	0.987112	0.87801	0.000333	2.44428	0.994861	0.909502	0.001426
MRI-ESM2-0	2.430481	0.99084	0.911775	0.000386	2.754317	0.989183	0.880503	5.16×10^{-7}	2.472997	0.993823	0.907291	0.001735
NESM3	2.393744	0.988131	0.913526	0.00115	2.625448	0.989953	0.890104	8.44×10^{-5}	2.475125	0.992128	0.907585	0.002775

Note: MAE, S, SS, and TS are mean absolute error, Taylor diagram-based quantifiers, spatial skill scores, and temporal skill scores, respectively. MAE is measured in °C, and the others are dimensionless indicators. The closer MAE and TS are to 0, the better the simulation ability of the model; the closer S and SS are to 1, the better the simulation ability of the model.

5. Conclusions

Based on the 24 GCMs in CMIP6 and temperature data with high spatial and temporal resolution generated by the delta statistical downscaling model, this study assessed the evolution of ET_0 in the YRB under four emission scenarios (SSP126, SSP245, SSP370, and SSP585) for the near (2022–2040), mid- (2041–2060), and long (2081–2100) term future. The major conclusions are as follows:

The regionally high-precision climate data generated by delta statistical downscaling based on multiple interpolation methods reduced the uncertainty in the GCM dataset. For the YRB, tas selected the climate models ACCESS-ESM1-5, CMCC-CM2-SR5, and INM-CM5-0; tasmax selected ACCESS-CM2, ACCESS-ESM1-5, and MRI-ESM2-0; and tasmin selected ACCESS-CM2, ACCESS-ESM1-5, and MPI-ESM1-2-LR. The equal-weighted multi-model ensemble had smaller mean absolute errors and higher correlation coefficients than single climate models, and CMIP6 efficiently simulated the temperature and ET_0 in the YRB.

Compared with that of the historical period (1901–2014), the annual ET_0 in the YRB under different emission scenarios (SSP126, SSP245, SSP370, and SSP585) in the future (2022–2100) substantially increased; the rate increased with the increase in emission concentration, and the ET_0 in 2100 under the SSP585 scenario reached 1170.39 mm. Morlet wavelet analysis revealed that ET_0 in the YRB had cyclic patterns of 34–38, 34, 39, and 27–32 years under the SSP126, SSP245, SSP370, and SSP585 scenarios, respectively.

Compared with that in the historical period, the ET_0 variation in the YRB in the near-, mid-, and long-term future under different future scenarios exhibited strong spatial heterogeneity. EOF analysis revealed that ET_0 had positive EOF1 values under all four emission scenarios, exhibiting a spatially consistent trend of ET_0 variation across the region. A maximum variation of 112.91% occurred in the western part of the YRB in the long-term future (2081–2100) under the SSP585 scenario. Without a scientific response, future increases in ET_0 could further reduce the shortage of water resources in the YRB.

Supplementary Materials: The following supporting information can be downloaded at: https://www.mdpi.com/article/10.3390/rs14225674/s1. Figure S1: Interannual variations in tas (a)/tasmax (b)/tasmin (c) in the Yellow River Basin over the historical period (1901–2014) and under different future emission scenarios; Figure S2: Spatial variations in the near (2022–2040; (a,d,g,j)), mid- (2041–2060; (b,e,h,k)), and long (2081–2100; (c,f,i,l)) term future difference between tasmax and tasmin of the Yellow River Basin relative to the historical period (1901–2014) under four SSP scenarios (SSP126, SSP245, SSP370, and SSP585); Figure S3: Spatial variations in the near (2022–2040; (a,d,g,j)), mid- (2041–2060; b,e,h,k), and long (2081–2100; (c,f,i,l)) term future annual precipitation of the Yellow River Basin relative to the historical period (1901–2014) under four SSP scenarios (SSP126, SSP245, SSP370, and SSP585).

Author Contributions: Conceptualization, S.J. and A.W.; methodology, C.S.; validation, S.J., A.W. and C.S.; investigation, A.W.; resources, S.J.; data curation, A.W.; writing—original draft, S.J. and A.W.; writing—review and editing, C.S. and K.W.; supervision, A.W. and K.W.; funding acquisition, S.J. and C.S. All authors have read and agreed to the published version of the manuscript.

Funding: This research was funded by the Training Program for Young Backbone Teachers in Colleges and Universities of Henan Province (2021GGJS003), the Henan Natural Science Foundation (212300410413), the Henan Youth Talent Promotion Project (2021HYTP030), the China Postdoctoral Science Foundation (2020M672247), the Key Science and Technology Project of Henan Province (No. 222102320108), and the First-class Project Special Funding of Yellow River Laboratory (No. YRL22IR12).

Institutional Review Board Statement: Not applicable.

Informed Consent Statement: Not applicable.

Data Availability Statement: Not applicable.

Conflicts of Interest: The authors declare no conflict of interest.

References

1. Nooni, I.K.; Hagan, D.F.T.; Wang, G.; Ullah, W.; Lu, J.; Li, S.; Dzakpasu, M.; Prempeh, N.A.; Lim Kam Sian, K.T.C. Future Changes in Simulated Evapotranspiration across Continental Africa Based on CMIP6 CNRM-CM6. *Int. J. Environ. Res. Public Health.* **2021**, *18*, 6760. [CrossRef] [PubMed]
2. Wang, T.; Tu, X.; Singh, V.P.; Chen, X.; Lin, K. Global data assessment and analysis of drought characteristics based on CMIP6. *J. Hydrol.* **2021**, *596*, 126091. [CrossRef]
3. Zhang, Y.; Fu, B.; Feng, X.; Pan, N. Response of Ecohydrological Variables to Meteorological Drought under Climate Change. *Remote Sens.* **2022**, *14*, 1920. [CrossRef]
4. Allen, R.G.; Pereira, L.S.; Raes, D.; Smith, M. *Crop Evapotranspiration-Guidelines for Computing Crop Water Requirements*; FAO Irrigation and Drainage Paper 56; FAO: Rome, Italy, 1998.
5. Liu, Y.; Yue, Q.; Wang, Q.; Yu, J.; Zheng, Y.; Yao, X.; Xu, S. A Framework for Actual Evapotranspiration Assessment and Projection Based on Meteorological, Vegetation and Hydrological Remote Sensing Products. *Remote Sens.* **2021**, *13*, 3643. [CrossRef]
6. Kundu, S.; Mondal, A.; Khare, D.; Hain, C.; Lakshmi, V. Projecting Climate and Land Use Change Impacts on Actual Evapotranspiration for the Narmada River Basin in Central India in the Future. *Remote Sens.* **2018**, *10*, 578. [CrossRef]
7. Shi, L.; Feng, P.; Wang, B.; Liu, D.L.; Zhang, H.; Liu, J.; Yu, Q. Assessing future runoff changes with different potential evapotranspiration inputs based on multi-model ensemble of CMIP5 projections. *J. Hydrol.* **2022**, *612*, 128042. [CrossRef]
8. Zeng, P.; Sun, F.; Liu, Y.; Feng, H.; Zhang, R.; Che, Y. Changes of potential evapotranspiration and its sensitivity across China under future climate scenarios. *Atmos. Res.* **2021**, *261*, 105763. [CrossRef]
9. Guo, D.; Westra, S.; Maier, H.R. Sensitivity of potential evapotranspiration to changes in climate variables for different Aus-tralian climatic zones. *Hydrol. Earth Syst. Sci.* **2017**, *21*, 2107–2126. [CrossRef]
10. Jhajharia, D.; Dinpashoh, Y.; Kahya, E.; Singh, V.P.; Fakheri-Fard, A. Trends in reference evapotranspiration in the humid region of northeast India. *Hydrol. Process.* **2012**, *26*, 421–435. [CrossRef]
11. Kuang, X.; Jiao, J.J. Review on climate change on the Tibetan Plateau during the last half century. *J. Geophys. Res. Atmos.* **2016**, *121*, 3979–4007. [CrossRef]
12. PascoliniCampbell, M.; Reager, J.T.; Chandanpurkar, H.A.; Rodell, M. Retraction Note: A 10 per cent increase in global land evapotranspiration from 2003 to 2019. *Nature* **2022**, *604*, 202. [CrossRef] [PubMed]
13. She, D.; Xia, J.; Zhang, Y. Changes in reference evapotranspiration and its driving factors in the middle reaches of Yellow River Basin, China. *Sci. Total Environ.* **2017**, *607–608*, 1151–1162. [CrossRef] [PubMed]
14. Yagob, D.; Deepak, J.; Ahmad, F.; Vijay, P.S.; Ercan, K. Trends in reference crop evapotranspiration over Iran. *J. Hydrol.* **2011**, *399*, 422–433.
15. Huang, J.; Zhai, J.; Jiang, T.; Wang, Y.; Li, X.; Wang, R.; Xiong, M.; Su, B.; Fischer, T. Analysis of future drought characteristics in China using the regional climate model CCLM. *Clim. Dyn.* **2018**, *50*, 507–525. [CrossRef]
16. Torres, R.R.; Marengo, J.A. Climate change hotspots over South America: From CMIP3 to CMIP5 multi-model datasets. *Theor. Appl. Climatol.* **2014**, *117*, 579–587. [CrossRef]
17. Chen, H.; Guo, J.; Xiong, W.; Guo, S.; Xu, C. Downscaling GCMs using the Smooth Support Vector Machine method to predict daily precipitation in the Hanjiang Basin. *Adv. Atmos. Sci.* **2010**, *27*, 274–284. [CrossRef]
18. Harding, K.J.; Snyder, P.K. Examining future changes in the character of Central U.S. warm-season precipitation using dynamical downscaling. *J. Geophys. Res.-Atmos.* **2014**, *119*, 13116–13136. [CrossRef]
19. Giorgi, F.; Mearns, L.O. Approaches to the simulation of regional climate change: A review. *Rev. Geophys.* **1991**, *29*, 191–216. [CrossRef]
20. Benestad, R.E. A comparison between two empirical downscaling strategies. *Int. J. Climatol.* **2001**, *21*, 1645–1668. [CrossRef]
21. Peng, S.; Ding, Y.; Wen, Z.; Chen, Y.; Cao, Y.; Ren, J. Spatiotemporal change and trend analysis of potential evapotranspiration over the Loess Plateau of China during 2011–2100. *Agric. For. Meteorol.* **2017**, *233*, 183–194. [CrossRef]
22. Xu, J.; Gao, Y.; Chen, D.; Xiao, L.; Ou, T. Evaluation of global climate models for downscaling applications centred over the Tibetan Plateau. *Int. J. Climatol.* **2017**, *37*, 657–671. [CrossRef]
23. Dosio, A.; Panitz, H.; Schubert-Frisius, M.; Lüthi, D. Dynamical downscaling of CMIP5 global circulation models over CORDEX-Africa with COSMO-CLM: Evaluation over the present climate and analysis of the added value. *Clim. Dyn.* **2015**, *44*, 2637–2661. [CrossRef]
24. Wang, L.; Chen, W. A CMIP5 multimodel projection of future temperature, precipitation, and climatological drought in China. *Int. J. Climatol.* **2014**, *34*, 2059–2078. [CrossRef]
25. Ahmadi, H.; Baaghideh, M. Assessment of anomalies and effects of climate change on reference evapotranspiration and water requirement in pistachio cultivation areas in Iran. *Arab. J. Geosci.* **2020**, *13*, 332. [CrossRef]
26. Das, S.; Datta, P.; Sharma, D.; Goswami, K. Trends in Temperature, Precipitation, Potential Evapotranspiration, and Water Availability across the Teesta River Basin under 1.5 and 2 °C Temperature Rise Scenarios of CMIP6. *Atmosphere.* **2022**, *13*, 941. [CrossRef]
27. Xing, W.; Wang, W.; Shao, Q.; Peng, S.; Yu, Z.; Yong, B.; Taylor, J. Changes of reference evapotranspiration in the Haihe River Basin: Present observations and future projection from climatic variables through multi-model ensemble. *Glob. Planet. Change* **2014**, *115*, 1–15. [CrossRef]

28. Liu, X.; Li, C.; Zhao, T.; Han, L. Future changes of global potential evapotranspiration simulated from CMIP5 to CMIP6 models. *Atmos. Ocean. Sci. Lett.* **2020**, *13*, 568–575. [CrossRef]
29. Nistor, M.; Mîndrescu, M.; Petrea, D.; Nicula, A.; Rai, P.K.; Benzaghta, M.A.; Dezsi, Ş.; Hognogi, G.; Porumb-Ghiurco, C.G. Climate change impact on crop evapotranspiration in Turkey during the 21st Century. *Meteorol. Appl.* **2019**, *26*, 442–453. [CrossRef]
30. Zhang, Q.; Zhang, Z.; Shi, P.; Singh, V.P.; Gu, X. Evaluation of ecological instream flow considering hydrological alterations in the Yellow River basin, China. *Glob. Planet. Change* **2018**, *160*, 61–74. [CrossRef]
31. Ding, Y.; Peng, S. Spatiotemporal change and attribution of potential evapotranspiration over China from 1901 to 2100. *Theor. Appl. Climatol.* **2021**, *145*, 79–94. [CrossRef]
32. Pan, S.; Xu, Y.; Xuan, W.; Gu, H.; Bai, Z. Appropriateness of Potential Evapotranspiration Models for Climate Change Impact Analysis in Yarlung Zangbo River Basin, China. *Atmosphere* **2019**, *10*, 453. [CrossRef]
33. Zhu, Y.; Chang, J.; Huang, S.; Huang, Q. Characteristics of integrated droughts based on a nonparametric standardized drought index in the Yellow River Basin, China. *Hydrol. Res.* **2016**, *47*, 454–467. [CrossRef]
34. Sheng, X.Y.; Wang, J.K.; Cui, Q.T.; Zhang, W.; Zhu, X.F. A feasible biochar derived from biogas residue and its application in the efficient adsorption of tetracycline from an aqueous solution. *Environ. Res.* **2022**, *207*, 112175. [CrossRef] [PubMed]
35. Yin, L.; Feng, X.; Fu, B.; Wang, S.; Wang, X.; Chen, Y.; Tao, F.; Hu, J. A coupled human-natural system analysis of water yield in the Yellow River basin, China. *Sci. Total Environ.* **2021**, *762*, 143141. [CrossRef] [PubMed]
36. Zhao, F.; Ma, S.; Wu, Y.; Qiu, L.; Wang, W.; Lian, Y.; Chen, J.; Sivakumar, B. The role of climate change and vegetation greening on evapotranspiration variation in the Yellow River Basin, China. *Agric. For. Meteorol.* **2022**, *316*, 108842. [CrossRef]
37. O'Neill, B.C.; Tebaldi, C.; van Vuuren, D.P.; Eyring, V.; Friedlingstein, P.; Hurtt, G.; Knutti, R.; Kriegler, E.; Lamarque, J.; Lowe, J.; et al. The Scenario Model Intercomparison Project (ScenarioMIP) for CMIP6. *Geosci. Model Dev.* **2016**, *9*, 3461–3482. [CrossRef]
38. Zhang, Z.H.; Deng, S.F.; Zhao, Q.D.; Zhang, S.Q.; Zhang, X.W. Projected glacier meltwater and river run-off changes in the Upper Reach of the Shule River Basin, north-eastern edge of the Tibetan Plateau. *Hydrol. Process.* **2019**, *33*, 1059–1074. [CrossRef]
39. Hay, L.E.; Wilby, R.; Leavesley, G.H. A comparison of delta change and downscaled GCM scenarios for three mountainous basins in the United States. *J. Am. Water Resour. Assoc.* **2000**, *36*, 387–397. [CrossRef]
40. Willmott, C.J.; Matsuura, K. Advantages of the mean absolute error (MAE) over the root mean square error (RMSE) in assessing average model performance. *Clim. Res.* **2005**, *30*, 79–82. [CrossRef]
41. Taylor, K.E. Summarizing multiple aspects of model performance in a single diagram. *J. Geophys. Res. Atmos.* **2001**, *106*, 7183–7192. [CrossRef]
42. Jiang, Z.; Song, J.; Li, L.; Chen, W.; Wang, Z.; Wang, J. Extreme climate events in China: IPCC-AR4 model evaluation and projection. *Clim. Change* **2012**, *110*, 385–401. [CrossRef]
43. Chen, W.; Jiang, Z.; Li, L. Probabilistic Projections of Climate Change over China under the SRES A1B Scenario Using 28 AOGCMs. *J. Clim.* **2011**, *24*, 4741–4756. [CrossRef]
44. Jiang, D.; Wang, H.; Lang, X. Evaluation of East Asian Climatology as Simulated by Seven Coupled Models. *Adv. Atmos. Sci.* **2005**, *22*, 479–495.
45. Ma, Z.; Sun, P.; Zhang, Q.; Zou, Y.; Lv, Y.; Li, H.; Chen, D. The Characteristics and Evaluation of Future Droughts across China through the CMIP6 Multi-Model Ensemble. *Remote Sens.* **2022**, *14*, 1097. [CrossRef]
46. Pelosi, A.; Medina, H.; Villani, P.; Urso, D.G.; Chirico, G.B. Probabilistic forecasting of reference evapotranspiration with a limited area ensemble prediction system. *Agric. Water Manag.* **2016**, *178*, 106–118. [CrossRef]
47. Yan, X.; Mohammadian, A. Estimating future daily pan evaporation for Qatar using the Hargreaves model and statistically downscaled global climate model projections under RCP climate change scenarios. *Arab. J. Geosci.* **2020**, *13*, 938. [CrossRef]
48. Hargreaves, G.H. Moisture Availability and Crop Production. *Trans. ASAE* **1975**, *18*, 980–0984. [CrossRef]
49. Hargreaves, G.H.; Samani, Z.A. Estimating Potential Evapotranspiration. *J. Irrig. Drain. Div.* **1982**, *108*, 225–230. [CrossRef]
50. Hargreaves, G.H.; Samani, Z.A. Reference Crop Evapotranspiration from Temperature. *Appl. Eng. Agric.* **1985**, *2*, 96–99. [CrossRef]
51. Li, H.; Wang, Y.; Jia, L.; Wu, Y.; Xie, M. Runoff Characteristics of the Nen River Basin and its Cause. *J. Mt. Sci.* **2014**, *11*, 110–118. [CrossRef]
52. Naren, A.; Maity, R. Modeling of local sea level rise and its future projection under climate change using regional information through EOF analysis. *Theor. Appl. Climatol.* **2018**, *134*, 1269–1285. [CrossRef]
53. North, G.R.; Bell, T.L.; Cahalan, R.F.; Moeng, F.J. Sampling Errors in the Estimation of Empirical Orthogonal Functions. *Am. Meteorol. Soc.* **1982**, *110*, 699–706. [CrossRef]
54. Singh, V.P.; Xu, C.Y. Evaluation and generalization of 13 mass-transfer equations for determining free water evaporation. *Hydrol. Process.* **1997**, *3*, 311–323. [CrossRef]
55. Blaney, H.F.; Criddle, W.D. Determining Consumptive Use and Irrigation Water Requirements. *Tech. Bull.* **1962**, *4*, 369–373.
56. Rohwer, C. *Evaporation from a Free Water Surface*; US Department of Agriculture: Washington, DC, USA, 1931.
57. Penman, H.L. Natural Evaporation from Open Water, Bare Soil and Grass. *Proc. R. Soc. Lond. Ser. A Math. Phys. Sci.* **1948**, *193*, 120–145.
58. Priestley, C.H.B.; Taylor, R.J. On the Assessment of Surface Heat Flux and Evaporation Using Large-Scale Parameters. *Mon. Weather Rev.* **1972**, *100*, 81–92. [CrossRef]
59. Jensen, M.E.; Haise, H.R. Estimating Evapotranspiration from Solar Radiation. *J. Irrig. Drain. Div.* **1963**, *89*, 15–41. [CrossRef]

60. Peng, S.; Gang, C.; Cao, Y.; Chen, Y. Assessment of climate change trends over the Loess Plateau in China from 1901 to 2100. *Int. J. Climatol.* **2018**, *38*, 2250–2264. [CrossRef]
61. Yang, X.; Zhou, B.; Xu, Y.; Han, Z. CMIP6 Evaluation and Projection of Temperature and Precipitation over China. *Adv. Atmos. Sci.* **2021**, *38*, 817–830. [CrossRef]
62. Er-Raki, S.; Chehbouni, A.; Khabba, S.; Simonneaux, V.; Jarlan, L.; Ouldbba, A.; Rodriguez, J.C.; Allen, R. Assessment of reference evapotranspiration methods in semi-arid regions: Can weather forecast data be used as alternate of ground meteorological parameters? *J. Arid. Environ.* **2010**, *74*, 1587–1596. [CrossRef]
63. Zhao, C.; Nan, Z.; Feng, Z. GIS-assisted spatially distributed modeling of the potential evapotranspiration in semi-arid climate of the Chinese Loess Plateau. *J. Arid. Environ.* **2004**, *58*, 387–403.
64. Bourletsikas, A.; Argyrokastritis, I.; Proutsos, N. Comparative evaluation of 24 reference evapotranspiration equations applied on an evergreen-broadleaved forest. *Hydrol. Res.* **2018**, *49*, 1028–1041. [CrossRef]
65. Lang, D.; Zheng, J.; Shi, J.; Liao, F.; Ma, X.; Wang, W.; Chen, X.; Zhang, M. A Comparative Study of Potential Evapotranspiration Estimation by Eight Methods with FAO Penman–Monteith Method in Southwestern China. *Water* **2017**, *9*, 734. [CrossRef]
66. Yan, X.; Mohammadian, A. Forecasting daily reference evapotranspiration for Canada using the Penman-Monteith model and statistically downscaled global climate model projections. *Alex. Eng. J.* **2020**, *59*, 883–891. [CrossRef]
67. Omer, A.; Elagib, N.A.; Ma, Z.; Saleem, F.; Mohammed, A. Water scarcity in the Yellow River Basin under future climate change and human activities. *Sci. Total Environ.* **2020**, *749*, 141446. [CrossRef]
68. Jiang, Z.; Yang, Z.; Zhang, S.; Liao, C.; Hu, Z.; Cao, R.; Wu, H. Revealing the spatio-temporal variability of evapotranspiration and its components based on an improved Shuttleworth-Wallace model in the Yellow River Basin. *J. Environ. Manag.* **2020**, *262*, 110310. [CrossRef]
69. Roderick, M.L.; Farquhar, G.D. The cause of decreased pan evaporation over the past 50 years. *Science* **2002**, *298*, 1410–1411. [CrossRef]
70. Li, Z.; Zheng, F.; Liu, W. Spatiotemporal characteristics of reference evapotranspiration during 1961–2009 and its projected changes during 2011–2099 on the Loess Plateau of China. *Agric. For. Meteorol.* **2012**, *154–155*, 147–155. [CrossRef]
71. Wang, W.; Shao, Q.; Peng, S.; Xing, W.; Yang, T.; Luo, Y.; Yong, B.; Xu, J. Reference evapotranspiration change and the causes across the Yellow River Basin during 1957-2008 and their spatial and seasonal differences. *Water Resour. Res.* **2012**, *48*, 113–122. [CrossRef]
72. Cook, P.A.; Black, E.C.L.; Verhoef, A.; Macdonald, D.M.J.; Sorensen, J.P.R. Projected increases in potential groundwater recharge and reduced evapotranspiration under future climate conditions in West Africa. *J. Hydrol. Reg. Stud.* **2022**, *41*, 101076. [CrossRef]
73. Gharbia, S.S.; Smullen, T.; Gill, L.; Johnston, P.; Pilla, F. Spatially distributed potential evapotranspiration modeling and climate projections. *Sci. Total Environ.* **2018**, *633*, 571–592. [CrossRef] [PubMed]
74. Dallaire, G.; Poulin, A.; Arsenault, R.; Brissette, F. Uncertainty of potential evapotranspiration modelling in climate change impact studies on low flows in North America. *Hydrol. Sci. J.* **2021**, *66*, 689–702. [CrossRef]
75. Liu, J.; Lu, Y. How Well Do CMIP6 Models Simulate the Greening of the Tibetan Plateau? *Remote Sens.* **2022**, *14*, 4633. [CrossRef]
76. Stouffer, R.J.; Eyring, V.; Meehl, G.A.; Bony, S.; Senior, C.; Stevens, B.; Taylor, K.E. CMIP5 Scientific Gaps and Recommendations for CMIP6. *Bull. Amer. Meteorol. Soc.* **2017**, *98*, 95–105. [CrossRef]
77. Le, T.; Bae, D. Response of global evaporation to major climate modes in historical and future Coupled Model Intercomparison Project Phase 5 simulations. *Hydrol. Earth Syst. Sci.* **2020**, *24*, 1131–1143. [CrossRef]
78. Gao, C.; Booij, M.J.; Xu, Y. Assessment of extreme flows and uncertainty under climate change: Disentangling the uncertainty contribution of representative concentration pathways, global climate models and internal climate variability. *Hydrol. Earth Syst. Sci.* **2020**, *24*, 3251–3269. [CrossRef]
79. Meinshausen, M.; Lewis, J.; McGlade, C.; Gütschow, J.; Nicholls, Z.; Burdon, R.; Cozzi, L.; Hackmann, B. Realization of Paris Agreement pledges may limit warming just below 2 °C. *Nature* **2022**, *604*, 304–309. [CrossRef]

sustainability

MDPI

Review

Sustainable Water Resources Management Assessment Frameworks (SWRM-AF) for Arid and Semi-Arid Regions: A Systematic Review

Badir S. Alsaeed [1,2,*], Dexter V. L. Hunt [1] and Soroosh Sharifi [1]

1 Department of Civil Engineering, School of Engineering, University of Birmingham, Edgbaston, Birmingham B15 2TT, UK
2 Department of Civil Engineering, College of Engineering, Qassim University, Buraydah 52571, Saudi Arabia
* Correspondence: bsa864@bham.ac.uk

Abstract: Sustainable water resources management assessment frameworks (SWRM-AF) with associated indicators and benchmarks have appeared widely during the last decades to improve or maintain water resources. Examination or evaluation of their appropriateness and refinement for particular arid and semi-arid regions is a relatively unexplored area. To fill this gap in knowledge, a systematic review of relevant 21st century studies identified within two extensive databases, Scopus and Engineering Village, and in grey literature, is undertaken in this study. Therein, 17 studies are identified and thoroughly explored to identify their focus, application, and framework construction. The results of the comparative analysis among these frameworks show that the average numbers of components and indicators are 4.5 and 17.6, respectively. Meanwhile, categorical rescaling (47.1%), equal weighting (47.1%), arithmetic technique (82.35%), local scale (52.8%), and interval of the final index value of [0–100] (41.2%) are the most commonly used normalization methods and elements. The paper concludes that none of the existing tools reviewed is 100% applicable for arid and semi-arid regions, and therefore the case is made for developing a new bespoke SWRM-AF. The outcomes of this paper provide some useful insights into what should be included therein (e.g., stakeholder engagement and specific indicators to fit the context).

Keywords: water resources management; sustainable assessment; water sustainable index; stakeholder; framework; indicator

Citation: Alsaeed, B.S.; Hunt, D.V.L.; Sharifi, S. Sustainable Water Resources Management Assessment Frameworks (SWRM-AF) for Arid and Semi-Arid Regions: A Systematic Review. *Sustainability* **2022**, *14*, 15293. https://doi.org/10.3390/su142215293

Academic Editors: Luis Garrote, Alban Kuriqi and Hossein Bonakdari

Received: 29 June 2022
Accepted: 11 November 2022
Published: 17 November 2022

1. Introduction

As a result of significant agricultural and industrial advancements in parallel with the peace and security afforded after the second world war, the global population has almost tripled from 2.7 billion to 7.5 billion in just seventy years [1]. This increase, accompanied by the changes in lifestyle (including eating habits) seen in many regions, is now placing significant stress on various natural resources (including but not limited to water) vital for human requirements. These requirements are categorized into basic, psychological, and self-fulfilment needs, based on Maslow's hierarchy of needs [2,3]. This research focuses on the water requirement, which might be considered the most important yet basic requirement for humans to survive. Nevertheless, the demand for this resource has never been greater than in the last few decades [4–6], especially in developing countries, where exceptional population growth, increased urbanization [7], and expansion in industrial and agricultural sectors have resulted in extreme water demand and water stress. The previous conditions could highly exacerbate the situation in some arid and semi-arid regions (ASAR) where limited natural water resources (WR) are available. Therefore, special attention and preparation should be given to this issue to ensure the longevity of these crucial resources, especially in regions with difficult climatic and weather-related issues, such as ASAR.

The term arid is typically used to describe the climate of regions that suffer from very high temperature and receive less than 100 mm of rainfall per year [8]. In contrast, the term semi-arid describes regions where the annual rainfall is between 250 and 500 mm/year [9]. Both types of regions feature evapotranspiration rates that are higher than the precipitation rate, with the potential for frequent severe droughts and infrequent but considerable floods [10]. Moreover, these regions are globally characterized as the most water-stressed areas, where the groundwater (GW), stored in aquifers, is the primary water source [11]. However, since some ASAR are characterized by low rainfall rates, and rain is essential to the speed and recharge time of aquifers, the use of GW is not very sustainable [12,13]. Furthermore, high dependence on GW with intensive pumping makes it prone to pollution, such as salinity intrusion [14,15]. Conversely, in coastal regions, water supply from desalination plants with many of the current technologies is unsustainable, given the high energy, environmental impact, and economic cost [16–18]. Therefore, water resources management (WRM) in such regions requires careful planning and assessment of sustainability, and thus requires appropriate tools.

Furthermore, global warming phenomena and the impacts of climate change are further pressurizing WR over the globe [5,19–22], not least in ASAR, requiring new solutions and approaches on both the demand and supply sides. Thus, the scientific community has conducted several meetings and studies during the last decades to address the consequences of such a trend [23–27]. One of the early attempts to deal with this issue was in 1992 during the International Conference on Water and the Environment [28], which ended with the declaration of the four Dublin principles, the third one stating clearly that any "development and management" in regard to water "should be based on a participatory approach ... at all levels". Hence, this principle informed one of the main strategies to enhance WRM and ensure the continuity of WR.

Assessing and managing WR in ASAR in a way that usefully informs decision-making is fraught with difficulty, especially with what appears to be a lack of region-specific frameworks, a lack of data collection and in the context of the natural and socio-economic (i.e., Sustainability) settings in which this needs to happen. A research gap exists in terms of identifying what general sustainable water resources management (SWRM) assessment frameworks exist, and whether they are applicable to ASAR. This is a key underlying philosophy behind this paper, the findings of which will be used to identify whether (a) existing frameworks are fit-for-purpose in ASAR; or (b) a bespoke framework should be derived. Moreover, if the latter outcome is found to be true, and in order to avoid reinventing the wheel, the systematic review and analysis of existing frameworks can be used to inform its derivation.

1.1. Sustainability and Sustainable Water Resources Management (SWRM)

The water cycle and its impact on related ecosystems represent a great example of a sustainable process that has existed for millions of years. However, current water demands and global climatic changes are impacting its ability to remain so [29,30].

The use of the terms "sustainability" and "sustainable development" has become ever more popular since Bruntland's [31] definition: *"to ensure that the current development meets the needs of current generation's without negatively impacting the capability of future generations to meet their needs"*. This has never been more important than for SWRM in ASAR, where GW is becoming depleted, negatively impacting the ability of future generations to draw down water and meet their needs—which due to growing populations, will be greater than today.

Another definition or principle for sustainability was introduced by Elkington [32] as: *"sustainability aims to ensure that the range of economic, social, and environmental options would stay open and not limited for the future generations because they were not hindered by the current human actions."* This has paved the way for the introduction of 17 sustainable development goals (SDG), the sixth of which is to *"ensure availability and sustainable management of water and sanitation for all"* [33]. This study is significant and motivated by such a global goal, and has never been more relevant in ASAR.

Sustainability itself has been widely recognized to stand on three common pillars or dimensions: the environment, the economy, and the society [34–38]. In other words, to obtain a sustainable system, its environment should be protected, the economy should be viable, and social equity and acceptance should be considered as much as possible.

Meanwhile, the importance of achieving a balance (rather than a trade-off) between these dimensions of sustainability has been a catalyst for much discussion [39–42]. For example, selling water in plastic bottles is both profitable for companies (economic) and satisfies the needs of many people (Social). However, the impact of this business on the environment is harmful if the bottles are not recycled. Therefore, to enhance the sustainability of any system, all three pillars need to be in balance. Moreover, for ASAR, the points at which the pillars interact for SWRM need to be considered ever more readily.

1.2. Assessment Frameworks for Sustainable Water Resources Management

To improve the sustainability of any WRM system, it is crucial to have an appropriate amount of different related indicators (i.e., quantitative and qualitative), metrics, and benchmarks contained within an assessment framework or index in order to help decision-makers and concerned stakeholders determine the current level (or performance) of their SWRM and improve it accordingly, should it be underperforming [43,44]. (N.B. The terms framework or index are used interchangeably within the literature; however, in this paper, they are considered to be one and the same.) The advantage of forming an indicator-based framework is its ability to help evaluate and elucidate multi-dimensional factors or thoughts that cannot be measured directly [45] and cannot be understood by only one component or indicator [46].

Indeed, collaboration among different stakeholders in developing a WR index is (and should always be) significant to ensure the index is acceptable [45]. By developing and using a suitable framework, all interested parties can understand the main issues that threaten sustainability in their system, and work co-operatively toward mitigating them. These issues can be simplified within the framework to a single number representing the general sustainability level of the whole WRM system. In most cases, having a quantifiable number would have a more substantial effect on the ability of the public/decision-makers to understand and therefore act in a more helpful way [47].

Furthermore, it is both beneficial and necessary to build any indicator-based framework based on a wide array of indicators [41] that have been widely vetted and endorsed and that can guide the assessment and improvement of the sustainability credentials for WRM systems [48]. Moreover, from a policy-making and management perspective, considering both water availability and access indicators is likely to be more emphasized (and therefore carry a higher weighting) for frameworks adopted in developing and water-poor countries than those in developed and water-rich countries [49]. Similarly, this would apply in countries in ASAR where appropriate "bespoke" frameworks are needed to improve or reform their WRM systems.

On the other hand, this study aims to review research published in the last two decades related to assessment frameworks for SWRM, focusing on checking to what extent they can be applicable for ASAR. Key objectives in the form of questions for the research include:

- Since the turn of the century, what indicator-based frameworks and/or indices have been used to assess the sustainability of WRM?
- What similarities and differences exist amongst indicator-based sustainability assessment frameworks of WRM, such as the number of components (and indicators) and the scaling, aggregating, and weighting methods?
- How effective are the current water resource indices or frameworks in assessing the sustainability of WRM in ASAR?

By answering these questions, it would be possible to ascertain whether a bespoke SWRM framework were needed within the context of ASAR.

The paper is divided into six sections. In Section 2, the methodology used to answer these questions is outlined. In Section 3, some general definitions of SWRM, along with criteria and related guidelines for making indicator-based frameworks are subsequently presented. In Section 4, the main elements of the indicator-based sustainability assessment framework of the WRM system are briefly illustrated. Section 5 provides the search results based on the criteria given in Section 2. These results include overviewing and analyzing the existing Sustainable Water Resources Management Assessment Framework(s) (SWRM-AF) developed since the turn of the century. A critique is provided that includes the advantages and disadvantages of each framework, followed by a brief comparative analysis. Section 6 discusses the results with a final evaluation of all frameworks included in this review to check their applicability for ASAR. Finally, conclusions are provided in Section 7, along with recommendations for future research.

2. Methodology

To answer the previous questions posed in Section 1.2, a systematic literature search using the two well-known databases Scopus and Engineering Village was conducted to check relevant studies. In the first stage, a group of pertinent keywords were identified and used to search databases using the title/abstract/keywords included in the papers. The first step required a filter, since the area of sustainability is extensive within the literature. Moreover, looking through a confined yet credible quantity for a literature review paper is crucial.

Therefore, the scope of this search was exclusive to peer-reviewed articles and peer-reviewed conference papers. Additionally, the search had two conditions for all included documents: (a) documents should have been produced in the period from 2000 to 2021 and (b) documents should be written in the English language only. This period was selected because several frameworks for assessing the WRM system were produced after 2000. Furthermore, this is consistent with the method applied by other authors, such as Topal et al. [50]. This method uses a four-step clustering algorithm (i.e., Scope, Target Group, Subject Domain, and Methods) to narrow the research area. This narrowing process would mean excluding, to some degree, any unrelated studies by using the OR operator within each category's keywords and the AND operator within each cluster [50]. The idea of this process is straightforward, requiring all studies covered in this review to be included in the intersection area of all four clusters.

2.1. Keyword Selection

In the Scope cluster, many terms mainly related to sustainability and WRM were used to define the largest frame with which the search should start. These specific terms and their derivatives were *"#water resources management"*, *"#water management"*, *"#water shortage"*, *"#water assessment"*, *"#SWRM"*, *"#sustainable assessment"*, *"#sustainable measurement"*, *"#water sustainable index"*, *"#sustainability principles"*, *"#sustainable development"*.

The Target Group of this study concerned the primary sectors that received water or were affected by any decisions related to its supply and demand. The main terms used for the Target Group cluster were: *"domestic water"*, *"municipal"*, and *"stakeholder"*.

The Subject Domain keywords were specific for the required method and its main parts that could evaluate the combination of the Scope and the Target Group and the geographic areas that needed to be investigated. The terms used in this search for these purposes were *"indicator"*, *"indicator-based"*, *"framework"*, *"criteria"*, *"index"*, *"component"*, *"arid"*, and *"semi-arid"*. It is worth mentioning that this category (i.e., Subject Domain) was used twice in the exact search. The first one included all required fields (i.e., Subject/Title/Abstract in the Engineering Village database, and Title/Abstract/Keywords in the Scopus database). The second one was only in the title, that is, one of the keywords needed to be in the article's title. This action was essential to reduce the enormous number of unrelated studies.

The fourth group, the methods of data collection or treatment based on the participatory approach, was assigned. The terms included in this cluster were *"survey"*, *"interview"*, *"questionnaire"*, and *"participatory"*.

2.2. Database Search

The search through the Scopus and Engineering Village databases was undertaken on the 14th of October 2021 and returned with 1428 and 1316 articles, respectively. However, the Engineering Village database was a combination of three databases: (1) Compendex, (2) GEOBASE, and (3) Inspec. For this reason, many of the 1316 articles were duplicated in the search output. Fortunately, the search engine had a feature to remove these duplications, and the number was subsequently reduced to 721 articles. This result, plus that from Scopus (i.e., 2149 articles) were merged in EndNote Library, which also has the advantage of automatically removing duplications, reducing the total number to 1627 articles. Among these papers, 174 were conference papers, while the remaining were peer-reviewed articles.

Before starting the manual search, inclusion and exclusion criteria needed to be assigned and followed generally. Under these criteria, any articles unrelated to the main scope (i.e., both WRM and sustainability), whether directly or indirectly, would be excluded right away. For example, many articles related mainly to the medical, education, and energy sectors were removed. In addition, if this criterion were applicable, another specific check was required to ensure that these studies had considered a framework or index by mentioning that clearly in either the title or the keywords. Consequently, both conditions were applied in the first screening stage by checking each title and all keywords of the 1627 papers. This stage resulted in a reduction in the number of articles to 400.

In the second round, abstracts were investigated concerning the target group and main elements of the subject domain (i.e., indicator, indicator-based, and component). The results dropped to 45. This round was supposed to be the last round, but after checking some articles among the 45, it appeared they lacked an applicable framework or index that included specific indicators. Therefore, a final round was added to skim-read each of the 45 papers and ensure that they contained these essential elements to be included in the full-text review. Consequently, 23 studies were selected to be included in the analysis. All these screening stages were summarized and illustrated below in Figure 1.

At the end of the systematic review, key methodological steps were applied to help meet underlying objectives, namely:

- Identification of SWRM definitions, guidelines and criteria (Section 3);
- Establishment of the main elements of indicator-based frameworks (Section 4);
- Provision of an overview of existing sustainable water resources management assessment frameworks (SWRM-AF) (Section 5).

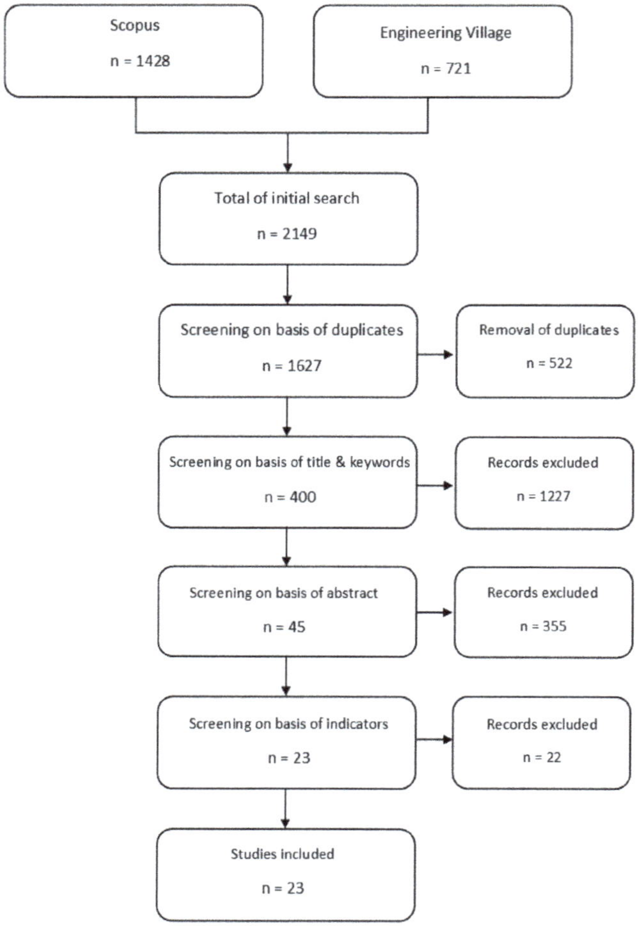

Figure 1. The selection process for articles.

3. Sustainable Water Resources Management (SWRM): Definitions, Guidelines, and Criteria

3.1. Definitions

While the definition of sustainability was previously mentioned in Section 1.1, it is essential to clarify further definitions used within this paper—not least WRM. Firstly, WR can be defined as any shape or state of natural waters that exist on the planet, whether above (e.g., rainwater in clouds), on (e.g., oceans and rivers), or under the ground (e.g., GW), that has the potential to be used by humans [51]. Secondly, management can be defined simply as the way to manage something. In terms of WRM, these definitions pertain to the supply of and demand for water and all matters related to them.

Furthermore, it can be considered that the definition of WR includes both the natural freshwater and saltwater that usually react to or are affected by the processes of the hydrological cycle and other species' activities. Humans are one of the species that can impact WR in their use of them, but what does it mean to make this process sustainable? Gleick et al. [52] defined sustainable water use as:

> *"the use of water that supports the ability of human society to endure and flourish into the indefinite future without undermining the integrity of the hydrological cycle or the ecological systems that depend on it."* [52] (p. 24)

However, humans have the most significant impact on the environment in general and on WR in particular due to their activities [19,53,54]. These impacts on WR are expected to expand in the future and cause more uncertainty in terms of water availability, more extreme weather events of droughts and floods, and quicker evaporation of surface water resources [53]. Hence, it is important to prepare carefully for these risks before they happen or increase to improve the sustainable management of WR systems.

Accordingly, the three definitions for the three terms (i.e., sustainability, water resources, and management) can be combined to present a possible explanation for SWRM. The function of such a definition is to help stakeholders from different backgrounds understand the target in a simple way, which would assist in the communication process, thereby gaining their trust and cooperation.

Pertinently to this matter and its purpose, the term integrated water resources management (IWRM) is defined as:

> *"a process which promotes the coordinated development and management of water, land and related resources, in order to maximise the resultant economic and social welfare in an equitable manner without compromising the sustainability of vital ecosystems".* [23] (p. 1)

Although the previous definition is widely known and broadly accepted by the scientific community, the understandable main aim is to maximize the benefits to the economy and society without harming the ecosystem or the environment. Meanwhile, it can be argued that the purpose of sustainability is slightly different, being more about obtaining the best result (i.e., optimizing) for all three aspects (i.e., economy, society, and environment) in as balanced a way as possible.

Therefore, the suggested definition for SWRM used in this paper is *"to ensure that the current management of water resources meets the need of the present generation in a way that balances between social, economic, and environmental factors avoiding negatively impacting future generations' capability to meet their water needs"*, accepting that future needs are not always easy to identify and require a range of foresight methods to predict. This definition requires a breakdown into several objectives or components that constitute indicators and sub-indicators to measure the performance of SWRM.

3.2. *Guidelines for the Development of the SWRM Framework*

Sustainability frameworks and their indicators, in general, could (and should) have different interpretations based on the perspective, context, and local conditions they are used for. For example, frameworks assigned for business or construction *per se* would be different than those for WRM. Indeed, each sector should have specific guidelines and criteria for any suggested indicator that matched its context [44,55].

First of all, the consideration and linkage of the three dimensions of sustainability (i.e., environmental, economic, and social) [56], in addition to the technical side in the criteria overall, are crucial to handling the complexity and uncertainty of water-related issues [57]. Hence, a sustainable system would not only facilitate the management of the infrastructure of water utilities with the supply and demand sides, but would also assure integration and fairness among the previously mentioned three core areas. Thus, it is essential in the developing stage of an SWRM framework to check whether any suggested indicator belongs (or not) to one of these four categories (i.e., technical or physical, environmental, economic, and social) before considering it.

The second general guideline can be elicited from one of the Dublin principles [28] (i.e., the third), which emphasizes the importance of a participatory approach for any development for WR. Thus, the involvement of stakeholders in developing an SWRM-AF, or at least the process of indicator selection, is necessary.

Another guideline is that the number of indicators should be appropriate. In other words, they should not be too numerous, since this would complicate the process of application and interpretation [58,59] and challenge the capacity of the financial and human resources in collection and analysis. Conversely, too small a number could result in inaccurate conclusions that would lead to weak policy decisions—not least because they would be based on inadequate data [44]. Hence, it is instrumental during the selection process to focus on just the right number of indicators whose details (i.e., data) are available, unambiguous, and comprehensive. Nevertheless, following the above guidelines as a first stage would require more specific criteria for the selection process of each indicator, as outlined in Section 3.3.

3.3. Criteria for Selection of SWRM Indicators

Specific criteria must be considered in order to select appropriate indicators for assessing SWRM-AF. One of these criteria is that these sustainable indicators should (1) work as a set; (2) be both simple and clear, and (3) contain sufficient information to help decision-makers provide efficient actions [60].

Moreover, Bell and Morse [61] identified other criteria as conditions for selecting indicators. Indicators must:

- Be relevant to the purpose for which they are used;
- Be comprehensive in the field of sustainability in parallel with the definition used;
- Have data available for all regions pertaining to the framework, and these data should be available from public sources, scientific or institutional.

Therefore, sustainability indicators should be filtered by the previous criteria to decide whether they are applicable and relevant to the system and whether they fit its definition. Additionally, data availability is significant; data must be authentic and from open sources, allowing access for all stakeholders. Furthermore, the United Nations [59] suggested additional selection criteria for sustainable development indicators, some of which could benefit the development of the SWRM framework, as follows:

- Designed on a national scale;
- Able to assess the progress of sustainability;
- Clear and understandable;
- Part of a conceptual foundation;
- Representative of an international consensus as much as possible, based on the context;
- Within the capacity of the government with regard to development;
- Reliant on cost-effective data of recognized quality.

Thus, it can be considered that any indicator should have specific features to be considered, such as being measurable, understandable, conceptual, and adaptable based on the function for which it is used.

Next, in Section 4, a brief overview of the main elements of the indicator-based assessment framework or index is outlined and briefly illustrated.

4. Main Elements of an Indicator-Based Assessment Framework

Before establishing or developing any assessment framework, it is vital to recognize and identify its main pillars. This process would ensure that the framework or index would be built clearly on a solid foundation. Therefore, the seven main elements of the indicator-based assessment framework, expressed explicitly and implicitly based on the literature analysis, are presented briefly below.

Overall, it can be said that any sustainability framework (or index) is constituted of several key parts: (1) a set of headline categories (components); (2) a set of underpinning indicators for each component, and (3) a set of second-order and possibly third-order sub-indicators [43]. To illustrate, a visual example for one of the SWRM-AFs included in this review (i.e., West Java Water Sustainability Index (WJWSI)- See Section 5.1) is presented in Figure 2, where the components are represented by the blue boxes, the indicators by the green ones, and the sub-indicators by the orange boxes.

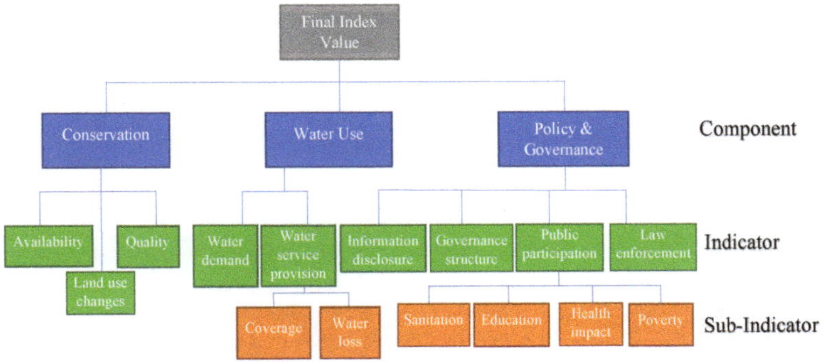

Figure 2. A visual example of the main parts that form an index/framework.

It can be observed that the index in Figure 2 is in a hierarchy shape, where the aggregating direction is a bottom-up process. A particular calculation method (i.e., rescaling or normalization) to have an equivalent value for each indicator and sub-indicator would be applied. Then, the aggregation and the weight of the output value of each sub-indicator would produce the indicator value. The same process is applicable for the resultant values of indicators and components in obtaining the final value of the index (i.e., the top grey box in Figure 2).

4.1. Indicator

The first element in forming an assessment framework or index is the indicator itself, which has the feature of being able to:

- combine with other indicator(s) to produce a component, and/or
- split to create more sections related to the same indicator, with each branch called a sub-indicator.

At the same time, a question might arise about what is meant by an indicator and what is the purpose of using it? Indicators present data about the case of a phenomenon [56], used mainly to measure/assess progress toward sustainability [62]. Moreover, indicators can reveal how countries (or regions) are coping with internal and external goals (e.g., SDGs) and conditions in terms of their sustainability obligations [60].

Indicators and sub-indicators are often objective and quantitative—representing a quantity or change based on metrics (e.g., water leakage rate [%], litres of water per person [l/p]). They may also include other aspects, such as area (e.g., [l/m^2]) or time periods [l/p/d] [l/p/yr]). On the other hand, they can be qualitative and subjective—dealing with cases that cannot be measured by a number, such as opinions, which differ from one person to another [44]. For example, they may be elicited by such questions as "How happy are you with your water provider (5 being very happy and 0 being very unhappy)"? However, in several SWRM frameworks in the literature, the value of a qualitative indicator or sub-indicator is converted into a number based on a pre-defined conditions or criteria to simplify the aggregation process, enabling the calculation of a final equivalent score for each component [63–65]. In general, combining the two types or classifications of indicators in the SWRM frameworks is not uncommon [66], although using only one or the other is more popular [63,65,67,68].

4.2. Benchmark

Furthermore, the second element of the indicator-based assessment framework is the benchmark or target (i.e., an aspired level of performance) with which any indicator and sub-indicator is usually measured or compared [44,68,69]—for example, domestic potable water consumption of 160 l/p/day. Thus, a baseline and specific range (i.e., roadmap and timeline) of values can be developed from any related benchmark to achieve this end goal, which might be to reach 120 l/p/day by 2025 and 80 l/p/day by 2040. This process is considered helpful for stakeholders and decision-makers to gain more comprehensive knowledge about the output of these indicators and enhance their contributions to moving towards, rather than away from, such an end goal.

4.3. Application Scale

Another vital element of the indicator-based assessment framework that needs to be carefully dealt with is the application scale. The scales assigned in the literature for SWRM indices in descending order are usually global, territorial (or regional), local, and community scales, in addition to river basins. Indeed, it is important to understand that the application of each scale might require different criteria and specific guidelines for the selection process of indicators that would form a suitable framework. For example, the Water Poverty Index (WPI) [67] (see Section 5.1) has two different versions/values of the same indicator because of the scale change. In other words, the original version was on a global scale with specific indicators of commonly available data among countries that can serve for this scale [67], while the second version was on a community scale, adding and removing some indicators to fit with the requirements of the case studies [70]. Therefore, knowing the appropriate application scale is essential.

4.4. Normalization Method

The fourth element of the indicator-based framework is the method of calculating sub-index values, or the normalization method (i.e., obtaining equivalent component values for each set of indicators and their following sub-indicators if applicable, as shown above in Figure 2). Before going further, it is essential to note that many indicators under the same index or framework would have different unit values. To illustrate, the water coverage or access indicator, which is common to numerous sustainable water indices, is usually measured as a percentage (%) of people who already have (or are connected to) the water service. On the other hand, the water quality indicator, which is also popular, is typically quantified by a unique summation of different sub-indicators. For instance, water turbidity, which refers to the solutions spectral light absorbance property, or "transparency", and is measured in nephelometric turbidity units (NTU), while another sub-indicator is the concentration of total suspended solids (TSS), measured in (Mg/L) [71]. Furthermore, if these indicators (i.e., water coverage and water quality) are categorized under one component with different unit values, they cannot be aggregated or compared directly. Therefore, a particular method to combine and compare their values as a normalization process should be chosen based on the features of the data and the goal of creating such a framework [43,46].

There are two widely used normalization methods in the literature for sustainable water indices addressing the issue of calculating the sub-index values:

(a) continued re-scaling [67,68], and
(b) categorical scaling [63,65].

The first method is also referred to as empirical normalization [72]. This method is proposed to re-scale the actual values of indicators by converting them mathematically into comparable numbers belonging to an identical interval of numbers ranging from either 0 to 1 or 0 to 100, based on the Equations (1) and (2), respectively [43]:

$$S_i = \frac{X_i - X_{\min}}{X_{max} - X_{\min}} \tag{1}$$

$$S_i = \frac{X_i - X_{\min}}{X_{max} - X_{\min}} \times 100 \tag{2}$$

where S_i is the component value for indicator i, X_i is the actual value for indicator i, and X_{min} and X_{max} are the minimum and maximum threshold values of the indicator, respectively; or in some cases, it can be said that X_{min} is the least-preferred value and the X_{max} is the most-preferred value, which means that to be able to use this method, the threshold values including the minimum and maximum should be identified for each indicator [43]. The advantage of this method is that it is easy and efficient in comparing the initial state of the indicator with alternatives [72]. Overall, this method might be more applicable when the assessment framework has a majority of quantitative indicators in terms of their data.

The second method for obtaining equivalent indicator values is categorical scaling, where the values of indicators are categorized and assigned based on pre-defined criteria [43]. These categories can be numbers, such as from 1 to 10, or descriptions and opinions, such as "low", "medium", or "high".

The general Equation (3) for using this method is presented below [43]:

$$S_i = \begin{cases} Z_j & \text{if} \quad X_i \text{ meets criteria 1} \\ Z_j & \text{if} \quad X_i \text{ meets criteria 2} \\ \vdots & \quad \dots \\ Z_n & \text{if} \quad X_n \text{ meets criteria } n \end{cases} \tag{3}$$

where S_i is the component value for the indicator i, X_i is the actual value for indicator i, Z_j *is* the category for X_i that meets criteria j, and n is the number of categories. Overall, this method has the advantage of providing the ability to work on both quantitative and qualitative data. For instance, because of the diversity of scales and units in their indicator-based system, Silva et al. [65] used a quali–quantitative scale working as a normalization step to aggregate and compare contrasting model elements.

4.5. Weighting Scheme

The fifth element of the indicator-based framework is the weighting scheme that should be considered before doing any aggregation for the product of the previous element (i.e., the normalization method). The weighting scheme is a process of multiplying each part of the indicator-based framework or index by a value representing its importance or weight during each calculation stage to get the final index number. These weighting techniques are classified in general, according to Nardo et al. [46], into two broad categories: (a) statistical-based methods, where weights are given based on the analysis of the indicator data (e.g., [73–76]), and (b) participatory-based methods, where weights are assigned based on the preference of expert decision-makers or stakeholders [43].

However, since the first approach is more complex and not used in most frameworks covered in this study, it is considered outside of the scope of this current paper. In addition, the participatory-based methods are preferred for use in SWRM because they match the Dublin principles' requirements and the definition of the IWRM. Moreover, participatory processes in these assessment types proved valuable and tended to lead to system change through cooperation [77,78]. Nevertheless, it is mandatory prior to using the participatory-based methods to consider providing appropriate justifications for the type of experts or people who have been selected [43], not least because this process might involve subjective judgment [43] and bias.

Furthermore, the weighting distribution scheme can be classified based on the literature of sustainable water indices, particularly in the participatory-based methods, into two schemes:

(a) the equal weights scheme, and
(b) the non-equal weights scheme.

According to Nardo et al. [46], most of the composite indicators, in general, have historically relied on equal weighting, and this also applies to some WR sustainable indices [63,65,67,68]. Indeed, it might be argued that a truly sustainable assessment system should equally balance the main elements of sustainability without introducing bias toward one aspect. For example, carbon and the race to achieve carbon neutrality is one key aspect here.

4.6. Aggregation Technique

The sixth element of the indicator-based framework is the aggregating method for the values of sub-indicators, indicators, and components. There are two common aggregating techniques, which are usually linked to the weighting schemes.

(a) Arithmetic (or linear) method
(b) Geometric method

The first one is the arithmetic (or linear) method, where all the output values of the indicators (or sub-indicators) are added together, then divided by their total number to obtain an equivalent value for each component (or indicator). This method is commonly called the mean or the average, which has the advantage of being simple, and the disadvantage of being sensitive to outlier values. The general expression for this method is shown in Equation (4) [79]:

$$I = \sum_{i=1}^{N} w_i S_i \tag{4}$$

where I is the aggregated component (or indicator), N is the total number of indicators (or sub-indicators) that needs to be calculated, S_i is the sub-index for the indicator i, and w_i is the weight of indicator i. Another feature of this method is that it can ensure perfect substitutability and compensability among sub-index values [46]. However, this method has been criticized, since it might hide or compensate for poor (or low) indicator quality if combined with a high-quality one [43,46,79].

The second method is the geometric aggregation method, where all the weighted sub-index values are multiplied instead of being added as in the arithmetic. Then, the result is powered by the inverse of their total numbers. Moreover, the geometric aggregation method does not have the feature of creating perfect substitutability and compensability among the sub-index values [43]. The general Equation (5) for using this method is given below [79]:

$$I = \prod_{i=1}^{N} {S_i}^{w_i} \tag{5}$$

where the symbols for Equation (4) are the same as for Equation (5); meanwhile, the weights w_i in both equations reflect the relative significance of S_i, and the summation of these weights should always equal one [79].

4.7. Final Index Value

The seventh element of the indicator-based framework is the final index value, which is the final goal of having an index. This element is usually represented by one number, and it is the final score of the standardized procedures of the fourth, fifth, and sixth elements of the indicator-based framework (i.e., normalization method, weighting scheme, and aggregation technique, respectively) [80]. This number is most likely to be from 0 to 100 or 0 to 1. The benefit of having such a number is to make the result of the whole framework easy to understand, not least by a range of different stakeholders, without the need for a more detailed assessment. Furthermore, classified interpretations for the overall sustainability level are sometimes given based on specific ranges of the final index value. For example, in a framework where the final index value is from 0 to 1, the low, intermediate, and high level of sustainability are interpretations for any final value lower than 0.5, from 0.5 to 0.8, and higher than 0.8, respectively [63].

5. Existing Sustainable Water Resources Management Assessment Frameworks (SWRM-AF): An Overview

After the previous brief exploration and explanation of the main elements of the indicator-based assessment framework, it would be helpful to provide an overview of the existing SWRM-AFs and check whether they are applicable to ASAR. Those presented in this section represent the result of the systemic literature review. This section is vital to finding any limitation or knowledge gap(s) in their respective application(s), and to ascertaining whether they would be suitable for application in different local contexts and conditions. For this reason, a specific search was conducted in this paper for every SWRM-AF available in two literature databases since the year 2000 (See Section 2).

Before going further, it is important to remember that this study focuses on the participatory method for the development of an SWRM-AF. This method is a critical process recommended by the principles of IWRM [81], where it is emphasized that stakeholders should be involved in the planning and implementation process [82]. However, in reality, the application of IWRM has faced different issues ranging between the complexity in measuring its effects and the difficulty in applying prescriptive ideals to the decision-making process [83]. Thus, considering that any indicator-based framework relies on a participatory technique would overcome the flaws of the application of IWRM. Additionally, this technique could gain the public's trust and would likely ensure their cooperation with any developed future plans and interventions after assessing their WRM system's sustainability.

5.1. Results of Systematic Literature Review

As illustrated previously in Figure 1 and discussed in Section 2.2, the final number of studies that matched the systematic review requirements from the two databases was narrowed in the final stages to only 23 studies. Of these 23 studies, which were supposed to be taken to the full review stage, 17 original frameworks were identified (Table 1). Inevitably, each of these frameworks has different purposes, uses different assessment techniques, and was made for a specific application at different scales and within diverse local contexts and conditions. Nevertheless, each of them was presented as a supportive tool to either measure or improve the level of sustainability of the WRM system, individually or collectively.

The other six studies were excluded for several reasons. One of these is that they applied one of the other 17 frameworks but with only minor changes. For example, by varying only the case study, which happened with a journal article [30] that applied the same Watershed Sustainability Index (WSI) [63] to a different region. Therefore, it was decided to only include the paper that introduced the original index in this review. In addition, a conference paper that suggested the application of the Canadian Water Sustainable Index (CWSI) to evaluate a specific case study had very few details about the index itself [84]. This was consequently replaced by the original framework published in a previous report [68]. Likewise, a conference paper [85] about some procedures used in developing the Water Needs Index (WNI) was excluded because the same index was provided in full detail in another paper [86] that was included in the review.

Another reason for excluding other papers was when their research served either as guidance on how to make indicators and frameworks with examples [58], or as criticism of the indicators assigned for the SDG number 6 [87].

The last reason for not including some studies in the final comparison, even though they had a framework and indicators, was that their purpose and indicators were not sufficiently focused on improving/assessing the sustainability of WRM. The first study of this type was a conference paper focused on evaluating the United States' infrastructure performance related to the water sector, without careful consideration of other dimensions of sustainability [88]. Similarly, to some degree, another study concentrated to some degree on evaluating the already existing performance indicators related to the water supply network that targeted the issue of water losses [89]. There were three main issues with the previous study: (1) the final product was not compatible with the definition of an index/framework; (2) it had too much technical detail in its indicators that were not all specifically related to sustainability, and (3) the final number of performance indicators reached 117, which did not comply with the guidance with regard to having a simple sustainable framework. Thus, this study was excluded. The remaining studies, ordered from the oldest to newest, are shown in Tables 1 and 2. Further comparative analysis among all frameworks included in Tables 1 and 2 is provided in Section 5.2.

Table 1. Summary and comparison of main elements of existing SWRM-AFs.

SWRM-AF Name	Acronym	Author(s), Year	Number of Indicators			Benchmark	Scale [Location]	Normalization Method	Weighting Scheme	Aggregation Tech.	Final Index Value
			Component	Indicator	S. Indicator						
Water Poverty Index	WPI	(Lawrence et al., 2002) [67]	5	17	15	yes	Global	Continuous rescaling	Equal	Arithmetic	0–100
Canadian Water Sustainability Index	CWSI	(Policy Research Initiative, 2007) [68]	5	15	×	yes	Community [2] [Canada]	Continuous rescaling	Equal	Arithmetic	0–100
Watershed Sustainability Index	WSI	(Chaves and Alipaz, 2007) [63]	4	15	×	yes	Local & regional [2] [Brazil]	Categorical rescaling	Equal	Arithmetic	0–1
West Java Water Sustainability Index	WJWSI	(Juwana et al., 2010) [90,91]	3	9	6	yes	Territorial (regional) [2] [Indonesia]	Continuous + Categorical rescaling	Equal + non-equal	Geometric	0–100
Water Needs Index	WNI	(Moglia et al., 2012) [86]	6	9	×	yes	Local (ward & district) [Vietnam]	Continuous rescaling	Non-equal (user defined)	Arithmetic	0–100
Water & Sanitation Sustainability Index	WASSI	(Iribarnegaray et al., 2015) [48]	9	15	2	yes	Local (urban & peri-urban) [Argentina]	Continuous + categorical rescaling	Equal	Arithmetic	0–100
Global Water Security Index	GWSI	(Gain et al., 2016) [7]	4	10	×	yes	Global	Continuous rescaling	Non-equal (authors defined)	Arithmetic	0–1
Hybrid Triple Bottom Line & Multi-criteria Decision Analysis	TBL-MCDA [1]	(Cole et al., 2018) [92]	3	44	×	yes	Local & community [USA]	Categorical rescaling	Equal	Arithmetic	1–5 [3]
Freshwater Health Index	FHI	(Vollmer et al., 2018) [93]	3	11	31	yes	Local & regional [2] [China]	Continuous + categorical rescaling	Equal + non-equal	Geometric+ Arithmetic	0–100 [3]
Assessing Water Security & Water–Energy–Food Nexus	WEF nexus [1]	(Marttunen et al., 2019) [49]	4	17	×	yes	National [Finland]	Categorical rescaling	×	×	×
Municipal Environmental Management	MEM	(Criollo et al., 2019) [94]	4	40	×	yes	Local & regional [Colombia]	Continuous rescaling	Non-equal (user defined)	Arithmetic	0–1

Table 1. *Cont.*

SWRM-AF Name	Acronym	Author(s), Year	Number of Indicators			Benchmark	Scale [Location]	Normalization Method	Weighting Scheme	Aggregation Tech.	Final Index Value
			Component	Indicator	S. Indicator						
River Basin Water Sustainability Index	RBWSI	(Silva et al., 2020) [65]	3	8	19 (54)	yes	Territorial regional [2] [N/A]	Categorical rescaling	Equal	Arithmetic	0–1
Water Sensitive Cities Index	WSC	(Rogers et al., 2020) [78]	7	34	×	yes	Local (metropolitan/municipal) [Australia]	Categorical rescaling	×	Arithmetic	1–5 [3]
Malaysia Manufacturing Industry Water Benchmarking System	MIWABS	(Bahar et al., 2020) [80]	4	9	×	yes	Factories level [Malaysia]	Proximity-to-target + categorical rescaling	Non-equal (user defined)	Arithmetic	0–100
Indicators of Integrated Water Resource Management	IIWRM [1]	(Ben-Daoud et al., 2021) [95]	4	12	×	yes	Local [2] [Morocco]	Categorical rescaling	Equal	Arithmetic	1–5
Sustainability Index	SI	(Najar and Persson, 2021) [96]	3	14	82	yes	Local [Sweden]	Survey (categorical rescaling)	Equal	Arithmetic	0–2
Rural Water Sustainability Index	RWSI	(Crispim et al., 2021) [97]	5	21	58	yes	Rural & community [Brazil]	Categorical rescaling	Non-equal (user defined)	Arithmetic	0–10
		Average	4.5	17.6	30.3						

[1] Indicates a suggested acronym; [2] designed for river basin scale; [3] does not have a final index value but a final value for each component only.

Table 2. Summary of why and how the existing SWRM-AFs have been developed with pros and cons.

Acronym [Reference]	Purpose	Selection Process for Indicators	Stakeholders Involved	Advantage	Disadvantage
WPI [67]	To find the relation between the water availability or scarcity impacts on the welfare level of human populations among 147 countries	Literature review then stakeholder opinion	Physical & social experts, academics, practitioners, others	Good range of stakeholders, helpful for general comparisons	General nature (or base) of indicators can neglect internal important issues related to the context of specific regions
CWSI [68]	To evaluate water sustainability and well-being in Canadian communities concerning freshwater	Literature review then stakeholder workshop	Government officials, academics, consultants	Participatory method with stakeholders in refining the selected indicators	Developed only for communities that depend on river basins
WSI [63]	To combine the treatment of the three pillars of sustainability within an integrated and dynamic process	Literature review	None	Equal weighting of indicators to ensure mutual respect among all sectors	No stakeholder engagement, developed only for river basins
WJWSI [90,91]	To identify main factors help improving WR, to assist in prioritize issues of WRM, and to communicate current condition of WR to community	Literature review then conceptual framework, then Delphi application & stakeholders' interview	Academics, consultants, government officials, community representatives	Participatory method with stakeholders in refining the selected indicators, good range of stakeholders	Developed for river basins particularly in Indonesia, unclear way of combination of normalization methods
WNI [86]	To pinpoint persistent water problems and hotspots that local water authorities should address	Literature review then stakeholder workshop	Academics, government officials	Participatory method with stakeholders in refining the selected indicators & assigning weights for components only	Indicator weightings assigned by researchers alone, component of aquatic ecosystems is specific for surface water
WASSI [48]	Developed as a tool to support governance procedures for more SWRM, applied to four cities in northern Argentina	Developed in collaboration with the provincial water company	Government officials, water Company	Helpful in comparing level of SWRM among cities, new information/data easily uploaded to web-interface	Website in Spanish, only one stakeholder group involved in the indicator selection process
GWSI [7]	To integrate physical and socio-economic aspects of security within a SWRM index	Literature review	None	Helpful for general comparisons, water security evaluation maps are well developed	General nature (or base) of indicators because of global scale, no stakeholder engagement
TBL-MCDA [92]	To evaluate the pillars (lenses) of sustainability related to using alternative water supply strategies versus maintaining the conventional system.	Developed in collaboration with technical experts & stakeholders	Technical experts, city departments, non-profit organization	Good range of stakeholders, performance indicators used with stakeholder preferences to support decision-making	Unclear if literature review used, Indicator number too large to be implemented in practical way, no final index value calculated.
FHI [93]	To integrate the multiple social, ecological, and governance dimensions toward the sustainability of freshwater management.	Literature review then scientific workshops & stakeholder opinion	Scientific experts, local stakeholders	Stakeholder engagement—include for indicator selection and partially in weightings	No final index value calculated. Developed for river basins
WEF nexus [49]	To evaluate water security and its trends in the future through a participatory process, and to analyse connections with water, energy, and food security in Finland	Literature review then stakeholder workshop	Academics, government officials, security organizations	Stakeholder engagement, high-level interviews, excel tool with different sheets	Highly qualitative, missed three main elements, difficult to use in other contexts/settings

Table 2. *Cont.*

Acronym [Reference]	Purpose	Selection Process for Indicators	Stakeholders Involved	Advantage	Disadvantage
MEM [94]	To create, as a bottom-up approach, a WRM that can measure local government administrations' dedication to sustainability	Literature review & stakeholder opinion then Interviews and online surveys	Academics, government & municipal officials, social organizations	Participatory method with stakeholders in refining the selected indicators and weights, Results published in a website	Environmental focus, large number of indicators that needed aggregation
RBWSI [65]	To evaluate and guide the decision-making process in promoting water sustainability as part of integrated river basin management (IRBM)	Literature interrogation	None	Literature reviewed using an inductive approach	No stakeholder engagement, large number of sub-indicators that needed aggregation, developed for river basins
WSC [78]	To evaluate a city's water sensitivity, create aspirational goals, and guide management actions to enhance water-sensitive processes	Literature review then consultation with stakeholders	Industry experts, academics	Participatory method for developing indicators and scoring system	High number of indicators, weightings seem ambiguous, no final index value
MIWABS [80]	To evaluate the industrial sector's water performance within a factory-level scale in Malaysia	Literature review then stakeholder workshop to screen & filter	Industry experts, academics	Weighting used analytical hierarchy process (AHP) applied to questionnaire output	Method for aggregation not reported, scale applicable to factory alone
IIWRM [95]	To produce an indicator-based framework to evaluate the application of IWRM within Meknes city, Morocco	Literature review then survey of stakeholder via questionnaires	Government officials (water sector actors), practitioners	Easy to interpret radar diagram used for displaying results	No evidence/justification for calculations or weighting scheme provided
SI [96]	To evaluate and guide Sweden's municipal water and wastewater sectors to be more sustainable	Swedish Water and Wastewater Association (SWWA) developed framework	Members of SWWA, water utilities of the municipalities	Annual survey—rigorously developed and well-written, simply to use/understand, results published in a web-based database	High number of sub-indicators, yearly application would have huge time, resource implications
RWSI [97]	To help decision-makers in the process of finding and prioritizing rural communities that need state intervention with regard to water provision	Literature review then Delphi method via questionnaires to stakeholders	Policymakers, technicians, experts, others	Participatory method with stakeholders in refining the selected indicators and weights	High number of indicators, mostly applicable to rural communities

5.2. Comparative Analysis of Existing SWRM-AFs

After the brief illustration of all the frameworks obtained from the systematic literature review (see Tables 1 and 2) a comparative analysis is performed in order to collectively get valuable observations and insights. The comparative analysis is undertaken using the aspects previously detailed in Section 4 and the key headings shown in Tables 1 and 2.

5.2.1. Number and Type of Components

The first observation was in regard to the number of components (Figure 3), where their total number was 76, while the different investigated frameworks used an average number of 4.5 components. Moreover, thirteen frameworks (76.5% of the total) opted for three to five components, with four being the most widely adopted featuring within six studies (35.3% of the total), whilst three and five components were featured in four and three frameworks (i.e., 29.4% and 17.6% of total), respectively. The other frameworks adopted six, seven, or nine components (23.5% of total). The highest number of components (9) was found in WASSI [48] and the least numbers of components (3) were found in RBWSI [65], FHI [93], TBL-MCDA [92] and WJWSI [90,91]. Based on this observation, it can be suggested that for any new SWRM-AF being developed, the number of components should preferably stay within the threshold of three to five, with a preference of four, since it was the most repeated number.

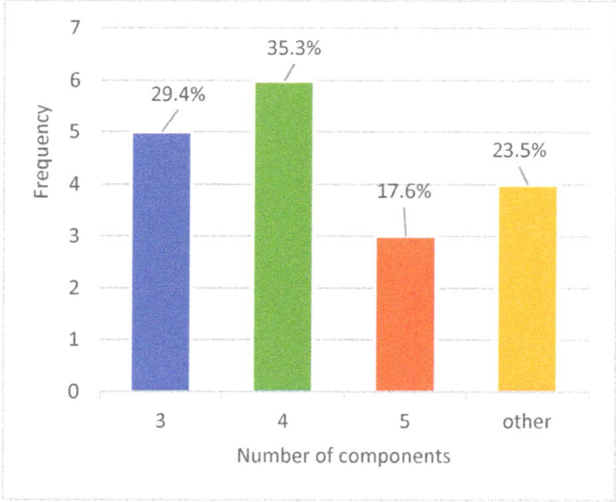

Figure 3. Total number of components used in each framework.

Regarding the types of components, a thematic analysis was conducted to categorize them in two steps. The first step was to check the common words in the title of the components that were repeated based on their numbers. A criterion was suggested to eliminate any word repeated less than three times. Therefore, only 63 components distributed among 14 main words were included in this analysis, as seen in Figure 4. The most-repeated words were "resource" and "water" (i.e., seven times for each), followed by "environment" and "access", which were mentioned six times. In contrast, "capacity", "social", "infrastructure", "quality", and "service" were the least-repeated words, with only three repetitions for each.

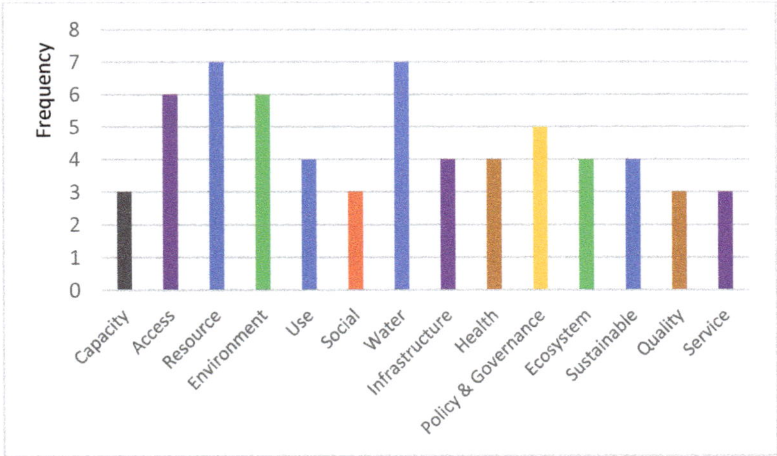

Figure 4. Number of most repeated-words in the titles of components.

Further investigation, which was the second step, highlighted that thematic categorization was possible by combining those categories in Figure 4 that served the same theme, as shown in Figure 5.

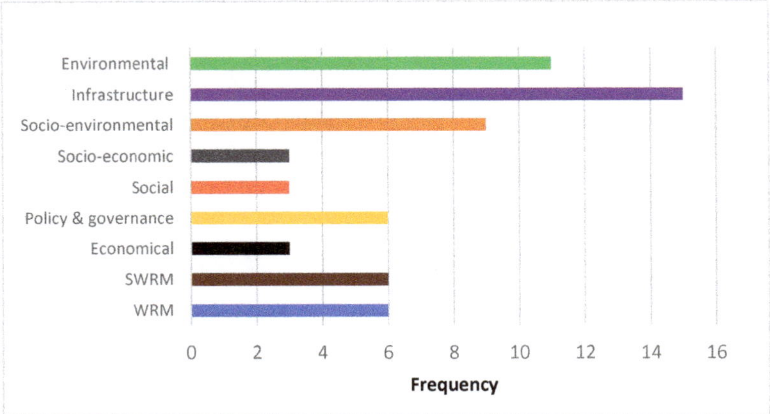

Figure 5. Main themes of components based on their repeated number.

Overall, it can be seen that the infrastructure, environmental, and socio-environmental components are critical in any SWRM-AF, since they have the biggest shares.

5.2.2. Number of Indicators

The second observation concerned the number of indicators. From the interrogation of Table 1, it can be seen that the average number of indicators in all included frameworks was 17.6 indicators. However, it can also be seen that most frameworks (twelve–70.6% of total) had a total number of indicators ranging between 9 [73,83,88] and 17 [49,67] (inclusive), leading to an average of 12.75 in this discrete group. The most repeated number of indicators therein were nine [80,86,91] and fifteen [48,63,68], where each of these numbers was found in three of the seventeen frameworks. Four of the remaining frameworks (i.e., 23.5% of total) had a higher number of indicators, 21 in RWSI [97], 34 in WSC [78], 40 in MEM [94], and 44 in TBL-MCDA [92], respectively, while only one study (i.e., RBWSI [65]) had a lower number, with eight indicators. The lower number was not typical; however, this framework had a unique design, with two orders of sub-indicators.

5.2.3. Number of Sub-Indicators and Benchmarks

In terms of sub-indicators, Table 1 shows that they were not always available. In other words, only seven frameworks (41.1% of total) included them. The average number was 30.3 sub-indicators, with a minimum of 2 in WASSI [48] and a maximum of 82 in SI [96]. In terms of benchmarking, all frameworks reviewed contained these (see Table 1)

5.2.4. Scale of Application

Various scales can be seen within the frameworks reviewed (Figure 6).

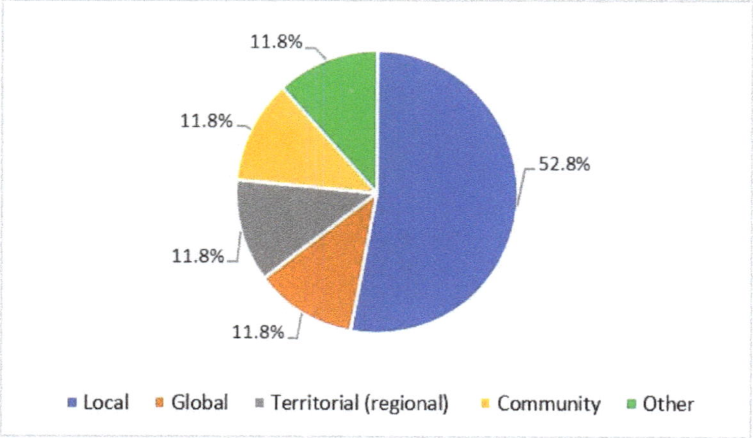

Figure 6. Scale of Application.

The global scale appeared only twice in WPI [67] and GWSI [7], likely because the amount of time, effort, and required data are extensive. The scale with the most significant share (9 studies or 52.8%) tended toward the local (mainly city) scale whilst the remaining six studies were evenly split between the community scale [68,92] and territorial (regional) scale [63,94], which refers to large areas, such as those with several cities. The last of these is the "other" category, with two frameworks, which included the national and factory scales [80,96]. It is also worth noting that six studies (i.e., 35.3% of total) considered areas with river basins [63,65,68,90,93,95].

5.2.5. Normalization Process

All percentages for the process of normalization are shown in Figure 7. Eight studies used categorical rescaling [49,63,65,78,92,95–97], and five studies used continuous rescaling [7,67,68,86,94], with 47.1% and 29.4% of the total share, respectively. Three studies (17.6% of total) used a combination of both [48,90,93]. This option is not common, because this task would be confusing for non-expert stakeholders. On the other hand, only one framework (i.e., MIWABS [80]) used a different approach: the Proximity-to-Target [80], which happened to be a very close match to continuous rescaling, albeit with subtle, nuanced differences.

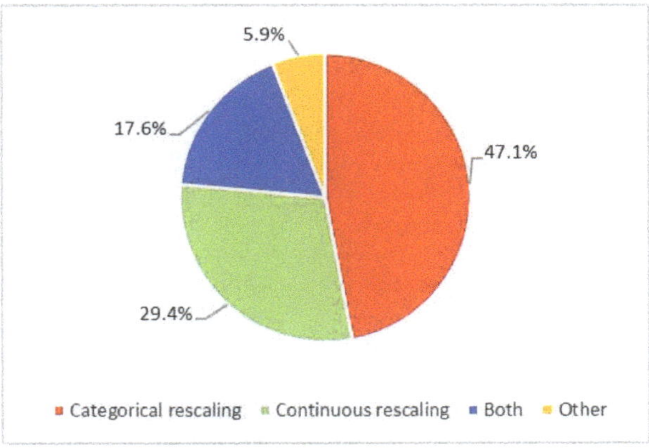

Figure 7. Percentage of Normalization Method.

5.2.6. Weighting Process

The process of weighting indicators and components was seen in 15 of the 17 (90%) frameworks reviewed (Figure 8). Out of all the frameworks reviewed, the preference for allocating equal weights was dominant in eight studies [48,63,65,67,68,92,95,96] with a percentage of 47.1%. This aligns with the ethos of sustainability, which is about balancing, rather than trading off, respective pillars. Five studies [7,80,86,94,97] considered the non-equal weights (user-defined), with a percentage of 29.3% of the total. Only two frameworks [90,93] (11.8% of total) adopted a combination of both approaches.

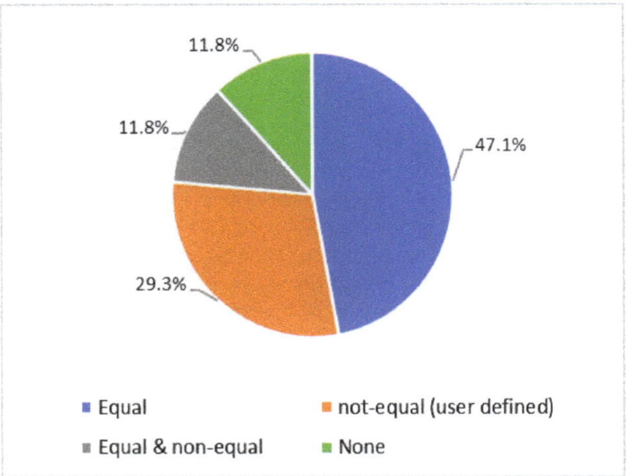

Figure 8. Percentage of Weighting Scheme.

5.2.7. Aggregation Technique and Final Index Value

The next element is the aggregation technique, which is used in combination with the weighting scheme in order to reach a final index value. Most frameworks [7,48,63,67,68, 78,80,86,92,94–97] (i.e., 14 or 82.35%) relied on the arithmetic technique—calculating the average rescaling value of indicators. The geometric technique was used twice: one time alone [90] and the other in combination with the arithmetic technique [93]. In contrast, the WEF nexus framework [49] used neither aggregation technique nor final index value.

In Figure 9, it can be seen that the most widely adopted interval for the final index value (with 41.2% of total) was 0 and 100 [48,67,68,80,86,90,93]. Therefore, it can be suggested that this interval was the most preferred choice for both experts and stakeholders within the frameworks reviewed. The second most widely adopted interval for final index value (with 23.5% of total) was 0 and 1 [7,63,65,94]. The third most widely adopted interval for final index value (with 17.6% of total) was 1 to 5 [78,92,95]. A category called "other" was used to combine any final index value with a unique range that appeared once in the frameworks reviewed. This happened in only two indices [96,97] (11.8% of total). The last category called "without" for the WEF nexus framework [49] (5.9% of total), which does not have any final index value. Meanwhile, only the final assessment for each indicator (or criterion as they called it) is provided with a qualitative description in an individual assessment card without aggregating all indicators or components to get a single final value.

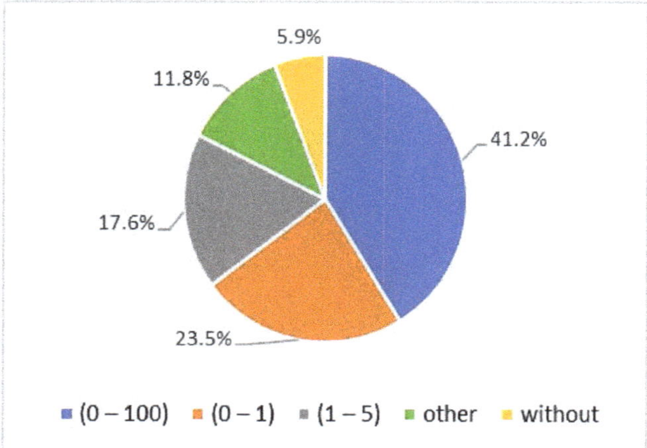

Figure 9. Interval of Final Index Value.

5.3. SWRM-AF for ASAR

After going through the systematic literature review, it was clear that no dedicated SWRM-AF had been explicitly made for or applied to ASAR—these areas lack any permanent rivers or river basins. That said, the frameworks reviewed in Sections 5.1 and 5.2 had considerable use in being developed (in whole or part) for such purposes.

By furthering the scope of the review to the grey literature outside the two databases that were checked previously, another two additional frameworks were identified, namely the Arab Water Sustainability Index (AWSI) [98] and the Abu Dhabi Water Index (ADWI) [99]. Thus, an overview and brief analysis is provided in the following sections to check their effectiveness.

5.3.1. Arab Water Sustainability Index (AWSI)

The AWSI is presented as a monitoring tool to address the water sustainability state relative to a base condition or period [98]. The scale of its application could be considered a national scale. In this index, 22 Arab countries, where 82.2% of their weather is either arid or semi-arid, were evaluated through four main components that were divided into only eight indicators. These components can be classified by checking their indicators and main themes or categories from Figure 5, as follows:

1. water crowding (related to WRM category);
2. water dependency (related to SWRM category);
3. water scarcity (related to SWRM category);
4. environmental sustainability (related to socio-environmental category).

Based on our comparative analysis above, it can be said that four components are an adequate number; however, the number of indicators (eight only) is lower than the average number of indicators, which was found to be approximately 18. Other main elements of indicator-based framework were used, such as the benchmark, the aggregation technique (i.e., arithmetic), and the final index value ranging between 0 to 100%. The normalization method of AWSI is based on a statistical method (i.e., principal components analysis (PCA)), which was also used to assign weights, which were not equal, for each component and indicator. A unique advantage of AWSI is the consideration of conventional and non-conventional WR (e.g., GW and desalination water, respectively), which is crucial for ASAR.

Meanwhile, the continuous rescaling method as a normalization method was mentioned, but whether that was for application or just information was unclear. Overall, even though the pillars of sustainability were considered, the stakeholders' participation in all phases did not exist in AWSI, which does not match the general guidance in developing such a framework. Therefore, to avoid such limitations, it is still required to have a more helpful framework that can gain public trust and cooperation.

5.3.2. Abu Dhabi Water Index (ADWI)

The other framework is the ADWI, which was developed through the adoption of the cause–effect approach (DSR—Driving force, State, Response) to deal with the challenging context of WRM of the United Arab Emirates (UAE) (i.e., very much at a local scale) [99]. The selection of indicators was based on a review of the literature, followed by checking of the availability of their data and whether they were relevant to the UAE environment. This process resulted in four categories (i.e., components), nineteen indicators, and twelve sub-indicators. Then, the benchmark for most of the indicators was obtained from the literature. These components with our main themes or categories from Figure 5 are as follows:

1. water availability (related to WRM category);
2. water quality (related to socio-environmental category);
3. water use efficiency (related to SWRM category);
4. policy and governance (related to policy & governance category).

Overall, the methodology for building ADWI was well-organized and systematically illustrated. In addition, taking the conventional and non-conventional WR into account is essential for the context of ASAR, which is another advantage similar to AWSI.

On the other hand, while considering all sustainability pillars in any SWRM-AF is significant, little attention was given here to the economic pillar. Additionally, ADWI seemed to lack any stakeholder participation or involvement. However, an indicator to measure the public participation in water activities existed, but it was based only on the researchers' evaluation. Moreover, the normalization method seems to equate to categorical rescaling. Still, the scoring criteria were not entirely clear (i.e., all scores were either good, represented by happy face, or poor, represented by sad face, while only one seemed neutral).

Furthermore, the weighting scheme, aggregation technique, and final index value did not exist in this methodology, except for the calculation of sub-indicators. Therefore, it can be said that ADWI was an attempt to develop a particular framework for ASAR, but with many limitations. Hence, it is important to build or develop an SWRM-AF that could avoid these flaws and is suitable to fit the main requirements and contexts of ASAR by considering stakeholder participation. A summary of the main elements that form the above two SWRM-AFs is presented in Table 3 to make the process of comparing them simpler.

Table 3. Summary and comparison of existing SWRM-AF for ASAR.

SWRM-AF	Number of indicators			Benchmark	Scale [Location]	Normalization	Weighting Scheme	Aggregation Tech.	Final Index Value
	Component	Indicator	S. Indicator						
AWSI	4	8	×	yes	National [Arab countries]	Principal components analysis	Non-equal	Arithmetic	0–100
ADWI	4	19	12	yes	Local [UAE]	Categorical rescaling	×	×	×

6. Discussion

This research sought to identify whether any existing SWRM-AF would be suitable for application in arid or semi-arid regions; by way of Section 6.1, this is explored further. Section 6.2 identifies the shortfalls of this research before the next steps of research are determined in Section 6.3.

6.1. Existing SWRM-AFs and Their Applicability for ASAR

The review has helped identify six key requirements that a framework would need in order for it to be considered appropriate for application in ASAR. In other words, they should:

1. adopt a participatory approach (i.e., stakeholder engagement) during the selection process of indicators and assigning weights;
2. have appropriate numbers of indicators;
3. include all seven primary elements of the indicator-based framework (Sections 4.1–4.7).
4. include water scarcity (WS) as a key theme;
5. consider all Water Resources (WR)—conventional and non-conventional;
6. fit with an ASAR context.

With this in mind, Table 4 provides a synthesis of the analysis to evaluate (by way of grading) the 19 frameworks, including those from the systematic review and the previous two SWRM-AFs found in the grey literature. All of the checking aspects are based on the six requirements mentioned above. The first three aspects (i.e., 1 to 3) are considered general but essential for inclusion in any SWRM-AF. The last three aspects (i.e., 4 to 6) are specific and considered vital to any SWRM-AF for ASAR. In Table 4, one point was assigned for each aspect included—based on its existence, except for the participatory approach, where a point was equally divided between the selection and weighting. Additionally, half of the maximum point was given if the aspect was either partially fulfilled or partially existed. This meant a maximum value of 6 could be achieved where a framework met all six criteria fully.

Table 4. Evaluation of the applicability of each SWRM-AF for ASAR.

SWRM-AF	Participatory Approach		Number of Indicators	7 Main Elements	Water Scarcity	All WR	Fit ASAR	Total
	Selection	Weighting						
WPI	0.5	0	1	1	1	0	0.5	4
CWSI	0.5	0	1	1	1	0	0	3.5
WSI	0	0	1	1	0	0	0	2
WJWSI	0.5	0	1	1	1	0	0	3.5
WNI	0.5	0.25	1	1	0	0	0	2.75
WASSI	0.5	0	1	1	1	0	0	3.5
GWSI	0	0.25	1	1	1	0	0.5	3.75

Table 4. *Cont.*

SWRM-AF	Participatory Approach		Number of Indicators	7 Main Elements	Water Scarcity	All WR	Fit ASAR	Total
	Selection	Weighting						
TBL-MCDA	0.5	0.25	0	0.5	0	0	0	1.25
FHI	0.5	0.25	1	0.5	0	0	0	2.25
WEF nexus	0.5	0	1	0	0	0	0	1.5
MEM	0.5	0.5	0	1	0	0	0	2
RBWSI	0	0	0	1	1	0	0	2
WSC	0.5	0.25	0	0.5	0	0	0.5	1.75
MIWABS	0.5	0.5	1	1	0	0	0	3
IIWRM	0.5	0	1	1	0	0	0	2.5
SI	0.5	0	1	1	0	0	0.5	3
RWSI	0.5	0.5	0	1	1	0	0.5	3.5
AWSI	0	0	0	1	1	1	1	4
ADWI	0	0	1	0	1	1	1	4

Table 4 shows that the highest total points was 4 out of 6, found in three frameworks (i.e., WPI, AWSI, and ADWI). While two of these frameworks were developed mainly for ASAR (i.e., AWSI and ADWI), there were some general requirements identifiable by a zero in the respective columns. In other words, this research showed that there is no SWRM-AF that could be considered fully fit-for-purpose for application in ASAR. Hence, steps should be taken to address this gap in knowledge (See Section 6.3).

6.2. Shortfalls of this Research

This review paper goes some way towards filling the gap in knowledge with respect to identifying whether an SWRM-AF for ASAR exists. However, it should be noted that the review was restricted to two well-known academic databases (i.e., Scopus and Engineering Village) in addition to the search terms and filtering process adopted herein. Broadening the review to other databases (e.g., Google Scholar and Research Gate, to name just two) may have identified more literature (including grey literature) beyond the two most applicable papers found. In addition, this research was very much focused on the derivation of the frameworks themselves, and not on the detailing (and usefulness) of individual indicators or the data availability enabling the actual measurement of their values. Hence, whilst the need for a new framework was identified by this review, more stages of research are required during its derivation (See Section 6.3).

6.3. Next Stage of Research

The next area of research will seek to develop a SWRM-AF for ASAR that satisfies all six aspects outlined in Section 6.1 In order to ensure it is both practical and meaningful for application, a conceptual framework for ASAR will be developed. This will involve some key steps:

- providing a detailed map of all components, indicators and sub-indicators;
- developing the methodology for selecting important indicators for each component;
- justifying (by way of stakeholder engagement) indicators and weights adopted;
- applying the framework to case studies (likely Kingdom of Saudi Arabia and elsewhere).
- refining SWRM-AF based on user feedback.

7. Conclusions

The sustainability of water supply to match proper demand is crucial for any future planning for the WRM system. This strategy became more significant in areas with limited WR and located in ASAR with challenging water conditions. During the last few decades, many scientific meetings and recommendations were conducted and presented to tackle the WRM issues, such as the Brundtland's definition, the Dublin principles, and the IWRM definition and principles. These efforts were the foundation for introducing guidance and criteria that led to the creation of several SWRM-Afs, such as those manifested above. However, it is essential to remember that sustainability does not mean focusing only on one pillar. Attaining a balance between the three pillars (i.e., environmental, social, economic) would generate the best results. This consideration should be accounted for during the development process of any tool that aims to improve and monitor sustainability progress. One of these tools is the indicator-based framework for assessing sustainability. Therefore, having specific and clear SWRM-AF to measure the level of SWRM would undoubtedly help improve the longevity of such vital resources.

Whilst many SWRM-AFs were developed for this purpose in the past, such as those described briefly in this review, it has been shown that they are insufficient to assess some ASAR. Moreover, even where frameworks have been developed specifically for ASAR, many shortfalls exist. That said, this review helps recognize the primary elements required to establish this type of framework. Moreover, detailed investigation and comparison among SWRM-AFs have helped identify similarities, differences, and limitations/knowledge gaps. As such, several recommendations are suggested based on the results of this review:

- Before establishing or developing an SWRM-AF, it is important to consider and comply with the specific guidelines and criteria for having one. Otherwise, the output of this process would not be practical and rigorous enough.
- Then, having all seven standard main elements of SWRM-AF clearly defined and justified during both the development and application stages will make the SWRM-AF less challenging to reapply in general. This includes its adoption by the scientific community and water authorities in regions with similar conditions. In contrast, ignoring some of these elements could reduce the whole benefit of the framework and make it obsolescent.
- For any new SWRM-AF, it would be preferred to select elements and normalization methods with a higher application rate, such as those highlighted in Section 5.2. For example, while the application of local scale (52.8%) and the final index value of [0–100] (41.2%) seem more popular in many frameworks, categorical rescaling (47.1%), equal weighting (47.1%), and the arithmetic technique (82.35%) are the most commonly used normalization methods. Thus, choosing them might ensure more confidence in both decision-makers and the public in the output of such a framework.
- The participation of stakeholders in developing SWRM-AF is essential, and helps reveal their main concerns. Their involvement could occur during the process of indicator selection, in the weighting stage, or in both stages. Hence, bias in the output of SWRM-AF can be eliminated or at least reduced, while motivation and awareness of SWRM among stakeholders would be higher. More importantly, this participation, especially by expert stakeholders, could be part of the validation for the SWRM-AF, making it more credible.
- Finally, it was found that the SWRM-AF for ASAR for particular countries without any permanent rivers or lakes is needed, since water shortage conditions are a serious threat for these countries, and little or inadequate research has been conducted to develop such a tool.

Therefore, a conceptual SWRM-AF for ASAR is recommended to tackle this issue, and its development is currently underway.

Author Contributions: Conceptualization, B.S.A.; methodology, B.S.A.; investigation, B.S.A.; resources, B.S.A.; data curation, B.S.A.; writing—original draft preparation, B.S.A.; writing—review and editing, B.S.A., D.V.L.H., and S.S.; visualization, B.S.A.; supervision, D.V.L.H. and S.S.; project administration, D.V.L.H. All authors have read and agreed to the published version of the manuscript.

Funding: D.V.L.H. received the financial support of the UK EPSRC under grant EP/J017698/1 (Transforming the Engineering of Cities to Deliver Societal and Planetary Wellbeing, known as Livable Cities) and EP/P002021 (From Citizen to Co-innovator, from City Council to Facilitator: Integrating Urban Systems to Provide Better Outcomes for People, known as Urban Living Birmingham).

Institutional Review Board Statement: Not applicable.

Informed Consent Statement: Not applicable.

Data Availability Statement: Not applicable.

Acknowledgments: B.S.A. gratefully acknowledges the financial support of Qassim University in Saudi Arabia given to him during his doctoral studies.

Conflicts of Interest: The authors declare no conflict of interest.

Abbreviations

Abbreviations	
ADWI	Abu Dhabi Water Index
AHP	Analytical hierarchy process
ASAR	Arid and semi-Arid regions
AWSI	Arab Water Sustainability Index
CWSI	Canadian Water Sustainability Index
CVI	Climate Vulnerability Index
DSR	Driving force, state, response
FHI	Freshwater Health Index
GW	Groundwater
GWSI	Global Water Security Index
IIWRM	Indicators of integrated water resources management
IRBM	Integrated river basin management
IWRM	Integrated water resources management
MEM	Municipal environmental management
MIWABS	Malaysia manufacturing industry water benchmarking system
NTU	Nephelometric turbidity units
PCA	Principal components analysis
PTT	Proximity-to-target
RBWSI	River basin water sustainability index
RWSI	Rural water sustainability index
SI	Sustainability Index
SDG	Sustainable development goal
SWRM	Sustainable water resources management
SWRM-AF	Sustainable water resources management assessment framework
SWWA	Swedish Water and Wastewater Association
TBL-MCDA	Hybrid triple bottom line & multicriteria decision analysis
TSS	Total suspended solids
UAE	United Arab Emirates
WASSI	Water & Sanitation Sustainability Index
WEF nexus	Water–energy–food nexus
WJWSI	West Java Water Sustainability Index
WNI	Water Needs Index
WPI	Water Poverty Index
WR	Water resources
WRM	Water resources management
WSC	Water Sensitive Cities Index
WSI	Watershed Sustainability Index

Notations	
i	indicator (or component)
I	aggregated indicator (or component)
j	criteria
N	total number of indicators
S	sub-index value
w	weight
X	actual value
X_{max}	maximum threshold value
X_{min}	minimum threshold value
Z	category

References

1. UNDESA. *World Population Prospects 2019: Highlights*; Department of Economic and Social Affairs, Population Division: New York, NY, USA, 2019.
2. Maslow, A.H. A Theory of Human Motivation. *Psychol. Rev.* **1943**, *50*, 370–396. [CrossRef]
3. Maslow, A.H. *Motivation and Personality*; Motivation and personality; Harpers: Oxford, UK, 1954; p. xiv, 411.
4. Falkenmark, M.; Lundqvist, J.; Widstrand, C. Macro-Scale Water Scarcity Requires Micro-Scale Approaches. *Nat. Resour. Forum* **1989**, *13*, 258–267. [CrossRef] [PubMed]
5. Strayer, D.L.; Dudgeon, D. Freshwater Biodiversity Conservation: Recent Progress and Future Challenges. *J. North Am. Benthol. Soc.* **2010**, *29*, 344–358. [CrossRef]
6. Mekonnen, M.M.; Hoekstra, A.Y. Four Billion People Facing Severe Water Scarcity. *Sci. Adv.* **2016**, *2*, e1500323. [CrossRef] [PubMed]
7. Gain, A.K.; Giupponi, C.; Wada, Y. Measuring Global Water Security towards Sustainable Development Goals. *Environ. Res. Lett.* **2016**, *11*, 124015. [CrossRef]
8. Şen, Z.; Al-Harithy, S.; As-Sefry, S.; Almazroui, M. Aridity and Risk Calculations in Saudi Arabian Wadis: Wadi Fatimah Case. *Earth Syst. Environ.* **2017**, *1*, 26. [CrossRef]
9. Cirilo, J.A. Public Water Resources Policy for the Semi-Arid Region. *Estud. Avançados* **2008**, *22*, 61–82. [CrossRef]
10. Jolly, I.D.; McEwan, K.L.; Holland, K.L. A Review of Groundwater–Surface Water Interactions in Arid/Semi-Arid Wetlands and the Consequences of Salinity for Wetland Ecology. *Ecohydrology* **2008**, *1*, 43–58. [CrossRef]
11. Almazroui, M.; Islam, M.N.; Balkhair, K.S.; Şen, Z.; Masood, A. Rainwater Harvesting Possibility under Climate Change: A Basin-Scale Case Study over Western Province of Saudi Arabia. *Atmos. Res.* **2017**, *189*, 11–23. [CrossRef]
12. Shomar, B.; Darwish, M.; Rowell, C. What Does Integrated Water Resources Management from Local to Global Perspective Mean? Qatar as a Case Study, the Very Rich Country with No Water. *Water Resour. Manag.* **2014**, *28*, 2781–2791. [CrossRef]
13. Odhiambo, G.O. Water Scarcity in the Arabian Peninsula and Socio-Economic Implications. *Appl. Water Sci.* **2017**, *7*, 2479–2492. [CrossRef]
14. Alyamani, M.S. Salinity Problem of Groundwater in the Wadi Tharad Basin, Saudi Arabia. *GeoJournal* **1999**, *48*, 291–297. [CrossRef]
15. Kinzelbach, W.; Brunner, P.; von Boetticher, A.; Kgotlhang, L.; Milzow, C. Sustainable Water Management in Arid and Semi-Arid Regions. In *Groundwater Modelling in Arid and Semi-Arid Areas*; Wheater, H., Mathias, S., Li, X., Eds.; Cambridge University Press: Cambridge, UK, 2010; pp. 119–130. ISBN 978-051-176-028-0.
16. Roberts, D.A.; Johnston, E.L.; Knott, N.A. Impacts of Desalination Plant Discharges on the Marine Environment: A Critical Review of Published Studies. *Water Res.* **2010**, *44*, 5117–5128. [CrossRef] [PubMed]
17. Ghaffour, N.; Missimer, T.M.; Amy, G.L. Technical Review and Evaluation of the Economics of Water Desalination: Current and Future Challenges for Better Water Supply Sustainability. *Desalination* **2013**, *309*, 197–207. [CrossRef]
18. Davies, P.A.; Yuan, Q.; Richter, R. de Desalination as a Negative Emissions Technology. *Environ. Sci.: Water Res. Technol.* **2018**, *4*, 839–850. [CrossRef]
19. Vörösmarty, C.J.; Green, P.; Salisbury, J.; Lammers, R.B. Global Water Resources: Vulnerability from Climate Change and Population Growth. *Science* **2000**, *289*, 284–288. [CrossRef]
20. Lobell, D.B.; Burke, M.B.; Tebaldi, C.; Mastrandrea, M.D.; Falcon, W.P.; Naylor, R.L. Prioritizing Climate Change Adaptation Needs for Food Security in 2030. *Science* **2008**, *319*, 607–610. [CrossRef]
21. Blair, A.C.; Sanger, D.M.; Holland, A.F.; White, D.L.; Vandiver, L.A.; White, S.N. Stormwater Runoff—Modeling Impacts of Urbanization and Climate Change. In Proceedings of the American Society of Agricultural and Biological Engineers Annual International Meeting 2011, ASABE 2011, Louisville, KY, USA, 7–10 August 2011; Volume 6, pp. 4918–4933.
22. Mujumdar, P.P. Climate Change: A Growing Challenge for Water Management in Developing Countries. *Water Resour. Manag.* **2013**, *27*, 953–954. [CrossRef]
23. GWP-TAC. *Background Paper No. 4. Integrated Water Resources Management*; Technical Advisory Committee of Global Water Partnership: Stockholm, Sweden, 2000; p. 71.
24. Hanjra, M.A.; Qureshi, M.E. Global Water Crisis and Future Food Security in an Era of Climate Change. *Food Policy* **2010**, *35*, 365–377. [CrossRef]
25. Allouche, J. The Sustainability and Resilience of Global Water and Food Systems: Political Analysis of the Interplay between Security, Resource Scarcity, Political Systems and Global Trade. *Food Policy* **2011**, *36*, S3–S8. [CrossRef]
26. Vörösmarty, C.J.; Pahl-Wostl, C.; Bunn, S.E.; Lawford, R. Global Water, the Anthropocene and the Transformation of a Science. *Curr. Opin. Environ. Sustain.* **2013**, *5*, 539–550. [CrossRef]
27. Cuthbert, M.O.; Gleeson, T.; Moosdorf, N.; Befus, K.M.; Schneider, A.; Hartmann, J.; Lehner, B. Global Patterns and Dynamics of Climate–Groundwater Interactions. *Nat. Clim. Change* **2019**, *9*, 137–141. [CrossRef]
28. ICWE. *The Dublin Statement on Water and Sustainable Development*; International Conference on Water and the Environment: Dublin, Ireland, 1992.
29. Ni, X.; Wu, Y.; Wu, J.; Lu, J.; Wilson, P.C. Scenario Analysis for Sustainable Development of Chongming Island: Water Resources Sustainability. *Sci. Total Environ.* **2012**, *439*, 129–135. [CrossRef] [PubMed]
30. de Castro, C.O.; Loureiro, Ó.C.S.; Santos, A.V.; Silva, J.; Rauen, W.B. Water sustainability assessment for the region of Curitiba. *Int. J. Sustain. Build. Technol. Urban Dev.* **2017**, *8*, 184–194. [CrossRef]

31. WCED. *Our Common Future*; World Commission on Environment and Development; Oxford University Press: Oxford, UK, 1987; p. 300.
32. Elkington, J. *Cannibals with Forks: The Triple Bottom Line of 21st Century Business*; Capstone: Oxford, UK, 1997; ISBN 978-190-096-127-1.
33. United Nations. *Transforming Our World: The 2030 Agenda for Sustainable Development*; United Nations: New York, NY, USA, 2015.
34. Basiago, A.D. Economic, Social, and Environmental Sustainability in Development Theory and Urban Planning Practice. *Environmentalist* **1998**, *19*, 145–161. [CrossRef]
35. Stirling, A. The Appraisal of Sustainability: Some Problems and Possible Responses. *Local Environ.* **1999**, *4*, 111–135. [CrossRef]
36. Lehtonen, M. The Environmental–Social Interface of Sustainable Development: Capabilities, Social Capital, Institutions. *Ecol. Econ.* **2004**, *49*, 199–214. [CrossRef]
37. Gibson, R.B. Beyond the Pillars: Sustainability Assessment as a Framework for Effective Integration of Social, Economic and Ecological Considerations in Significant Decision-Making. *J. Env. Assmt. Pol. Mgmt.* **2006**, *08*, 259–280. [CrossRef]
38. Kuhlman, T.; Farrington, J. What Is Sustainability? *Sustainability* **2010**, *2*, 3436–3448. [CrossRef]
39. Purvis, B.; Mao, Y.; Robinson, D. Three Pillars of Sustainability: In Search of Conceptual Origins. *Sustain. Sci.* **2019**, *14*, 681–695. [CrossRef]
40. UN. *Overall Progress Achieved since the United Nations Conference on Environment and Development*; Commission on sustainable development fifth session; United Nations: New York, NY, USA, 1997.
41. Moldan, B.; Janoušková, S.; Hák, T. How to Understand and Measure Environmental Sustainability: Indicators and Targets. *Ecol. Indic.* **2012**, *17*, 4–13. [CrossRef]
42. Maiolo, M.; Pantusa, D. Sustainable Water Management Index, SWaM_Index. *Cogent Eng.* **2019**, *6*, 1603817. [CrossRef]
43. Juwana, I.; Muttil, N.; Perera, B.J.C. Indicator-Based Water Sustainability Assessment—A Review. *Sci. Total Environ.* **2012**, *438*, 357–371. [CrossRef] [PubMed]
44. Bertule, M.; Bjørnsen, P.K.; Costanzo, S.D.; Escurra, J.; Freeman, S.; Gallagher, L.; Kelsey, R.H.; Vollmer, D. *Using Indicators for Improved Water Resources Management: Guide for Basin Managers and Practitioners*; 2017; ISBN 978-879-063-405-6. Available online: https://ian.umces.edu/site/assets/files/11189/using-indicators-for-improved-water-resources-management.pdf (accessed on 28 March 2022).
45. Sullivan, C. Calculating a Water Poverty Index. *World Dev.* **2002**, *30*, 1195–1210. [CrossRef]
46. Nardo, M.; Saisana, M.; Saltelli, A.; Tarantola, S.; Hoffman, A.; Giovannini, E. *Handbook on Constructing Composite Indicators: Methodology and User Guide*; OECD Publishing: Paris, France, 2005; No. 2005/03. [CrossRef]
47. Streeten, P. Human Development: Means and Ends. *Bangladesh Dev. Stud.* **1994**, *21*, 65–76.
48. Iribarnegaray, M.A.; D'Andrea, M.L.G.; Rodriguez-Alvarez, M.S.; Hernández, M.E.; Brannstrom, C.; Seghezzo, L. From Indicators to Policies: Open Sustainability Assessment in the Water and Sanitation Sector. *Sustainability* **2015**, *7*, 14537–14557. [CrossRef]
49. Marttunen, M.; Mustajoki, J.; Sojamo, S.; Ahopelto, L.; Keskinen, M. A Framework for Assessing Water Security and the Water–Energy–Food Nexus—The Case of Finland. *Sustainability* **2019**, *11*, 2900. [CrossRef]
50. Topal, H.F.; Hunt, D.V.L.; Rogers, C.D.F. Urban Sustainability and Smartness Understanding (USSU)—Identifying Influencing Factors: A Systematic Review. *Sustainability* **2020**, *12*, 4682. [CrossRef]
51. Encyclopedia Britannica. *Water Resource*; Encyclopedia Britannica: Chicago, IL, USA, 2021.
52. Gleick, P.; Gomez, S.; Loh, P.; Morrison, J. *California Water 2020: A Sustainable Vision*; Pacific Institute Report; Pacific Institute for Studies in Development, Environment, and Security: Oakland, CA, USA, 1995; p. 113.
53. Cosgrove, W.J.; Loucks, D.P. Water Management: Current and Future Challenges and Research Directions. *Water Resour. Res.* **2015**, *51*, 4823–4839. [CrossRef]
54. Ivanova, D.; Stadler, K.; Steen-Olsen, K.; Wood, R.; Vita, G.; Tukker, A.; Hertwich, E.G. Environmental Impact Assessment of Household Consumption. *J. Ind. Ecol.* **2016**, *20*, 526–536. [CrossRef]
55. Joumard, R.; Nicolas, J.-P. Transport Project Assessment Methodology within the Framework of Sustainable Development. *Ecol. Indic.* **2010**, *10*, 136–142. [CrossRef]
56. Bradley Guy, G.; Kibert, C.J. Developing Indicators of Sustainability: US Experience. *Build. Res. Inf.* **1998**, *26*, 39–45. [CrossRef]
57. Loucks, D.P.; Gladwell, J.S. (Eds.) *Sustainability Criteria for Water Resource Systems*; UNESCO International Hydrology Series; Cambridge University Press: Cambridge, UK, 1999.
58. Lundie, S.; Peters, G.; Ashbolt, N.; Lai, E.; Livingston, D. A Sustainability Framework for the Australian Water Industry. *Water J. Aust. Water Assoc.* **2006**, *33*, 83–88. [CrossRef]
59. United Nations (Ed.) *Indicators of Sustainable Development: Guidelines and Methodologies*, 3rd ed.; United Nations: New York, NY, USA, 2007; ISBN 978-921-104-577-2.
60. da Silva, J.; Fernandes, V.; Limont, M.; Rauen, W.B. Sustainable Development Assessment from a Capitals Perspective: Analytical Structure and Indicator Selection Criteria. *J. Environ. Manag.* **2020**, *260*, 110147. [CrossRef]
61. Bell, S.; Morse, S. *Sustainability Indicators: Measuring the Immeasurable*, 2nd ed.; Earthscan: London, UK, 2008; ISBN 978-185-383-498-1.
62. Hiremath, R.B.; Balachandra, P.; Kumar, B.; Bansode, S.S.; Murali, J. Indicator-Based Urban Sustainability—A Review. *Energy Sustain. Dev.* **2013**, *17*, 555–563. [CrossRef]
63. Chaves, H.M.L.; Alipaz, S. An Integrated Indicator Based on Basin Hydrology, Environment, Life, and Policy: The Watershed Sustainability Index. *Water Resour. Manag.* **2007**, *21*, 883–895. [CrossRef]

64. Juwana, I.; Muttil, N.; Perera, B.J.C. Application of West Java Water Sustainability Index to Three Water Catchments in West Java, Indonesia. *Ecol. Indic.* **2016**, *70*, 401–408. [CrossRef]
65. da Silva, J.; Fernandes, V.; Limont, M.; Dziedzic, M.; Andreoli, C.V.; Rauen, W.B. Water Sustainability Assessment from the Perspective of Sustainable Development Capitals: Conceptual Model and Index Based on Literature Review. *J. Environ. Manag.* **2020**, *254*, 109750. [CrossRef]
66. Juwana, I. Development of a Water Sustainability Index for West Java, Indonesia. Ph.D. Thesis, Victoria University, Footscray, VIC, Australia, 2012.
67. Lawrence, P.; Meigh, J.; Sullivan, C. *The Water Poverty Index:An International Comparison;* Keele Economics Research Papers; Centre for Economic Research, Keele University: Keele, UK, 2002; revised March 2003.
68. Policy Research Initiative. *Canadian Water Sustainability Index (CWSI): Project Report*; Policy Research Initiative: Ottawa, ON, Canada, 2007; ISBN 978-066-245-045-0.
69. Liverman, D.M.; Hanson, M.E.; Brown, B.J.; Merideth, R.W. Global Sustainability: Toward Measurement. *Environ. Manag.* **1988**, *12*, 133–143. [CrossRef]
70. Sullivan, C.A.; Meigh, J.R.; Giacomello, A.M. The Water Poverty Index: Development and Application at the Community Scale. *Nat. Resour. Forum* **2003**, *27*, 189–199. [CrossRef]
71. Parnian, A.; Furze, J.N.; Parnian, A.; Mayad, E.H. Water Purification Plantations for Oil and Gas Industries in Iran. *Environ. Sci. Pollut. Res.* **2021**, *28*, 64193–64198. [CrossRef] [PubMed]
72. Shilling, F. *The California Water Sustainability Indicators Framework: Draft Final Report*; 2013; p. 309. Available online: https://aquadoc.typepad.com/files/cwp_sif_phase-ii_draft_final_report_dec2013.pdf (accessed on 28 March 2022).
73. Hashimoto, T.; Stedinger, J.R.; Loucks, D.P. Reliability, Resiliency, and Vulnerability Criteria for Water Resource System Performance Evaluation. *Water Resour. Res.* **1982**, *18*, 14–20. [CrossRef]
74. Loucks, D.P. Quantifying Trends in System Sustainability. *Hydrol. Sci. J.* **1997**, *42*, 513–530. [CrossRef]
75. McMahon, T.A.; Adeloye, A.J.; Zhou, S.-L. Understanding Performance Measures of Reservoirs. *J. Hydrol.* **2006**, *324*, 359–382. [CrossRef]
76. Sandoval-Solis, S.; McKinney, D.C.; Loucks, D.P. Sustainability Index for Water Resources Planning and Management. *J. Water Resour. Plan. Manag.* **2011**, *137*, 381–390. [CrossRef]
77. Rijke, J.; Farrelly, M.; Brown, R.; Zevenbergen, C. Configuring Transformative Governance to Enhance Resilient Urban Water Systems. *Environ. Sci. Policy* **2013**, *25*, 62–72. [CrossRef]
78. Rogers, B.C.; Dunn, G.; Hammer, K.; Novalia, W.; de Haan, F.J.; Brown, L.; Brown, R.R.; Lloyd, S.; Urich, C.; Wong, T.H.F.; et al. Water Sensitive Cities Index: A Diagnostic Tool to Assess Water Sensitivity and Guide Management Actions. *Water Res.* **2020**, *186*, 116411. [CrossRef]
79. Swamee, P.K.; Tyagi, A. Describing Water Quality with Aggregate Index. *J. Environ. Eng.* **2000**, *126*, 451–455. [CrossRef]
80. Bahar, N.S.; Noor, Z.Z.; Aris, A.; Kamaruzaman, N.A.B. An Indicator Framework Approach on Manufacturing Water Assessment towards Sustainable Water Demand Management. *J. Environ. Treat. Tech.* **2020**, *8*, 875–883.
81. Xie, M. Integrated Water Resources Management (IWRM)—Introduction to Principles and Practices. In Proceedings of the Africa Regional Workshop on IWRM, Nairobi, Kenya, 29 October 29–3 November 2006.
82. Mostert, E.; Craps, M.; Pahl-Wostl, C. Social Learning: The Key to Integrated Water Resources Management? *Water Int.* **2008**, *33*, 293–304. [CrossRef]
83. Giordano, M.; Shah, T. From IWRM Back to Integrated Water Resources Management. *Int. J. Water Resour. Dev.* **2014**, *30*, 364–376. [CrossRef]
84. Attari, J.; Mojahedi, S.A. Water Sustainability Index: Application of CWSI for Ahwaz County. In Proceedings of the World Environmental and Water Resources Congress 2009, Kansas City, MO, USA, 17–21 May 2009; Volume 342, pp. 1664–1670.
85. Moglia, M.; Nguyen, M.N.; Neumann, L.E.; Cook, S.; Nguyen, T.H. Integrated Assessment of Water Management Strategies: Framework and Case Study. In Proceedings of the 20th International Congress on Modelling and Simulation, MODSIM 2013, Adelaide, SA, Australia, 1–6 December 2013; pp. 2262–2268.
86. Moglia, M.; Neumann, L.E.; Alexander, K.S.; Nguyen, M.N.; Sharma, A.K.; Cook, S.; Trung, N.H.; Tuan, D.D.A. Application of the Water Needs Index: Can Tho City, Mekong Delta, Vietnam. *J. Hydrol.* **2012**, *468–469*, 203–212. [CrossRef]
87. Guppy, L.; Mehta, P.; Qadir, M. Sustainable Development Goal 6: Two Gaps in the Race for Indicators. *Sustain. Sci.* **2019**, *14*, 501–513. [CrossRef]
88. Lee, J.; McNeil, S.; Li, Q. "Joshua" Development of a Water Infrastructure Performance Index for the USA. In Proceedings of the World Environmental and Water Resources Congress 2019, Pittsburgh, PA, USA, 19–23 May 2009; pp. 424–431. [CrossRef]
89. Kanakoudis, V.; Tsitsifli, S.; Samaras, P.; Zouboulis, A.; Banovec, P. A New Set of Water Losses-Related Performance Indicators Focused on Areas Facing Water Scarcity Conditions. *Desalination Water Treat.* **2013**, *51*, 2994–3010. [CrossRef]
90. Juwana, I.; Perera, B.J.C.; Muttil, N. A Water Sustainability Index for West Java. Part 1: Developing the Conceptual Framework. *Water Sci. Technol.* **2010**, *62*, 1629–1640. [CrossRef]
91. Juwana, I.; Perera, B.J.C.; Muttil, N. A Water Sustainability Index for West Java—Part 2: Refining the Conceptual Framework Using Delphi Technique. *Water Sci. Technol.* **2010**, *62*, 1641–1652. [CrossRef]
92. Cole, J.; Sharvelle, S.; Fourness, D.; Grigg, N.; Roesner, L.; Haukaas, J. Centralized and Decentralized Strategies for Dual Water Supply: Case Study. *J. Water Resour. Plan. Manag.* **2018**, *144*, 05017017. [CrossRef]

93. Vollmer, D.; Shaad, K.; Souter, N.J.; Farrell, T.; Dudgeon, D.; Sullivan, C.A.; Fauconnier, I.; MacDonald, G.M.; McCartney, M.P.; Power, A.G.; et al. Integrating the Social, Hydrological and Ecological Dimensions of Freshwater Health: The Freshwater Health Index. *Sci. Total Environ.* **2018**, *627*, 304–313. [CrossRef]
94. Criollo, R.; Malheiros, T.; Alfaro, J.F. Municipal Environmental Management Indicators: A Bottom-Up Approach Applied to the Colombian Context. *Soc. Indic. Res.* **2019**, *141*, 1037–1054. [CrossRef]
95. Ben-Daoud, M.; Moumen, A.; Sayad, A.; Elbouhadioui, M.; Essahlaoui, A.; Eljaafari, S. Indicators of Integrated Water Resources Management at the Local Level: Meknes as a Case (Morocco). In Proceedings of the E3S Web of Conferences, Kenitra, Morocco, 25–27 December 2021; Volume 234.
96. Najar, N.; Persson, K.M. A Sustainability Index within Water and Wastewater Management in Sweden: An Evaluation of Eight Case Studies. *Water* **2021**, *13*, 1879. [CrossRef]
97. Crispim, D.L.; Pimentel Da Silva, G.D.; Fernandes, L.L. Rural Water Sustainability Index (RWSI): An Innovative Multicriteria and Participative Approach for Rural Communities. *Impact Assess. Proj. Apprais.* **2021**, *39*, 320–334. [CrossRef]
98. Ali, H.M.M. Development of Arab Water Sustainability Index Using Principal Component Analysis. In Proceedings of the IWTC13, Hurghada, Egypt, 12–15 March 2009.
99. Alsalmi, H.; Elkadi, H.A.; Leao, S. Urban Growth in Arid Environments: Developing Water Sustainability Indicators for Abu Dhabi (UAE). In Proceedings of the 6th Making Cities Liveable Conference, in Conjunction with the Sustainable Transformation Conference, Melbourne, VIC, Australia, 17–19 June 2013.

remote sensing

MDPI

Article

Analyzing Driving Factors of Drought in Growing Season in the Inner Mongolia Based on Geodetector and GWR Models

Bowen Ji [1], Yanbin Qin [1,*], Tingbin Zhang [1,2], Xiaobing Zhou [3], Guihua Yi [4], Mengting Zhang [1] and Menglin Li [1]

1 College of Earth Science, Chengdu University of Technology, Chengdu 610059, China
2 State Environmental Protection Key Laboratory of Synergetic Control and Joint Remediation for Soil & Water Pollution, Chengdu University of Technology, Chengdu 610059, China
3 Geological Engineering Department, Montana Technological University, Butte, MT 59701, USA
4 College of Tourism and Urban-Rural Planning, Chengdu University of Technology, Chengdu 610059, China
* Correspondence: qinyanbin@cdut.cn

Abstract: As an important ecological security barrier in northern China, the Inner Mongolia Autonomous Region (hereinafter referred to as Inner Mongolia) is seriously affected by drought. It is of great significance to characterize the spatial distribution of drought and identify the influencing factors of drought. Due to complex interactions among drought driving factors, it is difficult to quantify the contribution of each driving factor to drought using linear correlation analysis alone. In this study, we used the Standardized Precipitation Evapotranspiration Index (SPEI) as a quantitative indicator of drought to discuss the spatiotemporal variation of drought during growing seasons in the Inner Mongolia from 2000 to 2018. We quantitatively characterized mode, scope, and intensity of changes in SPEI caused by drought-influencing factors such as weather, water, topography, soil, and human activities using the Geodetector and Geographically Weighted Regression (GWR) models. We concluded that about 20.3% of the region showed a downward trend in SPEI, with the fastest rate of decline in the central and western Inner Mongolia. Air temperature, precipitation, elevation, and distance to rivers are the main controlling factors in drought change, and the factor interactions showed nonlinear enhancement. The drought driving effect was obvious in Alxa League, Wuhai City, Ulanqab City, and Baotou City. The results will help us to understand the effects of the driving factors on drought and eventually help policymakers with water-resource management.

Keywords: growing season; driving factors of drought; Geodetector; GWR model; Inner Mongolia

Citation: Ji, B.; Qin, Y.; Zhang, T.; Zhou, X.; Yi, G.; Zhang, M.; Li, M. Analyzing Driving Factors of Drought in Growing Season in the Inner Mongolia Based on Geodetector and GWR Models. *Remote Sens.* **2022**, *14*, 6007. https://doi.org/10.3390/rs14236007

Academic Editor: Giorgio Baiamonte

Received: 15 October 2022
Accepted: 23 November 2022
Published: 27 November 2022

1. Introduction

As one of the most serious meteorological and environmental disasters, drought can severely impact the natural environment, crop production, social economy, and human life, but its impact mode is not easy to be quantified [1]. The Sixth Assessment Report (AR6) of the Intergovernmental Panel on Climate Change (IPCC) points out that in the past 30 years, the global average temperature has increased by 1.5 °C, extreme climate events occur frequently, and the degree of drought will continue to increase in the future. Drought impacts species and structure of vegetation. It is an important factor affecting vegetation growth, vegetation restoration, and soil desertification [2–4]. Changes in hydrothermal conditions can lead to biomass loss and ecosystem destruction. Therefore, investigating the spatiotemporal variation of drought during the growing season (from April to September) in Inner Mongolia, identifying causes of drought, and separating and quantifying relative contributions of the controlling factors of drought are of practical significance for drought remediation and ecosystem restoration.

Due to uncertainties in starting and ending times, spatial scale, time lag effects, and other factors of drought events, researchers mainly monitor and analyze drought effects through a series of drought indicators [1,5]. The Palmer Drought Severity Index (PDSI)

is the most widely used water-balance-based meteorological drought index, which comprehensively considers water supply and demand. However, it has limitations in judging short-term droughts [6]. The Standardized Precipitation Index (SPI) calculates the probability of precipitation distribution; however, it is difficult to handle the task of meteorological drought monitoring under the context of global change [7]. The Standardized Precipitation Evapotranspiration Index (SPEI) leverages the advantages of PDSI and SPI [8,9]. It not only considers the balance of water and energy, but also reflects the deficit and accumulation process of surface water. Therefore, it is widely used in climate studies [7,10], agriculture [11], hydrology [12,13], and so on. At the same time, SPEI can be calculated at multiple time scales. SPEI-3, used to characterize drought in a seasonal scale, reflects short-term regional meteorological drought. It has a direct correlation with grassland biomass and vegetation growth [14,15], and is an important index of vegetation to study drought in a growing season.

The China-Mongolia Arid and Semiarid Area (CMASA) is one of the eight major arid regions in the world, and it is also an inland arid region with the highest latitude. Inner Mongolia is located in the transition region between the arid and semi-arid areas in the east of CMASA. Due to the perennial influence of the westerly wind system, atmospheric circulation, and pressure field of Qinghai-Tibet Plateau, temperature rise in the west of Inner Mongolia is significantly higher than that in the inland and surrounding areas of China, and is particularly sensitive to climate change [16,17]. Inner Mongolia is China's main grassland for pasture and agriculture. It is an important ecological barrier to the North of China. For a long time, Inner Mongolia has suffered from frequent regional and local droughts, which have significantly impacted the local economy. The intensification of desertification caused by droughts has become the primary ecological and environmental concern in Inner Mongolia [18]. There are many research activities on long-term drought monitoring in Inner Mongolia: An et al. [19] analyzed the spatiotemporal variation of droughts in Inner Mongolia in the past 60 years; Pei et al. [20] compared the differences and applicability of SPI and SPEI drought indexes at different time scales; Tong et al. [21] used linear regression and wavelet analysis to identify drought changes and drought patterns; however, few studies have quantitatively explained the causes of the droughts. In the past, drought analysis and regional water resource planning were mainly based on linear correlation between factors [22–24]. However, drought is a complex regionalization event. It is generally hard to refine intensity and interaction among various factors in different regions using just the traditional linear regression analysis [25,26]. At the same time, drought is closely related to natural conditions, human activities, and their interactions. However, interactions among these factors have not been well-investigated [27]. Different land cover types, soil conditions, topography, and other factors may cause spatial differentiation of local drought. Geodetector and GWR are statistical models considering spatial nonstationarity and the modeling process is simple but intuitive. A combination of the two can accurately describe the action, path, and intensity of the influencing factors and has a good application prospect [28–30].

In this study, the seasonal SPEI-3 index (SPEI for short) was calculated based on the data at the meteorological stations in Inner Mongolia; the spatiotemporal variation trend of SPEI in the growing season from 2000 to 2018 was obtained using univariate linear trend analysis. The main controlling factors of drought change were identified through Geodetector. The GWR model was used to quantitatively evaluate the effect of various driving factors on SPEI change during the growing season, and to explain the interaction between the main controlling factors for spatial heterogeneity.

2. Materials and Methods

2.1. Study Area

Inner Mongolia is located in the northern border of China, spanning three major regions of northwest, north, and northeast, spreading along a long and narrow belt. The region covers a total area of about 1.183 million km^2 that accounts for about 12.1% of

China's total area. The region is rich in resources, with grasslands, forests, and arable land per capita ranking first in China. The Greater Khingan Range runs through the east of the study area in the north-south orientation. The Yin Mountains extend east-west in the south. Large deserts such as the Badain Jaran Desert, Tengger Desert, and Mu Us Desert are located in the west. The study area has an average altitude of about 1000 m. The climate in the region varies from arid-semiarid monsoon climate to humid-semi humid climate. It is often affected by cyclones on the Mongolian Plateau with strong wind in spring and by Lake Baikal, the world's largest freshwater lake by volume containing 22–23% of the world's fresh surface water. The climate is often controlled by prevailing westerlies belts or subtropical high-pressure belts, with high temperature and little rain in summer. Annual rainfall showed a decreasing trend from east to west and from north to south [22,31].

2.2. Data Sources and Preprocessing

2.2.1. Meteorological Data

Daily mean air temperature, monthly cumulative precipitation, daily mean wind speed, and daily mean sunshine duration at 110 meteorological stations (Figure 1a) in and around Inner Mongolia from 2000 to 2018 were selected as meteorological data, which were provided by the China Meteorological Data Service Centre (http://data.cma.cn/). A homogeneity test of the meteorological data was carried out to fill in unavailable data. A statistical analysis on the mean air temperature (°C), accumulated precipitation (mm), mean wind speed (m·s^{-1}), and mean sunshine duration (h) at stations was performed for the growing seasons from 2001 to 2018. Spatial resolution for ordinary Kriging interpolation was set to 1 km × 1 km; the geographic reference was set as WGS84/UTM zone 48°N.

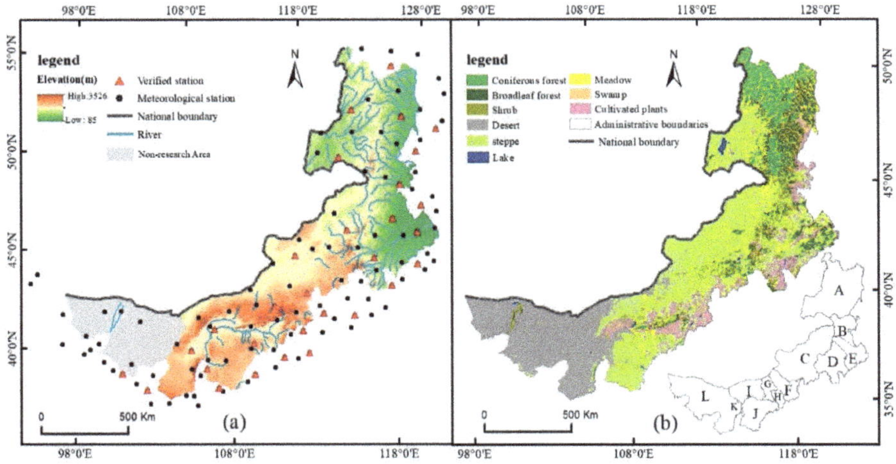

Figure 1. (**a**) The Digital Elevation Model (DEM) and meteorological station distribution, and (**b**) the vegetation types in Inner Mongolia (A—Hulunbuir, B—Hinggan League, C—Xilingol League, D—Chifeng, E—Tongliao, F—Ulangab League, G—Baotou, H—Hohhot, I—Bayannur League, J—Erdos, K—Wuhai, L—Alxa League).

2.2.2. DEM Data

DEM (Digital Elevation Model) was the Shuttle Radar Topography Mission (SRTM) data with a resolution of 90 m downloaded from the United States Geological Survey (USGS) data portal (http://earthexplorer.usgs.gov). After preprocessing, such as mosaicking and void-filling, the accuracy of the input topographic data had a standard error of 1 m. The DEM data were resampled to 1 km × 1 km and slope and aspect were then derived from the DEM data.

2.2.3. Other Data Sets

Population density data was from the population dataset produced by the Landscan Global team (https://landscan.ornl.gov/), with a spatial resolution of 0.01° × 0.01° (about 1 km). The data were produced according to the community standard of global population distribution data constructed from multivariable geographic dasymetric modeling and remote sensing image analysis [32]. Land use conversion type can effectively reflect intensity of human activities [33]. It is also an important surface condition for drought research. The land cover data in 2000 and 2018 were from the Resource and Environment Science Data Center of Chinese Academy of Sciences (http://www.resdc.cn). The data set was obtained by visual interpretation of Landsat TM/ETM+ images for different periods and was widely used [25]. Land cover was classified into six basic categories: cultivated land, forest, grassland, water area, construction land, and unused land. A land use conversion map from 2000 to 2018 was generated. Soil sediment contents were from the 1:1 million soil type map and the soil profile data was obtained from the second soil survey (http://www.resdc.cn). Soil texture was classified according to the content of sand, silt, and clay. The content of particles in different soils was shown as a percentage. Spatial data of main rivers and county stations were derived from the 1:4 million vector database provided by the National Geographic Center of China (http://ngcc.sbsm.gov.cn). Distance to Rivers (DTR) and Distance to Cities (DTC) were obtained through buffer zone analysis, and the spatial resolution was set to 1 km × 1 km. The 1:1 million vegetation type data released by the Resource and Environment Science Data Center of the Chinese Academy of Sciences (http://www.resdc.cn) were used. The vegetation in Inner Mongolia was reclassified into 8 classes: coniferous forest, broad-leaved forest, shrub, grassland, meadow, swamp, desert, and cultivated vegetation.

2.3. Methods

2.3.1. Calculation of the Standard Precipitation Evapotranspiration Index (SPEI)

The difference between precipitation and potential evapotranspiration (*PET*), *PD*, is a key parameter for SPEI calculation. *PET* was calculated using the Thornthwaite method because fewer meteorological elements are required [9], as follows:

$$PD_i = P_i - PET_i \tag{1}$$

$$PET = 16.0 \times \left(\frac{10Ti}{H}\right)^A \tag{2}$$

$$A = 6.75 \times 10^{-7}H^3 - 7.71 \times 10^{-5}H^2 + 1.79 \times 10^{-2}H + 0.492 \tag{3}$$

where P_i is the monthly precipitation of the *i*-th month, PET_i the monthly evapotranspiration, A a constant, and H the annual heat index. The log-logistic function based on three parameters (α, β, γ) was used to perform the normal fitting to the time series of PD_i and compute the probability distribution function $F(x)$. The log-logistic probability distribution function is given as below:

$$F(x) = [1 + (\frac{\alpha}{x-\gamma})^{\beta}]^{-1} \tag{4}$$

where α is the scale parameter, β the shape parameter, and γ the position parameter; all are obtained by the linear-moment method.

The probability distribution function was standardized to obtain the cumulative probability Q (Equation (5)):

$$Q = 1 - F(x) \tag{5}$$

and the SPEI value was then calculated as:

$$SPEI = \begin{cases} w - \frac{a_0 + a_1 w + a_2 w^2}{1 + d_1 w + d_2 w^2 + d_3 w^3}, w = \sqrt{-2\ln(Q)} (Q \le 0.5) \\ -(w - \frac{a_0 + a_1 w + a_2 w^2}{1 + d_1 w + d_2 w^2 + d_3 w^3}), w = \sqrt{-2\ln(1-Q)} (Q \ge 0.5) \end{cases} \tag{6}$$

where the constants $a_0 = 2.515517$, $a_1 = 0.802853$, $a_2 = 0.010328$, $d_1 = 1.432788$, $d_2 = 0.189269$, and $d_3 = 0.001308$. The degree of drought (Table 1) was classified according to the local climate conditions [21,34].

Table 1. Drought classification based on SPEI.

SPEI Value	Drought
>1	Severe wet
(0.5, 1]	Moderate wet
(0, 0.5]	Light wet
(−0.5, 0]	Light drought
(−1, −0.5]	Moderate drought
<−1	Severe drought

2.3.2. Trend Analysis

The univariate linear regression equation (Equation (7)) was used for trend analysis to calculate the variation trend of SPEI during the growing seasons from 2000 to 2018:

$$\theta_{slope} = \frac{n \times \sum_{i=1}^{n} (i \times SPEI_i) - \sum_{i=1}^{n} i \sum_{i=1}^{n} SPEI_i}{n \times \sum_{i=1}^{n} i^2 - (\sum_{i=2}^{n} i)^2} \quad (7)$$

where n ($n = 19$) is the length of time series and θ_{slope} is the slope in the linear regression equation. $\theta_{slope} > 0$ indicates that the drought trend is reduced; otherwise, the drought is aggravated. The variation trend of SPEI was divided into five levels based on the standard deviation (STD), i.e., significant degradation ($\theta_{slope} < -STD$), slight degradation ($-STD < \theta_{slope} < -0.5STD$), substantially unchanged ($-0.5STD < \theta_{slope} < 0.5STD$), slight improvement ($0.5STD < \theta_{slope} < STD$), and significant improvement ($\theta_{slope} > STD$).

2.3.3. Geodetector

Geodetector is a spatial statistical model based on spatial autocorrelation theory to reveal the spatial differentiation of geographic elements and their driving factors [30]. We mainly used the factor detector, ecological detector, and interactive detector within the model. The factor detector quantifies the contribution of influencing factors to dependent variables, and it is calculated as follows:

$$q = 1 - \frac{SSW}{SST} \quad (8)$$

$$SSW = \sum_{h=1}^{l} N_h {\sigma_h}^2, SST = N\sigma^2 \quad (9)$$

where SSW is the sum of factor variances over all layers and SST is the total sum of variance, where $h = 1, \ldots\ l$ is the number of strata of the dependent variable or independent variable; N_h and N are the number of units in class h and the whole region, respectively; and ${\sigma_h}^2$ and σ^2 are the variances of the dependent variable for the units in class h and the whole region, respectively. The larger the q-value is, the stronger the explanatory power of the factor to the drought phenomenon. The effective range of q is [0, 1].

The ecological detector uses an F test to measure the significant difference of the impact of different influencing factors on the spatial distribution of drought. The F value is determined as follows:

$$F = \frac{N_{n=1}(N_{n=2} - 1)\sigma_{n-1}^2}{N_{n=2}(N_{n=1} - 1)\sigma_{n-2}^2} \quad (10)$$

where $N_{n=1}$ and $N_{n=2}$ refer to the sample size of two random factors, and F reflects the significance level.

The interaction detector was used to identify whether two driving factors, x_1 and x_2, increase or decrease the explanatory power of the drought index SPEI when they work together (Table 2).

Table 2. Independent variable interaction type.

Judgement Condition	Interaction
$q(x_1 \cap x_2) < Min(q(x_1), q(x_2))$	Non-linearly weaken
$Min(q(x_1), q(x_2)) < q(x_1 \cap x_2) < Max(q(x_1), q(x_2))$	Non-linearly weaken by one factor
$q(x_1 \cap x_2) > Max(q(x_1), q(x_2))$	Mutually enhanced
$q(x_1 \cap x_2) = q(x_1) + q(x_2)$	Independent effect
$q(x_1 \cap x_2) > q(x_1) + q(x_2)$	Non-linearly enhanced

In addition to the influences of meteorological variables, droughts are also affected by other factors including geographic location, topography, soil, land cover type, human activities, etc. Land cover type affects runoff, infiltration, and evapotranspiration of surface water through water absorption [35]. We selected 12 potential drought driving factors as follows: Mean Air Temperature (MAT), Mean Precipitation (MP), Mean Wind Speed (MWS), and Mean Sunshine Duration (MSD) during the growing season, representing the meteorological conditions; Percent of Sand (POS) in soil, representing soil texture; elevation, slope, and slope aspect, representing topographic conditions; Distance to rivers (DTR), representing potential water availability; Distance to Prefecture Cities (DTC), Land-Use and Land-Cover Change (LUCC), and Average of population Density (AOPD), representing human factors that can transform and regulate the local environment [36]. These factors are easy to be quantified [25,37,38]. Since Geodetector can only handle discrete variables, the 12 variables need to be discretized individually. The LUCC factors were divided into 36 grades according to the land use type conversion maps from 2000 to 2018, the slope and aspect were divided into 9 grades, and each of the other 10 factors was divided into 6 grades by the Jenks Natural Breaks Classification Method (Figure 2).

2.3.4. The GWR Model

The GWR model is an extension of the ordinary linear regression analysis method [39], which can effectively estimate the data with spatial autocorrelation and reflect the spatial heterogeneity of parameters in different regions. The multi-variate linear regression equation is given by:

$$y_i = \beta_0(u_i, v_i) + \sum_{j=1}^{n} \beta_j(u_i, v_i)x_{ij} + \varepsilon_i \quad (11)$$

where β_0 represents the intercept; (u_i, v_i) represent the coordinates of the i-th sampling point; β_j (u_i, v_i) the j-th regression parameter on the i-th sampling point, which has geographic significance; $x_{i1}, x_{i2}, x_{i3}, \cdots, x_{in}$ are n regression variables at this point; and ε represents random error. Finally, the revised Akaike Information Criterion (*AIC*) was compared with the ordinary least squares (OLS) results. *AIC* is defined as:

$$AIC = -2InL(\hat{e}_L, y) + 2c \quad (12)$$

where y represents the sample set of the fitting value of the dependent variable SPEI, $L(\hat{e}_L, y)$ is the likelihood function, $\hat{e}_L$ is the maximum likelihood estimate of e_L, and c is the number of unknown parameters. The smaller the *AIC* is, the higher the fitting degree will be.

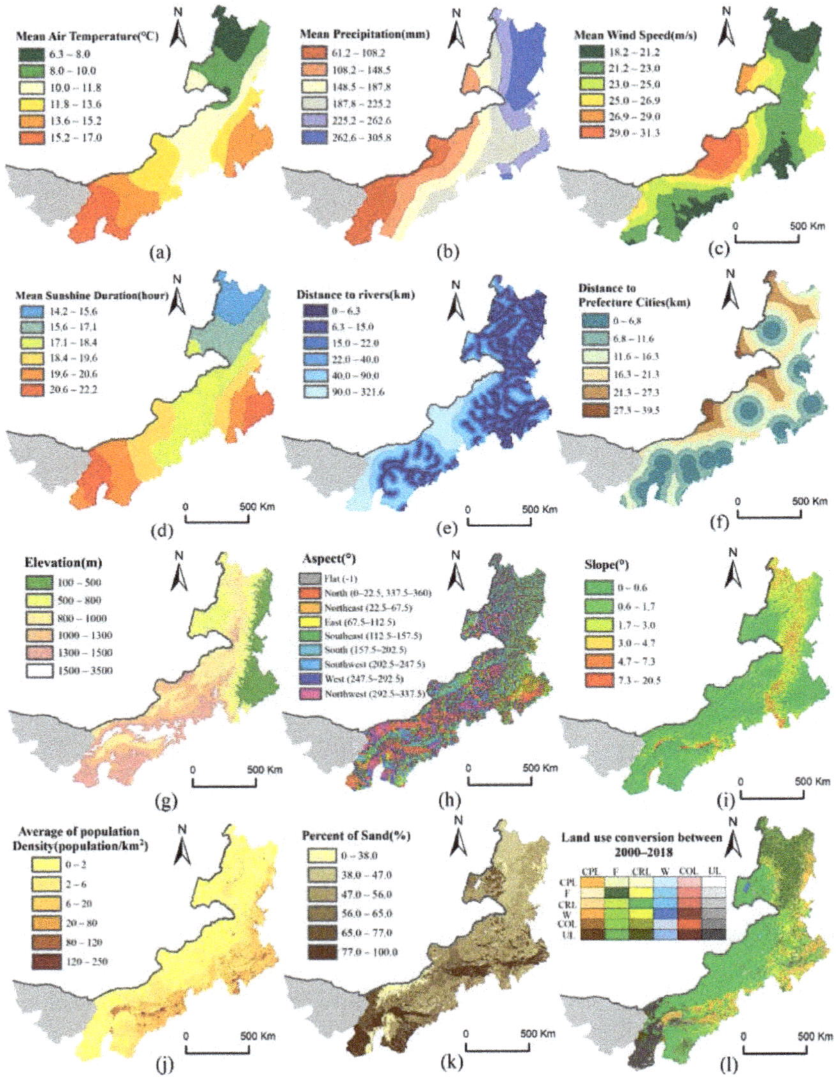

Figure 2. Factor grading (MAT(**a**), MP (**b**), MWS (**c**), MSD (**d**), DTR (**e**), DTC (**f**), Elevation (**g**), Aspect (**h**), Slope (**i**), AOPD (**j**), POS (**k**), LUCC (**l**) (CPL: CropLand; F: Frost; CRL: Crass Land; W: Water Area; COL: Construction Land; UL: Unused Land)).

3. Results and Analysis

3.1. Spatiotemporal Variation Characteristics of SPEI

The result of SPEI interpolation cross-validation shows a Pearson correlation coefficient of r = 0.85 and root-mean-square error RMSE = 1.15, indicating that the SPEI interpolation result has high accuracy. The statistical results of SPEI in the study area over the years show that (Table 3) the average annual SPEI of the growing season in Inner Mongolia from 2000 to 2018 is −0.03, representing a mild drought. The area in mild drought during the growing season reached 532,600 km^2, accounting for 52.39% of the total study area. The area in mild drought was the largest in 2000, accounting for 99.60% of the total, followed by 2001 and 2017, and that in 2012 was the smallest, accounting for only 6.01% of the total. Among the various types of droughts, the average annual area of mild drought accounted for about 74.34%, the highest proportion. The largest areas of moderate drought and severe drought occurred in 2000 and 2005, accounting for 73.10% and 36.17%, respectively.

Table 3. Change of drought area and proportion of various types of drought area in the study area from 2000 to 2018.

Year	Drought Area (Ten Thousand Km^2)	Percentage of Study Area	Percentage of Light Drought	Percentage of Moderate Drought	Percentage of Severe Drought	SPEI
2000	101.25	99.6%	23.28%	73.10%	3.62%	−0.62
2001	96.49	94.98%	20.44%	68.36%	11.21%	−0.68
2002	55.60	54.79%	73.00%	20.60%	6.39%	−0.01
2003	7.74	7.62%	100%	0	0	0.66
2004	51.40	50.59%	60.17%	39.82%	0.02%	0.03
2005	69.74	68.65%	33.45%	30.38%	36.17%	−0.43
2006	65.55	64.52%	98.63%	1.37%	0	−0.06
2007	84.12	82.80%	25.98%	40.67%	33.35%	−0.59
2008	16.74	16.48%	96.58%	3.42%	0	0.29
2009	73.59	72.43%	35.96%	53.26%	10.78%	−0.39
2010	66.59	65.54%	99.68%	0.32%	0	−0.03
2011	62.57	61.59%	69.74%	30.26%	0	−0.16
2012	6.11	6.01%	90.94%	9.06%	0	0.77
2013	29.32	28.86%	52.46%	26.51%	21.03%	0.51
2014	11.19	11.02%	100%	0	0	0.36
2015	17.87	17.59%	99.99%	0.01%	0	0.37
2016	34.37	33.80%	100%	0	0	0.19
2017	88.27	86.88%	28.08%	49.06%	22.86%	−0.61
2018	73.46	72.31%	68.36%	31.64%	0	−0.21
Annual average	53.26	52.39%	74.34%	18.89%	6.7%	0.03

The study area has high elevation in the west and low in the east (Figure 1a), and high in the south and low in the north. The spatial distribution of SPEI shows an increasing pattern from west to east with a rate of change of 0.008/degree and an increase of 0.01/degree from south to north (Figure 3). SPEI is highly sensitive to elevation gradients. Areas with high SPEI were mainly distributed in the 40~52°N area below 800 m in elevation, including Hinggan League, Hulunbuir City, Bairin Left Banner of Chifeng City, and other areas (Figure 1b); low SPEI appeared in areas with elevation between 1100 m~1400 m, in longitude between 105~115°E, and latitude between 40~45°N, mainly including Bayannur City, Baotou City, Ulanqab City, and West Ujimqin Banner of Xilingol League. In the Banner area, land covers are mainly grasslands, meadows, and deserts (Figure 1b).

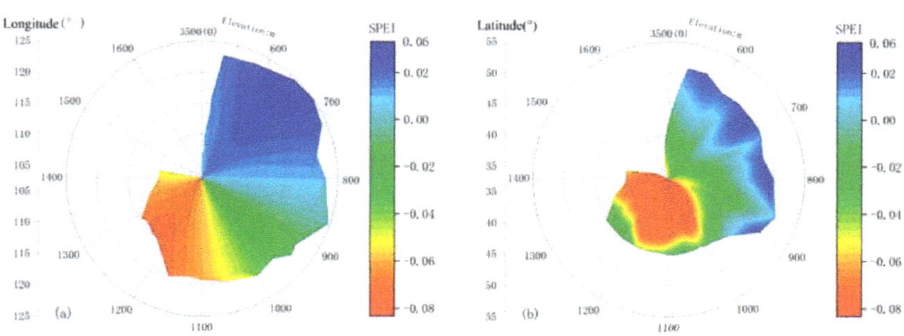

Figure 3. Spatial distribution of multi-year mean values of SPEI during the growing season in the study area from 2000 to 2018. (**a**) Longitude and elevation statistics. (**b**) Latitude and elevation statistics.

Trend analysis results show that (Figure 4a) there are significant differences in SPEI changes between the east and west of the study area. SPEI decreased with time significantly at a rate of $-0.40 \sim -0.25 \cdot (10a^{-1})$ in the west including Alxa Left Banner in Alxa League, Dorbod Banner in Ulanqab City, Darhan Muminggan United Banner in Baotou City, and Wuhai City, while SPEI increased significantly with time at a rate of change of $0.25 \sim 0.75 \cdot (10a^{-1})$ in Hulunbuir City, Hinggan League, Tongliao City, and the eastern part of Xilingol League. On the whole, the area with elevated SPEI was about 819,190 km^2, accounting for 79.70% of the study area. The land cover in the study area was relatively high in grassland, desert, and cultivated vegetation, reaching 43.68%, 11.94%, and 11.07%, respectively. The area of marsh was the smallest, accounting for only 3.69%. There are significant differences in the spatial distribution of vegetation (Figure 4b). The results of the SPEI variation trend in different land cover types showed that the area with a significantly higher SPEI (SPEI > STD) accounted for about 44.10% of the study area. The increasing trends of SPEI in swamp, coniferous forest, and broad-leaved forest were the most obvious, accounting for more than 85%. These land covers were located in a high-latitude, low-altitude forest area. The area has a large amount of precipitation, abundant water resource, and a low probability of drought. About 93.12% of the area where SPEI dropped significantly was located in the desert, accounting for about 61.53% of the total desert area. In the past 20 years, the mean annual precipitation in the desert areas of Inner Mongolia was less than 150 mm. Under the high-temperature and high-evaporation climatic conditions, water loss became severe and terrestrial carbon productivity was restricted, leading to an increased risk of drought [40].

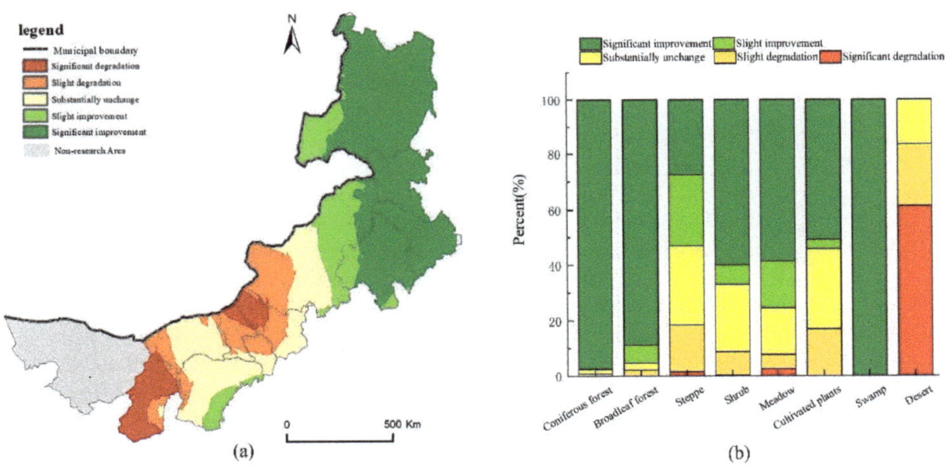

Figure 4. (**a**) The distribution of SPEI trend and (**b**) area percent in different land cover type.

3.2. Identification of Main Control Factors

The factor detection and ecological detection show that (Table 4) precipitation was the most explanatory factor ($q = 0.73$). From the q-values (Table 4), we can see the top four impact factors in decreasing order were MP > Elevation > MAT > DTR, and all passed the significance test ($p < 0.05$). Slope, aspect, LUCC, AOPD, and DTC have lower explanatory power for drought.

Table 4. *q* statistics and ecological detector.

Factor	Tag	*p* Value	*q*-Value	Rank
MAT	X_1	0.05	0.43	3
MP	X_2	0.05	0.73	1
MWS	X_3	0.05	0.13	
MSD	X_4	0.05	0.22	
DTR	X_5	0.05	0.42	4
DTC	X_6	0.05	0.03	
Elevation	X_7	0.05	0.53	2
Aspect	X_8	>0.1	0.01	
Slope	X_9	0.05	0.11	
AOPD	X_{10}	>0.1	0.06	
POS	X_{11}	0.05	0.23	
LUCC	X_{12}	0.05	0.26	

SPEI rates of change (θ_{slope}) versus factor level for various factors were shown in Figure 5. We can see that θ_{slope} shows a decreasing trend (Figure 5a), with increasing factor level for factors MAT (X_1), MSD (X_4), DTR (X_5), and elevation (X_7). The elevation factor has the greatest influence on θ_{slope} when the factor level is low (≤ 2), while factors MAT and MSD have a larger impact at a high factor level (>4). The elevation increases from the first level (100~500 m) to the fifth level (1500~3500 m), and the θ_{slope} decreases from $0.58 \cdot (10\ yr^{-1})$ to $-0.02 \cdot (10\ yr^{-1})$. Compared with other factors, the influence of DTR on θ_{slope} is approximately linear, and θ_{slope} drops to the lowest value ($0.08 \cdot (10\ yr^{-1})$) in the DTR interval of 90~321.6 km. On the contrary, θ_{slope} shows an increasing trend (Figure 5c) with the increasing factor level of MP (X_2), MWS (X_3), and Slope (X_9). Precipitation has the greatest impact on θ_{slope}. The amount of precipitation increases from level 1 (61.2~108.2 mm) to level 5 (262.6~305.8 mm), and θ_{slope} rises from $-0.075 \cdot (10\ yr^{-1})$ to $0.53 \cdot (10\ yr^{-1})$. The relationship between θ_{slope} and any of the following factors, DTC (X_6), aspect (X_8), AOPD (X_{10}), and POS (X_{11}), does not show a significant linear trend (Figure 5b). When POS is at the second level (38~47%), θ_{slope} reaches a peak value. This may imply that appropriate amount of sand is conducive to the respiration of plant roots, retaining soil moisture, and transportation of nutrients. However, a percentage of sand that is too high can easily cause surface degradation, soil moisture loss, and soil erosion. DTC, aspect, and AOPD have much smaller variation ranges of θ_{slope}, which indicates that these factors have little influence on the change of SPEI rate. SPEI over most land cover types (X_{12}) increased from 2000 to 2018, and SPEI decreased over only two land covers, in which cases water area was converted to construction area and unused land (Figure 5d). The θ_{slope} of the two land covers was $-0.09 \cdot (10\ yr^{-1})$ and $-0.03 \cdot (10\ yr^{-1})$, accounting for 0.1% and 11.5% of the study area, respectively. Among all land conversion types, the SPEI of unaltered forest land increased at the fastest rate $0.51 \cdot (10\ yr^{-1})$.

Overall, the factors with strong explanatory power in the factor detector have a larger fluctuation range of θ_{slope}. The θ_{slope} values of factors such as MP, elevation, MAT, and DTR are in four ranges of −0.07~0.55, −0.02~0.58, −0.04~0.53, and -0.09~$0.44 \cdot (10\ yr^{-1})$, respectively. The change range of SPEI rate influenced by natural factors, such as meteorology and topography in the study area, was larger than that influenced by human factors.

The ranking of the influence by interacting pair of factors was given in Table 5. Only the precipitation ∩ DTC ($X_2 \cap X_7$) and wind speed ∩ elevation ($X_1 \cap X_2$) pairs of the first 15 interacting pairs showed nonlinear enhancement; the others were dual-factor enhancement. Among them, the influence by the interaction between precipitation and elevation is the strongest, with a *q*-value of 0.870, followed by that between temperature and precipitation; the interaction between wind speed and elevation has the lowest explanatory power, with a *q*-value of 0.686. As expected, precipitation is an important source of water and a crucial driving factor in the process of drought changes. The difference in precipitation between the east and west of Inner Mongolia contributes mainly to the spatial differentiation of drought condition. Elevation also has a strong explanatory power for drought. It is an important topographical factor for driving drought in Inner Mongolia, and it is also an important factor in combining other factors to form a drought spatial pattern.

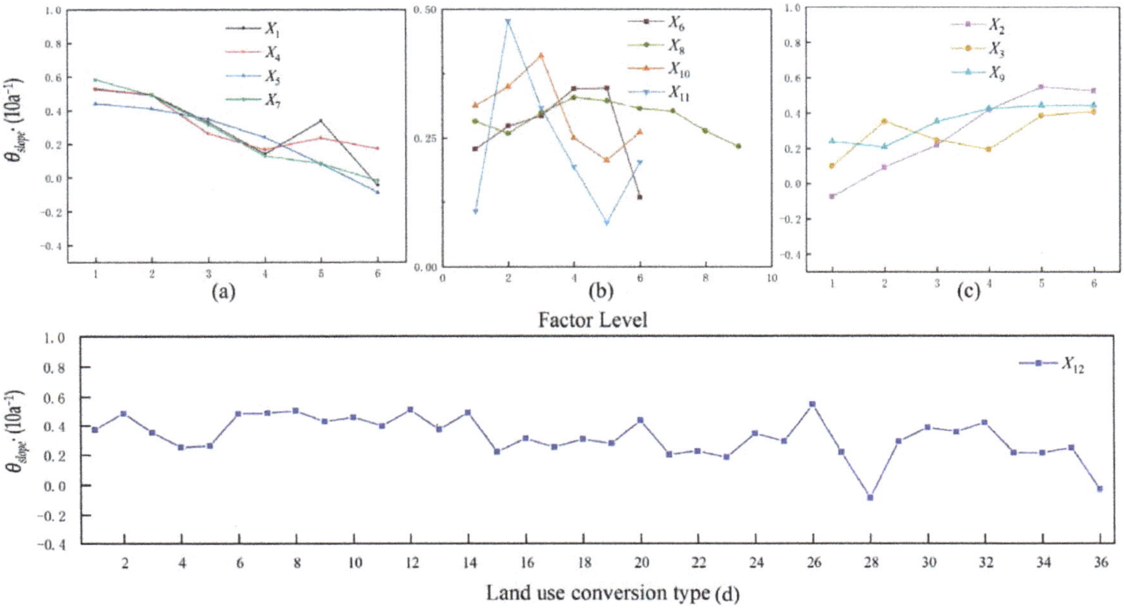

Figure 5. SPEI's variation trend with the influence of (**a**) MAT (X_1), MSD (X_4), DTR (X_5), Elevation (X_7); (**b**) DTC (X_6), Aspect (X_8), AOPD (X_{10}), POS (X_{11}); (**c**) MP (X_2), MWS (X_3), Slope (X_9); (**d**) LUCC (X_{12}) with their levels. The meanings of the land cover and land use codes for land use conversion types can be found in Figure 2.

Table 5. Influence of the interacting pairs of factors.

$q = A \cap B$	Results Comparison	Interaction Type	Rank
$X_1 \cap X_2 = 0.852$	$X_1 + X_2 > Max(X_1, X_2)$	Double-factor Enhance	2
$X_1 \cap X_7 = 0.753$	$X_1 + X_7 > Max(X_1, X_7)$	Double-factor Enhance	8
$X_2 \cap X_3 = 0.846$	$X_2 + X_3 > Max(X_2, X_3)$	Double-factor Enhance	3
$X_2 \cap X_4 = 0.836$	$X_2 + X_4 > Max(X_2, X_4)$	Double-factor Enhance	4
$X_2 \cap X_5 = 0.742$	$X_2 + X_5 > Max(X_2, X_5)$	Double-factor Enhance	11
$X_2 \cap X_6 = 0.770$	$X_2 + X_6 < X_2 \cap X_6$	Nonlinear Enhance	5
$X_2 \cap X_7 = 0.870$	$X_2 + X_7 > Max(X_2, X_7)$	Double-factor Enhance	1
$X_2 \cap X_9 = 0.745$	$X_2 + X_9 > Max(X_2, X_9)$	Double-factor Enhance	10
$X_2 \cap X_{10} = 0.756$	$X_2 + X_{10} > Max(X_2, X_{10})$	Double-factor Enhance	7
$X_2 \cap X_{11} = 0.751$	$X_2 + X_{11} > Max(X_2, X_{11})$	Double-factor Enhance	9
$X_2 \cap X_{12} = 0.762$	$X_2 + X_{12} > Max(X_2, X_{12})$	Double-factor Enhance	6
$X_3 \cap X_7 = 0.686$	$X_3 + X_7 < X_3 \cap X_7$	Nonlinear Enhance	15
$X_4 \cap X_7 = 0.737$	$X_4 + X_7 > Max(X_4, X_7)$	Double-factor Enhance	12
$X_5 \cap X_7 = 0.695$	$X_5 + X_7 > Max(X_5, X_7)$	Double-factor Enhance	14
$X_7 \cap X_{10} = 0.703$	$X_7 + X_{10} > Max(X_7, X_{10})$	Double-factor Enhance	13

Note: Only the first 15 combinations are ranked.

3.3. Spatial Difference of Main Control Factors

The GWR model was used to perform spatial regression analysis on the four main controlling factors, i.e., MAT, MP, Elevation, and DTR, and local adjusted R^2 and Akaike Information Criterion (*AIC*) as the evaluation indexes of the model fitting. Results show that the adjusted R^2 of the GWR model is 0.88, and the *AIC* value is −540.58. The action direction of the factor is reflected by the sign of the coefficient of the fitting equation. A negative coefficient in an area and the absolute value of the coefficient indicate that the area is drought-stricken and the strength of the driving effect, respectively.

The effects of the two meteorological factors, i.e., temperature and precipitation, have significant spatial differences. The overall fluctuation ranges of the two are relatively large, and the regression coefficient intervals are (−1.20, 0.60) and (−1.4, 0.55), respectively. The area in drought driven by temperature accounted for about 70.2%, of which the areas with strong temperature driving (−0.6~−1.20) were mainly located in Alxa Left Banner of Alxa League, Wuhai City, Hanggin Banner and Otog Banner of Ordos City, Urad Rear Banner of Bayannur City, and parts of Xilingol League (Figure 6a). Due to the large temperature difference between the east and west of the study area (up to 12.7 °C), the surface vegetation in the western hot area had strong transpiration and respiration, and the dry matter consumption and soil water loss were larger, which further expanded the arid area [41,42]. The difference of the driving results between precipitation and temperature factors is mainly in the semi-arid grasslands (Figure 6b), such as Xilinhot City, West Ujimqin Banner and East Ujimqin Banner in Xilingol League. The average precipitation in the growing season in this region was greater than 150 mm, which was enough for the growth of vegetation such as grassland, shrubs, and other vegetation [43].

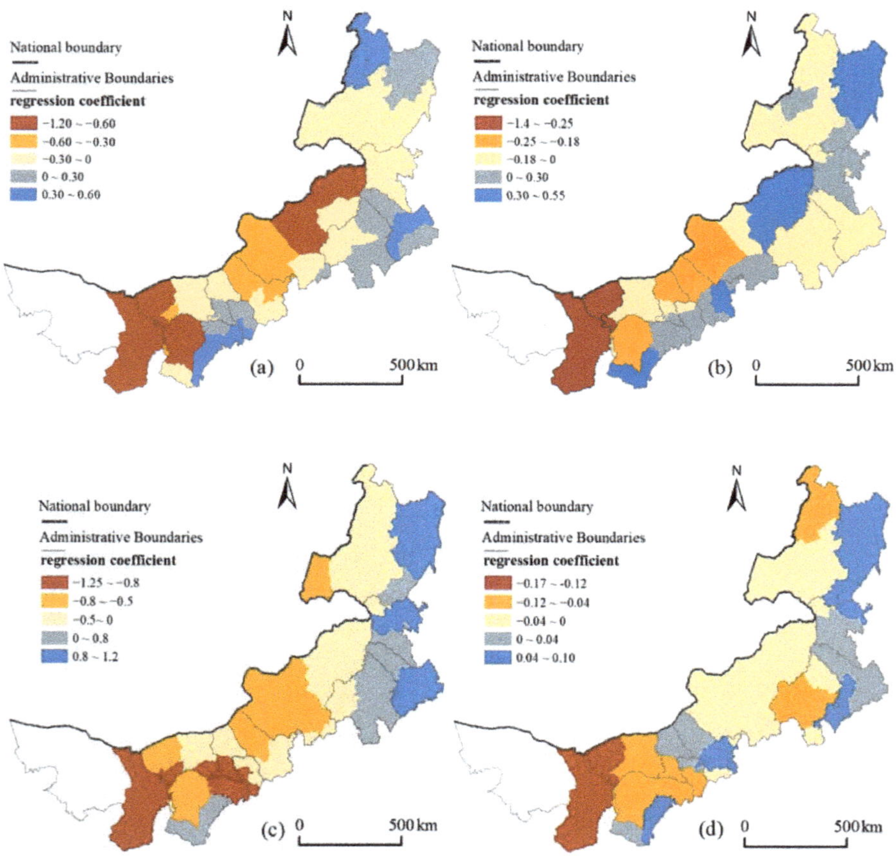

Figure 6. Distribution of the regression coefficient of SPEI with (**a**) MAT, (**b**) MP, (**c**) Elevation, and (**d**) DTR.

The influence of elevation mainly reflects the effect of the terrain. The GWR regression coefficient interval is (−1.25, 1.2), with the Great Khingan Range-Yin Mountains-Helan Mountains range as the boundary, and the SPEI driving coefficients on both sides are obviously different (Figure 6c). The central and western regions of Inner Mongolia (Alxa Left Banner, Dalate Banner, Zhungeer Banner, etc.) dominated by the Mongolian Plateau are generally higher than 1000 m in elevation, which has a significant effect on drought. The average GWR coefficient is −0.75. The eastern foothills of the Great Khingan Range-Yin Mountains and the southern foothills of Helan Mountains have lower average elevations, ranging from 100 to 500 m, and the average GWR coefficient is 0.5. The DTR factor reflects the water conservation within the basin, and the range of coefficient is the smallest (−0.17, 0.12). Due to low water conservation in Alxa League, Wuhai City, Ordos City, and Erguna in Hulunbuir City, the impact by DTR is shown in Figure 6d: the drought mitigation area driven by DTR is mainly located at the tributary of the Yellow River in Ordos City in the southwest of Inner Mongolia, Tabu River in Ulanqab City, and Dahei River (River inflow area), and the areas where Liaohe River, Songhua River, Nenjiang River, and other rivers adjacent to the Northeast Plain flow through.

4. Discussion

4.1. Driving Analysis of Drought in the Inner Mongolia

Drought is mainly caused by the imbalance of regional precipitation and evapotranspiration. We found that the change rate of SPEI during the growing season in Inner Mongolia from 2000 to 2018 ranged from −0.4 to 0.75·(10 yr^{-1}), and the area with decreasing SPEI accounted for 20.3% of the total area. The areas where the SPEI increased significantly (θ_{slope} is between 0.25 and 0.75·(10 yr^{-1})) are mainly located in Hulunbuir City, Hinggan League, and Tongliao City. The distribution characteristics of SPEI shown in this study are similar to the results of previous drought monitoring based on long-term series [19,20]. The differences are mainly manifested in the areas such as Hinggan League and Tongliao City, where drought changes increased significantly. Around 2000, the SPEI time series of the Mongolian Plateau showed a significant turning point from increasing to decreasing [34,44]. The results of the Geodetector modeling showed that the SPEI change was driven by four main controlling factors: air temperature, precipitation, DTR, and elevation (Table 4) during the growing season in the study area. Precipitation is a direct factor in drought (q = 0.73). Relevant studies have shown that in arid and semi-arid areas, vegetation growth and ecosystem health status depend directly on atmospheric precipitation [45]. The interaction detection results show that the joint effect of precipitation and elevation has the most explanatory power (q = 0.87). Conclusions about attribution analysis of drought agreed with a previous study [36]. However, we considered the special climatic background of Inner Mongolia in this study of meteorological drought. The impact of elevation reflects the influence of topography on mass and energy transportation and distribution of temperature and water availability that affect regional ecosystems through altering vegetation species and distribution and the formation and evolution of regional climate [46]. According to the drought trend (Figure 4a) and the spatial distribution of the regression coefficient of SPEI with GWR model factors (Figure 6), it was found that significantly reduced SPEI at a rate of −0.40~−0.25·(10 yr^{-1}) occurred in Alxa Left Banner in Alxa League, Dorbod Banner chain Ulanqab City, Darhan Muminggan United Banner in Baotou City, and Wuhai City in the western part of the study area. The drought in Alxa League and Wuhai City was caused by a synergy of hot air temperature, lack of precipitation, high elevation, and high DTR, while the drought in Ulanqab City and Baotou City was mainly caused by hot air temperature, lack of precipitation, and high elevation.

4.2. Variation of Explanatory Power of Factors in Different Elevations

Due to the large and high terrain environment of the Qinghai-Tibet Plateau in the region of the China-Mongolia Arid and Semiarid Area (CMASA), the lack of water vapor transported over the central and western Inner Mongolia has resulted in scarce precipitation [47]. The central and western part of the study area is dominated by plateaus and mountains. Due to the barrier and uplifting effects of the Great Khingan Range-Yin Mountains-Helan Mountains on water vapor, the eastern and southern piedmont of the mountains are the East Asian monsoon zone (elevation is about 150~500 m) and the west piedmont of the mountains is the non-monsoon zone (the elevation is generally higher than 1000 m).

As an important terrain factor, elevation has a significant impact on the spatial correlation of factors. We found an interesting pattern to speculate the relative importance of environmental and anthropogenic factors in our study area by elevation gradients. As the statistical results of q-values of various factors in different elevation intervals (Figure 7a), in the 100–500 m elevation interval, the average precipitation (X_1) in the growing season is 261.5 mm, and the q-value is the smallest. When the precipitation reaches a certain level, the impact of precipitation on SPEI decreases [48]. For the forest area, the ecological water storage is sufficient and the correlation between SPEI and sunshine duration (X_4) is stronger. In the elevation range of 800~1000 m, the q-values of factors such as air temperature (X_1), precipitation (X_2), sunshine duration (X_4), slope (X_9), POS (X_{11}), and LUCC (X_{12}) increased significantly, indicating that the change in SPEI was mainly affected by natural factors and some human activities. When the elevation increases to 1000~1300 m, the q-values of POS (X_{11}) and LUCC (X_{12}) reach the maximum of 0.38 and 0.42, respectively, and the q-values of other factors show a downward trend. In the area above 1300 m in elevation, the explanatory power of all factors decreases significantly with increasing elevation.

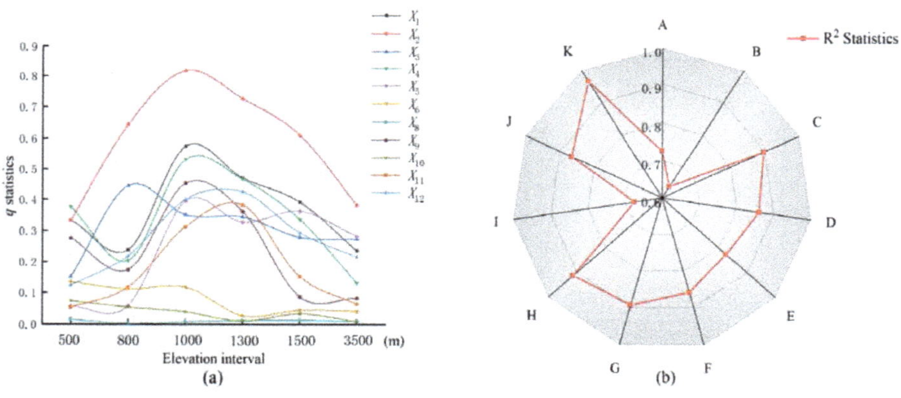

Figure 7. The change in q statistics of influencing factors on SPEI along the gradient of Elevation (**a**) and R^2 statistics of prefecture cities (**b**).

It is worth mentioning that the conversion of land use types is an important influencing factor reflecting human activities, as well as an important explanatory factor for drought changes (Table 3). Between 2000 and 2018, in the 800~1300 m elevation interval, SPEI was significantly enhanced by the land use conversion. According to statistics, the land use types that account for the largest area in this elevation interval are unaltered woodland and unaltered grassland (Table 6), which account for 48.3% and 15.5% of the total area of the region, respectively. The growth rates of SPEI are 0.52·(10 yr^{-1}) and 0.018·(10 yr^{-1}), respectively, indicating that the series of ecological restoration projects such as "closing hills for afforestation and reforestation, retiring grazing and raising grass" implemented by the Chinese government in Inner Mongolia since 2000 have played an important role [49]. In addition, in the conversion of land use from unused land to grassland and from farmland to grassland, SPEI increased by 0.24 and 0.58, respectively, which alleviated the drought conditions in the area to a large extent (Figure 5d). The SPEI of the unchanged farmland increased by 0.57, and the SPEI growth rate was 0.019·(10 yr^{-1}). This may be related to improvement in irrigation. Modern irrigation technology has improved the utilization rate of water resources and increased the field water holding capacity [50]. At the same time, grassland degradation caused by overgrazing and long-term abandonment of land aggravated the degree of drought [51,52]. For example, the grassland in the 800~1300 m elevation range was converted to unused land; as a result, the SPEI decreased by 0.42, while the SPEI of unaltered unused land decreased by 0.21.

Table 6. The area of specific land use conversion in 800~1300 m.

2000/2018 Unit: Km^2	Croplands	Forests	Grasslands	Water Areas	Construction Lands	Unused Lands
Croplands	14,143.289 (3.5)	948.394 (0.2)	7180.634 (1.8)	464.368 (0.1)	1530.627 (0.4)	1758.953 (0.4)
Forests	1357.162 (0.3)	61,982.534 (15.5)	17,312.724 (4.3)	0	0	811.536 (0.2)
Grasslands	5566.708 (1.4)	5677.182 (1.4)	193,072.543 (48.3)	1278.544 (0.3)	735.711 (0.2)	13,980.681 (3.5)
Water areas	477.401 (0.1)	0	952.564 (0.2)	2410.992 (0.6)	0	986.008 (0.2)
Construction lands	1372.649 (0.3)	159.97 (0.1)	1920.498 (0.5)	0	1092.653 (0.3)	492.902 (0.1)
Unused lands	626.973 (0.2)	3200.702 (0.8)	14,295.369 (3.6)	871.862 (0.2)	117.173 (0.1)	43,698.632 (10.9)

Note: The numbers in parentheses are the percentage of specific land use conversion to the total area (%).

4.3. Advantages and Limitations of GWR

The adjusted R^2 value from the GWR model is 0.88, and the *AIC* value is −540.58. Compared with the OLS model (R^2 = 0.54, *AIC* = −504.22), the *AIC* value is reduced by 36.36 and the degree of fit is higher, but there are regional differences. The regional statistics of R^2 shows that (Figure 7b) the largest value occurred in Wuhai City, Inner Mongolia (K) R^2 = 0.96, followed by Hohhot City (H) R^2 = 0.92, and the two prefecture-level cities with the smallest R^2 were Hinggan League (B) and Bayan. In Bayannur City (I), R^2 is 0.64 and 0.67, respectively. The difference in the accuracy of regional fitting may be related to the influence of elevation changes on the distribution of other factors in the large east-west span of the study area. Areas with small R^2 (A, B, I) have an average elevation of less than 800 m, and the average single factor *q*-value is 0.17. The average elevation of Wuhai City (K) and Hohhot City (H) are 1193 m and 1379 m, respectively, and the average *q*-value is 0.32 and 0.28, respectively. On the other hand, the area of each city is quite different, so is the statistical sample size, and the collinearity of the factors within the region may be another reason for the low fit in Hinggan League (B) and Bayannur City (I) [11,53,54].

4.4. Future Directions

Compared with traditional statistical models, we quantified the non-linear responses of independent variables and their interactions to SPEI change, without input of complex parameters. Further research may include: (1) using long-term SPEI data and more accurate PET calculation methods, such as the Penman and Hargreaves–Samani formula to produce more generalizable drought-driven results; (2) refining the spatial scale both horizontally and vertically, especially in eastern Inner Mongolia and western Mongolia, to generate results at higher resolutions.

5. Conclusions

Based on the multi-source data at the 110 meteorological stations, DEM, and vegetation types in Inner Mongolia and its surrounding areas, this study investigated the spatiotemporal variation of SPEI during the growing season in Inner Mongolia from 2000 to 2018. Through the introduction of time rate of change in SPEI, we used Geodetector and GWR models to screen the main controlling factors and then effectively quantified the impact of the factors on drought changes and the results are of great significance for drought-driven research. We made the following conclusions. (1) The SPEI in the growing season from 2000 to 2018 in Inner Mongolia showed a spatial variation pattern from dry west to wet east. The area with light drought accounts for the largest proportion in the whole region. (2) The inter-annual variation of SPEI shows an upward trend and the area of elevated SPEI accounted for 79.70% of the study area. These results indicate that the drought condition became alleviated with time during the growing season in Inner Mongolia. (3) The drought changes in Inner Mongolia were generally controlled by natural factors, with nonlinear interaction between factors enhancing drought impact. The aggravated drought in the central and western regions of the study area, such as Alxa League, Ulanqab City, Baotou City, and Wuhai City, were mainly driven by a synergy of hot air temperature, scarce precipitation, and high elevation, with significant impact from soil and LUCC at an elevation of 800~1300 m. The results from this study should be helpful for decision-making and management of regional water resources.

Author Contributions: Conceptualization, B.J.; formal analysis, Y.Q.; funding acquisition, G.Y.; methodology, M.Z.; software, M.L.; supervision, T.Z. and G.Y.; validation, X.Z.; visualization, Y.Q.; writing—original draft, B.J.; writing—review & editing, Y.Q., X.Z. and T.Z. All authors have read and agreed to the published version of the manuscript.

Funding: This research was funded by the National Natural Science Foundation of China (41801099), the second Tibetan Plateau Scientific Expedition and Research Program (2019QZKK0307), and the Key Research and Development Program of Sichuan (2022YFS0491).

Data Availability Statement: The meteorological data are available at http://data.cma.cn, accessed on 12 June 2022. The DEM are available at https://earthexplorer.usgs.gov/, accessed on 11 June 2022. The Population density data are available at https://landscan.ornl.gov/, accessed on 12 June 2022. The vegetation types data, land cover data, soil texture are available at https://www.resdc.cn/, accessed on 12 June 2022. Main rivers and county stations are available at http://ngcc.sbsm.gov.cn, accessed on 13 June 2022.

Conflicts of Interest: The authors declare no conflict of interest.

References

1. Mishra, A.K.; Singh, V.P. A review of drought concepts. *J. Hydrol.* **2010**, *391*, 202–216. [CrossRef]
2. Liu, D.; Ogaya, R.; Barbeta, A.; Yang, X.; Peñuelas, J. Contrasting impacts of continuous moderate drought and episodic severe droughts on the aboveground-biomass increment and litterfall of three coexisting Mediterranean woody species. *Glob. Chang. Biol.* **2015**, *21*, 4196–4209. [CrossRef] [PubMed]
3. Ma, X.; Zhao, C.; Yan, W.; Zhao, X. Influences of 1.5 °C and 2.0 °C global warming scenarios on water use efficiency dynamics in the sandy areas of northern China. *Sci. Total. Environ.* **2019**, *664*, 161–174. [CrossRef]
4. Ren, S.; Yi, S.; Peichl, M.; Wang, X. Diverse Responses of Vegetation Phenology to Climate Change in Different Grasslands in Inner Mongolia during 2000–2016. *Remote Sens.* **2018**, *10*, 17. [CrossRef]
5. Wang, L.; Chen, W. Applicability Analysis of Standardized Precipitation Evapotranspiration Index in Drought Monitoring in China. *Plateau Meteorol.* **2014**, *33*, 423–431. [CrossRef]
6. Steinemann, A. Drought indicators and triggers: A stochastic approach to evaluation. *JAWRA J. Am. Water Resour. Assoc.* **2003**, *39*, 1217–1233. [CrossRef]
7. Musei, S.K.; Nyaga, J.M.; Dubow, A.Z. SPEI-based spatial and temporal evaluation of drought in Somalia. *J. Arid Environ.* **2020**, *184*, 104296. [CrossRef]
8. Cook, B.I.; Smerdon, J.E.; Seager, R.; Coats, S. Global warming and 21st century drying. *Clim. Dyn.* **2014**, *43*, 2607–2627. [CrossRef]
9. Vicente-Serrano, S.M.; Beguería, S.; López-Moreno, J.I. A Multiscalar Drought Index Sensitive to Global Warming: The Standardized Precipitation Evapotranspiration Index. *J. Clim.* **2010**, *23*, 1696–1718. [CrossRef]
10. Rodrigues, M.; Jiménez-Ruano, A.; Peña-Angulo, D.; de la Riva, J. A comprehensive spatial-temporal analysis of driving factors of human-caused wildfires in Spain using Geographically Weighted Logistic Regression. *J. Environ. Manag.* **2018**, *225*, 177–192. [CrossRef]
11. Caetano, J.M.; Tessarolo, G.; De Oliveira, G.; Souza, K.D.S.E.; Diniz-Filho, J.A.F.; Nabout, J.C. Geographical patterns in climate and agricultural technology drive soybean productivity in Brazil. *PLoS ONE* **2018**, *13*, e0191273. [CrossRef] [PubMed]
12. Vicente-Serrano, S.M.; Beguería, S.; Lorenzo-Lacruz, J.; Camarero, J.J.; Lopez-Moreno, I.; Azorin-Molina, C.; Revuelto, J.; Morán-Tejeda, E.; Sanchez-Lorenzo, A. Performance of Drought Indices for Ecological, Agricultural, and Hydrological Applications. *Earth Interact.* **2012**, *16*, 1–27. [CrossRef]

13. Drumond, A.; Gimeno, L.; Nieto, R.; Trigo, R.M.; Vicente-Serrano, S.M. Drought episodes in the climatological sinks of the Mediterranean moisture source: The role of moisture transport. *Glob. Planet. Chang.* **2017**, *151*, 4–14. [CrossRef]
14. Hao, C.; Zhang, J.; Yao, F. Combination of multi-sensor remote sensing data for drought monitoring over Southwest China. *Int. J. Appl. Earth Obs. Geoinf.* **2015**, *35*, 270–283. [CrossRef]
15. Jin, L.; Wang, Y. The Impact of Drought on Biomass of Forage Grass in Hulunbuir Grassland. *Chin. J. Grassl.* **2020**, *42*, 80–90. [CrossRef]
16. Li, J.; He, Q.; Yao, J.; Hu, W. The characteristics of climate change and the impact factors analysis in the western part of Inner Mongolia. *J. Arid. Land Resour. Environ.* **2014**, *28*, 186–191. [CrossRef]
17. Tang, Q.; Liu, Y.; Zhang, C.; Su, F.; Li, Y.; Gao, Y.; Li, W.; Chen, D. Research progress on moisture source change of precipitation over the Tibetan Plateau and its surrounding areas. *Trans. Atmos. Sci.* **2020**, *43*, 1002–1009. [CrossRef]
18. Bailing, M.; Zhiyong, L.; Cunzhu, L.; Lixin, W.; Chengzhen, J.; Fuxiang, B.; Chao, J. Temporal and spatial heterogeneity of drought impact on vegetation growth on the Inner Mongolian Plateau. *Rangel. J.* **2018**, *40*, 113. [CrossRef]
19. An, Q.; He, H.; Nie, Q.; Cui, Y.; Gao, J.; Wei, C.; Xie, X.; You, J. Spatial and Temporal Variations of Drought in Inner Mongolia, China. *Water* **2020**, *12*, 1715. [CrossRef]
20. Pei, Z.; Fang, S.; Wang, L.; Yang, W. Comparative Analysis of Drought Indicated by the SPI and SPEI at Various Timescales in Inner Mongolia, China. *Water* **2020**, *12*, 1925. [CrossRef]
21. Tong, S.; Lai, Q.; Zhang, J.; Bao, Y.; Lusi, A.; Ma, Q.; Li, X.; Zhang, F. Spatiotemporal drought variability on the Mongolian Plateau from 1980–2014 based on the SPEI-PM, intensity analysis and Hurst exponent. *Sci. Total. Environ.* **2018**, *615*, 1557–1565. [CrossRef] [PubMed]
22. Qin, Y.; Zhang, T.; Yi, G.; Wei, P.; Yang, D. Remote sensing monitoring and analysis of influencing factors of drought in Inner Mongolia growing season since 2000. *J. Nat. Resour.* **2021**, *36*, 459–475. [CrossRef]
23. Yang, Z.; Ning, L.; Jidong, W. Analysis of Drought and its Possible Causes in Inner Mongolia Region for Nearly 30 Years. *J. Catastrophol.* **2013**, *28*, 67–73.
24. Psilovikos, A.; Tzimopoulos, C. Comparison of quadratic and non-linear programming (QP and NLP) optimization models in groundwater management. *J. Hydroinformatics* **2004**, *6*, 175–185. [CrossRef]
25. Zhu, L.; Meng, J.; Zhu, L. Applying Geodetector to disentangle the contributions of natural and anthropogenic factors to NDVI variations in the middle reaches of the Heihe River Basin. *Ecol. Indic.* **2020**, *117*, 106545. [CrossRef]
26. Wen, Q.; Sun, P.; Zhang, Q.; Yao, R. A multi-scalar drought index for global warming: The non-stationary stand-ardized precipitation evaporation index (NSPEI) and spatio-temporal patterns of future drought in China. *Acta Geogr. Sin.* **2020**, *75*, 1465–1482. [CrossRef]
27. Zhu, X.; Huang, C.; Wu, B.; Su, H.; Jiao, W.; Zhang, L. Research on remote sensing drought monitoring by considering spatial non-stationary characteristics. *Natl. Remote Sens. Bull.* **2019**, *23*, 487–500. [CrossRef]
28. Zhang, P.; Yang, D.; Zhang, Y.; Li, Y.; Liu, Y.; Cen, Y.; Zhang, W.; Geng, W.; Rong, T.; Liu, Y.; et al. Re-examining the drive forces of China's industrial wastewater pollution based on GWR model at provincial level. *J. Clean. Prod.* **2020**, *262*, 121309. [CrossRef]
29. Yang, Y.; Yang, X.; He, M.; Christakos, G. Beyond mere pollution source identification: Determination of land covers emitting soil heavy metals by combining PCA/APCS, GeoDetector and GIS analysis. *CATENA* **2019**, *185*, 104297. [CrossRef]
30. Wang, J.; Xu, C. Geodetector: Principle and prospective. *Acta Geogr. Sin.* **2017**, *73*, 219–231. [CrossRef]
31. He, D.; Yi, G.; Zhang, T.; Miao, J.; Li, J.; Bie, X. Temporal and Spatial Characteristics of EVI and Its Response to Climatic Factors in Recent 16 years Based on Grey Relational Analysis in Inner Mongolia Autonomous Region, China. *Remote Sens.* **2018**, *10*, 961. [CrossRef]
32. Dobson, J.E.; Bright, E.A.; Coleman, P.R.; Durfee, R.C.; Worley, B.A. LandScan: A Global Population Database for Estimating Populations at Risk. Photogramm. *Eng. Remote Sens.* **2000**, *66*, 849–857.
33. Yin, H.; Pflugmacher, D.; Li, A.; Li, Z.; Hostert, P. Land use and land cover change in Inner Mongolia—Under-standing the effects of China's re-vegetation programs. *Remote Sens. Environ.* **2018**, *204*, 918–930. [CrossRef]
34. Ming, B.; Guo, Y.; Tao, H.; Liu, G.; Li, S.; Wang, P. SPEIPM-based research on drought impact on maize yield in North China Plain. *J. Integr. Agric.* **2015**, *14*, 660–669. [CrossRef]
35. Luong, N.; Hiep, N.; Bui, T. Investigating the Spatio-Temporal Variation of Soil Moisture and Agricultural Drought towards Supporting Water Resources Management in the Red River Basin of Vietnam. *Sustainability* **2021**, *13*, 4926. [CrossRef]
36. Yang, S.; Quan, Q.; Liang, W.; Liu, T. Characteristics of Agricultural Droughts and Spatial Stratified Heterogeneity and Dependence of Dominant Factors in Inner Mongolia Autonomous Region, China. *Atmosphere* **2021**, *12*, 1249. [CrossRef]
37. Reiche, M.; Funk, R.; Zhang, Z.; Hoffmann, C.; Reiche, J.; Wehrhan, M.; Li, Y.; Sommer, M. Application of satellite remote sensing for mapping wind erosion risk and dust emission-deposition in Inner Mongolia grassland, China. *Grassl. Sci.* **2012**, *58*, 8–19. [CrossRef]
38. Jia, Y.; Cui, X.; Liu, Y.; Liu, Y.; Xu, C.; Li, T.; Ran, Q.; Wang, Y. Drought vulnerability assessment in Inner Mongolia. *Acta Ecol. Sin.* **2020**, *40*, 9070–9082. [CrossRef]
39. Fotheringham, A.S.; Charlton, M.E.; Brunsdon, C. Geographically Weighted Regression: A Natural Evolution of the Expansion Method for Spatial Data Analysis. *Environ. Plan. A* **1998**, *30*, 1905–1927. [CrossRef]

40. Alsafadi, K.; Al-Ansari, N.; Mokhtar, A.; Mohammed, S.; Elbeltagi, A.; Sammen, S.S.; Bi, S. An evapotranspiration deficit-based drought index to detect variability of terrestrial carbon productivity in the Middle East. *Environ. Res. Lett.* **2022**, *17*, 014051. [CrossRef]
41. Piao, S.; Friedlingstein, P.; Ciais, P.; Zhou, L.; Chen, A. Effect of climate and CO_2 changes on the greening of the Northern Hemisphere over the past two decades. *Geophys. Res. Lett.* **2006**, *33*, L23402. [CrossRef]
42. Zhang, R.; Zhao, X.; Zuo, X.; Degen, A.A.; Li, Y.; Liu, X.; Luo, Y.; Qu, H.; Lian, J.; Wang, R. Drought-induced shift from a carbon sink to a carbon source in the grasslands of Inner Mongolia, China. *CATENA* **2020**, *195*, 104845. [CrossRef]
43. Qiu, G.Y.; Xie, F.; Feng, Y.C.; Tian, F. Experimental studies on the effects of the "Conversion of Cropland to Grassland Program" on the water budget and evapotranspiration in a semi-arid steppe in Inner Mongolia, China. *J. Hydrol.* **2011**, *411*, 120–129. [CrossRef]
44. Jin, L.; Zhang, J.; Wang, R.; Bao, Y.; Guo, E. Analysis for Spatio-Temporal Variation Characteristics of Droughts in Different Climatic Regions of the Mongolian Plateau Based on SPEI. *Sustainability* **2019**, *11*, 5767. [CrossRef]
45. Brueck, H.; Erdle, K.; Gao, Y.; Giese, M.; Zhao, Y.; Peth, S.; Lin, S. Effects of N and water supply on water use-efficiency of a semiarid grassland in Inner Mongolia. *Plant Soil.* **2010**, *328*, 495–505. [CrossRef]
46. Pan, T.; Wu, S.; He, D.; Dai, E.; Liu, Y. Ecological Effects of Longitudinal Range-Gorge Land Surface Pattern and Its Regional Differentiation. *Acta Geogr. Sin.* **2012**, *67*, 13–26. [CrossRef]
47. An, C.-B.; Chen, F.-H.; Barton, L. Holocene environmental changes in Mongolia: A review. *Glob. Planet. Chang.* **2008**, *63*, 283–289. [CrossRef]
48. Ukkola, A.M.; Prentice, I.C.; Keenan, T.F.; Van Dijk, A.I.; Viney, N.R.; Myneni, R.B.; Bi, J. Reduced streamflow in water-stressed climates consistent with CO_2 effects on vegetation. *Nat. Clim. Chang.* **2016**, *6*, 75–78. [CrossRef]
49. Mu, S.; Li, J.; Chen, Y.; Gang, C.; Zhou, W.; Ju, W. Spatial Differences of Variations of Vegetation Coverage in Inner Mongolia during 2001–2010. *Acta Geogr. Sin.* **2012**, *67*, 1255–1268.
50. Shi, H.; Yang, S.; Li, R.; Li, X.; Li, W.; Yan, J.; Miao, Q.; Li, Z. Water-Saving Irrigation and Utilization Efficiency of Water and Fertilizer in Hetao Irrigation District of Inner Mongolia: Prospect for Future Research. *J. Irrig. Drain.* **2020**, *39*, 1–12. [CrossRef]
51. Dregne, H.E. Desertification of Arid Lands. *Econ. Geogr.* **1977**, *53*, 322–331. [CrossRef]
52. Julich, S.; Moorcroft, M.-A.; Feger, K.; van Tol, J. The impact of overgrazing on water fluxes in a semi-arid watershed—The suitability of watershed scale modeling in a data scarce area. *J. Hydrol. Reg. Stud.* **2022**, *43*, 101178. [CrossRef]
53. Shrestha, A.; Luo, W. Analysis of Groundwater Nitrate Contamination in the Central Valley: Comparison of the Geodetector Method, Principal Component Analysis and Geographically Weighted Regression. *ISPRS Int. J. Geo-Inf.* **2017**, *6*, 297. [CrossRef]
54. Zhao, R.; Zhan, L.; Yao, M.; Yang, L. A geographically weighted regression model augmented by Geodetector analysis and principal component analysis for the spatial distribution of PM2.5. *Sustain. Cities Soc.* **2020**, *56*, 102106. [CrossRef]

agriculture

MDPI

Article

Assessing the Representativeness of Irrigation Adoption Studies: A Meta-Study of Global Research

Ruchie Pathak * and Nicholas R. Magliocca

Department of Geography, University of Alabama, Tuscaloosa, AL 35401, USA
* Correspondence: rpathak@crimson.ua.edu

Abstract: For decades, nations around the world have been promoting irrigation expansion as a method for improving agricultural growth, smoothing production risk, and alleviating rural poverty. Despite its apparent advantages, suboptimal adoption rates persist. According to the existing literature, determinants of irrigation adoption are often highly dependent on cultural, contextual, and/or local institutional factors. Yet, studies from diverse geographies identify a consistent set of factors. Thus, to be able to make generalizable inferences from such studies, a global geographic representativeness assessment of irrigation adoption studies was conducted to determine whether identified factors influencing irrigation were the result of geographic, epistemological, or disciplinary biases. The results indicate that multiple geographic biases exist with respect to studying farmers' irrigation adoption decision-making. More research on this topic is being conducted in regions that have little to a high percentage of irrigation (>1%), are readily accessible, receive moderate amounts of average annual rainfall, and have moderate amounts of cropland cover. The results suggest the need to expand research efforts in areas with little to no irrigation to identify constraints and help accelerate economic growth, poverty reduction, and food and livelihood security for rural communities in these regions.

Keywords: agriculture technology; diffusion and adoption; farmers; climate change adaptation; systematic review

Citation: Pathak, R.; Magliocca, N.R. Assessing the Representativeness of Irrigation Adoption Studies: A Meta-Study of Global Research. *Agriculture* **2022**, *12*, 2105. https://doi.org/10.3390/agriculture12122105

Academic Editors: Robert J. Lascano and Alban Kuriq

Received: 26 October 2022
Accepted: 30 November 2022
Published: 8 December 2022

1. Introduction

One of the major global environmental issues confronting us today is climate change, which threatens our ability to meet the growing population demands for basic resources like food and water [1,2]. Due to its inherent link to natural resources, agriculture is highly sensitive to changing climatic conditions [3] and is among the most vulnerable sectors to climate change risks and impacts [4]. Changes in temperature and rainfall patterns will have direct and indirect impacts on our food systems, ranging from reduced crop production to volatility in markets and food prices [5,6]. Even though food production trends of the last 40 years have more or less kept pace with the rising food demands [7], pressure on our food systems will only intensify with changing consumption patterns, lifestyles, and diets in the coming years [1,8]. Additionally, in most developing countries, agriculture provides the main livelihood and employment opportunities for rural populations and contributes significantly to the national GDP [9]. Therefore, any reductions in production will impact agricultural economies and challenge the resilience of agricultural-dependent communities as well [9,10]. Hence, there is a need to strengthen local capacity to deal with forecasted and/or unexpected climatic changes [3], and this requires adaptation [11].

Adaptation is considered a vital component of any policy response to climate change in addition to mitigation [4], and often involves changes in processes, practices, or structures to reduce potential adverse impacts [3]. Sakschewski et al. (2014) in their assessment of agricultural production argued that production increases can be accomplished either by increasing land productivity or by increasing land resources, but since cropland expansion is limited, engineered or technological adaptive responses remain the most common in this sector [12]. One such adaptation strategy is to augment rainfed production with the use of irrigation [13]. Irrigation has the potential to buffer climate stress and increase production on existing agricultural lands, smooth production risks, and improve the growth of agricultural economies [9,13–15]. According to the UN Food and Agriculture Organization (FAO), the global area equipped for irrigation worldwide increased from 184 million ha in 1970 to 324 million ha in 2012 [16]. Much of this expansion has occurred in developing Asian countries [17], with China having the largest irrigated area in the world, followed by countries like India, the United States of America, Pakistan, and Iran in the top five [18,19].

Despite the multiple benefits, irrigation adoption among farming communities has been slow or the long-term investments needed delay its adoption [20]. This is because adoption of any technology, in general, is a complex sociological phenomenon [21] that involves a large number of factors affecting the adoption decision [21] and is seldom rapid [20]. Globally, many attempts from different disciplinary backgrounds have been made to identify the factors that act as barriers to irrigation technology adoption by farmers [22–27]. Studies from diverse geographies identify a consistent set of factors, with the cost of technology cited as the most common barrier to its adoption/uptake [28]. However, the existing literature also asserts that the determinants of irrigation adoption are often dependent on local culture, context, and/or policies [29]. For example, Alabama in the south-eastern U.S. receives an average of 55 inches of precipitation annually which allows for a long growing season in the state. However, the recent increase in flash drought instances within the state is a cause of worry for those practicing rainfed agriculture, especially the small farm owners, making them the most vulnerable to these changing climatic conditions (For more details see the U.S. Drought Monitor for Alabama from the year 2000–Present available at: https://www.drought.gov/states/alabama, accessed on 15 October 2022). Accordingly, this identification of factors influencing irrigation adoption across a wide range of geographic contexts will be useful when climate change necessitates adaptation in such unprecedented areas.

One explanation for this disconnect concerning the different factors affecting irrigation adoption, which we explore in this paper, is that the geographic contexts in which irrigation adoption studies are often conducted might be biased, and this bias has influenced the set of factors identified as having explanatory power. If such a bias exists, it would not be unique to irrigation adoption studies. For instance, Martin et al. (2012) found the global distribution and context of ecological field study sites to be biased toward more accessible locations with limited human influence. According to the authors, the geographical context of selected study sites greatly influenced the observations made within these locations [30]. Therefore, to better understand the reasons as to why a farmer chooses to adopt or not adopt irrigation, it is first essential to recognize the global extent and context-dependency of irrigation adoption. This can be achieved through a geographic representativeness analysis. Using this analytical approach, the representativeness of studies examining factors affecting the diffusion and adoption of irrigation by farmers from around the world will be assessed to determine whether the identified factors (influencing irrigation adoption) from a set of case studies selected through a systematic review were the result of certain geographic biases or not. Accordingly, to identify these potential biases, we test the following two hypotheses:

H$_1$. *The geographic context of irrigation adoption studies is biased towards locations with substantial levels of existing irrigation, relatively low annual precipitation, and greater accessibility to markets.*

H$_2$. *The same factors (affecting farmers' decision-making) are observed regardless of the geographic context of these studies.*

Thus, the goal of this review is to understand whether the apparent consistency of factors influencing irrigation adoption is the result of the geographic contexts in which it is studied. Given the emerging challenges presented by climate change, we suspect that there are settings in which irrigation (and the study of its adoption) is currently limited but would be beneficial (i.e., improved yields, profits). If this is the case, then the set of factors influencing the irrigation adoption process may be different than in contexts with established irrigation practices. To answer this question, we narrowed our review to articles that explicitly addressed the irrigation adoption process, rather than broader investigations of the adoption of climate-smart agriculture or best management practices, e.g., [31] or those that assessed the benefits of irrigation adoption, e.g., [32].

This paper is structured as follows. Sections 1.1–1.3 give a brief overview of the motivations behind irrigation adoption and the technological and theoretical perspectives commonly used to study the adoption process. Section 2 describes the procedure followed for this systematic review, followed by the presentation and discussion of the results of the geographic representativeness and adaptation factors' analyses in the subsequent sections and some concluding remarks.

1.1. Why Irrigation?

Irrigation refers to the systematic and artificial application of water to plants at regular intervals to assist in the growing of crops and maintenance of landscapes [33–35] and is usually classified as surface, sprinkler, and micro-irrigation [35,36]. Irrigated agriculture, which accounts for more than 70% of total global freshwater withdrawals [15,37], provides for about 40% of the world's agricultural production [38] from less than 20% of its area [39–41]. Asia continues to contribute the largest share of total irrigated area, followed by Latin America, while sub-Saharan Africa only contributes 6% of its cultivated area to irrigation [18,38].

Irrigation use increases and stabilizes crop production in areas that do not receive enough precipitation [42–44], and has helped shape the economies of many semi-arid and arid regions around the world [45,46]. It also contributes toward income stabilization of dependent communities by improving agricultural growth and smoothing production risk [47–49]. According to a study by Bhattarai et al. (2007) [50], irrigation use can provide direct benefits like increased crop production that go to individual farm owners and/or entire community(s), and indirect benefits that are accrued to the wider sectors of the economy. There can also be spillover effects, which is brought by the increased household spending in the local economy due to enhanced income and employment as a result of increased land productivity made possible by irrigation [48].

1.2. Technology Adoption and Related Theories

Adoption is the decision to use a particular technology or innovation by an individual [51,52], which then leads to its diffusion or dissemination within a social system [52,53]. There exists a plethora of literature on factors that determine the adoption of a technology. Various researchers even define 'technology' itself in different ways and based on their definitions and disciplinary backgrounds use different theories or models to study its adoption [29,54–57]. For instance, in marketing research, the purchase of a technology is often the focus rather than its actual use [53]. Within agriculture, scholars have commonly used economic models and theories to explain individual technology adoption decisions [57,58], which allow for only rational and objective decision-making behaviors of farmers, rather than their perceptions, and assume that they adopt technology only for profit or utility maximization [59,60]. Alternatively, some research has shifted analytical focus to the role of individual knowledge, perceptions, and/or attitudes in the decision-making process, which in turn are conditioned by extrinsic factors, such as characteristics of the individual (~age, gender, education levels, etc.) and their external environment [29,61]. While others have taken a relatively more macro-perspective as they focus not only on the individual but also on the characteristics of the technology in question and the infrastructure needed for its successful diffusion to comprehensively understand its spread across the entire society (or market) over time [51,62,63]. A more recent strand of literature on agricultural technology adoption has also included the role of social networks in influencing the adoption of agricultural technologies [64–68]. Another key element associated with the adoption decisions is that of uncertainty or risk, which refers to the suitability of technology with an individual's characteristics including his/her experience or skillset, and with their local conditions (~agronomic, economic, and/or climatic) [57,69].

Since there exists no single model for understanding the decision-making processes in which an individual engages before adopting a certain technology, adoption is examined through a combination of research paradigms [55,57]. Moreover, these studies mostly utilize regression models to explain the uptake of technology as a function of several independent variables [70,71] including personal characteristics, preferences, individual attitudes, economic or institutional constraints, that are gathered either through census data, surveys, or personal interviews or a combination of it [29].

1.3. Factors Affecting Irrigation Adoption

Studies suggest that uptake of a new technology is rarely rapid, particularly among small farmers in developing countries [61,72,73], and with a wide range of factors acting as possible deterrents [72,74,75]. For example, a study in Burkina Faso (West Africa) of 629 farmers highlighted the need for farmers' training and information dissemination on irrigation to increase adoption [76]. Another study investigated factors affecting the adoption and application of sprinkler irrigation technology by farmers in the county of Famenin, Iran, and showed that the adoption was influenced by both environmental factors, such as farm size, access to water, water quality, and non-environmental factors, including workforce number in the family, employment diversity, and participation in extension education and courses on agricultural water management [77]. Another study investigated the sources of variation for on-farm irrigation systems across producer fields in Nebraska (USA) [42]. Their findings showed that biophysical factors such as soil, crop type, and weather, explained about half of the observed variations in field irrigation. However, the rest of the variations remained unexplained, suggesting that both producer behavior and skills played a significant role in shaping these decisions. Another study looked at the effect of production risk on irrigation technology choice among small-scale farmers in Chile and their results indicated that more educated farmers, with credit access, receiving extension services, and living in communes with other adopters were more likely to use modern irrigation techniques [78]. Another study of 1500 farmers from Henan Province in China found that the farmers who believed in climate change adopted adaptation measures like irrigation to respond to and mitigate its negative impacts [79]. Thus, understanding the kinds of factors influencing adoption decisions is crucial not only for the propagators of these technologies to increase the likelihood of their adoption but also for identifying the overall determinants of agricultural growth and development [80].

2. Materials and Methods

2.1. Data Collection

2.1.1. Literature Search Strategy

A literature search was conducted using Science Direct and Scopus databases. The search was limited to only research articles written in the English language and published in peer-reviewed journals between 2000 and 2021. Articles prior to 2000 were excluded as the global irrigation dataset used in this analysis is based on the nationally reported statistics from around the year 2000 (more details about this dataset are presented in Table 1). Moreover, this also reflects the broader trends in irrigation adoption globally because the percentage of reported data on irrigation use from around the world is largest from the year 2000 onwards compared to the earlier years [81]. The steps taken in the search and screening process are presented in Figure 1. First, a target set of 10 articles containing both 'true positives' and 'true negatives' was assembled from a wide range of disciplines to represent the full range of publications in this research domain and assemble a set of search keywords. Target set articles are listed in Appendix A. Different keywords such as irrigation, technology adoption, agriculture, farmer decisions, water management, and climate change adaptations were combined using Boolean operators to download relevant studies. The specific search terms used were: (("irrigation") AND ("technology" OR "adoption") AND (("reasons and constraints") OR ("attitudes") OR ("drivers") OR ("perception") OR ("barriers")) AND ("climate change adaptation" OR "climate smart agriculture" OR "climate change" OR "adaptive capacity") AND (("drought") OR ("water management practices")) AND (("farmers") OR ("farmer decisions"))).

Table 1. Description and sources of all the datasets used in this analysis.

Dataset Name	Description	Source
Global Administrative Areas (GADM)	A spatial database of the location of administrative areas of all countries, at all levels of sub-division.	GADM (2018–2022) [82]
GLOBE Land Units (GLUs)	GLUs are equal-area hexagonal cells that cover the Earth's land surface and are based on the geodesic Discrete Global Grid (DGG) system of Kevin Sahr (2003).	GLOBE (2012)
Average Annual Precipitation	Average annual precipitation (mm/year) from 1950–1999. Native resolution is 30 arcminutes projected in Geographic Coordinate System WGS 1984.	Willmott & Matsuura (2001) [83]
Percent Crop Area	Percent crop land cover area per grid cell derived from HYDE (History Database of the Global Environment) land cover data. Native resolution is 0.5° projected in Geographic Coordinate System WGS 1984.	Klein Goldewijk et al. (2011) [84]
Market Access Index	Global grid of a normalized market access index based on travel time to cities with populations of at least 50,000 and 750,000.	Verburg et al. (2011) [85]
Percent Area Equipped for Irrigation	Global map of irrigation areas showing the amount of area equipped for irrigation around the year 2000 in the percentage of the total area on a raster with a resolution of 5 min.	Siebert et al. (2005) [86]

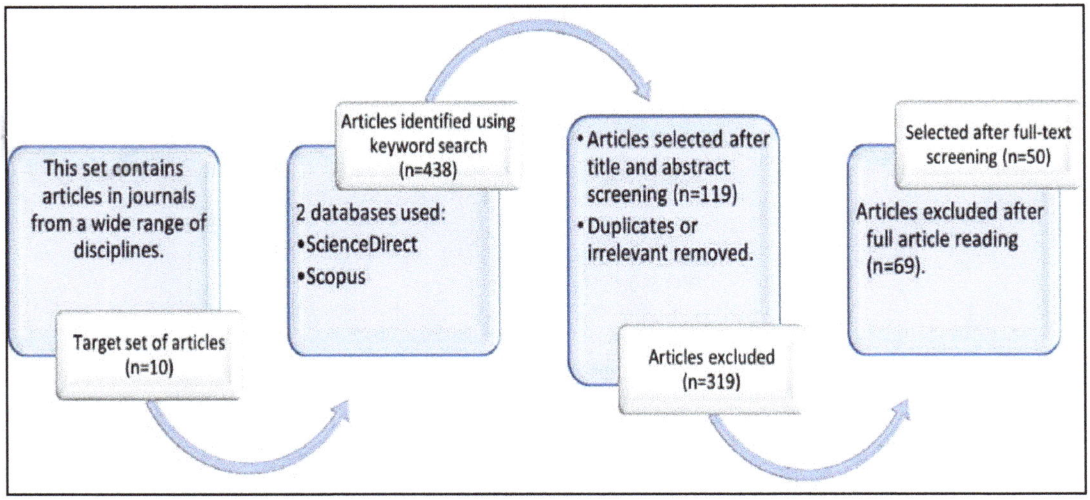

Figure 1. Steps involved to assemble research articles for this analysis.

2.1.2. Selection of Case Studies

After the literature search, the resulting dataset consisted of 438 publications. The next step was article screening to identify case studies that should be used in this meta-study. Both study titles and abstracts were checked and critically reviewed for suitability for this analysis. Articles were excluded if they did not (1) investigate the different factors/reasons affecting technology adoption within the agricultural sector, and (2) present an assessment of farmers' views or opinions. Conference proceedings, grey literature, reports, and duplicate articles were also excluded from the dataset. The initial screening reduced the number of eligible articles to 119. The second round of screening was performed using the full text of each remaining article. Articles were primarily screened to determine specifically if irrigation adoption by farmers was studied or not, irrespective of the type of irrigation system. For instance, many studies examined the adoption of several different agricultural practices together, in the form of climate change adaptation strategy, conservation agriculture, or as sustainable farming practices adopted by farmers including high-yielding crop varieties, different soil, and water management practices see, e.g., [87–89]. All the studies that did not include irrigation as one of the technologies or practices being studied were discarded. Moreover, studies that were conducted at a very large-scale and reported aggregated results (e.g., for entire U.S. mid-west region [90] or 11 African countries together [91]), were excluded to ensure comparability of results, since the goal was to examine the geographic contexts of these studies that would otherwise have been difficult to capture. Additionally, studies that investigated the benefits of irrigation adoption, assessed its impact on crop production under climate change, or estimated future adoption rates were also not considered, e.g., [92–95]. As a result, 50 case studies, which passed the inclusion and exclusion criterion were selected and used in this meta-study. A complete list of the studies included in this review is also provided in Appendix A.

2.2. Data Analysis

A representativeness analysis provides a robust statistical test to enable the user to investigate potential geographic biases within a collection of primary data observations (e.g., case studies) [96]. Using this analytical approach, for a given global variable of interest (e.g., average annual precipitation), the frequency distribution of the global variable within a user-specified geographic extent was compared with the frequency distribution of the observations in the sample collection, and the degree to which the sample collection's distribution is representative of the distribution of the global variable was quantified [96,97]. The null hypothesis for this analysis was that the frequency distributions of the global variable and sample collection are not statistically different. If the null hypothesis can be rejected with a low probability of type I error, then the sample can be declared as significantly biased. To enable comparability between values of the global variable and sample collection observations, which might include case study geographies of diverse extents, the standardized, hexagonal, and equal-area geographic units from the GLOBE system were used, known as GLOBE land units (GLUs). The degree of representedness (r) was then computed with a chi-squared ($\chi 2$) test and was characterized as follows:

$$\begin{aligned} r = 0 &\text{ if } f_e(g_v) = f_o(g_v) \\ -(1-p) &\text{ if } f_e(g_v) > f_o(g_v) \\ (1-p) &\text{ if } f_e(g_v) \le f_o(g_v) \\ \text{undefined} &\text{ if } f_e(g_v) = 0 \wedge f_o \neq 0 \end{aligned}$$

where $f_e(g_v)$ was the expected frequency of the bin to which GLU g belonged (calculated from the population set), $f_o(g_v)$ was the observed frequency of that bin (calculated from the sample set), and p was the p-value for the χ^2 test. The range of r is between [−1 to 1], with 0 indicating perfect representedness, negative numbers indicating under-representedness, and positive numbers indicating over-representedness [96].

Several data preparation steps were followed to produce the sample and global (~population) datasets. Table 1 describes all the datasets used for this analysis. After shortlisting the case studies, the locations of the study sites (total = 53) mentioned in each of the selected 50 articles were mapped using the shapefiles of administrative boundaries from the GADM dataset in ArcGIS Pro software (see Figure 2). Next, the global GLU feature layer obtained from GLOBE was filtered using several context variables (see Table 2) to restrict the global dataset to the expected geographic extent of agricultural areas. Case study locations were also intersected with the filtered GLU layer to form the sample dataset and to maintain a similar unit of analysis for both the layers. For each GLU, values of three variables—average annual precipitation (mm/year), percent crop area, and market access index were calculated. For the area equipped for irrigation (%) variable, mean values were computed using zonal statistics within each GLU for both the above feature layers. The extent/range of the selected four variables within both the global and sample layers are shown in Figures 3–6. For each of these four variables, these two datasets were divided into different intervals or bins. The binning strategy was kept the same as their source datasets (see Table 1 for dataset details) except for average annual precipitation variable for which a geometric interval was used. Finally, Pearson's χ^2 test for the independence of two datasets was conducted to compare the frequency distributions of the sample and global datasets for each of the selected four variables to determine the geographic representativeness of the assembled case studies on irrigation adoption and answer the first hypothesis.

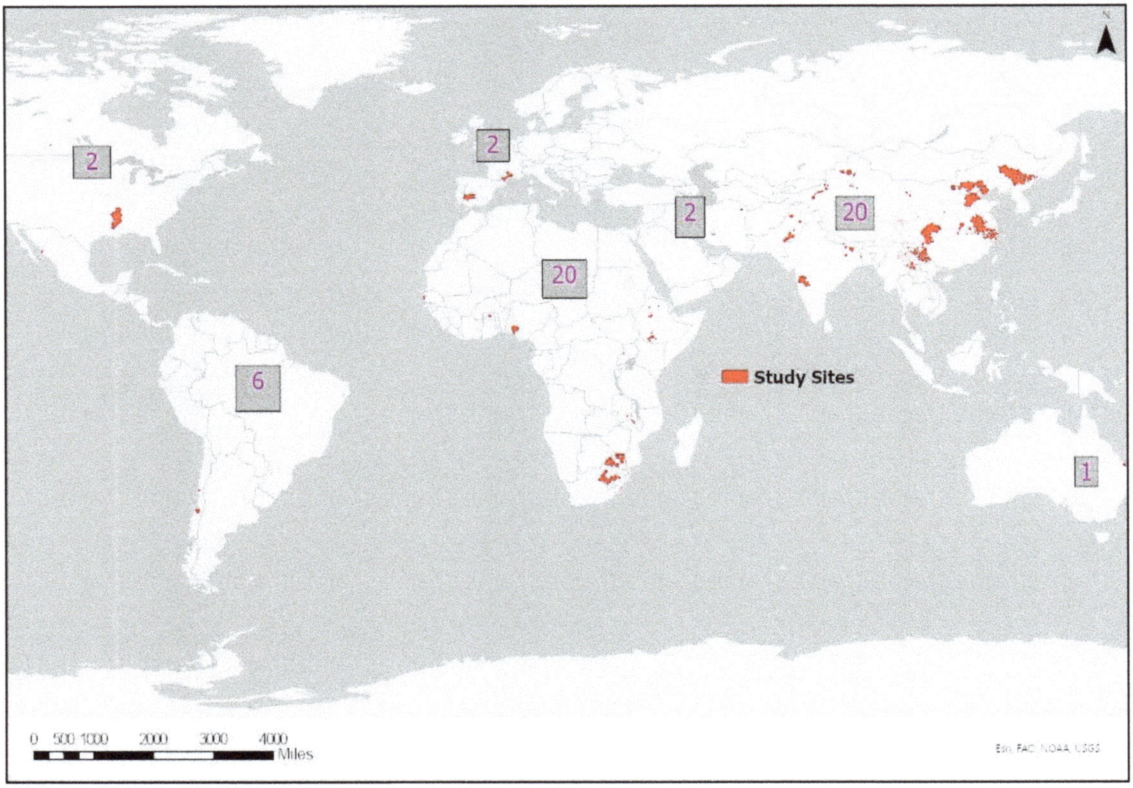

Figure 2. The map shows the location and distribution of selected cases.

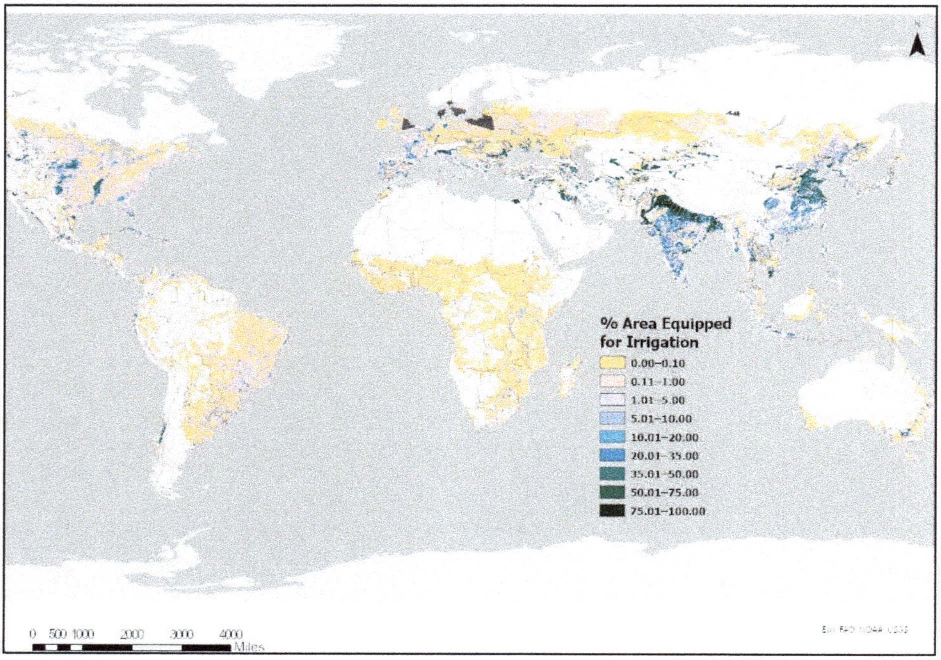

(a)

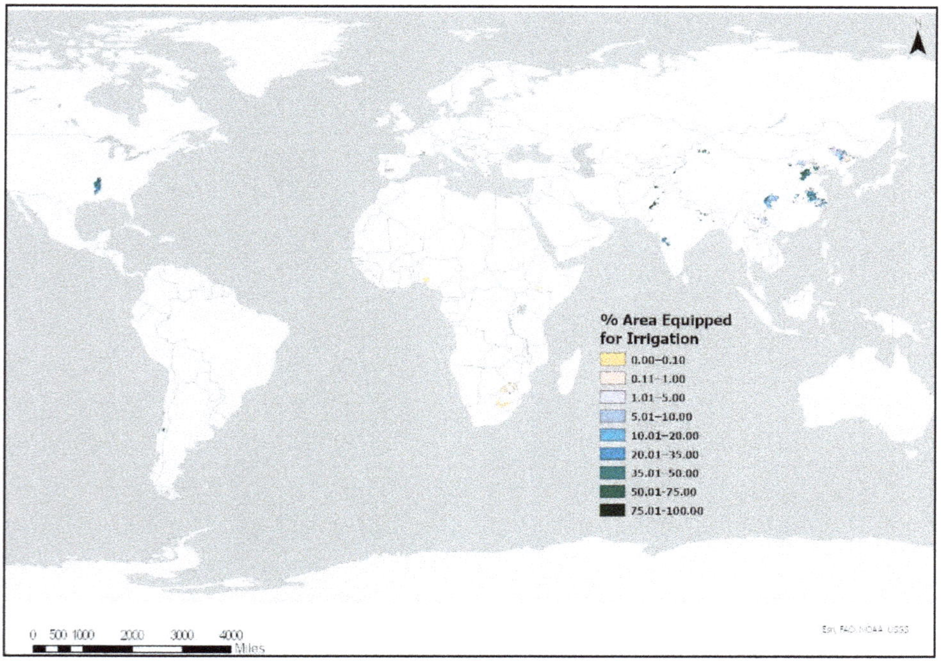

(b)

Figure 3. (**a**) Global extent for % Area Equipped for Irrigation variable. (**b**) Sample extent for % Area Equipped for Irrigation variable.

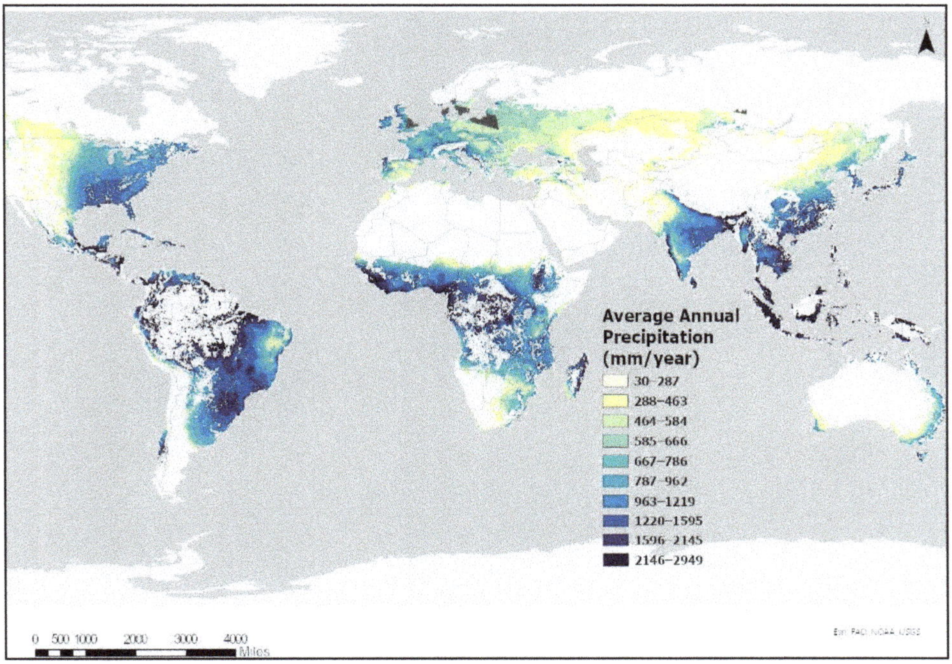

(a)

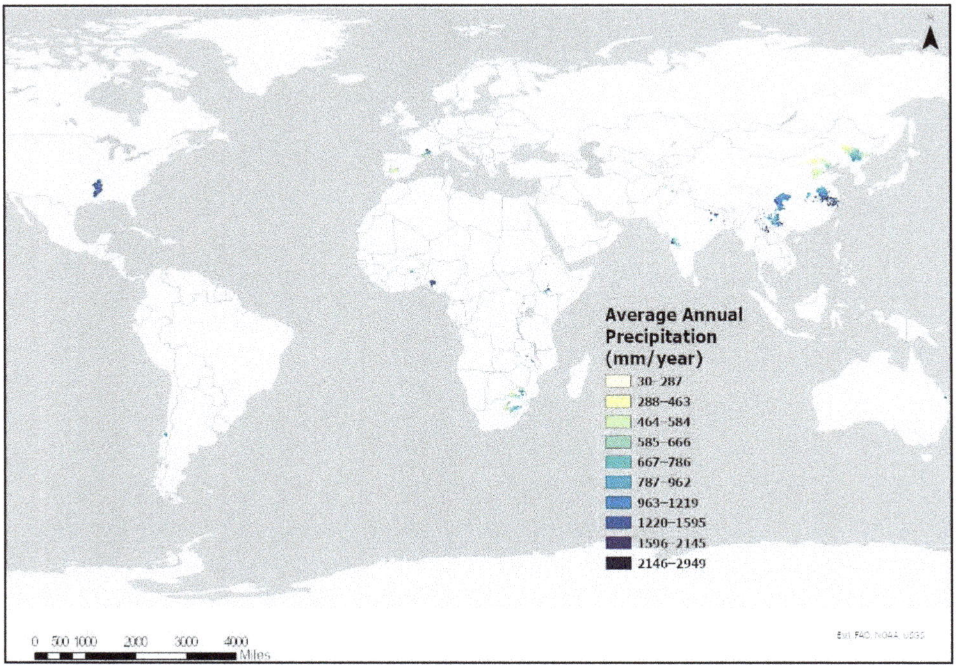

(b)

Figure 4. (**a**) Global extent for Avg Annual Precipitation variable. (**b**) Sample extent for Avg Annual Precipitation variable.

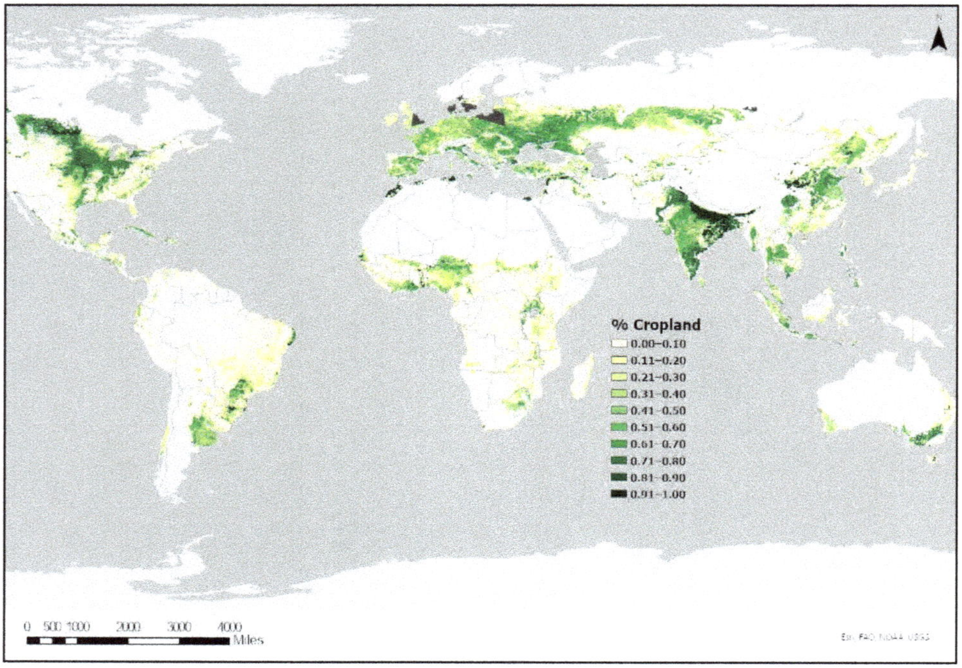

(**a**)

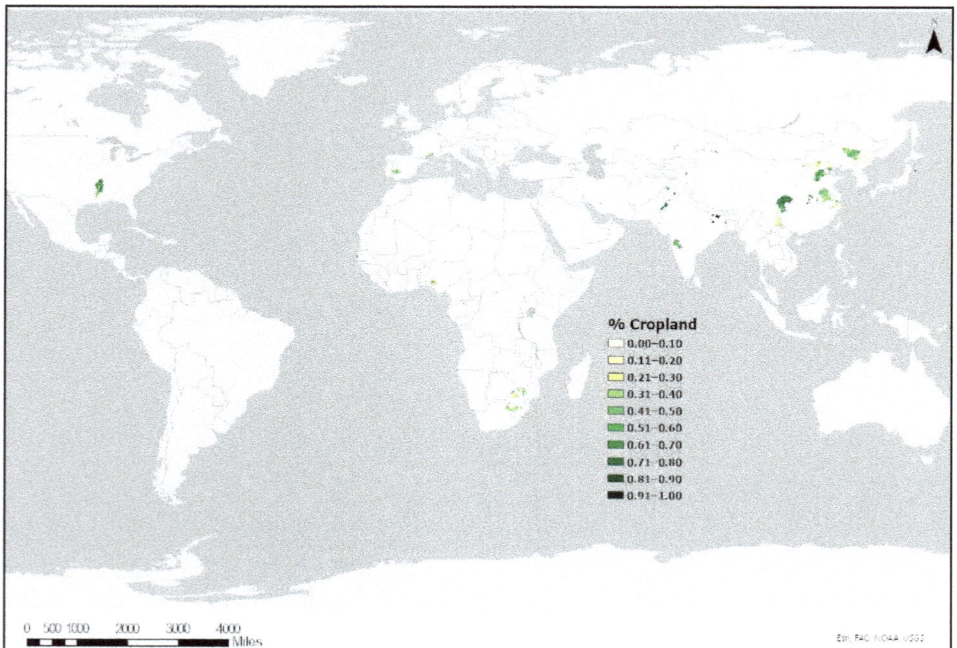

(**b**)

Figure 5. (**a**) Global extent for Percent Cropland variable. (**b**) Sample extent for Percent Cropland variable.

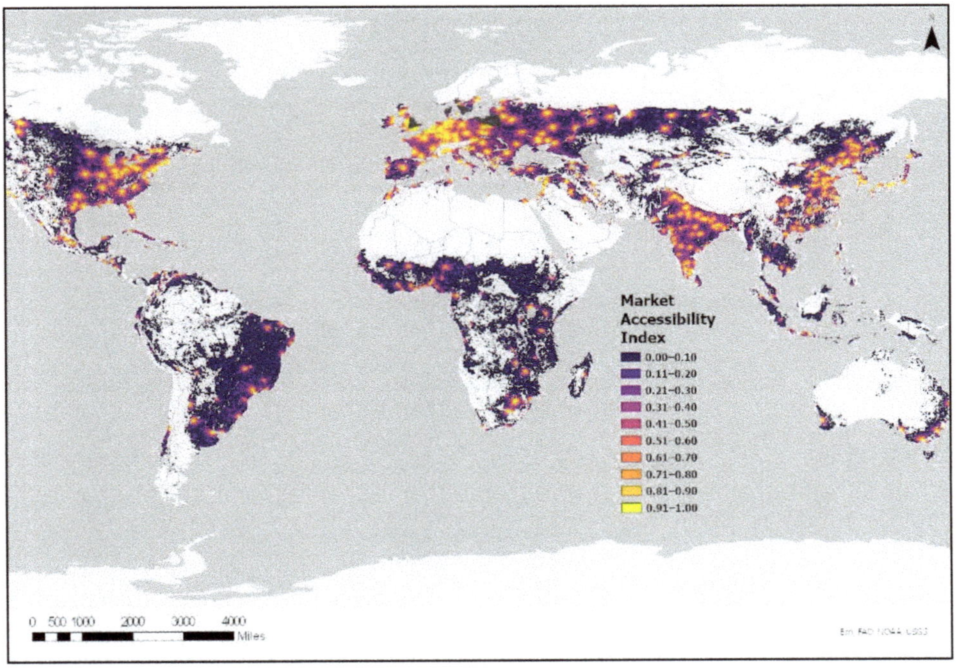

(**a**)

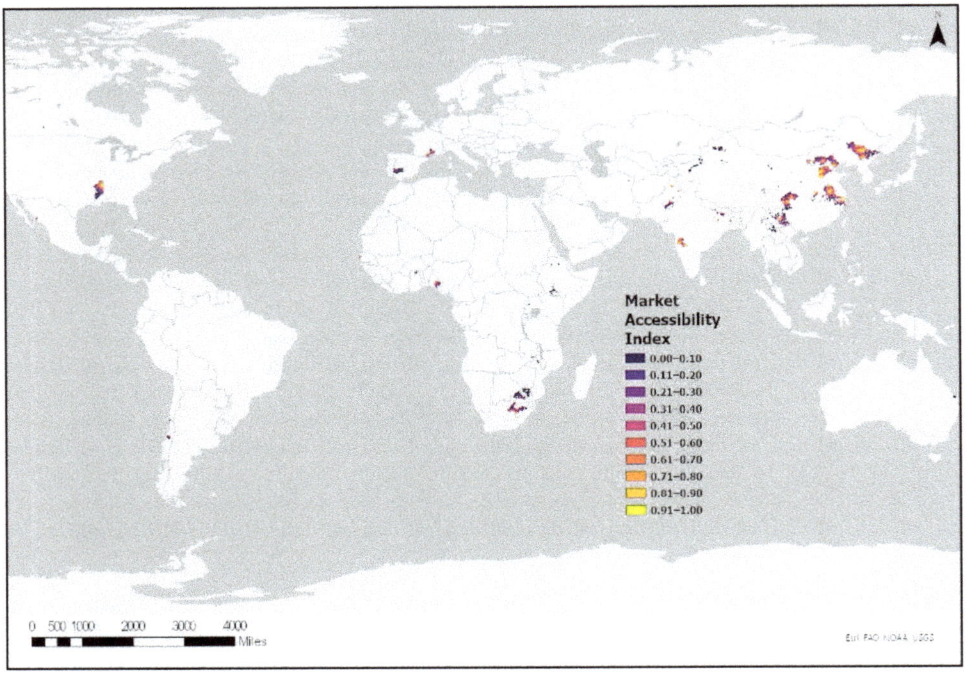

(**b**)

Figure 6. (**a**) Global extent for Market Access variable. (**b**) Sample extent for Market Access variable.

Table 2. Description of all the filters applied to the GLU layer obtained from GLOBE.

Variable Name	Description and Source	Filter(s) Applied
Olson Biomes	Terrestrial ecoregions of the world defined by climate, geology, and evolutionary history from Olson et al. (2001) [98]	Biomes—Boreal forests and Tundra removed.
Average Annual Temperature	Average annual temperature (°C) from 1951–2002. Values range from −28 °C to 31 °C. See [97] for more details.	Values greater than 28.57 °C and less than −12.2058 °C removed.
Average Annual Precipitation	Average annual precipitation (mm yr^{-1}) from 1950–1999 [83]. Values range from 0–10,572 mm/year.	Values greater than 2948.79 mm/yr and less than 30.0 mm/yr removed.
Population Density	Global model of population density from HYDE population model 2000 [84]. Values range from 0–62,018.	Values equal to '0' removed.
Percent Land Area	Percentage of land area contained within each GLU cell based on LandScan 2007 by Oak Ridge National Laboratory (2008). See [97] for details. Values range from 0–100%	Values less than 1 removed.
Percent Crop Area	Percent crop land cover area per grid cell derived from HYDE land cover data (2000) [84]. Values range from 0–100%	Values equal to '0' removed.
Slope Suitability Class	Global grid of land suitability for agriculture based on combined slope constraints [99]. Total 8 classes.	Classes 7 and 8 corresponding to 'Very Frequent Severe Constraints' and 'Unsuitable for Agriculture', respectively, removed.

To test the second hypothesis, first a list of factors reported to influence irrigation adoption decisions of farmers was compiled from the selected case studies. Factors affecting farmers' adoption decisions are often classified into broad clusters like—financial/economic, physical, institutional, and individual characteristics, but depending on the researchers' preferences and disciplinary backgrounds this categorization can vary [57,70]. For our study, based on the background literature, the different (influential) factors were clustered into seven broad categories—biophysical, demographic, geographic, technology-specific, social capital, farm enterprise, and institutional factors (Figure 7). Individual factors were coded using these broad categories for frequency analysis. Next, the relationships between these seven factor categories and their corresponding geographical contexts were examined using correspondence analysis. Correspondence analysis (CA) is a multivariate statistical technique and a useful visualization tool for summarizing, examining, and displaying the relationships between categorical data in a contingency table [100,101]. No underlying distributional assumptions are needed for this analysis and therefore, it accommodates any type of categorical variable—binary, ordinal, or nominal [102]. Moreover, the row and column points from the contingency table are shown together on a multi-dimensional map called biplot, which allows for easier visualization of the associations among variables [103,104]. CA uses the chi-square statistic to measure the distance between points on the map, but it does not reveal whether these associations are statistically significant and is therefore used only as an exploratory method [104].

All the above-mentioned statistical tests were conducted and developed in the PyCharm IDE (Integrated Development Environment) using pandas, Matplotlib, Prince, and Scipy Stats libraries.

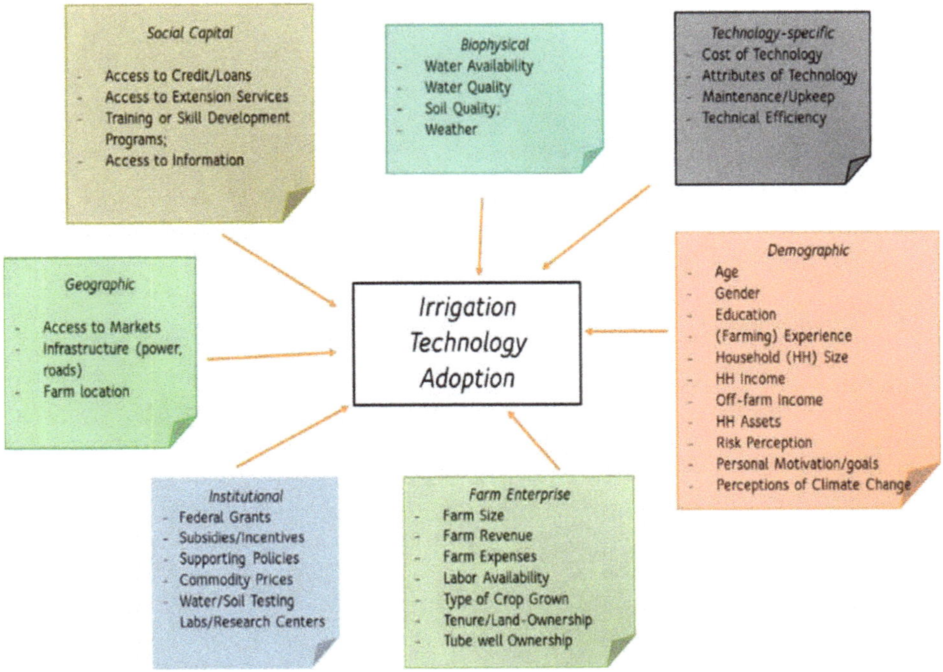

Figure 7. Categorization of different factors influencing farmers' irrigation adoption decision-making.

3. Results

3.1. Geographic Representativeness of Irrigation Adoption Studies

Geographic representativeness analyses were conducted for the percentage of GLU area equipped with irrigation, percentage of GLU area in cropland, average market accessibility, and average annual precipitation. Pearson's χ^2 tests for independence for each of the four variables (Tables 3–6) found that the observed (~sample) distributions were statistically different from the expected distributions.

Table 3. Pearson's χ^2 test results with percentage of area equipped for irrigation variable.

Bins	Frequency		χ^2 Statistic	*p*-Value *	Representedness Degree	
	Observed	Expected			r-Value **	Representedness
0.0–0.1	11469	20997	4951.870159	0.0	−1	**Highly under**
0.1–1.0	3634	5705	524.5160253	4.41×10^{-116}	−1	**Highly under**
1.0–5.0	5375	4079	203.1462568	4.30×10^{-46}	1	**Highly over**
5.0–10.0	3571	1769	654.2752213	2.63×10^{-144}	1	**Highly over**
10.0–20.0	3875	1710	906.3700068	4.05×10^{-199}	1	**Highly over**
20.0–35.0	3589	1236	1225.836112	1.48×10^{-268}	1	**Highly over**
35.0–50.0	2482	719	1013.405549	2.19×10^{-222}	1	**Highly over**
50.0–75.0	2536	694	1096.828508	1.62×10^{-240}	1	**Highly over**
75.0–100	749	369	130.4361647	3.29×10^{-30}	1	**Highly over**
Total Frequency	**37280**	**37280**	**Dist.χ^2= 24137.36522**	**Dist. *p*-value = 0.0**	**Diagnosis: Highly biased**	

* At 0.05 significance level; ** r-value calculation based on criteria defined in Section 3.2.

Table 4. Pearson's χ^2 test results with percentage of cropland variable.

Bins	Frequency		χ^2 Statistic	*p*-Value *	Representedness Degree	
	Observed	Expected			r-Value **	Representedness
0.0–0.1	6219	15900	5962.41	0.0	−1	**Highly under**
0.1–0.2	4595	5556	104.6993983	1.42×10^{-24}	−1	**Highly under**
0.2–0.3	5366	3863	278.0312679	2.02×10^{-62}	1	**Highly over**
0.3–0.4	6522	3245	1259.874418	5.93×10^{-276}	1	**Highly over**
0.4–0.5	6063	2856	1304.727198	1.06×10^{-285}	1	**Highly over**
0.5–0.6	3990	2551	345.7343162	3.60×10^{-77}	1	**Highly over**
0.6–0.7	2870	1954	185.2504582	3.46×10^{-42}	1	**Highly over**
0.7–0.8	2002	1319	146.4197305	1.05×10^{-33}	1	**Highly over**
0.8–0.9	203	416	73.20047746	1.17×10^{-17}	−1	**Highly under**
0.9–1.0	372	543	31.96753669	1.57×10^{-08}	−1	**Highly under**
Total Frequency	**38202**	**38202**	**Dist.χ^2= 15313.11155**	**Dist. *p*-value = 0.0**	**Diagnosis: Highly biased**	

* At 0.05 significance level; ** r-value calculation based on criteria defined in Section 3.2.

Table 5. Pearson's χ^2 test results with market accessibility variable.

Bins	Frequency		χ^2 Statistic	*p*-Value *	Representedness Degree	
	Observed	Expected			r-Value **	Representedness
0.0–0.1	10902	20736	5215.919285	0.0	−1	**Highly under**
0.1–0.2	4340	3907	25.36746844	4.74×10^{-07}	1	**Highly over**
0.2–0.3	4618	3186	292.2496414	1.61×10^{-65}	1	**Highly over**
0.3–0.4	4448	2677	484.9267588	1.81×10^{-107}	1	**Highly over**
0.4–0.5	3919	2075	614.9158442	9.54×10^{-136}	1	**Highly over**
0.5–0.6	3187	1481	663.2799333	2.89×10^{-146}	1	**Highly over**
0.6–0.7	2398	1335	317.6479106	4.71×10^{-71}	1	**Highly over**
0.7–0.8	2228	1172	342.6063765	1.73×10^{-76}	1	**Highly over**
0.8–0.9	1503	994	106.8413592	4.82×10^{-25}	1	**Highly over**
0.9–1.0	659	640	0.253736593	0.61	0.4	**Well-represented**
Total Frequency	**38202**	**38202**	**Dist.χ^2 = 12191.37033**	**Dist. *p*-value = 0.0**	**Diagnosis: Highly biased**	

* At 0.05 significance level; ** r-value calculation based on criteria defined in Section 3.2.

Table 6. Pearson's χ^2 test results with average annual precipitation (mm/year) variable.

Bins	Frequency		χ^2 Statistic	*p*-Value *	Representedness Degree	
	Observed	Expected			r-Value **	Representedness
30–287	1680	3293	558.91	1.45×10^{-123}	−1	**Highly under**
287–463	3449	5289	437.02	4.83×10^{-97}	−1	**Highly under**
463–584	6537	4397	488.33	3.28×10^{-108}	1	**Highly over**
584–666	3444	2457	178.54	1.01×10^{-40}	1	**Highly over**
666–786	3819	2999	108.02	2.66×10^{-25}	1	**Highly over**
786–962	6864	3658	1132.16	3.39×10^{-248}	1	**Highly over**
962–1219	6137	4917	157.17	4.71×10^{-36}	1	**Highly over**
1219–1595	5319	5660	12.30	4.54×10^{-04}	−1	**Highly under**
1595–2145	835	3261	1517.03	0.0	−1	**Highly under**
2145–2949	118	2270	2000.03	0.0	−1	**Highly under**
Total Frequency	**38202**	**38202**	**Dist.χ^2= 10070.33756**	**Dist. *p*-value = 0.0**	**Diagnosis: Highly biased**	

* At 0.05 significance level; ** r-value calculation based on criteria defined in Section 3.2.

The observed frequencies of the two lowest percent areas of irrigation were significantly lower than their expected frequencies (see Figure 8) and highly underrepresented (Table 3). Similarly, the remaining seven bins were highly over-represented in this collection as the observed frequencies of these bins were higher compared to their corresponding expected frequencies. Case studies of irrigation adoption were thus biased toward areas of existing agriculture, and studies were generally more over-represented as the area equipped for irrigation increased.

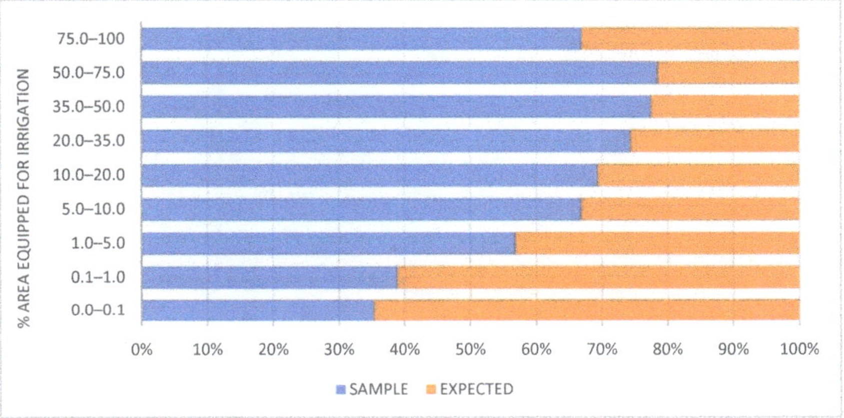

Figure 8. Percentage of Observed (~Sample) vs. Expected Counts for Irrigation Variable.

Similarly, in the case of the percent cropland variable (Table 4 and Figure 9), four out of ten bins (with very low and high cropland cover) were highly underrepresented. Irrigation adoption studies were more frequently conducted in areas with moderate extents of agricultural land use, and thus biased against areas of low or high cropland. This likely had implications for the irrigation adoption decisions studied. Locations that were dominantly or exclusively agricultural likely had better support services and infrastructure and did not compete with other land uses, which would presumably facilitate irrigation adoption. Conversely, farmers in low agricultural areas face the opposite conditions and may experience more barriers to irrigation adoption.

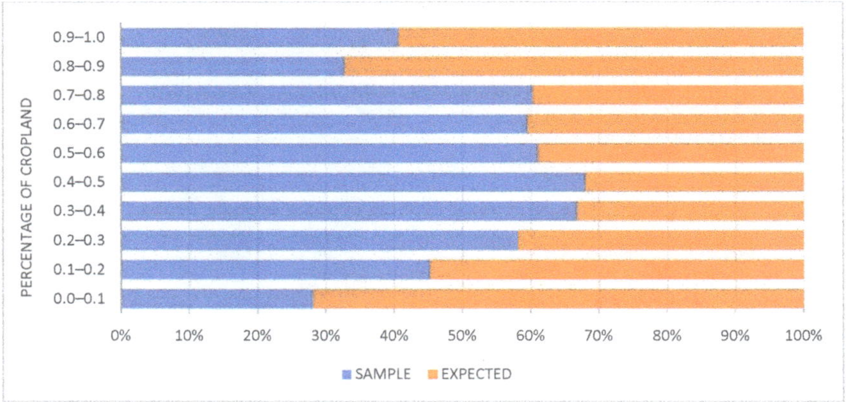

Figure 9. Percentage of Observed (~Sample) vs. Expected Counts for Cropland Variable.

In the case of the market access index, most of the bins (8 out of 10) were highly over-represented (Table 5 and Figure 10) with a bias toward areas having moderate-high market access. Market signals that might favor irrigation adoption were likely dampened in low market accessibility areas, which may not have been enough to overcome economic barriers to irrigation adoption. Additionally, remote areas are generally understudied due to access difficulties for researchers [30]. As a result, irrigation adoption studies were skewed toward locations with greater accessibility, including a well-represented sample of the most accessible locations.

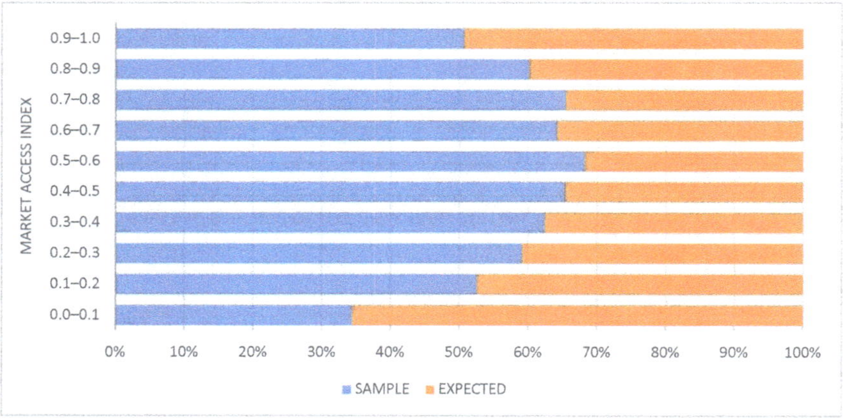

Figure 10. Percentage of Observed (~Sample) vs. Expected Counts for Market Accessibility Variable.

Finally, regions receiving moderate average annual rainfall (463–1219 mm/year) were highly over-represented, while regions with very low and high average annual rainfall were under-represented and understudied (Table 6 and Figure 11). The underrepresentation of low rainfall areas was surprising, but these may be neglected by irrigation adoption studies due to the necessity of irrigation and limited variability in decision-making. The limited sampling of high precipitation areas was not surprising, since areas receiving high average annual precipitation were more likely associated with rainfed agriculture. However, such areas may also include those in which seasonal drought is a concern despite high aggregate rainfall (e.g., humid southeast United States) and which potentially have unique sets of adoption decision factors.

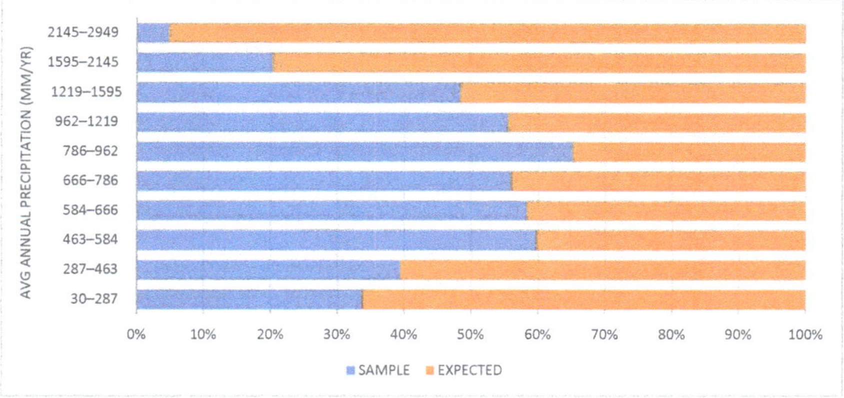

Figure 11. Percentage of Observed (~Sample) vs. Expected Counts for Average Annual Precipitation Variable.

3.2. Similarity of Irrigation Adoption Factors across Geographic Contexts

Most of the studies conducted in low irrigated regions of the world and that were highly underrepresented in this collection were from countries located in Africa and Latin America (see Table 7 and Figure 12). Further, Table 8 lists the different clusters of factors affecting irrigation adoption identified from the case studies, broken down by world regions. The frequency of each of the causal factors as reported in the case studies are provided in this table as an absolute number (this method of frequency analysis is based on the Geist & Lambin (2004) study). Only two case studies had a single variable (factor category) that explained farmers' decision-making regarding irrigation adoption, thus suggesting that the decision to adopt (or not) irrigation is best explained using a combination of factors (see Table 8). Dominating the broad clusters of factors affecting irrigation adoption decisions of farmers was the combination of—Biophysical, Demographic, Farm Enterprise, and Social Capital factors (B, D, F, S), followed by the cluster with Biophysical, Demographic, Farm Enterprise, Institutional, and Social Capital factors (B, D, F, I, S), with clear regional variations as both these clusters feature mainly in case studies from Asia and Africa. Cases from both these regions share a greater number of factors in common as compared to other regions. Demographic category that includes factors like age, gender, household size, and more (see Figure 7 for more details) featured the most, while both institutional and technology-related factor categories were least observed within these case studies. Further, demographic and social capital related factors together formed the most robust combination, although one that often occurred in combination with other clusters.

Table 7. Distribution of number of cases based on percentage of irrigation.

Percentage of Irrigation	No. of Cases	Degree of Representedness
0.0–0.1	4	Highly under
0.1–1.0	14	Highly under
1.0–5.0	7	Highly over
5.0–10.0	9	Highly over
10.0–20.0	6	Highly over
20.0–35.0	7	Highly over
35.0–50.0	2	Highly over
50.0–75.0	4	Highly over

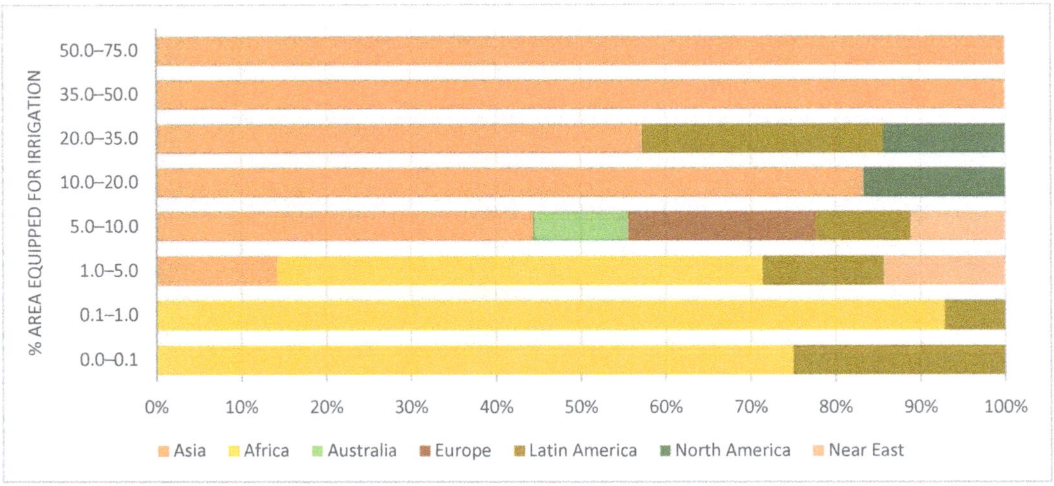

Figure 12. Distribution of study regions based on the percentage of area equipped for irrigation.

Table 8. Frequency of broad clusters of factors affecting irrigation adoption.

Study Sites / Factors	Asia (*n* = 20)	Africa (*n* = 20)	Australia (*n* = 1)	Europe (*n* = 2)	Latin America (*n* = 6)	North America (*n* = 2)	Near East (*n* = 2)	All Cases (*n* = 53)
SINGLE-FACTOR								
B	0	1	0	0	0	0	0	**1**
D	1	0	0	0	0	0	0	**1**
TWO FACTORS								
B, I	0	1	0	0	0	0	0	**1**
D, S	2	1	0	0	0	0	0	**3**
B, D	1	0	0	0	0	0	0	**1**
THREE FACTORS								
D, S, T	0	0	0	1	0	0	0	**1**
B, D, G	0	1	0	0	0	0	0	**1**
I, S, T	0	1	0	1	0	0	0	**2**
B, D, F	0	0	0	0	1	0	0	**1**
D, F, S	1	0	0	0	0	0	1	**2**
B, D, T	0	1	0	0	1	0	0	**2**
B, S, T	0	1	0	0	0	0	0	**1**
D, G, S	1	0	0	0	0	0	0	**1**
B, D, S	1	0	0	0	1	0	0	**2**
FOUR FACTORS								
B, D, F, S	3	2	0	0	0	0	1	**6**
B, D, F, G	1	0	0	0	0	0	0	**1**
B, D, I, S	0	0	0	0	0	1	0	**1**
D, G, I, S	0	1	0	0	0	0	0	**1**
D, G, S, T	0	1	0	0	0	0	0	**1**
B, F, G, S	0	1	0	0	0	0	0	**1**
D, F, I, S	0	1	0	0	0	0	0	**1**
D, F, S, T	0	0	1	0	1	0	0	**2**
B, D, G, T	0	1	0	0	0	0	0	**1**
D, F, G, S	1	1	0	0	1	0	0	**3**
FIVE FACTORS								
B, D, G, I, S	0	1	0	0	0	0	0	**1**
B, D, F, I, S	2	2	0	0	0	1	0	**5**
B, D, F, G, S	2	0	0	0	1	0	0	**3**
B, D, G, S, T	0	1	0	0	0	0	0	**1**
B, F, G, I, S	0	1	0	0	0	0	0	**1**
B, D, F, G, T	1	0	0	0	0	0	0	**1**
SIX FACTORS								
B, D, F, G, I, S	3	0	0	0	0	0	0	**3**
TOTAL CASES	**20**	**20**	**1**	**2**	**6**	**2**	**2**	**53**

B = Biophysical; D = Demographic; F = Farm Enterprise; G = Geographic; I = Institutional; S = Social Capital; T = Technology-specific.

Additionally, the CA biplot between the study regions and set of causal factors (Figure 13) was also prepared to visually identify and understand these regional variations. In this symmetric scatterplot, component 0 was represented by the horizontal axis and component 1 by the vertical axis. Together both the components explained about 45.68% of the variance/inertia in this dataset. Europe had high positive values along component 0 (horizontal axis), while Australia had high positive values along the vertical axis. Similarly, North America had high negative values and low positive values along vertical and horizontal axis, respectively. Moreover, from just visually inspecting this biplot it was evident that the set of factors influencing irrigation adoption (of farmers) in cases from Europe, Australia and North America were very different from each other as they were placed in separate quadrants and were also far from the origin. Australia and

Latin America study regions were placed in the same quadrant and thus, shared similar profiles, i.e., within both these regions similar combination of causal factors was observed as compared to say Europe or other regions (see Table 8 for more details). Further, the map also revealed that irrigation adoption by farmers from case studies in Europe was explained by a combination of only demographic, social capital, institutional, and technology-specific attributes. Whereas in case of North America, the strongest association was seen with factors like demographic, social capital, farm enterprise, institutional and biophysical.

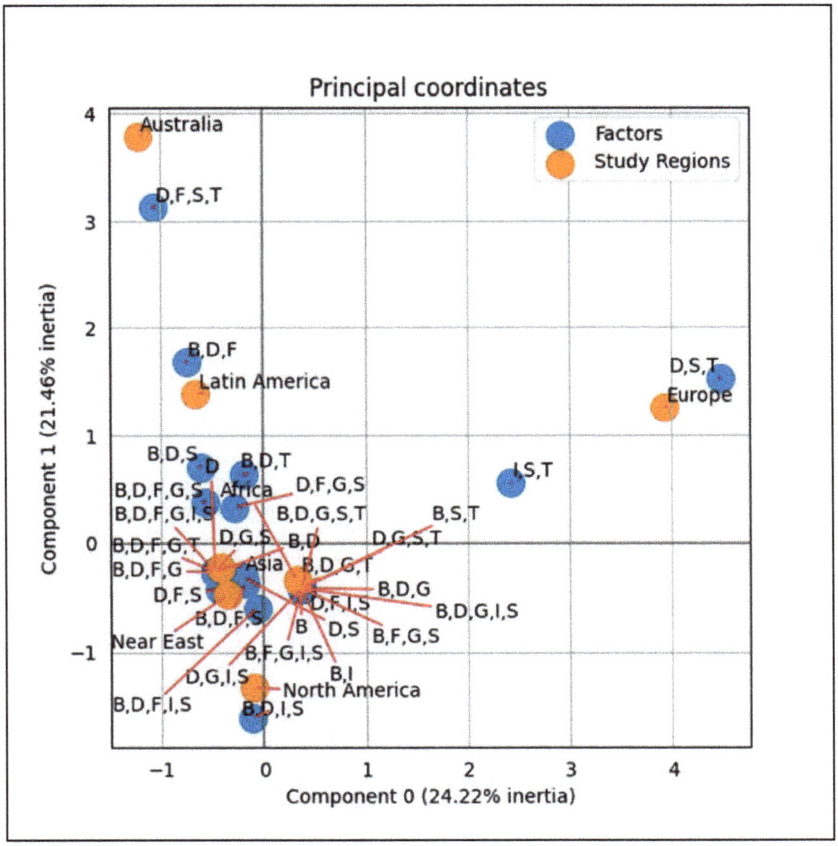

Figure 13. 2-D Correspondence Analysis biplot of Study Regions and Factors affecting Irrigation Adoption.

4. Discussion

In this paper, we explored the geographic contexts where irrigation adoption studies were conducted and the set of causal factors that were reportedly associated with irrigation adoption decisions. Based on the results of the systematic review, our first hypothesis held true. That is, the geographic contexts in which irrigation adoption studies were often conducted were biased. Geographic regions with less than 1% area equipped for irrigation, very low (less than 0.2%) and high (above 0.8%) percent of cropland, low market accessibility index (less than 0.1), and average annual precipitation with less than 463 mm/year and greater than 1219 mm/year, were highly underrepresented in this collection of case studies. In other words, these case studies were significantly biased toward areas where at least some amount of irrigation was already being practiced. An explanation for this bias towards irrigated areas could be that the research was motivated by the need to identify challenges and/or opportunities associated with further expansion. Additionally, low cropland areas were also understudied, because research might have been focused more on areas having a moderate or higher amount of cropland cover to encourage further agricultural growth and development. Usually, farmers in areas with a high percentage of cropland cover, because of the limited scope for further (land) expansion, are more likely be using intensive agricultural practices (like irrigation) to increase their crop productivity, hence the focus was towards areas with moderate amount of cropland. Further, highly accessible regions were over-represented in this collection, because research is often conducted in locations (and with communities) that are easily accessible (or reachable) as compared to remote or hard to reach locations [105]. There is also evidence that farmers with greater market access had stronger incentives to adopt irrigation for market production [106]. Hence, regions with low market accessibility were understudied and accordingly underrepresented. Similarly, regions with low and high average annual rainfall were also underrepresented and this might be due to the overall 'unsuitability' of this technology within these regions. For instance, if a region receives abundant rainfall, farmers might have a natural inclination to rely on rainfall for agricultural activities rather than investing in new technology, as irrigation is generally a substitute for rainwater [107]. For regions with low average annual rainfall, although irrigation technology can be very useful nevertheless, reliable access to water might hinder its widespread diffusion and subsequent adoption [108].

The second hypothesis that we tested in this paper held partially true as only the Demographic category of factors was observed as the most common among all the case study regions. This indicated that demographic factors such as a farmer's age, gender, household assets, income diversification options, and perceptions toward climate change (see Figure 3 for a complete list), significantly affected farmers' decisions to adopt (or not) irrigation irrespective of the geographic context. However, some distinct regional variations were also seen. For instance, studies from North America explained irrigation adoption behavior of farmers using a combination of only demographic, biophysical, social capital, farm-enterprise, and institutional factors. Factors related to place or technology did not feature in the case studies from this region. Similarly, for cases from Near East, only categories of factors such as demographic, farm enterprise, biophysical and social capital were observed. Both institutional and technology related factors were least observed among all these case studies. Further, the highest frequency was of the cluster with Biophysical, Demographic, Farm Enterprise, and Social Capital factors (B, D, F, S), followed by the cluster with Biophysical, Demographic, Farm Enterprise, Institutional, and Social Capital factors (B, D, F, I, S), suggesting that irrigation adoption decisions around the world are best explained by the combination of multiple and coupled factors instead of a single variable.

Moreover, majority of the case studies in this collection were from geographic regions of Asia and Africa and were clustered with a greater (and often similar) number of factors as compared to the rest. This suggests that some common challenges might possibly exist with regard to irrigation technology diffusion and adoption within these regions, even though the study sites within these regions (See Appendix A for more information on study locations) were different from each other in many other aspects beyond just percentage of irrigation or average annual precipitation (national wealth, population densities, etc.). A recent study on understanding sustainability challenges in three different rural landscapes, namely, Australia, central Romania, and southwestern Ethiopia, found similarities among these three different social-ecological systems, even though the systems examined appear to be very different on the surface [109], thus, highlighting the need for a comprehensive analysis to identify and better comprehend such common challenges.

Although a nearly similar set of factors were observed from case studies of Asia and Africa, many of the study sites from Africa with little to no irrigation (less than 1%) were understudied, while all those from Asia were over-studied and hence over-represented in this collection (Figure 12). One explanation for this research bias could be that the farmers in the study sites within Africa might still be in their early adoption phase. Given the low percentage of irrigated areas, one can argue that in these sites only a few individuals are taking the risk of investing in this technology. Moreover, this technology might not have been completely diffused within these sub-regions of Africa (east, west, and south), and as a result, this topic might be highly understudied within these sites because there is first a need to properly introduce this technology to the people, make them aware of its use and benefits, and only then can the adoption process be studied. Furthermore, based on the results of the frequency analysis, institutional and social capital related factors were most commonly observed in cases from this study region compared to others. These categories include factors like access to informational services, credit facilities, extension services, skill development programs, supporting policies, incentives, and subsidies. A study by Wozniak (1987) [110] highlighted the important role played by education and information on the new technology, particularly for early adoption. Another study by Diederen et al. (2003) [111] presented empirical evidence for explaining the differences in adoption behavior of innovators, early adopters, and laggards. Their findings suggested that innovators (~first or early users of technology) made more use of external sources of information. In a more recent study on the adoption of improved seed varieties by farmers in Ethiopia, the findings suggested that farmers' awareness about the available seed varieties is an important factor for the actual adoption to take place [112]. Teha & Jianjun (2021) [113] in their study on the adoption of small-scale irrigation found that 'government promotion' in the form of incentives and training positively affected a farmer's irrigation adoption decision. Thus, some kind of external support like extension and credit services are vital for farmers for enhancing the diffusion and adaptation of successful technologies and practices [114,115]. With limited information and support, a farmer's decision-making is primarily based on intuition and can be less efficient [116].

However, the results of this meta-study are limited in scope, since only peer-reviewed research articles that were available in the English language, in the two selected databases, and published on and after the year 2000 were considered for this analysis. Such a restriction on the publication date was imposed because the global irrigation dataset used in this analysis is based on the nationally reported statistics from around the year 2000. Further only articles that investigated the factors associated with irrigation adoption were selected for this analysis irrespective of the theoretical frameworks applied to examine a farmer's adoption behavior. Due to this, certain factors might be emphasized more than others. For instance, a social network analysis approach was used to assess the barriers to climate change adaptations in Spain [117]. Because of the specific framework used in this study, the barriers identified were mostly categorized within social capital and institutional categories (see Appendix A for study details). Similarly, another case study from Nepal, used risk perception and motivation theory to understand farmers preparedness to cope with the impacts of climate-change hazards [118], and as a result, only the factors characterized as demographic were identified from this case study. Moreover, conference proceedings and grey literature were also excluded from the dataset due to inconsistent methodology and results reporting. Such sources may have contained useful and unique insights, but issues of comparability with information gathered from peer-reviewed would have unduly complicated the analysis.

Despite the limitations mentioned above, the global representativeness analysis highlights the multiple (geographic) biases that exist with respect to studying farmers' irrigation adoption decision-making. More research on this topic is being conducted in regions that have little to high percentage of irrigation (>1%), are readily accessible, receive moderate amounts of average annual rainfall, and have moderate amounts of cropland cover. These results suggest the need to expand research efforts, particularly in areas with low irrigation and cropland cover to identify constraints to and help accelerate economic growth, poverty reduction, and food and livelihood security for rural communities in these regions.

5. Conclusions

Food production is still risky in many parts of the world, particularly in Sub-Saharan Africa, due to limited information about changing weather patterns, market access and demands, and unequal access to efficient technologies [116]. Additionally, this pressure on our global food systems will only intensify in the coming years with not only the changing consumption patterns but with the changing climatic conditions as well. For example, yield declines resulting from climate change (e.g., higher temperatures, increased seasonality, more frequent and severe hydroclimate events) have already occurred [119] and are expected to decrease the production of global consumable food calories by another 1% to 7% by the end of the century [120]. Irrigation currently remains one of the most critical inputs to farming today and is a key adaptation to variable precipitation and droughts resulting from changing climatic conditions [38]. New investments in irrigation infrastructure together with improved water management practices can not only minimize the impact of water scarcity but can also aid in meeting the water demands for global food production [121]. Further, managing and improving irrigation efficiency will, in turn, support global water, food, and energy goals [122]. Therefore, understanding the diverse reasons, motivations, and/or factors underlying the choices of producers regarding its adoption (or rejection), especially when climate change demands some kind of adaptation in unprecedented areas, will help better anticipate future food, energy, and water demands [123].

There is still much room left for improvements in both agricultural practices and water-use efficiency, but farmers' reluctance to adopt new technologies needs to be better understood if such sustainability targets are to be achieved [72], and societal resilience must be built to mitigate the impacts of future climatic changes [11]. In this study, we identified multiple geographic biases that exist with respect to studying farmers' irrigation adoption decision-making, thus, suggesting the need for extensive research even in areas with no irrigation and/or low cropland cover to identify opportunities for the implementation of other sustainable solutions to support agricultural development in these areas. Moreover, apart from these biases, some commonalities were observed in terms of constraints faced by farmers regarding irrigation technology adoption across different geographic landscapes. However, our findings also indicated that there may not be a 'standard set' of factors for understanding irrigation adoption, and nuances in the local context are just as important to identify as commonalities across settings. This suggests the need for more geographically comprehensive analyses that would enable comparative analysis of different landscapes, as well as studies that delve into the adoption process beyond individual technology adoption behaviors. Further, this kind of systems analysis will help unravel common challenges, drivers, and opportunities regarding agriculture development under changing climatic conditions across multiple systems, while also being attentive to local context offers the potential for co-learning [109,124].

Author Contributions: Both the authors (R.P. and N.R.M.) have made substantial, direct, and intellectual contribution to the work and approved it for publication. All authors have read and agreed to the published version of the manuscript.

Funding: The authors were supported by a National Science Foundation award INFEWS #1856054.

Institutional Review Board Statement: Not applicable.

Data Availability Statement: Publicly available datasets were analyzed in this study and their description and sources have already been listed above.

Conflicts of Interest: The authors declare that the research was conducted in the absence of any commercial or financial relationships that could be construed as a potential conflict of interest.

Appendix A

A1 Target Set

1. Alcon, F., Tapsuwan, S., Martínez-Paz, J. M., Brouwer, R., & de Miguel, M. D. (2014). Forecasting deficit irrigation adoption using a mixed stakeholder assessment methodology. *Technological Forecasting and Social Change*, *83*(1), 183–193. https://doi.org/10.1016/j.techfore.2013.07.003.
2. Asfaw, A., Simane, B., Bantider, A., & Hassen, A. (2019). Determinants in the adoption of climate change adaptation strategies: evidence from rainfed-dependent smallholder farmers in north-central Ethiopia (Woleka sub-basin). *Environment, Development and Sustainability*, *21*, 2535–2565.
3. Chandran, K. M., & Surendran, U. (2016). نصصخت سروي س آموزش ی کارگا ه های تیوضع آموزشی ملف های زکرم اطلاعا ت ملع بلا گ ویژ ه سروي س های STRS همجرت کنل های دنفم درخبرنام ه Study on factors influencing the adoption of drip irrigation by farmers in humid tropical Kerala, India. In *International Journal of Plant Production* (Vol. 10, Issue 3). www.SID.ir
4. Esham, M., & Garforth, C. (2013). Agricultural adaptation to climate change_ insights from a farming community in Sri Lanka. *Mitigation and Adaptation Strategies for Global Change*, *18*, 535–549.
5. Fan, Y., & McCann, L. (2020). Adoption of pressure irrigation systems and scientific irrigation scheduling practices by U.S. farmers: An application of multilevel models. *Journal of Agricultural and Resource Economics*, *45*(2), 352–375. https://doi.org/10.22004/ag.econ.302459
6. Greenland, S., Levin, E., Dalrymple, J. F., & O'Mahony, B. (2019). Sustainable innovation adoption barriers: water sustainability, food production and drip irrigation in Australia. *Social Responsibility Journal*, *15*(6), 727–741. https://doi.org/10.1108/SRJ-07-2018-0181
7. Huang, Q., Xu, Y., Kovacs, K., & West, G. (2017). ANALYSIS OF FACTORS THAT INFLUENCE THE USE OF IRRIGATION TECHNOLOGIES AND WATER MANAGEMENT PRACTICES IN ARKANSAS. *Journal of Agricultural and Applied Economics*, *49*(2), 159–185.
8. Mase, A. S., Gramig, B. M., & Prokopy, L. S. (2017). Climate change beliefs, risk perceptions, and adaptation behavior among Midwestern U.S. crop farmers. *Climate Risk Management*, *15*, 8–17. https://doi.org/10.1016/j.crm.2016.11.004
9. Mesfin, A. H., & Bekele, A. (2018). Farmers Perception on Climate Change and Determinants of Adaptation Strategies in Benishangul-Gumuz Regional State of Ethiopia. *International Journal on Food System Dynamics*, *9*(5), 453–469.
10. Ngigi, M. W., Mueller, U., & Birner, R. (2017). Gender Differences in Climate Change Adaptation Strategies and Participation in Group-based Approaches: An Intra-household Analysis From Rural Kenya. *Ecological Economics*, *138*, 99–108. https://doi.org/10.1016/j.ecolecon.2017.03.019.

A2 Shortlisted Case Studies

1. Abid, M., Scheffran, J., Schneider, U. A., & Ashfaq, M. (2015). Farmers' perceptions of and adaptation strategies to climate change and their determinants: the case of Punjab Province, Pakistan. *Earth System Dynamics*, *6*, 225–243.
2. Akrofi, N. A., Sarpong, D. B., Somuah, H. A. S., & Osei-Owusu, Y. (2019). Paying for privately installed irrigation services in Northern Ghana: The case of the smallholder Bhungroo Irrigation Technology. *Agricultural Water Management*, *216*, 284–293. https://doi.org/10.1016/j.agwat.2019.02.010
3. Alam, K. (2015). Farmers' adaptation to water scarcity in drought-prone environments: A case study of Rajshahi District, Bangladesh. *Agricultural Water Management*, *148*, 196–206. https://doi.org/10.1016/j.agwat.2014.10.011

4. Alauddin, M., Rashid Sarker, M. A., Islam, Z., & Tisdell, C. (2020). Adoption of alternate wetting and drying (AWD) irrigation as a water-saving technology in Bangladesh: Economic and environmental considerations. *Land Use Policy, 91*. https://doi.org/10.1016/j.landusepol.2019.104430
5. Ali, S., Ying, L., Nazir, A., Abdullah, Ishaq, M., Shah, T., Ye, X., Ilyas, A., & Tariq, A. (2021). Rural farmers perception and coping strategies towards climate change and their determinants: Evidence from Khyber Pakhtunkhwa province, Pakistan. *Journal of Cleaner Production, 291*. https://doi.org/10.1016/j.jclepro.2020.125250
6. Below, T. B., Mutabazi, K. D., Kirschke, D., Franke, C., Sieber, S., Siebert, R., & Tscherning, K. (2012). Can farmers' adaptation to climate change be explained by socio-economic household-level variables? *Global Environmental Change, 22*(1), 223–235. https://doi.org/10.1016/j.gloenvcha.2011.11.012
7. Budhathoki, N. K., Paton, D., A. Lassa, J., & Zander, K. K. (2020). Assessing farmers' preparedness to cope with the impacts of multiple climate change-related hazards in the Terai lowlands of Nepal. *International Journal of Disaster Risk Reduction, 49*. https://doi.org/10.1016/j.ijdrr.2020.101656
8. Bukchin, S., & Kerret, D. (2020). Character strengths and sustainable technology adoption by smallholder farmers. *Heliyon, 6*(8). https://doi.org/10.1016/j.heliyon.2020.e04694
9. Burnham, M., & Ma, Z. (2018). Multi-Scalar Pathways to Smallholder Adaptation. *World Development, 108*, 249–262. https://doi.org/10.1016/j.worlddev.2017.08.005
10. Chen, H., Wang, J., & Huang, J. (2014). Policy support, social capital, and farmers' adaptation to drought in China. *Global Environmental Change, 24*(1), 193–202. https://doi.org/10.1016/j.gloenvcha.2013.11.010
11. Chuchird, R., Sasaki, N., & Abe, I. (2017). Influencing factors of the adoption of agricultural irrigation technologies and the economic returns: A case study in Chaiyaphum Province, Thailand. *Sustainability (Switzerland), 9*(9). https://doi.org/10.3390/su9091524
12. Danso-Abbeam, G., Ojo, T. O., Baiyegunhi, L. J. S., & Ogundeji, A. A. (2021). Climate change adaptation strategies by smallholder farmers in Nigeria: does non-farm employment play any role? *Heliyon, 7*(6). https://doi.org/10.1016/j.heliyon.2021.e07162
13. Ebi, K., Padgham, J., Doumbia, M., Kergna, A., Smith, J., Butt, T., & McCarl, B. (2011). Smallholders adaptation to climate change in Mali. *Climatic Change, 108*, 423–436.
14. Esfandiari, M., Mirzaei Khalilabad, H. R., Boshrabadi, H. M., & Mehrjerdi, M. R. Z. (2020). Factors influencing the use of adaptation strategies to climate change in paddy lands of Kamfiruz, Iran. *Land Use Policy, 95*. https://doi.org/10.1016/j.landusepol.2020.104628
15. Esteve, P., Varela-Ortega, C., & Downing, T. (2018). A stakeholder-based assessment of barriers to climate change adaptation in a water-scarce basin in Spain. *Regional Environmental Change, 18*, 2505–2517.
16. Fagariba, C. J., Song, S., & Baoro, S. K. G. S. (2018). Climate change in Upper East Region of Ghana; Challenges existing in farming practices and new mitigation policies. *Open Agriculture, 3*(1), 524–536. https://doi.org/10.1515/opag-2018-0057
17. Fahad, S., Inayat, T., Wang, J., Dong, L., Hu, G., Khan, S., & Khan, A. (2020). Farmers' awareness level and their perceptions of climate change: A case of Khyber Pakhtunkhwa province, Pakistan. *Land Use Policy, 96*. https://doi.org/10.1016/j.landusepol.2020.104669
18. Funk, C., Raghavan Sathyan, A., Winker, P., & Breuer, L. (2020). Changing climate - Changing livelihood: Smallholder's perceptions and adaption strategies. *Journal of Environmental Management, 259*. https://doi.org/10.1016/j.jenvman.2019.109702
19. Graveline, N., & Grémont, M. (2021). The role of perceptions, goals and characteristics of wine growers on irrigation adoption in the context of climate change. *Agricultural Water Management, 250*. https://doi.org/10.1016/j.agwat.2021.106837

20. Herwehe, L., & Scott, C. A. (2017). Drought adaptation and development: small-scale irrigated agriculture in northeast Brazil. *Climate and Development*, *10*(4), 337–346.
21. Jamil, I., Jun, W., Mughal, B., Raza, M. H., Imran, M. A., & Waheed, A. (2021). Does the adaptation of climate-smart agricultural practices increase farmers' resilience to climate change? *Environmental Science and Pollution Research*, *28*, 27238–27249.
22. Jha, C. K., & Gupta, V. (2021). Farmer's perception and factors determining the adaptation decisions to cope with climate change: An evidence from rural India. *Environmental and Sustainability Indicators*, *10*. https://doi.org/10.1016/j.indic.2021.100112
23. Kabir, M. J., Cramb, R., Alauddin, M., Roth, C., & Crimp, S. (2017). Farmers' perceptions of and responses to environmental change in southwest coastal Bangladesh. *Asia Pacific Viewpoint*, *58*(3), 362–378.
24. Kalele, D. N., Ogara, W. O., Oludhe, C., & Onono, J. O. (2021). Climate change impacts and relevance of smallholder farmers' response in arid and semi-arid lands in Kenya. *Scientific African*, *12*. https://doi.org/10.1016/j.sciaf.2021.e00814
25. Kephe, P. N., Ayisi, K. K., & Petja, B. M. (2020). A decision support system for institutional support to farmers in the face of climate change challenges in Limpopo province. *Heliyon*, *6*(11). https://doi.org/10.1016/j.heliyon.2020.e04989
26. Keshavarz, M., & Moqadas, R. S. (2021). Assessing rural households' resilience and adaptation strategies to climate variability and change. *Journal of Arid Environments*, *184*. https://doi.org/10.1016/j.jaridenv.2020.104323
27. Khanal, U., Wilson, C., Hoang, V. N., & Lee, B. (2018). Farmers' Adaptation to Climate Change, Its Determinants and Impacts on Rice Yield in Nepal. *Ecological Economics*, *144*, 139–147. https://doi.org/10.1016/j.ecolecon.2017.08.006
28. Knapp, T., & Huang, Q. (2017). Do climate factors matter for producers' irrigation practices decisions? *Journal of Hydrology*, *552*, 81–91. https://doi.org/10.1016/j.jhydrol.2017.06.037
29. Koech, R., Haase, M., Grima, B., & Taylor, B. (2020). Barriers and measures to improve adoption of irrigation technologies: A case study from the Bundaberg region in Queensland, Australia.
30. Kumasi, T. C., Antwi-Agyei, P., & Obiri-Danso, K. (2019). Small-holder farmers' climate change adaptation practices in the Upper east region of Ghana. *Environment, Development and Sustainability*, *21*, 745–762.
31. Leroy, D. (2019). Farmers' Perceptions of and Adaptations to Water Scarcity in Colombian and Venezuelan Paramos in the Context of Climate Change. *Mountain Research and Development*, *39*.
32. Li, S., An, P. L., Pan, Z. H., Wang, F. T., Li, X. M., & Liu, Y. (2015). Farmers' initiative on adaptation to climate change in the Northern Agro-pastoral Ecotone. *International Journal of Disaster Risk Reduction*, *12*, 278–284. https://doi.org/10.1016/j.ijdrr.2015.02.002
33. Li, W., Ruiz-Menjivar, J., Zhang, L., & Zhang, J. (2021). Climate change perceptions and the adoption of low-carbon agricultural technologies: Evidence from rice production systems in the Yangtze River Basin. *Science of the Total Environment*, *759*. https://doi.org/10.1016/j.scitotenv.2020.143554
34. Mango, N., Makate, C., Tamene, L., Mponela, P., & Ndengu, G. (2018). Adoption of small-scale irrigation farming as a climate-smart agriculture practice and its influence on household income in the Chinyanja Triangle, Southern Africa. *Land*, *7*(2). https://doi.org/10.3390/land7020049
35. Marie, M., Yirga, F., Haile, M., & Tquabo, F. (2020). Farmers' choices and factors affecting adoption of climate change adaptation strategies: evidence from northwestern Ethiopia. *Heliyon*, *6*(4). https://doi.org/10.1016/j.heliyon.2020.e03867
36. Masud, M. M., Azam, M. N., Mohiuddin, M., Banna, H., Akhtar, R., Alam, A. S. A. F., & Begum, H. (2017). Adaptation barriers and strategies towards climate change:

Challenges in the agricultural sector. *Journal of Cleaner Production*, *156*, 698–706. https://doi.org/10.1016/j.jclepro.2017.04.060
37. Matewos, T. (2020). The state of local adaptive capacity to climate change in drought-prone districts of rural Sidama, southern Ethiopia. *Climate Risk Management*, *27*. https://doi.org/10.1016/j.crm.2019.100209
38. Mi, Q., Li, X., Li, X., Yu, G., & Gao, J. (2021). Cotton farmers' adaptation to arid climates: Waiting times to adopt water-saving technology. *Agricultural Water Management*, *244*. https://doi.org/10.1016/j.agwat.2020.106596
39. NGANGO, J., & HONG, S. (2021). Adoption of small-scale irrigation technologies and its impact on land productivity: Evidence from Rwanda. *Journal of Integrative Agriculture*, *20*(8), 2302–2312. https://doi.org/10.1016/S2095-3119(20)63417-7
40. Nguyen, N., & Drakou, E. G. (2021). Farmers intention to adopt sustainable agriculture hinges on climate awareness: The case of Vietnamese coffee. *Journal of Cleaner Production*, *303*. https://doi.org/10.1016/j.jclepro.2021.126828
41. Nigussie, Y., van der Werf, E., Zhu, X., Simane, B., & van Ierland, E. C. (2018). Evaluation of Climate Change Adaptation Alternatives for Smallholder Farmers in the Upper Blue-Nile Basin. *Ecological Economics*, *151*, 142–150. https://doi.org/10.1016/j.ecolecon.2018.05.006
42. Nyang'au, J. O., Mohamed, J. H., Mango, N., Makate, C., & Wangeci, A. N. (2021). Smallholder farmers' perception of climate change and adoption of climate smart agriculture practices in Masaba South Sub-county, Kisii, Kenya. *Heliyon*, *7*(4). https://doi.org/10.1016/j.heliyon.2021.e06789
43. Ojo, T. O., Adetoro, A. A., Ogundeji, A. A., & Belle, J. A. (2021). Quantifying the determinants of climate change adaptation strategies and farmers' access to credit in South Africa. *Science of the Total Environment*, *792*. https://doi.org/10.1016/j.scitotenv.2021.148499
44. Orduño Torres, M. A., Kallas, Z., & Ornelas Herrera, S. I. (2020). Farmers' environmental perceptions and preferences regarding climate change adaptation and mitigation actions; towards a sustainable agricultural system in México. *Land Use Policy*, *99*. https://doi.org/10.1016/j.landusepol.2020.105031
45. Pittman, J., Wittrock, V., Kulshreshtha, S., & Wheaton, E. (2011). Vulnerability to climate change in rural Saskatchewan: Case study of the Rural Municipality of Rudy No. 284. *Journal of Rural Studies*, *27*(1), 83–94. https://doi.org/10.1016/j.jrurstud.2010.07.004
46. Rico, L., Poblete, D., Meza, F., & Kerrigan, G. (2016). Farmers' Options to Address Water Scarcity in a Changing Climate: Case Studies from two Basins in Mediterranean Chile. *Environmental Management*, *109*, 958–971.
47. Sertse, S. F., Khan, N. A., Shah, A. A., Liu, Y., & Naqvi, S. A. A. (2021). Farm households' perceptions and adaptation strategies to climate change risks and their determinants: Evidence from Raya Azebo district, Ethiopia. *International Journal of Disaster Risk Reduction*, *60*. https://doi.org/10.1016/j.ijdrr.2021.102255
48. Shikuku, K. M., Winowiecki, L., Twyman, J., Eitzinger, A., Perez, J. G., Mwongera, C., & Läderach, P. (2017). Smallholder farmers' attitudes and determinants of adaptation to climate risks in East Africa. *Climate Risk Management*, *16*, 234–245. https://doi.org/10.1016/j.crm.2017.03.001
49. Udmale, P., Ichikawa, Y., Manandhar, S., Ishidaira, H., & Kiem, A. S. (2014). Farmers' perception of drought impacts, local adaptation and administrative mitigation measures in Maharashtra State, India. *International Journal of Disaster Risk Reduction*, *10*(PA), 250–269. https://doi.org/10.1016/j.ijdrr.2014.09.011
50. Zizinga, A., Kangalawe, R. Y. M., Ainslie, A., Tenywa, M. M., Majaliwa, J., Saronga, N. J., & Amoako, E. E. (2017). Analysis of Farmer's Choices for Climate Change Adaptation Practices in South-Western Uganda, 1980-2009. *Climate*, *5*(4), 89.

References

1. Sauer, T.; Havlik, P.; Schneider, U.; Schmid, E.; Kindermann, G.E.; Obersteiner, M. Agriculture and resource availability in a changing world: The role of irrigation. *Water Resour. Res.* **2010**, *46*, W06503. [CrossRef]
2. Yadav, P.; Jaiswal, D.K.; Sinha, R.K. Climate change. In *Global Climate Change*; Elsevier: Amsterdam, The Netherlands, 2021; pp. 151–174. [CrossRef]
3. Kurukulasuriya, P.; Rosenthal, S. *Climate Change and Agriculture A Review of Impacts and Adaptations*; World Bank: Washington, DC, USA, 2003.
4. Smit, B.; Skinner, M.W. Adaptation Options in Agriculture to Climate Change: A Typology. *Mitig. Adapt. Strateg. Glob. Chang.* **2002**, *7*, 85–114. [CrossRef]
5. Allouche, J. The sustainability and resilience of global water and food systems: Political analysis of the interplay between security, resource scarcity, political systems and global trade. *Food Policy* **2011**, *36*, S3–S8. [CrossRef]
6. Kogo, B.K.; Kumar, L.; Richard, K. Climate change and variability in Kenya: A review of impacts on agriculture and food security. *Environ. Dev. Sustain.* **2021**, *23*, 23–43. [CrossRef]
7. Cassman, K.G.; Dobermann, A.; Walters, D.T.; Yang, H. Meeting cereal demand while protecting natural resources and improving environmental quality. *Annu. Rev. Environ. Resour.* **2003**, *28*, 315–358. [CrossRef]
8. Sakschewski, B.; Von Bloh, W.; Huber, V.; Müller, C.; Bondeau, A. Feeding 10 billion people under climate change: How large is the production gap of current agricultural systems? *Ecol. Modell.* **2014**, *288*, 103–111. [CrossRef]
9. Calzadilla, A.; Rehdanz, K.; Betts, R.; Falloon, P.; Wiltshire, A.; Tol, R. Climate change impacts on global agriculture. *Clim. Change* **2013**, *120*, 357–374. [CrossRef]
10. Shiferaw, B.; Tesfaye, K.; Kassie, M.; Abate, T.; Prasanna, B.; Menkir, A. Managing vulnerability to drought and enhancing livelihood resilience in sub-Saharan Africa: Technological, institutional and policy options. *Weather Clim. Extrem.* **2014**, *3*, 67–79. [CrossRef]
11. Wilson, R.S.; Herziger, A.; Hamilton, M.; Brooks, J.S. From incremental to transformative adaptation in individual responses to climate-exacerbated hazards. *Nat. Clim. Change* **2020**, *10*, 200–208. [CrossRef]
12. Intergovernmental Panel on Climate Change. Adaptation Needs and Options. In *Climate Change 2014—Impacts, Adaptation and Vulnerability: Part A: Global and Sectoral Aspects: Working Group II Contribution to the IPCC Fifth Assessment Report: Volume 1: Global and Sectoral Aspects*; Intergovernmental Panel on Climate Change, Ed.; Cambridge University Press: Cambridge, UK, 2014; Volume 1, pp. 833–868.
13. Tack, J.; Barkley, A.; Hendricks, N. Irrigation offsets wheat yield reductions from warming temperatures. *Environ. Res. Lett.* **2017**, *12*, 114027. [CrossRef]
14. Troy, T.J.; Kipgen, C.; Pal, I. The impact of climate extremes and irrigation on US crop yields. *Environ. Res. Lett.* **2015**, *10*, 54013. [CrossRef]
15. Hejazi, M.I.; Edmonds, J.A.; Chaturvedi, V. Global Irrigation Demand? A Holistic Approach. *Irrig. Drain. Syst. Eng.* **2012**, *1*, 2–5. [CrossRef]
16. Angelakıs, A.N.; Zaccaria, D.; Krasilnikoff, J.; Salgot, M.; Bazza, M.; Roccaro, P.; Jimenez, B.; Kumar, A.; Yinghua, W.; Baba, A.; et al. Irrigation of world agricultural lands: Evolution through the Millennia. *Water* **2020**, *12*, 1285. [CrossRef]
17. Hussain, I.; Hanjra, M.A. Irrigation and poverty alleviation: Review of the empirical evidence. *Irrig. Drain.* **2004**, *53*, 1–15. [CrossRef]
18. International Commission on Irrigation & Drainage. ICID Database—World Irrigated Area. 2021. Available online: https://icid-ciid.org/Knowledge/world_irrigated_area (accessed on 30 March 2022).
19. Knoema. Total Area Equipped for Irrigation. 2019. Available online: https://knoema.com/atlas/topics/Land-Use/Area/Total-area-equipped-for-irrigation?type=maps (accessed on 18 March 2022).
20. Mottaleb, K.A. Perception and adoption of a new agricultural technology: Evidence from a developing country. *Technol. Soc.* **2018**, *55*, 126–135. [CrossRef]
21. McDonald, R.I.; Girvetz, E.H. Two Challenges for U.S. Irrigation Due to Climate Change: Increasing Irrigated Area in Wet States and Increasing Irrigation Rates in Dry States. *PLoS ONE* **2013**, *8*, e65589. [CrossRef]
22. United States Government Accountability Office (US GAO). *Irrigated Agriculture; Science, Technology Assessment, and Analytics, Natural Resources and Environment, Report to Congressional Requesters, Technologies, Practices, and Implications for Water Scarcity*; US GAO: Washington, DC, USA, 2019.
23. Combs, P. Evaluation of Factors Influencing Irrigation Adoption among Farmers in the Southeast. In *All Theses*; Clemson University: Clemson, SC, USA, 2019.
24. Patle, G.T.; Kumar, M.; Khanna, M. Climate-smart water technologies for sustainable agriculture: A review. *J. Water Clim. Chang.* **2019**, *11*, 1455–1466. [CrossRef]
25. Mbuli, C.S.; Fonjong, L.N.; Fletcher, A.J. Climate Change and Small Farmers' Vulnerability to Food Insecurity in Cameroon. *Sustainability* **2021**, *13*, 1523. [CrossRef]
26. Giannakis, E.; Bruggeman, A.; Djuma, H.; Kozyra, J.; Hammer, J. Water pricing and irrigation across Europe: Opportunities and constraints for adopting irrigation scheduling decision support systems. *Water Sci. Technol. Water Supply* **2016**, *16*, 245–252. [CrossRef]

27. Esham, M.; Garforth, C. Agricultural adaptation to climate change: Insights from a farming community in Sri Lanka. *Mitig. Adapt. Strateg. Glob. Chang.* **2013**, *18*, 535–549. [CrossRef]
28. Asare-Baah, L.; Zabawa, R.; Findlay, H.J.; Findlay, H. Participation in Selected USDA Programs by Socially Disadvantaged Farmers in Selected Black Belt Counties in Georgia. *J. Rural. Soc. Sci.* **2018**, *33*, 2.
29. Ruzzante, S.; Labarta, R.; Bilton, A. Adoption of agricultural technology in the developing world: A meta-analysis of the empirical literature. *World Dev.* **2021**, *146*, 105599. [CrossRef]
30. Martin, L.; Blossey, B.; Ellis, E. Mapping where ecologists work: Biases in the global distribution of terrestrial ecological observations. *Front. Ecol. Environ.* **2012**, *10*, 195–201. [CrossRef] [PubMed]
31. Prokopy, L.S.; Floress, K.; Klotthor-Weinkauf, D.; Baumgart-Getz, A. Determinants of agricultural best management practice adoption: Evidence from the literature. *J. Soil Water Conserv.* **2008**, *63*, 300–311. Available online: www.swcs.org (accessed on 15 November 2022).
32. Da Cunha, D.A.; Coelho, A.B.; Féres, J.G. Irrigation as an adaptive strategy to climate change: An economic perspective on Brazilian agriculture. *Environ. Dev. Econ.* **2015**, *20*, 57–79. [CrossRef]
33. Trapolino, M. Irrigation Technology in Agriculture: How New Technologies Overcome Challenges. 2019. Available online: https://www.agritechtomorrow.com/article/2019/01/top-article-from-2019-irrigation-technology-in-agriculture-how-new-technologies-overcome-challenges/11230 (accessed on 3 May 2021).
34. USDA Economic Research Service. Irrigation & Water Use. 2022. Available online: https://www.ers.usda.gov/topics/farm-practices-management/irrigation-water-use/ (accessed on 1 May 2022).
35. USGS. Irrigation Methods. 2018. Available online: https://www.usgs.gov/special-topics/water-science-school/science/irrigation-methods-quick-look (accessed on 1 May 2022).
36. Barta, R.; Broner, I.; Schneekloth, J.; Waskom, R. Farm Irrigation Systems. 2015. Available online: https://irrigazette.com/en/news/farm-irrigation-systems (accessed on 9 May 2022).
37. Michael, C.H.; Charles, H.U.; James, R.E.; Stefan, S.O.; Graham, V.M. World Atlas of Desertification, Publication Office of the European Union. 2018. Available online: https://wad.jrc.ec.europa.eu/ (accessed on 30 March 2022).
38. Rosegrant, M.W.; Ringler, C.; Zhu, T. Water for Agriculture: Maintaining Food Security under Growing Scarcity. *Annu. Rev. Environ. Resour.* **2009**, *34*, 205–222. [CrossRef]
39. Turral, H.; Svendsen, M.; Faures, J.M. Turral—Investing in Irrigation_Reviewing the past and looking to the future. *Agric. Water Manag.* **2010**, *97*, 551–560. [CrossRef]
40. FAO. The Irrigation Challenge. 2003. Available online: https://www.fao.org/publications/card/en/c/c72fbe14-0a6f-52d1-92b5-f29eea8e2ab7/ (accessed on 30 March 2022).
41. Khan, S.; Tariq, R.; Yuanlai, C.; Blackwell, J. Can irrigation be sustainable? In *Agricultural Water Management*; Elsevier: Amsterdam, The Netherlands, 2006; Volume 80, pp. 87–99.
42. Gibson, K.E.; Yang, H.S.; Franz, T.; Eisenhauer, D.; Gates, J.B.; Nasta, P.; Farmaha, B.S.; Grassini, P. Assessing explanatory factors for variation in on-farm irrigation in US maize-soybean systems. *Agric. Water Manag.* **2018**, *197*, 34–40. [CrossRef]
43. Grassini, P.; van Bussel, L.G.; Van Wart, J.; Wolf, J.; Claessens, L.; Yang, H.; Boogaard, H.; de Groot, H.; van Ittersum, M.K.; Cassman, K.G. How good is good enough? Data requirements for reliable crop yield simulations and yield-gap analysis. *Field Crop. Res.* **2015**, *177*, 49–63. [CrossRef]
44. Agriculture Victoria. Irrigation Management. 2022. Available online: https://agriculture.vic.gov.au/farm-management/water/irrigation/irrigation-management (accessed on 18 March 2022).
45. Lauer, S.; Sanderson, M. Irrigated agriculture and human development: A county-level analysis 1980–2010. *Environ. Dev. Sustain.* **2020**, *22*, 4407–4423. [CrossRef]
46. Mpanga, I.K.; Idowu, O.J. A decade of irrigation water use trends in Southwest USA: The role of irrigation technology, best management practices, and outreach education programs. *Agric. Water Manag.* **2021**, *243*, 106438. [CrossRef]
47. Evans, R.G.; Sadler, E.J. Methods and technologies to improve efficiency of water use. *Water Resour. Res.* **2008**, *44*. [CrossRef]
48. Fraiture, C.; Wichelns, D.C. Satisfying future water demands for agriculture. *Agric. Water Manag.* **2010**, *97*, 502–511. [CrossRef]
49. Zaveri, E.; Lobell, D.B. The role of irrigation in changing wheat yields and heat sensitivity in India. *Nat. Commun.* **2019**, *10*, 4144. [CrossRef] [PubMed]
50. Bhattarai, M.; Barker, R.; Narayanamoorthy, A. Who benefits from irrigation development in India? Implication of irrigation multipliers for irrigation financing. *Irrig. Drain.* **2007**, *56*, 207–225. [CrossRef]
51. Rogers, E.M. *Diffusion of Innovations*, 3rd ed.; Free Press: New York, NY, USA, 1962.
52. Kee, K.F. Adoption and Diffusion. In *The International Encyclopedia of Organizational Communication*; John Wiley & Sons, Inc.: Hoboken, NJ, USA, 2017; Volume 1, pp. 41–54.
53. Driessen, P.H.; Hillebrand, B. Adoption and Diffusion of Green Innovations. In *Marketing for Sustainability: Towards Transactional Policy-Making*; Bartels, G.C., Nelissen, W.J.A., Eds.; IOS Press: Amsterdam, The Netherlands, 2002; pp. 343–355. Available online: https://ssrn.com/abstract=2363527 (accessed on 5 April 2022).
54. Sunding, D.; Zilberman, D. The Agricultural Innovation Process: Research and Technology Adoption in a Changing Agricultural Sector. In *Handbook of Agricultural Economics*; Elsevier: Amsterdam, The Netherlands, 1999.
55. Straub, E.T. Understanding technology adoption: Theory and future directions for informal learning. *Rev. Educ. Res.* **2009**, *79*, 625–649. [CrossRef]

56. Kapoor, K.; Dwivedi, Y.; Williams, M. Innovation adoption attributes: A review and synthesis of research findings. *Eur. J. Innov. Manag.* **2015**, *17*, 327–348. [CrossRef]
57. Mwangi, M.; Kariuki, S. Factors determining adoption of new agricultural technology by smallholder farmers in developing countries. *J. Econ. Sustain. Dev.* **2015**, *6*, 208–217.
58. Feder, G.; Umali, D.L. The Adoption of Agricultural Innovations A Review. *Technol. Forecast. Soc. Chang.* **1993**, *43*, 215–239. [CrossRef]
59. Pham, H.-G.; Chuah, S.-H.; Feeny, S. Factors affecting the adoption of sustainable agricultural practices: Findings from panel data for Vietnam. *Ecol. Econ.* **2021**, *184*, 107000. [CrossRef]
60. Li, J.; Feng, S.; Luo, T.; Guan, Z. What drives the adoption of sustainable production technology? Evidence from the large scale farming sector in East China. *J. Clean. Prod.* **2020**, *257*, 120611. [CrossRef]
61. Meijer, S.S.; Catacutan, D.; Ajayi, O.C.; Sileshi, G.W.; Nieuwenhuis, M. The role of knowledge, attitudes and perceptions in the uptake of agricultural and agroforestry innovations among smallholder farmers in sub-Saharan Africa. *Int. J. Agric. Sustain.* **2015**, *13*, 40–54. [CrossRef]
62. Herrero, M.; Thornton, P.K.; Mason-D'Croz, D.; Palmer, J.; Benton, T.G.; Bodirsky, B.L.; Bogard, J.R.; Hall, A.; Lee, B.; Nyborg, K.; et al. Innovation can accelerate the transition towards a sustainable food system. *Nat. Food* **2020**, *1*, 266–272. [CrossRef]
63. Woltering, L.; Fehlenberg, K.; Gerard, B.; Ubels, J.; Cooley, L. Scaling—From "reaching many" to sustainable systems change at scale: A critical shift in mindset. *Agric. Syst.* **2019**, *176*, 102652. [CrossRef]
64. Genius, M.; Koundouri, P.; Nauges, C.; Tzouvelekas, V. Information transmission in irrigation technology adoption and diffusion: Social learning, extension services, and spatial effects. *Am. J. Agric. Econ.* **2013**, *96*, 328–344. [CrossRef]
65. Chen, H.; Wang, J.; Huang, J. Policy support, social capital, and farmers' adaptation to drought in China. *Glob. Environ. Chang.* **2014**, *24*, 193–202. [CrossRef]
66. Wossen, T.; Berger, T.; Di Falco, S. Social capital, risk preference and adoption of improved farm land management practices in Ethiopia. *Agric. Econ.* **2015**, *46*, 81–97. [CrossRef]
67. Hunecke, C.; Engler, A.; Jara-Rojas, R.; Poortvliet, P.M. Understanding the role of social capital in adoption decisions: An application to irrigation technology. *Agric. Syst.* **2017**, *153*, 221–231. [CrossRef]
68. Wang, J.; Klein, K.K.; Bjornlund, H.; Zhang, L.; Zhang, W. Adoption of improved irrigation scheduling methods in Alberta: An empirical analysis. *Can. Water Resour. J. Rev. Can. Des. Ressour. Hydr.* **2015**, *40*, 47–61. [CrossRef]
69. Chavas, J.P.; Nauges, C. Uncertainty, Learning, and Technology Adoption in Agriculture. *Appl. Econ. Perspect. Policy* **2020**, *42*, 42–53. [CrossRef]
70. Luu, T.D. Factors Influencing Farmers' Adoption of Climate-Smart Agriculture in Rice Production in Vietnam's Mekong Delta. *Asian J. Agric. Dev.* **2020**, *17*, 109–124. [CrossRef]
71. Amadu, F.O.; McNamara, P.E.; Miller, D.C. Understanding the adoption of climate-smart agriculture: A farm-level typology with empirical evidence from southern Malawi. *World Dev.* **2020**, *126*, 104692. [CrossRef]
72. de Witt, M.; de Clercq, W.P.; Velazquez, F.J.B.; Altobelli, F.; Marta, A.D. An in-depth evaluation of personal barriers to technology adoption in irrigated agriculture in South Africa. *Outlook Agric.* **2021**, *50*, 259–268. [CrossRef]
73. Hamdy, A.; Ragab, R.; Scarascia-Mugnozza, E. Coping with water scarcity: Water saving and increasing water productivity. *Irrig. Drain.* **2003**, *52*, 3–20. [CrossRef]
74. Annandale, J.G.; Stirzaker, R.J.; Singels, A.; van der Laan, M.; Laker, M.C. Irrigation scheduling research: South African experiences and future prospects. *Water SA* **2011**, *37*, 751–764. [CrossRef]
75. Garb, Y.; Friedlander, L. From transfer to translation: Using systemic understandings of technology to understand drip irrigation uptake. *Agric. Syst.* **2014**, *128*, 13–24. [CrossRef]
76. Zongo, B.; Diarra, A.; Barbier, B.; Zorom, M.; Yacouba, H.; Dogot, T. Farmers' Practices and Willingness to Adopt Supplemental Irrigation in Burkina Faso. *Int. J. Food Agric. Econ.* **2015**, *3*, 101–117.
77. Afrakhteh, H.; Armand, M.; Bozayeh, F. Analysis of Factors Affecting Adoption and Application of Sprinkler Irrigation by Farmers in Famenin County, Iran. *Int. J. Agric. Manag. Dev.* **2015**, *5*, 89–99. [CrossRef]
78. Salazar, C.; Rand, J. Production risk and adoption of irrigation technology: Evidence from small-scale farmers in Chile. *Lat. Am. Econ. Rev.* **2016**, *25*, 2. [CrossRef]
79. Zhai, S.-Y.; Song, G.-X.; Qin, Y.-C.; Ye, X.-Y.; Leipnik, M. Climate change and Chinese farmers: Perceptions and determinants of adaptive strategies. *J. Integr. Agric.* **2018**, *17*, 949–963. [CrossRef]
80. Hall, B.H.; Khan, B. Adoption of New Technology. 2003. Available online: http://www.nber.org/papers/w9730 (accessed on 5 August 2021).
81. Freydank, K.; Siebert, S. Towards Mapping the Extent of Irrigation in the Last Century: Time Series of Irrigated Area per Country. 2008. Available online: http://photogallery.nrcs.usda.gov/Index.asp (accessed on 29 April 2021).
82. GADM. GADM Maps and Data. 2022. Available online: https://gadm.org/data.html (accessed on 19 April 2021).
83. Willmott, C.; Matsuura, K. Terrestrial Air Temperature and Precipitation: Monthly and Annual Climatologies. 2001. Available online: http://climate.geog.udel.edu/~{}climate/html_pages/README.ghcn_clim2.html (accessed on 9 May 2022).
84. Klein Goldewijk, K.; Beusen, A.; Van Drecht, G.; De Vos, M. The HYDE 3.1 spatially explicit database of human-induced global land-use change over the past 12,000 years. *Glob. Ecol. Biogeogr.* **2011**, *20*, 73–86. [CrossRef]

85. Verburg, P.H.; Ellis, E.C.; Letourneau, A. A global assessment of market accessibility and market influence for global environmental change studies. *Environ. Res. Lett.* **2011**, *6*, 034019. [CrossRef]
86. Siebert, S.; Döll, P.; Hoogeveen, J.; Faures, J.-M.; Frenken, K.; Feick, S. Development and validation of the global map of irrigation areas. *Hydrol. Earth Syst. Sci.* **2005**, *9*, 535–547. [CrossRef]
87. Abebaw, D.; Haile, M.G. The impact of cooperatives on agricultural technology adoption: Empirical evidence from Ethiopia. *Food Policy* **2013**, *38*, 82–91. [CrossRef]
88. Abdulai, A.N.; Abdul-Rahaman, A.; Issahaku, G. Adoption and diffusion of conservation agriculture technology in Zambia: The role of social and institutional networks. *Environ. Econ. Policy Stud.* **2021**, *23*, 761–780. [CrossRef]
89. Abegunde, V.O.; Sibanda, M.; Obi, A. Determinants of the adoption of climate-smart agricultural practices by small-scale farming households in King Cetshwayo district municipality, South Africa. *Sustainability* **2020**, *12*, 195. [CrossRef]
90. Mase, A.S.; Gramig, B.M.; Prokopy, L.S. Climate change beliefs, risk perceptions, and adaptation behavior among Midwestern U.S. crop farmers. *Clim. Risk Manag.* **2017**, *15*, 8–17. [CrossRef]
91. Hassan, R.; Nhemachena, C.; Hassan, R.; Nhemachena, C. Determinants of African farmers' strategies for adapting to climate change: Multinomial choice analysis. *Afr. J. Agric. Resour. Econ.* **2008**, *2*, 83–104. Available online: http://ageconsearch.umn.edu (accessed on 15 November 2022).
92. Wheeler, S.; Zuo, A.; Bjornlund, H. Farmers' climate change beliefs and adaptation strategies for a water scarce future in Australia. *Glob. Environ. Chang.* **2013**, *23*, 537–547. [CrossRef]
93. Ortiz, R.; Sayre, K.D.; Govaerts, B.; Gupta, R.; Subbarao, G.V.; Ban, T.; Hodson, D.; Dixon, J.M.; Ortiz-Monasterio, J.I.; Reynolds, M. Climate change: Can wheat beat the heat? *Agric. Ecosyst. Environ.* **2008**, *126*, 46–58. [CrossRef]
94. Saadi, S.; Todorovic, M.; Tanasijevic, L.; Pereira, L.S.; Pizzigalli, C.; Lionello, P. Climate change and Mediterranean agriculture: Impacts on winter wheat and tomato crop evapotranspiration, irrigation requirements and yield. *Agric. Water Manag.* **2015**, *147*, 103–115. [CrossRef]
95. Zilberman, D.; Zhao, J.; Heiman, A. Adoption versus adaptation, with Emphasis on climate change. *Annu. Rev. Resour. Econ.* **2012**, *4*, 27–53. [CrossRef]
96. Schmill, M.D.; Gordon, L.M.; Magliocca, N.R.; Ellis, E.C.; Oates, T. GLOBE: Analytics for assessing global representativeness. In Proceedings of the 5th International Conference on Computing for Geospatial Research and Application, Washington, DC, USA, 25 September 2014. [CrossRef]
97. GLOBE. Global Representativeness Analysis. 2012. Available online: http://globe.umbc.edu/documentation-overview/representativeness/ (accessed on 4 March 2021).
98. Olson, D.M.; Dinerstein, E.; Wikramanayake, E.D.; Burgess, N.D.; Powell, G.V.; Underwood, E.C.; D'amico, J.A.; Itoua, I.; Strand, H.E.; Morrison, J.C.; et al. Terrestrial Ecoregions of the World: A New Map of Life on Earth: A new global map of terrestrial ecoregions provides an innovative tool for conserving biodiversity. *Bioscience* **2001**, *51*, 933–938. [CrossRef]
99. IIASA Global Agro-Ecological Zones (GAEZ). Slope Suitability Classes. 2001. Available online: https://gaez.fao.org/ (accessed on 9 May 2022).
100. Kabacoff, R. Correspondence Analysis. 2017. Available online: https://www.statmethods.net/advstats/ca.html (accessed on 29 August 2022).
101. TIBCO Software Inc. What Is Correspondence Analysis? 2022. Available online: https://www.tibco.com/reference-center/what-is-correspondence-analysis (accessed on 29 August 2022).
102. Sourial, N.; Wolfson, C.; Zhu, B.; Quail, J.; Fletcher, J.; Karunananthan, S.; Bandeen-Roche, K.; Béland, F.; Bergman, H. Correspondence analysis is a useful tool to uncover the relationships among categorical variables. *J. Clin. Epidemiol.* **2010**, *63*, 638–646. [CrossRef]
103. Bock, T. How Correspondence Analysis Works (A Simple Explanation). Available online: https://www.displayr.com/how-correspondence-analysis-works/ (accessed on 29 August 2022).
104. Doey, L.; Kurta, J. Correspondence Analysis applied to psychological research. *Tutor. Quant. Methods Psychol.* **2011**, *7*, 5–14. [CrossRef]
105. Pelletier, C.; Pousette, A.; Ward, K.; Fox, G. Exploring the perspectives of community members as research partners in rural and remote areas. *Res. Involv. Engagem.* **2020**, *6*, 3. [CrossRef] [PubMed]
106. Mwangi, J.K.; Crewett, W. The impact of irrigation on small-scale African indigenous vegetable growers' market access in peri-urban Kenya. *Agric. Water Manag.* **2019**, *212*, 295–305. [CrossRef]
107. Dhawan, B. Impact of Irrigation on Farm Economy in High Rainfall Areas: The Kal Project. *Econ. Polit. Wkly.* **1988**, *23*, A173–A175 + A177–A180.
108. Adeoti, A.I. Factors Influencing Irrigation Technology Adoption and its Impact on Household Poverty in Ghana. *J. Agric. Rural. Dev. Trop. Subtrop.* **2009**, *109*, 51–63.
109. Fischer, J.; Abson, D.J.; Dorresteijn, I.; Hanspach, J.; Hartel, T.; Schultner, J.; Sherren, K. Using a leverage points perspective to compare social-ecological systems: A case study on rural landscapes. *Ecosyst. People* **2022**, *18*, 119–130. [CrossRef]
110. Wozniak, G.D. Human Capital, Information, and the Early Adoption of New Technology. *J. Hum. Resour.* **1987**, *22*, 101–112. Available online: https://about.jstor.org/terms (accessed on 20 August 2022). [CrossRef]

111. Diederen, P.; Van Meijl, H.; Wolters, A.; Bijak, K. Innovation Adoption in Agriculture: Innovators, Early Adopters and Laggards Innovation Adoption in Agriculture: Innovators, Early Adopters and Laggards. 2003. Available online: https://hal.archives-ouvertes.fr/hal-01201041 (accessed on 20 August 2022).
112. Asfaw, S.; Shiferaw, B.; Simtowe, F.; Hagos, M. Agricultural technology adoption, seed access constraints and commercialization in Ethiopia. *J. Dev. Agric. Econ.* **2011**, *3*, 436–447.
113. Teha, D.; Jianjun, L. Factors Affecting Adoption of Small Scale Irrigation Technology: Insights from Sire Woreda, Oromiya Region, Ethiopia. *Am. J. Appl. Sci. Res.* **2021**, *7*, 84–101. [CrossRef]
114. Makate, C.; Makate, M.; Mutenje, M.; Mango, N.; Siziba, S. Synergistic impacts of agricultural credit and extension on adoption of climate-smartagricultural technologies in southern Africa. *Environ. Dev.* **2019**, *32*, 100458. [CrossRef]
115. Ajayi, T.; Fatunbi, O.; Yemi, A. *Strategies for Scaling Agricultural Technologies in Africa*; FARA Africa: Accra, Ghana, 2018.
116. Kim, J.; Shah, P.; Gaskell, J.C.; Prasann, A.; Luthra, A. *Scaling up Disruptive Agricultural Technologies in Africa*; The World Bank: Washington, DC, USA, 2020.
117. Esteve, P.; Varela-Ortega, C.; Downing, T. A stakeholder-based assessment of barriers to climate change adaptation in a water-scarce basin in Spain. *Reg. Environ. Chang.* **2018**, *18*, 2505–2517. [CrossRef]
118. Budhathoki, N.K.; Paton, D.; ALassa, J.; Zander, K.K. Assessing farmers' preparedness to cope with the impacts of multiple climate change-related hazards in the Terai lowlands of Nepal. *Int. J. Disaster Risk Reduct.* **2020**, *49*, 101656. [CrossRef]
119. Ray, D.K.; West, P.C.; Clark, M.; Gerber, J.S.; Prishchepov, A.V.; Chatterjee, S. Climate change has likely already affected global food production. *PLoS ONE* **2019**, *14*, e0217148. [CrossRef]
120. Ranganathan, J.; Waite, R.; Searchinger, T.; Zionts, J. *Regenerative Agriculture: Good for Soil Health, but Limited Potential to Mitigate Climate Change*; WRI: Washington, DC, USA, 2020.
121. Falkenmark, M.; Molden, D. Wake up to realities of river basin closure. *Int. J. Water Resour. Dev.* **2008**, *24*, 201–215. [CrossRef]
122. Lankford, B.; Closas, A.; Dalton, J.; Gunn, E.L.; Hess, T.; Knox, J.W.; van der Kooij, S.; Lautze, J.; Molden, D.; Orr, S.; et al. A scale-based framework to understand the promises, pitfalls and paradoxes of irrigation efficiency to meet major water challenges. *Glob. Environ. Chang.* **2020**, *65*, 102182. [CrossRef]
123. Graveline, N.; Grémont, M. The role of perceptions, goals and characteristics of wine growers on irrigation adoption in the context of climate change. *Agric. Water Manag.* **2021**, *250*, 106837. [CrossRef]
124. Andersson, J.A.; D'Souza, S. From adoption claims to understanding farmers and contexts: A literature review of Conservation Agriculture (CA) adoption among smallholder farmers in southern Africa. *Agric. Ecosyst. Environ.* **2014**, *187*, 116–132. [CrossRef]

water

MDPI

Article

Hydroclimate Impact Analyses and Water Management in the Central Rift Valley Basin in Ethiopia

Lemma Adane Truneh *, Svatopluk Matula and Kamila Báťková

Department of Water Resources, Faculty of Agrobiology Food and Natural Resources, Czech University of Life Sciences Prague, Kamýcká 129, Suchdol, 165 00 Prague, Czech Republic
* Correspondence: truneh@af.czu.cz

Abstract: This study explores the impacts of climate change on the major components of the water balance such as surface runoff (Q), water yield (WY), and evapotranspiration (ET) in the Central Rift Valley Basin (CRVB) in Ethiopia. Projected climate data from the climate emission scenarios were used for the analyses. Representative concentration pathway (RCP) data from the MIROC-RCA4 ensemble driving climate models were downscaled, bias-corrected, and applied for impact analyses. Climate scenario analyses for the near-term (2031–2060) and long-term (2070–2099) periods were used to assess the conditions of the water balance components. The endo hydrogenic CRVB was divided into three sub-basins, and their respective hydroclimatic impacts were simulated separately with calibrated Arc-SWAT models. The future impacts simulated on the annual average basis vary in their maximum ranges from −65.2% to +85.8% in Q, from −42.2% to +23.9% in WY, and from −4.1% to +17.3% in ET compared to the baseline data outputs in the individual sub-basin. Water management options according to the water balance sensitivities to the climate impacts were proposed for each of the sub-basins. SWAT-based studies aimed at balanced water resources management in combination with agricultural practices within the CRVB are recommended for future research.

Keywords: Arc-SWAT; climate change; climate scenario; water balance sensitivity; water management

Citation: Truneh, L.A.; Matula, S.; Báťková, K. Hydroclimate Impact Analyses and Water Management in the Central Rift Valley Basin in Ethiopia. *Water* **2023**, *15*, 18. https://doi.org/10.3390/w15010018

Academic Editor: Adriana Bruggeman

Received: 12 October 2022
Revised: 14 December 2022
Accepted: 15 December 2022
Published: 21 December 2022

1. Introduction

Sub-Saharan Africa is a region that is very sensitive to, and is highly affected by recurrent droughts, flooding, and untimely weather conditions. Floods and droughts have affected water supplies and have set a challenge for water management. At the same time, water management practices in these developing regions are not adequate for dealing with the challenges of significant changes in climate [1–4]. Increasing pressure on land and water resources due to population growth and human activities have also resulted in the degradation of vulnerable ecosystems and in reduced biodiversity [4–6]. Moreover, this degradation of ecosystems hinders the potential use of ecosystem services [7].

In addition, climate change is a driver of many societal and environmental problems of the 21st century [8,9]. Together with the impacts of population growth, it puts pressure on the management of natural resources such as water resources [5,10]. It can also alter the hydrological cycle, resulting in large-scale impacts on water availability. These impacts could be temporal or become permanent. Climate change can also affect the temporal conditions of the water balances [11]. Water balances are components of the water cycle that exist at different scales and in different conditions in each locality. They are highly affected by the state of the environment and by the climate. Climate change highly affects the water balance conditions both spatially and temporally at the local or regional scale. For instance, Africa is vulnerable to inter-annual climate variations due to the El-Niño southern oscillations [12,13]. To evaluate the conditions of water resources in a basin or region, it is essential to know the water balance conditions under certain circumstances. The water balance components may vary due to different spatial and temporal aggregations, reference periods, and climate change impacts, as well as the interventions of humans for the purpose of water use [14].

Climate change refers to changes in conditions such as temperature and rainfall over long periods of time in a region. It has been caused by the increasing concentration of greenhouse gases (GHGs) in the atmosphere since the pre-industrial era. The Intergovernmental Panel on Climate Change (IPCC) concluded that more than 90% of the accelerated warming of the past five decades has been caused by the industrial release of GHGs such as CO_2 into the atmosphere [15].

In the CRVB, there are high levels of rainfall variability, water scarcity, and weather variability, and it is a place where water resources planning and management are greatly challenged by the impacts of climate change [16]. For example, an increase in temperature and variability in rainfall affected the seasonal and total water supply and led to the occurrence of extreme hydrological events [17]. It is therefore essential to know the trend of climate change over a long period of time to manage possible extreme hydrological events, either droughts or flooding, in the region [15,18–20].

A climate impact study can also provide a reliable basis for water resources planning [21]. Nowadays, long-term water resources planning studies need to take into consideration ongoing and future global climate changes in order to curb the uncertainties in the management of water resources [22]. In such studies, the effects of climate change must be quantified with high spatial and temporal resolutions at basin scale [1,23–25].

Various studies have been carried out on the water resources of the Central Rift Valley Basin (CRVB) in an attempt to describe and evaluate the impact of climate change on existing water resources [16,26–30]. However, only a few of these studies have been aimed at analyzing the impacts of climate change based on various regional concentration pathway (RCP) simulations in different climate scenarios to evaluate the conditions of the components of the water balance in the sub-basins. For example, in Ethiopia, Legesse et al. (2003) used the Precipitation Runoff Modeling System (PRMS) model to simulate runoff, and they predicted a 30% decrease in runoff in response to a 10% decrease in the amount of precipitation [26,31]. A 1.5 °C increase in temperature resulted in a 15% decrease in runoff [32]. Similarly, it was indicated that a higher temperature leads to an increase in evaporation rates, reductions in stream flow, and an increase in the frequency of droughts [28]. In addition, a vast number of studies have been conducted to analyze the impacts of climate change on crop productions [17,18,33–35]. However, very little consideration has been given to the potential impact of climate change on the current and future water balance components in the region and on their management methods. Therefore, a deep understanding of the effects of climate change on the components of the water balance for identifying site-specific climate-smart agricultural water management measures is necessary. In this context, the findings of this study can contribute the input information for the purpose of agricultural water management in the CRVB to adapt to the impacts of climate change.

An analysis of the impact of climate change on the components of the water balance involves hydrological models and projected plausible future climate change variables from global circulation models (GCMs) [23,36–38]. The GCMs determine the effects of changing concentrations of greenhouse gases on global climate variables such as temperature, rainfall, evapotranspiration, humidity, and wind speed [38]. Similarly, global circulation models that predict long-term climate trends (rainfall, temperature, and humidity) are often unsuitable for regional scale studies because of their coarse grid-size resolution. It is therefore essential to downscale GCM data to the region-specific climate impact through the use of statistical or dynamical downscaling techniques [38,39].

Various hydrological models can be applied to analyze the impacts of changes in the climate [10]. These models investigate the degree to which observed changes in climate may affect the resources due to natural variability, human activity, or a combination of both [40]. The results and projections produced by such models provide essential information for making decisions of local, regional, and national importance on matters such as water resources management, agriculture, transportation, and urban planning [41]. However, hydrological models need to be calibrated to site-specific conditions before they are used for climate change impact analyses [22].

The general procedure for assessing the impacts of climate change on water resources and on watershed processes can be determined by physically-based distributed models. Due to its wider applicability and utility, different versions of SWAT have been used for several studies throughout the world [38]. SWAT has been used for hydrological modeling, soil erosion and sediment transport modeling, climate impact studies on stream flows, and modeling land use change and management impacts on sediment and stream flows. It can also be used for nutrient transport modeling in agricultural fields [38]. These studies have confirmed the successful use of the SWAT model across different watersheds on different scales and across different environmental, climatological, and hydrologic conditions [36,42,43].

The study presented here is therefore aimed at analyzing the impacts of climate change according to the regional RCP scenarios on the water balance components of the CRV sub-basins in Ethiopia. The results of the SWAT models integrating CMhyd, WGEN, and SWAT-CUP software packages, were used to identify possible sub-basin-wide water management options.

2. Materials and Methods

2.1. Description of the Study Location

The Central Rift Valley Basin (CRVB) is in Ethiopia between 38°15′ E and 39°30′ E longitude and 7°10′ N and 8°30′ N latitude, (Figure 1). It covers an area of approximately 9112.5 km^2. It is a hydrologically closed lakes region with no known outlets for its total basin [27]. The study basin is a vast closed area and thus was divided into smaller sub-basins with known outlets (Ketar, Meki, and Shalla sub-basins).

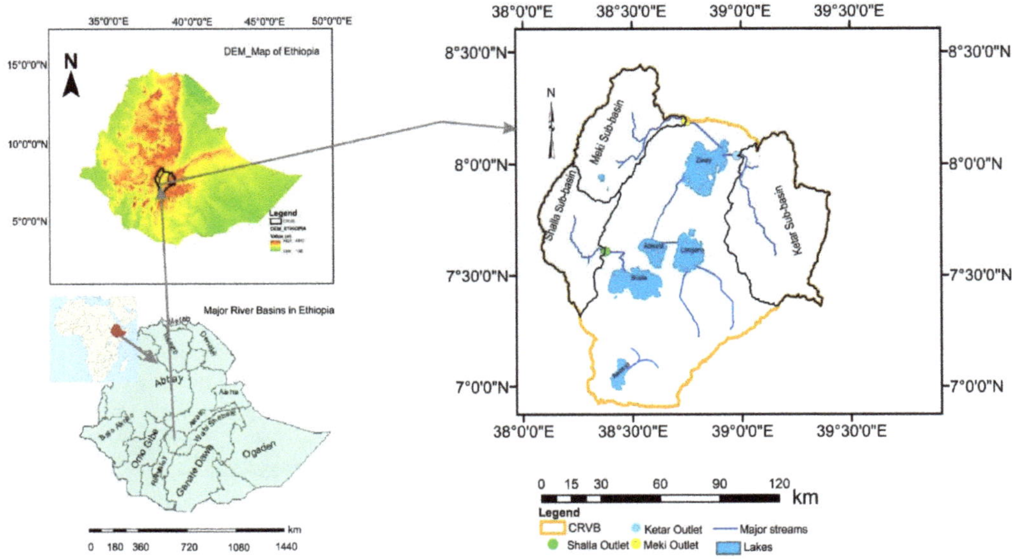

Figure 1. Location of Ethiopia in Africa, and the major river basins (bottom left); location of the study area within Ethiopia (top left), and the study sub-basins with their major stream outlets.

The mean annual rainfall of the study area varies between 600 mm near the lakes and 1200 mm–1600 mm in the highlands. The average minimum temperature is 10.5 °C, while the average maximum temperature is 24.3 °C [16]. CRVB comprises four major lakes:

Ziway, Shalla, Abiyata, and Langano. It also has perennial rivers, which include the Meki, the Ketar, and the Jidu rivers [16].

The CRVB has diverse soil types. It has varying infiltrability and associated runoff potential. Coarse-textured soils (LT Leptosols) with high infiltrability are dominant in the eastern and western highlands and in the valley floor around the lakes. Medium-textured soils (Euvertisols) with moderate infiltrability dominate the eastern and western mid-altitudes of the CRVB, whereas the lower reaches of the western highlands and some places in the central part of the eastern CRVB are dominated by fine-textured black soils (Vertisols) with lower infiltrability (Figure 2) [19].

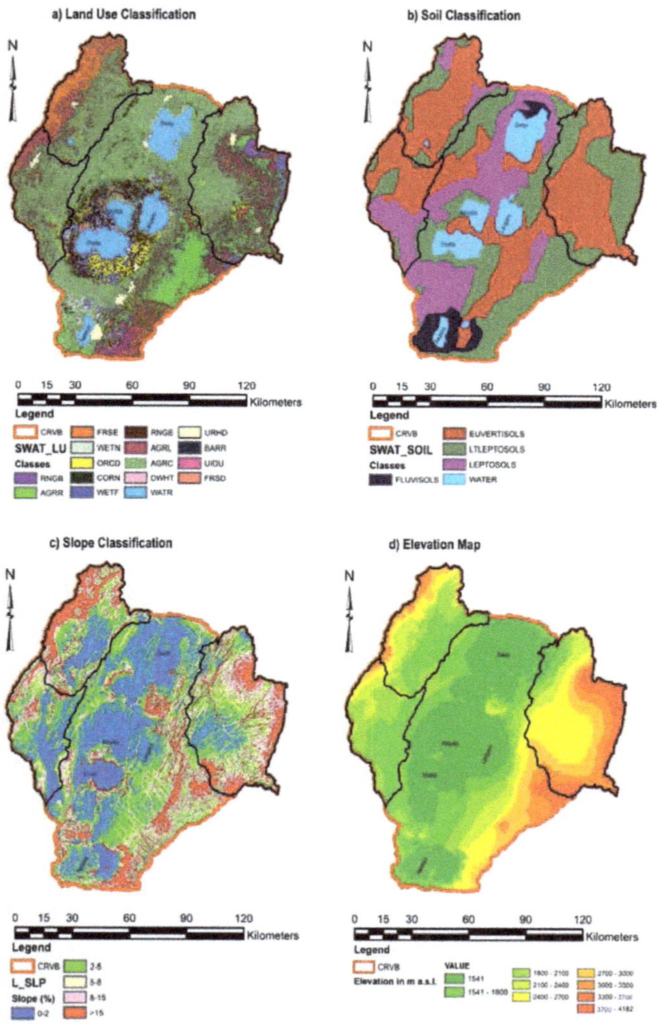

Figure 2. Distributions of land use, soil, slope, and elevation ranges in the CRVB (Note: the land use and soil codes are according to the SWAT classification standard as indicated in Tables S1 and S3 in the Supplementary File).

2.2. *Sub-Basin Selection Methods (Boundary Delineation)*

The hydrologically closed CRVB comprises many sub-basins. It was delineated and subdivided into major sub-basins in GIS according to their river systems, using the outlet points [16] as indicated in Figures 1 and 3. The DEM data were delineated in Arc SWAT and with the spatial analyst tool in ArcGIS. The total area of CRVB was delineated based on the watershed boundaries or water divide lines obtained from the Ministry of Water Resources of Ethiopia. The CRVB is an endo hydrogenic basin [27]. Since there is no single outlet for the CRVB, this study aims to investigate the hydroclimatic impacts via its major sub-basins with monitored outlets (Ketar, Meki, and Shalla). The selected sub-basins form parts of the CRVB with different characteristics which, when summed up, can generally characterize the climate impact conditions of the CRVB. The sub-basins were selected based on differences in agroecology, microclimate, and socio-environmental interactions. The analyses were performed for each of the sub-basins separately. The outlet locations of each sub-basin are indicated in Figures 1 and 3.

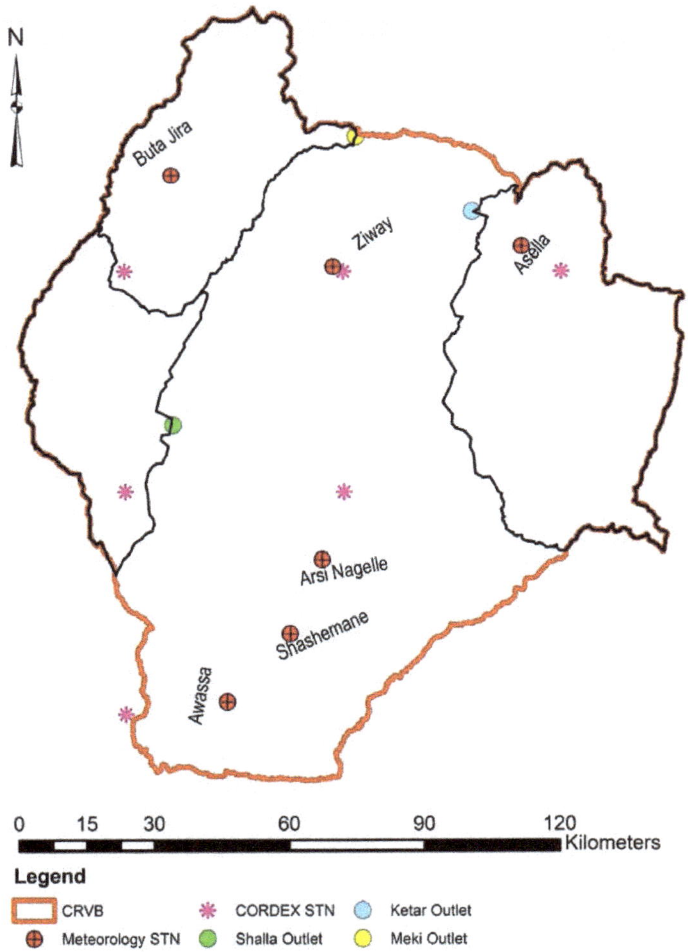

Figure 3. Locations of meteorology stations (Meteorology STN), CORDEX grid point (CORDEX STN), and outlets (Discharge monitoring stations).

2.3. Data Definition

2.3.1. Spatial Data

The spatial data used for the modeling were analyzed step-by-step. Initially, the digital elevation model (DEM) data of the CRVB was delineated with GIS into Ketar,

Meki, and Shalla. They were divided into sub-basins based on the topography and the river systems. Each sub-basin was consequently subdivided into hydraulic response units (HRUs) according to the land-use features, soil profile, and slope within SWAT. The major data inputs and their utilization are indicated in Table 1. The land uses characterize the hydrological process in the sub-basins. The land use map of the CRVB was obtained from the Ethiopian Geospatial and Information Institute (GSII).

The soil hydro-physical properties determine the existence and the quantity of each component of the water balance [44]. The soil physical properties and the area coverage of each of the soil types were classified based on the SWAT classification standards. The digitalized soil data for the study region with a resolution of 1ha was obtained from the Ministry of Agriculture and Natural Resources (MANR) of Ethiopia. The details are in the Supplementary File in Table S1. The spatial information maps of the study region including land use information, distribution of soil types, slope, and elevation information are indicated in Figure 2.

Table 1. Major input data used in the SWAT model.

	Data					
	Type	Format	Source	Year/scale	Resolution	Purpose
Weather data	Relative humidity	.xls	NMA	1984–2010	Daily	Analyze water balance (WB)
	Rainfall	.xls	NMA	1984–2010	Daily	Analyze rainfall trend and WB
	Sunshine hours	.xls	NMA	1984–2010	Daily	Analyze WB and solar radiation
	Temperature (Max and Min)	.xls	NMA	1984–2010	Daily	Analyze WB, and temp trend
	Wind	.xls	NMA	1984–2010	Daily	Analyze WB and wind trade
Spatial data	Land use	.shp	GSII	1996–2008	ha	Model land use and runoff
	Soil	.shp	MANR	NA	ha	Determine soil hydrology group
	DEM	.tiff	OBANR	2003–2008	30 m	Analyze location data sets
Hydrology data	River Discharge	.xls	MW	1900–2010	Monthly average	Analyze discharge trend, for model calibration and sensitivity

Note: NMA—National Meteorological Agency, GSII—Geospatial and Information Institute, MANR—Ministry of Agriculture and Natural Resources, OBANR—Oromia Bureau of Agriculture and Natural Resources, MW—Ministry of Water Resources.

2.3.2. Climate Data

Daily data on minimum and maximum temperature, hours of sunshine, relative humidity, wind speed, and precipitation from six meteorological stations, located in and near the sub-basins, were introduced into the model to simulate the water balances of the sub-basins (Table 1 and Figure 3). Hydrology data for stream flows were collected at the outlets indicated for each sub-basin. The CORDEX grid locations in the study area, based on which the climate data were downscaled and extracted, are also presented in Figure 3. The coordinate locations of the meteorology stations are indicated in Supplementary File in Table S2.

2.3.3. Baseline Data Processing with SWAT Weather Generator (SWAT-WGEN)

The weather data were statistically analyzed, and data qualities such as errors and outliers were assessed and adjusted by the weather database generator software (SWAT-WGEN). The data and their respective station coordinates (X, Y, and Z) were synchronized by the SWAT-WGEN. As a result, the SWAT model recognized the spatial distribution of the data supplied. SWAT-WGEN helps in statistical analyses, in data coding for SWAT use, and for data gap analyses as well as for spatial interpolation of the missed datasets. Special care was given to the input data within this study. The background data provided by the authorities were carefully checked and missing data were supplied if available. The data gaps in the collected baseline data were scattered, but on some days, they were sequential. These sequential data gaps ranged from one to only ten days maximum for some stations. The gaps were filled via interpolation by the software. These data gaps accounted for not more than 65 days out of the total 27 years per station, which is less than 0.66% of the data items. Simple arithmetic means (taking the averages of the values of the data series available before and after the missed data dates) were also applied to those stations where the gaps were scattered and not sequential to restore the missing values.

2.4. Arc SWAT Application

Arc SWAT 2012 was used for the hydroclimatic impact assessment of the CRVB. Arc SWAT 2012 is an Arc GIS extension program used for watershed modeling. The Soil and Water Assessment Tool (SWAT) is a widely used model for analyzing the water balances of a basin using long-term meteorological and spatial data of the area [45]. It is a physically-based, deterministic, continuous, watershed-scale simulation model developed by the U.S. Department of Agriculture—Agricultural Research Service (USDA) [45,46]. It is a model written in Fortran to analyze mainly water, nutrient, and sediment conditions in large basins and the behavior under climate changes [46]. It can also be applied to evaluate the impacts of various human, environmental, and infrastructural management interventions in basins. It involves systematic and interconnected spatial and weather data analyses to evaluate the intended goal at each hydraulic response unit (HRU).

In the application of the model, the Penman–Monteith method for evapotranspiration, the soil conservation service (SCS) curve number method for surface runoff determination, and the variable storage method to simulate channel water routing are employed to analyze the water balances.

The Water Balance Equations

In the analysis of the impacts of climate change on water balance components, the model operates based on the water balance equation indicated in Arnold et al. (2011) which is defined as:

$$SWt = SW_0 + \sum_{i}^{t}(Rday_i - Qsurf_i - Ea_i - Wseep_i - Qgw_i) \quad (1)$$

where SWt is soil water content (mm) at time t, SW_0 is initial soil water content (mm), t is simulation period (days), $Rday_i$ is amount of precipitation on the i-th day (mm), $Qsurf_i$ is amount of surface runoff on the i-th day (mm), Ea_i is amount of evapotranspiration on the i-th day (mm), $Wseep_i$ is amount of water entering the vadose zone from the soil profile on the i-th day (mm), and Qgw_i is amount of base flow on the i-th day (mm) [45].

Moreover, one of the critical parameters that are evaluated for sustainable water resource management of the study area is the water yield. The water yield is the aggregate sum of water leaving the HRU and entering the principal channel during a time step [45]. The water yield within a basin is evaluated by the model based on Equation (2). Considering the hydrological processes taking place continuously in the basin, the water yield, i.e., the net amount of water flowing past a given point on a stream during a given period, can be described by a basic model equation:

$$W_{yld} = Q_{sur} + Q_{lat} + Q_{gw} - T_{loss} \quad (2)$$

where W_{yld} is the water yield (mm), Q_{sur} is the surface runoff (mm), Q_{lat} is the contribution of the lateral flow to the stream (mm), Q_{gw} is the contribution of the groundwater to the streamflow (mm), and T_{loss} is the transmission losses (mm) from the tributary in the HRU by means of transmission through the bed.

2.5. Model Parameter Sensitivity Analysis

For a particular area of interest (CRVB), Arc-SWAT contains many hydrological parameters that need to be considered. However, not all the parameters may be contributing significantly to the model output, and it is therefore necessary to identify the input parameters that are significant [46]. In addition, the heterogeneity of the area makes it difficult for all SWAT parameters to be monitored simultaneously. Calibration and validation are required to identify the parameters to use for the specific area in a balanced way [47]. The parameter sensitivity scale developed by Lenhart et al. (2002) was used to classify the sensitivity of the parameters in the sub-basins [48]. It was scaled to the mean of index (I) values (Table 2).

Table 2. Parameter sensitivity scale classes assigned in SWAT as adapted from Lenhart et al. (2002) [48]).

Class	Mean of Index (I)	Category of Sensitivity
1	$0 \leq I \leq 0.05$	Small to negligible
2	$0.05 \leq I \leq 0.2$	Medium
3	$0.2 \leq I < 1$	High
4	$I \geq 1$	Very high

In addition, the most sensitive parameters used for stream flow analyses in the CRVB were selected on the basis of a tropical nature environment review recommendations [49]. The sensitivity ranking of the parameters (mean of index) is defined through an analysis of the values of the "t-stat" and "*p*-value" indexes in SWAT-CUP during calibration. The "t-stat" values are the t statistics. The t statistic is a measure of how extreme a statistical estimate is, and is calculated as:

$$t = \frac{M - \mu}{S_m} \tag{3}$$

Where t = t-stat, M = sample mean, μ = population mean and S_m = estimated standard error. The identified sensitive parameters are indicated in Table 3 with their descriptions.

Table 3. The most sensitive SWAT parameters identified in the CRV sub-basins, and their descriptions.

	Parameter	Description
1	CN2	SCS runoff curve number
2	ALPHA_BF	Base flow recession constant (days)
3	GW_DELAY	Ground water delay time for recharging the aquifer (days)
4	GWQMN	Water limit level in the aquifer for the occurrence of base flow (mm)
5	REVAPMN	Water limit level in the aquifer for revap to occur (mm)
6	GW_REVAP	Groundwater revap coefficient
7	ESCO	Soil evaporation compensation factor
8	EPCO	Plant uptake compensation factor
9	SURLAG	Delay time of direct surface runoff (days)
10	SOL_AWC	Available water capacity of the soil layer (mm mm^{-1})
11	SOL_K	Saturated hydraulic conductivity of the soil (mm h^{-1})
12	CH_K2	Effective hydraulic conductivity of the main channel (mm h^{-1})
13	SOL_Z	Depth from soil surface to the bottom of the layer (mm)
14	RCHRG_DP	Deep aquifer percolation fraction
15	HRU_SLP	Average slope steepness (m m^{-1})
16	BIOMIX	Bio-mixing efficiency

2.6. Model Calibration and Validation

Calibration and validation of the SWAT models were carried out using SWAT-CUP, a calibration uncertainty program for SWAT with the SUFI-2 algorithm, which is sequential uncertainty fitting, version 2. The program performed calibration, validation, sensitivity analysis (one at a time), and uncertainty analysis. In addition, the program links SUFI2, GLUE, ParaSol, MCMC, and PSO algorithms to SWAT [50]. The models were calibrated and validated using monitored stream flows from the outlets of the Ketar, Meki, and Jidu (Shalla) Rivers. The outlet locations were set at the flow gauging stations. The models were set to run for the baseline periods from 1984 to 2010 for each of the sub-basins (Ketar, Meki, and Shalla).

Calibration and validation help the model to resemble the study area in its operation by adjusting the sensitive model parameters. In this study, the observed stream flow data from 1990 to 2001, obtained from the Ministry of Water Resources of Ethiopia (MW), were used for calibration, and data from 2004 to 2010 were used for validation. The models of each of the sub-basins were calibrated and validated separately with their respective stream flow data from each sub-basin outlet (Figure 4). During calibration, the data from the first three years were kept as a warming-up period. These data allow the model to warm up, initialize, and approach reasonable initial values of the state variable of the model [50]. The adjusting values, as modified by SWAT-CUP to fit the values of the parameters to site-specific ranges, and the adjusting methods are presented in Table 4. The adjusting methods are indicated in the prefix of the parameter (V_, R_, and A_) and they are described in the table caption.

Table 4. Adjusting values and methods as adjusted by SWAT-CUP for the parameters.

Ketar		Meki		Shalla	
Parameter	Adjusting value	Parameter	Adjusting value	Parameter	Adjusting value
R__CN2.mgt	−0.44	R__CN2.mgt	−0.586	R__CN2.mgt	−0.155
V__ALPHA_BF.gw	0.629	V__ALPHA_BF.gw	0.348	R__ALPHA_BF.gw	−0.35
A__GW_DELAY.gw	12.251	A__GW_DELAY.gw	−17.291	A__GW_DELAY.gw	3.283
A__GWQMN.gw	336.23	A__GWQMN.gw	109.676	A__GWQMN.gw	−819.543
A__REVAPMN.gw	13.917	A__REVAPMN.gw	−126.446	A__REVAPMN.gw	213.915
A__GW_REVAP.gw	0.0403	A__GW_REVAP.gw	0.143	V__GW_REVAP.gw	0.18
V__ESCO.bsn	0.98	V__ESCO.bsn	0.43	V__ESCO.bsn	0.412
V__EPCO.bsn	0.221	R__EPCO.bsn	−0.662	V__EPCO.bsn	0.417
A__SURLAG.bsn	20.086	A__SURLAG.bsn	16.174	V__SURLAG.bsn	25.349
R__SOL_AWC(..).sol	1.29	R__SOL_AWC(..).sol	1.274	R__SOL_AWC(..).sol	NA*
R__SOL_K(..).sol	−0.661	R__SOL_K(..).sol	0.166	R__SOL_K(..).sol	0.149
V__CH_K2.rte	79.915	V__CH_K2.rte	NA*	A__CH_K2.rte	−74.91
R__SOL_Z(..).sol	0.665	R__SOL_Z(..).sol	NA*	R__SOL_Z(..).sol	NA*
R__RCHRG_DP.gw	−0.122	V__RCHRG_DP.gw	NA*	V__RCHRG_DP.gw	0.093
R__HRU_SLP.hru	NA*	R__HRU_SLP.hru	0.783	R__HRU_SLP.hru	NA*
R__BIOMIX.mgt	NA*	R__BIOMIX.mgt	0.205	R__BIOMIX.mgt	NA*

Note: R = relative, the parameter will be multiplied by the relative value as follows: value* (1 + R); V = replace, the parameter value will be replaced by the new values in the model; A = absolute, the parameter value will be added to the values in the model as follows: value + A; NA* = unchanged default values in the model.

2.7. Model Performance Evaluations

Before applying for analysis, the models' performances were assessed. Three main statistical parameters were used to evaluate the performance of the models: the coefficient of determination (R^2), the Nash–Sutcliffe efficiency (*NSE*), and the percentage of bias (*PBIAS*) [51]. R^2 is calculated as :

$$R^2 = \left[\frac{\sum_{i=1}^{N}(O_i - O)(S_i - S)}{\left[\sum_{i=0}^{N}(O_i - O)^2\right]^{0.5}\left[\sum_{i=0}^{N}(S_i - S)^2\right]^{0.5}} \right]^2 \quad (4)$$

R^2 ranges from 0.0 to 1.0. A higher value of R^2 indicates better performance of the model. The formula for calculating *NSE* is:

$$NSE = 1 - \frac{\sum_{i=1}^{N}(O_i - S_i)^2}{\sum_{i=1}^{N}(O_i - O)^2} \quad (5)$$

Nash–Sutcliffe Efficiency (*NSE*) is a normalized statistic, which measures the relative magnitude of the residual variance in comparison with the variance of the measured data. Like R^2, the higher the value of *NSE*, the better the performance of the model. *NSE* indicates the statistical relationship between simulated model values and observed values. It was stated that the "values of *NSE* vary from $-\infty$ to 1" [51,52].

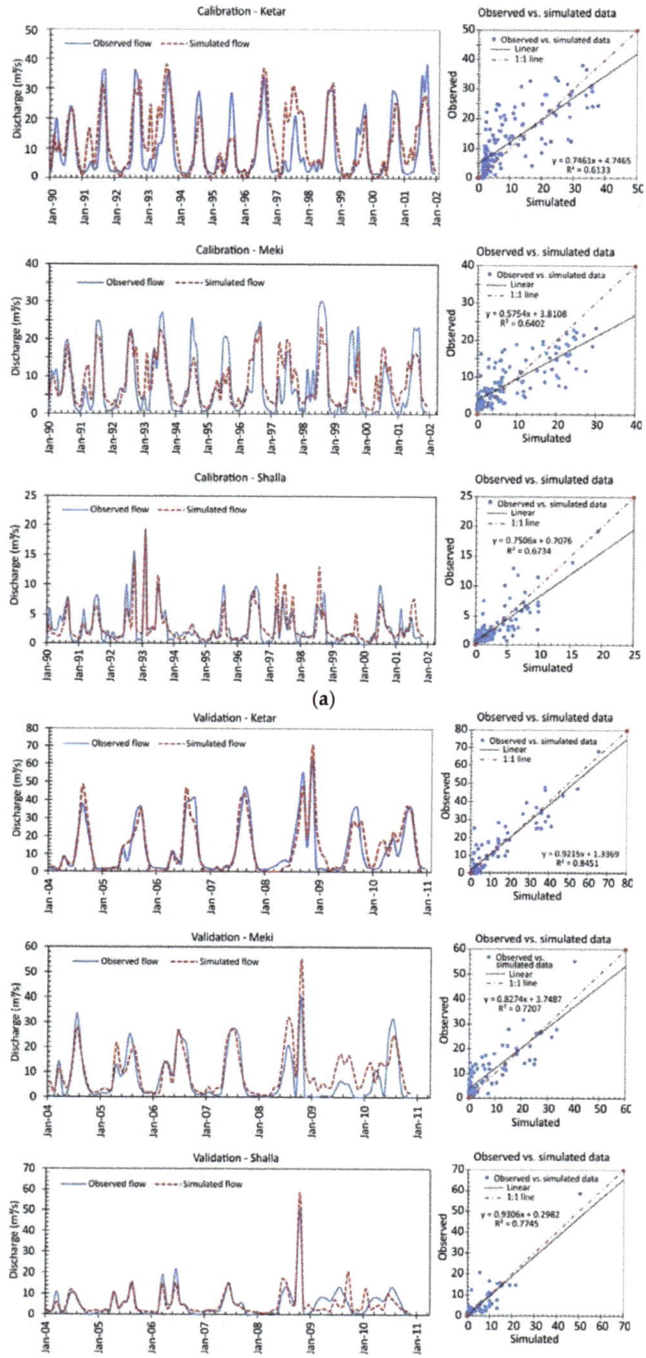

Figure 4. Calibration (**a**) and validation (**b**) results of the models for the CRV sub-basins.

PBIAS is calculated as:

$$\text{PBIAS} = \frac{\sum_{i=1}^{N}(S_i - O_i)}{\sum_{i=1}^{N} O_i} \times 100 \quad (6)$$

PBIAS measures the average tendency of the simulated values to be larger or smaller than their respective observed values. Positive *PBIAS* values indicate underestimation by the model, and negative values indicate overestimation. From the general statistics, the range within ±25% is acceptable [52].

In Equations (4)–(6), S is the mean of the simulated stream flows, O is the mean of the observed stream flows, S_i is the simulated stream flows, O_i is the observed stream flows, and N is the number of observations.

2.8. The Climate Scenario Application and Analyses Methods

2.8.1. Climate Scenario Analyses Setting and Simulation

An Arc-SWAT-based modeling approach to analyzing the impacts of climate change in the sub-basins of the CRV lakes region, and optimum agricultural water use and optimization strategies with respect to the identified impacts were carried out. Separate modeling for the selected sub-basins was performed. The climate scenarios (CSc) were set to analyze the impacts of climate change on the components of the water balance in the near-term (2031–2060) and in the long-term (2070–2099) periods for each of the regional concentration pathway (RCP) emission scenarios. The emission scenarios are RCP2.6 (low emission scenario), RCP4.5 (medium emission scenario), and RCP8.5 (high emission scenario). The simulations were categorized into seven CSc analyses, including the baseline data as listed in Table 5. The options for agricultural water use management are indicated based on the resulting water balance components affected by the changes in climate for each sub-basin.

The climate data were downscaled, bias corrected, analyzed, and simulated in an integrated manner with WGEN, CMhyd, and Arc SWAT. The WGEN software interlinks station coordinates and elevations with their respective data. All data statistics, such as average, standard deviation, mean, variance, etc., for each of the weather components downscaled were calculated and synchronized to their respective stations with WGEN. Rain Years, dew point, and other important variables useful for calculating the water balance components were also calculated and generated in WGEN. Finally, these climate data were imported into the SWAT models and simulated to see the changes in the components of the water balance that are especially useful for surface water sources.

2.8.2. Data Downscaling

Climate data stored in the World Climate Research Program (WCRP) databases were used. The data are from the experiments of CMIP5–RCP (RCP2.6-CMIP5, RCP4.5-CMIP5, and RCP8.5-CMIP5). These data were derived by the MIROC-RCA4 ensemble driving climate models under the GCM. The GCM data of these RCP data variables were regionalized to the regional climate model (RCM) with the Coordinated Regional Downscaling Experiment (CORDEX) for Africa, CORDEX-AFR-44. Both, historical data as well as the data of RCP2.6, RCP4.5, and RCP8.5 were downscaled by RCA4 models. RCA4 is the fourth version of the Rossby Center Regional Atmospheric model. It was originally developed by the Swedish Meteorological and Hydrological Institute within the CORDEX initiative. It is a dynamic downscaling method widely used with the CORDEX [23,53]. The downscaled datasets were daily precipitation, daily maximum near-surface air temperature, daily minimum near-surface air temperature, daily sunshine duration, near-surface relative humidity, and near-surface wind speed for future periods from 2006 to 2100. The duration of daily sunshine in units of seconds (s) was extracted from the model and adjusted to daily solar radiation with the units of kilowatt per square meter (KW/M^2) for SWAT use and to the SWAT input data standard units using Angstrom techniques [54].

2.8.3. Bias Correction

The data for precipitation and temperature were bias-corrected via linear scaling methods with CMhyd software, which is a SWAT community tool, before they were applied in the SWAT simulation. The need for bias correction is mainly due to onshore and offshore trade wind disturbances. The historical data from the model and the observed locational dataset from six stations in the study region were applied to the software. Data ranges from 1990 to 2006 were applied from the historical dataset of the climate model. Furthermore, observed datasets from the same periods were used to correct the biasedness created due to trade winds in the climate models. Parameters or correction factors for each month were developed in relation to the observed data range of the same time periods. Based on the parameters, the software adjusted the predicted rainfall and temperature values from the downscaled data. The corrected data values were applied to WGEN for statistical analyses and then to SWAT for simulation.

Table 5. Applied climate scenarios for analyzing the impacts of climate change on the major components of the water balance in the sub-basins.

Climate Scenario		
No.	Code	Description (Years)
1	NT-RCP2.6	RCP2.6 (2031–2060)
2	LT- RCP2.6	RCP2.6 (2070–2099)
3	NT-RCP4.5	RCP4.5 (2031–2060)
4	LT- RCP4.5	RCP4.5 (2070–2099)
5	NT-RCP8.5	RCP8.5 (2031–2060)
6	LT-RCP8.5	RCP8.5 (2070–2099)
7	BD	Observed baseline data (1984–2010)

Note: NT = Near term and LT-Long term.

3. Results and Discussion

3.1. Results of the Model Parameters Sensitivity Analyses

The parameter sensitivity analyses were carried out together with the calibration process, as it is necessary to include the flows estimated by SWAT and the monitored flows in the sub-basins. In general, a higher "t-stat" and a lower p-value indicate that the parameter is sensitive [55]. Based on the sensitivity scale developed by Lenhart et al. (2002), shown in Table 2, the following parameters were identified as highly sensitive in the Ketar sub-basin: EPCO, RCHRG_DP, SOL_K, GW_DELAY, CN2, REVAPMIN, and SURLAG. Similarly, ESCO, REVAPMIN, GWQMN, HRU_SLP, and GW-DEALY were very highly sensitive parameters in the Meki sub-basin, and ESCO, CH_K2, SOL_K, and GWQMN were very highly sensitive in the Shalla sub-basin. The description of the parameters is presented in Table 3. The differences in the sensitivity of the hydrological parameters in the sub-basins indicate that the sub-basins are heterogeneous, although they refer to a single, closed, lakes region. The differences are mainly due to land use, soil, hydrogeologic, and anthropogenic variations. The t-stat values of each of the selected parameters for each sub-basin are indicated in Table 6. The parameter description and their adjusting values are indicated in Tables 3 and 4.

3.2. Results of the Calibration and Validation of the Model

The calibration results indicate good agreement between the simulated and observed discharges in the sub-basins. The results for simulated and observed discharges in the sub-basins were evaluated against R^2, NSE, and $PBIAS$ during calibration and validation. The values in the Ketar sub-basin are in good agreement with $R^2 > 0.6$, $NSE > 0.5$, and $PBIAS \leq$ "$\pm$"25, (Figure 4a,b). Similarly, the results showed that the simulated and observed monthly discharges were in a good agreement during calibration and validation for the Meki and Shalla sub-basins (Table 7).

Table 6. Sensitivity or mean of index I values of the selected parameters for the sub-basins, according to their "t-stat" results as per the scale indicated in Table 2.

	Ketar		Meki		Shalla	
Parameter **	t-stat value	Sensitivity	t-stat value	Sensitivity	t-stat value	Sensitivity
R__CN2.mgt	1.408	Very high	−0.394	Negligible	−0.111	Negligible
V__ALPHA_BF.gw	0.046	Low	−0.997	Negligible	−1.643	Negligible
A__GW_DELAY.gw	1.206	Very high	1.951	Very high	−1.032	Negligible
A__GWQMN.gw	0.783	High	1.564	Very high	2.685	Very high
A__REVAPMN.gw	1.970	Very high	1.441	Very high	−1.116	Negligible
A__GW_REVAP.gw	0.710	High	0.844	High	NI*	NI*
V__ESCO.bsn	0.905	High	1.181	Very high	1.739	Very high
V__EPCO.bsn	1.013	Very High	−1.210	Negligible	−1.513	Negligible
A__SURLAG.bsn	2.329	Very high	−1.242	Negligible	0.744	High
R__SOL_AWC(..).sol	−1.034	Negligible	−3.957	Negligible	NI*	NI*
R__SOL_K(..).sol	1.202	Very high	−1.417	Negligible	1.197	Very high
V__CH_K2.rte	−0.551	Negligible	NI*	NI*	1.926	Very high
R__SOL_Z(..).sol	NI*	NI*	NI*	NI*	NI*	NI*
V__RCHRG_DP.gw	1.137	Very high	NI*	NI*	−1.986	Negligible
R__HRU_SLP.hru	NI*	NI*	1.799	Very high	0.084	Low
R__BIOMIX.mgt	NI*	NI*	1.669	Very high	0.798	High

Note: NI* = not identified, ** Parameter description is presented in Table 3.

Table 7. Model performance statistics for the Ketar, Meki, and Shalla sub-basins.

Sub-Basin	Calibration Statistics			Validation Statistics		
	R^2	*NSE*	*PBIAS*	R^2	*NSE*	*PBIAS*
Ketar	0.61	0.54	−22.5	0.85	0.84	−2.6
Meki	0.64	0.63	−4.81	0.72	0.64	−32.17
Shalla	0.67	0.66	0.2	0.77	0.74	1.34

Overall model performance statistics (R^2, *NSE*, and *PBIAS*) for the Ketar, Meki, and Shalla sub-basins are presented in Table 7.

3.3. *Climate Scenario Analyses Results and Discussion*

The results of the impacts of climate change on the major components of the water balance such as surface runoff (Q), water yield (WY), and evapotranspiration (ET) were evaluated in terms of their annual, seasonal, and monthly variations. The Q, WY, and ET were identified as the most sensitive elements of the water balance components in the CRVB. The simulated impacts of the climate scenarios on the water balance components are substantial. The percentage change in the Q, WY, and ET from their baseline simulated outputs for each sub-basin are presented in Table 8, together with the indication of the baseline annual rainfall data (averaged for years 1984–2010).

Table 8. The simulated mean annual changes, as a percentage, from the annual average values of the baseline outputs for the major components of the water balance in the sub-basins.

Sub-Basins	Ketar			Meki			Shalla		
Annual average rainfall (mm)		798.1			674.4			713.4	
Water balance components	Q	WY	ET	Q	WY	ET	Q	WY	ET
Baseline annual average output (mm)	103.8	492.2	282.5	53.5	257.5	393.1	44.2	326.7	363.8
			% of Δ						

Table 8. *Cont.*

Sub-Basins		**Ketar**			**Meki**			**Shalla**		
Scenarios	NT-RCP2.6	−62.2	−34.9	17.3	58.1	17.0	4.5	−3.5	0.9	12.2
	LT-RCP2.6	−55.0	−30.3	13.3	60.2	19.9	2.6	31.6	12.0	9.3
	NT-RCP4.5	−13.7	−35.9	−4.1	6.0	−1.1	5.6	−21.9	−10.1	9.2
	LT-RCP4.5	22.9	−28.7	−9.4	47.7	11.2	2.6	32.8	4.2	7.8
	NT-RCP8.5	−65.2	−42.2	7.4	58.3	13.0	6.4	−7.7	−2.4	10.8
	LT-RCP8.5	−60.5	−39.7	8.8	85.8	23.9	9.4	23.5	7.1	15.1

Note: % of Δ = Percentage of change of the component from its baseline output.

3.3.1. Ketar Sub-Basin

The resulting simulated ET, WY, and Q mean monthly values for the Ketar sub-basin are graphically displayed in Figure 5a. Changes in the Q pattern over the seasons in the Ketar sub-basin can be observed in Figure 5a. The highest Q season has shifted both in the near and long term of RCP4.5 to the months from March to May while it used to be between mid-June to the end of September in the baseline data outputs. The simulated annual variations from the base data are between −65.2% (LT-RCP8.5) and 22.9% (LT-RCP4.5). RCP 2.6 and RCP 8.5 analyses indicate that the expected runoff will decrease both in the near term and in the long term in relation to the baseline data simulation outputs. In all the seasons, for all RCPs, the runoff condition in the long term (LT) is higher than the runoff in the near-term (NT) period. However, the general trend indicates that the runoff is decreasing in this sub-basin in relation to the historical (baseline) period, but the rate of its reduction differs from one RCP to another and from one period to another.

In similar analyses, the WY in the Ketar sub-basin decreases for all RCPs, in both the NT and LT periods, except in the long-term periods of RCP4.5 for the months from April to June (Figure 5a). Generally, the impact is expected to reduce the WY in all projected scenarios, especially for the periods from July to October. However, the rate of reduction varies from RCP to RCP and varies from season to season. Nevertheless, the annual WY generation capacity of the Ketar sub-basin is higher than in the Meki and Shalla sub-basins, corresponding to the annual precipitation that is supplied. Almost half of the rainfall, 50% on an average, goes to the WY in all the scenarios, while the proportion is about 40% in the Meki sub-basin and about 44% in the Shalla sub-basin. The simulated WY in the RCPs follows a similar pattern to the observed base year simulations. It means that the seasonal change in WY is not disturbed in pattern but in quantity.

The ET in the sub-basin has bi-annual peaks between March and mid-May, and between July and September (Figure 5a). The ET is relatively low between mid-May and June. The rate of ET decreases between March and May in all the scenarios in relation to the observed data simulations except between June and September. ET will be higher in the Ketar sub-basin for RCP2.6 and RCP8.5, between June and September, than outputs from the base data. The significant change in ET mainly reflects the increase in temperature. Therefore, according to the RCP2.6 and RCP8.5 climate projections, the increase in ET will be higher than the RCP4.5 projections for ET. This is in line with the works of Musie et al. (2020) and Gadissa et al. (2019) in the Lake Ziway and CRV basins in Ethiopia, respectively [21,33]. Musie et al. (2020) used the SWAT model to evaluate the impacts of regional climate variabilities and land use change on the water resources in the Lake Ziway basin. They found an increase in surface runoff and water yield due to the climate scenarios from the year 2000 to 2017. Gadisa et al. (2019) used projected climate scenarios to evaluate stream flows for the medium-term (2040 to 2070) periods for the RCP4.5 and RCP8.5 scenarios.

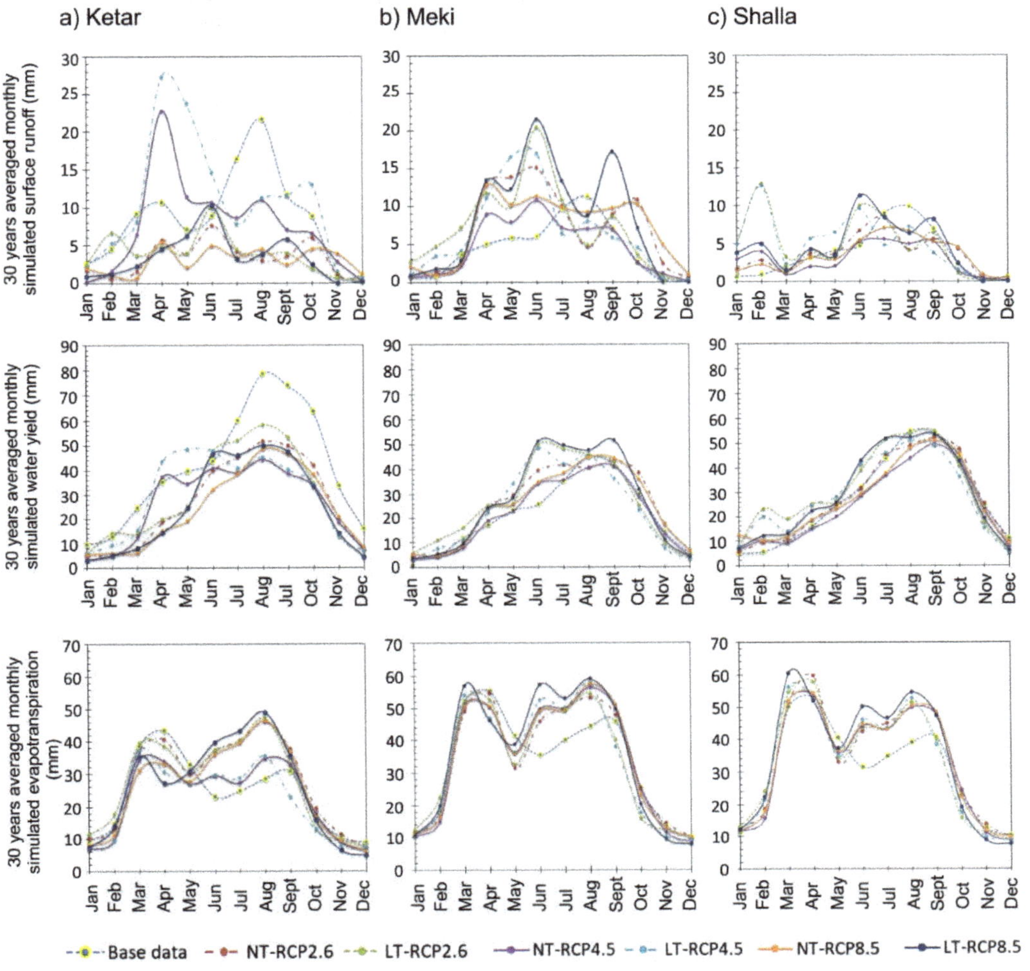

Figure 5. The simulated monthly distributions of Q, WY, and ET in the Ketar, Meki, and Shalla sub-basins for the applied climate scenarios. (**a**) Ketar, (**b**) Meki, (**c**) Shalla.

The results reported in both studies, and in Getnet et al. (2014), in the CRVB indicated that the hydrologic variations in water balance due to climate variability were highly significant [20,27,32]. However, in contrast to the study by Musie et al. (2020) [20], the hydroclimate in our study was more predominant in WY than ET in the Ketar sub-basin. Another study conducted in the CRVB in 2007 on climate change impacts on water availability with a SWAT model indicating an increase of averaged annual rainfall from 2001 to 2099 can also be found [56]. However, Gadissa et al. (2019) projected a reduction in precipitation by 7.97% and 2.55% under RCP4.5 and RCP8.5 respectively for the future period from 2040 to 2070 [32]. Reduction in precipitation has strong correlation with reduction in water yield and surface runoff. Our study is thus in line with the findings of Gadissa et al. (2019) [32] with minimal differences in the periods of occurrences. There are seasonal shifts in the pattern of occurrences of the components of the water balance when compared with the baseline data sets. These shifts are mainly from the changes in precipitation, temperature, and humidity patterns caused by greenhouse gases and other emissions.

3.3.2. Meki Sub-Basin

The Meki sub-basin is characterized by greater annual amounts of ET than in the Shalla and Ketar sub-basins. The annual surface runoff rises in all the RCP scenarios. There will be a seasonal shift of the peak runoff period from the usual July-to-September period to April-to-June in the sub-basin (Figure 5b). In the long-term periods of RCP2.6 and RCP8.5, the runoff will increase greatly in relation to the baseline data simulation outputs. However, RCP4.5 will create a moderate range of changes in relation to RCP2.6 and RCP8.5. The change in annual average runoff varies from 6% to 85% in reference to the baseline outputs. The projected monthly distribution shows that this water balance component varies significantly over the months in both the NT and LT period.

The change in averaged annual WY ranges from −1.1% to +23.9% in relation to the baseline data simulated. The scenario analysis also showed a remarkable increment in the WY amount between May and October for all RCP outputs. ET is the major water balance component of the sub-basin (Figure 5b). About 56% of the rainfall on average turns into ET. This indicates that the sub-basin water balance is highly sensitive to changes in temperature. Even though WY is good in the rainy seasons, most of it will be lost via ETs. Thus, for the Meki sub-basin, the impacts were more predominant in ET than in WY. This indicates the high seasonal weather variabilities in the sub-basin and its low hydroclimatic impact resilience. Similar findings were reported by Gadissa et al. (2019) and Musie et al. (2021) for this sub-basin. They used modeling approaches of RCM projections to assess the conditions of the Q, ET, and stream flows using the SWAT and WEAP models, respectively. In addition, Molla, (2014) has used physical assessment methods to indicate the sub-basin climate conditions [16,17,36]. These studies reported that the Meki sub-basin is the most hydroclimate-sensitive region. The strong weather variabilities in the sub-basin have resulted in wide ranges of changes in water resources similar to the findings of another study conducted by Getnet et al. (2014) in the CRVB [16,26,27]. The annual variations in this study are also relatively large for the sub-basin (Table 8). The modeling results in this study for the sub-basin are thus inconsistent with the above study findings.

3.3.3. Shalla Sub-Basin

The response of this sub-basin to the analysis in the model indicates a stronger range of variations in its water balance components. However, the Shalla sub-basin has a lower annual runoff amount than the Ketar and Meki sub-basins (Figure 5c). However, the changes in annual runoff vary between −21.9% and +32.8% from the baseline data simulation outputs. The average annual changes in WY vary from −10.1% to +12.0% because of the impacts. The changes in ET vary from +7.8% to +15.1%. The detail annual variations in percentage for each CSc and each component in each sub-basin are indicated in Table 8. ET increases significantly between June and September for all RCP projections. ET is the largest component, and most of the rainfall turns into ET. Because of the high ET and the small runoff, the entire sub-basin is characterized as a water-scarce region. The WY result for the Shalla sub-basin was moderate for all the CSc. Compared to other previous studies (for example Ayenew, 2007; Gadissa, et. al., 2018), Shalla has small WY output, but in the analyses conducted in this study, the sub-basin yielded a relatively higher amount [5,32]. The difference could possibly be due to its complex hydrogeologic setting that needs to be verified in further studies. However, there is agreement on the fact that its surface water availability will be depleted due to the high ET and the low Q occurrences.

The projected monthly average values of each of the water balance components in each sub-basin with their respective baseline monthly average output values for each of the scenarios are presented in Figure 6. It indicates that the hydroclimatic impacts in the future in the CRVB are very high. The baseline data outputs are indicated with yellow rings around their graphs.

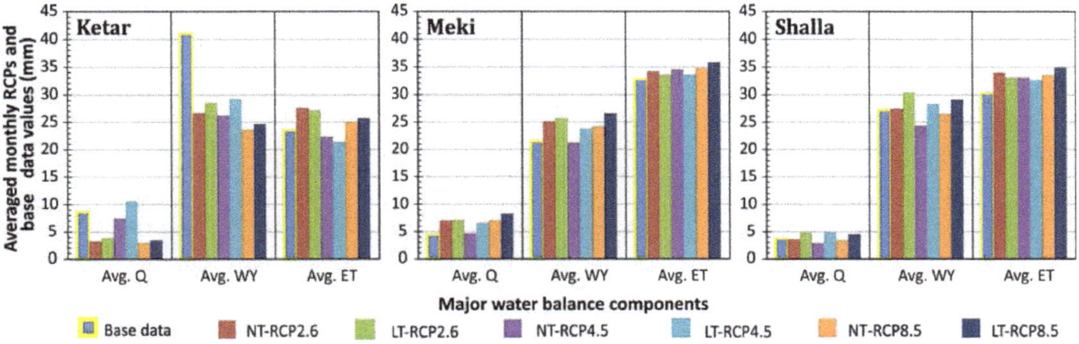

Figure 6. Monthly average values of 30 years of surface runoff (Q), water yield (WY), and evapotranspiration (ET) in the Ketar, Meki, and Shalla sub-basins for different climate scenario simulations in relation to the baseline data simulation outputs.

4. Discussion for Water Management Options

From the projected analyses of the impacts of climate change in the model, the major water balance components such as surface runoff and water yield are mainly expected to decrease, and evapotranspiration is projected to increase in the sub-basins. This will have an impact on the increasing demands for agricultural water in the sub-basins. Seasonal shifts in the patterns of the projected water balance distributions were also observed. Therefore, water management strategies that help mitigate the impacts should be identified and applied. Their application might help to face the food security challenge caused by the water shortage that would occur due to climate changes.

Based on the resulting projected water balances, agricultural water management in the Ketar sub-basin should, in the future, focus on the time modification of farm operations, and on water harvesting to store excess water occurring in the unusual months. Scarcity of water for agriculture is inevitable from the analyses (Figure 5a). Therefore, water saving, and water use optimization must be sought and applied in the future. The WY is the major water balance component of the Ketar sub-basin in all the scenarios, and its enhancement together with conservation, will make the basin rich enough in water to curb the impacts of climate change. In addition, irrigation water supply scheduling based on the modified climate pattern is the recommended method of agricultural water management for the Ketar sub-basin.

High water losses through ET in the Meki sub-basin can be mitigated by water management interventions such as crop mulching, farm operations during minimum evaporation seasons, favoring minimum tillage to reduce soil evaporation, selecting crops that are more resistant to high levels of evaporation, favoring efficient irrigation water application, and introducing regular soil and water conservation practices to reduce the high seasonal runoff and ET. In the Meki sub-basin, water harvesting and storage during periods of high runoffs can also reduce water scarcity during peaks in demand. High runoff management and protection infrastructures are also inevitable as there will be untimely and repeated higher runoff expected beyond the usual baseline trends, as per the analysis.

The high ET rates and low runoff makes the Shalla sub-basin a water-scarce region. The water scarcity problem in the sub-basin should be mitigated by improving WY via yield enhancement approaches that also help to reduce evaporation losses. These include soil and water conservation to improve subsurface storage, crop selection, farm operation scheduling based on the new climate pattern and minimum tillage to reduce soil evaporation, and the selection of highly ET-resistant crop varieties. Investigating afforestation for controlling ET losses, and controlled farm operations are also very crucial. Furthermore, inter-basin water transfers are recommended for adapting to the impacts on the sub-basin.

A study conducted by Kassie et al. (2015) applied an effective fertilizer with irrigation water as an adaptation measure to climate change for the maize crop in the CRVB. The study assessed the potential impacts of climate change on maize yield and explored specific adaptation options under climate change scenarios for the CRVB of Ethiopia by mid-century. They used GCM, RCPs, and crop models to search for adaptation options. The climate change impacts in their study are consistent with our study results. Their adaptation option offsets the severe impacts of yield loss in the area due to the climate impacts [34]. Thus, the effective application of fertilizer while producing the maize crop in the region together with irrigation water is crucial. In addition, the positive effects of changing the planting date were indicated in their study in offsetting the severe climate impacts on the maize crop [34].

Amare and Endalew (2016) assessed the importance of farm mechanization in rural Ethiopia for smallholder farmers. In their assessment, they indicated that mechanized farming helps in reducing water loss at the farms [57]. The study results showed that water distribution efficiencies in irrigated farms have been improved in the study regions, including the CRVB. This may be achieved by incorporating land use planning in a manner that its water allocations and use efficiencies will improve, for instance, farm mechanization and land leveling to minimize water loss and enhance even distribution [57]. Therefore, extensive farm mechanization and land leveling works are recommended as a means to improve water use and reduce its loss in the sub-basins' irrigated farm fields. These will help to increase the resilience capacity of the CRVB to the impacts of future climate changes.

Adaptation to climate impacts via water allocation planning based on weather, soil, and ecological characteristics and social benefit priorities can also reduce the unnecessary loss that may occur due to misallocation and weather variabilities. For instance, the cropping pattern alternatives that favor better gain based on the rainfall patterns of the rift valley region were adopted by some farmers, as indicated in the study conducted by Belay et al. (2017) [2]. The farmers applied a method of using different crop varieties of maize during long rainy seasons and during short rainy seasons. This has improved the gain in the worst water shortage seasons in the region, as reported in [2]. Accordingly, preparing alternative plans for seasonal climate change conditions for agricultural production, and for water use plans that can mitigate the dual impacts of climate and environmental changes while maximizing the benefits during the worst climate seasons are thus necessary. Hence, the possible alternative plans and the locally adopted measures by the farmers should be further assessed, tested, and applied in the worst seasons in the CRVB and in similar regions in the country. The plans need to be based on reliable data and on studies carried out for particular areas. This study aims to contribute to such a knowledge helping in the creation process of such adaptation plans for the CRVB in Ethiopia.

In addition, Kifle and Gebretsadikan (2016), conducted an experiment on the controlled application of irrigation water for potato production in the water-scarce region of Tigrai in Ethiopia [58]. They found positive effects of controlled irrigation water applications on potato production without losses for the deficit application of water with proper timing as means to curb water shortage due to climate changes. One of the best adaptation options for agricultural water uses in the sub-basins is thus the introduction of controlled irrigation that applies the water resources efficiently and that applies only the required amount of water at the proper time for effective use of the crops [58]. Controlled irrigation also helps avoid seepage and salinity problems via water applications to the required depth [58]. In addition, selecting fast-growing, highly productive quality seeds will help to save the resource for other economic and social uses. Controlled irrigation is thus recommended as a mitigating strategy for water scarcity and for environmental challenges that would occur due to the impacts of climate change and population growth.

For the CRVB, Musie et al., (2020) used SWAT models to assess the water conditions of terminal lakes in the CRVB and water management adaptation options. They recommended avoiding pollution of water sources and conserving the terminal lakes from pollution damages, both from sedimentation and other environmental pollutants. Thus, controlling the water level of the lakes, avoiding water quality degradations due to industrial and environmental wastes, and improving the storage capacity in the sub-basins will favor better use of the resources during peaks in demand [20].

Climate-based integrated development and use plan for the utilization of water resources according to its economic and social benefits, while safeguarding environmental sustainability, should be further assessed, modeled, and applied for its equitable use in equilibrium in the closed CRVB. Moreover, considering the response of the sub-basins to hydroclimatic impact while planning water use is crucial.

5. Conclusions

This paper investigated the impacts of future climate change on the major components of the water balance in the central rift valley basin in Ethiopia from the seasonal and spatial points of view. The evaluations are based on the magnitude of water yield, evapotranspiration, and surface runoff components change in relation to the baseline data outputs. Regional climate models (RCM) data in CORDEX—Africa were applied for the investigation. RCP data from the MIROC-RCA4 ensemble driving climate models were downscaled, bias-corrected, and used for the analyses. The methodology followed a calibrated Arc-SWAT modeling approach to search for basin-wide climate impacts on water resources and to indicate possible agricultural water management and adaptation strategies. The findings are solely based on model simulation outputs within the scope of its evaluations and error limitations.

Accordingly, the study identified a general decrease in water yield and surface runoff and a seasonal increase in evapotranspiration in the Ketar and Shalla sub-basins in both the near-term (2031–2060) and long-term (2070–2099) periods in comparison to the baseline period (1984–2010). However, all three water balance components projected were showing an increment in the Meki sub-basin for all the periods. The sub-basins were also found to be heterogeneous, and they showed variabilities in terms of their hydroclimatic reactions to the impacts of climate change even though they are in one endo hydrogenic region. In the sub-basins, some similarities were also found in the ways in which the pattern of the water balance components will be changed. However, the magnitudes of the impacts varied from sub-basin to sub-basin, between the RCPs, and between near-term and long-term periods due to the projected climate changes. These indicate that each of the sub-basin has a unique water balance environment.

The study also indicated the huge impacts of regional climate models (RCM) on surface components of the regional water cycle. These RCMs are a derivative of the Global Circulation Models (GCM).

The management interventions to mitigate the climate impacts should therefore be carried out according to the sub-basin water balance sensitivities while keeping the equilibrium in the closed CRVB water requirements. Finally, an investigated integrated watershed, agricultural water use, and farm management in the water–agriculture–land and climate nexus approaches following each sub-basin's climate responses, and other alternative resource management options for the closed CRVB must be determined and applied to cope with the hydroclimatic impacts.

The calibrated SWAT model has proved to be a useful tool for analyzing and identifying the temporal and spatial conditions of the water resources at a basin level under different climate change conditions in the CRVB. Therefore, further studies dealing with climate-based water resource management in combination with farming practices using the SWAT model would bring additional benefits.

Supplementary Materials: The following supporting information can be downloaded at: https://www.mdpi.com/article/10.3390/w15010018/s1, Table S1: Some physical properties of major soils in the CRV sub-basins; Table S2: Location of meteorological stations used for the analysis of the weather parameters in the CRVB; Table S3: SWAT land use code and their description.

Author Contributions: Conceptualization, L.A.T.; Methodology, L.A.T. and K.B.; Software, L.A.T. and K.B..; Validation, L.A.T., S.M. and K.B.; Formal Analysis, L.A.T.; Investigation, L.A.T. and K.B.; Resources S.M.; Data Curation, L.A.T. and K.B.; Writing—Original Draft Preparation, L.A.T.; Writing—Review & Editing, K.B.; Visualization, L.A.T., Svatopluk Matula; Supervision, Svatopluk Matula; Project Administration, S.M. and K.B.; Funding Acquisition, S.M. All authors have read and agreed to the published version of the manuscript.

Funding: This study was partly supported by the Czech National Agency for Agricultural Research; NAZV (project number QK1910086), and by the Czech University of Life Sciences Prague, Faculty of Agrobiology, Food and Natural Resources (project number SV21-20-21380).

Data Availability Statement: The source data and materials used in this study will be made available upon reasonable request from the corresponding author.

Acknowledgments: We acknowledge the financial support from Czech National Agency for Agricultural Research. We would also like to thank the Ethiopian Meteorological Agency, the Oromia Bureau of Agriculture and Natural Resources, the Ethiopian Ministry of Water Resources, the Ethiopian Geospatial and Information Institute, and the Ethiopian Ministry of Agriculture and Natural Resources for providing the data and for their support during data collection. We would also like to thank Robin Healey for his suggestions and comments relating to the text. We would also like to acknowledge the contribution of the anonymous reviewers who helped us to bring the manuscript into the present form.

Conflicts of Interest: We declare that we have no known competing financial interest or personal relationships that could appear to have influenced the work reported in this paper.

References

1. Worku, G.; Teferi, E.; Bantider, A.; Dile, Y.T. Prioritization of Watershed Management Scenarios under Climate Change in the Jemma Sub-Basin of the Upper Blue Nile Basin, Ethiopia. *J. Hydrol. Reg. Stud.* **2020**, *31*, 100714. [CrossRef]
2. Belay, A.; Recha, J.W.; Woldeamanuel, T.; Morton, J.F. Smallholder Farmers' Adaptation to Climate Change and Determinants of Their Adaptation Decisions in the Central Rift Valley of Ethiopia. *Agric. Food Secur.* **2017**, *6*, 24. [CrossRef]
3. Gebrechorkos, S.H.; Bernhofer, C.; Hülsmann, S. Climate Change Impact Assessment on the Hydrology of a Large River Basin in Ethiopia Using a Local-Scale Climate Modelling Approach. *Sci. Total Environ.* **2020**, *742*, 140504. [CrossRef] [PubMed]
4. Schmidt, E.; Zemadim, B. Expanding Sustainable Land Management in Ethiopia: Scenarios for Improved Agricultural Water Management in the Blue Nile. *Agric. Water Manag.* **2015**, *158*, 166–178. [CrossRef]
5. Ayenew, T. Water Management Problems in the Ethiopian Rift: Challenges for Development. *J. Afr. Earth Sci.* **2007**, *48*, 222–236. [CrossRef]
6. Gosain, A.K.; Rao, S.; Basuray, D. Climate Change Impact Assessment on Hydrology of Indian River Basins. *Curr. Sci.* **2006**, *90*, 346–353.
7. Alemayehu, T.; Furi, W.; Legesse, D. Impact of Water Overexploitation on Highland Lakes of Eastern Ethiopia. *Environ. Geol.* **2007**, *52*, 147–154. [CrossRef]
8. Ahmed, M.H. Climate Change Adaptation Strategies of Maize Producers of the Central Rift Valley of Ethiopia. *J. Agric. Rural Dev. Trop. Subtrop. JARTS* **2016**, *117*, 175–186.
9. Ngigi, S.N. *Climate Change Adaptation Strategies: Water Resources Management Options for Smallholder Farming Systems in Sub-Saharan Africa;* The Earth Institute at Columbia University: New York, NY, USA, 2009; ISBN 978-92-9059-264-8.
10. Hagemann, S.; Chen, C.; Clark, D.B.; Folwell, S.; Gosling, S.N.; Haddeland, I.; Hanasaki, N.; Heinke, J.; Ludwig, F.; Voss, F.; et al. Climate Change Impact on Available Water Resources Obtained Using Multiple Global Climate and Hydrology Models. *Earth Syst. Dyn.* **2013**, *4*, 129–144. [CrossRef]
11. Casagrande, E.; Recanati, F.; Rulli, M.C.; Bevacqua, D.; Melià, P. Water Balance Partitioning for Ecosystem Service Assessment. A Case Study in the Amazon. *Ecol. Indic.* **2021**, *121*, 107155. [CrossRef]
12. Conway, D.; Hanson, C.E.; Doherty, R.; Persechino, A. GCM Simulations of the Indian Ocean Dipole Influence on East African Rainfall: Present and Future. *Geophys. Res. Lett.* **2007**, *34*, 1–2. [CrossRef]
13. Khalilian, S.; Shahvari, N. A SWAT Evaluation of the Effects of Climate Change on Renewable Water Resources in Salt Lake Sub-Basin, Iran. *AgriEngineering* **2019**, *1*, 44–57. [CrossRef]
14. Müller Schmied, H.; Adam, L.; Eisner, S.; Fink, G.; Flörke, M.; Kim, H.; Oki, T.; Portmann, F.T.; Reinecke, R.; Riedel, C.; et al. Variations of Global and Continental Water Balance Components as Impacted by Climate Forcing Uncertainty and Human Water Use. *Hydrol. Earth Syst. Sci.* **2016**, *20*, 2877–2898. [CrossRef]
15. Uniyal, B.; Jha, M.K.; Verma, A.K. Assessing Climate Change Impact on Water Balance Components of a River Basin Using SWAT Model. *Water Resour. Manag.* **2015**, *29*, 4767–4785. [CrossRef]

16. Gadissa, T.; Nyadawa, M.; Behulu, F.; Mutua, B. The Effect of Climate Change on Loss of Lake Volume: Case of Sedimentation in Central Rift Valley Basin, Ethiopia. *Hydrology* **2018**, *5*, 67. [CrossRef]
17. Molla, M. The Effect of Climate Change and Variability on the Livelihoods of Local Communities: In the Case of Central Rift Valley Region of Ethiopia. *OALib* **2014**, *01*, 1–10. [CrossRef]
18. Shumet, A.G.; Mengistu, K. Assessing the Impact of Existing and Future Water Demand on Economic and Environmental Aspects (Case Study from Rift Valley Lake Basin: Meki-Ziway Sub Basin), Ethiopia. *Int. J. Waste Resour.* **2016**, *6*. [CrossRef]
19. Worqlul, A.W.; Dile, Y.T.; Schmitter, P.; Jeong, J.; Meki, M.N.; Gerik, T.J.; Srinivasan, R.; Lefore, N.; Clarke, N. Water Resource Assessment, Gaps, and Constraints of Vegetable Production in Robit and Dangishta Watersheds, Upper Blue Nile Basin, Ethiopia. *Agric. Water Manag.* **2019**, *226*, 105767. [CrossRef]
20. Musie, M.; Sen, S.; Chaubey, I. Hydrologic Responses to Climate Variability and Human Activities in Lake Ziway Basin, Ethiopia. *Water* **2020**, *12*, 164. [CrossRef]
21. Mengistu, D.; Bewket, W.; Dosio, A.; Panitz, H.-J. Climate Change Impacts on Water Resources in the Upper Blue Nile (Abay) River Basin, Ethiopia. *J. Hydrol.* **2021**, *592*, 125614. [CrossRef]
22. Beyene, T.; Lettenmaier, D.P.; Kabat, P. Hydrologic Impacts of Climate Change on the Nile River Basin: Implications of the 2007 IPCC Scenarios. *Clim. Change* **2010**, *100*, 433–461. [CrossRef]
23. Musie, M.; Sen, S.; Srivastava, P. Application of CORDEX-AFRICA and NEX-GDDP Datasets for Hydrologic Projections under Climate Change in Lake Ziway Sub-Basin, Ethiopia. *J. Hydrol. Reg. Stud.* **2020**, *31*, 100721. [CrossRef]
24. Pascual-Ferrer, J.; Candela, L. 'Water Balance on the Central Rift Valley', in case studies for developing globally responsible engineers GDEE (eds), Global Dimension in Engineering Education Barcelona. 2015, p. 31. Available online: http://gdee.eu/index.php/resources.
25. Taye, M.T.; Dyer, E.; Hirpa, F.A.; Charles, K. Climate Change Impact on Water Resources in the Awash Basin, Ethiopia. *Water* **2018**, *10*, 1560. [CrossRef]
26. Legesse, D.; Vallet-Coulomb, C.; Gasse, F. Hydrological Response of a Catchment to Climate and Land Use Changes in Tropical Africa: Case Study South Central Ethiopia. *J. Hydrol.* **2003**, *275*, 67–85. [CrossRef]
27. Getnet, M.; Hengsdijk, H.; van Ittersum, M. Disentangling the Impacts of Climate Change, Land Use Change, and Irrigation on the Central Rift Valley Water System of Ethiopia. *Agric. Water Manag.* **2014**, *137*, 104–115. [CrossRef]
28. Tekle, A. Assessment of Climate Change Impact on Water Availability of Bilate Watershed, Ethiopian Rift Valley Basin. In Proceedings of the AFRICON 2015, Addis Ababa, Ethiopia, 14–17 September 2015; pp. 1–5.
29. Muluneh, A.; Biazin, B.; Stroosnijder, L.; Bewket, W.; Keesstra, S. Impact of Predicted Changes in Rainfall and Atmospheric Carbon Dioxide on Maize and Wheat Yields in the Central Rift Valley of Ethiopia. *Reg. Environ. Change* **2015**, *15*, 1105–1119. [CrossRef]
30. Feleke, H.G. Assessing Weather Forecasting Needs of Smallholder Farmers for Climate Change Adaptation in the Central Rift Valley of Ethiopia. *J. Earth Sci.* **2015**, *6*, 312. [CrossRef]
31. Jansen, H.C.; Hengsdijk, H.; Legesse, D.; Ayenew, T.; Hellegers, P.; Spliethoff, P.C. Land and Water Resources Assessment in the Ethiopian Central Rift Valley: Project: Ecosystems for Water, Food and Economic Development in the Ethiopian Central Rift Valley. 2007. Available online: https://library.wur.nl/WebQuery/wurpubs/fulltext/19397 (accessed on 19 March 2021).
32. Gadissa, T.; Nyadawa, M.; Behulu, F.; Mutua, B. Chapter 13 - Assessment of Catchment Water Resources Availability under Projected Climate Change Scenarios and Increased Demand in Central Rift Valley Basin. In *Extreme Hydrology and Climate Variability*; Melesse, A.M., Abtew, W., Senay, G., Eds.; Elsevier: Amsterdam, The Netherlands, 2019; pp. 151–163. ISBN 978-0-12-815998-9.
33. Legesse Belachew, D.; Abiye, T.; Vallet-coulomb, C.; H, A. Streamflow Sensitivity to Climate and Land Cover Changes: Meki River, Ethiopia. *Hydrol. Earth Syst. Sci.* **2010**, *14*, 2277–2287. [CrossRef]
34. Kassie, B.T.; Asseng, S.; Rotter, R.P.; Hengsdijk, H.; Ruane, A.C.; Van Ittersum, M.K. Exploring Climate Change Impacts and Adaptation Options for Maize Production in the Central Rift Valley of Ethiopia Using Different Climate Change Scenarios and Crop Models. *Clim. Change* **2015**, *129*, 145–158. [CrossRef]
35. Asefa Bogale, G.; Temesgen, T. Environment Pollution and Climate Change Impacts and Challenges of Seasonal Variabilities of El Niño and La Niña on Crop and Livestock Production in The Central Rift Valley of Ethiopia: A Review. *Env. Pollut Clim. Chang.* **2021**, *5*, 2.
36. Musie, M.; Momblanch, A.; Sen, S. Exploring Future Global Change-Induced Water Imbalances in the Central Rift Valley Basin, Ethiopia. *Clim. Change* **2021**, *164*, 47. [CrossRef]
37. Mortezapour, M.; Menounos, B.; Jackson, P.L.; Erler, A.R. Future Snow Changes over the Columbia Mountains, Canada, Using a Distributed Snow Model. *Clim. Change* **2022**, *172*, 6. [CrossRef]
38. Leta, O.T.; El-Kadi, A.I.; Dulai, H.; Ghazal, K.A. Assessment of Climate Change Impacts on Water Balance Components of Heeia Watershed in Hawaii. *J. Hydrol. Reg. Stud.* **2016**, *8*, 182–197. [CrossRef]
39. Kotamarthi, R.; Hayhoe, K.; Wuebbles, D.; Mearns, L.O.; Jacobs, J.; Jurado, J. *Downscaling Techniques for High-Resolution Climate Projections: From Global Change to Local Impacts*; Cambridge University Press: Cambridge, UK, 2021; ISBN 978-1-108-47375-0.
40. Desta, H.; Lemma, B. SWAT Based Hydrological Assessment and Characterization of Lake Ziway Sub-Watersheds, Ethiopia. *J. Hydrol. Reg. Stud.* **2017**, *13*, 122–137. [CrossRef]

41. Liu, J.; Zhou, Z.; Yan, Z.; Gong, J.; Jia, Y.; Xu, C.-Y.; Wang, H. A New Approach to Separating the Impacts of Climate Change and Multiple Human Activities on Water Cycle Processes Based on a Distributed Hydrological Model. *J. Hydrol.* **2019**, *578*, 124096. [CrossRef]
42. Gadissa, T.; Nyadawa, M.; Mutua, B.M.; Behulu, F. Comparative Assessment of the Effect of Climate Change and Human Activities on Streamflow Regimes in Central Rift Valley Basin, Ethiopia. 2019. Available online: http://erepository.kibu.ac.ke/handle/123456789/950 (accessed on 15 December 2021). [CrossRef]
43. Baker, T.J.; Miller, S.N. Using the Soil and Water Assessment Tool (SWAT) to Assess Land Use Impact on Water Resources in an East African Watershed. *J. Hydrol.* **2013**, *486*, 100–111. [CrossRef]
44. Bát'ková, K.; Miháliková, M.; Matula, S. Hydraulic Properties of a Cultivated Soil in Temperate Continental Climate Determined by Mini Disk Infiltrometer. *Water* **2020**, *12*, 843. [CrossRef]
45. Arnold, J.G.; Kiniry, J.R.; Srinivasan, R.; Williams, J.R.; Haney, E.B.; Neitsch, S.L. *Soil and Water Assessment Tool Input/Output File Documentation Version 2009*; Texas Water Resources Institute: College Station, TX, USA, 2011; (accessed on 5 December 2021).
46. Abbaspour, K.C.; Rouholahnejad, E.; Vaghefi, S.; Srinivasan, R.; Yang, H.; Kløve, B. A Continental-Scale Hydrology and Water Quality Model for Europe: Calibration and Uncertainty of a High-Resolution Large-Scale SWAT Model. *J. Hydrol.* **2015**, *524*, 733–752. [CrossRef]
47. Teshager, A.D.; Gassman, P.W.; Secchi, S.; Schoof, J.T.; Misgna, G. Modeling Agricultural Watersheds with the Soil and Water Assessment Tool (SWAT): Calibration and Validation with a Novel Procedure for Spatially Explicit HRUs. *Environ. Manage.* **2016**, *57*, 894–911. [CrossRef]
48. Lenhart, T.; Eckhardt, K.; Fohrer, N.; Frede, H.-G. Comparison of Two Different Approaches of Sensitivity Analysis. *Phys. Chem. Earth Parts ABC* **2002**, *27*, 645–654. [CrossRef]
49. Setegn, S.G.; Srinivasan, R.; Dargahi, B. Hydrological Modelling in the Lake Tana Basin, Ethiopia Using SWAT Model. *Open Hydrol. J.* **2008**, *2*, 49–62. [CrossRef]
50. SWAT: Model Use, Calibration, and Validation. Available online: https://doi.org/10.13031/2013.42256 (accessed on 25 October 2021). [CrossRef]
51. Nash, J.E.; Sutcliffe, J.V. River Flow Forecasting through Conceptual Models Part I — A Discussion of Principles. *J. Hydrol.* **1970**, *10*, 282–290. [CrossRef]
52. Das, B.; Jain, S.; Singh, S.; Thakur, P. Evaluation of Multisite Performance of SWAT Model in the Gomti River Basin, India. *Appl. Water Sci.* **2019**, *9*, 134. [CrossRef]
53. Bc, H.; Rg, C. Climate Downscaling: Techniques and Application. *Clim. Res.* **1996**, *07*, 85–95. [CrossRef]
54. ARGAW, N. ESTIMATION OF SOLAR RADIATION ENERGY OF ETHIOPIA FROM SUNSHINE DATA. *Int. J. Sol. Energy* **1996**, *18*, 103–113. [CrossRef]
55. Moreira, L.L.; Schwamback, D.; Rigo, D.; Moreira, L.L.; Schwamback, D.; Rigo, D. Sensitivity Analysis of the Soil and Water Assessment Tools (SWAT) Model in Streamflow Modeling in a Rural River Basin. *Rev. Ambiente Amp Água* **2018**, *13*, 1–12. [CrossRef]
56. Zeray, L.; Roehrig, J.; Chekol, D. *Climate Change Impact on Lake Ziway Watershed Water Availability, Ethiopia*; Institute for Technology in the Tropics, University of Applied Science: Cologne, Germany, 2006.
57. Amare, D.; Endalew, W. Agricultural Mechanization: Assessment of Mechanization Impact Experiences on the Rural Population and the Implications for Ethiopian Smallholders. *Eng. Appl. Sci.* **2016**, *1*, 39–48.
58. Kifle, M.; Gebretsadikan, T.G. Yield and Water Use Efficiency of Furrow Irrigated Potato under Regulated Deficit Irrigation, Atsibi-Wemberta, North Ethiopia. *Agric. Water Manag.* **2016**, *170*, 133–139. [CrossRef]

water

MDPI

Review

How Can We Adapt Together? Bridging Water Management and City Planning Approaches to Climate Change

Vítor Vinagre [1,*], Teresa Fidélis [2] and Ana Luís [3]

[1] Department of Environment and Planning, Campus Universitário de Santiago, University of Aveiro, 3810-193 Aveiro, Portugal
[2] Research Unit on Governance, Competitiveness and Public Policies (GOVCOPP), Department of Environment and Planning, Campus Universitário de Santiago, University of Aveiro, 3810-193 Aveiro, Portugal
[3] AdP Valor, Rua Visconde de Seabra, 3, 1700-421 Lisboa, Portugal
* Correspondence: vcvinagre@ua.pt

Abstract: Different dynamics of climate change, population growth, and urbanisation challenge water service providers (WSPs) and those managing urban planning. The scientific community has been evidencing the concept of sustainable urban water management (SUWM) as a driver to foster the integration of the urban water cycle with its environmental, economic, and social sustainability dimensions. This article studies the approaches addressed by recent research on sustainable urban water management, focusing on the attention given by the scientific community to the way WSPs and city planners address the new challenges brought by climate change. A systematic review of existing literature shows how emergent challenges address the articulation between urban water cycle management and city planning. The results underline the need for the technical and economic evaluation of the overarching concept of SUWM systems, integrating values that go beyond financial issues; the need to address water scarcity not only from the supply side but also from the demand point of view; and the deepening of the relationship between new sources of water, such as the reuse, with the city planning in a context of climate change. Nevertheless, strategies for collaboration are still poorly addressed. The insights and gaps emerging from the analysis suggest new paths for research and practice in the field.

Keywords: climate change; adaptation; sustainable urban water management; city planning; urban planning; urban water management

Citation: Vinagre, V.; Fidélis, T.; Luís, A. How Can We Adapt Together? Bridging Water Management and City Planning Approaches to Climate Change. *Water* **2023**, *15*, 715. https://doi.org/10.3390/w15040715

Academic Editors: Alban Kuriqi and Luis Garrote

Received: 11 January 2023
Revised: 6 February 2023
Accepted: 9 February 2023
Published: 11 February 2023

1. Introduction

The world is rapidly urbanising. From 1950 to 2020, the population residing in cities increased from 0.8 billion (29.6%) to 4.4 billion (56.2%), and recent projections point towards it could reach 6.7 billion (68.4%) by 2050 [1].

The latest IPCC report highlights that global net anthropogenic GHG emissions in 2019 were 12% higher than in 2010 and 54% higher than in 1990 [2]. The implied global emissions by 2030 exceed pathways consistent with 1.5 °C and are near the upper end of the modelled pathways range, which keeps temperatures likely to limit warming to 2 °C [2]. In urban environments, observed climate changes impact human health, livelihoods, and critical infrastructure systems, which will be increasingly vulnerable if their design does not consider changing climate conditions [3].

Controlling greenhouse gas emissions and conserving dwindling water resources while feeding and serving a growing population is, in fact, a daunting task [4].

Whilst in the last century, the population grew three times, water consumption increased six times, following the average level of income, the evolution of habits, and a different demand for food [5], increasingly dependent on water, which represents, on a global average, about 70% of water consumption [6]. The area needed for irrigated agriculture increased, and consequently, so did the water needed for its production, which in turn is a competitor of the water required for other uses, such as industry, hydroelectric power production [7], and urban.

Furthermore, managing water resources, essential for human life, economic activities, and ecosystem functioning faces enormous challenges in a changing climate. It is known that water availability is not evenly distributed in the territory or in time [8]. The effects of climate change, namely through extreme phenomena such as droughts or floods, make the management process even more complex. Extreme hydrological events such as prolonged droughts and floods are increasingly frequent, creating great uncertainty about cities' water security [9]. This context can compromise the objectives of the United Nations Sustainable Development Goals (SDGs), especially the SDG11 Sustainable Cities and Communities and the SDG6 Clean Water and Sanitation [10].

In the "excess of water" dimension, it can be seen [11] that urban areas are particularly vulnerable to strong rainfall episodes due to their impermeable surfaces (such as roads, parking lots, and roofs) that prevent rainwater infiltration and, consequently, increase surface runoff and the risk of rain flooding [12]. Urban sprawl and a potential lack of water storage capacity in rainfall peaks lead to an insufficient drainage capacity of the water system, resulting in rain floods [12,13].

In the "water scarcity" dimension, i.e., when demand exceeds availability, the health and wellbeing of citizens, the quality of the urban environment, and socio-economic development are put at risk [8]. Related to the phenomena of scarcity due to climate change, the adoption of water reuse or rainwater harvesting efforts is critical, especially when considering urban development needs in warmer climates, the decline of water resources, the difficulty of transporting water between basins, and efforts to increase sustainability in urban planning and management [14–17].

On the one hand, water service providers need help with new problems concerning adaptation to climate change and the simultaneously evolving context, such as population growth, increasing urbanisation, and changes in consumption patterns [18,19]. On the other hand, critical aspects of managing the supply/demand balance are related to the dynamics of the territory and how the spatial planning and demography introduce new needs and consumption patterns [20], as well as new threats. For example, spatial distribution and consumption habits in the Barcelona region led to water consumption about ten times higher in peripheral areas (typically houses with lawns) compared to the urban core area with multifamily buildings [21]. Another challenge has to do with how these drivers call land planning for new solutions that integrate not only the necessary resilience to extreme drought and flood phenomena but also contribute to positive externalities at the level, for example, of blue and green infrastructures, enabling a better urban environment and improving the quality of life of populations [22–25]. Adapting cities to the effects of climate change on the water cycle is, therefore, a pressing issue. This involves assessing the adequacy of existing (often obsolete) infrastructures and their resizing and adaptation, whether in terms of the asset or how it is operated. Given the complexity of these challenges and the issues that must be addressed, including how and when they should be tackled, it requires the involvement of different actors in urban planning and water governance as well as risk management [26–28].

Entities responsible for managing the urban water cycle and associated social and ecological needs (water services, regulators, legislators) are thus called upon to rethink their decision-making processes [29]. However, they cannot act alone. The challenges of climate change reinforce the importance of the interrelationship between the management of water management services and the entities responsible for the planning and management of the urban territory [20,30]. The relative location of the economic activities that consume/reject water and the socio-economic relationships are aspects to consider for sound management of water resources, considering the supply, demand, and sustainability of the entire urban water cycle.

Faced with emerging water management challenges in cities, Marlow et al. ([31], p. 2) propose the overarching concept of sustainable urban water management (SUWM) "as an aspiration, SUWM reflects a generalised goal to manage the urban water cycle to produce more benefits than traditional approaches have delivered". Hurlimann and Wilson ([30], p. 1) consider that even if the concept of SUWM is not definitively enshrined, it "implies the consideration of climate change and the inclusion of both supply and demand side initiatives".

From what has been said, given the context of rapid change that is approaching, reaching "sustainable urban water management" is necessary [30].

While several studies have covered different aspects of the relationship between climate change, spatial planning, and the water cycle over the last few years, to our knowledge, there has yet to be a study that identifies, catalogues, and integrates consolidated expertise in these fields. Thus, this article aims to clarify and systematise existing knowledge, to systematise learnings and gaps, and to point out approaches that need further development. It undertakes a literature review focusing on two major questions: (i) What are the main themes addressed by contemporary research on sustainable urban water management? (ii) How is the scientific community addressing the collaboration between water management and urban planning agencies, and how is the relationship between climate change and the urban water cycle considered?

In this context, research and related dissemination become increasingly essential to support decision makers, water service providers, and communities for more robust climate change adaptation, infrastructure design, and operation in a potential new urban landscape. The article is organised into five sections. Following this introduction, Section 2 describes the methodology used to undertake the literature review. Section 3 presents the results, and Section 4 discusses the findings, insights, and gaps. Section 5 presents the main conclusions.

2. Methodology

A systematic literature review was chosen to identify, analyse, and interpret all the available research in this domain. This section describes the methodological process to initiate the search and to collect, screen, and analyse selected papers from the existing literature.

This review was conducted based on the Preferred Reporting Items for Systematic Review and Meta-Analysis (PRISMA) guidelines to guarantee the review process's reproducibility, traceability, and transparency. The review's objective is to find and analyse scientific literature framing urban adaptation in the context of climate change and its articulation with the urban water cycle and, with this, to respond to the research questions previously presented. Chronologically these steps were followed:

(a) The search string used in this review was first initiated by selecting an initial list of 15 relevant articles based on the expertise of the authors in the field, which were also chosen in the final set considered for analysis;
(b) Out of these articles, the first set of keywords was chosen and considered in the first search. Next, several test searches were performed with alternative combinations between keywords and their variants. The results from the test searches were discussed among the authors to refine the search strings until we were fully accomplished with the capability of the string to detect as much of the initial set of relevant and related publications as possible. The search strategy and results are presented in Table 1;
(c) Following this iterative strategy and after a series of test executions and reviews, which led to the selection of articles considered to be more relevant, we obtained the selected and unique set of search terms and keywords: climate change, sustainable urban water management, urban planning, and city planning. This step led to identifying 328 articles (from an initial universe of 524 items, from which we excluded the non-articles). The articles identified by the search engine were directly extracted into an Excel file offered by Scopus;
(d) For the quality evaluation, that is, relevance to the response to the research questions, the PRISMA tool was used for each article, providing an objective comparison between the articles and their classification, which resulted in a universe of 39 articles;

Table 1. Results of the combination of keywords in the iterative process.

Search Strategy for Scopus Conducted on 18 October 2022 Period: until 18 October 2022 Language: English Document Type Limits: Scientific Articles in Journals Search within the Title, Abstract, and Keywords		
Iteration	**Query**	**Records Retrieved**
a	(TITLE-ABS-KEY (climate AND change) AND TITLE-ABS-KEY (urban AND planning) AND TITLE-ABS-KEY (water AND reuse))	33
b	(TITLE-ABS-KEY (climate AND change) AND TITLE-ABS-KEY (water AND utilities) AND TITLE-ABS-KEY (municipalities))	25
c	(TITLE-ABS-KEY (climate AND change) AND TITLE-ABS-KEY (urban AND planning) AND TITLE-ABS-KEY (water AND management))	867
d	(TITLE-ABS-KEY (risk) AND TITLE-ABS-KEY (urban AND planning) AND TITLE-ABS-KEY (water AND management))	828
e	(TITLE-ABS-KEY (climate AND change) AND TITLE-ABS-KEY (sustainable AND urban AND water AND management))	667
f	(TITLE-ABS-KEY (climate AND change) AND TITLE-ABS-KEY (sustainable AND urban AND water AND management) AND TITLE-ABS-KEY (urban AND planning) OR TITLE-ABS-KEY (city AND planning))	328

The next step of filtering was performed to select additional relevant papers through the snowballing process. This step added 40 other articles and reports to the 15 previously identified.

Figure 1 synthesises the screening process and the number of articles excluded from the initial database and those that were added later. After concluding this screening process, the resultant set of articles was extracted into a final Excel file.

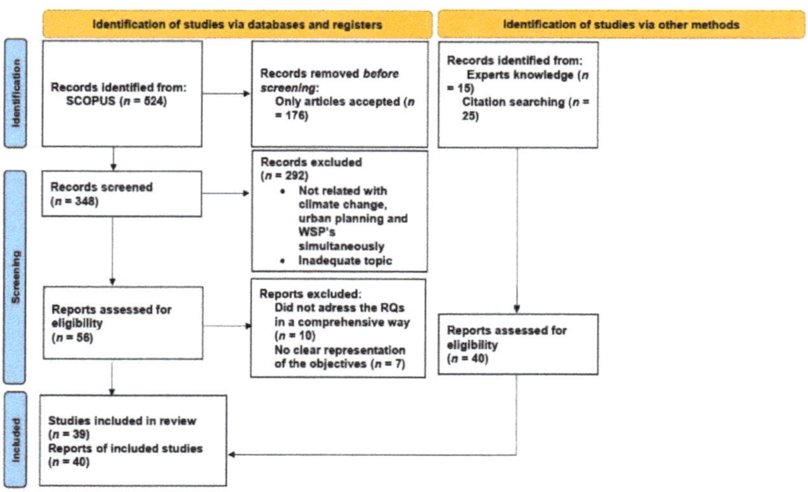

Figure 1. Methodological flow diagram summarising the steps to retrieve the articles (PRISMA).

3. Results and Content-Based Analysis

This section analyses the content of the selected articles, searching for how water service providers and urban planners respond to the challenges of climate change and how they respond, when they do, together. The information obtained is systematised and integrated into figures and tables. The presentation of results is first concentrated on the bibliometric analysis, then on the scientific literature, and finally on the grey literature.

3.1. Bibliometric Analysis

The evolution of articles published in recent years related to the selected keywords is represented in Figure 2, which shows a very sharp growth and attests to the growing scientific interest in the relationship between climate change, sustainable urban water management, and urban planning.

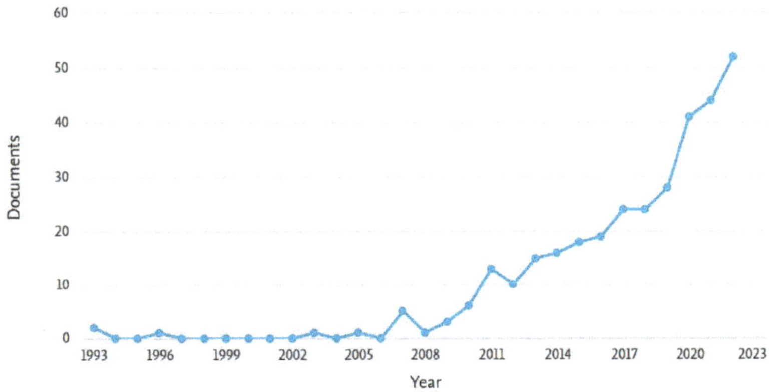

Figure 2. Annual distribution of the number of articles after Scopus search for the selected keywords and criteria. Figure source: Scopus.

The geographical distribution of articles and case studies, represented in Figure 3a,b, shows a prevalence of the Netherlands, USA, Australia, China, and the UK, which confirms the pancontinental nature of interest in this theme. That said, a joint analysis of those figures also suggests that most of the articles focus on the authors' territory, being the majority from developed countries, which makes it possible to infer that underdeveloped countries still need to be subject to such an in-depth analysis. The contribution of authors from Israel and Singapore should also be underlined, especially considering the perspective of the size of each of these countries, although it is known that these two countries are among the ones that faced severe water scarcity.

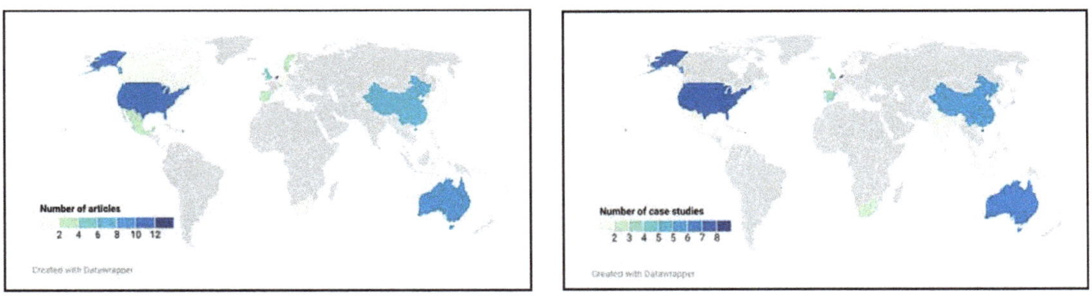

(a) (b)

Figure 3. (**a**) Geographical distribution of the articles considered in this article (1st author). Figure source: Datawrapper; (**b**) Geographical distribution of the case studies considered in the article. Figure source: Datawrapper.

3.2. Concepts of Sustainable Water Management in Cities

3.2.1. From the Scientific Literature

With the concept of SUWM, some authors [31,32] associate decentralisation, resource efficiency, and sustainability as critical factors. Also associated with this concept are technical configurations, such as the collection and reuse of rainwater [33], "grey water" recycling [34,35], design of "blue" and "green" infrastructures [36,37], and the optimisation of water consumption by the final consumer [38]. Within the concept of SUWM, other urban water cycle management concepts/frameworks are included, such as integrated urban water management (IUWM), water-sensitive cities, low-impact development (LID), sustainable urban drainage systems (SUDS), and Sponge Cities, to mention the most relevant. Despite their complementarity, other innovative city concepts primarily based on digital development and technology, such as those related to smart cities, will not be studied here since water management, sustainability, and urban planning are not at their core [39].

All these concepts, presented in Table 2, seek to respond to the new challenges by integrating the management of water resources with the drivers that most affect their availability and ecological status: climate change, population growth, and increasing urbanisation.

Table 2. Key sustainability concepts associated with water management and the new paradigms in the 21st century.

Concept	Definition	Origin
Low-Impact Development (LID) 1977 [40]	- *The original intent of LID was to achieve a 'natural' hydrology by use of site layout and integrated control measures. Natural hydrology referred to a site's balance of pre-development runoff, infiltration, and evapotranspiration volumes, achieved through a "functionally equivalent hydrologic landscape"* ([41], p. 3) - *The LID practice is an integrated watershed management strategy, which provides natural retention, treatment, and source protection capabilities.* - *It utilises natural processes to capture, treat, absorb, and infiltrate stormwater runoff that has increased in peak rate and volume with more pollutant contents* ([42], p. 1).	USA New Zealand (LIDUD)
Integrated Urban Water Management (IUWM) 1995 [43]	- *IUWM promotes a coordinated planning approach to drinking water, wastewater, and stormwater services that takes into consideration the broader implications of sustainable development, including energy demand, greenhouse gas emissions, solid waste generation, nutrient losses, life cycle costs, and community acceptability* ([44], p. 2) - *IUWM provides cities with a new framework for planning, designing, and managing urban water systems. An IUWM perspective enables all stakeholders to look at the urban water system holistically, as an integrated, cooperative venture, and together supply the capacity to predict the impacts of interventions across broad resource management units. By doing so, the framework facilitates the development of innovative solutions for urban water management and the prioritisation of resources* ([45] p. 58).	-

Table 2. *Cont.*

Concept	Definition	Origin
Water-Sensitive Urban Design (WSUD) 1996 [46]	- Lloyd et al. ([47], p. 2) describe WSUD as a *"philosophical approach to urban planning and design that aims to minimise the hydrological impacts of urban development on the surrounding environment. Stormwater management is a subset of WSUD directed at providing flood control, flow management, water quality improvements and opportunities to harvest stormwater to supplement mains water for non-potable uses"*. - *Water-sensitive urban design (WSUD) is supported by an underlying value of providing urban water services in a manner that considers the site-specific opportunities and limitations of development to provide water services in a way that protects and enhances local hydrological and ecological integrity. WSUD considers all aspects of the urban water cycle as a valuable resource. Incorporating WSUD in urban developments can also improve resilience to reduced yield from conventional water supply catchments due to potential climate change impacts* ([44], p. 2).	Australia
Sustainable Urban Drainage Systems (SUDS) 2000 [48]	- *SUDS consist of a range of technologies and techniques used to drain stormwater/surface water in a manner that is (arguably) more sustainable than conventional solutions. They are based on the philosophy of replicating, as closely as possible, the natural pre-development drainage from a site, consistent with the previously described principles behind LID.* - *Typically, SUDS are configured as a sequence of stormwater practices and technologies that work together to form a management train* ([41], p. 5). - *(. . .) SUDS can improve the sustainable management of water by replicating natural drainage patterns; reducing the peak flows, volume, and frequency of flows into watercourses from developed sites; removing pollutants from diffuse pollutant sources; and increasing the potential for rainwater harvesting. Consequently, SUDS can reduce downstream flooding risks, improve water quality, recharge groundwater and maintain base flows, reduce potable water demand, and improve local amenities through the provision of public open space and wildlife habitat* ([23], p. 7).	UK
Sustainable Urban Water Management (SUWM) 2008 [49,50]	- *SUWM concepts can be considered the next step in this co-evolution and reflect growing concerns over community wellbeing (rather than just public health), ecological health, and sustainable development, all of which can be collectively labelled as 'green' issues (. . .) SUWM reflects a generalised goal to manage the urban water cycle to produce more benefits than traditional approaches have delivered* ([31], p. 2). - *SUWM is advocated by an increasing number of scholars as an alternative paradigm to traditional water infrastructure and approaches, which can address the complex challenges facing urban water management. (. . .) SUWM is an umbrella concept that encapsulates the concepts of 'integrated urban water management' and "water-sensitive urban design" (WSUD)* ([51], p. 1).	-

Table 2. *Cont.*

Concept	Definition	Origin
Sponge City 2014 [52]	- *The Sponge City concept aims to (i) adopt and develop LID concepts, which improve effective control of urban peak runoff, and to temporarily store, recycle, and purify stormwater; (ii) upgrade the traditional drainage systems using more flood-resilient infrastructure (e.g., construction of underground water storage tanks and tunnels) and to increase current drainage protection standards using LID systems to offset peak discharges and reduce excess stormwater; and (iii) to integrate natural water bodies (such as wetlands and lakes) and encourage multi-functional objectives within drainage design (such as enhancing ecosystem services) whilst providing additional artificial water bodies and green spaces to provide higher amenity value* ([22], p. 2). - *(...) "sponge city" concept (...) represents a new urban development mode that is intended to manage effectively urban rainwater. This concept gives priority to protection and remediation of natural environments in urban planning and construction to ensure their ecosystem service function of water conservation. "Sponge city" vividly describes an urban environment that is devoted to finding ecologically suitable alternatives to transform urban infrastructures into green infrastructures so they could capture, control and reuse precipitation in a useful, ecologically sound way* ([53], p. 1).	China

3.2.2. From Grey Literature

The most structured recommendations in grey literature on how to address climate change in the design of cities, their relationship to water, and in some cases with risk, have been promoted through organisations such as the International Water Association (IWA (International Water Association), 2016), the World Bank [26,54], or the United Nations [28], among others. They are presented in the form of multi-stage frameworks that help build the response of cities and territories to the challenges of climate change, particularly in water systems. The methodologies and practices that stand out from the grey literature are shown in Table 3:

Table 3. Structured recommendations from grey literature.

Publication/ References	Main Recommendations	Case Studies Referred to in the Publications
United Nations Educational, Scientific and Cultural Organization (UNESCO) Climate Risk Informed Decision Analysis (CRIDA) [28]	Publication extract: *"(...) the UNESCO International Hydrological Programme presents, therefore, the Climate Risk Informed Decision Analysis (CRIDA). This approach provides a crucial framework to enable water managers and policy makers to assess the impact of climate uncertainty and change on their water resources and work towards effective adaptation strategies.* *This multi-step process embraces a participatory, bottom-up approach to identify water security hazards, and is sensitive to indigenous and gender-related water vulnerabilities.* *By engaging local communities in the design of the analysis, the information provided by scientific modelling and climate analysis can be tailored and thus provide more useful answers to the challenges they are facing. They are also providing a more informed starting point to assess the different options for adaptation, and design robust adaptation pathways, in line with the local needs.* *The CRIDA approach advocates hereby to move away from the "one size fits all" approach, and to pursue locally embedded solutions to the specific threats to water insecurity due to climate and other global changes (...)"* ([28], p. 9). Synthesis: It is a framework that considers a risk analysis and how it should be managed when a given system is confronted with climate change. It seeks to develop participatory adaptation methodologies involving different stakeholders and adapted to each location from a *bottom-up* perspective.	Colombo Bangkok Philippines Udon Thani Colombia Chile Mexico Guayaquil Zambia Sweden Rhine river Lake Ontario California

Table 3. *Cont.*

Publication/ References	Main Recommendations	Case Studies Referred to in the Publications
World Bank Water in Circular Economy and Resilience (WICER) [54]	Publication extract: "*WICER aims to promote a paradigm shift in the water sector. The shift involves moving away from linear thinking in the way we plan, design, and operate water infrastructure in urban settings towards a circular and resilient approach (...) Applying the WICER framework provides environmental benefits, as well as social, economic, and financial benefits. It is also a condition for achieving several of the global Sustainable Development Goals (SDGs)*" ([54], p. 43). Synthesis: This report aims to promote a common understanding of the definition and applications of circular economy principles and resilience in the urban water sector. It presents a framework to guide practitioners who are incorporating the principles in policies and strategies, planning, investment prioritisation, and design and operations to achieve three main outcomes: (1) deliver resilient and inclusive services; (2) design out waste and pollution, and (3) preserve and regenerate natural systems. These will ultimately improve livelihoods while valuing water resources and the environment. These outcomes are then deployed into three action plans each. It also states that cities and water utilities will only achieve a fully circular and resilient water system with the appropriate policy, institutional, and regulatory framework in place. It shows examples that investments in circular and resilient systems yield economic and financial payoffs and that the WICER framework could help utilities attract private-sector finance. To avoid being locked into linear and inefficient systems, low- and middle-income countries should also consider applying the WICER framework to design and implement circular and resilient water systems from the outset.	Durban Bogota River Chennai São Paulo Monclova Mostar Cali Ridgewood Santiago Atotonilco Indonesia Phnom Penh New Cairo S. Luis Potosi Nagpur Dakar Lingyuan Arequipa North Gaza
International Water Association The IWA Principles for Water-Wise cities [55]	Publication extract: "*Water-wise" behaviour means that leadership culture, governance arrangements, professional capacity, and innovative technology are all aligned with the objective of maximising sustainable urban water outcomes. Sustainable urban water management means that all water within the city (including reservoir and aquifer water, desalinated water, recycled water, and stormwater) is managed in a way that recognises the connection between services, urban design, and the basin, with an approach that maximises the achievement of urban liveability outcomes, and resilience to unexpected social, economic, or bio-physical shocks, while replenishing the environment.*" ([55], p. 2) Synthesis: The ultimate goal of the principles presented above is to encourage collaborative action, underpinned by a shared vision, so that local governments, urban professionals, and individuals actively engage in addressing and finding solutions for managing all waters of the city. From 5 building blocks and 4 levels of action, the 17 Principles are grouped into four categories: regenerative water services, water-sensitive urban design, basin-connected cities, and water-wise communities. Water-wise communities will use the building blocks to put the principles into action.	Amsterdam Berlin Brisbane Copenhagen Dakar Gothenburg Kampala Kunshan Lyon Melbourne Perth Shenzen Singapore Sydney Xi'an

This sub-section presented the status quo of the world's best-structured practical references in this field. It aimed to outline if their approaches complement those of the scientific literature referred to in the previous section. Further research could also help to assess how the scientific literature validates the grey literature.

3.3. Approaches to Climate Change Adaptation of Water Utilities and City Planning

Since the 19th century, water infrastructure has been centrally built to address hygiene and health issues, significantly reducing diseases and increasing health [56]. Centralised systems, the norm in cities, are characterised by extensive treatment, distribution, and collection facilities for treatment that connect distant points of origin/rejection and their final consumers [57]. Most developed countries spend between 1% and 6% of their annual GDP on centralised systems [58], resulting in substantial "sunken costs" and total dependence on water services (by nature with great inertia). Consequently, they face a blocking situation in which transforming alternatives for water management encounter barriers to entry [59].

These very centralised systems are based on mainly buried infrastructures whose main objective is to reach the water to citizens in quality and quantity, drain and treat the effluents generated, and drive rainwater as quickly as possible out of urban areas. Often, these infrastructures lead water from distant regions to the populations that need it through systems that favour the reliability and quality of the water supplied [31]. Similarly, treated effluents are often rejected from their place of production in large wastewater treatment plants. The dimensioning of these infrastructures is made for a distant design horizon with high initial sunken costs [10], which, in most cases, implies large tariffs for the first generations that use them and an idle capacity for at least the first years of activity [10]. This means a consequent waste of financial and operational resources, considering the need to "move" high water resources out of their natural "habitat", with impacts that go far beyond your place of consumption [31].

On the one hand, floods often involve the routing (undue if in separative networks) of rainwater to wastewater collectors, thus implying discharge into the water environment and in an uncontrolled form of crude effluent, more or less diluted. Besides this impact on the environment, floods also entail increasing human and property damage to which the current paradigm cannot respond, not only in developing countries (for instance, Mozambique and Pakistan) but also in developed countries, of which the 2021 floods in Germany are a recent example [60]. On the other hand, it is neither technical nor economically feasible [10,19]) to size wastewater and rainwater infrastructure for all extreme situations that potentially occur.

Concerning water scarcity, despite growing awareness of the effects of climate change, the transformative process of water management to include new sources (such as water reuse, rain harvesting, or desalination) or new conceptions (such as decentralised systems, green and blue infrastructures, etc.) is confronted with several obstacles that still lead to some inertia [32]. It should be said, however, that this latent stagnation is primarily presented by the incumbents, which present, among others, barriers to the implementation of innovative measures such as [32]: greater reliability (general) of centralised water systems, potentially lower costs of centralised systems (more significant economies of scale), perception of greater risk to public health by consumers, legislation not yet fully adequate for water reuse, uncertainties regarding the governance of the different systems in the future, and the lack of motivation in the entities that manage the status quo. This "lock-in" effect [61,62] is associated with apparent economies of scale, progress in the "learning curve", confidence in existing technologies, and network economies (agents using the same technology as their peers) [31], which translates into a barrier to innovation and the entry of more sustainable systems, perpetuating the incumbent.

However, there are also crucial motivations that lead to transformative processes in the relationship between urban water management and urban planning, such as [32,63,64]: the need to resize cities according to the variation of their population and consequent increase in consumption; public perception of the waste of the use of drinking water for irrigation, flushing of toilets, and washes; climate change, with a particular focus on capturing/deferring rainwater runoff and managing water scarcity; food security, as the lack of water, together with the degradation of agricultural land, leads to a reduction in agricultural productivity, which in turn leads to lower incomes and food availability [6,65]; increased consumption and decreased availability motivated by the average and "peak" increase in temperatures; sensitivity to phenomena such as self-sufficiency and the circularity of the economy. Naturally, in very concrete geographies, extreme phenomena of lack or excess of water are already the biggest catalysts for this paradigm shift, such as the cases of Israel, China, Australia, California, and Singapore, to report the most studied [66–68]. Thus, even in well-established and proven systems, the need for reinvestment, the urgent response to climate change, and the dynamics of urban expansion force a paradigm shift, which becomes necessary, both in underdeveloped countries and in developed countries.

For their part, in underdeveloped countries, in addition to drivers related to climate change and population growth, high rates of urban growth, poor trust in institutions [69], and uncertainty about city planning, combined with a lack of initial capital and high discount rates, lead to the trend of investing in rapid implementation solutions and in turn to a strong tendency to avoid significant investments in infrastructure [70].

The progressive hybridisation of centralised and decentralised systems has been reported as the most likely trend of implementation, combining the reliability and financial sustainability of centralised systems, so-called conventional, with the need to adapt cities to climate and demographic change, thus ensuring greater resilience. In this context, there is room for more consolidated studies, particularly about both levels' systemic and parallel functioning [32]. As such, the challenges presented above require different approaches and paradigms.

The way literature faces these challenges can unfold systematically in the following vectors: operational, organisational, institutional, behavioural, economical, technological, and urban planning.

Concerning the operational vector, the scientific literature points towards the definition of strategies to save water, reduce losses [10], minimise undue inflows to urban systems, and separate the sanitation of wastewater and rainwater [33,42] and the use of stored rainwater in periods of lower rainfall [71], either in a single-family management analysis [33] or in a city-level or basin-level approach [71,72]. Of course, some of these operational interventions must be integrated with the necessary investments corresponding to the economic and technological vectors.

At the organisational vector, the tendency referred to in the literature is for decentralising infrastructures and systems, corresponding to their greater spraying. The challenge arises regarding their management—in the local community, municipal, or WSPs—and, in any case, how centralised and open to citizen participation is. Questions are raised, as to how the "water decentralised infrastructures" should be created, given the technologically premature state of the proposed solutions and information regarding exploration costs, monitoring of their performance, and diffuse responsibility regarding their current management [31,44].

At the institutional level, efforts focus on sharing objectives and knowledge, usually with very flexible approaches, involving various stakeholders at national, regional, and local decision-making levels. At the economic vector, no direct savings in technical solutions related to sustainable water management are evident [31], especially considering the energy and operating and maintenance costs accompanying solutions such as desalination [35], rainwater harvesting [33,73], or reused water [32,74], where the "scale" factor is essential.

Concerning the behaviour of the final consumer, studies have been presented in Israel that correlate their perception of those reuses with their level of treatment, their possibility of use, and other variables, such as education and age [68,75]. The desire to consume alternative sources of water and the way the message is passed are fundamental aspects of its implementation [68,76]. However, some 13% of the consumers in a study conducted in the United States rejected the use of recycled water, depreciatingly called "from the toilet to tap" by those opposing it [77].

At the technological vector, the main trends concerning the challenges in the urban water sector for the 21st century are related to (i) the increased use of alternative sources of water, namely the reuse of rainwater [33,62,73,78], the reuse of water (direct or indirect), and desalination and the new technologies related to it that arise (Larsen et al., 2016); (ii) the "buffering" of extreme phenomena (usually related to floods) [8,42,62,79] in the search for more sustainable solutions with positive environmental externalities [35,68,80–82], such as Sponge Cities in China, greenfield expansions in Australia or redevelopment in the Netherlands [8,22,53,83]; and (iii) the application of information technologies to the planning of the urban cycle of water and cities [79,84,85]. Considering the various drivers referred to above, which influence the relationship between the sustainability of the management of urban water resources in the face of the challenges of climate change, population growth, and increasing urbanisation, growing literature is addressing the use of artificial intelligence to integrate the diversity of inputs. This literature seeks to integrate more technical and socio-economic baseline data, such as spatial planning, localisation of water infrastructures, impermeable surfaces, green areas, and green roof areas, among many others [38,84] to understand the practical implications that the future provides, depending on the simulated scenario. In a more focused way, several studies model and project the various possibilities of water reuse [86], rainwater reuse [73], or the behaviour of watersheds in extreme situations [87], among many others.

Finally, the scientific literature related to urban planning focuses on how positive externalities can be obtained in the pursuit of sustainable solutions that allow cities to tackle climate change ([20,22,30,71,88]. These analyses have focused on the preparation of cities for the management of water retaining [18,20] by defining the constructive details to be implemented in public/private infrastructures (porous pavements, green roofs, etc.) [42,64], by xeriscaping [87], through integrative interventions at the neighbourhood level [83,84], by the adaptation of the blueprint of cities to help landscape management in prioritising urban development strategies in the water-energy nexus [87] and the significant transformations of expansion or adaptation of cities considering rainwater management [71]. London is a paradigmatic example in how it defines water neutrality as a concept to frame the water stress in cities, integrating spatial data with an integrated urban water management model; this holistic, systemic design framework is designated CityPlan-Water [38]. In Table 4 we summarize the vectors presented in Figure 4 with the approaches to tackle climate change and urbanisation in the water sector.

Table 4. Synthesis of the approaches to tackle climate change and urbanisation in the water sector.

Vectors	Approaches	Examples
Operational	Strategies to save water and reduce losses	[10]
	Minimise undue runoff to sewer systems	[73]
	Separate sanitation of rainwater and wastewater	[33,42,76]
	Mindset towards the use of reserved rainwater in periods of lower rainfall	[33,71,72]
Organisational	Lack of knowledge	[24,25]
	Adaptation needs (WSP)	[31,44]
	Decentralisation vs Centralisation	[29,31,76]
	(Diffused) Responsibility	[44]
	Circular Economy	[15]
Institutional	Inter-organisational practices	[89–91]
	Water and land management communication	[20]
	Governance of adaptation	[90,92,93]
Economical	Financial (dis)advantages	[31]
	Costs and scale factor	[33,73,74]
	Last resort systems cost	[76,94]
	Food security	[6,65]
Behavioural	The context for the acceptance of water reuse	[68,75]
	Communication strategies for water reuse	[68,76]
Technological	Reuse of rainwater	[33,61,62,73,78]
	Reuse of water (direct or indirect)	[35,74,80,86,95–97]
	Watershed behaviour	[87]
	Buffering of extreme phenomena (floods)	[33,62,79,85]
	Positive externalities	[22,53,76,98]
	Information technology	[79,84,85,87,99]
Urban planning	Symbiosis of adaptation to climate change in cities and the water sector	[20,22,30,71,88]
	Interventions at the neighbourhood level	[83,84]
	Landscape management	[87]
	Major adaptations in cities	[38,71]
	Multilevel adaptation	[93]

The table above shows the main references found for each vector, outlining the limited number of those dealing with the relationship between the entities that manage the water services and the territory.

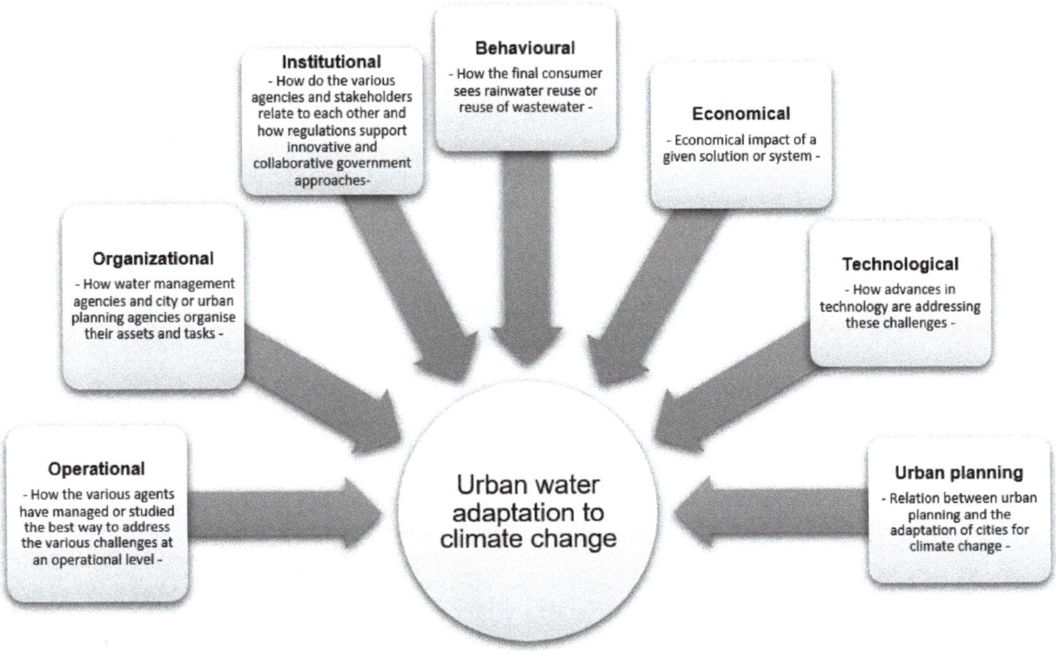

Figure 4. Main vectors contributing to the urban water management adaptation to climate change, according to the literature review.

4. Discussion

Considering the results mentioned above, several issues can be highlighted concerning the relationship between sustainable water management, urban planning, and climate change. In fact, despite the lack of practical implementation that does not yet follow the diversity of existing scientific literature on the sustainable management of the urban water cycle [38], it already presents a set of learnings/outcomes and gaps that allow us to perceive the main insights and the gaps to be filled.

One conclusion to be withdrawn from the outset is that the most significant innovations or need for innovation are mainly at the organisational and economic vectors and in the relationship between the various stakeholders and citizens/consumers and not so much in terms of technological development since the main drivers for change still arise in the paradigm shift from centralised to decentralised systems and how to share their management with the other stakeholders, including entities that manage the territory. Although there is a trend in the literature towards responding to climate change through decentralised systems, some of the best examples of success in adapting to climate change in the water supply sector, especially in terms of water reuse, occur using concepts of centralised systems, such as Singapore, Israel, or Southern California in the US.

Of course, in situations where redundancy exists, i.e., where centralised infrastructure remains a "last resort", there may be double pricing to sustain the sunken costs related to that system and the capital and operating cost tariffs associated with more sustainable methods. There is a need for a broader cost–benefit analysis involving not only the financial aspects but also the positive/negative externalities resulting from the implementation of more "sustainable systems" [76,94].

It is important to remark that the relationship between urban planning, WSPs, and climate change has also focused on flood control and less on water supply. There is, therefore, a gap in the need for scientific development [30,100]. This gap significantly increases when it comes to the integrated management of both "too much" or "too little" water, i.e., flood control and water supply.

For that matter, WSPs are facing increasing challenges in terms of water availability, management of consumption patterns, and the need for increased efficiency, which are alternatives to be developed to address the problem of lack of water [62], here still in the context of water directly collected from the water environment.

The planning of water infrastructures tends to be subordinate dweller to the planning of the territory [20] in a way that, in addition to being technically challenging, has also demonstrated other types of problems, such as complex collaboration in the face of more controversial situations of land use. On the other hand, the unavailability of staff in smaller locations and a level of diffuse responsibilities within and between each side, urban planning and WSPs, tend to hinder the necessary convergence. [20]. Adapting to a changing climate requires the collaboration of the disciplines of spatial planning and urban water supply management [30].

Many arguments and practices associated with concrete cases of articulation between WSPs and those that manage the territory can also be applied to the water supply strand. Consolidating a projection of the future—a practice to which urban planning is dedicated—with the projection of climate change is pointed out as being the way forward, to which the necessary articulation with the drainage and water supply strands of the WSPs is added.

Some of the barriers to overcome in the water sector are related to the "lock-in" effect related to the already mentioned inertia derived from the sector, often resonating on buried infrastructure with an extensive lifetime and high capital costs. On the other hand, it is a sector traditionally averse to innovation [10,76], both technical and operational [31]. The main insights and gaps that stand out from the literature are presented in Table 5, following Figure 4.

Table 5. Synthesis of the main insights and gaps that arise from the literature review.

Vectors	Insights and Gaps
Operational	The pluri-functionality of some installations (e.g., flood control and management, reuse of waters) contributes to a dispersion of the objectives to be achieved, often competing with each other or not taking advantage of their synergies [44].
	Regulatory changes and poor anticipation of operating costs are some of the risks most evidenced by experts dealing with water reuse [81].
	Grey water constitutes 50–80% of the total household wastewater produced, which enhances its future use after treatment [74,101].
	The decentralisation of systems can present great advantages in areas where, under "business as usual" conditions, it would be necessary to expand a centralised system, thus contributing to a more resilient system with less investment in capital, thus enhancing greater naturalisation of the same [31].
	Given the lock-in effect, the trend will be the coexistence of centralised and decentralised systems, thus operating a gradual change between both philosophies and a path leading to their hybridisation. The management of water demand, and not only the increase in its supply, can contribute decisively to the minimisation of the risk of water scarcity [31,83,99].

Table 5. *Cont.*

Vectors	Insights and Gaps
Organisational	Theoretical studies are presented that present ways for the adaptation of WSPs in different contexts and how their adaptability and learning are essential to meet the challenge of climate change [102].
	This adaptation can be facilitated by political and legislative measures [103].
	Reference [103] also demonstrated through an intersectoral study conducted in the UK that the water supply and flood control sectors are those in which at the institutional level, there was more significant activity to adapt to climate change, often from a top-down perspective (above the local level), with climate change triggers (actual or perished) and legislation, despite the fact that the need for interventions on the ground have greater difficulty in implementing if they are only motivated by climate change.
	There is a lack of deepening between sustainable urban water management measures, citizens' perception and socio-economic issues, and the use of the territory [30,63,76].
	There is a need for analyses that realise the possibilities of a relationship between sectoral measures of water resilience with the components of business and political decisions and studies relating to the relationship between conventional and decentralised water resource management systems [104].
	There is a gap in the knowledge about the implementation of the resilience of water systems more qualitatively and less quantitatively towards a more in-depth risk assessment and the relationship between water resilience systems—designated at flood level—and water supply, transport, energy, and waste collection systems [105].
	There is a need for more consolidated studies and reports on how to operate and maintain hybrid systems (centralised and decentralised) and how to define the attribution of responsibilities [76].
Institutional	The planning of the territory will tend to be more challenging and complex in the future, not only motivated by an entity—municipality or managing body of the water service provider—but more integrated towards closer interests, which may even converge to new structures or allocation of responsibilities in the management of public services [104]. Bearing in mind the need to create sustainable urban and regional planning practices, the articulation between institutions and spatial planning policies and water management tends to become one of the central concerns [93]. Given the constraints of articulating rigid institutional structures, informal networking structures are beginning to appear between different interests that fill the gaps between the different, and often conflicting, official organisations [91].
Economical	In most cases, a financial analysis, pure and complex, that is, intending to obtain a net present value (NPV) as a central element for the viability of a given project, will not result in an advantage of SUWM systems, and there is a need to converge to a more holistic approach that takes into account the minimisation of the risk inherent to phenomena of scarcity or excess water, the creation of leisure spaces, the minimisation of heat islands phenomena, increased resilience and the potential reuse of captured/reused water [31,83].
	There is a need for a deepening of knowledge that further characterises the vectors that contribute to a more comprehensive "value" of a given SUWM solution [31,106].

Table 5. *Cont.*

Vectors	Insights and Gaps
	Several analyses are confronted with the need to consider social, environmental, and economic factors together, and not just a financial analysis that can often not be shown to be favourable in determining the solution whose positive externalities are too evident [76,80].
	The use of rainwater harvesting systems can also contribute to a decrease in costs with public drainage systems. A cost–benefit analysis that highlights the scale of these benefits is necessary [33].
Behavioural	Among the reasons for some resistance on the part of consumers to the use of recycled water are real or assumed health risks, mistrust of authorities responsible for managing water and minimising health risks, and disgust with the idea often referred to as the "yuck factor". There are even cases where, although it is proven that treated water is purer than bottled water or tap water, due to the "yuck factor" acceptance is nil by some consumers. There is also resistance on the part of consumers to have direct contact with reused water, especially if there is a perception of health risks [75].
	The level of resistance often has to do with the availability of water and its cost [66].
	In Singapore, a positive press speech and a well-founded sense of safety in the face of water reuse, given the use of state-of-the-art and redundant technology, were essential aspects for the community's good acceptance of reused water, called NEWater, which is used for indirect potable use, to be introduced into raw water reservoirs. The blended water undergoes naturalisation and further treatment in conventional waterworks to create drinking water. A similar situation occurs in Southern California [68,107]
Technological	Water treatment for reuse purposes converges on increasingly advanced technologies, including membrane bioreactors (MBRs) or MBRs combined with forward osmose (FO) towards greater energy efficiency, of which Singapore is the gold standard in integrating water reuse at the scale of a large city [68,74].
	Typically, the most economical reuse of grey water is that associated with showers and washbasins, and less than associated with kitchen stalls and washing machines, considering its higher content in fats and detergents. This knowledge can be necessary in the definition of internal drainage networks of buildings considering the reuse of grey waters [23,74].
	The main forms of water reuse occur through its direct reuse (after treatment), adequate discharge in water medium with characteristics adjusted to the receiving medium, or through the recharge of the water environment through which water will be obtained again for consumption. In addition to saving the water balance, they also make it possible to know the reused water available for future use (depending on consumption) and the possibility of using nutrients in agriculture/irrigation (depending on the type of treatment) [72,80].
	There is still ignorance about the effects and amount of micropollutants of direct recycled water in humans, so the tendency will still be to avoid its use for consumption or cooking [35].
	The reuse of rainwater is not (directly) economically viable on a small scale (isolated dwellings or small condominiums), as concluded from studies carried out in Spain and the Netherlands [33,73].
	Desalination is even more expensive than water reuse, and both are more expensive than direct capture in the water environment [35,68]. Even so, technological development, both in terms of performance and energy consumption, has been presenting solutions and systems that are very promising in the fields of desalination and water reuse [95–97,108].

Table 5. *Cont.*

Vectors	Insights and Gaps
	There is a panoply of technical solutions that contribute in a proven way. For flood damnation and uncontrolled runoff in cities, such as (bioretention) system sites, artificial wetland sites, infiltration-only systems, permeable pavements, green roofs, artificial ponds, bioswales). Some solutions are even quick-fix implementations contributing to a fast and economic control of the flow, such as draining pavements, as studied in the cases of Sponge Cities and Parma [22,44].
	There is a much greater reference to solutions for flood control than to the reuse of water or rainwater collection [30].
	There is a need to deepen knowledge of the relationship between urban planning, climate change, and water use using IT tools [99].
Urban planning	The instruments of urban planning and governance of cities are essential for implementing a strategy of greater water resilience, bringing both conceptually and physically closer to the various actors that can contribute to its realisation. Territory planning can even act as an instrument to facilitate the implementation of these measures [30,63].
	The articulation between urban planning and urban water management can be materialised, for example, by imposing water reuse measures in multi-purpose projects, careful location of treatment infrastructures, and reuse of "grey" waters; the location of "blue" and "green" infrastructures in a manner reconciled with the rainwater network and the conditioned approximation of industries with high consumption of service water to WWTP, among many others [32,63,74,83].
	Attention has been paid to the potential role of spatial planning in adapting to climate change in the urban water supply sector. The land use policy plays a vital role in influencing water use (demand) through planning mechanisms such as urban shape control, density, and space, as well as the recognised impact that urban development has on the water quality of the natural environment. It is not too much to stress the role that water demand should play in the planning of the territory to ensure sustainable water supply in the medium and long term [20,30,104,109,110]

The table above systematises the existing insights and gaps, constituting a basis for future integrated or vector-focused studies.

5. Conclusions

Through a literature review, this paper systematised the main concepts involving urban planning and the sustainable management of urban water (SUWM) in a context of demographic, urban, and climate change, as well as the way the scientific community interprets and tackles these challenges.

It noted an increasing concern for climate change in the context of the urban water cycle and urban management, mainly concerning flood control and not so much about cities' preparation for scarcity and water savings. Studies addressing the maximisation of water resources were also noticeable but fewer about control and management of demand. It is also perceived that the growth and adaptation of urban water systems cannot continue to be done incrementally, as it has been so far.

Knowledge deepening is required in the technical and economic evaluation of the overarching concept of SUWM systems in a way that integrates values beyond financial matters and introduces an accurate cost–benefit analysis of the solutions for society. New forms of growth, contemplating a hybridisation of systems (centralised systems that grow in a decentralised way), imply new paradigms of assessment, management, and collection of tariffs for which more consolidated knowledge is required.

Achieving synergies and economies of scale, in the panorama of cities, for systems of rainwater harvesting and water reuse are presented as themes in need of development, in particular in the way they can involve the planning of cities and their stakeholders, not only from a design perspective but also in its management, decision making, and in the preparation of the final consumer for the "new water" that can be used in a context increasingly focused on the circular economy.

The grey literature produced by international organisations has complemented the scientific literature by presenting frameworks for some of these measures that will allow the various stakeholders to consider infrastructure planning in the context of climate change according to risk.

There is also a clear need for further studies and practice on the relationship between the various actors, particularly those managing the territory and water services, towards a collaborative response to the challenges of climate change. Despite the evident constraints, yes, adapting together is possible and desirable. Further research is required, though, to clarify the design of the new institutional bridges, necessary steps, and means.

Author Contributions: Conceptualisation, V.V., T.F. and A.L.; methodology, V.V.; validation, V.V., T.F. and A.L.; formal analysis, V.V.; investigation, V.V.; writing—original draft preparation, V.V.; writing—review and editing, V.V., T.F. and A.L.; supervision, T.F. and A.L. All authors have read and agreed to the published version of the manuscript.

Funding: This research received no external funding.

Data Availability Statement: All the data is available within this manuscript.

Acknowledgments: The authors are grateful to Águas de Portugal Group and the University of Aveiro for supporting this research.

Conflicts of Interest: The authors declare no conflict of interest.

References

1. United Nations (UN). *Revision of World Urbanization Prospects*; United Nations Department of Economic and Social Affairs: New York, NY, USA, 2018.
2. Shukla, P.R.; Skea, J.; Slade, R.; Al Khourdajie, A.; van Diemen, R.; McCollum, D.; Pathak, M.; Some, S.; Vyas, P.; Fradera, R.; et al. IPCC, 2022: Annex V: Expert Reviewers of the IPCC Working Group III Sixth Assessment Report. In *Climate Change 2022: Mitigation of Climate Change*; Contribution of Working Group III to the Sixth Assessment Report of the Intergovernmental Panel on Climate Change; Cambridge University Press: Cambridge, UK; New York, NY, USA, 2022; Available online: www.ipcc.ch (accessed on 21 November 2022).
3. Pörtner, H.-O.; Roberts, D.C.; Poloczanska, E.S.; Mintenbeck, K.; Tignor, M.; Alegría, A.; Craig, M.; Langsdorf, S.; Löschke, S.; Möller, V.; et al. IPCC, 2022: Summary for Policymakers. In *Climate Change 2022: Impacts, Adaptation and Vulnerability*; Contribution of Working Group II to the Sixth Assessment Report of the Intergovernmental Panel on Climate Change; Cambridge University Press: Cambridge, UK; New York, NY, USA, 2022. [CrossRef]
4. Godfray, H.C.J.; Beddington, J.R.; Crute, I.R.; Haddad, L.; Lawrence, D.; Muir, J.F.; Pretty, J.; Robinson, S.; Thomas, S.M.; Toulmin, C. Food security: The challenge of feeding 9 billion people. *Science* **2010**, *327*, 812–818. [CrossRef]
5. Seelen, L.M.S.; Flaim, G.; Jennings, E.; de Senerpont Domis, L.N. Saving water for the future: Public awareness of water usage and water quality. *J. Environ. Manag.* **2019**, *242*, 246–257. [CrossRef] [PubMed]
6. van Leeuwen, C.J.; Frijns, J.; van Wezel, A.; van de Ven, F.H.M. City Blueprints: 24 Indicators to Assess the Sustainability of the Urban Water Cycle. *Water Resour. Manag.* **2012**, *26*, 2177–2197. [CrossRef]
7. Kumar, P. Hydrocomplexity: Addressing water security and emergent environmental risks. *Water Resour. Res.* **2015**, *51*, 5827–5838. [CrossRef]
8. He, C.; Liu, Z.; Wu, J.; Pan, X.; Fang, Z.; Li, J.; Bryan, B.A. Future global urban water scarcity and potential solutions. *Nat. Commun.* **2021**, *12*, 4667. [CrossRef] [PubMed]
9. Tortajada, C.; Biswas, A.K. Water management in post-2020 world. *Int. J. Water Resour. Dev.* **2020**, *36*, 874–878. [CrossRef]
10. Larsen, T.A.; Hoffmann, S.; Lüthi, C.; Truffer, B.; Maurer, M. Emerging solutions to the water challenges of an urbanising world. *Science* **2016**, *352*, 928–933. [CrossRef] [PubMed]
11. Hartmann, T. Clumsy Floodplains. In *Clumsy Floodplains*; Routledge: London, UK, 2016. [CrossRef]
12. Zhou, Q. A Review of Sustainable Urban Drainage Systems Considering the Climate Change and Urbanization Impacts. *Water* **2014**, *6*, 976–992. [CrossRef]

13. Mikovits, C.; Rauch, W.; Kleidorfer, M. Dynamics in Urban Development, Population Growth and their Influences on Urban Water Infrastructure. *Procedia Eng.* **2014**, *70*, 1147–1156. [CrossRef]
14. Bohman, A.; Glaas, E.; Karlson, M. Integrating Sustainable Stormwater Management in Urban Planning: Ways Forward towards Institutional Change and Collaborative Action. *Water* **2020**, *12*, 203. [CrossRef]
15. Fidélis, T.; Cardoso, A.S.; Riazi, F.; Miranda, A.C.; Abrantes, J.; Teles, F.; Roebeling, P.C. Policy narratives of circular economy in the EU—Assessing the embeddedness of water and land in national action plans. *J. Clean. Prod.* **2021**, *288*, 125685. [CrossRef]
16. Grant, S.B.; Saphores, J.-D.; Feldman, D.L.; Hamilton, A.J.; Fletcher, T.D.; Cook, P.L.; Stewardson, M.; Sanders, B.F.; Levin, L.A.; Ambrose, R.F.; et al. Taking the 'waste' out of 'wastewater' for human water security and ecosystem sustainability. *Science* **2012**, *337*, 681–686. [CrossRef] [PubMed]
17. Opher, T.; Friedler, E. Comparative LCA of decentralised wastewater treatment alternatives for non-potable urban reuse. *J. Environ. Manag.* **2016**, *182*, 464–476. [CrossRef]
18. Treuer, G.; Koebele, E.; Deslatte, A.; Ernst, K.; Garcia, M.; Manago, K. A narrative method for analysing transitions in urban water management: The case of the Miami-Dade Water and Sewer Department. *Water Resour. Res.* **2017**, *53*, 891–908. [CrossRef]
19. Li, E.; Endter-Wada, J.; Li, S. Characterising and Contextualising the Water Challenges of Megacities. *J. Am. Water Resour. Assoc.* **2015**, *51*, 589–613. [CrossRef]
20. Gober, P.; Larson, K.; Quay, R.; Polsky, C.; Chang, H.; Shandas, V. Why Land Planners and Water Managers Don't Talk to One Another and Why They Should! *Soc. Nat. Resour.* **2013**, *26*, 356–364. [CrossRef]
21. Domene, E.; Saurí, D. Urbanisation and water consumption: Influencing factors in the metropolitan region of Barcelona. *Urban Stud.* **2006**, *43*, 1605–1623. [CrossRef]
22. Chan, F.K.S.; Griffiths, J.A.; Higgitt, D.; Xu, S.; Zhu, F.; Tang, Y.-T.; Xu, Y.; Thorne, C.R. 'Sponge City' in China—A breakthrough of planning and flood risk management in the urban context. *Land Use Policy* **2018**, *76*, 772–778. [CrossRef]
23. Dolman, N.; Savage, A.; Ogunyoye, F. Water-sensitive urban design: Learning from experience. In Proceedings of the Institution of Civil Engineers: Municipal Engineer; Thomas Telford Ltd: London, UK, 2013; Volume 166, No. 2. pp. 86–97. [CrossRef]
24. Ferguson, B.C.; Brown, R.R.; Deletic, A. Diagnosing transformative change in urban water systems: Theories and frameworks. *Glob. Environ. Change* **2013**, *23*, 264–280. [CrossRef]
25. Ferguson, B.C.; Frantzeskaki, N.; Brown, R.R. A strategic program for transitioning to a Water Sensitive City. *Landsc. Urban Plan.* **2013**, *117*, 32–45. [CrossRef]
26. Ray, P.A.; Brown, C.M. *Confronting Climate Uncertainty in Water Resources Planning and Project Design the Decision Tree Framework*; World Bank Publications: Washington, DC, USA, 2015. [CrossRef]
27. Luís, A.; Garnett, K.; Pollard, S.; Lickorish, F.; Jude, S.; Leinster, P. Fusing strategic risk and futures methods to inform long-term strategic planning: Case of water utilities. *Environ. Syst. Decis.* **2021**, *41*, 523–540. [CrossRef]
28. Mendoza, G.; Jeuken, A.; Matthews, J.H.; Stakhiv, E.; Kucharski, J.; Gilroy, K.; Unesco; International Center for Integrated Water Resources Management; International Hydrological Programme. *Climate Risk Informed Decision Analysis (CRIDA): Collaborative Water Resources Planning for an Uncertain Future*; UNESCO Publishing: Paris, France, 2019.
29. Azhoni, A.; Jude, S.; Holman, I. Adapting to climate change by water management organisations: Enablers and barriers. *J. Hydrol.* **2018**, *559*, 736–748. [CrossRef]
30. Hurlimann, A.; Wilson, E. Sustainable Urban Water Management under a Changing Climate: The Role of Spatial Planning. *Water* **2018**, *10*, 546. [CrossRef]
31. Marlow, D.R.; Moglia, M.; Cook, S.; Beale, D.J. Towards sustainable urban water management: A critical reassessment. *Water Res.* **2013**, *47*, 7150–7161. [CrossRef]
32. van Duuren, D.; van Alphen, H.J.; Koop, S.H.A.; de Bruin, E. Potential transformative changes in water provision systems: Impact of decentralised water systems on centralised water supply regime. *Water* **2019**, *11*, 1709. [CrossRef]
33. Hofman, J.; Paalman, M. *Rainwater Harvesting, a Sustainable Solution for Urban Climate Adaptation?* KWR Watercycle Research Institute: Nieuwegein, The Netherland, 2014.
34. Campisano, A.; Butler, D.; Ward, S.; Burns, M.J.; Friedler, E.; DeBusk, K.; Fisher-Jeffes, L.N.; Ghisi, E.; Rahman, A.; Furumai, H. Urban rainwater harvesting systems: Research, implementation and future perspectives. *Water Res.* **2017**, *115*, 195–209. [CrossRef]
35. Rygaard, M.; Binning, P.J.; Albrechtsen, H.J. Increasing urban water self-sufficiency: New era, new challenges. *J. Environ. Manag.* **2011**, *92*, 185–194. [CrossRef] [PubMed]
36. Demuzere, M.; Orru, K.; Heidrich, O.; Olazabal, E.; Geneletti, D.; Orru, H.; Bhave, A.G.; Mittal, N.; Feliú, E.; Faehnle, M. Mitigating and adapting to climate change: Multi-functional and multi-scale assessment of green urban infrastructure. *J. Environ. Manag.* **2014**, *146*, 107–115. [CrossRef] [PubMed]
37. Jiang, Y.; Zevenbergen, C.; Ma, Y. Urban pluvial flooding and stormwater management: A contemporary review of China's challenges and 'sponge cities' strategy. *Environ. Sci. Policy* **2018**, *80*, 132–143. [CrossRef]
38. Puchol-Salort, P.; Boskovic, S.; Dobson, B.; van Reeuwijk, M.; Mijic, A. Water neutrality framework for systemic design of new urban developments. *Water Res.* **2022**, *219*, 118583. [CrossRef]
39. Miranda, A.C.; Fidélis, T.; Roebeling, P.; Meireles, I. Assessing the Inclusion of Water Circularity Principles in Environment-Related City Concepts Using a Bibliometric Analysis. *Water* **2022**, *14*, 1703. [CrossRef]
40. Barlow, D.; Burrill, G.; Nolfi, J. Research Report on Developing a Community Level Natural Resource Inventory System. Center for Studies in Food Self-Sufficiency, 1977. Available online: http://vtpeakoil.net/docs/NR_inventory.pdf (accessed on 21 November 2022).

41. Fletcher, T.D.; Shuster, W.; Hunt, W.F.; Ashley, R.; Butler, D.; Arthur, S.; Trowsdale, S.; Barraud, S.; Semadeni-Davies, A.; Bertrand-Krajewski, J.-L.; et al. SUDS, LID, BMPs, WSUD and more—The evolution and application of terminology surrounding urban drainage. *Urban Water J.* **2015**, *12*, 525–542. [CrossRef]
42. Huang, J.J.; Li, Y.; Niu, S.; Zhou, S.H. Assessing the performances of low impact development alternatives by long-term simulation for a semi-arid area in Tianjin, northern China. *Water Sci. Technol.* **2014**, *70*, 1740–1745. [CrossRef]
43. Geldof, G.D. Adaptive water management: Integrated water management on the edge of chaos. *Water Sci. Technol.* **1995**, *32*, 7–13. [CrossRef]
44. Sharma, A.K.; Pezzaniti, D.; Myers, B.; Cook, S.; Tjandraatmadja, G.; Chacko, P.; Chavoshi, S.; Kemp, D.; Leonard, R.; Koth, B.; et al. Water Sensitive Urban Design: An Investigation of Current Systems, Implementation Drivers, Community Perceptions and Potential to Supplement Urban Water Services. *Water* **2016**, *8*, 272. [CrossRef]
45. Bahri, A. Integrated Urban Water Management. 2012. Available online: www.gwptoolbox.org (accessed on 21 November 2022).
46. Mouritz, M.J. Sustainable Urban Water Systems: Policy and Professional Praxis. PhD Thesis, Murdoch University, Perth, Australia, 1996. Available online: https://researchrepository.murdoch.edu.au/view/author/Mouritz,Mike.html (accessed on 25 November 2022).
47. Lloyd, S.D.; Wong, T.H.F.; Chesterfield, C.J. Water Sensitive Urban Design—A Stormwater Management Perspective. Melbourne, September 2002. Available online: https://www.researchgate.net/publication/260400236_Water_Sensitive_Urban_Design_-_A_Stormwater_Management_Perspective (accessed on 20 November 2022).
48. Construction Industry Research and Information Association; Sustainable Urban Drainage Scottish Working Party. *Sustainable Urban Drainage Systems: Design Manual for Scotland and Northern Ireland*; Construction Industry Research and Information Association: London, UK, 2000; p. 114.
49. Makropoulos, C.K.; Memon, F.; Shirley-Smith, C.; Butler, D. Futures: An exploration of scenarios for sustainable urban water management. *Water Policy* **2008**, *10*, 345–373. [CrossRef]
50. Novotny, V. Sustainable urban water management. In *Water and Urban Development Paradigms: Towards an Integration of Engineering, Design and Management Approaches, Proceedings of the International Urban Water Conference, Heverlee, Belgium, 15–19 September 2009*; CRC Press: Boca Raton, FL, USA, 2008; pp. 19–31. [CrossRef]
51. Barron, N.J.; Kuller, M.; Yasmin, T.; Castonguay, A.C.; Copa, V.; Duncan-Horner, E.; Gimelli, F.M.; Jamali, B.; Nielsen, J.S.; Ng, K.; et al. Towards water sensitive cities in Asia: An interdisciplinary journey. *Water Sci. Technol.* **2017**, *76*, 1150–1157. [CrossRef]
52. Ministry of Housing and Urban-Rural Development. *The Construction Guideline of Sponge City in China—Low Impact Development of Stormwater System (Trail)*; Ministry of Housing and Urban-Rural Development: Beijing, China, 2014.
53. Liu, H.; Jia, Y.; Niu, C. 'Sponge city' concept helps solve China's urban water problems. *Environ. Earth Sci.* **2017**, *76*, 473. [CrossRef]
54. Delgado, A.; Rodriguez, D.; Amadei, C.; Makino, M. Water in Circular Economyand Resilience (Wicer). 2021. Available online: www.worldbank.org (accessed on 21 November 2022).
55. IWA (International Water Association). 2nd Edition for Urban Stakeholders to Develop a Shared Vision and Act towards Sustainable Urban Water in Resilient and Liveable Cities Principles Water Wise Cities for. 2016. Available online: https://iwa-network.org/wp-content/uploads/2016/10/IWA_Brochure_Water_Wise_Communities_SCREEN-1.pdf (accessed on 21 November 2022).
56. Geels, F.W. The dynamics of transitions in socio-technical systems: A multi-level analysis of the transition pathway from horse-drawn carriages to automobiles (1860–1930). *Technol. Anal. Strateg. Manag.* **2005**, *17*, 445–476. [CrossRef]
57. Domènech, L. Rethinking water management: From centralised to decentralised water supply and sanitation models. *Doc. Anal. Geogr.* **2011**, *57*, 293–310. [CrossRef]
58. Cashman, A.; Ashley, R. Costing the long-term demand for water sector infrastructure. *Foresight* **2008**, *10*, 9–26. [CrossRef]
59. Marshall, G.R.; Alexandra, J. Institutional path dependence and environmental water recovery in Australia's Murray-Darling Basin. *Water Altern.* **2016**, *9*, 679–703.
60. Lehmkuhl, F.; Schüttrumpf, H.; Schwarzbauer, J.; Brüll, C.; Dietze, M.; Letmathe, P.; Völker, C.; Hollert, H. Assessment of the 2021 summer flood in Central Europe. *Environ. Sci. Eur.* **2022**, *34*, 107. [CrossRef]
61. Delgado-Ramos, G.C. Water and the political ecology of urban metabolism: The case of Mexico City. *J. Polit. Ecol.* **2015**, *22*, 98–114. [CrossRef]
62. Suleiman, L.; Olofsson, B.; Saurí, D.; Palau-Rof, L. A breakthrough in urban rain-harvesting schemes through planning for urban greening: Case studies from Stockholm and Barcelona. *Urban Urban Green* **2020**, *51*, 126678. [CrossRef]
63. Leigh, N.G.; Lee, H. Sustainable and resilient urban water systems: The role of decentralisation and planning. *Sustainability* **2019**, *11*, 918. [CrossRef]
64. Koop, S.H.A.; van Leeuwen, C.J. The challenges of water, waste and climate change in cities. *Environ. Dev. Sustain.* **2017**, *19*, 385–418. [CrossRef]
65. Reznik, A.; Feinerman, E.; Finkelshtain, I.; Fisher, F.; Huber-Lee, A.; Joyce, B.; Kan, I. Economic implications of agricultural reuse of treated wastewater in Israel: A statewide long-term perspective. *Ecological. Econ.* **2017**, *135*, 222–233. [CrossRef]
66. Craddock, H.A.; Rjoub, Y.; Jones, K.; Lipchin, C.; Sapkota, A.R. Perceptions on the use of recycled water for produce irrigation and household tasks: A comparison between Israeli and Palestinian consumers. *J. Environ. Manag.* **2021**, *297*, 113234. [CrossRef]
67. Hamin, E.M.; Gurran, N. Urban form and climate change: Balancing adaptation and mitigation in the U.S. and Australia. *Habitat. Int.* **2009**, *33*, 238–245. [CrossRef]

68. Lefebvre, O. Beyond NEWater: An insight into Singapore's water reuse prospects. *Curr. Opin. Environ. Sci. Health* **2018**, *2*, 26–31. [CrossRef]
69. UNESCO. World Water Assessment Programme Water for People Water for Life the United Nations World Water Development Report Executive Summary. 2003. Available online: https://unesdoc.unesco.org/ark:/48223/pf0000129556.locale=en (accessed on 31 October 2022).
70. Maurer, M.; Scheidegger, A.; Herlyn, A. Quantifying costs and lengths of urban drainage systems with a simple static sewer infrastructure model. *Urban Water J.* **2013**, *10*, 268–280. [CrossRef]
71. Schuetze, T.; Chelleri, L. Integrating Decentralised Rainwater Management in Urban Planning and Design: Flood Resilient and Sustainable Water Management Using the Example of Coastal Cities in The Netherlands and Taiwan. *Water* **2013**, *5*, 593–616. [CrossRef]
72. Ranatunga, T.; Tong, S.T.Y.; Sun, Y.; Yang, Y.J. A total water management analysis of the Las Vegas Wash watershed, Nevada. *Phys. Geogr.* **2014**, *35*, 220–244. [CrossRef]
73. Angrill, S.; Segura-Castillo, L.; Petit-Boix, A.; Rieradevall, J.; Gabarrell, X.; Josa, A. Environmental performance of rainwater harvesting strategies in Mediterranean buildings. *Int. J. Life Cycle Assess* **2017**, *22*, 398–409. [CrossRef]
74. Li, F.; Wichmann, K.; Otterpohl, R. Review of the technological approaches for grey water treatment and reuses. *Sci. Total Environ.* **2009**, *407*, 3439–3449. [CrossRef] [PubMed]
75. Portman, M.E.; Vdov, O.; Schuetze, M.; Gilboa, Y.; Friedler, E. Public perceptions and perspectives on alternative sources of water for reuse generated at the household level. *Water Reuse* **2022**, *12*, 157–174. [CrossRef]
76. Moglia, M.; Cook, S. Transformative approaches for sustainable water management in the urban century. *Water* **2019**, *11*, 1106. [CrossRef]
77. Rozin, P.; Haddad, B.M.; Slovic, P.; Haddad, B.; Nemeroff, C. Psychological Aspects of the Rejection of Recycled Water: Contamination, Purification and Disgust. 2015. Available online: https://www.researchgate.net/publication/279316792 (accessed on 25 November 2022).
78. Fisher-Jeffes, L.; Carden, K.; Armitage, N.P.; Winter, K. Stormwater harvesting: Improving water security in South Africa's urban areas. *S. Afr. J. Sci.* **2017**, *113*, 4. [CrossRef] [PubMed]
79. González-García, A.; Palomo, I.; González, J.A.; López, C.A.; Montes, C. Quantifying spatial supply-demand mismatches in ecosystem services provides insights for land-use planning. *Land Use Policy* **2020**, *94*, 104493. [CrossRef]
80. Garcia, X.; Pargament, D. Reusing wastewater to cope with water scarcity: Economic, social and environmental considerations for decision-making. *Resour. Conserv. Recycl.* **2015**, *101*, 154–166. [CrossRef]
81. West, C.; Kenway, S.; Hassall, M.; Yuan, Z. Expert opinion on risks to the long-term viability of residential recycled water schemes: An Australian study. *Water Res.* **2017**, *120*, 133–145. [CrossRef] [PubMed]
82. Asano, T. Urban water recycling. *Water Sci. Technol.* **2005**, *51*, 83–89. [CrossRef] [PubMed]
83. Salinas Rodriguez, C.N.A.; Ashley, R.; Gersonius, B.; Rijke, J.; Pathirana, A.; Zevenbergen, C. Incorporation and application of resilience in the context of water-sensitive urban design: Linking European and Australian perspectives. *Wiley Interdiscip. Rev. Water* **2014**, *1*, 173–186. [CrossRef]
84. Dada, A.; Urich, C.; Berteni, F.; Pezzagno, M.; Piro, P.; Grossi, G. Water Sensitive Cities: An Integrated Approach to Enhance Urban Flood Resilience in Parma (Northern Italy). *Climate* **2021**, *9*, 152. [CrossRef]
85. Xiang, X.; Li, Q.; Khan, S.; Khalaf, O.I. Urban water resource management for sustainable environment planning using artificial intelligence techniques. *Environ. Impact Assess. Rev.* **2021**, *86*, 106515. [CrossRef]
86. Li, Q.; Wang, W.; Jiang, X.; Lu, D.; Zhang, Y.; Li, J. Analysis of the potential of reclaimed water utilisation in typical inland cities in northwest China via system dynamics. *J. Environ. Manag.* **2020**, *270*, 110878. [CrossRef] [PubMed]
87. Yang, J.; Wang, Z.H. Planning for a sustainable desert city: The potential water buffering capacity of urban green infrastructure. *Landsc Urban Plan.* **2017**, *167*, 339–347. [CrossRef]
88. Zhang, Q.; Liu, S.; Wang, T.; Dai, X.; Baninla, Y.; Nakatani, J.; Moriguchi, Y. Urbanisation impacts on greenhouse gas (GHG) emissions of the water infrastructure in China: Trade-offs among sustainable development goals (SDGs). *J. Clean. Prod.* **2019**, *232*, 474–486. [CrossRef]
89. de Groot, B.; Leendertse, W.; Arts, J. Co-Evolution of Organizations in Infrastructure Planning: The Role of Communities of Practice as Windows for Collective Learning Across Project-Oriented Organizations. *Adm. Soc.* **2022**, *54*, 1328–1356. [CrossRef]
90. Bloemen, P.; van der Steen, M.; van der Wal, Z. Designing a century ahead: Climate change adaptation in the Dutch Delta. *Policy Soc.* **2019**, *38*, 58–76. [CrossRef]
91. Innes, J.E.; Booher, D.E.; di Vittorio, S. Strategies for Megaregion Governance. *Taylor Fr. J.* **2010**, *77*, 55–67. [CrossRef]
92. Huntjens, P.; Lebel, L.; Pahl-Wostl, C.; Camkin, J.; Schulze, R.; Kranz, N. Institutional design propositions for the governance of adaptation to climate change in the water sector. *Glob. Environ. Change* **2012**, *22*, 67–81. [CrossRef]
93. Wissink, B. Towards sustainable urban and regional planning: Experiences from the Netherlands. *Res. Urban Sociol.* **2014**, *14*, 59–79. [CrossRef]
94. Voskamp, I.M.; van de Ven, F.H.M. Planning support system for climate adaptation: Composing effective sets of blue-green measures to reduce urban vulnerability to extreme weather events. *Build Environ.* **2015**, *83*, 159–167. [CrossRef]
95. Li, Z.; Zhang, P.; Guan, K.; Yoshioka, T.; Matsuyama, H. Water flux enhancement of PVDF membrane by a facile coating method for vacuum membrane distillation. *Desalination* **2022**, *536*, 115818. [CrossRef]

96. Li, Z.; Das, S.; Sekine, T.; Mabuchi, H.; Kaneko, R.; Sakai, J.; Irie, T.; Kamio, E.; Yoshioka, T.; Suo, J.; et al. Control over the Hydrophilicity in the Pores of Covalent Organic Framework Membranes for High-Flux Separation of Dyes from Water. *ACS Appl. Nano Mater.* **2022**, *5*, 17632–17639. [CrossRef]
97. Jia, Y.; Guan, K.; Zhang, P.; Shen, Q.; Li, Z.; Istirokhatun, T.; Matsuyama, H. Asymmetric superwetting Janus structure for fouling- and scaling-resistant membrane distillation. *J. Memb. Sci.* **2022**, *657*, 120697. [CrossRef]
98. van den Bosch, M.; Sang, Å.O. Urban natural environments as nature-based solutions for improved public health—A systematic review of reviews. *Environ. Res.* **2017**, *158*, 373–384. [CrossRef]
99. Willuweit, L.; O'Sullivan, J.J. A decision support tool for sustainable planning of urban water systems: Presenting the Dynamic Urban Water Simulation Model. *Water Res.* **2013**, *47*, 7206–7220. [CrossRef] [PubMed]
100. Stead, D. Urban planning, water management and climate change strategies: Adaptation, mitigation and resilience narratives in the Netherlands. *Int. J. Sustain. Dev. World Ecol.* **2014**, *21*, 15–27. [CrossRef]
101. Friedler, E.; Hadari, M. Economic feasibility of on-site greywater reuse in multi-storey buildings. *Desalination* **2006**, *190*, 221–234. [CrossRef]
102. Larsen, T.A.; Udert, K.M.; Lienert, J. *Source Separation and Decentralization for Wastewater Management*; IWA Publishing: London, UK, 2013; Available online: www.iwappublishing.com (accessed on 25 November 2022).
103. Tompkins, E.L.; Adger, W.N.; Boyd, E.; Nicholson-Cole, S.; Weatherhead, K.; Arnell, N. Observed adaptation to climate change: UK evidence of transition to a well-adapting society. *Glob. Environ. Change* **2010**, *20*, 627–635. [CrossRef]
104. Trapp, J.H.; Kerber, H.; Schramm, E. Implementation and diffusion of innovative water infrastructures: Obstacles, stakeholder networks and strategic opportunities for utilities. *Environ. Earth Sci.* **2017**, *76*, 154. [CrossRef]
105. Bruaset, S.; Sægrov, S. Using the multiple scenario approach for envisioning plausible futures in long-term planning and management of the urban water pipe systems. *Eur. J. Futures Res.* **2018**, *6*, 7. [CrossRef]
106. Daniels, P.; Porter, M.; Bodsworth, P. Externalities in Sustainable Regional Water Strategies: Application of a Simple Methodology. 2012. Available online: http://www.griffith.edu.au/ (accessed on 27 October 2022).
107. Hoo, R. Managing water demand in Singapore through a systems perspective. *Int. J. Water Resour. Dev.* **2019**, *36*, 879–887. [CrossRef]
108. Quon, H.; Sperling, J.; Coughline, K.; Greene, D.; Miara, A.; Akar, S.; Talmadge, M.; Stokes-Draut, J.R.; Macknick, J.; Jiang, S. Pipe Parity Analysis of Seawater Desalination in the United States: Exploring Costs, Energy, and Reliability via Case Studies and Scenarios of Emerging Technology. *ACS ES T Eng.* **2002**, *2*, 434–445. [CrossRef]
109. Gober, P.; Kirkwood, C.W.; Balling, R.C.; Ellis, A.W.; Deitrick, S. Water Planning Under Climatic Uncertainty in Phoenix: Why We Need a New Paradigm. *Assoc. Am. Geogr.* **2010**, *100*, 356–372. [CrossRef]
110. Yong, S.T.Y.; Chen, W. Modeling the relationship between land use and surface water quality. *J. Environ. Manag.* **2002**, *66*, 377–393. [CrossRef]

water

MDPI

Article

Shallow Groundwater Responses to Rainfall Based on Correlation and Spectral Analyses in the Heilonggang Region, China

Chaoyue Wang [1,2,3], Fenggang Dai [1,2,3], Yang Liu [4], Yunmeng Wang [1,2,3], Hui Li [5,6] and Wenjing Qu [1,2,3,*]

1 Hebei Province Collaborative Innovation Center for Sustainable Utilization of Water Resources and Optimization of Industrial Structure, School of Water Resources and Environment, Hebei GEO University, Shijiazhuang 050031, China; wangchaoyue08@163.com (C.W.); daifenggang@126.com (F.D.); wangym2105@163.com (Y.W.)
2 Hebei Province Key Laboratory of Sustained Utilization and Development of Water Resources, School of Water Resources and Environment, Hebei GEO University, Shijiazhuang 050031, China
3 Hebei Center for Ecological and Environmental Geology Research, School of Water Resources and Environment, Hebei GEO University, Shijiazhuang 050031, China
4 Hebei Geo-Environment Monitoring, Shijiazhuang 050000, China; liuyhky@163.com
5 The Sixth Geological Brigade of Geological Mineral Exploration and Development Bureau of Hebei Province, Shijiazhuang 050085, China; 13731140891@163.com
6 Aerial Measurement Technology Application Research Center under the Geologial Mineral Exploration and Development Bureau of Hebei Province, Shijiazhuang 050085, China
* Correspondence: quwenjing90@163.com

Abstract: To clarify the long-term dynamics of groundwater and its response to precipitation in the Heilonggang region, China, variation trends of the monitored groundwater were studied according to the Mann–Kendall statistics. Based on observations from four monitoring wells, the persistence and periodicities of the loose porous aquifers, and the interrelation between precipitation and groundwater levels was examined based on a number of tools including autocorrelation, cross-correlation, spectral analysis, and wavelet transform. The results show that the groundwater levels at W1 and W2 present a downward trend while those at W3 and W4 show an upward trend. The dominant time period increases from 2.1 years (upstream) to 3.7 years (downstream). The time lags between aquifers and rainfall at W1, W2, W3 and W4 are 139.14 ± 59.76 days (2008–2020), 23.27 ± 12.03 days (2005–2014), 145.01 ± 68.00 days (2007–2020), and 59.22 ± 26.14 days (2005–2019), respectively. The lags for the pumping years are 1.2~2.1 times of those during the years without pumping. The recharge ratio overestimated by the water table fluctuation method decreases from 0.32 at W2 to 0.17 at W4, suggesting that the site W2 has a good potentiality for groundwater recharge. This research helps us better understand the local groundwater circulation and provide references for groundwater management.

Keywords: groundwater; precipitation; correlation; spectral analyses; wavelet transform

Citation: Wang, C.; Dai, F.; Liu, Y.; Wang, Y.; Li, H.; Qu, W. Shallow Groundwater Responses to Rainfall Based on Correlation and Spectral Analyses in the Heilonggang Region, China. *Water* **2023**, *15*, 1100. https://doi.org/10.3390/w15061100

Academic Editors: Alban Kuriqi and Luis Garrote

Received: 14 February 2023
Revised: 7 March 2023
Accepted: 9 March 2023
Published: 13 March 2023

1. Introduction

With the increase in extreme weather events induced by climate change, understanding the response mechanism of groundwater systems to precipitation is of critical importance [1–4]. To date, two valuable correlation methods including the auto-correlation and cross-correlation function have been typically applied to assess the responses of aquifers to precipitation [5–7]. The auto-correlation analysis characterizes the degree of "inertia" [8,9], or "persistence" [3], or "memory effect" [10] of an individual time series. The cross-correlation analysis can provide the response time between rainfall and groundwater levels (GWLs). In addition, the correlation methods are usually combined with spectral analysis, such as Fourier analysis, to detect the periodicity of a signal.

Jenkins and Watts [11] and Box and Jenkins [12] explained the basic principles of correlation and spectral analyses. Mangin [13] applied these methods to three karstic systems in the Pyrenean Mountains (France–Spain), and their work indicated that these analyses provide an excellent method for the investigation of a hydrological system. After that, correlation and spectral analyses have been widely used in different aquifer systems, including karstic systems [14–16], alluvial aquifers [3,5], and coastal aquifers [17,18], etc.

Although correlation and spectral analysis have been widely used in hydrology and hydrogeology, these methods cannot describe how the frequency of a signal changes with time. Instead, the wavelet transform method is an effective tool for detecting the periodicity of a nonlinear system and can provide localized intermittent periodicities [19–21]. Based on the cross wavelet and wavelet coherence methods, Yu and Lin [22] found that the temporal lag from rainfall to groundwater was about 3.71–72.07 days for the Pingtung Plain in Taiwan. Zhang et al. [23] found that the time lag of groundwater depth to precipitation in the Yellow River Delta during 2006–2010 ranged from 35. 51 to 178. 36 days, and the relationship between groundwater depth and precipitation is largely affected by land use types, soil texture, and micro-geomorphic types. Cai et al. [24] show that the response time of groundwater levels to rainfall during 2006–2018 extended from 80 to ~190 days in Puyang, Henan, China, and that it increases with the burial depth of groundwater. Although there are many studies investigating the relationship between rainfall and aquifers, seldom does research focus on areas facing severe water shortage such as the Heilonggang region, China.

The Heilonggang region is located in the southeast of Hebei Province (Figure 1), a part of the North China Plain, and serves as an important agricultural planting area for the nation. Agricultural water accounts for 76% of water resource consumption in this area, and more than 80% of agricultural irrigation water comes from groundwater [25]. The increasing demand for groundwater resources has led to the decline in GWLs and caused a series of ecological problems such as land subsidence, etc. Previous studies of this area mainly focus on the optimization of irrigation and planting regimes [26–28], and few researchers have paid attention to the long-term groundwater dynamics and associated responses to rainfall in this region. Clarifying the relationship between groundwater and precipitation can help us better understand the local groundwater circulation, provide references for groundwater resource management, and be conducive to ecological protection [29,30]. Therefore, in this study, we aim to (1) estimate the persistence of aquifers and their responses to rainfall through the auto-correlation and cross-correlation functions, (2) identify groundwater periodicity through spectral analyses, (3) determine the groundwater response time through cross-correlation and cross spectral analyses, and (4) explore the influences of rainfall intensity, humidity index, and groundwater pumping on the response time.

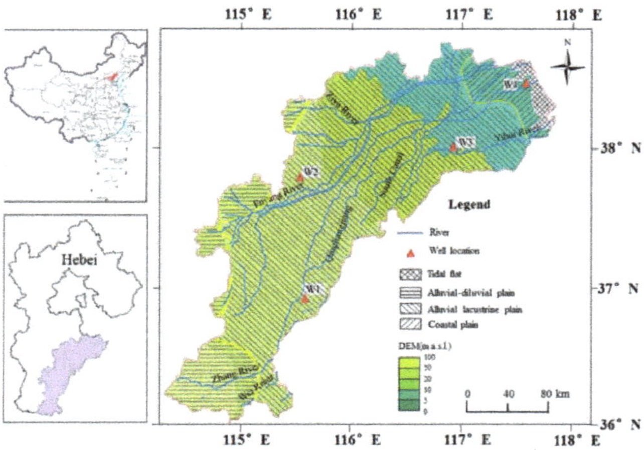

Figure 1. Location of the study area.

2. Materials and Methods

2.1. Study Site

The Heilonggang region covers an area of 34,700 km^2. The terrain in this region is gentle and inclines slightly from southwest to northeast with a topographic slope of 0.2~0.1‰. From west to east, the geomorphic type is dominated by the mountain alluvial-diluvial plain, the central alluvial-lacustrine plain and the coastal plain, respectively. Accordingly, from west to east, sediments change from gravel in front of the mountain, to medium coarse and medium fine sand in the middle, then to fine sand in the coastal area.

The study site is mainly affected by the warm temperate semi-arid and semi-humid continental monsoon with four distinct seasons. The annual average precipitation is 500~600 mm, mostly concentrated in July and August. The river system in the Heilonggang area comprises the Zhangweinan Canal system, the Heilonggang Canal system, and the Ziya River system. All rivers finally drain into the Bohai Sea (Figure 1). Reservoirs have been built at the upper reaches of almost all rivers, and more than 80% of the surface runoff is detained by the reservoirs, resulting in long-term drying of the rivers flowing through the area and greatly reducing the amount of river leakage.

Aquifers in the Heilonggang area are mainly composed of Quaternary strata. The Quaternary aquifer system can be divided into four aquifer groups from top to bottom (Figure S1). Among them, the first aquifer group is the aquifer we focused on in this study. It is composed of Holocene and Upper Pleistocene loose sediments, with a bottom depth of 20~50 m [31]. The vadose zone lithology in this area is mainly composed of silt and silty clay (Figure 2).

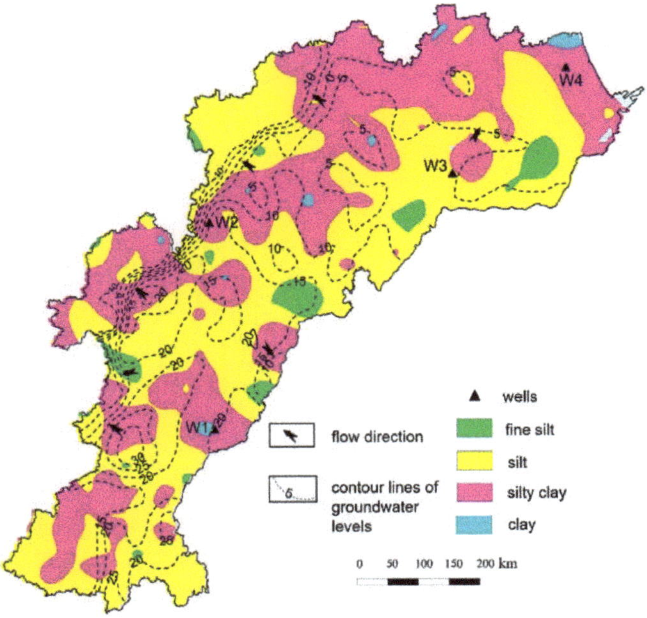

Figure 2. Vadose zone lithology and the flow direction map (2020).

In the 1950s, the degree of groundwater exploitation in this area was very low, and the whole groundwater flow system was in a natural state. Shallow groundwater generally flows from southwest to northeast (Figure S2a). Under the influence of human activities, local groundwater depression cones have formed in the north part of the region (Figure S2b). At present, the groundwater depression cones have further expanded towards the west (Figure 2), and the southwest–northeast hydraulic gradient has decreased greatly compared to the natural state.

Shallow fresh groundwater mainly occurs in the west and north of the research area, and shallow brackish water mainly occurs in the east of the region. The distribution area of the shallow fresh groundwater accounts for 51.6% of the total Heilonggang area [32]. Rainfall serves as the main source for the river and shallow groundwater.

2.2. Datasets

The groundwater data were obtained from the "Groundwater Almanacs of Geological Environment Monitoring in China". Four boreholes monitoring a substantial length of shallow groundwater were selected in this study, as shown in Figure 1. Detail characteristics of each monitoring well are presented in Table 1. The rainfall data before 2018 was extracted daily from the "China meteorological forcing dataset (1979–2018)" [33], and the data after 2018 was collected from China Meteorological Administration (http://data.cma.cn/, accessed on 1 June 2022).

Table 1. Detail information of the monitoring wells. The thickness of unsaturated zones was calculated as the average depth over the observation period.

Well Number	Surface Elevation (m a.s.l.)	Monitoring Depth (m)	Unsaturated Zone Thickness (m)	Observation Period (dd-mm-yyyy)	Frequency (Day^{-1})
W1	34.44	11~48	5.6	5 February 2006—30 December 2020	5
W2	25.23	0~10	3.3	10 January 2005—30 October 2015	10
W3	8.51	0.5~4.33	2.2	5 January 2005—30 December 2020	5
W4	2.16	6.4~8.1	1.1	5 January 2005—30 December 2019	5

2.3. Mann–Kendall Trend Analysis

The Mann–Kendall trend test is a nonparametric statistical test [34,35]. It is not necessary to assume that samples obey a certain distribution and are not disturbed by a few outliers.

The trend of time series is determined using Z values:

$$Z = \begin{cases} (S-1)/\sqrt{n(n-1)(2n+5)/18} & S > 0 \\ 0 & S = 0, \\ (S+1)/\sqrt{n(n-1)(2n+5)/18} & S < 0 \end{cases} \quad (1)$$

where S is the testing statistic, and n is the sample size. A positive value of S indicates an increasing trend, and vice versa. If $|Z| \geq 1.96$, it indicates the time series passes the significance test with 95% confidence.

2.4. Auto-Correlation and Cross-Correlation Functions

The auto-correlation functions were used to depict the persistence degree of the time series. The auto-correlation coefficient $r_{xx}(k)$ is expressed as [12]:

$$r_{xx}(k) = \frac{\frac{1}{N}\sum_{i=1}^{N-k}(x_i-\overline{x})(x_{i+k}-\overline{x})}{\sigma_x^2} = \frac{\sum_{i=1}^{N-k}(x_i-\overline{x})(x_{i+k}-\overline{x})}{\sum_{i=1}^{N}(x_i-\overline{x})^2}, \quad (2)$$

where N is the length of the series; k is the time lag; and $\overline{x}$ is the arithmetic mean of the series.

The cross-correlation coefficient between series x and y is defined as [12]:

$$r_{xy}(k) = \frac{\frac{1}{N}\sum_{i=1}^{N-k}(x_i-\overline{x})(y_{i+k}-\overline{y})}{\sigma_x\sigma_y} = \frac{\sum_{i=1}^{N-k}(x_i-\overline{x})(y_{i+k}-\overline{y})}{\sqrt{\sum_{i=1}^{N}(x_i-\overline{x})^2\sum_{i=1}^{N}(y_i-\overline{y})^2}}, \quad (3)$$

where N is the length of the series; k is the time lag; and σ_x and σ_y are the standard deviations.

The cross-correlation functions characterize the relationships between the input and output signals. Here, we take the rainfall as input signals and the GWLs as output signals. If the rainfall can be considered random, the cross-correlation functions give the impulse response of the aquifer.

The sliding-window cross-correlation method followed by Delbart et al. [2] was also carried out to investigate the influences of rainfall intensity on the response time. The sliding-window cross-correlation method consists of slicing the x (precipitation) and y (GWLs) series with partially superposed windows. For each window, the r_{xy} values between rainfall and GWLs are computed, and the corresponding response time is identified. Then, a time series of response times is obtained for different windows. In this study, the window length is 6 months, and the sliding interval is 1.5 months. Note that the correlation coefficients should be not lower than the standard error of ~$2/\sqrt{N}$, where N is the sample size, and "2" is the critical value for the 0.95 probability of the normal distribution. That is, values for which the r_{xy} peak was not significant at a 95% confidence level were left out.

2.5. Wavelet Transform

Hydrologic time series are usually nonstationary with temporal variations in both frequency and amplitude. Through the continuous wavelet transform, a complete time-scale representation of localized and transient phenomena occurring at different time scales can be obtained [36]. The wavelet spectrum is defined as the modulus of wavelet coefficients. This wavelet spectrum can also be averaged in time, known as the global averaged wavelet spectrum [20], which helps to identify the characteristic periods within a single time series. Here, the Morlet wavelet serves as the wavelet mother function [37].

After the continuous wavelet transform, the cross wavelet transform (XWT) is used to determine the cross wavelet spectrum of two time series, and to examine relationships between the two series. The XWT is defined as:

$$W_t^{XY}(s) = W_t^X(s)W_t^{Y*}(s), \tag{4}$$

where $W_t^X(s)$ is the wavelet transform of time series x_t (rainfall) at frequency scale s; and $W_t^{Y*}(s)$ is the complex conjugate of wavelet transform $W_t^Y(s)$ for y_t (GWLs). The XWT can be represented using polar coordinates:

$$W_t^{XY}(s) = \left|W_t^{XY}(s)\right|e^{t\varphi_t(s)}, \tag{5}$$

where $\left|W_t^{XY}(s)\right|$ is the power of the cross wavelet; and $\varphi_t(s)$ is the phase angle, which denotes the delay between the two series at time t and scale s.

The time lag ΔT at a scale s between the two signals is calculated by:

$$\Delta T = T(s) \times \frac{\varphi_t(s)}{2\pi}, \tag{6}$$

where $T(s)$ is the period relative to the scale s.

The distribution of the cross wavelet power is:

$$P\left(\frac{\left|W_t^{XY}(s)\right|}{\sigma_X\sigma_Y} < p\right) = \frac{Z_v(p)}{v}\sqrt{P_k^X P_k^Y}, \tag{7}$$

where P_k^X and P_k^Y are Fourier background spectra of the two series x_t and y_t; and $Z_v(p)$ is the confidence level associated with the probability p, and $p = \int_0^{Z_v(p)} f_v(z)dz$. The 5% significance level is determined using $Z_2(95\%) = 3.999$.

This study performed XWT on the original time series of rainfall and GWLs to understand their relationships. To exclude the edge effects, the cone of influence is introduced for all wavelet transforms. Codes used for wavelet analyses were based on those originally written by Grinsted et al. [19] and finished by Matlab 2014b (MathWorks, Natick, MA, USA).

3. Results

3.1. Observed Time Series and Trend Analysis

Piezometric levels of all boreholes shown in Figure 3 characterize the behavior of unconfined aquifers. The average GWLs decrease from upstream to downstream, i.e., from 28.77 m at W1, to 21.90 m at W2, to 6.42 m at W3, and to 1.09 m at W4. The standard deviation of the GWLs also declines from 1.25 m at W1 in the west, to 0.70 m at W2 in the middle, and to ~0.5 m for the other two wells in the east of the region. According to the Pearson Type-III distribution curve, the wet year and dry year are identified and denoted by the blue and orange bands, respectively. For example, in Figure 3d, the years 2010 and 2014 are identified as the wet year and dry year, respectively.

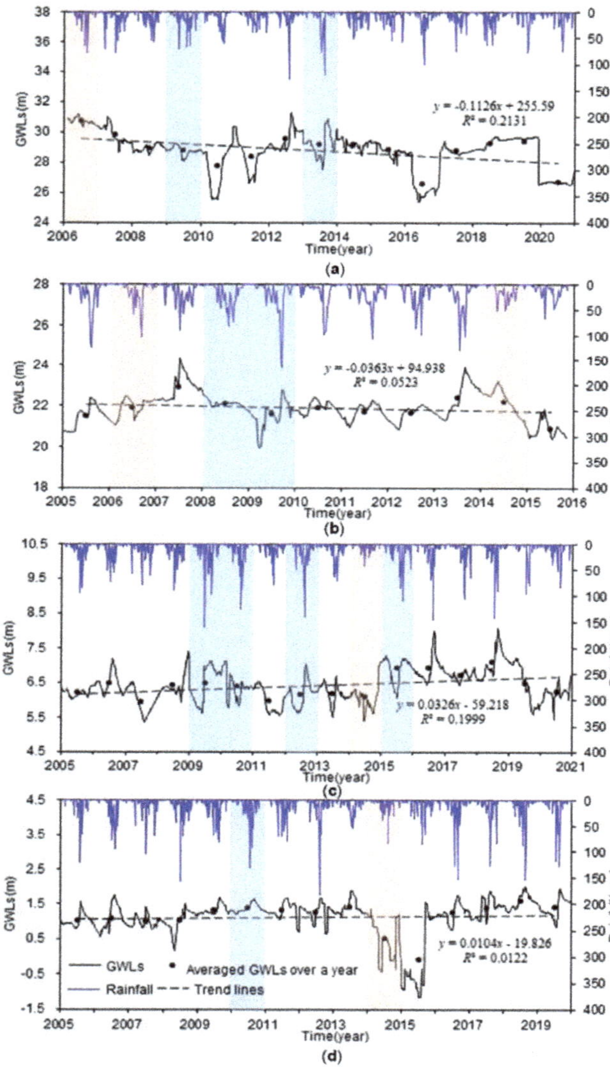

Figure 3. Observed time series of rainfall and GWLs at (**a**) W1, (**b**) W2, (**c**) W3, and (**d**) W4. The blue and orange bands indicate the wet year and dry year, respectively.

According to the Mann–Kendall statistics (Table 2), the GWLs at W1 and W2 present a downward trend while the GWLs at W3 and W4 show an upward trend (all at the 95% confidence level). For the annual averaged GWLs, piezometric levels of W1 and W2 decrease at a rate of 0.11 m/y and 0.04 m/y, respectively. Annual averaged GWLs of W3 and W4 increase at a rate of 0.03 m/y and 0.01 m/y, respectively. Both W3 and W4 are located in Cangzhou City. In 2008, Cangzhou City was listed as a national pilot area for groundwater protection, and the amount of exploitation decreased rapidly. Yan et al. [38] also reported that after 2008, the amount of artificial mining decreased and the depth of groundwater decreased. Our study result is in line with them.

Table 2. Test results of the Mann–Kendall trend analysis.

	W1	W2	W3	W4
S	−114,083	−7903	124,953	127,118
Z	−9.7	−3.1	9.6	10.7

3.2. *Auto-Correlation and Cross-Correlation Analyses*

Figure 4a shows that the auto-correlation coefficients of rainfall decay quickly close to zero, and all of the correlograms become null within 3 months, implying that the rainfall is relatively random. By comparison, GWLs present a long memory effect relative to rainfall (Figure 4b). The auto-correlation functions of GWLs show that the order of increasing inertia ranks as follows: W2→W1→W3→W4. For example, the auto-correlation slope (the slope of the auto-correlation coefficient before the curve becomes flat) increases from -12.7×10^{-2} month^{-1} at W2 to -5.0×10^{-2} month^{-1} at W4 (Table 3), and the time lag required for auto-correlation coefficients to reach 0.2 ($k_{0.2}$ values) also rises from W2 (5.7 months) to W4 (12.3 months). Note that without considering W2, there is an upward trend in persistence from upstream to downstream, which has also been identified by Duvert et al. [3] in a subtropical agricultural catchment dominated by alluvial aquifers in southeast Queensland, Australia.

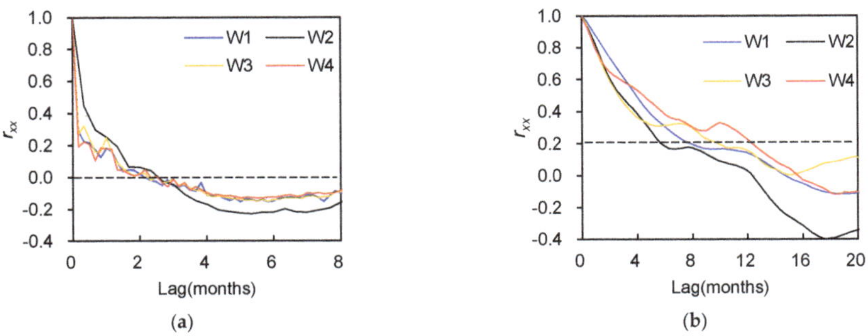

Figure 4. Auto-correlation functions for (**a**) rainfall, and (**b**) GWLs.

Table 3. Parameters of the auto-correlation functions.

	W1	W2	W3	W4
Slope ($\times 10^{-2}$ month^{-1})	−9.3	−12.7	−5.0	−5.0
$k_{0.2}$ (months)	7.8	5.7	9.8	12.3

Figure 5 shows that the peak value of r_{xy} between precipitation and GWLs is the maximum of 0.52 at W2, followed by 0.45 at W1, 0.41 at W3, and 0.40 at W4. It is interesting that this order is consistent with the above ranking result from the auto-correlation functions of GWLs. That is, the shorter the memory time, the greater the correlation coefficient. The time lags corresponding to the peak values are 0.67 months at W2and W3, and 1.33 months at W1 and W4.

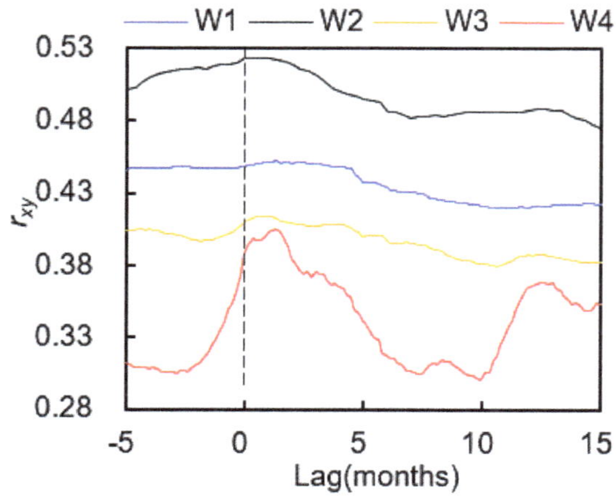

Figure 5. Cross-correlation diagrams between rainfall and GWLs.

3.3. Continuous Wavelet Spectra

Wavelet power spectra for rainfall and GWLs were plotted in Figure 6. Warmer colors denote higher power. It is statistically significant that the rainfall spectrum has a clear annual periodicity throughout the study period, which is mainly caused by the annual wet/dry cycle. For groundwater, this annual periodicity was identified during 2009–2014 and 2015–2017 for W1, 2008–2015 for W2, 2006–2011 and 2014–2016 for W3, and 2014–2016 for W4.

It can be seen that high-power frequencies in the rainfall spectra are absorbed and filtered by the aquifer to produce the groundwater signals. Therefore, aquifers serve as low-pass filters, which is consistent with the research of Imagawa et al. [5] and Duvert et al. [3]. It is interesting that the period when the maximum value in the global wavelet spectrum is achieved increases gradually from 2.1 years at W1 (upstream) to 3.7 years at W4 (downstream). Gómez et al. [39] also identified "longer aquifer regulation times in larger basins". The increasing time period from upstream to downstream we observed here further demonstrates the impacts of regional water circulation.

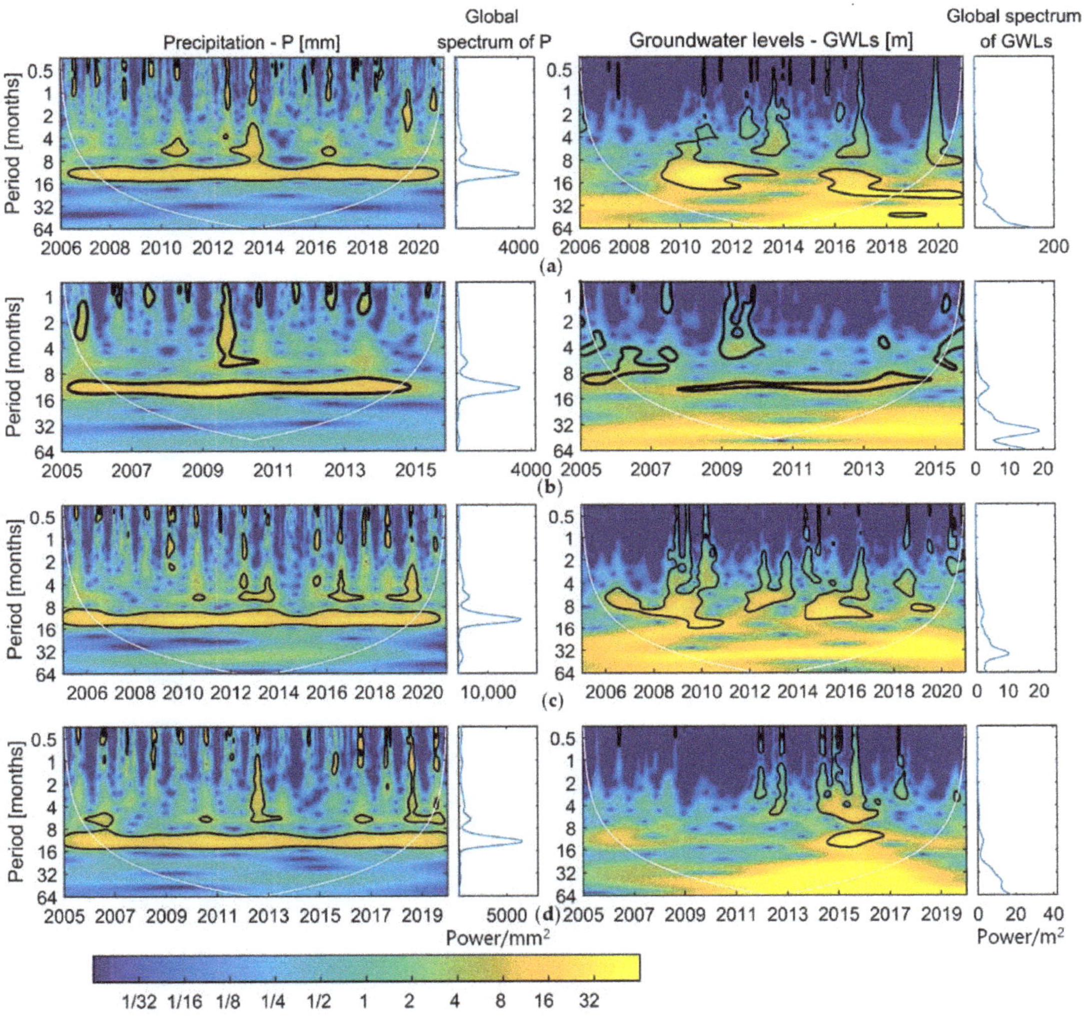

Figure 6. Continuous wavelet spectra for both precipitation and GWLs at (**a**) W1, (**b**) W2, (**c**) W3, and (**d**) W4, with the global wavelet spectrum right side of each subplot. Zones surrounded by black lines have significant wavelet power at the 95% confidence level. White lines denote the cone of influence.

3.4. Cross Wavelet Spectral Analysis

Cross wavelet spectra are given in Figure 7. The averaged phase angles are 2.40 rad, 0.40 rad, 2.50 rad, and 1.02 rad for W1, W2, W3, and W4, respectively. That is, the groundwater lags behind precipitation by 139.14 days at W1, 23.27 days at W2, 145.01 days at W3, and 59.22 days at W4, respectively (Table 4).

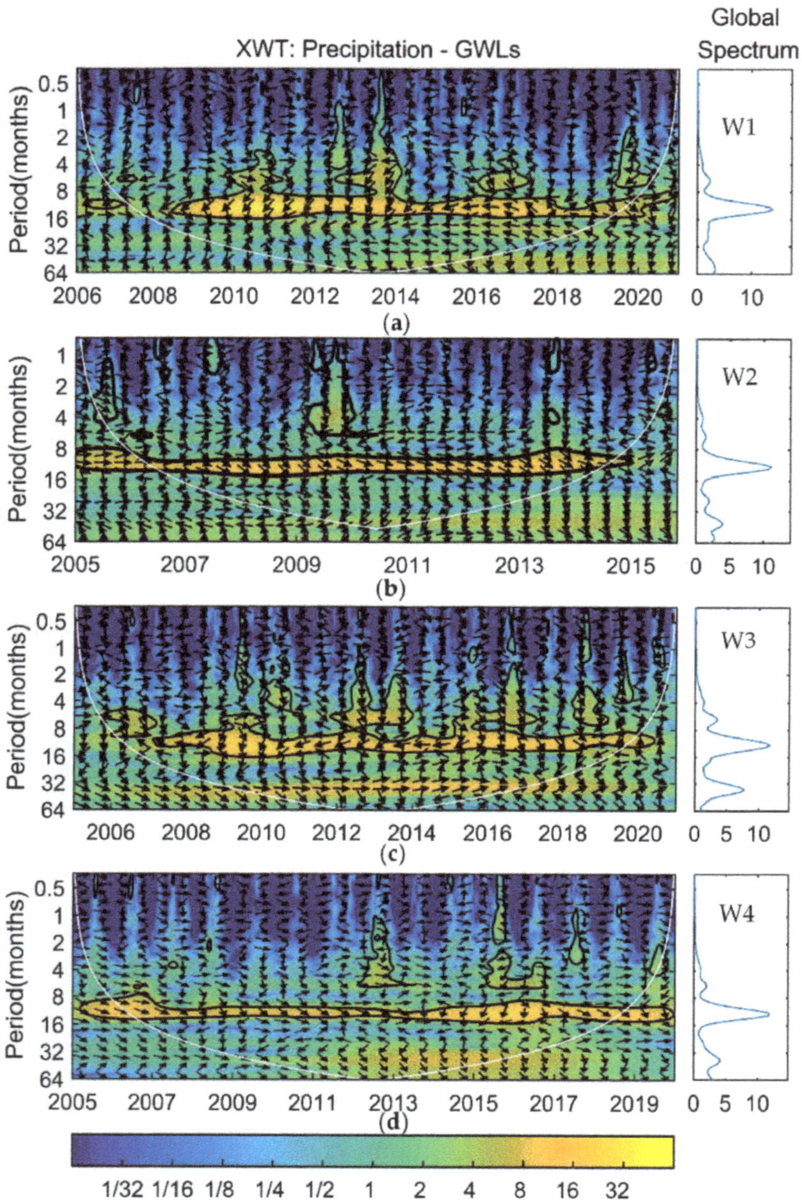

Figure 7. Cross wavelet spectra (left) with global wavelet spectra (right) between rainfall and GWLs at (**a**) W1, (**b**) W2, (**c**) W3, and (**d**) W4. Zones surrounded by black lines have significant wavelet power at the 95% confidence level. White lines denote the cone of influence. The phase angles are indicated by the black arrows.

Table 4. Time lags from the cross wavelet spectra between rainfall and GWLs (period of 365 days band).

Title 1	W1	W2	W3	W4
All significant periods Lags (days)	2008–2020 139.14 ± 59.76	2005–2014 23.27 ± 12.03	2007–2020 145.01 ± 68.00	2005–2019 59.22 ± 26.14
Wet years Lags (days)	2009, 2013 126.77 ± 11.07	2008–2009 39.07 ± 5.97	2009–2010, 2012, 2015 132.48 ± 20.30	2010 49.40 ± 2.37
Years of pumping Lags (days)	2009–2011, 2013–2017, 2020 156.10 ± 38.79	2006, 2008–2009 2014 39.07 ± 6.05	2007, 2010–2020 159.19 ± 74.44	2008, 2014–2016, 2019 88.74 ± 18.25
Years of no or little pumping Lags (days)	2008, 2012, 2018–2019 112.06 ± 50.74	2005, 2007, 2010–2013 23.31 ± 8.68	2008, 2009 123.38 ± 23.85	2005–2007, 2009–2013, 2017–2018 44.46 ± 14.29

The temporal lags for the wet years are also calculated. The time lags for the wet years at W1, W3, and W4 have been shortened by 12 days, 13 days, and 10 days, respectively. This further strengthens the conclusion that high rainfall shortens aquifer response time [2,3]. However, the time lags for wet years at W2 have been prolonged by 16 days. This is mainly caused by human pumping activities, which will be further discussed in Section 4.2.

Compared to the results of other studies shown in Table 5, the response times of W1 and W3 are comparable to those of most wells located in the Yellow River Basin [6,23,40,41]. The response time of W2 is close to the minimum value observed in the Yellow River Delta [23]. The response time of W4 is close to the shortest lags of Jinan Baiquan Spring Watershed and the largest ones of Pingtung Plain. There are many factors that can affect the groundwater response time. Here, we only considered the effects of rainfall intensity, pumping activities, and humidity index in the next section.

Table 5. Time lags of groundwater to rainfall at different study areas.

References	Study Sites	Study Periods	Depth(m)	Time Lags (Days)
Yu and Lin [22]	Pingtung Plain, Tainwan	2005–2010	-	3.71–72.07
Zhang et al. [23]	Yellow River Delta	2006–2010	1.2–2.2	35.51–178.36
Qi et al. [40]	Baiquan Spring Watershed, Jinan	1990–2011	~20–70	80.8–185.37
Cai et al. [6]	Puyang area, Henan	2006–2018	1–35.1	128–175
Feng et al. [41]	Xiongan New Area, China	1991–2016	30–152	147.56–177.20

4. Discussion

4.1. Rainfall Intensity

The time series of response times obtained from the sliding-window cross-correlation method is shown in Figure 7. It can be seen that the fluctuation of groundwater at W2 is almost consistent with that of precipitation: when the rainfall intensity becomes smaller, the GWLs become lower, and vice versa. The fast response leads to a short response time, which is within one month through the year. The same is true for W4 under wet and normal conditions, during which the response time is no more than 1.7 months. However, under dry conditions such as the year of 2014, the response time becomes larger, reaching up to 3 months. Generally speaking, aquifers at W2 and W4 react quickly to local rainfall. In contrast, wells one and three respond slowly to the rainfall with visible time lags as shown in Figure 7a,c. The variation range of the response time is 0~3.7 months for W1 and 0~3.5 months for W3. These values further verified the time lags as shown in Table 4.

4.2. Pumping

As we have mentioned above, agricultural development in this area relies heavily on groundwater. To ensure the winter wheat production, groundwater has to be extracted from March to June if there is not sufficient rainwater. For W2, we can see a significant decline in the water level from 2008 to 2009 despite the wet year, during which the maximum time lag could reach 45 days. This phenomenon is also observed at W4: the maximum response time over the drought period of 2014–2015 was prolonged to 113 days, which was very close to the response time obtained from the sliding-window cross-correlation method as shown in Figure 8.

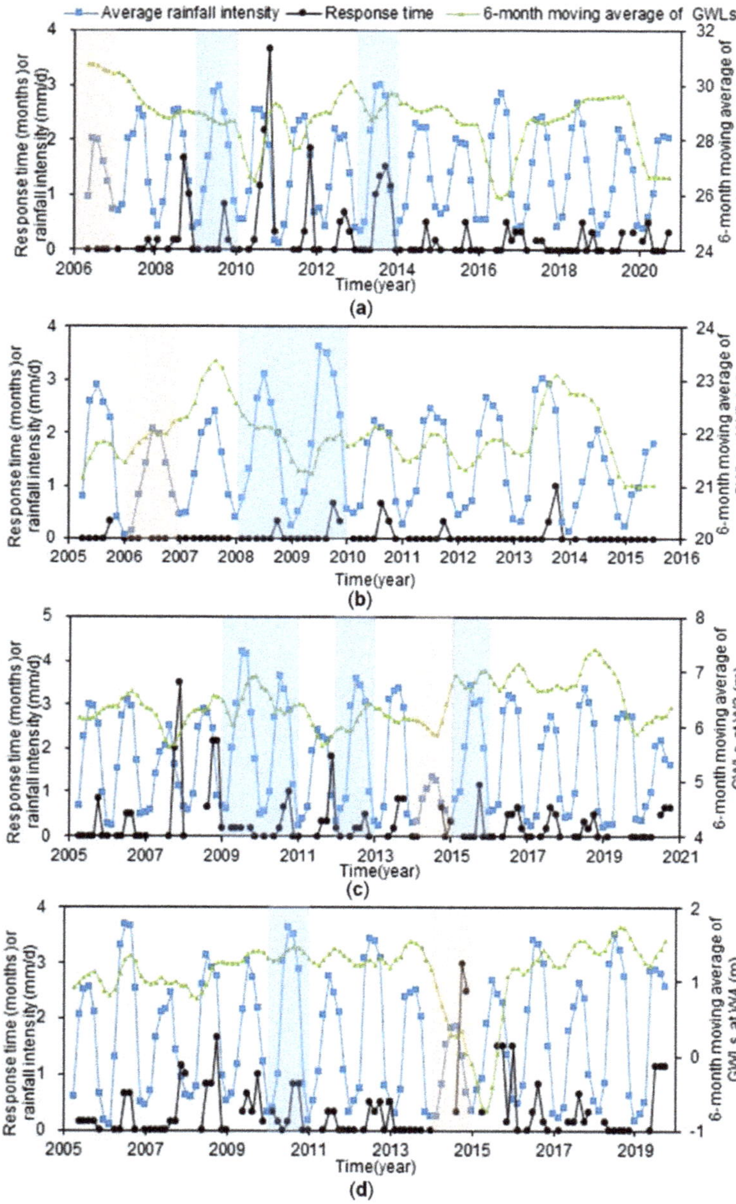

Figure 8. Time series of rainfall intensity, response time, and 6-month moving average of GWLs at (**a**) W1, (**b**) W2, (**c**) W3, and (**d**) W4. The blue and orange bands indicate the wet and dry years, respectively.

To further explore the effect of pumping on the time lags, we counted the lags in the years of pumping and those in the years of no or little pumping for each well (Table 4). The results show that the temporal lags for the pumping years are 1.3~2.0 times those during the years without pumping. Pumping can lead to a dropdown of the GWLs, with an increasing unsaturated zone thickness, and thus a longer time is needed for the aquifer to receive the infiltrated rainfall signal. Therefore, the time lags between groundwater and precipitation will be enlarged by groundwater pumping activities.

4.3. Humidity Index

The potential evapotranspiration (PET) data were obtained from the "1 km monthly potential evapotranspiration dataset in China (1990–2021)" [42], based on the Hargreaves method [43]. The annual average PET at the four stations did not vary greatly, and ranges from 1190 mm at W1, to 1214 mm at W2, to 1187 mm at W3, and 1069 mm at W4. Seasonal variations in PET have also been observed (Figure S3). Generally, the variation pattern of PET is consistent with that of precipitation, though sometimes the peak of PET arrives ~1 month before the precipitation.

The humidity index (the aridity index in UNEP, [44]) can be used to assess the surface-water stress or the arid degree at a given location [45]. It is defined as the ratio of precipitation P to the PET:

$$HI = \frac{P}{PET} \tag{8}$$

The results from the cross wavelet spectra between HI and GWLs are given in Table 6 and Figure S4. Compared to the time lags between P and GWLs as shown in Table 4, the lags between HI and GWLs are shortened by 14% at W1, 36% at W2, 28.5% at W3, and 19.6% at W4, respectively. This can be understood since not all rainfall can recharge aquifers. Only the effective rainfall can produce surface runoff or infiltrate into the subsurface. Compared to the precipitation, the humidity index reflects comprehensive influences of both P and PET, and to some extent the effective rainfall. Therefore, the GWLs response more quickly to the HI index than to the rainfall.

Table 6. Time lags from the cross wavelet spectra between HI and GWLs (period of 365 days band).

	W1	W2	W3	W4
All significant periods	2009–2017	2005–2014	2008–2012, 2014–2018	2005–2007, 2014–2019
Lags (days)	119.11 ± 39.41	14.85 ± 10.07	103.65 ± 40.20	47.60 ± 22.61

4.4. Comprehensive Analysis

To further investigate the controlling factors of the time lags between aquifers and rainfall identified using the XWT method, changes in lags induced by different factors are given in Table 7. It can be seen that variations in the time lags at W2 and W4 are most sensitive to pumping. In fact, Figure 3 and Table 4 indicate that compared to W1 and W3, W2 and W4 are in areas less affected by human activities with few pumping years. Therefore, they are very sensitive to the pumping under the regional background of groundwater level recession. The large percentage change in time lags induced by pumping at these areas highlights the fragility of local aquifer systems.

On the other hand, variations in the time lags at W1 and W3 are less affected by pumping, but most sensitive to HI. They are not sensitive to pumping because things could hardly become worse at these areas that have suffered from pumping to a certain extent. These unhealthy aquifers that have been affected by pumping, or these somewhat thirsty aquifers, will try to recover as long as they are replenished by rainfall. Therefore, they are most sensitive to HI. This highlights the resilience of local aquifers.

Table 7. Variation in time lags due to different factors.

Factors	Variation	W1	W2	W3	W4
Wet years	Absolute changes (days)	−12.37		−12.53	−9.82
	Percentage change	−8.89%		−8.64%	−16.58%
Pumping	Absolute changes (days)	16.96	15.8	14.18	29.52
	Percentage change	12.19%	67.90%	9.78%	49.85%
HI	Absolute changes (days)	−20.03	−8.42	−41.36	−11.62
	Percentage change	−14.40%	−36.18%	−28.52%	−19.62%

Assuming that all rises in water level are due to recharge from precipitation, the water table fluctuation (WTF) method [46,47] tell us the recharge ratio (α) can be estimated as:

$$\alpha = S_y \frac{\sum h}{\sum P} \tag{9}$$

where S_y is the specific yield; Σh is cumulative rise in water-level; and ΣP is the total precipitation in the period corresponding to the water level rise.

Here, the hydrologic year during which the water table rose most significantly is chosen, and the maximum empirical values of S_y provided by [48] are also used. As a result, the recharge ratio we estimated as shown in Table 8 is larger than or close to the maximum value of the empirical values. As a whole, the recharge ratio decreases from upstream to downstream. Specially, it decreases from W2 to W4, which is in line with the inertia ranks as shown in Figure 4. Meanwhile, considering the smallest time lags indicated by the cross wavelet spectra, the site of W2 has a good potentiality for groundwater recharge.

Table 8. Recharge ratio calculated using the WTF method.

	W1	W2	W3	W4
hydrologic year	2018	2013	2009	2018
Vadose zone lithology	Silty clay	Silty clay	Silt	Silty clay
S_y	0.05	0.05	0.074	0.05
Recharge ratio α	0.27	0.32	0.25	0.17
Empirical values of α	0.18–0.26	0.15–0.26	0.20–0.28	0.12–0.19

4.5. Limitations

There are many factors affecting the groundwater response time. Here, we only discussed the influences of rainfall intensity, evaporation, and groundwater pumping. Other factors, such as river water, were not taken into account. In fact, these factors are not isolated, but interact in various geological and geographical contexts. It is reported that the partial wavelet coherency method can detect localized and scale-specific bivariate relationships between predictor and response variables after removing the impact of other variables [30,49,50]. Therefore, in future research, the partial wavelet coherence method can be further implemented to distinguish the impacts of climate change and human activities on the GWLs, for better understanding their impacts on the groundwater flow system.

5. Conclusions

This research provides analyses of the long-term dynamics, the persistence, and the periodicity of shallow groundwater located in the Heilonggang region, China. The interrelation between precipitation and groundwater levels is also examined based on correlation and spectral analyses. The results of this research provide a more complete understanding of the local groundwater circulation system. The major conclusions are as follows.

Firstly, trend analysis shows that the GWLs at W1 and W2 present a downward trend while the GWLs at W3 and W4 show an upward trend over the observation period. Therefore, more attention should be paid to the upstream of the aquifer system.

Secondly, auto-correlation analysis indicates a rising trend in the memory time for aquifers from upstream to downstream. The cross-correlation analysis stresses that the shorter the memory time, the greater the correlation coefficient between rainfall and GWLs. The continuous wavelet spectra shows that the dominant period increases gradually from

2.1 years at W1 to 3.7 years at W4, further demonstrating longer regulation times from upstream to downstream.

Thirdly, both the cross wavelet spectra and the sliding-window cross-correlation method display that wells two and four respond quickly while wells one and three respond slowly to the local rainfall. The time lags between aquifers and rainfall at W1, W2, W3, and W4 are 139.14 ± 59.76 days (2008–2020), 23.27 ± 12.03 days (2005–2014), 145.01 ± 68.00 days (2007–2020), and 59.22 ± 26.14 days (2005–2019), respectively. The temporal lags of groundwater to precipitation are shortened by 10~13 days during the wet year conditions, and the lags during the pumping years are 1.3~2.0 times those during the years without pumping. The time lags between HI and GWLs are reduced by 14~36%, compared to those between rainfall and GWLs.

Further analysis shows that variations in the time lags at W2 and W4 are most sensitive to pumping, while variations in the time lags at W1 and W3 are less affected by pumping, but most sensitive to HI. The overestimated recharge ratio decreases from 0.32 at W2 to 0.17 at W4, suggesting that the site of W2 has a good potentiality for groundwater recharge.

Finally, although this research provides a new insight into relations between precipitation and groundwater in the study area, there are still some other factors, such as river water, which were not considered. Future researchers could possibly use the partial wavelet coherence method to distinguish the effects of human activities and climate change on the GWLs.

Supplementary Materials: The following supporting information can be downloaded at: https://www.mdpi.com/article/10.3390/w15061100/s1, Figure S1. Aquifer groups in the Heilonggang region. Figure S2. Shallow groundwater flow field in (a) 1959 and (b) 2005. Figure S3. Monthly precipitation (P) and monthly potential evapotranspiration (PET) at (a) W1, (b) W2, (c) W3 and (d) W4. Figure S4. Cross wavelet spectra between HI and GWLs at (a) W1, (b) W2 (c) W3 and (d) W4.

Author Contributions: Methodology, C.W.; formal analysis, C.W.; investigation, H.L.; data curation, Y.L.; writing—original draft preparation, C.W. and W.Q.; Writing—Review & Editing, W.Q.; funding acquisition, F.D.; visualization, Y.W. All authors have read and agreed to the published version of the manuscript.

Funding: This work was financially supported by the S&T Program of Hebei (No: 22373601D), the Water Conservancy Science and Technology Project of Hebei Province, China (2021-39), the Open Fund for Hebei Province Collaborative Innovation Center for Sustainable Utilization of Water Resources and Optimization of Industrial Structure (XTZX202111), the Scientific Research Projects of the Higher University in Hebei (BJK2022007), the Natural Science funds Project in Hebei Province (D2020403022, E2021403001), National Pre-research Funds of Hebei GEO University in 2023 (KY202314), the Youth Project of Hebei GEO University (QN202118 and QN202141).

Institutional Review Board Statement: Not applicable.

Informed Consent Statement: Not applicable.

Data Availability Statement: The data are contained within the article.

Acknowledgments: We are very grateful to the three anonymous reviewers for their constructive comments.

Conflicts of Interest: The authors declare no conflict of interest.

References

1. Taylor, R.G.; Scanlon, B.; Doll, P.; Rodell, M.; van Beek, R.; Wada, Y.; Longuevergne, L.; Leblanc, M.; Famiglietti, J.S.; Edmunds, M.; et al. Ground water and climate change. *Nat. Clim. Change* **2013**, *3*, 322–329. [CrossRef]
2. Delbart, C.; Valdes, D.; Barbecot, F.; Tognelli, A.; Richon, P.; Couchoux, L. Temporal variability of karst aquifer response time established by the sliding-windows cross-correlation method. *J. Hydrol.* **2014**, *511*, 580–588. [CrossRef]
3. Duvert, C.; Jourde, H.; Raiber, M.; Cox, M.E. Correlation and spectral analyses to assess the response of a shallow aquifer to low and high frequency rainfall fluctuations. *J. Hydrol.* **2015**, *527*, 894–907. [CrossRef]
4. Cuthbert, M.O.; Gleeson, T.; Moosdorf, N.; Befus, K.M.; Schneider, A.; Hartmann, J.; Lehner, B. Global patterns and dynamics of climate–groundwater interactions. *Nat. Clim. Change* **2019**, *9*, 137–141. [CrossRef]
5. Imagawa, C.; Takeuchi, J.; Kawachi, T.; Chono, S.; Ishida, K. Statistical analyses and modeling approaches to hydrodynamic characteristics in alluvial aquifer. *Hydrol. Process.* **2013**, *27*, 4017–4027. [CrossRef]

6. Cai, Z.; Ofterdinger, U. Analysis of groundwater-level response to rainfall and estimation of annual recharge in fractured hard rock aquifers, NW Ireland. *J. Hydrol.* **2016**, *535*, 71–84. [CrossRef]
7. Meng, Q.; Xing, L.; Liu, L.; Xing, X.; Zhao, Z.; Zhang, F.; Li, C. Time-lag characteristics of the response of karst springs to precipitation in the northern China. *Environ. Earth Sci.* **2021**, *80*, 348. [CrossRef]
8. Delbart, C.; Valdés, D.; Barbecot, F.; Tognelli, A.; Couchoux, L. Spatial organization of the impulse response in a karst aquifer. *J. Hydrol.* **2016**, *537*, 18–26. [CrossRef]
9. Geng, X.; Boufadel, M.C. Spectral responses of gravel beaches to tidal signals. *Sci. Rep.* **2017**, *7*, 40770. [CrossRef]
10. Duy, N.L.; Nguyen, T.V.K.; Nguyen, D.V.; Tran, A.T.; Nguyen, H.T.; Heidbüchel, I.; Merz, B.; Apel, H. Groundwater dynamics in the Vietnamese Mekong Delta: Trends, memory effects, and response times. *J. Hydrol. Reg. Stud.* **2021**, *33*, 100746. [CrossRef]
11. Jenkins, G.M.; Watts, D.G. *Spectral Analysis and Its Applications*; Holden Day Inc.: San Francisco, CA, USA, 1968; p. 525.
12. Box, G.E.P.; Jenkins, G.M. *Time Series Analysis: Forecasting and Control*; Prentice Hall: Upper Saddle River, NJ, USA, 1976; p. 575.
13. Mangin, A. Pour une meilleure connaissance des systèmes hydrologiques à partir des analyses corrélatoire et spectrale. *J. Hydrol.* **1984**, *67*, 25–43. [CrossRef]
14. Larocque, M.; Mangin, A.; Razack, M.; Banton, O. Contribution of correlation and spectral analyses to the regional study of a large karst aquifer (Charente, France). *J. Hydrol.* **1998**, *205*, 217–231. [CrossRef]
15. Labat, D.; Ababou, R.; Mangin, A. Rainfall-runoff relations for karstic springs. Part II: Continuous wavelet and discrete orthogonal multiresolution analyses. *J. Hydrol.* **2000**, *238*, 149–178. [CrossRef]
16. Massei, N.; Dupont, J.; Mahler, B.; Laignel, B.; Fournier, M.; Valdes, D.; Ogier, S. Investigating transport properties and turbidity dynamics of a karst aquifer using correlation, spectral, and wavelet analyses. *J. Hydrol.* **2006**, *329*, 244–257. [CrossRef]
17. Shi, D.C.; Lin, G. Application of spectral analysis to determine hydraulic diffusivity of a sandy aquifer (Pingtung County, Taiwan). *Hydrol. Process.* **2004**, *18*, 1655–1669. [CrossRef]
18. Kim, J.; Lee, J.; Cheong, T.; Kim, R.; Koh, D.; Ryu, J.; Chang, H. Use of time series analysis for the identification of tidal effect on groundwater in the coastal area of Kimje, Korea. *J. Hydrol.* **2005**, *300*, 188–198. [CrossRef]
19. Grinsted, A.; Moore, J.C.; Jevrejeva, S. Application of the cross wavelet transform and wavelet coherence to geophysical time series. *Nonlinear Proc. Geoph.* **2004**, *11*, 561–566. [CrossRef]
20. Labat, D. Cross wavelet analyses of annual continental freshwater discharge and selected climate indices. *J. Hydrol.* **2010**, *385*, 269–278. [CrossRef]
21. Li, J.; Pu, J.; Zhang, T.; Wang, S.; Huo, W.; Yuan, D. Investigation of transport properties and characteristics of a large karst aquifer system in southern China using correlation, spectral, and wavelet analyses. *Environ. Earth Sci.* **2021**, *80*, 84. [CrossRef]
22. Yu, H.; Lin, Y. Analysis of space–time non-stationary patterns of rainfall–groundwater interactions by integrating empirical orthogonal function and cross wavelet transform methods. *J. Hydrol.* **2015**, *525*, 585–597. [CrossRef]
23. Zhang, C.; Huang, C.; He, Y.; Liu, Q.; Li, H.; Wu, C.; Liu, G. An analysis of the space-time patterns of precipitation-shallow groundwater depth interactions in the Yellow River Delta. *Hydrogeol. Eng. Geol.* **2020**, *47*, 21–30. (In Chinese)
24. Cai, Y.; Huang, R.; Xu, J.; Xing, J.; Yi, D. Dynamic Response Characteristics of Shallow Groundwater Level to Hydro-Meteorological Factors and Well Irrigation Water Withdrawals under Different Conditions of Groundwater Buried Depth. *Water* **2022**, *14*, 3937. [CrossRef]
25. Yang, M. Current Situation Research on Groundwater Irrigation Management in Heilonggang Region. Master's Thesis, Hebei Agriculture University, Baoding, China, 2018. (In Chinese)
26. Jin, M.; Zhang, R.; Sun, L.; Gao, Y. Temporal and spatial soil water management: A case study in the Heilonggang region, PR China. *Agr. Water Manag.* **1999**, *42*, 173–187. [CrossRef]
27. Pan, D.; Ren, L.; Liu, Y. Application of distributed hydrological model for optimizing irrigation regime in Heilong-gang and Yundong plain II. Water production function and irrigation regime optimization. *J. Hydraul. Eng.* **2012**, *43*, 777–784. (In Chinese) [CrossRef]
28. Zhang, X.; Ren, L. Simulating and assessing the effects of seasonal fallow schemes on the water-food-energy nexus in a shallow groundwater-fed plain of the Haihe River basin of China. *J. Hydrol.* **2021**, *595*, 125992. [CrossRef]
29. Alfio, M.R.; Balacco, G.; Rose, M.D.; Fidelibus, C.; Martano, P. A Hydrometeorological Study of Groundwater Level Changes during the COVID-19 Lockdown Year (Salento Peninsula, Italy). *Sustainability* **2022**, *14*, 1710. [CrossRef]
30. Gu, X.; Sun, H.; Zhang, Y.; Zhang, S.; Lu, C. Partial Wavelet Coherence to Evaluate Scale-dependent Relationships Between Precipitation/Surface Water and Groundwater Levels in a Groundwater System. *Water Resour. Manag.* **2022**, *36*, 2509–2522. [CrossRef]
31. Bai, G. Analysis of Groundwater Regulation and Storage Conditions in Heilonggang Region, Hebei. Master's Thesis, Shijiazhuang University of Economics, Shijiazhuang, China, 2015. (In Chinese)
32. Wang, B. Study on Comprehensive Quality Assessment and Coupling-Coordination Relationsihp of Water-Land Resources in Heilonggang District. Ph.D. Thesis, Chinese Academy of Geological Sciences, Beijing, China, 2011. (In Chinese)
33. He, J.; Yang, K.; Tang, W.; Lu, H.; Qin, J.; Chen, Y.; Li, X. The first high-resolution meteorological forcing dataset for land process studies over China. *Sci. Data* **2020**, *7*, 25. [CrossRef]
34. Mann, H.B. Non-parametric tests against trend. *Econometrica* **1945**, *13*, 245–259. [CrossRef]
35. Kendall, M.G. *Rank Correlation Methods*, 4th ed.; Griffin, C., Ed.; Bucks: London, UK, 1975.
36. Torrence, C.; Compo, G.P. A practical guide to wavelet analysis. *Bull. Am. Meteorol. Soc.* **1998**, *79*, 61–78. [CrossRef]

37. Morlet, J.; Arens, G.; Fourgeau, E.; Giard, D. Wave propagation and sampling theory-Part II, Sampling theory and complex waves. *Geophysics* **1982**, *47*, 222–236. [CrossRef]
38. Yan, B.; Li, X.; Hou, J.; Bi, P. Study on the dynamic characteristics of shallow groundwater level under the influence of climate change and human activities in Cangzhou, China. *Water Supply* **2020**, *21*, 797–814. [CrossRef]
39. Gómez, D.; Melo, D.C.D.; Rodrigues, D.B.B.; Xavier, A.C.; Guido, R.C.; Wendland, E. Aquifer Responses to Rainfall through Spectral and Correlation Analysis. *J. Am. Water Resour. Assoc.* **2018**, *54*, 1341–1354. [CrossRef]
40. Qi, X.; Li, W.; Yang, L.; Shang, H.; Yi, F. The Lag Analysis of Groundwater Level Anomaliesto Precipitation Anomaly of Jinan Baiquan Springs Watershed. *Earth Environ.* **2015**, *43*, 619–627. (In Chinese)
41. Feng, W.; Qi, X.; Li, H.; Li, W.; Yin, X. Wavelet analysis between groundwater level regimes and precipitation, North Pacific Index in the Xiongan New Area. *Hydrogeol. Eng. Geol.* **2017**, *44*, 1–8. (In Chinese)
42. Peng, S.Z.; Ding, Y.X.; Wen, Z.M.; Chen, Y.M.; Cao, Y.; Ren, J.Y. Spatiotemporal change and trend analysis of potential evapotranspiration over the Loess Plateau of China during 2011–2100. *Agr. Forest Meteorol.* **2017**, *233*, 183–194. [CrossRef]
43. Hargreaves, G.H.; Samani, Z.A. Reference crop evapotranspiration from temperature. *Appl. Eng. Agric.* **1985**, *1*, 96–99. [CrossRef]
44. United Nations Environment Programme. *World Atlas of Desertification*, 2nd ed.; Arnold, Hodder Headline, PLC: London, UK, 1997.
45. Hirata, A.; Kominami, Y.; Ohashi, H.; Tsuyama, I.; Tanaka, N.; Nakao, K.; Hijioka, Y.; Matsui, T. Global estimates of stress-reflecting indices reveal key climatic drivers of climate-induced forest range shifts. *Sci. Total Environ.* **2022**, *824*, 153697. [CrossRef]
46. Healy, R.W.; Cook, P.G. Using groundwater levels to estimate recharge. *Hydrogeol. J.* **2002**, *10*, 91–109. [CrossRef]
47. Moon, S.K.; Woo, N.C.; Lee, K.S. Statistical analysis of hydrographs and water-table fluctuation to estimate groundwater recharge. *J. Hydrol.* **2004**, *292*, 198–209. [CrossRef]
48. Research Group of "57-01-03". National Key Science and Technology Project during the Seventh Five-Year Plan Period. Assessment of groundwater resources in north China. *Bull. Inst. Hydrogeol. Eng. Geol. CAGS* **1993**, *9*, 33–53. (In Chinese)
49. Hu, W.; Si, B. Technical note: Improved partial wavelet coherency for understanding scale-specific and localized bivariate relationships in geosciences. *Hydrol. Earth Syst. Sci.* **2021**, *25*, 321–331. [CrossRef]
50. Zhou, Z.; Liu, S.; Ding, Y.; Fu, Q.; Wang, Y.; Cai, H.; Shi, H. Assessing the responses of vegetation to meteorological drought and its influencing factors with partial wavelet coherence analysis. *J. Environ. Manag.* **2022**, *311*, 114879. [CrossRef] [PubMed]

sustainability

MDPI

Article

Relationship between Indigenous Knowledge Development in Agriculture and the Sustainability of Water Resources

Ali Sardar Shahraki [1], Thomas Panagopoulos [2,*], Hajar Esna Ashari [3] and Ommolbanin Bazrafshan [4]

1 Department of Agriculture Economics, University of Sistan and Baluchestan, Zahedan 9816745785, Iran
2 Faculty of Science and Technology, University of Algarve, Campus de Gambelas, 8000 Faro, Portugal
3 Department of Agriculture Economics, University of Jiroft, Jiroft 7867161167, Iran
4 Department of Natural Resources Engineering, Faculty of Agriculture and Natural Resources Engineering, University of Hormozgan, Bandar Abbas 7916193145, Iran
* Correspondence: tpanago@ualg.pt; Tel.: +351-289800900; Fax: +351-289818419

Abstract: The relationship between agricultural knowledge and water management is very important. Indigenous knowledge in agriculture can improve the water crisis situation and alleviate water stress from dry and semi-arid areas. Therefore, the combination of these two impacts can improve the agricultural sector and reduce the effects of drought. The purpose of this study was to investigate the factors affecting indigenous knowledge and the sustainable management of water resources for optimal water use in agriculture in the Sistan region of Iran. Alongside field research and interviews with 40 indigenous experts and experts from the Jihad-e-Agriculture sector of the Sistan region, the required information was collected by means of a questionnaire. Using the fuzzy hierarchy process (FAHP), the factors affecting indigenous knowledge and the sustainable management of water resources for optimal water use in the Sistan region were ranked. The final rankings of the factors influencing indigenous knowledge for optimal agricultural use of water resources indicate that the educational-extensional factor, with a final weight of 0.37, is the first priority, while social factors, government support, economics, farmers' knowledge, and information, with weights of 0.24, 0.21, 0.13, and 0.03, respectively, are the next priorities. It is recommended that the indigenous knowledge of local authorities be augmented, and that farmers be encouraged to use modern irrigation techniques to optimize the agricultural irrigation of water.

Keywords: indigenous knowledge; sustainable water resources management; optimal consumption; fuzzy hierarchy technique (FAHP)

Citation: Shahraki, A.S.; Panagopoulos, T.; Ashari, H.E.; Bazrafshan, O. Relationship between Indigenous Knowledge Development in Agriculture and the Sustainability of Water Resources. *Sustainability* **2023**, *15*, 5665. https://doi.org/10.3390/su15075665

Academic Editors: Alban Kuriqi and Luis Garrote

Received: 19 February 2023
Revised: 8 March 2023
Accepted: 20 March 2023
Published: 23 March 2023

1. Introduction

On a planet where more than two thirds of its surface is covered by water, there is a misconception that water resources cannot be scarce. In order for the use of water to continue, it cannot be withdrawn from reservoirs or other water sources at a rate faster than its natural cycle is capable of restoring and reproducing [1]. Hydrologists and specialists are of the opinion that whenever the ratio of population to volume of renewable freshwater resources exceeds a certain limit, noticeable increases in water scarcity and the pressure and stress caused by water shortage are inevitable. Over the last few decades, this ratio has reached or exceeded the critical limit in more than 24 countries across the world [2,3]. The realities of today's world show that a scarcity of any resource increases tensions and creates new ones. Water is not humane, and today, with a deficiency of water resources, along with the rapid increase in consumption demand in the urban, industrial, and agricultural sectors, water shortage has become the biggest nightmare and challenge for governments. Today, tension over the distribution of freshwater resources across the world, which spans all regions of the world, comes in various forms, creating of conflict between the urban and agricultural need for water and the environment [4,5]. With the growing population, increasing living standards, and increasing attention paid to environmental issues, the attention paid to water resource management has increased. On the other hand, in many countries, water has become a scarce resource. This has been due to population growth on the one hand and to the rapid growth of economic, commercial, and developmental activities on the other [6]. Therefore, when striving for water sustainability, there is a continuous need for long-term agricultural policies, and the correct and optimal use of renewable resources is one of the goals of many governments, including that of Iran [7]. In the current situation, across the world, the issue of water has become one of the main problems and crises and has played an important role in human life compared to previous years. In the vast and rainy country of Iran, farmers face the challenge of dealing with water scarcity. The water factor, its economic role, and its effect on social formation are some of the most important factors affecting the complexities of rural Iran. How water resources should be managed mainly depends on economic indicators, but the idea of sustainable development and the sustainable use of water resources has been considered, which plays a major role in planning and legislation. Therefore, across the world, different methods regarding the exploitation of water resources have been defined, according to the socio-economic and indigenous status of communities [8].

Some analysts and experts consider food and job security to be the main prerequisite for the realization of "national security"; while these two factors are dependent on "water security", this issue is much more prominent in countries and regions whose economies are dependent on agriculture. In terms of national security, the management and planning of water resources in border areas is more important than in other areas [9]. The most important economic sector for villages is the agricultural sector, which depends on the natural potential of the environment for development, progress, and both quantitative and qualitative increase. For sustainable rural development to take place, any development program must be based on a mutual understanding of local needs and resources. In addition, water resources are the most important and basic environmental potential for the development of the agricultural sector. On the other hand, water security improves the economic and social situation of farmers. This means that if the role of human resources in society has been the main axis of development, the role and position of water resources among natural resources is the center of development and has the highest effect on the development of human society and rural areas [10].

Part of the national capital of any ethnic group is indigenous knowledge, which encompasses indigenous beliefs, values, and ecological knowledge of their living environment, which has been the result of years of trial and error in the natural, economic, and social environment. Indigenous knowledge is highly vulnerable to extinction due to its oral nature; however, there are still ambiguities in the knowledge of odor in each region that must be examined to help overcome the problems inflicted by water shortage and drought in a region [11]. The drying up of water sources in big cities is considered to be one of the reasons for migration, which causes many jobs to disappear in water-scarce regions. The lack of a systemic approach, inefficient policy making, inappropriate laws, and pressure put on the agricultural sector under the pretext of increasing employment are among the main causes of the water crisis. The most important advantage of modern knowledge is the ability to create new technologies and transfer them from one environment to another, and one of the important characteristics of indigenous knowledge is its simplicity and compatibility with the environment. Therefore, the two sources of knowledge can be complementary, and a combination of the two can be effective in achieving success and progress. Now it seems that, for several reasons, addressing indigenous knowledge is not a choice but a necessity [12]. Due to its biological nature, agriculture is the largest consumer of water resources. In Iran, most water extraction (87 billion out of 95 billion cubic meters) is consumed by the agricultural sector, a large volume of which (63 billion cubic meters) is wasted due to improper irrigation methods [13].

The agricultural sector in Iran is in a good position with regard to its potential compared to other countries. In terms of irrigated land, Iran ranks fifth after India, China, America, and Pakistan, and in terms of the total area of agricultural land (rain-fed and irrigated) it ranks twentieth. The per-capita area of irrigated land (8 million hectares) is equal to 115 hectares per 1000 people, which is 2.5 times more than the per-capita irrigated land in the world, which is about 45 hectares per 1000 people. On the other hand, the low yield per unit area has caused the efficiency and productivity to decrease despite the land and water resources. For example, the yield of wheat in America is 14 tons per hectare, while in Iran, it is 5.3 tons. The main driver of agriculture across the world is water, so in order to increase the production of agricultural products, irrigation projects have played an essential role throughout the last half century. The production of agricultural products in countries located in arid and semi-arid regions is highly dependent on water, which accounts for more than 90% of the raw production of agricultural products. The agricultural sector is considered to be the most important and main source of food supply in the world; therefore, it plays a significant role in maintaining the balance of food, social, and political security of countries. In past years, agriculture has faced many fluctuations in the area of cultivated land and the yield of crops. Many factors, such as lack of water and salinity, poor management, lack of knowledge and awareness, the existence of competition between different sectors (environment, industry, and household) with the agricultural sector, the wear and tear of water facilities, slow development of grain-cultivated land, and land use change, cause a decrease agricultural product have been produced [2].

In the Sistan region in the southeast of Iran, about 90% of the residents are employed in agriculture. Due to frequent droughts over the last 20 years, the agricultural sector in this region has been almost destroyed. Therefore, the study and analysis of water input as a strategic resource is of great interest to researchers, who can take important steps in improving the conditions of this region by modeling and predicting the future.

Omani [14] studied the effective factors in the sustainable management of agricultural resources in the northern part of the Modaress watershed in the Khuzestan province. The results indicate that five factors, including economic characteristics, variables of educational and extension activities, the social, knowledge and information, and government support, together account for 71% of changes in the level of sustainable management of agricultural water resources. Arfai [15] studied the factors affecting indigenous knowledge in the optimal use of water in the agricultural sector. Their results indicate that there is a positive and significant relationship between independent variables—educational-promotional, cultural-social, economic, and managerial factors—, and the use of indigenous knowledge for optimal water consumption. Additionally, indigenous knowledge of efficient water use showed that educational-extension factors and economic factors have a positive role in optimal water use. Panahi [7] analyzed the factors affecting the optimal management of water resources in the Iranian agricultural system. In explaining the component of optimal management of agricultural water using structural equation modeling, it was found that 37% of the total variable dependent changes for the optimal management of agricultural water resources could be explained by four factors: government activities, extension services, individual and physical factors, and the use of management mechanisms by the producers. Bandani et al. [16] studied and analyzed the role of indigenous knowledge in the sustainable rural development of the Ghaemabad rural district of the Sistan region. The results show that being involved in the culture of the villagers and the participatory aspect of indigenous knowledge has been more effective than other factors in this village. Indigenizing and organizing various resources, including manpower and material capital, along with other experiences in the villages, is a complex and difficult task, but a possible one. Bouzarjomehri et al. [11] studied the local role and local traditions of women in the production and management of livestock products in the village of Abu Nasr, Bavanat city. The results indicate that there is a rich local knowledge and local traditions among rural women in the region, which is due to the traditional (rich) milking system and milk and dairy management mechanism. Rahimian [17], in a study, investigated the factors affecting the sustainable management of water resources among irrigated wheat farmers in the city of Koohdasht. The results, based on a correlation test, indicated that the relationship between sustainable water resource management and annual income varies according to the agricultural sector, the area under total wheat land, training provided to farmers in water management, and farmers' perception of direct and significant water shortage crisis, with variables such as the farmers' plots and percentage of farm slope being reversed and significant. Akhavan and Behbahaninia [18] studied the economic factors affecting the sustainable development and management of water resources in the agricultural sector. The collective effect of independent variables on the dependent variable through multiple regression indicated that from the perspective of experts on economic factors, farmers' knowledge and experts' insights have had a positive effect on the sustainable development and management of water resources in agriculture. Afshari et al. [19] investigated the determinants of farmers' attitudes towards sustainable water resource management in Komijan city. The results show that there was a significant difference between farmers' attitudes towards water resource management according to the type of water source ownership, so that farmers with private property had a more desirable attitude than farmers with common ownership. Hassani et al. [20] studied the factors affecting the water resource management behaviors of farmers in the Hamadan-Bahar plain. Based on their results, the perception of vulnerability, perception of severity of degradation, perception of barriers to water resource protection and responsibility had a significant correlation with water resource management behaviors at a level of one percent. Iglesia and Garot [21] examined adaptation strategies for managing agricultural water under climate change in Europe. The results indicate that the greatest opportunities to take action to improve compliance capacity and respond to change according to water needs are to reform water policy, provide adequate training for farmers, and provide effective financial instruments. Valipour [22] examined land use policy and agricultural water management in the first half of the present century in Africa. The results show that Africa needs government policies to encourage farmers to use irrigation systems and increase cultivation intensity for the irrigated area. In the same study by Valipour et al. [23], the results showed that trial and error policies should be avoided, and expert opinion applied to irrigation systems for each crop. Jacob et al. [6] investigated the relationship between knowledge and practice in sustainable water resource management. Their research shows that participatory processes in short-term decisions, such as water allocation decisions, are modifiable and are not suitable for long-term infrastructure decisions. Another important result indicated that capacity construction costs for stakeholders in the water management decision-making process are not known. Thus, a lack of understanding of the costs and associated complexities may contribute to citizens' lack of successful acceptance of infrastructure decisions. Kernecker et al. [24] examined women's local knowledge of water resources and adaptation to changing landscapes in the Veracruz Mountains in Mexico. This study shows that women in the study area have acquired their water management by relying on their local knowledge about landscape, climate and social networks. These results suggest that women's local knowledge can play an important role in planning development projects and helping women to adapt to sudden changes. Greenland et al. [25] examined in a study the improvement of agricultural water sustainability using farm water management strategy and encouraging drip irrigation. Research findings show that farmers do not accept drip irrigation due to cost and a lack of understanding, as well as problems with installation and maintenance. The solution proposed was to promote effective training programs for drip irrigation. Shahbakhsh et al. [1] conducted a piece of research titled "assessment of modern approach of water governance in the development of water exploitation systems in Sistan region". The simple additive weighting (SAW) method that is used here is one of the multiple attribute decision-making (MADM) methods. The indicators of water governance principles were derived from the opinions of 30 water experts, faculty members, and water users in the Sistan region using the SAW method, and weights were assigned to them to form MADM matrices. According to the results, six indicators were derived as the indicators determining the principles of water governance. 'Traditional users' was selected as the strongest system and 'irrigation and drainage networks exploitation companies' as the weakest system. Additionally, according to the results obtained from water experts, the first rank was assigned to 'irrigation and drainage networks exploitation companies' (A2) with a final crisp score of 6.818, followed by 'water user cooperatives' (A4) with a final crisp score of 6.515 in the second rank, and 'private firms' (A6) with a final crisp score of 6.308 in the third rank. Farrokhzadeh et al. [2] studied sustainable water resource management in an arid area using a coupled optimization-simulation modeling technique. In the study, a multi-objective optimization model was linked with the water evaluation and planning (WEAP) software to optimize water allocation decisions over multiple years. The results were analyzed by comparing purely economic versus multi-objective scenarios on the Pareto front. Finally, the disadvantages and advantages of these scenarios were also qualitatively described to assist the decision process for water resource managers. Abbasian et al. [3] examined a research paper about the economic management of water using valuation policy in mango orchards, with an emphasis on environmental inputs in Chabahar County. The study used cross-sectional data for the 2018–2019 crop years in order to estimate the price of water for mango, and to also estimate its demand, with an emphasis on environmental inputs. To this end, the real price of water was determined using the residual method, and the demand function was estimated using the translog cost function and the equations of the contribution of inputs to cost. The results support the good fit of the model used for the cost function of mango in the studied county. The results of the coefficients in Chabahar County indicate that water cost has a positive relationship with the prices of manure, water, seedling, and crop yield, and a negative relationship with the prices of pesticides and chemical fertilizers. Based on the results of the water demand function, water is a substitute for manure, chemical fertilizer, and seedling with partial elasticities of >1, revealing the impact of water use management and economic valuation on improving the use of other environmental inputs (pesticides, manure, and chemical fertilizers) and seedling, as well as the water itself, in mango production in this region. It is recommended that policies such as optimal pricing of inputs including pesticides, manure, chemical fertilizers, and seedling be adopted in order to curb the resulting environmental pollution.

This literature review shows that proper management is essential to deal with the water crisis. The objectives of water resource management include improving the allocation of water resources, improving consumer behavior, promoting methods to reduce water losses and prevent drought in agricultural activities, and improving the efficiency of water resource capacity and facilities. On the other hand, indigenous knowledge systems have much power in the field of sustainable development in rural areas. Looking at the characteristics of indigenous knowledge systems indicates that indigenous knowledge will be able to use water more effectively, as it is systematic, preserves biodiversity, relies on needs, is participatory, accessible and multi-dimensional, and adapts to people's culture. Indigenous knowledge can play an important role in the process of the sustainable development of a village. For this purpose, one of the objectives of this study was to investigate the factors affecting indigenous knowledge in water consumption in agriculture, while our other purpose was to investigate the factors affecting the sustainable management of agricultural water resources in the Sistan region. Figure 1 shows the geographical location of the study area.

This paper aims to study the factors affecting indigenous knowledge in the optimal use of water in the agricultural sector and the factors affecting the sustainable management of agricultural water resources, and to apply a multi-indicator decision-making modeling under fuzzy logic.

Figure 1. The location of the Sistan region in the southeast of Iran [1].

2. Materials and Methods

When using the fuzzy analytic hierarchy process (FAHP), the first step is to determine the pairwise comparison matrix in FAHP, as follows [26]:

$$A = \begin{bmatrix} 1 & a_{12} & \cdots & a_{1n} \\ a_{21} & 1 & \cdots & a_{2n} \\ \vdots & \vdots & \ddots & \vdots \\ a_{n1} & a_{n2} & \cdots & 1 \end{bmatrix} = \begin{bmatrix} 1 & a_{12} & \cdots & a_{1n} \\ 1/a_{12} & 1 & \cdots & a_{2n} \\ \vdots & \vdots & \ddots & \vdots \\ 1/a_{1n} & 1/a_{2n} & \cdots & 1 \end{bmatrix} \quad (1)$$

The pairwise comparison matrix is based on the opinion of the decision maker and the elements of each level, considered separately. Matrices include pairwise comparisons between the criteria considered and pairwise comparison matrices of the options under consideration for each criterion.

In general, if the number of options and criteria are M and N, respectively, the pairwise comparison matrices of the options will be M × M and the pairwise comparison matrix of the criteria will be an N × N matrix. The elements of the pairwise comparison matrices with A_{ij} are shown. In the FAHP method, it is assumed that $A_{ij} = 1/A_{ji}$. To perform the relative measurement (or degree of importance) of option M, the comparison operation is performed in pairs. This means that each specific option is not compared to the other available options at the same time; at any given time, it can only be compared to another option [26].

In the present study, the Chang (1996) method was used to perform the FAHP technique. In this method, each criterion is assigned to a fuzzy set of 4, Mi (Li Mi, Ui) is converted, and after drawing a hierarchical tree, the target levels, criteria, and options are determined. In the next step, the matrices are agreed upon according to the decision tree, and, using the opinions of experts in the form of fuzzy triangular numbers in the form of matrices, Equation (2) is formed [26].

$$A = \begin{bmatrix} (1,1,1) & \begin{Bmatrix} \tilde{a}_{121} \\ \tilde{a}_{122} \\ \vdots \\ \tilde{a}_{12p_{12}} \end{Bmatrix} & \cdots & \cdots & \begin{Bmatrix} \tilde{a}_{1n1} \\ \tilde{a}_{1n2} \\ \vdots \\ \tilde{a}_{1np_{1n}} \end{Bmatrix} \\ \begin{Bmatrix} \tilde{a}_{211} \\ \tilde{a}_{212} \\ \vdots \\ \tilde{a}_{21p_{21}} \end{Bmatrix} & (1,1,1) & \cdots & \cdots & \begin{Bmatrix} \tilde{a}_{2n1} \\ \tilde{a}_{2n2} \\ \vdots \\ \tilde{a}_{2np_{2n}} \end{Bmatrix} \\ \vdots & \vdots & \vdots & \vdots & \vdots \\ \vdots & \vdots & \vdots & \vdots & \vdots \\ \begin{Bmatrix} \tilde{a}_{n11} \\ \tilde{a}_{n12} \\ \vdots \\ \tilde{a}_{n1p_{n1}} \end{Bmatrix} & \begin{Bmatrix} \tilde{a}_{n21} \\ \tilde{a}_{n22} \\ \vdots \\ \tilde{a}_{n2p_{n2}} \end{Bmatrix} & \cdots & \cdots & (1,1,1) \end{bmatrix} \tag{2}$$

In the next step, the arithmetic mean of decision makers' opinions is calculated as relation (2):

$$\tilde{A} = \begin{bmatrix} (1,1,1) & \tilde{a}_{12} & \cdots & \tilde{a}_{1n} \\ \tilde{a}_{21} & (1,1,1) & \cdots & \tilde{a}_{2n} \\ \vdots & \vdots & \vdots & \vdots \\ \tilde{a}_{n1} & \tilde{a}_{n2} & \cdots & (1,1,1) \end{bmatrix} \tag{3}$$

The arithmetic mean matrix of decision makers' opinions (*jia*) is calculated according to the following equation [26].

$$\tilde{S}_i = \sum_{j=1}^{n} \tilde{a}_{ij} \; i = 1,2,\ldots,n \tag{4}$$

The next step is to calculate the sum of the row elements of the matrix obtained from the arithmetic mean [26]. In the the next step, Equation (4) is used to normalize the matrix of rows:

$$\tilde{M}_i = \tilde{S}_i \otimes \left[\sum_{i=1}^{n} \tilde{S}_i\right]^{-1} \quad i = 1,2,\ldots,n \tag{5}$$

Now, according to the mathematical relations governing triangular fuzzy numbers, if it is represented as (Li, Mi, Ui), the above relation is adjusted in the following order:

$$\tilde{M}_i = \left(\frac{l_i}{\sum_{i=1}^{n} u_i}, \frac{m_i}{\sum_{i=1}^{n} m_i}, \frac{u_i}{\sum_{i=1}^{n} l_i}\right) \tag{6}$$

Determining the degree of probability as being larger, (degree of probability) $M_2 = (L_2, M_2, U_2) \geq M_1 = (L_1, M_1, U_1)$ is calculated as follows:

$$VM_2 \succ M_1 = Suby \geq x[Min(\mu M_1, \mu M_2)] \tag{7}$$

where Y and X are the values of the membership functions of each criterion on the fuzzy function axis.

$$V(M_2 \geq M_1) = \mu(d) = \begin{cases} 1 \; if \; m_2 \geq m_1 \\ 0 \; if \; l_1 \geq u_2 \\ \frac{l_1 - u_2}{(m_2 - u_2) - (m_1 - l_1)} \; otherwise \end{cases} \tag{8}$$

Here, d is the maximum distance between the two membership functions 1 M and 2 M. To compare M_1 and M_2, the value of V ($M_2 \geq M_1$); V ($M_1 \geq M_2$) is needed and K is estimated as follows:

$$\begin{aligned} V(M \geq M_1, M_2, M_3, \ldots, M_k) \\ = V[(M \geq M_1)\&(M \geq M_2)\&(M \geq M_3)\&\ldots\&(M \geq M_k)] = \\ \Rightarrow MinV(M \geq M1) \end{aligned} \quad (9)$$

It can thus be written as follows:

$$d'(A_i) = MinV(M_i \geq M_k)k = 1, 2, 3, \ldots, n; k \neq i \quad (10)$$

Thus, the weight of each vector is obtained as follows:

$$W' = \left(d'(A_1), d'(A_2), d'(A_3), \ldots, d'(A_n)\right)^T \quad (11)$$

Here, W′ is a non-fuzzy number. By normalizing the obtained weights, the final weights can be obtained:

$$W' = \left[\frac{d'(A_1)}{\sum_{i=1}^{n} d'(A_i)}, \frac{d'(A_2)}{\sum_{i=1}^{n} d'(A_i)}, \ldots, \frac{d'(A_n)}{\sum_{i=1}^{n} d'(A_i)}\right]^T \quad (12)$$

The above weights are definite (non-fuzzy). The last step is the combination of weights; by combining each of the weights of the options and criteria, the final weight of the options is obtained [26].

$$\tilde{U}_i = \sum_{j=1}^{n} \tilde{W}_i \tilde{r}_{ij} \quad (13)$$

In multi-criteria decision-making models (MCDM), questionnaires and surveys completed by experts and top experts are used to complete the data. Therefore, in the first stage, experts related to the subject under study were selected and the questionnaires were distributed and completed by them. In this study, according to the objectives, 40 experts from the Agriculture and Water Resource Management Organization were identified, and a questionnaire was distributed among them to complete the information. FUZZY AHP software was used for data analysis. The opinions of experts regarding the identification of factors affecting indigenous knowledge in the optimal use of water in the agricultural sector and the study of factors affecting the sustainable management of agricultural water resources are given in Table 1.

Table 1. The factors affecting indigenous knowledge and the level of sustainable management of agricultural water resources in the optimal use of water in agriculture.

Factors Affecting Indigenous Knowledge in Optimal Water Consumption	Indicators
Visiting programs for farmers regarding indigenous knowledge	Educational-promotional
Training classes on the use of indigenous knowledge	Educational-promotional
Practical projects about indigenous knowledge projects	Educational-promotional
Using publications	Educational-promotional
Codified training packages, e.g., movies, books, tapes, etc.	Educational-promotional
Special seminars for promoting the development of indigenous knowledge	Educational-promotional
Internet network and eLearning	Educational-promotional
Beliefs about indigenous knowledge	Cultural-social
Farmers' interest in using indigenous knowledge	Cultural-social
Positive attitude towards indigenous knowledge	Cultural-social
Communication with neighbors regarding the use of indigenous knowledge	Cultural-social
Interest in joining social groups	Cultural-social
Membership with the water cooperatives association	Cultural-social
Providing special human resources for indigenous knowledge projects	Managerial

Table 1. *Cont.*

Factors Affecting Indigenous Knowledge in Optimal Water Consumption	Indicators
Organizing financial mechanisms for officials regarding to indigenous knowledge	Managerial
Managers' attitude toward indigenous knowledge	Managerial
Practical participation of managers in indigenous knowledge	Managerial
Assigning powers and decisions to farmers and bottom-up decision making	Managerial
Transfer of powers and decisions to farmers in decision making up and down	Economic
Farmers' income	Economic
Insurance for agricultural products at risk of drought	Economic
Extension of drought loan repayment	Economic
Drought loans	Economic
Bank facilities for purchasing pomp motors and machines	Economic
Using publications	Educational-promotional
The rate of use of radio and television	Educational-promotional
Holding classes and training courses	Educational-promotional
Distribution of educational magazines and publication promoting	Educational-promotional
Using educational workshops	Educational-promotional
Communication with specialists (agriculture promoters)	Educational-promotional
Social participation	Social
Social status	Social
Delivery of subsidized inputs to recipients of water resources management	Government support
Material and spiritual incentives for water resource recipients	Government support
Farmer's crop area	Economic
Type of exploitation system	Economic
Farmers' income	Economic
Mechanization level	Economic

3. Discussion

3.1. Identifying the Factors Affecting Indigenous Knowledge of the Optimal Use of Water in the Agricultural Sector

According to Table 2, it is observed that in the matrix of pairwise comparison of indices of factors affecting indigenous knowledge in optimal water consumption in the agricultural sector of Sistan, according to experts of Jihad Keshavarzi, the area of relations with neighbors regarding the use of indigenous knowledge, membership in the aquifer organization, farmers' incomes, and drought loans (11, 13, 19, and 23) are superior to other indicators. Meanwhile, visitation programs for farmers on indigenous knowledge, training classes on the use of indigenous knowledge, practical projects around indigenous knowledge projects, and the provision of specialized human resources in terms of indigenous knowledge projects (1, 2, 3, and 14) have a weak advantage over other indicators.

Table 2. Paired comparison matrix of indicators of factors affecting indigenous knowledge in optimal water consumption in the Sistan agricultural sector.

Comparison	Criterion 1	Criterion 2	Criterion 3	Criterion 4	Criterion 5	Criterion 6	Criterion 7	Criterion 8	Criterion 9	Criterion 10	Criterion 11	Criterion 12
Criterion 1	1, 1, 1	3, 5, 7	1, 3, 5	1, 3, 5	2, 4, 6	1, 2, 4	2, 4, 6	3, 5, 7	5, 7, 9	4, 6, 8	5, 7, 9	1, 2, 4
Criterion 2	0.2, 0.33, 0.14	1, 1, 1	1, 1, 1	2, 4, 6	1, 1, 1	1, 2, 4	1, 3, 5	1, 2, 4	1, 3, 5	1, 3, 5	1, 1, 1	1, 2, 4
Criterion 3	0.33, 1, 0.2	1, 1, 1	1, 1, 1	3, 5, 7	1, 1, 1	3, 5, 7	3, 5, 7	1, 2, 4	1, 2, 4	2, 4, 6	2, 4, 6	1, 1, 1
Criterion 4	0.33, 1, 0.2	0.2, 0.5, 0.16	0.2, 0.33, 0.14	1, 1, 1	2, 4, 6	4, 6, 8	1, 1, 1	1, 3, 5	3, 5, 7	3, 5, 7	1, 1, 1	3, 5, 7
Criterion 5	0.2, 0.5, 0.16	1, 1, 1	1, 1, 1	0.2, 0.5, 0.16	1, 1, 1	1, 3, 5	3, 5, 7	1, 1, 1	1, 1, 1	1, 1, 1	2, 4, 6	1, 1, 1
Criterion 6	0.5, 1, 0.25	0.5, 1, 0.25	0.2, 0.33, 0.14	0.16, 0.25, 0.12	0.33, 1, 0.2	1, 1, 1	1, 3, 5	1, 2, 4	1, 1, 1	1, 1, 1	4, 6, 8	1, 1, 1
Criterion 7	0.2, 0.5, 0.16	0.33, 1, 0.2	0.2, 0.33, 0.14	1, 1, 1	0.2, 0.33, 0.14	0.33, 1, 0.2	1, 1, 1	3, 5, 7	1, 1, 1	1, 1, 1	1, 1, 1	2, 4, 6
Criterion 8	0.2, 0.33, 0.14	0.5, 1, 0.25	0.5, 1, 0.25	0.33, 1, 0.2	1, 1, 1	0.5, 1, 0.25	0.2, 0.33, 0.14	1, 1, 1	1, 1, 1	1, 1, 1	7, 9, 11	1, 1, 1
Criterion 9	0.14, 0.2, 0.11	0.33, 1, 0.2	0.5, 1, 0.25	0.2, 0.33, 0.14	1, 1, 1	1, 1, 1	1, 1, 1	1, 1, 1	1, 1, 1	1, 2, 4	5, 7, 9	1, 1, 1
Criterion 10	0.16, 0.25, 0.12	0.33, 1, 0.2	0.2, 0.5, 0.16	0.2, 0.33, 0.14	1, 1, 1	1, 1, 1	1, 1, 1	1, 1, 1	0.5, 1, 0.25	1, 1, 1	5, 7, 9	4, 6, 8
Criterion 11	0.14, 0.2, 0.11	1, 1, 1	0.2, 0.5, 0.16	1, 1, 1	0.2, 0.5, 0.16	0.16, 0.25, 0.12	1, 1, 1	0.11, 0.14, 0.09	0.14, 0.2, 0.11	0.14, 0.2, 0.11	1, 1, 1	7, 9, 11
Criterion 12	05, 1, 0.25	0.5, 1, 0.25	1, 1, 1	0.2, 0.33, 0.14	1, 1, 1	1, 1, 1	0.25, 0.5, 0.16	1, 1, 1	1, 1, 1	0.11, 0.14, 0.09	0.11, 0.14, 0.09	1, 1, 1
Criterion 13	0.16, 0.25, 0.12	0.11, 0.14, 0.09	0.5, 1, 0.25	1, 1, 1	0.12, 0.16, 0.1	0.11, 0.14, 0.09	0.16, 0.25, 0.12	0.11, 0.14, 0.09	0.14, 0.2, 0.11	0.2, 0.33, 0.14	0.11, 0.14, 0.09	1, 1, 1
Criterion 14	1, 1, 1	1, 1, 1	1, 1, 1	0.33, 1, 0.2	1, 1, 1	0.11, 0.14, 0.09	0.16, 0.25, 0.12	0.2, 0.33, 0.14	1, 1, 1	0.11, 0.14, 0.09	0.11, 0.14, 0.09	1, 1, 1
Criterion 15	0.33, 1, 0.2	0.5, 1, 0.25	1, 1, 1	0.5, 1, 0.25	0.16, 0.25, 0.12	0.14, 0.2, 0.11	1, 1, 1	0.14, 0.2, 0.11	0.14, 0.2, 0.11	0.16, 0.25, 0.12	1, 1, 1	0.2, 0.33, 0.14
Criterion 16	0.333, 1, 0.2	0.5, 1, 0.25	1, 1, 1	0.5, 1, 0.25	1, 1, 1	0.16, 0.25, 0.12	0.14, 0.2, 0.11	0.14, 0.2, 0.11	0.14, 0.2, 0.11	0.14, 0.2, 0.11	1, 1, 1	0.14, 0.2, 0.11
Criterion 17	0.5, 1, 0.25	0.33, 1, 0.2	0.33, 1, 0.2	0.14, 0.2, 0.11	1, 1, 1	1, 1, 1	1, 1, 1	1, 1, 1	1, 1, 1	1, 1, 1	1, 1, 1	1, 1, 1
Criterion 18	0.11, 0.14, 0.09	0.11, 0.14, 0.09	1, 1, 1	0.33, 1, 0.2	0.11, 0.14, 0.09	0.16, 0.25, 0.12	1, 1, 1	0.16, 0.25, 0.12	0.14, 0.2, 0.11	0.16, 0.25, 0.12	0.14, 0.2, 0.11	1, 1, 1
Criterion 19	0.11, 0.14, 0.09	0.12, 0.16, 0.1	1, 1, 1	1, 1, 1	0.11, 0.14, 0.09	0.14, 0.2, 0.11	0.14, 0.2, 0.11	0.16, 0.25, 0.12	0.16, 0.25, 0.12	0.11, 0.14, 0.09	0.11, 0.14, 0.09	0.12, 0.16, 0.1
Criterion 20	0.5, 1, 0.25	0.5, 1, 0.25	1, 1, 1	0.2, 0.33, 0.14	0.25, 0.5, 0.16	0.2, 0.33, 0.14	1, 1, 1	0.2, 0.33, 0.14	0.16, 0.25, 0.12	0.11, 0.14, 0.09	1, 1, 1	0.11, 0.14, 0.09
Criterion 21	1, 1, 1	0.5, 1, 0.25	0.33, 1, 0.2	0.12, 0.16, 0.1	1, 1, 1	1, 1, 1	1, 1, 1	1, 1, 1	1, 1, 1	1, 1, 1	1, 1, 1	1, 1, 1
Criterion 22	0.33, 1, 0.2	0.11, 0.14, 0.09	1, 1, 1	0.25, 0.5, 0.16	0.16, 0.25, 0.12	0.16, 0.25, 0.12	0.5, 1, 0.25	0.16, 0.25, 0.12	0.14, 0.2, 0.11	0.11, 0.14, 0.09	1, 1, 1	0.16, 0.25, 0.12
Criterion 23	0.11, 0.14, 0.09	0.11, 0.14, 0.09	1, 1, 1	0.16, 0.25, 0.12	0.11, 0.14, 0.09	0.14, 0.2, 0.11	1, 1, 1	0.16, 0.25, 0.12	0.14, 0.2, 0.11	0.14, 0.2, 0.11	1, 1, 1	0.14, 0.2, 0.11
Criterion 24	0.33, 1, 0.2	1, 1, 1	1, 1, 1	1, 1, 1	0.33, 1, 0.2	0.33, 1, 0.2	0.11, 0.14, 0.09	0.11, 0.14, 0.09	0.11, 0.14, 0.09	0.11, 0.14, 0.09	0.33, 1, 0.2	0.11, 0.14, 0.09
Comparison	**Criterion 13**	**Criterion 14**	**Criterion 15**	**Criterion 16**	**Criterion 17**	**Criterion 18**	**Criterion 19**	**Criterion 20**	**Criterion 21**	**Criterion 22**	**Criterion 23**	**Criterion 24**
Criterion 1	4, 6, 8	1, 1, 1	1, 3, 5	1, 3, 5	1, 2, 4	2, 4, 6	7, 9, 11	1, 2, 4	1, 1, 1	1, 3, 5	7, 9, 11	1, 3, 5
Criterion 2	7, 9, 11	1, 1, 1	1, 2, 4	1, 2, 4	1, 3, 5	2, 4, 6	6, 8, 10	1, 2, 4	1, 2, 4	2, 4, 6	2, 4, 6	1, 1, 1
Criterion 3	1, 2, 4	1, 1, 1	1, 1, 1	1, 1, 1	1, 1, 1	1, 1, 1	1, 1, 1	1, 1, 1	1, 3, 5	1, 1, 1	1, 1, 1	1, 1, 1
Criterion 4	1, 1, 1	1, 3, 5	1, 2, 4	1, 2, 4	5, 7, 9	1, 3, 5	1, 1, 1	3, 5, 7	6, 8, 10	2, 4, 6	4, 6, 8	1, 1, 1
Criterion 5	6, 8, 10	1, 1, 1	4, 6, 8	1, 1, 1	1, 1, 1	2, 4, 6	7, 9, 11	2, 4, 6	1, 1, 1	3, 5, 7	2, 4, 6	4, 6, 8
Criterion 6	7.9, 11	2, 4, 6	5, 7, 9	4, 6, 8	1, 1, 1	4, 6, 8	5, 7, 9	3, 5, 7	1, 1, 1	4, 6, 8	5, 7, 9	1, 3, 5
Criterion 7	4, 6, 8	4, 6, 8	1, 1, 1	5, 7, 9	1, 1, 1	1, 1, 1	5, 7, 9	1, 1, 1	1, 1, 1	1, 2, 4	1, 1, 1	2, 4, 6

Table 2. *Cont.*

Comparison	Criterion 13	Criterion 14	Criterion 15	Criterion 16	Criterion 17	Criterion 18	Criterion 19	Criterion 20	Criterion 21	Criterion 22	Criterion 23	Criterion 24
Criterion 8	7, 9, 11	3, 5, 7	5, 7, 9	6, 8, 10	1, 1, 1	4, 6, 8	4, 6, 8	3, 5, 7	1, 1, 1	4, 6, 8	4, 6, 8	7, 9, 11
Criterion 9	5, 7, 9	1, 1, 1	6, 8, 10	5, 7, 9	1, 1, 1	5, 7, 9	4, 6, 8	4, 6, 8	1, 1, 1	5, 7, 9	6, 8, 10	7, 9, 11
Criterion 10	3, 5, 7	2, 4, 6	4, 6, 8	6, 8, 10	1, 1, 1	4, 6, 8	7, 9, 11	4, 6, 8	1, 1, 1	7, 9, 11	5, 7, 9	7, 9, 11
Criterion 11	7, 9, 11	7, 9, 11	1, 1, 1	1, 1, 1	1, 1, 1	5, 7, 9	7, 9, 11	1, 1, 1	1, 1, 1	1, 1, 1	1, 1, 1	1, 3, 5
Criterion 12	1, 1, 1	1, 1, 1	3, 5, 7	5, 7, 9	1, 1, 1	1, 1, 1	6, 8, 10	7, 9, 11	1, 1, 1	4, 6, 8	5, 7, 9	7, 9, 11
Criterion 13	1, 1, 1	1, 1, 1	1, 1, 1	1, 1, 1	1, 1, 1	3, 5, 7	6, 8, 10	1, 1, 1	1, 1, 1	1, 1, 1	1, 1, 1	1, 1, 1
Criterion 14	1, 1, 1	1, 1, 1	3, 5, 7	2, 4, 6	1, 1, 1	2, 4, 6	7, 9, 11	1, 1, 1	1, 1, 1	2, 4, 6	1, 2, 4	4, 6, 8
Criterion 15	1, 1, 1	0.2, 0.33, 0.14	1, 1, 1	1, 1, 1	1, 1, 1	1, 1, 1	7, 9, 11	1, 1, 1	1, 1, 1	1, 1, 1	1, 1, 1	2, 4, 6
Criterion 16	1, 1, 1	0.25, 0.5, 0.16	1, 1, 1	1, 1, 1	1, 1, 1	1, 1, 1	6, 8, 10	1, 1, 1	1, 1, 1	1, 1, 1	1, 1, 1	3, 5, 7
Criterion 17	1, 1, 1	1, 1, 1	1, 1, 1	1, 1, 1	1, 1, 1	5, 7, 9	7, 9, 11	7, 9, 11	1, 1, 1	4, 6, 8	4, 6, 8	7, 9, 11
Criterion 18	0.2, 0.33, 0.14	0.25, 0.5, 0.16	1, 1, 1	1, 1, 1	0.14, 0.2, 0.11	1, 1, 1	7, 9, 11	1, 1, 1	1, 1, 1	1, 1, 1	1, 1, 1	3, 5, 7
Criterion 19	0.12, 0.16, 0.1	0.14, 0.11, 0.09	0.14, 0.11, 0.09	1, 1, 1	0.14, 0.11, 0.09	0.14, 0.11, 0.09	1, 1, 1	1, 1, 1	1, 1, 1	1, 1, 1	1, 1, 1	1, 1, 1
Criterion 20	1, 1, 1	1, 1, 1	1, 1, 1	1, 1, 1	0.14, 0.11, 0.09	1, 1, 1	1, 1, 1	1, 1, 1	1, 1, 1	4, 6, 8	1, 2, 4	3, 5, 7
Criterion 21	1, 1, 1	1, 1, 1	1, 1, 1	1, 1, 1	1, 1, 1	1, 1, 1	1, 1, 1	1, 1, 1	1, 1, 1	7, 9, 11	1, 1, 1	7, 9, 11
Criterion 22	1, 1, 1	0.25, 0.5, 0.16	1, 1, 1	1, 1, 1	0.25, 0.5, 0.16	1, 1, 1	1, 1, 1	0.25, 0.5, 0.16	0.14, 0.11, 0.09	1, 1, 1	7, 9, 11	2, 4, 6
Criterion 23	1, 1, 1	0.5, 1, 0.25	1, 1, 1	1, 1, 1	0.25, 0.5, 0.16	1, 1, 1	1, 1, 1	0.5, 1, 0.25	1, 1, 1	0.14, 0.11, 0.09	1, 1, 1	2, 4, 6
Criterion 24	1, 1, 1	0.16, 0.25, 0.12	0.25, 0.5, 0.16	0.2, 0.33, 0.14	0.14, 0.11, 0.09	0.2, 0.33, 0.14	1, 1, 1	0.2, 0.33, 0.14	0.14, 0.11, 0.09	0.25, 0.5, 0.16	0.25, 0.5, 0.16	1, 1, 1

3.2. Investigating the Factors Affecting the Sustainable Management of Agricultural Water Resources for the Optimal Use of Water in the Agricultural Sector

The results of Table 3, which were obtained from the pairwise comparison matrix of the studied indicators, show the factors affecting the sustainable management of agricultural water resources in the optimal use of water in the agricultural sector. Acceptors of water resources management, the type of exploitation system, farmers' income, and the level of mechanization (criteria 4, 9, 12, 13, and 14) were preferable to other criteria.

According to the results related to the pairwise comparison of visit program indicators for farmers regarding indigenous knowledge, the prevalence of training classes on the use of indigenous knowledge, practical projects on indigenous knowledge projects, the use of publications, written training packages (film, books, tapes, etc.), specialized seminars on promotional projects for the development of indigenous knowledge, and the use the internet and e-learning, which are related to the educational-promotional factor, show that the socio-cultural factor has a relative superiority over other indicators.

According to the results of a pairwise comparison of belief options towards indigenous knowledge, farmers' interest in using indigenous knowledge, communication with neighbors regarding the use of indigenous knowledge, interest in membership in social groups (baneh) and membership in the water organizations branch were related to the socio-cultural factor. The managerial factor is superior over other criteria.

The results of pairwise comparison of criteria related to the managerial factor—providing specialized manpower for indigenous knowledge projects, organizing financial mechanisms of officials regarding indigenous knowledge, the attitudes of managers towards indigenous knowledge projects, the practical participation of officials and managers in indigenous knowledge promotion projects, transferring powers and decisions to farmers, bottom-up decision-making—show that the economic factor is relatively superior over other factors.

The results of a pairwise comparison of options related to the economic factor—farmers' income, insurance for agricultural products at risk of drought, drought grants, the extension of drought loan repayment, drought loans, bank facilities for the purchase of pump motors and machines, show that the socio-cultural factor is relatively superior.

According to the opinions of experts in the Sistan region, indicators have been identified to investigate the factors affecting indigenous knowledge in the optimal use of water in the agricultural sector and have been fitted using the analytic hierarchy process (FAHP) model. Considering that the purpose of this study was to investigate the factors affecting indigenous knowledge in the optimal use of water in the agricultural sector, prioritization has been conducted and, as can be seen in Figure 2, the educational-extension factor, with a final weight of 0.46, has had the greatest impact on indigenous knowledge in the optimal use of agricultural water, followed by socio-cultural (0.33), managerial (0.15), and economic (0.06) factors.

According to the results obtained from a pairwise comparison of educational-promotional factors—the use of publications, the use of radio and television, holding classes and training courses, the distribution of magazines and educational publications, the use of training workshops, and communication with relevant specialists (agricultural promoters)—the factor (option) of farmers' knowledge and information was superior. A pairwise comparison of the criteria for government support—the delivery of subsidized inputs to the recipients of water resources management and the material and spiritual incentives of the recipients of water resources—showed that the knowledge and information of farmers was superior over other factors.

Table 3. Paired comparison matrix of indicators for determining the factors affecting the sustainable management of agricultural water resources in the optimal use of water in the agricultural sector of Sistan.

Comparison	Criterion1	Criterion2	Criterion3	Criterion4	Criterion5	Criterion6	Criterion7	Criterion8	Criterion9	Criterion10	Criterion11	Criterion12	Criterion13	Criterion14	Criterion15	Criterion16
Criterion 1	1, 1, 1	1, 1, 1	1, 1, 1	1, 3, 5	1, 1, 1	2, 4, 6	1, 1, 1	1, 1, 1	3, 5, 7	1, 1, 1	1, 1, 1	4, 6, 8	6, 8, 10	7, 9, 11	1, 1, 1	5, 7, 9
Criterion 2	1, 1, 1	1, 1, 1	1, 1, 1	1, 2, 4	1, 1, 1	3, 5, 7	1, 1, 1	1, 1, 1	3, 5, 7	1, 1, 1	1, 1, 1	4, 6, 8	5, 7, 9	7, 9, 11	1, 1, 1	5, 7, 9
Criterion 3	1, 1, 1	1, 1, 1	1, 1, 1	1, 3, 5	1, 1, 1	2, 4, 6	1, 1, 1	1, 1, 1	4, 6, 8	1, 1, 1	1, 1, 1	5, 7, 9	6, 8, 10	7, 9, 11	1, 1, 1	6, 8, 10
Criterion 4	0.33, 1, 0.2	0.5, 1, 0.25	0.33, 1, 0.2	1, 1, 1	1, 1, 1	1, 1, 1	1, 1, 1	1, 1, 1	1, 3, 5	1, 1, 1	1, 1, 1	1, 3, 5	3, 5, 7	3, 5, 7	1, 1, 1	1, 3, 5
Criterion 5	1, 1, 1	1, 1, 1	1, 1, 1	1, 1, 1	1, 1, 1	4, 6, 8	1, 1, 1	1, 1, 1	3, 5, 7	1, 1, 1	1, 1, 1	3, 5, 7	5, 7, 9	7, 9, 11	1, 1, 1	5, 7, 9
Criterion 6	0.25, 0.16, 0.12	0.33, 1, 0.2	0.25, 0.16, 0.12	1, 1, 1	0.25, 0.16, 0.12	1, 1, 1	1, 1, 1	1, 1, 1	1, 1, 1	1, 1, 1	1, 1, 1	1, 3, 5	7, 9, 11	1, 2, 4	1, 1, 1	1, 3, 5
Criterion 7	1, 1, 1	1, 1, 1	1, 1, 1	1, 1, 1	1, 1, 1	1, 1, 1	1, 1, 1	1, 1, 1	4, 6, 8	1, 1, 1	1, 1, 1	5, 7, 9	3, 5, 7	7, 9, 11	1, 1, 1	6.8, 10
Criterion 8	1, 1, 1	1, 1, 1	1, 1, 1	1, 1, 1	1, 1, 1	1, 1, 1	1, 1, 1	1, 1, 1	4, 6, 8	1, 1, 1	1, 1, 1	4, 6, 8	4, 6, 8	7, 9, 11	1, 1, 1	6, 8, 10
Criterion 9	0.33, 0.2, 0.14	0.33, 0.2, 0.14	0.25, 0.16, 0.12	0.33, 1, 1, 0.2	0.25, 0.16, 0.12	1, 1, 1	0.25, 0.16, 0.12	0.25, 0.16, 0.12	1, 1, 1	1, 1, 1	1, 1, 1	1, 2, 4	1, 3, 5	3, 5, 7	1, 1, 1	1, 1, 1
Criterion 10	1, 1, 1	1, 1, 1	1, 1, 1	1, 1, 1	1, 1, 1	1, 1, 1	1, 1, 1	1, 1, 1	1, 1, 1	1, 1, 1	1, 1, 1	5, 7, 9	5, 7, 9	7, 9, 11	1, 1, 1	7, 9, 11
Criterion 11	1, 1, 1	1, 1, 1	1, 1, 1	1, 1, 1	1, 1, 1	1, 1, 1	1, 1, 1	1, 1, 1	1, 1, 1	1, 1, 1	1, 1, 1	6, 8, 10	5, 7, 9	7, 9, 11	1, 1, 1	7, 9, 11
Criterion 12	0.25, 0.16, 0.12	0.25, 0.16, 0.12	0.2, 0.14, 0.11	0.33, 1, 0.2	0.33, 0.2, 0.14	0.33, 1, 0.2	0.2, 0.14, 0.11	0.25, 0.16, 0.12	0.5, 1, 0.25	0.2, 0.14, 0.11	0.2, 0.14, 0.11	1, 1, 1	1, 2, 4	1, 2, 4	1, 1, 1	1, 1, 1
Criterion 13	0.2, 0.14, 0.11	0.2, 0.14, 0.11	0.2, 0.14, 0.11	0.33, 0.2, 0.1	0.2, 0.14, 0.11	0.14, 0.11, 0.09	0.33, 0.2, 0.14	0.2, 0.14, 0.11	0.33, 1, 0.2	0.2, 0.14, 0.11	0.2, 0.14, 0.11	0.5, 1, 0.25	1, 1, 1	7, 9, 11	1, 1, 1	1, 1, 1
Criterion 14	0.14, 0.11, 0.09	0.14, 0.11, 0.09	0.14, 0.11, 0.09	0.33, 0.2, 0.1	0.14, 0.11, 0.09	0.2, 0.14, 0.11	0.2, 0.14, 0.11	0.14, 0.11, 0.09	0.33, 0.2, 0.14	0.14, 0.11, 0.09	0.14, 0.11, 0.09	0.5, 1, 0.25	0.14, 0.11, 0.09	1, 1, 1	1, 1, 1	1, 1, 1
Criterion 15	1, 1, 1	1, 1, 1	1, 1, 1	1, 1, 1	1, 1, 1	1, 1, 1	1, 1, 1	1, 1, 1	1, 1, 1	1, 1, 1	1, 1, 1	1, 1, 1	1, 1, 1	1, 1, 1	1, 1, 1	7, 9, 11
Criterion 16	0.2, 0.14, 0.11	0.2, 0.14, 0.11	0.2, 0.14, 0.11	0.33, 1, 0.2	0.2, 0.14, 0.11	0.33, 1, 0.2		0.14, 0.11, 0.09	1, 1, 1	0.14, 0.11, 0.09	0.14, 0.11, 0.09	1, 1, 1	1, 1, 1	1, 1, 1	0.14, 0.11, 0.09	1, 1, 1

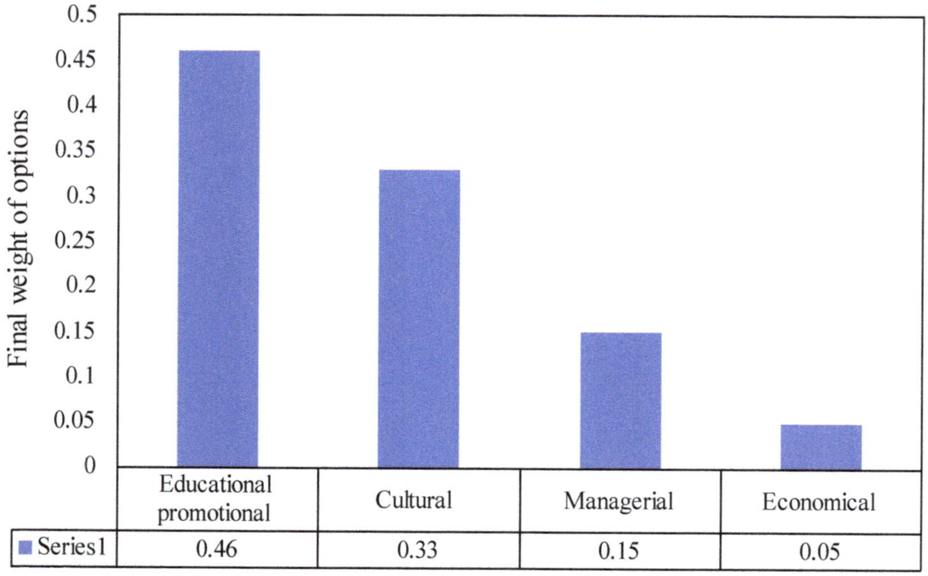

Figure 2. Weight of options based on indicators of factors affecting indigenous knowledge in optimal water consumption in agriculture.

The results of pairwise comparison of economic factors—the level of farmers' crop cultivation, the type of exploitation system, farmers' income, and the level of mechanization—showed that farmers' knowledge and information is superior to other options. Parallel comparison of options according to the criteria of farmers' technical knowledge of the stages up to harvest and their level of education, which are related to the factor of knowledge and information of farmers, showed that all education-extension, social, and government support factors are equally superior.

Scientific research is set up to discover the truth, find solutions to challenges, and improve human lives. Communication with neighboring countries and the use of their knowledge contributes to the optimal use of water in agriculture, and, by joining a aquifer organization, this creates the ground for correct and optimal consumption, as well as increasing farmers' incomes and the loans given to farmers during droughts. The above can be one of the most important factors affecting the indigenous knowledge of the region.

The results of evaluating the factors affecting the sustainable management of agricultural water resources in the optimal use of water in the agricultural sector showed that the distribution of educational magazines and publications, the delivery of subsidized inputs to recipients of water resources management, the type of exploitation system, farmers' income, and the level of mechanization were the preferred criteria. The results of this research confirm those of Shahbakhsh et al. [1], Abbasian et al. [2], and Farrokhzadeh et al. [3]. The existence of promotional activities, such as the distribution of magazines and publications, increases farmers' awareness of the sustainability of water resources, and authorities create incentive amongst designers to promote the sustainability and optimal use of water resources. Based on the factors affecting the sustainable management of agricultural water resources in the optimal use of water in the agricultural sector, the educational-extension factor, with a final weight of 0.37 is defined here as the first priority (Figure 3). The reason for the superiority of the criteria was due to the use of publications, the use of radio and television, holding classes and training courses, distributing educational-promotional magazines and publications, using training workshops, and communicating with specialists (agricultural promoters). Meanwhile, social factors, government, economic support, and farmers' knowledge and information, with final weights of 0.24, 0.21, 0.13, and 0.03, are the next priorities.

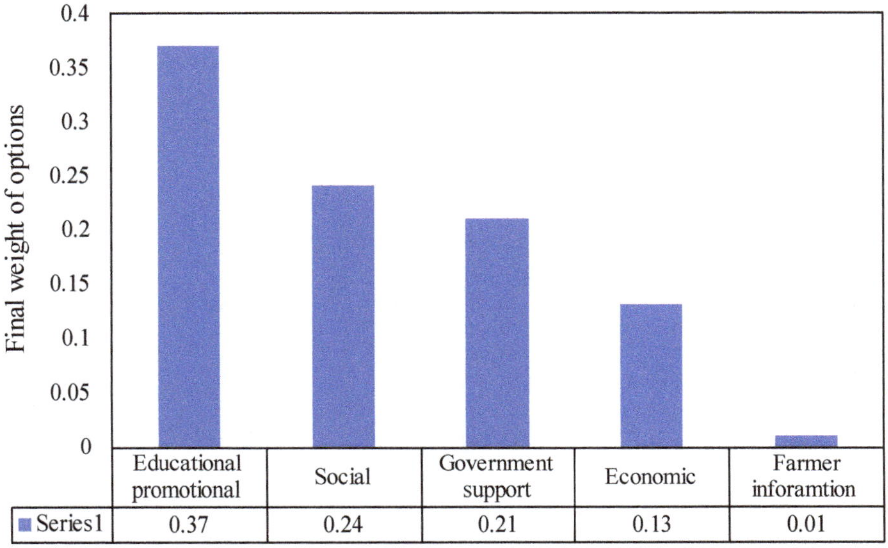

Figure 3. Final weight of options based on the indicators of factors affecting the sustainable management of agricultural water resources in the optimal use of water in agriculture.

4. Conclusions

Water shortage is one of the main limiting factors for the development of economic activities in the coming years. Iran is located in one of the driest regions in the world. Increasing population and the limitations of water resources make the targeted management of water resources even more necessary. Therefore, maintaining water resources and applying proper management in the exploitation of the above resources, especially in the agricultural sector, is of particular importance. The reason for this is that the production of agricultural products is intended to meet food needs, and in the future, due to population growth on the one hand and restrictions on arable land to limit water on the other hand, meeting the country's food needs will be a major problem.

Indigenous knowledge, which includes local beliefs, values and awareness, and their social knowledge, is a part of the national capital of any ethnic group and is of particular importance in the optimal use of water in the agricultural sector. In the present study, fuzzy hierarchical analysis (FAHP) was applied separately to the factors affecting local knowledge and sustainable management of agricultural water resources for the optimal use of water in the agricultural sector. The criteria and study options were grouped according to previous studies and using the opinions of experts, and the relationship between them was examined. The results of evaluation of factors affecting indigenous knowledge in optimal water consumption in agriculture indicated that, among the indicators used, indices of relations with neighbors regarding the use of indigenous knowledge, membership in water collectors, farmers' incomes and drought loans, were superior to other indicators.

Based on results of this research, the following suggestions are presented:

1. To increase the level of indigenous knowledge of regional officials in relation to the optimal and correct use of water in agriculture.
2. To use the experiences of neighboring countries for indigenous knowledge and optimal use of agricultural water in the region.
3. To establish research organizations on activities and studies of indigenous knowledge in the region with optimal water consumption.

4. To hold training classes related to the sustainability of water resources.
5. To encourage and support farmers in using new irrigation methods.

Author Contributions: Conceptualization, A.S.S. and O.B.; methodology, O.B., H.E.A.; Validation, T.P., H.E.A.; formal analysis, A.S.S., data curation, A.S.S.; writing—original draft preparation, A.S.S.; writing—review and editing, supervision, O.B. and T.P.; funding acquisition, T.P. All authors have read and agreed to the published version of the manuscript.

Funding: Foundation for Science and Technology–Portugal grant: UIDB/04007/2020.

Informed Consent Statement: Informed consent was obtained from all subjects involved in the study.

Data Availability Statement: Data is available on demand.

Acknowledgments: The study was supported by a grant from the Foundation for Science and Technology–Portugal (UIDB/04007/2020).

Conflicts of Interest: The authors declare no conflict of interest.

References

1. Shahbakhsh, F.; Safdari, M.; Sardar Shahraki, A. Assessment of modern approach of water governance in the devel-opment of water exploitation systems in Sistan region. *J. Appl. Res. Water Wastewater* **2020**, *7*, 127–136.
2. Farrokhzadeh, S.; Hashemi Monfared, S.A.; Azizyan, G.; Sardar Shahraki, A.; Ertsen, M.W.; Abraham, E. Sustainable Water Resources Management in an Arid Area Using a Coupled Optimization-Simulation Modeling. *Water* **2020**, *12*, 885. [CrossRef]
3. Abbasian, M.; Shahraki, A.S.; Ahmadi, N.A. Economic Management of Water by Using Valuation Policy in Mango Orchards with an Emphasis on Environmental Inputs in Chabahar County. *Iran. Econ. Rev.* **2021**, *26*, 727–738. [CrossRef]
4. Trenčiansky, M.; Štěrbová, M.; Výbošťok, J. The Influence of the Transition to Ecological Farming on the Quality of Runoff Water. *Sustainability* **2022**, *14*, 15412. [CrossRef]
5. Trenčiansky, M.; Štěrbová, M.; Výbošťok, J.; Lieskovský, M. Impacts of forest cover on surface runoff quality in small catchments. *Bioresources* **2021**, *16*, 7830–7845. [CrossRef]
6. Jacobs, K.; Lebel, L.; Buizer, J.; Addams, L.; Matson, P.; McCullough, E.; Garden, P.; Saliba, G.; Rijsberman, F.; Manning, N.; et al. Increasing green and blue water productivity to balance water for food and environment. In Proceedings of the Final Report of the 4th World Water Forum Water, Food, and Environment Theme Baseline document, Mexico City, Mexico, 16–22 March 2006; pp. 139–143.
7. Panahi, F. Analysis of factors affecting the optimal management of water resources in the iranian agricultural system. *Agri-Cult. Ext. Educ. Res.* **2012**, *1*, 110–117.
8. Giupponi, C.; Mysiak, J.; Fassio, A.; Cogan, V. MULINO-DSS: A computer tool for sustainable use of water resources at the catchment scale. *Math. Comput. Simul.* **2004**, *64*, 13–24. [CrossRef]
9. Saffari, H. Identifying and explaining the factors affecting water resources management technology in agriculture and natural resources. *Res. Technol.* **2015**, *2*, 112–128.
10. Rashidipour, L.; Kalantari, K.; Rezvanfar, A. Investigating the problems and limitations of water resources and its impact on the socio-economic status of wheat farmers in the central part of Saqez. *Agric. Econ. Develop.* **2011**, *19*, 183–204.
11. Bouzarjomehri, K.A.; Mahsoumi Jashni, M.; Jahantigh, H.A. Investigating the indigenous role and local traditions of women in the production and management of livestock products (study case: Abu Nasr village, Bavanat city). *Bi-Q. J. Iran. Indig. Knowl.* **2016**, *6*, 91–120.
12. Bouzarjomehri, K. Handbook of strategies for optimizing water consumption in agriculture with emphasis on some indige-nous technologies in iran and other countries. *Iran. J. Agric. Econ. Dev. Res.* **2009**, *42*, 80–84.
13. CWWEC. *Country Water and Wastewater Engineering Company*; Information Center for Electricity, Water and Energy Engineering Services: Saba, The Netherlands, 2011.
14. Omani, A.R. Investigating the effective factors in sustainable management of agricultural resources in the northern part of Modares watershed in Khuzestan province. *Watershed Manag. Res.* **1950**, *88*, 27–34.
15. Arfai, M. Investigating the factors affecting indigenous knowledge in optimal water consumption in agriculture. *Agric. Ext. Educ. Res.* **2011**, *3*, 93–103.
16. Bandani, M.; Mir Lotfi, M.R.; Seibanishad, A. Study and analysis of the role of indigenous knowledge in sustainable rural development of Ghaemabad rural district of Sistan region. In Proceedings of the National Conference on Civil Engineering and Architecture with an Approach to Sustainable Development, Tehran, Iran, 1 October 2015.
17. Rahimian, M. Investigating the factors affecting the sustainable management of water resources among irrigated wheat farmers in Koohdasht. *Iran. Agric. Ext. Educ. Sci.* **2015**, *12*, 233–247.
18. Akhavan, M.; Behbahaninia, A. Investigating the economic factors affecting sustainable development and management of water resources in agriculture. *Sustain. Dev. Environ.* **2017**, *4*, 31–41.

19. Afshari, S.; Rezaei, R.A.; Gholizadeh, H.; Shaban Alifami, H. Factors determining farmers' attitudes towards sustainable management of water resources (case study: Komijan County). *Q. J. Environ. Educ. Sustain. Dev.* **2017**, *6*, 101–113.
20. Hassani, N.; Yadalahi, P.; Mortazavi, A.A. Investigating the factors affecting farmers' water resources management behaviors (case study: Hamadan-Bahar Plain). *J. Water Resour. Eng.* **2017**, *10*, 1–10.
21. Iglesias, A.; Garrote, L. Adaptation strategies for agricultural water management under climate change in Europe. *Agric. Water Manag.* **2015**, *155*, 113–124. [CrossRef]
22. Valipour, M. Land use policy and agricultural water management of the previous half of century in Africa. *Appl. Water Sci.* **2014**, *5*, 367–395. [CrossRef]
23. Valipour, M.; Ziatabar Ahmadi, M.; Raeini-Sarjaz, M.; Gholami Sefidkouhi, M.A.; Shahnazari, A.; Fazlola, R.; Darzi-Naftchali, A. Agricultural water management in the world during past half century. *J. Arch. Agron. Soil Sci.* **2015**, *61*, 657–678. [CrossRef]
24. Kernecker, M.; Vogl, C.R.; Meléndez, A.A. Women's local knowledge of water resources and adaptation to landscape change in the mountains of Veracruz. *Mexico. Ecol. Soc.* **2017**, *22*, 37. [CrossRef]
25. Greenland, S.J.; Dalrymple, J.; Levin, E.; O'Mahony, B. Improving Agricultural Water Sustainability: Strategies for Effective Farm Water Management and Encouraging the Uptake of Drip Irrigation. In *The Goals of Sustainable Development*; David, C., Shahla, S., Abdul, M., Eds.; Springer: Berlin/Heidelberg, Germany, 2018; pp. 111–123.
26. Ataie, M. *Fuzzy Multi Criteria Decision Making*; Shahrood University of Technology: Shahrood, Iran, 2009.

water

MDPI

Article

High Resolution Estimation of Ocean Dissolved Inorganic Carbon, Total Alkalinity and pH Based on Deep Learning

Charles Galdies [1,*] and Roberta Guerra [2,3]

1 Institute of Earth Systems, University of Malta, MSD 2080 Msida, Malta
2 Department of Physics and Astronomy (DIFA), Alma Mater Studiorum—Università di Bologna, 40126 Bologna, Italy
3 Interdepartmental Research Centre for Environmental Sciences (CIRSA), University of Bologna, 48123 Ravenna, Italy
* Correspondence: charles.galdies@um.edu.mt

Abstract: This study combines measurements of dissolved inorganic carbon (DIC), total alkalinity (TA), pH, earth observation (EO), and ocean model products with deep learning to provide a good step forward in detecting changes in the ocean carbonate system parameters at a high spatial and temporal resolution in the North Atlantic region (Long. −61.00° to −50.04° W; Lat. 24.99° to 34.96° N). The in situ reference dataset that was used for this study provided discrete underway measurements of DIC, TA, and pH collected by M/V Equinox in the North Atlantic Ocean. A unique list of co-temporal and co-located global daily environmental drivers derived from independent sources (using satellite remote sensing, model reanalyses, empirical algorithms, and depth soundings) were collected for this study at the highest possible spatial resolution (0.04° × 0.04°). The resulting ANN-estimated DIC, TA, and pH obtained by deep learning shows a high correspondence when verified against observations. This study demonstrates how a select number of geophysical information derived from EO and model reanalysis data can be used to estimate and understand the spatiotemporal variability of the oceanic carbonate system at a high spatiotemporal resolution. Further methodological improvements are being suggested.

Keywords: ocean acidification; ocean carbonate system; dissolved inorganic carbon; total alkalinity; pH; North Atlantic; spatiotemporal variability; earth observation; deep learning

Citation: Galdies, C.; Guerra, R. High Resolution Estimation of Ocean Dissolved Inorganic Carbon, Total Alkalinity and pH Based on Deep Learning. *Water* **2023**, *15*, 1454. https://doi.org/10.3390/w15081454

Academic Editors: Luis Garrote and Alban Kuriqi

Received: 25 February 2023
Accepted: 3 April 2023
Published: 7 April 2023

1. Introduction

The global oceans constitute an important component in the global carbon cycle. They are also a major sink of human-induced emissions of CO_2. When CO_2 dissolves under typical ocean surface conditions, 90% of this CO_2 is formed as HCO_3^-, 9% as HCO_3^{2-}, and only 1% as undissociated CO_2 *(aq)* and H_2CO_3 [1]. The four important parameters that are needed to understand the ocean carbonic acid system include the dissolved inorganic carbon (DIC), the total alkalinity (TA), the pH, and the pCO_2 in surface water.

In the past decades, most of our understanding of the ocean carbonate system is derived from in situ observations. Now, thanks to global networking programs, observations have increased widely and consistently, due to ship surveys, the ARGOS project, and mooring and autonomous platforms; furthermore, due to the availability of ever more complex biogeochemical models, the understanding of ocean global and regional carbonate system has advanced considerably. These activities provide accurate, long-term time series fCO_2 datasets, such as those found in the Surface Ocean CO_2 Atlas—SOCAT—[2,3] and the Global Ocean Data Analysis Project (GLODAPv2.2022), consisting of data products of biogeochemical data collected through the chemical analysis of water samples, including TA, DIC, and many others [4]. This information now shows that surface ocean waters show around a 26% increase in concentration of hydrogen ions since 1860, which is equivalent to a drop in pH from 8.2 to 8.1 [5]. This change has been mainly attributed to the rising anthropogenic emissions of CO_2 [5].

From a measurement point of view, changes in pH occur on a large spatial scale and can be influenced by different environmental parameters, especially at the local scale. Due to their very nature, direct field measurements are inherently limited in spatial (time series, moored stations) and/or temporal resolution (ship surveys). Earth observation (EO), on the other hand, offers an avenue for expanding observations and analyzing the temporal and spatial variability of the global ocean and its properties. While EO has proved to be a difficult tool for the direct monitoring of seawater pH and its impact on marine organisms, satellite remote sensing can indirectly measure this by providing us with a range of related physico-chemical and biological processes occurring at the ocean surface at an unprecedented spatiotemporal scale. In addition, even though in situ surface measurements offer a geographically limited representation of the entire oceanic volume and its contents, remote sensing observations of the global ocean become very important for the study of the carbonate system, due to the fact that the change in ocean chemistry arises first in the ocean surface. Thus, environmental satellites have great potential in this field.

At the local level, coastal communities are most vulnerable to a lowering pH, especially where the ocean chemistry is changing most rapidly due to multiple stressors. These communities have the potential of being the worst hit, both economically and socially, especially those who derive benefits from calcifying organisms and other vulnerable species [6]. This explains the need for the rapid monitoring of such coastal waters.

This study asks the following research questions: (1) how can we provide information on the state of ocean carbonate information (such as pH and other important carbonate chemistry parameters) at suitable geographical scales that are useful for the management of marine resources? and (2) how can a more robust monitoring of the ocean carbonate system be made available; one that is chemically, biologically, and physically linked to a good number of environmental drivers instead of a much smaller number of parameters, such as salinity, temperature, and chlorophyll? [7].

In seeking to address these research questions, this study moves away from others that have modeled ocean carbonate parameters at coarse temporal [8] and spatial scales (around 500–1500 km; [9]). Instead, it aims to provide ocean carbonate system parameter information at an unmatched high spatial (4 km) and temporal (such as daily) level via gridded ocean maps, with the opportunity of assimilating this into daily operational monitoring and forward the modeling that is used by a wide variety of ocean end users. This goes perfectly in line with NOAA-SOCAN's top research priorities, i.e., "to monitor key ocean parameters across various spatial and temporal scales that will provide information on mechanistic drivers of acidification and input parameters for predictive model algorithm development" (known as 'priority 1′) by developing "operational and qualitative models that can transition to end users and adapting existing models to understand acidification" (known as 'priority 3′) from a "regional perspective as well as in specific systems" [10]. The end-user sectors of this data may range from artisanal and small-scale or semi-industrial fisheries and bivalve aquaculture [11] to coastal managers and policy makers whose actions need to become more adaptive in the short term.

To resolve this challenging aim, this study uses the artificial neural network (ANN) method to fix those specific, inter-related environmental conditions that can lead to particular states of the ocean carbonate system. It does so by following the approach that has been taken by the latest ocean research that uses time-finite, individual-ship-based transect measurements that cross extended oceanic areas such as the North Atlantic Ocean [12], the northwest European shelf seas [13], and the North Pacific Ocean [14], among others.

The calculations that have been carried out in this study were performed at a very high spatiotemporal resolution of a so-far unique list of environmental drivers that, in combination, are able to describe and model the much-needed detailed spatiotemporal variation of surface DIC, TA, and pH. This approach can lead to the prediction of a unique set of high-resolution, daily DIC, TA, and pH regional ocean surface grid maps, with potential applications in future studies focused on the local dynamics of the carbonate systems in both coastal and oceanic areas.

Now, the vast availability of daily EO data and related ancillary data are ideally suited for the ANN's model-free estimators and for predictive data mining. In this study, the ANN allows the processing of different chemical, biological, and physical ocean values by estimating the most probable field values on the basis of their previous patterns, as observed out in the field. Depending on the algorithmic architecture, the ANN is able to perform its estimations through association, clustering, and prediction of the required output variables. While keeping in mind the practicality and the feasibility of this study, it is very important to create an ANN architecture that is able to learn, and ultimately model, the association between the ocean carbonate parameters and the largest possible number of oceanic physicochemical and biological processes. The potential use of such a tool can be extremely important for the validation of numerical ocean modeling and the prediction of changes in ocean carbonate chemistry.

2. Materials and Methods

2.1. Study Area

The study area covers part of the Atlantic Ocean, comprising part of the Iberian Plain, with the Canary basin on the east side and the North American basin on the western side, reaching to the Puerto Rico trench.

Time series measurements show that the North and Central Atlantic constitutes the largest reservoir of anthropogenic CO_2 [15–17] and displays a surface ocean pH decline [18]. Moreover, a strong correlation between the pCO_2 and the surface water pH was identified by Bates et al., 2012 [19], with the latter showing a definitive decrease in the North Atlantic Ocean between 1984 and 2012. Furthermore, in its 2015 and 2016 State of the Climate, NOAA reported a world record in terms of large sea surface temperature and upper ocean heat content anomalies in large swaths of the western North Atlantic Ocean [20,21]. This extreme event can offer an interesting opportunity to continue studying the changes in DIC, with respect to pH, TA, and sea temperature [22], whilst making use of the novelty of this study.

2.2. Field Data

This study made use of the best surface underway data available over the study area for the period of 2015–2016. The Ocean Carbon and Acidification Data Portal of the National Centers for Environmental Information provides only one set of surface underway data (NCEI Accession 0154382) that contains the three core study variables of DIC, TA, and pH over the study area covering the period of analysis (https://www.ncei.noaa.gov/data/oceans/ncei/ocads/metadata/0154382.html (accessed on 20 February 2023)). Additional surface underway datasets are available; however, these consist of an increasingly limited number of observations (such as NCEI Accession 0157237, 0157352, 0157312, and 0110259), for which suitable co-located and co-temporal satellite and model reanalysis data are not available.

2.2.1. In Situ Observations of the Carbonate System

From 7 March 2015 to 6 November 2016, the M/V Equinox (ID: MLCE) sailed across the North Atlantic Ocean three times. Discrete surface underway measurements of seawater DIC, TA, and pH were performed on all cruises (Figure 1). The details of the laboratory methods onboard the M/V Equinox are well documented [23] as NCEI Accession 0154382. This research was conducted in support of the coastal monitoring and research objectives of the NOAA Ocean Acidification Program (OAP) and the Climate Program Office. The research cruise covered an area from $-78.9797°$ W to $-10.3998°$ E and from 38.4622° N to 19.2893° S.

In addition to DIC, TA, and pH, M/V Equinox also collected sea surface temperature and sea surface salinity measurements with a documented uncertainty of ± 0.001 °C and $\pm 0.005\%$, respectively (see https://www.ncei.noaa.gov/data/oceans/ncei/ocads/metadata/0154382.html (accessed on 20 February 2023)). The range of the values collected during the cruise mission is shown in Table 1.

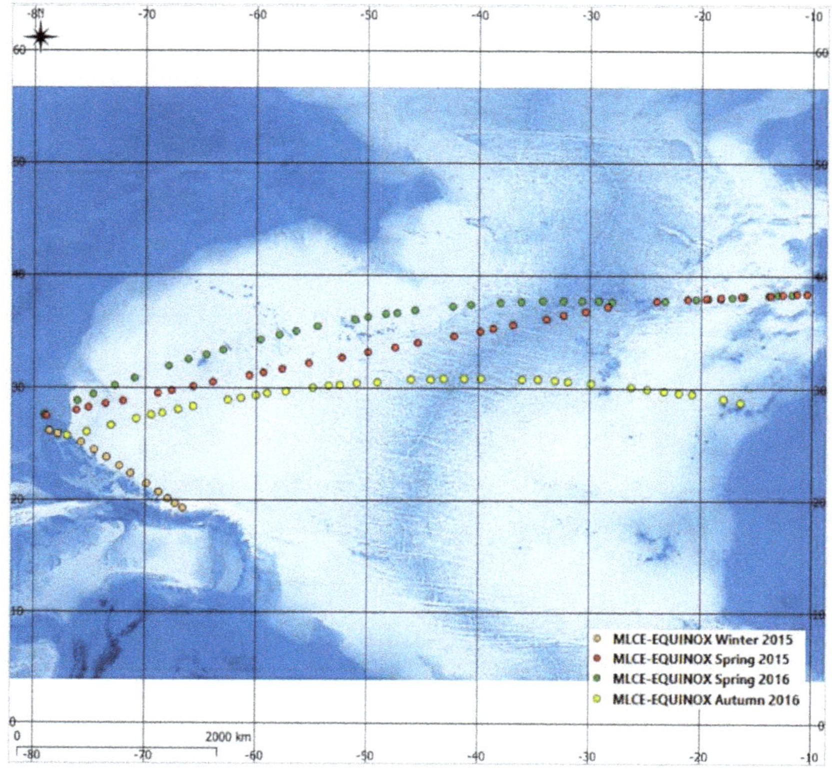

Figure 1. Sampling periods of in situ discrete underway samples of DIC, TA, and pH measured by M/V Equinox (source: NCEI Accession 0154382) overlaid over bathymetry (source: GEBCO) for Longitude −80° to −10° and Latitude +18° to +40°. Inset: Winter 2015: Validation dataset 1; Autumn 2016: Validation dataset 2; Spring 2015: Validation dataset 3; Spring 2016: ANN training dataset. The observations along the red transect were used to train the ANN for the prediction of DIC, TA, and pH. The surface underway measurements shown in brown, yellow, and green were used to validate the ANN algorithm against other independent datasets.

Table 1. Data value range and difference Δ along the transects M/V Equinox (ID: MLCE NCEI Accession 0154382) for the entire cruise period.

Parameter	Range	Δ
SST (°C)	15.2–27.5	12.3
SSS (PSU)	35.46–36.95	1.49
DIC (μmol.kg^{-1})	2025–2126	101
TA (μmol.kg^{-1})	2350–2439	89
pH	7.964–8.142	0.178

2.2.2. Remote Sensing Data and Reanalysis Data

Co-temporal and co-located met-ocean parameters that are considered to be somehow connected with the ocean carbonate system were derived from independent sources using earth observation satellite remote sensing (BD 1–7; PD 1–5), model reanalyses (PD 10), and empirical algorithms (PD 6–8) (Table 2). The GEBCO bathymetry (PD 9) was derived from a mix of ship track soundings, with the interpolation between soundings guided by satellite-derived gravity data.

Table 2. Co-temporal and co-located environmental drivers derived from independent sources that range from satellite remote sensing and model analyses to empirical algorithms were collected.

Parameter	Code	Source	Resolution	Reference
Biological drivers				
Water-leaving surface reflectance (Rrs) at 412, ... 555 nm)	BD 1	MODIS (Aqua, Terra)	0.042°, daily, global	[24,25]
Rrs 443/555	BD 2	MODIS (Aqua, Terra)	0.042°, daily, global	[24,25]
Rrs 531/555	BD 3	MODIS (Aqua, Terra)	0.042°, daily, global	[24,25]
Rrs 443/488	BD 4	MODIS (Aqua, Terra)	0.042°, daily, global	[24,25]
Chlorophyll-a	BD 5	MODIS (Aqua, Terra)	0.042°, daily, global	[26]
Particulate Inorganic Carbon (PIC)	BD 6	VIIRS	0.042°, daily, global	[27]
Particulate Organic carbon (POC)	BD 7	VIIRS	0.042°, daily, global	[28,29]
Physical drivers				
Sea surface salinity	PD 1	SMOS	0.05°, daily, global	[30]
Sea surface temperature	PD 2	OISST	0.25°, daily, global	[26]
Wind speed	PD 3	ASCAT	0.25°, daily, global	[26]
Wind direction	PD 4	ASCAT	0.25°, daily, global	[26]
Wind stress	PD 5	ASCAT	0.25°, daily, global	[31]
Transfer velocity (w)	PD 6	Based on ASCAT	0.25°, daily, global	[32]
Transfer velocity	PD 7	Based on ASCAT	0.25°, daily, global	[33]
Transfer velocity	PD 8	Based on ASCAT	0.25°, daily, global	[34]
Bathymetry	PD 9	GEBCO	0.083°, global	[35]
Mean layer depth	PD 10	Global ocean 1/12° physics analysis and forecast updated daily. Copernicus marine environment monitoring service.	0.083°, daily mean, global analyses, 50 depth levels	[36,37]

2.2.3. Justification for the Selection and Use of Environmental Drivers

Figure 2 shows the linkage between the various environmental drivers used in this study and how these were used to model the target ocean surface DIC, TA, and pH. The environmental drivers can be seen to represent the following three proxies of oceanic processes:

1. Kinetic forcing, by looking at atmospheric stability (proxies, such as transfer velocity, that affect the partial pressure of CO_2 (pCO_2), wind speed, wind direction, and wind stress on the ocean surface);
2. Thermohaline forcing, by looking at proxies such the sea surface temperature and the sea surface salinity;
3. Biological forcing, by looking at proxies such as chlorophyll-a, surface reflectance and its ratios, and particulate organic and inorganic carbon and its ratios;
4. Water-side convection and upwelling, by looking at proxies such as mixing layer depth and bathymetry.

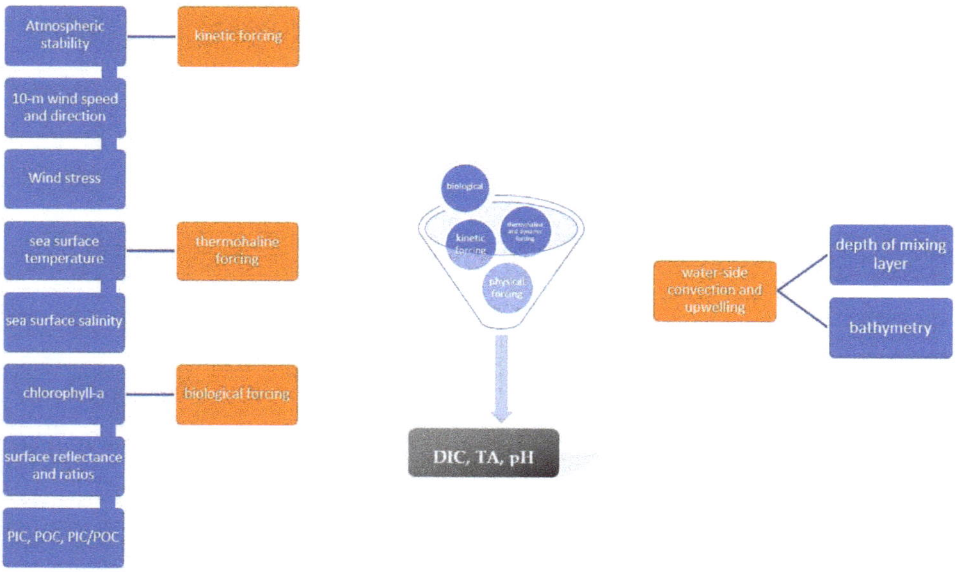

Figure 2. Linkage between the various groups of environmental drivers and how these were used to model or predict the three target parameters of surface DIC, TA, and pH. The environmental drivers can be seen as representing some of the main met-ocean processes influencing these three target variables (based on [38]).

These four processes were used to closely represent as much as possible the forcing that leads to the derivation of DIC, TA, and pH using our algorithm. Native resolution grids of all of the environmental drivers considered for this study, including PD1, were resampled to a common 0.04° × 0.04° global raster grid for a suitable retrieval of all co-located data. Table 3 provides a summarized justification for the inclusion of these drivers into the predictive algorithm.

Table 3. Justification of the use of the biological and physical drivers of surface DIC, TA, and pH used for this study.

Environmental Driver	Summary	Reference
Transfer velocity	The transfer velocity describes the efficiency exchange of CO_2 across the air–sea interface and dissolution in water on the basis of ΔpCO_2 between the water and the atmosphere.	[32–34,39–45]
Wind speed (U_{10}) and direction (DD)	The wind speed determines the structure and fluxes at the air–sea interface. It has an important effect on the magnitude and direction of the CO_2 flux across the air–sea interface, which differs according to the prevalent wind and turbulence regimes.	[46–53]
Mean layer depth	This is the depth at which the density difference from the surface reaches 0.02 kg m^{-3}. Within this layer, the properties of density, temperature, and salinity are more uniform, due to the mixing. When this layer is well-defined, a significantly enhanced transfer velocity within it is observed.	[36,37,54–57]

Table 3. *Cont.*

Environmental Driver	Summary	Reference
Wind stress	Wind stress is able to affect the vertical transport of dissolved gases, such as CO_2.	[31]
Sea surface salinity	Sea surface salinity has been used as a proxy indicator for pCO_2 using statistical analysis and artificial neural networks. CO_2 solubility is a function of temperature and salinity.	[31,58–61]
Sea surfacetemperature	Sea surface pCO_2 depends on the SST, such that when the SST increases by 1 °C, the surface pCO_2 increases 4-fold.	[26,62–68]
Depth	The depth and structure of the sea bottom can influence the intensity of upwelling. High levels of CO_2 from deep water can be brought to the surface through upwelling and released into the atmosphere. This can be enhanced in the case of an existing deep-water circulation.	[69]
Biological activity	Photosynthesis acts to bind CO_2 into organic matter and can affect DIC concentration. Studies show that chlorophyll-a correlates well with pCO_2.	[26,67,70]
Particulate Organic carbon (POC)	POC is a proxy of coccolithophore production, which in turn is often used as a measure of net productivity. The phenomenon of sinking POC is part of the biological pump, which provides a mechanism for the sequestration of carbon in the deep ocean.	[25,71]
Particulate Inorganic Carbon (PIC)	PIC is used as a measure of net calcification by coccolithophores. The PIC:POC ratio is considered to be an important term for modeling carbon cycling in the oceans and, therefore, is a good indicator of changes in seawater CO_2.	[72–74]

2.3. *Algorithm Development and Validation*

2.3.1. Training of the ANN

For this study, a back propagation neuron (BPN) algorithm was trained by supervised learning by providing it with values of the co-located and co-temporal environmental drivers (Figure 3) and the corresponding DIC, TA, and pH (Figure 2) that constitute the final output for this study. Since the BPN algorithm is central to much current work on learning in NN, and has been independently invented several times (e.g., [75,76]), we used this algorithm to perform our desired task. The BPN algorithm feeds forward the input training pattern, which is then followed by the back propagation of the associated error, and which is finally expressed as a weight adjustment.

In order to supply training power to the BPN algorithm, the in situ Spring 2016 dataset (i.e., 16–24 April 2016) measurements (Table 1) were used as the values of the output neurons, while their corresponding (i.e., co-located and co-temporal) physico-chemical and biological drivers (Table 2), which were obtained independently, were used as the values of the input neurons. The location of the sampling points spanned across the entire North Atlantic Ocean, and thus presented the desired wide-ranging variability in both the physico-chemical and the biological conditions, which in turn led to the value range of DIC, TA, and pH observed during that period (Table 1). This process was carried out to optimize the BPN weights, such that the error function became minimal. The choice of the input (predictors) and output (predictands) dataset was targeted towards having a BPN algorithm that was able to model the output variables under different physical environmental conditions within the area of interest.

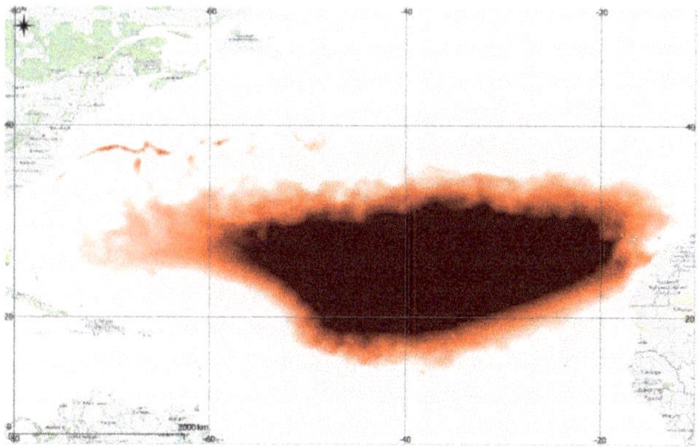

PD 1:

Salinity (PSU)

36.5
36.609
36.718
36.827
36.936

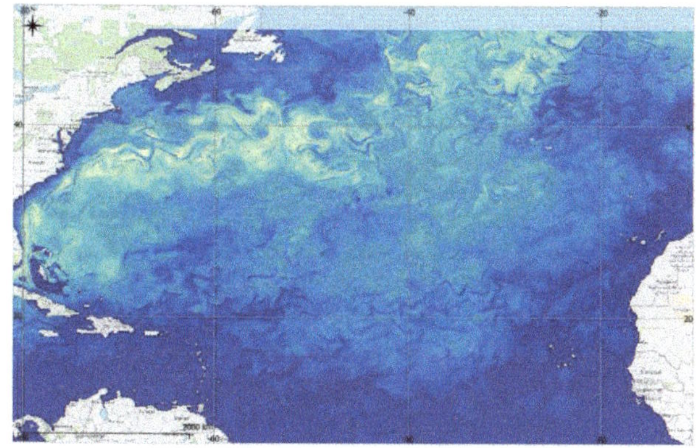

PD 10: Ocean mixed layer thickness defined by sigma theta

5.1881
32.6548
60.1215
87.5881
115.0548

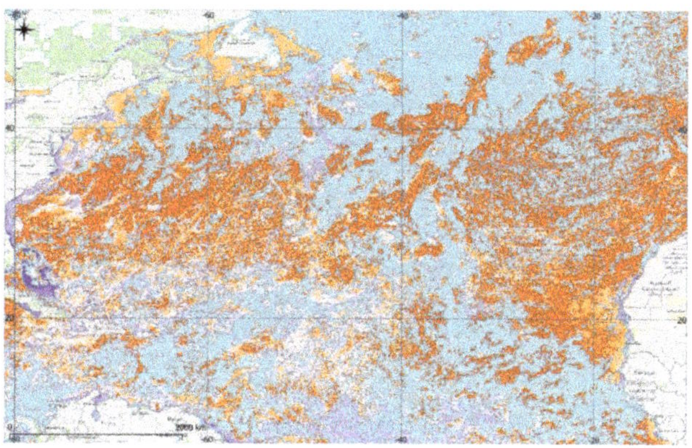

BD 6: PIC (log) (mol.m^{-3}) Scale factor: 0.2

0.0000
0.0030
0.0059
0.0088
0.0118

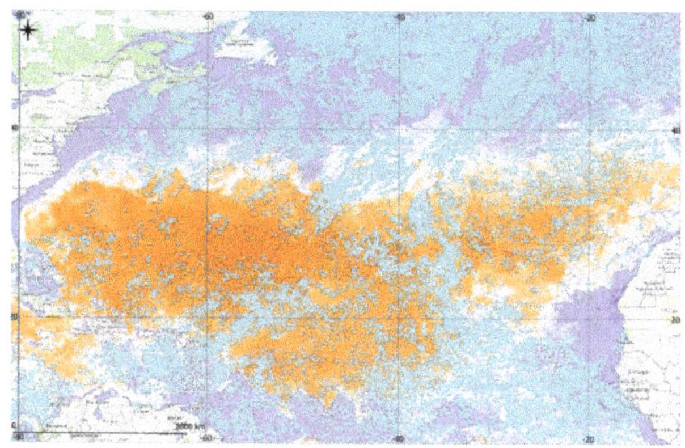

BD 7: POC (log) (mg.m^{-3})

Scale factor: 0.2

17.5999
32.1975
39.6025
61.2014

Figure 3. *Cont.*

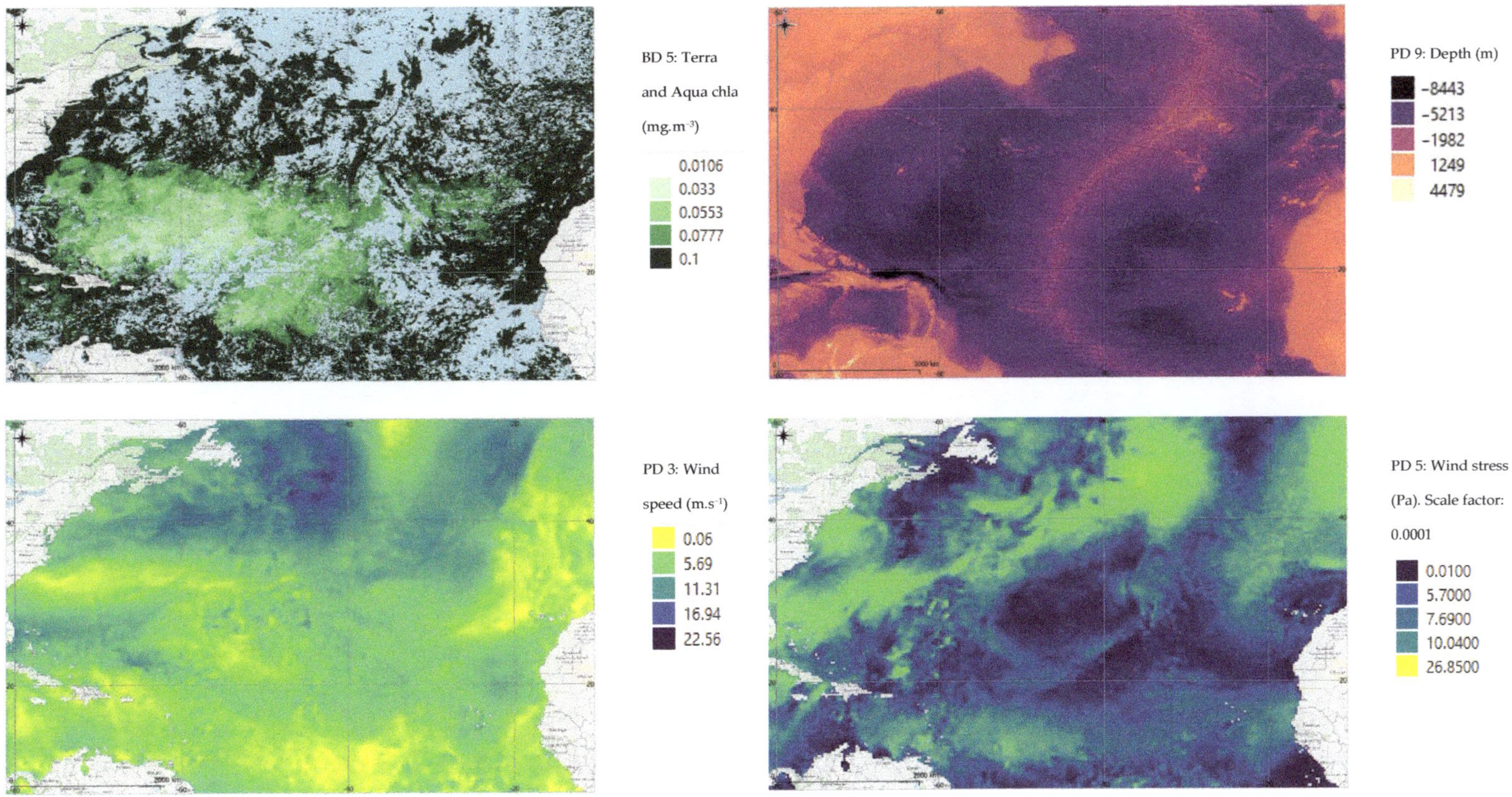

Figure 3. *Cont.*

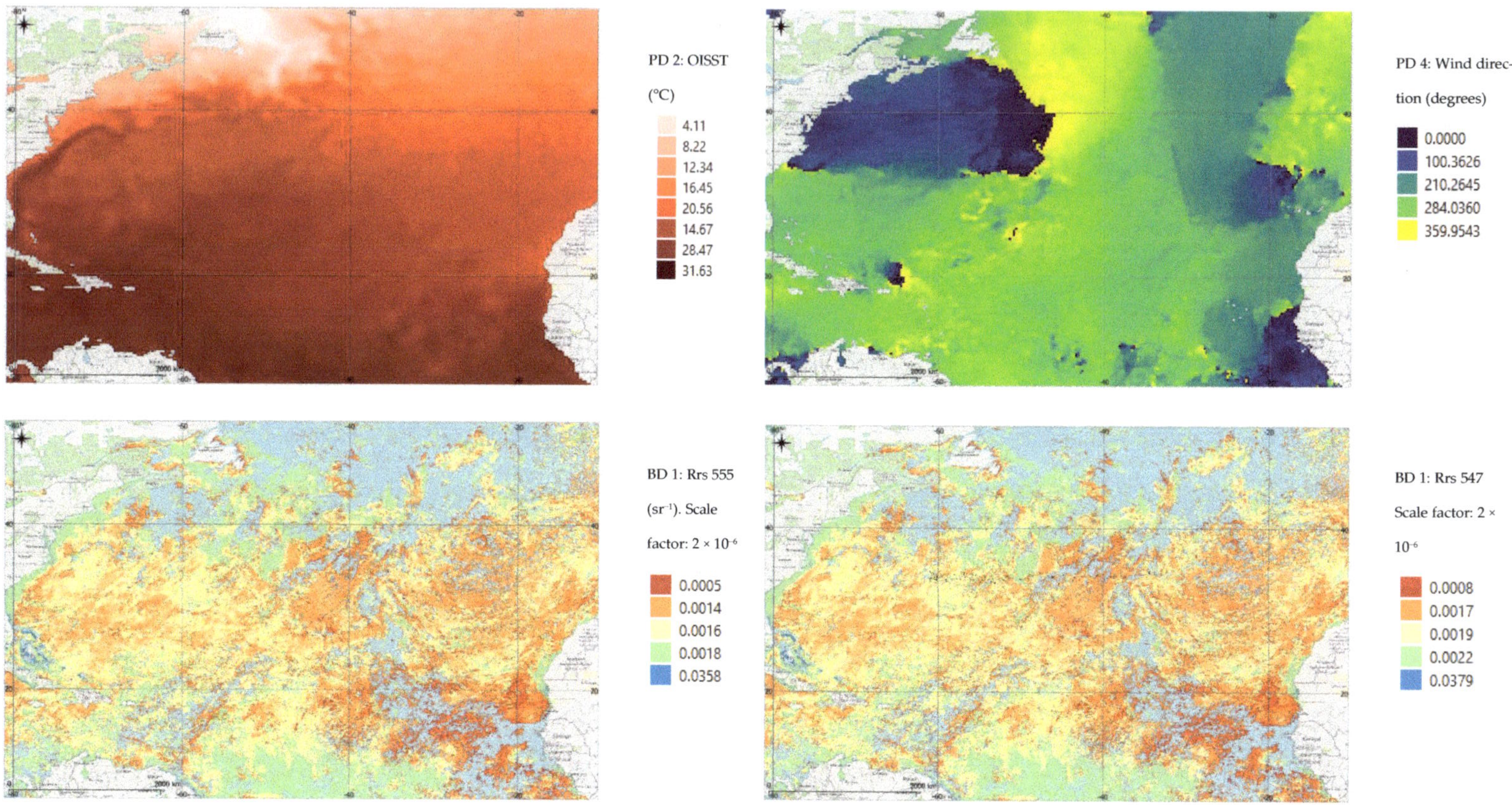

Figure 3. *Cont.*

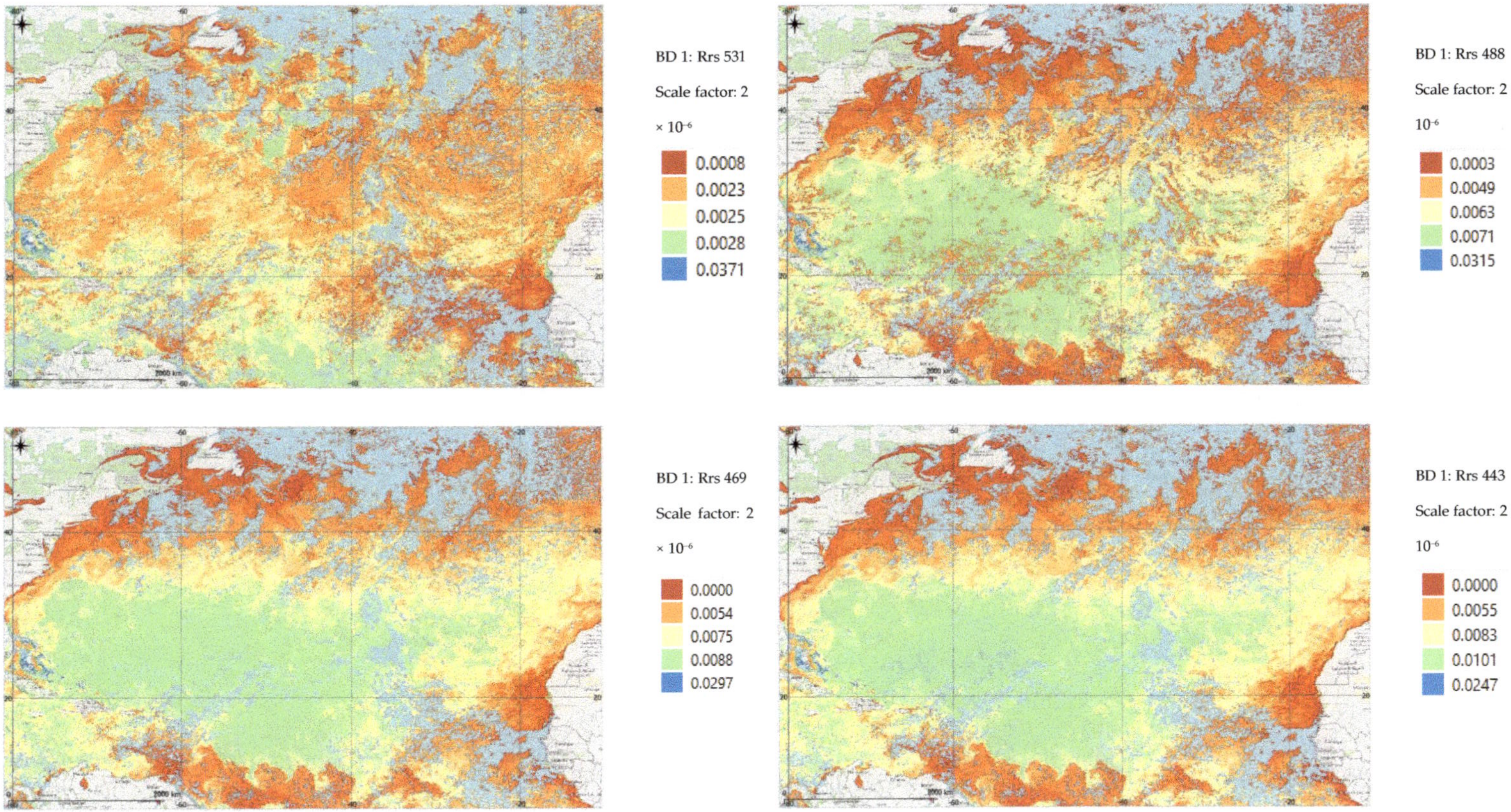

Figure 3. *Cont.*

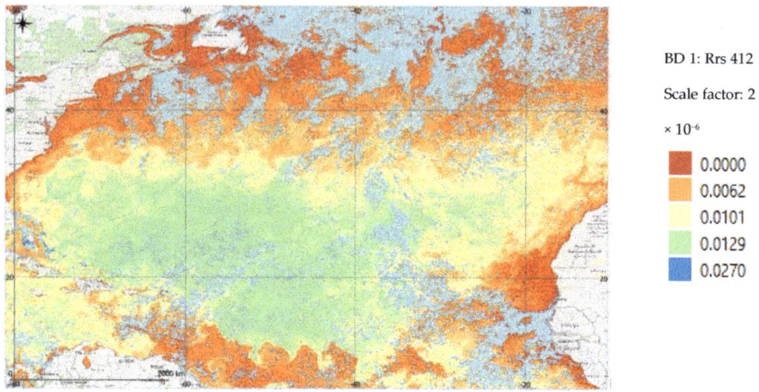

Figure 3. Multi-source, composite physico-chemical and biological data covering the period of Spring 2016, from which co-located and co-temporal values at points shown in Figure 1 were extracted to train the BPN algorithm. The sources and codes of these multi-source, independent datasets are listed in Table 2.

During this algorithm training, the net output was compared with the target value and the resultant error was calculated. It was here that the error factor was distributed back to the hidden layer, and the weights were updated accordingly. The error factor was calculated in a similar manner for all of the units, and their weights were updated simultaneously.

The ultimate objective here was to reduce this training error for the BPN algorithm until the ANN learned on the basis of the training data. The weights were gradually adjusted by means of a learning rule until they were capable of optimizing the predictive modeling of DIC, TA, and pH, as shown in Equation (1), as follows:

$$f(\text{DIC, TA, pH}) = (U_{10}, \text{DD, wind stress, transfer velocity, depth, SSS, chlorophyll-a, SST,} \\ R_{rs}\,412, R_{rs}\,443/555, R_{rs}\,531/555, R_{rs}\,443/48, \text{PIC, POC, MLD}) \quad (1)$$

Multi-source, geo-located EO and model reanalysis datasets (Level 4, SMI format) covering the period of 16–24 April 2016 were derived (Figure 3) from the co-located and co-temporal values (corresponding to BD 1–7 and PD 1–10) at the points shown in red (i.e., Spring 2016: ANN training set) in Figure 1. Choosing the right number of hidden neurons is usually performed through trial and error [77]. The ANN optimal topology hinges on the complexity of the relations between inputs and outputs. In this study, two sets of hidden neurons were tested: $n = 5$ and $n = 10$, where the assumption was that the greater the number of nodes, the smaller the error on the training set. However, at a certain point, the generalization began to increase, and the first structure (i.e., $n = 5$) was chosen on the basis of the smallest value for RMSE that was achieved during the training phase. The best topology found in this study consisted of an input layer with 17 neurons, 5 neurons in the hidden layer, and an output layer consisting of 3 neurons whose output gave the scaled DIC, TA, and pH (Figure 4). The training algorithm adjusted the bias and weighting factors according to the negative gradients of the error cost function [58] for the final training pattern.

The ANN training process algorithm for DIC, TA, and pH is shown in Figure 4 as follows:

1. The collection of co-located and co-temporal input (i.e., independent environmental drivers) and co-located and co-temporal output (i.e., cruise measurements of DIC, TA, and pH) datasets;
2. The data were normalized and scaled to the range of 0 to 1 to suit the transfer function in the hidden (sigmoidal, discrete; logistical implementation) and output layer (linear): $\hat{A} = (A - A_{min})/(A_{max} - A_{min})$, where $\hat{A}$ is the normalized value and A_{min} and A_{max} are the minimum and maximum values of A, respectively;
3. Neural network designing and training;
4. The testing of the ANN topology.

The training of the BPN algorithm started by using a small, random weight. It propagated each input pattern to the output layer, compared the pattern in the output layer with the correct one, and adjusted the weights according to the back propagation learning algorithm. After the presentation of around 10,000 patterns, the weights converged, i.e., the network picked up the correct pattern, and the error-correction learning stopped. In so doing, the network systematically reduced and/or reinforced the weights of the connection architecture and all of the 'knowledge' in the BPN was then contained in the weights. Naturally, the magnitude of this error depended on the choice, relation, quality, and accuracy of the inputs (predictors).

The predictive power of the BPN algorithm was maximized by means of the following steps:

1. A large number of iterations was used (circa 10,000) in order to minimize the processing error of the training set as much as possible. The training was stopped when a very small and stable training error was achieved (circa 0.0007);
2. The number of learning samples consisted of entire sets of measurements spanning the northwestern Atlantic, with its inherent physical (including bathymetry, surface

salinity, winds, wind stress, temperature, and mixed layer depths) and biological (chlorophyll-a, dissolved organic carbon, and surface-leaving reflectance) parameters, in order to model the highest possible scenario for appropriate learning under a wide range of variability. This training procedure can be further improved by including input and output variables with a greater degree of variability, such as measurements covering other regional areas and time periods;
3. An optimal number of hidden units ($n = 5$) was found with the sigmoid activation function and a liner output unit to derive an optimal 'expressive' power of the network. The present training set presented a 'smooth' function and therefore the number of hidden units needed was kept to a minimum ($n = 5$). For strongly fluctuating functions, more hidden units are generally needed, which does not seem to be a requirement for our study.

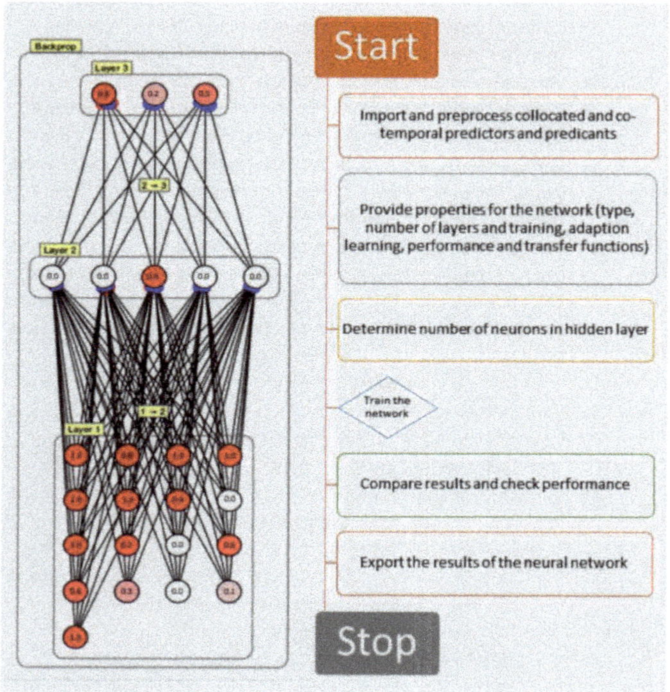

Figure 4. The network architecture of the BPN model (left) showing 5 neurons in the hidden layer and the respective weights (in red and blue) of each connection. The values in each of the neurons is a scaled down value (1 decimal place) of the input, hidden, and output neurons, corresponding to one possible solution between the proxy environmental drivers (predictors) and the values for DIC, TA, and pH (predictands). Layer 1 (input): 17 neurons (see Table 1 for a list of input neurons). Layer 2 (hidden): 5 neurons. Layer 3 (output): 3 neurons: DIC, TA, and pH. Steps involved in the development of the BPN model (right).

2.3.2. Performance of the BPN Algorithm

Entire M/V Equinox cruise transect datasets were reserved and used as independent datasets to validate the performance of the BPN training method. This is a common practice that ensures that the model can produce reliable estimates outside the range of the learning data (generalization capabilities) [78]. Thus, by assigning the trained BPN algorithm with the values of the fixed set of co-located and co-temporal input neurons as the physico-chemical and biological drivers, the resulting ANN-output-modeled DIC, TA, and pH were validated against the assigned datasets (Table 1).

The precision of the machine learning approach was evaluated, which was based on the trained BPN model, through a comparison with the M/V Equinox dataset using the mean bias (MB; Equation (2)) and the root mean square error (RMSE; Equation (3)), as well as the slope of the linear regression between the ANN-retrieved values and the corresponding in situ measured values, as follows:

$$MB = \Sigma_{n,i=1}(x_i - y_i)/n \tag{2}$$

$$RMSE_{fo} = \left[\frac{\sum_{i=1}^{N}\left(z_{f_i} - z_{oi}\right)^2}{N}\right]^{1/2} \tag{3}$$

where the mean bias is how far the model is from the ground truth data and RMSE determines the error on the test set (or generalization error). The objective of the best BPN model topology was based on the lowest possible metrics for the entire test data.

2.4. Construction of Gridded DIC, TA, and pH Gridded Data for 30 October 2016

Finally, the ability of the trained BPN algorithm to process and generate a huge number of DIC, TA, and pH data points was applied to a 1.1 million km^2 subset area located in the mid-North Atlantic Ocean, represented by a total number of 63,360 gridded data points (each encompassing the full set of 17 environmental drivers when available). In view of the extensive retrieval and processing requirements, these data points were based on the validated ANN algorithm and initiated by the physico-chemical and biological drivers that were retrieved on 30 October 2016.

The geographical extent of this area was west −61.00°; east −50.04°; west–east 10.96°; south 24.99°; north 34.96°; and south–north 9.96°. This area was chosen on the basis of its interesting hydrodynamics, as well as on its inter-annual trends in CO_2 concentrations. The large temporal and spatial gradients of pCO_2, as well as its variability driven by a diversity of physical and biological processes, make the analysis of the carbonate chemistry over the region both interesting and challenging [79]. The study's region of interest is influenced by the North Atlantic gyre and has a seasonal surface temperature variation of about 8 to 10 °C, occurring alongside a fluctuation in the MLD between the Northern Hemisphere's winter and summer seasons. On average, the MLD deepens to 200 m in winter up to about 10 m in summer. Generally, nutrients remain below the euphotic zone for most of the year, resulting in low primary production. During winter convective mixing, nutrients penetrate the euphotic zone, causing a short-lived phytoplankton bloom in the spring. All of these seasonal changes ultimately influence the total amount of CO_2 in the seawater.

All of the grid-point predictor variables were inserted in the BPN algorithm and the values of DIC, TA, and pH were modeled for that day for the entire area, with a native grid size of 0.04167°. On 30 October, there was a total of 7897 empty grid cells in this area that were attributed to cloud cover and, therefore, the lack of optically retrieved remotely sensed predictors (i.e., chlorophyll-a, R_{rs}, PIC, and POC).

3. Results and Discussion

3.1. Validation between Remotely Sensed- and Cruise-Derived SST and SSS Data

Table 4 shows a strong correlation between SST and SSS derived from the full cruise-segmented datasets (see Figure 1) and the remotely sensed PD1 and PD2.

3.2. Performance of the BPN Algorithm

By means of the independent validation datasets, we evaluated the performance of the algorithm by comparing the BPN-retrieved values of DIC, TA, and pH with the measurements that were taken by M/V Equinox (NCEI Accession 0154382) elsewhere, during the different time periods.

Table 4. Correlation between same variables obtained remotely and by M/V Equinox [ID: MLCE; 7 March 2015 to 6 November 2016] cruise-segmented datasets. Their p-value is <0.00001 and all of the correlations are significant at $p < 0.05$.

Sampling Period	Pearson Correlation R–Sea Surface Temperature	Pearson Correlation R–Sea Surface Salinity
7–8 March 2015	0.78	0.93
28 April–6 May 2015	0.99	0.69
16–24 April 2016	0.98	0.90

3.2.1. M/V Equinox—7–8 March 2015

Figure 5 shows the distribution and the statistical significance of the data points within the range that is shown by both in situ and ANN-estimated values, as well as the existence of outliers. The co-located, ANN-estimated DIC, TA, and pH values were in very good agreement with the surface underway measurements given that the BPN algorithm was trained on the data that were collected during 28 April–6 May 2015 along the entire North Atlantic width. The results show that the mean biases for DIC, TA, and pH are −2.5 μmol.kg^{-1}, −3.2 μmol.g^{-1}, and 0.0048, respectively. Compared to the range of DIC, TA, and pH that is shown by the surface underway measurements along all of the transects (Table 1), the values for the mean bias show low variations and a good ANN algorithm performance. Importantly, apart from the fact that no outliers were detected, the overall dispersion of the ANN-estimated values is well within the range of those shown by the M/V data. Some skewness is shown by the ANN-estimated pH and, to a lesser extent, for DIC. The similarity between these three sets of data is statistically significant at the 99% confidence level.

These results point to an effective BPN algorithm that is able to capture the information provided by the chosen environmental drivers. It is important to note that for oceanic and coastal regions with a different matrix of environmental drivers (such as for areas with high chlorophyll-a, where the net productivity is likely to perturb the carbonate system more, or in areas where there are river inputs), further learning of the BPN algorithm is therefore recommended.

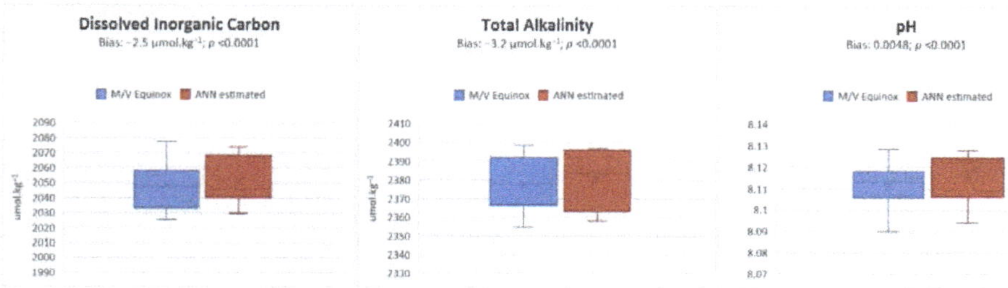

Figure 5. Median and variability of ANN-estimated values fall within those shown by discrete underway measurements (Winter 2015 cruise transect). The means of the two datasets are similar at the 99% C.L.

3.2.2. M/V Equinox—30 October to 6 November 2016, North Atlantic Ocean (20° N to 40° N; −80° W to −10° W)

Similarly, Figure 6 shows the resultant statistical evaluation when the ANN-estimated values were compared against the corresponding in situ data. As for the previous validation set, the predictions for the October–November 2016 dataset were in good agreement with the co-located and co-temporal M/V Equinox data. Overall, the ANN-estimated data show less dispersion than the in situ values and that the spread of the former is well within that shown by the data from M/V Equinox. The few ANN-estimated outliers are well within the interquartile range of the M/V Equinox data.

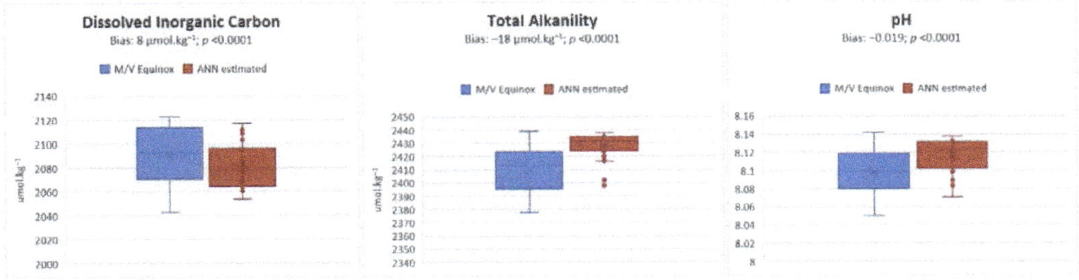

Figure 6. Median and variability of ANN-estimated values fall within those shown by discrete underway measurements (Autumn 2016 cruise transect). The means of the two datasets are similar at the 99% C.L.

The performance indicators between the modeled and the validation dataset 1 (i.e., the 7–8 March 2015 in situ dataset) point to a stronger estimation than in the case of the second validation dataset. This is most likely because dataset 1 is based on the same seasonal variations of the carbonate chemistry when compared to the second validation sample that was collected during the Autumn of 2016. The mean bias values generally show a non-Gaussian distribution and spread, with the exception of TA for both of the validation datasets, and pH for the Spring 2015 dataset (Figure 7). In the latter case, the residuals are skewed toward lower modeled values.

The uncertainties that were inherent in the in situ measurements were not included in the metadata information within NCEI Accession 0154382, and therefore this element of uncertainty attributed to the surface underway observation could not be evaluated. Overall, however, the results' metrics are very comparable to the validation metrics that were obtained by Fourrier et al., for their neural network estimation of pH and total alkalinity in the Mediterranean [80]. It is rather complex to identify the main sources of the observed metric errors in view of (1) the procedure that was used by this study and (2) the uncertainty embedded in the in situ data that were used for both the BNP algorithm training and its validation; however, this bias could be expected to decrease if the following steps are taken:

1. The further training of the BNP algorithm. In so doing, the training process of the BNP algorithm should allow for further 'learning' from the local/regional variability of both the predictors and predictands;
2. Although the neural networks have the ability to 'generalize', the additional retrieval of in situ measurements of surface DIC, TA, and pH from cruises can be carried out during other seasons over the same area, and combining this with the training set that was used for the BNP algorithm might prove useful;
3. Expand the range of predictors (i.e., environmental drivers; see Section 3.3.2 below).

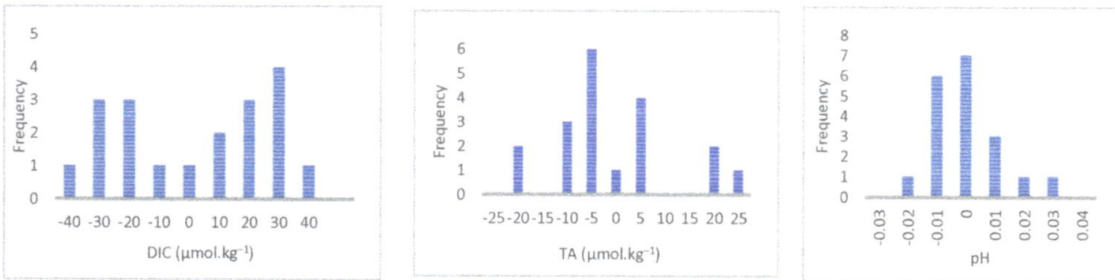

(**a**) M/V Equinox: 7–8 March 2015, off Bahamas and Dominican Republic.

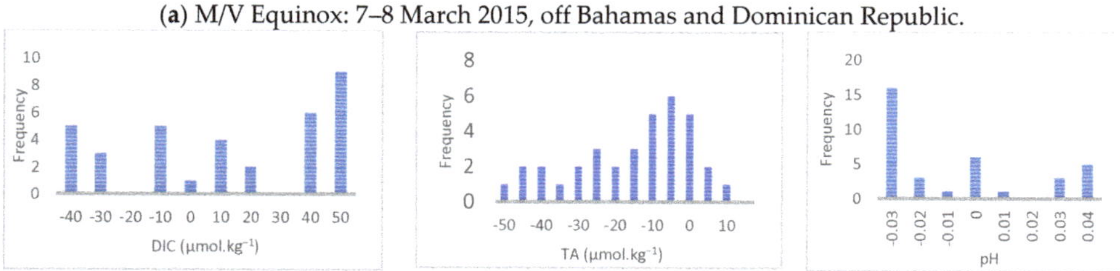

(**b**) M/V Equinox: 30 October–6 November 2016, North Atlantic Ocean.

Figure 7. Histogram reporting the distribution of the mean bias values for DIC, TA, and pH.

3.3. *Model Applications: ANN-Derived Ocean Variability of DIC, TA, and pH over the Mid-North Atlantic Ocean*

Based on the previous two validation studies that span different time periods and geographical areas (where each area manifests its own variability in terms of the magnitude of the environmental drivers), we were able to apply the validated ANN topology to model DIC, TA, and pH within the ROI described in Section 2.4 at a resolution of 0.04167°. The final product was a set of gridded, time-specific geophysical maps of these predictands (i.e., surface DIC, TA, and pH). The resolution of these maps took on the native resolution of the input (i.e., predictor) datasets (i.e., 17 environmental drivers). If needed, these raster outputs can be subsequently re-gridded to coarser resolutions in order to (1) further understand the spatiotemporal variability of the carbonate system over specific oceanic regions, (2) comprehensively map the carbonate system components in support of the cruise data, and (3) input the predicted values into numerical modeling systems (such as ocean forecasting models).

Figure 8a–d represents the gridded output of DIC, TA, and pH maps for the area of interest that were produced by the ANN algorithm. The data gaps represent that no ocean surface data are available whenever clouds obstruct part of the field of view of the optical satellite sensors, at which points the ANN algorithm nullifies the predictions. These high-resolution data representing the carbonate system of the area can be exploited by other modeling activities, including data assimilation for general circulation models [81] and improved model reanalyses [82], as well as the identification of daily trends over sensitive marine areas [83].

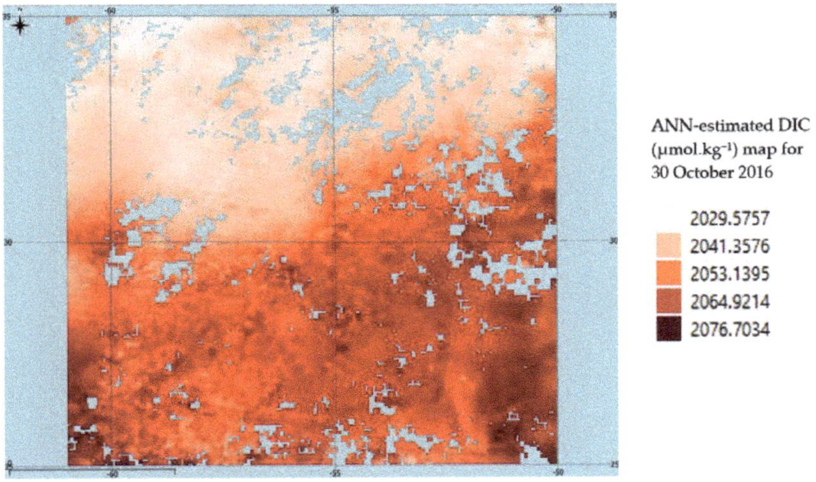

(**a**)

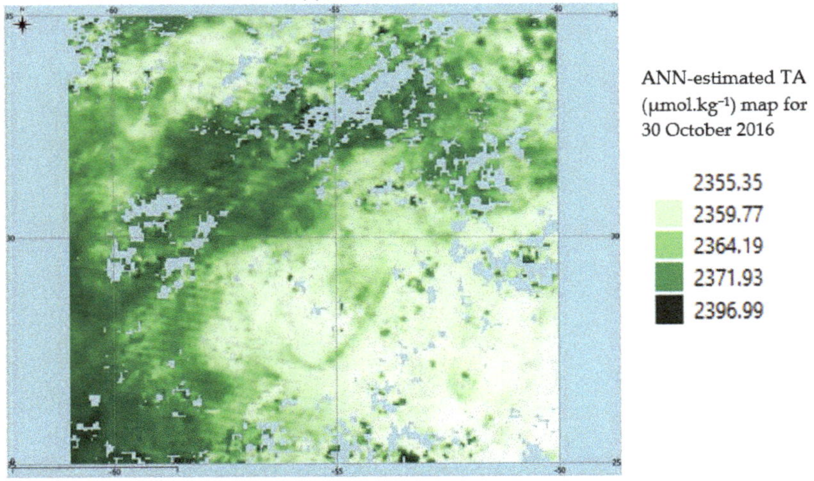

(**b**)

Figure 8. *Cont.*

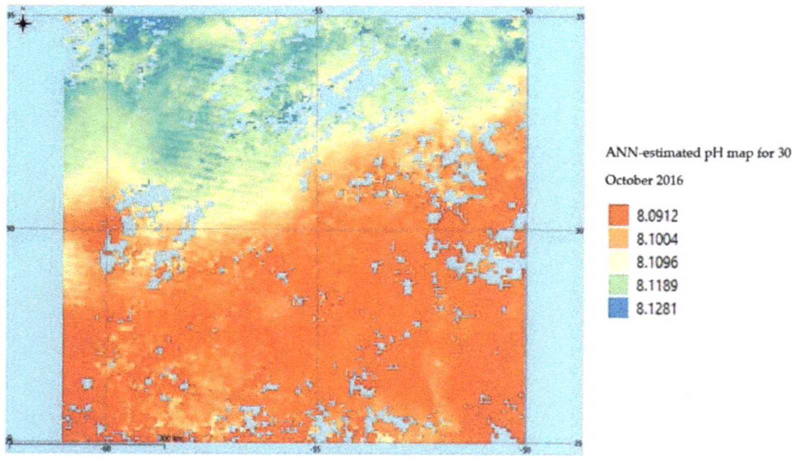

(c)

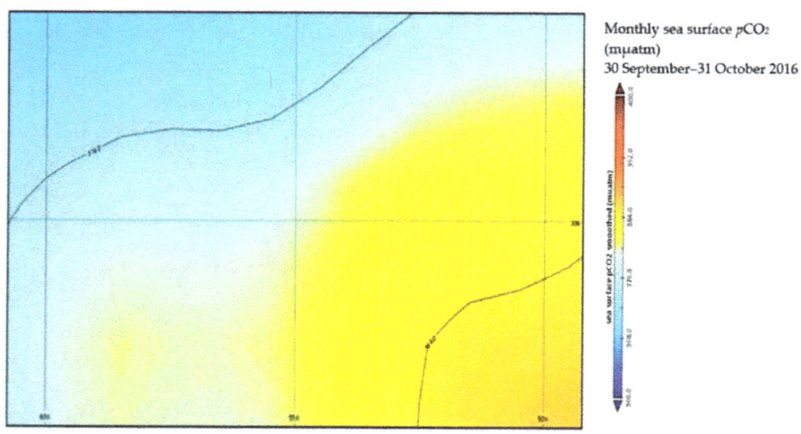

(d)

Figure 8. Modeled, gridded (**a**) DIC, (**b**) TA, and (**c**) pH maps for the ROI produced by the ANN algorithm. The spatial resolution is 0.04° × 0.04°, which corresponds to the native spatial resolution of some of the predictands. The (**d**) pCO_2 map valid for 30 September until 31 October 2016 has been inserted for reference [78]).

Figure 8d shows how the co-temporal spatial distribution of pCO_2 that has been derived by the Landschützer et al., dataset [84] and grid-resampled over our exact area of study is similar to the way that the ANN-estimated pH is distributed. It clearly shows higher pCO_2 levels over areas with a lower pH estimate (Figure 8c). This relationship corresponds with the results that were obtained by Sutton et al., (2014) and by Bates et al., (2012) when they studied the variability between pCO_2 and pH over the Pacific Ocean and the Atlantic Ocean surface, respectively [19,85]. In our study, the subtle gradient in pCO_2 from east to west at around 27° N in Figure 8d is well captured by the modeled spatial variation of the pH high resolution field over the same area (Figure 8c, including the relatively lower pH values corresponding to the northerly pCO_2 'tongue' originating from around −58° W, 26° N (Figure 8d).

3.3.1. Validation of the Modeled Data over the Mid-North Atlantic Ocean

In Situ Cruise Data

The results of the data validation against the in situ datasets available over the same area using validation dataset 3 are shown in Table 5. The in situ cruise transect (comprising StationIDs 1120000–1160000) did not include any pH measurements along the way. The linear regression analysis shows that the correlation between the TA datasets is statistically significant ($p < 0.05$; 95% C.L.). Moreover, the regressed observations and ANN-estimated DIC and TA values fall within the predicted 95% confidence level of the regression line.

Table 5. Corresponding ship-based and ANN-estimated values for DIC, TA, and pH. In situ pH measurements were not collected by M/V Equinox during part of the transect of 28 April–6 May 2015 (Validation dataset 3). (n/a: not available). The location of the individual StationIDs is as follows: 1120000: (31.1390° N, −60.5765° W); 1130000: (31.3795° N, −59.3347° W); 1140000: (31.7085° N, −57.6472° W); 1150000: (32.1818° N, −55.2020° W); 1160000: (32.7458° N, −52.2730° W); 1200000: (34.0460° N, −45.4433° W); and 1330000: (27.5105° N, −78.8207° W).

	Discrete Underway Measurements			ANN Estimation			Mean Bias		
StationID	**DIC ($\mu mol \cdot kg^{-1}$)**	**TA ($\mu mol \cdot kg^{-1}$)**	**pH**	**DIC ($\mu mol \cdot kg^{-1}$)**	**TA ($\mu mol \cdot kg^{-1}$)**	**pH**	**DIC ($\mu mol \cdot kg^{-1}$)**	**TA ($\mu mol \cdot kg^{-1}$)**	**pH**
1120000	2074	2387	n/a	2064	2385	8.111	10	2	n/a
1130000	2078	2397	n/a	2066	2378	8.105	12	19	n/a
1140000	2076	2404	n/a	2071	2368	8.099	5	36	n/a
1150000	2081	2400	n/a	2066	2375	8.104	15	25	n/a
1160000	2083	2392	n/a	2062	2384	8.112	21	7	n/a
1200000	2078	2382	8.073	2075	2378	8.096	3	4	−0.023
1330000	2095	2390	8.073	2073	2388	8.099	21	1	−0.025

Hindcast Biochemistry Data

In order to extend the validation of our BPN algorithm, additional independent sources of daily and/or monthly 2016 oceanic surface pH maps were sought; however, this type of dataset proved to be scarce, whilst 2016 daily/monthly gridded oceanic TA and DIC data are non-existent. As of now, the Copernicus Marine Service (CMEMS) makes available the following three datasets: (1) the Global Ocean Biochemistry Hindcast, which consists of both daily and monthly gridded maps (however, the geographic information about pH is only available at a monthly temporal resolution at 0.25° by 0.25° grid resolution); (2) the Global Ocean—in situ reprocessed carbon observations—SOCATv2021, which provides point measurements of DIC, TA, and pH, such as NCEI Accession 0154382; and (3) the Global Ocean Surface Carbon database, which provides pH data on a monthly basis at 1° by 1° grid resolution.

The correlation between the modeled pH (for 30 November 2016) and that derived from the Global Ocean Biochemistry Hindcast (16 October–15 November 2016 at 00:00 h UT) over the area of study is shown in Figure 9. This hindcast database provides monthly data starting as of the 16th day of each month, and therefore this data represents the average value for an entire month. In spite of their slight temporal difference, the two datasets are shown to be strongly correlated together, with an R^2 of 0.81 (Figure 9c), indicating a good statistical similarity, as well as an impressive spatial similarity for pH (Figure 9a,b). From an atmosphere–ocean dynamical point of view, this correlation points to a slowly changing pH distribution for the study area over a monthly scale.

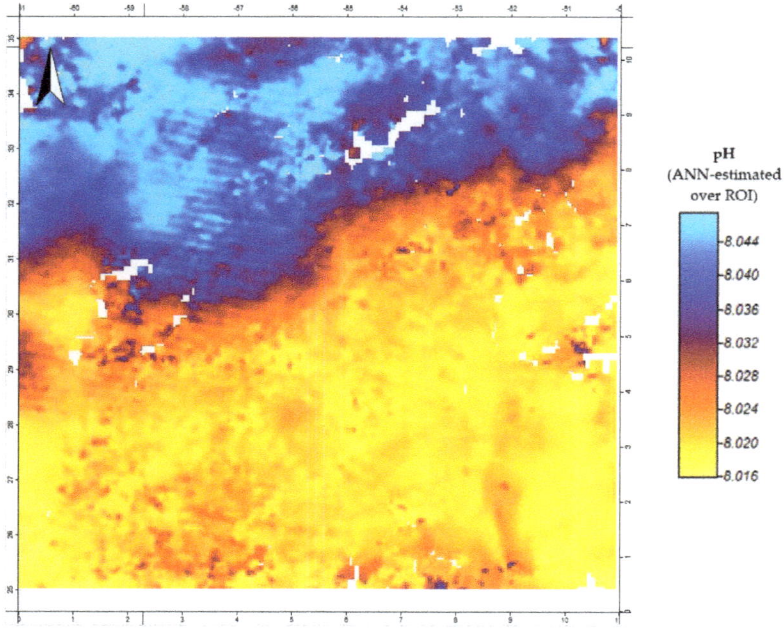

(a)

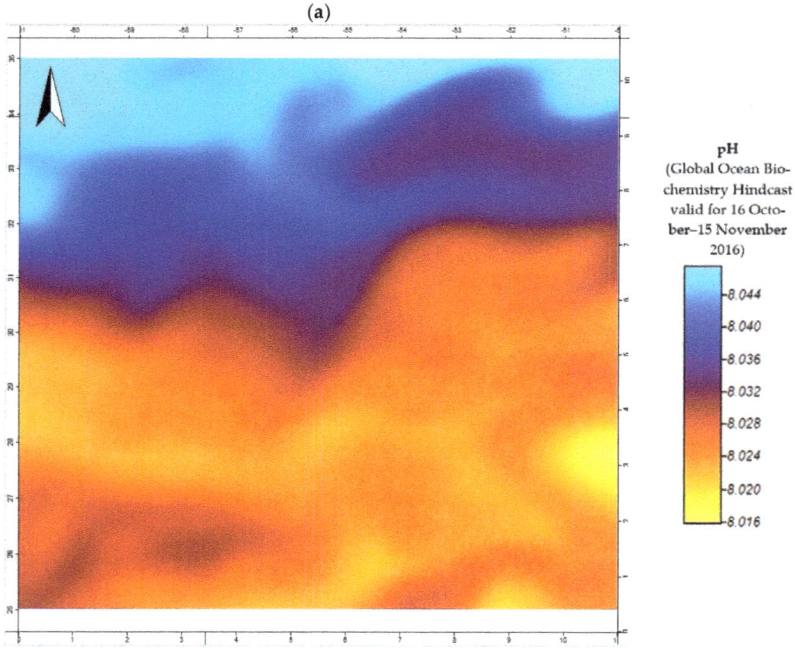

(b)

Figure 9. *Cont.*

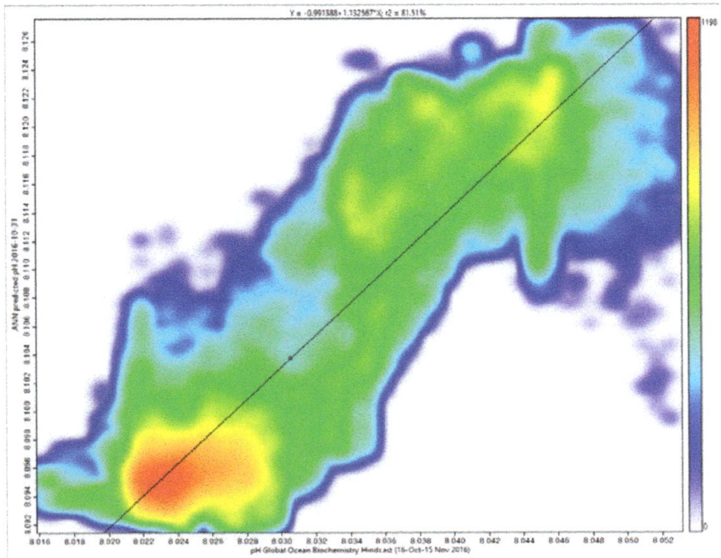

(**c**)

Figure 9. pH distribution map (**a**) ANN-estimated pH valid for 30 October 2016; (**b**) extracted from the Global Ocean Biochemistry Hindcast valid for 16 October–15 November 2016, and (**c**) scatterplot between (**a**) and (**b**) (R^2 = 0.81).

3.3.2. Caveats and Recommendations

This study is limited to the estimation of some elements of the carbonate chemistry for the mid-latitude of the North Atlantic Ocean based on their variability during the late winter, spring, and autumn of 2015 and 2016. Whether this neural network algorithm is applicable to other regions of the global oceans and/or for other time periods needs further investigation. The further development and training of the ANN algorithm is therefore recommended. This can be carried out by incorporating (1) a larger scalar variability of the same environmental drivers that are used at the highest spatiotemporal resolution possible in order to improve the learning of the BPN model, and (2) new environmental drivers, such as daily air–sea surface heat fluxes, 2 m air temperature, and air pressure at the highest spatial resolution possible. These may include freshwater influx through precipitation and conditions of the air–sea interface, such as heat fluxes (latent and sensible) and related physical values (such as the sub-layer depth [46]). The atmospheric conditions at sea level are an important parameter that influence the solubility of CO_2 in a unit volume of liquid [86]. Increasing the range of EO-based environmental drivers is now becoming more technically feasible, thanks to cloud servers and computing. Equally important would be the derivation of pCO_2 as another predictand from our artificial neural network algorithm [87]. Due to the limited time available in obtaining high resolution atmospheric and ocean modeled data, the inclusion of these additional environmental drivers was beyond the scope of the present study. The incorporation of (3) dynamical adjustments made to numerical ocean models [88] on the basis of chosen environmental drivers may further enhance the accuracy of the BPN algorithm. For example, it is necessary to take time-dependent temperature variations into account whenever the wind stress is estimated since it varies by more than a factor of two between 0° and 30 °C because of its dependence on temperature (the Schmidt number).

It is expected that the demand for high resolution DIC, TA, and pH maps, as estimated by deep learning, will, for many reasons, increase in the future. One important use is their support in the monitoring of proposed Ocean Acidification Refugia (OAR), such as the likes of extensive seagrass meadows and dense algal beds [89,90], and algal boundary layers [91,92], slow-flow habitats [93], deep-sea mounts [94], and areas that are isolated from ocean upwelling [95,96]. These are examples of highly localized areas that can vary dramatically across spatial scales from few millimeters (in the case of algal boundary layers) to hundreds of meters squared (such as in the case of extensive seagrass beds), with no clear criteria as to what makes each area a potential OAR other than the observed transient increases in seawater pH relative to the surrounding waters. Kapsenberg and Cyronak (2019) point out the lack of clear, agreed-upon functional criteria for OAR in the context of climate change, which makes it difficult for managers, legislators, and scientists to assess where to invest management efforts [97]. In this regard, this study becomes promising as a way to provide a means by which the daily determination of carbonate chemistry can be made available across multiple spatial scales down to at least a 4 km^2 horizontal resolution. In doing so, new target refugia can be proposed for research and management purposes.

4. Conclusions

Changes in ocean carbonate chemistry are a large spatiotemporal scale phenomenon that certainly needs to be monitored at the local scale. This study addresses its first research question by showing a way to produce high resolution, accurate, gridded maps of DIC, TA, and pH that are ideally suited for more localized ocean carbonate studies and applications.

Ship-based sampling remains subjected to limited ship time and human resources, costs, and weather conditions that prevent sampling in specific areas or at certain times of the year. Yet, they remain fundamental for numerical model validation and initialization tasks. This study shows a way to generate very-high-resolution gridded maps of ocean surface DIC, TA, and pH using an ANN approach in a robust and efficient way. This was carried out by addressing the second research question of this study. The future availability of more EO products hosted by cloud-serving computing environments and deep learning will soon be a determining factor towards the future automation of the synthesis of similar, highly detailed, daily carbonate chemistry maps for the global oceans. This technology will definitely help various ocean-related communities to better mitigate and adapt to the expected long-term changes. This is why we feel that high resolution EO products, coupled with deep learning, will provide us with an indirect way to monitor the chemical changes in seawater at an unprecedented resolution.

Author Contributions: Conceptualization, C.G.; Data curation, C.G.; Formal analysis, C.G.; Investigation, C.G.; Methodology, C.G.; Resources, C.G.; Software, C.G.; Validation, C.G.; Visualization, C.G.; Writing—original draft, C.G.; Writing—review and editing, C.G. and R.G. All authors have read and agreed to the published version of the manuscript.

Funding: This research was funded by COST Action CA 15217 Ocean Governance for Sustainability—Challenges, Options, and the Role of Science through short-term scientific missions (STSM) on the 'Modeling Ocean Acidification in the Gulf of Cadiz (MOsAiGC)' [CA15217-STSM-39764], and 'Contemporary Acidification Trends in the coastal NorThEaStern Atlantic (ATlaNTES)' [CA15217-STSM 40669].

Institutional Review Board Statement: IES-2022-00027. University of Malta.

Informed Consent Statement: Not applicable.

Data Availability Statement: Not applicable.

Acknowledgments: The authors would like to acknowledge Barbero, Leticia, Wanninkhof, Rik, Pierrot, and Denis for the collection of dissolved inorganic carbon, total alkalinity, pH, and other variables collected from surface and discrete observations using a flow-through pump and other instruments from M/V Equinox in the North Atlantic Ocean from 7 March 2015 to 6 November 2016 (NCEI Accession 0154382). The US DOC, NOAA, OAR, and the Atlantic Oceanographic and Meteorological Laboratory are also being acknowledged for hosting and making available this information: https://www.ncei.noaa.gov/access/metadata/landing-page/bin/iso?id=gov.noaa.nodc:0154382 (accessed on 20 February 2023). We would also like to acknowledge COST Action CA 15217 Ocean Governance for Sustainability—Challenges, Options, and the Role of Science—for supporting short-term scientific missions (STSM) on the 'Modeling Ocean Acidification in the Gulf of

Cadiz (MOsAiGC)' DOI: 10.13140/RG.2.2.32657.99681, and 'Contemporary Acidification Trends in the coastal NorThEaStern Atlantic (ATlaNTES)'. COST is supported by the EU Framework Program Horizon 2020.

Conflicts of Interest: The authors declare no conflict of interest.

References

1. Feely, R.A.; Orr, J.; Fabry, V.J.; Kleypas, J.A.; Sabine, C.L.; Langdon, C. Present and Future Changes in Seawater Chemistry Due to Ocean Acidification. In *Carbon Sequestration and Its Role in the Global Carbon Cycle*; American Geophysical Union: Washington, DC, USA, 2009; Volume 183, pp. 5–8. [CrossRef]
2. Bakker, D.C.E.; Pfeil, B.; O'Brien, K.M.; Currie, K.I.; Jones, S.D.; Landa, C.S.; Lauvset, S.K.; Metzl, N.; Munro, D.R.; Nakaoka, S.; et al. Surface Ocean CO_2 Atlas (SOCAT) V4. *Pangaea* 2016. [CrossRef]
3. Sabine, C.L.; Hankin, S.; Koyuk, H.; Bakker, D.C.E.; Pfeil, B.; Olsen, A.; Metzl, N.; Kozyr, A.; Fassbender, A.; Manke, A.; et al. Multispectral Remote Sensing Algorithms for Particulate Organic Carbon (POC): The Gulf of Mexico. *Remote Sens. Environ.* **2009**, *113*, 50–61. [CrossRef]
4. Lauvset, S.K.; Lange, N.; Tanhua, T.; Bittig, H.C.; Olsen, A.; Kozyr, A.; Álvarez, M.; Becker, S.; Brown, P.J.; Carter, B.R.; et al. An updated version of the global interior ocean biogeochemical data product, GLODAPv2.2021. *Earth Syst. Sci. Data* **2021**, *13*, 5565–5589. [CrossRef]
5. Tanhua, T.; Orr, J.C.; Lorenzoni, L.; Hansson, L. Monitoring ocean carbon and ocean acidification. *WMO Bull.* **2015**, *64*, 48–51.
6. NRDC. Ocean Acidification Hotspots. Natural Resources Defence Council. 2022. Available online: https://www.nrdc.org/resources/ocean-acidification-hotspots (accessed on 24 September 2022).
7. Zeng, J.; Matsunaga, T.; Saigusa, N.; Shirai, T.; Shin-Ichiro, N.; Zheng-Hong, T. Technical note: Evaluation of three machine learning models for surface ocean CO_2 mapping. *Ocean Sci.* **2017**, *13*, 303–313. [CrossRef]
8. Jiang, Z.; Song, Z.; Bai, Y.; He, X.; Yu, S.; Zhang, S.; Gong, F. Remote Sensing of Global Sea Surface pH Based on Massive Underway Data and Machine Learning. *Remote Sens.* **2022**, *14*, 2366. [CrossRef]
9. Bittig, H.C.; Steinhoff, T.; Claustre, H.; Fiedler, B.; Williams, N.L.; Sauzède, R.; Körtzinger, A.; Gattuso, J.P. An alternative to static climatologies: Robust estimation of open ocean CO_2 variables and nutrient concentrations from T, S, and O_2 data using bayesian neural networks. *Front. Mar. Sci.* **2018**, *5*, 328. [CrossRef]
10. Hall, E.R.; Wickes, L.; Burnett, L.E.; Scott, G.I.; Hernandez, D.; Yates, K.K.; Barbero, L.; Reimer, J.J.; Baalousha, M.; Mintz, J.; et al. Acidification in the U.S. Southeast: Causes, Potential Consequences and the Role of the Southeast Ocean and Coastal Acidification Network. *Front. Mar. Sci.* **2020**, *7*, 548. [CrossRef]
11. Ekstrom, J.; Suatoni, L.; Cooley, S.; Pendleton, L.H.; Waldbusser, G.G.; Cinner, J.E.; Ritter, J.; Langdon, C.; Van Hooidonk, R.; Gledhill, D.; et al. Vulnerability and adaptation of US shellfisheries to ocean acidification. *Nat. Clim. Change* **2015**, *5*, 207–214. [CrossRef]
12. Burgos, M.; Sendra, M.; Ortega, T.; Ponce, R.; Gómez-Parra, A.; Forja, J.M. Ocean-Atmosphere CO_2 Fluxes in the North Atlantic Subtropical Gyre: Association with Biochemical and Physical Factors during Spring. *J. Mar. Sci. Eng.* **2015**, *3*, 891–905. [CrossRef]
13. Rérolle, V.M.C.; Achterberg, E.P.; Ribas-Ribas, M.; Kitidis, V.; Brown, I.; Bakker, D.C.; Lee, G.A.; Mowlem, M.C. High-resolution pH measurements using a lab-on-chip sensor in surface waters of northwest European shelf seas. *Sensors* **2018**, *18*, 2622. [CrossRef]
14. Palevsky, H.I.; Quay, P.D. Influence of biological carbon export on ocean carbon uptake over the annual cycle across the North Pacific Ocean. *Glob. Biogeochem. Cycles* **2017**, *31*, 81–95. [CrossRef]
15. Liu, M.; Tanhua, T. Water masses in the Atlantic Ocean: Characteristics and distributions. *Ocean Sci.* **2021**, *17*, 463–486. [CrossRef]
16. Vázquez-Rodríguez, M.; Touratier, F.; Lo Monaco, C.; Waugh, D.; Padin, X.A.; Bellerby, R.G.J.; Goyet, C.; Metzl, N.; Ríos, A.F.; Pérez, F.F. Anthropogenic carbon in the Atlantic Ocean: Comparison of four data-based calculation methods. *Biogeosciences* **2009**, *6*, 439–451. [CrossRef]
17. Khatiwala, S.; Tanhua, T.; Mikaloff Fletcher, S.; Gerber, M.; Doney, S.C.; Graven, H.D.; Gruber, N.; McKinley, G.A.; Murata, A.; Ríos, A.F.; et al. Global ocean storage of anthropogenic carbon. *Biogeosciences* **2013**, *10*, 2169–2191. [CrossRef]
18. Lauvset, S.K.; Gruber, N.; Landschützer, P.; Olsen, A.; Tjiputra, J. Trends and drivers in global surface ocean pH over the past 3 decades. *Biogeosciences* **2015**, *12*, 1285–1298. [CrossRef]
19. Bates, N.R.; Best, M.H.P.; Neely, K.; Garley, R.; Dickson, A.G.; Johnson, R.J. Detecting Anthropogenic Carbon Dioxide Uptake and Ocean Acidification in the North Atlantic Ocean. *Biogeosciences* **2012**, *9*, 2509–2522. [CrossRef]
20. NOAA. State of the Climate: Ocean Heat Storage. 2015. Available online: https://www.climate.gov/news-features/featured-images/2015-state-climate-ocean-heat-storage (accessed on 24 September 2022).
21. NOAA. Reporting on the State of the Climate in 2016. Available online: https://www.ncei.noaa.gov/news/reporting-state-climate-2016 (accessed on 24 September 2022).
22. Clark, D.R.; Flynn, K.J. The Relationship between the Dissolved Inorganic Carbon Concentration and Growth Rate in Marine Phytoplankton. *Proc. Biol. Sci.* **2000**, *267*, 953–959. Available online: https://www.jstor.org/stable/1571440 (accessed on 22 August 2022). [CrossRef]

23. Barbero, L.; Wanninkhof, R.; Pierrot, D. *Dissolved Inorganic Carbon, Total Alkalinity, pH, and Other Variables Collected from Surface and Discrete Observations Using Flow-Through Pump and Other Instruments from M/V Equinox in the North Atlantic Ocean from 2015-03-07 to 2016-11-06 (NCEI Accession 0154382)*; Dataset; NOAA National Centers for Environmental Information: Washington, DC, USA, 2016. [CrossRef]
24. Ioannou, I.; Gilerson, A.; Gross, B.; Moshary, F.; Ahmed, S. Deriving ocean color products using neural networks. *Remote Sens. Environ.* **2013**, *134*, 78–91. [CrossRef]
25. Stramska, M. Particulate organic carbon in the global ocean derived from SeaWIFS ocean color. *Deep. Sea Res. I* **2009**, *56*, 1459–1470. [CrossRef]
26. Zhu, Y.; Shang, S.; Shai, W.; Dai, M. Satellite-derived surface water pCO_2 and air-sea CO_2 fluxes in the northern South China Sea in summer. *Prog. Nat. Sci.* **2009**, *19*, 775–779. [CrossRef]
27. Balch, W.M.; Gordon, H.R.; Bowler, B.C.; Drapeau, D.T.; Booth, E.S. Calcium carbonate budgets in the surface global ocean based on MODIS data. *J. Geophys. Res.* **2005**, *110*, C07001. [CrossRef]
28. Stramski, D.; Reynolds, R.A.; Kahru, M.; Mitchell, B.G. Estimation of particulate organic carbon in the ocean from satellite remote sensing. *Science* **1999**, *285*, 239–242. [CrossRef] [PubMed]
29. Stramska, M.; Stramski, D. Variability of particulate organic carbon concentration in the north polar Atlantic based on ocean color observations with Sea-viewing Wide Field-of-view Sensor (SeaWiFS). *J. Geophys. Res.* **2005**, *110*, C10018. [CrossRef]
30. Boutin, J.; Vergely, J.L.; Marchand, S.; D'Amico, F.; Hasson, A.; Kolodziejczyk, N.; Reul, N.; Reverdin, G.; Vialard, J. New SMOS Sea Surface Salinity with reduced systematic errors and improved variability. *Remote Sens. Environ.* **2018**, *214*, 115–134. [CrossRef]
31. Wanninkhof, R. Relationship between wind speed and gas exchange over the ocean revisted. *Limnol. Oceanogr.* **2014**, *12*, 351–362. [CrossRef]
32. Wanninkhof, R. Relationship between wind speed and gas exchange over the ocean. *J. Geophys. Res.* **1992**, *97*, 7373–7382. [CrossRef]
33. Wanninkhof, R.; McGillis, W.R. A Cubic Relationship between Air-Sea CO_2 Exchange and Wind Speed. *Geophys. Res. Lett.* **1999**, *26*, 1889–1892. [CrossRef]
34. Iwano, K.; Takagaki, N.; Kurose, R.; Komori, S. Mass transfer velocity across the breaking air–water interface at extremely high wind speeds. *Tellus B Chem. Phys. Meteorol.* **2013**, *65*, 1. [CrossRef]
35. Hall, J.K. GEBCO centennial special issue—Charting the secret world of the ocean floor: The GEBCO project 1903–2003. *Mar. Geophys. Res.* **2006**, *27*, 1–5. [CrossRef]
36. Dickson, A.G.; Sabine, C.L.; Christian, J.R. *Guide to Best Practices for Ocean CO_2 Measurements*; PICES Special Publication; NOAA: Washington, DC, USA, 2007; Volume 3, pp. 1–191.
37. Turk, D.; Malacic, V.; De Grandpre, M.D.; McGillis, W.R. Carbon dioxide variability and air-sea fluxes in the northern Adriatic Sea. *J. Geophys. Res.* **2010**, *115*, C10. [CrossRef]
38. Wanninkhof, R.; Asher, W.E.; Ho, D.T.; Sweeney, C.S.; McGillis, W.R. Advances in quantifying air-sea gas exchange and environmental forcing. *Ann. Rev. Mar. Sci.* **2009**, *1*, 213–244. [CrossRef]
39. Macintyre, S.; Wanninkhof, R.; Chanton, J.P. Trace gas exchange across the air-water interface in freshwaters and coastal marine environments. In *Biogenic Trace Gasses: Measuring Emissions from Soils and Waters*; Mattson, P.A., Harris, R.C., Eds.; Blackwell: New York, NY, USA, 1995; pp. 52–57.
40. Sweeney, C.; Gloor, E.; Jacobson, A.R.; Key, R.M.; McKinley, G.; Sarmiento, J.L.; Wanninkhof, R. Constraining global air–sea gas exchange for CO_2 with recent bomb ^{14}C measurements. *Glob. Biogeochem. Cycles* **2007**, *21*, GB2015. [CrossRef]
41. Ho, D.T.; Law, C.S.; Smith, M.J.; Schlosser, P.; Harvey, M.; Hill, P. Measurements of air-sea gas exchange at high wind speeds in the Southern Ocean: Implications for global parameterizations. *Geophys. Res. Lett.* **2006**, *33*, L16611. [CrossRef]
42. Fairall, C.W.; Hare, J.E.; Edson, J.B.; McGillis, W. Parameterization and Micrometeorological Measurement of Air–Sea Gas Transfer. *Bound. Layer Meteorol.* **2000**, *96*, 63–106. [CrossRef]
43. Krakauer, N.Y.; Randerson, J.T.; Primeau, F.W.; Gruber, N.; Menemenlis, D. Carbon Isotope Evidence for the Latitudinal Distribution and Wind Speed Dependence of the Air–sea Gas Transfer Velocity. *Tellus Ser. B Chem. Phys. Meteorol.* **2006**, *58*, 390–417. [CrossRef]
44. Suzuki, N.; Doneloan, M.A.; Takagaki, N.; Komori, S. Comparison of the global air-sea CO_2 gas flux on the difference of transfer velocity model. *J. Adv. Mar. Sci. Technol. Soc.* **2015**, *21*, 59–63. [CrossRef]
45. Boutin, J.; Quilfen, Y.; Merlivat, L.; Piolle, J.F. Global average of air-sea CO_2 transfer velocity from QuikSCAT scatterometer wind speeds. *J. Geophys. Res.* **2009**, *114*, C04007. [CrossRef]
46. Woolf, D. Bubbles and their role in gas exchange. In *The Sea Surface and Global Change*; Liss, P., Duce, R., Eds.; Cambridge University Press: Cambridge, UK, 1997; pp. 173–206. [CrossRef]
47. Wanninkhof, R.; Knox, M. Chemical enhancement of CO_2 exchange in natural waters. *Limnol. Oceanogr.* **1996**, *41*, 689–698. [CrossRef]
48. McGillis, W.R.; Edson, J.B.; Zappa, C.J.; Ware, J.D.; McKenna, S.P.; Terray, E.A.; Hare, J.E.; Fairall, C.W.; Drennan, W.; Donelan, M.; et al. Air-sea CO_2 exchange in the equatorial Pacific. *J. Geophys. Res.* **2004**, *109*, C08S02. [CrossRef]
49. Zappa, C.J.; McGillis, W.R.; Raymond, P.A.; Edson, J.B.; Hintsa, E.J.; Zemmelink, H.J.; Dacey, J.W.; Ho, D.T. Environmental turbulent mixing controls on air-water gas exchange in marine and aquatic systems. *Geophys. Res. Lett.* **2007**, *34*, 28790. [CrossRef]

50. Monahan, E.C.; Dam, H.G. Bubbles: An estimate of their role in the global oceanic flux of carbon. *J. Geophys. Res.* **2001**, *106*, 9377–9383. [CrossRef]
51. McNeil, C.; D'Asaro, E. Parameterization of air–sea gas fluxes at extreme wind speeds. *J. Mar. Syst.* **2007**, *66*, 110–121. [CrossRef]
52. Vogelzang, J.; Stoffelen, A.; Verhoef, A.; Figa-Saldaña, J. On the quality of high-resolution scatterometer winds. *J. Geophys. Res.* **2011**, *116*, C10033. [CrossRef]
53. Galdies, C.; Donoghue, D.N.M. A first attempt at assimilating microwave-derived SST to improve the predictive capability of a coupled, high-resolution Eta-POM forecasting system. *Int. J. Remote Sens.* **2009**, *30*, 6169–6197. [CrossRef]
54. Rutgersson, A.; Smedman, A. Enhancement of CO_2 transfer velocity due to water-side convection. *J. Mar. Syst.* **2010**, *80*, 125–134. [CrossRef]
55. Brostrom, G. The role of the seasonal cycles for the air–sea exchange of CO_2. *Mar. Chem.* **2000**, *72*, 151–169. [CrossRef]
56. Luger, H.; Wallace, D.W.R.; Kortzinger, A.; Nojiri, Y. The pCO_2 variability in the midlatitude North Atlantic Ocean during a full annual cycle. *Glob. Biochem. Cycles* **2004**, *18*, GB3023. [CrossRef]
57. Lellouche, J.-M.; Legalloudec, O.; Regnier, C.; Levier, B.; Greiner, E.; Drevillon, M. Quality Information Document for Global Sea Physical Analysis and Forecasting Product 2016. GLOBAL_ANALYSIS_FORECAST_PHY_001_024. Issue 2.0 Provided by the Copernicus Marine Environment Monitoring Service. Available online: https://hpc.niasra.uow.edu.au/dataset/550d2a5a-b66c-4318-aac2-c0fcf64370c0/resource/2d66a089-fe71-47ea-8245-6e1f1d469f59/download/global-analysis-forecast-phy-001-0241551608429013.nc (accessed on 24 September 2022).
58. Zeng, J.; Nojiri, Y.; Landschützer, P.; Telszewski, M.; Nakaoka, S. A global surface ocean fCO_2 climatology based on a feedforward neural network. *J. Atmos. Ocean Technol.* **2014**, *31*, 1838–1849. [CrossRef]
59. Chen, S.; Hu, C.; Barnes, B.B.; Wanninkhof, R.; Cai, W.-J.; Barbero, L.; Pierrot, D. A machine learning approach to estimate surface ocean pCO_2 from satellite measurements. *Remote Sens. Environ.* **2019**, *228*, 203–226. [CrossRef]
60. Cosca, C.E.; Feely, R.A.; Boutin, J.; Etcheto, J.; McPhaden, M.J.; Chavez, F.P.; Strutton, P.G. Seasonal and interannual CO_2 fluxes for the central and eastern equatorial Pacific Ocean as determined from fCO_2-SST relationships. *J. Geophys. Res.* **2003**, *108*, 3278. [CrossRef]
61. Landschützer, P.; Gruber, N.; Bakker, D.C.E.; Schuster, U.; Nakaoka, S.; Payne, M.R.; Sasse, T.P.; Zeng, J. A neural network-based estimate of the seasonal to inter-annual variability of the Atlantic Ocean carbon sink. *Biogeosciences* **2013**, *10*, 7793–7815. [CrossRef]
62. Takahashi, T.; Olafsson, J.; Goddard, J.; Chipman, D.W.; Sutherland, S.C. Seasonal variation of CO_2 and nutrients in the high-latitude surface oceans: A comparative study. *Glob. Biogeochem. Cycles* **1993**, *7*, 843–878. [CrossRef]
63. Hales, B.; Takahashi, T.; Bandstra, L. Atmospheric CO_2 uptake by a coastal upwelling system. *Glob. Biogeochem. Cycles* **2005**, *19*, GB1009. [CrossRef]
64. Stephens, M.P.; Samuels, G.; Olson, D.B.; Fine, R.A.; Takahashi, T. Sea-air flux of CO_2 in the North Pacific using shipboard and satellite data. *J. Geophys. Res.* **1995**, *100*, 13571–13583. [CrossRef]
65. Ono, T.; Saino, T.; Kurita, N.; Sasaki, K. Basin-scale extrapolation of shipboard pCO_2 data by using satellite SST and Chla. *Int. J. Remote Sens.* **2003**, *25*, 3803–3815. [CrossRef]
66. Bates, N.R.; Takahashi, T.; Chipman, D.; Goddard, J.G.; Howse, F.; Knap, A.H. Diurnal to Seasonal Variability of pCO_2 in the Sargasso Sea. Variability of PCO2 on Diel to Seasonal Timescales in the Sargasso Sea near Bermuda. *J. Geophys. Res. Ocean.* **1998**, *1031*, 15567–15586. [CrossRef]
67. Watson, A.; Robinson, C.; Robinson, J.; William, P.I.B.; Fasham, M. Spatial variability in the sink for atmospheric carbon dioxide in the North Atlantic. *Nature* **1991**, *350*, 50–53. [CrossRef]
68. Banzon, V.; Smith, T.M.; Chin, T.M.; Liu, C.; Hankins, W. A long-term record of blended satellite and in situ sea-surface temperature for climate monitoring, modeling and environmental studies. *Earth Syst. Sci. Data* **2016**, *8*, 165–176. [CrossRef]
69. NOAA. What Is Upwelling? Available online: https://oceanservice.noaa.gov/facts/upwelling.html (accessed on 24 September 2022).
70. Vargas, C.A.; Cuevas, L.A.; Broitman, B.R.; San Martin, V.A.; Lagos, N.A.; Gaitán-Espitia, J.D.; Dupont, S. Upper environmental pCO_2 drives sensitivity to ocean acidification in marine invertebrates. *Nat. Clim. Change* **2022**, *12*, 200–207. [CrossRef]
71. Duforêt-Gaurier, L.; Loisel, H.; Dessailly, D.; Nordkvist, K.; Alvain, S. Estimates of particulate organic carbon over the euphotic depth from in situ measurements. Application to satellite data over the global ocean. *Deep. Sea Res. Part I Oceanogr. Res. Pap.* **2010**, *57*, 351–367. [CrossRef]
72. Findlay, H.S.; Calosi, P.; Crawfurda, K. Determinants of the PIC:POC response in the coccolithophore *Emiliania huxleyi* under future ocean acidification scenarios. *Limnol. Oceanogr.* **2011**, *56*, 1168–1178. [CrossRef]
73. Ridgwell, A.; Schmidt, D.N.; Turley, C.; Brownlee, C.; Maldonado, M.T.; Tortell, P.; Young, J.R. From laboratory manipulations to Earth system models: Scaling calcification impacts of ocean acidification. *Biogeosciences* **2009**, *6*, 2611–2623. [CrossRef]
74. Hu, C.; Lee, Z.; Franz, B. Chlorophyll-a algorithms for oligotrophic oceans: A novel approach based on three-band reflectance difference. *J. Geophys. Res.* **2012**, *117*, C01011. [CrossRef]
75. Bryson, A.E.; Ho, Y.-C. *Applied Optimal Control*; Blaisdell Publishing Co.: Waltham, MA, USA, 1969.
76. Rumelhart, D.E.; Hinton, G.E.; Williams, R.J. Learning representations by back-propagating errors. *Nature* **1986**, *323*, 533536. [CrossRef]
77. Svozil, D.; Kvasnicka, V.; Pospichal, J. Introduction to multi-layer feed-forward neural networks. *Chemom. Intell. Lab. Syst.* **1997**, *39*, 43–62. [CrossRef]
78. Bishop, C.M. *Neural Networks for Pattern Recognition*; Oxford University Press: New York, NY, USA, 1995.

79. Chau, T.T.T.; Gehlen, M.; Chevallier, F. A seamless ensemble-based reconstruction of surface ocean pCO_2 and air–sea CO_2 fluxes over the global coastal and open oceans. *Biogeosciences* **2022**, *19*, 1087–1109. [CrossRef]
80. Fourrier, M.; Coppola, L.; Claustre, H.; D'Ortenzio, F.; Sauzède, R.; Gattuso, J.-P. A Regional Neural Network Approach to Estimate Water-Column Nutrient Concentrations and Carbonate System Variables in the Mediterranean Sea: CANYON-MED. *Front. Mar. Sci.* **2020**, *7*, 620. [CrossRef]
81. Campos Velho, H.F.; Furtado, H.C.M.; Sambatti, S.B.M.; Osthoff Ferreira de Barros, C.B.; Welter, M.E.S.; Souto, R.P.; Carvalho, D.; Cardoso, D.O. Data Assimilation by Neural Network for Ocean Circulation: Parallel Implementation. *Supercomput. Front. Innov.* **2022**, *9*, 74–86. [CrossRef]
82. Irrgang, C.; Saynisch-Wagner, J.; Thomas, M. Machine Learning-Based prediction of spatiotemporal uncertainties in global wind velocity reanalyses. *J. Adv. Model. Earth Syst.* **2020**, *12*, e2019MS001876. [CrossRef]
83. United Nations. Ocean Acidification Research on Local Scales. 2022. Available online: https://sdgs.un.org/partnerships/ocean-acidification-research-local-scales (accessed on 24 September 2022).
84. Landschützer, P.; Gruber, N.; Bakker, D.C.E. Decadal variations and trends of the global ocean carbon sink. *Glob. Biogeochem. Cycles* **2016**, *30*, 1396–1417. [CrossRef]
85. Sutton, A.J.; Feely, R.A.; Sabine, C.L.; McPhaden, M.J.; Takahashi, T.; Chavez, F.P.; Friederich, G.E.; Mathis, J.T. Natural variability and anthropogenic change in equatorial Pacific surface ocean pCO_2 and pH. *Glob. Biogeochem. Cycles* **2014**, *28*, 131–145. [CrossRef]
86. Kraus, E.B.; Businger, J.A. *Atmosphere-Ocean Interaction*; Oxford University Press: Oxford, UK, 1994; 352p.
87. Galdies, C.; Garcia-Luque, E.; Guerra, R. *Variability CO_2 Parameters in the North Atlantic Subtropical Gyre: A Neural Network Approach*; Ocean Carbon and Biogeochemistry (OCB) Summer Workshop; Woods Hole Oceanographic Institution: Falmouth, MA, USA, 2018.
88. Lu, F.; Harrison, M.J.; Rosati, A.; Delworth, T.L.; Yang, X.; Cooke, W.F.; Jia, L.; McHugh, C.; Johnson, N.C.; Bushuk, M.; et al. GFDL's SPEAR seasonal prediction system: Initialization and ocean tendency adjustment (OTA) for coupled model predictions. *J. Adv. Model. Earth Syst.* **2020**, *12*, e2020MS002149. [CrossRef]
89. Hendriks, I.E.; Olsen, Y.S.; Ramajo, L.; Basso, L.; Steckbauer, A.; Moore, T.S.; Howard, J.; Duarte, C.M. Photosynthetic activity buffers ocean acidification in seagrass meadows. *Biogeosciences* **2014**, *11*, 333–346. [CrossRef]
90. Young, C.S.; Gobler, C.J. The ability of macroalgae to mitigate the negative effects of ocean acidification on four species of North Atlantic bivalve. *Biogeosciences* **2018**, *15*, 6167–6183. [CrossRef]
91. Cornwall, C.E.; Hepburn, C.D.; McGraw, C.M.; Currie, K.I.; Pilditch, C.A.; Hunter, K.A.; Boyd, P.W.; Hurd, C.L. Diurnal fluctuations in seawater pH influence the response of a calcifying macroalga to ocean acidification. *Proc. R. Soc. B Biol. Sci.* **2013**, *280*, 20132201. [CrossRef]
92. Noisette, F.; Hurd, C. Abiotic and biotic interactions in the diffusive boundary layer of kelp blades create a potential refuge from ocean acidification. *Funct. Ecol.* **2018**, *32*, 1329–1342. [CrossRef]
93. Hurd, C.L. Slow-flow habitats as refugia for coastal calcifiers from ocean acidification. *J. Phycol.* **2015**, *51*, 599–605. [CrossRef]
94. Tittensor, D.P.; Baco, A.R.; Hall-Spencer, J.M.; Orr, J.C.; Rogers, A.D. Seamounts as refugia from ocean acidification for cold-water stony corals. *Mar. Ecol.* **2010**, *31*, 212–225. [CrossRef]
95. Chan, F.; Barth, J.A.; Blanchette, C.A.; Byrne, R.H.; Chavez, F.; Cheriton, O.; Feely, R.A.; Friederich, G.; Gaylord, B.; Gouhier, T.; et al. Persistent spatial structuring of coastal ocean acidification in the California Current System. *Sci. Rep.* **2017**, *7*, 2526. [CrossRef]
96. Kapsenberg, L.; Hofmann, G.E. Ocean pH time-series and drivers of variability along the northern Channel Islands, California, USA. *Limnol. Oceanogr.* **2016**, *61*, 953–968. [CrossRef]
97. Kapsenberg, L.; Cyronak, T. Ocean acidification refugia in variable environments. *Glob. Change Biol.* **2019**, *25*, 3201–3214. [CrossRef] [PubMed]

water

MDPI

Article

Water Balance Trends along Climatic Variations in the Mediterranean Basin over the Past Decades

Zaib Unnisa [1,2,3,*], Ajit Govind [2], Bruno Lasserre [3] and Marco Marchetti [3]

1 Department for Innovation in Biological, Agri-Food and Forest Systems (DIBAF), University of Tuscia, 01100 Viterbo, Italy
2 International Center for Agricultural Research in Dry Areas (ICARDA), Cairo 11435, Egypt
3 Department of Biosciences and Territory, University of Molise, 86090 Pesche, Italy
* Correspondence: zaib.unnisa@unimol.it

Abstract: The heterogeneous ecosystems in the Mediterranean Basin (MB) are becoming sensitive to water stress. To investigate the climatic stress, a water budget study was conducted over the basin using TerraClimate simulations for a long temporal range (1990–2020). According to the budget accounting, forested regions received the highest precipitation (P) on average compared to other land use types (annual mean ~≈ 633 mm yr^{-1}), and even then, they were in a water deficit state (−0.42 mm yr^{-1}). Tree plantations in North Africa (Libya and Morocco) were also in a water deficit state; however, their average P was very low (≈12 mm yr^{-1}) compared to that of northern parts of the MB, and the average Actual Evapotranspiration (AET) was ≈15 mm yr^{-1}. Also, the water balance in other land use systems (rain-fed, irrigated croplands, and rangelands) was either negative or near zero. As a whole, the basin's average annual P was ≈538 mm yr^{-1}, the annual average AET was ≈415 mm yr^{-1}, and the runoff (Q) was equivalent to 123 mm yr^{-1}, which shows a strong influence of ET over the region. Since runoff was negligible in most dry areas, the AET's large contribution was notable in the North African base of the Atlas Mountains including the Nile delta region. This indicates that precipitation and evaporation are the principal mechanisms of the water balance in the MB. The result shows strong climate variability over Southern Europe, Turkey, and the western Balkans in the recent years, affecting the AET and making their land use systems more vulnerable to water stress. This benchmark study signifies the consistent need for water storage in the Mediterranean vegetation systems of the basin. It also indicates two distinct climate clusters for water balance modeling.

Keywords: Mediterranean basin; climate fluctuations; water deficit; Mediterranean ecosystems; Budyko curve

Citation: Unnisa, Z.; Govind, A.; Lasserre, B.; Marchetti, M. Water Balance Trends along Climatic Variations in the Mediterranean Basin over the Past Decades. *Water* **2023**, *15*, 1889. https://doi.org/10.3390/w15101889

Academic Editors: Alban Kuriqi and Luis Garrote

Received: 9 January 2023
Revised: 7 March 2023
Accepted: 12 March 2023
Published: 16 May 2023

1. Introduction

In the Mediterranean climate, changes to freshwater availability directly affect total annual precipitation volumes. A rise in temperature also increases the evaporative demand. These effects bring variations in the water balance of the MB, and their frequent monitoring is significantly important to maintain socio-economic balance in water use and distribution. Such long-term monitoring also helps in understanding the nature of hydro-climatic fluctuations in the basin, which can also be human evolved. However, climatic effects are the prime contributor to the hydrological imbalances in the arid and semi-arid production systems of the MB, as these systems are more vulnerable to freshwater decline. Recent studies envisaged that the MB would experience an aridity of more than twice its present extent with future climate changes [1,2]. An increase in drought frequency and warm spells has also been documented in the MB in the last 40 years (1979–2018) [3,4]. These dry weather events are predicted to be more pronounced in hyper-arid regions like North Africa, southern Spain, and the Middle East [5]. Frequent dry spells, especially during the crop growing season, can increase water demand and can make basin ecosystems more sensitive to water stress [6]. Changes such as late springs, coupled with freshwater decline, are already affecting the net ecosystem productivity of the basin [7]. By 2050, the MB will experience more water deficit with a 30–50% decline in freshwater resources [8]. In such a case, an accurate assessment of the water budget is extremely important to conduct and control supply and demand imbalances.

Many studies found Evapotranspiration (ET) as an important hydrological flux in context with surface and deep surface responses to climatic variations [9,10]. It is also a principal water-extracting component in the most arid regions, and its accurate quantification along with precipitation is becoming more critical with the growing water scarcity in the basin. Several studies assessed the Mediterranean water budget using ET at various scales with a variety of projections and methodologies, i.e., Med-CORDEX [11], the Weather Research and Forecasting model [12], the bucket with a bottom hole (BBH) model with RCMs SMH-E and SMH-B [13], the LPJ model [14], and the transient model [15]. A few studies used the water budget equations with additional parameters, or the runoff or ET estimation at basin scale, while some modified water balance methods for region-specific studies in order to carry out the water budget explicitly [16–21]. Usually, in the majority of water budget estimations, the interpolated or gridded datasets of precipitation (P), actual ET (AET), and runoff (Q) are used, which are either simulated with climate or hydrological models or derived from reanalysis climate data. Some studies compared the budget inferences from different data sources and compared their associated uncertainties or absolute errors in budget estimates. Since the accuracy in estimations is a priority in hydrological assessments in order to set reliable benchmarks for long-term water management, it is necessary to account for and fix large errors in budget closures with associated uncertainties. For this study, we used the recently released monthly TerraClimate product for the water budget evaluation and Budyko curve analysis to investigate the water-limited state in different parts of the basin. We also assessed the water stress conditions in various land uses (wetland, agriculture, forest, and urban) at a plausible resolution (1 km). The TerraClimate product has already gained confidence in various hydrological assessment studies, i.e., in closing a water budget [18], freshwater flux differences [22], the impact of extreme events on the budget [23] and has also been validated with in situ data in various locations.

The terrestrial water budget is mainly constrained by energy and water limitations. These limitations are determined by the simple Budyko hypothesis. This technique is robust for evaluating water resources [24] in data scarce regions at various scales, i.e., global [25,26], national or large river [27], and basin-scale [28,29]. It relates the long-term average of a basin's actual evapotranspiration (AET) to its potential evapotranspiration (PET) and precipitation (P) in a semi-empirical way. It has many applications in hydrological studies; for instance, Li and Quiring (2021) [30] used the Budyko framework to investigate spatial heterogeneity in the factors that control the water balance, while looking at the importance of forest coverage as a function of climate. Wang and Hejazi (2011) [31] developed a decomposition method using the Budyko framework, to link discharge variability with hydrological components. Xu (2011) [32] also used a Budyko-like framework to develop a regression model and decompose the contribution of anthropogenic activities to the reduction of water resources in a basin. Our study quantifies the current nature of the water balance at the annual time step, using a first-order water balance approach and a Budyko framework to derive clusters with terrestrial water balance components. We also accounted for potential water storage differences among the existing land use patterns in the MB. The results may be used as a benchmark for tracking the changes in the water budget after the 1991–2020 period, and may assist decision-makers in setting future targets for water storage in the basin.

2. Materials and Methods

This study used monthly TerraClimate datasets for computing the 30-year average water budget using three principal hydrological components: P, AET, and Q. TerraClimate is mainly derived from Climate Research Unit CRU 4.0 [33], in substitution with Japanese reanalysis JRA55 data, which are particularly used for sparse station regions such as Africa. The hydrological flux components (AET and Q) were calculated via a modified Thornthwaite-Mather climatic water balance model, wherein storage or drainage is constrained with the defined soil water storage capacity with the lookup table method. The Thornthwaite-Mather climatic water balance model uses the relation of P, PET (calculated using the Penman-Monteith equation), soil moisture, and snowpack water storage to compute the water balance. In this case, some of P is intercepted by the canopy and lost through evaporation. The remainder, i.e., throughfall, is added to the soil water pool. When the resulting volumetric soil water content exceeds the capacity of the soil, this excess is deemed to be lost as runoff or drainage (Q), which is used in this analysis.

All of these components were validated from station data and streamflow gauges [34]. The product has better spatial realism (4 km) and quality in terms of average absolute error than its older versions. The details on the TerraClimate product development are in [35].

The entire 30-year budget calculation (1991–2020) was conducted in the google earth engine (GEE) platform (source code and datasets are available). Potential and actual evapotranspiration, runoff, and precipitation products were rescaled with 0.1, and all were resampled to 1 km resolution using the bilinear resampling method embedded in GEE. To quantify the water surplus and deficit state of basin's land use systems, the most recent Mediterranean Land System Map (2017) was used. The details of the dataset are given in the Table 1 below and the land cover classification methodology is discussed in [36].

Table 1. Datasets used for trend analysis and the water budget calculation for the study period (1991–2020).

Dataset	Variable	Spatial Resolution	Source
Terraclimate, University of Idaho	Precipitation Potential Evapotranspiration (ASCE Penman-Montieth) Actual Evapotranspiration Runoff	4.6 km resampled 1 km	Abatzoglou et al. 2018 [35]
Mediterranean land systems	26 land cover types	2 km resampled 1 km	Malek and Verburg, 2017 [36]

The PER method used in this study provides temporal variability of water budget changes using P, E, and Q [21]. In this method, the continuity equation gives the change in the storage within a specific area by subtracting the input precipitation from output sources (evapotranspiration and runoff), over the territory only.

$$P - Q - E = \Delta S \tag{1}$$

where P = precipitation, Q = runoff, E = evapotranspiration, and ΔS = water storage.

The annual averages were computed for these three water balance components within the basin domain. Then, the class-wise average value of each component was retrieved to find out which regions and land use land cover systems were in a water deficit state. To understand regional water trends, the basin was subdivided into four regions to compare annual budget time series; the sub-division is shown in Figure 1. Later, a simple Mann-Kendall statistical test was applied on annual datasets to assess whether the budget and its components were increasing over time or decreasing, and whether the trend in either direction was statistically significant. In last, the Budyko framework [37] was applied at the mean annual time scale to find aridity clusters in the basin.

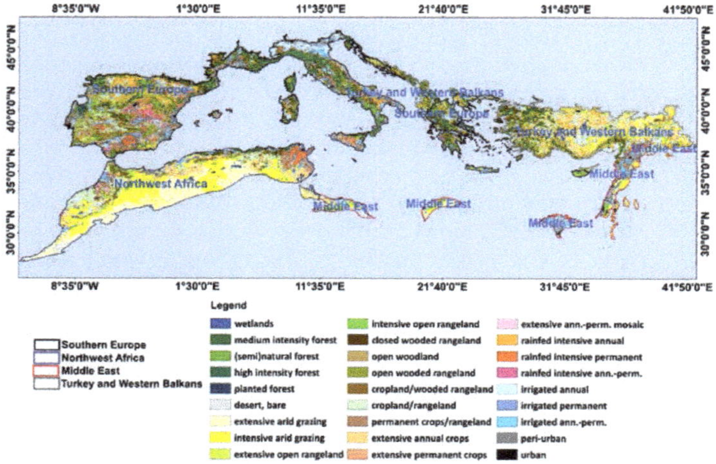

Figure 1. Diverse land systems in Mediterranean region with 26 thematic classes representing intensive and extensive typologies of forests, crops, wetland, agro-pastoral, and grazing systems lying in the four sub-regions differentiated with colored outlines, i.e., Southern Europe (black), Northwest Africa (blue), Middle East (red), and Turkey and Western Balkans (grey).

2.1. Budyko Curve Theory

The Budyko technique sets a semi-empirical relation between the evaporative index, AET/P, and the aridity index, PET/P. This relation can be termed as the Budyko AET/P versus PET/P curve, and is a conventional metric of climate aridity, often used to identify regions under climatic water deficits. In this technique, the annual water balance can be expressed as a function of the available water and energy, and the ratio of the mean annual evaporation to the mean annual precipitation (Ea/P, evaporative index) is manifested by the ratio of the mean annual potential evaporation to the mean annual precipitation (E_0/P, dryness index). If the dryness index is less than one, it means that the energy supply is the limiting factor for evaporation; if it is greater, then the water supply is the limiting factor. These indexes are typically affected by climate seasonality, topography, soil types, and vegetation. Yu et al. 2021 [38] revealed that vegetation coverage had a dominant effect on AET changes, while climate seasonality that has a negative effect on AET had a relatively lower effect. This indicates that vegetation dynamics can change the water budget situation in the basin, but at this broad region scale, our focus was to investigate the storage and consumption of water in the current land use system considering the atmospheric supply only. Landscape dynamics can only be assessed with the availability of a dynamic land use land cover (LULC) with a similar typology, which is missing in this study. Since the region is so diverse, 90% accuracy in LULC assessment is difficult to achieve.

2.2. Land Use Systems in the MB

In this study, the recent land use land cover map used represents the vegetation of 27 countries, covering an area of 2.3 million km^2. The basin surrounding the Mediterranean Sea has similar climatic and biophysical traits. In previous studies, the Nile River basin was usually not considered as being part of the MB, with the reason being that the source of this transboundary basin is in the tropics, and its tropical hydrological regime and water management issues go beyond the Mediterranean context [8]. Now in the new classification map, the Nile Delta and Apennines forests of Italy are included, with the justification of their sharing of common Mediterranean ecoregions falling in the Middle East and Southern European regions.

These 26 land cover classes shown in the map (Figure 1) were derived from land cover, livestock density, irrigation extent, and different intensity proxies' information from existing databases, using expert-based hierarchal classifications; the detailed methodology is given in [36]. This map is specifically used because of its significantly improved thematic typology at a higher resolution (2 km) compared to existing land cover products.

It sufficiently represents variations in the land systems across the region and gives a realistic view of climatic impacts on landscapes, compared to other classified maps. As the distribution indicates, a majority of intensive rain-fed systems exist in Spain, intensive and extensive arid grazing systems in North Africa, and a majority of irrigated croplands exist in Italy and Egypt. Forests are more common in Italy and Scandinavian countries, while Turkey, Spain, and the Balkans have a majority of rain-fed cropping systems and fewer forests. Their areal densities are given in Table 2, and the sub-regions shown in colored borders are used for further investigation on the water budget.

Table 2. Annual average of hydrological components for land use land cover classes.

Major Class	Area %	P (mm yr^{-1})	AET (mm yr^{-1})	Q (mm yr^{-1})	ΔS (mm yr^{-1})
Bare and Open Grazing systems	22.6%	486.96 ± 110	359.32 ± 58	127.61 ± 69	0.18 ± 28
Cropland Systems	37.9%	540.56 ± 111	429.60 ± 63	110.91 ± 65	0.00 ± 36
Forest systems	10.1%	633.63 ± 125	463.56 ± 62	170.88 ± 80	−0.42 ± 36
Agro-silvo pastoral mosaics	23.3%	633.41 ± 133	479.99 ± 68	153.62 ± 86	−0.41 ± 41

3. Results

This study examined the water balance of diverse land use systems in the MB, in which a 30-year average trend was determined for the Mediterranean ecosystems in the spatial domain of four major regions surrounding the Mediterranean Sea: Europe, North Africa, the Middle East, and Turkey and the Balkans (Figure 1). The geographical location, climatic zone of the sub-region, and various spatial and temporal processes can all affect the overall water balance of the basin. In areas where rainfall is scarce, the major inflows come from the sea (desalinated water) into the terrestrial ecosystem. However, only P could be accounted here as main inflow. On the other hand, evaporation from water bodies is also considered negligible, which is often true in arid and semi-arid areas at the coarser resolution; thereby, AET from vegetation was taken as the main outflow component. In this study, we analyzed the basin's average budget state in the current climate conditions, while exploring the significance of ET in the arid and semi-arid land systems through trend analysis. Such analysis is essential to understand a basin's behavior, priorities, and needs.

As can be seen in Figure 2, the average P was spatially distributed between 2413 and 12 mm yr^{-1}, with the lowest in Egypt and entire North African belt (less than 200 mm yr^{-1}). The latter is comparable to the 1979–1999 period-averaged precipitation map found in [39]. While the maximum amount of rain was confined to small areas in Italy, the Balkans, and coastal areas of North Africa and Spain, as expected, the basin showed contrasting humid and arid climate compositions. In the northern regions, the P was >1000 mm yr^{-1} and PET < 1000 mm yr^{-1}; in contrast, the southern regions had a P of <300 mm yr^{-1} and a PET of >1000 mm yr^{-1}. The surplus water was available mostly in the coastal regions of Lebanon and Israel, and parts of the Middle East, while the majority of the basin had a storage of nearly −0.5 – 2 mm yr^{-1}, which indicates a critical water deficit at an annual time step in a large part of the region.

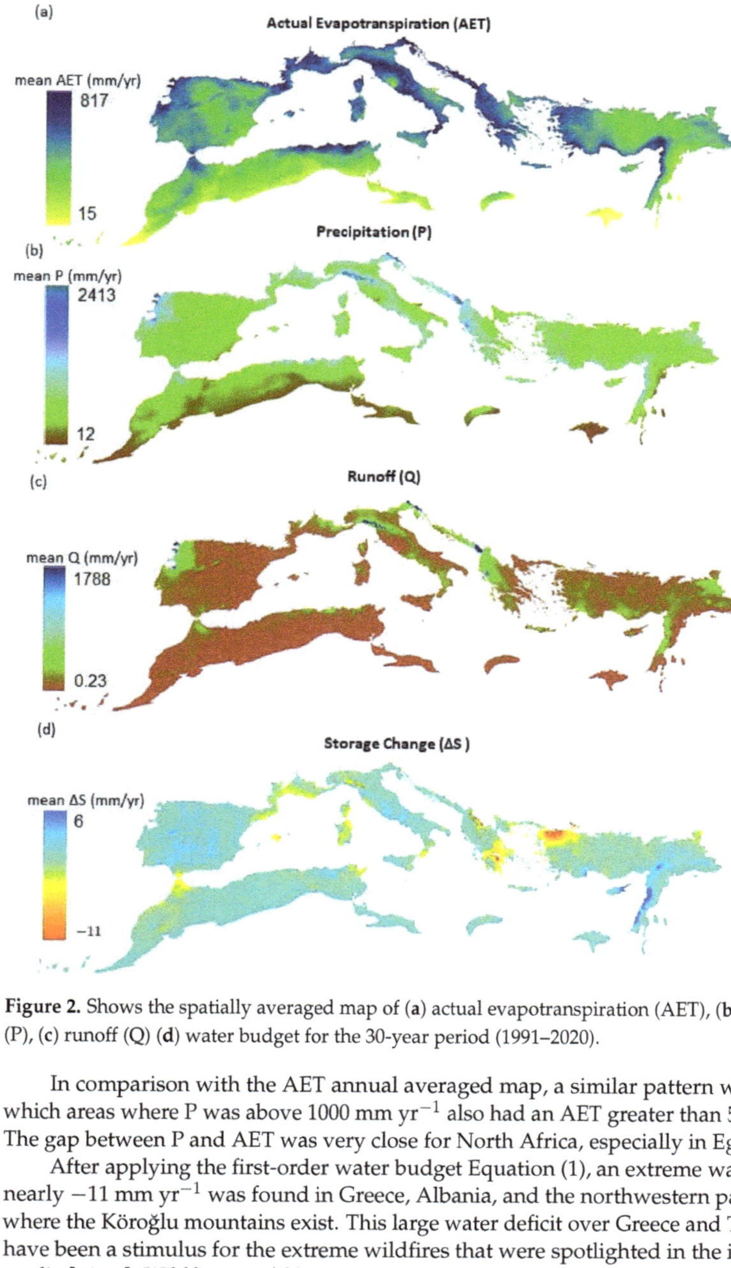

Figure 2. Shows the spatially averaged map of (**a**) actual evapotranspiration (AET), (**b**) precipitation (P), (**c**) runoff (Q) (**d**) water budget for the 30-year period (1991–2020).

In comparison with the AET annual averaged map, a similar pattern was found, in which areas where P was above 1000 mm yr^{-1} also had an AET greater than 500 mm yr^{-1}. The gap between P and AET was very close for North Africa, especially in Egypt.

After applying the first-order water budget Equation (1), an extreme water deficit of nearly -11 mm yr^{-1} was found in Greece, Albania, and the northwestern part of Turkey where the Köroğlu mountains exist. This large water deficit over Greece and Turkey could have been a stimulus for the extreme wildfires that were spotlighted in the international media [40,41]. Wildfires could be a consequence of increasingly drier conditions over the Mediterranean region. We also identified the Balkans, Maghreb, North Adriatic, Central Spain, and Turkey as the most affected regions from frequent wildfire events, and consider them to be under a climate crisis. Reduction in the stream flows are noticeably impacted by climate change, this observed deeply at the catchment scale in Catalonia, Spain [42]. Another scientific reason for the water deficit could be orographic effects in the mountainous region, where increased evapotranspiration rates at the peaks makes more water available for vegetation at foothills, and simultaneously less water available for transpiration in the mountains, making the water deficit more prominent over the peaks. As the annual average of ET and Q cannot justify the reasoning fully, neither lateral distributions are accounted for in this analysis. To justify this deficit, available water in the form of infiltration and groundwater is also required in the budget analysis, as Freund and Kirchner (2017) [43] considered P + net lateral transfer to quantify its effect on average ET; they found that ET increases when the dryness index increases with altitude, due to the lateral movement of water from more humid uplands to more arid lowlands. It shows that lateral water transfers will strongly affect the average ET if the source (or recipient) location is energy limited and the recipient (or source) location is water limited, which is not accountable with the existing Thornthwaite water balance approach. Overall, the runoff (Q) and P have similar spatial patterns, which indicates areas where P is more allocated to the mountainous systems which have greater runoff, and also highlights the near-zero runoff condition in the majority of the downstream basin areas. This situation will increase in the future, and the zero runoff condition will markedly increase with a decreasing P influx, as a notable decrease in streamflow is also suspected for the period 2076–2100 in the Mediterranean region, as predicted via climate modelling [40].

This spatial pattern (Figure 2) was further analyzed with a recent and detailed high-resolution land use land cover map, and it indicated that all types of vegetation on average were under a water-deficit state (Table 2). Surprisingly, forest and pastoral systems spreading over the entire basin showed on average a negative water balance. Medium intensity forests, followed by bare areas, received the highest precipitation on annual average among all land cover types. The class-wise statistics also showed the prominence of ET in the water budget of the basin. Cropland systems that had the highest distribution of nearly 38% in the basin had an average water budget close to zero, and were especially negative in the irrigated systems. Even forests covering just 10% of the land had a negative water balance. This clearly shows that Mediterranean vegetation systems are under climate-induced water stress. In fact, in all land use systems, the water storage situation was either negative or zero in average conditions, which can be seen in the minor classes as well (Appendix A). This was further observed in the graph of LULC systems that showed the minimal difference between inflow and outflow (Figure 3).

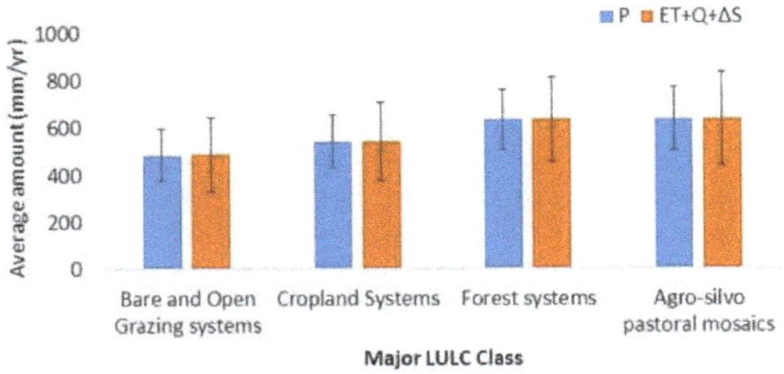

Figure 3. Inflow and outflow budget differences close to zero, among major land use land cover class of the MB.

This water-limited state was further verified with the Budyko framework that represents a simple first-order relationship among hydrological components. In this regard, each region's annual averaged evaporative and dryness index was plotted over the Budyko curve. Some interesting clusters formed with the sub-regional study that clearly justifies regional climatic differences in the Budyko curve. The majority index values found in the water-limited region of the Budyko curve in the horizontal direction indicate the dominant climatic influence on the hydrological cycle in the basin, as the basin is getting warmer and drier over time. Southern Europe (SE), Turkey, and the west Balkans showed more similarity in their chronological patterns for PET while showing a diagonal position on the curve; these regions are indicating a relatively less constrained state by available water supply, compared to the Middle East and Northwest Africa, where climate variability is more pronounced with a dispersed pattern and is highly constrained by water limitations.

North Africa's closeness to its water limit is shown in Figure 4, which depicts its fragility and vulnerability to a dry climate far greater than other regions. Nevertheless, this analysis covers only the Mediterranean region under thick or sparse vegetation. The climatic effects on vegetative parts are obvious from declining productivity and growing food insecurities in the region.

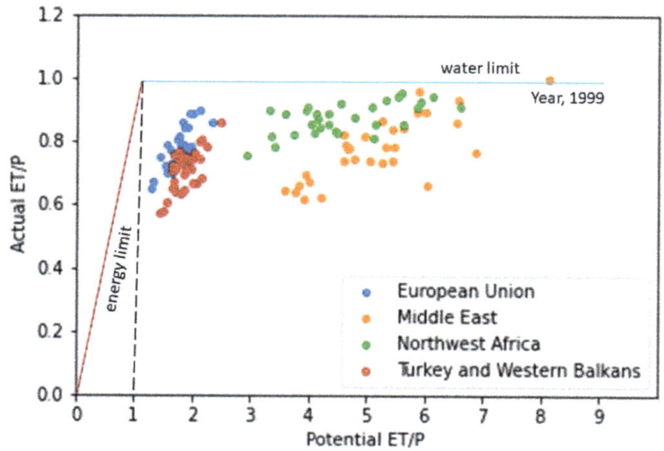

Figure 4. The dots represent the distribution of the mean annual evaporative index (E/P) versus the mean annual dryness index (PET/P), accounting for the TerraClimate averages over the 30-year period for the MB. The red and blue lines are the hypothetical water and energy limits of the Budyko framework.

The time series of hydrological components for the thirty years (1990–2020) were computed to observe the climate variability in each region. The region-wise stacked time series of P, AET, Q, and ΔS revealed that synchronicity exists between temporal averages of AET and P because this basin area is mostly covered with vegetation. However, the P and AET ranges vary with the regions, as the Middle East and northwest Africa have both a lower P and AET in the comparable range of 200–400 mm yr^{-1}. The gap between the average P and AET trend lines is lesser than the difference gap found in SE and the Turkey and the Balkans regions, where runoff is more prominent and generally highest (Figure 5). Davraz et al. (2014) [44] also found rainfall to be a prominent cause for the surface water level increase in Turkey, with direct and indirect recharges through runoff. It also indicates that the runoff varies by region in the basin, which is also linked to the AET distribution. The average precipitation peaks for SE and NA are chronologically similar (1996, 2003, 2010, 2018), indicating that extreme precipitation events happen after every seven or eight years. The years 2003 and 2018 are also important in terms of the peak found in the Middle East, the Balkans, and Turkey. The difference gap between AET and P in SE and Turkey and the Balkans is larger because of the prominent runoff trend ranges between 0 and 300 mm yr^{-1}, their AET average ranges between 400 and 500 mm yr^{-1}, and P between 600 and 800 mm yr^{-1}. This trend analysis indicates the regional differences in the water budget components, with more water storage fluctuations in the Balkans and Turkey followed by Southern Europe. The water budget in these regions is also close to zero and negative in some places, which means that much of the precipitation consumed in the basin and annual variability in storage is greater than that of the MENA region. By noticing temporal signals, more climate fluctuations in terms of precipitation can be observed in the recent decades, especially for Southern Europe, Turkey, and the Balkans. Ajjur and Baalousha (2021) [45] also observed substantial changes in the components of the hydrological cycle, which has increased the imbalance between the water supply and demand across the MENA region. According to climate projections for the middle and end of century (2021–2050, 2071–2100), temperature rise is the principal reason for PET losses that amplified the AET, which in result reduced the water availability in the MENA region [10] (Ajjur and Al Ghamdi 2021). In Europe, Deszi et al. (2018) [46] identified AET as an important driver of water availability, and their future projections over Europe postulated a pronounced water deficit in Southern Europe with reference to an ET increase. This also supports budget inferences, and testifies growing hydrological constraints in the current climatic conditions.

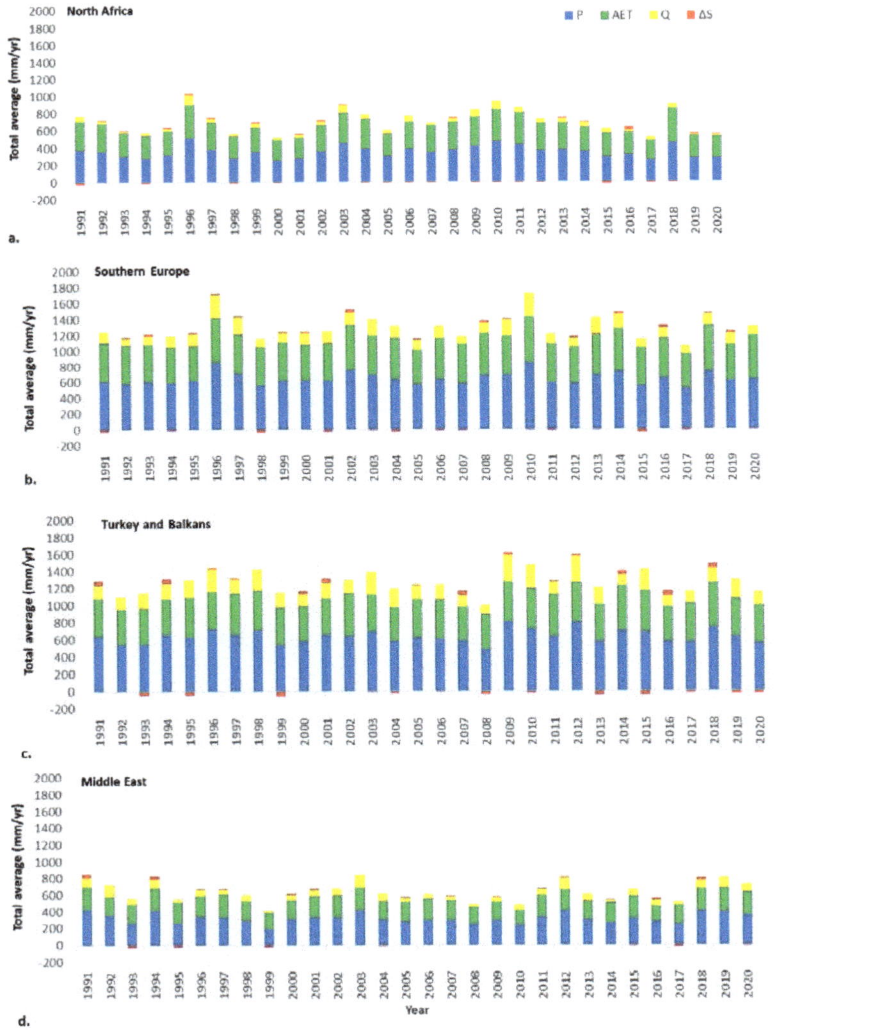

Figure 5. Annual average moisture budget over four regions for 1990 – 2020. Given are annual averaged trend of precipitation (blue), evapotranspiration (green), runoff (yellow) and minus change in terrestrial water storage (red), for (**a**) North Africa (**b**) Southern Europe (**c**) Turkey and the Balkans (**d**) Middle East.

Figure 6 shows a clear demarcation of latitudinal similarity among the four sub-regions in the temporal trends of the budget components. SE and Turkey and the Balkans have shown similar magnitudes in P and AET with some variations in peak years. Trendlines of all budget components showed a similar trend of a slight uplift in AET and P, while also a statistically insignificant decreasing trend found in SE, via the Mann Kendall test (Table 3). No trend was observed in most cases, which indicates only a very slight change in the behavior of the basin.

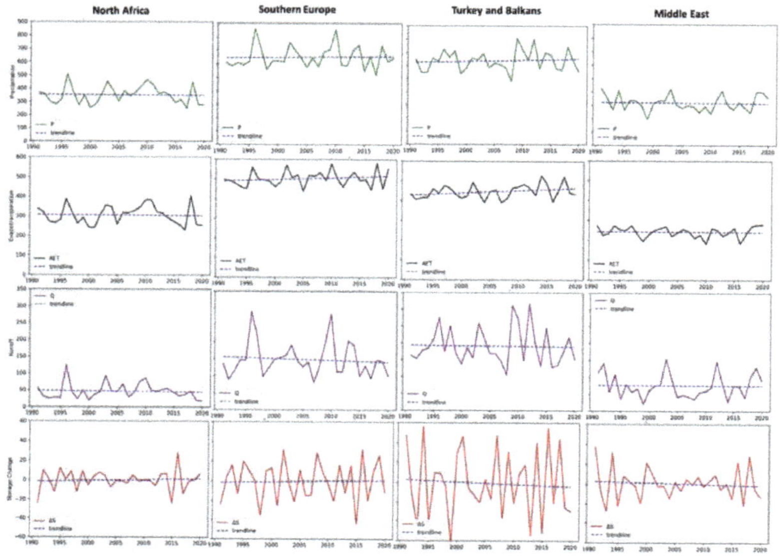

Figure 6. Long-term time series of water budget components over four sub-regions, i.e., North Africa, Southern Europe, Turkey and the Balkans, and the Middle East, with their averaged trend behavior.

The entire region observed a peak in the P and a storage change in the recent year 2018, which was also noticed as a year of climate extremes (drought and heavy precipitation) in North Africa [47]. Overall, the time series graph shows greater hydro-climatic fluctuations over SE, Turkey and the Balkans, indicating that climate predictions could be more uncertain and challenging here. As AET accounting is based on the water balance approach, wherein a significant part of irrigation and external water sources is missing and is not visible in the Middle Eastern Nile region as well, it could affect budget studies with conventional techniques. Remote sensing-based ET computations should be included in the budget assessments in future budgeting to account for irrigated ET. However, this study is primarily based on the terrestrial water budget, in which P is the main inflow and AET is computed following a one-dimensional simplified water balance approach explained in the methodology section. This first-order water balance calculation could reveal the storage condition on an annual basis in the major land use systems in the region, which are elaborated in this study in different ways.

Table 3. Mann-Kendall test results.

Sub-Region	P				AET				Q				ΔS			
	Trend	S Statistic	Z	*p*-Value	Trend	S Statistic	Z	*p*-Value	Trend	S Statistic	Z	*p*-Value	Trend	S Statistic	Z	*p*-Value
North Africa	no trend	−6	−1.22	0.22	no trend	−6	−1.22	0.22	no trend	−6	−1.22	0.22	no trend	4	0.73	0.46
Southern Europe	no trend	2	0.24	0.8	decreasing	−10	−2.2	0.02	no trend	4	0.73	0.46	no trend	6	1.22	0.22
Turkey & Balkans	no trend	0	0	1	no trend	4	0.73	0.46	no trend	8	1.714	0.086	no trend	−2	−0.24	0.8
Middle East	no trend	−4	−0.73	0.46	no trend	0	0	1	no trend	−6	−1.22	0.22	no trend	−4	−0.73	0.46

4. Discussion

Like in other studies, we also found the highest precipitation (more than 1200 mm) to be over the Adriatic coast, the Alpine region, and over the coast of Turkey and the Atlantic Iberian Peninsula. The lowest precipitation was found in the southeast region of the basin, over the southern Iberian Peninsula, and over the northern coast of Africa, with less than 400 mm of precipitation. Past studies determined that precipitation changes in the MB are partially influenced by North Atlantic Oscillations (NAO), upon which their vegetation dynamics and compositions are based. Precipitation is significantly correlated with the NAO, which is known to be the primarily responsible atmospheric event for budget fluctuations; however, no significant correlation has been found for evaporation yet [48]. Besides that, there could be multiple factors that act together and contribute to the basin's hydrological responses. They could be region-specific characteristics, topography, geography, location, or precipitation regimes; however, their proportional influence has not yet been accounted for in the basin. A recent study indicated that the incoming P flux mostly comes from sources inside the Mediterranean. This proportion is only 35%, while 10% is from ET over nearby land in continental Europe, and 25% originates in the North Atlantic. The remaining 30% comes from the tropical Pacific or the Southern Hemisphere, indicating its direct connections with multiple locations on earth due to a global terrestrial energy redistribution. Therefore, fluxes have a more global than regional influence that links climate change effects with the water cycle in the basin [49,50].

As also noticed over Turkey and the west Balkans (Figures 5c and 6), hydro-climatic fluctuations are projected to increase, which would decrease annual precipitation in the southern Europe–Turkey region and the Levant, while in the Arabian Gulf area precipitation may increase. Besides that, daytime maximum temperatures appear to increase most rapidly in the northern part of the region, i.e., the Balkan Peninsula and Turkey, which means more evaporation. This will have marked effects on the ecosystems' productivity and functioning, as seen in the analysis. Moutahir [51] also noticed negative trends in the different water balance components, although they focused on the pine stands in the sub-humid belt of Spain; according to their budget analysis, the native pine population is likely to disappear in the future under extreme climate scenarios of water stress. The water-limiting conditions in the region are threatening for native tree plantations. Even the small quantity of water used by trees is important in the hydrologic budget of the Mediterranean areas where rainfall is limited; this indicates how significant the accurate computation of AET is in the budget [52]. In these regions, AET is mostly found to be higher: more than 54% of total precipitation, and mostly where tree density was lower [53]. This also shows how crucial precipitation is for forest growth and how it can control regional forest expansions or declines. Since many projected studies revealed that rainfall frequency is likely to be lower in the future, it would reduce the chances for aquifer recharge, despite their increasing size with extreme rainfall events. Changes in the precipitation intensity, size, and temporal distribution are expected in this region, and will have different effects on the water balance. Even in older studies, after precipitation, evaporation is recorded as the largest term in the Mediterranean freshwater budget in the 50 years of one study period (1948–1998) [48]. They also found that the Mediterranean region was under a freshwater deficit at the annual scale. The decrease in open water evaporation under future climate scenarios as a result of increased relative humidity will have a positive effect on the water cycle, but this trend would be seasonal and common in the winter only.

The goal of this paper was to provide a picture of the mean annual water budget along with the long-term variability of hydrological fluxes in the MB. It also justifies that complex morphology and climate variability contribute to significant annual differences in total precipitation and its geographical distribution. Rainfall and surface inflows (streams and rivers) are the major inflows, whereas evapotranspiration from different land uses and drains from the region to sea are the major outflows. A recent in situ study in Italy also validated our conclusion about the deficit state of the basin. They found a negative trend in the estimated infiltration for the consecutive five hydrological years (2017–2022). This infiltration decrease was associated with a decrease in precipitation; however, they found AET in a less significant negative trend in the same period. This affirms that climate stresses are dominant in the MB, and that reduced surface water input will affect the groundwater input as well [54,55]. Scarascia-Mugnozza et al. [56] measured the hydrological balance at different integration times and found evapotranspiration was correlated to the water status of soils and plants. The indicated water stress can affect carbon metabolism, the water relations of forest trees, and ecosystem stability. All in all, AET was found to be an effective measure. This study highlights that both the accounting of the water budget at multi-scale and multi-functional systems are very important in considering the significant climatic influence over the water cycle and the budget. These findings could facilitate future climate preparedness and monitoring tasks.

5. Conclusions

This study attempted to investigate long-term water budget trends in the basin using only TerraClimate estimates, with which the annual averaged water storage in the major land distributions is assessed. The long-term water budget assessment has proven the hydrological constraints in the MB region under the current climate. These constraints found spatial variations due to an unequal distribution of precipitation over the basin.

From the Budyko curve analysis, we found that AET in the basin is mainly controlled by evaporative demand. This increase in demand makes the basin more arid, and with dwindling freshwater influx, causes the region to be water limited. The long-term climate statistics showed that its spatially averaged annual precipitation is $\approx$538 mm yr^{-1}, while water consumption through AET is $\approx$415 mm yr^{-1} and runoff is 123 mm yr^{-1}, which indicates that the major contribution of AET in the basin is in response to the climate. It is noteworthy to mention that the temporal trend of AET is getting closer to precipitation in the North African base of the Atlas Mountains, where the water budget is $\approx$ 0 mm yr^{-1}, including the Nile delta region, which shows pressure on external water resources. In contrast, the gap between AET and P is wider for Turkey, the Balkans, and the EU, with pronounced runoff patterns. The strong climate variability found in the recent decades for Turkey, the Balkans, and Southern Europe clearly indicates them as more vulnerable to climate effects and affected by irregular patterns.

As a whole, the basin is in a water deficit state due to the strong effect of evaporative demand and the limited water supply to the basin. Since these budget inferences are only based on TerraClimate estimates, which has gained confidence in recent studies, storage biases have not been verified with other global climate products and model-based estimations. The purpose of this study was to show the nature of water balance by already peer-reviewed and published data (TerraClimate) from a first-order perspective. With this, the study opens the opportunity to add more inputs from different model sources for hydrological sub-components, to contribute to the storage calculation over the basin.

Our results suppress the need for continuing serious actions on basin water storage at a large scale, and alarms about escalating climate fluctuations in the MB, especially in Southern Europe, Turkey, and the Balkans. Middle Eastern and North African ecosystems are fragile to water stress where remote sensing-based dynamic water budgeting is preferred for its irrigated, rain-fed, and pastoral land use systems.

Author Contributions: Conceptualization, Z.U. and A.G.; methodology, Z.U. and A.G.; data curation, Z.U.; writing—Z.U.; writing—review and editing, Z.U. and B.L.; supervision, M.M. and A.G. All authors have read and agreed to the published version of the manuscript.

Funding: This work is funded by (1) the CGIAR Initiative on Climate Resilience, ClimBeR and (2) Regional Water Harvesting Potential Mapping Project under SIDA and FAO.

Data Availability Statement: Source code is available on GitHub (https://github.com/zaib19/Water-Balance-in-the-Mediterranean-Basin-between-1990-and-2020).

Acknowledgments: We acknowledge the doctoral scholarship to Z.U. from the Department of Innovation in Biology, Agri-food and Forest systems (DIBAF), University of Tuscia, Viterbo, Italy. We are also grateful to the International Center for Agricultural Research in the Dry Areas (ICARDA) for hosting the traineeship to Z.U. Thanks to the National Center for Atmospheric Research at the University of California for sharing the TerraClimate product publicly.

Conflicts of Interest: The authors declare no conflict of interest.

Appendix A

Table A1. Class-wise 30-year averaged budget.

Major Class	Minor Class	Area %	P	ET	Q	ΔS
Bare and Open Grazing systems	bare	1.00	910.75	573.36	335.57	0.27
	ext. open rangeland	0.70	444.12	347.12	98.72	0.17
	ext. arid grazing	10.20	243.65	210.90	32.38	0.34
	int. open rangeland	1.80	495.93	383.30	113.22	0.08
	int. arid grazing	8.90	340.33	281.93	58.17	0.05
	Average	Σ 22.6%	486.96	359.32	127.61	0.18
Cropland Systems	ext. annual	10.40	460.19	359.67	100.53	−0.18
	ext. permanent	1.30	608.68	477.60	131.23	−0.22
	ext. annual permanent	2.70	548.40	442.63	105.59	0.11
	rainfed int. annual	5.80	569.68	443.52	126.11	0.07
	rainfed int. permanent	2.20	521.70	435.03	86.37	0.17
	rainfed int. ann. -perm.	1.30	543.74	436.50	107.06	0.26
	irrigated annual	8.70	569.70	430.66	139.38	−0.25
	irrigated permanent	2.20	531.43	440.53	91.00	−0.09
	irri ann. -perm	3.30	511.55	400.23	110.95	0.10
	Average	Σ 37.9%	540.56	429.60	110.91	0.00
Forest systems	medium intensity forest	6.20	837.39	593.49	244.69	−1.03
	semi (natural)	2.60	760.82	549.04	212.35	−0.22
	high intensity	1.00	769.99	561.11	210.02	−0.81
	planted forests	0.30	166.35	150.59	16.45	0.38
	Average	Σ 10.1%	633.63	463.56	170.88	−0.42
Agro-silvo pastoral mosaics	cropland/rangeland	6.50	509.44	405.95	103.49	−0.16
	open woodland	3.20	702.44	514.28	188.24	−0.28
	open wooded rang.	3.50	688.44	505.79	182.84	−0.50
	cropland/wooded rang.	6.80	656.81	501.44	155.72	−0.40
	perm. crops/wooded ran.	1.70	479.15	386.31	92.62	0.00
	closed wooded ran.	1.60	764.19	566.18	198.85	−1.11
	Average	Σ 23.3%	633.41	479.99	153.62	−0.41

References

1. Barredo, J.I.; Mauri, A.; Caudullo, G.; Dosio, A. Assessing shifts of Mediterranean and arid climates under RCP4. 5 and RCP8. 5 climate projections in Europe. In *Meteorology and Climatology of the Mediterranean and Black Seas*; Birkhäuser: Cham, Switzerland, 2019; pp. 235–251.
2. Zdruli, P. Land resources of the Mediterranean: Status, pressures, trends and impacts on future regional development. *Land Degrad. Dev.* **2014**, *25*, 373–384. [CrossRef]
3. Vogel, J.; Paton, E.; Aich, V.; Bronstert, A. Increasing compound warm spells and droughts in the Mediterranean Basin. *Weather Clim. Extrem.* **2021**, *32*, 100312. [CrossRef]
4. Hoerling, M.; Eischeid, J.; Perlwitz, J.; Quan, X.; Zhang, T.; Pegion, P. On the increased frequency of Mediterranean drought. *J. Clim.* **2012**, *25*, 2146–2161. [CrossRef]
5. Tramblay, Y.; Llasat, M.C.; Randin, C.; Coppola, E. Climate change impacts on water resources in the Mediterranean. *Reg. Environ. Chang.* **2020**, *20*, 83. [CrossRef]
6. Cardell, M.F.; Amengual, A.; Romero, R. Future effects of climate change on the suitability of wine grape production across Europe. *Reg. Environ. Chang.* **2019**, *19*, 2299–2310. [CrossRef]
7. Milano, M.; Ruelland, D.; Fernandez, S.; Dezetter, A.; Fabre, J.; Servat, E. Facing climatic and anthropogenic changes in the Mediterranean basin: What will be the medium-term impact on water stress? *C. R. Geosci.* **2012**, *344*, 432–440. [CrossRef]
8. Milano, M.; Ruelland, D.; Fernandez, S.; Dezetter, A.; Fabre, J.; Servat, E.; Fritsch, J.M.; Ardoin-Bardin, S.; Thivet, G. Current state of Mediterranean water resources and future trends under climatic and anthropogenic changes. *Hydrol. Sci. J.* **2013**, *58*, 498–518. [CrossRef]
9. Avanzi, F.; Rungee, J.; Maurer, T.; Bales, R.; Ma, Q.; Glaser, S.; Conklin, M. Climate elasticity of evapotranspiration shifts the water balance of Mediterranean climates during multi-year droughts. *Hydrol. Earth Syst. Sci.* **2020**, *24*, 4317–4337. [CrossRef]
10. Ajjur, S.B.; Al-Ghamdi, S.G. Evapotranspiration and water availability response to climate change in the Middle East and North Africa. *Clim. Chang.* **2021**, *166*, 1–28. [CrossRef]
11. Raymond, F.; Ullmann, A.; Tramblay, Y.; Drobinski, P.; Camberlin, P. Evolution of Mediterranean extreme dry spells during the wet season under climate change. *Reg. Environ. Chang.* **2019**, *19*, 2339–2351. [CrossRef]

12. Abdelwares, M.; Lelieveld, J.; Hadjinicolaou, P.; Zittis, G.; Wagdy, A.; Haggag, M. Evaluation of a regional climate model for the Eastern Nile Basin: Terrestrial and atmospheric water balance. *Atmosphere* **2019**, *10*, 736. [CrossRef]
13. Bargaoui, Z.; Foughali, A.; Tramblay, Y.; Houcine, A. Evaluation of two bias-corrected regional climate models for water budget simulations in a Mediterranean basin. *IAHS-AISH Publ.* **2013**, *359*, 73–79.
14. Santini, M.; Collalti, A.; Valentini, R. Climate change impacts on vegetation and water cycle in the Euro-Mediterranean region, studied by a likelihood approach. *Reg. Environ. Chang.* **2014**, *14*, 1405–1418. [CrossRef]
15. Oroud, I.M. Water budget assessment within a typical semiarid watershed in the eastern Mediterranean. *Environ. Process.* **2015**, *2*, 395–409. [CrossRef]
16. Falalakis, G.; Gemitzi, A. A simple method for water balance estimation based on the empirical method and remotely sensed evapotranspiration estimates. *J. Hydroinformatics* **2020**, *22*, 440–451. [CrossRef]
17. Wong, J.S.; Zhang, X.; Gharari, S.; Shrestha, R.R.; Wheater, H.S.; Famiglietti, J.S. Assessing water balance closure using multiple data assimilation–and remote sensing–based datasets for canada. *J. Hydrometeorol.* **2021**, *22*, 1569–1589. [CrossRef]
18. Zhang, Y.; Pan, M.; Sheffield, J.; Siemann, A.L.; Fisher, C.K.; Liang, M.; Beck, H.E.; Wanders, N.; MacCracken, R.F.; Houser, P.R.; et al. A Climate Data Record (CDR) for the global terrestrial water budget: 1984–2010. *Hydrol. Earth Syst. Sci.* **2018**, *22*, 241–263. [CrossRef]
19. Sanchez-Gomez, E.; Somot, S.; Mariotti, A. Future changes in the Mediterranean water budget projected by an ensemble of regional climate models. *Geophys. Res. Lett.* **2009**, *36*, L21401. [CrossRef]
20. Sheffield, J.; Ferguson, C.R.; Troy, T.J.; Wood, E.F.; McCabe, M.F. Closing the terrestrial water budget from satellite remote sensing. *Geophys. Res. Lett.* **2009**, *36*, 07403. [CrossRef]
21. Zeng, N.; Yoon, J.H.; Mariotti, A.; Swenson, S. Variability of basin-scale terrestrial water storage from a PER water budget method: The Amazon and the Mississippi. *J. Clim.* **2008**, *21*, 248–265. [CrossRef]
22. Zhao, Y.; Lu, Z.; Wei, Y. An assessment of global precipitation and evapotranspiration products for regional applications. *Remote Sens.* **2019**, *11*, 1077. [CrossRef]
23. Susanti, I.; Sipayung, S.B.; Siswanto, B.; Maryadi, E.; Latifah, H.; Nurlatifah, A.; Supriatin, L.S.; Witono, A.; Suhermat, M. Implications of extreme events on the water balance in Java. *AIP Conf. Proc* **2021**, *2331*, 030008. [CrossRef]
24. Gunkel, A.; Lange, J. Water scarcity, data scarcity and the Budyko curve—An application in the Lower Jordan River Basin. *J. Hydrol. Reg. Stud.* **2017**, *12*, 136–149. [CrossRef]
25. Luo, Y.; Yang, Y.; Yang, D.; Zhang, S. Quantifying the impact of vegetation changes on global terrestrial runoff using the Budyko framework. *J. Hydrol.* **2020**, *590*, 125389. [CrossRef]
26. Fang, K.; Shen, C.; Fisher, J.B.; Niu, J. Improving Budyko curve-based estimates of long-term water partitioning using hydrologic signatures from GRACE. *Water Resour. Res.* **2016**, *52*, 5537–5554. [CrossRef]
27. Singh, R.; Kumar, R. Vulnerability of water availability in India due to climate change: A bottom-up probabilistic Budyko analysis. *Geophys. Res. Lett.* **2015**, *42*, 9799–9807. [CrossRef]
28. Abera, W.; Formetta, G.; Borga, M.; Rigon, R. Estimating the water budget components and their variability in a pre-alpine basin with JGrass-NewAGE. *Adv. Water Resour.* **2017**, *104*, 37–54. [CrossRef]
29. Fernandez, R.; Sayama, T. Comparison of future runoff projections using Budyko framework and global hydrologic model: Conceptual simplicity vs process complexity. *Hydrol. Res. Lett.* **2015**, *9*, 75–83. [CrossRef]
30. Li, Z.; Quiring, S.M. Identifying the dominant drivers of hydrological change in the contiguous United States. *Water Resour. Res.* **2021**, *57*, e2021WR029738. [CrossRef]
31. Wang, D.; Hejazi, M. Quantifying the relative contribution of the climate and direct human impacts on mean annual streamflow in the contiguous United States. *Water Resour. Res.* **2011**, *47*, W00J08. [CrossRef]
32. Xu, L. The land surface water and energy budgets over the Tibetan Plateau. *Nat. Preced.* **2011**, 1. [CrossRef]
33. Harris, I.P.D.J.; Jones, P.D.; Osborn, T.J.; Lister, D.H. Updated high-resolution grids of monthly climatic observations–the CRU TS3. 10 Dataset. *Int. J. Climatol.* **2014**, *34*, 623–642. [CrossRef]
34. Wang-Erlandsson, L.; Bastiaanssen, W.G.; Gao, H.; Jägermeyr, J.; Senay, G.B.; Van Dijk, A.I.; Guerschman, J.P.; Keys, P.W.; Gordon, L.J.; Savenije, H.H. Global root zone storage capacity from satellite-based evaporation. *Hydrol. Earth Syst. Sci.* **2016**, *20*, 1459–1481. [CrossRef]
35. Abatzoglou, J.T.; Dobrowski, S.Z.; Parks, S.A.; Hegewisch, K.C. TerraClimate, a high-resolution global dataset of monthly climate and climatic water balance from 1958–2015. *Sci. Data* **2018**, *5*, 1–12. [CrossRef]
36. Malek, Ž.; Verburg, P. Mediterranean land systems: Representing diversity and intensity of complex land systems in a dynamic region. *Landsc. Urban Plan.* **2017**, *165*, 102–116. [CrossRef]
37. Budyko, M.I. *Climate and Life*; Academic Press: Orlando, FL, USA, 1974.
38. Yu, Y.; Zhou, Y.; Xiao, W.; Ruan, B.; Lu, F.; Hou, B.; Wang, Y.; Cui, H. Impacts of climate and vegetation on actual evapotranspiration in typical arid mountainous regions using a Budyko-based framework. *Hydrol. Res.* **2021**, *52*, 212–228. [CrossRef]
39. Adam, J.C.; Clark, E.A.; Lettenmaier, D.P.; Wood, E.F. Correction of global precipitation products for orographic effects. *J. Clim.* **2006**, *19*, 15–38. [CrossRef]
40. Liakos, C.; Labropoulou, E.; Woodyatt, A. Greece Faces 'Disaster of Unprecedented Proportions' as Wildfires Ravage the Country. CNN. Available online: https://edition.cnn.com/2021/08/09/europe/greece-wildfire-warning-climate-intl/index.html (accessed on 10 August 2021).

41. Smith, P. Greek Wildfires Are the 'Harsh Reality of Climate Change,' Experts Warn. NBCNews. Available online: https://www.nbcnews.com/news/world/greek-wildfires-are-harsh-reality-climate-change-experts-warn-n1276311 (accessed on 9 August 2021).
42. Pascual, D.; Pla, E.; Lopez-Bustins, J.A.; Retana, J.; Terradas, J. Impacts of climate change on water resources in the Mediterranean Basin: A case study in Catalonia, Spain. *Hydrol. Sci. J.* **2015**, *60*, 2132–2147. [CrossRef]
43. Rouholahnejad Freund, E.; Kirchner, J.W. A Budyko framework for estimating how spatial heterogeneity and lateral moisture redistribution affect average evapotranspiration rates as seen from the atmosphere. *Hydrol. Earth Syst. Sci.* **2017**, *21*, 217–233. [CrossRef]
44. Davraz, A.; Sener, E.; Sener, Ş.; Varol, S. Water Balance of the Eğirdir Lake and the Influence of Budget Components, Isparta, Turkey. *Süleyman Demirel Üniversitesi Fen Bilim. Enstitüsü Derg.* **2014**, *18*, 27–36.
45. Ajjur, S.B.; Baalousha, H.M. A review on implementing managed aquifer recharge in the Middle East and North Africa region: Methods, progress and challenges. *Water Int.* **2021**, *46*, 578–604. [CrossRef]
46. Dezsi, Ş.; Mîndrescu, M.; Petrea, D.; Rai, P.K.; Hamann, A.; Nistor, M.M. High-resolution projections of evapotranspiration and water availability for Europe under climate change. *Int. J. Climatol.* **2018**, *38*, 3832–3841. [CrossRef]
47. Nangombe, S.; Zhou, T.; Zhang, W.; Wu, B.; Hu, S.; Zou, L.; Li, D. Record-breaking climate extremes in Africa under stabilized 1.5 C and 2 C global warming scenarios. *Nat. Clim. Chang.* **2018**, *8*, 375–380. [CrossRef]
48. Mariotti, A.; Struglia, M.V.; Zeng, N.; Lau, K.M. The hydrological cycle in the Mediterranean region and implications for the water budget of the Mediterranean Sea. *J. Clim.* **2002**, *15*, 1674–1690. [CrossRef]
49. Lionello, P.; Malanotte-Rizzoli, P.; Boscolo, R.; Alpert, P.; Artale, V.; Li, L.; Luterbacher, J.; May, W.; Trigo, R.; Tsimplis, M.; et al. The Mediterranean climate: An overview of the main characteristics and issues. *Dev. Earth Environ. Sci.* **2006**, *4*, 1–26.
50. Lelieveld, J.; Hadjinicolaou, P.; Kostopoulou, E.; Chenoweth, J.; El Maayar, M.; Giannakopoulos, C.; Hannides, C.; Lange, M.A.; Tanarhte, M.; Tyrlis, E.; et al. Climate change and impacts in the Eastern Mediterranean and the Middle East. *Clim. Chang.* **2012**, *114*, 667–687. [CrossRef]
51. Moutahir, H.; Casady, G.; Manrique-Alba, A.; Ruiz-Yanetti, S.; Maturano, A.; Zeramdini, A.; Bellot Abad, J. Can we better understand land surface phenology changes using hydrological variables instead of climate variables. In Proceedings of the XIV MEDECOS & XIII AEET Meeting, Seville, Spain, 31 January–4 February 2017; Volume 31.
52. Llorens, P.; Latron, J.; Gallart, F. Dinámica espacio-temporal de la humedad del suelo en un área de montaña mediterránea. Cuencas experimentales de Vallcebre (Alto Llobregat). *Estud. Zona No Saturada Suelo* **2003**, *6*, 71–76.
53. Ungar, E.D.; Rotenberg, E.; Raz-Yaseef, N.; Cohen, S.; Yakir, D.; Schiller, G. Transpiration and annual water balance of Aleppo pine in a semiarid region: Implications for forest management. *For. Ecol. Manag.* **2013**, *298*, 39–51. [CrossRef]
54. Mollema, P.; Antonellini, M.; Gabbianelli, G.; Laghi, M.; Marconi, V.; Minchio, A. Climate and water budget change of a Mediterranean coastal watershed, Ravenna, Italy. *Environ. Earth Sci.* **2012**, *65*, 257–276. [CrossRef]
55. Delle Rose, M.; Martano, P. Datasets of Groundwater Level and Surface Water Budget in a Central Mediterranean Site (21 June 2017–1 October 2022). *Data* **2023**, *8*, 38. [CrossRef]
56. Scarascia-Mugnozza, G.; Callegari, G.; Veltri, A.; Matteucci, G. Water balance and forest productivity in mediterranean mountain environments. *Ital. J. Agron.* **2010**, *5*, 217–222. [CrossRef]

MDPI
St. Alban-Anlage 66
4052 Basel
Switzerland
www.mdpi.com

MDPI Books Editorial Office
E-mail: books@mdpi.com
www.mdpi.com/books

www.ingramcontent.com/pod-product-compliance
Lightning Source LLC
Chambersburg PA
CBHW061450220926
51850CB00004B/58
* 9 7 8 3 0 3 6 5 9 3 4 8 7 *